Genomics of Crucifer's Host-Pathosystem

Govind Singh Saharan • Naresh K. Mehta • Prabhu Dayal Meena

Genomics of Crucifer's Host-Pathosystem

Springer

Govind Singh Saharan
Department of Plant Pathology
CCS Haryana Agricultural University
Hisar, Haryana, India

Naresh K. Mehta
Department of Plant Pathology
CCS Haryana Agricultural University
Hisar, Haryana, India

Prabhu Dayal Meena
Plant Protection Unit
ICAR-Directorate of Rapeseed-Mustard Research
Bharatpur, Rajasthan, India

ISBN 978-981-19-3811-5 ISBN 978-981-19-3812-2 (eBook)
https://doi.org/10.1007/978-981-19-3812-2

This Springer imprint is published by the registered company Springer Nature Singapore Pte Ltd.
The registered company address is: 152 Beach Road, #21-01/04 Gateway East, Singapore 189721, Singapore

Foreword

The family Brassicaceae include 338 genera with 3709 species of which the genus *Brassica* comprises 37 species. A number of *Brassica* species are of global economic importance for yielding quality edible oil, nutritious vegetables, and animal forage and as with most crops, Brassicas are subject to considerable losses caused by biotrophic, necrotrophic, and hemibiotrophic pathogens. Recent advances in molecular techniques have greatly expanded our understanding of plant pathology, disease resistance, and immunity, enabling their application through plant breeding to agricultural crop systems. Within the Brassicas, host-pathosystems of various crops are being studied genomically using multiomic research technologies.

Plant immune/defense systems rely on their ability to recognize the presence of the pathogen, to carry out signal transduction, and to respond defensively through pathways involving many genes and/or their products. The plant immune systems have provided both increased evolutionary opportunity for pathogen resistance and additional mechanisms for pathogen inhibition of such defense responses. However, sequencing the genomes of plant pathogens has shown variability in genome size and its structure. Genomic data generated so far has provided insightful information to understand the mechanisms and the changes in genome size and adaptive evolution in plant pathogens. The pathogens provide prominent advantageous associations between rapidly evolving transposable elements and virulence genes that cause variation in virulence of the phenotypes. The genome architecture, pathogenicity factors, and determinants of host specificity of some pathogens are yet to be known.

The outstanding book *Genomics of Crucifer's Host-Pathosystem* authored by Prof. (Dr.) G. S. Saharan, Prof. (Dr.) Naresh K. Mehta, and Dr. Prabhu Dayal Meena is the ninth book in the series on diseases of crucifers being published by Springer, Nature, after the books on *Albugo*, *Alternaria*, *Erysiphe*, *Hyaloperonospora*, *Plasmodiophora*, *Sclerotinia*, Crucifer's host resistance, and *Molecular mechanism of crucifer's host resistance*. The book is an annotated compilation that explains the characteristics of the recent development in genome assembly of important pathogens of crucifers to gain insights into host-pathosystem. The manuscript has been prepared after critical analysis of the world literature on all the major diseases of crucifers with encyclopedic treatments for better comprehension by readers. The book provides helpful background and current information that enhances our insight on host-pathosystem of crucifers.

My heartiest congratulations to the authors for bringing out their lifelong professional interest, and expertise with comprehensive treatise at the most needed time. It will be quite useful for researchers, teachers, students, extension workers, and all those who are concerned with the growth and development of cruciferous crops.

Department of Plant Pathology,
University of Wisconsin
Madison, WI, USA

Paul H. Williams

Preface

Crucifer encompasses one of the major groups of crops belonging to the family Brassicaceae yielding high-quality edible oil and nutritious vegetables. Major diseases of these crops cause heavy yield and quality losses at global level which are increasing with the climate change threatening food security. The use of omics technologies allows us to understand the genomics of crucifers' host-pathosystem at molecular level to get deeper insights into the interacting host and pathogen to reveal systems biology. The integrated multiomics and systems biology data will enable breeding of high-quality disease-resistant *Brassica* crops in a more holistic, targeted, and precise way for better management of long-term co-evolutionary process of host and pathogen and in desired direction. Application of omics approaches has allowed getting high-quality genome assemblies of all the six *Brassica* species of "U" triangle in addition to the model plant *Arabidopsis thaliana* and major pathogens of *Brassica* crops. It has greatly facilitated the discovery of candidate gene for effectors and virulence factors of the pathogens. Transcriptomics of virulence-related genes has been applied to study pathogen gene expression during host invasion in order to monitor molecular mechanisms and events involved in pathogenesis. Secretomics revealed effector proteins and specific enzymes crucial for host colonization by the pathogens. Interactomics studied networks of genes and proteins interactions in biological host-pathosystem. Biometabolomics provided an insight into metabolic changes in the host during host-pathogen interactions and its effects on host-pathosystem.

The book *Genomics of Crucifer's Host-Pathosystem* is a compilation of latest research achievements of *Brassica* scientists using omics approaches in understanding host-pathogen interaction, molecular detection, identification, and functional characterization of effectors/genes including pathogenomics and biometabolomics, genomics of pathogenic variability, and genomics modulation of crucifers' host-pathosystem through nine chapters on all the important pathogens of crucifers. The book is a source of information which has been vividly illustrated with photographs, graphs, figures, histogram, tables, and colored plates, which makes it stimulating, effective, and easy to comprehend by readers. The headings and subheadings of each chapter have been arranged in numbered series to make the subject matter contiguous. A chapter on standardized, reproducible protocols in studying the host-pathosystem has been included for the researchers of cruciferous crops for developing resistant cultivars after understanding the host-pathosystem. The last section deals with the gaps in understanding, knowledge of genomics, and offers suggestions for future research priorities in order to initiate advanced research programs in conducting studies to foster success in *Brassica* production and productivity through the development of improved disease-resistant varieties for the benefit of farmers.

The authors are confident that this book will be immensely useful to researchers especially *Brassica* breeders, molecular biologists, plant pathologists, teachers, extension specialists, students, industrialists, farmers, and all others who are interested to grow healthy and profitable cruciferous crops all over the world. Any shortcomings, lacunae, and flaws in the book are the responsibility of the authors. Any suggestions by readers are always a source of inspiration for the authors, and suggestions for its improvement are most welcome.

Hisar, Haryana, India Govind Singh Saharan
Hisar, Haryana, India Naresh K. Mehta
Bharatpur, Rajasthan, India Prabhu Dayal Meena

Acknowledgments

The authors are highly grateful to all the persons, scientists, publishers, societies, journals, institutes, websites, and all others whose valuable materials such as photographs (macroscopic, microscopic, electron micrographs, scanning electron micrographs), drawings, figures, histograms, graphs, tables, and flowcharts, etc. have been used through reproduction in this book. The address of the author(s)/ source(s) from where material was adopted/produced from institutes, websites, societies, etc. can be obtained through the reference which has been cited in the reference section of the book in each chapter. We, the authors, duly thankfully acknowledge and put on record all the authors whose excellent and valuable publications/articles find their place in this manuscript by citing their names against tables, figures, plates, dendrograms, etc. We have sought the necessary permission to reproduce their material in this book, and majority of them have granted the permission to reuse them. The authors are sincerely thankful to all the scientists, publishers, journals, institutes, societies, and websites whose materials have been reproduced in one or the other form in the present manuscript but forgot to acknowledge their name(s) inadvertently. Their prompt help in providing permission, information, photographs, figures, and consents for reproduction in the present manuscript has been duly acknowledged, and the authors are grateful to all of them. There might be some errors, mistakes, and shortcomings in this manuscript. We shall greatly appreciate healthy criticism and suggestions for the improvement of this publication in future.

Contents

About the Authors

Govind Singh Saharan, Ex. Professor and Head, Department of Plant Pathology, CCS HAU, Hisar, has conducted research in diverse fields of Plant Pathology and has published 250 articles in national and international journals. He has been editor/author of several books, monographs, and Crop Production Compendium (CABI). He is on the panel of Experts of SAU, ICAR, IARI, CSIR, UGC, and Department of Biotechnology. Dr. Saharan has been a visiting Professor at the University of Alberta, Edmonton, Canada; Agriculture and Agri-Food Canada, Saskatoon, Canada; and Rothamsted Research, IACR, Harpenden, UK. He has been President (NZ) of the Indian Phytopathological Society, Editor-in-Chief of a few journals, and President of the Indian Society of Mycology and Plant Pathology. He has been awarded with the Y. L. Nene Outstanding Plant Pathology Teacher Award by ISMPP, Udaipur, and Life Time Achievement Award by the Society for Rapeseed-Mustard Research, Bharatpur, India.

Naresh K. Mehta, Former Associate Dean, Professor and Consultant Faculty, Department of Plant Pathology, CCS HAU, Hisar, has been teaching Plant Pathology courses to UG and PG students. He has conducted research in diverse fields of Plant Pathology especially on rapeseed-mustard. He has guided several M.Sc. and Ph.D. students. He has published about 200 research articles and editor/author of several books, book chapters, review articles, and teaching manuals. Dr. Mehta has been President of INSOPP, Ludhiana, and Editor-in-Chief of a journal and served as a member of the editorial board of various phytopathological societies in India. Dr. Mehta has been awarded the

Dr. Y. L. Nene Outstanding Plant Pathology Teacher Award by ISMPP, Udaipur; Dr. T. S. Thind Distinguish Plant Pathologist Award by INSOPP, Ludhiana, and Ms. Manju Utereja Memorial Gold Medal, HAU, Hisar. He is on the panel of Experts of SAU, ICAR, UGC, and IARI and a member of various national and international committees. Dr. Mehta has been a visiting scientist to the University of Alberta, Edmonton, Canada.

Prabhu Dayal Meena, Principal Scientist (Plant Pathology) at ICAR-Directorate of Rapeseed-Mustard Research, Bharatpur, Rajasthan, India, has conducted research on different aspects of rapeseed-mustard diseases, including host resistance, management, epidemiology, and biology of crucifer's pathogens. He has associated in the development of six high yielding varieties and registered seven trait-specific genotypes. He has published 120 research papers, reviews, and book chapters and 13 books. Dr. Meena is Fellow of Indian Society of Mycology and Plant Pathology, Plant Protection Association of India, Indian Society of Oilseed Research, Şociety for Rapeseed-Mustard Research, and Indian Phytopathological Society. He has been honored with the Dr. P.R. Kumar Outstanding Brassica Scientist Award and Gold Medal of Society for Rapeseed-Mustard Research. He is founder secretary, managing editor, and Chief Editor of Society for Rapeseed-Mustard Research. He has visited the UK under Indo-UK Collaborative Research on Oilseed Brassica crops. He has been guiding several M.Sc. and Ph.D. students.

Abbreviations

AAFC	Agriculture and Agri-Food Canada
AB	Alternaria blight
ABA	Abscisic acid
ABC	ATP-binding cassette
ABP1	Auxin binding protein 1
AFLP	Amplified fragment length polymorphism
AG	Anastomosis groups
AIP	Aminoindan phosphonic acid
AM	Arbuscular mycorrhizal
ANOVA	*Analysis of variance*
AOC3	*Allene oxide cyclase 3*
AOS	Allene oxide synthase
ap	Appressorium
APX	Ascorbate peroxidase
ARF-GEF	ADP ribosylation factor-GTP exchange factor
AT	Associative transcriptomics
AtSTP4	*Arabidopsis sugar transport protein 4*
AUDPC	Area Under Disease Progress Curve
Avr	Avirulence
BA	Benzoic acid
BABA	β-amino butyric acid
BC	Backcross
Bgt	*Blumeria graminis* f. sp. *tritici*
BiFC	Bimolecular fluorescence complement
BLAST	Basic Local Alignment Search Tool
BLAT	BLAST-like alignment tool
BLUPs	Best linear unbiased predictions
BSA	Bulked Segregant Analysis
BSMV	Barley stripe mosaic virus
CA	Constitutively activated
CAC	Clathrin adaptor complex
CAD	Cinnamyl alcohol dehydrogenase

CaM	Calmodulin
CaMV	Cauliflower mosaic virus
CAT	Catalase
CBD	Chitin binding domain
CBls	Calcineurin-B-like protein
CC	Coiled coil
CDPKs	Calcium-dependent protein kinases
CF	Culture/cultural filtrate
ChIP	Chromatin immune-precipitation
CHS	Chalcone synthase
CI	Cylindrical inclusion protein
CK	Cytokinins
cM	Centi Morgans
CMLs	Calmodulin-like protein
CNGCs	Cyclic nucleotide gated ion channels
COIP	Co-immune precipitation
CW	Cell wall
CWAs	Cell wall apposition
CWDE	Cell wall-degrading enzymes
DAI	Days after inoculation
DAMPs	Damage-associated molecular patterns
DAPC	Discriminant analysis of principal components
DAS	Days after sowing
DEGs	Differentially expressed genes
DGE	Differential gene expression
DH	Doubled haploid
DI	Disease indices
DM	Downy mildew
DNA	Deoxyribonucleic acid
dpi	Days post-inoculation
DTT	Dithiothreitol
Ec	*Erysiphe cruciferarum*
ECD	European clubroot differential
EDS1	Enhanced disease susceptibility 1
EHM	Extrahaustorial membrane
ELISAs	Enzyme-linked immunosorbent assays
EMS	Ethyl methanesulfonate
ENU	Ethyl nitrosourea
Ep	*Erysiphe pisi*
e-PCR	Electronic PCR
ERK	Extracellular signal-regulated kinase
ESTs	Expressed sequence tags
ET	Ethylene
ETI	Effector-triggered immunity
ETS	Effector-triggered susceptibility

f. sp.	Fungal forma specialis
FCF	Fungal culture filtrate
FDR	False discovery rate
Fig	Figure
FPKM	Fragments Per Kilobase of transcript per Million mapped
FTIR	Fourier-transform infrared spectroscopy
GA	Gibberellins
GBS	Genotyping by sequencing
GC	*Golovinomyces cichoracearum*
GC-MS	Gas chromatography-mass spectrometry
GCRMA	Guanine cytosine robust multi-array analysis
GDIs	Guanine nucleotide dissociation inhibitors
GDP	Guaracol-dependent peroxidase
GFP	Green fluorescent protein
GISH	Genomic in situ hybridization
GLM	Generalized linear models
GM	Genetically modified
GO	Gene ontology
Go	*Golovinomyces orontii*
GPCR	G-protein-coupled receptors
GSS	Genomic survey sequences
GWAS	Genome-wide association analysis
HGP	Human Genome Project
HIGS	Host-induced gene silencing
HMM	Hidden Markov Model
HNRT	Homoeologous nonreciprocal transposition
Hp	*Hyaloperonospora parasitica*
Hpa	*Hyaloperonospora arabidopsidis*
hpi	Hours post inoculation
HR	Hypersensitive response
HRMS	High-resolution mass spectrometry
HS	Highly susceptible
HSPs	Heat shock proteins
HTGs	High-throughput genome sequences
HTS	Host targeting signal
IAA	Indole-3-acetic acid
IAN	Indole-3-acetonitrile
IAOx	Indole-3-acetaldoxime
IC	Isochorismate
ICAR	Indian Council of Agricultural Research
ICIM	Inclusive composite interval mapping
ICM	Composite interval mapping
ICS	Isochorismate synthases
iGS	Indole glucosinolates
ILs	Introgression lines

IM	Interval mapping
INA	Isonicotinic acid
IP	Intron polymorphic
IPG	Immobilized pH gradient strips
ISSRs	Inter-simple sequence repeats
ITCs	Isothiocyanates
ITS	Internal transcribed spacer
JA	Jasmonic acid
KASP	Kompetitive allele-specific PCR
LD	Linkage disequilibrium
LGs	Linkage groups
LIF	Lignification inducing factor
Lm	*Leptosphaeria maculans*
LOD	Logarithm of odds difference
LRR	Leucine-rich repeat
LRR-RLKs	Leucine-rich repeat receptor like kinase
LRR-RLPs	Leucine-rich repeat receptors–like protein
LRRs	Leucine-rich repeats
LysM	Lysine motif
LZ	Leucine zipper
MAB	Marker-assisted backcross breeding
MAMPs	Microbe-associated molecular patterns
MAP	Mitogen-activated protein
MAPK	Mitogen-activated protein kinase
MAPKKK	MAP kinase kinase kinase
MAS	Marker-assisted selection
MeSA	Methyl salicylate
MET	Multi-environment trials
Mips	Macrophage infectivity potentiators
MKK	MAP kinase kinase
ML	Maximum likelihood
MLM	Mixed linear model
MLO	Mildew resistance locus O
MPK	MAP kinase
MPTO	Methylthiopentanaldoxime
MQM	Multiple QTL mapping
MR	Moderately resistant
MS	Mass spectrometry
MSA	Multiple sequence alignment
MT	Microtubule
MTAs	Marker trait associations
MYA	Million years ago
MYB	Myeloblastosis
NBS	Nucleotide binding site
NBS-LRR	Nucleotide-binding site leucine-rich repeat

NBT	Nitro blue tetrazolium
NC	Nucleotidyl cyclase
NCBI	National Center for Biotechnology Information
NGS	Next generation sequencing
NHR	Nonhost resistance
NIa	Nuclear inclusion protein a
NILs	Near isogenic lines
NIRS	Near-infrared reflectance spectroscopy
NLRs	Nucleotide-binding site leucine-rich repeats
NMR	Nuclear magnetic resonance
NO	Nitric oxide
NPR1	Nonexpressor of pathogenesis-related genes 1
NWCVT	National winter canola variety trials
OA	Oxalic acid
OD	Optical density
OGs	Oligogalacturonides
On	*Oidium neolycopersici*
OST1	Open stomata 1
PA	Phosphatidic acid
PAD4	Phytoalexin-deficient 4
PAL	Phenylalanine ammonia lyase
PAMP	Pathogen-associated molecular pattern
PCA	Principal component analysis
PCD	Programmed cell death
PCR	Polymerase chain reaction
PDA	Potato dextrose agar
PDI	Percent disease index
PDR	Pleiotropic drug resistance
PEIs	Pectinesterase inhibitors
PEN2	Peroxisome-associated myrosinase penetration 2
PFK	Phosphofructokinase
PGA	Polygalacturonase
PGIPs	Polygalacturonase inhibitor proteins
PIC	Polymorphic information content
PIPs	Phosphatidyl inositol phosphates
PM	Plasma membrane
PM	Powdery mildew
PMEI	Pectin methylesterase inhibitors
PMSF	Phenylmethylsulphonyl fluoride
PO	Peroxidase
pp	Penetration peg formation
PPO	Polyphenol oxidase
PPT	Phosphinothricin
PR	Pathogenesis related
PRRs	Pattern recognition receptors

PTI	PAMPs-triggered immunity
PTI	Pattern-triggered immunity
pv.	Pathovar
QDR	Quantitative disease resistance
qRT-PCR	Real-time quantitative PCR
QTL	Quantitative trait loci
R	Resistance
RAPD	Random amplification of polymorphic DNA
RFLP	Restriction fragment length polymorphism
RGL	Resistant genes like
RIN	RNA integrity numbers
RLCK	Receptor-like cytoplasmic kinase
RLKs	Receptor-like kinases
RLPs	Receptor-like proteins
RNA	Ribonucleic acid
RNAi	RNA interference
ROI	Reactive oxygen intermediates
ROP	Rho of plants
ROS	Reactive oxygen species
RT-PCR	Reverse transcription and quantitative reverse transcription-polymerase chain reaction
SA	Salicylic acid
SAG	Salicylic acid glycoside
SAG101	Senescence-associated gene 101
SAM	*S*-adenosine-L-methionine
SAR	Systemic acquired resistance
SARF	Sum of adjacent recombination fractions
SCAR	Sequence characterized amplified region
SD	Standard deviation
Si	Silicon
SMA	Single marker analysis
SMRT	Single-molecule real-time
SNAP	Soluble *N*-ethylmaleimide-sensitive factor adaptor protein
SNP	Single nucleotide polymorphism
SOD	Superoxide dismutase
SRAP	Sequence-related amplified polymorphism
SSIL	*Sclerotinia sclerotiorum integrin-like*
SSR	Sclerotinia stem rot
SSR	Simple sequence repeat
STK	Serine-threonine kinase
STS	Sequence tagged sites
TAIR	The *Arabidopsis* Information Resource
TBIAs	Tissue blot immunoassays
TBS	Tris-buffered saline
TDFs	Transcript-derived fragments

TFs	Transcription factors
TGS	TRIS-Glycine-SDS
TIGS	Transient-induced gene silencing
TIR	Toll/Interleukin-1 receptor
TM	Trans-membrane
TMM	Trimmed mean of means
TNL	TIR-NBS-LRR
TuMV	Turnip mosaic virus
TuYV	Turnip yellows virus
UPGMA	Unweighted pair group method with arithmetic mean
UPS	Ubiquitin proteasome system
USA	United States of America
USDA	United States Department of Agriculture
UTR	Untranslated region
VAMPs	Vesicle-associated membrane proteins
VIGS	Virus-induced gene silencing
VPg	Genome-linked viral protein
WAKL	Wall-associated kinase-like
WGT	Whole genome triplication

Symbols

$	Dollar
%	Percent
≈	Is approximately equal to
/	Per
:	Ratio
~	Approximately
£	Pound
+	Plus
<	Less than
=	Is equal to
>	More than
±	Plus-minus sign
×	Multiply
≤	Less than or equal to
≥	More than or equal to
µg	Microgram
µl	Microliter
µm	Micrometer
µmol	Micromole
AU$	Australian dollar
B	Boron
bp	Base pair

Ca	Calcium
Ca^{2+}	Calcium cation
CFU	Colony forming unit
cM	Centi Morgans
cm	Centimeter
cm^2	Square centimeter
cm^{-3}	Per cubic centimeters
cvs.	Cultivars
d	Day
dpi	Days post-inoculation
dS/m	DeciSiemens per meter
e.g.	For example
et al.	"et alia" meaning "and others"
g	Gram
g^{-1}	Per gram
Gy	Gray
h	Hours
h^2	Heritability
H_2O_2	Hydrogen peroxide
i.e.	id est, meaning "that is"
kb	Kilobase
KCl	Potassium chloride
kV	Kilovolt
L	Liter
L^{-1}	Per liter
Log	Logarithm
M	Million
m	Meter
M	molar
m^2	Square meter
Mb	Million bases
Mb	Megabyte
Mg	Magnesium
mg	Milligram
$MgCl_2$	Magnesium chloride
Mha	Million hectare
min	Minute
ml	Milliliter
mL^{-1}	Per milliliter
mm	Millimeter
mM	Millimolar
mmole	Millimoles
mmt	Metric million tonnes
MW	Molecular weight
N	Nitrogen

nm	Nanometer
nmoles	Nanomole
OC	Degree Celsius
P	Potassium
pH	Potential of hydrogen
rpm	Revolutions per minute
S	Sulfur
S	Susceptible
s	Second
s^{-1}	Per second
sdH_2O	Sterile deionized water
SE	Standard error
Subsp.	Subspecies
V	Volts
v/v	Volume by volume
v:w	Volume equal to weight
var.	Variety
viz.	Videlicet meaning "namely" or "which is"
w/v	Weight by volume
β	Beta
μM	Micrometers
Σ	Sigma means "sum up"

Genomics of Crucifer's Host-Pathosystem: Prologue

1

Abstract

Crucifer's belong to the family Cruciferae containing over 3660 species which includes very important crops of human and animal needs yielding quality edible oil, industrial oilseeds, vegetables, and fodder crops. The largest group of crops contains 39 species of Brassicaceae grown all over the world including *Arabidopsis thaliana* which has been used widely for dissecting molecular mechanisms of crucifer's host-pathosystem through multiomics approaches. Under natural and cultivated conditions, crucifer's are challenged by several abiotic and biotic stresses viz., *Albugo* (White rust), *Alternaria* (Alternaria blight), *Colletotrichum* (Anthracnose), *Erysiphe* (Powdery mildew), *Fusarium* (Fusarium wilt), *Hyaloperonospora* (Downy mildew), *Leptosphaeria* (Blackleg), *Plasmodiophora* (Clubroot), *Pseudocercosporella* (White leaf spot), *Pyrenopeziza* (Light leaf spot), *Sclerotinia* (Stem rot), Turnip mosaic virus (*TuMV*), *Verticillium* (Verticillium wilt), *Xanthomonas* (Black rot), and *Heterodera* (Cyst nematode). These pathogens have been used as model host-pathosystem to reveal genomics of crucifer's host-pathosystem. The genomics of plants was initiated after the sequencing of *A. thaliana* genome for the first time in 1990. The genome size of *A. thaliana* is 125 Mbp containing 25,498 genes encoding proteins from 11,000 families. Now, the genome of all the six *Brassica* species has been sequenced. The sequence analysis has revealed genome size of *B. carinata* 642 Mb, *B. juncea* 922 Mb, *B. napus* 925 Mb, *B. nigra* 591 Mb, *B. oleracea* 584.60 Mb, and *B. rapa* 485 Mb. The genome of major pathogens of crucifer's has also been sequenced. The availability of genome sequence analysis of both host and pathogen has allowed rapid identification of candidate genes during different events of host–pathogen interaction to understand genes governing pathogenesis and host resistance. The application of omics technologies like NGS, Pangenomics, SNP, In Silico, BSA, Ren Seq, Effectoromics, Transcriptomics, Proteomics, Secretomics, Interactomics, and

G. Singh Saharan et al., *Genomics of Crucifer's Host-Pathosystem*, https://doi.org/10.1007/978-981-19-3812-2_1

Metabolomics has benefitted greatly in revealing the complex biological, genetical, and molecular mechanisms of crucifer's host-pathosystem. Crucifers have developed molecular mechanisms in response to multiple stresses which are activated through complex signaling pathways for the expression of genes to overcome these stresses. Several genes are differentially expressed in response to multiple stresses. At global level, 16 pathogens causing crucifers diseases are considered as of major consequences based on their geographical distribution, host range, losses caused, and resources spend to manage them. Using crucifer's host-pathosystem with these pathogens, several genes have been identified, mapped on *Brassica* chromosomes with functional characterization, isolation, and cloning to produce resistant cultivars of *Brassica*. In recent years, QRT-PCR, a powerful and efficient technique has become the first choice in quantitative gene expression in crucifer's host-pathosystem for various biological and molecular functions. However, data normalization is essential for reliable output of QRT-PCR assays to avoid unsuitable choice of reference genes.

Keywords

Economic importance · Crucifers' · *Brassica* genomics · *Arabidopsis* · Omics approaches · Host-pathosystem · Secretomics · Metabolomics · Biotic and abiotic stress · Expression of genes · Major pathogens at global level · Their morphology · Disease symptomatology · Host range · Epidemiology · Fine structures · Pathogenic variability · Infection and pathogenesis · Host resistance · Disease management · Differential genes to overcome biotic stress · Signaling system of crucifers' · Genomics of crucifers host resistance

1.1 Introduction

The family Cruciferae is comprised of most diverse plants containing over 3660 species including important edible, industrial oilseeds, vegetable, and fodder crops. The largest group of crops belonging to family Brassicae contains 39 species, and includes six U's triangle crops grown all over the world for oilseeds, vegetables, and as forage crops. The *Arabidopsis thaliana* is also a member of this family which has become the most important crop to dissect molecular mechanisms and molecular genetical, interactions using *Arabidopsis* host-pathosystem as a model in the scientific community of the world.

Cruciferous crop plants like all other living organisms do not live in isolation but are under the influence of interactions among themselves as well as biotic (pathogen, pests, weeds), abiotic (temperature, salinity, alkalinity, drought), and environmental factors (above and below grounds) where they are cultivated or wild. Under natural conditions, they are always challenged simultaneously by several biotic, abiotic, and internal defense reactions against these stresses. To defend themselves against all kinds of stresses, plants employ pre- and post defense mechanisms triggered by signaling molecules. Multiple stress responses are coordinated by complex signaling network molecules which converge at point of defense. The application of omics and genomics approaches has facilitated to have deeper insight into these complex interactions and use the modern genomics technology to breed stress resistant idiotype crucifers' crops. More than 50 pathogens are reported to cause diseases in crucifers' crops all over the world. Out of these, 16 pathogens are widely distributed causing heavy yield losses in cruciferous crops. The genomics of crucifer's host-pathosystem has been investigated at global level using these pathogens as model system. Each of the following factors has great significance and bearing on comprehension of this complex biological system and all events during each stage of interaction for better management of crucifers crops against all kinds of stresses with higher yield and quality of crops.

1.2 Economic Importance of Crucifers

Crucifers include large number of vegetables and oil yielding crops of family Cruciferae. The largest group of crops belongs to family Brassicae. The *Brassica* genus is a member of Brassicaceae (Cruciferae) and contains 39 species (http://www.theplantlist.org/). Among the *Brassica* species, six constitute U'sTriangle2: three diploid species, namely *Brassica rapa* (AA genome: $2n = 2x = 20$), *Brassica nigra* (BB: $2n = 2x = 16$), and *Brassica oleracea* (CC: $2n = 2x = 18$), and three allotetraploid species, namely *Brassica juncea* (AABB: $2x = 4x = 36$), *Brassica napus* (AACC: $2n = 4x = 38$), and *Brassica carinata* (BBCC: $2n = 4x = 34$). The triangle model provides the fundamental relationships among these *Brassica* species and is used as an important guideline for both evolutionary research and the improvement of *Brassica* crops via interspecies crossing to facilitate gene exchanges.

The crucifers' crops have great economic and morphological importance in daily life. Crops of the genus *Brassica* (tribe Brassiceae), which are in the same taxonomic family as *Arabidopsis thaliana*, are widely used in the cuisine of many cultures, due, in part, to the many choices of edible forms in the genus. Economically, *Brassica* is loosely categorized into oilseed, vegetable, and condiment crops. The *B. napus*, *B. rapa* (formerly *campestris*), *B. juncea*, and *B. carinata* provide about 12% of the world-wide edible vegetable oil supplies. The *B. oleracea* and *B. rapa*, the so-called "cole crops," comprise many of the vegetables in our daily diet. Several of these vegetables have extreme morphological characteristics. Examples of such morphologies include the enlarged inflorescence of cauliflower (*B. oleracea* subspecies *botrytis*) and broccoli (*B. oleracea* subspecies *italica*); the enlarged stem of kohlrabi (*B. oleracea* subsp. *gongylodes*) and marrow stem kale (*B. oleracea* subspecies *medullosa*); the enlarged root of turnip (*B. rapa* subspecies *rapifera*); the enlarged and twisted leaves of Pak-choi (*B. rapa* subspecies *chinensis*), Chinese cabbage (*B. rapa* subspecies *pekinensis*); and the enlarged single apical bud of cabbage (*B. oleracea* subspecies *capitata*) or the many auxiliary buds of Brussels sprout (*B. oleracea* subspecies *gemmifera*). Finally, the seed of *B. nigra* is utilized as a condiment-mustard. *Brassica* species are a valuable source of dietary fiber, vitamin C, and other possible salubrious factors such as anticancer compounds. Estimates of the economic importance of *Brassica* species are conservative, because several cole crops such as collards are cultivated primarily for local or home use, but are none the less a dietary main stay in low-income communities where other fresh vegetables can be prohibitively expensive.

Crops belonging to the *Brassica* genus are among the ten most economically important vegetable crops in global agriculture and markets. Due to their wide adaptation and ability to thrive under varying agroclimatic conditions, *Brassica* crops are grown throughout the world for food, animal forage and fodder and also for industrial applications. They also have an allelopathic use for sustainable agriculture and are grown as phytore mediators against heavy metals such as cadmium. As far as food is concerned, now a day, consumers are demanding products that are rich in nutrients for optimal health benefits. In this respect, the popularity of *Brassica* products is increasing because of their nutritional value, and anticancer, antioxidant, and anti-inflammatory properties. Nutritionally, these vegetables are low-fat, have a high vitamin (C and E) content and contain minerals (P, S, Cl, Ca, Fe, Sr, K, Cr, Mn, Se, and Zn) and fiber. In addition, they contain important phytochemicals that are beneficial for human health, such as anthocyanins, flavonoids, terpenes, *S*-methyl cysteine sulfoxide, coumarins, and other small compounds. However, the most characteristic compounds are glucosinolates, which are a group of secondary metabolites that are only present in Brassicaceae and immediate families. They have various functions within the plant, being especially important in the defense against pathogens and herbivores (Poveda et al. 2020).

Many *Brassica* crops are of great economic significance, as they are cultivated as vegetables, oilseed sources, condiments, and forages. Climate change, pathogen variation, and inappropriate farming methods, such as continuous and high-intensity cropping, contribute to disease outbreaks, which pose threats to crucifers "production." Various biotic and abiotic stresses cause production losses (Saharan et al. 2017). Traditional approaches for disease prevention include agricultural, physical, chemical, and biological controls, and integrated pest management (IPM) strategies. Physical approaches, such as high-temperature treatment and light trapping, chemicals, such as fungicides and bactericides, and biological agents, such as *Bacillus subtilis* and arbuscular mycorrhiza, are frequently used. IPM has been extensively studied and can achieve some effect for certain diseases. However, the approaches are often complicated, costly, and/or environmentally damaging. In contrast, natural resistance in crucifer's hosts is the most desirable strategy and could be integrated with other approaches for high-efficiency disease control. Two types of plant immunity have been identified to date: pathogen/microbe-associated molecular pattern (PAMP/MAMP)-triggered immunity, which is activated by cell surface-localized pattern recognition receptors by the recognition of PAMPs/MAMPs, and effector-triggered immunity activated by host resistance (R) genes through the recognition of pathogen-specific effector molecules, which is in accord with the gene-for-gene theory. Most R genes identified to date encode nucleotide-binding leucine-rich repeats (NB-LRRs), including coiled-coil NB-LRRs (CC-NB-LRRs) and Toll interleukin 1 receptor NB-LRRs (TIR-NB-LRRs). Moreover, some

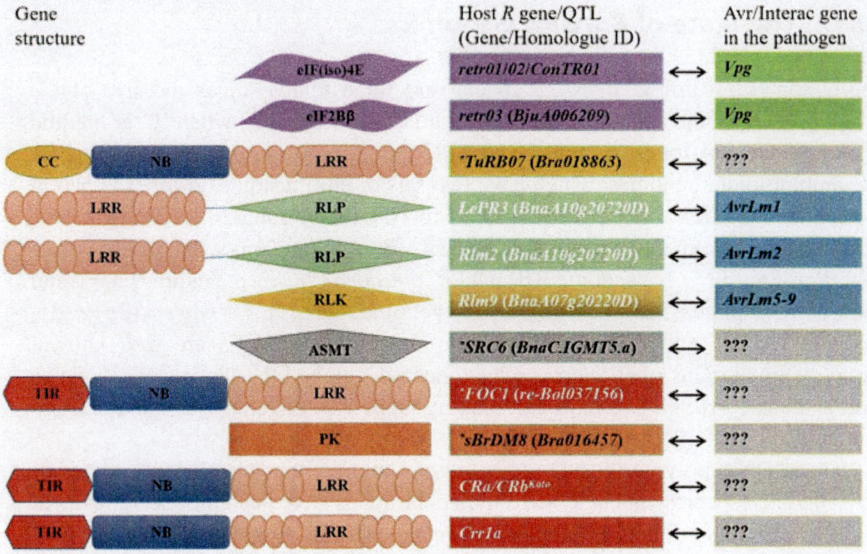

Fig. 1.1 Resistance genes identified in *Brassica* crops and their avirulence/interactor genes in the pathogens. ASMT, *N*-acetylserotonin Omethyltransferase; CC, coiled-coil domain; eIF (iso) 4E, eukaryotic translation initiation factor isoform 4E; eIF2Bβ, eukaryotic translation initiation factor 2Bβ; LRR, leucine-rich repeat; NB, nucleotide-binding domain; PK, protein kinase; RLP, receptor-like protein. *Putative genes that have not been functionally validated. ??? The avirulence or interaction genes in the pathogens that have not yet been characterized (Lv et al. 2020)

R genes encode receptor-like kinases (RLKs), transmembrane receptor-like proteins (RLPs), cytoplasmic kinases, and proteins with a-typical molecular motifs. Various R genes with flexible molecular mechanisms provide powerful weapons that protect the plant host from pathogens.

Many R genes have been identified and successfully applied to improve *Brassica* crop resistance against various diseases, which not only ensures *Brassica* production but also facilitates the discovery of host–pathogen interactions. Moreover, the genomic era characterized by massive genome and omic data has made fast and accurate R gene studies possible. The release of the reference genome data of the six *Brassica* species in addition to *B. carinata* has provided vital information for determining the genetic and molecular basis of disease resistance. Since the 2010s, researchers have performed extensive, high-quality genomic, postgenomic, and omics studies in *Brassica* species and have discovered a variety of R genes and closely related genes, which not only provide further insight into the resistance molecular mechanism and host–pathogen co-evolutionary arms race but also facilitate accurate molecular breeding at the whole-genome level (Lv et al. 2020; Fig. 1.1).

1.3 The State of *Brassica* Genomics

The genomes of diploid *Brassica* species are three to five times the size of the *Arabidopsis* genome, ranging from 0.97 pg/2C (468 Mb/1C; where C is haploid DNA per nucleus), for *B. nigra* to 1.37 pg/2C (662 Mb/1C) for *B. oleracea* 696 Mb. The amphidiploid genomes, which have two sets of chromosomes from each parent species, range from 2.29 pg/2C (1105 Mb/1C) to 2.56 pg/2C (1235 Mb/1C). Relationships among the three diploid species—*B. rapa* (syn. rapa; $2n = 20$, genome AA), *B. nigra* ($2n = 16$, genome BB), and *B. oleracea* ($2n = 18$, genome CC)—and three amphidiploids *B. napus* ($2n = 38$, genome AACC), *B. juncea* ($2n = 36$, genome AABB), and *B. carinata* ($2n = 34$, genome BBCC) are well known. Rapid-cycling strains of *B. oleracea* have been developed that have a life cycle as short as that of *Arabidopsis thaliana*, making them easier to study (Nagaharu 1935; Williams and Hill 1986; Paterson et al. 2001).

Neutral DNA polymorphisms are common among coeno-specific genotypes of most *Brassica* species, and at least 15 molecular maps have been produced, using at least 900 different publicly available *Brassica* and *Arabidopsis* DNA probes (which are each used with similar efficacy). Virtually all the probes hybridize to two or more loci, even in diploid *Brassica* species; however, only a subset of loci segregates for allelic variation in any one mapping population. The subset of genetically mapped loci can, therefore, be used to infer the locations of many additional sequence tagged sites (STSs) (Lan et al. 2000). The comparative organization of the chromosomes of *Brassica* and *Arabidopsis* is well studied. On the basis of 186 corresponding loci, about 19 chromosome structural rearrangements differentiate *B. oleracea* and *A. thaliana* orthologs, suggesting that chromosomal tracts of about 20–25 cM remain largely co-linear in the two taxa (Lan et al. 2000). Microsynteny studies involving comparative genetic and physical mapping of specific chromosome segments have shown largely conserved gene order in *Arabidopsis* and *Brassica*, but some disruption in gene content by deletions or insertions. Furthermore, duplicated partial gene clusters are commonly found in both *Arabidopsis* and *Brassica* species. Comparative sequencing has permitted orthology assignment of some duplicated segments in *Arabidopsis* and *Brassica* species (Wroblewski et al. 2000).

1.4 Crucifers' Host Genome

The members of the crucifer family are too diverse and the genomic structures are very complicated. In general, the cultivated *Brassica* species as per the triangle of U including diploid as well as allotetraploids species, the diploid possess the most variability while there is less variability, as compared to the diploid, in the amphidiploid species, as it is derived from the diploid species with the few chance crosses followed by the amphidiplodization. Also the selection pressure on the amphidiploid species is more due to its nucleo-cytoplasm and environmental instability. But still a considerable genetic diversity is present within these three amphidiploid species (Song et al. 1998). Based upon studies of genetic diversity, *B. napus* may be

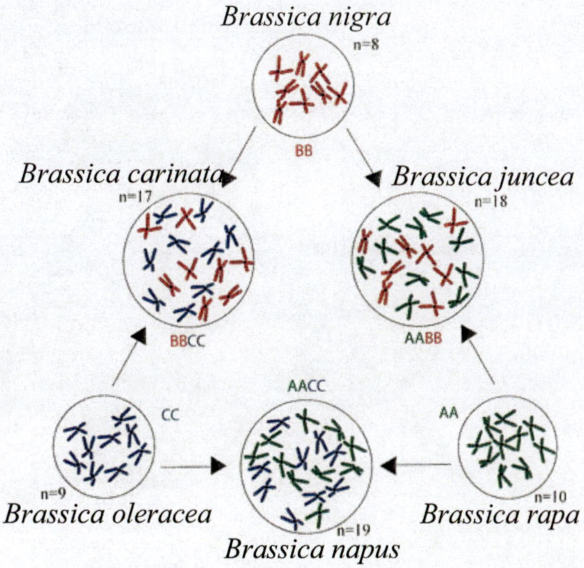

Fig. 1.2 Representing interrelationship among the diploid and amphidiploid species in triangle of U by Nagaharu (1935)

considered as the most ancient amphidiploid, succeeded by *B. juncea* and *B. carinata*. Two major factors are responsible for general diversity within amphidiploids, multiple hybridizations with different diploid parents and genome modifications following polyploidization. Early comparative studies conducted at the level of genetic linkage maps revealed extensive duplication within *Brassica* genomes (Lagercrantz and Lydiate 1996) and tracts of co-linearity disrupted by multiple rearrangements between the genomes of *B. nigra* and *A. thaliana* (Lagercrantz 1998). The genome of different U-triangle species along with *Arabidopsis thaliana* (Figs. 1.2 and 1.3) is as follows.

1.4.1 *Arabidopsis thaliana*

The flowering plant *Arabidopsis thaliana* is an important model system for identifying genes and determining their functions. So, *A. thaliana* made a large contribution to our molecular understanding of key concepts in biology by successful integration of different fields of molecular biology research. The chromosome number of *Arabidopsis thaliana* is $2n = 10$ and it has a polyploid origin, although there has been a dramatic genome reconstruction and loss of genes from the duplicated genome segments (Arabidopsis Genome Initiative 2000; Blanc et al. 2000; Ku et al. 2000; Mayer et al. 2001). The *Brassica* and *Arabidopsis* lineages diverged 20 MYA (Yang et al. 1999). Phylogenetic analysis groups of the *Brassica* species into the *nigra* and *rapa/oleracea* lineages (Warwick and Black 1991), which diverged 8 MYA (Lysak et al. 2005). With the progression of the *Arabidopsis* genome project during the late 1990s, thousands of "interesting" genes were

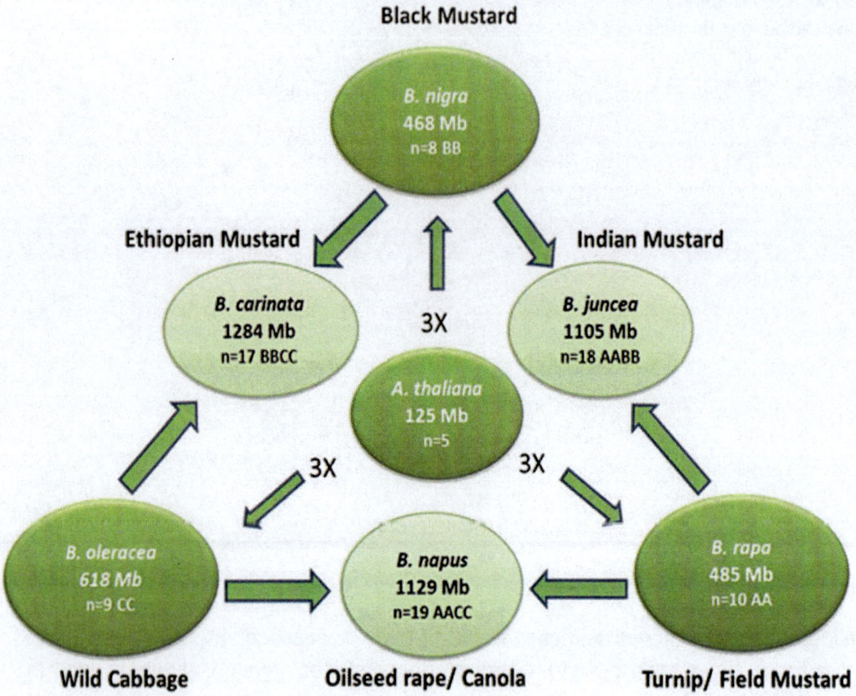

Fig. 1.3 U's triangle of *Brassica* species with their estimated genome size. The inferred relationship with the model species *A. thaliana* is indicated, but it should be noted that each of the species shown share a common ancestor with an increased chromosomes number

discovered at an unprecedented pace, fueling the desire by plant researchers to test their favorite hypothesis about these genes' potential functions. The genome size of *Arabidopsis thaliana* is 125 Mbp and genome contains 25,498 genes encoding proteins from 11,000 families (Arabidopsis Genome Initiative 2000).

1.4.2 *Brassica carinata*

Brassica carinata (BBCC, $n = 17$) was formed from *B. oleracea* (CC, $n = 9$) and *B. nigra* (BB, $n = 8$). This species is characterized by the slow steady growth of *B. oleracea* and the mustard oil content of *B. nigra*. Wild forms of *B. carinata* are not known, but primitive domesticated types are cultivated in upland areas of Ethiopia and further south into Kenya. This hybrid may have originated from kale land races of *B. oleracea* types hybridizing with wild or semi-domesticated forms of *B. nigra*. Both kale and *carinata* crops thrive in cool environments typical of the Ethiopian plateau. The local farmers grow them in "kale gardens." This term, translated into many languages and dialects, is commonly found throughout those rural areas using vegetables derived from *B. oleracea* types. *Carinata* crops themselves are alternatively named "guomin," Abyssinian mustard or Ethiopian cabbage, and provide leafy vegetables and sources of oils. Cabbage itself can be a ubiquitous term used to describe cole *Brassicas* and not necessarily synonymous with the sophisticated heads seen on today's super market shelves. The flow cytometry estimation of *B. carinata* gave a picture of genome size around 642 Mb (Johnston et al. 2005).

1.4.3 *Brassica juncea*

Brassica juncea ($n = 18$) is a hybrid between *B. rapa* ($n = 10$) and *B. nigra* ($n = 8$) (Frandsen 1943) producing large leaves and with the rapid growth of *B. rapa* and the mustard oil of *B. nigra*. Use of *B. juncea* as a source of vegetable oil is gaining importance in India; while throughout Asia, especially in China and Japan, the plant has a great diversity of cultivated forms used as staple vegetables of immense dietary importance. Reputably, wild forms are still found on the Asia Minor plateau and in southern Iran. The genome size of *B. juncea* variety Tumida is around 922 Mb (Yang et al. 2016).

1.4.4 *Brassica napus*

Brassica napus ($n = 19$), has wild forms in Sweden, Denmark, The Netherlands, and the UK. It developed from hybridization between *B. rapa* ($n = 10$) and *B. oleracea* ($n = 9$) followed by chromosome doubling. This hybrid may have formed as *B. oleracea* types expanded their range along the coasts of northern Europe and *B. rapa* extended from the Irano-Turanian regions. Alternatively, *B. napus* may have Mediterranean origins or, as seems likely, there were several centers of evolution. The wild populations of *B. napus* have acquired major scientific significance recently as they present means of determining the potential for gene flow to and from genetically modified cultivars of oilseed rape. The sequence analysis reveals the genome size of winter *B. napus* is 925 Mb (Lee et al. 2020).

1.4.5 *Brassica nigra*

Brassica nigra (black mustard), itself the ancestor of culinary mustards, is found widely and distributed as annual herbs growing in shallow soils around major rocky Mediterranean coasts. The *B. nigra* is the diploid species and also progenitor of the two allotetraploids species viz., *B. juncea* and *B. carinata*. The chromosome numbers of *B. nigra* is 16 with BB genome. The estimated genome size as per the flow cytometry analysis is 632 Mb (Johnston et al. 2005). The sequence of *B. nigra* variety Sangam using Illunima HiSeq platform depict the genome size of *B. nigra* as 591 Mb (Yang et al. 2016).

1.4.6 *Brassica oleracea*

Natural populations of *Brassica oleracea* and associated types are seen as potential progenitors of many European cole vegetables. These inhabit rocky cliffs in cool damp coastal habitats. They have slow, steady growth rates and are capable of conserving water and nutrients. This is an important vegetable crop with diploid in nature with CC genome having 18 chromosomes. Also, this is the one of the progenitor species of *B. napus* and *B. carinata*. The flow cytometry study reveals that the genomic size of *B. oleracea* is around 696 Mb (Johnston et al. 2005). The draft sequence analysis gave a picture of the genome of *B. oleracea* with genome size of 584.60 Mb which contained 47,772 genes and about 56.65% of genome with repetitive sequences (Sun et al. 2019).

1.4.7 *Brassica rapa*

The putative ancestor of many Oriental *Brassica* vegetables, *B. rapa* originates from the fertile crescent in the high plateaux regions in today's Iran–Iraq–Turkey. Here these plants grow rapidly in the hot, dry conditions, forming copious seed. This is one of the progenitors of the amphidiploid *Brassica juncea* and *Brassica napus* with AA genome and 20 chromosomes. This includes several important vegetable and oil crops. The flow cytometric estimation suggested that the genome size of *Brassica rapa* genome is around 529 Mbp (Johnston et al. 2005). The *B. rapa* genome was the first to be sequenced among the *Brassica* species with genome size of around 485 Mb (Zhang et al. 2018a; Wang et al. 2011).

1.4.8 Crucifer's as Ideal Host-Pathosystem for Genomics Research

Brassica host-pathosystem research has benefitted greatly from the close relationship between the *Brassica* crops and crucifer's model plant *Arabidopsis thaliana*, which was the first plant species to have a fully sequenced and annotated genome. The availability of the *A. thaliana* sequence allowed the rapid identification of candidates genes expressed during different crucifer's host–pathogens interactions governing different functions of pathogenesis/virulence for host susceptibility and resistance/avirulence for host resistance. The unique genetic relationship has been observed between the six most commonly grown *Brassica* crop species with model species *A. thaliana* which share a common ancestor for host-pathosystem (Fig. 1.3). This genetic relationship has provided researchers ample opportunities to track the genes functioning during *Brassica* host-pathosystem to understand molecular mechanisms and systems biological functions. The use of different omics technologies has further strengthened the avenues for wider scope of utilizing the knowledge in breeding for high yielding and resistance to biotic and abiotic stresses of cruciferous crops. The application and use of integrated omics technologies from

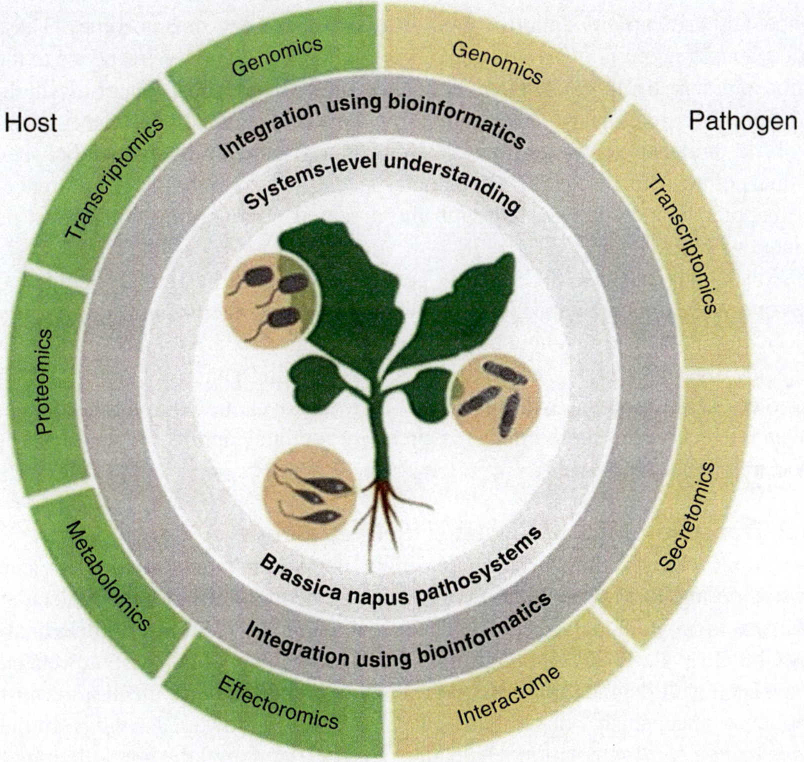

Fig. 1.4 The integration of omics technologies in *Brassica* host-pathosystem (Neik et al. 2020)

both the host and pathogen has opened ways in breeding *Brassica* cultivars that are better to resist pathogens attack (Fig. 1.4).

1.4.9 Utilization of Omics Approaches in Crucifer's Host-Pathosystem

1.4.9.1 High-Quality Genome Assemblies

High-quality reference genome assemblies for the major pathogens of *Brassica* are currently available, where long-read sequencing approaches such as ONT MinIon sequencing and PacBio sequencing were applied to assemble the genomes of *Leptosphaeria maculans*, *Sclerotinia sclerotiorum*, *Alternaria brassicae,* and *A. alternata*. The genome of *Plasmodiophora brassicae* was assembled using Illumina Hiseq2500 technology, while the *A. candida* genome was assembled using Roche/454; both are short-read sequencing technologies.

These high-quality genome assemblies of the *Brassica* pathogens revealed that the fungal pathogens contain high genomic variation, including mutations largely induced by transposable elements (TE), large-scale chromosomal re-arrangements, presence–absence variation, and the gain or loss of accessory chromosomes. These genetic events are continuously and actively evolving in the adaptive response to the selection pressure by the host plant resistance mechanism, thus generating high-genome-plasticity regions, which are often found distributed in compartments where most of the virulence genes are housed. The extent of genetic divergence between individuals of the same fungal plant pathogen species is also high; hence, reference-based mapping is a challenge, although the diverged regions may not always be associated with virulence.

The availability of these high-quality genome assemblies has greatly facilitated the discovery of candidate genes for effectors and virulence factors and significantly advanced our understanding about the pathogen in relation to its evolutionary pattern and species diversity through comparative and population genomics studies. Such deep molecular information will also allow us to uncover the complex *Brassica*–pathogen interactions, as these omics resources are routinely applied in plant pathology research (Neik et al. 2020).

1.4.9.2 Transcriptomics of Virulence-Related Genes

Transcriptome analysis has been increasingly applied to study pathogen gene expression during host invasion, allowing the real-time monitoring of the molecular mechanisms involved in pathogenesis. The recent genome-wide transcriptome analysis by Chittem et al. (2020) who highlighted the involvement of peroxisome-related pathways, cell wall degradation by various enzymes, and the detoxification of host metabolites as mechanisms of virulence by *S. sclerotiorum* toward *B. napus*. In the *B. napus–L. maculans* interaction, Haddadi et al. (2019) reported the upregulation of genes for fungal toxin biosynthesis during the necrotrophic stage of infection. Furthermore, several *L. maculans* effectors have been predicted from RNA-seq data, consistent with the results of Sonah et al. (2016), who detected different sets of genes coding for several effector proteins, where expression was correlated to *L. maculans* lifestyle transition from a biotrophic to necrotrophic stage. This observation provides an additional clue in deciphering the arsenal of virulence mechanisms employed by *L. maculans*, one of the most notoriously adaptive disease-causing pathogens of the *Brassica* family. In Clubroot, small secreted proteins (SSPbPs) have been identified that were assumed to play critical functions in primary and secondary infections, leading to hypertrophic tissue development (Chen et al. 2019). For bacterial pathogens, such as *X. campestris* and *P. carotovorum*, a variety of mechanisms are deployed to evade host resistance including the release of extracellular enzymes such as cellulase, mannanase, pectinase, protease, polygalacturonases (PGs) and pectate lyase (Type II secretion system), the injection of effector proteins (Type III secretion system), as well as the production of exopolysaccharides and biofilm formation. These virulence mechanisms can be manipulated by various techniques such as genome editing or developing cultivars that can undermine such mechanisms (Büttner and Bonas 2010; Charkowski et al. 2011). Through, the in silico analysis of comparative genomic and transcriptomic data of *S. sclerotiorum*, 80 putative secondary metabolite gene clusters implicated in virulence in *B. napus* were identified in subtelomeric regions close to transposable elements, with the upregulation of 12 polyketidesynthases (PKSs) and enzymes during *S. sclerotiorum* infection of *B. napus*, revealing clues about the virulence pathway in the *B. napus* host. Enzymes associated with secondary metabolites production in *S. sclerotiorum* to suppress host defense mechanisms, such as PKS, non-ribosomal peptidesynthase (NRPS), and chalcone synthase (CHS), were upregulated in an RNA-seq experiment studying *B. napus*–*S. sclerotiorum* disease resistance (Graham-Taylor et al. 2020).

1.4.9.3 Secretomics of Host–Pathogen Interaction

The fungal plant pathogen secretes a whole suite of proteins, collectively known as the secretome, during its interaction with the host. The secretome comprises effector proteins and specific enzymes crucial for host colonization, and the composition of each of these secreted proteins may vary between pathogen types based on the pathogen's mode of nutrition and lifestyle. The identification of these secreted proteins is key to understanding the pathogenicity process of the fungal plant pathogen in the host plant. The availability of rich omics resources and advanced bioinformatics pipelines for diverse fungal plant pathogen species have enabled the quick prediction of effector proteins across kingdom-wide fungal species with different lifestyles and have accelerated the cloning and functional characterization of candidate effectors (Van de Wouw and Idnurm 2019).

An understanding of the structural features of effector proteins, the diversity of the effector genes, and how these genes play a role in the pathogenicity and the evolutionary patterns of the genes are important for uncovering the complexity of the resistance mechanism in *Brassica*–pathogen interactions to support breeding-resistant *Brassica* varieties. Due to sequence diversity of effector for avoiding recognition by the host immune system, the specific function and mechanism of effector in inducing pathogenicity in the host are difficult to determine. However, the majority of effector genes can be predicted or identified more accurately and effectively based on the known characteristics of cloned effectors using improved computational, bioinformatics software combined with machine learning. Examples include Effector P 2.0, Signal P, and Apoplast P. Bioinformatics tools specific for the identification of transposable elements (TEs) have also been developed. Besides proteinaceous effector molecules, nonproteinaceous effectors in fungal pathogens, such as secondary metabolites, small noncoding RNAs, and their biological roles in pathogenicity, have also been studied in plant–fungus interactions (Depotter and Doehlemann 2020; Carreón-Anguiano et al. 2020).

The conventional secretion pathway for the proteins during plant–pathogen interaction involves the endoplasmic reticulum–Golgi pathway. However, increasing evidence has shown that some secretomes of plant or fungal/oomycete proteins are secreted independently of the classical pathway during plant–pathogen interactions. Hence, it is important that proteins lacking signal peptides within the fungal/oomycete secretome are not overlooked when identifying candidate effectors (Ruano and Scheuring 2020; Vincent et al. 2020). Some signaling molecules produced by phytopathogenic fungal species play a part in virulence resemble homologous signaling molecules in the host, acting as mimics to evade the plant immune system for successful disease development. For instance, oxylipins, which are important signaling molecules commonly found in animals, plants, and fungi, play a role in growth, development, and the defense response, with one of the examples being jasmonate. In *Brassica*, oxylipins were found to display fungicidal activity against *A. brassicae*, *L. maculans*, *S. sclerotiorum,* and *Verticillium longisporum*. In phytopathogenic fungi, oxylipins have been found to be involved in disease progression through the modification of the plant host defense mechanisms, an example being *F. oxysporum* hijacking the oxylipin JA signaling pathway in *A. thaliana* (Deboever et al. 2020; Genva et al. 2019; Patkar and Naqvi 2017). The gene expression profile from the *P. brassicae* Pb3 genome assembly revealed that the pathogen contains genes that are associated with the biosynthesis of the plant hormones cytokinin and auxin, suggesting a potential role for these hormones in virulence activity in the host plant, while gene clusters for the synthesis of the ABA hormone were detected in *L. maculans*, suggesting a putative role of ABA production in disease progression in *B. napus*. One of the well-characterized effectors for *P. brassicae* is the benzoic acid (BA)/SA methyltransferase protein (PbBSMT), which suppresses host SA signaling during plant defense. The functional role of PbBSMT is similar to that of the SABATH methyltransferase gene family in *A. thaliana*, AtBSMT1, where the genes play a role in converting SA into methyl salicylate (MeSA), which is the inactive form of SA, thereby compromising the SAR defense response. A transcriptomic study of *P. brassicae* infection in Kohlrabi (*B. oleracea* var. *gongylodes*) showed that PbBSMT was one of the highest-expressed pathogen genes in the root gall tissue, playing a role in the local reduction of SA via PbBMST-mediated methylation. It was found from another cloning experiment with AtBSMT1 vs. PbBSMT in *P. brassicae*-infected *Arabidopsis* (host) and *A. candida*-infected *Arabidopsis* (non-host), comparing the level of SA inactivation in *Arabidopsis*, that PbBSMT resulted in higher levels of SA inactivation, meaning PbBSMT suppressed the host and non-host SAR defense mechanisms at a greater level. Multiomics approaches combining genomics, transcriptomics, proteomics, and metabolomics using computational strategies will allow us to identify suitable mimicking molecules in the fungal and/or host species that trigger stronger plant defense systems during plant–pathogen interactions (Spallek et al. 2018; Darma et al. 2019; Ciaghi et al. 2019).

Beneficial bacterial endophytes found in the apoplast region of *B. napus* have been shown to inhibit the growth of *X. campestris*, *S. sclerotiorum,* and *L. maculans*, thus acting as natural biological control against *B. napus* diseases. The high-through put sequencing of the fungal endophytes obtained from healthy roots of tumorous stem mustard (*B. juncea* var. *tumida*) and *P. brassicae*-infected roots of the same plant species showed a more diverse composition of the fungal endophytes in the healthy roots compared to in the diseased roots. This suggests a strong interaction network in the fungal endophyte community that contributes toward the optimum health of the host plant. A combination of secretomic and proteomic analysis of the apoplast fluids will allow us to identify and characterize the diverse apoplast proteins and further elucidate their role in protecting *B. napus* from pathogen invasion (Romero et al. 2018; Tian et al. 2019).

1.4.9.4 Interactomics of Host–Pathogen Interaction

Interactomics is the study of networks of gene and protein interactions in biological systems. Understanding the biological process and pathogenicity mechanisms of the fungal pathogens in *Brassica* crops is crucial for the identification of disease resistance targets. A web-based database called the Pathogen–Host Interactions database (PHI-base) has been set up that stores curated experimental data obtained from host–pathogen studies, encompassing phenotypic data and biological data on pathogenicity, virulence, and effector gene functions from fungal, oomycete, and bacterial pathogens from animal, plant, fungal, and insect host species, with embedded search links including BLAST, PubMed, UniProt Knowledge base, and others. Complementing PHI-base, PHI-Nets provides information related to networks of molecular and biological protein–protein interactions for the understanding of pathogenicity and virulence mechanisms in host–pathogen relationships (Ahmed et al. 2018; Janowska-Sejda et al. 2019).

1.4.9.5 Metabolomics of Crucifers' Host–Pathosystem

Metabolomics in plant pathology refers to the study of host plant metabolism changes in response to pathogen infection that provides an understanding of how host–pathogen interaction, through a (de)activation of metabolites and associated signaling pathways, could lead toward a resistant or susceptible outcome for the host. Metabolites associated with black rot infection in *B. oleracea* were identified using liquid chromatography-quadrupole time-of-flight (LC-QTOF)-based metabolite profiling, which revealed that metabolic changes in the host occurred 48 h after infection and implicated photosynthesis, alkaloids, coumarins, and sphingolipids as involved during the infection process. Systems biology was constructed to model the metabolic pathway for JA-signaling in the *Brassica–Alternaria* patho-physiology to identify important elements in the regulation of resistance mechanisms and to pin point molecular targets for engineering enhanced resistance in *Brassica* crops (Castro-Moretti et al. 2020; Tortosa et al. 2018).

The huge number of data collected from multiomics technologies will be useful in the construction of network biology, where mathematical models and computational approaches are implemented to predict the pathogenicity and virulence mechanisms in plant–pathogen interactions. Metabolic pathways supported by mathematical modeling can be used to study how cells within a multicellular organism work cooperatively to carry out a particular function. An example of systems biology was carried out for *A. thaliana–S. sclerotiorum* pathosystem, where a genome-scale metabolic model of *S. sclerotiorum* based on global gene expression was constructed to assess the metabolic activity in different parts of the hyphal cells, supporting the hypothesis that cooperation in *S. sclerotiorum* hyphal cells is necessary for virulence and host colonization (Peyraud et al. 2019).

1.5 Genomics of Crucifers Biotic and Abiotic Stresses

The concept of stresses on plant is "unfavorable and environmental constraints in plants" or "all agents can act as stressors, producing both stress and specific action," and "there exist stressor-specific responses and nonspecific general responses." *Brassica* crops experience many different types of stresses. These are biotic stresses like fungal, bacterial, viral diseases, and insect pests, while abiotic stresses are salinity, extremes temperature (cold, heat), water shortage (drought or dehydration), water logging, etc. The plant can respond to abiotic stresses not only by fast acclimations but also by particular long-term adaptations such as leaf shape, size and thickness, stomata density and distribution, chloroplast structure and function by increasing the levels of photo protecting enzymes, and stress metabolites (Obidiegwu et al. 2015). These types of adaptation may take place within few days or in a week depending on plant nature. Biotic stress is the result of damage of plants by other living organisms, such as bacteria, viruses, fungi, parasites, insects, weeds, and cultivated or native plants. Abiotic stress is the lifeless factors on the living organisms in a specific environment. The lifeless variable must control the environment beyond its normal range of variation to unfavorably affect the population performance or individual physiology of the organism in a significant way. The biotic and abiotic stresses are very importance for crop production worldwide; it's significantly reducing the yield of crops. These stresses are major constraints for crop growth, which consequently hampers the productivity of crop plants. Crops yield are reduced about 25% worldwide due to diseases and insect infestation (Savary et al. 2012). Globally, abiotic stress is the key sources of crop loss, reducing more than 50% average yields for most major crop plants (Acquaah 2007). Particularly, drought and salinity are becoming devastating in many regions of the world, and may cause serious salinization of more than 50% of all arable lands by the year 2050 (Wang et al. 2003). However, the biotic stress is one of the important features that have significant effect on crop growth and development and ultimately responsible for huge losses of crop yield. Moreover, abiotic stress is another confronts inducing a strong pressure on crop production. Now-a-days, research has generally centralized on understanding plant responses to abiotic or biotic stresses (Qin et al. 2011; Todaka et al. 2012; Thakur and Sohal 2013; Stotz et al. 2014). In the field condition, plants have to face more than one stress simultaneously, and the response cannot be predicted based on the plant's response to the individual single stresses (Atkinson and Urwin 2012). Plants can demonstrate different levels of sensitivity depending on the natural condition and the developmental stage of the plant (Mittler and Blumwald 2010). Different types of interactions can take place between the defense mechanism after sensitivity of the stresses. It is confusing, whether biotic and abiotic stresses are rather antagonistic, synergistic, or additive, inducing more or less susceptibility to a specific kind of stress. In most of the times, exposed on different type of stresses can also lead to antagonistic responses in the plants (Yasuda et al. 2008; Ton et al. 2009), for examples, under drought stress condition, common beans show more symptoms when infected by *Macrophomina phaseolina* (Suleman et al. 2001) and applied

ABA on tomato leaves increases the susceptibility to *Botrytis cinerea* (Audenaert et al. 2002).

The *Brassica* is a very important vegetable group worldwide which comprises some 100 species, including rapeseed (*Brassica napus* L.), mustard (*Brassica juncea* L.), cabbage (*Brassica oleracea* L.), and turnip rape (*Brassica rapa* L.) that are mainly grown for oil, condiments, vegetables, or fodder (Ashraf and McNeilly 2004). These crops are sternly challenged by biotic stresses (fungal and bacterial pathogens, viral diseases) and abiotic stresses including drought, heat, cold, or salinity are the major factors that cause enormous losses of yield. Sixteen pathogens have been considered as of major consequences at global level (Table 1.1; Saharan et al. 2021b). *Fusarium oxysporum* is the plant pathogen that infects *Brassica* crops causing the wilt disease leading to great loss in quantity and quality of oilseed *Brassica* (Gaetan 2005). The blackleg or stem cankers causing pathogen *Leptosphaeria maculans* is the most devastating fungal diseases for oilseed *Brassica*, particularly canola (*B. napus* and *B. rapa*) (Howlett et al. 2001; Fitt et al. 2006). *Brassica* vegetables are highly susceptible to bacterial soft rot disease which is the most severe and destructive disease that causes serious damage and economic losses across the members of *Brassica*. Many abiotic stresses directly or indirectly affect the *Brassica* crops, due to the fact that they are mainly grown in arid and semiarid areas. However, its growth and yield have greatly decreased owing to abiotic stresses. Most of the *Brassica* species were grown normally at 15–22 °C temperature. The temperature above 35 °C and below 10 °C was injurious to reproductive organs at different developmental stages of *Brassica* crops (Angadi et al. 2000). The *Brassica rapa* was more sensitive to heat stress than *B. napus* and *B. juncea*. However, generally the *Brassica* spp. of tropical and subtropical origins is sensitive to chilling (0–15 °C) and freezing (<0 °C) stress and lacks this mechanism of cold acclimation (Sanghera et al. 2011).

Stresses could be managed in two ways in cruciferous crops: (1) Crop management; (2) Biotechnological approaches. Abiotic stress effects can be managed by agronomic practices such as irrigation by quality water, soil amendment, altering sowing time, etc. This strategy is not sustainable and economic for vegetable production (Farooq et al. 2009). Therefore, the permanent solution for crucifers' production against stresses needs to be undertaken. Plants have their own defense mechanism that regulate the expression of stress-related genes which will be necessary for the stress tolerance crop variety development (Dita et al. 2006). Exposure of plants to biotic and abiotic stress leads to the adjustment of stresses using their genetical mechanism and produced high yield during stress condition. On the other hand, when plants face up to adverse conditions, few genes are active against such stresses; researchers have used these active genes to develop stress tolerance crop varieties (Kayum et al. 2016b).

Table 1.1 Global ranking of crucifer's major pathogens/diseases based on geographical distribution (GD), host range (HR), yield losses (YL), and genomics research (GR) (updated 2020)

S. No.	Pathogen	Disease	GD in number of countries	Host range number	Average/ (range) yield loss (%)	GR no. of papers	Rank				Average (Rank)	References
							GD	HR	YL	GR		
1.	*Albugo candida*	White rust	100	300	34	122	6	79	4	33	30.5 (VI)	Saharan et al. (2014)
2.	*Alternaria brassicae*	Alternaria blight	96	4000	45.1	183	12	114	11	46	45.75 (II)	Saharan et al. (2016)
3.	*Colletotrichum higginsianum*	Anthracnose	100	100	50	9	2	2	1	4	2.25 (–)	Cannon et al. (2012)
4.	*Erysiphe cruciferarum*	Powdery mildew	44	100	27.3	134	18	31	5	80	33.5 (V)	Saharan et al. (2019)
5.	*Fusarium oxysporum*	Wilt	100	100	15.9	20	1	0	1	18	5.0 (–)	Matic et al. (2018); Lange et al. (2007)
6.	*Hyaloperonospora parasitica*	Downy mildew	102	99	37.3	15	3	4	1	7	3.75 (–)	Saharan et al. (2017)
7.	*Leptosphaeria maculans*	Blackleg	102	35	42	142	14	37	3	88	35.5 (IV)	Fitt et al. (2006)
8.	*Plasmodiophora brassicae*	Clubroot	144	3700	28.5	273	21	127	17	108	68.25 (I)	Saharan et al. (2021a, b, c)
9.	*Pseudocercosporella capsellae*	White leaf spot	29	22	15	4	1	3	0	0	1.0 (–)	Gunasinghe et al. (2020)
10.	*Pyrenopeziza brassicae*	Light leaf spot	100	33	22	9	5	2	1	1	2.25 (–)	Fitt et al. (1997)
11.	*Rhizoctonia solani*	Seedling rot/Collar rot	70	200	30	5	0	1	0	4	1.25 (–)	Verma (1996b)

12.	*Sclerotinia sclerotiorum*	Stem rot	102	600	40	166	22	70	3	71	41.5 (III)	Saharan and Mehta (2008)
13.	*Turnip mosaic virus*	TuMV	106	8		77	5	18	2	52	19.25 (VII)	Spence et al. (2007)
14.	*Verticillium longisporum*	Wilt	10	9		27	5	8	0	14	6.75 (–)	Tyvaert et al. (2014); Novakazi et al. (2015); Depotter et al. (2016)
15.	*Xanthomonas campestris pv. campestris*	Black rot	157	10		73	5	39	1	28	18.25 (VIII)	Gupta et al. (2013)
16.	*Heterodera*	Nematode	91			1				1		–
						3135[a]						

Figure in the bracket are the ranking of the disease
– Data not available
[a] Total publications

1.6 Major and Devastating Stresses of Crucifer's Crops

Globally, crucifer vegetable crops yield is reduced more than 50% due to abiotic stresses (Bray et al. 2000). Climatic changes will influence the severity of abiotic stress imposed on vegetables crops. The genera *Brassica* are an important and diverse vegetable crops grown worldwide, include a number of economically important crops such as *B. rapa*, *B. oleracea*, *B. napus*, etc. (Cardoza and Stewart 2004) and these crop plants are affected by biotic and abiotic stresses. The *Brassica* vegetables are soft and succulent and generally consist of more than 85% water. Therefore, water scarcity as well as excessive water significantly influences the yield and quality of these crops; abiotic stress particularly cold, salinity, and drought stresses dramatically reduce productivity of these crops. Different types of stresses are faced by *Brassica* vegetables during their growth cycle and try to overcome those stresses by adapting or producing different mechanisms (Table 1.2).

Cold is one of the devastating stresses for *Brassica* vegetable crops. Under cold stress condition, water in plant body become ice and rupture the plant cell, ultimately significant yield losses with great risks for future global food security. Cold stress is even more destructive stress when it imposed in seedling and reproductive stages. Drought stress causes an increase of solute concentration in the soil, leading to an osmotic flow of water out of plant cells. Extreme drought stress conditions will

Table 1.2 Mechanisms of stress responses by crucifers (Kayum et al. 2016b)

Stresses	Agent	Probable response
Biotic	Bacteria	Plant produce Collagen and tyloses which is responsible against Bacterial pathogens
	Fungi	Formation of ROS which is activated against fungi
Abiotic	Low temperature (Chilling injury)	Activate DREB1/CBF genes and produce detoxification of ROS and alter glyoxalate pathway that developed low temperature tolerance plants.
	High salinity	WRKY, NAC, MYB TFs activate salt resistant genes that alter glyoxalate pathway and waxy layer formation; finally produce salinity resistant plants
	Draught/water shortage	Stomatal closure reduces transpiration that manages the drought stress
	Heat	Efficient protein repair systems and general protein stability support
	Natural mineral deficiency (nitrogen shortage)	survival, temperature can lead to acclimation
	Water logging	Development of cavities mostly in the roots that facilitate the exchange of oxygen and ethylene between shoot and root (parenchyma)
Hormonal factors	SA	SA regulate the stomatal closure that is directly involved in drought stress
	JA and ABA	JA uses as stress signaling

ROS reactive oxygen species, *SA* salicylic acid, *JA* Jasmonic acid, *ABA* abscisic acid negatively impact on crop yield. Due to soft and succulent in nature of the cabbage family, majority numbers of the vegetable crops are sensitive to drought stress, mainly during flowering and seed development stages.

Salinity is also the similar trends of problem like drought for vegetables crops. It is projected that about 17% and 30% of cultivated lands and irrigated agricultural land of the worldwide are affected by high salinity, respectively (Vijayvargiya and Kumar 2011). In tropical areas, high evapo-transpiration results in substantial water loss due to hot and dry environments, thus leaving salt around the plant roots which interferes with the plants ability to uptake water. Under salt stress, plant is reflected in loss of turgor, growth reduction, wilting, leaf curling, leaf abscission, decreased photosynthesis, respiratory changes, loss of cellular integrity, tissue necrosis, and potentially death of the plant (Cheeseman 1988).

Most of the *Brassica* crops are sensitive to salinity stress during seedling and early growing stages. Therefore, it is well characterized that the response of plants to abiotic stresses depends on the plant growing stage and the duration and severity of the stress (Bray 2002). Like abiotic stresses, *Brassica* are also highly susceptible to several biotic stresses. *Fusarium* wilt, Clubroot, soft rot, and Black rot are the important diseases (Ahmed et al. 2012b; Kayum et al. 2015a). Two pathogens, *Pectobacterium carotovorum* subsp. *carotovorum* and *F. oxysporum* f. sp. *conglutinans*, are specifically attacked *Brassica* crops and cause soft rot and *Fusarium* wilt diseases. Bacterial soft rot and Fusarium wilt are the most severe and destructive disease, which causes huge damage and enormous losses across the members of *Brassica*. Control of this pathogen is difficult due to its wide range of host's selectivity and to survive on plant debris. These two diseases break out devastatingly when attack at seedling stage (Ahmed et al. 2012a). Many plant defense or defense-related genes contain W-box sequences in their promoter's region, including pathogenesis-related (PR) genes and the regulatory *NPR1* gene (Yu et al. 2001). The W-box sequences are necessary for the inducible expression of these defense genes.

1.7 Expression of Genes to Overcome the Abiotic Stresses

Plants have developed different defense mechanism through activation of complex signaling cascade in varying stress conditions (Abuqamar et al. 2009). Plants exposed to abiotic and biotic stresses, promote to activate protein kinase cascades and specific ion channels keep turn on, or producing reactive oxygen species (ROS), phytohormones such as salicylic acid, jasmonic acid, abscisic acid (ABA), etc. (Fujita et al. 2006). A simple model has been proposed in by Kayum et al. (2016a) (Fig. 1.5), where different mechanisms of responses to abiotic stresses by plants along with their corresponding are presented for better understanding.

The transcriptional regulatory networks of abiotic stress signals and gene expression have been studied (Chaves et al. 2003; Yamaguchi-Shinozaki and Shinozaki 2006; Perez-Clemente et al. 2013). The major findings include: The signal perception, signal transduction, transcription factors activation, activation of cis-acting

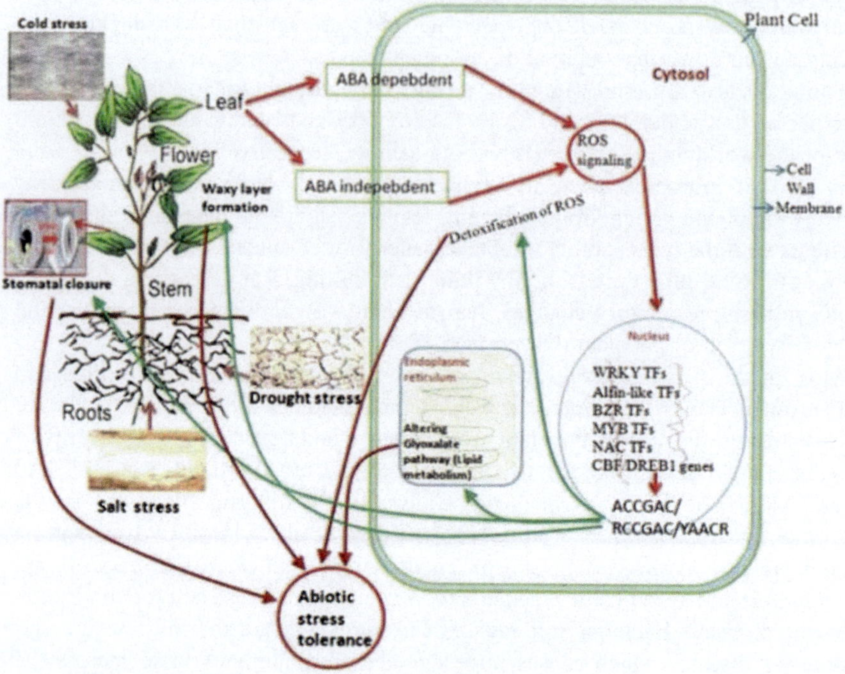

Fig. 1.5 Simple model of different signaling pathway involved in plants to overcome the abiotic stresses. *ABA* abscisic acid, *ROS* reactive oxygen species (Kayum et al. 2016a)

elements, and stress responsive gene expression. Plant cell sensors or receptors located in the cell wall or membrane perceive stress stimuli. Abiotic stress signal transduction is through (1) ABA-dependent and (2) ABA-independent pathways. In the ABA-dependent pathway, ABRE is the main ABA responsive elements that activate the stress responsive genes. On the other hand, in the ABA-independent pathway, dehydration responsive element (DRE) is involved in drought, cold, and salt stress responsive genes regulation (Fig. 1.5). These signals are received by cell membrane sensors that pass from plants when they face the abiotic stresses, accordingly cells try to response the signals through producing ROS, calcium ions, inositol phosphate, etc. inside the cells. However, high level of ROS in intracellular space may cause damage to the lipid, protein, and finally DNA. Moreover, phosphorylation mediated by protein kinase helps to activate or suppress different numbers of transcription factors (TFs) and bind specifically to cis-acting elements in the stress-responsive genes promoters and control their transcriptional level for combat against abiotic stresses (Danquah et al. 2014).

In the meantime, other upstream components control TFs at the transcription level (Hirayama and Shinozaki 2010) and subject to different levels modifications at the post-transcription stage, such as ubiquitination and sumoylation, thus forming a complex regulatory network that control the stress responsive genes expression level. These genes are involved in detoxification of ROS, altering glyoxalate pathway, stomatal closure, and waxy layer formation (Dong et al. 2006; Miura et al. 2007; Mizoi et al. 2013). Therefore, breeder should keep in consideration of the above mentioned components in the Fig. 1.5, from signal receptors to the downstream stress responsive genes that constitute the generic pathway for plant abiotic stress signal transduction (Fig. 1.5) for developing abiotic stress tolerance cultivars.

1.8 Differential Gene Expression in *Brassica* Under Abiotic Stresses

The abiotic stresses are becoming alarming situation for *Brassica* crop production worldwide due to quickly changes of global environment. The different genes are differentially expressed under such type of stresses in *B. rapa* and *B. oleracea*. WRKY and Alfin-like transcription factors have been well documented in response to many abiotic stresses in *B. rapa* (Kayum et al. 2015a, b). Among the *WRKY* genes, *BrWRKY22*, *BrWRKY44*, *BrWRKY70*, and *BrWRKY72* in Chiifu showed about 175-fold, 32-fold, 54-fold, and 42-fold higher expressions, respectively, during cold stress. Therefore, *BrWRKY22*, *BrWRKY44* genes could be introgressed or over-expressed in *B. rapa* vegetables for getting cold tolerant lines. Additionally, under salt stress condition, *BrWRKY 17*, *BrWRKY 57*, and *BrWRKY 58* showing about five-, two-, and three folds higher expression, respectively and *BrWRKY 51*, *BrWRKY 65*, *BrWRKY 98*, and *BrWRKY 104* showed their highest expression after drought stress treatment and were about 85-fold, 15-fold, 18-fold, and 38-fold higher than the control, respectively (Kayum et al. 2015a). On the other hand, *BrAL 2*, *BrAL 3*, *BrAL 7*, *BrAL 9*, *BrAL 12*, *BrAL 13*, *BrAL 14*, and *BrAL 15* showed higher expression in response to all abiotic stresses (cold, salt, and drought) (Kayum et al. 2015b). Lee et al. (2012) found that all eight CBF/DREB1 genes responses against cold stress and some of them also responsive to salt, drought, and ABA treatment. Several numbers of NAC TFs are involved in various abiotic stress of Chinese cabbage (Liu et al. 2014). Several numbers of BZR transcription factor and longevity assurance gene of *B. rapa* showed responsiveness against cold, drought, and ABA stresses (Ahmed et al. 2012b; Saha et al. 2015).

The Alfine-like transcription factors in *B. oleracea* (BoAL 1, 4, 5, 6, 7, 8, 9, 10, and 12) have been reported in response to abiotic stresses like cold, salinity, drought, and ABA (Kayum et al. 2016a), and Wang et al. (2015) showed that *BrMYB 210*, *BrMYB 137*, *BrMYB 88*, *BrMYB 154*, and *BrMYB 222* significantly responses against cold and osmotic stress and *BrMYB 261* (an ortholog of *AtMYB28*) had no response to cold treatment, but was expressed dramatically by osmotic stress. Among the MYB genes, six genes (*BrMYB 140*, *BrMYB 172*, *BrMYB 229*, *BrMYB 208*, *BrMYB 137*, and *BrMYB 210*) were significantly upregulated by ABA treatment but down-regulated by auxin treatment. However, *BrMYB 210* (an ortholog of *AtMYB96*) showed down-regulation against auxin treatment in leaves (Wang et al. 2015). Ahmed et al. (2015) showed that 12 BoCRGs were expressed differentially after cold stress treatment in two contrasting cabbage lines, and *BoCRG54*, *BoCRG*

56, *BoCRG* 59, *BoCRG* 62, *BoCRG* 70, *BoCRG* 72, and *BoCRG* 99 were predicted to be involved in cold regulatory pathways (Kayum et al. 2016a).

1.9 Major Pathogens of Crucifers at Global Level

Several pathogens of fungal, bacterial, viral, MLO's, and nematode origin are known to cause infection and damage to cruciferous crops at their different stages of crop growths. However, at global level, some of them can be considered of major consequences based on their geographical distribution, host range, losses caused, and resources spend by the scientists to investigate them in depth for planning strategies to manage (Table 1.1). Following pathogens have been exploited by the molecular biologists and molecular plant pathologists to investigate molecular mechanisms of host–pathogen interaction and genomics of crucifers' host-pathosystem using different *Brassica* host-pathosystem at different levels of transcriptomics analysis. Several genes have been identified, mapped on chromosomes of *Brassica* species along with functional characterization, isolation, and cloning for producing effective, durable disease resistance cultivars for better management of crucifers' diseases (Saharan et al. 2021a, b).

1.9.1 *Albugo* (White Rust)

White rust (WR; white colored pustules), white blister or white blister rust (release white powder or form galls) and stag heads (malformation of inflorescence) are the common names of the disease caused by *Albugo* spp. on more than 400 species of plants worldwide. It is known to cause infection on plants of genera belonging to families of Acanthaceae, Aizoaceae, Allionaceae, Amaranthaceae, Ambrosiaceae, Apiaceae, Asteraceae, Boraginaceae, Brassicaceae, Carduaceae, Capparidaceae, Caryophyllaceae, Chenopodiaceae, Cichoraceae, Cleomaceae, Compositae, Convolvulaceae, Crassulaceae, Cruciferae, Fabaceae, Fumariaceae, Gentianaceae, Nyctaginaceae, Orchidaceae, Papaveraceae, Peperomiaceae, Portulacaceae, Resedaceae, Scrophulariaceae, Solanaceae, and Urticaceae (Saharan and Verma 1992; Saharan et al. 2014).

1.9.1.1 The Causal Organism

On the basis of symptoms on the host plants, the disease is known as white rust (WR), white blister, stag heads, and white blister rust. Symptoms of the disease can be observed in the form of two categories. First, the general symptoms described in book chapters, reviews, monographs, research papers, etc. Second, specific symptoms and symptom variability recorded on different genotypes of *Brassica* crops sown on different dates, with different cultural conditions like, soil type, nutrition, variety, environment, and locations. Isolate-specific symptoms have also been observed (Kiran and Saharan 2002). Biga (1955) re-examined the morphology of the sporangia and constructed a key, in Italian, to differentiate species of genus

Albugo. Wilson (1907) described the North American species, Savulescu (1946a, b) the Romanian species, and Wakefield (1927) gave a historical note on the genus in her account of the species in South Africa. Morphological details of *Albugo* on crucifers were given by Togashi and Shibasaki (1934), and Togashi et al. (1930, 1931). Baker (1955) upheld the validity of *Albugo* against *Cystopus*. Waterhouse (1975) described the species of *Albugo* of the Aizoaceae in the United Kingdom. An account of Indian species was given by Butler and Bisby (1934) and Safeefulla (1952a, b). Burdyukova (1980) described the Russian species of *Albugo*. Choi and Priest (1995) have given a key to the genus *Albugo*. A key to *Albugo* species parasitic to Brassicales based on oospore characteristics has also been prepared by Choi et al. (2011a, b, c). A revision of genus *Albugo* has been presented by Thines and Spring (2005).

A detailed historical account of genus *Albugo* through research developments after its discovery as a pathogen has been nicely given by Thines and Spring (2005) in chronological order. The genus *Albugo* (Pers.) Roussel (1806) comprises ca. 50 species that are obligate biotrophic parasites of Dicotyledonae (Biga 1955; Choi and Priest 1995). Some species, like *Albugo occidentalis*, *A. portulacearum*, *A. candida*, and *A. tragopogonis*, have economic relevance, causing diseases known as "white rusts" (WR) in various crop plants. The genus *Albugo* had first been placed in the *Protomyceae* by Gray (1821). Later, de Bary (1863) transferred *Albugo* to the *Peronosporaceae* under the generic name *Cystopus* Lév., which had been introduced by Leveilla (1847) after the sexual form of *Albugo* had been discovered. Through the use of *Cystopus* by de Bary, this generic name became widespread and was adopted in the *Sylloge Fungorum* (Berlese and de Toni 1888 and proceeding volumes), and also by Wakefield (1927). Biga (1955) pointed out that the name *Albugo* was in consent with the International Code of Botanical Nomenclature, which allows the further use of *Albugo*, although the sexual form had been described under the generic name *Cystopus*.

In 1893, Schroter introduced the family *Albuginaceae* to accommodate the genus *Albugo*. The monogeneric *Albuginaceae* has traditionally been placed within the order *Peronosporales* (downy mildews) (Kirk et al. 2001; Dick 2001). However, molecular phylogenetic studies do not support this placement (Petersen and Rosendahl 2000; Cooke et al. 2000; Riethmüller et al. 2002), but placed *Albugo* basal to the *Pythiales*. Moreover, Hudspeth et al. (2003) even doubted a closer relationship between the WR (which were found to be basal to the *Rhipidiales* in their survey), and the DM, and thus suggested that *Albugo* should be raised to ordinal level. Wolf and Wolf (1947) probably noticed the distinctness of this group as they mentioned an order *Albuginales*, but provided no description and, in addition, did not include the order in their key to the orders of the Oomycetes.

Due to the seemingly morphological uniformity of *Albugo*, the phylogenetic studies have been mainly restricted to molecular phylogenetics, *A. candida* being the only species examined. Although characters like the structure of sporangia or oospore have been used to distinguish between species or species groups within *Albugo*, they have mostly been applied to construct identification keys (Migula 1910; Biga 1955; Choi and Priest 1995), and have not been critically evaluated as possible tools for separating monophyletic groups within *Albugo*. Other useful characters, like metabolic pathways, chemical characters or ultra-structural features, which have been brought into consideration (Hall 1996; Spring and Thines 2004), have not yet been taken into account. Since the studies of de Bary (1863), the presence or absence of an equatorial wall thickening of sporangia has been noticed and used to divide *Albugo* into the sections *Annulatae* and *Aequales* for species with or without equatorial wall thickening. The first section consisted of *A. tragopogonis* and *A. bliti*, the latter of the rest of *Albugo* species recognized at that time. These two sections have proven to be polyphyletic by the molecular phylogenetic analyses of Riethmüller et al. (2002).

Another feature, which has been used to distinguish between groups of *Albugo* species, is the presence or the absence of a distinctive primary or terminal sporangium. Such sporangia, differing from the other sporangia in size, type of germination, color, wall thickness, and wall structure, are reported from *Albugo* parasitic on Asterids and Caryophyllids, but it was claimed that they are lacking in *Albugo* parasitic on *Brassicaceae*, where all sporangia are reported to be similar (de Bary 1863; Berlese and de Toni 1888). However, this dimorphism is encountered in all species of *Albugo*. This genus contains at least three groups of species that deserve generic level. Taking into account the unique features of these organisms, the subclass Albuginomycetidae and the order *Albuginales* are introduced. The genus *Albugo* is amended, and two new genera, *Pustula* and *Wilsoniana*, are introduced for the WR of Asteridae and Caryophyllidae, respectively. Based on morphological, ultra-structural, and molecular grounds, eight new combinations are proposed: *Pustula chardiniae*, *P. hydrocotyles*, *P. tragopogonis*, *Wilsoniana achyrantis*, *W. bliti*, *W. platensis*, *W. amaranthi,* and *W. portulacae*.

Albuginaceae J. Schröt., Nat. Pflanzenfamilien 1(1): 110 (1893); ***Typus:*** *Albugo* (Pers.) Roussel, Flore du Calvados, 2e Ed.: 30 (1806).

Albugo (Pers.) Roussel, **emend.**

Species typica: *Albugo candida* (Pers.) Roussel, Flore du Calvados, 2e Ed.: 41 (1806).

Chromista, *Albuginales*. Hyphae intercellular, haustoria intracellular, small, mostly globose vesicles. Sori formed beneath the epidermis of the host plant, white to yellowish, first covered by the epidermis of the host plant. Mature sporangia dispersed through ruptured epidermis of the host plant. Sporangiophores colorless, agglomerated in large fascicles or sori, unbranched, clavate to cylindrical. Sporangia produced basipetally, in chains, at the top of the sporangiophore. Sporangia first formed similar to the others, in contrast to *Wilsoniana* and *Pustula*, sporangia colorless, globose to subglobose, wall of uniform thickness, or slightly thickened in the equatorial part, commonly forming a broad annulus, in contrast to *Pustula*, where the annulus is usually thicker and narrow; surface appearing verrucose in scanning electron microscopy, the warts forming lined to reticulate structures; on germination usually producing zoospores. Oogonia are globose to irregular; oospores globose, yellowish to brownish, on germination producing zoospores on the infected tissues.

1.9.1.2 Taxonomy and Morphology

The systematic position of the pathogen described by Ainsworth et al. (1973) is as: Kingdom: Mycota; Division: Eumycota; Subdivision: Mastigomycota; Class: Oomycetes; Order: Peronosporales; Family: Albugonacae; Genus: *Albugo*; Species: *candida*.

The pathogen is *Albugo candida* (Pers. Ex Lev.) Kuntze = (*Albugo cruciferarum* S.F. Gray). It is an obligate parasite. However, Thines and Spring (2005) while revising *Albugo*, has elevated it to Chromista with the class: Peronosporomycetes, and order: Albuginales. Historical and current phylogenetic taxonomy of *Albugo* causing WR is as follow: Traditionally, species in the *Albuginaceae* were thought to be host family specific, especially in the *Brassicaceae* and *Asteraceae*. This view is, however, asserted by molecular phylogenetic studies (Thines and Spring 2005; Voglmayr and Riethmüller 2006; Choi et al. 2006), which revealed that several distinct lineages are present on the same host family and that at least some species are possibly host genus specific.

Several species of *Albugo* parasitic to *Brassicaceae* have been described, but are mostly treated as synonyms of *A. candida*, among which the most widely known binominals are *U. cruciferarum* DC., *Cystopus sphaericus* Bonord., *Caeoma candidum* Schltdl., *A. wasabiae* Hara, *A. macrospora* (Togashi) S. Ito, *A. lepidii* N.A.S. Rao, *U. cheiranthi* Pers., and *U. thlaspi* Sowerby. Listing all binominals, including combinations and varieties would not serve practical purpose. This is due to the fact that *A. candida s.l.* is parasitic to a wide range of *Brassicaceae*, and that WR incidence is very common on some of the most widely distributed *Brassicaceae*. After listing a few synonyms, Fischer (1892) mentioned that there are numerous synonyms of this "*Vielbenannten Pilzes*" [fungus named many times]. Also in the brassicaceous hosts formerly assigned to the *Capparaceae*, two species of *Albugo* have been described. The type material of one of these, *A. chardonii* W. Weston, is indistinguishable from *A. candida* in molecular phylogenetic reconstructions, and therefore, needs also to be considered a synonym of or at least a species closely related to *A. candida s.str*. The slight wall thickening at the base and sides given in the key of Choi and Priest (1995) is a feature present in some sporangia of almost every collection of *A. candida* (Thines and Spring 2005). The other species described in *Capparaceae* is *A. capparis* (de Bary) Kuntze, which has been doubted to be a distinct species almost since its description. de Bary (1863), who formally described the species, which was given as a variety of *A. candida* by Rabenhorst (1844), noted this species to be "*in speciminibus C. candida omnino similia.*" Later also Pirotta (1884, as cited in Fischer 1892) was not able to distinguish between *A. candida* and *A. capparis*. This finding was also confirmed in Saccardo's Sylloge *Fungorum* (Berlese and de Toni 1888), who stated this species to be "*A C. candida non v. vix distinguendus.*" Biga (1955) included this species in his key, which is starting at the host family level, but the features given by him are identical to *A. candida* var. *macrospora*, comparing both sporangia and oospore features. Therefore, it is interesting that Choi and Priest (1995), who unfortunately did not mention any material they might have studied, cited the two references mentioned above (Biga 1955; Saccardo 1888) to give the oospore ornamentation as "tuberculate," a feature that would indicate that this species might be related to *A. lepidii*, and also not to be included in *A. candida s.str*. Specimens of *Albugo* from *Capparaceae* sequenced so far are either placed within *A. candida s.str*. or in a sister clade to *A. lepidii*, among other specimens from *Brassicaceae*. Whether this clade contains several distinct species, which could include *A. capparis* and *Uredo cheiranthi*, or not, has to be elucidated by thorough investigation of Rabenhorst's material. *Albugo* from *Capsella* in Korea is placed sister to the clade comprising *A. lepidii*, and several other distinct lineages, indicating that describing a new species for the pathogen on *Capsella* from Korea, which is distinct from the *Capsella* infecting *Albugo candida s.str*. is highly warranted (Choi et al. 2007).

1.9.1.3 The Disease and Symptomatology

All the above ground parts of the plant may show infection. In the initial stage of infection, localized isolated spiny white pustules or blisters up to 1–2 mm in diameter are formed on the aerial parts preferentially leaves and cotyledons (Plate 1.1a). The upper area of the leaf becomes tan yellow. Numerous pustules ultimately coalesce to form larger pustules (Plate 1.1e). Often pustules appear in a circular arrangement around one big central pustule (Plate 1.1c). Mature pustules rupture to release a mass of white sporangia (Plate 1.1b, d). In some cases, pustules may be surrounded by a yellow halo. On leaves, initial infection is on the under surface, but under humid conditions, both sides may produce white minute pustules. Pustules are also common on inflorescence and floral parts. In the young stem and inflorescence, the pathogen becomes systemic in the infected tissues and stimulates deformities with pronounced hypertrophy. The disease is sometimes called the "cancer" of rapeseed as one phase of this disease shows grotesque malformations of the young shoots and inflorescence. The leaves on the systemically infected plants may become thickened, fleshy, pallid, and distorted or rolled. When systemic infection takes place in seedling stage, the entire plant may remain dwarf with small leaves. Stem may show various types of swellings. The pathogen stimulates hypertrophy and hyperplasia consequently resulting in deformation of floral parts (Plate 1.1f, g). Later in the season, the infected flowering shoots show very dramatic changes. The axis becomes swollen, curved, twisted giving a zig-zag appearance. The flowers and young fruits may also become malformed. Finally, the flowering shoot may consist only the swollen axis and spiny pedicels of the flowers and fruits (Plate 1.1h). Such structures are reminiscent of the antlers of stags and are therefore referred to as "stag heads." These deformities are attributed to hormonal imbalance induced by the pathogen. There is also evidence of changes in chlorophyll content of diseased leaves. Stimulations of normal buds result into production of lateral shoots. The floral parts show irregular enlargement with distortion of the tissues due to hypertrophy and hyperplasia. The flower stalk may be enormously thickened. The flower organs become wholly or in part swollen, fleshy, green or violet in color, and persist instead of petals and stamens falling off early. The petals may look like sepals, and stamens become leaf-like or occasionally like carpels; ovules and pollen grains are usually atrophied resulting in the sterile condition of the flowers. The pistil also becomes malformed and larger in size and sometimes the pods transform into leaf

Plate 1.1 (**a**) Circular creamy color pustule on leaf, (**b**) white circular pustules on lower surface of leaf, (**c**) yellowish brown patches on upper surface of the leaf having white pustules, (**d**) close up of white pustule, (**e**) heavily infected leaf showing white pustules, (**f**) hypertrophies of inflorescence–stag head, (**g**) stag head formation showing pustules running in row, (**h**) Stag head on diseased plants without pathogen growth

like structures. Such malformed pods lack seed development. Persistent, systemic, and asymptomatic infection of *Albugo* has been detected in wild crucifer species (Jacobson et al. 1998). The mixed infection of WR and DM is conspicuous at floral stage of crop growth. On the malformed floral parts, WR produces pustules and DM produces a coating of fine white mass of sporangia (Plate 1.5g). At maturity, under high humidity, malformed inflorescence may be covered with saprophytes turning into dark brown to black color. Plants may remain stunted due to heavy infection (Saharan 1992a, b; Saharan and Verma 1992; Saharan and Mehta 2002). The pathogen *Albugo* (Pers.) Gray 1821; =*Uredo albugo* Persoon 1801; =*Elysibe sensu* Walker 1833; =*Cystopus Leveille* 1847; =*Cystopus* deBary 1863.

As a unique feature within the *Peronosporales*, sporangia in *Albugo* are formed basipetally in chains beneath the epidermis of the host plant, whereas in all other obligate biotrophic pathogens within the *Peronosporales*, single sporangia are formed almost simultaneously on the ultimate branch lets of sporangiophores, which emerge through the stomata of the host plant. Furthermore, the way of sexual reproduction, especially as described for the multinucleate oospore formation of *A. portulacae* and *A. bliti* (Stevens 1899, 1901a, b, c, 1904), points to the unrelatedness of the WR and the DM. The unique mode of oospore germination, producing zoospores without prior formation of a germ tube (de Bary 1863; Schroter 1893), also clearly separates *Albugo* from the rest of the *Peronosporales*.

1.9.1.4 Infection, Pathogenesis, and Disease Cycle

The survival of a pathogen between cropping seasons and its effective dispersal to healthy plants are crucial aspects of the plant disease cycle. If either of these is prevented, the disease will not occur. Most pathogens possess mechanisms to survive intercrop periods or periods of unfavorable environmental conditions. The spread of inoculum can be air borne, soil borne, water borne, seed or clone borne, or vector borne. Air borne inoculum can travel for great distances, even across oceans, while soil-borne inoculum is rarely spread any great distance. Many pathogens are dispersed by more than one mechanism. These fungi are able to form resting structures that enable them to survive long periods without a suitable host or when environmental conditions are unfavorable.

In all species of *Albugo*, oospores are the primary source of inoculum. In perennial hosts like horseradish, the mycelium persists in the crowns, and occasionally in the lateral roots. Secondary infection and spread of disease during the growing seasons is by means of sporangia, which are readily carried short distances by splashed water droplets or, to a certain degree, by air currents. Moisture on the host surface is essential both for sporangial germination and infection by zoospores. The most likely primary infection sites from germinating overwintered oospores in soil are the emerging cotyledons. The zoospores derived from germinating oospores are also capable of causing infection, but no evidence exists of infection arising directly by germ tubes produced from germinating oospores. The process of infection and pathogenesis in susceptible, moderately resistant and resistant *Brassica* genotypes has been determined. Molecular mechanism of host–pathogen interaction is also known (Saharan et al. 2014).

1.9.1.5 Epidemiology

Incidence and severity of white rust (WR) varies from year to year on different crops. Factors affecting initiation and development of the disease have not been studied in detail. On spinach, sporangial production is profuse at low temperatures but oospore production is very abundant at higher temperatures. On water spinach in Hong Kong, during the main growing season (May to October), temperatures of 24–29 °C with heavy dew formation is ideal for infection and disease development. On sunflower in Australia, WR development is greatest between 20 and 25 °C. It is reported that the optimum temperatures for the development of WR on potted radish plants were 16–18 °C, with a range of 12–21 °C; the disease did not develop below 6–7 °C or above 28–29 °C. Relative humidity (RH) of 60–80% was more favorable than a saturated environment, CO_2 between 70 and 80 mm, and Hg (9.6–11%) inhibited disease development. The fungus was highly sensitive to the effect of ultraviolet light; the optimum dose for inhibition of disease was 25–30 min. WR of horseradish is most favored at 15–20 °C. On *Amaranthus* species, ideal condition for the spread of WR around Delhi (India) occurs from October to the end of February. During this period, warm days with low humidity facilitate desiccation of sporangia while cool temperatures and heavy dew deposition at night provide optimum conditions for sporangial germination and infection (Saharan et al. 2014).

1.9.1.6 Pathogenic Variability

Physiologic specialization has long been known in *Albugo candida.* Initially two specialized groupings of *Albugo,* one attacking *Capsella, Lepidium,* and *Arabis,* and the other attacking *Brassica, Sinapis,* and *Diplotaxis were recognized.* However, there was hesitation to use the phrase "biological forms" for this fungus. Later, it was demonstrated that this fungus on *Capsella bursa-pastoris* should be considered a distinct form. Several biological forms in *Albugo tragopogonis have been observed.* The distinct races occur on *Ipomoea batatas* and on *Ipomoea* spp. in the USA and West Indies. The *Albugo on Ipomoeae-panduratae* is divided into two biologic species. Eight morphological forms within *A. candida were distinguished.* Later, ten varieties of *A. candida* based on host specialization and morphology were made. Three biologic forms of *A. candida,* the first on *Raphanus sativus,* the second on *Brassica juncea,* and the third on *Brassica rapa var. Chinensis were recognized.* Twenty races of *A. candida* in Britain were described. The sporangia of *Albugo* from *Brassica* and *Raphanus* were 20 × 18 μm in size, while those from *Cardamine, Capsella, Draba,* and *Arabis* measured 15.5 × 14.5 μm; these were classified as macrospora and microspora, respectively. At global level, more than 115 pathotypes of *A. candida* have been reported from eight countries of the world (Saharan 2010; Saharan et al. 2014).

1.9.1.7 Host Resistance

The resistance in host is governed by one, two, or more dominant genes (AC-7-1, AC-7-2, AC-7-3), additive genes with epistatic effects, and single recessive genes (WPr) along with a single gene (WPR 4) conferring broad spectrum resistance to reces, AC-2, 4, 7, and 9. The inheritance of virulence in *Albugo–Brassica* system is controlled by a single-dominant avirulence gene AC-2. Resistant genes effective against one or more reces of *A. candida* have been mapped and identified on the chromosome of *B. juncea*, *B. rapa*, *B. napus*, and *A. thaliana*. Slow white rusting, induced systemic resistance and plant defense resistant genes have been identified (Saharan et al. 2014).

1.9.1.8 Disease Management

The information on the management of white rust (WR) of cultivated crops have been generated on the aspects of fungicidal control, cultural control, host resistance, biological control, and their use in an integrated way. Early work on the chemical control of white rust (WR) diseases was focused on the use of copper based fungicides to control the leaf phase (LP) of the disease. It was reported that *Brassica* diseases can be controlled by frequent sprayings with copper fungicides. The use of Bordeaux mixture or perenox for the control of WR of *Brassicas* and other diseases was recommended. With the progress in the development of dithiocarbamate, control of WR was attained with multiple applications of protectant fungicides. However, these fungicides provided little protection from the stag head phase of the disease. Acylalanines, which are specifically active against Peronosporales, have now been proven useful to control both the leaf phase (LP) and stag head phase of white rust, either with a seed dressing or soil drenching, and with one or two foliar applications. Three sprays of polyram M at the rate of 0.2% at 15-day intervals were most effective in controlling WR of *Brassica rapa* var. Sarson in Pakistan. Many fungicides have been evaluated for efficacy against WR of *Brassica* crops in India. The fungicides such as Benlate 0.1%, Calixin 0.1%, Difolatan 0.2%, Dithane Z-78 0.2%, Miltox 0.3%, Thiovit 0.3%, mancozeb 0.2%, mancozeb + metalaxyl (Ridomil-MZ) 0.05%, and Ridomil-MZ 0.2% are effective in controlling both leaf (LP) and SP of the disease and increasing crop yield (Saharan and Verma 1992; Saharan et al. 2014).

1.9.2 *Alternaria* (Alternaria Blight)

The Alternaria is a ubiquitous fungal genus that includes saprobic, endophytic, and pathogenic species. It is associated with a wide variety of substrates including seeds, plants, agricultural products, animals, soil, and atmosphere. Species of *Alternaria* are known as serious plant pathogens, causing major losses on a wide range of crops. There are over 4000 *Alternaria*/host associations recorded in the USDA Fungal Host Index ranking the genus 10th among nearly 2000 fungal genera based on the total number of host records. While very few species appear to have a sexual stage (*Pleospora*) to their life cycle, the most lack sexuality altogether. Several species of *Alternaria* with their taxonomic and morphological characteristics, pathogenic nature, and synonymous along with distribution have been described extensively (Saharan et al. 2016).

1.9.2.1 The Disease and Causal Organism

Alternaria blight caused by four species of *Alternaria* is known by several names (synonyms) on the basis of color, shape, size, and location of lesions on different parts of cruciferous host species. Species pathogenic on cruciferous crops are *A. brassicae*, *A. brassicicola,* and *A. raphani*. Among the four species of *Alternaria*, *A. brassicae* is more common on oil-yielding crops, while other three species are major pathogens of vegetable crops. Phylogeny, taxonomy, morphology, classification, infection process, pathogenesis, disease cycle, and identification characteristics along with synonymous of all four species have been well documented. Identification, cloning, and sequencing of virulence genes of *Alternaria* infecting crucifers have resolved some doubts about their relationship with cruciferous hosts. Pathogenicity factor and transcription factor *Amr 1* have been identified in *A. brassicicola.* A nonribosomal peptide synthase gene (*AbNPs2*) is important for cell wall integrity, conidial viability, and virulence of aged spores of *A. brassicicola.* More than 100 genes have been functionally analyzed through various techniques like gene knockout and overexpressions making *A. brassicicola* the species of choice for functional genomics research. The pathogen survives through infected seed, crop residue, cruciferous hosts, weeds, microsclerotia, and chlamydospores and all acts as a source of primary inoculum. Secondary infection during the crop season is through conidia produced on infected host plants (under high humidity), which spread through rain splashes and wind to complete the disease cycle (Saharan et al. 2016).

General symptoms are similar on all plant hosts with minor variations influenced by environmental conditions, host species, genotypes, nutritional conditions, and pathotypes. On oil-yielding *Brassica* crops, all three species of *Alternaria* (*A. brassicae*, *A. brassicicola*, and *A. raphani*) are reported to cause symptoms at the seedling stage on cotyledons, and at the adult stage on leaves, leaf petiole, stem, inflorescence, siliquae, and seeds (Plate 1.2a–g). There may be variations in shape, size, color, and intensity of lesions on different host tissues and species under different agro-ecological zones. In general, lesions produced by *A. brassicae* are gray compared to black sooty velvety by *A. brassicicola.* All three species produce distinct lesions with yellow halos around them. The disease first appears on the cotyledons and hypocotyls in the form of small light brown lesions which soon turn black due to the appearance of spore masses (under humid conditions) which act as a source of infection for the other parts of the plant (Plate 1.2a). Damping-off of rapeseed seedlings due to *A. brassicicola* is common in Finland (Tahvonen 1979).

The initial infection by *A. brassicae* on the lower leaves/cotyledonary leaves starts as minute brown to blackish lesions (Plate 1.2a), which multiply rapidly and later spread to the upper leaves, stem, and siliquae. On the leaves, lesions may vary in size from 1 to 20 mm (Plate 1.2b). In some *Brassica* species, formation of concentric rings in the lesions, and a zone of yellow halo around (Plate 1.2b, c, d) the lesions is very prominent (Kadian and Saharan 1983). Several lesions on the leaves coalesce to cause blighting and defoliation under humid weather (Plate 1.2e). The lesions on the stem are at first linear and then expand but remain usually elongated with pointed ends (Plate 1.2f). In severe attacks, the upper part of the stem, and siliquae wither. Siliquae may show sunken, dark brown to black circular

Plate 1.2 (**a**) Circular brownish to black spots on cotyledons leaf, (**b**) brownish spots on leaves with yellow halo, (**c**) spots on leaf showing concentric rings, (**d**) spots may coalesce to form big spots, (**e**) several lesions on the leaves coalesce to cause blighting and defoliation, (**f**) pointed and elongated lesions on the stem, (**g**) on siliquae lesions are dark brown to black in color

lesions. Deep lesions on the siliquae cause infection in the seed. On siliquae of *B. rapa* var. Toria, Yellow and Brown Sarson, lesions are more prominent, enlarged, and dark brown to black in color than on *B. juncea* where they are light colored with distinct gray centers. The diseased seeds just beneath the black spot are small, shriveled, and gray to brown in color (Plate 1.2g). In years of severe outbreaks, infection of the stem may be sufficiently intense to cause the whole plant to wither before many of the pods have even been matured or formed (Kolte 1985; Saharan and Chand 1988; Vasudeva 1958). Detailed symptomatology of *Brassica* species infected with *A. brassicae* has been described by Conn et al. (1990) and Kadian and Saharan (1983).

In *A. brassicae* infections in rapeseed, the chloroplasts are more severely affected than the mitochondria. Changes in chloroplasts include swelling and eventual disintegration. The mitochondria, although, show swelling of intracristal spaces, but without disintegration. In severely affected cells, vesiculated plasma membranes with electron-dense deposits, intact grana from disintegrated chloroplasts, and mitochondria with swollen intracristal spaces are recognizable (Tewari 1991). On Brassicaceae, the lesions on leaves due to *A. brassicicola* are sooty black, velvety, and copiously covered with black conidiophores and conidia, whereas those caused by *A. brassicae* are gray, dense, and sparsely covered with brown conidiophores and conidia (Changsri 1961). The lesions caused by *A. raphani* are small with raised margins and are surrounded by a yellow translucent halo. Isolate-specific lesions have been recorded on *B. juncea* genotypes which vary in size, shape, color, number of concentric rings in each spot, and presence and absence of yellow halo around the spots (Kiran et al. 2004). Under field conditions on *B. juncea* leaves, symptoms of mixed infections of Alternaria blight, powdery mildew, and white rust can also be observed. *Alternaria alternata* causes root rot of rape (*R. rapa*) in Canada (Berkenkamp and Vaartonou 1972), and root and foot rot of *Eruca sativa* in India (Bhargava et al. 1980). The first evidence of the disease in the field is poor germination of seeds. The infected cotyledons first become chlorotic and then necrotic. The necrotic lesions extend on the stem up to 6–7 cm above the soil level. The root system of severely infected plants is completely destroyed (Bhargava et al. 1980). Pathotype-specific symptoms can be observed on *Brassica* genotypes. The disease is widely distributed all over the world wherever the congenial environmental conditions and host–pathogen interaction occur. Sometimes, under field conditions, symptoms of mixed infections of *Alternaria* blight, powdery mildew and white rust can be observed on *B. juncea* leaves. Its host range is very wide ranging from oil-yielding *Brassica* to vegetables and a large number of cruciferous weeds. The disease causes heavy yield losses in the form of quality and quantity of cruciferous crops (11–100%). The disease incidence and intensity can be accessed through various keys, scales and diagrams and using modern techniques including remote sensing, video image and PCR (Saharan et al. 2016).

1.9.2.2 Infection and Pathogenesis

Depending upon the growing season, *Alternaria* species in crucifers cause primary infection through the windborne spores produced from oversummered or overwintered primary inoculum in infected seeds, plant debris, collateral hosts, and weeds. Under congenial environmental conditions, spores (conidia) germinate readily in the presence of moisture by producing a germ tube, which penetrates the hosts directly or through stomata. Further, the process of infection and pathogenesis is governed by a number of external and internal factors of a specific host–pathogen interaction. Induction of cell wall-degrading enzymes, secretion of toxins, and enforcement of fungal cell wall architecture and detoxification of host defense molecules are essential functions for successful infection and pathogenesis of *Alternaria* species. Genes important for each of these functions have been identified in various fungal species. A transcription factor gene, *AbVF19*, has been discovered from *A. brassicicola*, which positively regulates hydrolytic enzyme-coding gene. Another transcription factor gene, *Amr1*, negatively regulates a subset of these genes during the late stage of pathogenesis and positively regulates melanin biosynthesis during conidiogenesis. It seems that the suppressive functions of *Amr 1* gene contribute to the specialized adaptation of *A. brassicicola* as efficient and successful facultative parasites of cruciferous plants (Saharan et al. 2016).

1.9.2.3 Epidemiology

Alternaria blight of crucifers occurs and develops in the epidemic form under the combination of a number of factors including availability of susceptible host cultivars, source of virulent pathogen inoculum in the vicinity, and favorable environmental and cultural conditions. Temperature range of 15–25 °C, relative humidity >90%, wind velocity of 2–5 km/h, and intermittent rains are the most abiotic conducive conditions for development of *Alternaria* blight under field conditions. In addition, closer spacing of 30 × 15 cm, high doses of nitrogen (>80 kg/ha), frequent irrigation and growing of susceptible cultivars further rapidly increase the disease severity in rapeseed–mustard. The disease progression under field conditions as influenced by environmental conditions and dates of sowing has been measured through the development of models. Disease forecasting models have been developed taking into account the leaf wetness period; number of rainy days; minimum, maximum and optimum temperatures and relative humidity; date of sowing; variety and species of *Brassica* crops grown under different agro-ecological conditions (Saharan et al. 2016).

1.9.2.4 Pathogenic Variability

Three categories of such populations are of direct interest to the Brassicalogists of the world. In *Alternaria–Brassica* host-pathosystem, the following categories of variability exist: although the genus *Alternaria* is known as an imperfect fungus, it shows genetic variability within a species, and this variability might be due to the existence of mutation, somatic hybridization, heterokaryosis, uniform host selection, extensive dispersal, and/or a cryptic sexual stage. Several determinant attributes have been assessed to observe variability in *Alternaria* species infecting crucifers. The variability in *Alternaria* has been expressed with respect to pathological, morphological, cultural, nutritional, symptomatological, biochemical, thermo-sensitivity, fungicidal sensitivity, genetical, molecular, and proteomic level by the isolates of *A. brassicae* collected from different sources (Saharan et al. 2016).

1.9.2.5 Fine Structures

Among the four *Alternaria* species infecting cruciferous hosts, information on fine structures through electron microscopy has been generated on *A. brassicicola*. The sequence of events in the spore production and maturation has been described. Changes in the internal organelles of hypha, conidium, and conidiophore have been observed. Conidiophores have a similar structure to mature hyphae, except that after spore production, they have pore in the tip and an annulus. There are variations in the number of nuclei in the cells of the vegetative hyphae, conidiophores, and conidia with six chromosomes in dividing nuclei of vegetative hyphae (Saharan et al. 2016).

1.9.2.6 Host Resistance

Attempts have been made to transfer resistance from wild Cruciferae to cultivated *Brassica*. While transferring the resistance into cultivated *Brassica*, the knowledge of pathogenic variability, sources of resistance and methods to determine nature, mechanism and stability of resistance is essential. Various techniques to incorporate the resistance in the cultivated *Brassica* as well as to evaluate the resistance in the field, in screen house and in the laboratory conditions have been developed. The pathogenic variability, especially in *A. brassicae*, has been identified and requires special attention. The mechanisms of resistance in rapeseed and mustard due to epicuticular wax, as well as for pre-existing rapid accumulation of biochemical components, have been studied. The role of phytoalexins in the resistance has been confirmed. The importance of phytotoxic compounds including destruxin B, destruxin B2, homodestruxin B, and desmethyldestruxin B in the pathogenesis of *A. brassicae* has been determined. The fact that the resistant *Brassica* varieties produce other metabolites including sesquiterpenes deoxyuvidin, β-albrassitriol, isoalbrassitriol, and brassicadiol has also been confirmed. Scanning electron microscopy (SEM) has also revealed the sequestration of calcium during pathogenesis by *A. brassicae* (Saharan 1992a, b, 1997; Saharan et al. 2003, 2005, 2016).

1.9.2.7 Disease Management

No single method or approach in current use is feasible, viable, stable, effective, and economical in dealing with any host–pathogen system. Therefore, it is necessary to integrate all the methods of plant disease control, which are available, including cultural, chemical, biological, and host resistance to effectively manage *Alternaria* diseases of brassicaceous plants (Eastburn 1989; Saharan 1992b; Verma and Saharan 1994; Saharan et al. 2016).

1.9.3 *Colletotrichum* (Anthracnose)

The *Colletotrichum higginsianum* causes anthracnose disease on a wide range of cruciferous plants, including agronomically important species, such as members of genera *Brassica* and *Raphanus* as well as the model plant *Arabidopsis thaliana* (Narusaka et al. 2004; O'Connell et al. 2004). The pathogen has long been considered as the causal agent on Chinese flowering cabbage (Scheffer 1950), Pak Choi (*B. chinensis* Kitam) (Mahmodi et al. 2013).

1.9.3.1 The Disease and Causal Organism

The hemibiotrophic ascomycete fungus *Colletotrichum higginsianum* Sacc. belongs to the family Glomerellaceae. It is parasitic or saprophytic on host plants by feeding initially on living cells as a biotroph before switching to a necrotrophic mode of growth. The fructification is subcuticular or sub-epidermal with or without setae. The stromatic tissues are restricted to the base of the acervulus forming conidia on the upper surface. Mycelium is immersed, branched, septate, and hyaline at first white pale brown or dark brown. Acervuli are formed on fruits, subcuticular or epidermal, rounded, or elongate, approximately 350 μm in diameter intra- and subepidermal with disrupting outer epidermal growth. Setae are abundant, brown, 1–5 septate, rigid, and hardly swollen at the base, slightly tapered toward the polar acute apex up to 250 μm and 5–8 μm wide. Conidia are falcate, fusiform, apicus, acute, narrow, truncated base, aseptate, hyaline, uninucleate, 16–30 × 2.5–4 μm from hyaline to faintly brown cylindrical with septate or aseptate phialidie conidiophores. Spores are in clusters in droplets of pale buff to salmon color. Appressoria are formed which are medium-brown, clavate to cylindrical, edge usually entire, 6–25 × 4–10 μm, often become complex and forming long closely branched chains. The perfect stage has been described under *Glomerella* in the Ascomycotina.

It causes anthracnose disease on many plants in the Brassicaceae family. *Colletotrichum* spp. cause anthracnose diseases that typically appear as small or large dark-colored spots or slightly sunken lesions on the foliage, stems, or fruit of a wide range of important crop and ornamental plants (Agrios 2005). The pathogen is responsible for typical sunken symptoms on leaves, stems and petioles of rapeseed-mustard. In natural environments, microorganisms normally produce toxic chemicals to compete with other organisms and are capable of tolerating toxicity of these natural chemicals (Meena et al. 2021).

1.9.3.2 Economic Importance

The *Colletotrichum higginsianum* was reported to cause typical anthracnose lesions on the leaves, petioles, and stems of turnip, mustard, and Chinese cabbage (Higgins 1917). In South China, this fungus usually causes typical water-soaked lesions on leaves of Chinese cabbage (*B. parachinensis*), leading to 30–40% yield loss yearly (Yang et al. 2008). To invade host tissue, conidia first attach to plant surfaces and germinate to form melanized appressoria. After that, *C. higginsianum* penetrates the plant cell with high turgor pressure generated in the melanized appressorium, and then large bulbous biotrophic hyphae form in the first infected cell. Finally, the fungus differentiates secondary hyphae to kill host tissues (Narusaka et al. 2004; Meena et al. 2021).

The pathogen uses a two-stage hemibiotrophic infection process, initially growing biotrophically inside a living host cell before switching to a destructive and invasive necrotrophic phase. The first step in the establishment of infection is the adhesion of spore to the surface of the plant. After connection to the plant surface, conidia germinate to form dome-shaped, melanized appressoria that pierce the cuticle and cell wall directly, by means of a narrow penetration peg. Swollen primary hyphae then invade living epidermal cells and invaginate the host plasma membrane, like the haustoria of obligate biotrophs. Both the biotrophic phase and the specialized primary hyphae of *C. higginsianum* are entirely restricted to the first infected epidermal cell, in contrast to other hemibiotrophic *Colletotrichum* species (O'Connell et al. 2004). Thereafter, the fungus switches to a necrotrophic mode of growth, associated with production of thin secondary hyphae which ramify within and between host cells and kill host cells ahead of infection. The asexual stage is completed by the production of acervuli on the surface of the dead host tissue. Acervuli consist of a mass of short conidiophores interspersed with long hair-like setae, which have darkly melanized cell walls like the appressoria. The *C. higginsianum–A. thaliana* pathosystem provides an attractive model for the molecular dissection of plant–fungal interactions.

1.9.3.3 Host Range

The disease has been reported on a wide range of cruciferous plants, including agronomically important species, such as members of genera *Brassica* and *Raphanus* as well as the model plant *Arabidopsis thaliana* (Narusaka et al. 2004; O'Connell et al. 2004), on Arugula (*Eruca sativa*) in Florida (Patel et al. 2014), on *Boehmeria nivea* (Wang et al. 2011) and *Rumex acetosa* (Zhang et al. 2018a, b) in China. *Colletotrichum higginsianum*, with straight conidia, has long been considered as the causal agent of anthracnose of crucifers including Chinese flowering cabbage (Scheffer 1950; Horie et al. 1988; Zhang et al. 1998, 2000; Zhou et al. 2002; Damm et al. 2014). Mahmodi et al. (2013) reported that *C. capsici* also could cause anthracnose on Pak Choi (*B. chinensis* Kitam), another member of Cruciferae. The *Arabidopsis–Colletotrichum* pathosystem provides an integrated system, with extensive information on the host plant and availability of genomes for both partners, to illustrate many of the important concepts governing fungal–plant interactions, and to serve as an excellent starting point for broad perspectives into issues in plant pathology (Yan et al. 2018). On the crucifer host, a set of specialized infection structures are generated by the fungus, consisting of germ tubes, appressoria, biotrophic hyphae, and necrotrophic hyphae (O'Connell et al. 2012). Functions of a CDC_{25} ortholog results implies that $ChCDC_{25}$ is involved in pathogenicity by regulation of cAMP signaling pathways in *C. higginsianum* (Yan et al. 2020).

1.9.3.4 Disease Management

The disease management practices for anthracnose is not available in respect to the crucifers' crops. However, the common fungicides used for the management of anthracnose on other crops can be used for this crop also. RCH1 a locus identified from *Arabidopsis* confers resistance to *C. higginsianum* (Narusaka et al. 2004).

1.9.4 *Erysiphe* (Powdery Mildew)

Powdery mildews (Erysiphales) are a very large group of biotrophic pathogens which grow principally on the aerial parts of angiosperms including crucifers, and cause qualitative, and quantitative damage on a wide variety of crops. Throughout, the more than 250 years since Linnaeus first gave a scientific name to a powdery mildew, they have figured prominently in the history of plant pathology.

1.9.4.1 The Disease and Causal Organism

The white, floury superficial mycelium, conidiophores, conidia, and pinhead dark brown chasmothecia (cleistothecia) of powdery mildews make them among the most easily recognized and best known plant pathogens. The taxonomy, and identification of powdery mildews are based on the characteristics of teleomorph (cleistothecial appendages, number of asci per cleistothecium), and anamorphs morphology of conidial stages on host including host range data, etc. With the use of scanning electron microscopy (SEM), light microscopy (LM), host range, and phylogenetic analysis, identification keys have been constructed for different powdery mildews genera, and species infecting different crops including crucifers. The major powdery mildew of crucifers is caused by *E. cruciferarum* under field conditions all over the world. However, on *Arabidopsis* under artificial inoculation conditions, nonadapted species of powdery mildew (not reported under natural conditions) like *Golovinomyces* (syn. *Erysiphe*) *cichoracearum*, *G. orontii,* and *Oidium neolycopersici* have been used for molecular and genetical studies. Powdery mildews can take epidemic form under field conditions within a very short period of time if congenial weather conditions prevail after landing of primary inoculum on the host surface (Saharan et al. 2019).

On the basis of morphological, pathological, host range, anatomical, phylogenetical, developmental features, characteristics of the powdery mildew fungus at anamorph, and teleomorph states of the causal organism of crucifer's powdery mildew has been identified as *Erysiphe cruciferarum* Opiz. ex L. Junell. Under artificial inoculation conditions, three other powdery mildew fungi viz., *Golovinomyces cichoracearum*, *G. orontii,* and *Oidium neolycopersici* are able to complete their anamorph state on a crucifers weed *Arabidopsis thaliana*. *Arabidopsis*–powdery mildew host-pathosystem has been used as a model system for molecular and genetical studies all over the world. The four powdery mildew species pathogenic on crucifers have been distinguished on the basis of conidial size, shape, appearance, number of conidia per conidiophore, conidiophores shape, and size, haustoria, hyphal branch angles, conidial germination with respect to temperature, relative humidity, light, and substrate, size of chasmothecia, asci, ascospores, number of ascospores per chasmothecium, and type of appendages. Phylogenetics relationship of crucifer's powdery mildew with other powdery mildew species has been determined through sequence analysis of the ITS rDNA, and sequence comparison of DNA encoding the 5.8s, rRNA, and ITS2. The powdery mildew pathogen genomes, and transcriptomes have been determined. Powdery mildew fungi have sizeable genomes, which are about four times larger than those of other ascomycetes.

The number of coding genes in the powdery mildew genomes is comparatively low. The structure of powdery mildew family, Erysiphaceae, the position of genus *Erysiphe*, its sections, and subsections *Golovinomyces*, key to the recognized genera, and species of the family, and relationship within the family have been exploited (Saharan et al. 2019).

The powdery mildew of crucifers shows its symptoms on aerial parts of host plants in the form of white to dirty white circular floury patches on leaves, stems, inflorescence, and siliquae/pods. These floury patches increase in size, and coalesce to cover entire plant with the increase in atmospheric temperature. It appears in the form of dirty white circular, floury patches on leaves, stems, and pods (Plate 1.3a–f). With the increase in temperature, white floury patches increase in size, and coalesce to cover entire plant. Severely diseased plants appear as if they are dusted with white, granular, chalk like powdery mass (Plate 1.3a–c). Such plants are poor in growth, and produce less siliquae. Stems are covered with mildew growth which appears purplish or bleached brown below the fungal growth (Plate 1.3d, e). The young green siliquae show white to dirty white circular mildew growth in the initial stage of infection (Plate 1.3f). Later, such siliquae are completely covered with white mass of mycelium and conidia. Severely diseased siliquae remain small in size, and produce small sized shriveled few seeds. At maturity stage of the crop, infected leaves, stem, and siliquae show minute, circular black or dark brown spherical, scattered, or concentrated bodies of perithecia, cleistothecia, or chasmothecia (Plate 1.3g; Saharan and Kaushik 1981; Saharan 1998; Saharan et al. 2005). Rutabagas (*B. napus* ssp. *rapifera*) plants infected with powdery mildew disease typically exhibit premature senescence of foliage. In addition, this disease predisposes rutabagas to other foliage and insect infections. In severe cases, the plant canopy is reduced drastically (Shattuck and Parry 1990). Under artificial inoculation conditions, symptoms development on *B. napus* crop is greatly influenced by the growth stages of host.

Mixed infection of Alternaria blight, and powdery mildew is very common on *B. juncea* leaves in the form of brown spots with concentric rings covered with dirty white floury patches specially when temperature is between 20 and 25 °C. Similarly, mixed infection of white rust and powdery mildew is also common on *B. juncea* leaves showing creamy to whitish pustules covered with dirty white floury patches of powdery mildew. Association of these (all three) diseases may be visible depending upon weather conditions prevailing in an area coinciding with infection and development of pathogens.

Under field conditions, severity of powdery mildew of mustard is visible in the form of white powdery mildew growth of conidia and conidiophores on leaves, stems, siliquae, and whole plants with deformed leaves. Similar symptoms appear on siliquae of *Brassica rapa* ssp. Yellow Sarson. Severely diseased leaves exhibit chlorotic or necrotic brown lesions, and leaf distortion followed by senescence. Severely diseased siliquae remain small in size, and produce small-sized shriveled few seeds. The disease is distributed in more than 25 countries of the world infecting more than 120 crucifers host plants including vegetables, and oil yielding economically important crops. Powdery mildew of crucifers causes 10–90% loss in yield of

Plate 1.3 (**a**) White to dirty white circular floury patches on leaves, (**b**) white floury patches increase in size, and coalesce to cover entire leaf, (**c**) close up of white floury patches showing fungal mycelial growth and conidia, (**d**) white floury patches on the stem, (**e**) white floury patches increase in size, and cover entire stem, (**f**) white floury patches on the silique, (**g**) severely infected plant

rapeseed–mustard crops with reduction in oil quality, and quantity up to 7% (Saharan et al. 2019).

1.9.4.2 Infection, Pathogenesis, and Disease Cycle

Powdery mildew fungi of crucifers are ectoparasites on their respective host plant species. Primary infection takes place through ascospores or conidia after landing on the host surface through germ-tube which develops an appressorium at the tip, a thickened infection structure. After cell wall penetration, specialized hyphae feeding structure called haustorium is formed which gets necessary nutrients from the host tissues for the growth, and development of the pathogen to produce mycelium, conidiophores, and conidia to continue its asexual state. Secondary spread of the disease is through air borne conidia causing infection, and pathogenesis under congenial environmental conditions. At the maturity stage of the crop, sexual state of the pathogen is visible in the form of small, pinhead, dark brown spherical chasmothecia which over summer or over winter to serve as a source of primary inoculum for the next season. Chasmothecia imbibe to burst, and release asci, and ascospores which germinate after landing on host surface, and cause primary infection. The pathogenesis of powdery mildew fungi is achieved through the suppression of the hosts preformed, and inducible penetration resistance mechanisms as well as maintenance of the compatible interaction after host cell penetration. After host cell penetration, the haustorial complex is established, which consists of the fungal haustorium, an extra haustorial modified plant plasma membrane, and the haustorial matrix. The haustorial complex is the site of the most intimate contact of the pathogen with its host cell, and functions in nutrient uptake, and provides a platform for the delivery of pathogens effector genes for pathogenesis. The expression of hundreds of host genes are altered within 8 hpi indicating that signaling between the pathogen, and individual host cells already occurs a few hours post contact of the powdery mildew conidia or ascospores with the leaf surface of about the stage of penetration, but certainly prior to establishment of the extra haustorial matrix. For the successful pathogenesis of crucifer's powdery mildew, there is great role of genetic factors (genes), MLO proteins, and host transcriptional changes. Powdery mildew of crucifers is activated by SA signaling pathways at the early stage of infection, and by the JA/ET signaling pathways at the later stage of pathogenesis. Infection is highly activated by chitinase in the genotypes. Adjustment of the plant host metabolism to support the growth of the powdery mildew pathogen is consistent with an increased ploidy level of the mesophyll cells underlying infected epidermal cells. Powdery mildew disease cycle operates at both asexual and sexual stages depending on its host plant species infected under prevailing environmental conditions (Saharan et al. 2019).

1.9.4.3 Fine Structures and Electron Microscopy

The formation of fine structures of powdery mildew pathogens during infection, and pathogenesis after artificial inoculation of *Arabidopsis* have been revealed through light, and scanning electron microscopy (SEM). The sequence of events has been observed from time of inoculation to production of asexual states, and symptoms production on the host leaf, and stem. The comparative observations on the morphological characteristics of different structures formed by three pathogens viz., *E. cruciferarum*, *E. orontii*, and *E. cichoracearum* of *Arabidopsis* have been distinguished as a compatible host–parasite interaction system for molecular, and genetical studies. The observations have been recorded on pathogens hypha, its branches; angles, mycelium growth; conidiophore shape, size; number of conidia formed; conidial shape, size, germ tubes number, shape, its tips shape; haustoria, its shape, number of infection pegs, their shape; appressoria, their shape, and formation of papillae. For each event, time period has been recorded including incubation period of powdery mildew pathogens on *Arabidopsis*. The fine structures revealed through SEM such as septa, haustorial sac, extra haustorial membrane, vacuolar membrane of the host cell, host cytoplasm, granular structures in the extra haustorial matrix, space between the haustorial sac, and the haustorium, and haustorial lobes are very crucial to understand the mechanisms of pathogenesis by powdery mildews of crucifers in susceptible and resistant hosts. The observation of encapsulation of haustoria, formation of papilla like structures, deposition of granular material adjacent to haustoria, and necrosis of host cells are potential resistant reactions of the invaded host plant. Now a day, SEM observations have become important, and essential criteria for all taxon of powdery mildews especially when only anamorph states are produced like in powdery mildews of *Arabidopsis* (Saharan et al. 2019).

1.9.4.4 Epidemiology and Forecasting

Epidemiology of crucifier's powdery mildews is influenced by numerous factors of host, and pathogen origin under the prevailing environmental, and physical factors of soil, nutrition, and cultivation practices. After landing of inoculum in the form of pathogens conidia/ascospores, their germination, and formation of appressoria to cause infection is influenced by environmental factors like temperature (minimum, maximum, averages), relative humidity (morning, evening, averages), sunshine hours, and wind velocity along with critical host growth stages. Host factors like age of the leaf, and stem under the influence of ambient temperature determines the infection, and spread of disease under field conditions. Interaction of pathogen with host at right growth stage coinciding with congenial environmental conditions favors disease development at faster rate. Stem infection is maximum, if they are exposed to the inoculum for longer duration especially at maturity stage of the crop. Infection on younger plants and newly emerging leaves is rare. Disease development is at faster rate if host–pathogen interaction coincides with favorable crop growth stages, and crucial environmental variables. Late planting is associated with high disease intensity. Host resistance curbs the area under disease progress curve (AUDPC), less disease intensity, longer incubation, and latent periods, low infection rate, less number of powdery mildew colony with less number of conidiophores, and conidial production. Disease forecasting system has been developed using historical and current data on maximum severity of disease, crop age at first appearance of the disease, and crop age at peak severity of disease for two susceptible cultivars sown at different location. Weather data have been analyzed to develop disease forecasting models (Saharan et al. 2019).

1.9.4.5 Host Resistance

Host resistance in crucifers to powdery mildews is multilayered, and multi-components both at pre- and postpenetration stages. The intricate immune responses are evolved through accumulation of ROS, H_2O_2, deposition of callose, pectin, cellulose, waxes, silicon, ion fluxes, formation of papilla, cell wall apposition, phenolic compounds, over expression of *R* genes, PR proteins, protein phosphorylation, biosynthesis of phytoalexins, fungal enzymes inhibiters, chito-octamers, triggering of HR induction of SAR, and nonhost resistance mechanisms. These cytoskeleton components have very important and crucial functional, and structural roles in host resistance to powdery mildew pathogens of crucifers. Defenses are activated either through SA signaling, and simultaneous perception of ethylene, and jasmonic acid (JA). The overexpression of several *R* genes in crucifers' powdery mildew host-pathosystem induces host resistance. MLO genes encoding seven-trans-membrane, calmodulin-binding protein confers broad spectrum resistance to adapted powdery mildews of *Arabidopsis.edr* mutants of *Arabidopsis* have a general link between SA mediated resistance, mitochondrial function, and programmed cell death. Powdery mildews mutants confer resistance to powdery mildew through altered cell wall composition of host. Increased SA enhances the expression of RPW 8.1, and RPW 8.2 leading to HR, or SHL, and resistance. Bj NPR1 gene activates SAR to confer broad spectrum resistance to powdery mildew of *B. juncea*. At ROP regulated At RLCK V1 A3 has a role in basal resistance to powdery mildews. The At MLO2, At MLO6, and At MLO12 triple mutants are resistance to powdery mildews, and induce PCD in response to infection by *E. cruciferarum*. There is a role of WRKY transcription factors, and over expression of *R* genes like PMR, MLO, PEN, EDR, MAPK, MAPK 65-3, MPR1, PAD3, PAD4, ED5, SNARE, RLCKs, and KDL (At CEP1) to confer R to powdery mildews of crucifers. Higher levels of camalexin contribute to the enhanced R to powdery mildew in Cyp83 a1-3 mutants of *Arabidopsis.* SR1 plays a critical role in powdery mildew resistance by regulating EIN3, and NDR1 expression. There is harmonious condition between transcriptional regulation, and resistance to powdery mildews. The application of *Trichoderma harzianum* and its CF induces (ISR) resistance in crucifers. Mechanisms of nonhost R in crucifers to powdery mildews have been unrevealed which is strong, and durable. Nonhost resistance is PEV-gene-mediated at preinvasion, and controlled by genes EDS1, PAD4, and SAG (101) at postinvasion of powdery mildew pathogens. In cabbage, R to powdery mildew is controlled by a single dominant gene with modifiers. A single R gene controls R to powdery mildew in HC-1, and PCC-2 with complete dominance. In *Arabidopsis*, R to PM is polygenic, and based on *R* gene *RPW8* or on combination of *RPW* 8 gene complex locus. Powdery mildew resistance genes of *Arabidopsis* have been mapped on chromosomes II (RPW1), III (RPW2, RPW3, RPW7, and RPW8), IV (RPW4), and V (RPW 5, RPW 6). In rapeseed gamma rays mutagenic plants exhibits R to powdery mildew due to an increase in concentration of unsaturated fatty acids with 18 carbon atoms. Induction of glucosinolates and camalexin plays important roles for resistance to powdery mildew of crucifers. Camalexin biosynthesis and accumulation are affected by WRKY 18, WRKY 40 transcription factor of *Arabidopsis*, and enhances upon *G. orontii* infection to confer resistance. Transfer of powdery mildew resistance to *B. oleracea* from *B. carinata* through embryo rescue has been confirmed. Major gene sources of resistance against powdery mildew of crucifers have been identified (Saharan et al. 2019).

1.9.4.6 Disease Management

Powdery mildew of crucifers is a compound interest disease (polycyclic) whose epidemic development within the growth cycle of the host is determined by the initial amount of inoculum or primary inoculum, and the rate at which the disease/ inoculum increases, i.e., the apparent infection rate. To reduce the severity of powdery mildew, the reproductive rate of the pathogen must be decreased. The major contributing factors which affect the reproduction rate of pathogen are reduction in the infection frequency or lesion, colony numbers, and size, longer latent period of the pathogen, and decreased spore production. These can be controlled by spraying of fungicides soon after the appearance of the disease symptoms, and repeating of second spray at an interval of 10 days. To escape the crop from powdery mildew infection, date of planting should be manipulated to avoid congenial weather conditions, and susceptible growth stage of the crop. Out of several chemicals tested, fungicides like sulfur (wettable sulfur), Karathane, Calixin, Dinocap, flusilazole, and carbendazim have shown their efficacy against powdery mildew with positive effect on total yield, and yield contributing attributes. Some biological control agents have shown promise to control powdery mildew when used as seed treatment and foliar sprays (*Allium sativum*, *Trichoderma harzianum,* and *Pseudomonas fluorescens*). Cultural practices like mixed cropping with pea, and wider spacing of crop reduces powdery mildew intensity under field conditions. Crop nutrition in the form of FYM is better than chemical fertilizers to manage powdery mildew of crucifers but higher yields are obtained with chemical fertilization. Integrated disease management using all means to control powdery mildew gives better results for high yield, and better quality of crucifers (Saharan et al. 2019).

1.9.5 *Fusarium* (Fusarium Wilt)

Fusarium species are well known pathogens of several crops but following three species are important on crucifers causing Fusarium wilt.

1.9.5.1 *Fusarium Oxysporum* f. sp. *Conglutinans*

The Disease and Causal Organism

The pathogen causing wilt of canola was first reported from Russia in 1996 and was identified as *Fusarium oxysporum* f. sp. *conglutinans*, race 1. Three years later, a new disease of canola called Fusarium wilt was reported by Lange et al. (2001) in Alberta, Canada. Fusarium wilt is a serious biosecurity threat to Australia's canola industry. This pathogen specifically attacks *Brassica* species and is morphologically indistinguishable from other *F. oxysporum* strains. Sporodochia and macro conidia

Plate 1.4 Plant showing infection of *Fusarium* sp.

are abundant on CLA (Carnation leaf agar). Macro conidia are short to medium in length and usually 3-septate. Micro conidia are usually 0-septate, oval, or reniform and are formed in false heads. Chlamydospores are formed abundantly on agar (Meena et al. 2021).

The pathogen causing canola wilt mainly affects canola and mustard. The infected plants may show wilting symptom and yellowish lesion may be seen at collar region of the plant (Plate 1.4). It is also known to cause "Fusarium Yellows" on a range of other crucifers in particular vegetable *Brassicas*. Fusarium wilt can reduce yield by up to 30%; however, in worst cases, complete yield loss can occur (Lange et al. 2007). The symptoms of the disease include poor seed set, premature senescence and premature shattering of siliques. Sometimes, symptoms may be present on one side of the stem, or affecting only one or a few branches, while other plant parts appear normal. Vascular discoloration of the main stem is the key diagnostic symptom.

Host Range and Distribution

Fusarium wilt of canola is reported from Argentina, Canada, India, and Russia. The pathogen causing Fusarium wilt (*Fusarium oxysporum* f. sp. *conglutinans* race 1), is known to widely occur in many countries. This pathogen has also been reported from Queensland, Australia in 1961 causing "Yellows disease" on cabbages (Meena et al. 2021).

Disease Management

Diagnosis is based upon fungal morphological characters on Special Minimal Medium (SNA). Control options are limited and use of resistant varieties is the best option. Crop rotation and good weed control is vital to minimize the inoculum levels.

1.9.5.2 *Fusarium Oxysporum*

The Disease and Causal Organism

Fusarium oxysporum is known as a wilt pathogen of many plant species worldwide. Examples include Fusarium wilt of oilseed mustard (*Brassica juncea*) in India, and wilt of *Brassica napus* rapeseed in Russia. The formae speciales that can attack canola is *conglutinans*, which includes all crucifer infecting isolates. Fusarium wilt was not identified in canola until recently, and was only reported for the first time in Alberta, and then in Saskatchewan and Manitoba of Western Canada in 1999 (Klassen et al. 2007; Meena et al. 2021).

The *Fusarium avenaceum* is less well documented as a *Brassica* pathogen. *F. roseum* has been implicated several times in Canada and France as a seedling blight and root rot pathogen of *Brassica napus* and *Brassica rapa. Fusarium roseum* is an obsolete epithet that has since been divided into several species, including *F. avenaceum. Fusarium avenaceum* probably over winters as mycelium in infested plant debris. Host range studies indicate that isolates that caused Fusarium wilt of canola can infect cereals, and may cause disease in those crops. *F. avenaceum* has long been known to cause root rots and seedling blights on canola. It also causes wilt-like symptoms of canola, although whether it infects the xylem vessels is currently unknown. It is not clear about the relationship between canola seedling blight and root rots, and Fusarium wilt caused by *F. avenaceum*.

Fusarium oxysporum usually winters as soil borne chlamydospores. Two other spore types (macro- and microconidia) are also formed. Spores germinate, and penetrate roots directly or through cracks formed by emerging lateral roots. The expanding mycelium then infects the vascular tissue, which becomes discolored and blocked. This blockage of the vascular tissue prevents water movement up the stem, resulting in turn in wilting of the plant. Spores are also formed on aerial plant parts, which fall to the ground and re-infest the soil. Occasionally, seed is infected via the vascular tissue, but usually the seed aborts before this can occur.

Disease Management

Avoid continuous cropping in the same field and follow field sanitation and crop rotation practices. To eradicate seed borne inoculum, seed treatment with Bavistin @ 2 g/kg or Thiram or Captan @ 3 g/kg seed is essential (Meena et al. 2021).

1.9.5.3 *Fusarium equiseti* (Corda) Sacc.

Fusarium equiseti (Corda) Sacc. is considered to be a weak pathogen on cereals and is sporadically established to be associated with Fusarium head blight-infected kernels (Xue et al. 2006) and on roots of *B. napus* and Indian mustard (Meena et al. 2019).

The Disease and Causal Organism

This species generally occurs in tropical and subtropical regions (Booth 1978; Bosch and Mirocha 1992), but it has also been recovered from cereals in temperate areas, including the Soviet Union, Europe, and North America (Kosiak et al. 2003; Stack et al. 1997; Tekauz et al. 2005; Wing et al. 1993; Xue et al. 2006; Meena et al. 2021).

Fusarium equiseti is a soil inhabitant and can transmit a disease to seeds, roots, tubers, and fruit of numerous crop plants. This new species is unique in having colonies on PDA at 25 ± 2 °C, fast growing, rosy pink to vanacous, reverse pell leuteus to light ochraceous. A dense white mycelium develops, that turns first beige, finally buff brown. Micromorphology shows that the hyphae are hyaline, smooth, pigmented and 2.75–4.77 μm wide. Single-spore pure culture produce chlamydospores (intercalary), fusoid, hyaline, and smooth walled. While, macroconidia are abundantly produced, fusoid, smooth walled, hyaline, foot cell present and sickle shaped, 2-septate with 14.59–45.48 × 2.88–4.10 μm in size. Though, microconidia are fusoid, 1-septate and 2.5–2.67 × 2.67–13.7 μm in size (Meena et al. 2019).

Host Range

It has been formerly concerned as a causal agent of disease on different plant species, such as cotton (*Gossypium hirsutum* L.; Chimbekujwo 2000), cowpea (*Vigna unguiculata* (L.) Walp.; Rodrigues and Menezes 2005), lentils (*Lens culinaris* Medik.; Chaudhary and Kaur 2002), sugar beet (*Beta vulgaris* L.; Stojsin et al. 2001), potato (*Solanum tuberosum* L.; Rai 1979; Theron and Holtz 1989) and pine (*Pinus* spp.; Ocamb and Juzwik 1995). Seed decay and reddish brown to black lesions were observed on hypocotyls and roots of kidney bean (*Phaseolus vulgaris*), bush bean (*Phaseolus lunatus*), broad bean (*Vicia faba*), chickpea (*Cicer arietinum*), and pea (*Pisum sativum*). A brownish discoloration and water-soaking symptoms developed on roots of alfalfa (*Medicago sativa*), canola (*Brassica napus*), wheat (*Triticum aestivum*), barley (*Hordeum vulgare*), and oat (*Avena sativa*) seedlings. Tomato (*Lycopersicon esculentum*), pepper (*Capsicum annum*), carrot (*Daucus carota*), and cucumber (*Cucumis sativus*) plants did not exhibit any visible symptoms (Goswami et al. 2008). The disease is distressing farmers to cultivate Indian mustard crop (Meena et al. 2019, 2021).

1.9.6 *Hyaloperonospora* (Downy Mildew)

Downy mildew as a disease of crucifers was reported for the first time on *Capsella bursa-pastoris* by Persoon (1796). The third largest group of downy mildews, which is mostly restricted to one plant family, is the brassicolous downy mildews (BDM), even though a few species of this group are parasitic to other plant families, such as Capparaceae, Resedaceae, Limnanthaceae, Cistaceae, and Zygophyllaceae. *Hyaloperonospora* is characterized by dichotomously branching and tree like sporangiophores, comparatively thin-walled oospores, and globose to lobate haustoria. Like *Perofascia*, this genus was segregated from *Peronospora* only after molecular phylogenetic analysis, a modern technique which could prove distinctiveness from *Peronospora*. The genus *Hyaloperonospora* is the third largest genus of downy mildews containing more than 100 species, which can infect economically important Brassicaceae crops. The downy mildew of *Arabidopsis thaliana* and *Hyaloperonospora arabidopsidis* has become a model organism to dissect plant-pathogen interactions. The downy mildew of crucifers is a widely prevalent and is a very destructive disease all over the world wherever cruciferous plant species, cultivated or wild, are found. The disease is very devastating causing yield losses in oil-yielding *Brassica* crops and cruciferous vegetables ranging from 50% to 100% depending upon the amount of pathogen inoculum present in the soil or near the vicinity of host, favorable environmental conditions present for infection, and development, cultural practices adopted, and disease management practices followed (Saharan et al. 2017).

1.9.6.1 The Disease and Causal Organism

The downy mildew disease is caused by the fungus *Hyaloperonospora parasitica* (Gaum.) Goker. The nomenclature of the disease is based on symptoms on the infected leaves with downy fungal growth. Downy mildew of crucifers is caused by an obligate pathogen, *Hyaloperonospora parasitica* (Gaum.) Goker synonymy *Peronospora parasitica* (Pers. ex. Fr.) Fr. The genus *Peronospora s. lat.* is one of the largest genera of downy mildews (Peronosporales, Peronosporomycetes, phylum Heterokonta). The phylogenetic analyses of DNA sequences have received increasing attention in Peronosporales, opening a new era in the investigation of their phylogenetic relationships. Both genera *Hyaloperonospora* and *Perofascia* almost exclusively infect hosts in the Brassicaceae and related families. The genera were segregated from *Peronospora* due to pronounced morphological and molecular differences. Researchers presented molecular and morphological evidence to split *Peronospora* into three separate genera, *Peronospora s.str.*, *Hyaloperonospora*, and *Perofascia*, and the latter two genera were found to be entirely restricted to a single host family, Brassicaceae (Saharan et al. 2017).

The disease appears on all above-ground plant parts but its symptoms are usually more conspicuous on leaves, stems, and inflorescences. The upper surface of affected young and older leaves has ill-defined, irregular, pale yellow necrotic lesions, whereas the lower surface is covered by white-gray mycelium. The cotyledons and older leaves may be killed prematurely when single lesions coalesce to form large blotches. At the seedling stage on cotyledons (Plate 1.5a), and the first few true leaves, small angular translucent light-green lesions appear. These lesions later enlarge, and develop into grayish-white, irregular necrotic patches on the upper surface of the leaf (Plate 1.5b) while downy fungal growth appear on the under surface (Plate 1.5c). In a severe attack, diseased leaves dry up, and shrivel. On cotyledons, necrotic lesions are more pronounced in *B. rapa*, whereas on true leaves, lesions are conspicuous on *B. juncea*.

Symptoms of mixed infection of downy mildew, and white rust are common on leaves, and inflorescence of *B. rapa*, and *B. juncea*. On leaves, downy growth of the fungus appears in or around the white rust pustules. On malformed inflorescences, sporulation of the downy mildew fungus is predominant in the form of white granular conidia, and conidiophores (Plate 1.5d) (Saharan 1992a). According to Butler (1918) owing to very frequent co-existence of white rust, and downy mildew,

Plate 1.5 (**a**) Downy mildew growth on cotyledon leaves of rapeseed-mustard; (**b**) yellowish flecks on the upper surface of the leaf of mustard; (**c**) downy mildew growth on the lower surface of mustard leaf; (**d**) initial growth of downy mildew on stag head; (**e**) inflorescence showing fresh conidial growth of pathogen, (**f**) stag head showing full conidial growth of *Hyaloperonospora*; (**g**) white conidial growth on stag head

it is not easy to separate their effects, but white rust produces the greatest deformities in the stem, and flowers (Awasthi et al. 1995, 1997). Stem swellings may be limited often with abrupt bending of the stalk, or swelling may be several inches long (Plate 1.5d). The axis of the inflorescence is equally susceptible to deformity. The leaves, and flowers are not often swollen, except for the young ovary, which may be transformed to a twisted body about 2 or 3 in. in length (Plate 1.5e, f). More often, the floral buds are atrophied with all the parts (i.e., sepals, petals, stamens, and pistil) being shrunken, and almost colorless. If the attack is late, the buds/silique may be partly normal, partly deformed or atrophied, and a single bud may similarly be affected in part only. Diseased pods may be visible with angular brown lesions, and under high humidity, a sparse white-grayish mycelium with numerous sporangia and sporangiophores may be visible. Severe attacks may lead to premature ripening and drying of the siliquae. Grayish white growth of conidia of pathogen is quite visible at the later stage of the crop growth on stag head (Plate 1.5g). There is never any trace of the violet color produced in downy mildew infections which often occur with white rust.

Systemically mixed infected plants with *H. parasitica,* and *A. candida* have stunted, and thickened growth of the whole plant which bears profuse sporulation of both pathogens. Hypertrophy of the affected cells, which is mainly attributed to infection with *A. candida*, causes thickening of the stem, and inflorescence. The hypertrophied tissues tend to attract infection by *H. parasitica* because their relative susceptibility to this pathogen is much higher than normal tissues (Awasthi et al. 1995, 1997). The pith of the stem has more hypertrophied tissue than the cortex. The affected inflorescence either bears no silique, or produces abnormal silique, which are often curled without seeds. In the initial stages, an affected inflorescence does not show typical symptoms of infection such as the presence of oospores, and downy growth on the surface. But at the later stages, conidial fungal growth of downy mildew, and sporangiophores blisters of white rust occurs on the surface of affected tissue, and formation of oospores takes place in the tissue as it dries. Necrotic lesions bearing downy growth of the fungus may also be observed on well-developed silique (Vasudeva 1958; Kolte 1985; Saharan 1992a; Awasthi et al. 1995, 1997; Saharan et al. 2017).

The internal changes due to infection by *Hyaloperonospora* differ from those caused by *Albugo* in many respects. With *Hyaloperonospora* infections, the palisade cells of the leaf are not changed. In the deeper layers of the cortex, endoderm, and pericycle, new cell layers may be formed by normal cell division. Only the cells around the vascular bundles become enlarged, and thin-walled; the rest of the inter-fascicular sclerenchyma remains unaltered. There is no inter-fascicular cambium in these host plants. The cambium of the vascular bundles remains active whereas the xylem, and phloem vessels become enlarged, and separated by radial bands of parenchyma when hyphae have penetrated. There are no accessory bundles. In general, the effect on the cells seems to be more destructive than in *Albugo*, the chlorophyll content is diminished, and the cell contents more rapidly used up. There is no tendency for chlorophyll accumulation in unusual places as with white rust. In general, the effect of downy mildew on the cell seems to be more destructive than white rust (Butler 1918). The protoplast of epidermal cells responds differently to haustorial development than protoplast of the mesophyll cells (Chou 1970). The epidermal cells result in a severe disruption of the protoplast. The central vacuoles contract, and probably undergo fragmentation, the plasmalemma is broken down or detached from the wall, and numerous vesicles are formed from it. The cytoplasm is either dislocated, and aggregated into a vacuolated blob or completely dispersed to the extent that its identity cannot be discerned.

Consequently, haustoria in epidermal cells are not, in most cases, surrounded by a clearly defined layer of host cytoplasm. Haustoria formation in a mesophyll cell cause less disruption. The host cytoplasm is merely invigilated by the invading haustorium, while the tonoplast and plasmalemma apparently remain intact. This pathogen, alone or in combination with *Albugo candida* (white rust), is responsible for causing severe losses in the yield of several temperate and tropical Brassicaceae crops, particularly rapeseed and mustard (Saharan et al. 2017).

1.9.6.2 Fine Structures and Electron Microscopy

Electron microscopy in association with physiological, biochemical, and genetical studies has provided information which helps in understanding the complex host-parasite relationship of this disease. With the use of light electron microscopy and transmission electron microscopy, fine structures of *Hyaloperonospora parasitica* have been observed during the process of host penetration, haustorium development, host–pathogen interface, conidia and conidiophore formation, and disease development along with host response, cytology, and genomics of the pathogen (Saharan et al. 2017).

1.9.6.3 Pathogenic Variability

Specificity in the downy mildew fungus on crucifers is very complex, since it occurs on a wide range of wild hosts as well as agricultural and horticultural species. However, there has been little sustained effort to introduce resistance to the disease; hence there has been less selection pressure exerted on the pathogen population than is the case with many other obligate parasites. Further impetus has been added by the exponential growth in research on the wild crucifer *Arabidopsis thaliana*, as a host for *Hyaloperonospora parasitica* and serving as a model system for genetic and molecular analysis. Discontinuities in the host range of isolates from different host genera and species suggest that the fungus may exist as a series of pathotypes adapted to each host of origin, although some cross infections may occur. There is also growing evidence that within host species, specificity may be determined by genotype-specific interactions consistent with a gene-for-gene recognition system. Specificity might therefore be expressed at several levels including family, genus, species, and cultivar or accession. In view of the close cytogenetic relationship between the major *Brassica* species, coupled with the strongly out breeding nature of several of these crops, some overlap in the host range of species-adapted isolates is perhaps predictable. Physiologic specialization/pathogenic variability have long been recorded in *H. parasitica* infecting cruciferous plants all over the world (Saharan et al. 2017).

1.9.6.4 Infection, Pathogenesis, and Disease Cycle

Oospores formed in malformed and senescence host tissues constitute an important means of survival of *H. parasitica* over periods under unfavorable environmental conditions. It is also known to survive through mycelium and conidia for a short period of time. Pathogen is also known to be active during over-summering and over-wintering under unfavorable environmental conditions in the diseased debris. Germinating oospores, intact oospores, mycelia, and conidia of *H. parasitica* were observed in the severely infected macerated host tissues. The maximum conidial discharge of *H. parasitica* from kohlrabi leaves is between 5 and 6 a.m. (258 conidia/cm^2). The conidial discharge decreases greatly from 12 noon to 8 p.m. Maximum conidial germination of *H. parasitica* from *B. juncea* was at 20 °C temperature. The germination of oospores of *H. parasitica* is dependent on temperature, light, pH of the medium, and age of oospores.

In crucifers, the downy mildew infection may be either general or local. In the former case, all or most of the leaves, and inflorescence, may bear conidiophores. Although some parts (especially the stem) may show no external injury, microscopic examination shows the mycelia are in the tissues. Generalized infection is restricted to young tissues, and seedlings may show completely infected leaves. Localized infection also occurs in young tissues, especially those still in active division. In the hypertrophy caused by *Albugo*, the cells of the epidermis and cortex are dividing and may readily give entrance to *Hyaloperonospora*. Young inflorescence may wholly or partly be infected, while normal tissues of older stems and leaves below the initial site of infection may remain free (Saharan et al. 2017).

Downy mildew of crucifers is a polycyclic disease under natural conditions. The pathogen requires certain specific nutrients which are produced or become available during pathogenesis to proliferate and continue in life cycle. The disease cycle involves the changes in the plant and the plant symptoms as well as those in the pathogen and spans periods within a growing season and from one growing season to the next. The main events in a disease cycle include inoculation (landing of pathogen propagules on the plant in the nature or otherwise), penetration, infection, growth and reproduction of the pathogen, dissemination of the pathogen, and overwintering or over summering of the pathogen (Saharan et al. 2017).

1.9.6.5 Epidemiology

In epidemics of downy mildews, the pathogen population starts from a low level of initial inoculum which then increases exponentially through successive cycles on the host during the growing season. The seasonal increase of the pathogen population has been investigated much more thoroughly than that of the initial inoculum. Information has been generated on the multiplication phase of the disease, which relates to the sequence of events in the life of the pathogen on its host, which are infection, colonization, and sporulation. The studies on forecasting of crucifers' downy mildew disease are limited to prediction models (Saharan et al. 2017).

1.9.6.6 Mixed Infection or Association of Downy Mildew and White Rust

The association of downy mildew (DM) and white rust (WR) infection on oilseed *Brassicas*, vegetable *Brassicas*, wall flowers, and stocks has been observed. The association or mixed infections, or DM-WR disease complex, or simultaneous occurrence of *A. candida* and *Hyaloperonospora parasitica* on leaves, stems, inflorescence, and siliquae of crucifers in nature is very common. *A. candida* appears first on leaves, which probably predisposes the host to infection by *H. parasitica*. The intensity of mixed infections varies from 0.5% to 35%, depending on local weather conditions. Severity of mixed infections on leaves is not related to infections on inflorescence. It seems that greater susceptibility of new inflorescence and their availability over extended periods of time are associated with this phenomenon. The hypertrophied and malformed inflorescences of *B. juncea* infected with *A. candida* are usually covered heavily with the white powdery growth of conidia and conidiophores. *Albugo candida* alone, on artificial inoculation of flower buds, induces typical hypertrophy of the inflorescence (Saharan et al. 2017).

1.9.6.7 Biometabolomics

Biochemical studies on the growth and survival of a pathogen and the changes it induces in its host can ultimately lead to a better understanding of the disease development, its epidemiology, and its control. Ideal prerequisites for meaningful studies of the biochemistry of the host-parasite interaction are (a) a clear understanding of the genetic control of virulence and avirulence in the parasite and of susceptibility and resistance in the host, (b) precise histological and cytological descriptions of spore germination and infection and the establishment and development of the host–pathogen interaction, and (c) the availability of methods for maintaining the pathogen alone and in combinations with its host, under controlled conditions. Unfortunately, these criteria have not yet been satisfactorily met for downy mildew of crucifers. The biosynthesis of the defense-related compounds like camalexin, salicylic acid, and lignin in *Arabidopsis* provides resistance to crucifers against downy mildew pathogen (Saharan et al. 2017).

1.9.6.8 Host Resistance

Genetic resistance is the most important attribute of the host defense against *H. parasitica*. Host resistance provides an economical, environmentally benign, and widely accepted method of managing downy mildew of crucifers. Present day's emphasis on organic agriculture relies mainly on large-scale deployment of disease-resistant crop cultivars in addition to the use of techniques such as cultural practices (sowing time, soil solarization, crop rotation, use of organic manures, plant population, clean cultivation, soil and plant moisture regulation) and biological disease control. *Brassica* host–pathogen interaction can be classified into two types, compatible and incompatible interaction, leading to the phenotypes of susceptibility and resistance to certain pathogens. Obviously, the incompatible interaction has been extensively exploited by the *Brassica* breeders to develop resistant cultivars for increasing crop production. Large-scale deployment of elite cultivars carrying an R gene (monoculture) imposes higher selection pressure on the pathogen carrying the cognate *Avr* gene to survive, resulting in the modification or deletion of the recognized *Avr* gene or generation of novel effector genes that can escape the recognition of the old "*R*" gene. This *R-Avr* interactive co-evolution explains why many resistant cultivars in field lose their resistance in a relatively short period of time. Hence, knowledge on the molecular mechanisms of *Brassica* host–pathogen interaction and use of horizontal, durable, and broad-spectrum resistance has become very important in disease resistance breeding programs. The interplay of host resistance and pathogen virulence in the latter system is likely to be quite different to the major gene-based system. Major gene resistance in crucifers to downy mildew has been exploited and used for breeding resistant cultivars (Saharan et al. 2017).

1.9.6.9 Disease Management

To manage downy mildew of crucifers, no single method or approach is considered feasible, effective, environmentally safe, and economical. It is always essential to integrate the available methods for disease control/management. Various practices like cultural practices, seed treatment, use of resistant varieties, biological control, soil treatment, and compost treatment with an application of fungicides at the time of disease appearance have been advocated. Downy mildew of crucifers can be managed effectively under field conditions using integrated disease management strategy (Saharan et al. 1997, 2017).

1.9.7 *Leptosphaeria* (Blackleg)

Blackleg disease caused by the ascomycete fungus, *Leptosphaeria maculans,* is an important disease on oilseed rape (canola, rapeseed, *Brassica napus*, *Brassica juncea*, *Brassica rapa*) causing seedling death, lodging or early senescence.

1.9.7.1 The Disease and Causal Organism

Phoma lingam (Tode. Fr.) Desmaz a anamorphic stage of *L. maculans* causes symptoms on a large number of plant species from the family Brassicaceae (Punithalingam and Holliday 1972). The pathogen causes spots on cotyledons, leaves and pods, canker on stems and root crowns and lesions on upper stems (Gabrielson 1983; Petrie 1979; Paul and Rawlinson 1992). During primary infection in the fall, the fungus releases ascospores from mature pseudothecia (Huang et al. 2003; Marcroft et al. 2003). Ascospore release lasts for 3–4 months, but it is most intensive in first 2 months (McGee 1977; Salam et al. 2003). In addition to ascospores, pycnospores too may cause disease symptoms (Gosende et al. 2003).

Initial symptoms of seed-borne infection are seen on cotyledons as distinct circular lesions with large numbers of pycnidia at growth stage (GS) 1.0–1.4 (Plate 1.6a) (Sylvester-Bradley and Makepeace 1985). The fungus is able to grow from cotyledon lesions (Bokor et al. 1975) and leaf lesions (Hammond et al. 1985; Hammond and Lewis 1986a) biotrophically in the lamella and petiole of the leaf to colonize the hypocotyl and stem (Hammond and Lewis 1987). Leaf lesions or phoma leaf spots may vary in appearance depending on host resistance,

Plate 1.6 (**a**) Initial symptoms on cotyledonary and first leaves, (**b**) lesion on the leaf, (**c**) lesion causing blight and producing pycnidia on the leaf. (Photos: Courtesy-Venkat R Chapara, USA)

L. maculans group and the stage of lesion development, but appear to develop similarly in Australia, Canada and Europe. Lesions caused by the A group of *L. maculans* first appear as pale green spots, which enlarge to 1–2 cm in diameter, often turning pale brown and containing tiny dark specks, pycnidia, that produce conidia (Plate 1.6b, c). Eventually the center of the lesion may break or fall out completely. Smaller, darker leaf spots with few or no pycnidia tend to be caused by the B group of *L. maculans* (Brun et al. 1997; Ansan-Melayah et al. 1997). Hypocotyl infection produces a constriction in the stem above the ground and below the first leaves. The black/brown blackleg lesions on hypocotyls can cause severe seedling blight and up to 70% of seedlings have been killed in individual crops (Barbetti and Khangura 1999). Even up to the six-leaf stage, blackleg lesions may completely sever the tender stem base of seedlings (Barbetti and Khangura 1999). Blackleg seems an appropriate term with which to describe these black/brown lesions, when they form at the base of green hypocotyls of seedlings or on green stems of older plants during flowering. Blackleg lesions at the stem base (crown or root collar) often have a distinct dark brown or purple margin, and are typically associated with leaf scars; such lesions originate from lesions on leaves early in the season (Hammond et al. 1985).

During the pod development and seed ripening stage, these lesions may spread or coalesce and crack open to form the dry rots or cankers (Paul and Rawlinson 1992) that generally form the damaging phase of epidemics. The terminology used to describe these stem base disease symptoms includes terms such as blackleg, crown canker, canker, collar rot and basal canker; these symptoms on the dry, woody stem base seem most appropriately called "crown canker" since they are no longer black. The severe epidemics is probably less severe cankers developed on new, more resistant cultivars such as Jet Neuf and Rafal (Gladders and Symonds 1995; Gladders et al. 1998), although Humpherson-Jones (1986) noted that large numbers of pseudothecia were still produced on their stubble residues. Severe crown canker epidemics are common in Western Europe but these stem base symptoms are rare in eastern Europe (Jedryczka et al. 1999). Both crown cankers and phoma stem lesions may girdle the stem, causing premature ripening of the pods due to disrupted water transport (Davies 1986), and in severe cases the stem is weakened enough to cause lodging and death of the plant. Late infections on the pods (siliques) are rare but phoma pod spots can cause premature ripening, splitting and decreased yield (Petrie and Vanterpool 1974). More importantly, pod infections can spread to the seed inside (Wood and Barbetti (1977b); Kharbanda and Stevens 1993).

1.9.7.2 Economic Importance

Blackleg has been reported in all canola-growing regions except China (Fitt et al. 2008; Zhang et al. 2014). China only has the closely related species, *L. biglobosa* brassicae, which although it can cause yield losses of up to 10–30%, but not as devastating as *L. maculans* (Liu et al. 2014; Cai et al. 2018). In Australia, Canada and the United Kingdom, annual yield losses associated with blackleg disease caused by *L. maculans* are 10–20% (West et al. 2001; Fitt et al. 2006; Van de Wouw et al. 2016; Zhang and Fernando 2018), but can be much higher in some regions in some years. Epidemics causing significant levels of disease occurred in regions of France in 2000 and in Lower Eyre Peninsula, Australia, in 2003 (Rouxel et al. 2003; Sprague et al. 2006a, b). Pedras et al. (1996) stated that annual crop losses due to *P. lingam* in Canada exceed 30 million dollars. In Great Britain, yield reductions with susceptible cultivars reach up to 50% in years when stem canker attack is intensive form (Gladders and Musa 1979). These epidemics in Australia have resulted in up to 90% yield losses (Sprague et al. 2006b).

1.9.7.3 Distribution

The *Leptosphaeria maculans* (Desm.) Ces. and de Not, the teleomorph of *Phoma lingam* (Tode. Fr.) Desmaz, is present for a long time in some European countries (Great Britain (UK), Germany, Denmark, France, etc.), Canada, USA and Australia. Stem canker is the major rapeseed disease on the global scale (Gosende et al. 2003; Howlett et al. 2001). Stem canker is an economically important disease of rapeseed in the main production regions of Australia, Canada and Europe (West et al. 2001; Salam et al. 2003).

1.9.7.4 Pathogen Morphology

Depending on isolate and location of origin, on PDA, the fungus forms a white to grayish-white mycelium with round, brown pycnidia (Pound 1947). Two types of pycnidia are formed: type 1 is brown scleroids with a narrowed down ostiole, and type 2 are round black pycnidia (Punithalingam and Holliday 1972).

On PDA medium, the mycelium develops slowly, forming round to wedge-shaped colonies. The aerial mycelium is sparse, loose, whitish to gray. The substrate mycelium is gray-whitish. The pathogen release dark brown pigment into the medium. After a week or so, the fungus forms round, dark brown to black pycnidia whose size is 185–384 μm × 220–510 μm. The pycnidia occur individually at first, forming stromata later on. Pycnospores are single-celled, hyaline, short cylindrical, mostly strait but some slightly bent, with oil drops on both end, their size range around 1.25–2.36 × 2.60–5.70 μm. On oat-agar medium, the colonies are round-shaped. After 15 days, they fill the entire Petri dish. The aerial mycelium is white at the beginning, somewhat more compact than that developing on PDA, turning light brown later. Pigment color in this medium is somewhat lighter. Pycnidia occur individually and are typically round and light brown. Pycnospores correspond in shape, color and size to those growing on PDA. However, the disease is rare in Scotland and in the large area of oilseed rape grown in Asia, mostly in India (as *B. juncea* and *B. rapa*) and China (as *B. napus*). Within Australia, Canada or Europe, the severity of epidemics differs greatly between seasons, between regions and between crops since where the disease occurs, total destruction of the crop due to seedling death is rare and usually yield losses at harvest are <10%, although that can reach 30–50% (Hall et al. 1993; Zhou et al. 1999; Barbetti and Khangura 1999).

1.9.7.5 Epidemiology

Both ascospores and conidia adhere to the cotyledons and leaves of new crops and germinate in humid or wet conditions to produce hyphae that cause infection. Infection is predominantly via stomatal pores (Hammond et al. 1985; Chen and Howlett 1996) but also via wounds. Although *L. maculans* produces a wide range of cell-wall degrading enzymes, no cutinase was detected by Annis and Goodwin (1996). Hall (1992) reported that ascospores germinated within 4 h at 4–28 °C in laboratory experiments, suggesting that infection is normally limited by wetness rather than by temperature. This was confirmed by Biddulph et al. (1999), who found that 4 h was the minimum wetness period required to produce leaf lesions by ascospores of both A and B group *L. maculans* and that most lesions were produced following a leaf wetness duration of 48 h at 12–20 °C. Compatible leaf infections by A group ascospores produced leaf lesions (phoma leaf spot) after 5 days at 20 °C and after 2 weeks at 8 °C; these controlled environment results were validated by observing progress of phoma leaf spot epidemics in relation to weather in UK winter oilseed rape crops in several seasons (Biddulph et al. 1999). However, the incubation period between infection and leaf lesion formation may differ between cultivars and between leaves at different positions (Poisson and Pérès 1999). Wood and Barbetti (1977b) reported that only one or two ascospores were needed to initiate a lesion in optimal temperature and wetness conditions. However, in controlled environment experiments, conidia were able to infect only wounded leaves, petioles and stem (Hammond 1985) and were unlikely to infect unwounded leaves unless applied at very high concentrations to older leaves (Vanniasingham and Gilligan 1989). Secondary infections caused by conidia are rare in Canada and Europe but more common in Australia (Alabouvette et al. 1974; Barbetti 1976; Wood and Barbetti (1977b); Gladders and Musa 1980; Thurwachter et al. 1999). In Western Australia, conidia can spread the disease up to 1 m from original foci (Barbetti 1976) in crops of very susceptible cultivars. This may reflect the greater aggressiveness of Australian isolates (Kucher et al. 1993) but secondary infections are not generally considered to affect yield (Hall 1992).

The relationship between climatic factors, such as temperature and rainfall, and differences between continents in the epidemiology and severity of phoma stem canker epidemics is complex, with different factors affecting survival of inoculum, maturation of pseudothecia, timing of ascospore release, infection conditions and host resistance. Climatic factors affect the persistence of pseudothecia on crop debris because the rate of degradation of debris is affected by wetness and soil temperature, with survival favored by dry summers or cold winters. Residues remain an inoculum source for up to 4 years in western Australia, due to the hot, dry summers (Barbetti and Khangura 1997), and for several years in western Canada, with its very cold winters and dry hot summers (Petrie 1986). In contrast, the mild, wet climate in the UK promotes rapid decomposition of debris, which generally degrades within 2 years (West et al. 1999). In south-east Australia, oilseed rape residues declined in volume by 90% within 1 year and produced fewer ascospores in the second year than in the first year after harvest. Pseudothecial maturation depends on both temperature and wetness, with an optimum at 14–15 °C (Petrie 1994; Poisson 1997; Peres et al. 1999).

Symptom severity on cotyledons, leaves and stems appears to be increased by high temperatures. On plants inoculated at the cotyledon growth stage, crown canker was more severe at temperatures above 12 °C than at lower temperatures (Barbetti 1975; McGee 1977). Cotyledons inoculated with A group isolates at 18 °C produced incompatible reactions but at 27 °C the reaction was compatible (Badawy et al. 1992). This supports the suggestion by Boudart (1982) that resistance genes of juvenile plants may be temperature sensitive, producing more disease at 24 °C than at 14 °C. On plants inoculated at the green bud stage, phoma stem canker symptoms were less severe at 12 °C than at 18 °C (McGee and Petrie 1979). The most severe epidemics, in Australia, are associated with a climate in which temperatures are typically 25–30 °C during the period when phoma stem canker lesions are developing. Severe epidemics can also occur in Canada and Eastern Europe, where crops experience high summer temperatures. Although such severe epidemics do not occur on oilseed rape in China, despite the high summer temperatures, this could be due to the scarcity or absence of the A group (West et al. 2000). Additionally, the severity of phoma stem canker epidemics decreases in the United Kingdom with distance northwards and the disease is rare on oilseed rape in Scotland, possibly because host resistance operates well at the low summer temperatures found there. However, higher temperatures may also increase phoma stem canker severity because water movement restrictions, caused by stem infection, could prevent increased transpiration rates, leading to water-stress. This would result in the plant senescing early and could explain why drought-induced water-stress, occurring around harvest, exacerbates the yield loss caused by crown canker in Western Australia (Barbetti and Khangura 1997). Water stress induced by application of a desiccant to facilitate harvesting of a crop can also cause *L. maculans* to spread rapidly in stems as they senesce.

Physical damage can increase infection; wind damage at the early seedling stage, sometimes combined with resultant sand-blasting, can increase susceptibility to infection in Western Australia (Barbetti and Khangura 1999) and damage caused by hail in Canada was thought to enhance disease symptoms. These climatic and population factors, which influence the severity of epidemics, need to be considered when devising the best strategy for managing phoma stem canker epidemics in Australia, Canada or Europe.

1.9.7.6 Infection, Pathogenesis, and Disease Cycle

The infection process begins with the release of sexual spores (ascospores) from stubble (crop debris) left from the previous season following a rainfall event. These ascospores land on the cotyledons or leaves of seedlings, germinate and then invade via wounds or the stomatal aperture. The fungus then colonizes the intercellular spaces, causing necrotic cotyledon and leaf lesions before entering a biotrophic phase whilst growing down the petiole and into the stem. The necrotic cotyledon and leaf lesions then produce asexual pycnidiospores, which are dispersed via rain-splash and can give rise to secondary infection events. At plant maturity, the fungus cause blackening of the stem and cankering thus restrict the nutrient flow up the plant, and in severe situations, completely killing the plant (Hammond et al. 1985; Hammond and Lewis 1987; Chen and Howlett 1996).

After harvest, the senescent stem tissues are rapidly colonized by *L. maculans* and pycnidia are produced abundantly. Additionally, conidia are able to colonize stubble saprophytically and may increase inoculum levels and subsequent numbers of pseudothecia. Pseudothecia develop in these residues and, when mature, ascospores are released over an extended period of time. Ascospores from infested residue were considered a risk to crops several kilometers away in Australia (Bokor et al. 1975), Canada (Petrie 1978) and Europe (Gladders and Musa 1980). However, a separation of only 2 km is now recommended in Western Australia, reflecting the current use of more resistant cultivars and the practical difficulties of avoiding proximity to stubble-based inoculum (Barbetti and Khangura 1999). The greatest risk of infection is within 500 m of the source with spore numbers decreasing with distance due to deposition and dilution, as indicated by experiments in Australia (Barbetti et al. 2000). Ascospores can remain viable for about 6 weeks (Paul and Rawlinson 1992) and it is probable that a small number of ascospores travel considerable distances. Although rare, seed-borne infection can be very important in spreading the disease to completely new area as *L. maculans* can be found in crucifer seed as dormant mycelium within the seed coat or even within the embryo (Jacobsen and Williams 1971).

In experiments in Europe, pathogenicity to oilseed rape stems differed between A and B group isolates (Johnson and Lewis 1994). On susceptible cultivars, A group isolates caused damaging brown, cortical lesions but B group isolates penetrated the leaf gap to enter the stem pith, rarely causing externally visible phoma stem cankers (Hammond and Lewis 1986b, 1987; Johnson and Lewis 1994). However, as stems senesce and turn brown, abundant pycnidia of both groups are produced on their surfaces. Furthermore, whereas the A group is generally associated with damaging crown cankers at the base of plants, the B group seems to be more prevalent on upper parts of the stems (Johnson and Lewis 1994; Thurwachter et al. 1999; Wang 1999). Nevertheless, severe phoma stem lesion epidemics have been associated with the B group in Poland (Jedryczka et al. 1999).

1.9.7.7 Disease Management

Differences in the epidemiology of phoma stem canker and economics of oilseed production between different regions have considerable implications for the management of the disease by cultural, chemical and other methods. The impact of blackleg disease is minimized through the combination of cultural, chemical and genetic control strategies (Fitt et al. 2006; Van de Wouw et al. 2016; Zhang and Fernando 2018). Cultural strategies to minimize blackleg include rotation of canola with other crop species and avoidance of *L. maculans*-infested stubble. Stubble management and good crop rotation (4 years breaks between oilseed rape crops are usually recommended) decrease the risks of infection by ascospores released from colonized residue. Different combinations of fungicide seed treatments, soil fungicides or foliar fungicide sprays are used for control of phoma stem canker in different regions, depending on the epidemiology of the disease and the economics of the crop. Fungicides are applied as fertilizer-amended seed dressing and/or by spraying (Van de Wouw et al. 2016).

Genetic resistance is a major tool in the fight against blackleg disease. Generally, quantitative resistance is assessed at maturity, as it is expressed in the stem, leading to the term adult plant resistance. Although little is known about quantitative resistance in the *L. maculans–Brassica* interaction, the combination of quantitative resistance and major-gene resistance can lead to more durable disease control in the field (Brun et al. 2010; Delourme et al. 2014). Qualitative or major-gene resistance is much better understood. It occurs in a gene-for-gene interaction whereby for each resistance gene in the *Brassica* host, there is a corresponding avirulence gene in *L. maculans* (Balesdent et al. 2005). These avirulence genes belong to a larger class of genes, known as effectors, which are a general class of genes involved in disease progression. The avirulence effector genes, although involved in the specific recognition of the corresponding resistance genes, presumably also play a general role in disease progression. Major-gene resistance is usually assessed at the seedling stage of development, using cotyledon assays; however, it has also been shown phenotypically to be expressed in true leaves, branches and pods (Balesdent et al. 2005; Marcroft et al. 2012; Elliott et al. 2016; Sprague et al. 2018).

1.9.8 *Plasmodiophora* (Clubroot)

The clubroot of crucifers was observed for the first time in 1736 on turnip. During the last 284 years, it has become a very serious devastating disease in more than 88 countries of the world on economically and nutritionally very important precious crops of oilseeds and vegetables like rapeseed (Canola), mustard, cauliflower, cabbage, Chinese cabbage, Brussels sprouts, Kale, Kohlrabi, radish, rutabaga, and turnip. It is causing very heavy losses in these crops ranging from 10% to 100% with an average loss of 10–15% at global level with decrease in 2–6% in oil content, and 50% increase in chlorophyll content in oil. Its host range is very wide with infection on more than 330 genera and 3700 species of crucifers besides several weeds, and nonhost plants where it survives in their rhizosphere. It is a consistent and potential threat to cruciferous crops covering 2.69% of total cropped area of the world with an area of 1384.8 million ha and 147.69 million metric tons production. Its causal organism was identified and established as *P. brassicae* Woronin in 1875 after 140 years of first report in 1736. Its biology, ecology, *Brassica* host–pathogen interaction, and management strategies were unveiled only in the last few decades. The pathogen is obligate, biotrophic, protists, nonaxenic, microscopic, single celled, intracellular, soil borne, highly variable in virulence, produces well protected, and robust resting spores, resilient to harsh soil environment, and can survive in soil for more than 20 years. It has two stages of infections after resting spores germination by primary zoospores as primary root hairs infection, and secondary zoospores as secondary cortical cells infection resulting in clubroots as characteristics symptoms of the disease. *Plasmodiophora brassicae* genome is compact, small in size (25 Mb), with high gene density, and have 13 chitin synthesis genes. Potential pathogenicity genes have been identified. In the soil environment, surviving resting spores germinate to release mobile biflagellate primary zoospores which exist in the soil moisture film for very short time to cause primary root hair infection following multiplication and causing secondary stage infection of root cortical cells through secondary zoospores to induce massive host cells hypertrophy by growth regulators production exhibiting clubroot symptoms. Such clubroots are full of resting spores (inoculum load) which acts as source of primary inoculum and liberate back into the soil to continue life cycle of the pathogen. The different components of pathogens life cycle viz., zoospores, sporangia, and sporangiospore develop, and grow well protected within the host cells. Clubroot has a very complex disease cycle, and secretive life style of *P. brassicae* where even the occurrence, and position of meiosis is still unclear. Fine structures by electron microscopy, rapid identification, and detection of *P. brassicae* from seed, soil, host tissues, and water, identification, isolation, cloning, and mapping of CR genes; detection of metabolic pathways, and modification of plant hormone homeostasis by the pathogen; identification of pathotypes during host–pathogen interaction using molecular approaches of omics, genomics, proteomics, and transcriptomics have been revealed. However, there are still several gaps to fill in for fully comprehension and effective management of clubroot of crucifers (Saharan et al. 2021c).

1.9.8.1 The Disease and Causal Organism

The disease is known by name clubroot due to its characteristics symptoms of infected hypertrophied root cortex cells leading to the formation and development of root galls. There is direct and strong relationship between disease incidences, degree of soil infestation, and crop yield loss. Host resistance helps to reduce yield losses. All cultivated *Brassica* vegetables and oil yielding crops including several cruciferous weeds are susceptible to this disease. The pathogen can survive in the root cortex of noncrucifer's plants also. Symptoms may vary depending on the growth stage of the crop when infection occurs. The characteristics symptoms of the disease appear on the roots as spindle like shredded, knobby or club shaped swellings. The clubroot has been assessed in the form of incidence and intensity using different rating scales, descriptive as well as pictorial to assess yield losses, host resistance, and efficacy of technology. The disease is usually monocyclic, and zoospore movement in the soil is limited. However, any activity that results in transport of infested soil from one point to another has the potential to disseminate clubroot (Saharan et al. 2021c).

The pathogen *Plasmodiophora brassicae* Woronin is a very serious, soil borne, intracellular, obligate, nonaxenic, endoparasites, wide spread, and destructive to cruciferous crops all over the world causing clubroot disease which is nutrient sink to the host plant. The pathogen is highly variable, and exists as mixture of pathotypes in the soil, and even in single-clubbed roots. The isolates differ in preferences, and aggressiveness toward different cruciferous hosts. The survival of the pathogen in soil leads to infection, development, and multiplication in host tissues, and in soil. It produces morphological structures like resting spores, zoospores, and plasmodia. The pathogen has a very complex life cycle consisting of different zoosporic stages (primary and secondary), formation of primary and secondary plasmodia, and resting spores inside the host cells. Plasmodia provoke hypertrophy, and hyperplasia leading to the development of galls that obstruct nutrient, and water transport to the infected plants. The only host free stages are the short-lived biflagellate zoospores, and resting spores. Resting spores are extremely resilient to harsh environmental conditions, and survive in the soil for more than 20 years. Although, traditionally, *Plasmodiophora* was classified as fungus but now molecular phylogenetic analysis of the small subunit (SSU) ribosomal RNA, actin, and polyubiquiten genes suggest that it belongs to the protists phylum *Cercozoa* in the kingdom Rhizaria, and Plasmodiophorida order. During the sexual reproduction, meiosis appears to occur in the plasmodium before resting spore formation but how and when karyogamy takes place is not exactly known. The *P. brassicae* genome is compact with small size (24.2–25.5 Mb), high gene density, and with scarcity of repeats. In its genome, there is reduction, and overlapping of intergenic distances. High degree of variation in field populations indicates sexual recombination. The molecular biology techniques have revealed *P. brassicae* genome size, recombination pattern, chromosome number, number of genes, gene structure, gene functions during pathogenesis, disease developmental stages, pattern of gene expression, and signaling for metabolic pathways (Saharan et al. 2021c).

The disease is known by several common names in different countries of the world. In countries such as Belgium, and Germany, the multiplicity of names probably indicates a considerable degree of regionality, and localization reflecting the distinctly separate dialects used by farmers, and growers living in historically separate states. The naming also reflects the relative importance of the crops affected. In northern Europe, cabbage was the major crop grown, and hence diseased by clubroot, whereas further in southern France, cauliflower was of greater importance. Colonists from Europe took their names for this disease with them to the New World, and at the same time the pathogen was exported in animal fodder such as swedes (*Brassica napus*), and turnips (*B. rapa*) used as nutrition for their beasts. It is possible that *P. brassicae* reached China, and Japan much earlier as ancestral forms of *B. rapa* moved from southeastern Europe, establishing Chinese cabbage, and all its associated variant forms (Dixon 2007). In southern Europe the disease was viewed as a form of plant "*hernia*," whereas further north as "*club foot*," a term exported by German, and Scandinavian colonists with their crops. In recent years, the disease has generally been referred to as "*clubroot*" as opposed to "*club foot*." This reflects a desire for uniformity in the nomenclature of diseases stemming from European Union rules relating to crop cultivar assessment.

Very often the symptoms of clubroot infection initiation are not seen above ground until after the host passes the seedling stage. Since infection initiates below ground, and above ground symptoms are expressed only when gall formation interrupts movement of water, and nutrients in the host plant foliage. Unless the roots of such infected plants are examined carefully after pulling, disease diagnosis is ignored. The infected roots quite often appear to be healthy which at later stage show characteristically swollen, malformation, and clubs (Plate 1.7a). Above ground symptoms consists of a flogging or wilting which may be confused with any other soil borne disease or deficiency or abiotic stress till it is confirmed from observation of clubroots. It is a very typical disease which requires experience, and knowledge of identification by the observer for proper diagnosis of the disease. Clubroot galls are a nutrient sink, so they tie up nutrients, and severely infected roots of crucifers cannot transport sufficient water, and nutrients for aboveground plant parts. Symptoms may vary depending on the growth stage of the crop when infection occurs. Early infection at the seedling stage can result in pale green to yellowish leaves than wilting, stunting, and yellowing of infected plants in the late rosette to early podding stages. Such symptoms may be wrongly attributed to heat stress during periods with high temperatures or to other diseases such as blackleg or Fusarium wilt or other root rots or even to water, and salt stresses. In such cases, proper diagnosis includes digging up wilted plants to check for gall formation on roots (Plate 1.7a–c).

1.9.8.2 Fine Structures and Electron Microscopy

During infection process, small galls emerge from epidermal cells in the elongation zone of host roots, and abundant secondary plasmodia, and resting spores can be seen within the parenchymal cells of such galls. At early infection stages in host cells, primary plasmodia contain complex cellular components including nuclei, lipid droplets, an endo-membrane system and visible in epidermal cells of the root

Plate 1.7 (**a**) Clubroot formation initial symptoms on roots, (**b**) advances in clubroot formation, (**c**) severely damaged roots with pathogen. (Photos: Courtesy-Venkat R Chapara, USA)

elongation zone. Cortical invasion by *P. brassicae* shows disruption of cell wall, presence of vesicles or inclusion bodies inside the cell wall, cell wall thickness in combination with plasmodesmata and elongation and/or disorganization of host nuclei. Multinucleate, intracellular plasmodia are visible within the clubs in the cortex and pericycle cells. Gall development amplifies merismatic activity within the vascular cambium, and phloem parenchyma cells in the region of the hypocotyl. It also perturbs vascular development with reduction in xylem, and increase in phloem parenchyma cells in infected plants. Scanning electron microscopy of infected host galls revealed initial plasmodia of the pathogen as spherical, smooth surfaces with average diameter of 3.7 μm increasing to average size of 4.2 μm in diameter. Young plasmodia are round with partial rough surfaces and average 8 μm in diameter. Vegetative plasmodia are smooth surface, spherical and 45 μm in diameters which almost fill the host cells. Young resting spores are connected by fine filamentous materials, and mature resting spores have spiny surfaces. In resting sporangia during primary zoosporogenesis, single zoospores are differentiated within each sporangium being initiated by the emergence of two flagella in the tight space formed by invagination of plasma membrane within the sporangium. A deep grove is formed on the zoospore cell body by further invagination of the plasma membrane. Two flagella appear to coil around the zoospore cell body in parallel along this groove. A nucleolus like region is visible in the mature zoospores. A partially opened germ pore is seen in the sporangium containing the mature zoospores. Zoospores have electron dense mitochondria, and well developed endoplasmic reticulum. Mitotic nuclear division is characterized by a persistent nucleolus, bipolar centerioles paired end–to-end, densely stained chromatin, and a complex array of membranes that surrounded and ramify through the spindle. Chromatin begins to condense in prophase, and is aligned at metaphase in a reticulate plate on the nuclear equator. Large amounts of membrane cisternae within the spindle are visible at metaphase. Several concentric sheets of perinuclear endoplasmic reticulum surround the spindle, and give the mitotic nucleus the superficial appearance of having an intact nuclear envelop (Saharan et al. 2021c).

1.9.8.3 Pathogenic Variability

Pathogenic variability in *P. brassicae* was detected long back during 1930s on the basis of differential interactions with cruciferous species. First generation widely used host differentials all over the world included genotypes from *B. napus* var. *napobrassicae*, *B. napus* var. *rapifera*, *B. oleracea* var. *capitata*, *B. rapa* var. *pekinensis*, *B. rapa* ssp. *rapifera,* and *B. napus* var. *napus*. The occurrence, distribution, and pattern of *P. brassicae* virulence/pathotypes have been characterized from Australia, Canada, China, Czech Republic, Finland, France, GDR, Iceland, India, Japan, Korea, New Zealand, Norway, Poland, Sweden, Turkey, UK, USA, and USSR. Molecular techniques like PCR, RAPD, and RFLP/AFLP have also been used to characterize *P. brassicae* isolates. Although these studies have revealed high degree of polymorphisms among as well as within *P. brassicae* populations but so far no correlations have been established with pathogenicity. Modified differentials sets have been used at Canada, China, Japan, Korea, and Turkey to get clearer picture of pathotypes in their countries by including genotypes of *Brassica* having local/national importance. Numerous pathotypes of *P. brassicae* have been identified from more than 20 countries of the world with designations as serial numbers (1, 2, . . .*n*), triplicates (16/14/15, 16/2/14, . . . *n*), and letters (A, B, C, . . . X1, P1, P2, . . . *n*). However, the characterization of pathotypes with respect to virulence spectrum, and erosion of clubroot cultivars is from very limited countries. For rapid identification, and detection of pathotypes infecting *Brassica* species, molecular markers have been developed. A *CR 811* gene has been identified in pathotype 3, and 5 which plays an important role in host resistance, and *P. brassicae* pathogenicity. Isolate/pathotype specific molecular markers are very useful tools to identify genes to test, and differentiate unknown pathotypes. Differential sets development, and used so far have several limitations to detect, and identify new pathotypes/distribution of virulence all over the world. Since the sets consist of genetically heterogeneous *Brassica* species which is against to the International standard differentials host with near isogenic lines. Similarly, isolates used so far are composed of genetically heterogeneous populations instead of genetically pure single spore isolates. Similarly, designation and nomenclature of pathotypes have limitations since different systems have been followed (serial number), triples (binary notation) which makes difficult to compare, analyze, and detection of new pathotypes. Following this system, similar pathotypes have been reported from different countries which is not consistent with the biology of the pathogen. The intermediate and fluctuating results of certain deferential hosts following inoculation with the population of *P. brassicae* indicates importance of single spore isolates for virulence analysis prevalent world over (Saharan et al. 2021c).

1.9.8.4 Infection, Pathogenesis, and Disease Cycle

The infection process by *P. brassicae* is initiated soon after the germination of resting spores triggered by host exudates under favorable temperature conditions of 20–25 °C. Resting spores in the soil germinate, and release primary zoospores that penetrate root hairs to form primary plasmodia. The primary plasmodia form multinucleate zoosporangia that produce uninucleate secondary zoospores. These secondary zoospores are released into the soil and infect root cortical cells and develop into motile secondary plasmodia. Cortical infection results in hyperplasia and hypertrophy of the roots cells which produces swollen, rounded clubroot symptoms. Clubroot development and severity of the pathogenesis is induced, and regulated by several metabolites, growth regulators, sugars, hormones, enzymes, proteins, lipids, glucosinolates, etc. during host–pathogen interactions. The involvement of auxin, and cytokinin as pathogenicity factors cause increased cell division, and cell elongation during late infection stages. Molecular mechanism of pathogenesis revealed a serine protease (Pro1) gene important for resting spore germination. During infection, and pathogenesis *P. brassicae* members of the TF families, MYB, bHLH, and APZ/EREBP are highly up or down regulated in both shoot and root of host plants. *P. brassicae* infection triggers active sugars translocation between the sugars producing tissues, and the clubbed tissues, and SWEET family genes are involved in this process. *Brassica* specific genes (BSGs) play important roles in *P. brassicae* stress responses. Genes Pb-Bripa, and PbPGA are expressed in early stages of infection, mi RNAs are expressed during the process of clubroot interaction, and development. G protein-coupled receptors (GPCR) signal transduction pathways play important roles in the growth, development, and pathogenesis of *P. brassicae* to their cruciferous host species extensively (Saharan et al. 2021c).

Plasmodiophora brassicae causing clubroot disease of cruciferous crops very successfully completes its disease cycle through three stages of life cycle; first long time survival of resting spores in the soil, second root hairs infection by the primary zoospores, and third cortical infection by secondary zoospores. In the infected root cells, the pathogen develop into secondary plasmodia which cause hypertrophy followed by gall formation and generation of new resting spores and root exudates. Calcium ions trigger the germination of the resting sores. The germination of mature resting spores is affected by pH, humidity, temperature, and other inorganic ions as well as biological factors. The zoospores released from resting spores survive for short periods of time. In dual culture, the development of *P. brassicae* in hairy roots is similar to that found in intact plants. The secondary plasmodia in the inner cortical cells are distributed by cytoplasmic streaming and host cell division. The role of effectors/gene has been determined in the disease cycle. The serine protease Pro1 is involved in the germination of resting spores. During root hair infection by primary zoospores, primary plasmodia, secondary zoospore, secondary plasmodia, proteases, and methyltransferases are produced to suppress plant immunity for the success of infection process (Saharan et al. 2021c).

1.9.8.5 Epidemiology and Disease Forecasting

Clubroot of crucifers is strictly a disease of soil environment; therefore, its development is influenced by environmental parameters of soil, and their interactions with all abiotic, and biotic soil inhabitants. The disease development is affected by factors such as temperature, soil pH, soil type, rainfall, and drainage; micro nutrients like boron, calcium, and nitrogen; inoculum potential, host resistance, weeds, cultural practices, and interactions of all these factors in a soil complex environment. Temperature below 17 °C inhibits the development of clubroot at all stages of its disease cycle. Alkaline pH also reduces infection, and symptoms development. However, alkaline pH does not eliminate clubroot when other conditions are conducive for infection. In primary and secondary infections, nutrient availability cannot limit clubroot development in the rhizosphere. Both calcium and nitrate form of nitrogen, and calcium cyanamide inhibits the development of plasmodium into sporangia in root hairs, and reduces the variability of resting spores, and the ability of primary zoospores to invade the host. Root hairs infection, cortical infection, and disease severity is high at 25 °C temperature and pH of 5.0–6.0 but decreases as pH increases, and temperature decreases. Clubroot severity is positively correlated with inoculum density in the field. Host resistance inhibits cortical infection by the pathogen. Disease forecasting is based on congenial factors for disease development (Saharan et al. 2021c).

1.9.8.6 Biochemistry of Host–Pathogen Interaction

The life cycle of *P. brassicae* is entirely different from other biotrophic pathogens. The intimate relationship between host, and pathogen leads to change in hormonal status, cellular development of pathogen, and sources sink relations of clubbed roots and shoot of host plant. In the infected plants growth-promoting hormones such as auxins, cytokinin, and changes in secondary metabolites are associated with club development which exhibits their effect either directly on growth hormones or indirectly via their bioactive properties. The change in status, and metabolism of hormones, and compounds like auxins, cytokinin, glucosinolates, camalexin, brasinisteroids, gibberellin, jasmonic acid, ethylene, flavoids, and sugars have effects on clubroot phenotypes. The metabolic sink results in the accumulation of starch grains in clubbed tissues. The different pathways and hormonal signals result in a homeostasis between the pathogen, and the host plant resulting galls formation and/or suppression of gall growth. A protein (PbGH3) can modify auxins, and jasmonic acid produced in infected root galls. The resilient resting spores of the pathogen contain chitin in their cell walls. A trehalase encoding gene is primarily expressed in spores to release glucose as an energy source in the germination process. The *PLeff* genes are expressed in plasmodia, the *Heff* genes are expressed in the transcriptomes of the host, and the *PLeff* genes are expressed late in the life cycle during spore formation. A complete pathway for the spermidine biosynthesis has been predicted in the *P. brassicae* (Saharan et al. 2021c).

1.9.8.7 Host Resistance

Genetical and molecular mechanisms of host resistance in the form of qualitative and quantitative genes have been expressed against *P. brassicae* in the economically important and widely grown cruciferous species viz., *B. rapa*, *B. oleracea*, *B. napus*, *R. sativus,* and *Arabidopsis thaliana*. Mapping of quantitative and qualitative traits loci (QTL's) have revealed several (12) dominant loci in *B. rapa* more than 22 QTL's in *B. oleracea,* and more than 19 QTL's in *B. napus* for clubroot resistance to different pathotypes of *P. brassicae*. Clubroot resistance in *Brassica* species is inherited by many dominant, recessive, additive and complementary genes. Substantial progress has been made in breeding clubroot resistance genotypes/cvs. of *B. rapa*, *B. oleracea*, *B. napus*, *R. sativus,* and *A. thaliana*. Molecular mechanisms to identify host resistance have been extensively exploited through genetic mapping of CR loci, genomics, transcriptomics, and proteomics approaches. Multiple Pathotypes specific QTL's conferring resistance to *P. brassicae* have been identified and functionally characterized by RNA sequencing and genotyping-by-sequencing to facilitate cloning of R genes. To counteract the role of mitogen in modulation of CR QTL's, new sources of resistance have been found among the oilseed rape accessions under low and high N conditions. Marker assisted selection of CR genes have ensured accumulation of more than one R gene into a single CR cultivar for more durability under field conditions against different pathotypes of *P. brassicae*. To trigger the pathogen associated molecular pattern (PAMP's) immunity 33 chitinase genes have been identified in *B. rapa* with their structure, evolution, and chromosome location. Map-based cloning and characterization of *Crr1A* genes which confers resistance to *B. rapa* encodes a Toll-Interlinked-1 receptor/nucleotide binding set/Leucine-rich repeat (TIR-NB-LRR) protein expressed in the stele and cortex of hypocotyl and root, where secondary infection of pathogen occurs but not in root hairs. Differential *R* genes expression and their functional mechanisms during life cycle of pathogen to confer resistance at different stages of infection, and pathogen development have been elucidated. Phenyl-propanoid pathway genes (Lignin biosynthesis) are upregulated in CR *B. rapa*. Clubroot resistance has been transferred in *Brassica* through molecular marker assisted selection, interspecific hybridization, distant hybridization, and mutational approaches. Effective management of R cultivars of *Brassica* can enhance durability, and cost effective control of clubroot. The nomenclature of CR genes has not been standardized in *Brassica* species. A basal defense mechanism is activated by a virulent pathogen through induction of trehalase gene. Cellular resistance is expressed through accumulation of ROS or lignification of epidermal or cortical cells in the resistant cvs. Small spheroid galls of clubroot acts as an indicator of host resistance. Histological resistance is expressed through deposition of callose in response to pathogen invasion. Systematic acquired resistance is associated with up regulation of pathogenesis related genes PR-1, and PR-2 (Saharan et al. 2021c).

1.9.8.8 Disease Management

Clubroot disease management has been aimed to target all the three stages of disease cycle. The first most important step is to bring inoculum load in the soil below minimum threshold level of resting spores. The second and third steps are to suppress primary root hairs infection, and secondary cortical infections, respectively which results in club formations to continue life cycle of pathogen, *P. brassicae*. Any activity that results in transport of soil or crop residues infested with *P. brassicae* from one point to another has the potential to disseminate the pathogen. The most effective cultural control is through removal of diseased plants, improved drainage, deep ploughing, ridging, long crop rotation (>3 years), using clean and treated seed, avoiding application of raw manures, sanitization (cleaning and disinfecting of machinery, equipment, vehicle, tools, footwear, seed drill, plant materials, irrigation water etc.), and manipulating date of sowing to sow the crop seed when soil temperature are too cold for clubroot pathogen development. Use of clean, treated seeds, crop rotation, and bait crops with resistant, susceptible, and non-*Brassica* plants reduce clubroot inoculum in the infected fields. Soil solarization in combination with low rates of fumigation (Dazomet) can be a means of clubroot control. Postharvest weed management along with susceptible host plants is very crucial factor to reduce inoculum potential in the fields. Several simple low cost cultural strategies may be useful for effective management of disease in the endemic areas. Intercropping of *Brassica* and flaxseed reduces clubroot with increase in yield. Soil amendment with lime, wood ash or calcium cyanamide increases soil pH, and is an important clubroot management strategy. Judicious use of nutrients like N, P, K, Ca, B, Molybdenum, and organic matter helps in management of clubroot through maintenance of pH with conducive effects on host growth, and suppressive soil environment for pathogen infection and development. Several chemicals used as soil drench, surfactant, and soil fumigants control clubroot through inhibition of resting spores germination, and suppression of root hairs, and cortical infection. Clubroot has been managed with the use of biological agents of bacterial, fungal, and plant products origin which produce anti-microbial metabolites against *P. brassicae*, and colonize host roots to induce resistance. Commercial biofungicides like Serenade, Prestop, Mycostop, granulate formulation Plasma soil, and Root Shield Plus with fungicides like fluazinam, and cyazofamid are very effective against clubroot. Disease management through host resistance is the most economical, and ecofriendly strategy. Disease resistant cultivars of *Brassica* crops have been developed, and are being used to manage clubroot in the countries like Canada, China, Australia, Japan, Germany, and Korea. Integrated approaches using cultural, chemical, biological, and host resistance has been extremely effective clubroot management strategy along with high yield of cruciferous crops. In Latin American countries, clubroot has been managed with modification of cultural operations depending on soil environment, pH, and inoculum load in each country, in addition to the growing of resistance cultivars and using biological, and chemical control strategies (Saharan et al. 2021c).

1.9.9 *Pseudocercosporella* (White Leaf Spot)

1.9.9.1 The Disease and Causal Organisms

White leaf spot disease caused by *Pseudocercosporella* (Syn. *Neopseudocercosporella capsellae*) (Ellis & Everhart) Videira & Crous is an important disease on many Brassicaceae including oilseed, wild, vegetable, condiment, and fodder *Brassica* species (Petrie and Vanterpool 1978; Sumner et al. 1978; Barbetti and Sivasithamparam 1981; Cerkauskas et al. 1998; Ocamb 2014; Gunasinghe et al. 2020). *Pseudocercosporella capsellae* has both sexual and asexual life cycles. The teleomorph stage (*Mycosphaerella capsellae*) is produced at the end of the crop season. It is primarily responsible for survival of the pathogen over the period between crops and for production of primary inoculum for the next season (Inman et al. 1991). It is assumed that its survival in Australia, over the summer is by means of stromatic hyphae on infected stubble with subsequent production of conidia providing the primary inoculum for the next season (Barbetti and Khangura 2000; Meena et al. 2021).

Taxonomically *Neopseudocercosporella capsellae* belongs to the *Cercospora* complex (Cercosporoid), where differentiation within the complex is based on morphological characteristics of spores and reproductive structures (Braun 2001). Within the complex, *N. capsellae* is in the class Dothideomycetes, the family Mycosphaerellaceae (Deighton 1973), and the genus *Neopseudo-cercosporella* (Videira et al. 2016). Twenty-six synonyms are documented for this pathogen previous to the recent name changed to *Neopseudocercosporella capsellae* (Videira et al. 2016). On artificial media, the pathogen's growth rate is noticeably slow, taking 3 weeks for a colony to reach 1–2 cm in diameter (Crossan 1954). Slow growth is a likely reason for under-reporting of the pathogen as the disease causal agent. On a variety of culture media, it produces dark to olivaceous-gray stromatic colonies with dentate margins (Inman 1992). Optimum growth from hyphae occurs between 20 and 24 °C and at pH 5.5–7.0 on potato dextrose agar (PDA) (Okullo'kwany 1987). Young colonies produce thin and hyaline hyphae becoming thick-walled, septate, brown hyphae with stroma-like or sclerotia-like structures, which give rise to conidia (Crossan 1954). However, *N. capsellae* does not sporulate on commonly used artificial media including PDA (Miller and McWhorter 1948; Crossan 1954), but produces conidia when the growing colonies on V8 or distilled water agar are exposed to near-UV light around 365–370 nm (Inman et al. 1991). Petrie and Vanterpool (1978) observed and extracted a red/purple/pink pigment produced by *N. capsellae* mycelial mats. Now this pigment has been confirmed as the mycotoxin cercosporin (Gunasinghe et al. 2016a; Meena et al. 2021).

White leaf spot lesions first appear on older lower leaves, and severe infections may induce leaf fall (Barbetti and Khangura 2000). The symptoms on Brassicas can vary depending upon the degree of leaf surface waxiness or the host species (Crossan 1954). In particular, on oilseed rape, *N. capsellae* initiates infections on leaves observable as numerous brown, elongated spots that later become white or pale beige. These leaf lesions can be up to 1 cm in diameter and subsequently coalesce to form large irregular shaped lesions. The mature lesions, usually surrounded by a chestnut-brown margin with a distinct delimitation between healthy and diseased tissues, are visible on upper and lower leaf surfaces. The pathogen produces evenly distributed tufts of conidiophores on upper and lower surfaces (amphigenous) of the diseased leaf (Deighton 1973).

Related *Brassica* pathogens can also develop similar symptoms on leaves. Ringspot disease symptoms caused by *N. brassicae* could be misleading particularly when developed under dry conditions in the absence of spermogonia and/or perithecia in typical zonate arrangement (Petrie and Vanterpool 1978; Corlett 1999). On oilseed rape, the early development stage of blackleg caused by *L. maculans* can resemble white leaf spot, as leaf lesions are first visible as brown discolored areas. However, white leaf spot lesions are easily distinguished from mature blackleg lesions, which have small black pycnidia within the lesions and exhibit shot-holes when they age (Brun et al. 1997). Light leaf spot disease of *Brassicas* caused by *P. brassicae* causes irregular lesions with white spore pustules (Claassen 2016) that can be misidentified as white leaf spot disease. However, older light leaf spot lesions are much larger than *N. capsellae* lesions and have a bleached and papery center, and the infection of the leaf can cause leaf distortion (McCartney and Lacey 1990).

The leaf spot symptoms in *Brassica* species with waxy leaves (cabbage) or turnip, mustard, and Chinese cabbage varieties, differ from each other and may not resemble the white leaf spot lesions described on other hosts (Crossan 1954). On cabbage, initial dark gray or black dendritic lesions may resemble downy mildew symptoms caused by *Hyaloperonospora parasitica* (Pers.), but mature white leaf spot lesions become more or less rectangular or rounded with well-defined margins with ashy black center (Miller and McWhorter 1948). In contrast, typically round, semi-transparent larger lesions with brownish-gray centers with well-defined brown or tan margins are typical symptoms on turnip and wild mustard (Miller and McWhorter 1948). The pathogen can colonize stems and pods as the crop matures. The infection of stems or pods initially results in gray to black lesions. Geographic areas in which the sexual stage occurs, numerous tiny dark specks appear within the gray-speckled areas due to the formation of the sexual stage (Petrie and Vanterpool 1978). These stem lesions are somewhat superficial and do not damage the pith (Inman 1992). Pod lesions can be mistaken for *Alternaria* spp. lesions. However, *N. capsellae* lesions have dark reticulation within the brown pod-spot and a less well-defined margin (Lane and Gladders 2008). These pod lesions can expand rapidly to cover large areas with depressed centers (Carmody 2017). By harvest, plants can become completely discolored with the entire field of plants appearing purple or gray (Anonymous 2020).

1.9.9.2 Distribution

The *Neopseudocercosporella capsellae* has been recorded from all subcontinents except Antarctica, over a wide range of climatic conditions from temperate to tropical, including the principal oilseed rape or mustard producing countries of the European Union, China, India, Canada, Australia, and Japan (CMI 1986). However, the disease severity and incidence differ geographically due to variation in pathogen populations, host species, cultivars grown, different agricultural practices adopted, and prevailing local climatic conditions (West et al. 2001), in particular, temperature, humidity, and rainfall (Siebold and Von Tiedemann 2012). White leaf spot disease of oilseed Brassica occur across the oilseed *Brassica* growing countries including France (Penaud 1987; Perron and Souliac 1990), the UK (Inman 1992), Canada (Petrie and Vanterpool 1978), Germany (Amelung and Daebeler 1988; Sochting and Verreet 2004) and Australia (Barbetti and Sivasithamparam 1981; Howlett et al. 1999; Eshraghi et al. 2005; Meena et al. 2021).

1.9.9.3 Host Range

The *Pseudocercosporella capsellae* has a wide host range (Boerema and Verhoeven 1980; Gudelj et al. 2004), infecting diverse wild and cultivated crucifers, including oilseed, forage (Sumner et al. 1978) and vegetable Brassicas (Deighton 1973; Cerkauskas et al. 1998). In Western Australia, *P. capsellae* was first recorded in 1942 on *Brassica napus* (Shivas 1989), subsequently on turnip rapeseed (*B. rapa*), Chinese cabbage (*B. rapa* var. *chinensis*), broccoli (*B. oleracea* var. *italica*), swede (*B. napus* var. *napobrassica*) (Shivas 1989), and more recently on *B. juncea* (Eshraghi et al. 2005). Some species of the genus *Cercospora* are known as producers of cercosporin, visible as a purple-pink pigment adjacent to the margins of colonies growing on artificial media (Petrie and Vanterpool 1978; Daub and Ehrenshaft 2000).

The pathogen *N. capsellae* has been isolated from leaf lesions on "wild" or "weedy" crucifers such as wild radish (*Raphanus raphanistrum*) and wild turnip (*Brassica rapa* ssp. *sylvestris*) (Marchionatto 1947; Deighton 1973; Morris and Crous 1994; Francis and Warwick 2003; Maxwell and Scott 2008). It has also been recorded causing disease on false flax (*Camelina sativa*), a recently introduced oilseed crop in Europe (Föller and Paul 2002), Chinese cabbage, mustard type *Brassica* vegetables, and cauliflower (Lancaster 2006). Moreover, *N. capsellae* can produce leaf spots interspersed with symptoms caused by other *Brassica* pathogens.

1.9.9.4 Economic Importance

Severe losses due to white leaf spot disease can occur at the seedling stage (Ocamb 2014) or in older plants when susceptible varieties are grown under environmental conditions favorable for disease development (Reyes 1979; Penaud 1987; Barbetti and Khangura 2000). In both these situations, and particularly when the environment is conducive, white leaf spot causes significant yield losses of at least 30% in oilseed *Brassicas* predominantly through defoliation and the development of pod lesions (Penaud 1987; Barbetti and Khangura 2000). Pod infections by *N. capsellae* can cause 15% yield losses in France (Penaud 1987).

Variation of pathogen populations can relate to differences in pathogenicity and/or virulence of isolates of the same pathogen (McDonald and Linde 2002), and such differences can be attributed to differences in genetic makeup of the fungal isolates (Kuninaga et al. 1997). Understanding this variation within populations of agronomically significant pathogens is an essential criteria toward developing and applying management strategies, and in understanding the pathogen's ability to respond to implemented management practices.

1.9.9.5 Epidemiology

Under favorable environmental conditions, *N. capsellae* conidia or ascospores produce lesions on the leaves, while only conidia are responsible for stem and pod lesions that appear later in the crop season (Petrie and Vanterpool 1978; Penaud 1987; Barbetti and Khangura 2000; Inman et al. 1991, 1999). Optimal conidial germination occurs at 20–24 °C and is inhibited below 8 °C or above 28 °C (Crossan 1954). Germination is usually from apical cells, or less frequently from basal cells (Morris and Crous 1994). The conidia can produce multiple germ tubes from each conidial cell (Petrie and Vanterpool 1978; Gunasinghe et al. 2016b) and cleavage at a septum of a conidium can produce multiple conidia, each producing a germ tube to infect the host (Gunasinghe et al. 2016b). Hyphae of germinating spores invade the host tissue through natural openings such as stomata (Crossan 1954; Gunasinghe et al. 2016b). Temperatures of 18–19 °C and high humidity (100%) with at least 8 h of continuous leaf wetness are ideal for white leaf spot infections (Inman 1992; Meena et al. 2021).

1.9.9.6 Pathogenic Variability

Variation of pathogen populations can relate to differences in pathogenicity and/or virulence of isolates of the same pathogen (McDonald and Linde 2002), and such differences can be attributed to differences in genetic makeup of the fungal isolates (Kuninaga et al. 1997). Differences observed across isolates of *P. capsellae* on potato dextrose agar (PDA), such as the production of diffusible purple-pink pigment and growth habits, suggested possible variation within pathogen populations. Furthermore, in previous studies by Gunasinghe et al. (2013) and Eshraghi et al. (2007), where plants were sprayed with a mixture of isolates of *P. capsellae*, different reactions were observed across diverse *Brassica* species, including the oilseed species *B. napus*, *B. carinata* and *B. juncea*. The cotyledon screening tests of *B. napus* (oilseed rape), *B. juncea* (Indian mustard), and *R. raphanistrum* (wild radish), under controlled environmental conditions, were employed for 52 isolates of *P. capsellae* to assess their pathogenicity and virulence differences. Fifty-four *P. capsellae* isolates were characterized using ITS sequencing and phylogenetic analysis. Growth rate and ability to produce the purple-pink pigment indicative of cercosporin for the same 54 isolates were evaluated on three different media: potato dextrose, V8 juice, and malt extract agars. Gunasinghe et al. (2015) reported that *Pseudocercosporella capsellae* reference isolates from other countries generally grouped into a single separate cluster, highlighting the genetic distinctiveness of Western Australian isolates.

1.9.9.7 Disease Management

The host resistance is generally the preferred and most cost-effective option for disease management, cultural controls, such as residue removal, crop rotation, good hygiene practices, and fungicides can all be utilized to manage white leaf spot disease. Recommended cultural practices to prevent white leaf spot epidemics are typically aimed at reducing initial levels of inoculum or the spread of the established pathogen. These cultural disease control methods are preventive and indirect in action against the pathogen, and therefore, awareness of the biology of the pathogen facilitates the disease management practices by identifying the most vulnerable stages of the pathogen to attack (Ogle and Dale 1997). Destroying crop residues and crop rotation are recommended to reduce the inoculum levels of necrotrophic pathogens with similar biology and survival mechanisms on crop residues (Bokor et al. 1975; Ogle and Dale 1997). This practice has been highly effective when the pathogen's survival period on the susceptible host or host residues is shorter than the rotation intervals between susceptible crops (Ogle and Dale 1997). The *N. capsellae* has a survival period on crop residue of balanced supply of nitrogen decreases the disease potential (Duff et al. 2006). Generally, fungicide treatment is not recommended to control white leaf spot disease in the absence of any other disease, as fungicides may not be cost-effective (Inman 1992). However, in severe epidemics or when multiple diseases co-occur such as blackleg together with white leaf spot, application of fungicides can be beneficial and economic. If conducive weather continues after initial infections, fungicide applications are recommended every 2–3 weeks to prevent disease spread (Ocamb 2016). If white leaf spot disease was present in the previous season, foliar fungicide application before the main rains early in the season can reduce or even prevent the disease, and monthly fungicide applications throughout the season will reduce the carryover of the pathogen in infected residues (Ocamb 2016). In general, fungicides effective on other foliar pathogens will also reduce *N. capsellae* (Ocamb 2016; Meena et al. 2021).

1.9.10 *Pyrenopeziza* (Light Leaf Spot)

1.9.10.1 The Disease and Causal Organism

Light leaf spot (LLS) disease of *Brassicas*, caused by the hemibiotrophic fungal pathogen *Pyrenopeziza brassicae* (anamorph *Cylindrosporium concentricum*), is known as one of the most damaging diseases of winter oilseed rape (*B. napus*) in Northern Europe (Boys et al. 2007). In Ireland, *P. brassicae* was observed for the first time during the 1964–1965 season, causing light leaf spot disease on broccoli, cabbage and Brussels sprouts (Staunton 1967). Several severe epidemics have been reported in winter oilseed rape in the UK since the first major epidemic recorded in 1974 (Simons and Skidmore 1988). In France, the disease was first reported in 1978 and there were severe epidemics in the 1980s and 2000s (Pilet et al. 1998; Karolewski et al. 2006). In Germany, occurrence of light leaf spot was widespread in the late 1980s (Pilet et al. 1998). Light leaf spot also occurs in Poland, with severe damage during mild winters (Karolewski 1999; Koike et al. 2007). *Pyrenopeziza brassicae* is also prevalent on Brassicas in the wet, cool climate of New Zealand, with severe outbreaks of light leaf spot reported on vegetable Brassicas (Cheah et al. 1980; Vegetables New Zealand 2016; Meena et al. 2021).

1.9.10.2 Economic Importance

It has been estimated that pre-harvest pathogens cause yield losses of 9–15% in crop production each year. The losses can be a much greater percentage of seed yield for certain crops (Teng et al. 1984; Oerke 2006).

1.9.10.3 Host Range

Light leaf spot also occurs on different types of *B. oleracea* and other related *Brassica* species or subspecies. These include Brussels sprouts (*B. oleracea* var. *gemmifera*), cabbage (*B. oleracea* var. *capitata*), cauliflower (*B. oleracea* var. *botrytis*), broccoli (*B. oleracea* var. *oleracea* var. *italica*), kale (*B. oleracea* var. *acephala*), turnip (*B. rapa* ssp. *rapa*), swede (*B. rapa* ssp. *rapifera*), Chinese cabbage (*B. rapa* ssp. *pekinensis*) and black mustard (*B. nigra*) (Maddock and Ingram 1981; Simons and Skidmore 1988; Boys 2009; Karolewski 2010). Spread of light leaf spot between different *Brassica* host species has been suggested (Wafford et al. 1986) but there has been little work on this disease (Meena et al. 2021).

1.9.10.4 Epidemiology

Light leaf spot epidemics are usually initiated by airborne ascospores of *P. brassicae*, which are forcibly released from apothecia (cup-shaped fruiting bodies) produced on diseased plant stem, pod or leaf debris (Gilles et al. 2001a, b). Ascospores produced on crop debris at other times may be important in initiating epidemics on crops of vegetable Brassicas or secondary epidemics on oilseed rape (McCartney and Lacey 1990; Gilles et al. 2000, 2001b; Karolewski et al. 2012). These airborne ascospores may be sampled by volumetric spore samplers as a means of forecasting risk of severe light leaf spot epidemics. Light leaf spot epidemics are favored by wet weather, which encourages production and dispersal of conidia (Meena et al. 2021).

1.9.11 *Rhizoctonia* (Seedling Rot; Collar Rot; Blight)

1.9.11.1 The Disease and Causal Organism

Rhizoctonia solani Kuhn (teleomorph = *Thanatephorus cucumeris* Donk) is a ubiquitous soil-borne plant pathogenic fungus belongs to the order Agonomycetales, family Agonomycetaceae (Mycelia sterilia) in which fungus lack even the perfect state and reproduce by the fragmentation of the mycelium. It is plurivorous fungus causing charcoal rot, damping off of various crop plants. Colonies on potato dextrose agar (PDA) are light brown to brown, with concentric zones of oppressed and aerial mycelium. Sclerotia are produced in various sizes and numbers, and are distributed in the zones of aerial mycelia. Infection cushions are formed on both susceptible and resistant hypocotyls, but are more frequent and form earlier on susceptible than on resistant hypocotyls. In culture, sclerotia are most common. The fungus is facultative necrotrophs. Mycelium is superficial or immersed, hyaline to brown, branched, septate, often tree like in form. The basic characters of the genus are formation of sclerotia. Sclerotia are dark brown to black, smooth, hard, and are formed of thick walled cells of irregular size but of uniform texture. The cells of the hyphae are barrel shaped, anatomizing frequently, branching more or less at right angles and pale brown to brown color. Perfect stages are known as *Thanatephorus* or *Ceratobasidium* and belong to basidiomycotina. *R. bataticola* also produces imperfect state (pycnidial state) i.e., *Macrophomina phaseolina* belonging to Sphaeropsidales. In case of perfect state of *Rhizoctonia solani (Thanatephorus cucumeris)* pycnidia are formed which are separate, globose, dark brown, immersed within one cavity, thick walled, erumpent, 100–200 μm in diameter (Meena et al. 2021).

The pathogen *R. solani* is associated with stem and root rot diseases in dicotyledonous crop species belonging to Brassicaceae (Gugel et al. 1987; Sneh et al. 1991; Tewoldemedhin et al. 2006). *Rhizoctonia solani* AG2-1 is the principal pathogenic strain causing damping-off and seedling and mature plant root rot (brown girdling root rot) in oilseed rape and canola (*Brassica napus* and *B. rapa*) in western Canada and the United States (Verma 1996b). Several brassicaceous weeds, including bail mustard [*Neslia paniculata* Desv.], stinkweed (*Thlaspi arvense* L.) and shepherd's purse [*Capsella bursa-pastoris* (L.) Medic], also are affected by the brown root rot caused by *R. solani*. Besides root and collar rot, *R. solani* also causes a severe leaf blight of oilseed rape and mustard in India (Chauhan and Saksena 1974). In Finland, damping-off due to *R. solani* occurred in 90% of spring rape (*B. napus* and *B. rapa*) fields surveyed during 1981–1882 (Tahvonen et al. 1984; Meena et al. 2021).

The pathogen *Rhizoctonia solani* causes damping-off can be pre-emergent when seeds fail to germinate or young plants fail to emerge, or postemergent when the stem of the seedling is infected, causing a water-soaked lesion at the soil line, wilting, and eventual plant collapse. Damping-off usually develops in patches within the seedbed or flat. The pathogen initially causes brown to dark brown lesions at the collar region of the seedlings (Plate 1.8a), the legion advances and causing shriveling of the collar region (Plate 1.8b). The stems of crucifers are frequently attacked when plants are mature and the lesion further enlarges and girdles the entire root region at the soil level of the plant (Plate 1.8c, d). The infected stems darken above and below the soil line and the outside tissue decays and sloughs off, resulting in a thin and wiry stem especially in cruciferous vegetables. Affected plants may stay erect, but remain stunted, yield poorly, and may eventually die. The infected mustard plants become yellowish due to restricted movement of nutrients etc. At the maturity, the infected roots may breakdown at the soil level showing yellowish and weak roots (Plate 1.8e) and root of the infected plant may looks like dry and shriveled. The infection in the field can be observed in the early stages of the crop as yellowish and wilted plants in patches in the field (Plate 1.8f).

1.9.11.2 Host Range and Disease Distribution

The wide host range of *Rhizoctonia* spp., which includes roots of volunteer plants, grassy weeds and rotation crops, allows this pathogen to survive in the soil from season to season (Smiley et al. 1992). The fungi generally grouped as *Rhizoctonia* types occur in all parts of the world and are probably indigenous to uncultivated

Plate 1.8 (**a**) Initial brown to black lesions on the collar region; (**b**) entire collar region become brown and shriveled; (**c**) the lesion enlarges and girdle the entire collar region; (**d**) lesions advances and restricted the upward movement of the nutrients infected roots become dark brown to black; (**e**) severely infected plants root may break down at the soil level; (**f**) infected field

areas and multiply and spread through agricultural operations. The host plants infected by *Thanatephorus cucumeris* reported in Japan alone belong to 35 orders, 52 families, and 125 genera and comprise more than 142 species from the Cycadopsida to the Monocotyledons.

1.9.11.3 Pathogenic Variability

In *Rhizoctonia solani* and binucleate *Rhizoctonia* spp. AGs strains have been identified in relation to many diseases. Isolates of certain AGs were found to infect many host plants, while those of other AGs have a rather limited host ranges. AG 1 and its subgroups infect many different crops. The most important disease caused by isolates of this AG is the rice sheath blight. Isolates of AG 2 and its subgroups infect rice, sugar beet and com. Isolates of AG 3 cause the black scurf disease on potato tubers. Isolates of AG 4 have a wide host range, including cotton and onion whereas AG 8 cause the severe "bare patch" disease on wheat and barley. Isolates of other anastomosis groups may be weakly pathogenic on several plants or not pathogenic at all such as AG 6, which is considered to be mainly nonpathogenic. Certain isolates are capable to form mycorrhizal associations with orchids (Masuhara et al. 1993).

Anastomosis groups of *R. solani*, binucleate *Rhizoctonia* spp. or *Waitea* species have been reported in 47 countries or districts; but more likely at least some AGs are present in every country. Distribution of anastomosis groups depends for pathogenic groups on the host plant distribution and on the climate and vegetation types for saprophytic groups. However, they are basically cosmopolitans, since they have a relatively wide host range and a flexible capability for adaptation. *Rhizoctonia* spp. has been extensively studied in recent years, due to the introduction of new techniques, research approach. As opposed to other groups, isolates of AGs 5, 6, 7, 9, 10, and BI of *R. solani* have been reported to be weakly pathogenic or not pathogenic. Isolates of AG 7 have been reported as a severe pathogen on carnation (Lo et al. 1990) in Taiwan and on soybean in Indonesia (Naito et al. 1991). Therefore, the possibility that isolates of AGs 6, 9, 10, and BI are severe pathogens on some plants under certain environmental conditions should not be overruled. The main mycorrhizal fungi of orchids are binucleate *Rhizoctonia* (*R. repens*, *R. anaticula*, and others). *R. solani* has also been recorded to form mycorrhizal associations with orchids. However, relationships between AGs of *R. solani* and orchids have now been studied in detail. AG 5 and AG 6 have been isolated from orchids and it has been shown that all AGs of *R. solani* can potentially form mycorrhiza1 associations (Masuhara et al. 1993). Furthermore, several new isolates of multinucleate *Rhizoctonia* have been found to be involved in orchid mycorrhiza1 associations. In Australia, AG 8 has been divided into five groups according to the zymogram patterns of the isolates (MacNish et al. 1993). It is quite possible that additional groups will be found in AG 8 in future. Cultural appearances are diverse among isolates from USA, Australia, and Scotland and may deserve special attention in this respect. AG 8 is known as a bridging group. Isolates of this group were found to be capable to fuse with isolates of AGs 2-1, 2-2, and BI (Rovira et al. 1986). The bridging behavior has not been sufficiently studied. In South Africa, the "Crater rot" disease of wheat has been known for a long time (Scott et al. 1979). It should be determined to which AG the isolates causing this disease, belong. Researchers in South Africa claim that this pathogen is *R. solani*, but the isolates do not fuse with any representative tester isolates of the known AGs (Lubbe and Meyer 1995). International distribution of *Rhizoctonia* isolates is basically not acceptable for quarantine reasons. Over the years, Ogoshi (1996) sent isolates to 34 countries, 77 locations, and 85 researchers. In a sense this is dangerous with regard to dissemination and infestation of the pathogens. It is suggested to establish a system, where at least one *Rhizoctonia* researcher in every country, district or continent will assume the responsibility to maintain and distribute *Rhizoctonia* spp. isolates. The key researchers should keep isolates of *R. solani*, binucleate *Rhizoctonia* spp. and *Waitea* AGs from their own countries or districts and determine their AGs with the known authentic isolates carefully. The local isolates could then be made available to researchers in the country, district or continent as tester strains in the country, district or continent.

The wide host range of *Rhizoctonia* spp., which includes roots of volunteer plants, grassy weeds and rotation crops, allows this pathogen to survive in the soil from season to season (Smiley et al. 1992). *R. solani* is associated with stem and root rot diseases in dicotyledonous crop species belonging to Brassicaceae (Gugel et al. 1987; Sneh et al. 1991; Tewoldemedhin et al. 2006). *Rhizoctonia solani* AG2-1 is the principal pathogen causing damping-off and seedling and mature plant root rot (brown girdling root rot) in oilseed rape and canola (*Brassica napus* and *B. rapa*) in western Canada and the United States (Verma 1996b). Several brassicaceous weeds, including bail mustard [*Neslia paniculata* Desv.], stinkweed (*Thlaspi arvense* L.) and shepherd's purse [*Capsella bursa-pastoris* (L.) Medic], also are affected by the brown root rot caused by *R. solani*. Besides root and collar rot, *R. solani* also causes a severe leaf blight of oilseed rape and mustard in India (Chauhan and Saksena 1974). In Finland, damping-off due to *R. solani* occurred in 90% of spring rape (*B. napus* and *B. rapa*) fields surveyed during 1981–1882 (Tahvonen et al. 1984; Meena et al. 2021).

1.9.11.4 Economic Importance

The predominant population of *R. solani* causing severe seedling diseases associated with establishment losses of up to 80–100% and final yield loss of up to 30% of oil seed rape (OSR, *Brassica napus*) worldwide belongs to AG2-1 strain (Tahvonen et al. 1984; Kataria and Verma 1992; Khangura et al. 1999).

1.9.11.5 Epidemiology

Incidence and severity of pre- and postemergence damping-off and root rot in rapeseed fields is influenced by weather and soil conditions and by inoculum density of the anastomosis groups (AGs) of *R. solani* in the soil. Teo et al. (1988) showed that early planting of rapeseed, before the soil had warmed to 15 °C, was more favorable to damping-off by AG 2-1 than AG 4 strain, while damping-off due to AG 4 increased as the soil warmed to 20 °C. Temperatures in the range of 18–30 °C support luxuriant growth of both AGs, and AG 4 isolates have a higher growth rate than AG 2-1 isolates at these temperatures (Verma 1996a; Meena et al. 2021).

1.9.11.6 Disease Management

The lack of adequate genetic resistance to *R. solani* in currently grown rapeseed and canola cultivars, and the absence of cultural methods for suppressing *R. solani* populations in the field, caused research to be directed toward investigating chemical control. Carboxin (Vitavax RS) and iprodione (Rovral ST) in mixtures with insecticide gamma-HCH (Lindane) are although recommended seed treatments to control damping-off and seedling root rot, but they do not provide control against the mature plant root rot.

1.9.12 *Sclerotinia* (Stem Rot)

The Sclerotinia is recognized as an important plant pathogen due to its worldwide distribution, its wide host range and the difficulties encountered in controlling the diseases it causes. Although, pathogen is most frequently observed in the regions tending to be cool and moist. It is also reported to occur in some semi-arid regions where conditions would seem unfavorable for disease development. Disease outbreaks in these drier areas occur during the summer months in irrigated fields since irrigation provides favorable microclimatic conditions for disease even though the macroclimate is unfavorable, or the disease occurs during the cooler winter months. Yield losses in susceptible crops vary and may be as high as 100%. *Sclerotinia* is generally considered to be a simple interest pathogen utilizing ascospores as the primary inoculum. Although micro conidia are produced by this pathogen, their role, if any in its life cycle has not yet been shown. Secondary spread of disease by plant to plant infection can occur by mycelial growth between stem bases or by contact between aerial parts of infected plants with neighboring healthy plants late in the growing season.

The Sclerotinia is one of the most devastating and cosmopolitan plant pathogens. More than 60 names have been used to refer to diseases caused by this fungal pathogen. The fungus infects over 500 species of plants worldwide. *S. sclerotiorum* has been reported from many countries located in all continents. The occurrence and distribution of *S. sclerotiorum*, *S. trifoliorum,* and *S. minor* on various crops in the UK and somewhere in almost every country has been recorded (Saharan and Mehta 2008). The *Sclerotinia* diseases are known under a variety of names and the most common of these along with some of the host plants most seriously affected are the cottony rot, white mold, or watery soft rot of beans, cabbage, carrot, eggplant, citrus, peanut, potato, stock, tobacco, etc.; stem rot and timber rot of cucumber, squash, beans, artichoke, asparagus, chrysanthemum, dahlia, delphinium, peony, potato, tomato, soybean, sweet potato, etc.; lettuce drop, broad bean, beet, cabbage, etc.; damping off of celery, lettuce, rapeseed-mustard, etc.; crown rot or wilt of sunflower, safflower, columbine, snapdragon, etc.; blossom blight of narcissus, camellia, etc.; pink joint of red pepper, stem canker of hollyhock, root and crown rot of clover; white blight, white rot, stem blight, stalk break and stem canker of rapeseed-mustard, head rot of sunflower, safflower, castor and dollar spot of turf grass (causal fungus previously known as *Sclerotinia homoeocarpa* now considered as species of *Lenzia* and *Moellerodiscus*).

1.9.12.1 *The Disease and Causal Organism*

Despite the continued interest of mycologists and plant pathologists, the taxonomic position of the economically important, plant pathogenic species of *Sclerotinia* has remained unresolved. Over 250 species of diverse relationships, both pathogenic and nonpathogenic, have been assigned to the genus *Sclerotinia* with consequent controversy and confusion over generic limits (Kohn 1979). Attempts to delimit the pathogenic species assigned to the genus using a few traditional characters have resulted in recognition of too many or too few species in relation to what is now known about the biology and micro-anatomy of this group to satisfy the practical need of the plant pathologist to name the pathogen in hand. Examination of type specimens of preserved material as well as observation of living isolates in the light of micro-anatomical and cultural characters employed by contemporary discomycotina taxonomists has resulted in the delimitation of plant pathogenic species of *Sclerotinia* in the revised and more limited circumscription of the genus presented.

The life cycle of *Sclerotinia* spp. can be completed on a variety of artificial and synthetic media. As a result, numerous articles reporting various aspects of growth and morphogenesis of *Sclerotinia* spp. have been published. The formation and germination of dark-colored, hardened compact masses of fungal tissue known as sclerotia have been studied in numerous fungi. From such studies, certain similarities and differences have been noted. Depending on the manner in which hyphae aggregate, the main types of sclerotium development are classified as terminal, strand or loose. Electron microscopy in association with physiological, biochemical and genetical studies has provided information which helps in understanding the complex host-parasite relationship of the disease (Saharan and Mehta 2008).

Based on the symptoms, the disease has been named as white blight, white rot, stem blight, stalk break, stem canker and rape canker. Under natural conditions, the stem of the plant is seen affected more frequently, though all above-ground parts are subject to attack by the disease. Symptoms on the stem become visible as elongated, water-soaked lesions, which later on are covered by a cottony mycelial growth of the fungus (Plate 1.9a, b). When the stem is completely girdled by such lesions, the plant wilts and dries. Sometimes, the infection is restricted to a smaller area of pith, which results in slow stunting of the plant and premature ripening rather than the sudden collapse of the affected plants (Plate 1.9c, d). Such plants under field conditions can be easily identified from a distance because of premature ripening (Plate 1.9e). The affected stem tends to shred and numerous grayish-white to black, spherical sclerotia appear either on the surface or in the pith of the affected stem (Plate 1.9f). When the crop is at seed maturation stage, the plants tend to lodge, touching the siliquae to the soil level. Such plants, though remaining free from stem or aerial infection throughout, show rotting of the siliquae with profuse fungal growth, along with sclerotial bodies just above the soil level. In an early stage of infection, damping-off, root rot and death of the whole plant may be observed. Lesions on leaves are grayish, irregularly shaped and often associated with adhering petals. Occasionally, gray mold (*Botrytis cinerea*) is associated with stem rot, covering infected areas with brown to gray fluffy mycelium (Saharan and Mehta 2008).

1.9.12.2 Economic Importance

Yield losses due to *Sclerotinia* diseases in susceptible crops vary and may be as high as 100%. In vegetable and fruit crops, losses are due to the infected produce rotting in the field before harvest and also due to rotting during storage. In seed crops, yields are reduced by both as reduction in seed size from the premature ripening of infected plants and by the loss of seed during harvesting. Seed loss is attributable to the replacement of seed by sclerotia in crops like sunflower and beans. The shattering of

Plate 1.9 (**a**)-Mycelia growth of the pathogen at the base of the stem, (**b**) carpogenic infection at axis of the leaf, (**c**) growth of the pathogen due to aerial infection at the stem, (**d**) stem dried due to aerial infection, (**e**) severely damaged stem showing the growth of the pathogen, (**f**) sclerotia are producing in the pith of the stem, (**g**) severely white stem rot infected field

prematurely ripened seed pods before harvest and loss of quality in the form of smaller, shrunken and chaffy seed in crops like rapeseed and sunflower has been observed. The fungal sclerotia may be mixed with seeds of sunflower, rapeseed and beans. Losses of some crops from diseases caused by *Sclerotinia sclerotiorum* and other species of *Sclerotinia* are millions of dollars annually. The losses are directly from loss of yield and indirectly from lessened quality (loss in grade). There also is another form of loss caused by this pathogen, the production lost due to abandonment of fields for growing preferred crops for less lucrative ones, or to noncrop plants, weeds or fallow (Purdy 1979).

1.9.12.3 Pathogenic Variability

Considerable variation in morphology and pathogenicity has been observed among isolates of *S. sclerotiorum*. Variation in pathogenicity has been associated with the production of pectolytic enzymes, cellulase, hemicellulase, phosphatidase, and oxalic acid. However, variation in these factors does not appear to be the primary determinant of pathogenicity as correlations have not always been apparent. In *S. trifoliorum*, a degenerate nonvirulent isolate of the pathogen produces more protopectinase than a normal strain but only the normal isolate secretes a toxin and is considered virulent. Little information is available on additional mechanisms that affect virulence among isolates of *S. sclerotiorum*. In *S. sclerotiorum* mycelial compatibility groups (MCGs) have been identified where early MCG has unique DNA fingerprint and two MCGs include strains with one of two fingerprints and interpreted as two clones (Saharan and Mehta 2008).

1.9.12.4 Infection, Pathogenesis, and Disease Cycle

The pathogen over winters as mycelium in dead or living plants and as sclerotia on or within infected tissues or as sclerotia that have fallen on the ground or through infected seed and propagating materials. Being a wide host–pathogen various hosts provide good opportunity for successful survival of the pathogen under all possible adverse conditions. Under favorable environmental conditions, the sclerotia germinate and produce one to many slender stalks terminating at a small, 5–15 mm in diameter, disc or cup-shaped apothecium in which asci and ascospores are produced. Large numbers of ascospores are discharged from the apothecia into the air over a period of 2–3 weeks. The ascospores are blown away and if land on susceptible plant parts, germinate and cause infection. Very frequently, the sclerotia cause infection by producing mycelial strands, which attack and infect young plant stems directly. Under moist conditions, the latter method of infection is probably more common than the one by ascospores (Abawi and Grogan 1975; Agrios 2005; Sherf and Macnab 1986; Walker 1969). According to Tu (1988), *Sclerotinia sclerotiorum* survives in infected seeds of white bean as dormant mycelium in testa and cotyledons over a 3 year period up to 85–89%. Under favorable conditions, sclerotia formed on infected seeds may be capable of producing apothecia later in the same season (Hungerford and Pitts 1953).

The life cycle of *S. sclerotiorum* is a relatively simple one as compared to certain other fungi like rusts. The sclerotia are the perpetuating structures of *S. sclerotiorum*.

Sclerotia may survive for 3–5 years in soil, assuring pathogen's availability when a host crop is planted. These survival structures become distributed throughout the tilled levels of soil. Although some sclerotia are destroyed by other organisms, a substantial number remain viable near the soil surface each year regardless of the host crop being grown. When conditions are favorable, these sclerotia germinate to form either a mycelium or apothecia. Mycelial production by sclerotia is negligible unless an exogenous source of energy is supplied and the infection of host plants can only occur if this energy source is available. Mycelial infection occurs at or below the soil line and has been reported in sunflower (Huang and Hoes 1980) and in beans (Saharan and Mehta 2008).

1.9.12.5 Epidemiology and Forecasting

Species of the genus *Sclerotinia* can function either as soil borne or airborne pathogens. Infections of above-ground plant parts result from ascosporic inoculum, whereas soil line infection may result either from ascospores or sclerotia. Below-ground infection, however, results from mycelial germination of soil borne sclerotia. Accordingly, the epidemiology of these two types of infections incited by *Sclerotinia* spp. is quite different and the effect of weather factors on their incidence and development differs considerably. Abawi and Grogan (1979) proposed that generally, the large sclerotia-type isolates of *Sclerotinia* (represented by *S. sclerotiorum* and *S. trifoliorum*) function primarily by producing apothecia and that mycelial germination from sclerotia contributes minimally, if at all, to the development of epidemics. In contrast, infections incited by the small sclerotial type isolates (*S. minor*) originate primarily through the eruptive mycelial germination of sclerotia, whereas production of apothecia under natural conditions occurs very rarely and thus is of minor importance. The different modes of infection exhibited by the large- and small sclerotia type isolates of *Sclerotinia* probably results from continued adaptation of these species to their ecological niches. Only a limited amount of detailed epidemiological information is available on a few important diseases.

Diseases caused by *Sclerotinia sclerotiorum* and *Sclerotinia minor* are responsible for economically important losses on several crops including canola, cabbage, carrots, celery, lettuce, snap beans, soybean and white beans. Crops such as cabbage and celery, the etiology of these diseases are known, but little information is available on the epidemiology. In these crops, disease avoidance and cultural practices are the primary methods of disease management, although fungicides are sometimes applied after symptoms are observed. For other crops such as beans, canola, carrots, lettuce, and soybean, the epidemiology has been described and at least partially quantified. Based on these epidemiological studies, disease-forecasting systems have been developed for canola, lettuce and beans. Epidemics in snap beans are associated with ascospores infecting petals as the primary inoculum and forecasting is based on soil moisture, rainfall, crop flowering, canopy enclosure and apothecia. Epidemics in carrots are bicyclic and represent a different situation. Epidemics in the field are associated with infection on senescing leaves in contact with moist soil under the carrot canopy. Forecasting is based on soil moisture, canopy enclosure, senescing leaves, air and soil temperature and the presence and number of apothecia. Epidemics in storage are associated with air temperature, rate of cooling, surface wetness and pre-existing infection. Despite the availability of forecasting systems for diseases caused by *Sclerotinia* spp. on several crops, there are no examples of organized monitoring or forecasting programs for these diseases. Anecdotal comments suggest that the reasons for the lack of development and implementation of forecasting models include the variable severity of epidemics, a lack of registered fungicides, little or no infrastructure to deliver disease-forecasting systems, and declining prevalence of integrated pest management (IPM) programs (McDonald and Boland 2004).

1.9.12.6 Host Resistance

Biotechnological approaches to enhance disease resistance involves either exploitation of natural forms of resistance or genetic engineering approaches, such as the introduction of chitinases, glucanases and other antifungal proteins. The former approach may involve the introduction of novel resistance genes from wild species and the subsequent introgression of genes through the use of molecular markers, or attempts to clone resistance genes. The approach to gene cloning is most likely to be successful to exploit *Arabidopsis*. Chitinases, which hydrolyze the β-(1-4)-glycoside in chitin (a major component of fungal cell walls), are often induced in plants following fungal attack and it is thought that these are involved in plant defense. Lines which contain chitinases that are specifically induced in *Brassica* by pathogens may give a higher degree of protection. Another strategy has been the introduction of a gene for oxalate oxidase in order to reduce susceptibility to infection by *Sclerotinia sclerotiorum* which relies upon the production of oxalic acid in the infection process (Saharan and Mehta 2008).

1.9.12.7 Disease Management

Most diseases caused by *Sclerotinia* have not been managed consistently and economically. The explosive pathogenicity of *Sclerotinia* species under favorable conditions and the ability of their sclerotia to withstand adverse conditions allow them to be successful pathogens on many crops. In field crops management practices should be applied to increase attainable yield despite their association with high disease risk. Integration of approaches like crop rotation with nonhost crop, clean cultivation, deep summer ploughing, balanced fertilization, timely sowing, seed treatment with carbendazim at 2 g/kg seed followed by two sprays of carbendazim at 1% 45–50, and 65–70 days after sowing significantly controls the disease (Meena et al. 2014).

1.9.13 *Turnip Mosaic Virus-TuMV* (Turnip Mosaic)

1.9.13.1 The Disease and Causal Organism

The *Turnip mosaic virus* (*TuMV*) is a member of the genus Potyvirus of family Potyviridae. The *TuMV* is a positive strand RNA virus with a genome of 9830–9833 nucleotides (Ohshima et al. 1996) and well characterized molecular biology (Walsh and Jenner 2002). *TuMV* is one of the most prevalent viruses, which was first discovered in the United States, is threatening *Brassica* vegetables around the world, especially in Europe, Asia and North America (Tomlinson 1987; Walkey and Pink 1988; Walsh and Jenner 2002). The *TuMV* disease was first described in *B. rapa* in the USA (Gardner and Kendrick 1921; Schultz 1921), while it was later found in *B. oleracea* in the UK (Smith 1935), and in *B. napus* in China (Ling and Yang 1940). Currently, *TuMV* is one of the most prevalent viruses and is threatening *Brassica* vegetables around the world, especially in Europe, Asia and North America (Tomlinson 1987; Walkey and Pink 1988; Walsh and Jenner 2002; Meena et al. 2021).

The first symptoms to develop on *Brassica* seedlings are chlorotic spots on inoculated leaves, mottling followed by systemic vein clearing, mosaic and/or necrosis, leaf distortion and often stunting. Infected plants are often stunted (Plate 1.10a–d). Some *Brassica* cultivars develop progressive necrosis of leaves, petioles and stem with some virus isolates, leading to plant death, particularly in *B. napus* (Walsh and Tomlinson 1985). A number of biotic and abiotic factors affect symptom expression, particularly temperature and virus isolate (pathotype/'strain'). The symptoms can become progressively less pronounced when the plants become infected as they mature, to the point of not being easily visible in late infections (Plate 1.10e, f). The effect of temperature varies considerably for different host and virus isolate interactions. Pound and Walker (1945) noted that mottle symptoms of cabbage plants infected by *TuMV* were more pronounced and virus titer was higher in plants grown at 28 °C than those grown at 16 °C. Seed pods of infected *Brassica* can be reduced in number and size, some are malformed and seedless, and the size of individual seeds and seed yield are reduced. Seed viability is also affected (Walsh and Tomlinson 1985). The cytoplasm of infected epidermal, mesophyll and phloem cells contains cylindrical cytoplasmic inclusions consisting of pinwheels, bundles, scrolls and laminated aggregates (Edwardson 1974; Edwardson and Christie 1986).

Infection by *TuMV*-infected plants may allow secondary infection by destructive pathogens. In field situations, mixed infections of *TuMV* and other viruses are often encountered. In China, mixed infections often occur with *Cucumber mosaic virus* and a tobamo-virus (Chiu and Chang 1982). In Europe, *TuMV* often occurs with *CaMV* (Jenner and Walsh 1996), *Broccoli necrotic yellows virus* (Walsh and Tomlinson 1985) and *Beet western yellows virus* (Hardwick et al. 1994). The symptoms are usually more severe where plants are simultaneously infected with other plant viruses (Pound and Walker 1945; Sano and Kojima 1989).

1.9.13.2 Economic Importance

The *Turnip mosaic virus* caused a loss of 30% in *B. napus* production in Canada (Shattuck and Stobbs 1987), as well as seed yield losses of up to 70% in *B. napus* in the UK (Hardwick et al. 1994) and 50% reductions in *B. oleracea* var. *capitata* (cabbage) head production in Kenya (Spence et al. 2007).

Plate 1.10 (**a**) Initial symptoms of mosaic on leaves, (**b**) severely infected leaf, (**c**) *TuMV*-infected plants, (**d**) dwarf plant due to initial infection, (**e**) severely infected plants, (**f**) severely infected plants

1.9.13.3 Host Range

The *Turnip mosaic virus* possesses an exceptionally broad host range; it infects 156 plant genera representing many arable, vegetable, and ornamental crops including oilseed and vegetable Brassicas, peas, lettuce, rhubarb, and stocks (Edwardson and Christie 1991). Natural infection by *TuMV* is common in Brassicaceae weeds. Infected crops include mustard (*Brassica juncea*), field pea (*Pisum sativum*), chickpea (*Cicer arietinum*), coriander (*Coriandrum sativum*) and forage (turnip, *Brassica rapa*) (Schwinghamer et al. 2014). It infects mainly Brassicaceae hosts, including Brassicas (*Brassica*) and non-Brassicas such as radish (*Raphanus sativa* L.). Canola is the Brassicaceae crop of most concern to grain growers, but *TuMV* also infects at least 39 other dicotyledonous plant families and three mono-cotyledonous families either naturally or by controlled inoculation (Edwardson and Christie 1991; Walsh and Jenner 2002; Tomimura et al. 2004).

Although, *TuMV* is naturalized in all states of Australia (McLean and Price 1984), it is reported to cause significant disease only in vegetable crops and one ornamental species (Brassicaceae: *Matthiola incana* L.). Affected hosts from the eastern states of NSW, Tas, Queensland, Victoria, and South Australia include a range of *Brassica* vegetables, rocket (Brassicaceae: *Eruca vesicaria* L. subsp. *sativa*), lettuce, and rhubarb (Polygonaceae: *Rheum rhabarbarum* L.). In NSW, *TuMV* was first recorded causing serious disease in cabbage, broccoli, and cauliflower (*Brassica oleracea* L. varieties) in the Sydney metropolitan area in 1957 and subsequently in irrigation areas in other parts of the state (Letham et al. 1975; Conroy 1959).

1.9.13.4 Pathogenic Variability

The *Turnip mosaic virus* is a member of the genus Potyvirus (type species Potato virus Y) in the family Potyviridae. The *TuMV* is the only potyvirus known to infect Brassicas (Walsh and Tomlinson 1985). The *TuMV* genome consists of a positive-sense single-strand RNA molecule of about 9830 nucleotides in length. The gene contains a single open reading frame (ORF) flanked by two untranslated regions (Basso et al. 1994). The ORF is translated into a single polyprotein, which is co-and post-translationally processed by three virus-encoded proteases (Walsh and Jenner 2002). The proteins released comprise P1, helper component protease (HCPro), P3, 6K1, cytoplasmic/cylindrical inclusion (CI), 6K2, virus-encoded genome linked (VPg) protein, nuclear inclusion protein a (NIa), nuclear inclusion protein b (NIb) and the coat protein (CP) (Walsh and Jenner 2002). *Turnip mosaic virus* probably originated from a virus of wild orchids in Germany and, while adapting to wild and domestic Brassicas, spread to Asia Minor via southern Europe no more than 700 years ago (Yasaka et al. 2017). Various strains and pathotypes of *TuMV* have been defined based on their ability to infect certain plant species or cultivars (Shattuck 1992). An early scheme used symptom type on cabbages and *Nicotiana glutinosa* to distinguish *TuMV* strains (Yoshii et al. 1998). Liu et al. (1996) also used symptom types and disease severity indices on a variety of *Brassica* species to define seven strains (Tu1–7). Other schemes were based on differential lines of a single species, either *B. rapa* (strains C1–C4; Provvidenti 1980; strains C1–C5; Green and Deng 1985) or *B. napus* (pathotypes 1–12) (Walsh 1989; Jenner and Walsh 1996).

Some resistance genes to *TuMV* are known to be temperature-sensitive with resistance breaking down at temperatures greater than 30 °C (Walsh 1989). In other studies on other species (*Armoracia rusticana* and *Cichorium intybus*), lower temperatures of 15–16 °C have been shown to induce more severe symptoms than higher temperatures of 28 °C (Pound 1948; Provvidenti et al. 1979). Symptoms in Brassicaceae can be confused with those caused by *Cauliflower mosaic virus* (*CaMV*) (Tomlinson 1970).

1.9.13.5 Host Resistance

Extreme forms of "race"-specific resistance to the virus have been identified in swede (Tomlinson and Ward 1978, 1982; Shattuck and Stobbs 1987; Doucet et al. 1990; Shattuck 1992) and rape (Walsh and Tomlinson 1985; Walsh 1989; Stobbs et al. 1989), forms of *B. napus* and in *B. rapa* (Chen 1980; Provvidenti 1980; Niu et al. 1983; Green and Deng 1985; Yoon et al. 1993; Suh et al. 1995). Some of these sources of resistance may be immunity where the virus fails to replicate (Provvidenti

1980; Tomlinson and Ward 1982; Shattuck and Stobbs 1987; Walsh 1989). However, no such major sources of resistance have been found in *B. oleracea* types (Walkey 1982; Pink et al. 1986; Pink and Walkey 1988; Walkey and Pink 1988). Where known, the *B. napus* resistance genes are in the A genome of *B. napus*, suggesting that there are no genes for extreme forms of resistance to *TuMV* in the C genome of *B. oleracea*.

Some sources of resistance have been identified to *TuMV* in *B. rapa* appear to have resistance to a broad spectrum of pathotypes (Suh et al. 1995; Jenner and Walsh 1996). For one of these resistances, no isolate of *TuMV* so far tested has been found to overcome the resistance. The resistance is not immunity, replication is detected in inoculated leaves, but the viruses do not appear to spread systemically indicating that the resistance is effective against long-distance movement of the virus. One *Brassica* resistance gene (TuRB01) has been mapped and others are currently being mapped culture (Jenner and Walsh 1996). This will facilitate rapid incorporation of resistance to *TuMV* into desirable genetic backgrounds via marker-assisted breeding. Interspecies transfer from *B. rapa* to *B. oleracea* is also being attempted using these techniques.

Sources of resistance to the virus have also been identified in non-*Brassica* species. The dominant Turnip gene in lettuce confers resistance to infection by *TuMV* and this gene has been mapped (Robbins et al. 1994). Most accessions of chicory (*Cichorium intybus*) tested by Provvidenti et al. (1979) were found to be resistant to *TuMV* and were proposed as a valuable source of resistance for endive and escarole (*Cichorium endivia*), all accessions of which tested were susceptible to the virus. Resistance to one strain of the virus has been reported in *Impatiens balsamina* (Provvidenti 1982).

1.9.13.6 Disease Management

It is difficult to control *TuMV* because it is mainly transmitted in a nonpersistent mode by at least 89 aphid species. Specifically, *TuMV* is introduced into plant cells via the stylet of aphids in a typical nonpersistent transmission mode during aphid probing or feeding. The wide host range, high genetic variability and transmission mode of TuMV make the virus difficult to control through traditional methods such as chemicals. Chemical insecticides may control one or multiple aphid species but, as there are more than 89 aphid species capable of transmission, these aphids are soon replaced by other species leading to a continuation of host infection. Additional methods of TuMV control include cultivation of resistant host varieties, which can provide a more effective and environmentally friendly approach. In particular, virus-resistant cultivars can be important in controlling TuMV epidemics, although resistance-breaking *TuMV* pathotypes that overcome single gene resistance may arise (Jenner and Walsh 1996; Hughes et al. 2002; Walsh et al. 2002; Gładysz and Hanus-Fajerska 2009). It is necessary to combine control measures (including identification of symptoms and virus by molecular, serological and electron microscopical methods) with host resistance to ensure effective TuMV-induced disease management (Jones 2004, 2006; Meena et al. 2021).

1.9.14 *Verticillium* (*Verticillium* Wilt or Verticillium Stem Striping)

The *Verticillium longisporum* (e.g., *V. dahliae* var. *longisporum* Stark; comb. nov. Karapapa) is a soil borne fungal pathogen causing vascular diseases of cruciferous plants (Karapapa et al. 1997; Zeise and von Tiedemann 2001, 2002a, b). Verticillium wilt is a disease on oilseed rape (*Brassica napus* L. sp. *oleifera*), threatening its production particularly in the northern European countries (Dunker et al. 2008). Oilseed rape colonized by *V. longisporum* does not develop wilting symptoms, and therefore the common name of Verticillium wilt is unsuitable for this crop. Therefore, Depotter et al. (2016) proposed "Verticillium stem striping" as the common name for *Verticillium* infections of oilseed rape (Meena et al. 2021).

1.9.14.1 The Disease and Causal Organism

Verticillium longisporum is a member of the Flavno nexudans lineage and thus lacks the ability to produce yellow hyphal pigmentation. The pathogen produces durable microsclerotia which accumulate in the soil and from which plant roots are attacked (Schnathorst 1981). The germination of resting structures, which is triggered by root exudates, fungal hyphae grow toward the root surface and penetrate the root epidermal cells near the root tips (Eynck et al. 2007; Zhou et al. 2006). Before entering into the xylem, the fungus traverses the root cortex inter- as well as intracellularly. During most of its life cycle, *V. longisporum* is confined to the vascular system, a nutrient-poor environment to which the fungus is well adapted (Pegg 1985; Van Alfen 1989). The pathogen spreads with growing hyphae and/or conidiophores conveyed with the transpiration stream into upper parts of the plant vascular system. As the host tissue turns to senescence, the pathogen enters a final saprophytic growth stage in which microsclerotia are abundantly formed in the dying stem parenchyma (Meena et al. 2021).

1.9.14.2 Economic Importance

Although, the impact of *V. longisporum* infection on oilseed yield remains unclear because Verticillium stem striping on oilseed rape appears toward the end of the cropping season. Significant yield reductions upon *V. longisporum* infection only occurred in a single field trial on all tested oilseed rape cultivars. These preliminary data suggest that Verticillium stem striping does not consistently impact oilseed rape yield, despite the occurrence of abundant disease symptoms (Depotter et al. 2019). Yield losses by Verticillium stem striping have been anticipated to range from 10% to 50%, yet experimental verifications of such estimations are lacking (Dunker et al. 2008). In contrast, Verticillium stem striping did not impact yield significantly in previous field studies despite the presence of stem striping symptoms (Dunker et al. 2008).

1.9.14.3 Infection, Pathogenesis, and Host Resistance

The breeding for resistance has been hampered by the lack of sufficient resistance in commercially available breeding material as well as in the different varieties. However, a promising level of resistance to *V. longisporum* has been identified in cabbage (*B. oleracea*) (Dixelius et al. 2005; Happstadius et al. 2003; Rygulla et al. 2007a). The significant improvement of resistance in *B. napus* against *V. longisporum* has been achieved by hybridization of resistant *B. oleracea* with *B. rapa* (Rygulla et al. 2007b). Resistance to wilt pathogens such as *Verticillium* is supposed to depend predominantly on the physical restriction and chemical inhibition of the pathogen during the systemic phase of colonization (Nicholson and Hammerschmidt 1992). Unlike many biotrophic and some necrotrophic interactions, complete resistance to vascular infection has been considered to be unlikely (Beckman 1987). Resistance to vascular pathogens is thus expressed "internally," representing an exceptional phenomenon in the realm of plant–pathogen interactions. Usually, this type of resistance is based on the rapid build-up of mechanical barriers, vascular gels or tyloses delaying or even preventing the spread of the pathogen in the vascular system (Sinha and Wood 1968; Meena et al. 2021).

Phenolic compounds delivered along the phenylpropanoid pathway play an important role in defense to pathogen infection either as preformed or postinfectional defense factors. They have been assigned to various important biological functions in defense such as cell wall reinforcement and antimicrobial activity (Tuncel and Nergiz 1993), as modulators of plant hormones in defense signaling or as scavengers of reactive oxygen species (Dixon and Paiva 1995). Particularly the phenolic polymer lignin is an important principal structural component of secondary vascular tissue and fibers in higher plants (Humphreys et al. 2002; Whetten and Sederoff 1995). It is known to play a fundamental role in mechanical support, solute conductance and disease resistance (Harakava 2005). Deposition of lignin, lignin-like polymers, and other wall-bound phenolic materials is reported to be a response to mechanical damage, wounding, or microbial infection (Boudet et al. 1995). In addition to cell wall strengthening and increased cell wall rigidity (Wardrop 1971), lignin deposition is supposed to decrease the diffusion of enzymes and toxins released from pathogenic fungal hyphae to the host, and of water and nutrients from the host to the fungus, thus essentially starving the intruder (Vance et al. 1980). This data on histochemical and physiological generated during the elucidation of mechanisms involved in the genotype-specific internal resistance of a *B. napus* genotype to *V. longisporum*. In particular, resistance is related to changes in the phenolic composition of root and stem tissues, to alterations in the vascular cell walls and to vessel occlusions occurring in the vascular tissue during particular stages of infection and systemic colonization. Pathogen spread in a susceptible and a resistant *B. napus* cultivar was followed by quantitative polymerase chain reaction (qPCR). The first histological and biochemical characterization of quantitative resistance in a rapeseed genotype against *V. longisporum* was documented by Eynck et al. (2009).

1.9.14.4 Disease Management

The pathogen responsible for disease in oilseed rape cannot be controlled efficiently with fungicides as well as their ability for survival of microsclerotia in the soil (Heale and Karapapa 1999) restricted the control measures of the disease on either cultural practices such as wider crop rotations or the use of resistant cultivars (Meena et al. 2021).

1.9.15 *Xanthomonas* (Black Rot)

Black rot of crucifers incited by *Xanthomonas campestris* pv. *campestris* (Pam.) Dowson (Xcc), is one of the most widespread and devastating worldwide. The black rot bacteria cause systemic infection in the susceptible plants and penetrate the plants through the hydathodes or wounds. Typical disease symptoms are "V"-shaped necrotic lesions appearing from the leaf margins with blackened veins.

1.9.15.1 The Disease and Causal Organism

It is a small, rod shaped, aerobic, gram negative, nonspore forming bacterium. The bacterium has a single polar flagellum and it is catalase positive, hydrogen sulfide positive, oxidase negative and does not produce nitrate or indole. It produces a yellowish extracellular polysaccharide (EPS) called xanthan on media containing glucose (Gupta et al. 2013). Black rot caused by the vascular, gram-negative bacterium, *X. campestris* pv. *campestris*, is one of the most devastating seed borne epidemic disease worldwide (Vicente et al. 2001; Taylor et al. 2002). The bacteria caused systemic infection in the susceptible host plants and bacterium penetrate the host plants species via natural openings (hydathodes on the leaf margins) or wounds and replicates in the vascular tissues (Cook et al. 1952b; Ignatov et al. 1999b). Once the bacteria invades the plant through hydathodes, *Xcc* multiply and colonize the vascular system, and extracellular polysaccharide (EPS) called xanthan are produced, which can obstruct the xylem vessels, lead to necrosis of tissue and severe leaf wilting (Williams 1980).

The initial symptoms of the disease occurs as V-shaped chlorotic to necrotic foliar lesions on the margin of the leaves. The point of the V-shaped lesions is directed toward a vein (Plate 1.11a–c) as the infection advances. Eventually, the pathogen affected larger area of the leaf and infected plants become necrotic and the veins turn black or brown (Plate 1.11d–g). Plant metabolic activities are severely affected with the advancement of the disease which results in stunted growth, wilting and stem rot where infected roots become black and shriveled (Plate 1.11h). As the pathogen proceeds from the leaf margins toward the veins, water stress and chlorotic symptoms develop due to occlusion of water-conducting vessels by bacteria and components of degraded plant cell walls. The darkening of vascular tissues following bacterial invasion gives the black rot symptoms. The disease can also be spread through infected seedlings, infested soil, crop debris and other host weeds species (Schaad and Dianese 1981; Vicente and Holub 2013; Meena et al. 2021).

1.9.15.2 Economic Importance

The disease has a wide geographical distribution and is destructive to crucifer vegetables causing yield (50–65%) and quality loss due to cultivar susceptibility (Williams 1980; Dhar and Singh 2014).

1.9.15.3 Infection, Pathogenesis, and Disease Cycle

The pathogen moves systemically through the vascular bundles turning them black and produces main symptoms on leaves as described by various workers. The

Plate 1.11 (**a**) V-shaped initial symptoms on the leaves; (**b**) advances of the V-shaped lesions on leaves; (**c**) symptoms further advances with yellow hallo along with veins; (**d**) Infected areas become brown and start dying; (**e**, **f**) infection further advances and cover larger areas along the veins; (**g**) brown to black area increases as the disease advances; (**h**) roots become black and shriveled in severely infected plants

pathogen produces two types of symptoms. Initially, the marginal chlorotic spots appear on the leaves followed by darkening or blackening of mid rib and veins (Cook et al. 1952a). However, Sutton and Williams (1970) reported that vein blackening was the first visible symptom which was due to the accumulation of melanin among xylem cells. The pathogen colonizes the vascular system after its entry into the plant and produces plentiful extracellular polysaccharide called xanthan (Jeanes 1973) which along with the bacterial cells plug the xylem vessels, restricting the water flow and resulting in the characteristic V-shaped chlorotic lesions originating from the margins of the leaves.

The bacteria causes' systemic infection in the susceptible host plants and bacterium penetrate the host plants species via natural openings (hydathodes on the leaf margins) or wounds and replicates in the vascular tissues (Cook et al. 1952b; Ignatov et al. 1999a, b). Once the bacteria invades the plant through hydathodes, *Xcc* multiply and colonize the vascular system, and extracellular polysaccharide (EPS) called xanthan are produced, which can obstruct the xylem vessels, lead to necrosis of tissue and severe leaf wilting (Williams 1980). The point of the V-shaped lesions directed toward a vein. Eventually, the pathogen affected regions on the plants become necrotic and the veins turn black or brown. Plant metabolic activities are severely affected with the advancement of the disease which results in stunted growth.

1.9.15.4 Host Range

Black rot was first reported from Iowa, North America, on rutabaga by Pammel (1895). In 1898, the disease was reported on cabbage from Wisconsin by Russel. Since then, there has been a growing recognition of seriousness of black rot on cruciferous crops throughout the world. In India, the disease was recorded for the first time on cauliflower by Patel et al. (1949). This disease is reported to cause huge loss from different regions of world such as Brazil, Ethiopia, South Africa, Belgium, Germany, Sweden, Hungry, France, the Netherlands, Italy, the USA, the UK, Nepal, China, Taiwan, and Canada along with Australia and India (Saharan 1992b; Mulema et al. 2012; Singh et al. 2016; Akhtar et al. 2017; Meena et al. 2021).

The *Xanthomonas campestris* pv. *campestris* (*Xcc*) belongs to the genus that causes diseases on at least 124 monocotyledonous and 268 dicotyledonous plant species including all major crop plants (Leyns et al. 1984). Several other plant species in the Brassicaceae family including radish, turnip, rutabaga, vegetable mustard, black mustard, wild mustard, ornamentals and cruciferous weeds also serve as a host of this pathogen. The disease was recorded on cabbage in the USA by Garman (1894). The bacterium is found to attack all members of Cruciferae family including rutabaga, cabbage, savoy cabbage, broccoli, turnip, radish, kohlrabi, *Lepidium sativum*, *Raphanus sativus* var. *caudatus*, *Brassica juncea*, fodder cabbage, *Capsella bursa-pastoris,* and *Centella asiatica* (Mahiar and Khalaif 1999).

1.9.15.5 Disease Management

Sanitation and management practices, including crop rotation, weed control and the use of assayed clean seed, can provide significant control of the disease (Schaad and Alvarez 1993). Seed treatments, including hot water, antibiotics, antibiotics and sodium hypochlorite, hydrogen peroxide and hot acidified cupric acetate or zinc sulfate, are available, but no treatment is totally effective. Several methods can be used to reduce the spread of the disease during transplant raising, including the use of web and flow irrigation systems instead of overhead irrigation and of chlorine dioxide in the irrigation water (Krauthausen et al. 2011; Meena et al. 2021).

1.10 Expression of Genes to Overcome the Biotic Stresses

Plants have two levels of immune system that protects them against different types of pathogens. At the first, a pattern recognition receptor (PRR) recognizes pathogens attacks on the surface of the plant cell. The PRR produces signaling and pass to nucleus that activate defense related genes. Thereby, the cells try to stop further invasions of the pathogens through producing huge amount of callose, tyloses and ROS. Plant cells are deposited with callose and tyloses just inside cell wall, resulting in cell walls thickness which protect from the pathogens (Fig. 1.6). Secondly, the plant cells start secretion of high amount of ROS and there is activation of pathogenesis related proteins (PRs) which threatening pathogens (Kumar 2013). At the same time, pathogens also are injecting sets of affecter molecule inside the cell that try to damage or defeat the plant immunity system. The affecter molecules damage transduction signaling and in response to that disrupt plant defense system and ultimately plants are infected by pathogens in susceptible plants (Fig. 1.6; Oppel et al. 2009).

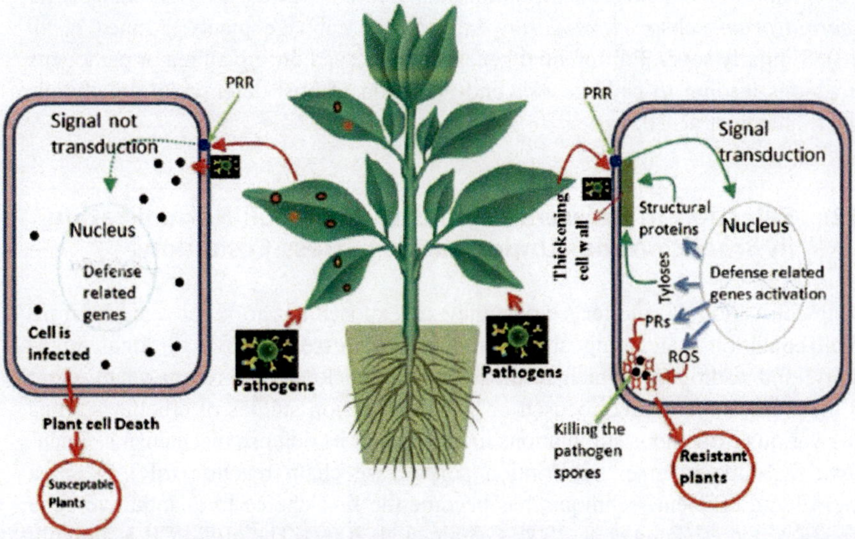

Fig. 1.6 Schematic model of plant–pathogen interaction and molecular mechanisms involved in resistance and susceptible reactions to pathogens attacks. *PRR* pattern recognition receptor, *PRs* pathogenesis related, *ROS* reactive oxygen species (Kayum et al. 2016a)

1.11 Differential Genes Expression in *Brassica* Under Biotic Stresses

Biotic stress is very devastating stress for cruciferous crops. Several numbers of genes have well documented that they are differentially expressed against different biotic stresses. However, when infection with *F. oxysporum* f. sp. *conglutinans*, six *Brewery* genes (*BrWRKY4*, *BrWRKY465*, *BrWRKY472*, *BrWRKY497*, *BrWRKY4133*, and *BrWRKY4141*) showed significantly higher expression and were about eight-, sixfold, sixfold, threefold, and fivefold higher, respectively (Kayum et al. 2015a, b). *BrWRKY141* showed about 180-fold higher expression after *P. carotovorum* subsp. *carotovorum* infection (Kayum et al. 2015a). After infection with *F. oxysporum* f. sp. *conglutinans*, *BrAL*2, *BrAL*23, *BrAL*24, *BrAL*27, *BrAL*29, *BrAL*210, *BrAL*213, *BrAL*214, and *BrAL*215 showed several fold higher expression, and *BrAL* 2, *BrAL* 3, *BrAL* 7, *BrAL* 9, *BrAL* 13, *BrAL* 14, and *BrAL* 15 were expressed in response to both biotic and abiotic stresses (Kayum et al. 2015b). Ahmed et al. (2012a) reported that three chitinase genes (*BrCLP1*, *BrCLP2*, and *BrCLP3*) responses against *P. carotovorum* subsp. *carotovorum* infection. The defensin-like family protein (BrDLFP) expression significantly increased after infection with *P. carotovorum* subsp. *carotovorum* in Chinese cabbage (Ahmed et al. 2012a). On the other hand, Alfin-like two genes (*BoAL8* and *BoAL12*) were induced after infection with *P. carotovorum* subsp. *carotovorum* in *B. oleracea* (Kayum et al. 2016a). Among the 12 thaumatin-like proteins, three expressed differentially after *P. carotovorum* subsp. *carotovorum* infection in cabbage plants (Ahmed et al. 2013). Similarly several differential genes are expressed during all major pathogens interactions leading to pathogenesis and induction of host defenses in cruciferous crops (Saharan et al. 2021a).

1.12 Selection of Reference Genes for qRT-PCR Normalization in *Brassica Napus* Under Various Stress Conditions

During cultivation of crucifer's crops, they are exposed to various adverse environmental conditions, including abiotic and biotic stresses, such as drought, high-salinity, and pathogens, which result in production losses. In recent years, more and more researchers have focused on gene expression studies of crucifer's crops under various exogenous stimulations to understand its defense mechanisms. Quantitative real-time reverse transcription polymerase chain reaction (qRT-PCR), a powerful and efficient technique, has become the first choice in quantitative gene expression studies (Fang et al. 2004; Guenin et al. 2009). The qRT-PCR is a highly sensitive technique; however, the accuracy of its results can be affected by certain uncertainties, such as RNA quantity, purity, reverse-transcription efficiency, and PCR amplification (Zhu et al. 2013). To control these variables, the careful selection of reference genes as internal controls for normalizing gene expression is essential (Guenin et al. 2009; Kundu et al. 2013). Ideal reference genes should be stably expressed regardless of the experimental conditions under study (Goossens et al.

2005; Schmittgen and Zakrajsek 2000). Traditional reference genes are mostly "housekeeping" genes that are thought to have stable expression patterns, such as 18S ribosomal RNA, actin, polyubiquitin, glyceraldehyde-3-phosphate dehydrogenase, and elongation factor 1-α (Wang et al. 2013). However, to date, more and more studies have suggested that the expressions of these genes change significantly under different experimental treatments (Kim et al. 2003; Nicot et al. 2005; Stürzenbaum and Kille 2001). Therefore, for accurate normalization in gene expression studies, it is necessary to evaluate the expression stability of reference genes and to select the most suitable reference genes under different experimental conditions before real-time PCR amplification. Several statistical algorithms have been developed to calculate gene expression stability and to determine the appropriate reference genes and their numbers required for normalization under specific experimental conditions (Kundu et al. 2013). In terms of geNorm algorithm, the *M* value of each gene, which is a stability measure, is calculated on the basis of the pair wise variation between it and all other control genes. The genes with the lowest *M* values have the most stable expression, and geNorm determines these genes by increasingly eliminating less stable genes from the candidate reference genes pool (Vandesompele et al. 2002).

NormFinder, another algorithm, employs a statistical and mathematical model that not only estimates the overall expression variation of a candidate gene, but also considers the variation between the chosen subgroups (Huis et al. 2010). To date, a number of studies focusing on reference gene evaluation and validation have been undertaken in various plants, including *Arabidopsis* (Czechowski et al. 2005). To evaluate the stability of nine candidate reference genes for normalization across a large set of biological samples representing *B. napus* under different stress conditions to study was conducted by Wang et al. (2014). Statistical algorithms implemented in geNorm and NormFinder were used. The results identified the best-performing reference genes in this model, and will benefit future gene expression studies in *B. napus* and other species of the genus *Brassica* (Wang et al. 2014).

Data normalization is essential for reliable output of quantitative real-time reverse-transcription polymerase chain reaction (qRT-PCR) assays, as the unsuitable choice of reference gene(s), whose expression might be influenced by exogenous treatments in plant tissues, could cause misinterpretation of results. Wang et al. (2014) investigated the expression variations of nine candidate reference genes in 40 samples of *B. napus* leaves subjected to various exogenous treatments. Parallel analyses by geNorm and NormFinder revealed that optimal reference genes differed across the different sets of samples. The best-ranked reference genes were *PP2A* and *TIP41* for salt stress, *TIP41* and *ACT7* for heavy metal (Cr6+) stress, *PP2A* and *UBC21* for drought stress, *Fbox* and *SAND* for cold stress, *F-box* and *ZNF* for salicylic acid stress, *TIP41*, *ACT7*, and *PP2A* for methyl jasmonate stress, *TIP41* and *ACT7* for abscisic acid stress, and *TIP41*, *UBC21*, and *PP2A* for *Sclerotinia sclerotiorum* stress. Two newly employed reference genes, *TIP41* and *PP2A*, showed better performances, suggesting their suitability in multiple conditions. To further validate the suitability of the reference genes, the expression patterns of *BnWRKY40* and *BnMKS1* were studied in parallel (Wang et al. 2014).

1.13 Signaling System of Crucifers' Host-Pathosystem in Response to Multiple Stresses

Cruciferous plants have developed numerous molecular mechanisms to respond to various environmental and multiple biotic and abiotic stresses. Whenever plant encounters any kind of stress, first of all there is perception of the signals by the potential sensors. This further generates secondary messengers (e.g., ABA, ROS, Ca^{2+}, inositol phosphates etc.). The secondary messengers can alter the levels of intracellular Ca^{2+}. Any change in cytoplasmic Ca^{2+} is sensed by Ca^{2+} sensors i.e. calcium binding proteins and after interacting with corresponding interacting partners they initiate a protein phosphorylation cascade. This signaling cascade ultimately regulates the transcription factors that manage specific suite of stress-regulated genes and the functional proteins involved in protecting the cell against stress. The expression of stress-regulated genes leads to the production of various kinds of regulatory molecules like the phytohormones ABA, ET, and SA and transcription factors etc. A second round of signaling can get started by these regulatory molecules which may further enhance the initial signaling pathway. These signaling networks are very complex and interconnected at many levels for transferring information to generate robust host resistance. Moreover, during signaling other components are also involved. A successful signal transduction pathway demands proper coordination of all signaling molecules. This is brought about by specific molecules whose functions are to transport, bring together or modify the signaling molecules. They themselves are not involved in relaying the signal. These proteins consist of scaffolding proteins, protein modifiers (e.g., enzymes needed for protein glycosylation, methylation, ubiquitination and lipidation) and adaptors. Cruciferous interactions with multiple stresses trigger a wide range of defense reactions at molecular and cellular levels to ensure that host cells are insulated from these challenges they encounter. The production of ROS, MAP Kinase cascades and hormone signaling are triggered by multiple stresses and play crucial roles as common components of the host immune system. *GbRLK*, a receptor like kinase (RLK) gene is involved in response to multiple stresses as positive regulator. Transcription factors (TFs) are the central core of the genes as defense signaling pathways to multiple stresses. MYB TFs play essential roles in multiple stress responses. ATAF1 is a multifaceted TF with versatile capabilities and plays an important role in mediating crosstalk between multiple stress responses. HSPs gene regulate the function of *NBS-LRR* genes in response to multiple stresses. The miRNAs may target only one locus or several loci in response to multiple stresses. Cell signaling in response to multiple stresses largely depends on the SnRK family of protein kinases in plants. Calcium is a key signaling pathway in host responses to multiple stresses through calcium sensor proteins like calcineurin-B-like protein (CBls), the calmodulin (CaM), and calmodulin-like protein (CMLs).

Transcriptome analyses based on microarrays and proteomic studies have given rise to rapid progress in the field of plant signal transduction and gene regulation. Gene Chip and cDNA microarrays together with massive whole genome sequencing proved to be useful in identifying novel signaling determinants on a whole-genome scale in response to various stresses. This has increased our understanding toward the complex mechanisms involved in various aspects of plant responses to environmental cues. Proteomics approach is useful in studying post-translational modifications of the proteins. It can also be used to clone unique genes by means of differential analysis which will expedite our understanding of stress signaling mechanisms in plants. Therefore, stress-tolerant plants can be genetically engineered by the combined efforts of plant molecular biologists, physiologists, and breeders (Saharan et al. 2021b).

1.14 Genomics of Crucifers' Host Resistance

The brighter side of crucifers' host-pathosystem is that *Brassica* scientists have generated enormous data on genetical and molecular aspects of host–pathogen interaction using major pathogens of seven Brassica species as the model biological system. The data generated on identification, isolation, molecular mapping and functional characterization of several genes are of very practical use and has been used for breeding disease resistance genotypes in *Brassica* crops at global level. The detailed description on genomics of crucifers host resistance has been covered separately in two books viz., Genomics of crucifers' host resistance and Molecular mechanisms of crucifers' host resistance published by Springer, Nature, the Netherlands (Saharan et al. 2021a, b).

References

Abawi GS, Grogan RG (1975) Source of primary inoculum and effects of temperature and moisture on infection of beans by *Whetzelinia sclerotiorum*. Phytopathology 65:300–309

Abawi GS, Grogan RG (1979) Epidemiology of diseases caused by Sclerotinia species. Symposium on Sclerotinia. Am Phytopathol Soc 69(8):899–904

Abuqamar S, Luo H, Laluk K, Mickelbart MV, Mengiste T (2009) Crosstalk between biotic and abiotic stress responses in tomato is mediated by the *AIM1* transcription factor. Plant J 58:347–360

Acquaah G (2007) Principles of plant genetics and breeding. Blackwell, Oxford, p 385

Agrios G (2005) Plant pathology, 5th edn. Elsevier Academic Press, Burlington, VA, p 952. https://doi.org/10.1016/C2009-0-02037-6

Ahmed NU, Park JI, Seo MS, Kumar TS, Lee IH, Park BS (2012a) Identification and expression analysis of chitinase genes related to biotic stress resistance in *Brassica*. Mol Biol Rep 39:3649–3657

Ahmed NU, Park JI, Jung HJ, Lee IH, Song IJ, Yang SY (2012b) Identification and characterization of longevity assurance gene related to stress resistance in *Brassica*. Afr J Biotechnol 11:12721–12727

Ahmed NU, Park JI, Jung HJ, Chung MY, Cho YG, Nou IS (2013) Characterization of thaumatin-like gene family and identification of Pectobacterium carotovorum subsp. carotovorum inducible genes in Brassica oleracea. Plant Breed Biotech 1:111–121

Ahmed NU, Jung HJ, Park JI, Cho YG, Hur Y, Nou IS (2015) Identification and expression analysis of cold and freezing stress responsive genes of *Brassica oleracea*. Gene 554:215–223

Ahmed H, Howton TC, Sun Y, Weinberger N, Belkhadir Y, Mukhtar MS (2018) Network biology discovers pathogen contact points in host protein-protein interactomes. Nat Commun 9:2312

Ainsworth GC, Sparrow FK, Sussman AS (1973) The fungi: an advance treatise, vol IV. Academic Press, New York. (A&B)

Akhtar J, Singh B, Kandan A, Kumar P, Chand D, Maurya AK, Dubey SC (2017) Interception of pathogens during quarantine processing: an effort towards safe introduction of oilseed and vegetable *Brassicas* germplasm in India. J Oilseed Brassica 8:120–130

Alabouvette C, Brunin B, Louvet J (1974) Recherches sur la maladie du colza due à*Leptosphaeria maculans* (Desm.) Ces. et de Not. IV. Pouvoir inffectieux des pycnidiospores et sensibilité variétale. Ann Phytopathol 6:265–275. (in French)

Amelung D, Daebeler F (1988) White spot (*Pseudocercosporella capsellae*/Ell. et Ev./Deighton)-a new disease of winter rape in the German Democratic Republic. Nachricht Pflanzensch DDR 42:73–74

Angadi SV, Cutforth HW, Miller PR, McConkey BG, Entz MH, Brandt SA (2000) Response of three *Brassica* species to high temperature stress during reproductive growth. Can J Plant Sci 80: 693–701

Annis SL, Goodwin PH (1996) Comparison of cell wall-degrading enzymes produced by highly and weakly virulent isolates of *Leptosphaeria maculans* in culture. Microbiol Res 151:401–406

Anonymous (2020) Canola encyclopedia: white leaf spot and gray stem. Canola Council of Canada, Winnipeg, MB

Ansan-Melayah D, Rouxel T, Bertrandy J, Letarnec B, Mendes-Pereira E, Balesdent M-H (1997) Field efficiency of *Brassica napus* specific resistance correlates with *Leptosphaeria maculans* population structure. Eur J Plant Pathol 103:835–841

Arabidopsis Genome Initiative (2000) Analysis of the genome sequence of the flowering plant *Arabidopsis thaliana*. Nature 408:796–815

Ashraf M, McNeilly T (2004) Salinity tolerance in *Brassica* oilseeds. Crit Rev Plant Sci 23:157–174

Atkinson NJ, Urwin PE (2012) The interaction of plant biotic and abiotic stresses: from genes to the field. J Exp Bot 63:3523–3543

Audenaert K, De Meyer GB, Hofte MM (2002) Abscisic acid determines basal susceptibility of tomato to *Botrytis cinerea* and suppresses salicylic acid-dependent signaling mechanisms. Plant Physiol 128:491–501

Awasthi RP, Nashaat NI, Kolte SJ (1995) Interaction between *Peronospora parasitica* and *Albugo candida* in relation to development of stag heads in *Brassica juncea*. In: Proc, National symposium on detection of plant pathogens and their management, Faizabad, India, January 18–20, p 36

Awasthi RP, Nashaat NI, Heran A, Kolte SJ, Singh U (1997) The effect of *Albugo candida* on the resistance to *Peronospora parasitica* and vice versa in rapeseed-mustard. In: Proceeding, ISHS Symposium on Brassicas, Tenth Crucifer Genetics Workshop, September 23–27, 1997, Rennes, France

Badawy HMA, Kakau J, Hoppe HH (1992) Temperature and ageing of host tissue affect the interactions between different oilseed rape cultivars and pathotype groups of *Leptosphaeria maculans*. J Phytopathol 134:255–263. https://doi.org/10.1111/j.1439-0434.1992.tb01234.x

Baker SD (1955) The genus Albugo in New Zealand. Trans R Soc N Z 82:987–993

Balesdent MH, Barbetti MJ, Li H, Sivasithamparam K, Gout L, Rouxel T (2005) Analysis of *Leptosphaeria maculans* race structure in a worldwide collection of isolates. Phytopathology 95:1061–1071

Barbetti MJ (1975) Effects of temperature on development and progression in rape of crown canker caused by *Leptosphaeria maculans*. Aust J Exp Agric Ani Husb 15:705–708

Barbetti MJ (1976) Late blackleg infections in rape are important. Aust Plant Pathol Soc Newsl 4:3–4

Barbetti MJ, Khangura RK (1997) Developments for better management of blackleg disease in Western Australia. In: Proceedings of the 11th Australian Research Assembly on *Brassicas*, Perth, WA, 1997, Perth, Australia: Agriculture Western Australia, pp 11–14

Barbetti MJ, Khangura RK (1999) Managing blackleg in the disease-prone environment of Western Australia. In: Proc 10th Int Rapeseed Congress, (1999) Canberra, Australia. http://www.regional.org.au/papers/index.htm

Barbetti MJ, Khangura RK (2000) Fungal diseases of canola in Western Australia. In: Agriculture Western Australia, Bulletin no. 4406. Agriculture Western Australia, Perth, WA

Barbetti MJ, Sivasithamparam K (1981) *Pseudocercosporella capsellae* and *Myrothecium verrucaria* on rapeseed in Western Australia. Australas Plant Pathol 10:43–44. https://doi.org/10.1071/APP9810043

Barbetti MJ, Carmody P, Khangura RK, Sweetingham M, Walton G (2000) Managing blackleg in Western Australia. In: Agriculture Western Australia, Bulletin no. 4400. Agriculture Western Australia, Perth, WA

de Bary A (1863) Recherches sur le developement quelques champignons parasites. Ann Sci Nat Bot 20(4):5–148

Basso J, Dallaire P, Charest PJ, Devantier Y, Laliberté JF (1994) Evidence for an internal ribosome entry site within the 5′ non-translated region of turnip mosaic potyvirus RNA. J Gen Virol 75: 3157–3165

Beckman CH (1987) The nature of wilt diseases of plants. American Phytopathological Society, St. Paul, MN

Berkenkamp B, Vaartonou H (1972) Fungi associated with rape root rot in Alberta. Can J Plant Sci 52:973–976

Berlese AN, de Toni JB (1888) Phycomycetae. In: Saccardo PA (ed) Sylloge Fungorum, vol 7, pp 181–322

Bhargava SN, Shukla DN, Singh N (1980) Root- and foot-rot of *Eruca sativa* caused by *Alternaria alternata* (Fr) Keissler. Curr Sci India 49:452

Biddulph JE, Fitt BDL, Leech PK, Welham SJ, Gladders P (1999) Effects of temperature and wetness duration on infection of oilseed rape by ascospores of *Leptosphaeria maculans* (stem canker). Eur J Plant Pathol 105:769–781

Biga MLB (1955) Review of the species of the genus *Albugo* based on the morphology of the conidia. Sydowia 9:339–358

Blanc G, Barakat A, Guyot R, Cooke R, Delseny M (2000) Extensive duplication and reshuffling in the *Arabidopsis* genome. Plant Cell 12:1093–1101

Boerema G, Verhoeven A (1980) Check-list for scientific names of common parasitic fungi. Series 2b: Fungi on field crops: cereals and grasses. Neth J Plant Pathol 83:165–204. https://doi.org/10.1007/BF01976647

Bokor A, Barbetti MJ, Brown AGP, MacNish GC, Wood PM (1975) Blackleg of rapeseed. J Dept Agric W A 16:7–10

Booth V (1978) *Fusarium equiseti*. In: IMI description of fungi and bacteria. CABI Biosciences, Surrey. No. 58: Sheet 571

Bosch U, Mirocha CJ (1992) Toxin production by *Fusarium* species from sugar-beets and natural occurrence of zearaenone in beets and beet fiber. Appl Environ Microbiol 58:3233–3239

Boudart G (1982) The black-leg disease: some aspects of the host–parasite relationship. Crucif Newsl 7:63–64

Boudet AM, Lapierre C, Grima-Pettenati J (1995) Biochemistry and molecular biology of lignification. New Phytol 129:203–236

Boys EF (2009) Resistance to *Pyrenopeziza brassicae* (light leaf spot) in *Brassica napus* (oilseed rape). Ph D Thesis, University of Nottingham, UK

Boys EF, Roques SE, Ashby AM, Evans N, Latunde-Dada AO, Thomas JE, West JS, Fitt BDL (2007) Resistance to infection by stealth: *Brassica napus* (winter oilseed rape) and *Pyrenopeziza brassicae* (light leaf spot). Eur J Plant Pathol 118:307–321. https://doi.org/10.1007/s10658-007-9141-9

Braun U (2001) Taxonomic notes on some species of the *Cercospora* complex (VII). Fungal Divers 8:41–71

Bray EA (2002) Classification of genes differentially expressed during water-deficit stress in *Arabidopsis thaliana*: an analysis using microarray and differential expression data. Ann Bot 89:803–811

Bray EA, Bailey-Serres J, Weretilnyk E (2000) Responses to abiotic stresses. In: Gruissem W, Buchannan B, Jones R (eds) Biochemistry and molecular biology of plants. ASPP, Rockville, MD, pp 1158–1249

Brun H, Levivier S, Eber F, Renard M, Chevre AM (1997) Electrophoretic analysis of natural populations of *Leptosphaeria maculans* directly from leaf lesions. Plant Pathol 46:147–154

Brun H, Chevre AM, Fitt BD, Powers S, Besnard AL, Ermel M, Huteau V, Marquer B, Eber F, Renard M, Andrivon D (2010) Quantitative resistance increases the durability of qualitative resistance to *Leptosphaeria maculans* in *Brassica napus*. New Phytol 185:285–299

Burdyukova LI (1980) Albuginacea fungi. Taxonomy, morphology, biology, and specialization. Ukrainskyi Botanichnyi Zhumal 37:65–74. (Russian)

Butler EJ (1918) Fungi and diseases in plant. Thaker Spink & Co, Calcutta, pp 297–300

Butler EJ, Bisby GR (1934) The fungi of India. Sci. Manager No. 1. ICAR, New Delhi, India

Büttner D, Bonas U (2010) Regulation and secretion of *Xanthomonas* virulence factors. FEMS Microbiol Rev 34:107–133

Cai Q, Qiao L, Wang M, He B, Lin FM, Palmquist J, Huang SD, Jin H (2018) Plants send small RNAs in extracellular vesicles to fungal pathogen to silence virulence genes. Science 360:1126–1129. https://doi.org/10.1126/science.aar4142

Cannon PF, Damm U, Johnston PR, Weir BS (2012) *Colletotrichum* – current status and future directions. Stud Mycol 73:181–213. https://doi.org/10.3114/sim0014

Cardoza V, Stewart CN (2004) *Brassica* biotechnology: progress in cellular and molecular biology. In Vitro Cell Dev Biol Plant 40:542–551

Carmody S (2017) Light leaf spot and white leaf spot of Brassicaceae in Washington State. M Sc Thesis, Washington State University

Carreón-Anguiano KG, Islas-Flores I, Vega-Arreguín J, Sáenz-Carbonell L, Canto-Canché B (2020) Eff Hunter: a tool for prediction of effector protein candidates in fungal proteomic databases. Biomolecules 10:712. https://doi.org/10.3390/biom10050712

Castro-Moretti FR, Gentzel IN, Mackey D, Alonso AP (2020) Metabolomics as an emerging tool for the study of plant-pathogen interactions. Metabolites 10:52

Cerkauskas R, Stobbs L, Lowery D, Van Driel L, Liu W, Van Schagen J (1998) Diseases pests and abiotic problems associated with oriental cruciferous vegetables in southern Ontario in 1993-1994. Can J Plant Pathol 20:87–94. https://doi.org/10.1080/07060669809500449

Changsri W (1961) Studies of *Alternaria* spp. pathogen on Cruciferae. Dis Abstr 21:1698

Charkowski A, Blanco C, Condemine G, Expert D, Franza T, Hayes C, Hugouvieux-Cotte-Pattat N, Solanilla EL, Low D, Moleleki L, Pirhonen M, Pitman A, Perna N, Reverchon S, Palenzuela PR, Francisco MS, Toth I, Tsuyumu S, van der Waals J, van der Wolf J, Gijsegem FV, Yang CH, Yedidia I (2011) The role of secretion systems and small molecules in Soft-Rot Enterobacteriaceae pathogenicity. Annu Rev Phytopathol 50:425–449

Chaudhary RG, Kaur A (2002) Wilt disease as a cause of shift from lentil cultivation in Sangod Tehsil of Kota (Rajasthan). Indian J Pulses Res 15:193–194

Chauhan LS, Saksena HK (1974) A new *Rhizoctonia* leaf blight of rapes and mustard. Indian J Farm Sci 2:98–99

Chaves MM, Maroco JP, Pereira JS (2003) Understanding plant responses to drought: from genes to the whole plant. Funct Plant Biol 30:239–264

Cheah LH, Hartill WFT, Corbin JB (1980) First report of the natural occurrence of *Pyrenopeziza brassicae* Sutton et. Rawlinson, the apothecial state of *Cylindrosporium concentricum* Greville, in *Brassica* crops in New Zealand. N Z J Bot 18:197–202

Cheeseman JM (1988) Mechanisms of salinity tolerance in plants. Plant Physiol 87:547–550

Chen QY (1980) Varietal resistance of Chinese cabbage varieties to turnip mosaic virus disease. Acta Phytothol Sin 7:201–208

Chen CY, Howlett B (1996) Rapid necrosis of guard cells is associated with the arrest of fungal growth in leaves of Indian mustard (*Brassica juncea*) inoculated with avirulent isolates of *Leptosphaeria maculans*. Physiol Mol Plant Pathol 48:73–81

Chen W, Li Y, Yan R, Xu L, Ren L, Liu F, Zeng L, Yang H, Chi P, Wang X, Chen K, Ma D, Fang X (2019) Identification and characterization of *Plasmodiophora brassicae* primary infection effector candidates that suppress or induce cell death in host and nonhost plants. Phytopathology 109:1689–1697. https://doi.org/10.1094/PHYTO-02-19-0039-R

Chimbekujwo IB (2000) Frequency and pathogenicity of Fusarium wilts (*Fusarium solani* and *Fusarium equiseti*) of cotton (*Gossypium hirsutum*) in Adamawa in Nigeria. Rev Biol Trop 48: 1–5

Chittem K, Yajima WR, Goswami RS, del Rio Mendoza LE (2020) Transcriptome analysis of the plant pathogen *Sclerotinia sclerotiorum* interaction with resistant and susceptible canola (*Brassica napus*) lines. PLoS One 15(3):e0229844. https://doi.org/10.1371/journal.pone.0229844

Chiu WF, Chang YH (1982) Advances of science of plant protection in the People's Republic of China. Annu Rev Phytopathol 20:71–92

Choi D, Priest MJ (1995) A key to the genus *Albugo*. Mycotaxon 53:261–272

Choi YJ, Hong SB, Shin HD (2006) Genetic diversity within the *Albugo candida* complex (Peronosporales, Oomycota) inferred from phylogenetic analysis of ITS rDNA and COX2 mtDNA sequences. Mol Phylogenet Evol 40:400–409

Choi YJ, Shin HD, Hong SB, Thines M (2007) Morphological and molecular discrimination among *Albugo candida* materials infecting *Capsella bursa-pastoris* world-wide. Fungal Divers 27:11–34

Choi YJ, Park MJ, Park JH, Shin HD (2011a) White blister rust caused by A. candida on Oilseed rape in Korea. Plant Pathol J 27:192

Choi YJ, Shin HD, Ploch S, Thines M (2011b) Three new phylogenetic lineages are the closest relatives of the widespread species Albugo candida. Fungal Biol 115:598–607

Choi YJ, Thines M, Shin HD (2011c) A new perspective on the evolution of white blister rust: *Albugo* s.str. (Albugonales; Oomycota) is not restricted to Brassicales but also present on Fabales. Org Divers Evol 11:193–199

Chou CK (1970) An electron-microscope study of host penetration and early stages of haustorium formation of *Peronospora parasitica* (Fr) Tul. on cabbage cotyledons. Ann Bot 34:189–204

Ciaghi S, Schwelm A, Neuhauser S (2019) Transcriptomic response in symptomless roots of clubroot infected kohlrabi (*Brassica oleracea* var. *gongylodes*) mirrors resistant plants. BMC Plant Biol 19:288. https://doi.org/10.1186/s12870-019-1902-z

Claassen B (2016) Investigations of Black leg and Light leaf spot on Brassicaceae hosts in Oregon. M Sc Thesis, Oregon State University

CMI (1986) Distribution map of plant diseases, pathogen: *Pseudocercosporella capsellae* (EII. and Ev.) Deighton: Commonwealth Mycological Institute. CAB, Wallingford

Conn KL, Tewari JP, Awasthi RP (1990) A disease assessment key for Alternaria black spot in rapeseed and mustard. Can Plant Dis Surv 70:19–22

Conroy RY (1959) Black ring spot disease of Crucifers. J Aust Inst Agric Sci 25:64–67

Cook AA, Larson RH, Walker JC (1952a) Relation of the black rot pathogen to cabbage seed. Phytopathology 42:316–320

Cook AA, Walker JC, Larson RH (1952b) Studies on the disease cycle of black rot of crucifers. Phytopathology 42:162–167

Cooke DEL, Drenth A, Duncan JM, Wagels G, Brasier CM (2000) A molecular phylogeny of *Phytophthora* and related Oomycetes. Fungal Genet Biol 30:17–32

Corlett M (1999) Mycosphaerella brassicicola. Can J Plant Pathol 21:52–54. https://doi.org/10.1080/07060661.1999.10600133

Crossan D (1954) *Cercosporella* leaf spot of crucifers. NC Agric Exp Stn Tech Bull 109:23

Czechowski T, Stitt M, Altmann T, Udvardi MK, Scheible WR (2005) Genome-wide identification and testing of superior reference genes for transcript normalization in *Arabidopsis*. Plant Physiol 139:5–17. https://doi.org/10.1104/pp.105.063743

Damm U, O'Connell RJ, Groenewald JZ, Crous PW (2014) The *Colletotrichum destructivum* [italicize] species complex–hemibiotrophic pathogens of forage and field crops. Stud Mycol 79:49–84. https://doi.org/10.1016/j.simyco.2014.09.003

Danquah A, de Zelicourt A, Colcombet J, Hirt H (2014) The role of ABA and MAPK signaling pathways in plant abiotic stress responses. Biotechnol Adv 32:40–52

Darma R, Lutz A, Elliott C, Idnurm E (2019) A Identification of a gene cluster for the synthesis of the plant hormone abscisic acid in the plant pathogen *Leptosphaeria maculans*. Fungal Genet Biol 130:62–71

Daub ME, Ehrenshaft M (2000) The photo activated *Cercospora* toxin cercosporin: contributions to plant disease and fundamental biology. Annu Rev Phytopathol 38:461–490. https://doi.org/10.1146/annurev.phyto.38.1.461

Davies JML (1986) Diseases of oilseed rape. In: Scarisbrick DH, Daniels RW (eds) Oilseed rape. Collins, London, pp 195–236

Deboever E, Deleu M, Mongrand S, Lins L, Fauconnier M-L (2020) Plant–pathogen interactions: underestimated roles of phyto-oxylipins. Trends Plant Sci 25:22–34

Deighton F (1973) Studies on *Cercospora* and allied genera. IV. *Cercosporella* Sacc., *Pseudocercosporella* gen. nov. and *Pseudocercsporidium* gen. nov. Mycol Pap 133:1–62

Delourme R, Bousset L, Ermel M, Duffe P, Besnard AL, Marquer B, Fudal I, Linglin J, Chadoeuf J, Brun H (2014) Quantitative resistance affects the speed of frequency increase but not the diversity of the virulence alleles overcoming a major resistance gene to *Leptosphaeria maculans* in oilseed rape. Infect Genet Evol 27:490–499

Depotter JRL, Doehlemann G (2020) Target the core: durable plant resistance against filamentous plant pathogens through effector recognition. Pest Manag Sci 76:426–431. https://doi.org/10.1002/ps.5677

Depotter JR, Deketelaere S, Inderbitzin P, Von Tiedemann A, Höfte M, Subbarao KV, Wood TA, Thomma BPHJ (2016) *Verticillium longisporum*, the invisible threat to oilseed rape and other Brassicaceous plant hosts. Mol Plant Pathol 17(7):1004–1016. https://doi.org/10.1111/mpp.12350

Depotter JRL, Thomma BPHJ, Wood TA (2019) Measuring the impact of *Verticillium longisporum* on oilseed rape (*Brassica napus*) yield in field trials in the United Kingdom. Eur J Plant Pathol 153:321–326. https://doi.org/10.1007/s10658-018-1537-1

de Roussel HFA (1806) Flore du Calvados, 2nd edn, Caen

Dhar S, Singh D (2014) Performance of cauliflower genotypes for yield and resistance against black rot (*Xanthomonas campestris* pv. *campestris*). Indian J Hortic 71(2):197–201

Dick MW (2001) *Straminipilous* fungi. Kluwer, Dordrecht

Dita MA, Rispail N, Prats E, Rubiales D, Singh KB (2006) Biotechnology approaches to overcome biotic and abiotic stress constraints in legumes. Euphytica 147:1–24

Dixelius C, Happstadius I, Berg G (2005) Verticillium wilt on *Brassica* oilseed crops – a Swedish perspective. J Swed Seed Assoc 115:36–48

Dixon GR (2007) Vegetable Brassicas and related crucifers. CABI, Wallingford, pp 1–2. Chapter 1

Dixon RA, Paiva NL (1995) Stress-induced phenylpropanoid metabolism. Plant Cell 7:1085–1097

Dong CH, Agarwal M, Zhang Y, Xie Q, Zhu JK (2006) The negative regulator of plant cold responses, *HOS1*, is a *RING E3* ligase that mediates the ubiquitination and degradation of ICE1. Proc Natl Acad Sci U S A 103:8281–8286

Doucet R, Shattuck VI, Stobbs LW (1990) Rutabaga germplasm *TuMV*-R possessing resistance to *Turnip mosaic virus*. Hortic Sci 25:583–584

Duff J, Sermon D, Walton G, Mangano P, Newman C, Walden K et al (2006) Growing Western Canola, an overview of canola production in Western Australia. Oilseeds Industry Association of Western Australia, Perth, WA

Dunker S, Keunecke H, Steinbach P, Tiedemann AV (2008) Impact of *Verticillium longisporum* on yield and morphology of winter oilseed rape (*Brassica napus*) in relation to systemic spread in the plant. J Phytopathol 156:698–707

Eastburn D (1989) Disease management of cabbage and broccoli-an IPM approach. Trans II State Hortic Soc 123:32–35

Edwardson JR (1974) Some properties of the potato virus V-group. FL Agric Exp Stn Monogr Ser 4:398

Edwardson JR, Christie RG (1986) *Turnip mosaic virus*. In: Viruses infecting forage legumes. Vol. II. Agri Exp Inst Food Agr Sci Monograph, 14. Univ Florida, Gainesville, Gainesville, FL, pp 438–453

Edwardson JR, Christie RG (1991) The Potyvirus Group. University of Florida, Monograph 16, vol 3. University of Florida, Gainesville, FL, pp 699–712

Elliott VL, Marcroft SJ, Howlett BJ, Van de Wouw AP (2016) Gene-for-gene resistance is expressed in cotyledons, leaves and pods, but not during late stages of stem colonization in the *Leptosphaeria maculans-Brassica napus* pathosystem. Plant Breed 135:200–207

Eshraghi L, You M, Barbetti M (2005) First report of white leaf spot caused by *Pseudocercosporella capsellae* on *Brassica juncea* in Australia. Plant Dis 89:1131. https://doi.org/10.1094/PD-89-1131B

Eshraghi L, Barbetti M, Li H, Danehloueipour N, Sivasithamparam K (2007) Resistance in oilseed rape (*Brassica napus*) and Indian mustard (*Brassica juncea*) to a mixture of *Pseudocercosporella capsellae* isolates from Western Australia. Field Crop Res 101:37–43. https://doi.org/10.1016/j.fcr.2006.09.006

Eynck C, Koopmann B, Grunewaldt-Stoecker G, Karlovsky P, Tiedemann AV (2007) Differential interactions of *Verticillium longisporum* and *V. dahliae* with *Brassica napus* detected with molecular and histological techniques. Eur J Plant Pathol 118:259–274

Eynck C, Koopmann B, Karlovsky P, von Tiedemann A (2009) Internal resistance in winter oilseed rape inhibits systemic spread of the vascular pathogen *Verticillium longisporum*. Phytopathology 99:802–811

Fang X, Willis RC, Hoang Q, Kelnar K, Xu W (2004) High-throughput sample preparation for gene expression profiling and *in vitro* target validation. JALA 9(3):140–145

Farooq M, Wahid A, Kobayashi N, Fujita D, Basra SMA (2009) Plant drought stress: effects, mechanisms and management. Agron Sustain Dev 29:185–212

Fischer A (1892) Phycomycetes. In: Fischer A, Hauck F, Limpricht G, Luerssen C, Richter P, Winter G (eds) Dr L Ranenhorst's Kryptogamen-Flora von Deutschland, Österreich und der Schweiz, vol 1, 2nd edn. von Eduard Kummer, Leipzig

Fitt BDL, Gladders P, Turner JA, Sutherland KG, Welham SJ, Davies JML (1997) Prospects for developing a forecasting scheme to optimize use of fungicides for disease control on winter oilseed rape in the UK. Asp Appl Biol 48:135–142

Fitt BDL, Brun H, Barbetti MJ, Rimmer SR (2006) Worldwide importance of phoma stem canker (*Leptosphaeria maculans* and *L. biglobosa*) on oilseed rape (*Brassica napus*). Eur J Plant Pathol 114:3–15

Fitt BDL, Hu BC, Li ZQ, Liu SY, Lange RM, Kharbanda PD, Butterworth MH, White RP (2008) Strategies to prevent spread of *Leptosphaeria maculans* (phoma stem canker) on to oilseed rape crops in China: costs and benefits. Plant Pathol 57:652–664

Föller I, Paul V (2002) Investigations on diseases of false flax (*Camelina sativa* (L.) Crtz.) with special regard to downy mildew (*Peronospora parasitica* (Pers.) Fr.). Soest Germany 2:43–52

Francis A, Warwick SI (2003) The biology of Canadian weeds *Neslia paniculata* (L.) Desv. Can J Plant Sci 83:441–451. https://doi.org/10.4141/P02-076

Frandsen KJ (1943) Experimental formation of *Brassica juncea*, (L.) Czern & Coss. Dansk Bot Arkiv Bd 11(4):Irl7

Fujita M, Fujita Y, Noutoshi Y, Takahashi F, Narusaka Y, Yamaguchi-Shinozaki K (2006) Crosstalk between abiotic and biotic stress responses: a current view from the points of convergence in the stress signaling networks. Curr Opin Plant Biol 9:436–442

Gabrielson RL (1983) Blackleg disease of cabbage caused by *Leptosphaeria maculans* (*Phoma lingam*) and its control. Seed Sci Technol 11:749–780

Gaetan SA (2005) Occurrence of Fusarium wilt on canola caused by *Fusarium oxysporum* f. sp. *conglutinans* in Argentina. Plant Dis 89:432

Gardner MW, Kendrick JB (1921) Turnip Mosaic. J Agric Res 22:123–124

Garman H (1894) A bacterial disease of cabbage. Kentucky Agric Exp Stn Rep 3:43–46

Genva M, Obounou Akong F, Andersson MX, Deleu M, Lins L, Fauconnier ML (2019) New insights into the biosynthesis of esterified oxylipins and their involvement in plant defense and developmental mechanisms. Phytochem Rev 18:343–358

Gilles T, Evans N, Fitt BDL, Jeger MJ (2000) Epidemiology in relation to methods for forecasting light leaf spot (*Pyrenopeziza brassicae*) severity on winter oilseed rape (*Brassica napus*) in the UK. Eur J Plant Pathol 106:593–605. https://doi.org/10.1023/a:1008701302853

Gilles T, Ashby AM, Fitt BDL, Cole T (2001a) Development of *Pyrenopeziza brassicae* apothecia on agar and oilseed rape debris. Mycol Res 105:705–714. https://doi.org/10.1017/s0953756201003902

Gilles T, Fitt BDL, Jeger MJ (2001b) Effects of environmental factors on development of *Pyrenopeziza brassicae* (light leaf spot) apothecia on oilseed rape debris. Phytopathology 91: 392–398. https://doi.org/10.1094/phyto.2001.91.4.392

Gladders P, Musa TM (1979) The development of *Leptosphaeria maculans* in winter oilseed rape and its implications for disease control. Pests Dis 89:129–136

Gladders P, Musa TM (1980) Observations on the epidemiology of *Leptosphaeria maculans* stem canker in winter oilseed rape. Plant Pathol 29:28–37

Gladders P, Symonds BV (1995) Occurrence of canker (*Leptosphaeria maculans*) in winter oilseed rape in eastern England 1977–93. Inter Org Biol Cont Bull 18:1–11

Gladders P, Symonds BV, Hardwick NV, Sansford CE (1998) Opportunities to control canker (*Leptosphaeria maculans*) in winter oilseed rape by improved spray timing. Inter Org Biol Cont Bull 21:111–120

Gładysz K, Hanus-Fajerska E (2009) Evaluation of the infectivity of selected turnip mosaic virus isolates towards white cabbage cultivars. Folia Hortic Ann 21(1):129–138

Goossens K, Poucke MV, Soom AV, Vandesompele J, Zeveren AV, Peelman LJ (2005) Selection of reference genes for quantitative real-time PCR in bovine pre-implantation embryos. BMC Dev Biol 5:27. https://doi.org/10.1186/1471-213X-5-27

Gosende S, Penaud A, Aubertat NJ, Schnieder O, Pinochet X (2003) Evolution of soil surface oilseed rape stubbles and their ability to produce spores of *Leptosphaeria maculans*: preliminary results. In: Proc. 11th Inter Rapeseed Congress, 4: AP 11 Denmark, vol 14, pp 1166–1168

Goswami RS, Dong Y, Punja ZK (2008) Host range and mycotoxin production by *Fusarium equiseti* isolates originating from ginseng fields. Can J Plant Pathol 30:155–160

Graham-Taylor C, Kamphuis LG, Derbyshire MC (2020) A detailed in silico analysis of secondary metabolite biosynthesis clusters in the genome of the broad host range plant pathogenic fungus *Sclerotinia sclerotiorum*. BMC Genome 21:7

Gray SF (1821) A natural arrangement of British plants: according to their relations to each other as pointed out by Jussieu, De Candolle, Brown: with an introduction to botany. Baldwin, Cradock and Joy, London, p 72

Green SK, Deng TC (1985) *Turnip mosaic virus* strains in cruciferous hosts in Taiwan. Plant Dis 69:28–31

Gudelj I, Fitt BDL, Van den Bosch F (2004) Evolution of sibling fungal plant pathogens in relation to host specialization. Phytopathology 94:789–795. https://doi.org/10.1094/PHYTO.2004.94.7.789

Guenin S, Mauriat M, Pelloux J, Van Wuytswinkel O, Bellini C, Gutierrez L (2009) Normalization of qRT-PCR data: the necessity of adopting a systematic, experimental conditions-specific, validation of references. J Exp Bot 60(2):487–493. https://doi.org/10.1093/jxb/ern305

Gugel RK, Yitbarek SM, Verma PR, Morrall RAA, Sadasivaiah RS (1987) Etiology of the *Rhizoctonia* root-rot complex of canola in the Peace River region of Alberta. Can J Plant Pathol 9:119–128

Gunasinghe N, You MP, Banga SS, Barbetti MJ (2013) High level resistance to *Pseudocercosporella capsellae* offers new opportunities to deploy host resistance to effectively manage white leaf spot disease across major cruciferous crops. Eur J Plant Pathol 138:873–890

Gunasinghe N, You MP, Barbetti MJ (2015) Phenotypic and phylogenetic studies associated with the crucifer white leaf spot pathogen, *Pseudocercosporella capsellae*, in Western Australia. Plant Pathol 65(2):205. https://doi.org/10.1111/ppa.12402

Gunasinghe N, You MP, Cawthray G, Barbetti M (2016a) Cercosporin from *Pseudocercosporella capsellae* and its critical role in white leaf spot development. Plant Dis 100:1521–1531. https://doi.org/10.1094/PDIS-10-15-1192-RE

Gunasinghe N, You M, Clode P, Barbetti M (2016b) Mechanisms of resistance in *Brassica carinata*, *B. napus* and *B. juncea* to *Pseudocercosporella capsellae*. Plant Pathol 65:888–900. https://doi.org/10.1111/ppa.12484

Gunasinghe N, Barbetti MJ, You MP, Burrell D, Neate S (2020) White leaf spot caused by *Neopseudo-cercosporella capsellae*: a re-emerging disease of Brassicaceae. Front Cell Infect Microbiol 10:588090. https://doi.org/10.3389/fcimb.2020.588090

Gupta M, Vikram A, Bharat N (2013) Black rot-a devastating disease of crucifers: a review. Agric Rev 34:269. https://doi.org/10.5958/j.0976-0741.34.4.012

Haddadi P, Larkan NJ, Borhan MH (2019) Dissecting R gene and host genetic background effect on the *Brassica napus* defense response to *Leptosphaeria maculans*. Sci Rep 9:6947. https://doi.org/10.1038/s41598-019-43419-9

Hall R (1992) Epidemiology of blackleg of oilseed rape. Can J Plant Pathol 14:46–55

Hall G (1996) Modern approaches to species concepts in downy mildews. Plant Pathol 45:1009–1026

Hall R, Peters RD, Assabgui RA (1993) Occurrence and impact of blackleg of oilseed rape in Ontario. Can J Plant Pathol 15:305–313

Hammond KE (1985) Systemic infection of *Brassica napus* L. spp. *oleifera* (Metzger) Sinsk. by *Leptosphaeria maculans* (Desm.) Ces. et de Not. PhD thesis, University of East Anglia, Norwich

Hammond KE, Lewis BG (1986a) The timing and sequence of events leading to stem canker disease in populations of *Brassica napus* var. *oleifera* in the field. Plant Pathol 35:551–564

Hammond KE, Lewis BG (1986b) Ultrastructure studies of the limitation of stem lesions caused by *Leptosphaeria maculans* on *Brassica napus* var. *oleifera*. Physiol Mol Plant Pathol 28:251–265

Hammond KE, Lewis BG (1987) Variation in stem infections caused by aggressive and non-aggressive isolates of *Leptosphaeria maculans* on *Brassica napus* var. *oleifera*. Plant Pathol 36:53–65

Hammond KE, Lewis BG, Musa TM (1985) A systemic pathway in the infection of oilseed rape plants by *Leptosphaeria maculans*. Plant Pathol 34:557–565

Happstadius I, Ljungberg A, Kristiansson B, Dixelius C (2003) Identification of *Brassica oleracea* germplasm with improved resistance to *Verticillium* wilt. Plant Breed 122:30–34

Harakava R (2005) Genes encoding enzymes of the lignin biosynthesis pathway in Eucalyptus. Genet Mol Biol 28:601–607

Hardwick NV, Davies JML, Wright DM (1994) The incidence of three virus diseases of winter oilseed rape in England and Wales in the 1991/92 and 1992/93 growing seasons. Plant Pathol 43:1045–1049

Heale JB, Karapapa KV (1999) The *Verticillium* threat to Canada's major oilseed crop: canola. Can J Plant Pathol 21:1–7

Higgins BB (1917) A *Colletotrichum* leaf spot of turnips. J Agric Res 10:157–165

Hirayama T, Shinozaki K (2010) Research on plant abiotic stress responses in the post-genome era: past, present and future. Plant J 61:1041–1052

Horie H, Sugata S, Abe Z (1988) Studies on anthracnose of komatsuna, *Brassica rapa*. Bull Tokyo Metro Agric Exp Stn 21:189–237

Howlett B, Ballinger D, Barbetti MJ (1999) Diseases. In: Salisbury PA, Potter TD, McDonald G (eds) Proc Canola in Australia. Canberra, Australia, 10th International Rapeseed Congress, Australia, pp 47–52

Howlett BJ, Idnurm A, Pedras MS (2001) *Leptosphaeria maculans*, the causal agent of blackleg disease of *Brassicas*. Fungal Genet Biol 33:1–14

Huang HC, Hoes JA (1980) Importance of plant spacing and sclerotial position to development of Sclerotinia wilt of sunflower. Plant Dis 64:81–84

Huang Y, Fitt B, Steed Y, Hall A (2003) Maturation of *Leptosphaeria maculans* pseudothecia in relation to forecasting phoma leaf spots (stem canker) on oilseed rape. In: Proc 11th International Rapeseed Congress, 4: AP 11 Denmark, vol 9, pp 1153–1156

Hudspeth DSS, Stenger D, Hudspeth MES (2003) A COX2 phylogenetic hypothesis of the downy mildews and white rusts. Fungal Divers 13:47–57

Hughes SL, Green SK, Lydiate DJ, Walsh JA (2002) Resistance to *Turnip mosaic virus*, in *B. rapa* and *B. napus* and the analysis of genetic inheritance in selected lines. Plant Pathol 51:567–573. https://doi.org/10.1046/j.1365-3059.2002.00755.x

Huis R, Hawkins S, Neutelings G (2010) Selection of reference genes for quantitative gene expression normalization in flax (*Linum usitatissimum* L.). BMC Plant Biol 10:71. https://doi.org/10.1186/1471-2229-10-71

Humpherson-Jones FM (1986) The occurrence of virulent pathotypes of *Leptosphaeria maculans* in *Brassica* seed crops in England. Plant Pathol 35:224–231

Humphreys TL, Schnizlein-Bick CT, Katz BP, Baldridge LA, Hood AF, Hromas RA, Spinola SM (2002) Evolution of the cutaneous immune response to experimental *Haemophilus ducreyi* infection and its relevance to HIV-1 acquisition. J Immunol 169:6316–6323

Hungerford W, Pitts R (1953) The *Sclerotinia* disease of beans in Idaho. Phytopathology 43:519–521

Ignatov A, Hida K, Kuginuki Y (1999a) Environmentally dependent change of disease symptoms caused by *Xanthomonas campestris* pv. *campestris* on *Brassicas*. Acta Phytopathol Entomol Hungarica 34(3):183–186

Ignatov A, Kuginuki Y, Hida K (1999b) Vascular stem resistance to black rot in *Brassica oleracea*. Can J Bot 77:442–446

Inman AJ (1992) The biology and epidemiology of white leaf spot *(Pseudocercosporella capsellae)* on oilseed rape. PhD Thesis, The University of London

Inman A, Sivanesan A, Fitt B, Evans R (1991) The biology of *Mycosphaerella capsellae* sp. nov., the teleomorph of *Pseudocercosporella capsellae*, cause of white leaf spot of oilseed rape. Mycol Res 95:1334–1342. https://doi.org/10.1016/S0953-7562(09)80586-8

Inman A, Fitt BDL, Todd A, Evans R (1999) Ascospores as primary inoculum for epidemics of white leaf spot (*Mycosphaerella capsellae*) in winter oilseed rape in the UK. Plant Pathol 48: 308–319. https://doi.org/10.1046/j.1365-3059.1999.00350.x

Jacobsen BJ, Williams PH (1971) Histology and control of *Brassica oleracea* seed infection by *Phoma lingam*. Plant Dis Rep 55:934–938

Jacobson DJ, Le Febvre SM, Ojerio RS, Berwald N (1998) Persistent, systemic, asymptomatic infections of *Albugo candida*, an oomycete parasite, detected in three wild crucifer species. Can J Bot 76(5):739–750

Janowska-Sejda EI, Lysenko A, Urban M, Rawlings C, Tsoka S, Hammond-Kosack KE (2019) PHI-nets: a network resource for ascomycete fungal pathogens to annotate and identify putative virulence interacting proteins and siRNA targets. Front Microbiol 10:2721. https://doi.org/10.3389/fmicb.2019.02721

Jeanes A (1973) Extracellular microbial polysaccharides: new hydrocolloides having both fundamental and practical importance. In: Bikales NM (ed) Polymer science and technology, vol 2. Plenum Press, New York, NY

Jedryczka M, Fitt BDL, Kachlicki P, Lewartowska E, Balesdent MH, Rouxel T (1999) Comparison between Polish and United Kingdom populations of *Leptosphaeria maculans*, cause of stem canker of winter oilseed rape. Zeitschrift für Pflanzenkrankheiten und Pflanzenschutz, (Phytopath Z) 106:608–617

Jenner CE, Walsh JA (1996) Pathotypic variation in turnip mosaic virus with special reference to European isolates. Plant Pathol 45:848–856

Johnson RD, Lewis BG (1994) Variation in host range, systemic infection and epidemiology of *Leptosphaeria maculans*. Plant Pathol 43:269–277

Johnston JS, Pepper AE, Hall AE, Chen ZJ, Hodnett G, Drabek J, Lopez R, Price HJ (2005) Evolution of genome size in Brassicaceae. Ann Bot 95:229–235. https://doi.org/10.1093/aob/mci016

Jones RAC (2004) Using epidemiological information to develop effective integrated virus disease management strategies. Virus Res 99:5–30

Jones RAC (2006) Control of plant virus diseases. Adv Virus Res 67:205–244

Kadian AK, Saharan GS (1983) Symptomatology, host range and assessment of yield losses due to *Alternaria brassicae* infection in rapeseed and mustard. Indian J Mycol Plant Pathol 13:319–323

Karapapa VK, Bainbridge BW, Heale JB (1997) Morphological and molecular characterization of *Verticillium longisporum* comb. nov., pathogenic to oilseed rape. Mycol Res 101:1281–1294

Karolewski Z (1999) The occurrence of light leaf spot on winter oilseed rape in Western Poland in 1991–1996 and the characteristics of *Pyrenopeziza brassicae* isolates. Phytopatol Pol 18:113–121

Karolewski Z (2010) Development of light leaf spot (*Pyrenopeziza brassicae*) on *Brassicas*. Phytopathologia 55:13–20

Karolewski Z, Fitt BDL, Latunde-Dada AO, Foster SJ, Todd AD, Downes K, Evans N (2006) Visual and PCR assessment of light leaf spot (*Pyrenopeziza brassicae*) on winter oilseed rape (*Brassica napus*) cultivars. Plant Pathol 55:387–400

Karolewski Z, Kaczmarek J, Jedryczka M, Cools HJ, Fraaije BA, Lucas JA, Latunde-Dada AO (2012) Detection and quantification of airborne inoculum of *Pyrenopeziza brassicae* in Polish and UK winter oilseed rape crops by real-time PCR assays. Grana 51:270–279. https://doi.org/10.1080/00173134.2011.653401

Kataria H, Verma PR (1992) *Rhizoctonia solani* damping-off and root rot in oilseed rape and canola. Crop Prot 11:8–13

Kayum MA, Jung HJ, Park JI, Ahmed NU, Saha G, Yang TJ (2015a) Identification and expression analysis of WRKY family genes under biotic and abiotic stresses in *Brassica rapa*. Mol Gen Genomics 290:79–95

Kayum MA, Park JI, Ahmed NU, Jung HJ, Saha G, Kang JG (2015b) Characterization and stress-induced expression analysis of Alfin-like transcription factors in *Brassica rapa*. Mol Gen Genomics 290:1299–1311

Kayum MA, Park JI, Ahmed NU, Saha G, Chung MY, Kang JG (2016a) Alfin-like transcription factor family: characterization and expression profiling against stresses in *Brassica oleracea*. Acta Physiol Plant 38:127

Kayum MA, Kim HT, Nath UK, Park JI, Kho KH, Cho YG, Nou IS (2016b) Research on biotic and abiotic stress related genes exploration and prediction in *Brassica rapa* and *B. oleracea*: a review. Plant Breed. Biotech 4(2):135–144. https://doi.org/10.9787/PBB.2016.4.2.135

Khangura RK, Barbetti MJ, Sweetingham MW (1999) Characterization and pathogenicity of *Rhizoctonia* species on canola. Plant Dis 83:714–721

Kharbanda PD, Stevens RR (1993) Seed testing for blackleg of canola. Alberta Environmental Centre, Vegreville, AB. AECV93–E1

Kim BR, Nam HY, Kim SU, Kim SI, Chang YJ (2003) Normalization of revere transcription quantitative-PCR with housekeeping genes in rice. Biotechnol Lett 25:1869–1872. https://doi.org/10.1023/A:1026298032009

Kiran G, Saharan GS (2002) Identification of pathotypes of *Albugo candida* with stable characteristic symptoms on Indian mustard. J Mycol Plant Pathol 32:46–51

Kiran G, Saharan GS, Mehta N, Sangwan MS (2004) Identification of pathotypes of *Alternaria brassicae* from Indian mustard [*Brassica juncea* (L) Czern and Coss]. J Mycol Plant Pathol 34: 15–19

Kirk PM, Cannon PF, David JC, Stalpers JA (2001) Ainsworth & Bisby's dictionary of the fungi, 9th edn. CABI, Egham

Klassen J, Yu C, Lange R, Fernando WGD (2007) Genetic variation in *Fusarium oxysporum* f. sp. *conglutinans* strains causing Fusarium-wilt of canola in Western Canada. Genetics and breeding: breeding for stress resistance. In: Proc International Rapeseed Congress, Wuhan, pp 433–435

Kohn LM (1979) Delimitation of the economically important plant pathogenic *Sclerotinia* species. Phytopathology 69(8):881–886

Koike ST, Gladders P, Paulus AO (2007) Vegetable diseases: a color handbook. Academic Press, San Diego, CA

Kolte SJ (1985) Diseases of annual edible oilseed crops. Vol. II. Rapeseed-mustard and sesame diseases. CRC Press, Boca Raton, FL, p 135

Kosiak B, Torp M, Skjerve E, Thrane U (2003) The prevalence and distribution of *Fusarium* species in Norwegian cereals: a survey. Acta Agric Scandi B Soil Plant Sci 53:168–179

Krauthausen HJ, Laun N, Wohanka W (2011) Methods to reduce the spread of the black rot pathogen, *Xanthomonas campestris* pv. *campestris*, in *Brassica* transplants. J Plant Dis Prot 118:7–16

Ku HM, Vision T, Liu J, Tanksley SD (2000) Comparing sequenced segments of the tomato and *Arabidopsis* genomes: large-scale duplication followed by selective gene loss creates a network of synteny. Proc Natl Acad Sci U S A 97:9121–9126

Kucher HR, Vandenberg CGJ, Rimmer SR (1993) Variation in pathogenicity of *Leptosphaeria maculans* on *Brassica* spp. based on cotyledon and stem reactions. Can J Plant Pathol 15:253–258

Kumar S (2013) Defence mechanism in plants: Phase I. Google Blogger. http://sanjeetbiotech.blogspot.kr/2013/03/defence-mechanism-in-plants-phase-i.html

Kundu A, Patel A, Pal A (2013) Defining reference genes for qPCR normalization to study biotic and abiotic stress responses in *Vigna mungo*. Plant Cell Rep 32:1647–1658. https://doi.org/10.1007/s00299-013-1478-2

Kuninaga S, Natsuaki T, Takeuchi T, Yokosawa R (1997) Sequence variation of the rDNA ITS regions within and between anastomosis groups in *Rhizoctonia solani*. Curr Genet 32:237–243

Lagercrantz U (1998) Comparative mapping between *Arabidopsis thaliana* and *Brassica nigra* indicates that *Brassica* genomes have evolved through extensive genome replication accompanied by chromosome fusions and frequent rearrangements. Genetics 150:1217–1228

Lagercrantz U, Lydiate DJ (1996) Comparative genome mapping in *Brassica*. Genetics 144:1903–1910

Lan HY, Wang CH, Zhang LH, Liu GZ, Wan LL, Chen ZH, Tian YC (2000) Studies on transgenic oilseed rape (*Brassica napus*) plants transformed with beta-1, 3-glucanase and chitinase genes and its resistance to *Sclerotinia sclerotiorum*. Chin J Biotech 16:142–146

Lancaster R (2006) Diseases of vegetable brassicas. Department of Agriculture, Government of Western Australia, Perth, WA

Lane A, Gladders P (2008) Pests and diseases of oilseeds, *Brassica* seed crops and field beans. In: Alford DV (ed) Pest and disease management handbook. Blackwell, Oxford, pp 52–83. https://doi.org/10.1002/9780470690475.ch3

Lange RM, Harrison LM, Hall M, MacDonald K, Kharbanda PD (2001) The first report of Fusarium wilt of canola in Alberta, 1999. Abstracts, Alberta Regional Meeting, The Canadian Phytopathological Society 1999 in Canadian Journal of Plant Pathology 23:185

Lange RM, Gossmann M, Büttner C (2007) Yield loss in susceptible cultivars of spring rapeseed due to Fusarium wilt caused by *Fusarium oxysporum*. Commun Agric Appl Biol Sci 72(4): 723–734

Lee SC, Lim MH, Yu JG, Park BS, Yang TJ (2012) Genome-wide characterization of the CBF/DREB1 gene family in *Brassica rapa*. Plant Physiol Biochem 61:142–152. https://doi.org/10.1016/j.plaphy.2012.09.016

Lee H, Chawla HS, Obermeier C, Dreyer F, Abbadi A, Snowdon R (2020) Chromosome-scale assembly of winter oilseed rape *Brassica napus*. Front Plant Sci 11:496. https://doi.org/10.3389/fpls.2020.00496

Letham DB, Daincs D, Hennissey J (1975) Turnip mosaic-a serious problem in hybrid cabbage crops. Agric Gaz N S W 86:52–53

Leveilla JH (1847) On the methodical arrangement of the Uredineae. Ann Sci Nat Ser 3:371

Leyns F, de Cleene M, Swings JG, de Ley J (1984) The host range of the genus *Xanthomonas*. Bot Rev 50(3):308–356

Ling L, Yang JY (1940) A mosaic disease of rape and other cultivated crucifers in China. Phytopathology 30:338–342

Liu XP, Lu WC, Liu YK, Wei SQ, Xu JB, Liu JB, Liu ZR, Zhang HJ, Li JL, Ke GL, Yao WY, Cai YS, Wu FY, Cao SC, Li YH, Xie SD, Lin BX, Zhang CL (1996) Occurrence and strain differentiation of turnip mosaic potyvirus and sources of resistance in Chinese cabbage in China. Acta Hortic 407:431–440

Liu T, Song X, Duan W, Huang Z, Liu G, Li Y (2014) Genome-wide analysis and expression patterns of NAC transcription factor family under different developmental stages and abiotic stresses in Chinese cabbage. Plant Mol Biol Rep 32:1041–1056

Lo CT, Tu CC, Tsai WH (1990) Anastomosis groups and pathogenicity of carnation. Plant Prot Bull 32:158–161

Lubbe EM, Meyer L (1995) Crater disease of wheat in South Africa. In: An International Symposium of *Rhizoctonia*, The Netherlands, p 59

Lv H, Fang Z, Yang L, Zhang Y, Wang Y (2020) An update on the arsenal: mining resistance genes for disease management of *Brassica* crops in the genomic era. Hortic Res 7:34. https://doi.org/10.1038/s41438-020-0257-9

Lysak MA, Koch MA, Pecinka A, Schubert I (2005) Chromosome triplication found across the tribe Brassiceae. Genome Res 15:516–525

MacNish GC, Carling DE, Brainard KA (1993) Characterization of *Rhizoctonia solani* AG-8 from bare patches by pectic isozyme (zymogram) and anastomosis techniques. Phytopathology 83: 22–27

Maddock SE, Ingram DS (1981) Studies of survival and longevity of the light leaf spot pathogen of Brassicas, *Pyrenopeziza brassicae*. Trans Br Mycol Soc 77:153–159. https://doi.org/10.1016/s0007-1536(81)80189-1

Mahiar M, Khalaif H (1999) Black rot of crucifers in Jordan: sources of inoculum. Dirasat Agric Sci 26:329–337

Mahmodi F, Kadir JB, Wong MY, Nasehi A, Puteh A, Soleimani N (2013) First report of anthracnose caused by *Colletotrichum gloeosporioides* on soybean (*Glycine max*) in Malaysia. Plant Dis 97:841

Marchionatto J (1947) Parasitic fungi of plants, new or little known in Argentina. Min Agric 3:11

Marcroft S, Sprague S, Salisburg P, Howlett B (2003) Survival and dissemination of *Leptosphaeria maculans* in southern Australia. In: Proc 11th. International Rapeseed Congress, 4: AP 11 Denmark, vol 11, pp 1157–1159

Marcroft SJ, Elliott VL, Cozijnsen AJ, Salisbury PA, Howlett BJ, Van De Wouw AP (2012) Identifying resistance genes to *Leptosphaeria maculans* in Australian *Brassica napus* cultivars based on reactions to isolates with known avirulence genotypes. Crop Past Sci 63:338–350. https://doi.org/10.1071/CP11341

Masuhara G, Katsuya K, Yamaguchi K (1993) Potential for symbiosis of *Rhizoctonia solani* and binucleate *Rhizoctonia* with seeds of *Spiranthes sinesis* var. *amoana in vitro*. Mycol Res 97: 746–752

Matic S, Gilardi G, Gullino ML, Garibaldi A (2018) Evidence for an expanded host range of *Fusarium oxysporum* f. sp. *chrysanthemi*. J Plant Pathol 100(1):97. https://doi.org/10.1007/s42161-018-0021-2

Maxwell A, Scott JK (2008) Pathogens on wild radish, *Raphanus raphanistrum* (Brassicaceae), in South-Western Australia–implications for biological control. Australas Plant Pathol 37:523–533. https://doi.org/10.1071/AP08040

Mayer K, Murphy G, Tarchini R, Wambutt R, Volckaert G, Pohl T, Düsterhöft A, Stiekema W, Entian KD, Terryn N, Lemcke K, Haase D, Hall CR, van Dodeweerd AM, Tingey SV, Mewes HW, Bevan MW, Bancroft I (2001) Conservation of micro-structure between a sequenced region of the genome of rice and multiple segments of the genome of *Arabidopsis thaliana*. Genome Res 11(7):1167–1174. https://doi.org/10.1101/gr.161701

McCartney H, Lacey M (1990) The production and release of ascospores of *Pyrenopeziza brassicae* on oilseed rape. Plant Pathol 39:17–32. https://doi.org/10.1111/j.1365-3059.1990.tb02471.x

McDonald MR, Boland GJ (2004) Forecasting diseases caused by *Sclerotinia* spp. in eastern Canada: fact or fiction. Can J Plant Pathol 26:480–488

McDonald BA, Linde C (2002) Pathogen population genetics, evolutionary potential and durable resistance. Annu Rev Phytopathol 40:349–379

McGee DC (1977) Blackleg (*Leptosphaeria maculans* (Desm) Ces. et de Not.) of rapeseed in Victoria: sources of infection and relationships between inoculum, environmental factors and disease severity. Aust J Agric Res 28:53–62

McGee DC, Petrie GA (1979) Seasonal patterns of ascospore discharge by *L. maculans* in relation to blackleg of oilseed rape. Phytopathology 69:586–589

McLean GD, Price LK (1984) Virus, viroid, mycoplasma and rickettsial diseases of plants in Western Australia. Tech Bull Dept Agric W A 68:22

Meena PD, Chattopadhyay C, Meena PS, Goyal P, Kumar VR (2014) Shelf life and efficacy of talc-based bio-formulations of *Trichoderma harzianum* isolates in management of Sclerotinia rot of Indian mustard (*Brassica juncea*). Ann Plant Prot Sci 22:127–135

Meena PD, Riteka G, Sharma P, Rai PK, Meena HS, Singh SK (2019) Isolation and characterization of a new fungal species, *Fusarium equiseti* (Corda) Sacc., from *Brassica juncea* in Bharatpur. J Oilseed Brassica 10(2):130–133

Meena PD, Mehta N, Saharan GS (2021) Minor pathogens: a worldwide challenge to cultivation of crucifers. Agric Res J 58(4):557–580. https://doi.org/10.5958/2395-146X.2021.00081.8

Migula W (1910) Familie Albuginaceae. In: Kryptogamen-Flora von Deutschland. Deutsch Österreich und der Schweiz. 3, Pilze, vol 1, pp 153–155

Miller P, McWhorter F (1948) A disease of cabbage and other crucifers due to *Cercosporella brassicae*. Phytopathology 38:893–898

Mittler R, Blumwald E (2010) Genetic engineering for modern agriculture: challenges and perspectives. Annu Rev Plant Biol 61:443–462

Miura K, Jin JB, Lee J, Yoo CY, Stirm V, Miura T (2007) SIZ1-mediated sumoylation of ICE1 controls CBF3/DREB1A expression and freezing tolerance in *Arabidopsis*. Plant Cell 19:1403–1414

Mizoi J, Ohori T, Moriwaki T, Kidokoro S, Todaka D, Maruyama K (2013) GmDREB2A;2, a canonical Dehydration-Responsive Element-Binding Protein2-type transcription factor in soybean, is post translationally regulated and mediates dehydration-responsive element-dependent gene expression. Plant Physiol 161:346–361

Morris MJ, Crous PW (1994) New and Interesting records of South African fungi. XIV. Cercosporoid fungi from weeds. S Afr J Bot 60:325–332. https://doi.org/10.1016/S0254-6299(16)30587-7

Mulema JK, Vicente JG, Pink DAC, Jackson A, Chacha DO, Wasilwa L, Kinyua ZM, Karanja DK, Holub EB, Hand P (2012) Characterization of isolates that cause black rot of crucifers in East Africa. Eur J Plant Pathol 133:427–438

Nagaharu U (1935) Genome-analysis in Brassica with special reference to the experimental formation of *B. napus* and peculiar mode of fertilization. Jpn J Bot 7:389–452

Naito S, Djaeni M, Nasution A, Purwanti H, Sutoyo AM, Kardin K, Achmad NH, Lidya B (1991) Soil-borne diseases affecting yield loss of soybean. In: Proc Final Seminar of the Strengthening of Pioneering Research for PALAWIJA Crop Indonesia, pp 24–37

Narusaka Y, Narusaka M, Park P, Kubo Y, Hirayama T, Seki M, Shiraishi T, Shida T, Nakashima M, Enju A, Sakurai T, Satou M, Kobayashi M, Shinozaki K (2004) RCH1, a locus in *Arabidopsis* that confers resistance to the hemibiotrophic fungal pathogen *Colletotrichum higginsianum*. Mol Plant-Microbe Interact 17:749–762. https://doi.org/10.1094/MPMI.2004.17.7.749

Neik TX, Amas J, Barbetti M, Edwards D, Batley J (2020) Understanding host–pathogen interactions in *Brassica napus* in the omics era. Plants 9:1336. https://doi.org/10.3390/plants9101336

Nicholson RL, Hammerschmidt R (1992) Phenolic compounds and their role in disease resistance. Annu Rev Phytopathol 30:369–389

Nicot N, Hausman JF, Hoffmann L, Evers D (2005) Housekeeping gene selection for real-time RT-PCR normalization in potato during biotic and abiotic stress. J Exp Bot 56:2907–2914. https://doi.org/10.1093/jxb/eri285

Niu X, Leung H, Williams PH (1983) Sources and nature of resistance to downy mildew and turnip mosaic virus in Chinese cabbage. J Am Soc Hortic Sci 108:775–778

Novakazi F, Inderbitzin P, Sandoya G, Hayes RJ, Tiedemann AV, Subbarao KV (2015) The three lineages of the diploid hybrid *Verticillium longisporum* differ in virulence and pathogenicity. Phytopathology 105:662–673

O'Connell R, Herbert C, Sreenivasaprasad S, Khatib M, Esquerré-Dugayé MT, Dumas B (2004) A novel *Arabidopsis-Colletotrichum* pathosystem for the molecular dissection of plant-fungal interactions. Mol Plant-Microbe Interact 17:272–282

O'Connell RJ, Thon MR, Hacquard S, Amyotte SG, Kleemann J, Torres M, Damm U, Buiate EA, Epstein L, Alkan N, Altmüller J, Alvarado-Balderrama L, Bauser CA, Becker C, Birren BW, Chen Z, Choi J, Crouch JA, Duvick JP, Farman MA, Gan P, Heiman D, Henrissat B, Howard RJ, Kabbage M, Koch C, Kracher B, Kubo Y, Law AD, Lebrun MH, Lee YH, Miyara I, Moore N, Neumann U, Nordström K, Panaccione DG, Panstruga R, Place M, Proctor RH, Prusky D, Rech G, Reinhardt R, Rollins JA, Rounsley S, Schardl CL, Schwartz DC, Shenoy N, Shirasu K, Sikhakolli UR, Stüber K, Sukno SA, Sweigard JA, Takano Y, Takahara H, Trail F, van der Does HC, Voll LM, Will I, Young S, Zeng Q, Zhang J, Zhou S, Dickman MB, Schulze-Lefert P, van Themaat EVL, Ma LJ, Vaillancourt LJ (2012) Lifestyle transitions in plant pathogenic *Colletotrichum* fungi deciphered by genome and transcriptome analyses. Nat Genet 44:1060. https://doi.org/10.1038/ng.2372

Obidiegwu JE, Bryan GJ, Jones HG, Prashar A (2015) Coping with drought: stress and adaptive responses in potato and perspectives for improvement. Front Plant Sci 6:542

Ocamb C (2014) Disease alert - white leaf spot and gray stem in crucifer seed crops in Western Oregon. Oregon State University, Corvallis, OR

Ocamb C (2016) A clinic close-up: blackleg, light leaf spot, and white leaf spot in Western Oregon. Oregon State University Extension Service, Corvallis, OR

Ocamb CM, Juzwik J (1995) *Fusarium* species associated with rhizosphere soil and diseased roots of eastern white pine seedlings and associated nursery soil. Can J Plant Pathol 17:325–330

Oerke EC (2006) Crop losses to pests. J Agric Sci 144:31–43

Ogle H, Dale M (1997) Disease management: cultural practices. In: Brown J, Ogle H (eds) Plant pathogens and plant diseases. Rockdale Publ, Armidale, pp 390–404

Ogoshi A (1996) Introduction—the genus *Rhizoctonia*. In: Sneh B, Jabaji-Hare S, Neate S, Dijst G (eds) *Rhizoctonia* species: taxonomy, molecular biology, ecology, pathology and disease control. Kluwer Academic Publishers, Dordrecht, pp 1–9

Ohshima K, Tanaka M, Sako N (1996) The complete nucleotide sequence of turnip mosaic virus RNA Japanese strain. Arch Virol 141:1991–1997

Okullo'kwany FS (1987) Studies on white leaf spot of turnips caused by *Cercosporella brassicae* (Fautr. and Roum.), Hoehnel. Master of Agricultural Science, University of Canterbury

Oppel CB, Dussourd DE, Garimella U (2009) Visualizing a plant defense and insect counter ploy: alkaloid distribution in Lobelia leaves trenched by a plusiine caterpillar. J Chem Ecol 35:625–634

Pammel LH (1895) Bacteriosis of rutabaga (*Bacillus campestris n. sp.*). IA Coll Agric Exp Stn Bull 27:130–134

Patel MK, Abhyankar SG, Kulkarni YS (1949) Black rot of cabbage. Indian Phytopathol 2:58–61

Patel JS, Costa de Novaes MI, Zhang S (2014) First report of *Colletotrichum higginsianum* causing anthracnose of Arugula (*Eruca sativa*) in Florida. Plant Dis 98(9):1269. https://doi.org/10.1094/PDIS-09-13-0926-PDN

Paterson AH, Lan T, Amasino R, Osborn TC, Quiros C (2001) *Brassica* genomics: a complement to, and early beneficiary of, the *Arabidopsis* sequence. Genome Biol 2(3):1011.1–1011.4

Patkar RN, Naqvi NI (2017) Fungal manipulation of hormone-regulated plant defence. PLoS Pathol 13(6):e1006334. https://doi.org/10.1371/journal.ppat.1006334

Paul VH, Rawlinson CJ (1992) Diseases and pests of rape. Verlag Th. Mann, Gelsenkirchen-Buer

Pedras CSM, Taylor LJ, Morales MV (1996) The blackleg fungus of rapeseed: how many species. Acta Hortic 407:441–446

Pegg GF (1985) Life in a black hole–the micro-environment of the vascular pathogen. Trans Br Mycol Soc 85:1–20

Penaud A (1987) La maladie des taches blanches du colza. Phytoma 95:23–26

Peres A, Poisson B, Le Sourne V, Maisonneuve C (1999) *Leptosphaeria maculans*: effect of temperature, rainfall and humidity on the formation of pseudothecia. In: Proc 10th International Rapeseed Congress, 1999, Canberra, Australia

Perez-Clemente RM, Vives V, Zandalinas SI, Lopez-Climent MF, Munoz V, Gomez-Cadenas A (2013) Biotechnological approaches to study plant responses to stress. BioMed Res Int 10: 654120. https://doi.org/10.1155/2013/654120

Perron G, Souliac L (1990) *Pseuocercosporella capsellae* (Et. et Ev) ou maladies des taches blanches du colza. Def Vegetaux 44:22–27

Persoon CH (1796) Observations mycologicae sur descriptions tan novorum quam notibilium *fungorum* exhibitae. Part 1:115

Petersen AB, Rosendahl S (2000) Phylogeny of the Peronosporomycetes (Oomycota) based on partial sequence of the large ribosomal subunit (LSU rDNA). Mycol Res 104:1295–1303

Petrie GA (1978) Occurrence of a highly virulent strain of blackleg (*Leptosphaeria maculans*). Can Plant Dis Surv 58:21–25

Petrie GA (1979) Blackleg of rape. Can Agric 24:22–25

Petrie GA (1986) Consequences of survival of *Leptosphaeria maculans* (blackleg) in canola stubble residue through an entire crop rotation sequence. Can J Plant Pathol 8:353. (Abstract)

Petrie GA (1994) Effects of temperature and moisture on the number, size and septation of ascospores produced by *Leptosphaeria maculans* (blackleg) on rapeseed stubble. Can Plant Dis Surv 74:141–151

Petrie GA, Vanterpool TC (1974) Infestation of crucifer seed in western Canada by the blackleg fungus *Leptosphaeria maculans*. Can Plant Dis Surv 54:119–123

Petrie GA, Vanterpool TC (1978) *Pseudocercosporella capsellae*, the cause of white leaf spot and gray stem of Cruciferae in Western Canada. Can Plant Dis Surv 58:69–72

Peyraud R, Mbengue M, Barbacci A, Raffaele S (2019) Intercellular cooperation in a fungal plant pathogen facilitates host colonization. Proc Natl Acad Sci U S A 116:3193

Pilet ML, Delourme R, Foisset N, Renard M (1998) Identification of QTL involved in field resistance to light leaf spot (*Pyrenopeziza brassicae*) and blackleg resistance (*Leptosphaeria maculans*) in winter rapeseed (*Brassica napus* L.). Theor Appl Genet 97:398–406. https://doi.org/10.1007/s001220050909

Pink DAC, Walkey DGA (1988) The reaction of summer- and autumn-maturing cauliflowers to infection by cauliflower and turnip mosaic viruses. J Hortic Sci 63:95–102

Pink DAC, Sutherland RA, Walkey DGA (1986) Genetic analysis of resistance in Brussels sprout to *Cauliflower mosaic* and *Turnip mosaic viruses*. Ann Appl Biol 109:199–208

Pirotta R (1884) Breve notizia sul Cystopus Capparidis. In: Nuovo Giornale Botanico Italiano

Poisson B (1997) Etudes relàtives à la maturation des péritheces de *Leptosphaeria maculans* sur les pailles de colza d'hiver nécrosées au collet. In: 5ème Conférence Sur les Maladies Des Plantes, Tours, France: (1997) ANPP1, pp 345–352. (in French)

Poisson B, Pérès A (1999) Study of rapeseed susceptibility to primary contamination of *Leptosphaeria maculans* in relation to plant vegetative stage. In: Proc 10th. International Rapeseed Congress, 1999, Canberra, Australia. http://www.regional.org.au/papers/index.htm

Pound SG (1947) Variability in Phoma lingam. J Agric Res 75(4):571

Pound GS (1948) Horseradish mosaic. J Agric Res 77:97–114

Pound GS, Walker JC (1945) Differentiation of certain crucifer viruses by the use of temperature and host immunity reactions. J Agric Res 71:255–278

Poveda J, Francisco M, Cartea ME, Velasco P (2020) Development of transgenic *Brassica* crops against biotic stresses caused by pathogens and arthropod pests. Plants 9:1664. https://doi.org/10.3390/plants9121664

Provvidenti R (1980) Evaluation of Chinese cabbage cultivars from Japan and the People's Republic of China for resistance to *Turnip mosaic virus* and *Cauliflower mosaic virus*. J Am Soc Hortic Sci 105:571–573

Provvidenti R (1982) A destructive disease of garden balsam caused by a strain of *Turnip mosaic virus*. Plant Dis 66:1076–1077

Provvidenti R, Robinson RW, Shail JW (1979) Chicory: a valuable source of resistance to Turnip mosaic for endive and escarole. J Am Soc Hortic Sci 104:726–728

Punithalingam E, Holliday P (1972) *Leptosphaeria maculans,* CMI Descriptions of pathogenic fungi and bacteria. No. 331. The Eastern Press Ltd, London

Purdy LH (1979) *Sclerotinia sclerotiorum*: history, diseases and symptomatology, host range, geographic distribution and impact. Phytopathology 69:875–880

Qin F, Shinozaki K, Yamaguchi-Shinozaki K (2011) Achievements and challenges in understanding plant abiotic stress responses and tolerance. Plant Cell Physiol 52:1569–1582

Rabenhorst GL (1844) Deutschlands Kryptogamen-Flora, Erster Band. Pilze, Leipzig

Rai RP (1979) *Fusarium equiseti* (Corda) Sacc. causing dry rot of potato tubers. New report. Curr Sci 48:1043–1045

Reyes A (1979) First occurrence of a severe white leaf spot on Chinese mustard in Canada. Can Plant Dis Surv 59:1–2

Riethmüller A, Voglmayr H, Göker M, Weiß M, Oberwinkler F (2002) Phylogenetic relationships of the downy mildews (Peronosporales) and related groups based on nuclear large subunit ribosomal DNA sequences. Mycologia 94:834–849

Robbins MA, Witsenboer H, Michelmore RW, Laliberté JF, Fortin MG (1994) Genetic mapping of turnip mosaic virus resistance in *Lactuca sativa*. Theor Appl Genet 89:583–589

Rodrigues AAC, Menezes M (2005) Identification and pathogenic characterization of endophytic *Fusarium* species from cowpea seeds. Mycopathologia 159:79–85

Romero FM, Rossi FR, Gárriz A, Carrasco P, Ruíz OA (2018) A bacterial endophyte from apoplast fluids protects canola plants from different phytopathogens via antibiosis and induction of host resistance. Phytopathology 109:375–383

Rouxel T, Penaud A, Pinochet X, Brun H, Gout L, Delourme R, Schmit J, Balesdent MH (2003) A 10-year survey of populations of *Leptosphaeria maculans* in France indicates a rapid adaptation towards the *Rlm1* resistance gene of oilseed rape. Eur J Plant Pathol 109:871–881

Rovira AD, Ogoshi A, McDonald HJ (1986) Characterization of isolates of *Rhizoctonia solani* from cereal roots in South Australia and New South Wales. Phytopathology 76:1245–1248

Ruano G, Scheuring D (2020) Plant cells under attack: unconventional endomembrane trafficking during plant defense. Plants 9(3):389. https://doi.org/10.3390/plants9030389

Rygulla W, Friedt W, Seyis F, Lühs W, Eynck C, Tiedemann AV, Snowdon RJ (2007a) Combination of resistance to *Verticillium longisporum* from zero erucic acid *Brassica oleracea* and oilseed *Brassica rapa* genotypes in resynthesized rapeseed (*Brassica napus*) lines. Plant Breed 126:596–602

Rygulla W, Snowdon RJ, Eynck C, Koopmann B, Tiedemann AV, Lühs W, Friedt W (2007b) Broadening the genetic basis of *Verticillium longisporum* resistance in *Brassica napus* by interspecific hybridization. Phytopathology 97:1391–1396

Saccardo PA (1888) *Cyctopus lepigoni*, *Cystopus bliti*, *Cystopus capparidis*, *Cystopus portulacae*. In: Sylloge Fungorum, vol 7, pp 235–237

Safeefulla KM (1952a) Morphological and cytological studies of *Albugo* species on *Ipomoea hederacea*. Curr Sci 21:287–288

Safeefulla KM (1952b) Species of the genus *Albugo* in India. Sci Cutt 18:282–283

Saha G, Park JI, Jung HJ, Ahmed NU, Kayum MA, Kang JG (2015) Molecular characterization of BZR transcription factor family and abiotic stress induced expression profiling in *Brassica rapa*. Plant Physiol Biochem 92:92–104

Saharan GS (1992a) Disease resistance. In: Labana KS, Banga SS, Banga SK (eds) Breeding oilseed Brassicas. Narosa Publ House, New Delhi, pp 181–205. Chapter 12

Saharan GS (1992b) Management of rapeseed-mustard diseases. In: Kumar D, Rai M (eds) Advances in oilseeds research. Sci Pub, Jodhpur, pp 152–188

Saharan GS (1997) Disease resistance. In: Kalia HR, Gupta SC (eds) Recent advances in oilseed Brassicas. Kalyani Pub, Ludhiana, pp 233–259. Chapter 12

Saharan GS (1998) Diseases of rapeseed and mustard and their management. In: Thind TS (ed) Diseases of field crops and their management. National Agricultural Technology Information Center, Ludhiana, pp 95–114

Saharan GS (2010) Analysis of genetic diversity in *Albugo*-crucifer system. J Mycol Plant Pathol 40:1–13

Saharan GS, Chand JN (1988) Diseases of oilseed crops (Hindi). Min. Human Resources, Govt of India. HAU Press, Hisar, pp 84–91. Chapter 3

Saharan GS, Kaushik JC (1981) Occurrence and epidemiology of powdery mildew of *Brassica*. Indian Phytopathol 34:54–57

Saharan GS, Mehta N (2002) Fungal diseases of rapeseed-mustard. In: Nagarajan S, Singh DP (eds) Role of resistance in intensive agriculture. Kalyani Pub, New Delhi, pp 98–108

Saharan GS, Mehta N (2008) Sclerotinia diseases of crop plants: biology, ecology and disease management. Springer, Dordrecht., 486 p. https://doi.org/10.1007/978-1-4020-8408-9

Saharan GS, Verma PR (1992) White rusts: a review of economically important species. International Development Research Centre (IDRC), Ottawa, ON

Saharan GS, Verma PR, Nashaat NI (1997) Monograph on downy mildew of crucifers. Tech Bull 1997-01, Cat No A5A-13/1997-01E. Agric and Agri-Food Canada, Saskatoon Research Centre, Saskatoon, SK. 197 p

Saharan GS, Mehta N, Sangwan MS (2003) Nature and mechanism of disease resistance to Alternaria blight in rapeseed-mustard system. Annu Rev Plant Pathol 2:85–125

Saharan GS, Mehta N, Sangwan MS (2005) Development of disease resistance in rapeseed-mustard. In: Saharan GS, Mehta N, Sangwan MS (eds) Diseases of oilseed crops. Indus Pub Co, New Delhi, pp 561–617

Saharan GS, Verma PR, Meena PD, Kumar A (2014) White rust of crucifers: biology, ecology and disease management. Springer Nature, Singapore. 244 p

Saharan GS, Mehta N, Meena PD (2016) Alternaria diseases of crucifers: biology, ecology and disease management. Springer, Singapore. 244 p

Saharan GS, Mehta N, Meena PD (2017) Downy mildew disease of crucifers: biology, ecology and disease management. Springer Nature, Singapore. 357 p

Saharan GS, Mehta N, Meena PD (2019) Powdery mildew disease of crucifers: biology, ecology and disease management. Springer Nature, Singapore. 362 p

Saharan GS, Mehta N, Meena PD (2021a) Molecular mechanism of crucifer's host-resistance. Springer Nature, Singapore., 835 p. https://doi.org/10.1007/978-981-16-1974-8/1

Saharan GS, Mehta N, Meena PD (2021b) Genomics of crucifer's host-resistance. Springer Nature, Singapore., 784 p. https://doi.org/10.1007/978-981-16-0862-9

Saharan GS, Mehta N, Meena PD (2021c) Clubroot disease of crucifers: biology, ecology and disease management. Springer Nature, Singapore., 757 p. https://doi.org/10.1007/978-981-16-2133-8

Salam MU, Khangura RK, Diggle AJ, Barbetti MJ (2003) Blackleg Sporacle. A model for predicting onset of pseudothecia maturity and seasonal ascospore showers in relation to blackleg of Canola. Phytopathology 93:1073–1081

Sanghera GS, Wani SH, Hussain W, Singh NB (2011) Engineering cold stress tolerance in crop plants. Curr Genomics 12:30–43

Sano Y, Kojima M (1989) Increase in cucumber mosaic virus concentration in Japanese radish plants co-infected with turnip mosaic virus. Annu Phytopathol Soc Jpn 55:296–302

Savary S, Ficke A, Aubertot JN, Hollier C (2012) Crop losses due to diseases and their implications for global food production losses and food security. Food Secur 4:519–537

Savulescu O (1946a) Study of the species of *Cystopus* (Pers.) Lev. Bucharest. Anal Acad Rous Mem Sect Stimtiface Soc 21:13

Savulescu O (1946b) A study on the European species of the genus *Cystopus* Lev. With special reference to the species found in Rumania. Thesis 213. University of Bucarest, Rumania. (Abstract in Rev Appl Mycol 27: 542, 1948)

Schaad NW, Alvarez A (1993) *Xanthomonas campestris* pv. *campestris*: cause of black rot of crucifers. In: Swings JG, Civerolo EL (eds) *Xanthomonas*. Chapman & Hall, London, pp 51–55

Schaad NW, Dianese JC (1981) Cruciferous weeds as sources of inoculum of *Xanthomonas campestris* in black rot of crucifers. Phytopathology 71:1215–1220

Scheffer RP (1950) Anthracnose leaf spot of crucifers. Tech Bull N C Agric Exp Stn 92:1–26

Schmittgen TD, Zakrajsek BA (2000) Effect of experimental treatment on housekeeping gene expression: validation by real-time, quantitative RT-PCR. J Biochem Biophys Methods 46:69–81

Schnathorst WC (1981) In: Mace ME, Bell AA, Beckman CH (eds) Life cycle and epidemiology of *Verticillium*. In: Fungal wilt diseases of plants. Academic Press, New York, NY, pp 81–111

Schroter J (1893) Peronosporinae. In: Engler A (ed) Die naturlichen Pflanzenfamilien, vol 1. Wilhelm Engelmann, Leipzig, pp 108–119

Schultz ES (1921) A transmissible mosaic disease of Chinese cabbage. J Agric Res 22:173–177

Schwinghamer MW, Schilg MA, Walsh JA, Bambach RW, Cossu RM, Bambridge JM, Hind-Lanoiselet TL, McCorkell BE, Cross P (2014) *Turnip mosaic virus*: potential for crop losses in the grain belt of New South Wales, Australia. Aust Plant Pathol 43:663–678

Scott DB, Visser CPN, Rufenacht EMC (1979) Crater disease of summer wheat in African drylands. Plant Dis Rep 63:836–840

Shattuck VI (1992) The biology, epidemiology, and control of turnip mosaic virus. Plant Breed Res 14:199–238

Shattuck VI, Parry R (1990) The occurrence of Powdery mildew on rutabagas in Southern Ontario. Can Plant Dis Surv 70:15–16

Shattuck VI, Stobbs LW (1987) Evaluation of rutabaga cultivars for turnip mosaic virus resistance and the Inheritance of resistance. Hortic Sci 22:935–937

Sherf AF, MacNab AA (1986) Vegetable diseases and their control. John Wiley & Sons, New York, NY

Shivas RG (1989) Fungal and bacterial diseases of plants in Western Australia. J R Soc WA 72:1–62

Siebold M, Von Tiedemann A (2012) Potential effects of global warming on oilseed rape pathogens in Northern Germany. Fungal Ecol 5:62–72. https://doi.org/10.1016/j.funeco.2011.04.003

Simons AJ, Skidmore DI (1988) Race-specific resistance to light leaf spot in *Brassica oleracea*. Trans Br Mycol Soc 90:431–435. https://doi.org/10.1016/s0007-1536(88)80152-9

Singh D, Rathaur PS, Vicente JG (2016) Characterization, genetic diversity and distribution of *Xanthomonas campestris* pv. *campestris* races causing black rot disease in cruciferous crops of India. Plant Pathol 65(9):1411–1418

Sinha AK, Wood RKS (1968) Studies on the nature of resistance in tomato plants to *Verticillium alboatrum*. Ann Appl Biol 59:143–154

Smiley RW, Ogg AG Jr, Cook RJ (1992) Influence of glyphosate on Rhizoctonia root rot, growth, and yield of barley. Plant Dis 76:937–942

Smith KM (1935) A virus disease of cultivated crucifers. Ann Appl Biol 22:239–242

Sneh B, Burpee L, Ogoshi A (1991) Identification of *Rhizoctonia* species. Am Phytopath Soc Press, St. Paul, MN

Sochting H, Verreet J (2004) Effects of different cultivation systems (soil management, nitrogen fertilization) on the epidemics of fungal diseases in oilseed rape (*Brassica napus* L. var. *napus*). J Plant Dis Prot 1:1–29. https://doi.org/10.1007/BF03356129

Sonah H, Zhang X, Deshmukh RK, Borhan MH, Fernando WGD, Bélanger RR (2016) Comparative transcriptomic analysis of virulence factors in *Leptosphaeria maculans* during compatible and incompatible interactions with canola. Front Plant Sci 7:1784. https://doi.org/10.3389/fpls.2016.01784

Song K, Osborn T, Williams P (1998) *Brassica* taxonomy based on nuclear restriction fragment length polymorphisms (RFLPs). Theor Appl Genet 75:784–794. https://doi.org/10.1007/BF00265606

Spallek T, Gan P, Kadota Y, Shirasu K (2018) Same tune, different song-cytokinins as virulence factors in plant-pathogen interactions? Curr Opin Plant Biol 44:82–87. https://doi.org/10.1016/j.pbi.2018.03.002

Spence NJ, Phiri NA, Hughes SL, Mwaniki A, Simons S, Oduor G, Chacha D, Kuria A, Ndirangu S, Kibata GN, Marris GC (2007) Economic impact of *Turnip mosaic virus, Cauliflower mosaic virus* and *Beet mosaic virus* in three Kenyan vegetables. Plant Pathol 56:317–323. https://doi.org/10.1111/j.1365-3059.2006.01498.x

Sprague SJ, Balesdent MH, Brun H, Hayden HL, Marcroft SJ, Pinochet X, Rouxel T, Howlett BJ (2006a) Major gene resistance in *Brassica napus* (oilseed rape) is overcome by changes in virulence of populations of *Leptosphaeria maculans* in France and Australia. In: Fitt BDL, Evans N, Cooke BM, Howlett BJ (eds) Sustainable strategies for managing *Brassica napus* (oilseed rape) resistance to *Leptosphaeria maculans* (phoma stem canker). Springer, New York, NY, pp 33–40

Sprague SJ, Marcroft SJ, Hayden HL, Howlett BJ (2006b) Major gene resistance to blackleg in *Brassica napus* overcome within three years of commercial production in southeastern Australia. Plant Dis 90:190–198

Sprague SJ, Marcroft SJ, Lindbeck KD, Ware AH, Khangura RK, Van de Wouw AP (2018) Detection, prevalence and severity of upper canopy infection on mature *Brassica napus* plants caused by *Leptosphaeria maculans* in Australia. Crop Past Sci 69:65–78

Spring O, Thines M (2004) On the necessity of new characters for classification and systematics of biotrophic Peronosporomycetes. Planta 219:910–914

Stack RW, Frohberg RC, Casper H (1997) Reaction of spring wheats incorporating Sumai#3-derived resistance to inoculation with seven *Fusarium* species. Cereal Res Commun 25:667–671

Staunton WP (1967) Studies on light leaf spot (*Gloeosporium concentricum*) of Brassicae. Irish J Agric Res 6(2):203–211

Stevens FL (1899) The compound oosphere of *Albugo bliti*. Bot Gaz 28:149–172. 225–245

Stevens FL (1901a) Gametogenesis and fertilization in *Albugo*. Contribution from the Hull Botanical Laboratory. XXIX. Bot Gaz 32:77–98

Stevens FL (1901b) Gametogenesis and fertilization in *Albugo*. Contribution from the Hull Botanical Laboratory. XXIX. Bot Gaz 32:177–169

Stevens FL (1901c) Gametogenesis and fertilization in *Albugo*. Contribution from the Hull Botanical Laboratory. XXIX. Bot Gaz 32:238–261

Stevens FL (1904) Oogenesis and fertilization in *Albugo ipomoeae-panduranae*. Bot Gaz 38:300–302

Stobbs LW, Hume D, Forrest B (1989) Survey of canola germplasm for resistance to *Turnip mosaic virus*. Phytoprotection 70:1–6

Stojsin V, Balaz F, Bagi F, Keljack I (2001) Pathogenicity of *Fusarium* spp. isolates from sugar beet root. Zaštita Bilja 52:241–249

Stotz HU, Mitrousia GK, de Wit PJ, Fitt BDL (2014) Effector-triggered defence against apoplastic fungal pathogens. Trends Plant Sci 19:491–500

Stürzenbaum SR, Kille P (2001) Control genes in quantitative molecular biological techniques: the variability of invariance. Compar Biochem Physiol B 130:281–289

Suh SK, Green SK, Park HG (1995) Genetics of resistance to five strains of *Turnip mosaic virus* in Chinese cabbage. Euphytica 81:71–77

Suleman P, Al-Musallam A, Menezes CA (2001) The effect of solute potential and water stress on black scorch caused by *Chalara paradoxa* and *Chalara radicicola* on date palms. Plant Dis 85: 80–83

Sumner D, Glaze N, Dowler C, Johnson A (1978) Foliar diseases of turnip grown for greens in intensive cropping systems. Plant Dis Rep 62:51–55

Sun D, Wang C, Zhang X, Zhang W, Jiang H, Yao X, Liu L, Wen Z, Niu G, Shan X (2019) Draft genome sequence of cauliflower (*Brassica oleracea* L var. *botrytis*) provides new insights into the C genome in *Brassica* species. Hortic Res 6:82

Sutton JC, Williams PH (1970) Comparison of extracellular polysaccharide of *Xanthomonas campestris* from culture and from infected cabbage leaves. Can J Bot 48(3):433–664. https://doi.org/10.1139/b70-089

Sylvester-Bradley R, Makepeace RJ (1985) Revision of a code for stages of development in oilseed rape (*Brassica napus* L.). Asp Appl Biol 10:395–400

Tahvonen R (1979) Seed-borne fungi on cruciferous cultivated plants in Finland and their importance in seedling raising. J Sci Agric Soc Finl 51:327–379

Tahvonen R, Hollo J, Hannukkala A, Kurppa A (1984) *Rhizoctonia solani* damping-off on spring turnip rape and spring rape (*Brassica* spp.) in Finland. J Sci Agric Soc Finl 56:143–154

Taylor JD, Conway J, Roberts SJ, Astley D, Vicente JG (2002) Sources and origin of resistance to *Xanthomonas campestris* pv. *campestris* in *Brassica* genomes. Phytopathology 92:105–111

Tekauz A, Mueller E, Beyene M, Stulzer M (2005) Fusarium head blight of winter wheat in Manitoba in 2004. Can Plant Dis Surv 85:47–48

Teng PS, Shane WW, MacKenzie DR (1984) Crop losses due to plant pathogens. Crit Rev Plant Sci 2:21–47. https://doi.org/10.1080/07352688409382187

Teo BK, Yitbarek SM, Verma PR, Morrall RAA (1988) Influence of soil moisture, seeding date, and *Rhizoctonia solani* isolates (AG 2–1 and AG 4) on disease incidence and yield in canola. Can J Plant Pathol 10:151–158

Tewari JP (1991) Structural and biochemical bases of the black spot disease of crucifers. Adv Struct Biol 1:325–349

Tewoldemedhin YT, Lamprecht SC, Mcleod A, Mazzola M (2006) Characterization of *Rhizoctonia* spp. recovered from crop plants used in rotational cropping systems in the Western Cape province of South Africa. Plant Dis 90:1399–1406. https://doi.org/10.1094/pd-90-1399

Thakur M, Sohal BS (2013) Role of elicitors in inducing resistance in plants against pathogen infection: a review. ISRN Biochem 2013:762412., 10 pages. https://doi.org/10.1155/2013/762412

Theron DJ, Holtz G (1989) *Fusarium* species associated with dry and stem-end rot of potatoes in South Africa. Phytophylactica 21:175–181

Thines A, Spring O (2005) A revision of *Albugo* (Chromista, Peronosporomycetes). Mycotaxon 92: 443–458

Thurwachter F, Garbe V, Hoppe HH (1999) Ascospore discharge, leaf infestation and variations in pathogenicity as criteria to predict impact of *Leptosphaeria maculans* on oilseed rape. J Phytopathol 147:215–222. https://doi.org/10.1046/j.1439-0434.1999.147004215.x

Tian X, Wang D, Mao Z, Pan L, Liao J, Cai Z (2019) Infection of *Plasmodiophora brassicae* changes the fungal endophyte community of tumourous stem mustard roots as revealed by high-throughput sequencing and culture-dependent methods. PLoS One 14:e0214975

Todaka D, Nakashima K, Shinozaki K, Yamaguchi-Shinozaki K (2012) Toward understanding transcriptional regulatory networks in abiotic stress responses and tolerance in rice. Rice 5:6

Togashi K, Shibasaki Y (1934) Biometrical and biological studies of *A. candida* (Pers.) O. Kuntze in connection with its specialization. Bull Imp Coll Agric For 18:88

Togashi K, Shibasaki Y, Kwanno Y (1930) Morphological studies of *Albugo candida*, the causal fungus of the white rust of cruciferous plants. Agric Hortic 7:859–882

Togashi K, Shibasaki Y, Sugana Y (1931) Morphological studies of white rust fungi in cruciferous plants. Jpn J Bot 5:82–83

Tomimura K, Spak J, Katis N, Jenner CE, Walsh JA, Gibbs AJ, Ohshima K (2004) Comparisons of the genetic structure of populations of *Turnip mosaic virus* in West and East Eurasia. Virology 330(2):408–423. https://doi.org/10.1016/j.virol.2004.09.040

Tomlinson JA (1970) *Turnip mosaic virus*. CMI/AAB descriptions of plant viruses, vol 8. CABI, Wallingford

Tomlinson JA (1987) Epidemiology and control of virus diseases of vegetables. Ann Appl Biol 110:661–681

Tomlinson JA, Ward CM (1978) The reactions of swede (*Brassica napus*) to infection by *Turnip mosaic virus*. Ann Appl Biol 89:61–69

Tomlinson JA, Ward CM (1982) Selection for immunity in swede (*Brassica napus*) to infection by *Turnip mosaic virus*. Ann Appl Biol 101:43–50

Ton J, van der Ent S, van Hulten M, Pozo M, van Oosten V, van Loon LC et al (2009) Priming as a mechanism behind induced resistance against pathogens; insects and abiotic stress. IOBC/WPRS Bull 44:3–13

Tortosa M, Cartea ME, Rodríguez VM, Velasco P (2018) Unraveling the metabolic response of *Brassica oleracea* exposed to *Xanthomonas campestris* pv. *campestris*. J Sci Food Agric 98: 3675–3683

Tu JC (1988) The role of white mold-infected white bean (*Phaseolus vulgaris* L) seeds in the dissemination of *Sclerotinia sclerotiorum* (Lib) de Bary. J Phytopathol 121:40–50

Tuncel G, Nergiz C (1993) Antimicrobial effect of some olive phenols in a laboratory medium. Lett Appl Microbiol 17:300–302

Tyvaert L, França SC, Debode J, Höfte M (2014) The endophyte *Verticillium* Vt305 protects cauliflower against Verticillium wilt. J Appl Microbiol 116:1563–1571

Van Alfen NK (1989) Molecular bases for virulence and avirulence of fungal wilt pathogens. In: Tjamos EC, Beckman CH (eds) Vascular wilt diseases of plants. Springer, Berlin, pp 19–32

Van de Wouw AP, Idnurm A (2019) Biotechnological potential of engineering pathogen effector proteins for use in plant disease management. Biotechnol Adv 37:1073–1087. https://doi.org/10.1016/j.biotechadv.2019.04.009

Van de Wouw AP, Idnurm A, Davidson JA, Sprague SJ, Khangura RK, Ware AH, Lindbeck KD, Marcroft SJ (2016) Fungal diseases of canola in Australia: identification of trends, threats and potential therapies. Aust Plant Pathol 45:415–423

Vance CP, Kirk TK, Sherwood RT (1980) Lignification as a mechanism of disease resistance. Annu Rev Phytopathol 18:259–288

Vandesompele J, De Preter K, Pattyn F, Poppe B, Van Roy N, De Paepe A, Speleman F (2002) Accurate normalization of real-time quantitative RT-PCR data by geometric averaging of multiple internal control genes. Genome Biol 3:RESEARCH0034. https://doi.org/10.1186/gb-2002-3-7-research0034

Vanniasingham VM, Gilligan CA (1989) Effects of host, pathogen and environmental factors on latent period and production of pycnidia of *Leptosphaeria maculans* on oilseed rape leaves in controlled environments. Mycol Res 93:167–174

Vasudeva RS (1958) Diseases of rape and mustard. In: Singh DP (ed) Rape and mustard. Indian Central Oilseed Committee, Hyderabad. 77 p

Vegetables New Zealand (2016) Vegetable *Brassica* IPM manual: pests, natural enemies, diseases and disorders of vegetable *Brassicas* in New Zealand. Horticulture New Zealand, Wellington. www.vegetablesnz.co.nz/research-and-development/current-research-projects/. Accessed 26 Feb 2017

Verma PR (1996a) Oilseed rape and canola diseases incited by *Rhizoctonia* species. In: Sneh B, Jabaji-Hare S, Neate S, Dijst G (eds) *Rhizoctonia* species: taxonomy, molecular biology, ecology, pathology and disease control. Springer, Dordrecht. https://doi.org/10.1007/978-94-017-2901-7_22

Verma PR (1996b) Biology and control of *Rhizoctonia solani* on rapeseed: a review. Phytoprotection 77:99–111

Verma PR, Saharan GS (1994) Monograph on Alternaria diseases of crucifers. Technical Bulletin 1994–6E. Saskatoon Research Centre, Agriculture and Agri-Food Canada, Saskatoon, SK

Vicente JG, Holub EB (2013) *Xanthomonas campestris* pv. *campestris* (cause of black rot of crucifers) in the genomic era is still a worldwide threat to Brassica crops. Mol. Plant Pathol 14(1):2–18. https://doi.org/10.1111/j.1364-3703.2012.00833.x

Vicente JG, Conway J, Roberts SJ, Taylor JD (2001) Identification and origin of *Xanthomonas campestris* pv. *campestris* races and related pathovars. Phytopathology 91:492–499

Videira S, Groenewald J, Braun U, Shin HD, Crous P (2016) All that glitters is not *Ramularia*. Stud Mycol 83:49–163. https://doi.org/10.1016/j.simyco.2016.06.001

Vijayvargiya S, Kumar A (2011) Influence of salinity stress on plant growth and productivity: salinity stress influences on plant growth. Lap Lambert Academic Publishers, Saarbrucken, p 180

Vincent D, Rafiqi M, Job D (2020) The multiple facets of plant-fungal interactions revealed through plant and fungal secretomics. Front Plant Sci 10:1626

Voglmayr H, Riethmüller A (2006) Phylogenetic relationships of *Albugo* species (white blister rusts) based on LSU rDNA sequence and oospore data. Mycol Res 110:75–85

Wafford JD, Gladders P, McPherson GM (1986) The incidence and severity of Brussels sprout diseases and the influence of oilseed rape. Asp Appl Biol 12:1–12

Wakefield EM (1927) The genus *Cystopus* in South Africa. Trans Br Mycol Soc 2:242–246

Walker JC (1969) Sclerotinia disease of vegetable and field crops. In: Plant pathology, 3rd edn. McGraw-Hill, New York, NY, p 417

Walkey DGA (1982) Reaction of spring cabbage cultivars to infection by *Turnip mosaic virus*. J Natl Inst Agric Bot 16:114–125

Walkey DGA, Pink DAC (1988) Reactions of white cabbage (*Brassica oleracea* var. *capitata*) to four different strains of turnip mosaic virus. Ann Appl Biol 112:273–284

Walsh JA (1989) Genetic control of immunity to *Turnip mosaic virus* in winter oilseed rape (*Brassica napus* ssp. *oleifera*) and the effect of foreign isolates of the virus. Ann Appl Biol 115:89–99

Walsh JA, Jenner CE (2002) *Turnip mosaic virus* and the quest for durable resistance. Mol Plant Pathol 3(5):289–300. https://doi.org/10.1046/j.1364-3703.2002.00132.x

Walsh JA, Tomlinson JA (1985) Viruses infecting winter oilseed rape (*Brassica napus* ssp. *oleifera*). Ann Appl Biol 107:485–495

Walsh JA, Rusholme RL, Hughes SL, Jenner CE, Bambridge JM, Lydiate DJ, Green SK (2002) Different classes of resistance to *Turnip mosaic virus* in *Brassica rapa*. Eur J Plant Pathol 108: 15–20

Wang G (1999) Evaluation of *Brassica napus* seed infection by *Leptosphaeria maculans/Phoma lingam*. M Sc thesis, University of Agriculture, Poznan

Wang W, Vinocur B, Altman A (2003) Plant responses to drought, salinity and extreme temperatures: towards genetic engineering for stress tolerance. Planta 218:1–14. https://doi.org/10.1007/s00425-003-1105-5

Wang TY, He F, Hu QW, Zhang Z (2011) A predicted protein-protein interaction network of the filamentous fungus *Neurospora crassa*. Mol BioSyst 7:2278–2285. https://doi.org/10.1039/c1mb05028a

Wang L, Wang Y, Zhou P (2013) Validation of reference genes for quantitative real-time PCR during Chinese wolfberry fruit development. Plant Physiol Biochem 70:304–310

Wang S, Wong D, Forrest K, Allen A, Chao S, Huang BE, Maccaferri M, Salvi S, Milner SG, Cattivelli L (2014) Characterization of polyploid wheat genomic diversity using a high-density 90 000 single nucleotide polymorphism array. Plant Biotechnol J 12:787–796

Wang Z, Tang J, Hu R, Wu P, Hou XL, Song XM (2015) Genome-wide analysis of the R2R3-MYB transcription factor genes in Chinese cabbage (*Brassica rapa* ssp. *pekinensis*) reveals their stress and hormone responsive patterns. BMC Genomics 16:17

Wardrop AB (1971) Occurrence and formation in plants. In: Sarkanen KV, Ludwig CH (eds) Lignins: occurrence, formation, structure, and reactions. Wiley Inter science, New York, NY, pp 19–41

Warwick SI, Black LD (1991) Molecular systematics of Brassica and allied genera (subtribe Brassicinae, Brassiceae)—chloroplast genome and cytodeme congruence. Theor Appl Genet 82:81–92. https://doi.org/10.1007/BF00231281

Waterhouse GM (1975) Species of *Albugo* on Aizoaceae. Trans Br Mycol Soc 65:504–507

West JS, Biddulph JE, Fitt BDL, Gladders P (1999) Epidemiology of *Leptosphaeria maculans* in relation to forecasting stem canker severity on winter oilseed rape in the UK. Ann Appl Biol 135:535–546

West JS, Evans N, Liu S, Hu B, Peng L (2000) *Leptosphaeria maculans* causing stem canker of oilseed rape in China. New Dis Rep. http://www.bspp.org.uk/ndr/2000/20003.htm

West JS, Kharbanda PD, Barbetti MJ, Fitt BDL (2001) Epidemiology and management of *Leptosphaeria maculans* (phoma stem canker) on oilseed rape in Australia. Can Eur Plant Pathol 50:10–27

Whetten R, Sederoff R (1995) Lignin biosynthesis. Plant Cell 7:1001–1013

Williams PH (1980) Black rot: a continuing threat to world crucifers. Plant Dis 64:736–742

Williams PH, Hill CB (1986) Rapid cycling populations of *Brassica*. Science 232:1385–1389

Wilson GW (1907) Studies in North American Peronasporales. I. The genus *Albugo*. Torry Bot Club Bull 34:61–84

Wing N, Bryden WL, Lauren DR, Burgess LW (1993) Toxogenecity of *Fusarium* species and subspecies in Section Gibbosum from different regions of Australia. Mycol Res 97:1441–1446

Wolf FA, Wolf FT (1947) The fungi, vol I. John Wiley and Sons, New York, pp 105–108

Wood PMc R, Barbetti M (1977a) The role of seed infection in the spread of blackleg of rape in Western Australia. Aust J Exp Agric 17:1040–1044

Wood PMc R, Barbetti MJ (1977b) A study on the inoculation of rape seedlings with ascospores and pycnidiospores of the blackleg disease causal agent *Leptosphaeria maculans*. J Aust Inst Agric Sci 43:79–80

Wroblewski T, Coulibaly S, Sadowski J, Quiros CF (2000) Variation and phylogenetic utility of the *Arabidopsis thaliana Rps2* homolog in various species of the tribe Brassiceae. Mol Phylogenet Evol 16:440–448

Xue AG, Ho KM, Butler G, Vigler BJ, Babcock C (2006) Pathogenicity of *Fusarium* species causing head blight in barley. Phytoprotection 87:55–61

Yamaguchi-Shinozaki K, Shinozaki K (2006) Transcriptional regulatory networks in cellular responses and tolerance to dehydration and cold stresses. Annu Rev Plant Biol 57:781–803

Yan Y, Yuan Q, Tang J, Huang J, Hsiang T, Wei Y, Zheng L (2018) *Colletotrichum higginsianum* as a model for understanding host–pathogen interactions: a review. Int J Mol Sci 19(7):2142. https://doi.org/10.3390/ijms19072142

Yan Y, Tang J, Yuan Q, Gu Q, Liu H, Huang J, Hsiang T, Zheng L (2020) ChCDC25 regulates infection-related morphogenesis and pathogenicity of the crucifer anthracnose fungus *Colletotrichum higginsianum*. Front Microbiol 11:763. https://doi.org/10.3389/fmicb.2020.00763

Yang YW, Lai KN, Tai PY, Li WH (1999) Rates of nucleotide substitution in angiosperm mitochondrial DNA sequences and dates of divergence between *Brassica* and other angiosperm lineages. J Mol Evol 48:597–604

Yang X, Feng HX, Yang YS (2008) Effects of silicon on flowering Chinese cabbage's anthracnose occurrence, flower stalk formation, and silicon uptake and accumulation. Chin J Appl Ecol 19(5):1006–1012

Yang J, Liu D, Wang X, Ji C, Cheng F, Liu B, Hu Z, Chen S, Pental D, Ju Y (2016) The genome sequence of allopolyploid *Brassica juncea* and analysis of differential homoeolog gene expression influencing selection. Nat Genet 48:1225–1232

Yasaka R, Fukagawa H, Ikematsu M, Soda H, Korkmaz S, Golnaraghi A, Katis N, Ho SYM, Gibbs AJ, Ohshima K (2017) The timescale of emergence and spread of turnip mosaic potyvirus. Sci Rep 7:4240. https://doi.org/10.1038/s41598-017-01934-7

Yasuda M, Ishikawa A, Jikumaru Y, Seki M, Umezawa T, Asami T (2008) Antagonistic interaction between systemic acquired resistance and the abscisic acid-mediated abiotic stress response in *Arabidopsis*. Plant Cell 20:1678–1692

Yoon JY, Green SK, Opena RT (1993) Inheritance of resistance to turnip mosaic virus in Chinese cabbage. Euphytica 69:103–108

Yoshii M, Yoshioka N, Ishikawa M et al (1998) Isolation of an *Arabidopsis thaliana* mutant in which accumulation of *Cucumber mosaic virus* coat protein is delayed. Plant J 13:211–219

Yu D, Chen C, Chen Z (2001) Evidence for an important role of WRKY DNA binding proteins in the regulation of *NPR1* gene expression. Plant Cell 13:1527–1540

Zeise K, von Tiedemann A (2001) Morphological and physiological differentiation among vegetative compatibility groups of *Verticillium dahlia* in relation to *V. longisporum*. J Phytopathol 149:469–475

Zeise K, von Tiedemann A (2002a) Host specialization among vegetative compatibility groups of *Verticillium dahlia* in relation to *Verticillium longisporum*. J Phytopathol 150:112–119

Zeise K, von Tiedemann A (2002b) Application of RAPD-PCR for virulence type analysis within *Verticillium dahlia* and *Verticillium longisporum*. J Phytopathol 150:557–563

Zhang X, Fernando WGD (2018) Insights into fighting against blackleg disease of *Brassica napus* in Canada. Crop Past Sci 69:40–47. https://doi.org/10.1071/CP16401

Zhang H, Zhou E, Liu Z, Yang M, Shen Y (1998) A technique for identification of resistance to anthracnose of *Brassica parachinensis* in seedling stage. J South China Agric Univ 19:47–50

Zhang H, Liu Z, Zheng Y, Huang H, Zhou E, Yang M (2000) The identification of resistance of Chinese flowering cabbage varieties to anthracnose. Guangdong Agric Sci 27:47–49

Zhang X, White RP, Demir E, Jedryczka M, Lange RM, Islam M, Li ZQ, Huang YJ, Hall M, Zhou G, Wang Z, Cai X, Skelsey P, Fitt BDL (2014) *Leptosphaeria* spp., phoma stem canker and potential spread of *L. maculans* on oilseed rape crops in China. Plant Pathol 63:598–612. https://doi.org/10.1111/ppa.12146

Zhang L, Cai X, Wu J, Liu M, Grob S, Cheng F, Liang J, Cai C, Liu Z, Liu B, Wang F, Li S, Liu F, Li X, Cheng L, Yang W, Li MH, Grossniklaus U, Zheng H, Wang X (2018a) Improved *Brassica rapa* reference genome by single-molecule sequencing and chromosome conformation capture technologies. Hortic Res 5:50. https://doi.org/10.1038/s41438-018-0071-9

Zhang YW, Xue LH, Li CJ (2018b) First report of anthracnose caused by *Colletotrichum higginsianum* on *Rumex acetosa* in China. Plant Dis 102(6):1174. https://doi.org/10.1094/PDIS-07-17-1082-PDN

Zhou Y, Fitt BDL, Welham SJ, Gladders P, Sansford CE, West JS (1999) Effects of severity and timing of stem canker (*Leptosphaeria maculans*) symptoms on yield of winter oilseed rape (*Brassica napus*) in the UK. Eur J Plant Pathol 105:715–728

Zhou E, Yang M, Zhang H, Liu Z (2002) Factors affecting the mycelia growth, sporulation and conidial germination of *Colletotrichum higginsianum* Sacc. J Nanjing Agric Univ 25:47–51

Zhou L, Hu Q, Johansson A, Dixelius C (2006) *Verticillium longisporum* and *V. dahliae*: infection and disease in *Brassica napus*. Plant Pathol 55:137–144

Zhu W, Wei W, Fu Y, Cheng J, Xie J, Li G, Yi X, Kang Z, Dickman MB, Jiang D (2013) A secretory protein of necrotrophic fungus *Sclerotinia sclerotiorum* that suppresses host resistance. PLoS One 8:e53901

Crucifer's Pathogens Genome

2

Abstract

The information on genome sequences, analysis and assembly of limited pathogens of crucifers has been generated and made available. It has given deeper insight into the molecular and biological functions of host–pathogen interactions in different ecological situations in nature and directed in the modern agricultural operations. The *Albugo candida* genome size is 45.3 Mb which is significantly smaller than a comparable biotrophs, *Hyaloperonospora arabidopsidis*. *A. candida* genome is deficient in enzymes required for nitrate and sulfate assimilation and has a reduced number of cell wall degrading enzymes. It has low number (26) of RXLR effector-like genes. *Alternaria brassicae* genome assembly consists of 17 contigs of 34.14 Mb sequences. The contigs ranges in length from 66.7 to 7.01 Mb. Of the 17 contigs, 1 represents the mitochondrial genome. The assembly contains nine complete chromosomes. The *Colletotrichum higginsianum* genome is 53.4 Mb in size and distributed among 12 chromosomes including two mini-chromosomes less than 1 Mb. The fungi causing powdery mildew disease has about four times larger genome size than those of other ascomycetes. The genome of *G. orontii* is 160 Mb in size. The fungi *Fusarium oxysporum* genome is 57.3 Mb in size with 583 scaffolds. The G:C content is 47.4%. The genome analysis of *Hyaloperonospora* has revealed phylogeny of *H. parasitica* with other species infecting cruciferous genera and species along with major species clusters in it. The final version of *H. arabidopsidis* genome of Emoy 2 isolates assembly consists of 81.6 Mb (V8.3.2) comprising of 42% of the *Hpa* genome. The genome assemblies for two isolates of *H. brassicae* differing for their virulence comprise of genome size of 72.762 and 76.950 Mb and 6470 and 6470 scaffolds. *Leptosphaeria maculans* genome is 45 Mb in size and organized in an unused patchwork fashion with G:C rich blocks known as isochore structure. There is reduction in intergenic space with multiple overlapping un-translated regions (UTRS) in *L. maculans*.

G. Singh Saharan et al., *Genomics of Crucifer's Host-Pathosystem*,
https://doi.org/10.1007/978-981-19-3812-2_2

Plasmodiophora brassicae genome is compact and small in size of 24.2–25.5 Mb with high gene density. Thirteen chitin syntheses gene have been identified. The white stem rot pathogen *Sclerotinia sclerotiorum* genome is 38.3 Mb in size and contains 679 sequence contigs giving 8X sequencing coverage of the genome. A complete assembly of *S. sclerotiorum* genome contains an additional 80S, 146 bp. It produces a number of extracellular pectin–degrading enzymes and biosyntheses genes of phytotoxins and other secondary metabolites. Bacterium *Xanthomonas campestris* comprises circular chromosomes of approximately 5,000,000 base pairs (bp), have a high G + C content and do not carry plasmids. Genome-wide comparisons have revealed three secretion system (type I, III, IV) associated with pathogenesis. The genome of *Turnip Mosaic Virus* (TuMV) isolates are 9798 nt in length, few are one to three nucleotide shorter in the terminal UTR's. Genome sequencing, comparative genome, transcriptome and secretome analysis of pathogenic species, clades and isolates facilitates determination of phylogeny, host specificity, pathogenicity factors/genes, evolution period, clonal reproduction, genetic exchange, and molecular genetical mechanisms and events in all host-pathosystem.

Keywords

Albugo genome · Evolution of obligate parasitism · Genome introgression · Assemblies · Intrarace diversity · Sequential infections · *Alternaria* genome · Comparative genomics · Phylogenomics · Bioinformatics analysis · Characterization of histidine kinase gene · *Colletotrichum* Genome · Sequencing · Transcriptome analysis · Phylogenetics · Sequence analysis · *Fusarium* genome · *Hyaloperonospora* Genome · Phylogeny · *Hyaloperonospora arabidopsidis* genome · *Leptosphaeria* genome · Identification of new subclade · *Plasmodiophora* genome · Nuclear genome analysis · Metabolic pathway · Distinguishing features · Phylogeny · cDNA cloning · *Sclerotinia* Genome · Phylogeny · Expression of genes · Sequences · Transcriptomics and secretomics analysis · Gene annotations · *Turnip Mosaic Virus* genome · Sequencing · Genome analysis · Phylogenetic analysis · Genetic diversity · *Xanthomonas* genome · Regulation of secretion systems · Type III effector genes · Virulence genes

2.1 Introduction

The development of genomics, proteomics, transcriptomics, molecular and biological techniques has revealed functional and evolutionary relationship of pathogen with each other including host-pathosystem of crucifers. Genome sequencing and analysis of crucifer's pathogenesis genome have given deeper insight into the evolutionary history, close relationship between necrotrophy and biotrophy, consequences of biotrophy, identification of defense suppressing and their host targets, host specificity genes, gain and loss of genes, phylogenetic relationship with other organism's species, races and strains. Genome sequencing has revealed clonal reproduction, genetic exchange between races, genome-wide similarity between races, levels of divergence between races, identification of virulence genes to understand population biology of pathogens. Genome sequencing and comparative analysis help to identify the shared and species-specific pathogenicity factors/genes in closely related species. It contributes immensely to clarify taxonomy of pathogens at species level. It also helps to determine secondary metabolites encoding gene clusters and effector coding genes. Absence and presence of species-specific regions on the chromosomes of the pathogen can be identified. Genome sequencing and functional genomics approaches have increased insight into molecular plant–pathogen interaction and help to assess genetic transformation efficiency using homologous recombination. It is possible to estimate divergence time and evolutionary status of related pathogenic species. A hybrid mapping and genome assembly approaches help to investigate gene content variation in different isolates of pathogens. The studies on genome reveal its organization, function of genes, gene inactivation, and modification as well as expression regulation. Genome sequences also facilitate comparative transcriptomic and secretomic analysis of pathogenic species, clades, and isolates. Comparative genomics is an important tool to mine pathogenicity determinants, host specificity factors and particularly type III effectors/gene in bacterial pathogens.

2.2 *Brassica–Albugo* Genome

2.2.1 Introduction

The *Albugo candida* (Pers.) Roussel is a key species for comparative genomics in the oomycetes as the archetypal crop pathogen in the Albuginales, which is an order that consists exclusively of obligate biotrophs. The Albuginales diverged early from the Peronosporales, which includes a wider spectrum of necrotrophs (*Pythium ultimum*), hemibiotrophs (*Phytophthora infestans* and *P. sojae*) and obligate biotrophic downy mildews (*Hyaloperonospora arabidopsidis*). White blister rust has been an important experimental model for investigating the molecular basis of disease resistance, pathogen virulence, host–parasite speciation, and the phenomenon of sustained defense suppression that is typically associated with "green islands" of compatible host tissue. The type species, *A. candida* (Pers.) Roussel was originally collected from a wild species *Capsella bursa-pastoris* but represents a highly variable complex of physiological races, each of which are pathogens in different oilseed, vegetable *Brassicas* or other wild crop relatives. More narrowly specialized species occur on other wild members of the Brassicaceae such as a common pathogen of *A. thaliana* that was named as *Albugo laibachii*. Rare accessions of *A. thaliana* are susceptible under laboratory conditions to *A. candida* races 2 and 7, derived from *Brassica juncea* and *Brassica rapa*, respectively, and have thus enabled genetic characterization of white rust resistance (WRR) genes in a nonhost crop relative. The first example, WRR4, encodes a TIR-NBS-LRR protein which confers broad spectrum resistance to at least four *A. candida* races in *A. thaliana*, and also confers resistance in transgenic oilseed crops of *B. juncea* and *B. napus*. The broad spectrum (race nonspecific) nature of WRR4 suggests that the multiple physiological races of *A. candida* from domestic and wild host species express a highly conserved avirulence protein (Saharan et al. 2014).

2.2.2 *Albugo candida* Genome

The genomic resources for *A. candida* have been developed to enable identification of candidate effector proteins expressed by the pathogen, including the predicted avirulence protein recognized by WRR4 in *A. thaliana*. The analysis of ESTs from two cDNA libraries and a draft genome sequence generated from the *A. candida* race 2 isolate (re-designated as Ac2VRR the "RR" suffix) was added to the pathotype name Ac2v to indicate the standard isolate chosen from the collection. The first cDNA library was constructed from susceptible *Brassica juncea* tissue infected with Ac2VRR, and a second library was made from Ac2VRR sporangiospores collected from susceptible seedlings. The predicted repertoire of genes from *Albugo candida* using transcript sequences and de novo sequence assembly of the pathogen genome has been summarized by Links et al. (2011). The historical reference isolate of *A. candida* (renamed Ac2VRR) used for the investigation represents a major destructive pathogen in oilseed and vegetable production of *Brassica* crops worldwide. Previous, pathology and molecular genetic studies using *A. candida* have been important for advancing in understanding of broad spectrum disease resistance and defense suppression in *A. thaliana* and the genetics of avirulence in *A. candida* (Cooper et al. 2008).

The *Albugo candida* is a member of the Albuginales, which is comprised exclusively of obligate biotrophs that have adapted to a wide diversity of hosts. This order is thought to have diverged early in oomycete evolution from other major plant pathogens including necrotrophic *Pythium* species which are the basal lineage in the Peronosporales, with respect to other hemi-biotrophic *Phytophthora* species and obligate biotrophic downy mildews. Thus, biotrophy has apparently evolved independently, in two major lineages of the oomycetes. Oomycete genomes range in size from the smallest 18 Mb genome of a necrotrophic *Pythium* to the largest >240 Mb genome of potato late blight pathogen *P. infestans*. The *A. candida* genome size (45.3 Mb) is significantly smaller than a comparable biotroph from the Peronosporales such as *H. arabidopsidis*, a common pathogen of *A. thaliana* in Europe. Similar to *H. arabidopsidis* and the obligate fungi, *A. candida* is deficient for enzymes required for nitrate and sulfate assimilation and has a reduced number of cell wall degrading enzymes (Spanu et al. 2010).

The most intriguing contrast found in the comparison of *A. candida* genome with other oomycetes is the low number of RXLR effector-like genes (26 were predicted as compared to 563 in *P. infestans*, 350 in the genomes of *P. sojae* and *P. ramorum* and 134 in the genome of *H. arabidopsidis* (Haas et al. 2009). This relatively small gene family in *A. candida* may be due to weak selection pressure from the slow evolution of white rust resistance genes in the host, and/or an apparently minimal repertoire of effectors in *A. candida* that have been more effective in suppressing and avoiding host defenses than in members of the Peronosporales. The lack of RXLR genes in the genome of *P. ultimum* and their discovery in a variant (Ac-RXL) form in the *A. candida* genome suggest that this class of effectors may have evolved independently along with biotrophy in the two major lineages of plant pathogenic oomycetes. The potential function of candidate CH × C proteins in *A. candida* also warrants further investigation as a potential novel class of effectors (Links et al. 2011).

The NEPs found in many plant pathogens, are capable of inducing cell death in plants. The *P. infestans* and *P. sojae* genomes contain genes that encode 27 NEPs and 39 NEPs, respectively. NEP expression levels in *P. sojae* correlates with a transition in disease development from biotrophy to necrotrophy. Links et al. (2011) were unable to identify any NEP orthologs in *A. candida* from the transcriptome or in the assembled genome sequence. The absence of NEP genes in *A. candida* is associated with necrotrophy. By contrast to the lack of NEPs in the *A. candida* genome, the *H. arabidopsidis* genome contains ten NEP-like genes. However, NEP proteins in *H. arabidopsidis* have high sequence divergence, probably to avoid triggering necrosis response in host (Baxter et al. 2010).

Albugo candida and *P. infestans* are also notably different in the respective number of elicitins and crinkler like proteins encoded in each genome. Links et al. (2011) identified nine elicit in and elicitin-like genes in Ac2VRR, far fewer than the 156 genes detected in *Phytophthora* species. Crinklers are also highly abundant in *Phytophthora* species with close to 200 genes reported for *P. infestans*. The *P. infestans* crinkler proteins are typified by an amino-terminal LFLAK domain and are capable of inducing necrosis in planta. While, Links et al. (2011) were able to identify six CRNs in the *A. candida* genome and none were detected as being highly expressed during infection or in sporangiospores, they did not induce necrosis in planta. Presence of tat-C orthologs and the conservation of tat-P containing proteins within the genomes of oomycetes are intriguing. In the case of *A. candida,* the tat-P containing complement of genes is reduced as compared to other oomycetes and the genome appears to harbor a tat-C ortholog. While these observations of a putative tat secretion system within oomycetes are preliminary, it forms an intriguing question for further investigation.

The differences in the repertoire of genes involved in pathogenicity, between *A. candida* and other oomycetes indicate that *A. candida* is highly adapted to life as an obligate biotroph. Among the canonical classes of pathogen effector molecules, many appear to be nonexistent in the *A. candida* genome or represent significantly smaller classes of genes. The analysis of the *A. candida* genome provides a basis for investigating the relationship between an obligate biotroph and its host. In general, these results portray a remarkably small eukaryotic genome for an organism that has evolved intimate biotrophic relationships independently from other oomycete lineages and has adapted to diverse host species (Links et al. 2011).

2.2.3 Evolution of Obligate Parasitism in *Albugo–Arabidopsis*

The obligate biotroph *A. laibachii* is a member of the Chromalveolata, which incorporates Dinophyta, Ciliophora, Heterokonts (containing the oomycetes), Haptophyta, Cryptophyta, and Apicomplexa which includes human parasites like *Plasmodium falciparum* and *Toxoplasma gondii*. Within the oomycetes, *A. laibachii* belongs to a lineage known as peronosporalean, which includes the hemibiotrophic pathogen of potato *P. infestans* and the necrotroph pathogen *P. ultimum*. Within this lineage, obligate biotrophy evolved twice independently in white rusts (Albuginales) and downy mildews (part of the Peronosporaceae). The downy mildew pathogen *H. arabidopsidis* and *A. laibachii* are both pathogens of the model plant *Arabidopsis thaliana*. While both show similar infection structures within the host, *A. laibachii* releases motile zoospores from asexual spores and sexual oospores, while *H. arabidopsidis* lacks all motile stages. Both pathogens are regularly found to co-infect plants and sporulate on the same leaf. A remarkable consequence of infection by *Albugo* sp. is enhanced host plant susceptibility to other parasites to which the host is resistant in the absence of *Albugo* infection, and also impairment of cell death mechanisms. *Albugo* sp. infect 63 genera and 241 species, including economically important *Brassica rapa* (canola), *B. juncea* (oilseed mustard), and *B. oleracea* (cabbage family vegetables) (Saharan et al. 2014). The analysis of oomycete evolutionary history suggest that *Albugo* is more closely related to necrotrophs such as *Pythium* than to downy mildews, and thus provides a unique system to study the evolution and consequences of biotrophy, and to identify new defense-suppressing effectors and their host targets (Thines and Kamoun 2010).

Plant pathogens that cannot grow except on their hosts are called obligate biotrophs. How such biotrophy evolves is poorly understood. Kemen et al. (2011) sequenced the genome of the obligate biotroph white rust pathogen (*A. laibachii*) of *Arabidopsis*. From comparisons with other oomycete plant pathogens, diatoms, and the human pathogen *Plasmodium falciparum*, it was revealed a loss of important metabolic enzymes. It also revealed the appearance of defense-suppressing "effectors," some carrying motifs known from other oomycete effectors, and discover and experimentally verify a novel class of effectors that share a CHXC motif within 50 amino acids of the signal peptide cleavage site. Obligate biotrophy involves an intimate association within host cells at the haustorial interface (where the parasite penetrates the host cell's cell wall), where nutrients are acquired from the host and effectors are delivered to the host. It was found that *A. laibachii*, like *H. arabidopsidis* and *P. falciparum*, lacks molybdopterin-requiring biosynthetic pathways, suggesting relaxed selection for retention of, or even selection against, this pathway. Kemen et al. (2011) proposed that when defense suppression becomes sufficiently effective, hosts become such a reliable source of nutrients that a free-living phase can be lost. These mechanisms leading to obligate biotrophy and host specificity are relevant not only to plant pathogenic but also to human pathogens.

The *A. laibachii* genome assembly sheds light on the evolution of biotrophy since it allows the first comparison of two oomycetes obligate biotroph pathogens (*A. laibachii* and *H. arabidopsidis*) that evolved biotrophy independently. In addition, *A. laibachii* shows the highest overall amino acid identity to the necrotroph pathogen *P. ultimum* and the hemibiotroph *P. infestans*. One of the striking results of this comparison is that all organisms able to build haustoria have lost their thiamine biosynthesis pathway, presumably because thiamine is easily obtained from hosts via the haustorial interface. A closer interface requires effective host defense suppression. Kemen et al. (2011) therefore, hypothesized that the evolution of biotrophy involves a series of steps: step 1, involving progressively more effective effectors to suppress defense, step 2, attenuated activation of defense by reduction in the inventory of cell wall hydrolyzing enzymes, resulting in, step 3, weak selection to maintain certain biosynthetic pathways if the products of the pathways can be directly obtained from the host. This results in progressively more comprehensive auxotrophy and culminates in irreversible biotrophy (Fig. 2.1).

2.2.4 Genome Introgression of *Albugo candida* Races

Some "generalist" parasite species have solved the dilemma by evolving multiple specialized races, each of which infects different hosts. The eukaryotic oomycete order Albuginales consists entirely of obligate biotrophic pathogens that cause disease on a broad range of plant hosts (Saharan et al. 2014). Its largest genus, *Albugo*, comprise ~50 (usually) specialist pathogens (Choi et al. 2009b, 2011; Thines et al. 2009a; Ploch et al. 2010), yet *A. candida* (Pers.) Roussel can infect over 200 species of plants in 63 genera from the families of Brassicaceae, Cleomaceae, and Capparaceae (Saharan and Verma 1992; Choi et al. 2009a, b). The *A. candida* infections are the causal agent of "white rust" disease resulting in 9–60% losses on economically important oilseed and vegetable *Brassica* crops (Harper and Pittman 1974; Barbetti and Carter 1986; Saharan and Verma 1992; Meena et al. 2002).

Albugo candida consists of different physiological races that each usually shows high host specificity (Hiura 1930; Pound and Williams 1963; Petrie 1988). More than 115 races of *A. candida* have been defined, based primarily on their host range (Saharan and Verma 1992; Saharan et al. 2014). *Albugo candida* is a diploid organism that reproduces both asexually and sexually, but the relative importance of both reproductive modes is not well established. During asexual reproduction, diploid zoospores are formed in a special propagule (the zoosporangium) on a thallus of hyphae beneath the leaf epidermis. White rust comprising large numbers of dehydrated sporangia eventually rupture the epidermis to release inoculum for dispersal. When reproducing sexually, fertilization between two isolates results in nonmotile diploid thick-walled oospores that can resist extreme temperatures and desiccation. Although the relative importance of different mechanisms of reproduction in the *Albugo* life cycle remains poorly understood, clonal reproduction enables rapid population expansion, especially in genetically uniform crop monocultures. The *A. candida* is considered a single species despite the fact it comprises several specialized physiological races that colonize different host plants. According to evolutionary and population genetic theory, adaptations and trade-offs associated

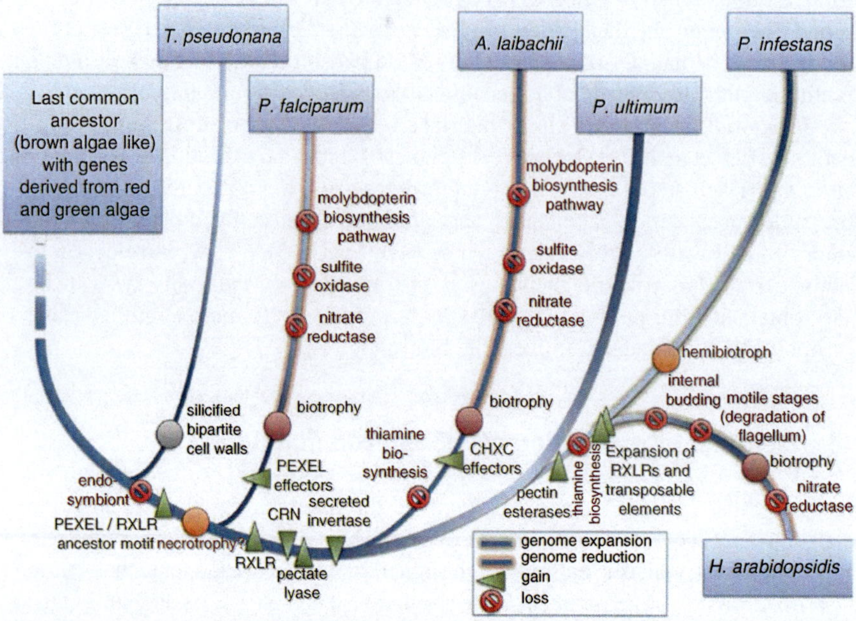

Fig. 2.1 Gain and loss of genes and pathways for selected Chromalveolata in comparison to *Albugo laibachii*. It was hypothesized that the last common ancestor of Chromalveolata was a brown-alga-like organism with genes from green and red algae integrated into the nuclear genome after primary and secondary endosymbiosis. While some heterokonts kept their secondary endosymbiont and, in the case of diatoms, acquired a silicated bipartite cell wall, others lost their secondary endosymbiont. The postulated after the loss of the endosymbiont, convergent evolution led to effector proteins like PEXEL and RXLR precursors. PEXEL effectors might have enabled *Plasmodium falciparum* to achieve more complex interactions with its host and establish intercellular growth. In addition to the RXLR effector proteins, oomycetes acquired or evolved another class of effectors, the CRNs and a secreted invertase that allows use of sucrose from host plants. Oomycetes that are biotrophs or hemibiotrophs lost their thiamine biosynthesis pathway and, in the case of *A. laibachii*, evolved a new "CHXC" effector class. After taking up the biotroph lifestyle, the genomes of *P. falciparum*, *Hyaloperonospora arabidopsidis*, and *A. laibachii* started a gene reduction that is exemplified by looking at enzymes that require molybdenum co-factors and the molybdopterin biosynthesis pathway. Hemibiotroph *Phytophthora infestans* instead shows a strong genome expansion. In this context, *H. arabidopsidis* showed a genome expansion and acquired biotrophy late, based on the loss of only one molybdenum dependent enzyme (Kemen et al. 2011)

with host specialization combined with strong population structuring can result in adaptive radiation and speciation (Abbott et al. 2013; Stukenbrock 2013). Conceivably, *A. candida* is an adaptive radiation "in progress," and the broad host range is realized by ongoing specialization of independent races that are on the road to speciation (Drès and Mallet 2002).

The *Albugo* spp. infection strongly suppresses host innate immunity. The *Albugo* spp. is unique compared to other microbial plant pathogens in enhancing host susceptibility to secondary infection by otherwise avirulent pathogens, including downy mildews (Cooper et al. 2008). It has been suggested that enhanced susceptibility imposed by *Albugo* might accelerate adaptation of other pathogen species to an *Albugo*-susceptible host (Thines 2014). However, no evolutionary rationale has been put forward to explain why it might be adaptive for *Albugo* spp. to render its hosts so susceptible to other pathogens that could compete for access to the same resources (Cooper et al. 2008). Arguably, suppression of host innate immunity could facilitate cohabitation of distinct physiological races and thus may enable genetic exchange between them. Introgression, here defined as the introduction by recombination of syntenic nucleotide variation from a parental donor race into the genome of a recipient race (Hedrick 2013), could slow down genetic divergence, and hence, retard speciation. On the other hand, introgression between races that are well-adapted to exploit different host plants could be maladaptive and strongly selected against because hybrids will inherit effector alleles derived from both parental races. Given that immune recognition of even a single effector is sufficient to trigger the immune response and stop an infection, hybrids that possess an expanded repertoire of effector alleles are likely to have a strong fitness disadvantage on most potential host plants.

Genome sequence assemblies of five isolates were collected from four host species (*Brassica oleracea*, *B. juncea*, *Capsella bursa-pastoris*, and *Arabidopsis thaliana*). One isolate of one race and two isolates of each of two additional races were generated by McMullan et al. (2015). It was shown that these races are not genetically isolated despite having non overlapping host ranges. Recombination analysis shows that there is widespread genetic exchange between *A. candida* races, and that hybridization leading to introgression has occurred numerous times, which include exchanges in the recent past. It was observed that whether preinfection of *Arabidopsis* and *Brassica* with virulent *A. candida* races results in enhanced host susceptibility, and found that preinfection with a virulent strain enables proliferation of an *A. candida* isolate that would otherwise not colonize that host. For the two races with two isolates, it was found that population expansion is by clonal reproduction (McMullan et al. 2015).

2.2.4.1 Single Race Isolates of *Albugo candida* are Host-Specific

The virulence of single race infections of AcNc2, Ac2V, and AcBoT was tested on different host species and cultivars. AcNc2 was propagated on *A. thaliana* Ws-2. Two *B. juncea* cultivars ("Cutlass" and "Czerniac") and 18 cultivars of *B. oleracea* were resistant to AcNc2. Screened 356 *A. thaliana* accessions for their susceptibility to AcNc2 and 38.5% were found susceptible to this race (Table 2.1). Race Ac2V was originally isolated in Canada on *B. juncea* (Rimmer et al. 2000). McMullan et al. (2015) confirmed virulence of Ac2V on *B. juncea* by spray inoculation of *B. juncea* "Cutlass" and "Czerniac." On 3-week-old *A. thaliana* plants, all 107 tested accessions were resistant. Two *B. oleracea* cultivars were also resistant to Ac2V (Table 2.1). Race AcBoT was virulent on all 15 tested cultivars of *B. oleracea*. In contrast, 34 accessions of *A. thaliana* were fully resistant to AcBoT, as were in *B. rapa* and *B. juncea* cultivars (Table 2.1). These tests confirm that *A. candida* races show pronounced host specificity to distinct host species. The races AcBoT and AcBoL only grow on *B. oleracea*, and AcNc2 and AcEm2 can only grow on a subset

Table 2.1 Virulence of the *Albugo candida* races on different plant host accessions (McMullan et al. 2015)

	Arabidopsis thaliana		*Brassica rapa*		*Brassica juncea*		*Brassica oleracea*	
A. candida race	+	–	+	–	+	–	+	–
AcNc2	137	219	0	1	0	1	0	18
AcBoT	1[a]	34	0	1	0	2	15	0
Ac2V	1[a, b]	107	1[c]	4[c]	6[c]	0[c]	0	2

+ Host–pathogen compatible interactions (number of susceptible accessions)
– Host–pathogen incompatible interactions (number of resistant accessions)
[a] *A. thaliana* Ws-eds1 (enhanced disease susceptibility) mutants were susceptible to all tested *A. candida* races
[b] In the laboratory conditions, the cotyledons of the *A. thaliana* accession Ws-3 were found to be susceptible to the Ac2V (Cooper et al. 2008)
[c] Data from (Rimmer et al. 2000) incorporated; one cultivar *B. rapa* (CrGC1-18, rapid-cycling accession) was infected by Ac2V race and four other tested cultivars ("Torch," "Colt," "Horizon," "Reward") were incompatible with Ac2V race. All analyzed cultivars of *B. juncea* (CrGC4-1S, "Burgonde," "Domo," "Cutlass") were susceptible to Ac2V

of *Arabidopsis* and *Capsella* genotypes. Genetic exchange between races is unlikely to occur unless they colonize the same host. The immune-compromised of *A. thaliana* Ws-2-eds1 mutant was susceptible to all races.

2.2.4.2 Genome Assemblies of *Albugo candida* Isolates

The AcNc2 *A. candida* assembly was used as the reference which comprises 34 Mb in 5212 contigs of ~160-fold coverage (Table 2.2). McMullan et al. (2015) assembled ~73% of an estimated 45 Mb genome of *A. candida* AcNc2 (Voglmayr and Greilhuber 1998; Links et al. 2011). A similar proportion (76%) of the *A. candida* Ac2V race genome was assembled by Links et al. (2011). The unassembled part of the genome (~11 Mb) is likely to include repeats and duplicated sequences. Repeat sequences constitute ~17.4% of the ~AcNc2 assembly. Approximately 8% of annotated repeats represent collapsed regions with coverage several times higher than average, so the real repeat content may be higher.

The *Ab* initio gene predictions were conducted with several gene prediction programs, resulting in 10,907 predicted gene models. About 90% (9830) of the predicted proteins have homologous sequences in the proteome of *A. candida* Ac2V race (15,824 genes) (Links et al. 2011). In about 1000 cases, McMullan et al. (2015) predicted a single-copy gene in the AcNc2 race. Links et al. (2011) have predicted multigene families in the Ac2V race, explaining the discrepancy in the predicted gene number for two assemblies. Only 37% of AcNc2 proteins showed significant sequence similarity to known proteins (BLASTP *E*-value $\leq 10^{-5}$). Using the Tribe-MCL algorithm, 3522 genes (32% of the predicted AcNc2 gene repertoire) were clustered into 1020 gene families. The largest gene tribes are protein kinases (187 members), kinesin (59 genes), and myosin (44 genes) like proteins and 35 genes homologous to the secreted "CHXC" proteins of *A. laibachii* (Kemen et al. 2011). A total of 915 genes in the AcNc2 assembly are predicted to encode

Table 2.2 Summary of the *Albugo candida* AcNc2, AcEm2, AcBoT, AcBoL and Ac2V genome assemblies (McMullan et al. 2015)

Parameters	AcNc2	AcEm2	AcBoT	AcBoL	Ac2V
Number of contigs	5212	11,581	11,929	11,143	12,210
N50 length (bp)	41,078	29,326	14,673	14,953	24,005
N50 number	231	284	584	581	353
Mean contig length (bp)	6610	2668	2781	2998	2770
Assembly size (bp)	34,454,169	33,409,146	33,184,526	33,409,856	33,823,601
GC content (%)	43.19%	43.09%	43.15%	43.15%	43.11%
CEGMA gene space coverage (%)[a]	93.55%	92.74%	93.55%	93.13%	92.34%
Average genome coverage	160	150	140	140	200
Repeat content (%)	17.4%	NA	NA	NA	NA
Predicted genes	10,907	NA	NA	NA	NA

[a] Completeness of the gene space in the different genome assemblies was estimated using CEGMA pipeline; the presence of over 90% of core eukaryotic genes in the assembly serves as an indication of an overall complete gene space

putative proteins with amino-terminal secretary signal peptides, but no transmembrane domain. Only 34% of the predicted secretome was functionally annotated (BLASTP, *E*-value $\leq 10^{-5}$), including 115 proteins (proteases, hydrolases, elicitin-like proteins, elicitors, protease inhibitors) that could be involved in plant cell wall degradation and protection against host defense enzymes. In addition to the 35 CHXC proteins (Tyler et al. 2006), further candidate virulence factors were identified including 19 homologs of the *Phytophthora* Crinkler effectors (Haas et al. 2009), and another 23 secreted proteins with "RXLR" and 35 proteins with similar "RXLQ" motifs. Both motifs are located in the N-terminal part of protein after the predicted signal peptide, thus, resembling the RXLR effectors of *P. infestans* and *H. arabidopsidis* (Haas et al. 2009; Baxter et al. 2010), but not having any other significant sequence similarity to these proteins. After carrying out several assemblies based on different k-mer lengths, the quality of each assembly was assessed with various parameters and one best assembly was chosen for each isolate (Table 2.2). The high similarity of the five *A. candida* isolates enabled to conclude that sequenced three "races," within which AcNc2 and AcEm2 were isolates of the same race and AcBoT and AcBoL were also isolates of the same race (Fig. 2.2). Therefore, genome comparisons were first conducted on one representative from each race (AcNc2, Ac2V and AcBoT), from each of which ~33–34 Mb of genome was assembled (Table 2.2).

2.2.4.3 Genome-Wide Similarity Between Races with Non-overlapping Host Range

The overall genome-wide similarity between races was assessed; McMullan et al. (2015) performed alignments of reads against the AcNc2 reference assembly. For the

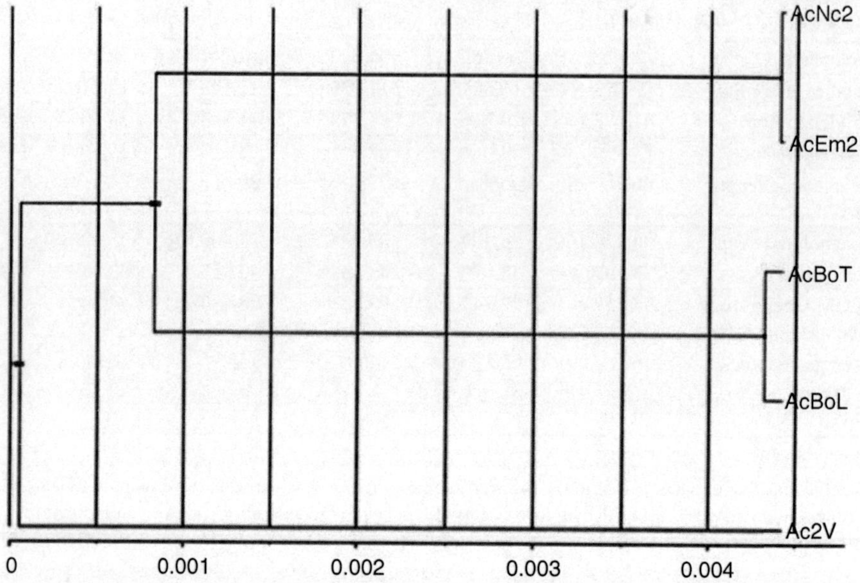

Fig. 2.2 Phylogeny shows that the five sequenced *Albugo candida* isolates fall into three divergent races. BEAST tree constructed based on using contig 1 (398,508 bp) shows the divergence among the three *A. candida* races (AcEm2 and AcNc2; AcBoT and AcBoL; Ac2V). Blue bars represent the 95% Higher Posterior Density (HPD). A strict molecular clock fixed at 1.0 in order to show the relationship in the scale of substitutions per site was used (McMullan et al. 2015)

majority of the AcNc2 genome, a significant positive correlation between read depth in the reference assembly and mapping depth of the Ac2V and AcBoT reads was found ($r = 0.65$, $p < 2.2e{-}16$; Fig. 2.3a). Some AcNc2 regions (3–4% of the assembly) showed low or zero coverage by Ac2V and/or AcBoT reads (Fig. 2.3), suggesting the presence of highly divergent or unique regions among the races. These are gene sparse regions (150 and 234 genes predicted in the AcNc2, respectively), without apparent enrichment for genes encoding for secreted proteins ($\chi^2 = 0.11$, d. f. $= 1$, $p > 0.7$). Amplification of randomly selected AcNc2 genes from these regions revealed that four of the selected five genes are indeed absent/or highly diverged in the AcBoT and Ac2V races, and present in the AcNc2 genome. The overall mean level of nucleotide identity in the homologous genomic regions among races is ~99% (Fig. 2.3b). The 25 polymorphic genomic regions were verified by Sanger sequencing. McMullan et al. (2015) used the longest of all contigs from AcNc2, "contig 1" (398,508 bp), to compare the levels of divergence between races vs. the number of heterozygous positions within each race. An extremely low proportion of sites (0.03% and 0.01%) on "contig 1" are heterozygous within AcNc2, AcEm2, and Ac2V races, respectively (Fig. 2.4). Races AcBoT and AcBoL are more heterozygous than Ac2V and AcNc2, with 0.65% of nucleotide positions in "contig 1" being heterozygous in AcBoT (Fig. 2.4). Importantly, >97% of all heterozygous positions are shared in AcBoL and AcBoT. In between-race

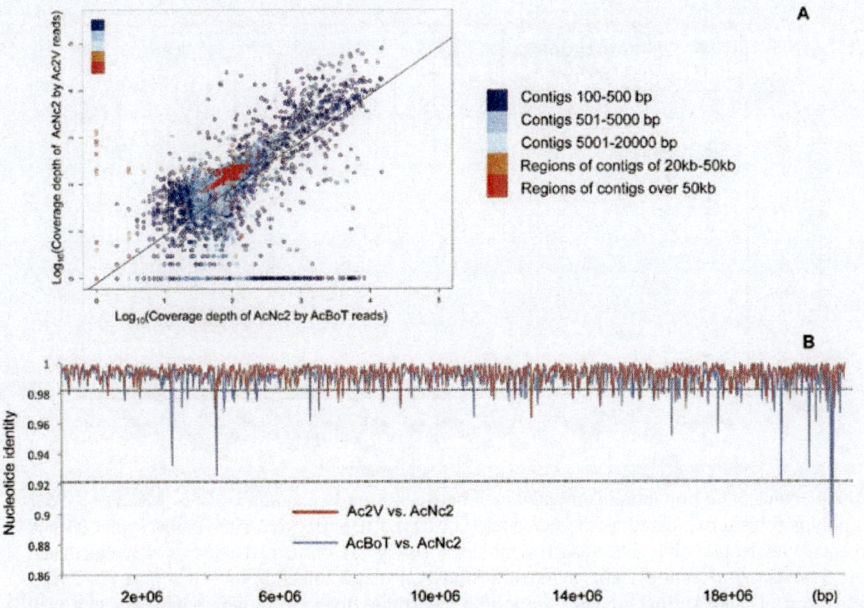

Fig. 2.3 Comparison of *Albugo candida* races using alignments of Illumina reads against the AcNc2 assembly. (**a**) Positive correlation between the depths of coverage of the reference assembly (AcNc2) by the Ac2V and AcBoT reads. For the reference contigs less than 20 kb, the mean coverage was calculated across the whole contig length and log-transformed. For the contigs over 20 kb, the mean coverage was calculated for the sliding window of 20 kb and log-transformed. *Y*-axis shows the log-transformed depth of the reference coverage by the Ac2V reads; *X*-axis shows the log-transformed depth of the reference coverage by the AcBoT reads. (**b**) Nucleotide identity among the homologous genomic regions of Ac2V, AcBoT and AcNc2. The mean identity was calculated for the sliding window of 20 kb (McMullan et al. 2015)

comparisons, ~1.0% of nucleotide positions on "contig 1" has diverged between AcBoT, Ac2V, and AcNc2.

2.2.4.4 Mosaic-Like Genome Structure of *Albugo candida* Races

Polymorphisms are not homogeneously distributed among *A. candida* races. Some regions of the genome are identical for up to 10,000 base pairs, whereas the local nucleotide identity is as low as 89% in other regions of up to 5 kb (Fig. 2.3b). McMullan et al. (2015) examined 133 contigs (12,373,253 bp), covering 38% of the reference assembly. Stretches of nucleotide similarity among races are distributed in a block-like structure; there are regions where AcNc2 is highly similar (or identical) to AcBoT and significantly (nucleotide divergence $\pi > 1\%$) diverged from Ac2V, and vice versa (Fig. 2.5). By using multiple different algorithms that can detect recombination in DNA sequence data incorporated in the software RDP3 (Martin et al. 2010). Recombination analysis of 133 contigs highlighted a total of 675 recombined blocks on 127 contigs that were significant (after Bonferroni correction) in three or more tests using RDP3.

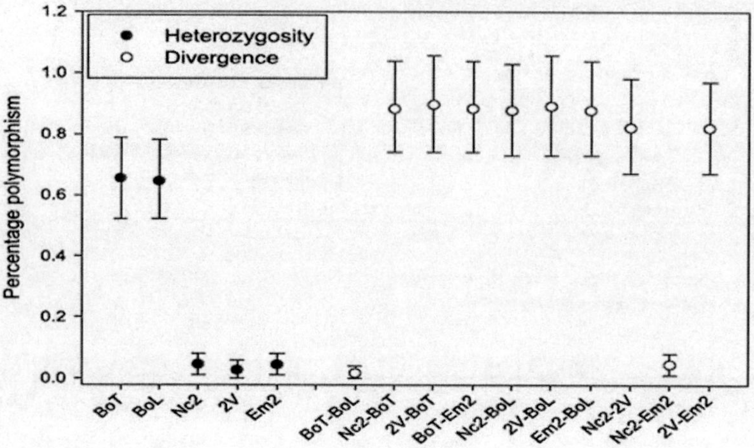

Fig. 2.4 Nucleotide polymorphism within and between *Albugo candida* isolates. Mean (±5–95% CI) polymorphism expressed as the percentage observed heterozygote sites (solid symbols) and percentage nucleotide divergence (open symbols) at contig 1. Confidence intervals were calculated using a bootstrap of contig 1 after removal of indels. Isolates infecting the same host plant (i.e., AcBoT-AcBoL and AcEm2-AcNc2) show little nucleotide divergence, which indicates that they are genotypically almost identical (i.e., diverged by less than 0.05%). Nevertheless, the *Brassica oleracea* infecting race (AcBoT and AcBoL) possess a relatively high heterozygosity compared to the isolates of the *Arabidopsis thaliana* infecting race. Moreover, most of this heterozygous polymorphism is shared (low nucleotide divergence) and presence of the majority of heterozygous sites is consistent with clonal reproduction (McMullan et al. 2015)

The combined length of all identified blocks is nearly 3 Mb or ~25% of the analyzed regions. These blocks indicate regions of genome in one race that derive from another race (or the ancestor of another race). Figure 2.5 illustrates the effect of genetic introgression on the pattern of nucleotide similarity between the three races in the largest contig of ~400 kb. The sequence (dis) similarity between the three races shows a mosaic-like genome structure with large regions where races AcNc2 and AcBoT show near sequence identity (yellow blocks in Fig. 2.5b), whilst other areas show a high level of sequence similarity between AcNc2 and Ac2V, and AcBoT and Ac2V (indicated by the purple and turquoise regions, respectively). Note that the presence of such well-defined blocks of high sequence similarity in an otherwise diverged genome is characteristic for rare introgression between organisms that show a high (yet incomplete) level of reproductive isolation. Noteworthy too is the fact that a high level of recombination (relative to the mutation rate) would homogenize the sequence divergence between the races, and hence, that this would not result in the observed mosaic-like structure.

Despite the fact that introgression between races is rare, it must have occurred multiple times between the ancestors of the three races given that the coalescence times varies markedly between the different blocks (Fig. 2.6).

Assuming a base mutation rate of $\mu = 10^{-8}$ per cell cycle, with 100 cell cycles per year (i.e., a combined mutation rate of 10^{-6} per year), analysis in the software

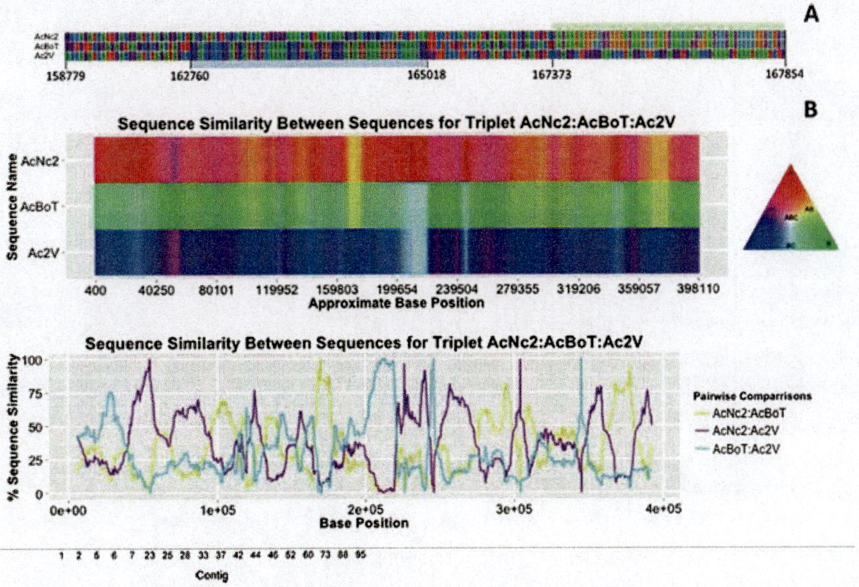

Fig. 2.5 Variation in sequence similarity between races. (**a**) Alignment of nucleotides in between positions 158,779 and 167,382 within "contig 1" of three *Albugo candida* races (AcNc2, AcBoT, and Ac2V) illustrating two recombination blocks colored blue and green. Both blocks show high sequence similarity between races. Also shown is the sequence divergence in between blocks. Alignment gaps and monomorphic sites have been removed. (**b**) The sequence similarity at "contig 1" among three *A. candida* races was visualized using the colors of a RBG color triangular in the software HybRIDS (http://www.norwichresearchpark.com/HybRIDS). Areas where two contigs have the same color (yellow, purple, or turquoise) are indicative of two races sharing the same polymorphisms. The linear plot of the proportion of SNPs shared between the three pair wise comparisons between the races. Shown on the *X*-axis is the actual base position. The graphs were made in the R package HybRIDS. http://www.norwichresearchpark.com/HybRIDS, McMullan et al. 2015)

HybRIDS shows that the most recent introgression event has occurred circa 220 years ago, whilst the oldest event occurred almost 200,000 years ago. The mean age calculated across all introgression events equals 6237 (±12,594) years (Fig. 2.6) (with a combined mutation rate of $\mu = 10^{-7}$ per base per year, the range in the age of introgression would span from 2200 to 2,000,000 years). Irrespective of the mutation rate, the principal finding is that genetic introgression among *A. candida* races is an ongoing evolutionary process occurring across a wide range of evolutionary times, and that it gives rise to mosaic genomes with the introgression blocks interspersed in the recipient genomic background. A total of 1655 predicted genes are located in the recombined regions, and among these, 125 are predicted to encode secreted proteins. In the introgression regions, McMullan et al. (2015) identified 14 genes encoding secreted proteases and hydrolases that in some pathogens act as virulence factors (Monod et al. 2002; Soanes et al. 2007; Lebrun et al. 2009). Thus, recombination between races, resulting in introgression can act as a mechanism for exchange of virulence gene alleles. However, neither gene density

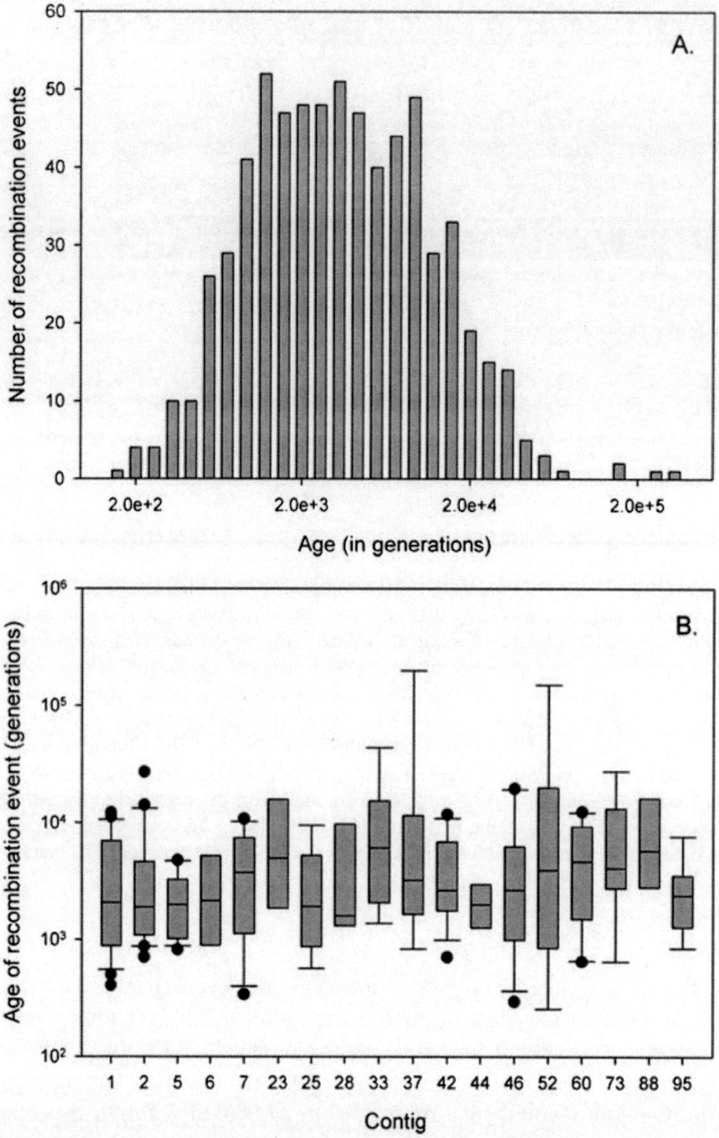

Fig. 2.6 Age of recombination blocks. (**a**) Age of the 675 recombination blocks (mutation rate of $\mu = 10^{-6}$) estimated using binomial mass function; (**b**) Box plot of the median (plus first nation blocks and third quartile) log-age of recombination events in contigs. Only contigs with eight or more events are shown. There is no significant difference in age of events between contigs (GLM: F22, 233 = 1.06, $p = 0.387$) (McMullan et al. 2015)

nor dN/dS are enriched or depleted in regions affected by introgression, compared to regions not affected (Paired t-test: $T = -0.05$, $p = 0.958$; Mann–Whitney test: $W = 3.15 \times 166$, $p = 0.152$, respectively).

2.2.4.5 Intrarace Diversity Suggests Clonal Propagation After Creation of Novel Adapted Allele Repertoires

The evolution and diversification of *A. candida* genomes to understand in a better way, McMullan et al. (2015) analyzed the pattern of recombination and nucleotide divergence with the inclusion of two additional *A. candida* isolates. The AcBoL is an additional isolate from the AcBoT race and AcEm2 is an additional isolate of the AcNc2 race (Fig. 2.2). McMullan et al. (2015) found evidence for 581 recombination events, of which 335 included AcBoT or AcBoL as recombinants. These two isolates shared a recombinant block in 97.3% of recombination events (AcBoT and AcBoL = 326; AcBoT = 6; AcBoL = 3). The AcNc2 and AcEm2 shared a recombinant block in 99.6% of events (AcNc2 and AcEm2 = 246; AcNc2 = 0; AcEm2 = 1). This demonstrates that the AcBoT/AcBoL and AcNc2/AcEm2 races have remained largely unchanged since their initial emergence. The nucleotide diversity within genomes (the observed heterozygosity) and the nucleotide divergence between genomes (genetic differentiation) can be used to further understand *A. candida* population biology.

Isolates AcBoT and AcBoL were both collected in Lincolnshire in 2009, and the observed heterozygosity in these isolates was at least 13 times higher than that of any of the other races (percentage heterozygous positions in contig 1: AcBoT = 0.653%; AcBoL = 0.644%; AcNc2 = 0.047%; AcEm2 = 0.044%; Ac2V = 0.028%) (Fig. 2.4). Remarkably, AcBoT and AcBoL are heterozygous for almost all of the same sites; 97.2% of the sites that are heterozygous in AcBoT are also heterozygous in AcBoL (and 98.5% vice versa). This is only consistent with clonal reproduction because after just one generation of sexual reproduction (or selfing with recombination), Mendelian segregation would eradicate this high level of genotypic similarity. Notably, such evidence for clonal propagation of a race would be difficult to obtain for haploid fungal pathogens. With little evidence for recombination and gene flow, most nucleotide divergence between AcBoT and AcBoL must have accumulated through mutation. The nucleotide divergence between these isolates is just 0.030% (121 polymorphisms in 398,508 bp in contig 1) (Fig. 2.4). Assuming a mutation rate of 1×10^{-8} per cell division, and 100 cell divisions per lineage per year, it was estimated that these isolates could have diverged 305 (262–353) years ago. The other pair of isolates, AcNc2 and AcEm2, were collected in Norfolk in 2007 and in Kent in 1993 (160 km apart), respectively. Similar to the former two isolates, AcNc2 and AcEm2 are also nearly identical (nucleotide divergence $\pi = 0.046\%$, that is, they are >99.95% identical) (Fig. 2.4). When assumed that their nucleotide divergence arose by mutation alone, their estimated divergence time is 890 (814–970) years. However, unlike AcBoT and AcBoL, AcNc2 and AcEm2 are much less heterozygous (Fig. 2.4), which suggests that another genetic mechanism, for example, loss of heterozygosity (Lamour et al. 2012), might be operating which has eradicated the gene diversity in the clonally propagating races over time.

2.2.4.6 Sequential Infections Abolish Host Specificity of Susceptibility to *Albugo candida* Infection

The *Albugo candida* infection compromises host resistance against otherwise avirulent pathogen species (Cooper et al. 2008). Conceivably, *A. candida* could suppress host defenses to otherwise avirulent races of *A. candida*, enabling co-infection and sexual exchange. To test this sequential inoculation experiments, identifying races using the genome sequences to create race-specific DNA markers were performed. Race-specific PCR of preinoculated plants shows preinfection by AcNc2 suppresses resistance in *A. thaliana* accession Ws-2 leaves toward the *B. juncea*-infecting race, Ac2V (Fig. 2.7a). Also, preinfection by AcBoT suppresses, *B. oleracea* resistance toward Ac2V (Fig. 2.7b). Furthermore, defense suppression was so effective that Ac2V was able to complete its life cycle on both *A. thaliana* Ws-2 and *B. oleracea* as observed by successful subsequent infection on *B. juncea* from the sequentially inoculated plants. In a reciprocal experiment, preinfection of *B. juncea* with Ac2V enabled AcNc2 growth on *B. juncea* (Fig. 2.7c). Therefore, AcNc2 not only can suppress Ac2V recognition on *A. thaliana*, but Ac2V is also capable of suppressing *B. juncea* resistance toward AcNc2. The TIR-NB-LRR resistance genes likely confer Ac2V resistance in *Arabidopsis* (McHale et al. 2006), as Ac2V grows on an eds1-1 mutant of *A. thaliana* Ws-2 (Parker et al. 1996). It has long been noted that *Albugo* sp. has a remarkable capacity to suppress immunity in their hosts (Cooper et al. 2008). McMullan et al. (2015) hypothesized that suppression of host innate immunity enables co-infection of hosts by races with otherwise non-overlapping host ranges, thus providing a remarkable mechanism to enable sexual genetic exchange between specialized *A. candida* races (McMullan et al. 2015).

2.3 *Brassica–Alternaria* Genome

2.3.1 Introduction

The *Alternaria brassicae* represents one of the six monotypic lineages within the *Alternaria* genus (Woudenberg et al. 2013). The *Alternaria* spp. infecting the Brassicaceae family includes *A. brassicae*, *A. brassicicola*, *A. raphani*, and *A. alternata* (Saharan et al. 2016; Dunbar et al. 2017). The primary infections are usually caused by the more invasive *A. brassicae*, while secondary infections are caused by *A. brassicicola* and *A. alternata*. The genomes of many *Alternaria* spp., including that of *A. brassicicola*, have been published (Belmas et al. 2018; Dang et al. 2015; Hu et al. 2012; Nguyen et al. 2016; Wolters et al. 2018; Woudenberg et al. 2015). Many of the *Alternaria* spp. has been sequenced using short-read technologies, and thus, only draft genomes are available. A gapless genome assembly of *A. solani* has been made available (Wolters et al. 2018). The use of third-generation sequencing technologies to obtain long reads offers a powerful approach to produce high-quality genome assemblies at affordable costs. Long reads are currently generated from Pacific Biosciences single-molecule real-time sequencing technology and Oxford Nanopore sequencing technology. A contiguous assembly is

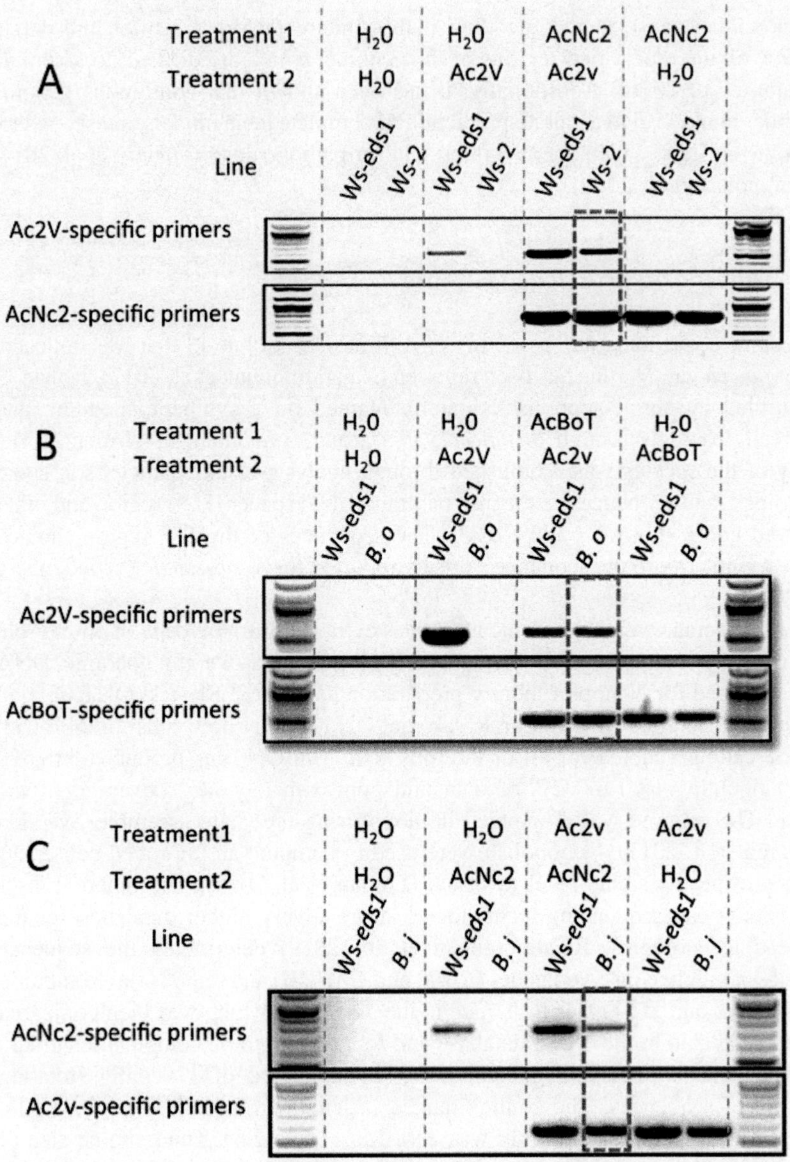

Fig. 2.7 PCR with race-specific markers on DNA prepared from plant tissue generated from co-infection assays. (**a**) Co-infection assay of AcNc2 followed by Ac2V onto Ws-eds1 and Ws-2. (**b**) Co-infection of AcBoT followed by Ac2V onto Ws-eds1 and *Brassica oleracea* (*Bo*). (**c**) Co-infection of Ac2V followed by AcNc2 onto Ws-eds1 and *B. juncea* (*B.j*). Bands highlighted in orange indicate amplification of secondary inoculum on usually nonhost plants upon primary inoculation with virulent *A. candida*. These experiments were repeated multiple times with similar results (McMullan et al. 2015)

an important because genes encoding pathogenicity factors are often present in rapidly evolving repeat-rich regions of the genome, which are difficult to assemble (Thomma et al. 2016). Additionally, it has been shown that contiguous genome assemblies could lead to accurate prediction of complete gene models and secondary metabolite clusters, which are also important for pathogenicity (Dallery et al. 2017; Rajarammohan et al. 2019).

2.3.2 *Alternaria brassicae* Genome

The genome sequencing and assembly of *A. brassicae* isolate J3 that was collected from *B. juncea* var. Varuna has been reported (Rajarammohan et al. 2017). Isolate J3 was purified by single-spore isolation, maintained on a synthetic medium, and periodically passage through *B. juncea* var. Varuna to maintain its virulence. The identity of the species was established through analysis of its characteristic spore morphology, and sequences of the internal transcribed spacer (ITS) region and other conserved genes (*GAPDH* and *RBP2*). The sequences of the ITS region showed complete similarity to the sequences already reported for *A. brassicae* (Woudenberg et al. 2013).

High-molecular-weight genomic DNA was extracted from mycelia of 15-day-old in vitro grown cultures of *A. brassicae*. High-molecular-weight genomic DNA (2 μg) was used for Nanopore library preparation using the LSK-108 ONT ligation protocol. The libraries were run on R 9.4 Spot ON MinION flow cells for 24 h with live base-calling enabled for all of the runs. The MinION runs produced 348,590 reads, amounting to 1.62 GB of data and approximately 46× coverage of the genome. The genome was assembled de novo using the Canu assembler, version 1.6 (Koren et al. 2017). Nanopolish were used to compute an improved consensus sequence using the signal-level raw data (Loman et al. 2015). The inherent error rates of long-read sequencing technologies are relatively higher than those of the short-read technologies. Rajarammohan et al. (2019) determined the sequence similarity of highly conserved genes (*RBP2* and *GAPDH*) and regions (large subunit, small subunit, and ITS) of *A. brassicae* in the assembly, which was then compared with that available in the NCBI database and found them to be completely similar.

The *Alternaria brassicae* genome assembly consisted of 17 contigs totaling 34.14 Mb of sequences. The contigs ranged in length from 66.7 kb to 7.01 Mb. The N50 of the genome assembly was 2.98 Mb (Table 2.3). The genome size of *A. brassicae* was found to be comparable with the genome size of other *Alternaria* spp. reported to date. Of the 17 contigs, 1 represented the mitochondrial genome. The circular nature of the contig representing the mitochondria could not be validated, hence, could not affirm the completeness of the mitochondrial genome. Its size of approximately 189 kb showed that the mitochondrial genome of *A. brassicae* was significantly larger than those reported for other *Alternaria* spp. (Liao et al. 2017). A comparison of mitochondrial genomes of *A. brassicae* and *A. alternata* revealed high levels of similarity; however, the expansion in the size in *A. brassicae* was due to extensive duplications. The presence of telomeric repeats at

Table 2.3 Summary of genome assembly of *Alternaria brassicae* (J3) and its comparison with other *Alternaria* spp. (Rajarammohan et al. 2019)

Parameters	*A. brassicae*	*A. solani*[a]	*A. brassicicola*[b]	*A. alternata*[c]	*A. alternata*[d]
Isolate name	J3	altNL03003	Abra43	RL-8442-2	SRC1 IrK2f
Sequencing technology[e]	Nanopore	PacBio SMRT	PacBio SMRT	PacBio SMRT	Illumina
Number of contigs	17	10	29	27	79
Number of contigs (>10,000 bp)	17	10	29	25	55
Largest contig	7,015,266	6,940,169	3,303,516	3,968,015	2,531,235
Total length	34,141,106	32,779,142	31,036,461	33,499,641	32,990,834
N50	2,988,132	2,866,566	2,101,997	2,828,041	1,097,884
GC (%)	50.7	51.32	50.85	50.95	51.4

[a] Wolters et al. (2018)
[b] Belmas et al. (2018)
[c] Nguyen et al. (2016)
[d] Zeiner et al. (2016)
[e] PacBio SMART = Pacific Biosciences single molecules real time

the ends of a contig has been considered to represent a complete chromosome (Seidl et al. 2015). Nine contigs with telomeric repeats on both ends were identified. One other contig larger than 1 Mb contained telomeric repeats on only one of the ends. Therefore, the current assembly contains nine complete chromosomes and one chromosome with telomeric repeats missing at one of the ends. Apart from these, there were six contigs, of which one of them was approximately 1 Mb in size. These contigs may represent dispensable chromosomes and need to be characterized further (Rajarammohan et al. 2019).

2.3.3 Comparative Genomics of *Alternaria* Species

Genome sequencing and comparative analysis can help identify shared and species-specific pathogenicity factors in closely related species. Genomic information for nearly 26 *Alternaria* spp. including *A. brassicae* is currently available and has contributed immensely to clarify the taxonomy of the *Alternaria* genus. However, comparative analysis to identify pathogenicity factors that confer the ability to infect a wide range of hosts have not been carried out. Most of the genomic information available for *Alternaria* spp. has been generated by shotgun sequencing approaches and hence is fragmented. A contiguous genome assembly is essential, especially when the aim is to identify and characterize pathogenicity factors or effectors, which are often present in rapidly evolving repeat-rich regions of the genome. Additionally, contiguous genome assemblies enable an accurate prediction of genes and gene clusters that are involved in various secondary metabolic processes, many of which are implicated to have an important role in pathogenicity. Long reads generated from Pacific Biosciences (PacBio) single-molecule real-time (SMRT) sequencing technology and Oxford Nanopore sequencing technology enable the generation of high-quality genome assemblies at affordable costs. Besides the recently announced near complete genome sequence of *A. brassicae*, three other near-complete genomes of *Alternaria* spp. have been reported recently.

Using the published high quality genome assembly of *A. brassicae*, Rajarammohan et al. (2019) annotated the genome and carried out comparative analysis of *A. brassicae* with other *Alternaria* spp. to discern unique features of *A. brassicae* vis-à-vis the other *Alternaria* species. They sequenced and annotated the genomes of two *A. alternata* isolates that were co-infecting *B. juncea*. The two *A. alternata* isolates had gene content, effector repertoire, and CAZyme profiles that were very similar to that of an earlier sequenced *A. alternata* isolate (ATCC34957; Table 2.3). This leads to conclude that these isolates are opportunistic pathogens with a limited ability to cause infection on their own but would contribute overall to the disease outcome of a primary *A. brassicae* infection. Additionally, it showed the presence of a dispensable chromosome in *A. brassicae*, a large-spored *Alternaria* species for the first time. The implications of a lineage-specific dispensable chromosome in *A. brassicae* toward pathogenesis remain to be unrevealed. The CAZyme profiles of nearly 30 Dothideomycetes described show that the CAZyme profiles of *A. brassicae* and *A. brassicicola* are different from the other *Alternaria* species.

Several important secondary metabolite gene clusters with putative roles in pathogenicity have also been identified. The identification of the biosynthetic cluster responsible for Destruxin B in *A. brassicae* paves the way for reverse genetics studies to conclusively determine the contribution of Destruxin B toward the pathogenicity of *A. brassicae*. The repertoire of effector side identified in the six *Alternaria* species was largely overlapping. It may thus, be hypothesized that host specificity in the *Alternaria* species may be conferred by the combined action of proteinaceous effectors and the secondary metabolite toxins (Rajarammohan et al. (2019).

2.3.4 Genomic Features of *Alternaria brassicae* and *Alternaria alternata* Isolates

The genomes of two isolates of *A. alternata* (PN1 and PN2) that were co-infecting *B. juncea* with *A. brassicae* were sequenced. The *A. brassicae* assembly has been described by Rajarammohan et al. (2019). Briefly, the assembly consisted of nine complete chromosomes and one chromosome with telomeric repeats missing at one of the ends. Apart from these chromosomes, there were six contigs of which one of them was ~1 Mb in size, which may together constitute a dispensable chromosome (Fig. 2.8). The N50 of the *A. brassicae* assembly was 2.98 Mb (Table 2.4). The two isolates co-infecting *B. juncea* were identified to be *A. alternata* based on their ITS and GAPDH sequences. The *A. alternata* assemblies Aat_PN1 and Aat_PN2 consisted of 14 contigs totaling to 33.77 Mb, and 15 contigs totaling to 33.53 Mb, respectively (Table 2.4). Six contigs in each of the two assemblies contained telomeric repeats on both ends and therefore, are most likely to represent full chromosomal molecules. Four other contigs in both the assemblies contained telomeric repeats on one end but were of similar size of full chromosome molecules as described in *A. solani*. Therefore, the genome assemblies for *A. alternata* isolates represented ten nearly complete chromosomes of each of the two isolates. Whole-genome alignments related with *Alternaria* spp. showed an overall synteny between the genomes with minor rearrangements (Fig. 2.9). Additionally, mitochondrial sequences were also obtained from the sequencing data for the two isolates of *A. alternata*. The mitochondrial genomes of the *A. alternata* strains were approximately 49,783 bp and 50,765 bp in size respectively, and showed high similarity with the previously published mitochondrial genome of *A. alternata* (Liao et al. 2017). Gene prediction following repeat masking resulted in the identification of 11,593, 11,495, and 11,387 genes in the *A. brassicae*, *A. alternata* PN1, and PN2 genome assemblies, respectively. This was comparable to the gene numbers estimated in other *Alternaria* spp. (Table 2.4). The BUSCO analysis showed that the gene models predicted in the three genomes covered 98% of the single-copy conserved fungal genes indicating near-completeness of the assemblies. The predicted genes were comprehensively annotated using a combination of databases as described in Fig. 2.8. In addition to the three genomes, the predicted genes de novo in the genome assemblies of three other *Alternaria* species which were

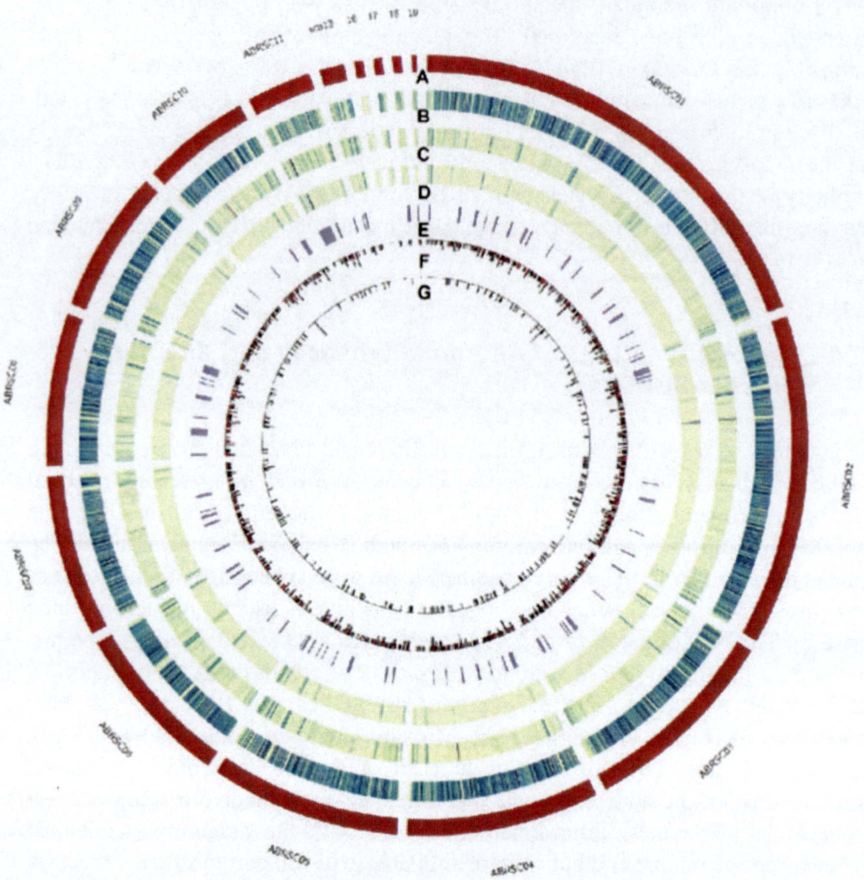

Fig. 2.8 Summary of *Alternaria brassicae* genome, (From outer to inner circular tracks). **(A)** pseudo chromosomes/scaffolds, **(B)** protein-coding genes, **(C)** repeat elements, **(D)** transposable elements (DNA and LTR), **(E)** predicted secondary metabolite clusters, **(F)** secreted proteins, **(G)** predicted effectors (Rajarammohan et al. 2019)

sequenced using long-read technologies viz. *A. brassicicola* (abra43), *A. alternata* (ATCC34957), and *A. solani* (altNL03003) (Table 2.4). These six genomes and their gene predictions were used for the comparative analysis of secondary metabolite encoding gene clusters and effector-coding genes.

2.3.5 Phylogenomics of *Alternaria brassicae* and *Alternaria brassicicola* Within *Alternaria* Genus

In order to accurately reconstruct the divergence and relationship between *A. brassicae*, the two *A. alternata* isolates (PN1 and PN2), and the other *Alternaria* species, phylogenomic analysis was conducted using 29 single-copy orthologs that

Table 2.4 Assembly statistics of the six near-complete *Alternaria* genome sequences (Rajarammohan et al. 2019)

Parameters	*A. brassicae* J3[a]	*A. alternata* PN1	*A. alternata* PN2	*A. solani* altNL03003[b]	*A. brassicicola* abra43[c]	*A. alternata* ATCC34957[d]
Assembly size (Mb)	34.14	33.77	33.53	32.78	31.04	33.48
No. of contigs	17	14	15	10	29	27
No. of contigs (>10,000 bp)	17	13	15	10	29	25
Largest contig (Mb)	7.1	6.86	6.76	6.94	3.3	3.96
N50	2.98	3.09	3.1	2.87	2.1	2.83
GC (%)	50.7	50.98	50.95	51.32	50.85	50.95
Repeat content (%)	9.33	2.43	2.64	5.71	9.3	2.71
Predicted genes	11,593	11,495	11,387	11,804	10,261	12,500

[a] Rajarammohan et al. (2019)
[b] Wolters et al. (2018)
[c] Belmas et al. (2018)
[d] Nguyen et al. (2016)

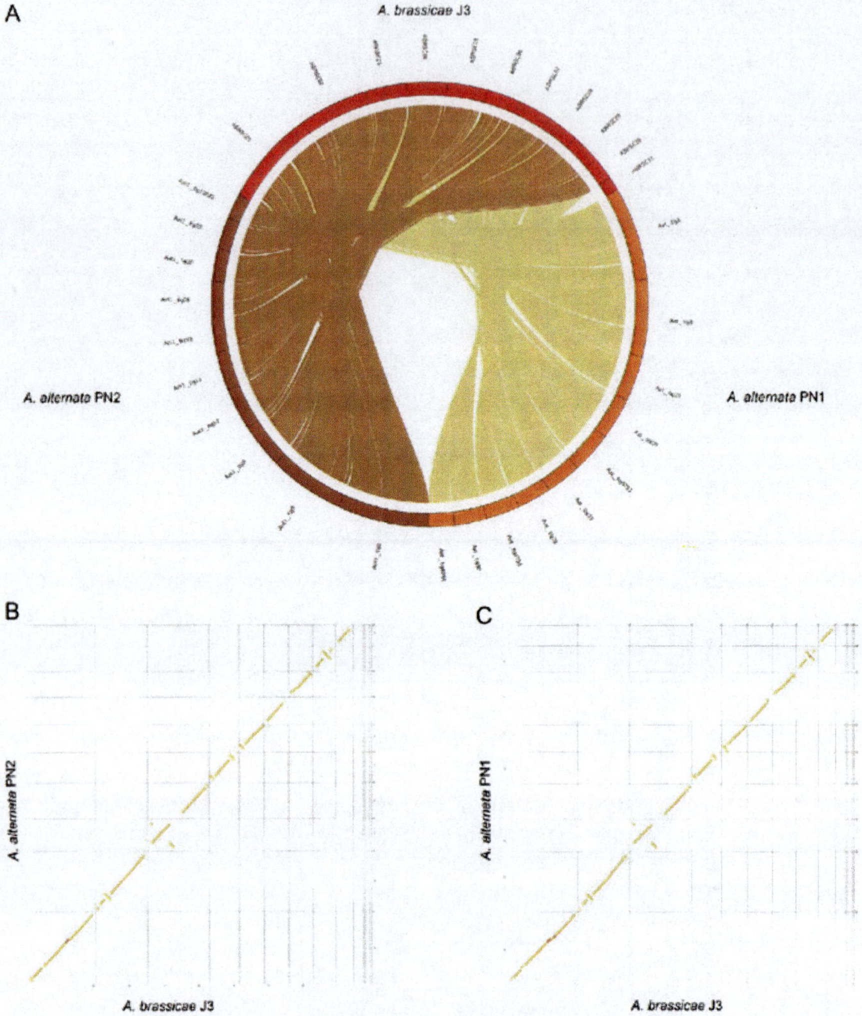

Fig. 2.9 Whole-genome alignments of *Alternaria alternata* PN1 and PN2 with *A. brassicae*. (**a**) Circos plot showing macrosynteny of *A. alternata* PN1 and PN2 with *A. brassicae* across all contigs except the dispensable contigs (ABRSC11, scaffold13,17,18,19), (**b**, **c**) Syntenic dot plots of *A. brassicae* with *A. alternata* PN1 and PN2 (Rajarammohan et al. 2019)

had the highest phylogenetic signal as calculated by the program Mirlo. Selection of genes with higher phylogenetic signals leads to phylogenies that are more congruent with the species tree. The resulting phylogeny showed that the large-spored *Alternaria* and small-spored *Alternaria* species clustered separately into two different clades (Fig. 2.10). The two major pathogens of the *Brassicas* viz. *A. brassicae* and *A. brassicicola* clustered separately from all the other *Alternaria* species,

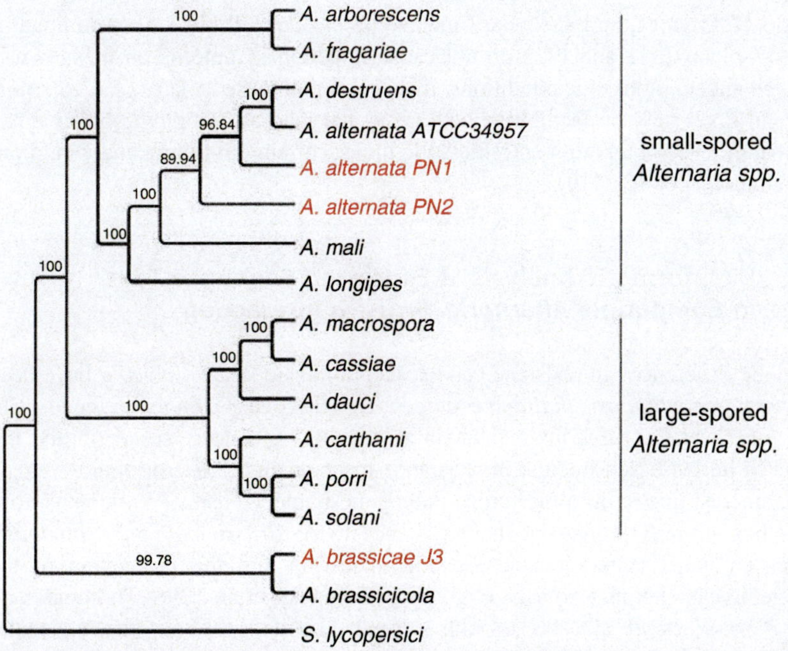

Fig. 2.10 Phylogenetic tree of *Alternaria* species with *Stemphylium lycopersici* as an out-group. The tree was constructed using 29 single-copy orthologs, which had the highest phylogenetic signal as calculated in Mirlo. Branch support values from 1000 bootstrap replicates are shown (Rajarammohan et al. 2019)

possibly indicating a different evolutionary trajectory based on the common host preferences of these two species.

2.3.6 Comparative Analysis of *Alternaria alternata* Isolates Obtained from Different Hosts

The genomes of *A. alternata* PN1 and PN2 (isolated from *B. juncea*) was compared to that of *A. alternata* ATCC34957 (isolated from sorghum) to identify any differences in their genomic content that might allow these to infect two very different species. Whole-genome alignments of *A. alternata* PN1 and PN2 to that of *A. alternata* ATCC34957 revealed very high levels of synteny and the absence of any species-specific regions. It was possible to identify 719, 152, and 586 isolate-specific genes between the three isolates of *A. alternata*, respectively. More than two-third of the isolate-specific genes in all the three isolates were uncharacterized proteins or had no annotations. Notably, all the three isolates did not contain any dispensable chromosomes which may confer pathogenicity, as has been reported for *A. alternata* isolates infecting many of the fruit crops such as citrus, pear, and apple. The gene repertoires of the three isolates also consisted of similar number and type of effectors, CAZymes, and secondary metabolite clusters (Table 2.5). Additionally, the two isolates PN1 and PN2 do not cause infection symptoms on their own in *B. juncea* under epiphytotic conditions. It suggests that these isolates of *A. alternata* (PN1 and PN2) may be facultative pathogens that lead a saprophytic lifestyle and may change over to a pathogenic life style under certain environmental conditions (Rajarammohan et al. 2019).

2.3.7 Bioinformatic Analysis of Expressed Sequence Tags in Compatible *Alternaria–Brassica* Interaction

The subtle differences in resistance and susceptibility to *A. brassicicola* have been discovered in *Arabidopsis thaliana* ecotypes. This discovery along with the identification of several *Arabidopsis* mutants displaying enhanced susceptibility has stimulated an increased interest in resistance mechanisms to necrotrophic fungi as well as interest in necrotrophic fungal pathogenesis mechanisms. Work by various groups has strongly suggested that resistance to *A. Brassicicola* in *Arabidopsis* requires an intact jasmonic acid signaling pathway and the production of the phytoalexin camalexin (Narusaka et al. 2003; Peninckx et al. 1996; Thomma et al. 1998; Van Wees et al. 2003). Still, current knowledge about the molecular mechanisms underlying the interactions between necrotrophic fungal pathogens and their hosts is extremely limited.

With the advent of whole-genome sequencing and functional genomics-based tools, the ability to understand molecular plant–microbe interactions have greatly increased. The power of functional genomics techniques, such as expressed sequence tags (ESTs) and microarray technologies lies in their ability to assess global expression of genes at specific time points in a specific environment. The EST sequencing and analysis has widely been used in the study of plant–microbe interactions (Skinner et al. 2001). The ESTs provide a relatively large, often whole-genome profile of genes expressed by an organism under a specific environmental condition, providing targets for in-depth analysis of the organism's response to the environmental condition in question. The ESTs are particularly powerful because they identify novel genes and are useful in annotating and analyzing whole-genome sequencing projects. The EST projects are often considered a logical starting point to study host–pathogen interactions where very little sequence data are publicly available for one or either of the organisms in question.

The cDNA subtraction and EST sequencing and clustering to gain a better understanding of the genes expressed during the *A. brassicicola–B. oleracea* (green cabbage) interaction has been utilized by Cramer et al. (2006). To identify the fungal genes in the EST collection produced from a subtracted cDNA library derived from infected plant tissue, a three-step bioinformatic approach for *A. brassicicola* a necrotrophic fungal pathogen that causes black spot disease on members of the Brassicaceae plant family were analyzed. In order to identify candidate fungal pathogenicity genes and characterize a compatible host response, a suppression subtractive hybridization (SSH) cDNA library enriched for

Table 2.5 Protein repertoires and functional classification of the six near-complete *Alternaria* genome sequences (Rajarammohan et al. 2019)

Parameters	*A. brassicae* J3	*A. alternata* PN1	*A. alternata* PN2	*A. brassicicola* abra43	*A. solani* altNL03003	*A. alternata* ATCC34957
Total proteins	11,593	11,495	11,387	10,261	11,804	12,500
CAZymes	508	542	550	484	538	571
Peptidases	277	259	253	262	283	299
Secreted proteins	1195	1284	1243	1052	1358	1414
Effectors	198	219	212	160	227	252

A. brassicicola and *B. oleracea* genes expressed during the interaction was created, along with a fungal cDNA library representing genes expressed during nitrogen starvation (NS). A total of 3749 and 2352 expressed sequence tags (ESTs) were assembled into 2834 and 1264 unisequences sets for the SSH and NS libraries, respectively. Cramer et al. (2006) compared two methods to identify the origins (plant vs. fungal) of ESTs in the SSH library using different classification procedures, with and without the availability of a database representing the *A. brassicicola* whole-genome sequence and Brassicaceae-specific genes. The BLASTX analysis of the 2834 unisequences set using the GenBank nonredundant database identified 114 fungal genes. Further BLASTN analysis of the genes with unidentifiable origin using a database consisting of the 1264 fungal unisequence set from the nitrogen-starved library identified 94 additional fungal genes. By contrast, BLASTN analysis of the same SSH unisequences set using a partially assembled *A. brassicicola* whole-genome draft sequence identified a total of 310 unisequences of fungal origin. The results indicated that even a small number of organism-specific EST sequences can be very helpful to identify pathogen genes in a library derived from infected tissue, partially overcoming the limitation of the public databases for little studied organisms. However, using the whole-genome draft sequence of *A. brassicicola*, Cramer et al. (2006) were able to identify approximately 30% more fungal genes in the SSH library than without utilizing this resource was adopted by utilizing public databases only, public databases and a database of 1264 unisequences assembled from an *A. brassicicola* EST collection and a Brassicaceae-specific database coupled with a database of the *A. brassicicola* whole-genome draft sequence. Data generated from these EST analyses provide the first profile of genes employed by *A. brassicicola* for pathogenesis of *Brassica* species, and a view of genes expressed by a susceptible host during fungal infection (Cramer et al. 2006).

2.3.8 Characterization of Histidine Kinase Gene *AbNIK1* from *Alternaria* Conferring High Dicarboximide and Phenylpyrrole Resistance

The dicarboximide fungicide iprodione is widely used as an active ingredient for seed dressing and spray programs to limit outbreaks of the disease on crucifers. The field isolates of *A. brassicicola* highly resistant to dicarboximides have been identified (Iacomi-Vasilescu et al. 2004). These isolates were also found to be highly resistant to phenylpyrroles. Laboratory mutants highly resistant to dicarboximides and phenylpyrroles have been reported for some fungal species (Fujimura et al. 2000; Yoshimi et al. 2003). However, the phenotype of the *A. brassicicola* isolates was uncommon since field isolates resistant to dicarboximides were usually phenylpyrrole-sensitive (Oshima et al. 2002). Moreover, in vitro tests revealed that, contrary to most laboratory mutants, *A. brassicicola* isolates were still pathogenic to host plants (Iacomi-Vasilescu et al. 2004).

The cellular targets of phenylpyrroles and dicarboximides have not yet been identified, but it was shown that they both stimulate glycerol synthesis in *Neurospora crassa* (Pillonel and Meyer 1997). This observation and the fact that osmo sensitive (os) mutants of *N. crassa* were also found to be resistant to dicarboximides and fludioxonil (Fujimura et al. 2000) suggest that the response to osmotic stress and fungicide resistance might be linked. In this model filamentous fungus, the *nik1/os-1* gene encodes a two-component histidine kinase (Alex et al. 1996; Schumacher et al. 1997) that was predicted to be the sensor in an osmotic response signal transduction cascade constituted by MAP kinases encoded by other osmotic loci such as os-2, os-4, and os-5 (Fujimura et al. 2003; Zhang et al. 2002). Ochiai et al. (2001) showed that mutations in the 90 N-terminal amino acid repeats of the histidine kinase Nik1p/Os-1p confer fungicide resistance and osmotic sensitivity. Similarly, mutations within the amino acid repeat domains of BcOs1p and AaHK1p, i.e., os-1p homologues from *Botrytis cinerea* and *A. alternata*, respectively, were found in iprodione-resistant field isolates (Oshima et al. 2002; Cui et al. 2004; Dry et al. 2004).

The phenotypic characterization of highly fungicide-resistant isolates (field and laboratory mutant strains) of *A. brassicicola* has been revealed by investigating their osmotic sensitivities (Avenot et al. 2005). The *AbNIK1*, a nik1/os1 orthologous gene have also been isolated from a wild-type fungicide-sensitive *A. brassicicola* strain and compared its sequence with corresponding histidine kinase genes from fungicide-resistant strains. Two strain groups were identified on the basis of their responses to high osmotic pressure. All the fungicide-resistant strains displaying a moderately osmosensitive phenotype were found to have null mutations in the *AbNIK1* gene. Other resistant strains had a unique phenotype characterized by their tolerance to osmotic stress and either they were likely mutated at a locus not corresponding to *AbNIK1*, or they produced altered forms of the *AbNIK1* protein.

Highly iprodione- and fludioxonil-resistant field and laboratory isolates of *A. brassicicola* were found to be either moderately sensitive or tolerant to osmotic stress. The *AbNIK1*, a two-component histidine kinase gene, was isolated from a fungicide-sensitive strain. The predicted protein possessed the six tandem amino acid repeats at the N-terminal end, which is a landmark of osmo-sensor histidine kinases from filamentous fungi. A comparison of the nucleic acid sequences of the Ab-*NIK1* gene from fungicide-sensitive and fungicide-resistant isolates revealed the presence of mutations in six of the seven resistant strains analyzed. Null mutants were all found to be moderately sensitive to osmotic stress, indicating that they are similar to *N. crassa* Type I os-1 mutants. Only one mutation, corresponding to a single amino acid change within the H-box of the kinase domain, was found in an osmo-tolerant strain. These results suggest that AbNIK1p participates in osmo regulation and that expression of the fully functional enzyme is essential for dicarboximide and phenylpyrrole antifungal activities (Avenot et al. 2005).

2.3.9 Phylogenetic Relationship of *Alternaria* species Determined With Nuclear Ribosomal DNA Sequences

The sequences coding for the nuclear 18s rRNA, 5-8s rRNA and the internal transcribed spacers (*ITS1* and *ITS2*) were amplified by the polymerase chain reaction and sequenced for one isolate each of *A. brassicae*, *A. brassicicola*, *A. raphani*, *A. alternata,* and *Pleospora herbarum.* The 5-8s rDNA sequences from the four *Alternaria* species are identical, and differed at only one base pair from that of *P. herbarum*. The internal transcribed spacer sequences especially *ITS1* are very variable in both base composition, and length. The 18s rDNA sequence are highly conserved, but enough variability is present to distinguish genera clearly. Phylogenetic analysis of the sequence data sets by both parsimony, and maximum-likelihood methods clearly separated genera and species. All of the *Alternaria* species are closely related to *Pleospora* also appeared to be more closely related to *Alternaria* than to *Leptosphaeria* (Jasalavich et al. 1995).

The taxonomists consider *Pleospora* and *Leptosphaeria* to be in separate loculoascomycetes families, the Pleosporaceae, and the Leptosphaeriaceae, respectively (Barr 1987; Eriksson and Hawksworth 1991), with the main differential criterion being the conidiomatal structure of the anamorphs. *Leptosphaeria* species have coelomycetous anamorphs, while genera in the Pleosporaceae have hyphomyceteous anamorphs (Barr 1987). The phylogenetic trees (Figs. 2.11, 2.12) based on rDNA sequences support this taxonomic view. All the Loculoascomycetes included in the analysis, i.e., *Leptosphaeria* and *Pleospora* as well as *Alternaria,* formed a phylectic group. Within this group, there are two major lineages: the Leptosphaeriaceae composed of *Leptosphaeria* and the Pleosporaceae composed of *Pleospora* and *Alternaria.* The two *Ophiosphaerella* species, which are included only in the analysis based on the 5-8s rDNA and flanking internal transcribed spacers, fall within the *Leptosphaeria* group (Fig. 2.11). *Ophiosphaerella korrae* (Walker and Smith) Shoemaker and Babc has been placed in *Leptosphaeria* (Walker and Smith 1972) while *O. herpotricha* (Fr.; Fr), Walker was previously in *Phaeosphaeria* (Holm 1957), a segregate of *Leptosphaeria* (Holm 1957).

Most species of *Alternaria* including the four species pathogenic to crucifers lack a known sexual stage. *Alternaria* and *Pleospora,* although distinct genera, are sister taxa (Fig. 2.12) based on the 18s rDNA sequence data. Simmons (1967) considered the *Alternaria*, *Stemphylium* and *Ulocladium* to be closely related on the basis of conidial development, as they all produce Dictyophoa spores. *P. herbarum* has a *Stemphylium* anamorphs, and provide the first molecular evidence that asexual species of *Alternaria* are indeed related to *Pleospora.* Therefore, the anamorphic *Alternaria* should be included in the Pleosporaceae. Wiltshire (1947) considered *A. raphani* and *A. brassicae* to be closely related. It would appear; however, that *A. raphani* is actually more closely related to *A. brassicicola* than to *A. brassicae* based on rDNA sequence data.

The pathogens *Alternaria alternata*, *A. brassicae*, *A. brassicicola,* and *A. raphani* formed a strong clade of very closely related sister taxa. The 18s rDNA resolved two subclades of species within *Alternaria* at a level of confidence of 95% (Fig. 2.12),

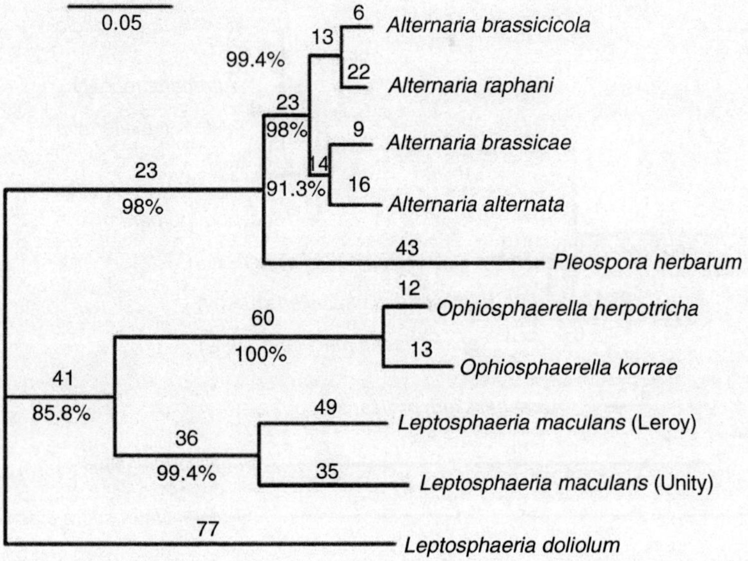

Fig. 2.11 The consensus tree generated by global parsimony bootstrap analysis of alignment of the 5-8s rDNA and flanking internal transcribed spaces (*ITS1* and *ITS* 2) sequences. The percentages represent the proportion of 1000 bootstrap replications in which the taxa to the right of the node were placed together by the program DNAPARS with randomization of the sequence input order. The other numbers represent the steps. Branch length (drawn in the horizontal dimension only) is the maximum-likelihood estimates made by the program DNAML. When the user tree was defined as the bootstrap consensus tree, the length of the vertical lines has no meaning and was adjusted arbitrarily for ease in labeling termini. *Leptosphaeria doliolum* was designated as the out group (Jasalavich et al. 1995)

based on five informative sites contained within the 300 bp at 3′ end of sequence alignment. Complete resolution of the species of *Alternaria* is achieved with the *ITS* sequence data (Fig. 2.11) which are much more variable, and contained more phylogenetically informative sites than the 18s rDNA. The *ITS* sequence data of *Leptosphaeria* are even more variable than those of *Alternaria*. Likewise, the range of genetic distance among the *Leptosphaeria* species is larger than those seen among the *Pleospora* and *Alternaria* species. The genetic distances between *P. herbarum* and any of the four *Alternaria* species examined are on the same order of magnitude as that between the highly virulent and weekly virulent isolates of *L. maculans*. These differences suggest that genus *Alternaria* encompasses less genetic variability than does the genus *Leptosphaeria*. This may reflect the differences in size and reproductive strategies of the two genera. *Alternaria* comprises roughly 60 species (Rao 1969; Rossman et al. 1987), while *Leptosphaeria* includes a heterogeneous mixture of about 1689 taxa (Crane and Shearer 1991), and is very likely polyphyletic. Perhaps the four *Alternaria* species have separated from each other only recently in evolutionary time and so have not yet accumulated many changes in their rDNA sequences. Alternatively, perhaps plant host association may exert a

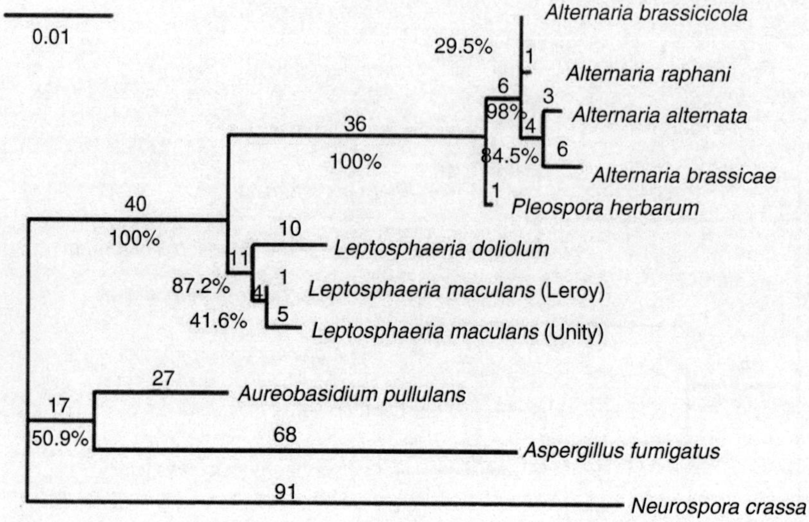

Fig. 2.12 The consensus tree generated by global parsimony bootstrap analysis of alignment of the 18s rDNA. The percentages represent the proportion of 1000 bootstrap replications in which the taxa to the right of the node were placed together by the program DNAPARS with randomization of the sequence input order. The other numbers represent the steps. Branch length (drawn in the horizontal dimension only) is the maximum-likelihood estimates made by the program DNAML. When the user tree was defined as the bootstrap consensus tree, the length of the vertical lines has no meaning and was adjusted arbitrarily for ease in labeling termini. *Neurospora crassa* was designated as the out group (Jasalavich et al. 1995)

pressure against further divergence. Along with phylogenetic relationship among the *Alternaria* species, portions of the rDNA sequence can have practical uses, e.g., production of diagnostic tools. The variability present in *ITS1* can be exploited to design species-specific probes and primers in order to identify rapidly species of *Alternaria* both in culture, and in plant materials (Jasalavich et al. 1995).

2.4 *Brassica–Colletotrichum* Genome

2.4.1 Introduction

The *Colletotrichum* is a large ascomycete genus comprising more than 190 species, many of which cause devastating diseases on a large range of agricultural and horticultural crops worldwide (Jayawardena et al. 2016). Among species of *Colletotrichum*, *C. higginsianum* is classified in a main phylogenetic clade within the *C. destructivum* complex, and causes anthracnose disease on a wide range of cruciferous plants, such as species of *Brassica* and *Raphanus* as well as the model plant *Arabidopsis thaliana*. Since most *A. thaliana* ecotypes are susceptible to *C. higginsianum*, the pathogen can be regarded as adapted for *A. Thaliana* (Shimada et al. 2006). As a typical hemibiotrophic fungus, *C. higginsianum* develops a series of specialized infection structures including germ tubes, appressoria, primary biotrophic hyphae (BH), and secondary necrotrophic hyphae (NH). Thus, *C. higginsianum* is one of the best-studied species within the genus *Colletotrichum* because of its interesting infection strategy, and the ease with which it can be cultured axenically and transformed with high efficiency by T-DNA transfer mediated by *Agrobacterium tumefaciens*. Furthermore, complete genome sequences and transcriptome data are available (Zampounis et al. 2016). For these reasons, the *C. higginsianum–Arabidopsis* pathosystem has become an attractive model for research on the molecular basis of fungal pathogenicity and plant–fungal interactions. *Colletotrichum higginsianum* is a hemibiotrophic ascomycetous fungus that causes economically important anthracnose diseases on numerous monocot and dicot crops worldwide. As a model pathosystem, the *Colletotrichum–Arabidopsis* interaction has the significant advantage that both organisms can be manipulated genetically. The genome sequence of *C. higginsianum* has provided insights into how genome structure and pathogen genetic variability has been shaped by transposable elements, and allows systematic approaches to longstanding areas of investigation, including infection, structure, differentiation and fungal–plant interactions. The *Arabidopsis–Colletotrichum* pathosystem provides an integrated system, with extensive information on the host plant and availability of genomes for both partners, to illustrate many of the important concepts governing fungal–plant interactions, and to serve as an excellent starting point for broad perspectives into issues in molecular plant pathology (Yan et al. 2018).

2.4.2 Genome Sequencing and Assembly

The first genome of *C. higginsianum* strain IMI 349063 was reported by O'Connell et al. (2012) using a multisource method, including short-read data from 454 GSFLX (350 bp) and Illumina GAII (100 bp) sequencing platforms together with a smaller number of longer Sanger reads. Optical mapping showed that the genome of *C. higginsianum* strain IMI 349063 was 53.4 Mb distributed among 12 chromosomes, including two mini chromosomes less than 1 Mb in size. However, the actual assembly was composed of over 10,000 contigs. This genome is smaller than other sequenced *Colletotrichum* genomes (88.3 Mb, 55.6 Mb, and 57.4 Mb such as *C. orbiculare* (88.3 Mb), *C. gloeosporioides* (55.6 Mb), and *C. graminicola* (57.4 Mb) (Gan et al. 2013). Surprisingly, more genes (16,172) were predicted from the *C. higginsianum* assembly in contrast to assemblies of *C. orbiculare* (13,479), *C. gloeosporioides* (15,469), and *C. graminicola* (12,006). One limitation of this assembly was that many of the predicted protein-coding genes were truncated or split between contigs, resulting in multiple gene calls. The fragmented nature of the assembly leading to incomplete gene calls was confirmed by a series of problematic experiments by Yan et al. (2018). The *C. higginsianum* strain IMI 349063 had been re-sequenced using the single-molecule real-time (SMRT) technique, and combined with previous optical mapping data, has achieved a gapless assembly of all 12 chromosomes except for the ribosomal DNA repeat cluster on chromosome 7 (Dallery et al. 2017). This assembly of nearly all chromosomes represents the most complete genome assembly to date of any *Colletotrichum* species, and becomes part of a short list of completely assembled genomes of phytopathogenic fungi, namely *Zymoseptoria tritici*, *Sclerotinia sclerotiorum*, *Botrytis cinerea*, *Verticillium dahlia,* and *Fusarium graminearum.* The final genome assembly of strain IMI 349063 contains 28 unitigs (chromosome 7 is represented by 13 small unitigs and the mitochondrial genome is represented by 3 unitigs) with a total length of 50.82 Mb. Based on the new gene annotation, a total of 14,651 protein-coding genes were predicted from the new genome assembly, 1521 fewer than the previous assembly.

Genome mining of *C. higginsianum* and *C. graminicola* for candidate secreted effector proteins (CSEPs), which serve as molecular weapons to evade or suppress plant immunity, revealed only 177 in *C. graminicola*, but 365 were found in *C. higginsianum.* The CSEPs are mostly small, cysteine rich proteins, averaging 110 residues in *C. higginsianum*. The larger, more diversified CSEP repertoire of *C. higginsianum* might be an adaptation to evade defenses and invade a broader range of host plants. The more accurate gene annotation from the new assembly revealed many secondary metabolism (SM) key genes and putative biosynthetic pathways. Interestingly, the annotation demonstrated that *C. higginsianum* encodes one of the largest repertoires of SM key genes and SM gene clusters of any sequenced ascomycete, suggesting a large capacity to produce diverse metabolites. Analysis of the mini-chromosomes showed that both are repeat-rich and AT-rich, gene-poor and highly enriched with genes encoding putative secreted effector proteins of unknown function, which are different in their content to the other 10 chromosome (Dallery et al. 2017). Surprisingly, a study revealed that strains lacking small chromosome 11 abort infection during biotrophy, while their ability to grow on artificial media was not affected, and that chromosome 12 can be lost without effects on virulence or growth on agar plates, indicating that a number of potential genes from chromosome 11 have critical functions in addressing plant host responses (Plaumann et al. 2018). The complete genome assembly allows for analysis of genomic features including transposable elements, telomeres, structural rearrangements and large gene clusters. Moreover, this assembly can be a reference for investigations of other isolates of *C. higginsianum* or other *Colletotrichum* species, and such data should facilitate future studies including those on functional genomics in this important model phytopathogen (Yan et al. 2018).

2.4.3 Transcriptome Analysis

Based on flow cytometric purification, intracellular biotrophic hyphae of *C. higginsianum* from infected *Arabidopsis* leaves were purified for biotrophic stage transcriptome analysis. Six fungal genes, namely homologues of NmrA, saccharopine dehydrogenase, CIH1 and three unigenes were specifically expressed in planta during the biotrophic phase, and the three unigenes (3, 125 and 143) are likely to encode small, soluble secreted proteins of unknown function that represent candidate fungal effectors (Takahara et al. 2009). RNA-Seq data for samples from infected *Arabidopsis* corresponding to prepenetration appressoria, the early biotrophic phase and the transition to necrotrophy have been released, and the transcription levels were found to be highly dynamic. At the appressorial phase, genes encoding CAZymes that were predicted to degrade cutin, cellulose, hemicellulose and pectin were upregulated, which may contribute to initial host penetration, together with a larger set of enzymes that potentially remodel the fungal cell wall. During early infection, the transcriptome was dominated by secondary metabolism genes, with the majority of expressed SM gene clusters being induced before penetration and during biotrophy, and not in vitro. Furthermore, the majority of CSEP-encoding genes were strongly induced during biotrophy, implying that effector production was especially prominent during the biotrophic stage. During the switch to nectrophy, there was induction of a wide variety of lytic enzymes, presumably as the fungus feeds on moribund and necrotic tissues to allow prolific growth and colonization leading to increased spore production. Among the lytic enzymes produced, there were CAZymes and putative-secreted proteases that may cut the various types of polysaccharides associated with the host cell walls. Furthermore, there was induction of many genes encoding plasma membrane transporters which may be required for movement and assimilation of metabolic products of this degradative enzymatic activity such as sugars, oligopeptides, and amino acids (O'Connell et al. 2012; Yan et al. 2018).

2.4.4 Genetic Transformation

A high efficiency transformation system using *Agrobacterium tumefaciens*-mediated transformation (ATMT) is available for *C. higginsianum*. It has become established as the method of random insertional mutagenesis and targeted gene disruption using homologous recombination. Random insertional mutagenesis is a powerful approach for discovering novel pathogenicity genes in fungi. Based on the first application of ATMT for insertional mutagenesis of *C. higginsianum*, a T-DNA insertion library was generated. By using a high-throughput infection assay on *A. thaliana* seedlings, among 8850 mutants, 40 mutants showed reproducible pathogenicity defects on *Arabidopsis* and *Brassica* plants, 6 were impaired in appressorial melanization, 15 had reduced penetration ability, 14 induced host papillae or hypersensitive cell death, and five were affected in the transition from biotrophy to necrotrophy. Flanking sequence analysis of the tagged genes led to the isolation of 14 putative pathogenicity genes (Table 2.6). Similarly, another T-DNA insertion library of *C. higginsianum* was also generated containing 5012 ATMT mutants, and six virulence-deficient mutants were acquired (Table 2.6). Identification and analysis of the T-DNA tagged loci of these mutants revealed several potential genes possibly related to virulence factors such as genes encoding a copper amine oxidase, an exosome component EXOSC1/CSL4, and other hypothetical proteins (Huser et al. 2009; Liu et al. 2013). Korn et al. (2015), from a collection

Table 2.6 Summary of *Colletotrichum higginsianum* genes identified from T-DNA flanking sequences (Yan et al. 2018)

Mutant	Insertion[a]	T-DNA insertion[b]	Putative function (NCBI accession)[c]	References
path-5	1	In predicted open reading frame (ORF)	Unknown	
path-7	2	In ORF	Hypothetical protein (FG06146.1)	
		1.5 kb upstream	Hypothetical protein (FG06145.1)	
path-8	1	In predicted ORF	Unknown	
path-9	1	1 kb downstream	Endo–1,3(4)–β–glucanase (AFUA_1G05290)	
path-12	1	In ORF	MFS transporter (NFIA_086030)	Huser et al. (2009)
path-16	1	In ORF	Ornithine decarboxylase (AY602214)	
path-19	1	In ORF	Arg–6 protein (EAA35492.1)	
path-23	2	620 bp upstream	Hypothetical protein (FG02446.1)	
		In predicted ORF	Unknown	
path-29	1	730 bp upstream	ATP–binding endoribonuclease (ACLA_048430)	
path-35	1	In ORF	Carbamoyl–phosphate synthetase (EAA36214.1)	
path-36	1	620 bp upstream	Importin 2 subunit (AFUA_1G15900)	
path-38	1	In ORF	Importin 2 subunit (AFUA_1G15900)	
T732	1	168 bp downstream	Copper amine oxidase (XP_001826965)	
T734	1	In ORF	Hypothetical protein (ELA33048)	
B30	2	In ORF	Exosome component EXOSC1/CSL4 (EFQ29835)	Liu et al. (2013)
		850 bp upstream	DUF221 domain protein (EFY94646)	
T45	Unknown		Hypothetical protein (EFQ29552)	
vir-2	2	supercontig_1.2671,583, RB	Phosphori bosylamino imidazole carboxylase (EFQ26499.1)	
vir-10	2	contig05930, 16777,LB	Kelch domain-containing protein (EFQ26610.1)	
vir-12	2	supercontig_1.3174,1154, LB	Plasma-membrane proton-efflux P-type ATPase	

(continued)

Table 2.6 (continued)

Mutant	Insertion[a]	T-DNA insertion[b]	Putative function (NCBI accession)[c]	References
vir-14	2	supercontig_1.6150,870, RB	ABC transporter (EFQ25092.1)	
		supercontig_1.903,6335, LB	Nucleoside-diphosphate-sugar epimerase	
vir-22	1	supercontig_1.3174,1748, RB	Plasma-membrane proton-efflux P-type ATPase	
vir-24	1	supercontig_1.3174,1422, RB	Plasma-membrane proton-efflux P-type ATPase	
vir-27	2	supercontig_1.6150,873, RB	ABC transporter (EFQ25092.1)	Korn et al. (2015)
		supercontig_1.826,7944, RB	STE like transcription factor	
vir-51	1	supercontig_1.1848,6585, LB	Unknown	
vir-52	2	contig 00557	Alanine dehydrogenase/ PNT domain containing protein (EFQ25467.1)	
		contig 11896	FAD dependent oxido reductase super family protein (XP_007280006)	
vir-53	2	supercontig_1.6692,RB	Unknown	
vir-56	3	supercontig_1.66,3878, LB	Peroxisomal membrane protein 24	
vir-76	2	supercontig_1.56,17248, LB	Spindle assembly checkpoint component MAD1	
vir-84	2	supercontig_1.3742,1175, LB	Sporulation protein RMD1 (ELA35952.1)	
vir-88	2	supercontig_1.5277,868–879	Mob1/phocein family protein (EFQ26211.1)	
vir-97	2	supercontig_1.3174,812, LB	Plasma-membrane proton-efflux P-type ATPase	
vir-102	1	supercontig_1.3174,793, LB	Plasma-membrane proton-efflux P-type ATPase	

[a] Number of insertion sites determined by Southern Blot analysis

[b] Locations of T-DNA insertion sites or position of the T-DNA border sequence in the *Colletotrichum* database Super contigs. Sequence names are shown with left border (LB) or right border (RB)

[c] Open reading frames (ORFs) predicted by Soft berry were used in the BLAST search

of 7200 insertion mutants, isolated 75 mutants with reduced symptoms. Among them, 19 were affected in host penetration, while 17 were affected in later stages of infection. The location of T-DNA insertions of only 16 mutants could be identified by polymerase chain reaction (PCR) for gene functional analysis (Table 2.6).

Among the T-DNA insertion library, two *C. higginsianum* mutants defective in the switch from biotrophy to necrotrophy showed high homology to conserved importin-β-2 proteins. This class of importins is known to mediate the nuclear importation of pre-mRNA processing proteins in mammals, yeast and plants. Importins are considered essential for the asexual/sexual development in *Trichoderma reesei* and pathogenicity in *Aspergillus nidulans*, *Phytophthora sojae*. Importantly, there have been no reports that importin proteins are involved in the transition from biotrophy to necrotrophy. Otherwise, many T-DNA tagged loci of these mutants were annotated as hypothetical proteins, and these genes have the potential to be involved in novel functional genes possibly related to virulence factors. There needs to be more attention and penetrating research on these genes to validate that the integrated T-DNA is responsible for the observed phenotypes, complementation experiments or the generation of targeted knockout mutants are important. Since targeted gene mutagenesis through homologous recombination occurs with relatively low frequency in *C. higginsianum*, an efficient gene knockout protocol, which could increase homologous recombination frequency to 60–90%, was established based on the inactivated Ku70 and Ku80 components of the nonhomologous end-joining (NHEJ) pathway in *C. higginsianum*.

The method of inactivated NHEJ pathway components to increase the homologous recombination frequency was previously reported in other filamentous fungi, and this method raised the transformation efficiency to 60–90% in *A. sojae* and *A. oryzae* and by over 80% in *M. oryzae*. Overall, these findings indicate that insertional mutagenesis by ATMT could be a valuable tool for genome-wide analysis of gene function in this important model pathogen (Yan et al. 2018).

2.5 *Brassica–Erysiphe* Genome

2.5.1 Introduction

On the basis of morphological, pathological, host range, anatomical, phylogenetically, developmental features, and characteristics of the powdery mildew fungus at anamorph, and teleomorph states, the major causal organism of crucifer's powdery mildew has been identified as *Erysiphe cruciferarum*. Under artificial inoculation conditions three other powdery mildew fungi viz., *Golovinomyces cichoracearum*, *G. orontii*, and *Oidium neolycopersici* are able to complete their anamorph state on a crucifers weed *Arabidopsis thaliana*. *Arabidopsis*–powdery mildew host-pathosystem has been used recently as a model system for molecular and genetical studies all over the world. Morphological features of all the four powdery mildew species pathogenic to crucifers', *Brassica* species, and *Arabidopsis* have been described as observed on the host species. Classification, taxonomy, and nomenclature of powdery mildew pathogens of crucifers have been updated on the basis of features on the surfaces of powdery mildew conidia revealed through scanning electron microscopy, and light microscopy at anamorph state where teleomorph state was absent. The four powdery mildew species pathogenic on crucifers have been distinguished on the basis of conidial size, shape, appearance, number of conidia per conidiophore, conidiophores shape, and size, haustoria, hyphal branch angles, conidial germination with respect to temperature, relative humidity, light, and substrate, size of chasmothecia, asci, ascospores, number of ascospores per chasmothecium, and type of appendages. Phylogenetics relationship of crucifer's powdery mildew with other powdery mildew species has been determined through sequence analysis of the ITS rDNA, and sequence comparison of DNA encoding the 5.8s, rRNA, and ITS2. The powdery mildew pathogen genomes and transcriptomes have been determined. Powdery mildew fungi have sizeable genomes, which are about four times larger than those of other ascomycetes. The number of coding genes in the powdery mildew genomes is comparatively low. The structure of powdery mildew family, Erysiphaceae, the position of genes *Erysiphe*, its sections, and subsections *Golovinomyces*, key to the recognized genera, and species of the family, and relationship within the family have been exploited (Saharan et al. 2019).

2.5.2 Present Status of Crucifers' Powdery Mildew Pathogen Genome

The fungi causing powdery mildew diseases have sizeable genomes, which are about four times larger than those of most other ascomycetes (average ascomycete genome size: 36.9 Mbp; Mohanta and Bae 2015). The genome of *G.o*, for example, is approximately 160 Mbp in size (Spanu et al. 2010). By contrast, the number of coding genes in the powdery mildew genomes is comparatively low (average number of ascomycete coding genes: 11,129; Mohanta and Bae 2015). Only ca. 6500 genes each have been annotated in the *B.gh*, and *B.gt* genomes, and ca. 7100 genes (on the basis of assembled transcript contigs) are expressed in *G.o* haustoria (Spanu et al. 2010; Weßling et al. 2012; Wicker et al. 2013; Kusch et al. 2014).

The biological reason for the surprisingly low gene number most likely lies in the biotrophic life style, due to the close association of parasite, and host, the fungus acquires its nutrients from the plant. As a result, the need for the maintenance of many complex biosynthesis pathways is low, whereas the requirement to control the host cell by secreted effectors is high (Spanu et al. 2010). Associated with this unusual ratio of genome size to gene number is the presence of numerous nested retro-transposons that cover most of the powdery mildew genomes. These retro-transposons are physically closely associated with effector protein-encoding genes, and are therefore thought to be involved in the rapid evolutionary adaptation of powdery mildews (Hacquard et al. 2013).

In a transcriptomic approach using a cDNA library obtained from mature *G.o* haustoria extracted from heavily infected *Arabidopsis* leaves, protein-coding genes for translation, and protein turnover were recognized to be most abundant (Weßling et al. 2012). This is in line with the finding that haustoria contain an abundance of cytoplasmic and endoplasmic reticulum-connected ribosomes, pointing at high levels of protein biosynthesis (Micali et al. 2011). Genes associated with mycelium development were also found to be highly represented in the *G.o* haustorial transcriptome. By contrast, the transcript levels of sugar and amino acid transporters are comparatively low (Weßling et al. 2012). A substantial proportion of the transcripts are predicted to encode secreted effector proteins: 115 *G.o* effector candidates (OECs) were discovered in the transcriptome of isolated haustoria (Weßling et al. 2012, 2014). Among them, 84 OECs were subject of a comprehensive protein interaction study with a subset of *Arabidopsis* host proteins. In this work, identification of an interspecies effector convergence network revealed common effector target proteins (hubs) for *Arabidopsis* pathogens from three kingdoms of life, i.e. *G.o* (a powdery mildew fungus), *H. arabidopsidis* (an oomycete), and *P. syringae* (a bacterium; Weßling et al. 2014). Interestingly, mutants of many of the respective host target genes show altered disease phenotypes (toward either increased resistance or higher susceptibility). This effector convergence suggests that biotrophic pathogens from different kingdoms manipulate the same host plant processes (Weßling et al. 2014). Among the common effector targets, proteins involved in cell cycle regulation/plant development are highly represented, e.g., the TFs TCP13, TCP14, and TCP19 (At5g51910). Since the TF MYB3R4 seems to be involved in powdery mildew-induced increase in polyploidy (Chandran et al. 2010), these findings may indicate that the manipulation of the host cell cycle is crucial for the *Go* infection process, and that of a range of other pathogens as well (Weßling et al. 2014).

2.5.3 Phylogenetics of Powdery Mildew Pathogens

Phylogenetics relationship of powdery mildew of crucifers with other powdery mildew species has been determined through sequence analysis of the ITS rDNA (Choi et al. 2009a), and sequence comparison of DNA encoding the 5.8S, rRNA, ITS1, and ITS2 (Adam et al. 1999).

2.5.3.1 Sequence Analysis of the ITS rDNA

The rDNA gene sequences covering both highly conserved ribosomal subunits, and variable internal transcribed spacer (ITS) regions are widely used to evaluate the polygenic relationships among related fungal species, and races within species (Saenz and Taylor 1999; Takamatsu and Kano 2001). The PCR amplification and sequencing of the ITS rDNA was performed by Choi et al. (2009a) using the following primer set: ITS 1: 5′TCCGTAGG-TGAACCTGCGG-3′, and ITS4: 5-′-TCCTCCGCTTATTGATATGGC-3′ (White et al. 1990). The resulting sequence of the complete ITS rDNA of *Erysiphe* sp. KUS-F23994 (accession no. FJ548627) has been deposited in the Gen Bank of National Centre for Biotechnology Information (NCBI, http://www.ncbi.nl.nih.gov/). The nearly complete ITS rDNA of *Erysiphe* sp. KUS-F23994 was aligned with other *Erysiphe* spp. nucleotide sequences from the NCBI Gen Bank database.

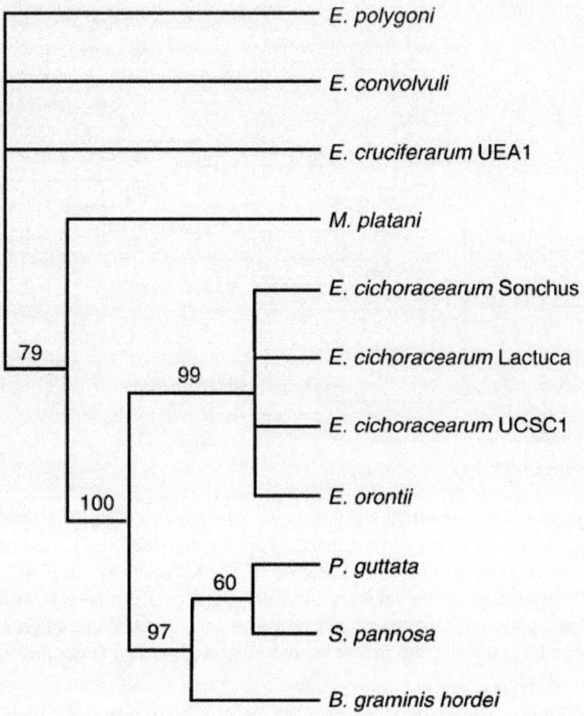

Fig. 2.13 Phylogenetic analysis of UEA1, UCSC1, and nine other powdery mildew fungi based on the sequences (Saenz and Taylor 1999). Strongly supported (95%) branches are shown in bold. Numbers shown are frequencies of branches in phylogenies estimated from bootstrap sampled data bases (Felsenstein 1985). Only values >60% are reported. *M. platani; Microsphaera platani*; *P. guttata; Phyllactinia guttata*; *S. pannosa; Sphaerotheca pannosa*; *B. graminis hordei; Blumeria graminis* f. sp. *hordei* (Adam et al. 1999)

Molecular phylogenetic reconstructions of these ITS rDNA sequences were done using MEGA4, version 4.0 (Tamura et al. 2007), for neighbor-joining analysis using Tajima-Nei distances. To test the reproducibility of results, 1000 bootstrap replications (Felsenstein 1985) were performed by the parameters in default values. As shown in ITS based phylogenetic tree (Fig. 2.13), the *Erysiphe* sp. KUS-F23994 formed a well-supported group (a highly bootstrap value of 97%) with three sequences of powdery mildew fungus *E. cruciferarum* which is able to infect *A. thaliana*, *B. campestris,* and *B. rapa*. However, sequence divergences between the *Erysiphe* sp. KUS-23994, and other *Erysiphe* species, except *E. cruciferarum,* was more than 3%. Taken together, the polygenetic analysis revealed that the *Erysiphe* sp. KUS-F23994 is indeed identical to *E. cruciferarum* (Choi et al. 2009a).

2.5.3.2 Sequence Comparison of DNA Encoding the 5.8S rRNA, ITS1, and ITS2

To distinguish the powdery mildew isolates of *E. cruciferarum* (UEA1), and *E. cichoracearum* (UCSC1), sequence comparison of DNA was performed by Adam et al. (1999). The rDNA gene cluster contains both highly conserved ribosomal subunits, and variable internal transcribed spacer (ITS) regions that can be compared to evaluate the phylogenetic relationships among related fungal species, and races within species (White et al. 1990; O'Donnell 1992). The nucleotide

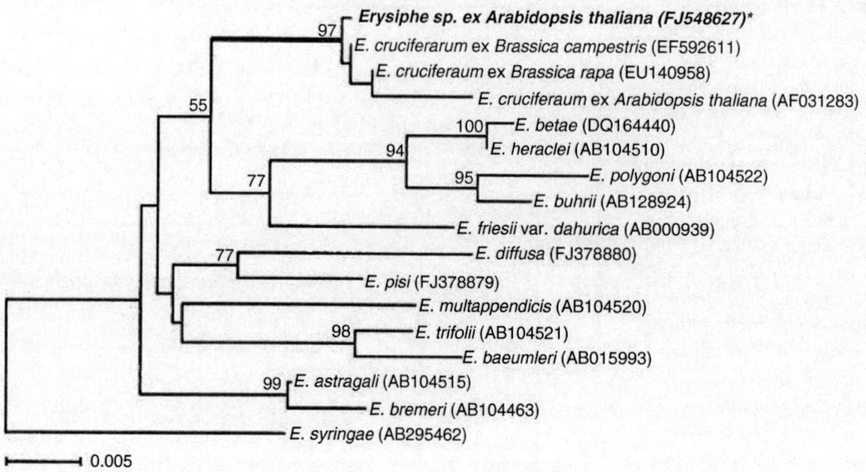

Fig. 2.14 Phylogenetic tree of *Erysiphe cruciferarum* KUS-F23994 (accession no. FJ548627), and other 16 *Erysiphe* spp., constructed by a neighbor-joining method based on the complete internal transcribed spacer (ITS) rDNA regions (ITS1, 5.8S rDNA, and ITS2). Numbers above the branches represent the bootstrap values of over 50% obtained from 1000 bootstrap replicates. Bar = Number of nucleotide substitutions per site. The GenBank accession numbers are represented in parentheses. Asterisk (*) denotes the isolate using in this study (Choi et al. 2009a)

sequences for both isolates were obtained from polymerase chain reaction (PCR) fragments defined by the primers NS7, and ITS4 (White et al. 1990). Each sequence included a portion (UCSC1, 384 bp; UEA1, 363 bp) of the 3′ end of the 18S rDNA, the complete 5.8S rDNA subunit (154 bp), and the two flanking internal transcribed spacers, ITS1, and ITS2, as well as a 57- to 60-bp portion of the nuclear large 28S rDNA. ITS1 varied in size from 223 bp to 187 bp between UEA1, and UCSC1, respectively. The ITS2 region was also smaller in UCSC1 (165 bp) than in UEA1 (184 bp). As expected, the 3′ end of 18S, the 5.8S, and 5′ end of 28S rDNA sequences displayed a high degree of conservation between the two *Erysiphe* isolates, with 4, 1, and 2 bp changes in the respective segments. The occurrence of base pair changes in the conserved rDNA subunit genes in addition to a number of base pair changes in ITS1, and ITS2 confirmed, at the molecular level, that both isolates were from distinct phylogenetic groups (Saenz et al. 1994; Hirata and Takamatsu 1996; Saenz and Taylor 1999). The sequences were aligned relative to each other, matching sequences of *B. graminis* f. sp. *hordei*, and eight other powdery mildews for a phylogenetic reconstruction (Saenz and Taylor 1999) (Fig. 2.14). The tree is a strict consensus of the eight most parsimonious trees, which were found by a heuristic search employing the random stepwise addition option of PAUP (Swofford 1993).

Branch support was determined by 1000 boot strapped data sets, which are displayed on the tree as numbers above the branches. The tree was 262 steps in length; the consistency index was 0.828, the retention index was 0.763, and the re-scaled consistency index was 0.632 (Farris 1989). This analysis showed that the

UCSC1 and the UEA1 isolates are distinct *Erysiphe* spp. The UCSC1 isolate grouped with other *E. cichoracearum* isolates, and *E. orontii*, while the UEA1 isolate grouped with *E. polygoni*, and *E. convolvuli* (Adam et al. 1999).

2.6 *Brassica–Fusarium* Genome

2.6.1 Introduction

The *Fusarium oxysporum*, a notorious plant pathogen, can infect hundreds of cultivated species, and at the same time, have distinct host specificity. This host specificity is used to classify *F. oxysporum* into formae speciales, and a forma specialis typically represents one to a few monophyletic clonal lineages that cause disease in a narrow range of taxonomically related plants (Kistler 1997). At the genomic level, host specificity corresponds to the presence of lineage-specific chromosomes (Ma et al. 2010). However, little is known about molecular mechanisms involved in these host-specific plant–fungus interactions. One of the three formae speciales that are pathogenic to the crucifer *Arabidopsis thaliana* (Diener and Ausubel 2005; Provart et al. 2016), the genome sequence of *F. oxysporum* f. sp. *matthiolae* will enable the genetic analysis of fungal pathogenesis and host immunity using the model plant *A. thaliana.*

2.6.2 *Fusarium oxysporum* Genome

Phylogenic analysis indicates that *F. oxysporum* f. sp. *matthiolae* isolates form a single clonal lineage (Bosland and Williams 1987; Kistler and Benny 1989; Kistler et al. 1987, 1991; O'Donnell et al. 2009), although two races of *F. oxysporum* f. sp. *matthiolae* are distinguished by the differential susceptibility of varieties of *Matthiola incana* (Bosland and Williams 1988). Natural variation of immunity is observed among wild accessions or ecotypes of *A. thaliana* toward Fusarium wilt (Diener and Ausubel 2005). Quantitative trait loci mapping in offspring of crosses between resistant and susceptible ecotypes has identified three resistance to *F. oxysporum* (*RFO*) genes—one receptor-like protein gene (*RFO2*) and two receptor-like kinase (RLK) genes (*RFO1* and *RFO3*)—from different RLK gene subfamilies (Cole and Diener 2013; Diener 2013; Diener and Ausubel 2005; Shen and Diener 2013). Because receptor-mediated immunity is reported to be the major determinant of disease resistance to *F. oxysporum* f. sp. *matthiolae* (Cole and Diener 2013), investigation of the interaction of *F. oxysporum* f. sp. *matthiolae* and *A. thaliana* should lead to a fundamental understanding of receptor-mediated plant immunity, especially against fungal pathogens. The genome sequence comes from DNA purified from *F. oxysporum* f. sp. *matthiolae* race 2, isolated from wilted garden stock (*M. incana*), a cultivated plant in the crucifer or mustard (Brassicaceae) family, prized for its colorful flowers (Baker 1948; Tatsuzawa et al. 2012). This strain was deposited in the American Type Culture Collection (ATCC 16603) and subsequently designated by P. H. Williams as PHW726 (Kistler et al. 1987).

The pipeline for genome assembly was adapted from Ayhan et al. (2018). Genomic DNA was purified from the mycelium of PHW726, and then sequenced by Illumina MiSeq and PacBio RS II plat forms with ×119 and ×21 coverage, respectively. Yu et al. (2020) used MiSeq paired-end sequencing with 150 cycles. The maximum size of the PacBio RS II reads was 59 kb while the mean size was 8.5 kb. Trimmomatic, version 0.32 (Bolger et al. 2014), was used to remove adaptors and trim ends of Illumina reads (parameters: ILLUMINACLIP: TruSeq3-PE.fa:2: 30:10 LEADING: 3 TRAILING: 3 SLIDING WINDOW: 4:15 MINLEN: 36). FastQC, version 0.11.5, was used to check the quality of all reads. SPAdes, version 3.9.1 (Antipov et al. 2016), was used to combine PacBio subreads and trimmed Illumina reads into an initial hybrid assembly with default parameters. BWA, version 0.7.12 (Li and Durbin 2009), was used to map the Illumina reads to the assembly. Further cleaning, fixing, and sorting of mapping reads was done with Picard, version 2.0.1, and Sam tools, version 1.3 (Li et al. 2009). A structural variant caller, GRIDSS, version 1.4.1 (Cameron et al. 2017), was used to identify links between scaffolds in the initial assembly. A custom script (scaffolding.m) was used for scaffolding. Minimap 2, version 2.17 (Li et al. 2018), was used to map PacBio subreads to new scaffolds, and links were manually inspected and, if necessary, fixed. Further polishing was performed by remapping Illumina reads to the assembly, during which Free Bayes, v0.9.10-3-g47a713e (Garrison and Marth 2012), was used to identify base variants between reads and the assembly (specially 70% support of minimal 10 alternate counts, with a minimal base mapping-quality greater than q30). Identified variants were used to correct the assembly by a custom script (FASTA edit With VCF. m). Repeat Marker 4.0.5 (Tarailo-Graovac and Chen 2009) was used to screen the repeats. Mummer 3.22 (Kurtz et al. 2004) was used to align the assembly with the reference genome assembly for the tomato pathogen *F. oxysporum* f. sp. *lycopersici* Fol4287 (Ma et al. 2010).

The final assembly was 57.3 Mb in total length and comprised 583 scaffolds with an N_{50} value of 0.77 Mb (Table 2.7). The largest scaffold size was 3.6 Mb. The GC

Table 2.7 Summary of the *Fusarium oxysporum* f. sp. *matthiolae* strain PHW726 genome assembly and a comparison with Fol4287 (Ayhan et al. 2018; Yu et al. 2020)

Variables	Statistics	
	PHW726	Fol4287
Assembly size (bp)	57,270,650	53,912,387
Core sequence size (bp)	43,818,233	42,239,438
Number of scaffolds	583	499
Number of core scaffolds	66	55
Size of largest scaffolds (bp)	3,557,637	5,733,288
Interspersed repeat content	5.35%	4.21%
N50 (bp)	774,050	1,338,693
N90 (bp)	47,752	49,310
L50 (bp)	18	11
GC Content	47.44%	47.68%

content was 47.4%. The size of total interspersed repeats was 3.1 Mb, which accounted for 5.4% of the assembly. A comparison with Fol 4287 assembly (Ayhan et al. 2018) suggested a larger assembly size and higher interspersed repeat content of PHW726. The size of sequence mapped to the core chromosomes of Fol4287 which, including 66 scaffolds (defined as core scaffolds), was 43.8 Mb. The assembly also included a scaffold of 52,365 bp that captured the entire mitochondrial DNA. This Whole Genome Shotgun project has been deposited at DNA Data Bank of Japan/European Nucleotide Archive/GenBank under the accession WJXY00000000. The version described is version WJXY01000000. This assembly for the genome of PHW726 should facilitate future molecular genetics and genomic studies. Candidate *F. oxysporum* f. sp. *matthiolae* genes that promote pathogenesis or elicit immune response in *A. thaliana* and *M. incana* can now be predicted, subcloned, and genetically characterized (Yu et al. 2020).

2.7 *Brassica–Hyaloperonospora* Genome

2.7.1 Introduction

The *Hyaloperonospora parasitica* causes downy mildew, a foliar disease of *Brassica* crops. The downy mildew causes considerable yield losses to all *Brassica* crops worldwide, especially in Europe, Asia, and Australia. Physiological races or pathotype variations have been described by various workers, but, few studies have achieved clear race differentiation (Saharan et al. 2017). Notably, Coelho et al. (2012) summarized six pathotypes and suggested five major-effect R loci corresponding to the observed phenotypes. The ideal control approach for downy mildew is to breed genetically resistant varieties. To date, several R loci have been mapped and applied inbreeding. Resistance to downy mildew is believed to be distinct at the *Brassica* seedlings and adult plant stages. Resistance mapping work has identified several R genes/loci. In *B. oleracea*, the first locus conferring resistance in the broccoli seedling stage was placed in a linkage group and was found to be located close to the glucosinolate related gene *BoGsl elong*. Another single dominant resistance gene expressed at the adult stage was identified in broccoli and was named *Pp523*. The genomic region containing this gene was further analyzed using SCAR and cleaved using amplified polymorphic sequence (CAPS) markers, as well as two bacterial artificial chromosome (BAC) libraries. In *B. rapa*, QTLs conferring seedling-stage resistance were discovered. Using a genetic linkage map generated with a DH population, the major effect locus *BraDM* was delimited to a region spanning 2.9 cM in the A08 linkage group. For adult-stage resistance, a monodominant gene, *BrRHP1*, was localized to a 2.2 Mb interval on the A01 linkage group. The mapping methods based on high-throughput resequencing have greatly promoted the identification of R genes. For example, using a high-density SNP-based map, a major locus, *sBrDM8*, was localized to a physical segment of ~228 kb, with one candidate kinase gene, *Bra016457* (Yu et al. 2016). The developed markers closely located with these R loci have been adopted for resistance breeding through MAS and have greatly contributed to resistance breeding. For example, Yu et al. (2011) converted the closely linked RAPD marker K14-1030 to a SCAR marker, which greatly improved selection efficiency in the progenies (Lv et al. 2020).

2.7.2 *Hyaloperonospora* Genome

The *Hyaloperonospora* genome was first sequenced and assembled in 2008 using *H. arabidopsidis* species through Sanger and Illumina sequencing approached by Baxter et al. (2010) with a genome size of 78 Mb with 9.5× coverage of the nuclear genome but did not assemble the mitochondrial genome in comparison to genomes of related, hemi-biotrophic *Phytophthora* species, the *Hpa* genome exhibits dramatic reductions in genes encoding: (1) RXLR effectors and other secreted pathogenicity proteins; (2) enzymes for assimilation of inorganic nitrogen and sulfur; (3) proteins associated with zoospore formation and motility. These attributes comprise a genomic signature of evolution toward obligate biotrophy.

2.7.2.1 Characteristics of *Hyaloperonospora arabidopsidis* Genome

Genome analysis was performed with *H. arabidopsidis* (*Hpa*) Emoy2 isolate (Holub 2008), using DNA from asexual spores. Sanger shotgun sequencing at 9.5-fold coverage, combined with 97 sequenced BAC inserts, yielded an assembly of 77.8 Mb. Illumina sequencing at 46-fold coverage yielded 3.8 Mb of additional sequence, which was integrated into the Sanger assembly to form the 81.6 Mb final version (v8.3.2). Forty-two percent of the *Hpa* genome is comprised of repetitive elements. Analysis of Sanger and Illumina read depth suggests that v8.3 contains approximately 12.7 Mb composed of tandem repeats compressed into reduced copies in the assembly, indicative of a genome size of around 100 Mb. The CEGMA pipeline, combined with manual examination, identified *Hpa* orthologs of 95% of the 248 conserved single-copy eukaryotic genes. Moreover, 94% of 31,759 EST reads aligned to the assembly, indicating that the draft genome assembly encompasses a very high percentage of the *Hpa* gene space (Baxter et al. 2010). The 14,543 genes were computationally predicted in v8.3, of which 80% are supported by ESTs and/or Illumina cDNA tags. This predicted gene content is similar to *P. ramorum* (65 Mb, 15,743 genes) and lowers than *P. sojae* (95 Mb, 19,027 genes) or *P. infestans* (240 Mb, 17,887 genes). Six thousand eight hundred and eighty-two predicted genes in *Hpa* had no identifiable ortholog in sequenced *Phytophthora* species, or similarity to known proteins, and as such represent potentially lineage-specific genes. Some of these genes may play roles that are specific to biotrophy. For example, a novel family of secreted small, cysteine-rich proteins exists in *Hpa* (PPAT12/24-like (Bittner-Eddy et al. 2003).

The distance between genes flanking the flagellar locus was smaller in *H. arabidopsidis* than *P. infestans* in 78% of cases; a similar conclusion was drawn from the prior study of eight loci (Baxter et al. 2010). Interestingly, this trend in intergenic distances is opposite to that observed for genes that share synteny but have functions unrelated to flagella. Of 120 gene pairs selected randomly from 40 different regions of their genomes, intergenic regions were smaller in *H. arabidopsidis* than *P. infestans* only 31% of the time, with median sizes of 310 nt and 510 nt, respectively. The loss of the flagella gene therefore appeared to facilitate the compaction of the *H. arabidopsidis* genome beyond what would be expected if only the flagella gene and its transcriptional regulatory sequences had been lost (Judelson et al. 2012).

Oomycete genomes contain a large amount of transposon and retro element-like sequences, but such elements did not appear to have the predominant role in eliminating flagellar genes. In a search of 32 intergenic regions that formerly contained a flagella gene, repetitive sequences were detected only in the case of IFT80-encoding gene PITG_18714. The 3.5 kb region containing the IFT80 gene and flanking noncoding DNA was reduced to 1.5 kb in *H. arabidopsidis*. This included about 400 nt of a Copia-like sequence having 46 relatives in the *H. arabidopsidis* genome, based on BLAST with a $E = 10^{-9}$ threshold. It is possible that transposable elements had a greater role than predicted by this analysis, which was biased by its focus on cases where gene order was conserved (Judelson et al. 2012). Judelson et al. (2012) observed compaction of the *H. arabidopsidis* genome at most affected loci. Such changes are estimated to account for a reduction of about 1 Mb; overall shrinkage is likely greater since some metabolic genes, protein kinases, and effectors are also reported to be absent in *H. arabidopsidis* (Baxter et al. 2010; Judelson and Ah-Fong 2010).

2.7.2.2 Genome Architecture of *Hyaloperonospora brassicae* Isolates for Virulence

The first draft genome assemblies for two isolates of *H. brassicae*, differing for their virulence was reported by You et al. (2021). They revealed genome sizes of 72.762 Mb and 76.950 Mb and 6438 and 6470 scaffolds respectively. In silico annotation allowed understanding of the genome architecture of *H. brassicae* in terms of genes for pathogenicity and virulence. The observed reduction in virulence or loss of pathogenicity in a larger number of genes in the sample with low virulence in comparison to sample with high virulence may reflect differential rates of mutation and selection during host–parasite co-evolution. Genomic resources development will aid in monitoring field prevalence of *H. brassicae* pathotypes and to detect early any virulence changes within pathogen populations (You et al. 2021). They sequenced the genomes of two *H. brassicae* isolates of high (Sample B) and low (Sample C) virulence. Sequencing reads were first assembled de novo with softwares SOAPdenovo2, ABySS V2.1, and Velvet V1.1 and later combined to create meta-assemblies with genome sizes of 72.762 Mb and 76.950 Mb and predicted gene densities of 1628/Mb and 1644/Mb, respectively. They annotated 12,255 and 13,030 genes with high proportions (91–92%) of complete BUSCOs for sample B and C, respectively. Comparative analysis revealed conserved and varied molecular machinery underlying the physiological specialization and infection capabilities. BLAST analysis against PHI gene database suggested a relatively higher loss of genes for virulence and pathogenicity in sample C compared to Sample B, reflecting pathogen evolution through differential rates of mutation and selection. It will enable identification and monitoring of *H. brassicae* virulence factors prevailing in-field (You et al. 2021).

The two isolates of pathogen samples were similar for molecular functions. However, both biological processes (biosynthetic and catabolic process, cell communication, and cellular component organization and localization) and cellular components (catalytic complex, coated membrane, endo-membrane system, endoplasmic reticulum membrane, and envelope) were distinctively different between the more virulent Sample B than the less virulent Sample C. Importantly, the genome sequences presented in the current study constitute an important resource for future functional genomics analysis, fostering DNA barcoding of *H. brassicae* isolates and the development of genetic markers (e.g., SNPs and short repeats) that could be used to assess the population diversity and to tag genes for virulence in the pathogen population by comparing highly virulent and nonvirulent isolates (You et al. 2021).

2.7.2.3 Genome Assemblies of *Hyaloperonospora brassicae* Isolates and Their Functional Characteristics

The genome assemblies for two samples of *H. brassicae* isolated from commercial rapeseed fields in Australia. Using SOAP denovo2, ABySS V2.1, and Velvet V1.1 to perform de novo assembly and create a meta-genome assembly, using software Met assembler was presented. You et al. (2021) generated 6438 scaffolds for sample B with an N50 of 23,533 with the longest scaffold being 154,210 bp. In contrast, 6470 scaffolds were produced for sample C with an N50 of 24,471 bp and highest scaffold length 197,135 bp. Met assembler has been widely used to develop metagenome assemblies. The genome size of sample C was 76.95 Mb compared to 72.16 Mb for sample B. The N50 value of both samples approximated 24 kb, indicating good quality for further downstream analysis. Using reference sequences for *H. arabidopsidis*, they predicted 36,819 and 40,346 protein-coding genes, for samples B and C, respectively, including those associated with plant–pathogen interactions. Of the predicted genes only 12,255 and 13,030 could be mapped and annotated for samples B and C, respectively. They achieved good assembly continuity using short-read data and annotated over 91–92% of BUSCO Alveolata dataset of single-copy orthologs, with assemblies relatively contiguous and with a negligible number of duplicated BUSCO genes. Comparative analysis of the genomic data obtained from two *H. brassicae* isolates revealed conserved and varied molecular machinery, underlying their physiological specialization and infection capabilities. Several gene classes were identified. These included necrosis-inducing proteins (NLPs) produced by fungal pathogens that cause cell death while stimulating the plant's immune reaction (Feng et al. 2014). Oome et al. (2014) showed that "downy mildew pathogen *H. arabidopsidis* encodes ten noncytotoxic NLPs (HaNLPs) that do not cause necrosis" but "act as potent activators of the plant immune system in *Arabidopsis thaliana*." Many gene classes they identified differed significantly between the model pathogen *H. arabidopsidis* and *H. brassicae*, in both identified class and numbers within each class. Of particular interest were the RXLR effectors that are primarily associated with the suppression of host immunity and play a major role in virulence for downy mildew species (Anderson et al. 2015). The predicted RXLR effectors and elicitors were more in the *H. brassicae* samples A and B compared with *H. arabidopsidis*. Zhang et al. (2019) also reported RXLR, effectors in *P. sojae* isolates, and of these, 42 core RxLR effectors were considered important for infection. The number of elicitors in both *H. brassicae* samples A and B were higher than in *H. arabidopsidis*. Elicitors are pathogen signal metabolites recognized by plant cells, and most constitute Pathogen-Associated Molecular Patterns (PAMPs). Plant pathogens must overcome host resistance to successfully infect plants and toward this outcome, fungal pathogens produce pectin lyases (PL) to catalyze the depolymerization of esterified pectin by a b-elimination mechanism (Lara-Márquez et al. 2011). The contribution of pectin lyases to *H. brassicae* infection remains largely unknown, similar to other oomycete pathogens, despite PL having been extensively studied in major fungal pathogens (Fu et al. 2015). Pectin-esterases are involved in plant tissue maceration and soft-rotting, and a few cellulases (glucanases/cellulases) of plant-pathogenic fungi have been shown to be involved in pathogenicity (Eshel et al. 2002). The pathogen *H. arabidopsidis* secretes proteins and fibrillar β-1, 3-glucans that binds its germ tubes to the substratum (Carzaniga et al. 2001), likely helping resist detachment by wind or rain and to help protect against desiccation (Judelson and Ah-Fong 2019).

2.7.2.4 Draft Genome Assemblies from Whole-Genome Sequences of Two *Hyaloperonospora brassicae* Isolate Samples Differing in Field Virulence on *Brassica napus*

The genome sequences, differing significantly to those of *H. arabidopsidis*, did show a large degree of consistency with those of *H. arabidopsidis*, at-least to the extent these find use in population genomic studies for the identification of virulence haplotypes and alleles. Comparative analysis of annotated protein-coding genes obtained from two *H. brassicae* isolates revealed conserved and varied molecular machinery underlying the physiological specialization and infection capabilities of this important pathogen. In particular, *H. brassicae* Samples A and B differed in terms of biological processes for their cellular components. Studies will pave the way for early detection of any virulence changes within the pathogen populations, and perhaps even for molecular identification of pathotypes of this pathogen. Such information is crucial for *H. brassicae* as there is not only virulence/pathotype diversity among its populations, but that such diversity is rapidly evolving. Further, genome sequencing or enrichment analysis may also be used to identify alleles associated with particular virulence/avirulence effects. However, while linkage mapping or genome-wide association mapping can enable more rapid generation of cultivars resistant to *H. brassicae*, more virulent newly evolving downy mildew genotypes will keep emerging. Hence, developing markers for virulence/pathotype classification would allow rapid molecular identification and monitoring of the prevalent virulence potentials existing within pathogen populations' in-field (You et al. 2021).

2.7.3 Phylogeny of *Hyaloperonospora*

The phylogenetic analysis of six clades and subclades of *Hyaloperonospora* species infecting different cruciferous species has been made by Goker et al. (2009). The best log likely hood value obtained with RA × ML on the complete dataset was 14,885.742; the alpha parameter of the gamma distribution was estimated at 0.293. The best tree is shown in Figs. 2.15 and 2.16 along with the ML BS values. Heuristic MP analysis of the complete dataset yielded 156 most parsimonious trees of length 2225, which were obtained in 182 of the 1 K replicates; the RI was 0.926. The strict consensus tree of the best trees was topologically similar to the ML tree. MP BS values, host names, and assignments of *Hyaloperonospora* specimens to species are also indicated on the ML tree (Figs. 2.15 and 2.16). Most of the groups were characterized by similar ML, and MP support values.

Using *Perofascia* as an out group, *Hyaloperonospora* was highly supported as monophyletic by a BS of 100% (Fig. 2.15). Even though, the sequences were sampled from a number of different *Lepidium* host species, *P. lepidii* appears as a single, and genetically rather uniform species. Within *Hyaloperonospora*, six diverse clades were recognized comprising more than one species supported with high BS, and a number of isolated species. Clade 1 is the sister group to a clade containing the remaining five clades plus all single-species lineages with low (65%) and high (98%), BS in ML, and MP analysis, respectively (Fig. 2.15). The resolution of the backbone connecting these remaining groups is low, in contrast to the usually highly supported terminal nodes. Clade 1 comprises six clearly distinct lineages that are parasitic on different European, and East Asian Cardamine or *Rorippa* species, European *Barbarea vulgaris* or *Arabis soyeri*. Except for the latter two, these host genera belong to Cardamininеae. Clade 2 comprises collections from European *Draba verna,* and East Asian *Draba nemorosa* in two distinct clusters, the genetic distance between which is considerable. The *H. arabidisalpinae* separates before clade 2, and *H. niessleana* after clade 2 on the same grade in the tree, but without support. The *H. isatidis*, *H. thlaspeos-arvensis*, and the parasite of *Biscutella auriculata* appear in close relationship to clade 3, but also without support. Clade 3 comprises parasites of a rather diverse assemblage of hosts, such as *Reseda*, which is one of the few host genera apart from Brassicaceae. The basal split within clade 3 between the East Asian *Sisymbrium luteum* pathogen, and the other specimens is moderately to highly supported (87–95% BS). Some of the sub clades of clade 3 are not easy to delineate; it is not clearly evident whether the parasite of *Arabisturrita* also belongs to *H. arabidopsidis* or whether *H. thlaspeosperfoliati* includes the specimens from *Noccaeaca erulescens*. Isolated lineages (Fig. 2.16) that are highly supported as monophyletic, but have unresolved inter-relationships, are represented by *H. cheiranthi*, *H. hesperidis*, *H. nesliae*, *H. camelinae*, and *H. parasitica*, as well as by the pathogens of *Lepidium draba*, *Descurainia sophia*, *Iberis sempervirens*, and *Helianthemum* (Cistaceae). Clade 4 comprises of *H. galligena,* and *H. berteroae*, whereas clade 5 consists of two highly distinct lineages of *Sisymbrium* parasites. The last clade, clade 6, contains a particularly large number of specimens, which is divided into two subclades: specimens that are mainly parasitic on

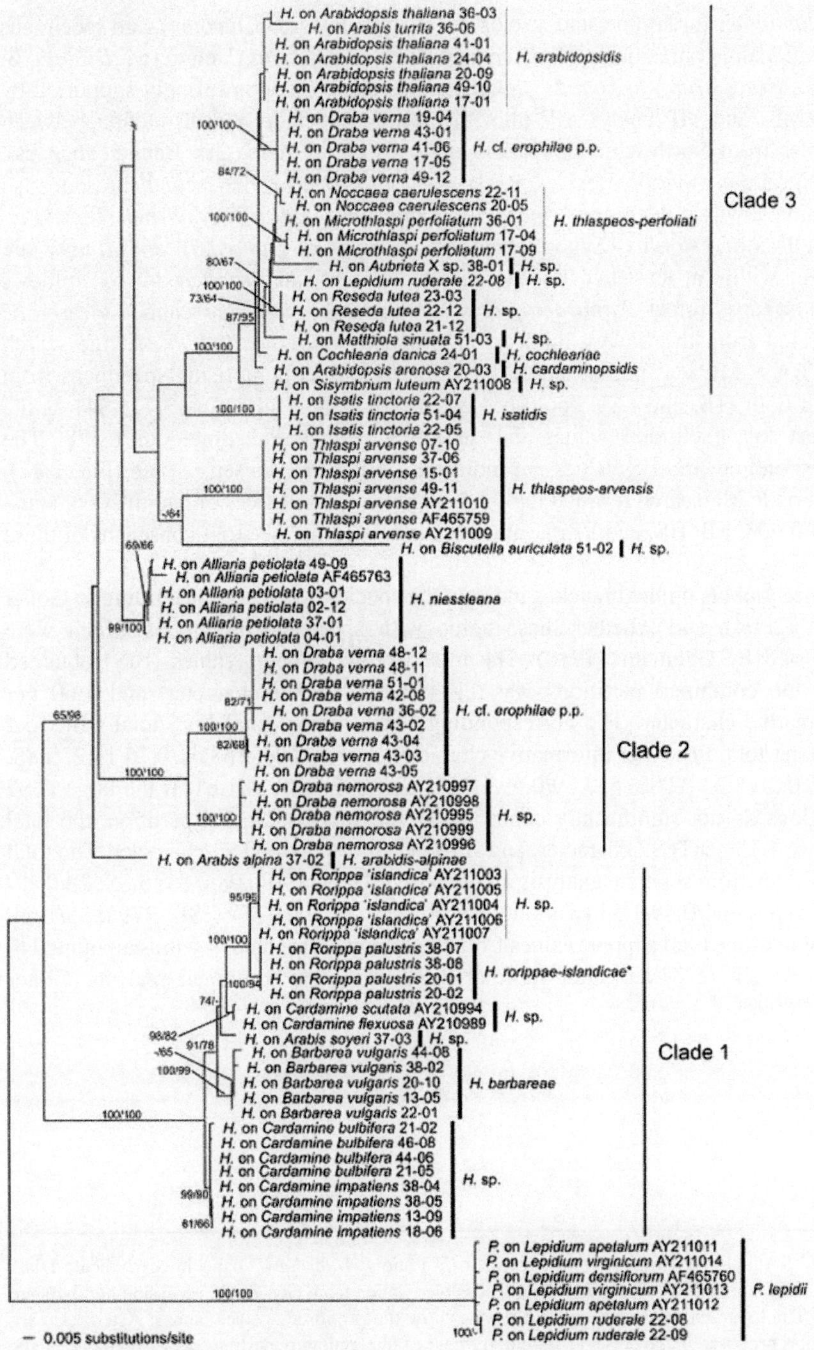

Fig. 2.15 Lower half of the phylogenetic tree inferred from the complete dataset with RA × ML under a GTRMIX nucleotide substitution model approximation and rooted with *Perofascia*. Branch lengths are scaled in terms of the expected number of substitutions per site. Numbers above branches represent BS values above 50% from ML (left) and MP (right) bootstrapping. Labels

Cardaminae (Cardamine, and its close relatives, such as *Nasturtium*), and specimens that are mainly parasitic on *Brassica,* and its close Relatives (Brassiceae; *Diplotaxis, Eruca, Erucastrum, Raphanus, Sinapis*). This subdivision is strongly supported by both ML, and MP analysis. Within the former subclade, the distinction between samples from North America (specimen 21-01), East Asia (Gen Bank sequences), and Europe is obvious. Besides parasites of Brassicaceae, the second subclade also contains downy mildews of *Armoracia, Lobularia, Lunaria, Sisymbrium, Teesdalia,* as well as of *Tribulus* (Zygophyllaceae). Parasites of *Diplotaxis,* and *Sinapis* are present within at least two lineages. Both *Sinapis,* and *Brassica* downy mildews seem to also infect *Armoracia. Hyaloperonospora lunariae* cannot clearly be separated from the downy mildew on *Erucastrumnas turtiifolium.* A tree search with RA × ML was performed on the reduced dataset comprising specimens from which both ITS, and LSU could be sequenced. This resulted in 13,776.841 as the highest log likelihood value, and an alpha parameter estimate of 0.278. The corresponding ML BS values are indicated on the MP consensus tree (Fig. 2.17). A heuristic MP search resulted in 404 most parsimonious trees of length 2066, and a RI of 0.904. MP BS, and PBrS values are indicated on the strict consensus of these trees (Fig. 2.17).

The symbols on the branches indicate the species boundaries according to Goker et al. (2004), and whether these agree with species delimitations which were suggested by Gaumann (1918). The total Bremer support values (BrS) obtained from the combined partitions was 1124, i.e. 0.48 per character, and 1.60 per informative character. The corresponding PbrS values (total PbrS/total PBrS per character/total PBrS per informative character) were: ITS1, 153.41/0.76/1.42; 5.8S, 59.86/0.36/2.22; ITS2, 537.39/0.59/1.61; LSU, 373.33/0.34/1.61. If the large ITS2 insertions do not significantly contribute to the support for that partition, the total BrS are 1.13 per ITS2 character, and 2.13 per informative ITS2 character. The total BrS values from separate analysis amounted to: ITS1, 196/0.98/0.81; 5.8S, 28/0.17/1.04; ITS2, 444/0.49/1.33 (without insertions: 444/0.93/1.76); LSU, 379/0.35/1.63. The sum of the total support values from the four separate analyses thus amounted to 1047, which are 77 less than the total BrS obtained in combined analysis (Goker et al. 2009).

Fig. 2.15 (continued) for *Hyaloperonospora* (*H.*) and *Perofascia* (*P.*) specimens indicate DNA isolation number and host species; if sequences were taken from Gen Bank, accession numbers are given. Thick vertical bars and adjacent names show the proposed species names. Asterisks mark binomials proposed in the present study. In the case of uncertainty regarding species boundaries, the bars are drawn in light gray. Thin vertical bars and adjacent numbers indicate clades apparently above species level as described in the text. See Fig. 2.16 for the upper half of this tree (Voglmayr 2003)

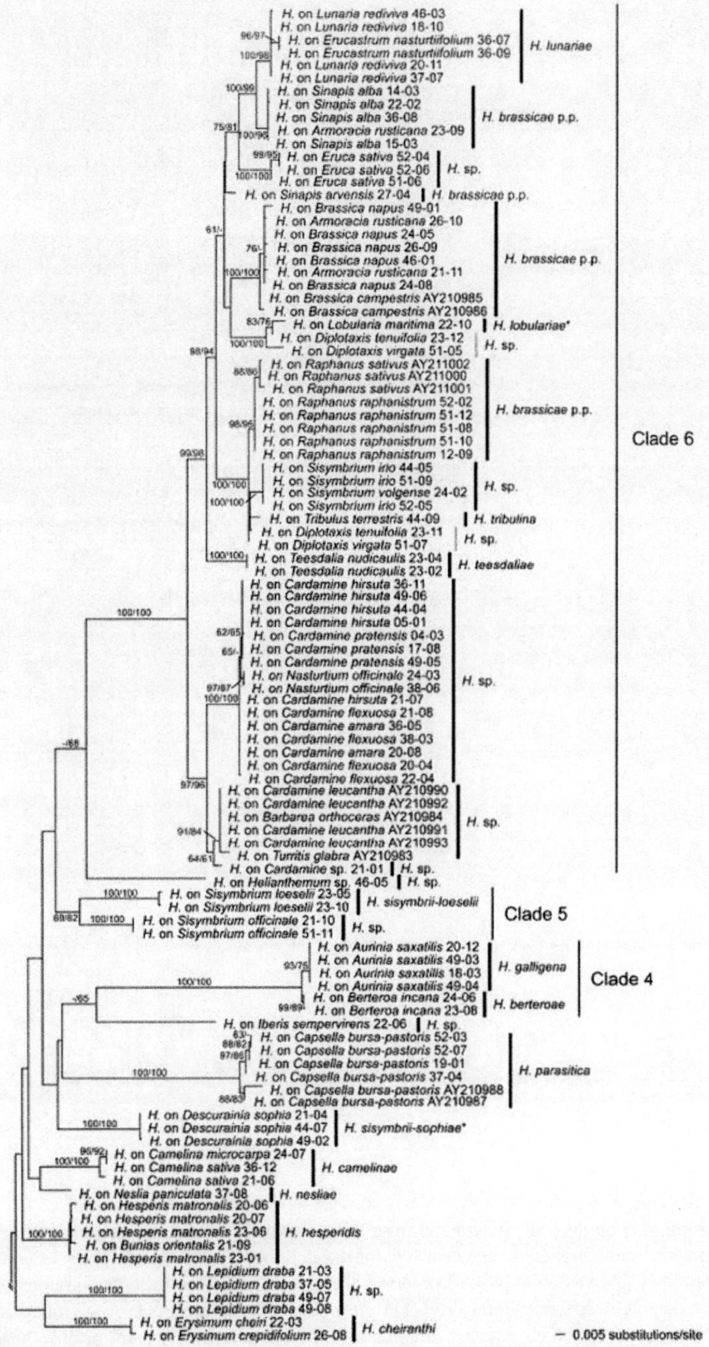

Fig. 2.16 Upper half of the phylogenetic tree depicted in Fig. 2.15, including clades 4–6. For a description, see Fig. 2.15 above (Voglmayr 2003)

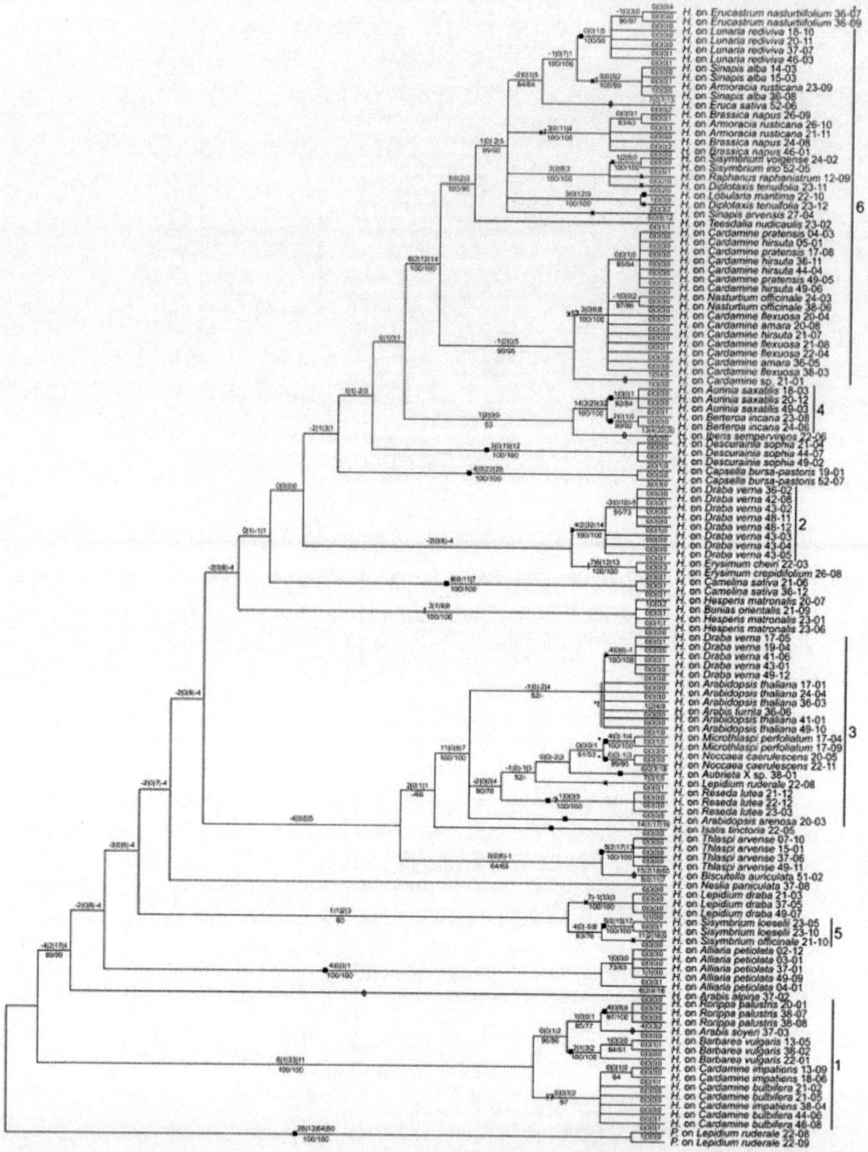

Fig. 2.17 Strict consensus of the 404 most parsimonious trees (length: 2066 bp) inferred from the reduced dataset that contains only specimens from which both ITS, and LSU sequences could be obtained. Numbers below branches represent RA × ML/GTRCAT (left), and MP (right) BS values above 50%. Numbers above branches (except terminal ones) are partitioned Bremer support values; the partitions examined were (from left to right) ITS1, 5.8S, ITS2, and LSU rDNA. Numbers above terminal branches represent their average lengths as inferred with DELTRAN optimization, as implemented in PAUP from the same partitions. Specimen labels are as in Figs. 2.15 and 2.16; affiliation of specimens to the clades used in these figures is indicated by vertical bars, and numbers on the right side. The following symbols are used to indicate the suggested species boundaries according to Goker et al. (2004), as well as in the present text, and whether these are in accordance with Gaumann's (1918, 1923, 1926) taxonomy: (–) in agreement with Gaumann; (!) not in

2.7.4 Major Species Clusters in *Hyaloperonospora*

Molecular phylogenies (Goker et al. 2004, 2009) so far conducted have revealed many distinct lineages in *Hyaloperonospora*, which mostly represent single species although the backbone of the tree could not be well resolved. However, six clades could so far be identified, which are highly supported, and harbor several distinct species from different hosts (Fig. 2.18). These are clade 1, mainly on *Cardamine*, and its close relatives; clade 2, on *Draba* species; clade 3, on a variety of host species including *A. thaliana*; clade 4, on *Aurinia*, and *Berteroa*; clade 5, on different *Sisymbrium* species, and clade 6, on variety of host genera but with a focus on *Brassica*, and its relatives (wild Brassicaceae) in one subclade, and on *Cardamine* and its relatives (subtribe Cardaminінae) in the other subclade.

Hyaloperonospora brassicae, which includes the economically important pathogens of *Brassica*, *Raphanus*, and *Sinapis*, is a polyphyletic construct, as specimens from these three host genera are phylogenetically distinct with maximum bootstrap support. The most noteworthy species in clade 6, which also includes *H. brassicae*, is *Hyaloperonospora tribulina*. As *Hyaloperonospora* is primarily parasitic to Brassicaceae (i.e., the earliest diverging lineages are parasites of Brassicaceae), the occurrence of *Hyaloperonospora* on a Zygophyllaceae is there suit of a host jump from Brassicaceae to Zygophyllaceae. From the molecular genetics perspective, *Hyaloperonospora* clade 3 is probably the most interesting because, although it does not include economically important species, this clade includes a parasite of the most intensely studied model plant *A. thaliana*, also, *Arabidopsis arenosa* is among the hosts of clade 3. The investigation of the species closely related to *Hyaloperonospora arabidopsidis* offers the unique possibility to identify key pathogenicity genes involved in host jumps, and pathogen establishment on a specific host. Thus, the general patterns of how plant pathogen evolution and adaptation to specific hosts takes place might be revealed. Similar to clade 6, a host jump across host families occurred within the clade 3, with *Hyaloperonospora crispula s.l.*, which is parasitic to *Reseda lutea* and *Reseda luteola* of the Resedaceae.

2.7.5 *Hyaloperonospora arabidopsidis* on *Arabidopsis*

The widespread application of the broad species concept advocated by Yerkes and Shaw (1959), it was commonly believed that *Hyaloperonospora* on *Arabidopsis* was the same species as *Hyaloperonospora* on *Brassica*, and *Hyaloperonospora* on *C. bursa-pastoris*, the type host of *H. parasitica sensu stricto*. This led to confusion

Fig. 2.17 (continued) agreement, with Gaumann's species being paraphyletic, and including additional hosts; (×) not in agreement, Gaumann's species polyphyletic; (0) host not examined by Gaumann; (?) type host not included in our sample. Asterisks point to molecular uncertainty with respect to species boundaries (Voglmayr 2003)

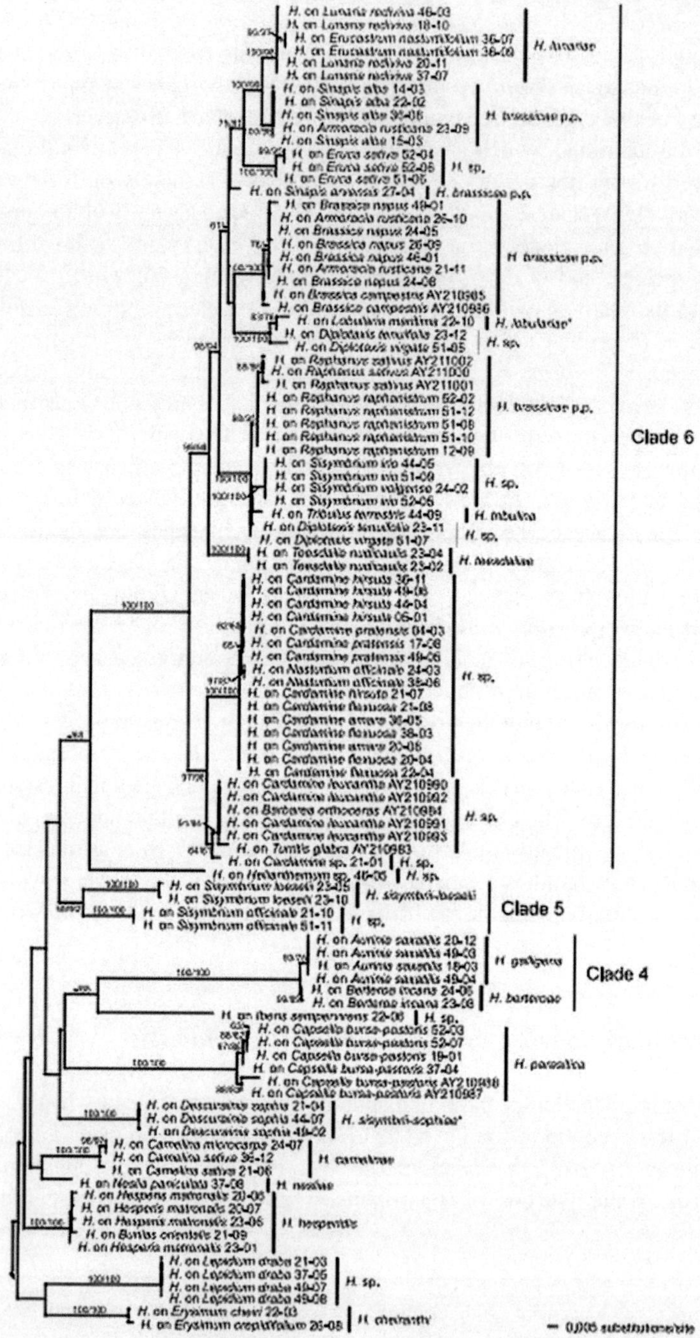

Fig. 2.18 Maximum-likelihood phylogenetic tree inferred with RA × ML from concatenated internal spacer region (ITS), and large ribosomal subunit (LSU) rDNA sequences. The dataset represents a subset of the one analyzed by Goker et al. (2009). Technical details on the inference of this tree as well on the files used, and on the origin of the sequences are provided in Goker et al.

and misconceptions over the taxonomy of the *Arabidopsis*-infecting downy mildew. Research on the genetics of the downy mildew pathogen of *Arabidopsis* has been justified by, among other arguments, the claims that this pathogen belongs to the same species as economically important lineages that infect *Brassica*, *Raphanus*, and *Sinapis* crops. In addition to the fact that *Arabidopsis* is the most extensively studied model plant, this claim, based on the approach of Yerkes and Shaw (1959). However, it was soon realized that the pathogens from different hosts (e.g., *Arabidopsis*, *Brassica*, and *Sinapis*) were not compatible regarding host range (Tor et al. 1994; Satou and Fukumoto 1996).

Therefore, different "races" of *Hyaloperonospora* on various hosts were named, assuming these would merely represent specialized forms of the same species. In the light of molecular phylogenetic reconstructions (Rehmany et al. 2000; Riethmuller et al. 2002; Voglmayr 2003; Choi et al. 2003; Goker et al. 2004, 2009), this perception needs to change, as these studies revealed that the genus *Hyaloperonospora* contains a multitude of distinct species, which have partly gone through a long independent evolution. Considering this, the problems in relating results regarding pathogenicity genes, and avirulence genes obtained from *H. arabidopsidis* to *H. brassicae* become understandable. Because of the high genetic divergence of *H. brassicae* and *H. arabidopsidis*, it is not surprising that pathogenicity genes, which are in common for *H. arabidopsidis* and *H. brassicae*, are hard to find.

The *Hyaloperonospora arabidopsidis* is the most intensely studied obligate biotrophic pathogen because the host genome was the first plant genome available whereas *H. brassicae* is still not well studied in many respects. For *H. parasitica*—a species so far known only from *C. bursa-pastoris* only very few data are available. As this pathogen neither occurs on an economically important nor a genetically well-studied host, it could be assumed that this species will never attract considerable attention. However, it should be borne in mind that *C. bursa-pastoris* is probably the most widely distributed host for a *Hyaloperonospora* species, and that *H. parasitica* is among the most often observed downy mildew diseases in natural populations in Europe, and therefore is well suited for investigations on ecology and population structure of a downy mildew species (Thines et al. 2009a).

Fig. 2.18 (continued) (2009). Note that the recognition of the clade numbers is based on the extended sampling used in Goker et al. (2009), whereas this figure only shows the specimens for which both ITS, and LSU rDNA could be amplified. The numbers above the branches are bootstrap support values equal to or larger than 60% from 100 replicates. *H.*, *Hyaloperonospora*; *P.*, *Perofascia* (Thines et al. 2009b)

2.8 *Brassica–Leptosphaeria* Genome

2.8.1 Introduction

The pathogen *Leptosphaeria maculans* causing blackleg or stem canker is a disastrous fungal disease for *Brassica* crops. The first epidemic was reported on cabbage in Wisconsin. However, blackleg has become a problem in terms of mass infection of oilseed rape only since the middle of the twentieth century, especially in Australia, North America, and Europe. The *L. maculans* exhibits a high level of diversification and has been assigned to different races/pathotypes. Resistance gene mapping work has been conducted since the 1990s and some cultivars with improved resistance to blackleg are available. Most blackleg resistance genes/QTLs originated from the *B. napus* A genome. Ferreira et al. (1995) first applied a double haploid (DH) population from *B. napus* to localize the major locus *LEM1* on N7. Using a similar method, Dion et al. (1995) identified another major gene, *LmFr1*. Mayerhofer et al. (1997) detected a major locus, *LmR1*, and co-segregating markers were developed. Delourme et al. (2004) reported the mapping of resistance loci in two genomic regions and a cluster consisting of five R genes was proposed as the candidate. Fine mapping work was conducted extensively after 2010. Long et al. (2011) identified two resistance genes, *BLMR1* and *BLMR2*, and fine mapping of *BLMR1* resulted in the closest marker distance of 0.13 cM. Jestin et al. (2011) used an association mapping method to characterize the molecular diversity using 128 oilseed rape accessions and identified five novel alleles. Raman et al. (2012) positioned a new major locus, *Rlm4*, and the deposited region was further analyzed, with several candidates being characterized. In addition, blackleg resistance loci have been transferred from wild relatives of *B. rapa* and *B. oleracea* to *B. napus*. Yu et al. (2012) mapped blackleg resistance derived from the wild relative and *LepR1*-*LepR3* was identified. Larkan et al. (2013, 2014) employed map-based cloning to isolate LepR3, which encoded an RLP, representing the first cloned blackleg disease resistance gene. They further isolated the *Rlm2* gene, which is an allelic variant of *LepR3*. Another blackleg resistance gene, *Rlm9*, which encodes a wall-associated kinase-like protein, a newly discovered class of race specific plant RLK resistance genes has been cloned. In addition to the major locus, some QTLs have also been characterized, including six and four that are stable under different environmental conditions (Larkan et al. 2016).

2.8.2 *Leptosphaeria maculans* Genome

Unlike most other dothideomycetes whose genomes have been sequenced, *L. maculans* has a high amount (33%) of repetitive DNA and a correspondingly large genome (45 Mb). The genome is organized in an unusual patchwork fashion with GC-rich blocks that are gene rich, and AT-rich blocks that are gene poor, with relatively sharp demarcation between blocks. Such an arrangement is known as an isochore structure (Rouxel et al. 2011). The AT-rich, gene-poor regions house mainly class I long terminal repeat (LTR) retro-transposons that have been degenerated by RIP mutation. Such regions contain 3.5% of the total genes in the genome, but 20% of these genes encode small secreted proteins, including effectors such as avirulence genes. The genomic location of effector genes embedded within and between degenerated transposable elements enables them to be easily lost or inactivated by RIP mutation during sexual reproduction, which occurs prolifically on canola stubble. This genomic plasticity appears to be the basis of the rapid evolution of virulence in populations of this fungus and the consequent propensity of the fungus to cause resistance breakdown. In view of the unusual isochore structure of the genome of *L. maculans*, it is essential to know whether the genomes of four other members of the *L. maculans/L. biglobosa* species complex had a similar structure. This species complex comprises lineages that infect cruciferous plants. At least two lineages also infect *B. napus*, *L. biglobosa* "brassicae" and *L. biglobosa* "Canadensis," but the symptoms are much less severe than that those caused by *L. maculans* "brassicae" (*L. maculans* "brassicae" is termed *L. maculans* for simplicity) and stem cankering is seldom reported (Mendes-Pereira et al. 2003; Voigt et al. 2005).

2.8.3 *Leptosphaeria* species Genomes

The genomes of *L. biglobosa* "brassicae," *L. biglobosa* "Canadensis," *L. biglobosa* "thlaspii," *L. maculans* "lepidii," and another isolate of *L. maculans* "brassicae" were sequenced (Grandaubert et al. 2014). The genomes of all species encode similar numbers of small secreted proteins (ca. 650–750), which include putative effectors. However, *L. maculans* "brassicae" is the only member with high amounts of repetitive DNA and an isochore structure. The genomes of the four other members are compact (30–32 Mb), with transposable elements (4% of the genome) mainly restricted to chromosome ends. The transposons identified across the Leptosphaeriacae contain both class I (LTR retrotransposons) and class II (DNA transposons) types, with a similar number of families in each class (Grandaubert et al. 2014). Analysis of a range of ascomycete genomes showed that these transposable element sequences are restricted to the dothideomycete order Pleosporales, with the exception of two families. Alignment of 19 protein sequences from 55 fungal taxa revealed that the *Leptosphaeria* species diverged from sampled members of Pleosporales about 73,000,000 years ago (MYA). The time required for acquisition of transposable elements during species divergence was determined (Fig. 2.19). Phylogenetic analysis of the transposable element families revealed a relatively constant rate of introduction of families during species divergence over the past 100,000,000 years. At some point after the divergence of *L. maculans* "brassicae" from *L. maculans* "lepidii" (5.1 MYA) four transposon families proliferated within the *L. maculans* "brassicae" genome (Fig. 2.19). If the repeat sequences are ignored, these two fungi have nearly perfect chromosomal synteny. It is unclear what triggered the rapid proliferation of transposons in *L. maculans* "brassicae"; three of these four families are found across the Pleosporales but in

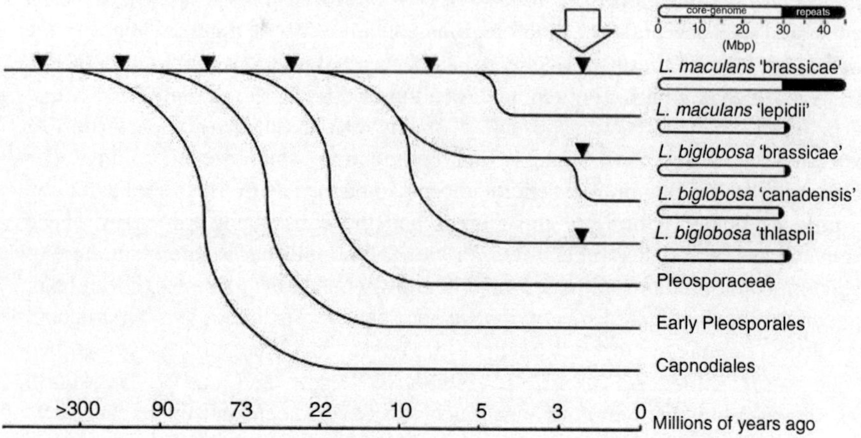

Fig. 2.19 Timing of divergence of species and invasion of transposons in five members of the *L. maculans/L. biglobosa* species complex. Each introduction of one or more transposon families is marked with a single triangle on the phylogeny diagram. A major invasion/proliferation (indicated by arrow) of four families of transposons into *L. maculans* "brassicae" occurred after divergence of this species from *L. maculans* "lepidii." The amount of repeat sequence in each genome is drawn below each species name; white shading indicates the nonrepetitive core regions of the genome and black shading represents repetitive sequences comprising mainly transposable elements. Approximate timing of divergence events is marked on a nonlinear scale bar. (All data are from Grandaubert et al. 2014 and Howlett et al. 2015)

relatively low numbers (Grandaubert et al. 2014). The differences in genome structure, particularly in terms of transposable element content and distribution, correlates with the ability of *L. maculans* "*brassicae*" to evolve, cause disease and adapt better to its hosts than other members of this species complex (Howlett et al. 2015).

2.8.4 Identification of a New Subclade of *Leptosphaeria* from *Brassica*

The *Leptosphaeria maculans* and *L. biglobosa* are two closely related fungal species that together forma species complex that causes blackleg or Phoma stem canker of crucifers, including *Brassica napus*, *B. juncea*, *B. oleracea*, and *B. rapa*. The *L. maculans* and *L. biglobosa* were previously described as virulent and weakly virulent, respectively, with genetic differences and distinct phenotypic (disease) expression on oilseed rape (*B. napus*) leaves or stems. During infection of *B. napus*, *L. maculans* produces large gray/green leaf lesions, and then grows down the vascular tissue to the stem, where the fungus causes necrotic stem cankers. In contrast, *L. biglobosa* causes small, dark leaf lesions and typically is restricted to infection of the upper stem. Consequently, *L. maculans* has been considered much more damaging than *L. biglobosa* (Fitt et al. 2006a).

The *Leptosphaeria biglobosa* and *L. maculans* have been estimated to have diverged approximately 22 million years ago based on comparative genome sequencing analysis of 19 conserved proteins. The *L. maculans* has been categorized into two subclades: "Brassicae" and "lepidii"; while *L. biglobosa* isolates have been categorized into six subclades: "Brassicae," "canadensis," "thlaspii," "erysimii," "australensis," and "occiaustralensis". The *L. biglobosa* "brassicae" is the most common of the *L. biglobosa* species and has been found in most oilseed rape growing regions. Although isolates of *L. biglobosa* are often found in association with *L. maculans*, *L. biglobosa* "brassicae" is the sole subspecies that has been identified in China to date.

The *Leptosphaeria biglobosa* "canadensis" has been isolated from both *B. napus* and *B. juncea* crops in both Canada and Australia, and is most closely related to *L. biglobosa* "brassicae." The remaining four *L. biglobosa* subspecies are less common and have only been reported in specific situations: *L. biglobosa* "thlaspii" was obtained from the cruciferous weed, *Thlaspi arvense*, in central Canada; *L. biglobosa* "erysimii" was obtained from a weed of an *Erysimum* sp., also in Canada; *L. biglobosa* "australensis" has been isolated from *B. napus* and *B. juncea* in the United States of America and Australia, and *L. biglobosa* "occiaustralensis" has been isolated mainly from oilseed rape cultivars with resistance derived from *B. rapa* sp. *sylvestris*, and wild radish (*Raphanus raphanistrum*) in western Australia, Chile, and Georgia. The *L. maculans–L. biglobosa* species complex had been separated into seven distinct groups using the sequence of the ITS region. In these studies, ITS rDNA, and *actin* and *b-tubulin* sequences have been widely used for the phylogeny analysis of the *Leptosphaeria* complex and new subclade identification. Parsimony and distance analysis separated these species with groups corresponding to specific host plants and geographic origin. In 2008, Vincenot et al. (2008) identified a new subclade of *L. biglobosa* (*L. biglobosa* "occiaustralensis") using ITS rDNA, *actin*, and *b-tubulin* sequences in Western Australia. The phylogeny of *L. biglobosa* isolates on the American continent has also been analyzed by using ITS rDNA, *actin*, and *b*-tubulin sequences. *L. maculans* has spread into many areas where only *L. biglobosa* was present, suggesting that *L. biglobosa* may have evolved earlier than *L. maculans* from a common ancestor (Zou et al. 2019).

The isolates of a proposed new subclade of *L. biglobosa* obtained from a certified organic *B. rapa* vegetable seed crop of the cv. Mizspoona, grown in the Willamette Valley of Oregon, USA in 2013–2014 that had developed classic symptoms of blackleg have been characterized by Zou et al. (2019). Pathogenicity and molecular characterization clearly identified these isolates as belonging to a distinct subclade compared to isolates of the six known subclades of this species. They proposed naming of the new subclade as *L. biglobosa* "americensis." Relationships between members of the two *Leptosphaeria* species, *L. maculans* and *L. biglobosa*, based on three conserved DNA regions yielded fairly consistent phylogenetic trees and were extremely comparable to earlier published studies (Fig. 2.20). Earlier studies found that isolates of the subclade, *L. biglobosa* "occiaustralensis," were most closely related to isolates of *L. biglobosa* "canadensis," while isolates of *L. biglobosa* "australensis" and "erysimii" were closely related to each other. The topologies of

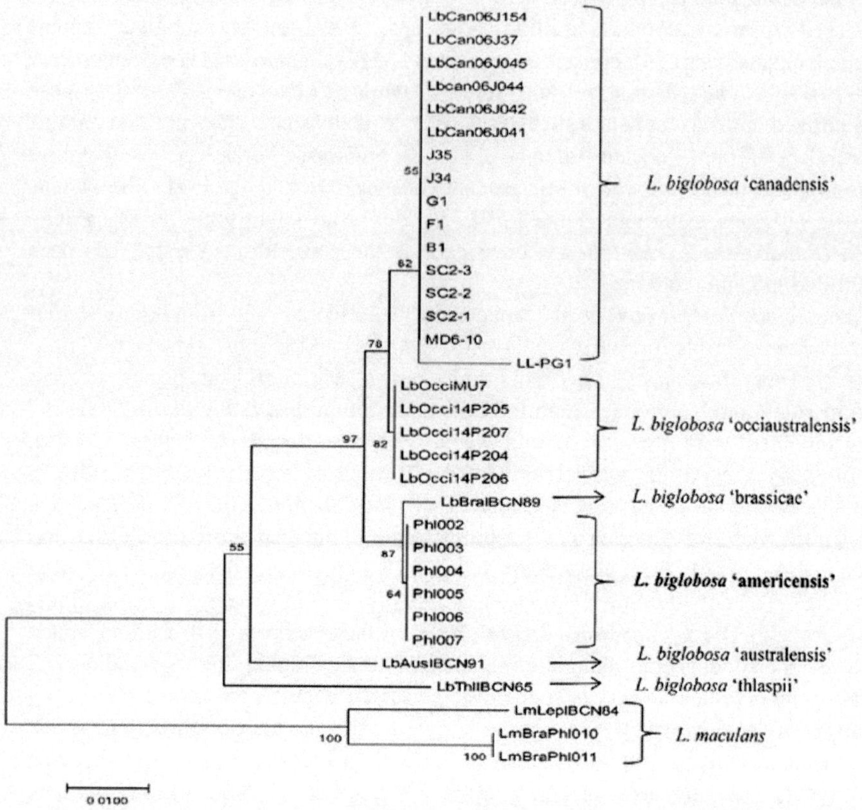

Fig. 2.20 Phylogenetic analysis of the *Leptosphaeria maculans–L. biglobosa* species complex based on maximum parsimony analysis of the concatenated sequences of the internal transcribed spacer (ITS) region of ribosomal DNA, *actin*, and *β-tubulin* (one of two parsimony trees). Bootstrap values (1000 replications) are indicated as a percentage above the node. The length of the tree is 449 steps (CI = 0.84, RI = 0.91). The tree was similar to trees constructed by neighbor-joining (NJ) or maximum-likelihood (ML) analysis. Three *L. maculans* isolates were included as out group control (Zou et al. 2019)

these subclades are consistent with the data generated. In addition, *L. biglobosa* "thlaspii" isolates were most closely related to *L. maculans* isolates in the phylogenetic trees constructed. The distance and parsimony analysis on concatenated sequences or a single gene sequence with high bootstrap value supported that subclades including the six *L. biglobosa* isolates is indeed a new subspecies. The *L. biglobosa* "canadensis" was close to *L. biglobosa* "occiaustralensis" through concatenated/single gene sequence analysis. The six *B. rapa* derived isolates formed an independent group and was mostly close to *L. biglobosa* "brassicae" when using concatenated sequences of three conserved genes and consistent with those trees constructed by single sequence of ITS rDNA or *actin*.

The inconsistent tree obtained by phylogeny analysis using the *b-tubulin* sequence, which showed the new six isolates as closer to *L. biglobosa* "canadensis," reinforced that these six isolates cannot be categorized into either "brassicae" or "canadensis" and formed a distinct group. To eliminate the bias of the phylogenetic tree constructed by the single gene sequence, concatenated sequences of three conserved genes were applied for parsimony analysis; these six isolates formed a separate subclade that was closer to *L. biglobosa* "brassicae." More genomic or transcription data will provide further insights on the characterization of the new subclade in the future for investigation of the difference between this new subclade and other subclades of *L. biglobosa*. Trees obtained through parsimony analysis were similar to those obtained by the NJ method, and sufficient bootstraps supported the hypothesis that the *B. rapa*-derived isolates formed amonophyletic clade. Phylogenic analysis using NJ, parsimony, maximum-likelihood, and minimum evolution analysis resulted in trees of similar topologies, with greater bootstrap values for the NJ trees compared with the other trees. It remains unclear why the different subspecies of *L. biglobosa* are so distinct from one another.

With the exception of *L. biglobosa* "brassicae" and *L. biglobosa* "canadensis," generally, few isolates have been cultured from each of the subspecies, making population surveys difficult. Therefore, limited information is known regarding gene flow across geographical regions. Isolates of *L. biglobosa*, which can coexist on host plants with isolates of *L. maculans*, typically are associated with upper stem lesions on infected *Brassica* plants, and are considered weakly virulent or even avirulent. The *L. biglobosa* isolates have only caused significant losses in areas with high summer temperatures, in Poland. Vincenot et al. (2008) demonstrated that *L. biglobosa* "occiaustralensis" isolates were more virulent than *L. biglobosa* "canadensis" or "australensis" isolates, and *L. biglobosa* "brassicae" isolates produced large leaf lesions on *B. napus* plants. The pathogenicity of *L. biglobosa* "americensis" isolates on *Brassica* spp. indicated that these isolates were intermediate in virulence on *B. napus* and caused more severe disease on *B. juncea* and *B. rapa*. Van de Wouw et al. (2008) reported that *L. biglobosa* "canadensis" isolates induced similar size lesions on *B. juncea* cotyledons as *L. maculans* isolates.

In contrast, *L. biglobosa* "americensis" isolates caused significantly larger lesions than *L. biglobosa* "canadensis" isolates on both resistant and susceptible lines of *B. napus*, and large lesions on *B. juncea* and *B. rapa* lines. These results were confirmed by the accumulation of lignified material observed in stained, infected cotyledons, with more diffuse and less lignified material in the cotyledons caused by *L. biglobosa* "americensis" and *L. maculans* isolates than isolates of other subclades of *L. biglobosa*. It remains unclear why the different subspecies of *L. biglobosa* show different reactions on different *Brassica* species, however, it raises the question of whether effector genes are present in these *L. biglobosa* subspecies and whether they may play a role in the pathogenicity of the various *Brassica* species. The genome sequencing of a subset of *L. biglobosa* species has shown different suites of effector-like proteins, thus supporting this hypothesis, however, the functions of these effector-like proteins remain unknown in these subspecies (Ghanbarnia et al. 2015).

In contrast to isolates of other *L. biglobosa* subclades, *L. biglobosa* "americensis" isolates characterized represent the first *L. biglobosa* subclade derived from *B. rapa* that caused severe symptoms on *B. rapa*, *B. oleracea*, and *B. juncea* cultivars. The evolution of *Leptosphaeria* isolates appears to be influenced by internal and external forces, including mutation, reproduction, gene flow, genetic drift, natural selection (temperature and geography), and host specificity. Unlike the well understood gene-for-gene relationship of R genes and corresponding avirulence genes in the *B. napus*–*L. maculans* pathosystem, it remains unclear whether a gene-for-gene interaction exists among *L. biglobosa* isolates and *Brassica* species. The study validated and enriched the known diversity of *L. biglobosa* isolates, and provides a resource not only to help determine whether the evolutionary process of *L. biglobosa* isolates is influenced by geographic location and host selection pressures, but also to underline the possible existence of plant defense responses among *Brassica* species and *L. biglobosa* isolates. As reflected by the more severe disease symptoms associated with the *L. biglobosa* "americensis" isolates on cultivars of three *Brassica* spp. evaluated compared to isolates of other subclades, further research is needed to investigate potential differences in resistance to *L. biglobosa* "americensis" isolates vs. isolates of other subclades, and to find sources of resistance to isolates of the different subclades of *L. biglobosa* (Balesdent et al. 2005; Rouxel et al. 2011; Zou et al. 2019).

2.8.5 Gene Loss in *Leptosphaeria maculans*

Genome-wide gene loss detection in three *L. maculans* isolates has been observed. Golicz et al. (2015) utilized a hybrid mapping and genome assembly approach to investigate gene content variation in three isolates of *L. maculans*, an important pathogen of canola. The results suggest that genes involved in a number of biological processes can be influenced by gene loss. These include genes involved in oxidative processes, DNA maintenance, cell signaling and sexual reproduction.

Gene content variation was analyzed by a combination of read mapping and assembly. The genome assemblies obtained for isolates 21 and 41 are smaller in size compared to the published reference sequence which is 45 Mbp in length (Rouxel et al. 2011). This may result from different assembly algorithms used. Roughly 30% of the *L. maculans* genome is composed of repetitive elements (transposons) (Rouxel et al. 2011) which are unlikely to be assembled by Velvet. The v23.1.3 isolate differs by a total of 28 genes in isolate 21 and 52 genes in isolate 41. The higher number of differences between isolate v23.1.3 and isolate 41 is consistent with the published observations, suggesting that isolate 41 is a more distant relative of v 23.1.3 than isolate 21 (Zander et al. 2013). Two classes of genes found in the *L. maculans* genome have previously been shown to display presence/absence variation: the mating type (MAT) locus and avirulence (*Avr*) genes, and these genes were used to verify results obtained. *L. maculans* contains a single mating type (*MAT*) locus, with two alternate forms, which must be different for two isolates to mate (Venn 1979). The *MAT1*-1 is 1368 bp long, containing a 45-bp intron and

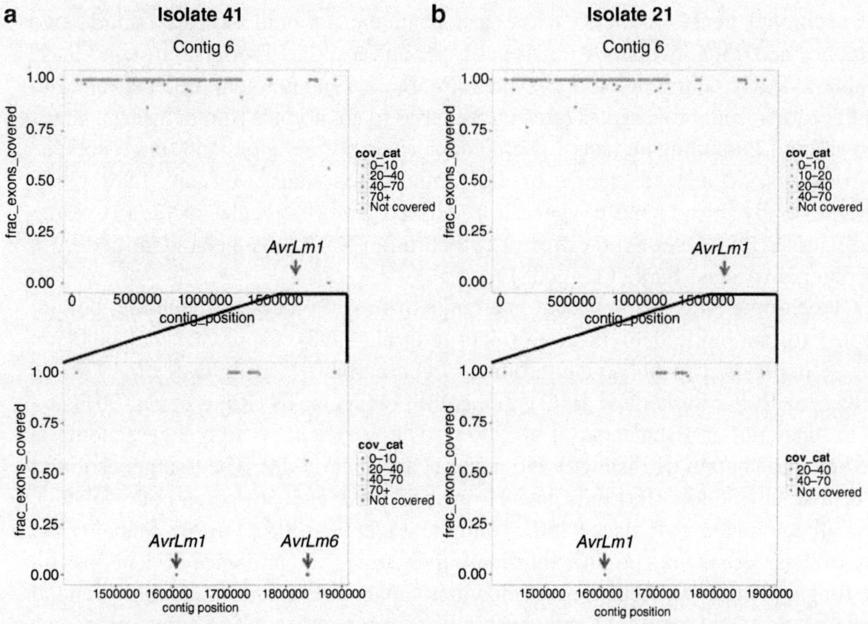

Fig. 2.21 *AvrLm* gene loss in *Leptosphaeria maculans*. Plots visualizing lost *AvrLm* genes in isolates 41 and 21. Each dot represents a gene. Coordinate on the *x*-axis is the position along the contig. Coordinate on the *y*-axis is the fraction of the total number of positions within exons of the gene covered by mapped reads. Coverage category (*cov_cat*) is the average coverage depth across exons of the gene. (**a**) Isolate 41 shows loss of two *AvrLm* genes present on super contig 6—*AvrLm1* and *AvrLm6*. (**b**) Isolate 21 shows loss of one *AvrLm* gene present on super contig 6—*AvrLm1* (Golicz et al. 2015)

encoding a protein of 441 amino acids. The *MAT1*-2 is 1246 bp, encoding a predicted protein of 397 amino acids (Cozijnsen and Howlett 2003). The reference sequence isolate v23.1.3 is known to contain *MAT1*-2 (Fitt et al. 2006a, b). The results found suggest that isolate 41 is the opposite *MAT1*-1 mating type. The results were validated using PCR. The PCR results confirmed in silico gene loss predictions. Two of the predicted lost genes are classified as avirulence genes, which have previously been shown to display presence/absence variation between isolates (Fudal et al. 2009; Gout et al. 2006, 2007; Van de Wouw et al. 2010; Zander et al. 2013). It suggests that both isolates 21 and 41 demonstrate differential *AvrLm* gene loss (Fig. 2.21), with *AvrLm1* predicted to be absent in both isolates, *AvrLm4*-7 and *AvrLm11* present in both isolates, while *AvrLm6* is present in isolate 21, but absent in isolate 41. The results are in full agreement with published observations of the virulence specificity of these isolates (Raman et al. 2012; Zander et al. 2013).

The loss of *AvrLm1* is most likely due to a larger deletion of a whole genomic region including the AvrLm1 locus (Fudal et al. 2009; Gout et al. 2007). The loss of *AvrLm6* might be due to a larger genomic deletion or repeat induced point (RIP) mutations, which can lead to functional loss of a gene (Fudal et al. 2009; Van de Wouw et al. 2010).

Additional genes predicted to be lost from one or both isolates include two cytochrome P450s, two DNA repair helicases and a serine threonine protein kinase (Table 2.8). Cytochrome P450 enzymes (P450s) are hemthiolate proteins found in all life forms from prokaryotes (archea, bacteria) to eukaryotes (fungi, insects, plants and animals, including humans). These enzymes catalyze region- and stereo-specific conversions of a wide range of lipophilic compounds to more hydrophilic derivatives by introducing oxygen atom derived from molecular oxygen (Cresnar and Petric 2011). Over 8700 different cytochrome P450s have been identified from 113 fungal species (Park et al. 2008).

Cytochrome P450 is implicated in a range of fungal processes including, but not limited to, antibiotic detoxification (George et al. 1998), naphthalene metabolism and detoxification of polycyclic aromatic hydrocarbons (Cerniglia et al. 1978; da Silva et al. 2004; Sutherland 1992), gibberellin biosynthesis (Rojas et al. 2001) and toxin biosynthesis (Bhatnagar et al. 2003). The *L. maculans* reference genome is predicted to contain 62 distinct P450 genes (Park et al. 2008). The two genes found to be lost, G085580.1 (lost in isolate 41 only) and G085620.1 (lost in both isolates), are both annotated as E class P450, group I (Park et al. 2008). Further characterization of these genes may provide information regarding the influence of gene loss on the fungal phenotype. Helicases are enzymes capable of separating double-stranded nucleic acid (DNA or RNA) and are involved in a number of biological processes including replication, recombination, transcription and translation. The number of helicase genes in eukaryotic organisms is relatively high, with approximately 1% of genes in eukaryotic genomes predicted to encode RNA or DNA helicases (Wu 2012). Furthermore, redundancy of genes coding for helicases has been observed. In rice, a total of 115 helicase genes have been identified, several of which were redundant (Umate et al. 2011) suggesting that gene loss may affect helicase genes without a pronounced effect on the phenotype. Genes coding for protein kinases belong to large families and their products possess diverse functions. They are also thought to exhibit a rapid birth–death cycle (Hardie 1999; Rudrabhatla et al. 2006; Zheng et al. 2011). Therefore, protein kinases are expected to be affected by gene loss. The avirulence (*AvrLm*) genes are subject to presence/absence variation. The gene loss influences additional genes in *L. maculans*, including genes possibly involved in oxidative processes, DNA maintenance, cellular signaling and sexual reproduction. The spatial distribution of genes lost suggests that genes can be lost either in isolation or as consecutive losses of multiple genes presumably as a part of larger deletion (Golicz et al. 2015).

Six blackleg isolates recovered from *Brassica rapa* cv. Mizspoona in the Willamette Valley of Oregon were characterized as *L. biglobosa* based on standard pathogenicity tests and molecular phylogenetic analysis. These isolates were compared to 88 characterized *L. biglobosa* isolates from western Canada, 22 isolates from Australia, and 6 *L. maculans* isolates from Idaho, USA using maximum parsimony and distance analysis of phylogenetic trees generated from the ITS rDNA (internal transcribed spacer rDNA) sequence, and the *actin* and β-*tubulin* gene sequences. The *Leptosphaeria biglobosa* isolates derived from *B. rapa*

Table 2.8 List of genes identified as lost in re-sequenced *Leptosphaeria maculans* isolates relative to the reference genome (Golicz et al. 2015)

Isolate 21	Protein description	Isolate 41	Protein description
		G030170.1[a]	Predicted protein
		G037480.1[a]	Predicted protein
		G043120.1[a]	Predicted protein
G046740.1[a]	Predicted protein		
G049660.1	M1 protein (*AvrLm1*)	G049660.1	M1 protein (AvrLm1)
		G049920.1[a]	Predicted protein
		G049930.1[a]	Predicted protein
		G049940.1	M6 protein (*AvrLm6*)
G076370.1[a]	Predicted protein		
		G070000.1[a]	Histidine triad
G079780.1[a]	Predicted protein	G079780.1[a]	Predicted protein
G081630.1[a]	Predicted protein	G081630.1[a]	Predicted protein
G081640.1[a]	Predicted protein	G081640.1[a]	Predicted protein
G082370.1[a]	Predicted protein		
G082540.1[a]	Glucose-6-phosphate isomerase	G082540.1[a]	Glucose-6-phosphate isomerase
G085580.1	Cytochrome p450	G085580.1	Cytochrome p450
G085590.1	Transferase family protein	G085590.1	Transferase family protein
G085610.1[a]	Predicted protein	G085610.1[a]	Predicted protein
G085620.1	Benzoate 4-monooxygenase cytochrome p450	G085620.1	Benzoate 4-monooxygenase cytochrome p450
		G085630.1	Carboxylesterase family protein
		G086520.1	DNA repair helicase
		G086530.1	DNA repair helicase
G090230.1[a]	Predicted protein	G090230.1[a]	Predicted protein
		G098080.1	Serine threonine protein kinase
		G098090.1	Predicted protein
		G103490.1[a]	Phosphotransferase family protein
G104510.1[a]	Predicted protein	G104510.1[a]	Predicted protein
		G109260.1	Het domain-containing protein
G110260.1[a]	Predicted protein	G110260.1[a]	Predicted protein
G113050.1	Tpr domain protein	G113050.1	Tpr domain protein
G113060.1	Predicted protein	G113060.1	Predicted protein
G113080.1	Predicted protein	G113080.1	Predicted protein
		G114280.1	Predicted protein
		G114380.1	Pyridoxamine phosphate oxidase
		G114390.1	Mating-type mat1-2 protein
G116710.1	Glycosyltransferase family 34 protein		
G123850.1[a]	Predicted protein	G123850.1[a]	Predicted protein

The *Avr* genes are shown in italics

[a] Genes deemed less confident by the authors (length <300 aa and without EST evidence at the time of prediction pipeline)

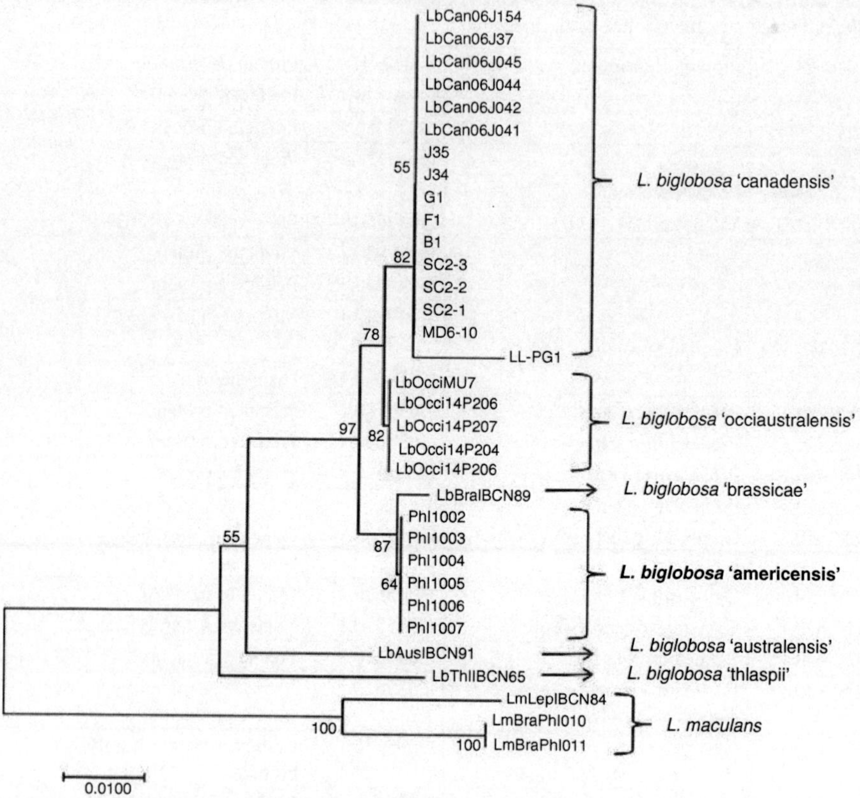

Fig. 2.22 Phylogenetic analysis of the *Leptosphaeria maculans–L. biglobosa* species complex based on maximum parsimony analysis of the concatenated sequences of the internal transcribed spacer (ITS) region of ribosomal DNA, *actin*, and β-*tubulin* (one of two parsimony trees). Bootstrap values (1000 replications) are indicated as a percentage above the node. The length of the tree is 449 steps (CI = 0.84, RI = 0.91). The tree was similar to trees constructed by neighbor-joining (NJ) or maximum-likelihood (ML) analysis. Three *L. maculans* isolates were included as out-group control isolates (Zou et al. 2019)

collected in Oregon formed a separate subclade based on concatenated gene sequences or a single gene sequence, regardless of the analysis (Fig. 2.22).

Pathogenicity tests showed that these isolates failed to infect either resistant or susceptible *B. napus* cultivars, but caused severe symptoms on three *B. rapa* cultivars (Accession number: UM1113, UM1112, and UM1161), a *B. oleracea* var. *capitata* (cabbage) cultivar (Copenhagen Market), and two *B. juncea* cultivars (CBM, a common brown Mustard, and Forage). These findings demonstrated that the *L. biglobosa* isolates derived from a *B. rapa* crop in Oregon were genetically distinct from existing species of *L. biglobosa*, and constitute a new subclade, herein proposed as *L. biglobosa* "americensis" by Zou et al. (2019).

2.8.6 Contrasting Genomic Organization of *Leptosphaeria* species

The *Leptosphaeria maculans* "*brassicae*" (*Lmb*) and *Leptosphaeria maculans* "*lepidii*" (*Lml*) are closely related phytopathogenic species that exhibit a large macrosynteny but contrasting genome structure. The *Lmb* has more than 30% of repeats clustered in large repeat rich regions, while the *Lml* genome has only a small amount of evenly distributed repeats. Repeat-rich regions of *Lmb* are enriched in effector genes, expressed during plant infection. The distinct genome structures of *Lmb* and *Lml* provide an excellent model for comparing the organization of pathogenicity genes in relation to the chromatin landscape in two closely related phytopathogenic fungi. The performed chromatin immune-precipitation (ChIP) during axenic culture, targeting histone modifications typical for hetero-chromatin or euchromatin, combined with transcriptomic analysis to analyze the influence of chromatin organization on gene expression was reported by Soyer et al. (2021). In both species, it was found that facultative heterochromatin is enriched with genes lacking functional annotation, including numerous effector and species-specific genes. Notably, orthologous genes located in H3K27me3 domains are enriched with effector genes. Compared to other fungal species, including *Lml*, *Lmb* is distinct in having large H3K9me3 domains associated with repeat-rich regions that contain numerous species-specific effector genes. Discovery of these two distinctive heterochromatin landscapes now raises questions about their involvement in the regulation of pathogenicity, the dynamics of these domains during plant infection and the selective advantage to the fungus to host effector genes in H3K9me3 or H3K27me3 domains (Table 2.9; Figs. 2.23 and 2.24; Soyer et al. 2021).

2.9 *Brassica–Plasmodiophora* Genome

2.9.1 Introduction

Clubroot, caused by the obligate telluric biotroph protist *Plasmodiophora brassicae* Wor., is one of the economically most important diseases affecting *Brassica* crops in the world. The pathogen leads to severe yield losses worldwide. Clubroot development is characterized by the formation of galls on the roots of infected plants. Above-ground symptoms include wilting, stunting, yellowing and premature senescence. The pathogen has a complex life cycle comprising three stages: survival in soil as resting spores, root hair infection, and cortical infection. First, resting spores can survive for a long period (more than 20 years) in the soil. Second, germination of haploid resting spores releases primary zoospores which infect root hairs, leading to intracellular haploid primary plasmodia. In the third stage, after nuclear division, secondary zoospores may be released into the soil and penetrate the cortical tissues. The step of released zoospores in the soil before invading the root cortex is not universally accepted among the research community. During this last stage, the pathogen develops into secondary multinucleate diploid plasmodia which cause the hypertrophy and hyperplasia of infected roots into characteristic clubs.

Table 2.9 Number of proteinaceous effector genes located in different genomic compartments in *Leptosphaeria maculans brassicae* and *Leptosphaeria maculans lepidii* (Soyer et al. 2021)

	Number of genes	Effector genes			Specific effector genes		
		Number	Proportion	*P* value[a]	Number	Proportion	*P* value[a]
L. maculans brassicae							
Genome	13,047	1080	8.2×10^{-2}	–	860	6.2×10^{-2}	–
TE associated genes[b]	2466	289	1.2×10^{-1}	5.6×10^{-10}	210	8.5×10^{-2}	1.4×10^{-6}
H3K4me2domains[c]	7373	433	5.8×10^{-2}	6.7×10^{-14}	319	4.3×10^{-2}	4.1×10^{-11}
H3K9me3domains[c]	104	38	3.6×10^{-1}	2.2×10^{-16}	35	3.4×10^{-1}	2.2×10^{-16}
H3K27me3domains[c]	2020	286	1.4×10^{-1}	2.2×10^{-16}	200	9.9×10^{-2}	3.6×10^{-12}
H3K9 + H3K27me3domains[c]	101	24	2.4×10^{-1}	1.6×10^{-8}	16	1.6×10^{-1}	5.5×10^{-5}
L. maculans lepidii							
Genome	11,272	892	7.9×10^{-2}	–	618	5.5×10^{-2}	–
TE associated genes[b]	641	79	1.2×10^{-1}	3.7×10^{-8}	66	1×10^{-1}	$3.4 \times .10^{-11}$
H3K4me2domains[c]	6065	266	4.4×10^{-2}	3.9×10^{-14}	163	2.7×10^{-2}	5.8×10^{-14}
H3K9me3domains[c]	70	14	2×10^{-1}	1.3×10^{-5}	14	2×10^{-1}	1.8×10^{-9}
H3K27me3domains[c]	1501	217	1.45×10^{-1}	2.2×10^{-16}	152	1×10^{-1}	2.2×10^{-16}
H3K9 + H3K27me3domains[c]	58	18	2.8×10^{-13}	1.6×10^{-8}	14	2.4×10^{-1}	2.2×10^{-16}

[a] A χ^2 test was applied to compare proportion of effector genes in the genome and in the genomic compartment analyzed

[b] Genes located up to 2 kb upstream or downstream of a transposable element sequence

[c] Genes located in a H3K4me2, H3K9me3, H3K27me3 or H3K9me3/H3K27me3 domain in vitro

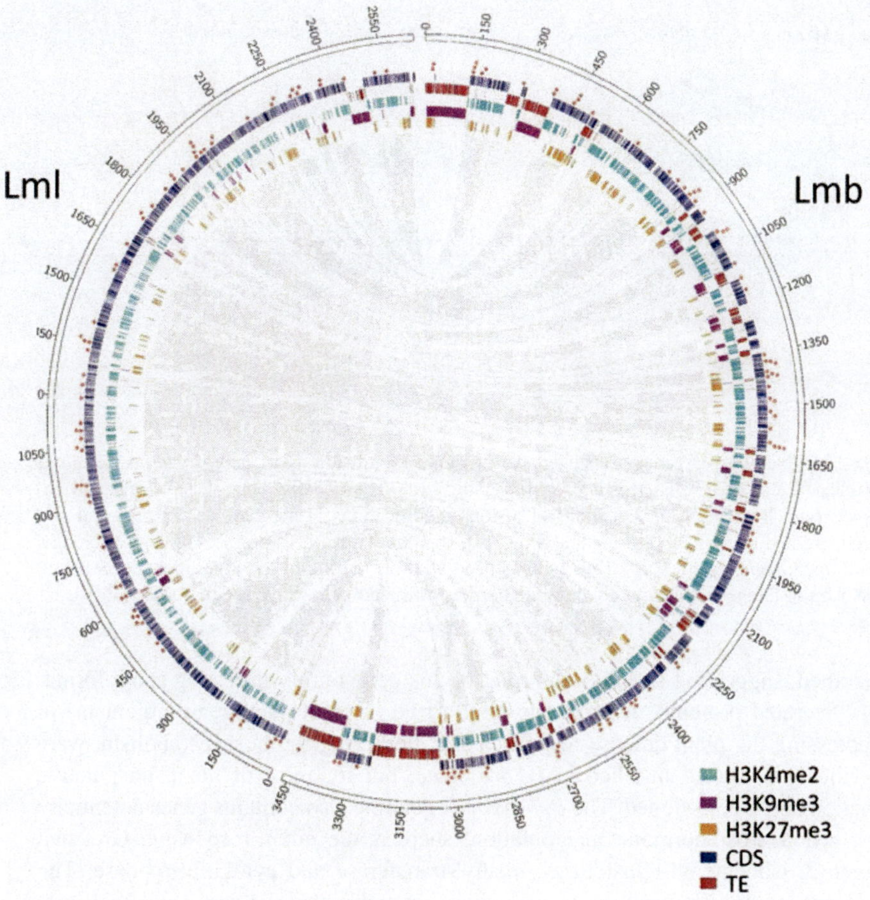

Fig. 2.23 Genome of *Leptosphaeria maculans* "*brassicae*" harbors large TE-rich, H3K9me3 domains compared to *Leptosphaeria maculans* "*lepidii*." Example of Super Contig 2 of *Lmb* and Scaffold 1 of *Lml*. ChIP-seq was performed with antibodies targeting H3K4me2 (light blue), H3K9me3 (purple) or H3K27me3 (orange); rectangles indicate location of significantly enriched domains identified using RSEG (Song and Smith 2011). Location of CDS (dark blue); transposable elements (TE, red); H3K4me2 (light blue); H3K9me3 (purple); H3K27me3 (orange); *Lmb*, *L. maculans* "*brassicae*"; *Lml*, *L. maculans* "*lepidii*." Genes encoding proteinaceous effectors are indicated with a red square (Soyer et al. 2021)

Secondary plasmodia finally release a new set of haploid resting spores. Several *P. brassicae* genomes are available, including one reference genome from a European isolate, and additional re-sequencing of five Canadian and one Chinese isolates.

The pathotype classification of these isolates is difficult to compare because they have been characterized using different differential hosts sets. In previous studies, the main characteristics of the *P. brassicae* genome were described, such as its small size (24.2–25.5 Mb) and high gene density. Thirteen chitin synthases were

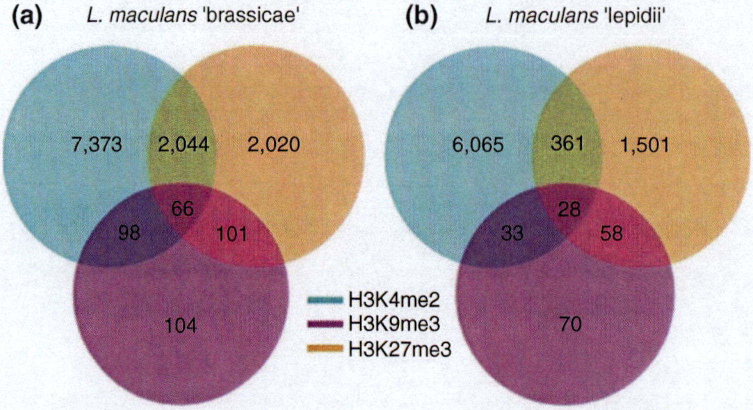

Fig. 2.24 Number of genes associated with histone modifications in *Leptosphaeria maculans* "brassicae" and *Leptosphaeria maculans* "lepidii." (**a**) *L. maculans* "brassicae," Lmb and (**b**) *L. maculans* "lepidii," Lml. Locations of histone modifications in the genomes of *Lmb* and *Lml* were identified using RSEG (Song and Smith 2011). Blue: H3K4me2; purple: H3K9me3; orange: H3K27me3. Genes were considered as associated with any of the histone modifications when at least 1 bp of the gene was found within the borders of the domain (Soyer et al. 2021)

described, suggesting an important role for this gene family in resting spore formation. Secreted proteins, described to play a role in plant pathogenic organisms in suppressing the plant defense responses and modifying the host metabolism, were previously predicted in silico in *P. brassicae*, but for most of them, no putative function could be assigned. The *P. brassicae* genome also contains genes potentially involved in host hormone manipulation, such as the auxin-responsive Gretchen Hagen 3, isopentenyl-transferases, methyl-transferase and cytokinin oxidase. The clubroot genome also appears to be lacking several metabolic pathways, a characteristic of the eukaryotic biotrophic plant pathogens. The missing genes encode proteins involved in sulfur and nitrogen uptake, as well as the arginine, lysine, thiamine and fatty acid biosynthesis pathways. In addition, only a few carbohydrate active enzymes (CAZymes), involved in the synthesis, metabolism and transport of carbohydrates, were found. Schwelm et al. (2016) summarized the current life stage specific transcriptomics data related to the molecular events in the lifecycle of *P. brassicae*. During spore germination and at the primary zoospore stage, enzymes involved in chitinous cell wall digestion are highly expressed. Several metabolism pathways such as the starch, citrate cycle, pentose phosphate pathway, as well as pyruvate and trehalose metabolism were also very active. Although the molecular mechanisms involved in primary root hair infection are unknown, the secondary phase has been better described.

During re-infection by the secondary zoospores in root hairs or the root cortex, genes involved in basal and lipid metabolism were highly expressed. The plant clubroot symptoms (hyperplasia and hypertrophy of the infected tissues leading to the gall phenotype) were found to result from the deregulation of the host plant primary and secondary metabolism as well as from the modification of plant hormone homeostasis by the pathogen. The main plant hormone pathways modified were cytokinin biosynthesis, auxin homeostasis, and salicylic acid and jasmonic acid metabolism. However, the mechanisms by which the pathogen may drive plant hormone homeostasis are still not completely deciphered. At the end of the *P. brassicae* life cycle, secondary plasmodia transform into resting spores. In these structures, trehalose could play an important role for their long-term survival capacity. This description of the current transcriptomics data in *P. brassicae* shows that some important steps in *P. brassicae* developmental stages are still not well-characterized at the molecular level. Daval et al. (2019) reported the genome sequence of the European eH *P. brassicae* isolate belonging to a pathotype widespread in France but different to e3, the reference clubroot isolate. For the first time, the mitochondrial genome of this clubroot pathogen was described. The second aim was to improve genome annotation, to predict new metabolic pathways and to develop Clubroot Cyc, a metabolic pathway data base for *P. brassicae*. This new tool gives researchers access to a repertoire of *P. brassicae* metabolic and transport pathways as well as to an Omics Viewer to analyze expression data, which can be updated with new data.

2.9.2 Early Phase of *Plasmodiophora brassicae* Genomics

During early phase of investigations, the number of chromosomes of *P. brassicae* has been a matter of debate for many years. Early light microscopy studies reported haploid numbers of $n = 4$ (Terby 1924), $n = 5$ (Tommerup and Ingram 1971), and $n = 8$ (Jones 1928). Therefore, karyotypic analysis based on serial sections of synaptonemal complexes in pachytene nuclei were carried out and revealed a haploid chromosomal number of 20 (Braselton 1982). Later studies using pulse field gel electrophoresis reduced the number of chromosomes to 6 (Bryan et al. 1996) and 13 (Ito et al. 1994). Both groups reported chromosomal sizes of ca 0.5–2.2 Mb.

During investigations on the genomic DNA of *P. brassicae*, the presence of host DNA in the resting spores and zoospores of *P. brassicae* was found (Bryngelsson et al. 1988). It was speculated that the continuous uptake of host DNA can be of importance for the evolution of *P. brassicae*. It may be the existence of a sequence which hybridized with *Brassica* DNA in the *P. brassicae* genomic DNA (Buhariwalla and Mithen 1995). The cloned sequence showed homology to a tandem repeat element from *B. nigra* (Gupta et al. 1990). However, there is no evidence at present that host DNA is stably integrated into the genome of *P. brassicae*, although the continuous uptake of DNA does provide a route by which horizontal DNA transfer may occur. Also the mechanism of DNA uptake by the pathogen remains to be elucidated. Plasmodia of *P. brassicae* have been observed to engulf host organelles and cytoplasm (Buczacki 1983), which may provide a means of host DNA uptake.

To detect resting spores by polymerase chain reaction (PCR), specific primers were developed for *P. brassicae* DNA. The PCR is a sensitive tool, which is increasingly used for the detection and identification of plant pathogens (Henson and French 1993). Several investigations have shown that it is possible to isolate genomic DNA from resting spores of *P. brassicae* (Bischoff 1994; Bryngelsson et al. 1988; Buhariwalla and Mithen 1995; Buhariwalla et al. 1995; Ito et al. 1994). The cloning of several putative repetitive/high copy number DNA elements led to the development of specific PCR primers for the amplification of polymorphic DNA from *P. brassicae* (Buhariwalla et al. 1995) and was the first step toward detection of fungal resting spores by molecular methods. Ito et al. (1997) have attempted to isolate ipt (isopentenyl transferase)-like sequences from *P. brassicae*, because it is known that plasmodia can convert adenine to transzeatin (Müller and Hilgenberg 1986). Although a *P. brassicae*-specific DNA sequence was obtained by using PCR primers based on the ipt sequence of *Agrobacterium tumefaciens*, it did not show any homology to ipt. The sequence was conserved among all *P. brassicae* isolates tested and, thus, can be used for the detection of the fungus. This set of primers can specifically amplify a single-copy DNA sequence from *P. brassicae* DNA, whereas attempts to amplify DNA from other fungi including *Polymyxa betae*, *Polymyxa graminis,* and *Olpidium brassicae* gave negative results and can be, thus, used as a diagnostic tool.

An approach for the isolation of differentially induced proteins during pathogenesis can be the separation of proteins from infected and control roots on two-dimensional gel electrophoresis and subsequent comparison of the pattern. Hansen et al. (1994) examined protein patterns of *Brassica oleracea* roots infected with *P. brassicae* and found that several proteins present in healthy roots were reduced in clubroots. Ito et al. (1996) investigated the two-dimensional protein pattern of a susceptible and resistant response in roots of Chinese cabbage infected with two different *P. brassicae* isolates. They observed proteins which were enhanced in the susceptible response, repressed in susceptible response, enhanced in the resistant response, and repressed in resistant response. N-terminal sequencing of a protein increased in the susceptible response revealed homology to pathogenesis-related protein group 5 (Ito et al. 1996). A previous study on PR-proteins during pathogenesis showed no clear correlation between the activity of peroxidase and chitinase isoenzymes in roots of a susceptible and a resistant cultivar of Chinese cabbage and resistance/tolerance to clubroot (Ludwig-Muller et al. 1994). However, in the susceptible interaction, a novel chitinase isoenzyme with a pi of 7.3 was induced (Ludwig-Muller et al. 1994).

It has been shown in the interaction of plants with arbuscular mycorrhizal fungi that it was possible to isolate differentially expressed genes from the host plant and the AM fungus (Martin-Laurent et al. 1997; Murphy et al. 1997). Using a PCR-coupled subtractive hybridization technique (Buchanan-Wollaston and Ainsworth 1997) differentially expressed genes during clubroot development in Chinese cabbage were isolated. Three independent differentially expressed clones were isolated from a subtractive cDNA library constructed from root tissue 28 days after inoculation and sequenced. One was shown to have homology to small heat shock proteins (Ludwig-Muller 1999).

2.9.3 Main Features of the Nuclear Genome of *Plasmodiophora*

For a better understanding of the pathogenesis and metabolic potential of *P. brassicae*, Daval et al. (2019) generated a de novo assembled genome of the eH isolate (belonging to pathotype 1. A paired-end shotgun library consisting of 73,082,688 reads (read length of 101 bp and inserts size of 430 bp) and approximately 225-fold genome sequence coverage was obtained. Two additional paired-end libraries were also generated. A 3 kb insert mate-pair Illumina library resulted in 97,629,456 reads (read length of 101 bp and insert size of 3200 bp) and a 5 kb insert mate-pair Illumina library resulted in 81,813,222 reads (read length of 101 bp and insert size of 5050 bp). Both libraries displayed also approximately 225-fold genome sequence coverage. The final assembly consisted of 136 gap-filled scaffolds with a N50 of 741, 293 bp and the total sequence length was 24,559,637 bp with a 59.29% GC content. Basic information about the assembled genome and predicted genes is shown in Table 2.10, in comparison with the assembly of e3 and ZJ-1 genomes, while the ZJ-1 resources corresponded to draft data, the assemblies of eH and e3 shared the same levels of quality. The main features of the eH genome were indeed similar to the reference e3 genome, particularly concerning GC content (Fig. 2.25), but the eH genome displayed a less fragmented assembly (Table 2.10) and a higher number of predicted genes (12,811 in eH and 9730 in e3). Figure 2.25 shows a visual representation of the eH genome's first 43 scaffolds (which represent 95% of the total genome sequence) compared to the other available *P. brassicae* sequences (Rolfe et al. 2016). The gene density, calculated as the number of genes every 20 kb, was nine on average. This gene density is relatively high as previously described, and contributes to the small genome size compared to the two known genomes of free-living Rhizaria, *Bigelowiella natans* (~100 Mb) and *Reticulomyxa filosa* (~320 Mb). Gene similarity between the eH and e3 isolates was studied by comparing the gene sequences of both genomes. Using a stringent query span of 80% (meaning that the percentage of alignment coverage of the query sequence should be at least 80%, making sure that genes with common regions), 60% of the genes found in the eH genome showed similarity with the e3 genes. Among the 6654 alignments displaying the query span of 80%, 4702 were with 100% gene identity and only 52 were with <60% gene identity (value often used to define genes as highly divergent). For the ZJ-1 isolate's genome, annotated regions are not available which made it unfeasible to carry out such accurate comparisons between genes and only contigs were compared which include noncoding regions. The scaffolds #44 to #242 (corresponding to 5% of the total genome) and in particular scaffolds #88 to #242 were more difficult to assemble and thus few genes were annotated in these regions.

Table 2.10 Genome statistics of *Plasmodiophora brassicae* assemblies. BUSCO analysis was done on genome sequences with the Ensembl Protists database (Daval et al. 2019)

Isolates	eH	e3 Schwelm et al. (2015)	ZJ-1 Bi et al. (2016)
Assembly accession	POCA00000000.1	GCA_001049375.1	GCA_002093825.1
Assembly assessment			
Genome assembly size	24,559,637 nt	24,050,826 nt	24,831,078 nt
Number of assembled scaffolds	136	165	667
Mean scaffold size	180,585 nt	145,762 nt	37,228 nt
Median scaffold size	11,351 nt	36,592 nt	2588 nt
Minimal scaffold size	509 nt	1868 nt	2001 nt
Maximal scaffold size	1,480,576 nt	1,189,627 nt	1,479,440 nt
N50	741,293 nt	472,887 nt	510,270 nt
Genomic G + C content	59.29%	59.37%	59.38%
Busco 3.0.2			
Complete and single-copy	192	191	184
Complete and duplicated	3	3	10
Fragmented	3	3	3
Missing	17	18	18
Annotation statistics			
Number of protein coding genes	12,590		
Average gene size	1398 nt		
Average exon size	335 nt		
Average exon number per gene	3.65		
Average intron size	66 nt		
Average intron number per gene	2.65		
Genes with 5′ UTR	25%		
Genes with 3′ UTR	35%		
Gene with transcript evidence	6736		
Gene without transcript evidence	6065		

2.9.4 Mitochondrial Genome, Structure, and Gene Content

In the circular genome representation (Fig. 2.25), scaffold #43 showed a low GC content and one match with a *P. brassicae* mitochondrial gene. In addition, as a high AT content was found in some Plasmodiophorid mitochondrial genomes, for example in *S. subterranea*, these observations suggested that the scaffold #43 could correspond to the *P. brassicae* mitochondrial genome. To ensure that no errors occurred in its assembly, a new assembly of this scaffold #43 was performed

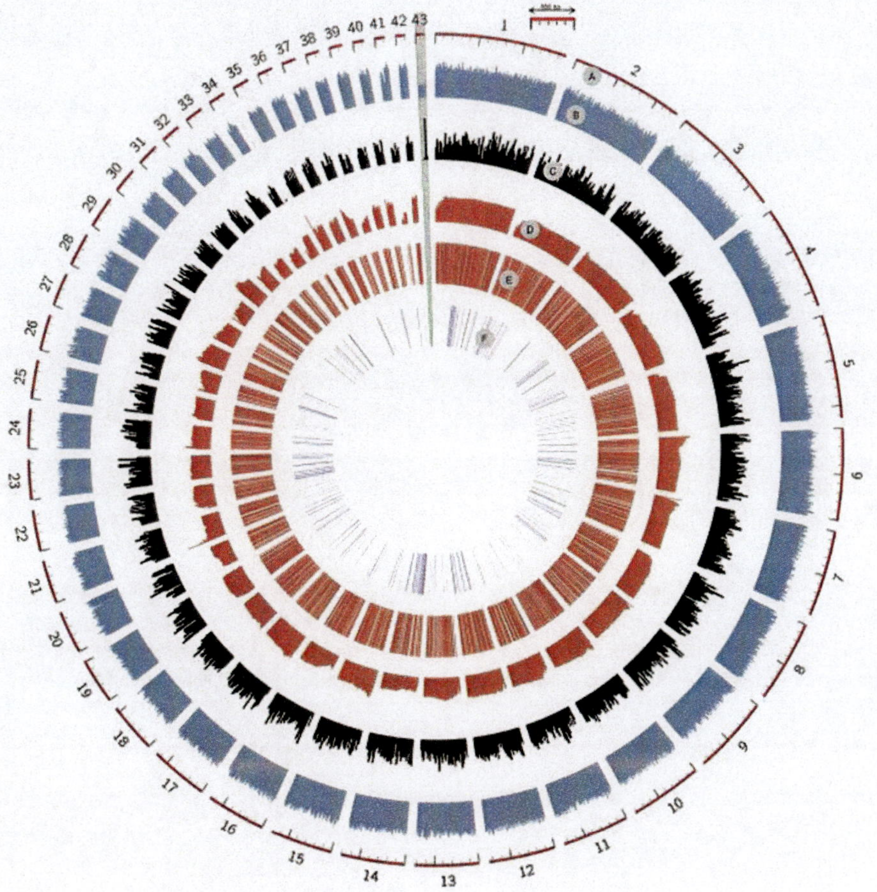

Fig. 2.25 Visual representation of the eH genome first 43 scaffolds. In the first outer circle (**A**) are represented the first 43 scaffolds of the assembled eH genome that correspond to 95% of the total genome sequence. In the second and third outer circles (**B** and **C**), the percentage in GC and the gene density (calculated as the number of genes every 20 kb) are indicated, respectively. In the two following inner circles, is represented the gene similarity between the isolates eH and e3: **D** circle pictures the relative rate every 10 kb of the eH genes number to the eH genes number that have similarity to e3 genes (80% threshold); **E** circle displays this similarity as lines when genes from eH and e3 have 80% similarity. The inner circle shows the gene similarity between the isolates eH and ZJ-1 (**E**, purple lines), and eH and Led09 (**E**, green lines). The gray scaffold 43 represented the mitochondrial genome (Daval et al. 2019)

using software specifically designed for organelle assembly (NOVOPlasty). Both assemblies were similar suggesting consistent results.

The newly assembled sequence of eH scaffold #43 is 102,962 bp, with a 24.67% GC content (Fig. 2.26) and was very similar (Max score = 91,831, Total score = 1.728e+05, Query cover = 90%, *E* value = 0.0, Identity = 99%) to scaffold #49 of the genome of the *P. brassicae* isolate ZJ-1. The size of the mitochondrial

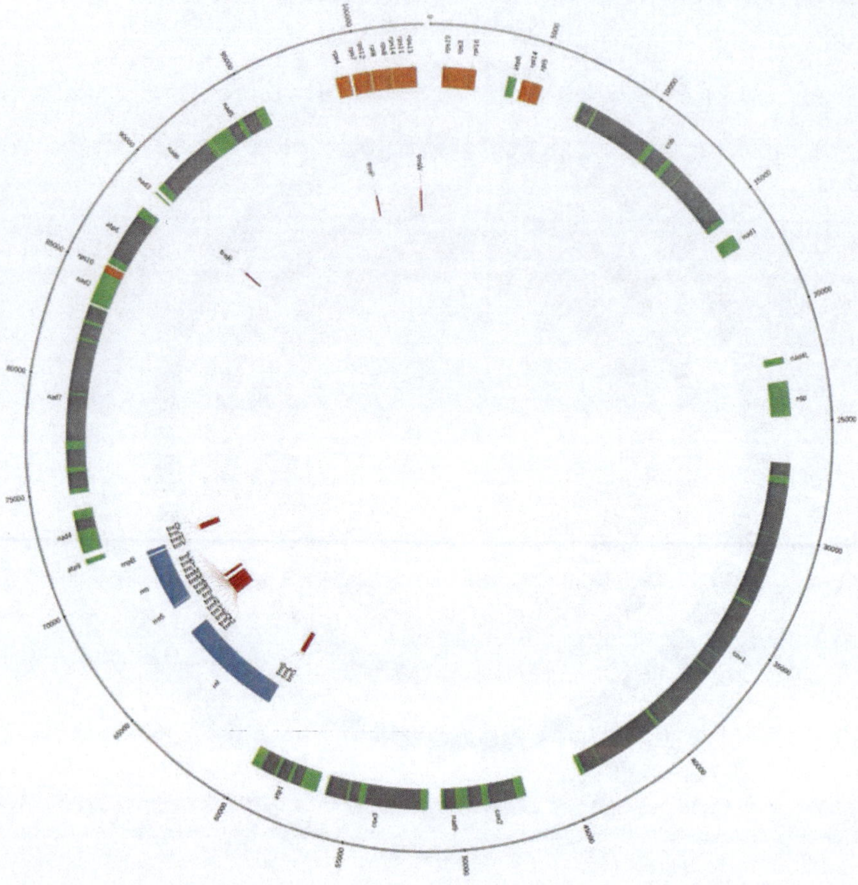

Fig. 2.26 Circular DNA map of the mitochondrial genome of *Plasmodiophora brassicae*. The outer circle shows the position of protein coding genes with exons represented in green and introns in gray, and the position of ribosomal protein coding genes in orange. The inner rings depict the position of structural RNAs: rRNA and one ncRNA (rnpB) in blue, and tRNA in red (Daval et al. 2019)

P. brassicae sequence (102,962 bp) was larger than that of *S. subterranean* (37,699 bp). Such diversity in genome size was previously described in Chlorarachniophytes, a small group of algae from the super group Rhizaria phylum Cercozoa, in which the mitochondrial genomes ranged from 34 to 180 kbp. Genome sizes vary considerably because of variation in the intergenic regions and intron content, which both contribute to mitochondrial genome plasticity. In *P. brassicae*, numerous long introns were found in mitochondrial genes in contrast to the *Lotharella oceanica* (Rhizaria group) and *B. natans* mtDNAs (Tanifuji et al. 2016). To assess the extent of mitochondrial genome rearrangements more closely, mitochondrial genomes of *P. brassicae* (eH and ZJ-1), *S. subterranea*, *B. natans*, *L. oceanica*, and *Paracercomonas marina* were aligned using the Simple Synteny

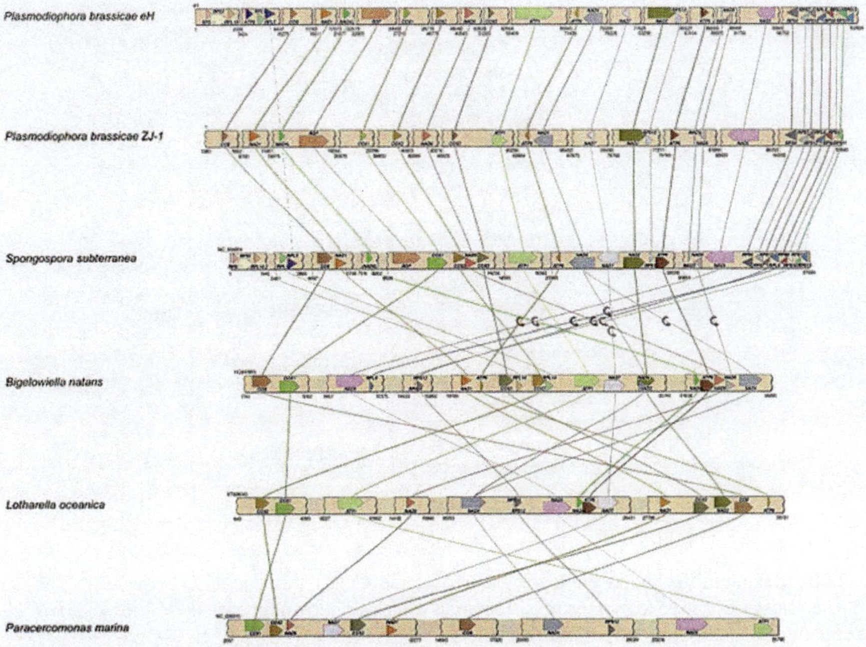

Fig. 2.27 Syntenic comparison between mitochondrial genomes. The names of the genes are indicated. Lines connecting regions of sequence similarity are drawn. Dotted lines show missing sequences (Daval et al. 2019)

alignment tool and the protein sequences. Agenome synteny map is shown in Fig. 2.27. Despite the significant difference in total size, mitochondrial genomes of *P. brassicae* eH and ZJ-1 and *S. subterranea* exhibited a similar architecture in which conservation of homologous genes and gene order was identified, but some genes are missing in ZJ-1. The organization of the three other mitochondrial sequences was also in part conserved in the content of genes but in a lesser extent concerning the gene order. The genome sizes varied considerably between these species, mainly because of the variations of the intergenic regions and the intron content, making impossible an accurate synteny's analysis using gene sequences. Both mitochondrial sequences were also quite similar between *P. brassicae* and *S. subterranea*, when the blast best-hit analysis of the *P. brassicae* mitochondrial genes were displayed for the "protein sequences" (32 protein-coding genes) and for the "complete gene sequences" (the 28 other types of genes), 60 genes were found among which 56 were also found among the 59 described in *S. subterranea*. Genes coding for proteins involved in respiration and oxidative phosphorylation (17 genes: ATP 1–7, 4 L, 9, COB, COX 1–3, ATP 1, 6, 8, 9), and for ribosomal subunit proteins (10 for small subunit: RPS, 4, 7, 8, 10, 11, 12, 13, 14, 19, and 4 for large subunit: RPL 5, 6, 14, 16) were identified. As in *S. subterranea*, 24 genes coding for tRNA were found in the *P. brassicae*, but tRNH described in *S. subterranea* was not found

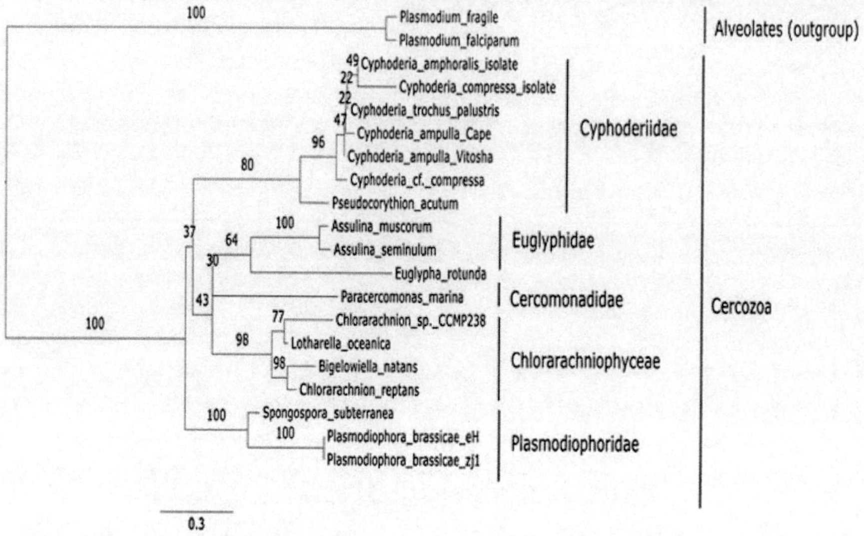

Fig. 2.28 Maximum-likelihood phylogenetic tree of the COX1 mitochondrial sequences. The Alveolates cluster was used to root the tree. Bootstrap values are shown for each node. The scale bar corresponds to 0.3 substitutions per site (Daval et al. 2019)

in *P. brassicae* and tRNQ was found only in *P. brassicae* (and not in *S. subterranea*). Interestingly, the protein-coding genes in *P. brassicae* displayed a mosaic structure with small parts spread across the scaffold, as described for mitochondrial sequences of the chlorarachniophyte *B. natans* (belonging to the Rhizaria group) and the model cryptophytes protist Guillardia theta; both algal nuclear genomes are mosaics of genes derived from the host and endo symbionts. No transposable elements were detected in the *P. brassicae* mitochondrial genome. In order to verify the *P. brassicae* taxonomic classification, mitochondrial sequences of the *COX1* gene were used in phylogenetic analysis. For this, 20 sequences of the *COX1* gene from diverse taxa were retrieved from NCBI (not found for ZJ-1 isolate), aligned, and used to create a maximum-likelihood phylogenetic tree (Fig. 2.28), knowing that analysis with the Bayesian method gave the same results. The *COX1* gene was chosen because it was retrieved from more species than the other mitochondrial genes. Both *P. brassicae* and *S. subterranea* strains were clustered in a branch with very high bootstrap and a posterior probability of 0.99, confirming that both were closely related and belonged to the order of Plasmodiophoridae. The different known orders inside the Rhizaria were also relatively well related since families of Cercozoa (Plasmodiophoridae, Chlorarachniophyceae, Cercomonadidae, Euglyphidae, and Cyphoderiidae) were clustered in a clade supported by a posterior probability superior to 0.9.

2.9.5 Nuclear Genome Annotation

For annotation, the RNAseq of resting spores was used and resulted in 23,540,954 reads. Other transcriptomes obtained in planta from other studies did not improve the annotation, showing that the transcriptional machinery was operational in resting spores. This result helps to compare the e3 reference genome for which seven transcriptomes were used, including those of resting and maturing spores (Schwelm et al. 2015). The genome annotation was improved compared to the e3 reference genome by interrogating several external databases. The annotation data predicted 12,590 protein coding genes in eH genome (Table 2.10). In total, 8989 genes were associated with GO terms, 4523 with EC numbers and 8231 with Inter Pro signatures. In rare cases of conflicting annotation, the results from manual annotation and Blast2Go were favored. Carbohydrate-active enzyme sequences of both the e3 and eH genomes were compared using dbCAN, a web server and Data Base for automated Carbohydrate-active enzyme ANnotation. For the eH isolate, 287 enzymes were predicted. Eighty-nine Glycoside Hydrolases (GH), 111 Glycosyl-transferases, 47 Carbohydrate Esterases (CE), 7 Polysaccharide Lyases, 23 Carbohydrate-Binding Modules (CBM), and 10 Auxiliary Activities. GH18, CE4, and CBM18 were the most abundant enzymes (respectively 19, 21, 7 were predicted) with in some of these groups. Among these 287 eH genes with predicted carbohydrate-active enzyme activity, 223 matched with e3 genes. The list of CAZymes was similar in both the eH and e3 genomes. Using Signal P and TMHMM to predict the secretome, 741 protein sequences were found to harbor a secretion signal prediction without transmembrane helix. Furthermore, 543 sequences corresponded to small proteins (<450 amino acids). Schwelm et al. (2015) analyzed three genome and predicted 553 secreted proteins including 416 with a secretome. An additional annotation was carried out by blasting eH data against clubroot published sequences. Ninety-nine percent and 100% identity matches were observed for the *P. brassicae* methyltransferase gene and the serine protease gene, respectively. Three hundred and eighty-eight out of the 426 genes described by Bi et al. (2016) involved in lipid metabolism were found in the eH genome. From the 118 genes identified by Feng et al. (2013) in the Led 09 isolate's genome, 23 matched the eH genome. Genes identified as being involved in pathogenicity (i.e., with functions such as P-type ATPase activity, Mps1 binder-like protein) were mapped to the eH genome. To complete the annotation, out of the 9730 genes in the e3 genome, 6626 were matched to the eH genome. The further comparison of eH and e3 genomes, a global comparison of orthologous gene clusters study was performed. Usually, orthology is the most accurate way of describing differences and similarities in the composition of genomes from different species, but it can be also used to compare strains intra species. The number of orthologous gene clusters was 8168, among which 7485 ortho groups were shared by e3 and eH isolates, and 683 clusters were specific to one of the two isolates, showing certain relatedness between both stains.

The annotations were added to the corresponding eH genes using python scripts querying the Uniprot database. Using this updated annotation, the genes were

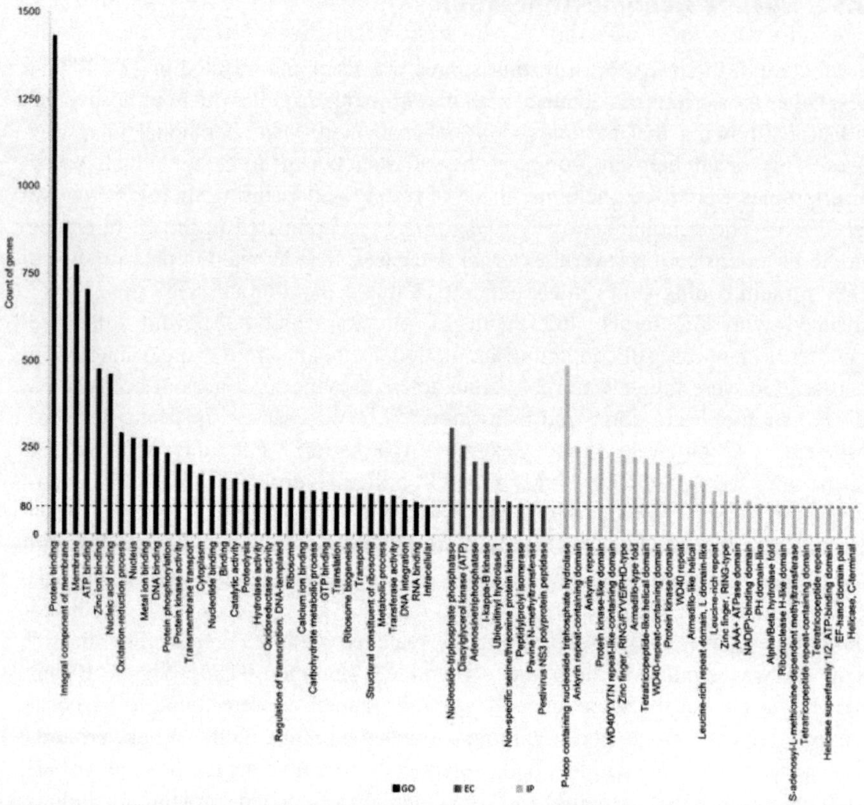

Fig. 2.29 Assignment of genes in different annotation categories. GO, Gene Ontology term (Blast2GO); EC, Enzyme Commission number (Blast2GO and EC Domain Miner); IP, Inter pro term (Blast2GO and Gene Ontology Database). The terms in which at least 80 genes were classified are represented (Daval et al. 2019)

classified into subsystems whose main terms are described in Fig. 2.29. Most of the annotated genes were classed in the protein-binding (1433) GO term category. This class, not accurately defined for their function, still remained high because of the lack of functional data in Plasmodiophorid genomics resources. Ninety-six genes encoding potential RNA-binding proteins having at least one RNA-binding domain were identified. This protein family could play a role in *P. brassicae* pathogenicity, as described in *Ustilago maydis*, the basidiomycete causative agent of corn smut disease. From the Inter Pro (IP) database, the ankyrin repeat (258 and 259) was found to be the major protein domain. The ankyrin repeat is the most common protein–protein interaction motif in nature. It is predominantly found in eukaryotic proteins, but also in many bacterial pathogens which employ various types of secretion systems to deliver ankyrin-containing proteins into eukaryotic cells where they mimic or manipulate various host functions. In eukaryotes, proteins with ankyrin repeats are involved in many cellular functions, such as transcriptional regulation, signal transduction, cell cycle but also in virulence in Amoebae, suggesting an adaptation to their intracellular localization. The high number of ankyrin repeats in the *P. brassicae* genome could suggest a role in its capacity to invade host plant cortical cells. To summarize the long list of GO terms by removing redundancy, asemantic similarity-based graph was constructed with REViGO. This allowed a representative subset of the terms to be found using a clustering algorithm that relies on semantic similarity measures. The Revigo Tree map showed a high representation of the GO terms linked to protein phosphorylation. Changes in the phosphorylation state have a broad impact on cells because they affect many processes such as the subcellular protein localization, protein–protein interactions or signaling cascades. Moreover, interference of the host phosphorylation machinery is a common strategy used by pathogens to promote infection and development in host tissues, particularly in plant pathogenic fungi. A better understanding of the regulation of protein phosphorylation could help in developing targets for regulation of *P. brassicae* infection processes. Another GO term found frequently in the *P. brassicae* genome annotation was related to lipid metabolism. This is consistent with Bi et al. (2016) who reported proteins associated with lipid droplets in resting spores of *P. brassicae* and showed that lipid conversion is high in this protist during cortex infection. Finally, the codon usage, a pattern that describes the degeneracy of the genetic code and illustrates differential usage of synonymous codons, was deciphered in nuclear genome versus mitochondrial genome.

Using the genetic invertebrate mitochondrial code from NCBI, the codon usage of the *P. brassicae* mitochondrial ORFs was determined. This showed that, as usually described in other species, the codon usage was not identical between nuclear and organelle in *P. brassicae*. As the major pattern in codon usage was a result of differences among the genes in the frequencies of G + C, which is coherent with the higher G + C content in mitochondrial sequence than in nuclear genome (Daval et al. 2019).

2.9.6 Metabolic Pathway Database

Taking advantage of the improved annotation, *P. brassicae* specific database was created using Pathway Tools Software. New genome annotation and available databases (Uniprot, Gene Ontology and EC domain Miner) served to create a Genbank entry. Using this file as input, the Patho Logic component of Pathway Tools could be used to create a new Pathway Genome Data Base (PGDB) containing the predicted *P. brassicae* metabolic pathways. The database is called Clubroot Cyc and predicts 115 pathways, 1938 enzymatic reactions, 71 transport reactions, 4468 enzymes, 123 transporters, and 1567 compounds (https://pathway-tools.toulouse.inra.fr/CLUBROOT). It links genomic data to additional information such as protein annotation, enzymatic reactions and pathways. To validate the relevance and functionality of the ClubrootCyc database, the metabolic reconstruction was tested with available published transcriptomics data described by Bi et al. (2016). Visual analysis of omics gene-expression showed that the major pathway was involved in fatty acid and lipid biosynthesis and degradation, consistent with the results of Bi et al. (2016) who showed that lipid metabolism is enriched in cortical infection stage of gall development. The development of this *P. brassicae*-specific database is thus a useful tool for integrating transcriptomics data, gene and metabolic networks and to gain further information than just a list of over- or under-expressed genes between experimental conditions.

2.9.7 Prediction of New Pathways

Clubroot Cyc also allowed to access new pathways and to complete annotation data. A complete pathway for the spermidine biosynthesis was predicted in eH *P. brassicae*. Particularly, the three enzymes (EC 2.7.2.11, EC 1.2.1.41, and EC2.6.1.13) required for conversion of glutamate to ornithine, were present in the eH genome although they were not found in the genomes described by Rolfe et al. (2016) and Schwelm et al. (2015). Ornithine is converted into putrescine, a precursor of the polyamine spermidine. The roles of polyamines synthesized by plants are well documented. They are involved in fundamental cellular processes, such as chromatin organization, cell proliferation, differentiation, and programmed cell death, but also in adaptive responses to abiotic and biotic stresses. Polyamines have been found being accumulated in the galls caused by *P. brassicae*. In turnip (*Brassica rapa* var. *rapa*), the levels of putrescine, spermidine and spermine were reported to be higher in the roots infected with clubroot than in the noninfected normal roots. In cabbage, broccoli, and Komatsuna clubroot galls, the levels of putrescine, spermidine, and spermine also increased after infection with *P. brassicae*. The high level of polyamines measured in the roots may be contributed by the host plant, as reported by Cao et al. (2008) who showed the upregulation of the *B. napus* polyamine biosynthetic enzyme spermidine synthase in response to *P. brassicae* infection. However, it could also be due to *P. brassicae* since it is difficult to distinguish between metabolites of plant or pathogen origin. Polyamine biosynthesis could be vital to the survival and pathogenesis of *P. brassicae* as described in other plant pathogens.

The chorismate/Shikimate pathway, which was predicted by Schwelm et al. (2015) but lacking the chorismate mutase (EC 5.4.99.5), was found to be complete in the eH *P. brassicae* genome suggesting that this pathway is operational in *P. brassicae*. This pathway is known in plants, bacteria, fungi, and apicomplexan parasites and is central for the biosynthesis of carbocyclic aromatic compounds. Chorismate is the common branch point for the production of these metabolites, and some intermediates in these pathways also have important physiological relevance. The salicylate, which is derived from chorismate, is involved in plant defense against *P. brassicae*. Anthranilate, another derivative of chorismate, is the precursor of the *Pseudomonas quinolone* signal, which regulates numerous virulence factors in *Pseudomonas aeruginosa*. The chorismate mutase enzyme of this pathway, newly identified in the eH genome, was described as being involved in gall formation in

Meloidogyne javanica and in *Ustilago maydis*. The potential role of this enzyme in *P. brassicae* pathogenicity would require functional studies.

Clubroot Cyc also predicted the L-Dopachrome biosynthesis pathway, which is involved in melanin synthesis. Melanin is a broad term for a group of natural pigments found in most organisms and localized in cell walls of different structures of fungi. They represent well-known virulence factors for several pathogenic fungi and for bacteria. So Clubroot Cyc, a freely accessible tool that can be updated with newly published results, provides a way to easily visualize omics data and represent metabolic pathways potentially involved in particular experimental conditions of interest. Metabolic pathway prediction in *P. brassicae* is still, however, incomplete and poorly annotated because there is still not enough information in the reference databases. To make further progress in metabolic network reconstruction and modeling, appropriate gap-filling methods are required. Software is being developed to identify and add lacking pathways, such as Meneco, a tool dedicated to the topological gap-filling of genome-scale draft metabolic networks (Prigent et al. 2017).

2.9.8 Advantages of Plasmodiophorid Genomics

One area where callus cultures appear to hold a great deal of promise is in the study of plasmodiophorid genomics, which have been severely limited by the lack of uncontaminated pathogen tissue (Siemens et al. 2009a, b). The few previous insights into plasmodiophorid genomics have come from clubroot galls on *Arabidopsis thaliana* (Bulman et al. 2006, 2007), because these allow for partial purification of large amounts of *P. brassicae* resting spores and the plant genes can be easily identified, owing to the availability of the complete host genome (Kaul et al. 2000). Plasmodiophorid member *Spongospora* genomic investigations on callus material will be much simpler than from field grown galls where multiple contaminating microbes may be present. The desiccation in particular, led to very little plant host DNA being retrieved from the callus, further simplifying access to *S. subterranea* DNA. In combination with powerful next generation sequencing techniques and increased plant genome resources (Visser et al. 2009), access to plasmodiophorid genomes will rapidly move forward in the near future. Fresh, dividing callus also seems an excellent, relatively homogenous substrate for matching "in planta" plasmodiophorid gene transcripts to a genomic scaffold.

2.9.9 Distinguishing Features of *Plasmodiophora brassicae* Genome

A draft genome of *P. brassicae* pathotypes 3, and 6 (Pb3, and Pb6) differs in their host range. Pb3 is highly virulent on *B. napus* (also infects other *Brassica* species) while Pb6 infects only vegetable *Brassica* crops. Both the Pb3, and Pb6 genomes are highly compact, each with a total size of 24.2 Mb, and contain less than 2% repetitive DNA. Clustering of genome-wide single nucleotide polymorphisms (SNP) of Pb3,

Pb6, and three additional re-sequenced pathotypes (Pb2, Pb5, and Pb8) shows a high degree of correlation of cluster grouping with host range. The Pb3 genome features significant reduction of intergenic space with multiple examples of overlapping un-translated regions (UTRs). Dependency on the host for essential nutrients is evident from the loss of genes for the biosynthesis of thiamine, and some amino acids, as well as the presence of a wide range of transport proteins, including some unique to *P. brassicae*. The annotated genes of Pb3 include those with a potential role in the regulation of the plant growth hormones cytokinin, and auxin. The *P. brassicae* genome sequence reveals a compact genome, a dependency of the pathogen on its host for some essential nutrients, and a potential role in the regulation of host plant cytokinin, and auxin. Genome annotation supported by RNA sequencing reveals significant reduction in intergenic space which, in addition to low repeat content, has likely contributed to the *P. brassicae* compact genome (Rolfe et al. 2016). In common with many pathogens, *P. brassicae* has been shown to synthesize trehalose. The Pb3 genome contained three genes encoding trehalose phosphate synthase (TPS) (PbPT3Sc00016_A_8.299_1, PbPT3Sc00018_Sm_4.147_1, PbPT3Sc00059_A_1.289_1), and single genes encoding trehalose phosphate phosphatase (PbPT3Sc00030_A_6.312_1), and trehalase (PbPT3Sc00031_A_12.282_1). The TPS gene PbPT3Sc 00059_A_1.289_1 matched the fragment isolated by Brodmann et al. (2002), and the corresponding complete gene has been reported (Bulman et al. 2007).

The *P. brassicae* genome differ due to the small size, and scarcity of repeats in comparison with most of other obligate biotrophs such as rust, and powdery mildew fungi along with the downy mildew oomycete *H. arabidopsidis*. Transposable and other repetitive elements are factors that contribute to genome expansion, evolution, and possible diversification of effector genes in biotrophic fungi. Another unique feature of the *P. brassicae* genome is the reduction (and in some occasions overlapping UTRs) of intergenic distances. This resembles the reduction of intergenic intervals in *Microsporidia*, another group of obligate intracellular pathogen in which the reduction of gene space has forced promoters, and terminators into adjacent genes, and consequently the presence of mRNAs that encode sequences from more than one gene (Spanu 2012).

Comparison of the SNP profile of Pb3 with four other Canadian pathotypes showed a clear difference between Pb3, and Pb6 while the SNP profile for Pb2, Pb5, and Pb8 were very similar. Interestingly, this SNP-based grouping correlated with host specificity of these isolates. Despite the marked differences between the genome-wide SNP profile of Pb3, and Pb6, these two pathotypes were very similar with respect to genome structure, and their effector profile. As a biotrophic pathogen, *P. brassicae* relies on host plant carbon sources for growth. Among the Pb3 predicted transporters were a monosaccharide, sucrose, and several major facilitate or super family (MSF) transporters that transport different compounds including simple sugars. A novel group of transporters among the predicted proteins of Pb3 with substrate binding motifs only found in prokaryotic ABC transporters. Seven proteins with these bacterial transporter motifs had 7 transmembrane sweet-taste receptor of 3 GCPR (7tm3) joined with either a bacterial extracellular solute-binding protein (SBP) or a bacterial periplasmic-binding (PBP) domain. Bacterial SBP motifs are essential for transmembrane transport irrespective of solute concentration, as they restrict substrate diffusion away from the cell, and improve transporter affinity. The evolutionary origin of these transporters remains to be established but they are likely to represent examples of horizontal gene transfer from prokaryotes to eukaryotes which have been documented in other protist genomes. Schwelm et al. (2015) also reported that a number of *P. brassicae* GH18 chitinases were closely related to bacterial chitinases which have not been reported in other eukaryotes before. Plant biotrophic pathogens are efficient in evading host recognition, and suppressing plant defense through secretion of effector proteins. A total of 221 small secreted proteins evaluated among the predicted in Pb3 proteins, out of which 60% (132 SSPs) showed no homology to the NCBI nr database. Twenty-eight SSPs that were differentially expressed during infection consisted of proteins with chitin binding domains, proteinase activity, and peptidase inhibitors. Salicylic acid (SA) is the main plant hormone for activating defense against biotroph pathogens. A putative Pb3 effector with homology to plant protein BSMT that function as SA methyl transferase has been identified. In planta, BSMT generates MeSA at the site of pathogen infection. MeSA acts as a SAR signal in distal leaves of infected plants. A BSMT from *P. brassicae* was reported by Ludwig-Müller et al. (2015). The function of this gene in induction of SAR, and it was differentially expressed during infection. One possible role for *P. brassicae* BSMT is to dampen the effect of SA in local tissue by converting it to MeSA with the potential benefit of protecting above-ground parts of the plant from other pathogens, and thus sustaining the life of the host plant which is required for obligate biotrophic pathogens. SA mediated plant defense is mainly described for foliar plant pathogens, and there is little information about involvement of the SA dependent defense pathway in roots. The possible involvement of SA against root colonizing pathogens is the induction of a SA marker gene encoding pathogenesis related protein 1 (PR-1) in barley roots during challenge with *Bipolaris sorokiniana*, a root pathogen of barley. Lemarié et al. (2015) reported that in *Arabidopsis* plants with partial resistance to clubroot SA was induced. Two *B. napus* orthologues of *Arabidopsis* PR1 (At2g14610) were expressed in *B. napus* roots infected with Pb3 but only one of these orthologues was differentially expressed. Expression of this gene started at week 3, and then increased at weeks 4, and was significantly upregulated at weeks 5, and 6 (sevenfold at week 5, and 12-fold at week 6). This pattern of PR1 gene expression suggests the suppression of the SA pathway at earlier stages of infection, as was also supported by HORMONOMETER analysis (Rolfe et al. 2016).

2.9.10 Evolution and Phylogeny of *Plasmodiophora brassicae*

Divided into five kingdoms, eukaryotes contain most of phagotrophic Protozoa, which are the ancestors of four distinct clades, including osmotrophic Fungi, epithelial phagotrophic Animalia, photosynthetic Plantae, and Chromista (Cavalier-Smith et al. 2018). Rhizaria, the main heterotrophic infrakingdom in the protist groups, are markedly different from those of animals, fungi or plants. The eukaryotic super group Rhizaria also contains important plant pathogens, including the obligate biotrophic Plasmodiophorida, which belongs to the *Phytomyxea* of Rhizaria (Schwelm et al. 2018). Based on the partial transcriptome sequencing of eight Rhizaria genomes, it has been revealed that Rhizaria is a sister of the Halvaria infrakingdom within the Harosa chromist subkingdom (Cavalier-Smith et al. 2018). Therefore, comparative genomics analysis may prove useful in determining the evolutionary status of the *Rhizaria* clade within the eukaryotic tree. A comparative genomics analysis was performed among *P. brassicae* and 18 other available eukaryotic model species. Ninety genes were selected for phylogenetic analysis based on the conservation of these single-copy orthologs across species. The phylogenomic analysis of this large dataset strongly supported the unique evolutionary status of the *Rhizaria*. This comprehensive study of the most poorly understood eukaryotic branch will undoubtedly fill an important gap in our understanding of the eukaryotic tree of life. Though sharing an independent branch of the phylogenetic tree with common ancestors, the genome sizes of the species in Rhizaria group vary dramatically (*P. brassicae* ~24 Mb, *B. natans* ~100 Mb, and *R. filose* ~320 Mb) (Table 2.11; Fig. 2.30). This may be a result of evolutionary genomic events (expansion or contraction). The genomes of *P. brassicae* exhibit a low level of similarity to known sequences from the other eukaryotic species (Perez-Lopez et al. 2018). Although, it is extremely difficult to annotate *P. brassicae* genes, determining the key proteins involved in the interactions between this pathogen and its host will improve our understanding of eukaryotic plant pathogens, which is so far predominantly from fungi, nematodes, and oomycetes. *P. brassicae* maintains high genetic diversity despite low repetitive content and a smaller genome. Certain biotrophs, including obligate biotrophic powdery mildew and rust fungi, display marked increases in overall genome size compared to that of their close relatives either nonpathogenic or nonbiotrophic. One possible explanation for this is that any increase is invariably the result of increased proliferation of transposable elements detectable as repetitive DNA (Spanu 2012); however, this is not the case with biotrophic *P. brassicae*. The genome size of *P. brassicae* (e3, Pb3, and ZJ-1) is small (24.1–25.5 Mb) (Schwelm et al. 2015; Rolfe et al. 2016), and smaller than that of most other members in the SAR super group, whereas the genome of *P. brassicae* has a higher gene density. Some of the predicted ORFs from the *P. brassicae* genome revealed a reduction in the length of the intergenic region. The overlapping of un-translated regions (UTRs) is one of the reasons for the compact genome structure of *P. brassicae* (Rolfe et al. 2016). The sequenced genome of *S. subterranea*, another *Plasmodiophorida* (*Phytomyxea*, *Rhizaria*) that causes potato powdery scab, also exhibits a similarly small size (28.08 Mb) (Ciaghi et al. 2018). The ZJ-1 genome only consists of approximately 2% repeated content. Transposable elements (TEs) are stretches of DNA that move throughout the genome and have significant influences on the organization and evolution of the genome (Medstrand et al. 2005). Inserting TEs into genes or gene regulatory regions can potentially disrupt the gene's function and trigger chromosome rearrangements. Additionally, the mutational activities of TEs contribute to genetic diversity and

Table 2.11 Genome comparison of *Plasmodiophora brassicae* and the other representative SAR super group (Bi et al. 2019)

	Stramenopiles	Alveolata	Rhizaria				
			Plasmodiophora brassicae				
Features	*Phytophthora infestans*	*Plasmodium falciparum*	ZJ-1	e3	Pb3	*Bigelowiella natans*	*Reticulomyxa filosa*
Genome size (Mb)	228.54	22.8	24.1	24	24.2	100	320
Gene number	18,179	5268	10,951	9730	10,851	21,708	40,433
G/C content (%)	50.97	19.4	59.4	58.5	59	45	35
Gene density[a]	79.54	231.05	454.40	405.42	448.39	217.08	126.35

[a] The unit of gene density is genes/Mb

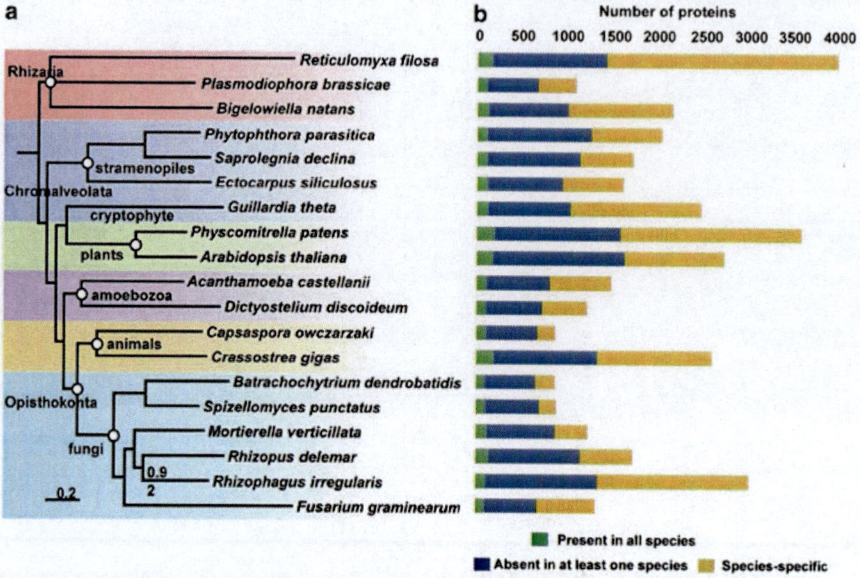

Fig. 2.30 Phylogeny and ortholog distribution in *Plasmodiophora brassicae* and other species. (**a**) The phylogeny was constructed by the maximum-likelihood method from a concatenated alignment of 90 single-copy orthologs conserved across species. The scale bar of branch length represents 0.2 amino acid-substitutions/site, and the bootstrap values are indicated on nodes (the value of 1 was omitted). Different subphylum branches are highlighted as follows: Rhizaria, red; Chromalveolata, dark blue; plants, light green; amoebozoa, light purple; animals, dark orange; fungi, light blue. (**b**) The colored bars represent orthologous proteins identified using OrthoMCL in the species shown in A. The bars are divided into three categories: present in all species (green), absent in at least one other species (blue), species-specific (orange) (Bi et al. 2019)

innovation among organisms (Biemont and Vieira 2006). Despite of a low rate of TEs, the key players in evolution, *P. brassicae* strains possess a marked genetic diversity.

This could be reflected in single-nucleotide polymorphisms (SNPs). The SNP cluster analysis of the available *P. brassicae* genomes revealed a high degree of correlation between SNPs, host ranges, and regional origins, and 48,578 SNPs were identified within the genomes of the Pbe3 and Pb3 strains of *P. brassicae* (Rolfe et al. 2016). The significant rearrangement events observed in both the ZJ-1 and Pbe3 strains, suggesting that a rapid adaptation via dynamic regions within the genomes is not a universal phenomenon. Certain species, such as *P. brassicae*, have developed alternative strategies for maintaining genetic diversity and evolution that are independent of transposable elements. This confirms that different organisms possess diverse mechanisms for creating genetic diversity (Bi et al. 2019).

2.9.11 Gene's Discovery of *Plasmodiophora brassicae* Isolates

Several DNA detection assays have been used for *P. brassicae* gene discovery using PCR primers designed in the sequence of ribosomal internal transcribed spacer (ITS) sequences (Faggian et al. 1999; Wallenhammar and Arwidsson 2001; Kageyama et al. 2003). These standard PCR-based assays tests have been found highly sensitive for spore detection but are not quantitative. A real-time PCR test based on the ribosomal ITS sequences have been designed in combination with measurements of the fatty acid biomarker arachidonic acid, quantification of *P. brassicae* (Sundelin 2008). Many DNA detection studies have been combined with surveys of molecular diversity among *P. brassicae* field or single-spore isolates. In particular, there is hope that molecular markers will eventually allow the rapid identification of virulence patterns among isolates. In some of the earliest studies, distantly related isolates of *Plasmodiophora* with divergent virulence were characterized by PCR amplification patterns using arbitrary primers (Buhariwalla et al. 1995; Moller and Harling 1996; Yano et al. 1997), and sequence-generated primers (Buhariwalla et al. 1995; Ito et al. 1999; Klewer et al. 2001). The number of distinguishing patterns was low, however, virulence patterns, and PCR patterns could not be correlated unequivocally. Early successes in the cloning of repetitive DNA elements of the pathogen were achieved when small *P. brassicae* DNA libraries were established (Buhariwalla and Mithen 1995; Buhariwalla et al. 1995; Ito et al. 1998; Klewer et al. 2001). Difficulties in obtaining high-quality genomic *P. brassicae* DNA, limited the library size in these studies to between 27 and 150 clones.

The application of 19 random primers, and 9 sequenced-derived primers to obtain 103 polymorphic bands among 46 related isolates demonstrated the very high level of genetic diversity among isolates from one field (Manzanares-Dauleux et al. 2001). Two RAPD markers (Manzanares-Dauleux et al. 2001), and a sequence-characterized amplified region (SCAR) marker (Manzanares-Dauleux et al. 2000) were correlated to isolates of pathotype group 1. In a similar set of experiments, Klewer et al. (2001) examined the molecular diversity of single-spore isolates originating from a German field isolate "e." Reliable differentiation of isolates was produced with sequence derived primers, and RFLP analysis using repetitive elements as hybridization probes. Using three highly repetitive *P. brassicae* probes, and combinations of restriction enzymes, the number of polymorphic bands in these fingerprint patterns was increased to more than 100, and the patterns of different single-spore isolates showed fewer common bands (Fahling et al. 2003). Isolate-specific banding patterns have been obtained using a combination of sequence-derived primers from the borders of highly repetitive elements of *P. brassicae* (sequence-anchored primer), and primers for genomic walking (random primer). Overall, the first simple molecular techniques to characterize isolates are now established but there is as yet no standard set of markers to describe an isolate are available. A widespread screening of pathotypes by standardized molecular techniques would make it easier to describe pathotypes, and map their distribution. Molecular techniques that rapidly describe the virulence characteristics of *P. brassicae* field populations would be highly valuable for studies of population structure, and dynamics in *Brassica* crops which thereby benefit the rapeseed growers and farmers (Siemens et al. 2009a, b).

2.9.12 The cDNA Cloning Approaches to Identify *Plasmodiophora brassicae* Genes

The isolation of a single-copy DNA sequence is unique to *P. brassicae* (Ito et al. 1997). While from subtractive cDNA libraries, Graf et al. (2004) isolated a few partial cDNAs that had no significant homology to GenBank sequences, whereas Ito et al. (1998) found a single-copy fragment with high similarity to the yeast RNA polymerase II gene. Brodmann et al. (2002) isolated a short partial sequence for a trehalose-phosphate synthase (PbTPS). Ando et al. (2006) designated a full-length *P. brassicae* cDNA clone of a single-copy gene PbSTKL1 due to a serine/threonine kinase-like domain at the C-terminal region. An acceleration of the hunt for *Plasmodiophora* genes was achieved when more than 70 cDNA clones of the pathogen were isolated by suppression subtractive hybridization (SSH) or by direct EST sequencing (Bulman et al. 2006). More than 40 of these cDNA sequences were full-length. The study showed that working with a pool containing both plant, and *P. brassicae,* cDNA was an acceptable starting point for plasmodiophorid gene discovery. Again using SSH, but starting with *Brassica rapa* rather than *Arabidopsis* clubroot galls. Sundelin (2008) isolated about 140 genes, half of these gene sequences originated from the pathogen, of these, ten clones were newly characterized *P. brassicae* genes. The remainder of the *P. brassicae* genes had been identified by Bulman et al. (2006), showing the consistency of results between these two SSH studies. Moreover, multiple copies of some *P. brassicae* genes (PbGST1, PbSUR2, and PbSHSP1) were sequenced in both studies, indicating a high transcription of these genes or perhaps preferential amplification of the transcripts (Siemens et al. 2009a, b).

2.9.13 *Plasmodiophora brassicae* Gene Structure

The presence of introns in plasmodiophorid genes was apparent during cloning of short *P. brassicae* gene fragments (Brodmann et al. 2002). The first full structure of *P. brassicae* protein-coding gene was provided when Ando et al. (2006) sequenced both the cDNA, and the genomic DNA of the PbSTKL1 gene. By comparing these sequences, the PbSTKL1 gene was observed to contain 14 intron sequences. Subsequently, Bulman et al. (2007) obtained a genomic DNA sequence corresponding with 24 *P. brassicae* cDNAs, whereas Sundelin (2008) confirmed the genomic DNA sequence corresponding with five newly characterized genes. These included the cDNA and genomic sequence of the first ATPase gene (PbATP1) cloned from a plant parasitic protozoan species. The PbATP1 protein contained at least eight transmembrane helixes and a number of protein domains characteristic for P-type ATPases. Although the total number of genomic DNA sequences is small, the structure of *P. brassicae* genes appears to be consistent between these studies. On the whole, *P. brassicae* genes are intron-rich. The great majority of the introns were bordered by the consensus border sequences 5′ GT, and AG 3′ which contained consensus branch point sequences, indicating that the splicing event in *P. brassicae* proceeds in a similar way to that in the majority of eukaryotes. Examination of the promoter, and transcription starter sites of genes indicated that *P. brassicae* transcription is likely to begin from initiator elements, as no TATA box-containing promoters were found. Where neighboring genes were confirmed, intergenic distances were found to be short, ranging from 44 to 470 bp, but a number of larger DNA fragments of up to 2.2 kb were also found, and these contained no obvious genes (Bulman et al. 2007). Calculations were made using a mean gene size of 1500 bp, and a mean intersection size of 500 bp mean that a *Plasmodiophora* genome of approximately has 20 Mbp with the potential to contain 10,000 genes. Consequently, 99% of the coding sequences on the genetic map of *P. brassicae* remain unknown (Siemens et al. 2009a, b).

2.10 *Brassica–Sclerotinia* Genome

2.10.1 Introduction

The *Sclerotinia sclerotiorum* causing stem rot is a worldwide catastrophe for *Brassica* production, especially in oilseed rape, in which yield losses can range from 10% to 80%, with low oil quality. The *S. sclerotiorum* was reported as the pathogen of stem rot in 1837 and is now found worldwide. The disease exhibits little host specificity and eight pathotypes have been identified. The *S. sclerotiorum* is able to persist for several years in the soil and the most desirable approach for its control to use resistant cultivars. Unfortunately, to date, no highly resistant resource has been characterized in *Brassica* crops, making breeding work for stem rot resistance difficult (Barbetti et al. 2012; Ge et al. 2012; Saharan and Mehta 2008).

The mapping of R-loci has been focused on *B. napus*; however, only partial resistance has been characterized in both the A and C genomes. Zhao and Meng (2003) first identified three QTLs for leaf resistance and three other QTLs for stem resistance in the seedling and adult stages, respectively, but no common QTLs were identified. Zhao et al. (2006) identified eight and one QTL involved in two segregating DH populations, with each explaining 6–22% of the observed variance, still with no common QTLs. Yin et al. (2010) detected ten, one, and ten QTLs in one DH population using three inoculation procedures, and only two common QTLs were detected. Wu et al. (2013) identified three QTLs at the seedling stage for leaf resistance and ten QTLs for stem resistance at the adult stage. Two major QTLs could be detected repeatedly and a candidate resistance gene, *BnaC.IGMT5*, was first identified. These studies revealed abundant QTLs but seldom common ones, indicating the complicated genetic structure of these plants. The release of the *B. napus* genome sequence has strongly facilitated mapping work. Fomeju et al. (2014) first adopted a genome-wide association study (GWAS) using 116 materials genotyped with 3228 SNPs and the results indicated that 64 genomic regions are involved in stem rot resistance. Wei et al. (2016) combined GWAS and SNP array analysis using 347 *B. napus* accessions and 17 significant regions were located on the A8 and C6 chromosomes. These SNPs on Chr. A8 were placed in a 409 kb segment, with candidate genes being suggested. Using a similar method, Wu et al. (2016) genotyped 448 accessions and 26 SNPs corresponding to three loci were associated with SR resistance. In total, 39 candidates were proposed. Gyawali et al. (2016) performed a GWAS using microsatellite markers in a global collection of 152 accessions and found that 34 loci were significantly associated. Now, many loci opposing stem rot have been characterized but none have been functionally characterized.

2.10.2 Early Phase of *Sclerotinia sclerotiorum* Genome

A 1380 bp intervening sequence within the mitochondrial small subunit rRNA (nst ssurRNA) gene of *S. sclerotiorum* has been sequenced and identified as group 1-intron. The intron shows a close similarity in secondary structure to the subgroup-IC2 introns from *Podospora* (ND 3i1, ND 5i2, and CO Ii5) and *Neurospora* (ND 5i1). Screening by means of Southern hybridization and PCR amplification detected the intron in the mt ssu rRNA gene of *S. minor*, *S. trifoliorum*, and *S. cepivorum*, but not in other members of the Sclerotineaceae (Carbone et al. 1995). The *S. sclerotiorum* can be transformed with polyethylene glycol-mediated protoplast transformation but the absence of reproductivity and stability of transformants combined with the multinucleate nature of hyphal cells have allowed relatively few molecular studies of the pathogen (Rolland et al. 2003). The first *Agrobacterium tumefaciens* mediated transformation of *S. sclerotiorum* was reported through co-cultivation with ascospores (Weld et al. 2005). Utilizing the benefits of *Agrobacterium* mediated transformation of bi-nucleate ascospores, future gene transfer, gene knock out and insertional mutagenesis studies on this pathogen should expedite. Nonetheless, the utility of reporter genes such as the green fluorescent protein (GFP) gene have already been exploited in *S. sclerotiorum* to gain more understanding on the biology of the fungus both in vitro and in planta (De Silva et al. 2005; Guimaraes and Stotz 2004; Lorang et al. 2001).

The 38-Mb genome of *S. sclerotiorum* was sequenced through support from the US department of Agriculture by the Broad Institute of Harvard and the Massachusetts Institute of Technology. Based on shotgun sequencing, the current assembly contains 679 sequence contigs giving 8X sequencing coverage of the genome (*Sclerotinia sclerotiorum* Sequencing Project: http://www.broad.mit.edu/annotation/fungi/*Sclerotinia*/*sclerotiorum*). Trace sequence has been deposited at the National Centre for Biotechnology Information (NCBI). The assembled sequence was released in the autumn of 2005. The publication of the full genome sequence opens a new chapter in research on this important pathogen (Bolton et al. 2006).

2.10.3 Phylogeny of *Sclerotinia* and Related Genera

Phylogenies have been constructed based on nuclear ribosomal internal transcribed spacer (ITS) DNA sequences from an in-group consisting of 50 isolates representing 24 species of the discomycete family Sclerotiniaceae and an out-group consisting of five related taxa of the same family. The ingroup taxa are: three *Botrytis* spp., two *Botryotinia* spp., one *Ciborinia* sp., one *Dumontinia* sp., one *Grovesinia* sp., six *Myriosclerotinia* spp., nine *Sclerotinia* spp., and one *Sclerotium* sp. The out-group taxa are: one *Ciboria* sp., one *Encoelia* sp., and three *Monilinia* spp. The type species is included for all taxa except for *Ciborinia* and *Encoelia*. Several of the included taxa are important plant pathogens. The resulting phylogenies are discussed with regard to morphology, life history and taxonomy. A suspected relationship between *Sclerotinia borealis* and *S. tetraspora*, and *Myriosclerotinia* is rejected, while a suspected relationship between *Ciborinia ciborium* and *Myriosclerotinia* is strongly supported. *Sclerotinia ulmariae*, previously synonymized with *Dumontinia tuberosa*, is reinstated as an independent species of *Dumontinia*. Two new combinations, *Dumontinia ulmariae* and *Myriosclerotinia ciborium*, are proposed. The imperfectly known taxon *Sclerotium cepivorum* seems most closely related to *Dumontinia*. It is concluded that *Dumontinia* and *Myriosclerotinia*, as currently conceived, are monophyletic and that *Botryotinia* along with *Botrytis* anamorphs probably also constitute a monophyletic lineage. The genus *Sclerotinia* is probably polyphyletic and characterized by simple isomorphies rather than synapomorphies. Two putatively new taxa, *Sclerotinia* sp.1 and *Sclerotinia* sp.2, are most closely related to *S. minor*, *S. sclerotiorum*, and *S. trifoliorum* and to *S. borealis*, respectively (Holst-Jensen et al. 1998).

2.10.4 Expression of Genes for Enzymes Activity

Matured sclerotia have weak but detectable β-1,3 glucanase and β-glucosidase activity, but there is no amylase or trehalase activity. With the beginning of germination, the activity of β-1,3 glucanase apparently increases, whereas β-glucosidase activity remains at the initial level. The increase in β-1,3 glucanase activity is most pronounced when apothecial stipes are actively produced. The activities of enzymes of carbohydrate metabolism have been compared between the different parts of germinating sclerotia where mature apothecia are seen, i.e., mature apothecia, stipes (immature apothecia) and sclerotia (Fig. 2.31). The specific activities of β-1,3 glucanase and β glucosidase are highest in the sclerotia. At later periods of germination, when most stipes had developed into mature apothecia, detectable amylase and trehalase activity is found in the respective parts. In this state, the specific activity of amylase is highest in mature apothecia, but the trehalase activity is almost equal in the three parts. The highest activity of glucose-6-phosphate dehydrogenase is found in apothecia. By contrast, in germinating sclerotia, activity is very low, though somewhat higher than in un-germinating sclerotia (Fig. 2.32). These results clearly show the difference of physiological states between apothecia and sclerotia; the

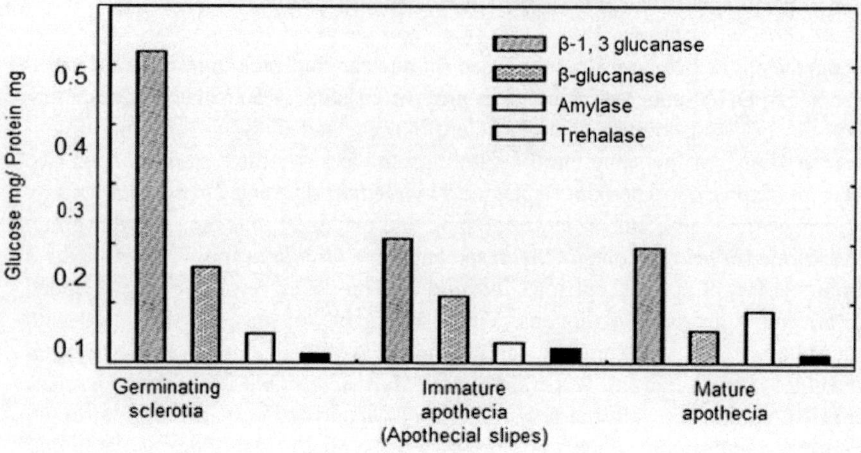

Fig. 2.31 Comparison between carbohydrase activities of germinating sclerotia, immature and mature apothecia (Saito 1977)

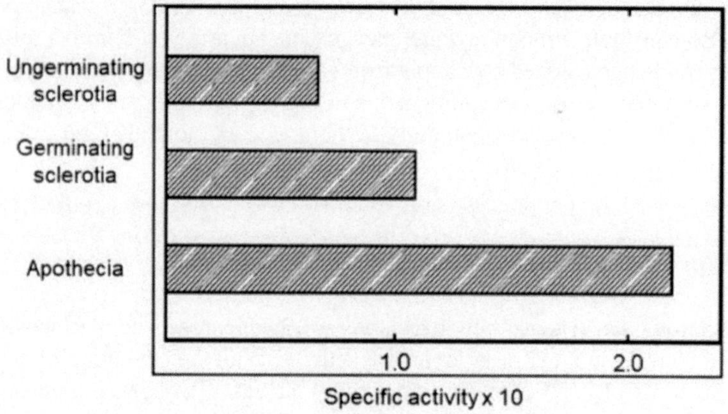

Fig. 2.32 Activities of glucose-6-phosphate dehydrogenase in un-germinating and germinating sclerotia and apothecia (Saito 1977)

biosynthesis of materials needed for cellular construction is promoted in stipes and apothecia. In contrast, the catabolism of endogenous reserves takes place mainly in the medullary cells of sclerotia (Saito 1977). The *Sclerotinia sclerotiorum* produces a number of extracellular pectin-degrading enzymes. The complete sequence of a gene (*pg1*) encoding an endo-polygalacturonase (*PG1*) has been cloned and determined. The coding region consisted of a noninterrupted 1143-bp open reading frame. *S. sclerotiorum pg1* was compared with other fungal PG-encoding genes. Basic transcription control sequences were identified in the five non coding regions.

The deduced amino acid (aa) sequence (380 aa) of the enzyme was compared with seven fungal PG sequences and showed a high level of identity (41.5–59.8%). Predicted secondary structures were compared, revealing a similar protein organization most probably in antiparallel beta sheets. Hybridization analysis using a pg1 0.65-kb *BamHI* fragment as a probe allowed the identification of seven different recombinant phages from a genomic library. Analysis of the hybridizing restriction fragments suggested that PG-encoding genes are organized as a family (Reymond et al. 1994).

2.10.5 Expression of Calcineurin Genes for Pathogenesis

The *Sclerotinia sclerotiorum* is a necrotrophic, omnivorous plant pathogen with worldwide distribution. Sclerotia of *S. sclerotiorum* are pigmented, multihyphal structures that play a central role in the life and infection cycles of this pathogen. Calcineurin, a Ser/Thr phosphatase linked to several signal-transduction pathways, plays a key role in the regulation of cation homeostasis, morphogenesis, cell-wall integrity and pathogenesis in fungi. Calcineurin expression in *S. sclerotiorum* is altered in a phase-specific manner during sclerotial development. Inhibition of calcineurin by FK506, cysclosporin A, or inducible antisense calcineurin expression impairs sclerotial development at the prematuration phase and increases germination of preformed sclerotia. Induction of antisense calcineurin expression in *S. sclerotiorum* results in reduced pathogenesis on tomato and *Arabidopsis*. However, secretion of oxalic acid, a key virulence factor of *S. sclerotiorum* is not altered. Inhibition of calcineurin conferred a reduction in cell wall β-1,3-glucan content and increased sensitivity to cell wall degrading enzymes and to the glucan synthase inhibitor caspofungin. Thus, calcineurin plays a major role in both sclerotial development and pathogenesis of *S. sclerotiorum* and most likely, other phytopathogens (Harel et al. 2006).

2.10.6 The Main Features of the *Sclerotinia sclerotiorum* Genome

The genome of *S. sclerotiorum* has been sequenced using isolates of the pathogen through analysis of transcriptome and secretome as follows.

2.10.6.1 Genome Sequences

The genome of *S. sclerotiorum* strain 1980, originally isolated from beans in western Nebraska USA, was first sequenced in 2011 using Sanger technology with 9.1× coverage. As approximately 1.6 Mb of sequences were not covered by the 38.0 Mb scaffolds, and thus missed from the assembly, the final size of *S. sclerotiorum* genome was estimated to be approximately 38.3 Mb, which consists of 16 linkage groups likely corresponding to 16 chromosomes (GenBank accession numbers: AAGT01000000). Most of the uncovered regions are located in the middle of chromosomes, which probably correspond to centromeres. The genome was predicted to contain 14,522 genes, with average GC contents of 41.8%. After removing genes encoding small proteins that are less than 100 amino acids in length or without evidence of expression (from ESTs and/or microarray signals), 11,860 predicted proteins with high confidence were deduced. Initial analysis of the genome sequences revealed a large number of virulence related genes that were possessed by the *S. sclerotiorum* 1980 strain, including genes encoding cell wall degrading enzymes (CWDEs) and biosynthesis genes of phytotoxins and other secondary metabolites. Amselem et al. (2011) also identified a significant amount of repetitive transposable elements (TEs), which comprises approximately 7.7% of the whole genome. This led to a postulation that the *S. sclerotiorum* genome experienced a recent major remodeling that was associated with a dramatic expansion of TEs. The TE expansion in *S. sclerotiorum* might also have an impact on its genome organization and function of gene inactivation, modification, or expression regulation (Amselem et al. 2011). Derbyshire et al. (2017) sequenced the genome of the same *S. sclerotiorum* 1980 strain to near completion in 2017 using PacBio technology. It was annotated using extensive RNA-Seq data and manual curation. Therefore, this version of the whole-genome sequence became the choice of standard for *Sclerotinia*. They identified 70 candidate effector genes, and found a significant association between the positions of these secreted proteins and regions with a high relative RIP (repeat-induced point mutation) index, which suggested that *S. sclerotiorum,* exhibits a subtly enhanced mutation rate of secreted proteins in specific genomic compartments as a result of transposition and RIP activity. Derbyshire et al. (2017) sequenced the genomes of 25 field isolates of *S. sclerotiorum* that were collected from four different continents—Australia, Africa (north and south), Europe, and North America (Canada and the Northern United States). They conducted SNP-based analysis on population structure and selective sweeps. These 25 isolates can be grouped into two major populations, where population 1 consists of 11 isolates from Canada, the USA, and France, and population 2 includes nine isolates from Australia and one from Morocco. A single candidate selective sweep was identified in the Australian and Moroccan group, which covers less than 0.001% of the genome. It is reasonable to speculate that a slow evolution rate and extensive self-fertilization of *S. sclerotiorum* may have led to a striking absence of strong selective sweeps, as there is only one gene within this 10 kb region, which encodes a major facilitator super family transporter that was only negligibly expressed at late stages of *Brassica napus* infection. In support of such genome conservation, *S. sclerotiorum* strains from even a tropical and a temperate region, although segregated in the populations, still exhibit similar genetic structure (Del Ponte et al. 2017; Xia et al. 2020).

2.10.6.2 Transcriptomic and Secretomic Analysis

The availability of the high-quality genome sequences facilitated comparative transcriptomic and secretomic analysis of this notorious pathogen, which in turn facilitated biological investigations. Seifbarghi et al. (2017) was able to delineate gene expression patterns that signified transitions between pathogenic phases of *S. sclerotiorum*, namely, host penetration, ramification, and necrotrophic stages, through RNA-Seq analysis focusing on events occurring through the early (1 h) to

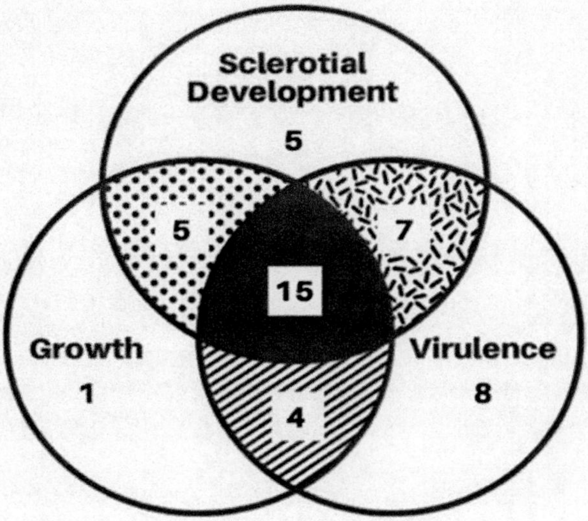

Fig. 2.33 A Venn diagram summary of the genes that have been studies with mutant analysis (detailed in Table 2.12 (Xia et al. 2020)

the middle (48 h) stages of infection. These expression data provide evidence for the occurrence of a brief biotrophic phase soon after host penetration. A total of 432 proteins were identified in *S. sclerotiorum* in an interspecies comparative analysis of the predicted secretomes of *S. sclerotiorum* and *Botrytis cinerea*. Among them, 16% of the encoding genes reside in small gene clusters that are distributed over 13 of the 16 predicted chromosomes. These candidate genes provide a reservoir for future reverse genetics analysis to delineate the pathogenesis mechanisms in the *S. sclerotiorum*–host plant interactions. Mutant analysis is paramount to establish a causal relationship between a gene and a studied biological process, as omics experiments only provide vague association clues. All of the genes studied in *S. sclerotiorum* so far using such convincing mutant analysis methods have been summarized. A Venn diagram of the functions of these encoded proteins is provided for readers' overview (Fig. 2.33; Table 2.12). Xia et al. (2020) also constructed a chromosomal map for these genes to detect possible clustering events that are often associated with virulence factors (Fig. 2.34). However, little clustering is observed, except for genes contributing to mating and melanization, reflecting the under-studied nature of this fungus.

2.10.7 Genome Architecture of *Sclerotinia sclerotiorum*

Predicted secretome and effector studies have been facilitated by the huge increase in availability of fungal genome sequences in recent years. However, although there are about 1509 fungal genomes that have been sequenced (http://www.ncbi.nlm.nih.gov/genome/browse/), many are fragmented and poorly annotated. The regions missing from these genomes are largely composed of repetitive sequence. This is because the technologies used to sequence them produce reads up to a maximum of

Table 2.12 *Sclerotinia sclerotiorum* genes with functional characteristics (the genes are ordered with regard to their chromosomal positions, Fig. 2.34) (Xia et al. 2020)

Gene code (new)	Gene code (old)	Strains	Mutant name	Gene full name	Mutant type	Mutant phenotypes					
						Hyphal growth	Sclerotial formation	Oxalate production	Virulence	Appressoria formation	Induce host HR/resistance
sscle_01g003850SS	1G_02068	Ep-1PNA367, Heilongjiang, China	ssvp1	Small secreted virulence-related protein	RNAi	–	NA	NA	+	NA	+
sscle_01g004990SS	1G_01919	1980, Nebraska, USA	svf1	Survival factor 1 homologue	RNAi	–	NA	–	+	+	NA
sscle_01g006030SS	1G_01788	1980, Nebraska, USA	cna1	Catalytic subunit calcineurin encoding gene	RNAi	+	+	–	+	–	NA
sscle_01g007450SS	1G_01614	Ep-1PNA367, Heilongjiang, China	sop1	Microbial opsin homolog gene	RNAi	+	+	NA	+	NA	NA
sscle_01g011030SS	1G_01151	1980, Nebraska, USA	sfh1	GATA-box and SNF5 domains containing transcription factor	RNAi	+	+	NA	+	+	NA
sscle_02g013550SS	1G_12694	HA61, Jiangsu, China	shk1	Histidine kinases	Deletion	+	+	NA	–	NA	NA
sscle_02g020240SS	1G_04003	1980, Nebraska, USA	mat1-1-5	Mating-type gene	Deletion	–	–	NA	–	NA	NA
sscle_02g020250SS	1G_04004	1980, Nebraska, USA	mat1-1-1	Mating-type gene	Deletion	–	–	NA	–	NA	NA
1G_04005	1G_04005	1980, Nebraska, USA	mat1-2-4	Mating-type gene	Deletion	–	–	NA	–	NA	NA
sscle_02g020270SS	1G_04006	1980, Nebraska, USA	mat1-2-1	Mating-type gene	Deletion	+	–	NA	–	NA	NA
sscle_03g025030SS	1G_00699	WMA1, Washington USA	sod1	Cu/Zn superoxide dismutase	T-DNA	–	–	–	+	NA	NA
sscle_03g025030SS	1G_00699	1980, Nebraska, USA	sod1	Cu/Zn superoxide dismutase	Deletion	+	+	+	+	NA	NA

sscle_03g028510SS	1G_00263	Isolated from an infected canola stem Alberta, Canada	v263	Hypothetical secreted protein	Deletion	–	NA	NA	+	NA	NA
sscle_03g031470SS	1G_13314	Isolated from diseased rapeseed, Alberta, Canada	scd1	Scytalone dehydratase	Deletion	+	+	NA	–	NA	NA
sscle_03g031480SS	1G_13315	Isolated from diseased rapeseed Alberta, Canada	thr1	Trihydroxynaphthalene reductase	Deletion	+	+	NA	–	NA	NA
sscle_03g031520SS	1G_13322	1980, Nebraska, U.S.A.; UF1 Florida; WMA, Washington USA	pks13	Polyketide synthase	Deletion by CRISPR	–	–	NA	–	+	NA
sscle_03g031670SS	1G_13339	274, Nebraska, USA	pth2	Peroxysomal carnitine acetyl transferase	Deletion	+	+	+	+	+	NA
sscle_04g034810SS	1G_02462	Isolated from an infected canola stem, Alberta, Canada	axp	Arabinofuranosidase/-xylosidase	Deletion	NA	NA	NA	+	NA	NA
sscle_04g034960SS	1G_02486	Sunf-M, Inner Mongolia, China	caf1	Secreted protein with a putative Ca^{2+}-binding EF-hand motif	RNAi, T-DNA	+	+	–	+	+	+
sscle_04g037170SS	1G_02784	1980, Nebraska, USA	scat1	Type A catalase	Deletion	+	+	NA	+	NA	NA
sscle_04g038020SS	1G_02904	Ep-1PNA367, Heilongjiang, China	cvnh	Secreted protein	RNAi	+	+	NA	+	NA	NA
sscle_05g046240SS	1G_05917	SUN-F-M, Inner Mongolia, China	sl2	Cell wall protein	RNAi	–	+	NA	–	NA	NA
sscle_05g046390SS	1G_05899	1980, Nebraska, USA	trr1	Thioredoxin reductase	RNAi	NA	+	NA	+	NA	NA

(continued)

Table 2.12 (continued)

sscle_05g046790SS	1G_05839	1980, Nebraska, USA	bi1	Bax inhibitor-1 protein	RNAi	+	–	–	+	NA	NA
sscle_05g046830SS	1G_05834	JRUF1, Florida, USA	foxe2	Forkhead-box transcription factor family gene	Deletion, T-DNA	–	–	NA	–	–	NA
sscle_05g048220SS	1G_05661	1980, Nebraska, USA	nox1	NADPH oxidase	RNAi	–	+	+	+	NA	NA
sscle_06g049430SS	1G_07404	1980, Nebraska, USA	rhs1	Rearrangement hot spot repeat-containing protein	RNAi	+	+	–	+	+	NA
sscle_06g049780SS	1G_07360	1980, Nebraska, USA	fkh1	Atypical forkhead (FKH)-box-containing protein	RNAi	+	+	NA	+	NA	NA
sscle_06g049830SS	1G_07355	1980, Nebraska, USA	pac1	pH-Responsive transcription factor	Deletion	–	+	+	+	NA	NA
sscle_06g049890SS	1G_07345	NGA4, Anhui, China	pemg1	Elicitor-homologous protein	RNAi	+	NA	–	+	+	NA
sscle_06g051560SS	1G_07136	UF1, Florida, USA	ste12	Downstream transcription factor of MAPK pathway	RNAi	+	+	NA	+	+	NA
sscle_07g055970SS	1G_03527	SUN-F-M, Inner Mongolia, China	hex1	Woronin body major protein	RNAi	–	+	NA	NA	NA	NA
sscle_07g058030SS	1G_03252	UF1, Florida, USA	ams2	Cell-cycle-regulated GATA transcription factor	RNAi	+	–	NA	+	+	NA
sscle_07g058620SS	1G_03171	1980, Nebraska, USA	pka1	Protein kinase A	Deletion	–	–	–	NA	NA	NA
sscle_07g059700	MK992913	1980, Nebraska, USA	qdo	Quercetin dioxygenase gene	Deletion	–	NA	NA	+	NA	NA
sscle_08g064670SS	1G_05163	1980, Nebraska, USA	cry1	Cryptochrome family CRY-DASH ortholog	Deletion	–	–	NA	–	NA	NA

sscle_08g065550SS	1G_05284	NGA4, Anhui, China	nac	Nascent polypeptide-associated complex—subunit	RNAi	–	–	–	+	NA	NA
sscle_08g066770SS	1G_05445	1980, Nebraska, USA	smk3	Slt2 ortholog	Deletion	+	+	NA	+	+	NA
sscle_08g067830SS	1G_05588	1980, Nebraska, USA	mads	MADS-box proteins	RNAi	+	–	NA	+	NA	NA
sscle_08g068500SS	1G_14133	Ep-1PNA367, Heilongjiang, China	itl	Integrin-like protein	RNAi	+	+	NA	+	NA	+
sscle_08g068530SS	1G_14127	1980, Nebraska, USA	ggt1	Glutamyl transpeptidase	Deletion	–	+	NA	–	+	NA
sscle_09g069850SS	1G_10796	1980, Nebraska, USA	odc2	Oxalate decarboxylases	Deletion	–	–	+	+	+	NA
sscle_10g075560SS	1G_08218	WMA1, Washington, USA	oah	Oxaloacetate acetylhydrolase	T-DNA	–	–	+	+	NA	NA
sscle_10g075560SS	1G_08218	1980, Nebraska, USA	oah1	Oxaloacetate acetylhydrolase	Deletion	–	+	+	+	+	NA
sscle_10g075560SS	1G_08218	1980, Nebraska, U.S.A.; UF1, Florida; WMA, Washington, USA	oah1	Oxaloacetate acetylhydrolase	Deletion by CRISPR	–	+	+	+	+	NA
sscle_10g077630SS	1G_08489	1980, Nebraska, USA	pph1	Type 2A Ser/Thr phosphatase catalytic subunit PP2Ac	RNAi	+	+	NA	NA	NA	NA
sscle_10g079320SS	1G_14065	1980, Nebraska, USA	ssp1	Development-specific protein	Deletion	–	–	NA	NA	NA	NA
sscle_11g082700SS	1G_07871	1980, Nebraska, USA	rgb1	Type 2A Ser/Thr phosphatase B subunit	RNAi	–	+	NA	+	+	NA
sscle_11g083230SS	1G_07798	SUN-F-M, Inner Mongolia, China	gpd	Glyceraldehyde-3-phosphate dehydrogenase	RNAi	NA	+	NA	NA	NA	NA

(continued)

Table 2.12 (continued)

sscle_11g083680SS	1G_07749	1980, Nebraska, USA	xyl1	Endo β-1, 4-xylanase	Deletion	+	+	NA	+	NA	NA
sscle_11g083950SS	1G_07715	1980, Nebraska, USA	sac1	Adenylate cyclase	Deletion	+	+	–	+	NA	NA
sscle_12g087830SS	1G_11172	1980, Nebraska, USA	nox2	NADPH oxidase	RNAi	–	+	–	–	NA	NA
sscle_12g090900SS	1G_11866	1980, Nebraska, USA	smk1	ERK-type MAP kinase	RNAi	+	+	NA	NA	NA	NA
sscle_14g099710SS	1G_08814	1980, Nebraska, USA	odc1	Oxalate decarboxylases	Deletion	–	NA	NA	–	NA	NA
sscle_16g107670SS	1G_10096	1980, Nebraska, USA	cp1	Cerato-platanin protein	Deletion	–	–	NA	+	NA	+
sscle_16g107670SS	1G_10096	FXGD2, Anhui, China	sm1	Cerato-platanin protein	RNAi	+	+	–	+	NA	NA
sscle_16g107930SS	1G_10135	1980, Nebraska, USA	fdh1	Formaldehyde dehydrogenase	Deletion	–	+	NA	+	+	NA
sscle_16g109570	ANQ80447	1980, Nebraska, USA	nsd1	GATA-type Ivb zinc-finger transcription factor	Deletion	–	–	NA	+	+	NA

Gene code (new)	Secretion signal	Tested hosts	Other functions	Reference
sscle_01g003850SS	Yes	Canola (Zhongyou 821)	HR induction	Lyu et al. (2016)
sscle_01g004990SS	No	*Arabidopsis* and canola (Zhongyou 821)	Cell wall integrity, ROS production	Yu et al. (2019)
sscle_01g006030SS	No	*Arabidopsis* and tomato (Bonny Best)	Hyphal elongation	Harel et al. (2006)
sscle_01g007450SS	No	*Arabidopsis*	Stress responses	Lyu et al. (2016)
sscle_01g011030SS	No	Soybean, common bean and tomato	ROS accumulation	Liu et al. (2018)
sscle_02g013550SS	No	Rapeseed, strawberry, tomato and cucumber	Oxidative stresses, glycerol accumulation	Duan et al. (2013)
sscle_02g020240SS	No	Tomato	Apothecial development	Doughan et al. (2016)
sscle_02g020250SS	No	Tomato	Apothecial development	Doughan et al. (2016)
1G_04005	No	Tomato	Apothecial development, ascospore production	Doughan et al. (2016)

(continued)

Table 2.12 (continued)

sscle_02g020270SS	No	Tomato	Apothecial development	Doughan et al. (2016)
sscle_03g025030SS	No	Pea (Guido)	Detoxification of host ROS	Xu and Chen (2013)
sscle_03g025030SS	No	Tomato (Garden Peach) and tobacco	Oxidative stress tolerance, repression of host ROS	Veluchamy et al. (2012)
sscle_03g028510SS	Yes	Canola		Liang et al. (2013)
sscle_03g031470SS	No	Rapeseed (Westar)	Hyphal branching	Liang et al. (2017)
sscle_03g031480SS	No	Rapeseed (Westar)	Hyphal branching	Liang et al. (2017)
sscle_03g031520SS	No	Soybean, canola, tomato (Better Boy), faba bean (Windsor) and pea (Sugar Daddy)	Pigmentation of compound appressoria	Li et al. (2018)
sscle_03g031670SS	No	Soybean	Oxalic acid accumulation	Liberti et al. (2013)
sscle_04g034810SS	Yes	Canola		Yajima et al. (2009)
sscle_04g034960SS	Yes	*Arabidopsis*, rapeseed, pak choi cabbage hot pepper, tomato, cucumber and soybean		Xiao et al. (2014)
sscle_04g037170SS	No	Tomato (Garden Peach)	Modulation of ROS	Yarden et al. (2014)
sscle_04g038020SS	Yes	*Arabidopsis*		Lyu et al. (2015)
sscle_05g046240SS	Yes	Rapeseed and *Arabidopsis*	Cellular integrity	Yu et al. (2012)

sscle_05g046390SS	No	*Arabidopsis* and tobacco	Oxidative stress tolerance	Zhang et al. (2019)
sscle_05g046790SS	No	*Arabidopsis* and tomato	Stress responses, hyphal tip branching	Yu et al. (2015)
sscle_05g046830SS	No	Tomato	Apothecial development	Wang et al. (2016)
sscle_05g048220SS	No	Tomato (Rutger)	ROS regulation	Kim et al. (2011)
sscle_06g049430SS	Yes	*Arabidopsis* and canola (Zhongshuang 9)		Yu et al. (2017)
sscle_06g049780SS	No	Tomato	Cellular integrity	Fan et al. (2017)
sscle_06g049830SS	No	*Arabidopsis* and tomato (Bonnie Best)		Rollins et al. (2003)
sscle_06g049890SS	No	Oilseed rape and soybean	Negative regulator of growth and virulence	Rollins et al. (2003)
sscle_06g051560SS	No	Bush bean and tomato		Xu et al. (2018)
sscle_07g055970SS	No		Cellular integrity	Yu et al. (2012)
sscle_07g058030SS	No	Common bean	Chromosome segregation, number and distribution of sclerotia	Liu et al. (2018)

(continued)

Table 2.12 (continued)

sscle_07g058620SS	No		PKA activity	Jurick et al. (2004)
sscle_07g059700	Yes	*Arabidopsis*	Flavanol degradation	Chen et al. (2019)
sscle_08g064670SS	No	*Arabidopsis* and tomato	Sclerotial mass, response to UV	Veluchamy and Rollins (2008)
sscle_08g065550SS	No	Oilseed rape and tobacco		Li et al. (2015)
sscle_08g066770SS	No	Canola (Westar)	Cuticle penetration, cell wall integrity	Bashi et al. (2016)
sscle_08g067830SS	No	Tomato		Qu et al. (2014)
sscle_08g068500SS	Yes	*Arabidopsis* and canola	Hyphal branching, suppression of host defense	Zhu et al. (2013)
sscle_08g068530SS	Yes	Tomato		Li et al. (2012)
sscle_09g069850SS	Yes	Common bean (Bush Blue Lake 47), soybean (Harosoy), tomato (Bonnie Best) and celery		Liang et al. (2015)
sscle_10g075560SS	No	Faba bean (Broad Windsor), pea (Guido), green bean (Great North Tara) and soybean (Skylla)		Xu et al. (2015)

sscle_10g075560SS	No	Tomato (Bonnie Best), common bean (Bush Blue Lake 47), soybean (Harosoy), canola, *Arabidopsis* and sunflower		Liang et al. (2015)
sscle_10g075560SS	No	Soybean, canola, tomato (Better Boy), faba bean (Windsor), and pea (Sugar Daddy)		Li et al. (2018)
sscle_10g077630SS	No			Erental et al. (2007)
sscle_10g079320SS	No		Resistance to glycoside—containing antibiotics	Li and Rollins (2010)
sscle_11g082700SS	No	*Arabidopsis* and tomato (Bonny Best)	MAPK pathway	Erental et al. (2007)
sscle_11g083230SS	No			Yu et al. (2012)
sscle_11g083680SS	Yes	Canola and *Arabidopsis*	Apothecia formation	Yu et al. (2016)
sscle_11g083950SS	No	Tomato (Bonnie Best)	Apothecia production cAMP-signaling	Ii and Rollins (2007)
sscle_12g087830SS	No	Tomato (Rutger)	ROS regulation	Kim et al. (2011)
sscle_12g090900SS	No		pH-dependent regulation	Chen et al. (2004)

(continued)

Table 2.12 (continued)

sscle_14g099710SS	Yes	Common bean (Bush Blue Lake 47), soybean (Harosoy) and tomato (Bonnie Best)		Liang et al. (2015)
sscle_16g107670SS	Yes	*Arabidopsis* and tobacco	Interaction with PR1	Yang et al. (2018)
sscle_16g107670SS	Yes	Tobacco, oilseed rape and soybean		Pan et al. (2018)
sscle_16g107930SS	No	Common bean and tomato	Osmotic oxidative stress resistance	Zhu et al. (2019)
sscle_16g109570	No	Tomato (Bonnie Best), celery and tomato	Ascogonia formation, Apothecium development	Pan and Rollins (2017)

The assembly genome for the new version gene code is ASM185786v1 published in 2017 and ASM14694v1 for the old version published in 2011. "+" in the table represents that the corresponding phenotype of mutant is altered compared to wild type strain, while "–" means that the phenotype remains unchanged. NA means not assessed. The secretion signals were found using "SignalP-5.0." Unless specified, all deletions were generated with homologous recombination. The Latin names of host species used for pathogenicity test of mutants are: *Apium graveolens* (celery), *Arabidopsis thaliana* (Arabidopsis), *Brassica napus* (canola, oilseed rape, rapeseed), *Brassica rapa* subsp. *chinensis* (pak choi cabbage), *Capsicum frutescens* (hot pepper), *Cucumis sativus* (cucumber), *Fragaria ananassa* (strawberry), *Glycine max* (soybean), *Helianthus annuus* (sunflower), *Lycopersicon solanum* (tomato), *Nicotiana benthamiana* (tobacco), *Phaseolus vulgaris* (bush bean, common bean, green bean), *Pisum sativum* (pea), and *Vicia faba* (faba bean). The ecotype of *A. thaliana* mentioned in the table is Columbia-0. The specific local cultivars of host species are shown in brackets

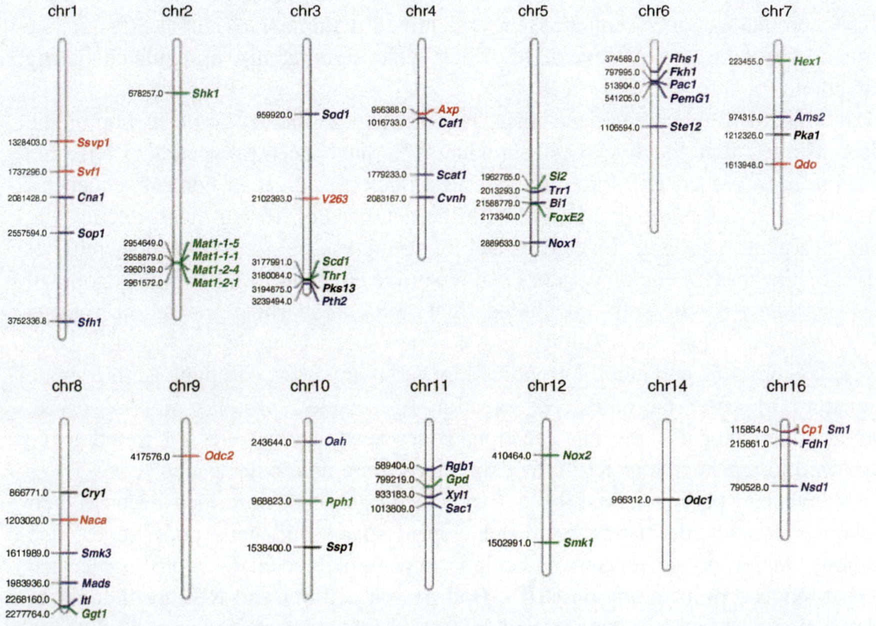

Fig. 2.34 Map positions of the genes in Table 2.12. The numbers on the left of each chromosome represent the locations of these genes. Genes labeled in green are involved in the development of *Sclerotinia sclerotiorum*, including hyphal growth, sclerotial formation, apothecial formation and etc. Genes labeled in red mainly play roles in virulence of *S. sclerotiorum*. Genes labeled in blue are involved in both while the ones labeled in black are involved in other biological processes. The chromosomal map was drawn using "Map Chart" software (Xia et al. 2020)

1000 bp (Sanger) and in many cases only 300 bp (Illumina) (Goodwin et al. 2016). Sequence assemblies based on reads that are substantially shorter than repeated elements do not produce a reliable contiguous sequence as the repeats are often collapsed into a single, short sequence during graph based assembly (Ekblom and Wolf 2014). This poses a particular problem to researchers interested in fungal effectors, which, as aforementioned, often cluster within repetitive genomic regions (Thomma et al. 2016). The Pacific Biosciences developed the single-molecule real-time sequencing (PacBio) platform, which produces reads of up to 60 kb (Goodwin et al. 2016). Thus far, two gapless fungal plant pathogen genomes have been assembled using PacBio reads, in combination with optical mapping—a process that uses fluorescent staining in conjunction with restriction enzymes to infer genomic structure based on restriction site patterns across chromosomes (Faino et al. 2015; van Kan et al. 2017). The existing *S. sclerotiorum* genome, although of high quality compared with many other fungal genomes, is not complete, with an estimated 1.6 Mb of missing sequence, based on comparison of the assembly with an optical map (Amselem et al. 2011). To improve upon the previous genome, its annotations, and subsequent effector analysis, Derbyshire et al. (2017) used the existing optical map in combination with PacBio and Illumina sequencing. Using this improved genome sequence, they identified a number of novel effector-like protein encoding genes, several of which were significantly upregulated during infection.

It was also recharacterized that TEs in the *S. sclerotiorum* genome and demonstrated that the five most abundant TEs have been subjected to RIP and exhibit an increased RIP index relative to a randomized set of control sequences. Additionally, comparative analysis with the genomes of several host-specific filamentous plant pathogens revealed a lack of strong association between secreted or effector-like protein encoding genes and repetitive elements or RIP in the genome of *S. sclerotiorum*. It was concluded that the broad host range necrotroph *S. sclerotiorum* does not exhibit a classic "two-speed genome" as found in host-specific biotrophic and hemibiotrophic fungi and oomycetes. Although a more subtle signature of potential enhanced selection of secreted proteins was observed, predicted effector-like protein encoding genes specifically were not found to be localized to repeat-rich or RIP-affected regions more than other genes.

Many plant pathogens exhibit a "two-speed" genome, described because their genomes contain alternating gene rich, repeat sparse and gene poor, repeat-rich regions. In fungi, the repeat-rich regions may be subjected to a process termed repeat-induced point mutation (RIP). Both repeat activity and RIP are thought to play a significant role in evolution of secreted virulence proteins, termed effectors. Derbyshire et al. (2017) presented a complete genome sequence of *S. sclerotiorum* generated using Single Molecule Real-Time Sequencing technology with highly accurate annotations produced using an extensive RNA sequencing data set. They have identified 70 effector candidates and have highlighted there in planta expression profiles. Furthermore, they characterized the genome architecture of *S. sclerotiorum* in comparison to plant pathogens that exhibit "two-speed" genomes. It has been shown that there is a significant association between positions of secreted proteins and regions with a high RIP index in *S. sclerotiorum* but they did not detect a correlation between secreted protein proportion and GC content. Neither did detect a negative correlation between CDS content and secreted protein proportion across the *S. sclerotiorum* genome. It was concluded that *S. sclerotiorum* exhibits subtle signatures of enhanced mutation of secreted proteins in specific genomic compartments as a result of transposition and RIP activity. However, these signatures are not observable at the whole-genome scale (Fig. 2.33; Derbyshire et al. 2017).

2.10.7.1 A Complete New Assembly of the *Sclerotinia sclerotiorum* Genome

A complete and accurately annotated genome of the destructive plant pathogenic fungus *S. sclerotiorum* has been presented by Derbyshire et al. (2017). The previous genome assembly was based on Sanger sequencing at a depth of 9.1× combined with optical mapping. Though this assembly was relatively good quality, it was estimated to have been missing approximately 1.6 Mb of sequence. The new assembly of *S. sclerotiorum* contains an additional 805,146 bp. Though this is only half the estimated missing sequence observed in the study. It was concluded that the new

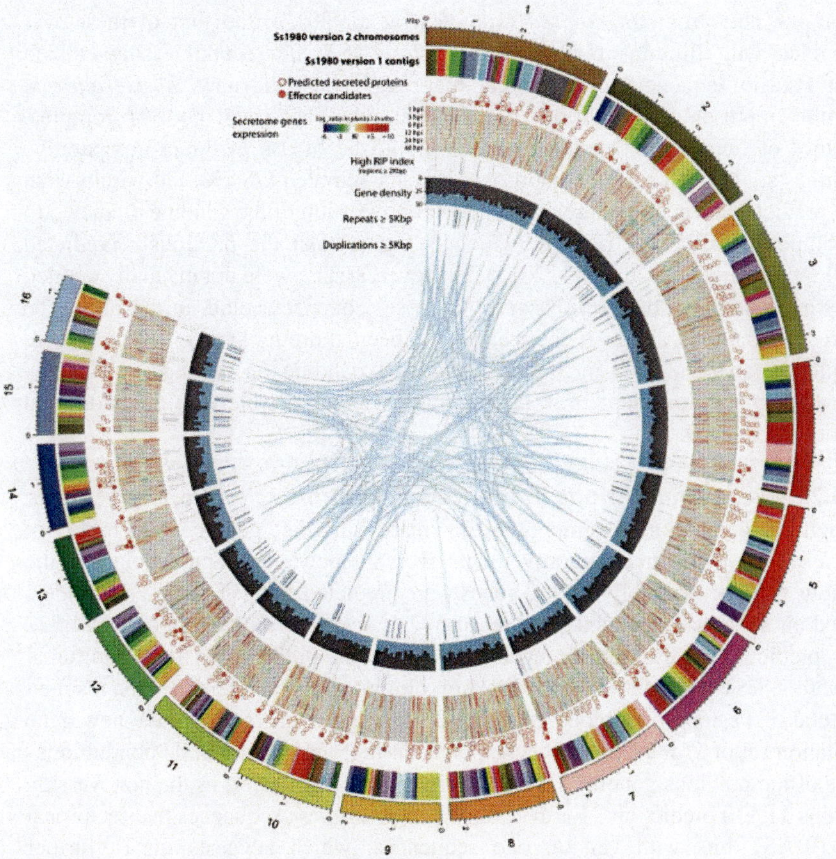

Fig. 2.35 Circos plot depicting features of the new *Sclerotinia sclerotiorum* assembly and positions of corresponding version one contigs. From outer to inner circular tracks: The 16 version two assembly (Ss1980 version 2) contigs colored to distinguish neighboring contigs; the contigs from the version one assembly (Ss1980 version 1) colored to distinguish neighboring contigs; circles depicting positions of secreted proteins and SsPEs—clear circles = secreted proteins, colored circles = SsPEs; log fold change in expression of secreted proteins from in vitro to each of the *B. napus* time points tested, from blue (downregulated) to red (upregulated); regions with a high RIP index as predicted by RipCal; repeat sequences from REPET de novo annotation of more than 5 kb. The lines crossing the center of the Circos plot join sequences of the same repeat element (Derbyshire et al. 2017)

S. sclerotiorum genome is practically complete, because 14 of the 16 chromosomes have been sequenced from telomere to telomere with virtually no gaps (Fig. 2.35).

The only gap present is a region of the nucleolar organizing complex (containing rDNA repeats). A similar result was found in the recently published PacBio-sequenced genome of the plant-pathogenic fungus *V. dahliae*. The only other two complete fungal plant pathogen genomes (apart from *V. dahliae*) are those of the wheat head blight fungus *Fusarium graminearum* (King et al. 2015), and the broad host-range necrotroph *B. cinerea* (van Kan et al. 2017); the first of these was assembled with Illumina mate pair technology and the second was assembled using PacBio sequencing and optical mapping. Thus, the new *S. sclerotiorum* assembly represents an addition to a thus far relatively small pool of complete genomes of fungal plant pathogens, and should be of use in future comparative genomics studies. Most of the additional bases assembled (78.7%) fall within what were predicted as repetitive regions. The total proportion of the genome predicted to be composed of TEs is 12.96%, which is higher than the previously predicted 7.7–9.5% (Amselem et al. 2011, 2015). However, as TEs were not manually curated following automated detection, several of these sequences could, in the future, be marked as dubious. Though they have not been thoroughly investigated, more complete analysis of TEs in *S. sclerotiorum*, expanding on findings by Santana et al. (2014) and Amselem et al. (2015) may now be possible with the additional repetitive sequence.

2.10.7.2 Gene Annotations in the New *Sclerotinia sclerotiorum* Genome

Though moderate improvements were made in the genome assembly of *S. sclerotiorum*, the main improvements were in new gene annotations. In the previous genome assembly, 14,522 genes were predicted using ab initio gene finders trained on a manually curated gene set of 542 *S. sclerotiorum* genes. These 14,522 gene predictions were further evaluated based on predictions from additional software's, EST and BLAST evidence, homology to TEs, and length. This resulted in a total of 11,860 nondubious gene calls (Amselem et al. 2011). In the new gene annotation set of *S. sclerotiorum*, when considering, only the 11,860 "nondubious" genes of the previous genome version, there are still 730 fewer, as the new version contains 11,130 predictions. Further, only 10,528 of these sequences had reciprocal best BLAST hits with version one sequences, which suggests that a further 602 sequences were either not present in the previous genome or had changed substantially enough in the version two prediction set for them to be unable to retrieve their corresponding loci as hits through BLAST analysis. Further inspection revealed that 1301 version two loci resulted from fusion of previous loci. This is in contrast to the 279 version two gene predictions that resulted from splitting of version one predictions. Thus, it can be surmised that the decrease in filtered gene predictions from the version one assembly to the version two is largely a result of the joining of previous separate loci.

Further analysis illustrated that the version two gene predictions were more accurate than the version one predictions. Blasting of amino acid sequences against the NCBI nr database indicated that version two sequences on average covered more of and had a higher identity to their best BLAST hits. This implies that the new gene predictions were able to obtain more BLAST hits from accurate protein predictions that represent truly conserved sequences. A similar analysis was performed to compare the *Parastagonospora nodorum* annotation set with a previous version (Syme et al. 2016). Accuracy at the protein level can also be inferred by the increased number of predicted functional domains and GO terms.

Additionally, version two predictions were more similar to their corresponding Cufflinks transcripts based on BLAST analysis. Though median coverage was only improved from 99% to 100% in the version two predictions relative to version one, a high interquartile range of 32% for the version one sequences, as opposed to 1% for the version two sequences, would indicate a decreased degree of variability in correspondence with Cufflinks transcripts in the version two predictions. That the version two predictions are more supported by the RNASeq data is also evident in the observation that the median coverage of version two CDSs is higher than those of version one (Derbyshire et al. 2017).

2.10.7.3 Identification of a Seven-Member Effector Gene Family in *Sclerotinia sclerotiorum*

In several plant pathogenic fungi, a clear link between the genomic positions of effector-like genes and TEs has been identified (Rouxel et al. 2011; Grandaubert et al. 2014; Amselem et al. 2015; Selin et al. 2016; Asman et al. 2016; Faino et al. 2016). It has been hypothesized that such positioning allows for expansion of effector gene families and subsequent diversification through mutation (Dong et al. 2015). This, in theory, could enhance the capacity of a fungus to rapidly adapt to newly evolved R genes or lost susceptibility loci in host plants. Additionally, RIP, a fungal defense mechanism against TEs, is thought to enhance mutation rates of effectors embedded within TE-rich regions. This process involves mutation of C to T bases in duplicated sequences. Repeat-rich regions containing effector-like proteins are usually gene-poor compared with the rest of the genome. This allows for housekeeping genes to remain unaffected by TE-associated evolutionary processes.

In the *B. napus,* pathogen *L. maculans*, it has been shown that most TEs are inactive. This is thought to be a result of an ancestral TE invasion followed by extensive RIP. Approximately 20% of effector-like sequences in *L. maculans* are associated with RIP-affected TE containing regions, whereas only 4.2% are associated with gene-rich regions. Though the sequences in repetitive regions generally do not appear to be in families that have expanded through TE activity, it has been shown that they are likely to have undergone RIP-like mutation (Rouxel et al. 2011). Derbyshire et al. (2017) carried out several analyses to test this hypothesis to determine whether SsPEs occurred in families, by conducting OrthoMCL analysis using the gene predictions of *S. sclerotiorum* and *B. cinerea* to detect recent paralogs. Further, to determine whether SsPE families were likely formed through TE activity, multiple sequence alignment of regions containing effectors and inspected them for the presence of TE sequences as predicted by REPET. It was identified a single SsPE family, which included seven proteins with more than 90% amino acid identity. The sequences in this family exhibited a single ortholog in *B. cinerea*, indicating that they possibly appeared subsequent to the divergence of *S. sclerotiorum* and *B. cinerea*. Six of the sequences in this family were already identified by Guyon et al. (2014), though a detailed investigation of their association with transposition activity has not been carried out. None of these sequences contained any predicted functional domains, underlining their possible specialized functions as effectors. Furthermore, multiple alignments of regions surrounding these sequences showed that they were within a 4–5 kb genomic segment that was repeated across six different chromosomes. Intriguingly, each of these SsPEs was embedded in either a partial or complete Gypsy LTR-retro element. This would indicate that these duplicated sequences arose as a result of TE activity. Thus, there is some evidence of effector evolution occurring through transposition in *S. sclerotiorum*, the first evidence of this phenomenon in this species. Based on this observation, it was hypothesized that *S. sclerotiorum* may harbor TE-rich, gene sparse, and potentially RIP affected regions that are enriched for effector proteins. To test this hypothesis, a comparative analysis of the genome of *S. sclerotiorum* with the "two-speed" genomes of *B. graminis*, *L. maculans*, and *P. infestans* was performed. It was found that out of these four microbial genomes, only that of *L. maculans* showed a bimodal GC content. This is consistent with results from the same analysis performed by Testa et al. (2016), who used the previous version of the *S. sclerotiorum* genome. This would indicate that GC content of *B. graminis*, *P. infestans*, and *S. sclerotiorum* is consistent across the whole genome.

There are two possible explanations (1) The genomes analyzed do not exhibit or only exhibit a limited level of RIP, and (2) the genomes exhibit a consistent level of RIP throughout, and do not harbor RIP-affected regions in distinct, GC-depleted compartments. It has been shown that *S. sclerotiorum* and *L. maculans* exhibit active RIP, whereas *B. graminis* does not; *P. infestans* is not thought to harbor RIP as RIP is a fungus-specific mechanism (Hane et al. 2015). The *S. sclerotiorum* displays an intermediate frequency of RIP and *L. maculans* displays a high frequency of RIP. *Blumeria graminis* shows an RIP-like signature in some TE copies but it is thought that this is the result of ancestral activity, as this species does not harbor the genes necessary for RIP. Unlike *L. maculans*, *S. sclerotiorum* has been shown to exhibit two kinds of RIP, both CpT → TpT and CpA → TpA transitions (Amselem et al. 2015). The results are consistent with this analysis as *S. sclerotiorum* was found to exhibit dominance of these two types of RIPs across copies of the most abundant TE. Additionally, the TpA/ApT RIP indices of the five most numerous TEs were significantly higher than random sets of control sequences. As *B. graminis* has only an extremely weak signature of ancestral RIP, it is unlikely to exhibit a bimodal GC content at the whole-genome scale. However, as RIP appears to be at a fairly significant level in *S. sclerotiorum*, the lack of bimodality of its genome GC content could indicate a lack of specific RIP-affected genome compartments such as those observed in *L. maculans*.

In *Leptosphaeria maculans*, there was a significant association between secreted proteins and TE sequences, but no significant association between effector-like proteins and TE sequences. However, adjusting a to 0.1 showed that there was possible evidence of a general trend toward association of effector sequences with repeats in *L. maculans*. These results are consistent with the previously proposed hypotheses that *B. graminis* and *P. infestans* do not exhibit active RIP but do compartmentalize effector sequences in TE-rich regions (Haas et al. 2009; Pedersen et al. 2012). It is also consistent with the hypothesis that *L. maculans* repeats have been subjected to extensive RIP, and were hence not as easily discovered by standardized repeat calling pipeline (RepeatMasker). Despite this, there was still evidence of association of secreted and effector-like proteins with repeat sequences in *L. maculans*. Conversely, in *S. sclerotiorum*, there was no significant association between secreted or effector-like proteins and TEs. This would suggest that TE activity is not a dominant mode of effector or secreted protein evolution in *S. sclerotiorum*. Indeed, the high degree of homology of SsPEs to proteins in other fungi (90% were homologous at e^{-10}) would indicate that they may be fulfilling conserved functions that are not under extensive selection pressure exerted by host species. To test whether secreted and effector-like proteins are collocated with RIP-affected sequence in *S. sclerotiorum*, the same analysis was performed as for the repeat sequences for regions with a high RIP index as specified by RipCal. This showed that *L. maculans* exhibited a significant association between both secreted proteins and effector-like proteins and RIP-affected sequences, whereas *B. graminis* exhibited a significant negative association between effector and secreted proteins and RIP-affected sequence.

Intriguingly, in *S. sclerotiorum*, there was a significant association between secreted proteins and RIP-affected sequence. However, there was no association between effector-like sequences and RIP-affected sequence. This would indicate that there may be some hot spots of RIP associated with secreted protein diversification in the *S. sclerotiorum* genome. Indeed, several regions can be readily identified in Fig. 2.35. At the distal ends of chromosomes 7, 15, and 16, there appear to be clusters of secreted and effector-like proteins that occur in proximity to regions with a high RIP index and several repeat sequences. However, the impact of these regions on evolution and host specificity remains to be elucidated.

Based on these observations, it was hypotheses that secreted proteins were associated with AT-rich regions in *S. sclerotiorum*. Furthermore, test whether they were associated with gene sparse regions, which would be indicative of a significant level of compartmentalization away from housekeeping genes. It was found that in *S. sclerotiorum*, *B. graminis*, and *P. infestans*, there was no correlation between secreted protein proportion and GC content. However, in *L. maculans*, there was a significant negative correlation between GC content and secreted protein proportion. This would indicate that in *S. sclerotiorum*, there is no significant compartmentalization of secreted proteins into ATrich (i.e., RIP-affected) sequence regions. Though it is possible that there are hotspots of this kind of activity in the genome, a 100-kb sliding window was too large to detect them. However, as it was able to detect an association in *L. maculans*, for which RIP-affected genome compartments with effectors have been characterized, it would at least demonstrate that in *S. sclerotiorum* this kind of phenomenon is not as extensive as in *L. maculans*. In addition, the genomes of *B. graminis*, *P. infestans*, and *L. maculans* exhibited a significant negative correlation between CDS content (used to infer gene richness) and secreted protein proportion. This supports previous observations that effector-like and secreted proteins are often positioned in gene sparse regions in these species. Conversely, there was no correlation between gene content and secreted protein proportion in *S. sclerotiorum*. Again, it is possible that there are small hotpots of secreted protein diversification that are gene poor; Fig. 2.35 illustrates such candidate regions (e.g., the distal end of chromosome 7). However, at a macro scale, this phenomenon was not identified in *S. sclerotiorum* as it was in organisms previously characterized (Raffaele and Kamoun 2012). Therefore, it was proposed that compartmentalization of secreted proteins in *S. sclerotiorum*, if at all present, is far subtler than it is in the host-specific biotrophic and hemibiotrophic organisms (Derbyshire et al. 2017).

2.11 *Brassica–Turnip Mosaic Virus* Genome

2.11.1 Introduction

The genus Potyvirus contains more than 200 members or possible members and belongs to the largest plant virus family, Potyviridae (Berger et al. 2000). Potyviruses have flexuous filamentous particles that contain an approximately 10 kb positive-sense single-stranded RNA that is covalently linked to a virus genome-linked protein (VPg) at the 59 end and polyadenylated at the 39 end (Revers et al. 1999). The RNA has a single open reading frame that is translated into a large polyprotein, which is proteolytically cleaved into mature proteins by three virus-encoded proteinases (Riechmann et al. 1992; Urcuqui-Inchima et al. 2001). The potyvirus P3 protein is proteolytically cleaved from polyprotein by HC-Pro and NIa-Pro, resulting in either the P3 protein or its precursor P3-6K1 (Riechmann et al. 1992). Among potyviruses, there is relatively little similarity in P3 proteins compared to other proteins (Urcuqui-Inchima et al. 2001). The P3 proteins are thought to be involved in virus replication (Merits et al. 1999), accumulation (Klein et al. 1994), symptomatology (Chu et al. 1997; Saenz et al. 2000), resistance breaking (Hjulsager et al. 2002; Jenner et al. 2002, 2003; Johansen et al. 2001), and cell-to-cell movement (Dallot et al. 2001; Johansen et al. 2001). The *Turnip mosaic virus* (*TuMV*) is a member of the genus Potyvirus (Berger et al. 2000) and infects cruciferous plants throughout the world (Walsh and Jenner 2002). Cruciferous plants include a large variety of economically important crops belonging to different genera (Gomez-Campo 1999). The *TuMV* has adapted to such diversified cruciferous crops and the many *TuMV* isolates in the world have been classified into several strains or pathotypes (Fujisawa 1990; Green and Deng 1985; Jenner and Walsh 1996; Provvidenti 1980; Stavolone et al. 1998; Stobbs and Shattuck 1989). Many *TuMV* isolates have been collected around the world and grouped into four lineages by sequencing and phylogenetic analysis of sequences encoding P1 and CP (Ohshima et al. 2002), CP (Sanchez et al. 2003), or the entire polyprotein (Tomimura et al. 2003). These classifications were based on host range, but the factors affecting the virus host range in cruciferous plants are not clear (Suehiro et al. 2004).

2.11.2 Genome Sequence of *Turnip Mosaic Virus*

The genomes of 48 *TuMV* isolates from Europe and five from USA were sequenced. These were analyzed along with 102 other genomic *TuMV* sequences, mostly of East

Asian isolates, obtained from the international sequence databases. Most were 9798 nt in length; a few were one to three nucleotides shorter in the terminal UTRs, whereas those of OM isolates were 6 nt (i.e., two codons) shorter in the polyprotein region. All of the motifs reported in potyvirus genes, encoded proteins, and the "Pretty Interesting Potyviridae ORF" (P3N-PIPO) were present. The P1 gene and its encoded protein were the most variable, and these had few totally conserved residues or compact motifs. The P3 gene and its encoded protein were only slightly less variable. The sequences are available in the GenBank, EMBL, and DDBJ databases with Accession Codes AB701690-AB701742 (Gibbs and Ohshima 2010; King et al. 2012).

2.11.3 The *Turnip Mosaic Virus* Genome Characteristics and the Virus Phylogeny

The full viral genome sequence of the Croatian isolate CRO184A was obtained and deposited in GenBank under the accession number KF595121. Phylogenetic analysis (Fig. 2.36) has placed it in the world-B phylogenetic lineage (Tomimura et al. 2003). The isolate shared the highest nucleotide identity (98–45%) with the French isolate FRA2 from *B. napus* (Jenner and Walsh 1996) belonging to the same lineage and the host type B (Tomimura et al. 2004). As it was previously confirmed for the French isolate FRA2 (Nguyen et al. 2013a), the recombination analysis program did not identify any inter lineage recombination events within CRO184A sequence either. Also, the CP-region analysis, including more sequences of European isolates available in GenBank, confirmed its highest homology (99%) to FRA2 and more distant relationship with isolates from Eastern Europe. The presence of DAG-motive (Atreya et al. 1990) was confirmed from translated CP gene portion of the CRO184A *TuMV* genome by using the appropriate options in the abovementioned sequence analysis programs (Seruga Musić et al. 2014).

2.11.4 Genome Analysis of *Turnip Mosaic Virus* Co-infecting *Brassica*

Molecular characterization of a pathogenic complex infecting winter oilseed rape (*Brassica napus* ssp. *oleifera* (DC.) Metzg.) plants showing typical rape phyllody symptoms along with some atypical changes revealed *Phytoplasma* ("*Candidatus Phytoplasma*") presence which was confirmed by PCR-RFLP and 16S rRNA gene sequencing. Phylogenetic analysis of phytoplasma *amp*, tufB, secY, groEL, and ribosomal protein genes confirmed its affiliation to the "Ca. *P. asteris*" species. However, in the *amp* gene encoding a specific protein crucial for insect transmission specificity, significant SNPs were found. Biological and serological tests revealed the co-infection with *Turnip mosaic virus* (*TuMV*). The phylogenetic analysis of full *TuMV* genome sequence, the first reported from the Balkans, classified it into the world-B phylogenetic lineage. A pathogenic complex consisting of "Ca., *P. asteris*" and *TuMV* have been found to co-infect oilseed rape plants for the first time has been

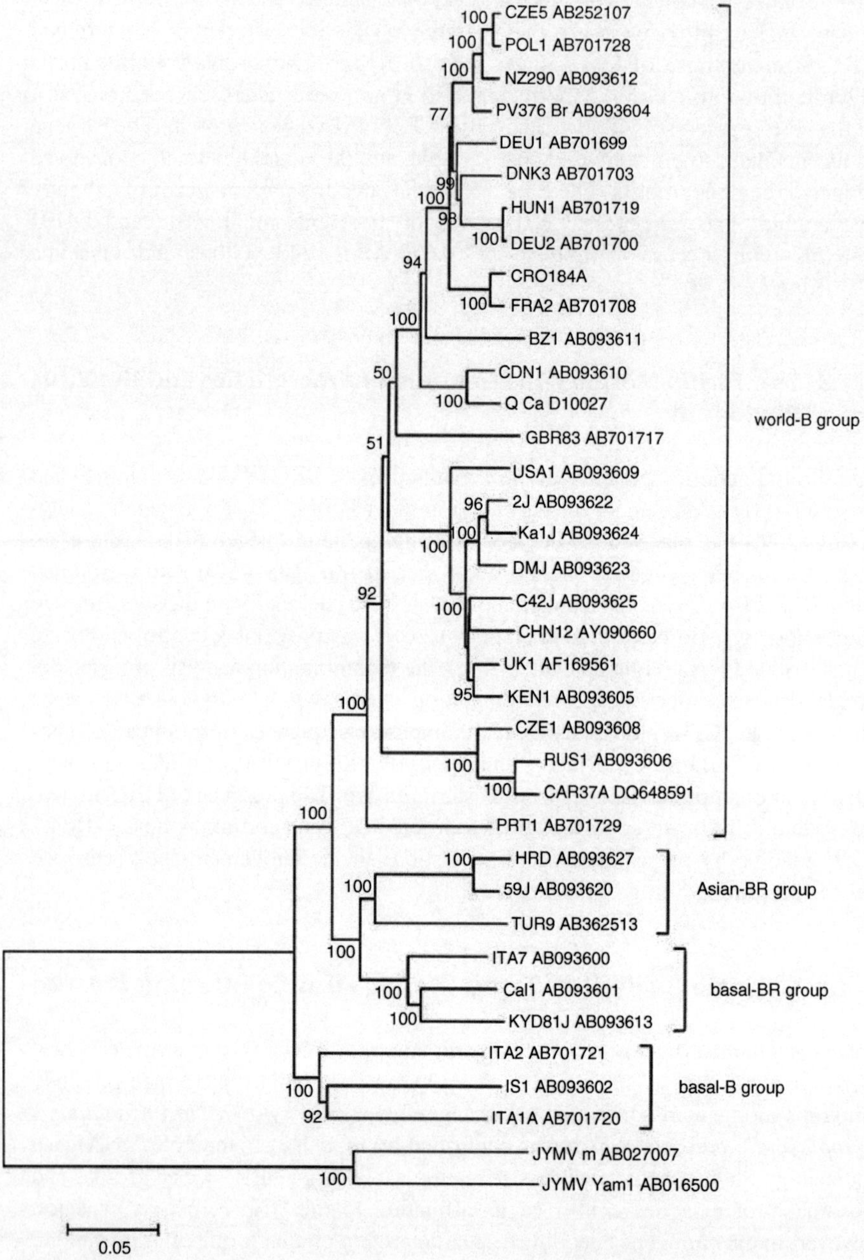

Fig. 2.36 Phylogenetic tree (NJ method, Kimura 2-parameter) based on the complete polyprotein sequences of nonrecombinant *TuMV* isolates. The Croatian oilseed rape isolate CRO184A shares 98.45% nucleotide identity with *TuMV* isolate FRA2. Clusters are delineated in accordance with the phylogenetic groups of Tomimura et al. (2003) and Nguyen et al. (2013a, b). (Seruga Musić et al. 2014)

molecularly characterized. Rape phyllody is a serious problem in rapeseed production. The molecular information from this first multigene analysis of "Ca. *P. asteris*" strain associated with rape phyllody as well as the first report of the complete sequence of *TuMV* isolate from the Balkans is a starting point for understanding the disease complexity and management (Seruga Musić et al. 2014).

2.11.5 Phylogenetic Analysis

Separate phylogenetic trees were estimated for the poly-proteins and for the individual HC-Pro, P3, NIb, and CP genes/regions of the 155 isolates. Inconsistent and poorly supported relationships among the resulting trees indicated that some isolates were recombinants, as found previously. Accordingly, Huson and Bryant (2006) checked the sequences for recombination using split decomposition. The OM isolates formed a single nonrecombinant lineage distinct from all the *TuMV*-BI lineages, and closest to the basal-B lineage (Fig. 2.37a). The extent of the reticulations at the base of the world-B and Asian-BR lineages suggests that most of these sequences are recombinants. Further analysis using recombination-detection methods confirmed that many of the sequences were recombinants, with only 37 of the sequences showing no significant evidence of recombination.

Fifty sequences were interlineage recombinants (had "parents" from different lineages; red names in Fig. 2.37a), whereas 68 sequences were intralineage recombinants (had "parents" from the same lineage; blue names in Fig. 2.37a). When the interlineage recombinant sequences were removed, and the remaining sequences were analyzed again by split decomposition, the branching patterns of the major lineages were resolved (Fig. 2.37b). Figure 2.38 shows a maximum-likelihood tree of the amino acid sequences encoded by the few genomic sequences that had no evidence of recombination. It confirms that the OM isolates forma monophyletic lineage that is sister to the *TuMV*-BI lineages and closest to the basal-B group. No recombination sites were detected in the 59 and 39 noncoding regions. However, they were found throughout the coding regions of many of the genomes, especially in the P1 gene and CI-VPg regions as reported earlier. Therefore, the HC-Pro, P3, NIb, and CP genes/regions of the genomes that were not intralineage recombinants, showed no intragenic recombination, selected for phylogenetic analysis using maximum likelihood in PhyML version 3 and neighbor joining in PHYLIP. The resulting trees grouped the *TuMV* BI sequences into the four major groups reported earlier, with the OM isolates grouped as a monophyletic sister lineage to all of the other *TuMV* lineages. An exception to this occurred in the HC-Pro trees, where the OM lineage was sister to the basal-B lineage.

All of these topologies were supported by high boots trap values (Ohshima et al. 2007). These results raised the question as to whether the OM isolates were closer to the TuMV-BIs or to the out group viruses. In a maximum-likelihood phylogeny of 37 nonrecombinant *TuMV*-BI genomes, the four *TuMV*-OM genomes, and the four out group genomes, the mean patristic distances between the out group sequences and the BI and OM genomes were 7.9761.06 and 7.6361.09 subs/site, respectively,

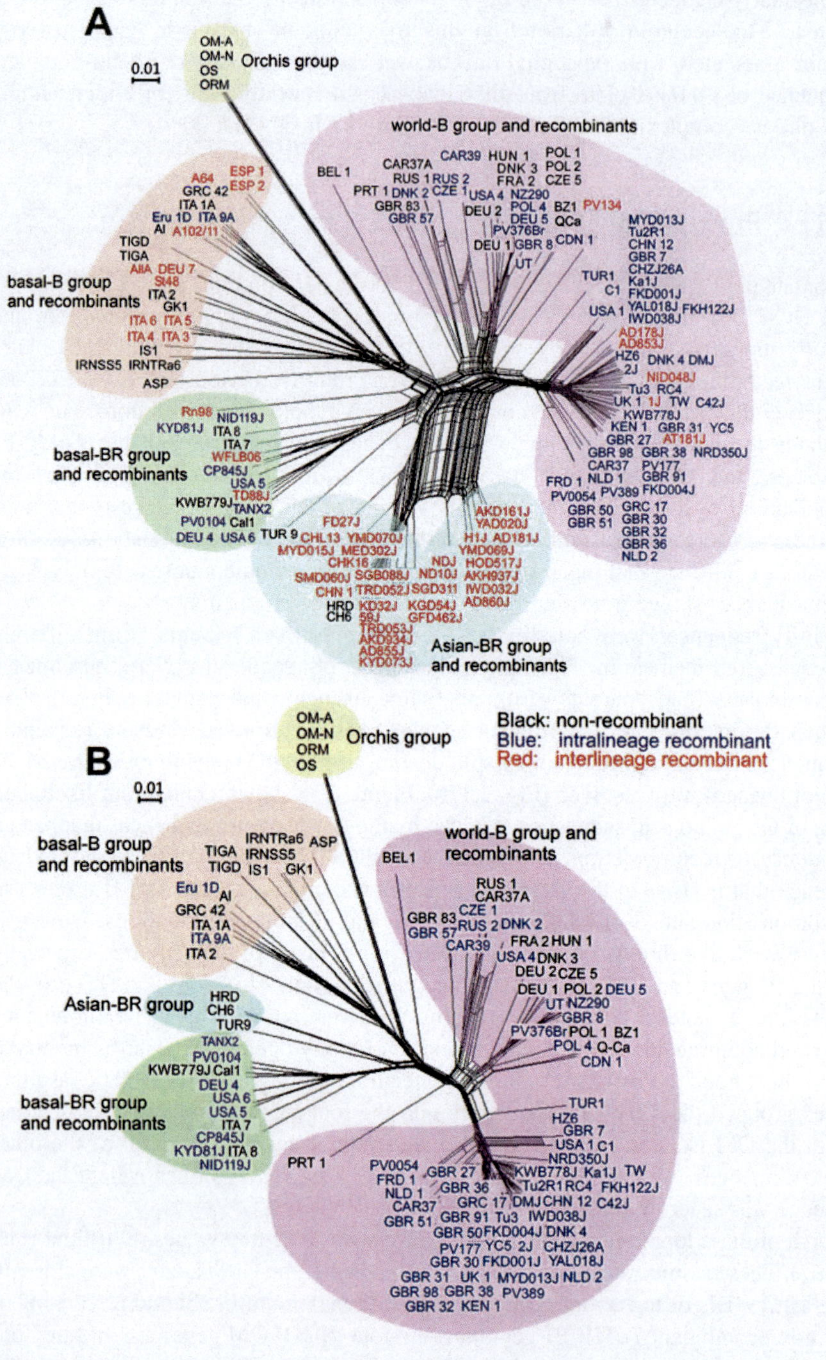

Fig. 2.37 Split-decomposition phylogenetic networks. Networks inferred from (**a**) polyprotein sequences of 155 *Turnip mosaic virus* (*TuMV*) isolates and (**b**) 105 sequences remaining after removing inter lineage recombinants. The isolates of nonrecombinants (acronyms in black), intrarecombinants (acronyms in blue) and inter-recombinants (acronyms in red) are separately listed. doi: https://doi.org/10.1371/journal.pone.0055336.g001 (Nguyen et al. 2013a, b)

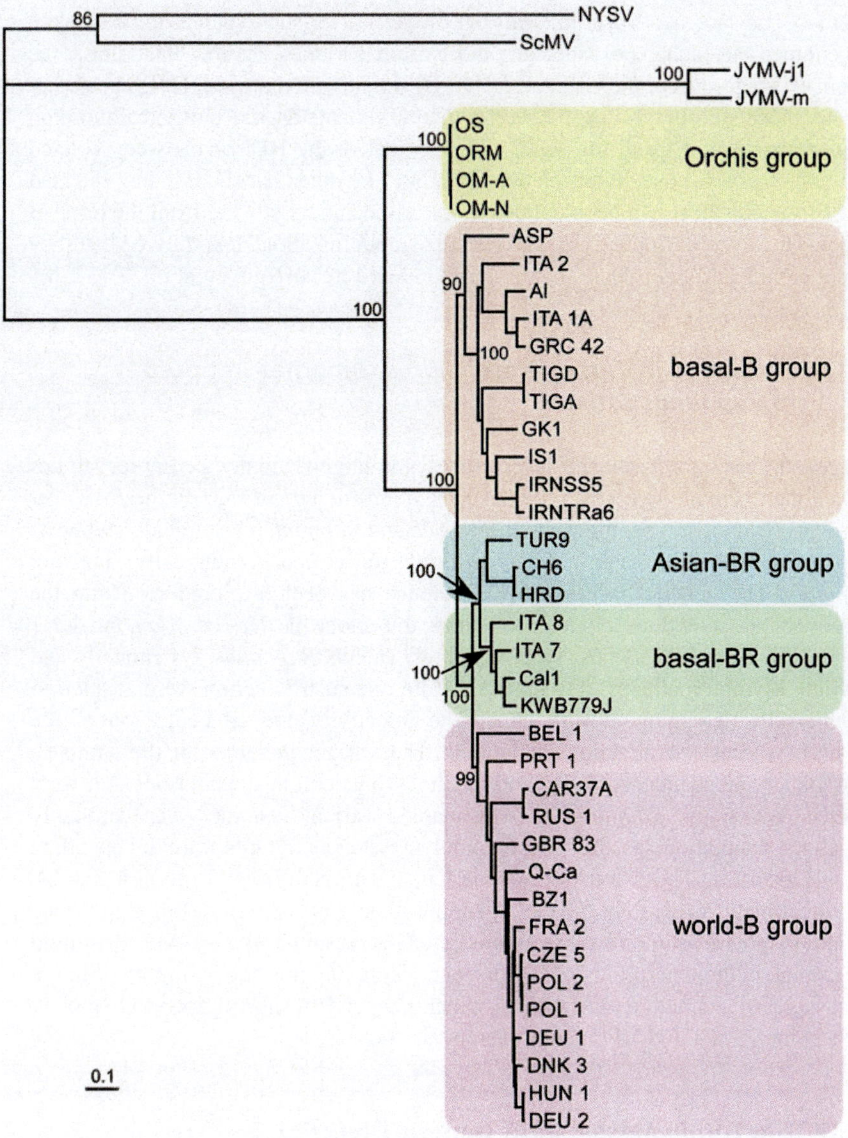

Fig. 2.38 Maximum-likelihood tree of the complete polyprotein sequences of 37 nonrecombinant *Turnip mosaic virus* isolates. Nodes are labeled with bootstrap support percentages (Nguyen et al. 2013a, b)

but only 1.0160.03 subs/site between the BI and OM genomes. Thus, the BI and OM genomes represent distinct populations that are much closer to one another than to the out group viruses. The sequences of the OM isolates are closer to those of *TuMV* isolates throughout the genome, and both are very different from the out group genomes. They are also closer in terms of other characteristics, such as the lengths of the genomes and each gene, especially those of the P1 and CP genes. In addition, the protein cleavage sites of the OM and *TuMV*-BI isolates are more similar to each other than to those of out groups. It seems that OM isolates are close to, although biologically distinct from, the *TuMV*-BIs. Several of the BI isolates were isolated from non-*Brassica* hosts other than orchids, but like other *TuMV*-BIs, they infected most Brassicas. Most had been collected in Europe and most were from the basal-B lineage. However, a Monte Carlo "provenance randomization" test showed that they did not significantly cluster in maximum-likelihood trees (Nguyen et al. 2013a, b).

2.11.6 Mutations in *Turnip Mosaic Virus* Genomes Adapted to *Raphanus sativus*

The genetic basis for virulence in potyviruses is largely unknown. Earlier studies showed that there are two host types of *Turnip mosaic virus* (*TuMV*); the *Brassica/Raphanus* (BR)-host type infects both *Brassica* and *Raphanus* systemically, whereas the *Brassica* (B)-host type infects *Brassica* fully and systemically, but not *Raphanus*. The genetic basis of this difference has been explored by using the progeny of an infectious clone, p35Tunos; this clone is derived from the UK1 isolate, which is of the B-host type, but rarely infects *Raphanus* systemically and then only asymptomatically. Two inocula from one such infection were adapted to *Raphanus* by passaging, during which the infectivity and concentration of the virions of successive infections increased. The variant genomes in the samples, 16 in total, were sequenced fully. Four of the 39 nucleotide substitutions that were detected among the *Raphanus sativus*-adapted variant genomes were probably crucial for adaptation, as they were found in several variants with independent passage histories. These four were found in the protein 1 (P1), protein 3 (P3), cylindrical inclusion protein (CI) and genome-liked viral protein (VPg) genes. One of four "parallel evolution" substitutions, $_{3430}$G-A, resulted in a$_{1100}$ Met-Ile amino acid change in the C terminus of P3. It seems likely that this site is important in the initial stages of adaptation to *R. sativus*. Other independent substitutions were mostly found in the P3, CI and VPg genes (Tan et al. 2005).

2.11.7 The *Turnip Mosaic Virus* Genetic Diversity

The *Turnip mosaic virus* is a member of the genus Potyvirus (type species Potato virus Y) in the family Potyviridae. The *TuMV* is the only potyvirus known to infect *Brassicas* (Walsh and Tomlinson 1985). The *TuMV* genome consists of a positive-sense single-strand RNA molecule of about 9830 nucleotides in length. The gene contains a single open reading frame (ORF) flanked by two untranslated regions (Basso et al. 1994). The ORF is translated into a single polyprotein, which is co- and post-translationally processed by three virus-encoded proteases (Walsh and Jenner 2002). The proteins released comprise P1, helper component protease (HCPro), P3,

6K1, cytoplasmic/cylindrical inclusion (CI), 6K2, virus-encoded genome linked (VPg) protein, nuclear inclusion protein a (NIa), nuclear inclusion protein b (NIb) and the coat protein (CP) (Walsh and Jenner 2002). The *Turnip mosaic virus* probably originated from a virus of wild orchids in Germany and, while adapting to wild and domestic Brassicas, spread to Asia Minor via southern Europe no more than 700 years ago (Yasaka et al. 2017). Various strains and pathotypes of *TuMV* have been defined based on their ability to infect certain plant species or cultivars (Shattuck 1992). An early scheme used symptom type on cabbages and *Nicotiana glutinosa* to distinguish *TuMV* strains (Yoshii et al. 1998). Liu et al. (1996) also used symptom types and disease severity indices on a variety of *Brassica* species to define seven strains (Tu 1–7). Other schemes are based on differential lines of a single species, either *B. rapa* (strains C1–C4; Provvidenti 1980; strains C1–C5; Green and Deng 1985) or *B. napus* (pathotypes 1–12) (Walsh 1989; Jenner and Walsh 1996; Li et al. 2019).

2.12 *Brassica–Xanthomonas* Genome

2.12.1 Introduction

Three isolates of *Xanthomonas campestris* pv. *campestris* (*Xcc)* have been fully sequenced, including the type strain ATCC33913 (NCPPB528; LMG568; ICMP13; DSM3586) (da Silva et al. 2002), isolate 8004 (a spontaneous rifampicin resistant mutant from NCPPB1145) (Qian et al. 2005) and the industrial high-xanthan-producing isolate B100 (Vorhoelter et al. 2008). The type strain has been identified as race 3 (Vicente et al. 2001), isolate 8004 has been assigned to race 9 (Fargier and Manceau 2007) and B100 belongs to race 1. The isolate 8004 has featured in several studies of phytopathological properties, such as the secretion of extracellular enzymes and exopolysaccharides (Tang et al. 1991), cell–cell signaling and biofilm formation (Torres et al. 2007). Further information about the *Xcc* genomes is summarized in Table 2.13.

An isolate of *X. campestris* pv. *raphani* (756C), a closely related pathogen that causes a nonvascular, leaf spot pathogen on Brassicaceae and other hosts such as tomato, has also been sequenced (Bogdanove et al. 2011). These *X. campestris* genomes comprise circular chromosomes of approximately 5,000,000 base pairs (bp), have a high G + C content and do not carry plasmids. The comparison of the sequences of *Xcc* type strain ATCC33913 and 8004 indicated that significant genomic-scale rearrangement across the replication axis between two IS1478 elements and the loss and acquisition of blocks of genes, rather than point mutations, constitute the main genetic variation between the strains (Qian et al. 2005). The sequence of B100 is extensively collinear to the strain 8004, but differs from the type strain ATCC33913 by the inversion of a large chromosomal fragment (Vorhoelter et al. 2008). This may indicate that the strains 8004 and B100 have originated via recent recombination events (Qian et al. 2005). He et al. (2007) have constructed a whole-genome microarray of the *Xcc* strain 8004 and have used this resource to

Table 2.13 *Xanthomonas campestris* pv. *campestris* strains that have been completely sequenced and features of the genome (Vorhoelter et al. 2008)

Isolate	ATCC 33913	8004	B100
Other designations	NCPPB (528), HRI5212, LGM 568, CFBP 5241	Rifampicin-resistant laboratory strain derived from NCPPB 1145, CFBP 6650	
Host of origin	Brussels sprouts (*Brassica oleracea* var. *gemmifera*)	Cauliflower (*B. oleracea* var. *botrytis*)	
Country of origin	UK	UK	
Year of isolation	1957	1958	
Race	3	9	1
Size (bp)	5,076,187	5,148,708	5,079,002
G + C content (%)	65.0	64.9	65.0
CDS—predicted number	4181	4273	4471
CDS—function assigned	2708	2671	2878
Ribosomal RNA operons	2	2	2
Transfer RNAs	54	54	54
Insertion sequence elements	109	115	59
Reference	da Silva et al. (2002)	Qian et al. (2005)	Vorhoelter et al. (2008)

CDS coding DNA sequence, *CFBP* Collection Française de Bactéries Phytopathogènes, *INRA* France, *HRI* School of Life Sciences (ex-Warwick HRI)

study the genetic diversity and host specificity of this pathogen by array-based comparative hybridization with genomes of 18 pathogenic isolates collected from different host plants and various geographical regions of China. A core set of 3405 conserved coding sequences was identified and 730 coding sequences that were absent or highly divergent among the isolates. Of the 304 proven or postulated pathogenicity genes currently known in *Xanthomonas*, 258 were conserved and 46 were highly divergent among the isolates.

These current reference genomes of *Xcc* are useful for genome-wide comparisons with publicly available genomes from other *Xanthomonas* species, including strain 306 of *X. axonopodis* pv. *citri* (da Silva et al. 2002), which represents the citrus canker pathogen, strain 85-10 of *X. axonopodis* pv. *vesicatoria* (Thieme et al. 2005), which represents the bacterial spot pathogen which specifically attacks pepper (and not tomato), strains of the rice bacterial blight pathogen *X. oryzae* pv. *oryzae*, KACC10331 (Lee et al. 2005), MAFF 311018 (Ochiai et al. 2005) and PXO99A (Salzberg et al. 2008), strain BLS256 of *X. oryzae* pv. *oryzicola* (Bogdanove et al. 2011), strain GPE PC73 of *X. albilineans* (Pieretti et al. 2009) and strains from *X. vasicola* pv. *vasculorum* and *musacearum*, the cause of banana wilt (Studholme et al. 2010). Lu et al. (2008) compared six gene clusters associated with pathogenicity across genomes of eight *Xanthomonas* isolates representing vascular and non-vascular diseases of rice, *Brassicas*, pepper and citrus. Interestingly, plasmid DNA was only observed in the reference isolates of *X. axonopodis* pv. *citri* and pv. *vesicatoria* (Vicente and Holub 2013).

2.12.2 Regulation of Secretion Systems of *Xanthomonas* Genome

Reference genomes of *Xcc* have begun to reveal an extensive inventory of genes that may be required for plant associations on the basis of homology with known genes from other *Xanthomonas* or *Pseudomonas* species. Most importantly, genome-wide comparisons have revealed three secretion systems (types II, III and IV) that have been associated with pathogenic bacteria of plants or animals. The type II secretion system (T2SS) mediates the transport of proteins into the extracellular space. This system can secrete plant cell wall-degrading enzymes, including cellulose, polygalacturonase, xylanase, and protease. Two classes of T2SS have been identified in *Xanthomonas*: *xps* and *xcs*. The *xps* cluster is present in all *Xanthomonas* genomes currently sequenced and in other genera (*Xylella*), whereas the *xcs* genes are only present in some of the *Xanthomonas* genomes, including *Xcc* and *X. campestris* pv. *raphani* (Lu et al. 2008). Qian et al. (2005) obtained six pathogenicity-deficient mutants (*xpsD*, *xpsE*, *xpsF*, *xpsK*, *xpsL*, and *xpsM*) related to the *xps* system, whereas no pathogenicity deficient mutants were found on the 12 annotated genes related to the *xcs* system; therefore, the *xps* genes are related to pathogenicity, whereas the *xcs* genes may have other roles not essential for pathogenicity or may be nonfunctional.

The mutation of DsbB, a protein involved in disulfide bond formation in the periplasm of *Xcc*, reduced virulence, cell mobility and bacterial growth in planta and resulted in ineffective T2SS and T3SS (Jiang et al. 2008b). The type IV secretion system (T4SS) is important for the release of macromolecules, and this system is used by Gram-negative bacteria to translocate protein and DNA substrates across the cell envelope into target cells (Souza et al. 2011). However, the role of T4SS in the pathogenicity of *Xanthomonas* is uncertain. The T4SS carried in the race 9 isolate 8004 is encoded mainly by genes in the *virD4* and the *virB* cluster that consists of nine open reading frames (ORFs). Qian et al. (2005) identified a mutated gene encoding the channel-forming protein *VirB8*, and concluded that the T4SS contributes to the virulence of the pathogen. However, He et al. (2007) have shown that deletion of *virD4* and the *virB* cluster of strain 8004 does not affect the virulence of the pathogen. The T3SS is generally thought to be essential for pathogenicity and for the initiation of disease in susceptible host plants, and can be involved in the elicitation of cell death and other defense responses in resistant plants.

The T3SS is encoded by the *hrp* cluster of genes and translocates effector proteins directly into the plant cells; it is also possible that effectors can exit the bacterial phytopathogens via the Hrp pathways (Lindgren 1997). Individual genes have been named *hrp*, *hrp*-conserved (*hrc*) or *hrp* associated (*hpa*). Most *hrp* gene sequences are only found in *Xanthomonas* and related genera, whereas *hrc* genes are conserved in many plant and animal pathogens. In *Xcc*, the *hrp* cluster consists of 26 genes extending from *hpa2* to *hrpF* (da Silva et al. 2002). The *Xcc* genomes do not contain *hpa3* and the N-terminal region of *xopF1*, although these genes are present in other *Xanthomonas* species, including *X. campestris* pv. *raphani* (Lu et al. 2008). Four mutants of genes encoding the T3SS machinery (*hrcJ*, *hrcU*, *hrcV*, *hrpE*) and one encoding a regulatory protein (*hrpG*) were nonpathogenic (Qian et al. 2005). Mutations in the regulator *hrpG* of strain 8004 can make all the *hrp* genes express constitutively and enhance the intensity of the HR reaction in pepper (Jiang et al. 2006). Wei et al. (2007) showed that *hpaR*, a putative *marR* family transcriptional regulator, is essential for the pathogenicity of strain 8004 on cabbage, is required for the HR response on the nonhost pepper and is under the positive control of the two key *hrp* gene regulators HrpG and HrpX. The GntR family is a frequent group of helix–turn–helix transcriptional regulators in bacteria. One of the six GntR regulators, HpaR1, positively and negatively affects the expression of HR and pathogenicity *hrp* genes that encode the type III secretion via hrpG (An et al. 2011). Several genes that are regulated in a HrpX-dependent manner possess a consensus nucleotide sequence TTCGB–N15–TTCGB (B can be base C, G, or T), which has been termed the plant-inducible promoter box, or PIP box. The detection of PIP boxes provides a refined tool for the identification of candidate genes in the hrpX regulon, as well as effector protein genes of the type III pathway. The 17 putative PIP box sequences have been described in the promoter regions of *Xcc* ATCC33913 (da Silva et al. 2002) and 56 genes have been predicted to have PIP boxes in strain 8004 (Jiang et al. 2009).

2.12.3 Function of Type III Effector Genes of *Xanthomonas*

Many studies have been conducted to identify and understand the function of type III effectors in *Xanthomonas* species. White et al. (2009) observed type III effectors from *Xanthomonas* and classified them into 39 different groups (called Xop) reflecting sequence similarity. From these, 21 appear to be present in the *Xcc* strain ATCC33913. Nine effectors, considered to be core effectors, are present in almost all *Xanthomonas* isolates, with the exception of *X. albilineans* and, in some cases, *X. campestris* pv. *raphani*; other effectors are found in a more limited number of species/isolates (Bogdanove et al. 2011; Ryan et al. 2011). From the extensive list of genes encoding putative T3SS proteins in *Xcc* and *X. campestris* pv. *raphani*, summarized in Table 2.14, effort has begun to identify potential effector proteins

Table 2.14 Occurrence, features, and function of putative and proven effector/avirulence genes of *Xanthomonas campestris* pv. *campestris* (Vicente and Holub 2013)

Effector class	Synonym	Gene ID in ATCC33913	Gene ID in 8004	Gene ID in B100	Gene ID in *Xcr* 756C	PIP box	Features and function on pathogenicity
Core effectors[a]							
AvrBs2		XCC0052	XC 0052	xcc-b100 0057	No	Yes	Glycerophosphoryl diester phosphodiesterase
							Avirulence in *Brassica juncea*, *B. carinata,* and *B. oleracea* and HR on pepper (Ignatov et al. 2002) or no effect (Castaneda et al. 2005)
XopK		XCC2899	XC 1210	xcc-b100 1254	No		Unknown
XopL	XopLR	XCC4186	XC 4273	xcc-b100 4400	No	Yes	Leucine-rich protein
							Reduced virulence on radish (Jiang et al. 2009)
XopN		XCC0231	XC 0241	xcc-b100 0253	No		ARM/HEAT repeat
							Reduced virulence on radish (Jiang et al. 2008a)
XopP		XCC1247	XC 2994	xcc-b100 3057	XCA_1500	Yes	Unknown
							Reduced virulence on radish (Jiang et al. 2009)
XopQ		XCC1072	XC 3177	xcc-b100 3274	No	Yes	Putative inosine–uridine nucleotide *N*-ribohydrolase
							Reduced virulence on radish (Jiang et al. 2009)
XopR		XCC0258	XC 0268	xcc-b100 0280	XCA_4254		Unknown
XopX		XCC0529	XC 0541	xcc-b100 0558	No		Methionine-rich protein
		XCC0530	XC 0542	and 0559			
XopZ		XCC1975	XC 2210	xcc-b100 2274	No		Unknown

(continued)

Table 2.14 (continued)

Effector class	Synonym	Gene ID in ATCC33913	Gene ID in 8004	Gene ID in B100	Gene ID in *Xcr* 756C	PIP box	Features and function on pathogenicity
Other effectors[b]							
AvrBs1		XCC2100	XC 2081	xcc-b100 2396	No	No	HR on pepper (He et al. 2007)
XopD		XCC2896	XC 1213	xcc-b100 1256	No		SUMO cysteine protease (C48 family), EAR motif, DNA binding, nuclear localization (Canonne et al. 2012)
XopE	AvrXccE1	XCC1629	XC 2602	No	No	Yes	Putative transglutaminase
							Avirulence on *B. rapa* (He et al. 2007)
XopF		XCC1218	XC 3024	xcc-b100 3087	XCA_1470		Unknown
XopG		XCC3258	XC 0967	xcc-b100 2655	No		M27 family peptidase (*Clostridium* toxin)
XopH	AvrBs1.1	XCC2099	XC 2082	xcc-b100 2395	No	No	Putative tyrosine phosphatase
XopJ	AvrXccB	XCC3731	XC 3802	No	No	Yes	Putative cysteine protease (C55 family) or serine/threonine acetyltransferase. Ubiquitin-like protease
							No effect on pathogenicity (Jiang et al. 2009)
XopAC	AvrAC	XCC2565	XC 1553	xcc-b100 1596	XCA_2914	Yes	Leucine-rich protein
							Avirulence recognized in vascular tissues of *Arabidopsis* (Xu et al. 2008)
XopAD		No	No	No	XCA_1464		SKWP repeat protein
XopAG		XCC3600	XC 0563	xcc-b100 0580	No		Unknown
XopAH	AvrXccC	XCC2109	XC 2004	xcc-b100 2071	No	Yes	Dual function. Avirulence in *B. juncea* and *B. rapa*; virulence in *B. oleracea* (Castaneda et al. 2005; He et al. 2007; Wang et al. 2007)

XopAL1	AvrPhpE	XCC1246	XC 2995	xcc-b100 3058	XCA_1499	Yes	Reduced virulence on radish (Jiang et al. 2009)
XopAL2		Downstream XCC3574	Downstream XC_3916	xcc-b100 0616	No		Unknown
XopAM	XopR1	XCC1089	XC 3160	xcc-b100 3256	No	Yes	Reduced virulence on radish (Jiang et al. 2009)
XopAT		No	No	No	XCA_1464a	Yes	Unknown. No similarity with other known proteins (Bogdanove et al. 2011)
XopA		XCC1240	XC 3002	xcc-b100 3065	XCA_1492		Harpin, may not be a t3e
HpaA		XCC1224	XC 3018	xcc-b100 3081	XCA_1476		T3 secretion control protein, may not be a t3e
HrpW		XCC1219	XC 3023	xcc-b100 3086	XCA_1471		Pectate lyase, may not be a t3e
AvrXccA1	AvrXca	XCC4229	XC 4318	xcc-b100 4450	XCA_4581	No	May not be a t3e. Virulence of *Xcr* in *Arabidopsis* (Parker et al. 1993)
AvrXccA2		XCC2396	XC 1716	xcc-b100 1770	XCA_2696	No	Unknown, may not be a t3e

t3e type III effector

Adapted from da Silva et al. (2002), White et al. (2009), Jiang et al. (2009), Ryan et al. (2011)

[a] Nine core effectors are found in most *Xanthomonas* spp., with the exception of *X. albilineans* and, in some cases, *X. campestris* pv. *raphani* (*Xcr*)

[b] Other effectors are found in a more limited number of *Xanthomonas* spp. isolates

(which are important for pathogenicity in a susceptible host) and avirulence determinants (which elicit defense in a resistant host). Effector proteins are generally thought to be involved in the suppression of the host defense system or in the alteration of host metabolism to favor nutrient uptake of the pathogen. However, experimental verification of effector function can be difficult. The deletion of individual effector genes can often result in little or no reduction in pathogenicity because of the presence of alternative functional (redundant) effector proteins (Cunnac et al. 2009). Some T3SS effectors may activate defenses (rapid host cell death) when infiltrated into plant tissue, and thus indicate a potential role as avirulence (Avr) proteins in a resistant host (Flor 1971; Keen 1990). Xu et al. (2006) showed that the mutagenesis of eight candidate genes from strain 8004 had no effect on pathogenicity in Chinese radish and cabbage. However, it was showed that mutants of *avrBs1* lost the ability to elicit an HR in the pepper line ECW10R that contains the corresponding R gene *Bs1* (Xu et al. 2006; He et al. 2007). The AvrBs1 effector is responsible for the elicitation of HR on pepper dependent on T3SS (Xu et al. 2006). Whereas, Castaneda et al. (2005) did not detect an HR variation between the *avrBs1* mutants and the wild-type 528T. He et al. (2007) suggested that it is possible that the function of *avrBs1* is redundantly encoded in the *Xcc* type strain and that the expression is different in the two strains.

An *avrBs2* homologue, designated *avrRxc1/3*, determined the avirulence of *Xcc* 512/2 (a natural mutant strain of a race 3 isolate) on *Brassica* plants (Ignatov et al. 2002) with the B genome, including *B. juncea* cv. Florida Mustard and *B. carinata* line PI199947, and on some *B. oleracea* cultivars (including the lines SR1/3, Badger Inbred-16/2 and cv. Miracle F1). This gene was also responsible for a mesophilic HR on leaves of Florida Mustard and nonhost pepper plants carrying *Bs2*. On Chinese radish and cabbage, Xu et al. (2006) showed that *avrBs2* is required for full virulence of strain 8004. In contrast, Castaneda et al. (2005) showed that the deletion of the entire putative *avrBs2* homologue in the type strain had no effect on the interaction with Florida Mustard and pepper. The strains employed by Ignatov et al. (2002) are probably derived from the type strain used by da Silva et al. (2002) and Castaneda et al. (2005); therefore, it is difficult to reconcile these results and the role of the *avrBs2* homologue on pathogenicity/race specificity still needs to be clarified. Genes from the AvrBs3 family have been found in a number of *Xanthomonas* species and pathovars, and in *Ralstonia solanacearum* (White et al. 2009). This large family of closely related T3 effectors constitutes the TAL effector family. All members of the family have a common structure with tandem repeats of a sequence of amino acids, typically of 34 amino acids in the central part of the proteins, nuclear localization signals (NLSs) and an acidic activation domain (AD); the avirulence and virulence specificities of an AvrBs3 member are dependent on the number and order of the repeats and the NLSs and AD (Boch and Bonas 2010; Gurlebeck et al. 2006).

Interestingly, the three *Xcc* strains sequenced do not contain genes from the *avrBs3/pthA* family, whereas three members of this family have been identified in an isolate named as *X. campestris* pv. *armoraciae*. These three genes were designated *hax2*, *hax3*, and *hax4* (homologue of *avrBs3* in *Xanthomonas*) (Kay et al. 2005). The three Hax proteins are translocated via the T3SS. The Hax3 and

Hax4 proteins have the typical structure of the AvrBs3-like effectors, with 34 amino acid repeats in the central part, but Hax2 has 35 amino acid repeats that contain an additional proline residue. The three *hax* genes have an additive effect on disease symptoms in radish, with hax2 having the strongest influence, and two *hax* genes (*hax3* and *hax4*) have a *Bs4*-dependent avirulence activity in tomato (Kay et al. 2005). Transconjugants of Chinese isolates containing *avrXccE1* (*xopE*) became avirulent on Chinese cabbage (*B. rapa*) cv. Zhongbai-83 (He et al. 2007). In contrast, Castaneda et al. (2005) did not notice any effect on pathogenicity on the plants tested, but it is possible that the lines tested by these authors did not possess the *R* gene responsive to *avrXccE1*. Jiang et al. (2008a) showed that XopN from strain 8004 (designated XopXccN) is required for full virulence. A mutant of the strain 8004 with an insertion in *xopXccN* was significantly weaker than strain 8004 when inoculated on Chinese radish (*Raphanus sativus* var. *radiculus*) cv. Manshenhong. The expression of *xopXccN* is regulated by the key *hrp* regulators HrpG and HrpX. The protein sequence of XopD has been identified in strain B100 (Canonne et al. 2012). This type III effector targets the *Arabidopsis* transcription factor MYB30, possibly as part of a virulence strategy that allows *Xcc* to suppress the plant cell responses to infection (Canonne et al. 2011).

Mutagenesis of the eight candidate effector genes from type strain ATCC33913 (syn. 528T) had no effect on pathogenicity when the mutants were used to inoculate six crucifer species, and none of the mutated genes singly or in any combination affected the nonhost HR elicited by the type strain in pepper. However, insertion or deletion mutants in a locus of a gene designated *avrXccFM* (XopAH) became virulent on Florida Mustard, and therefore changed the race specificity of the isolate (Castaneda et al. 2005). The T3SS effector AvrXccC of strain 8004, which belongs to the AvrB effector family of *Xanthomonas*, has been shown to have a dual effect: this effector is required for full bacterial virulence in a susceptible *B. oleracea* cv. Jingfeng 1 and for avirulence in a *B. napiformis* L.H. Bailey [syn. *B. juncea* (L.) Czern. var. *napiformis* (Pailleux and Bois) Kitam.] accession (un-specified) (Wang et al. 2007). The *avrXccC* gene was expressed in the race 2 strain HRI3849A (which causes symptoms on the mustard accession used by these authors) and the resulting strain was avirulent. He et al. (2007) also showed that an *avrXccC* mutant of 8004 became pathogenic on mustard (*B. juncea* var. *megarrhiza* Tsenet Lee) cv. Guangtou and Chinese cabbage (*B. rapa* ssp. *pekinensis*) cv. Zhongbai-83. The *avrXccC* gene of strain 8004, which confers avirulence on some mustard cultivars, is identical to *avrXccFM* of strain ATCC33913 (syn. 528T), which confers avirulence on Florida Mustard (He et al. 2007). Wang et al. (2007) showed that AvrXccC is anchored to the plant plasma membrane and its avirulence function for host recognition depends on its location.

The expression of *avrXccC* is *hrpG/hrpX* dependent. *Xcc* also contains genes encoding leucine-rich repeat (LRR) proteins. These motifs are commonly involved in protein–protein interactions and are found in several classes of plant disease resistance genes. Xu et al. (2008) showed that one of these genes, *xopAC*, named *avrACXcc8004*, encodes a protein containing LRRs; this gene is a type III effector that appears to be restricted to strains of *X. campestris*. The *A. thaliana* accession

Col-0 was resistant to the wild-type 8004 strain and to an *avrACXcc8004* mutant when the leaf mesophyll was infiltrated with bacterial suspensions, but the Col-0 ecotype became susceptible to the mutant when the bacterium was introduced into the vascular system by piercing the central vein of the leaves, indicating that the product of *avrACXcc8004* is recognized in vascular tissues and this gene might be related to the ability to colonize the xylem of a host.

Some *avr*/effector genes have also been identified in closely related pathovars, including isolates identified as *X. campestris* pv. *armoraciae* and *raphani* (possibly synonymous). Parker et al. (1993) cloned an avirulence gene of an isolate identified as *X. campestris* pv. *raphani* (isolate 1067). This isolate was avirulent in all the *A. thaliana* accessions tested by the authors. When the *avrXca* gene was transferred to *Xcc* strain 8004, it strongly reduced symptom development and bacterial growth in Columbia plants, but did not affect virulence to *Brassica* plants. The *avrXca* gene encodes a protein of 67 kDa that has no homology with known sequences and confers avirulence in a number of *A. thaliana* accessions, except one ("Kas-1"). The avirulence phenotype is not *Hrp* dependent and the interaction with *A. thaliana* did not lead to a characteristic HR, indicating that this may not be a type III effector. Recently, Corbett et al. (2005) have characterized a virulence factor, designated Svx, of *E. carotovora* ssp. a*troseptica* which shows homology to AvrXca. This protein is secreted by the type II secretion apparatus and the transcription of the *svx* gene is regulated by quorum sensing; the function of this protein is unknown, but the authors consider that it may play a role in the pathogenicity of *E. carotovora* pv. *atroseptica*. So far, there is some evidence that 12 of the putative effector genes from *Xcc* are phenotypically functional in a very restricted number of isolates. The avirulence genes presented in the model have not been proven and linked to the effectors of *Xcc*, although it is likely that A1 might correspond to XopAH and A4 to XopE. Some putative effectors could be pseudo genes or poorly expressed genes in the isolates studied, or the genes might have small effects on pathogenicity or fitness that have not been detected in the assays (Castaneda et al. 2005). Further research into the putative effectors using other isolates and hosts might still show that they have a function in pathogenicity and/or race specificity. It is also possible that some effectors were once pathogenicity genes that determined the host range of the pathogen; the function of some of these effectors could be gratuitous or even detrimental, but their structure might allow rapid adaptive selection for pathogenic function (Gabriel 1999).

2.12.4 Virulence and Pathogenicity Effectors/Genes of *Xanthomonas*

The surface structure and appendages of phytopathogenic bacteria are important for the attachment, colonization and infection of the host. Type IV pili may contribute to bacterial pathogenesis by affecting surface motility, micro colony and biofilm formation, adhesion, immune evasion and cell signaling (Craig et al. 2004). The comparison of genomes from isolates ATCC33913 and 8004 showed that at least 26 genes related to the pili assembly are highly conserved, and mutations in two of the assembly genes (*pilB* and *pilC*) have reduced virulence (Qian et al. 2005). Many phytopathogenic bacteria produce a large number of factors that might be essential or contribute to cause disease. The bacteria from the genus *Xanthomonas* typically produce yellow, membrane-bound pigments, called xanthomonadins. These pigments have a role in the maintenance of the ecological fitness of the bacteria, protecting the cells against photo oxidative stress. Xanthomonadin biosynthesis is encoded by the *pig* cluster of genes. Poplawsky and Chun (1998) have shown that the mutation of *pigB* results in a reduction in epiphytic survival and decrease in infection via the hydathodes. The biosynthesis of xanthomonadin is regulated by 3-hydroxybenzoic acid, a diffusible factor (DF) that is associated with a range of biological functions; DF-deficient mutants are nonpigmented, impaired in survival ability and less virulent (He et al. 2011).

The bacterium *Xcc* produces a range of extracellular enzymes (including proteases, pectinases and endoglucanase). The extracellular enzymes are capable of degrading the plant cell components and may be required to overcome plant defense responses, to allow bacteria to move into un-colonized plant tissues and to mobilize plant polymers for nutritional purposes (Torres et al. 2007). In *Xcc*, the synthesis of extracellular degrading enzymes and exopolysaccharides is regulated by products of the *rpf* genes (rpfABFCHGDIE) (da Silva et al. 2002). The *rpf* gene cluster of *Xcc* regulates the genes involved in the synthesis of extracellular hydrolytic enzymes and extracellular polysaccharides (EPSs) (including xanthan) and plays a role in motility, toxin production, oxidative stress resistance, aerobic respiration and biofilm formation (Dow et al. 2003; Wilson et al. 1998). A diffusible signal factor (DSF) responsible for cell–cell signaling is involved in the transcriptional control of these genes (Barber et al. 1997). Some genes under the control of the rpf/DSF system are required for the first stages of endophytic colonization (Gudesblat et al. 2009). Genes that encode proteins involved in the synthesis or degradation of cyclic-di-guanosine-3′,5′-monophosphate (cyclic-di-GMP), a second messenger involved in the regulation of cellular functions, have been shown to play significant roles in virulence to plants. Complex regulatory networks possibly allow the bacteria to adapt to different environments and to modulate virulence factors (Ryan et al. 2007). The expression levels of proteases and endoglucanases are reduced when the *rpfI* gene is inactivated in *Xcc*, suggesting that *rpfI* may have a function in the extensive tissue degeneration that is characteristic of black rot (Dow et al. 2000). RpfF is involved in the synthesis of DSF, and a two component system comprising RpfG and the complex sensor kinase RpfC is involved in sensing and responding to DSF (Ryan et al. 2010). Mutants deficient in RpfG, a phosphodiesterase that degrades cyclic-di-GMP, or the elevation of cyclic-di-GMP via the overexpression of a GGDEF domain protein WspR, caused the aggregation of cells, reduction in mobility and decrease in production of virulence factors, extracellular enzymes and exopolysaccharides (Hsiao et al. 2011). Mutations of *rpfF* or *rpfC* also lead to a reduction in the synthesis of virulence factors. The cell–cell interactions are highly regulated and dynamic; the interactions involving HD-GYP and GGDEF domain proteins are dependent on DSF signaling influence the localized expression of cyclic-di-GMP and mediate virulence (Ryan et al. 2007).

The Mips (macrophage infectivity potentiators) genes encode proteins reported as virulence factors in human pathogens. Zang et al. (2007) showed that a *mip*-like gene of *Xcc* is involved in pathogenicity through an effect on the production of exopolysaccharides and on the activity of extracellular proteases. EPSs play an important role in the pathogenicity and virulence of many bacteria, both in terms of direct interactions with host cells and in conferring protection against hostile environments (Coplin and Cook 1990). The most important EPS secreted by *Xcc* is xanthan gum, an EPS that has many industrial applications. Xanthan is a complex EPS with a cellulosic backbone and trisaccharide side chains of two mannose and one glucuronate residues that are attached to every second glucose in the backbone (Becker et al. 1998; Jansson et al. 1975; Vorhoelter et al. 2008; Yun et al. 2006). Yun et al. (2006) have shown that a xanthan minus mutant strain and a mutant strain that produces truncated xanthan fail to cause disease in *A. thaliana* and *Nicotiana benthamiana*, and induce the deposition of callose in these plants; pretreatment of the plant with xanthan restores the pathogenicity of both strains. The results indicate that xanthan induces susceptibility to *Xcc* by suppressing callose deposition. The bacterial cyclic β-(1,2)-glucan also has an effect on plant defense suppression and callose deposition, but, unlike xanthan, which acts locally, cyclic glucan acts systemically as part of a counter-defensive strategy that might facilitate the spread of the pathogen in the plants (Rigano et al. 2007).

The Xanthan has a broad range of applications in food and nonfood products from the oil, pharmaceutical, cosmetic, paper, paint and textile industries, and is mainly used as a thickening, stabilizing, gelling and emulsifying agent (Becker et al. 1998). The biosynthesis of xanthan production and the genes that encode for the enzymes involved have been studied (Ielpi et al. 1993; Becker et al. 1998). The Xanthan synthesis is encoded by 12 *gum* genes (*gumBCDEFGHIJKLM*), which are located in a single gene cluster of 12 kb that is mainly expressed as an operon from a promoter upstream of the first gene, *gumB* (Vojnov et al. 2001). The genes required for the synthesis of xanthan and its nucleotide sugar precursors are highly conserved among the three sequenced strains (ATCC33913, 8004 and B100), but are significantly different when compared with the other sequenced *Xanthomonas*; a model for the biosynthesis of xanthan based on the genome annotation has been proposed (Vorhoelter et al. 2008). The disruption of the *rmlA*, *xanA*, and *gumK* genes, which are responsible for the biosynthesis of intermediates for xanthan production, can lead to a loss of pathogenicity (Qian et al. 2005).

The lipopolysaccharide (LPS) is an essential component of the outer membrane of Gram-negative bacteria, and is an important virulence factor in *Xanthomonas* (Dow et al. 1995; He et al. 2007; Patil et al. 2007). Over 20 genes for LPS synthesis have been characterized in *Xcc*, including *xanAB*, *rmlABCB*, *rfaXY*, *lpsIJ* and 15 genes that constitute the *wxc* cluster (Vorholter et al. 2001). The *wxc* gene cluster is involved in the synthesis of the LPS O-antigen, the most variable part of LPS. The *wxc* genes are highly divergent between *Xcc* strains, indicating that different isolates might produce varied LPSs (He et al. 2007). The LPS gene cluster of *Xcc* B100 is significantly different from those of 8004 and ATCC33913, but is almost identical to the LPS cluster of *X. campestris* pv. *raphani* 756C. This could indicate that the LPS cluster of B100 was acquired recently via horizontal gene transfer from *X. campestris* pv. *raphani* 756C or, more likely, that the *X. campestris* pv. *raphani* strain diverged recently from an *Xcc* with a B100-type LPS locus (Patil et al. 2007). Qian et al. (2005) showed that the disruption of a number of *wxc* genes resulted in a significant reduction in virulence (Vicente and Holub 2013).

2.12.5 Characterization of *Xanthomonas* Genome for Virulence Genes

Extensive molecular genetics of *Xanthomonas campestris* has identified key pathogenicity determinants such as extracellular polysaccharides, lipo-polysaccharides, DSF-dependent quorum sensing, extracellular enzymes (proteases, cellulases, ...) exported by the type II secretion systems or proteins secreted by the type III secretion (T3S) system. Type III-secreted proteins (T3SP) include type III effector (T3E) proteins which are injected inside the plant cells where many of them interfere with cell physiology and plant immunity. Because the Hrp (hypersensitive response and pathogenicity) T3S system is essential for *Xanthomonas* virulence, it is assumed that the type III effectome is also globally essential for any interaction with the plant. Yet, individual effectors usually have limited or nonsignificant contribution to pathogenicity when studied alone, probably reflecting functional redundancy and/or additivity between effectors. *hrp* genes are not expressed in Xanthomonas cultivated in rich media. Yet, *hrp* gene expression can be induced in *Xanthomonas* in specific minimal media (XVM2, MME, MMX) that were thought to mimic in planta conditions. Two master regulators of the *hrp* systems which are both required for virulence and *hrp* gene expression in minimal medium have been identified: HrpX is an AraC-type transcriptional activator inducing the expression of all *hrp* operons but hrp A upon binding to the plant-inducible promoter (PIP) box (TTCGB-N15-TTCGB; B represents C, G, or T) in the promoter region. Expression of most T3Egenes is under the control of *hrpX*. HrpG, an OmpR-type transcriptional regulator and its putative cognate sensor kinase HpaS control the expression of *hrpX*, the *hrpA* operon and other genes. Interestingly, several point mutations in *hrp*G (hrpG*) can render its activity constitutive in the absence of inducing condition and result in an increased aggressiveness on plants (Koebnik et al. 2006; Li et al. 2014; Wengelnik et al. 1999).

Four complete genome sequences are available in the *X. campestris* species (pathovar *campestris*: 8004, ATCC33913, B100; pathovar *raphani*: 756C) and six draft genomes (pathovar *campestris*: Xca5, JX, B-1459, CN14, CN15 and CN16). These strains have in common a ca. 5-Mb circular chromosome with approx. 65% G + C content. The presence of plasmids has only been reported in strains B-1459, CN14, CN15 and CN16. No genomic information is available yet on the pathovars incanae and the nonpathogenic group of *X. campestris*. Comparative genomics has proven to be an important tool to mine for pathogenicity determinants, host specificity factors and in particular for T3E. Due to the lack of conserved secretion/translocation signals in T3SP, those can only be predicted by indirect means: in silico proteomes can be studied by homology searches with known T3E sequences, by searching for eukaryotic signatures (nuclear localization signals, myristoylation/palmitoylation signals, F-box motifs, . . .) and/or predicted genes can be scrutinized for the presence of a PIP box. Indeed, co-expression of T3SP with the Hrp T3S system is the rule and has proven to be a powerful tool for the identification of novel T3E in *Xanthomonas*, *Ralstonia solanacearum* and *Pseudomonas syringae*. To this end, the analysis of *Xanthomonas transcriptomes* in hrp-inducing conditions or in deregulated mutants expressing constitutively the hrp regulon has been a major source for the discovery of novel effectors (Roux et al. 2015).

The genome sequencing of four *X. campestris* strains of the pathovars incanae, *raphani,* and NP and their comparison with publicly available *X. campestris* genomes have been analyzed. This approach identified a very limited core type III effectome for this species of *Xanthomonas*. A transcriptomic analysis of the hrpG regulon has also been performed in a *X. campestris* pv. *raphani* strain CFBP 5828R and provides important information for the structural annotation of the genes and precious hints for the identification of candidate type III effector proteins as well as genes which could contribute to pathogenicity of *X. campestris*. The observed genomic coverage of the species by sequencing and annotating the genomes of strains from pathovar *incanae* (CFBP 1606R and CFBP 2527R), pathovar *raphani* (CFBP 5828R) and a pathovar formerly named *barbareae* (CFBP 5825R) was characterized. While comparative analysis identified a large core ORFeome at the species level, the core type III effectome was limited to only three putative type III effectors (XopP, XopF1 and XopAL1). In *Xanthomonas*, these effector proteins are injected inside the plant cells by the type III secretion system and contribute collectively to virulence. A deep and strand-specific RNA sequencing strategy was adopted in order to experimentally refine genome annotation for strain CFBP 5828R. This approach also allowed the experimental definition of novel ORFs and noncoding RNA transcripts. Using a constitutively active allele of hrpG, a master regulator of the type III secretion system, aHrpG-dependent regulon of 141 genes co-regulated with the type III secretion system was identified. Importantly, all these genes but seven are positively regulated by HrpG and 56 of those encode components of the Hrp type III secretion system and putative effector proteins. This dataset is an important resource to mine for novel type III effector proteins as well as for bacterial genes which could contribute to pathogenicity of *X. campestris* (Roux et al. 2015).

2.12.6 Genomic Diversity in the *Xanthomonas campestris* species

There are 13 genome sequences available for *Xanthomonas campestris*, which provide an overview of its four major genomic groups. With a complete T3S system and 19 predicted T3E, it may not be appropriate to consider strain CFBP 5825 to be nonpathogenic as proposed. The NP designation only indicates that none of the Brassicaceae tested at the time were appropriate hosts for these strains under the inoculation conditions tested. Future studies should determine if strain CFBP5825 possesses for instance a functional T3S system and if it is virulent on at least one Brassicaceae plant.

A great genomic diversity can be observed both at the intra- and interpathovar levels. Previous studies demonstrated that the pathovar campestris is composed of at least three clades (Guy et al. 2013). This study suggests that comparable diversity is expected for pathovars raphani and incanae. The lack of common CRISPR spacers between the two strains of the pathovar raphani is particularly striking. Yet, a polyphyletic origin of this pathovars is unlikely since the CRISPR repeats are identical between the two strains. Thus, it suggests that the two strains diverged a long time ago to allow the loss of all ancestral spacers in at least one strain and the likely acquisition of a significant number of new spacers since this event. Importantly, no CRISPR loci could be identified in the other strains so that genomic diversity within this species cannot be precisely determined using this tool and should thus rely on MLSA, AFLP, or MLVA (multilocus variable number of tandem repeats analysis) schemes as developed for other *Xanthomonas* species (Ngoc et al. 2009; Pruvost et al. 2011). The *P. syringae* core genome is composed of 3397 genes (genomes comprise from ca. 5000 to 8000 genes) which is close to the 3481 genes for *X. campestris*. In both instances, comparable numbers of strains were analyzed: 19 *P. syringae* genomes vs. 13 for *X. campestris*. While estimation of core genomes is rather insensitive to annotation quality or homogeneity, the pan genome size can be drastically affected. This could be one of the explanations for the rather small pan genome of the 13 *X. campestris* strains (6472 genes) compared to the 12,749 genes for the 19 *P. syringae* strains. Thus, *X. campestris* appears as a rather homogeneous genomic group compared to *P. syringae* which is considered as composed of several geno species. These observations are in agreement with the fact that *P. syringae* species has a larger host range compared to *X. campestris* that only infect plant of the Brassicaceae family (Bull and Koike 2015; Roux et al. 2015).

References

Abbott R, Albach D, Ansell S, Arntzen JW, Baird SJ, Bierne N, Boughman J, Brelsford A, Buerkle CA, Buggs R, Butlin RK, Dieckmann U, Eroukhmanoff F, Grill A, Cahan SH, Hermansen JS, Hewitt G, Hudson AG, Jiggins C, Jones J, Keller B, Marczewski T, Mallet J, Martinez-Rodriguez P, Möst M, Mullen S, Nichols R, Nolte AW, Parisod C, Pfennig K, Rice AM, Ritchie MG, Seifert B, Smadja CM, Stelkens R, Szymura JM, Väinölä R, Wolf JBW, Zinner D (2013) Hybridization and speciation. J Evol Biol 26:229–246. https://doi.org/10.1111/j.1420-9101.2012.02599.x

Adam L, Ellwood S, Wilson I, Saenz G, Xiao S, Oliver RP, Turner JG, Somerville S (1999) Comparison of *Erysiphe cichoracearum* and *E. cruciferarum* and a survey of 360 *Arabidopsis thaliana* accessions for resistance to these two powdery mildew pathogens. Mol Plant Microbe Interact 12(12):1031–1043

Alex LA, Borkovich KA, Simon MI (1996) Hyphal development in *Neurospora crassa*: involvement of a two-component histidine kinase. Proc Natl Acad Sci U S A 93:3416–3421

Amselem J, Cuomo CA, van Kan JAL, Viaud M, Benito EP, Couloux A, Coutinho PM, de Vries RP, Dyer PS, Fillinger S, Fournier E, Gout L, Hahn M, Kohn L, Lapalu N, Plummer KM, Pradier JM, Quévillon E, Sharon A, Simon A, ten Have A, Tudzynski B, Tudzynski P,

Wincker P, Andrew M, Anthouard V, Beever RE, Beffa R, Benoit I, Bouzid O, Brault B, Chen Z, Choquer M, Collémare J, Cotton P, Danchin EG, Da Silva C, Gautier A, Giraud C, Giraud T, Gonzalez C, Grossetete S, Güldener U, Henrissat B, Howlett BJ, Kodira C, Kretschmer M, Lappartient A, Leroch M, Levis C, Mauceli E, Neuvéglise C, Oeser B, Pearson M, Poulain J, Poussereau N, Quesneville H, Rascle C, Schumacher J, Ségurens B, Sexton A, Silva E, Sirven C, Soanes DM, Talbot NJ, Templeton M, Yandava C, Yarden O, Zeng Q, Rollins JA, Lebrun MH, Dickman M (2011) Genomic analysis of the necrotrophic fungal pathogens *Sclerotinia sclerotiorum* and *Botrytis cinerea*. PLoS Genet 7(8):e1002230. https://doi.org/10.1371/journal.pgen.1002230

Amselem J, Lebrun MH, Quesneville H (2015) Whole genome comparative analysis of transposable elements provides new insight into mechanisms of their inactivation in fungal genomes. BMC Genomics 16(1):141

An SQ, Lu GT, Su HZ, Li RF, He YQ, Jiang BL, Tang DJ, Tang JL (2011) Systematic mutagenesis of all predicted GntR genes in *Xanthomonas campestris* pv. *campestris* reveals a gntR family transcriptional regulator controlling hyper-sensitive response and virulence. Mol Plant-Microbe Interact 24:1027–1039

Anderson RG, Deb D, Fedkenheuer K, McDowell JM (2015) Recent progress in RXLR effector research. Mol Plant-Microbe Interact 28:1063–1072

Ando S, Yamada T, Asano T, Kamachi S, Tsushima S, Hagio T, Tabei Y (2006) Molecular cloning of *PbSTKLl* gene from *Plasmodiophora brassicae* expressed during club root development. J Phytopathol 154:185–189

Antipov D, Korobeynikov A, McLean JS, Pevzner PA (2016) HYBRIDSPADES: an algorithm for hybrid assembly of short and long reads. Bioinformatics 32:1009–1015

Asman AKM, Fogelqvist J, Vetukuri RR, Dixelius C (2016) *Phytophthora infestans* Argonaute 1 binds microRNA and small RNAs from effector genes and transposable elements. New Phytol 211:993–1007

Atreya CD, Raccah B, Pirone TP (1990) A point mutation in the coat protein abolishes aphid transmissibility of a potyvirus. Virology 178:161–165

Avenot H, Simoneau P, Iacomi-Vasilescu B, Bataille-Simoneau N (2005) Characterization of mutations in the two-component histidine kinase gene AbNIK1 from *Alternaria brassicicola* that confer high dicarboximide and phenylpyrrole resistance. Curr Genet 47:234–243. https://doi.org/10.1007/s00294-005-0568-2

Ayhan DH, Lopez-Díaz C, Di Pietro A, Ma LJ (2018) Improved assembly of reference genome *Fusarium oxysporum* f. sp. *lycopersici* strain Fol4287. Microbiol Resour Announc 7:e00910–e00918

Baker KF (1948) Fusarium wilt of garden stock (*Mathiola incana*). Phytopathology 38:399–403

Balesdent M, Barbetti M, Li H, Sivasithamparam K, Gout L, Rouxel T (2005) Analysis of *Leptosphaeria maculans* race structure in a worldwide collection of isolates. Phytopathology 95:1061–1071

Barber CE, Tang JL, Feng JX, Pan MQ, Wilson TJG, Slater H, Dow JM, Williams P, Daniels MJ (1997) A novel regulatory system required for pathogenicity of *Xanthomonas campestris* is mediated by a small diffusible signal molecule. Mol Microbiol 24:555–566

Barbetti MJ, Carter EC (1986) Diseases of rapeseed. In: Lawson JA (ed) Rapeseed in Western Australia. No. 4105. WA Department of Agriculture, Perth, WA, pp 14–19

Barbetti MJ, Banga SS, Salisbury PA (2012) Challenges for crop production and management from pathogen biodiversity and diseases under current and future climate scenarios—case study with oilseed Brassicas. Field Crop Res 127:225–240

Barr ME (1987) Prodromus to class Loculoascomycetes. Hamilton I. NewelL Inc., Amherst, MA

Bashi ZD, Gyawali S, Bekkaoui D, Coutu C, Lee L, Poon J, Rimmer SR, Khachatourians GG, Hegedus DD (2016) The *Sclerotinia sclerotiorum* Slt2 mitogen-activated protein kinase ortholog, *SMK3*, is required for infection initiation but not lesion expansion. Can J Microbiol 850:836–850. https://doi.org/10.1139/cjm-2016-0091

Basso J, Dallaire P, Charest PJ, Devantier Y, Laliberte JF (1994) Evidence for an internal ribosome entry site within the 59 non-translated region of turnip mosaic potyvirus RNA. J Gen Virol 75: 3157–3165

Baxter L, Tripathy S, Ishaque N, Boot N, Cabral A, Kemen E, Thines M, Ah-Fong A, Anderson R, Badejoko W, Bittner-Eddy P, Boore JL, Chibucos MC, Coates M, Dehal P, Delehaunty K, Dong S, Downton P, Dumas B, Fabro G, Fronick C, Fuerstenberg SI, Fulton L, Gaulin E, Govers F, Hughes L, Humphray S, Jiang RH, Judelson H, Kamoun S, Kyung K, Meijer H, Minx P, Morris P, Nelson J, Phuntumart V, Qutob D, Rehmany A, Rougon-Cardoso A, Ryden P, Torto-Alalibo T, Studholme D, Wang Y, Win J, Wood J, Clifton SW, Rogers J, Van den Ackerveken G, Jones JD, McDowell JM, Beynon J, Tyler BM (2010) Signatures of adaptation to obligate biotrophy in the *Hyaloperonospora arabidopsidis* genomed. Science 330: 1549–1551. https://doi.org/10.1126/Science1195203

Becker A, Katzen F, Puhler A, Ielpi L (1998) Xanthan gum biosynthesis and application: a biochemical/genetic perspective. Appl Microbiol Biotechnol 50:145–152

Belmas E, Briand M, Kwasiborski A, Colou J, N'Guyen G, Iacomi B, Grappin P, Campion C, Simoneau P, Barret M, Guillemette T (2018) Genome sequence of the necrotrophic plant pathogen *Alternaria brassicicola* Abra43. Genome Announce 6:e01559–e01517

Berger PH, Barnett OW, Brunt AA et al (2000) Family Potyviridae. In: van Regenmortel MHV, Fauquet CM, Bishop DHL et al (eds) Virus taxonomy: Seventh report of the international committee on taxonomy of viruses. Academic Press, San Diego, CA, pp 703–724

Bhatnagar D, Ehrlich KC, Cleveland TE (2003) Molecular genetic analysis and regulation of aflatoxin biosynthesis. Appl Microbiol Biotechnol 61:83–93. https://doi.org/10.1007/s00253-002-1199-x

Bi K, He Z, Gao Z, Zhao Y, Fu Y, Cheng J, Xie J, Jiang D, Chen T (2016) Integrated omics study of lipid droplets from *Plasmodiophora brassicae*. Sci Rep 6:36965

Bi K, Chen T, He Z, Gao Z, Zhao Y, Liu H, Fu Y, Xie J, Cheng J, Jiang D (2019) Comparative genomics reveals the unique evolutionary status of *Plasmodiophora brassicae* and the essential role of GPCR signaling pathways. Phytopathol Res 1:12

Biemont C, Vieira C (2006) Genetics: junk DNA as an evolutionary force. Nature 443(7111): 521–524

Bischoff M (1994) PCR- und RAPD-Analysis an *Plasmodiophora brassicae* Woronin und Vertretern der Gattung *Brassica* L.: Identifikation einer Nitrilasesequenz in Brassicaceen. Diploma Thesis, J.W. Goethe-Universität, Frankfurt am Main, Germany

Bittner-Eddy PD, Allen RL, Rehmany AP, Birch P, Beynon JL (2003) Use of suppression subtractive hybridization to identify downy mildew genes expressed during infection of *Arabidopsis thaliana*. Mol Plant Pathol 4(6):501–507. https://doi.org/10.1046/j.1364-3703.2003.00194.x

Boch J, Bonas U (2010) *Xanthomonas* AvrBs3 family-type III effectors: discovery and function. Annu Rev Phytopathol 48:419–436

Bogdanove AJ, Koebnik R, Lu H, Furutani A, Angiuoli SV, Patil PB, Van Sluys MA, Ryan RP, Meyer DF, Han SW, Aparna G, Rajaram M, Delcher AL, Phillippy AM, Puiu D, Schatz MC, Shumway M, Sommer DD, Trapnell C, Benahmed F, Dimitrov G, Madupu R, Radune D, Sullivan S, Jha G, Ishihara H, Lee SW, Pandey A, Sharma V, Sriariyanun M, Szurek B, Vera-Cruz CM, Dorman KS, Ronald PC, Verdier V, Dow JM, Sonti RV, Tsuge S, Brendel VP, Rabinowicz PD, Leach JE, White FF, Salzberg SL (2011) Two new complete genome sequences offer insight into host and tissue specificity of plant pathogenic *Xanthomonas* spp. J Bacteriol 193:5450–5464

Bolger AM, Lohse M, Usadel B (2014) Trimmomatic: a flexible trimmer for Illumina sequence data. Bioinformatics 30:2114–2120

Bolton MD, Thomma BPHJ, Nelson BD (2006) Pathogen profile: *Sclerotinia sclerotiorum* (Lib.) de Bary, biology and molecular traits of a cosmopolitan pathogen. Mol Plant Pathol 7:1–16

Bosland PW, Williams PH (1987) An evaluation of *Fusarium oxysporum* from crucifers based on pathogenicity, isozyme polymorphism, vegetative compatibility, and geographic origin. Can J Bot 65:2067–2073

Bosland PW, Williams PH (1988) Pathogenicity of geographic isolates of *Fusarium oxysporum* from crucifers on a differential set of crucifer seedlings. J Phytopathol 123:63–68

Braselton JP (1982) Karyotypic analysis of *Plasmodiophora brassicae* based on serial thin sections of pachytene nuclei. Can J Bot 60:403–408

Brodmann D, Schuller A, Ludwig-Muller J, Aeschbacher RA, Wiemken A, Boller T, Wingler A (2002) Induction of trehalase in *Arabidopsis* plants infected with the trehalose-producing pathogen *Plasmodiophora brassicae*. Mol Plant-Microbe Interact 15(7):693–700

Bryan RJ, Trese AT, Braselton JP (1996) Molecular karyotypes for the obligate, intracellular, plant pathogens, *Plasmodiophora brassicae* and *Spongospora subterranea*. Mycologia 88:358–360

Bryngelsson T, Gustafsson M, Green B, Lind C (1988) Uptake of host DNA by the parasitic fungus *Plasmodiophora brassicae*. Physiol Mol Plant Pathol 33:163–171

Buchanan-Wollaston V, Ainsworth C (1997) Leaf senescence in *Brassica napus*: cloning of senescence related genes by subtractive hybridization. Plant Mol Biol 33:821–834

Buczacki ST (1983) Plasmodiophora. An interrelationship between biological and practical problems. In: Buczacki ST (ed) Zoosporic plant pathogens. Academic Press, London, pp 161–191

Buhariwalla S, Mithen R (1995) Cloning of a Brassica repetitive DNA element from resting spores of *Plasmodiophora brassicae*. Physiol Mol Plant Pathol 47:95–101

Buhariwalla S, Greaves S, Magrath R, Mithen R (1995) Development of specific PCR primers for the amplification of polymorphic DNA from the obligate root pathogen *Plasmodiophora brassicae*. Physiol Mol Plant Pathol 47:83–94

Bull CT, Koike ST (2015) Practical benefits of knowing the enemy: modern molecular tools for diagnosing the etiology of bacterial diseases and understanding the taxonomy and diversity of plant pathogenic bacteria. Annu Rev Phytopathol 53:157–180. https://doi.org/10.1146/annurev-phyto-080614-120122

Bulman S, Siemens J, Ridgway HJ, Eady C, Conner AJ (2006) Identification of genes from the obligate intracellular plant pathogen, *Plasmodiophora brassicae*. FEMS Microbiol Lett 264: 198–204

Bulman S, Ridgway HJ, Eady C, Conner AJ (2007) Intron-rich gene structure in the intracellular plant parasite *Plasmodiophora brassicae*. Protist 158:423–433

Cameron DL, Schröder J, Penington JS, Do H, Molania R, Dobrovic A, Speed TP, Papenfuss AT (2017) GRIDSS: sensitive and specific genomic rearrangement detection using positional de Bruijn graph assembly. Genome Res 27:2050–2060

Canonne J, Marino D, Jauneau A, Pouzet C, Briere C, Roby D, Rivas S (2011) The *Xanthomonas* type III effector XopD targets the *Arabidopsis* transcription factor MYB30 to suppress plant defense. Plant Cell 23:3498–3511

Canonne J, Pichereaux C, Marino D, Roby D, Rossignol M, Rivas S (2012) Identification of the protein sequence of the type III effector Xop D from theB100 strain of *Xanthomonas campestris* pv *campestris*. Plant Signal Behav 7:184–187

Cao T, Srivastava S, Rahman MH, Kav NNV, Hotte N, Deyholos MK, Strelkov SE (2008) Proteome-level changes in the roots of *Brassica napus* as a result of *Plasmodiophora brassicae* infection. Plant Sci 174:97–115

Carbone I, Anderson JB, Kohn LM (1995) A group-I intron in the mitochondrial small subunit ribosomal RNA gene of *Sclerotinia sclerotiorum*. Curr Genet 27:166–176

Carzaniga R, Bowyer P, O'Connell RJ (2001) Production of extracellular matrices during development of infection structures by the downy mildew *Peronospora parasitica*. New Phytol 149:83–93

Castaneda A, Reddy JD, El-Yacoubi B, Gabriel DW (2005) Mutagenesis of all eight avr genes in *Xanthomonas campestris* pv. *campestris* had no detected effect on pathogenicity, but one avr gene affected race specificity. Mol Plant–Microbe Interact 18:1306–1317

Cavalier-Smith T, Chao EE, Lewis R (2018) Multigene phylogeny and cell evolution of chromist infra kingdom Rhizaria: contrasting cell organization of sister phyla Cercozoa and Retaria. Protoplasma 255(5):1517–1574

Cerniglia CE, Hebert RL, Szaniszlo PJ, Gibson DT (1978) Fungal transformation of naphthalene. Arch Microbiol 117:135–143. https://doi.org/10.1007/BF00402301

Chandran D, Inada N, Hather G, Kleindt CK, Wildermuth MC (2010) Laser micro dissection of *Arabidopsis* cells at the powdery mildew infection site reveals site-specific processes and regulators. Proc Natl Acad Sci U S A 107:460–465. https://doi.org/10.1073/pnas.0912492107

Chen C, Harel A, Gorovoits R, Yarden O, Dickman MB (2004) MAPK regulation of sclerotial development in *Sclerotinia sclerotiorum* is linked with pH and CAMP sensing. Mol Plant-Microbe Interact 17:404–413. https://doi.org/10.1094/MPMI.2004.17.4.404

Chen J, Ullah C, Reichelt M, Gershenzon J, Hammerbacher A (2019) *Sclerotinia sclerotiorum* circumvents flavonoid defenses by catabolizing flavonol glycosides and aglycones. Plant Physiol 180:1975–1987. https://doi.org/10.1104/pp.19.00461

Choi YJ, Hong SB, Shin HD (2003) Diversity of the *Hyaloperonospora parasitica* complex from core brassicaceous hosts based on ITS rDNA sequences. Mycol Res 107:1314–1322

Choi HW, Choi YJ, Kim DS, Hwang IS, Choi DS, Kim NH, Lee DH, Shin HD, Nam J, Hwang BK (2009a) First report of powdery mildew caused by *Erysiphe cruciferarum* on *Arabidopsis thaliana* in Korea. Plant Pathol J 25(1):86–90

Choi YJ, Shin HD, Hong SB, Thines M (2009b) The host range of *Albugo candida* extends from Brassicaceae through Cleomaceae to Capparaceae. Mycol Prog 8:329–335

Choi YJ, Thines M, Shin HD (2011) A new perspective on the evolution of white blister rust: *Albugo* s.str. (Albugonales; Oomycota) is not restricted to Brassicales but also present on Fabales. Org Divers Evol 11:193–199

Chu M, Lopez-Moya JJ, Llave-Correas C, Pirone TP (1997) Two separate regions in the genome of the tobacco etch virus contain determinants of the wilting response of tabasco pepper. Mol Plant-Microbe Interact 10:472–480

Ciaghi S, Neuhauser S, Schwelm A (2018) Draft genome resource for the potato powdery scab pathogen *Spongospora subterranea*. Mol Plant-Microbe Interact 31(12):1227–1229

Coelho PS, Vicente JG, Monteiro AA, Holub EB (2012) Pathotypic diversity of *Hyaloperonospora brassicae* collected from *Brassica oleracea*. Eur J Plant Pathol 134:763–771. https://doi.org/10.1007/s10658-012-0052-z

Cole SJ, Diener AC (2013) Diversity in receptor-like kinase genes is a major determinant of quantitative resistance to *Fusarium oxysporum* f.sp. *matthioli*. New Phytol 200:172–184

Cooper AJ, Latunde-Dada AO, Woods-Tor A, Lynn J, Lucas JA, Crute IR, Holub EB (2008) Basic compatibility of *Albugo candida* in *Arabidopsis thaliana* and *Brassica juncea* causes broad-spectrum suppression of innate immunity. Mol Plant-Microbe Interact 21:745–756. https://doi.org/10.1094/MPMI-21-6-0745

Coplin DL, Cook D (1990) Molecular genetics of extracellular polysaccharide biosynthesis in vascular phytopathogenic bacteria. Mol Plant–Microbe Interact 3:271–279

Corbett M, Virtue S, Bell K, Birch P, Burr T, Hyman L, Lilley K, Poock S, Toth I, Salmond G (2005) Identification of a new quorum-sensing-controlled virulence factor in *Erwinia carotovora* subsp. *atroseptica* secreted via the type II targeting pathway. Mol Plant-Microbe Interact 18:334–342

Cozijnsen AJ, Howlett BJ (2003) Characterization of the mating-type locus of the plant pathogenic ascomycete *Leptosphaeria maculans*. Curr Genet 43:351–357. https://doi.org/10.1007/s00294-003-0391-6

Craig L, Pique ME, Tainer JA (2004) Type IV pilus structure and bacterial pathogenicity. Nat Rev Microbiol 2:363–378

Cramer RA, Mauricio La Rota C, Cho Y, Thon M, Craven KD, Knudson DL, Mitchell TK, Lawrence CB (2006) Bioinformatic analysis of expressed sequence tags derived from a compatible *Alternaria brassicicola –Brassica oleracea* interaction. Mol Plant Pathol 7(2):113–124

Crane JL, Shearer CA (1991) A nomenclature of *Leptosphaeria* V. Cesati & G. de Notaris. IL Nat Hist Surv Bull 34:1–355

Cresnar B, Petric S (2011) Cytochrome P450 enzymes in the fungal kingdom. Biochim Biophys Acta Prot Proteom 1814:29–35. https://doi.org/10.1016/j.bbapap.2010.06.020

Cui W, Beever RE, Parkes SL, Templeton MD (2004) Evolution of an osmo sensing histidine kinase in field strains of *Botryotinia fuckeliana* (*Botrytis cinerea*) in response to dicarboximide fungicide usage. Phytopathology 94:1129–1135

Cunnac S, Lindeberg M, Collmer A (2009) *Pseudomonas syringae* type III secretion system effectors: repertoires in search of functions. Curr Opin Microbiol 12:53–60

Da Silva AC, Ferro JA, Reinach FC, Farah CS, Furlan LR, Quaggio RB, Monteiro-Vitorello CB, Van Sluys MA, Almeida NF, Alves LM, Do Amaral AM, Bertolini MC, Camargo LE, Camarotte G, Cannavan F, Cardozo J, Chambergo F, Ciapina LP, Cicarelli RM, Coutinho LL, Cursino-Santos JR, El-Dorry H, Faria JB, Ferreira AJ, Ferreira RC, Ferro MI, Formighieri EF, Franco MC, Greggio CC, Gruber A, Katsuyama AM, Kishi LT, Leite RP, Lemos EG, Lemos MV, Locali EC, Machado MA, Madeira AM, Martinez-Rossi NM, Martins EC, Meidanis J, Menck CF, Miyaki CY, Moon DH, Moreira LM, Novo MT, Okura VK, Oliveira MC, Oliveira VR, Pereira HA, Rossi A, Sena JA, Silva C, De Souza RF, Spinola LA, Takita MA, Tamura RE, Teixeira EC, Tezza RI, Trindade dos Santos M, Truffi D, Tsai SM, White FF, Setubal JC, Kitajima JP (2002) Comparison of the genomes of two *Xanthomonas* pathogens with differing host specificities. Nature 417:459–463

Dallery JF, Lapalu N, Zampounis A, Pigne S, Luyten I, Amselem J, Wittenberg AHJ, Zhou S, de Queiroz MV, Robin GP, Auger A, Hainaut M, Henrissat B, Kim KT, Lee YH, Lespinet O, Schwartz DC, Thon MR, O'Connell RJ (2017) Gapless genome assembly of *Colletotrichum higginsianum* reveals chromosome structure and association of transposable elements with secondary metabolite gene clusters. BMC Genomics 18:667

Dallot S, Quiot-Douine L, Saenz P, Cervera MT, Garcia J-A, Quiot J-B (2001) Identification of Plum pox virus determinants implicated in specific interactions with different *Prunus* spp. Phytopathology 91:159–164

Dang HX, Pryor B, Peever T, Lawrence CB (2015) The *Alternaria* genome database: a comprehensive resource for a fungal genus comprised of saprophytes, plant pathogens, and allergenic species. BMC Genomics 16:239

Daval S, Belcour A, Gazengel K, Legrand L, Gouzy J, Cottret L, Lebreton L, Aigu Y, Mougel C, Manzanares-Dauleux MJ (2019) Computational analysis of the *Plasmodiophora brassicae* genome: mitochondrial sequence description and metabolic pathway database design. Genomics 111(6):1629–1640. https://doi.org/10.1016/j.ygeno.2018.11.013

De Silva A, Bolton MD, Nelson BD (2005) Transformation of *Sclerotinia sclerotiorum* with the green fluorescent protein gene and expression of fluorescence of host tissues. Phytopathology 95:523

Del Ponte EM, Eduardo S, Lehner MS, Ju TJDP, Mizubuti G, Pethybridge SJ (2017) Independently founded populations of *Sclerotinia sclerotiorum* from a tropical and a temperate region have similar genetic structure. PLoS One 12:e0173915

Delourme R, Pilet-Nayel ML, Archipiano M, Horvais R, Tanguy X, Rouxel T, Brun H, Renard M, Balesdent MH (2004) A cluster of major specific resistance genes to *Leptosphaeria maculans* in *Brassica napus*. Phytopathology 94:578–583. https://doi.org/10.1094/PHYTO.2004.94.6.578

Derbyshire M, Denton-Giles M, Hegedus D, Seifbarghy S, Rollins J, van Kan J, Seidl MF, Faino L, Mbengue M, Navaud O, Raffaele S, Hammond-Kosack K, Heard S, Oliver R (2017) The complete genome sequence of the phytopathogenic fungus *Sclerotinia sclerotiorum* reveals insights into the genome architecture of broad host range pathogens. Genome Biol Evol 9(3): 593–618

Diener AC (2013) Routine mapping of Fusarium wilt resistance in BC1 populations of *Arabidopsis thaliana*. BMC Plant Biol 13:171

Diener AC, Ausubel FM (2005) *Resistance to Fusarium oxysporum 1*, a dominant *Arabidopsis* disease-resistance gene, is not race specific. Genetics 171:305–321

Dion Y, Gugel RK, Rakow GFW, Séguin-Swartz G, Landry BS (1995) RFLP mapping of resistance to the blackleg disease [causal agent, *Leptosphaeria maculans* (Desm.) Ces. et de Not.] in canola (*Brassica napus* L.). Theor Appl Genet 91:1190–1194. https://doi.org/10.1007/BF00220928

Dong S, Raffaele S, Kamoun S (2015) The two-speed genomes of filamentous pathogens: waltz with plants. Curr Opin Genet Dev 35:57–65

Doughan B, Rollins JA, Avery SV (2016) Characterization of *MAT* gene functions in the life cycle of *Sclerotinia sclerotiorum* reveals a lineage-specific *MAT* gene functioning in apothecium morphogenesis. Fungal Biol 120:1105–1117. https://doi.org/10.1016/j.funbio.2016.06.007

Dow JM, Osbourn AE, Wilson TJ, Daniels MJ (1995) A locus determining pathogenicity of *Xanthomonas campestris* is involved in lipopolysaccharide biosynthesis. Mol Plant Microbe Interact 8:768–777

Dow JM, Feng JX, Barber CE, Tang JL, Daniels MJ (2000) Novel genes involved in the regulation of pathogenicity factor production within the rpf gene cluster of *Xanthomonas campestris*. Microbiology 146:885–891

Dow JM, Crossman L, Findlay K, He YQ, Feng JX, Tang JL (2003) Biofilm dispersal in *Xanthomonas campestris* is controlled by cell–cell signaling and is required for full virulence to plants. Proc Natl Acad Sci U S A 100:10995–11000

Drès M, Mallet J (2002) Host races in plant–feeding insects and their importance in sympatric speciation. Philos Trans R Soc Lond Ser B Biol Sci 357:471–492. https://doi.org/10.1098/rstb.2002.1059

Dry IB, Yuan KH, Hutton DG (2004) Dicarboximide resistance in field isolates of *Alternaria alternata* is mediated by a mutation in a two-component histidine kinase gene. Fungal Genet Biol 41:102–108

Duan Y, Ge C, Liu S, Wang J, Zhou M (2013) A two-component histidine kinase shk1 controls stress response, sclerotial formation and fungicide resistance in *Sclerotinia sclerotiorum*. Mol Plant Pathol 14:708–718. https://doi.org/10.1111/mpp.12041

Dunbar M, Adhikari A, Kontz B, Varenhorst A, Nleya T, Byamukama E, Mathew F (2017) First report of Alternaria black spot caused by *Alternaria alternate* on *Brassica carinata* in South Dakota. Plant Dis 101:1951

Ekblom R, Wolf JBW (2014) A field guide to whole-genome sequencing, assembly and annotation. Evol Appl 7:1026–1042

Erental A, Harel A, Yarden O (2007) Type 2A phosphoprotein phosphatase is required for asexual development and pathogenesis of *Sclerotinia sclerotiorum*. Mol Plant-Microbe Interact 20:944–954. https://doi.org/10.1094/MPMI-20-8-0944

Eriksson OE, Hawksworth DL (1991) Outline of the ascomycetes -1990. Systema Ascomycetum 9: 39–271

Eshel D, Miyara I, Ailing T, Dinoor A, Prusky D (2002) pH regulates endo-glucanase expression and virulence of *Alternaria alternata* in *Persimmon* fruit. Mol Plant-Microbe Interact 15:774–779

Faggian R, Bulman SR, Lawrie AC, Porter IJ (1999) Specific polymerase chain reaction primers for the detection of *Plasmodiophora brassicae* in soil and water. Phytopathology 89:392–397

Fahling M, Graf H, Siemens J (2003) Pathotype separation of *Plasmodiophora brassicae* by the host plant. J Phytopathol 151(7–8):425–430. https://doi.org/10.1046/j.1439-0434.2003.00744.x

Faino L, Seidl MF, Datema E, van den Berg GCM, Janssen A, Wittenberg AHJ, Thomma BPHJ (2015) Single-Molecule Real-Time sequencing combined with optical mapping yields completely finished fungal genome. MBio 6:e00936–e00915

Faino L, Seidi MF, Shi-Kunne X, Pauper M, van den Berg GCM, Wittenberg AHJ, Thomma BPHJ (2016) Transposons passively and actively contribute to evolution of the two-speed genome of a fungal pathogen. Genome Res 26:1091–1100

Fan H, Yu G, Liu Y, Zhang X, Liu J, Zhang Y, Rollins JA, Sun F, Pan H (2017) An atypical fork head-containing transcription factor SsFKH1 is involved in sclerotial formation and is essential for pathogenicity in *Sclerotinia sclerotiorum*. Mol Plant Pathol 18:963–975. https://doi.org/10.1111/mpp.12453

Fargier E, Manceau C (2007) Pathogenicity assays restrict the species *Xanthomonas campestris* into three pathovars and reveal nine races within *X. campestris* pv. *campestris*. Plant Pathol 56:805–818

Farris JJ (1989) The retention index and the rescaled consistency index. Cladistics 5:417–419

Felsenstein J (1985) Confidence limits on phylogenies: an approach using the Boots trap. Evolution 39:783–791

Feng J, Hwang SF, Strelkov SE (2013) Assessment of gene expression profiles in primary and secondary zoospores of *Plasmodiophora brassicae* by dot blot and real-time PCR. Microbiol Res 168(8):518–524

Feng BZ, Zhu XP, Fu L, Lv RF, Storey D, Tooley P, Zhang XG (2014) Characterization of necrosis-inducing NLP proteins in *Phytophthora capsici*. BMC Plant Biol 14:126. https://doi.org/10.1186/1471-2229-14-126

Ferreira ME, Rimmer SR, Williams PH, Osborn TC (1995) Mapping loci controlling *Brassica napus* resistance to *Leptosphaeria maculans* under different screening conditions. Phytopathology 85:213–217

Fitt BDL, Brun H, Barbetti MJ, Rimmer SR (2006a) World-wide importance of Phoma stem canker (*Leptosphaeria maculans* and *L. biglobosa*) on oilseed rape (*Brassica napus*). Eur J Plant Pathol 114:3–15

Fitt BDL, Evans N, Howlett BJ, Cooke M (2006b) Sustainable strategies for managing *Brassica napus* (oilseed rape) resistance to *Leptosphaeria maculans* (phoma stem canker), vol 114. Springer, New York, NY

Flor HH (1971) Current status of gene-for-gene concept. Annu Rev Phytopathol 9:275–296

Fomeju FB, Falentin C, Lassalle G, Manzanares-Dauleux MJ, Delourme R (2014) Homoeologous duplicated regions are involved in quantitative resistance of *Brassica napus* to stem canker. BMC Genome 15:498

Fu L, Zhu C, Ding X, Yang X, Morris PF, Tyler BM, Zhang X (2015) Characterization of cell-death-inducing members of the pectate lyase gene family in *Phytophthora capsici* and their contributions to infection of pepper. Mol Plant-Microbe Interact 28:766–775

Fudal I, Ross S, Brun H, Besnard A-L, Ermel M, Kuhn M-L, Balesdent M-H, Rouxel T (2009) Repeat-induced point mutation (RIP) as an alternative mechanism of evolution toward virulence in *Leptosphaeria maculans*. Mol Plant-Microbe Interact 22:932–941. https://doi.org/10.1094/MPMI-22-8-0932

Fujimura M, Ochiai N, Ichiishi A, Usami R, Horikoshi K, Yamaguchi I (2000) Fungicide resistance and osmotic stress sensitivity in os mutants of *Neurospora crassa*. Pestic Biochem Physiol 67(2):125–133. https://doi.org/10.1006/pest.2000.2479

Fujimura M, Ochiai N, Oshima M, Motoyama T, Ichiishi A, Usami R, Horikoshi K, Yamaguchi I (2003) Putative homologs of SSK22 MAPKK kinase and PBS2 MAPK kinase of Saccharomyces cerevisiae encoded by os-4 and os-5 genes for osmotic sensitivity and fungicide resistance in *Neurospora crassa*. Biosci Biotechnol Biochem 67:186–191

Fujisawa I (1990) *Turnip mosaic virus* strains in cruciferous crops in Japan. Jpn Agric Res Q 23: 289–293

Gabriel DW (1999) Why do pathogens carry avirulence genes? Physiol Mol Plant Pathol 55:205–214

Gan P, Ikeda K, Irieda H, Narusaka M, O'Connell RJ, Narusaka Y, Takano Y, Kubo Y, Shirasu K (2013) Comparative genomic and transcriptomic analysis reveal the hemibiotrophic stage shift of *Colletotrichum* fungi. New Phytol 197:1236–1249. https://doi.org/10.1111/nph.12085

Garrison E, Marth G (2012) Haplotype-based variant detection from short-read sequencing. arXiv. http://arxiv.org/abs/1207.3907v2

Gaumann E (1918) Uber die Formen der *Peronospora parasitica* (Pers.) Fries. Ein Beitrag zur Speziesfrage bei den parasitischen Pilzen. Beih Bot Centralbl 34(1):395–533

Gaumann E (1923) Contributions toward a monograph of the genus *Peronospora* corda. Beitrage zur kryptogamenflora de schweiz 5:360

Gaumann E (1926) Uber die Spezialisierung des falschen Mehltaus (*Peronospora brassicae* Gm.) auf dem Kohl und seinen Verwandten. Landwirtschaftliches Jahrbuch der Schweiz 40:463–468

Ge XT, Li YP, Wan ZJ, You MP, Finnegan PM, Banga SS, Sandhu PS, Garg H, Salisbury PA, Barbetti MJ (2012) Delineation of *Sclerotinia sclerotiorum* pathotypes using differential resistance responses on *Brassica napus* and *B. juncea* genotypes enables identification of resistance to prevailing pathotypes. Field Crop Res 127:248–258. https://doi.org/10.1016/j.fcr.2011.11.022

George H, Hirschi K, Van Etten H (1998) Biochemical properties of the products of cytochrome P450 genes (PDA) encoding pisatin demethylase activity in *Nectria haematococca*. Arch Microbiol 170:147–154. https://doi.org/10.1007/s002030050627

Ghanbarnia K, Fudal I, Larkan NJ, Links MG, Balesdent MH, Profotova B, Fernando WGD, Borhan MH (2015) Rapid identification of the *Leptosphaeria maculans* avirulence gene AvrLm2 using an intraspecific comparative genomics approach. Mol Plant Pathol 16:699–709

Gibbs A, Ohshima K (2010) Potyviruses in the digital revolution. Annu Rev Phytopathol 48:205–223. https://doi.org/10.1146/annurev-phyto-073009-114404

Goker M, Riethmüller A, Voglmayr H, Weiss M, Oberwinkler F (2004) Phylogeny of *Hyaloperonospora* based on nuclear ribosomal internal transcribed spacer sequences. Mycol Prog 3:83–94

Goker M, Voglmayr H, Oberwinkler F (2009) Species delimitation in downy mildews: the case of *Hyaloperonospora* in the light of nuclear ribosomal ITS and LSU sequences. Mycol Res 113:308–325. https://doi.org/10.1016/j.mycres.2008.11.006

Golicz AA, Martinez PA, Zander M, Patel DA, Van De Wouw AP, Visendi P, Fitzgerald TL, Edwards D, Batley J (2015) Gene loss in the fungal canola pathogen *Leptosphaeria maculans*. Funct Integr Genomics 15:189–196

Gomez-Campo C (1999) Taxonomy. In: Gomez-Campo C (ed) Biology of *Brassica* coenospecies. Elsevier Science, Amsterdam, pp 3–32

Goodwin S, McPherson JD, Mc Combie WR (2016) Coming of age: ten years of next-generation sequencing technologies. Nat Rev Genet 17:333–351

Gout L, Fudal I, Kuhn ML, Blaise F, Eckert M, Cattolico L, Balesdent MH, Rouxel T (2006) Lost in the middle of nowhere: the *AvrLm1* avirulence gene of the Dothideomycete *Leptosphaeria maculans*. Mol Microbiol 60:67–80

Gout L, Kuhn ML, Vincenot L, Bernard-Samain S, Cattolico L, Barbetti M, Moreno-Rico O, Balesdent MH, Rouxel T (2007) Genome structure impacts molecular evolution at the *AvrLm1* avirulence locus of the plant pathogen *Leptosphaeria maculans*. Environ Microbiol 9:2978–2992

Graf H, Fahling M, Siemens J (2004) Chromosome polymorphism of the obligate biotrophic parasite *Plasmodiophora brassicae*. J Phytopathol 152:86–91

Grandaubert J, Lowe RGT, Soyer JL, Schoch CL, van de Wouw AP, Fudal I, Robbertse B, Lapalu N, Links MG, Ollivier B, Linglin J, Barbe V, Mangenot S, Cruaud C, Borhan H, Howlett BJ, Balesdent M-H, Rouxel T (2014) Transposable element-assisted evolution and adaptation to host plant within the *Leptosphaeria maculans-Leptosphaeria biglobosa* species complex of fungal pathogens. BMC Genomics 15:891. https://doi.org/10.1186/1471-2164-15-891

Green SK, Deng TC (1985) Turnip mosaic virus strains in cruciferous hosts in Taiwan. Plant Dis 69:28–31

Gudesblat GE, Torres PS, Vojnov AA (2009) *Xanthomonas campestris* overcomes *Arabidopsis* stomatal innate immunity through a DSF cell-to-cell signal-regulated virulence factor. Plant Physiol 149:1017–1027

Guimaraes RL, Stotz HU (2004) Oxalate production by *Sclerotinia sclerotiorum* deregulates guard cells during infection. Plant Physiol 136:3703–3711

Gupta V, Jagannathan V, Lakshmikumaran M (1990) A novel AT-rich tandem repeat of *Brassica nigra*. Plant Sci 68:223–229

Gurlebeck D, Thieme F, Bonas U (2006) Type III effector proteins from the plant pathogen *Xanthomonas* and their role in the interaction with the host plant. J Plant Physiol 163:233–255

Guy E, Genissel A, Hajri A, Chabannes M, David P, Carrère S, Lautier M, Roux B, Boureau T, Arlat M, Poussier S, Noël LD (2013) Natural genetic variation of *Xanthomonas campestris* pv. *campestris* pathogenicity on *Arabidopsis* revealed by association and reverse genetics. mBio 4(3):e00538–e00512. https://doi.org/10.1128/mBio.00538-12

Guyon K, Balague C, Roby D, Raffaele S (2014) Secretome analysis reveals effector candidates associated with broad host range necrotrophy in the fungal plant pathogen *Sclerotinia sclerotiorum*. BMC Genomics 15:336

Gyawali S, Harrington M, Durkin J, Horner K, Parkin IAP, Hegedus DD, Bekkaoui D, Buchwaldt L (2016) Microsatellite markers used for genome-wide association mapping of partial resistance to *Sclerotinia sclerotiorum* in a world collection of *Brassica napus*. Mol Breed 36:72

Haas BJ, Kamoun S, Zody MC, Jiang RHY, Handsaker RE, Cano LM, Grabherr M, Kodira CD, Raffaele S, Torto-Alalibo T, Bozkurt TO, Ah-Fong AMV, Alvarado L, Anderson VL, Bollmann M, Armstrong R, Avrova A, Baxter L, Beynon J, Boevink PC, Stephanie R, Bos JIB, Bulone V, Cai G, Cakir C, Carrington JC, Chawner M, Conti L, Costanzo S, Ewan R, Green NF, Fischbach MA, Fugelsad J, Gilroy EM, Gnerre S, Green PJ, Grenville-Briggs LJ, Griffith J, Grunwald NJ, Horn K, Horner NR, Hu CH, Huitema E, Jeong DH, Jones AME, Jones JDG, Jones RW, Karlsson EK, Kunjeti SG, Lamour K, Liu Z, Ma L, MacLean D, Chibucos MC, McDonald H, Mc Walters J, Meijer HJG, Morgan W, Morris PF, Munro CA, O'Neill K, Ospina-Giraldo M, Pinzón A, Pritchard L, Ramsahoye B, Ren Q, Restrepo S, Roy S, Sadanandom A, Savidor A, Schornack S, Schwartz DC, Schmann UD, Schwessinger B, Seyer L, Sharpe T, Silvar C, Song J, Studholme DJ, Sykes S, Thines M, van de Vondervoort PJI, Phuntumart V, Wawra S, Weide R, Win J, Young C, Zhou S, Fry W, Meyers BC, van West P, Ristaino J, Govers F, Birch PRJ, Whisson SC, Judelson HS, Nusbaum C (2009) Genome sequence and analysis of the Irish potato famine pathogen *Phytophthora infestans*. Nature 461:393–398

Hacquard S, Kracher B, Maekawa T, Vernaldi S, Schulze-Lefert P, van Themaat EVL (2013) Mosaic genome structure of the barley powdery mildew pathogen and conservation of transcriptional programs in divergent hosts. Proc Natl Acad Sci U S A 110(24):E2219–E2228. https://doi.org/10.1073/pnas.1306807110

Hane JK, Williams AH, Taranto AP, Solomon PS, Oliver RP (2015) Repeat induced point mutation: a fungal-specific, endogenous mutagenesis process. In: van den Berg MA, Maruthachalam K (eds) Genetic transformation systems in fungi, vol 2. Springer International Publishing, pp 55–68

Hansen CE, Flengsrud R, Kopperud C (1994) Two-dimensional gel electrophoresis of proteins in healthy roots and clubroots of *Brassica oleracea*. Acta Agric Sci B Soil Plant Sci 44:123–128

Hardie DG (1999) Plant protein serine/threonine kinases: classification and functions. Annu Rev Plant Physiol Plant Mol Biol 50:97–131. https://doi.org/10.1146/annurev.arplant.50.1.97

Harel A, Bercovich S, Yarden O (2006) Calcineurin is required for sclerotial development and pathogenicity of *Sclerotinia sclerotiorum* in an oxalic acid independent manner. Mol Plant-Microbe Interact 19:682–693

Harper FR, Pittman UJ (1974) Yield loss by *Brassica campestris* and *Brassica napus* from systemic stem infection by *Albugo cruciferarum*. Phytopathology 64:408–410

He YQ, Zhang L, Jiang BL, Zhang ZC, Xu RQ, Tang DJ, Qin J, Jiang W, Zhang X, Liao J, Cao JR, Zhang SS, Wei ML, Liang XX, Lu GT, Feng JX, Chen B, Cheng J, Tang JL (2007) Comparative and functional genomics reveals genetic diversity and determinants of host specificity among reference strains and a large collection of Chinese isolates of the phytopathogen *Xanthomonas campestris* pv. *campestris*. Genome Biol 8:R218

He YW, Wu J, Zhou L, Yang F, He YO, Jiang BL, Bai LQ, Xu YQ, Deng ZX, Tang JL, Zhang LH (2011) *Xanthomonas campestris* diffusible factor is3-hydroxybenzoic acid and is associated with xanthomonad in biosynthesis, cell viability, antioxidant activity, and systemic invasion. Mol Plant-Microbe Interact 24:948–957

Hedrick PW (2013) Adaptive introgression in animals: examples and comparison to new mutation and standing variation as sources of adaptive variation. Mol Ecol 22:4606–4618. https://doi.org/10.1111/mec.12415

Henson JM, French R (1993) The polymerase chain reaction and plant disease diagnosis. Annu Rev Phytopathol 31:81–109

Hirata T, Takamatsu S (1996) Nucleotide sequence diversity of rDNA internal transcribed spacers extracted from conidia and cleistothecia of several powdery mildew fungi. Mycoscience 37: 283–288

Hiura M (1930) Biologic forms of *A. candida* (Pers.) Kuntze on some cruciferous plants. Jpn J Bot 5:1–20

Hjulsager CK, Lund OS, Johansen IE (2002) A new pathotype of Pea seed borne mosaic virus explained by properties of the P3-6K1- and viral genome-linked protein (VPg)-coding regions. Mol Plant-Microbe Interact 15:169–171

Holm L (1957) Etudes taxonomiques sur les pleosporacees. Symb Bot Ups 14:1–188

Holst-Jensen A, Vaage M, Schumacher T (1998) An approximation to the phylogeny of *Sclerotinia* and related genera. Nor J Bot 18:705–719

Holub EB (2008) Natural history of *Arabidopsis thaliana* and oomycete symbioses. Eur J Plant Pathol 122:91–109. https://doi.org/10.1007/s10658-008-9286-1

Howlett BJ, Lowe RGT, Marcroft SJ, van de Wouw AP (2015) Evolution of virulence in fungal plant pathogens: exploiting fungal genomics to control plant disease. Mycologia 107(3): 441–451. https://doi.org/10.3852/14-317

Hsiao YM, Liu YF, Fang MC, Song WL (2011) XCC2731, a GGDEF domain protein in *Xanthomonas campestris*, is involved in bacterial attachment and is positively regulated by Clp. Microbiol Res 166:548–565

Hu J, Chen C, Peever T, Dang H, Lawrence C, Mitchell T (2012) Genomic characterization of the conditionally dispensable chromosome in *Alternaria arborescens* provides evidence for horizontal gene transfer. BMC Genomics 13:171

Huser A, Takahara H, Schmalenbach W, O'Connell R (2009) Discovery of pathogenicity genes in the crucifer anthracnose fungus *Colletotrichum higginsianum*, using random insertional mutagenesis. Mol Plant-Microbe Interact 22:143–156. https://doi.org/10.1094/MPMI-22-2-0143

Huson DH, Bryant D (2006) Application of phylogenetic networks in evolutionary studies. Mol Biol Evol 23:254–267

Iacomi-Vasilescu B, Avenota H, Bataille-Simoneau N, Laurent E, Guénard M, Simoneau P (2004) *In vitro* fungicide sensitivity of *Alternaria* species pathogenic to crucifers and identification of *Alternaria brassicicola* field isolates highly resistant to both dicarboximides and phenylpyrroles. Crop Prot 23:481–488

Ielpi L, Couso RO, Dankert MA (1993) Sequential assembly and polymeri-zation of the polyprenol-linked pentasaccharide repeating unit of the xanthan polysaccharide in *Xanthomonas campestris*. J Bacteriol 175:2490–2500

Ignatov AN, Monakhos GF, Djalilov FS, Pozmogova GV (2002) Aviru-lence gene from *Xanthomonas campestris* pv. *campestris* homologous to the avrBs2 locus is recognized in race-specific reaction by two different resistance genes in *Brassicas*. Russ J Genet 38:1404–1410

Ii WMJ, Rollins JVA (2007) Deletion of the adenylate cyclase (Sac1) gene affects multiple developmental pathways and pathogenicity in *Sclerotinia sclerotiorum*. Fungal Genet Biol 44:521–530

Ito S, Yano S, Tanaka S, Kameya-Iwaki M (1994) The use of resting spore spheroplasts in the DNA analysis of *Plasmodiophora brassicae*. Ann Phytopathol Soc Jpn 60:491–495

Ito S, Tanaka S, Miyanaga C, Takabayashi S, Yano S, Kameya-Iwaki M (1996) Constitutive and inducible proteins in the root of Chinese cabbage infected with *Plasmodiophora brassicae*. J Phytopathol 144:89–95

Ito S, Maehara T, Tanaka S, Kameya-Iwaki M, Yano S, Kishi F (1997) Cloning of a single-copy DNA sequence unique to *Plasmodiophora brassicae*. Physiol Mol Plant Pathol 50:289–300

Ito S, Nakamura T, Matsumoto T, Maehara T, Tanaka S, Kameya-Iwaki M, Kishi F (1998) Characterization of DNA sequences cloned from resting spores of *Plasmodiophora brassicae*. J Phytopathol 146(2–3):143–147. https://doi.org/10.1111/j.1439-0434.1998.tb04671.x

Ito S, Ichinose H, Yanagi C, Tanaka S, Kameya-Iwaki M, Kishi F (1999) Identification of an in planta-induced mRNA of *Plasmodiophora brassicae*. J Phytopathol 147:79–82

Jansson PE, Kenne L, Lindberg B (1975) Structure of extracellular polysac-charide from *Xanthomonas campestris*. Carbohydr Res 45:275–282

Jasalavich CA, Morales VM, Pelcher LE, Seguin-Swartz G (1995) Comparison of nuclear ribosomal DNA sequences from *Alternaria* species pathogenic to crucifers. Mycol Res 99(5): 604–614

Jayawardena RS, Hyde KD, Damm U, Cai L, Liu M, Li XH, Zhang W, Zhao W, Yan JY (2016) Notes on currently accepted species of *Colletotrichum*. Mycosphere 7:1192–1260. https://doi.org/10.5943/mycosphere/si/2c/9

Jenner CE, Walsh JA (1996) Pathotypic variation in *Turnip mosaic virus* with special reference to European isolates. Plant Pathol 45:848–856

Jenner CE, Tomimura K, Ohshima K, Hughes SL, Walsh JA (2002) Mutations in *Turnip mosaic virus* P3 and cylindrical inclusion proteins are separately required to overcome two *Brassica napus* resistance genes. Virology 300:50–59

Jenner CE, Wang X, Tomimura K, Ohshima K, Ponz F, Walsh JA (2003) The dual role of the potyvirus P3 protein of *Turnip mosaic virus* as a symptom and avirulence determinant in Brassicas. Mol Plant-Microbe Interact 16:777–784

Jestin C, Lodé M, Vallée P, Domin C, Falentin C, Horvais R, Coedel S, Manzanares-Dauleux MJ, Delourme R (2011) Association mapping of quantitative resistance for *Leptosphaeria maculans* in oilseed rape (*Brassica napus* L.). Mol Breed 27:271–287

Jiang B, Xu RQ, Li XZ, Wei HY, Bai F, Hu X, He YQ, Tang JL (2006) Construction and characterization of a *hrpG* mutant rendering constitutive expression of *hrp* genes in *Xanthomonas campestris* pv. *campestris*. Prog Nat Sci 16:480–485

Jiang BL, He YQ, Cen WJ, Wei HY, Jiang GF, Jiang W, Hang XH, Feng JX, Lu GT, Tang DH, Tang JL (2008a) The type III secretion effector xopXccN of *Xanthomonas campestris* pv. *campestris* is required for full virulence. Res Microbiol 159:216–220

Jiang BL, Liu J, Chen LF, Ge YY, Hang XH, He YQ, Tang DJ, Lu GT, Tang JL (2008b) DsbB is required for the pathogenesis process of *Xanthomonas campestris* pv. *campestris*. Mol Plant-Microbe Interact 21:1036–1045

Jiang W, Jiang BL, Xu RQ, Huang JD, Wei HY, Jiang GF, Cen WJ, Liu J, Ge YY, Li GH, Su LL, Hang XH, Tang DJ, Lu GT, Feng JX, He YQ, Tang JL (2009) Identification of six type III effector genes with the PIP box in *Xanthomonas campestris* pv. *campestris* and five of them contribute individually to full pathogenicity. Mol Plant-Microbe Interact 22:1401–1411

Johansen IE, Lund OS, Hjulsager CK, Laursen J (2001) Recessive resistance in Pisum sativum and potyvirus pathotype resolved in a gene-for-cistron correspondence between host and virus. J Virol 75:6609–6614

Jones PM (1928) Morphology and cultural history of *Plasmodiophora brassicae*. Arch Protistenkd 62:313–327

Judelson HS, Ah-Fong AM (2010) The kinome of *Phytophthora infestans* reveals oomycete-specific innovations and links to other taxonomic groups. BMC Genomics 11:700

Judelson HS, Ah-Fong AMV (2019) Exchanges at the plant-oomycete interface that influence disease. Plant Physiol 179:1198–1211

Judelson HS, Shrivastava J, Manson J (2012) Decay of genes encoding the oomycete flagellar proteome in the downy mildew *Hyaloperonospora arabidopsidis*. PLoS One 7(10):e47624. https://doi.org/10.1371/journal.pone.0047624

Jurick WM, Dickman MB, Rollins JA (2004) Characterization and functional analysis of a CAMP-dependent protein kinase a catalytic subunit gene (*Pka1*) in *Sclerotinia sclerotiorum*. Physiol Mol Plant Pathol 64:155–163. https://doi.org/10.1016/j.pmpp.2004.07.004

Kageyama K, Komatsu T, Suga H (2003) Refined PCR protocol for detection of plant pathogens in soil. J Gen Plant Pathol 69:153–160

van Kan JAL, Stassen JHM, Mosbach A, Theo AJ, Lee VD, Faino L, Farmer AD, Papasotiriou DG, Zhou S, Seidl MF, Cottam E, Edel D, Hahn M, Schwartz DC, Dietrich RA, Widdison S, Scalliet G (2017) A gapless genome sequence of the fungus *Botrytis cinerea*. Mol Plant Pathol 18:75–89

Kaul S, Koo HL, Jenkins J, Rizzo M, Rooney T, Tallon LJ, Feldblyum T, Nierman W, Benito MI, Lin X, Town CD, Venter JC, Fraser CM, Tabata S, Nakamura Y, Kaneko T, Sato S, Asamizu E, Kato T, Kotani H et al (2000) Analysis of the genome sequence of the flowering plant *Arabidopsis thaliana*. Nature 408:796–815

Kay S, Boch J, Bonas U (2005) Characterization of AvrBs3-like effectors from a Brassicaceae pathogen reveals virulence and avirulence activities and a protein with a novel repeat architecture. Mol Plant-Microbe Interact 18:838–848

Keen NT (1990) Gene-for-gene complementarity in plant–pathogen interactions. Annu Rev Genet 24:447–463

Kemen E, Gardiner A, Schultz-Larsen T, Kemen AC, Balmuth AL, Robert-Seilaniantz A, Bailey K, Holub E, Studholme DJ, Maclean D, Jones JD (2011) Gene gain and loss during evolution of obligate parasitism in the white rust pathogen of *Arabidopsis thaliana*. PLoS Biol 9:e1001094

Kim HJ, Chen C, Kabbage M, Dickman MB (2011) Identification and characterization of *Sclerotinia sclerotiorum* NADPH oxidases. Appl Environ Microbiol 77:7721–7729. https://doi.org/10.1128/AEM.05472-11

King AMQ, Adams MJ, Carstens EB, Lefkowitz EJ (2012) Virus taxonomy: classification and nomenclature of viruses. In: Andrew MQK, Adams MJ, Eric BC, Elliot JL (eds) Ninth report of the International Committee on Taxonomy of Viruses. International Union of Microbiological Societies Virology Division, Elsevier Academic, San Diego, CA

King R, Urban M, Hammond-Kosack MCU, Hassani-Pak K, Hammond-Kosack KE (2015) The completed genome sequence of the pathogenic ascomycete fungus *Fusarium graminearum*. BMC Genomics 16:544

Kistler HC (1997) Genetic diversity in the plant-pathogenic fungus *Fusarium oxysporum*. Phytopathology 87:474–479

Kistler HC, Benny U (1989) The mitochondrial genome of *Fusarium oxysporum*. Plasmid 22:86–89

Kistler HC, Bosland PW, Benny U, Leong S, Williams PH (1987) Relatedness of strains of *Fusarium oxysporum* from crucifers measured by examination of mitochondrial and ribosomal DNA. Phytopathology 77:1289–1293

Kistler HC, Momol EA, Benny U (1991) Repetitive genomic sequences for determining relatedness among strains of *Fusarium oxysporum*. Phytopathology 81:331–336

Klein PG, Klein RR, Rodrıguez-Cerezo E, Hunt AG, Shaw JG (1994) Mutational analysis of the tobacco vein mottling virus genome. Virology 204:759–769

Klewer A, Luerßen H, Graf H, Siemens J (2001) Restriction Fragment Length Polymorphism markers to characterize *Plasmodiophora brassicae* single-spore isolates with different virulence patterns. J Phytopathol 149:121–127

Koebnik R, Kruger A, Thieme F, Urban A, Bonas U (2006) Specific binding of the *Xanthomonas campestris* pv. *vesicatoria* Ara C-type transcriptional activator *HrpX* to plant-inducible promoter boxes. J Bacteriol 188(21):7652–7660. https://doi.org/10.1128/JB.00795-06

Koren S, Walenz BP, Berlin K, Miller JR, Bergman NH, Phillippy AM (2017) Canu: scalable and accurate long-read assembly via adaptive *k-mer* weighting and repeat separation. Genome Res 27:722–736

Korn M, Schmidpeter J, Dahl M, Müller S, Voll LM, Koch C (2015) A genetic screen for pathogenicity genes in the hemibiotrophic fungus *Colletotrichum higginsianum* identifies the plasma membrane proton pump Pma2 required for host penetration. PLoS One 10:e0125960. https://doi.org/10.1371/journal.pone.0125960

Kurtz S, Phillippy A, Delcher AL, Smoot M, Shumway M, Antonescu C, Salzberg SL (2004) Versatile and open software for comparing large genomes. Genome Biol 5:R12

Kusch S, Ahmadinejad N, Panstruga R, Kuhn H (2014) *In silico* analysis of the core signaling proteome from the barley powdery mildew pathogen (*Blumeria graminis* f.sp. *hordei*). BMC Genomics 15:843. https://doi.org/10.1186/1471-2164-15-843

Lamour KH, Mudge J, Gobena D, Hurtado-Gonzales OP, Schmutz J, Kuo A, Miller NA, Rice BJ, Raffaele S, Cano LM, Bharti AK, Donahoo RS, Finley S, Huitema E, Hulvey J, Platt D, Salamov A, Savidor A, Sharma R, Stam R, Storey D, Thines M, Win J, Haas BJ, Dinwiddie DL, Jenkins J, Knight JR, Affourtit JP, Han CS, Chertkov O, Lindquist EA, Detter C, Grigoriev IV, Kamoun S, Kingsmore SF (2012) Genome sequencing and mapping reveal loss of heterozygosity as a mechanism for rapid adaptation in the vegetable pathogen *Phytophthora capsici*. Mol Plant-Microbe Interact 25:1350–1360. https://doi.org/10.1094/MPMI-02-12-0028-R

Lara-Márquez A, Zavala-Páramo MG, López-Romero E, Calderón-Cortés N, López-Gómez R, Conejo-Saucedo U, Cano-Camacho H (2011) Cloning and characterization of a pectin lyase gene from *Colletotrichum lindemuthianum* and comparative phylogenetic/structural analysis with genes from phytopathogenic and saprophytic/opportunistic microorganisms. BMC Microbiol 11:260

Larkan NJ, Lydiate DJ, Parkin IAP, Nelson MN, Epp DJ, Cowling WA, Rimmer SR, Borhan MH (2013) The *Brassica napus* blackleg resistance gene LepR3 encodes a receptor-like protein triggered by the *Leptosphaeria maculans* effector AVRLM1. New Phytol 197:595–605. https://doi.org/10.1111/nph.12043

Larkan NJ, Lydiate DJ, Yu F, Rimmer SR, Borhan MH (2014) Co-localization of the blackleg resistance genes *Rlm2* and *LepR3* on *Brassica napus* chromosome A10. BMC Plant Biol 14:1595. https://doi.org/10.1186/s12870-014-0387-z

Larkan NJ, Yu F, Lydiate DJ, Rimmer SR, Borhan MH (2016) Single R gene introgression lines for accurate dissection of the *Brassica - Leptosphaeria* pathosystem. Front Plant Sci 7:1771. https://doi.org/10.3389/fpls.2016.01771

Lebrun I, Marques-Porto R, Pereira AS, Pereira A, Perpetuo EA (2009) Bacterial toxins: an overview on bacterial proteases and their action as virulence factors. Mini-Rev Med Chem 9:820–828. https://doi.org/10.2174/138955709788452603

Lee BM, Park YJ, Park DS, Kang HW, Kim JG, Song ES, Park IC, Yoon UH, Hahn JH, Koo BS, Lee GB, Kim H, Park HS, Yoon KO, Kim JH, Jung C, Koh NH, Seo JS, Go SJ (2005) The genome sequence of *Xanthomonas oryzae* pathovar oryzae KACC10331, the bacterial blight pathogen of rice. Nucleic Acids Res 33:577–586

Lemarié S, Robert-Seilaniantz A, Lariagon C, Lemoine J, Marnet N, Jubault M, Manzanares-Dauleux MJ, Gravot A (2015) Both the jasmonic acid and the salicylic acid pathways contribute to resistance to the biotrophic clubroot agent *Plasmodiophora brassicae* in *Arabidopsis*. Plant Cell Physiol 56(11):2158–2168

Li H, Durbin R (2009) Fast and accurate short read alignment with Burrows-Wheeler transform. Bioinformatics 25:1754–1760

Li M, Rollins JA (2010) The development-specific Ssp1 and Ssp2 genes of *Sclerotinia sclerotiorum* encode lectins with distinct yet compensatory regulation. Fungal Genet Biol 47:531–538. https://doi.org/10.1016/j.fgb.2010.03.008

Li H, Handsaker B, Wysoker A, Fennell T, Ruan J, Homer N, Marth G, Abecasis G, Durbin R, 1000 Genome Project Data Processing Subgroup (2009) The sequence alignment/map format and SAM tools. Bioinformatics 25:2078–2079

Li M, Liang X, Rollins JA (2012) *Sclerotinia sclerotiorum*-glutamyl transpeptidase (Ss-Ggt1) is required for regulating glutathione accumulation and development of sclerotia and compound appressoria. Mol Plant-Microbe Interact 25:412–420

Li RF, Lu GT, Li L, Su HZ, Feng GF, Chen Y, He Y-Q, Jiang B-L, Tang D-J, Tang J-L (2014) Identification of a putative cognate sensor kinase for the two-component response regulator

HrpG, a key regulator controlling the expression of the *hrp* genes in *Xanthomonas campestris* pv. *campestris*. Environ Microbiol 16(7):2053–2071. https://doi.org/10.1111/1462-2920.12207

Li XL, Guo M, Xu DF, Chen FX, Zhang HJ, Pan Y, Li MM, Gao Z (2015) The nascent-polypeptide-associated complex alpha subunit regulates the polygalacturonases expression negatively and influences the pathogenicity of *Sclerotinia sclerotiorum*. Mycologia 107:1130–1137. https://doi.org/10.3852/14-250

Li J, Zhang Y, Zhang Y, Yu PL, Pan H, Rollins JA (2018) Introduction of large sequence inserts by CRISPR-Cas9 to create pathogenicity mutants in the multinucleate filamentous pathogen *Sclerotinia sclerotiorum*. MBio 9(3):e00567–e00518. https://doi.org/10.1128/mBio.00567-18

Li G, Lv H, Zhang S, Zhang S, Li F, Zhang H, Qian W, Fang Z, Sun R (2019) *TuMV* management for *Brassica* crops through host resistance: retrospect and prospects. Plant Pathol 68:1035–1044. https://doi.org/10.1111/ppa.13016

Liang Y, Yajima W, Davis MR, Kav NNV, Stephen E (2013) Disruption of a gene encoding a hypothetical secreted protein from *Sclerotinia sclerotiorum* reduces its virulence on canola (*Brassica Napus*) disruption of a gene encoding a hypothetical secreted protein from *Sclerotinia sclerotiorum* reduces its virulence on Canola (*Brassica napus*). Can J Plant Pathol 35:46–55

Liang X, Liberti D, Li M, Kim Y, Hutchens A, Wilson RON, Rollins JA (2015) Oxaloacetate acetylhydrolase gene mutants of *Sclerotinia sclerotiorum* do not accumulate oxalic acid, but do produce limited lesions on host plants. Mol Plant Pathol 16:559–571. https://doi.org/10.1111/mpp.12211

Liang YUE, Xiong WEI, Steinkellner S, Feng JIE (2017) Deficiency of the melanin biosynthesis genes *SCD1* and *THR1* affects sclerotial development and vegetative growth, but not pathogenicity, in *Sclerotinia sclerotiorum*. Mol Plant Pathol 19:1–10. https://doi.org/10.1111/mpp.12627

Liao M, Chen C, Li Q (2017) The complete mitochondrial genome of *Alternaria alternata* (Hypocreales: Nectriceae). Mitochondrial DNA B 2:587–588

Liberti D, Rollins JA, Dobinson KF (2013) Peroxysomal carnitine acetyl transferase influences host colonization capacity in *Sclerotinia sclerotiorum*. Mol Plant–Microbe Interact 26:768–780

Lindgren PB (1997) The role of *hrp* genes during plant–bacterial interactions. Annu Rev Phytopathol 35:129–152

Links MG, Holub E, Jiang RH, Sharpe AG, Hegedus D, Beynon E, Sillito D, Clarke WE, Uzuhashi S, Borhan MH (2011) De novo sequence assembly of *Albugo candida* reveals a small genome relative to other biotrophic oomycetes. BMC Genomics 12:503

Liu XP, Lu WC, Liu YK, Wei SQ, Xu JB, Liu ZR, Zhang HJ, Li JL, Ke GL, Yao WY, Cai YS, Wu FY, Cao SC, Li YH, Xie SD, Lin BX, Zhang CL (1996) Occurrence and strain differentiation on Turnip mosaic potyvirus and sources of resistance in Chinese cabbage in China. Acta Hortic 407:431–440

Liu L, Zhao D, Zheng L, Hsiang T, Wei Y, Fu Y, Huang J (2013) Identification of virulence genes in the crucifer anthracnose fungus *Colletotrichum higginsianum* by insertional mutagenesis. Microbiol Pathol 64:6–17. https://doi.org/10.1016/j.micpath.2013.06.001

Liu L, Wang Q, Zhang X, Liu J, Zhang Y (2018) *Ssams2*, a gene encoding GATA transcription factor, is required for appressoria formation and chromosome segregation in *Sclerotinia sclerotiorum*. Front Microbiol 9:1–12. https://doi.org/10.3389/fmicb.2018.03031

Loman NJ, Quick J, Simpson JT (2015) A complete bacterial genome assembled de novo using only nanopore sequencing data. Nat Methods 12:733–735

Long Y, Wang Z, Sun Z, Fernando DWG, Mcvetty PBE, Li G (2011) Identification of two blackleg resistance genes and fine mapping of one of these two genes in a *Brassica napus* canola cultivar 'Surpass 400'. Theor Appl Genet 122:1223–1231. https://doi.org/10.1007/s00122-010-1526-z

Lorang JM, Tuori RP, Martinez JP, Sawyer TL, Redman RS, Rollins JA, Wolpert TL, Johnson KB, Rodriguez RJ, Dickman MB, Cluffetti LM (2001) Green fluorescent protein is lighting up fungal biology. Appl Environ Microbiol 67:1987–1994

Lu H, Patil P, Van Sluys M-A, White FF, Ryan RP, Dow JM, Rabinowicz P, Salzberg SL, Leach JE, Sonti R, Brendel V, Bogdanove AJ (2008) Acquisition and evolution of plant pathogenesis-associated gene clusters and candidate determinants of tissue-specificity in *Xanthomonas*. PLoS One 3:e3828

Ludwig-Muller J (1999) *Plasmodiophora brassicae*, the causal agent of clubroot disease: a review on molecular and biochemical events in pathogenesis/*Plasmodiophora brassicae*, the causative agent of cabbage hernia: an overview of molecular and biochemical processes during pathogenesis. J Plant Dis Prot 106(2):109–127

Ludwig-Muller J, Thermann P, Pieper K, Hilgenberg W (1994) Peroxidase and chitinase isoenzyme activities during root infection of Chinese cabbage with *Plasmodiophora brassicae* Woron. Physiol Plant 90:661–670

Ludwig-Müller J, Jülke S, Geiß K, Richter F, Mithöfer A, Šola I, Rusak G, Keenan S, Bulman S (2015) A novel methyltransferase from the intracellular pathogen *Plasmodiophora brassicae* methylates salicylic acid. Mol Plant Pathol 16(4):349–364. https://doi.org/10.1111/mpp.12185

Lv H, Fang Z, Yang L, Zhang Y, Wang Y (2020) An update on the arsenal: mining resistance genes for disease management of *Brassica* crops in the genomic era. Hortic Res 7:34

Lyu X, Shen C, Fu Y, Xie J, Jiang D, Li G, Cheng J (2015) Comparative genomic and transcriptional analysis of the carbohydrate-active enzymes and secretomes of phytopathogenic fungi reveal their significant roles during infection and development. Sci Rep 5:15565. https://doi.org/10.1038/srep15565

Lyu X, Shen C, Fu Y, Xie J, Jiang D, Li G, Cheng J (2016) A small secreted virulence-related protein is essential for the necrotrophic interactions of *Sclerotinia sclerotiorum* with its host plants. PLoS Pathol 12:e1005435. https://doi.org/10.1371/journal.ppat.1005435

Ma LJ, van der Does HC, Borkovich KA, Coleman JJ, Daboussi M-J, Di Pietro A, Dufresne M, Freitag M, Grabherr M, Henrissat B, Houterman PM, Kang S, Shim W-B, Woloshuk C, Xie X, Xu J-R, Antoniw J, Baker SE, Bluhm BH, Breakspear A, Brown DW, Butchko RAE, Chapman S, Coulson R, Coutinho PM, Danchin EGJ, Diener A, Gale LR, Gardiner DM, Goff S, Hammond-Kosack KE, Hilburn K, Hua-Van A, Jonkers W, Kazan K, Kodira CD, Koehrsen M, Kumar L, Lee Y-H, Li L, Manners JM, Miranda-Saavedra D, Mukherjee M, Park G, Park J, Park S-Y, Proctor RH, Regev A, Ruiz-Roldan MC, Sain D, Sakthikumar S, Sykes S, Schwartz DC, Turgeon BG, Wapinski I, Yoder O, Young S, Zeng Q, Zhou S, Galagan J, Cuomo CA, Kistler HC, Rep M (2010) Comparative genomics reveals mobile pathogenicity chromosomes in *Fusarium*. Nature 464:367–373

Manzanares-Dauleux MJ, Barret P, Thomas G (2000) Development of a pathotype specific SCAR marker in *Plasmodiophora brassicae*. Eur J Plant Pathol 106:781–787

Manzanares-Dauleux MJ, Divaret I, Baron F, Thomas G (2001) Assessment of biological and molecular variability between and within field isolates of *Plasmodiophora brassicae*. Plant Pathol 50:165–173

Martin DP, Lemey P, Lott M, Moulton V, Posada D, Lefeuvre P (2010) RDP3: a flexible and fast computer program for analyzing recombination. Bioinformatics 26:2462–2463. https://doi.org/10.1093/bioinformatics/btq467

Martin-Laurent F, van Tuinen D, Dumas-Gaudot E, Gianinazzi-Pearson V, Gianinazzi S, Franken P (1997) Differential display analysis of RNA accumulation in arbuscular mycorrhiza of pea and isolation of a novel symbiosis-regulated plant gene. Mol Gen Genet 256:37–44

Mayerhofer R, Bansal VK, Thiagarajah MR, Stringam GR, Good AG (1997) Molecular mapping of resistance to *Leptosphaeria maculans* in Australian cultivars of *Brassica napus*. Genome 40: 294–301. https://doi.org/10.1139/g97-041

McHale L, Tan X, Koehl P, Michelmore RW (2006) Plant NBS-LRR proteins: adaptable guards. Genome Biol 7:212

McMullan M, Gardiner A, Bailey K, Kemen E, Ward BJ, Cevik V, Robert-Seilaniantz A, Schultz-Larsen T, Balmuth A, Holub E, van Oosterhout C, Jones JDG (2015) Evidence for suppression of immunity as a driver for genomic introgressions and host range expansion in races of *Albugo candida*, a generalist parasite. elife 4:e04550. https://doi.org/10.7554/eLife.04550

Medstrand P, van de Lagemaat LN, Dunn CA, Landry JR, Svenback D, Mager DL (2005) Impact of transposable elements on the evolution of mammalian gene regulation. Cytogenet Genome Res 110(1–4):342–352

Meena PD, Awasthi RP, Chattopadhyay C, Singh F, Singh B, Gupta A (2002) Yield loss in Indian mustard due to white rust and effect of some cultural practices on Alternaria blight and white rust severity. J Oilseed Brassica 4:18–24

Mendes-Pereira E, Balesdent MH, Hortense B, Rouxel T (2003) Molecular phylogeny of the *Leptosphaeria maculans–L. biglobosa* species complex. Mycol Res 107:1287–1304

Merits A, Guo D, Jarveku LL, Saarma M (1999) Biochemical and genetic evidence for interactions between potato A potyvirus encoded proteins P1 and P3 and proteins of the putative replication complex. Virology 263:15–22

Micali CO, Neumann U, Grunewald D, Panstruga R, O'Connell R (2011) Biogenesis of a specialized plant–fungal interface during host cell internalization of *Golovinomyces orontii* haustoria. Cell Microbiol 13(2):210–226. https://doi.org/10.1111/j.1462-5822.2010.01530.x

Mohanta TK, Bae H (2015) The diversity of fungal genome. Biol Proced Online 17:8. https://doi.org/10.1186/s12575-015-0020-z

Moller M, Harling R (1996) Randomly amplified polymorphic DNA (RAPD) profiling of *Plasmodiophora brassicae*. Lett Appl Microbiol 22:70–75

Monod M, Capoccia S, Léchenne B, Zaugg C, Holdom M, Jousson O (2002) Secreted proteases from pathogenic fungi. Int J Med Microbiol 292:405–419. https://doi.org/10.1078/1438-4221-00223

Müller P, Hilgenberg W (1986) Isomers of zeatin and zeatin riboside in clubroot tissue: evidence for trans-zeatin biosynthesis by *Plasmodiophora brassicae*. Physiol Plant 66:245–250

Murphy PJ, Langridge P, Smith SE (1997) Cloning plant genes differentially expressed during colonization of roots of Hordeum vulgare by the vesicular arbuscular mycorrhizal fungus *Glomus intraradices*. New Phytol 135:291–301

Narusaka Y, Narusaka M, Seki M, Ishida J, Nakashima M, Kamiya A, Enju A, Sakurai T, Satoh M, Kobayashi M, Tosa Y, Park P, Shinozaki K (2003) The cDNA microarray analysis using an *Arabidopsis pad3* mutant reveals the expression profiles and classification of genes induced by *Alternaria brassicicola* attack. Plant Cell Physiol 44:377–387

Ngoc LBT, Verniere C, Jarne P, Brisse S, Guerin F, Boutry S, Gagnevin L, Pruvost O (2009) From local surveys to global surveillance: three high-throughput genotyping methods for epidemiological monitoring of *Xanthomonas citri* pv. *citri* pathotypes. Appl Environ Microbiol 75(4): 1173–1184. https://doi.org/10.1128/AEM.02245-08

Nguyen HD, Tomitaka Y, Ho SYW, Duchêne S, Vetten H-J, Lesemann D, Walsh JA, Gibbs AJ, Ohshima K (2013a) *Turnip mosaic* potyvirus probably first spread to *Eurasian Brassica* crops from wild orchids about 1000 years ago. PLoS One 8(2):e55336. https://doi.org/10.1371/journal.pone.0055336

Nguyen HD, Tran HTN, Ohshima K (2013b) Genetic variation of the *Turnip mosaic virus* population of Vietnam: a case study of founder, regional and local influences. Virus Res 171: 138–149. https://doi.org/10.1016/j.virusres.2012.11.008

Nguyen HD, Lewis CT, Levesque CA, Grafenhan T (2016) Draft genome sequence of *Alternaria alternata* ATCC 34957. Genome Announce 4:e01554–e01515

O'Connell RJ, Thon MR, Hacquard S, Amyotte SG, Kleemann J, Torres MF, Damm U, Buiate EA, Epstein L, Alkan N, Altmüller J, Alvarado-Balderrama L, Bauser CA, Becker C, Birren BW, Chen Z, Choi J, Crouch JA, Duvick JP, Farman MA, Gan P, Heiman D, Henrissat B, Howard RJ, Kabbage M, Koch C, Kracher B, Kubo Y, Law AD, Lebrun M-H, Lee Y-H, Miyara I, Moore N, Neumann U, Nordström K, Panaccione DG, Panstruga R, Place M, Proctor RH, Prusky D, Rech G, Reinhardt R, Rollins JA, Rounsley S, Schardl CL, Schwartz DC, Shenoy N, Shirasu K, Sikhakolli UR, Stüber K, Sukno SA, Sweigard JA, Takano Y, Takahara H, Trail F, van der Does HC, Voll LM, Will I, Young S, Zeng Q, Zhang J, Zhou S, Dickman MB, Schulze-Lefert P, van Themaat EVL, Ma L-J, Vaillancourt LJ (2012) Lifestyle transitions in plant pathogenic *Colletotrichum* fungi deciphered by genome and transcriptome analysis. Nat Genet 44:1060–1065. https://doi.org/10.1038/ng.2372

O'Donnell KO (1992) Ribosomal DNA internal transcribed spacers are highly divergent in the phytopathogenic ascomycete *Fusarium sambucinum* (*Gibberella pulicaris*). Curr Genet 22: 213–220

O'Donnell K, Gueidan C, Sink S, Johnston PR, Crous PW, Glenn A, Riley R, Zitomer NC, Colyer P, Waalwijk C, van der Lee T, Moretti A, Kang S, Kim H-S, Geiser DM, Juba JH, Baayen RP, Cromey MG, Bithell S, Sutton DA, Skovgaard K, Ploetz R, Corby Kistler H, Elliott M, Davis M, Sarver BAJ (2009) A two-locus DNA sequence database for typing plant and human pathogens within the *Fusarium oxysporum* species complex. Fungal Genet Biol 46: 936–948

Ochiai N, Fujimura M, Motoyama T, Ichiishi A, Usami R, Horikoshi K, Yamaguchi I (2001) Characterization of mutations in the two-component histidine-kinase gene that confer fludioxonil resistance and osmotic sensitivity in the os-1 mutants of *Neurospora crassa*. Pest Manag Sci 57:437–442

Ochiai H, Inoue V, Takeya M, Sasaki A, Kaku H (2005) Genome sequence of *Xanthomonas oryzae* pv. *oryzae* suggests contribution of large numbers of effector genes and insertion sequences to its race diversity. Jpn Agric Res Q 39:275–287

Ohshima Y, Yamaguchi R, Hirota T, Hamamoto K, Tomimura Z, Tan T, Sano F, Azuhata JA, Walsh J, Fletcher J, Chen A, Gera A, Gibbs A (2002) Molecular evolution of *Turnip mosaic virus*; evidence of host adaptation, genetic recombination and geographical spread. J Gen Virol 83:1511–1521

Ohshima K, Tomitaka Y, Wood JT, Minematsu Y, Kajiyama H, Tomimura K, Gibbs AJ (2007) Patterns of recombination in *Turnip mosaic virus* genomic sequences indicate hot spots of recombination. J Gen Virol 88:298–315

Oome S, Raaymakers TM, Cabral A, Samwel S, Böhm H, Albert I, Thorsten N, Van den Ackerveken G (2014) Nep1-like proteins from three kingdoms of life act as a microbe-associated molecular pattern in *Arabidopsis*. Proc Natl Acad Sci U S A 111:16955–16960

Oshima M, Fujimura M, Banno S, Hashimoto C, Motoyama T, Ichiishi A, Yamaguchi I (2002) A point mutation in the two component histidine kinase BcOS-1 gene confers dicarboximide resistance in field isolates of *Botrytis cinerea*. Phytopathology 92:75–80

Pan H, Rollins JA (2017) The GATA-Type IVb zinc-finger transcription factor SsNsd1 regulates asexual—Sexual development and appressoria formation in *Sclerotinia sclerotiorum*. Mol Plant Pathol 9:1679–1689

Pan Y, Wei J, Yao C, Reng H, Gao Z (2018) Plant science SsSm1, a cerato-platanin family protein, is involved in the hyphal development and pathogenic process of *Sclerotinia sclerotiorum*. Plant Sci 270:37–46. https://doi.org/10.1016/j.plantsci.2018.02.001

Park J, Lee S, Choi J, Ahn K, Park B, Park J, Kang S, Lee Y-H (2008) Fungal cytochrome P450 database. BMC Genomics 9:402. https://doi.org/10.1186/1471-2164-9-402

Parker JE, Barber CE, Fan MJ, Daniels MJ (1993) Interaction of *Xanthomonas campestris* with *Arabidopsis thaliana*: characterization of a gene from *X. c.* pv. *raphani* that confers avirulence to most *A. thaliana* accessions. Mol Plant-Microbe Interact 6:216–224

Parker JE, Holub EB, Frost LN, Falk A, Gunn ND, Daniels MJ (1996) Characterization of eds1, a mutation in *Arabidopsis* suppressing resistance to *Peronospora parasitica* specified by several different RPP genes. Plant Cell 8:2033–2046. https://doi.org/10.1105/tpc.8.11.2033

Patil PB, Bogdanove AJ, Sonti RV (2007) The role of horizontal transfer in the evolution of a highly variable lipopolysaccharide biosynthesis locus in Xanthomonads that infect rice, citrus and crucifers. BMC Evol Biol 7:243

Pedersen C, van Themaat EVL, McGuffin LJ, Abbott JC, Burgis TA, Barton G, Bindschedler LV, Lu X, Maekawa T, WeBling R, Cramer R, Christensen HT, Panstruga R, Spanu PD (2012) Structure and evolution of barley powdery mildew effector candidates. BMC Genomics 13:694

Peninckx IAMA, Eggermont K, Terras FRG, Thomma BPHJ, De Samblanx GW, Buchala A, Metraux JP, Manners JM, Broekaert WF (1996) Pathogen induced systemic activation of a plant defensin gene in *Arabidopsis* follows a salicylic acid- independent pathway. Plant Cell 8:2309–2323

Perez-Lopez E, Waldner M, Hossain M, Kusalik AJ, Wei Y, Bonham-Smith PC, Todd CD (2018) Identification of *Plasmodiophora brassicae* effectors-a challenging goal. Virulence 9(1):1344–1353

Petrie GA (1988) Races of *Albugo candida* (white rust and stag head) on cultivated cruciferae in Saskatchewan. Can J Plant Pathol 10:142–150

Pieretti I, Royer M, Barbe V, Carrere S, Koebnik R, Cociancich S, Couloux A, Darrasse A, Gouzy J, Jacques MA, Lauber E, Manceau C, Mangenot S, Poussier S, Segurens B, Szurek B, Verdier V, Arlat M, Rott P (2009) The complete genome sequence of *Xanthomonas albilineans* provides new insights into the reductive genome evolution of the xylem- limited *Xanthomonadaceae*. BMC Genomics 10:616

Pillonel C, Meyer T (1997) Effect of phenylpyrroles on glycerol accumulation and protein kinase activity of *Neurospora crassa*. Pestic Sci 49(3):229–236. https://doi.org/10.1002/(SICI)1096-9063(199703)49:3<229::AID-PS525>3.0.CO;2-T

Plaumann PL, Schmidpeter J, Dahl M, Taher L, Koch C (2018) A dispensable chromosome is required for virulence in the hemibiotrophic plant pathogen *Colletotrichum higginsianum*. Front Microbiol 9:1005. https://doi.org/10.3389/fmicb.2018.01005

Ploch S, Choi YJ, Rost C, Shin HD, Schilling E, Thines M (2010) Evolution of diversity in *Albugo* is driven by high host specificity and multiple speciation events on closely related Brassicaceae. Mol Phylogenet Evol 57:812–820

Poplawsky AR, Chun W (1998) *Xanthomonas campestris* pv. *campestris* requires a functional pig B for epiphytic survival and host infection. Mol Plant-Microbe Interact 11:466–475

Pound GS, Williams PH (1963) Biological races of *Albugo candida*. Phytopathology 53:1146–1149

Prigent S, Frioux C, Dittami SM, Thiele S, Larhlimi A, Collet G, Gutknecht F, Got J, Eveillard D, Bourdon J, Plewniak F, Tonon T, Siegel A (2017) Meneco, a topology-based gap-filling tool applicable to degraded genome-wide metabolic networks. PLoS Comput Biol 13(1):e1005276. https://doi.org/10.1371/journal.pcbi.1005276

Provart NJ, Alonso J, Assmann SM, Bergmann D, Brady SM, Brkljacic J, Browse J, Chapple C, Colot V, Cutler S, Dangl J, Ehrhardt D, Friesner JD, Frommer WB, Grotewold E, Meyerowitz E, Nemhauser J, Nordborg M, Pikaard C, Shanklin J, Somerville C, Stitt M, Torii KU, Waese J, Wagner D, McCourt P (2016) 50 years of *Arabidopsis* research: highlights and future directions. New Phytol 209:921–944

Provvidenti R (1980) Evaluation of Chinese cabbage cultivars from Japan and the People's Republic of China for resistance to *Turnip Mosaic Virus* and *Cauliflower Mosaic Virus*. J Am Soc Hortic Sci 105:571–573

Pruvost O, Verniere C, Vital K, Guerin F, Jouen E, Chiroleu F, Ah-You N, Gagnevin L (2011) Insertion sequence- and tandem repeat-based genotyping techniques for *Xanthomonas citri* pv. *mangiferae indicae*. Phytopathology 101(7):887–893. https://doi.org/10.1094/PHYTO-11-10-0304

Qian W, Jia Y, Ren SX, He YQ, Feng JX, Lu LF, Sun Q, Ying G, Tang DJ, Tang H, Wu W, Hao P, Wang L, Jiang BL, Zeng S, Gu WY, Lu G, Rong L, Tian Y, Yao Z, Fu G, Chen B, Fang R, Qiang B, Chen Z, Zhao GP, Tang JL, He C (2005) Comparative and functional genomic analysis of the pathogenicity of phytopathogen *Xanthomonas campestris* pv. *campestris*. Genome Res 15:757–767

Qu X, Yu B, Liu J, Zhang X, Li G, Zhang D (2014) MADS-Box transcription factor SsMADS is involved in regulating growth and virulence in *Sclerotinia sclerotiorum*. Int J Mol Sci 15:8049–8062

Raffaele S, Kamoun S (2012) Genome evolution in filamentous plant pathogens: why bigger can be better. Nat Rev Microbiol 10:417–430. https://doi.org/10.1038/nrmicro2790

Rajarammohan S, Kumar A, Gupta V, Pental D, Pradhan AK, Kaur J (2017) Genetic architecture of resistance to *Alternaria brassicae* in *Arabidopsis thaliana:* QTL mapping reveals two major resistance-conferring loci. Front Plant Sci 8:260

Rajarammohan S, Paritosh K, Pental D, Kaur J (2019) Comparative genomics of *Alternaria* species provides insights into the pathogenic lifestyle of *Alternaria brassicae* – a pathogen of the Brassicaceae family. BMC Genomics 20:1036. https://doi.org/10.1186/s12864-019-6414-6

Raman R, Taylor B, Marcroft S, Stiller J, Eckermann P, Coombes N, Rehman A, Lindbeck K, Luckett D, Wratten N, Batley J, Edwards D, Wang X, Raman H (2012) Molecular mapping of qualitative and quantitative loci for resistance to *Leptosphaeria maculans* causing blackleg disease in canola (*Brassica napus* L.). Theor Appl Genet 125:405–418. https://doi.org/10.1007/s00122-012-1842-6

Rao VG (1969) The genus *Alternaria* from India. Nova Hedwigia 17:219–258

Rehmany AP, Lynn JR, Tör M, Holub EB, Beynon JL (2000) A comparison of *Peronospora parasitica* (downy mildew) isolates from *Arabidopsis thaliana* and *Brassica oleracea* using amplified fragment length polymorphism and internal transcribed spacer 1 sequence analysis. Fungal Genet Biol 30:95–103

Revers F, Le Gall O, Candresse T, Maule AJ (1999) New advances in understanding the molecular biology of plant/potyvirus interactions. Mol Plant-Microbe Interact 12:367–376

Reymond P, Deleage G, Rascle C, Fevre M (1994) Cloning and sequence analysis of a polygalacturonase-encoding gene from the phytopathogenic fungus *Sclerotinia sclerotiorum*. Gene 146:233–237

Riechmann JL, Laın S, Garcıa JA (1992) Highlights and prospects of potyvirus molecular biology. J Gen Virol 73:1–16

Riethmuller A, Voglmayr H, Goker M, Weiß M, Oberwinkler F (2002) Phylogenetic relationships of the downy mildews (Peronosporales) and related groups based on nuclear large subunit ribosomal DNA sequences. Mycologia 94:834–849

Rigano LA, Payette C, Brouillard G, Marano MR, Abramowicz L, Torres PS, Yun M, Castagnaro AP, El Oirdi M, Dufour V, Malamud F, Dow JM, Bouarab K, Vojnov A (2007) Bacterial cyclic beta-(1,2)-glucanacts in systemic suppression of plant immune responses. Plant Cell 19:2077–2089

Rimmer SR, Mathur S, Wu CR (2000) Virulence of isolates of *Albugo candida* from western Canada to *Brassica* species. Can J Plant Pathol 22:229–235. https://doi.org/10.1080/07060660009500468

Rojas MC, Hedden P, Gaskin P, Tudzynski B (2001) The P450-1 gene of *Gibberella fujikuroi* encodes a multifunctional enzyme in gibberellin biosynthesis. Proc Natl Acad Sci U S A 98:5838–5843. https://doi.org/10.1073/pnas.091096298

Rolfe SA, Strelkov SE, Links MG, Clarke WE, Robinson SJ, Djavaheri M, Malinowski R, Haddadi P, Kagale S, Parkin IAP, Taheri A, Borhan MH (2016) The compact genome of the plant pathogen *Plasmodiophora brassicae* is adapted to intracellular interactions with host *Brassica* spp. BMC Genomics 17:272

Rolland S, Jobic C, Fevre M, Bruel C (2003) *Agrobacterium* mediated transformation of *Botrytis cinerea* simple purification of monokaryotic transformants and rapid conidia based identification of the transfer DNA host genomic DNA flanking sequences. Curr Genet 44:164–171

Rollins JA, Pathology P, Hall F (2003) The *Sclerotinia sclerotiorum pac1* gene is required for sclerotial development and virulence. Mol Plant-Microbe Interact 16:785–795. https://doi.org/10.1094/MPMI.2003.16.9.785

Rossman AY, Palm ME, Spielman LJ (1987) A literature guide for the identification of the plant pathogenic fungi. APS Press, St Paul. MN

Roux B, Bolot S, Guy E, Denancé N, Lautier M, Jardinaud M-F, Fischer-Le Saux M, Portier P, Jacques M-A, Gagnevin L, Pruvost O, Lauber E, Arlat M, Carrère S, Koebnik R, Noël L (2015) Genomics and transcriptomics of *Xanthomonas campestris* species challenge the concept of core type III effectome. BMC Genomics 16:975

Rouxel T, Grandaubert J, Hane JK, Hoede C, van de Wouw AP, Couloux A, Dominguez V, Anthouard V, Bally P, Bourras S, Cozijnsen AJ, Ciuffetti LM, Degrave A, Dilmaghani A, Duret L, Fudal I, Goodwin SB, Gout L, Glaser N, Linglin J, Kema GHJ, Lapalu N, Lawrence CB, May K, Meyer M, Ollivier B, Poulain J, Schoch CL, Simon A, Spatafora JW, Stachowiak A, Turgeon BG, Brett M, Tyler BM, Vincent D, Weissenbach J, Amselem J, Quesneville H, Oliver RP, Wincker P, Balesdent MH, Howlett BJ (2011) Effector diversification within compartments of the *Leptosphaeria maculans* genome affected by Repeat-Induced Point mutations. Nat Commun 2:202. https://doi.org/10.1038/ncomms1189

Rudrabhatla P, Reddy MM, Rajasekharan R (2006) Genome-wide analysis and experimentation of plant serine/threonine/tyrosine-specific protein kinases. Plant Mol Biol 60:293–319. https://doi.org/10.1007/s11103-005-4109-7

Ryan RP, Fouhy Y, Lucey JF, Jiang BL, He YQ, Feng JX, Tang JL, Dow JM (2007) Cyclic di-GMP signaling in the virulence and environmental adaptation of *Xanthomonas campestris*. Mol Microbiol 63:429–442

Ryan RP, McCarthy Y, Andrade M, Farah CS, Armitage JP, Dow JM (2010) Cell–cell signal-dependent dynamic interactions between HD-GYP and GGDEF domain proteins mediate virulence in *Xanthomonas campestris*. Proc Natl Acad Sci U S A 107:5989–5994

Ryan RP, Vorholter FJ, Potnis N, Jones JB, Van Sluys MA, Bogdanove AJ, Dow JM (2011) Pathogenomics of *Xanthomonas*: understanding bacterium–plant interactions. Nat Rev Microbiol 9:344–355

Saenz GS, Taylor JW (1999) Phylogeny of Erysiphales (powdery mildews) inferred from internal transcribed spacer (ITS) ribosomal DNA sequences. Can J Bot 77:150–168

Saenz GS, Taylor JW, Gargas A (1994) 18S rRNA gene sequences and supraordinal classification of the Erysiphales. Mycologia 86:212–216

Saenz P, Cervera MT, Dallot S, Quiot L, Quiot J-B, Riechmann JL, Garcıa JA (2000) Identification of a pathogenicity determinant of Plum pox virus in the sequence encoding the C-terminal region of protein P3+6K1. J Gen Virol 81:557–566

Saharan GS, Mehta N (2008) *Sclerotinia* diseases of crop plants: biology, ecology and disease management. Springer, Dordrecht, p 485

Saharan GS, Verma PR (1992) White rusts: a review of economically important species. IDRC-MR315e. International Development Research Centre, IV, Ottawa, ON. 65 p

Saharan GS, Verma PR, Meena PD, Kumar A (2014) White rust of crucifers: biology, ecology and management. Springer, New Delhi. 275 p

Saharan GS, Mehta N, Meena PD (2016) Alternaria blight of crucifers: biology, ecology and disease management. Springer, New Delhi. 291 p

Saharan GS, Mehta N, Meena PD (2017) Downy mildew disease of crucifers: biology, ecology and disease management. Springer, Singapore, p 347

Saharan GS, Naresh N, Meena PD (2019) Powdery mildew disease of crucifers: biology, ecology and disease management. Springer, Singapore. 362 p

Saito I (1977) Sclerotinia sclerotiorum (Lib.) de Bary. Studies on the maturation and germination of sclerotia of Sclerotinia sclerotiorum (Lib.) de Bary, a causal fungus of bean stem rot. Report Hokkaido prefectural Agric. Exp. Stn. No. 26, March, 1977. Hokkaido central agricultural experiment station, 069–13, Japan, 106 p

Salzberg SL, Sommer DD, Schatz MC, Phillippy AM, Rabinowicz PD, Tsuge S, Furutani A, Ochiai H, Delcher AL, Kelley D, Madupu R, Puiu D, Radune D, Shumway M, Trapnell C, Aparna G (2008) Genome sequence and rapid evolution of the rice pathogen *Xanthomonas oryzae* pv. *oryzae* PXO99A. BMC Genomics 9:204. https://doi.org/10.1186/1471-2164-9-204

Sanchez F, Wang X, Jenner CE, Walsh JA, Ponz F (2003) Strains of *Turnip mosaic* potyvirus as defined by the molecular analysis of the coat protein gene of the virus. Virus Res 94:33–43

Santana MF, Silva JCF, Mizubuti ESG, Arajo EF, Queiroz MV (2014) Analysis of Tc1-Mariner elements in *Sclerotinia sclerotiorum* suggests recent activity and flexible transposases. BMC Microbiol 14:256

Satou M, Fukumoto F (1996) The host range of downy mildew, *Peronospora parasitica*, from *Brassica oleracea*, cabbage and broccoli crops. Ann Phytopathol Soc Jpn 62:393–396

Schumacher MM, Enderlin CS, Selitrennikoff CP (1997) The osmotic-1 locus of *Neurospora crassa* encodes a putative histidine kinase similar to osmosensors of bacteria and yeast. Curr Microbiol 34:340–347

Schwelm A, Fogelqvist J, Knaust A, Jülke S, Lilja T, Bonilla-Rosso G, Karlsson M, Shevchenko A, Dhandapani V, Choi SR, Kim HG, Park JY, Lim YP, Ludwig-Muller J, Dixelius C (2015) The *Plasmodiophora brassicae* genome reveals insights in its life cycle and ancestry of chitin synthases. Sci Rep 5:11153

Schwelm A, Dixelius C, Ludwig-Müller J (2016) New kid on the block – the clubroot pathogen genome moves the plasmodiophorids into the genomic era. Eur J Plant Pathol 145:531–542

Schwelm A, Badstober J, Bulman S, Desoignies N, Etemadi M, Falloon RE, Gachon CMM, Legreve A, Lukes J, Merz U, Nenarokova A, Strittmatter M, Sullivan BK, Neuhauser S (2018) Not in your usual top 10: protists that infect plants and algae. Mol Plant Pathol 19(4): 1029–1044

Seidl MF, Faino L, Shi-Kunne X, van den Berg GC, Bolton MD, Thomma BP (2015) The genome of the saprophytic fungus *Verticillium tricorpus* reveals a complex effector repertoire resembling that of its pathogenic relatives. Mol Plant-Microbe Interact 28:362–373

Seifbarghi S, Borhan MH, Wei Y, Coutu C, Robinson SJ, Hegedus DD (2017) Changes in the *Sclerotinia sclerotiorum* transcriptome during infection of *Brassica napus*. BMC Genome 18: 266

Selin C, de Kievit TR, Belmonte MF, Fernando WGD (2016) Elucidating the role of effectors in plant-fungal interactions: progress and challenges. Front Microbiol 7:600. https://doi.org/10.3389/fmicb.2016.00600

Seruga Music M, Duc Nguyen H, Černi S, Mamula Đ, Ohshima K, Škorić D (2014) Multi-locus sequence analysis of 'Candidatus *Phytoplasma asteris*' strain and the genome analysis of *Turnip mosaic virus* co-infecting oilseed rape. J Appl Microbiol 117(8):774–785

Shattuck VI (1992) The biology, epidemiology and control of *Turnip mosaic virus*. Plant Breed Rev 14:199–238

Shen Y, Diener AC (2013) *Arabidopsis thaliana* resistance to *Fusarium oxysporum 2* implicates Tyrosine-sulfated peptide signaling in susceptibility and resistance to root infection. PLoS Genet 9:e1003525

Shimada C, Lipka V, O'Connell R, Okuno T, Schulze-Lefert P, Takano Y (2006) Non-host resistance in *Arabidopsis-Colletotrichum* interactions acts at the cell periphery and requires actin filament function. Mol Plant-Microbe Interact 19:270–279. https://doi.org/10.1094/MPMI-19-0270

Siemens J, Bulman S, Rehn F, Sundelin T (2009a) Molecular biology of *Plasmodiophora brassicae*. J Plant Growth Regul 28:245–251. https://doi.org/10.1007/s00344-009-9091-x

Siemens J, Graf H, Bulman S, In O, Ludwig-Müller J (2009b) Monitoring expression of selected *Plasmodiophora brassicae* genes during clubroot development in *Arabidopsis thaliana*. Plant Pathol 58:130–136

da Silva M, Esposito E, Moody JD, Canhos VP, Cerniglia CE (2004) Metabolism of aromatic hydrocarbons by the filamentous fungus *Cyclothyrium* sp. Chemosphere 57:943–952. https://doi.org/10.1016/j.chemosphere.2004.07.051

Simmons EG (1967) Typification of *Alternaria, Stemphylium*, and *Ulocladium*. Mycologia 59:67–92

Skinner W, Keon J, Hargreaves J (2001) Gene information for fungal plant pathogens from expressed sequences. Curr Opin Microbiol 4:381–386

Soanes DM, Richards TA, Talbot NJ (2007) Insights from sequencing fungal and oomycete genomes: what can we learn about plant disease and the evolution of pathogenicity? Plant Cell 19:3318–3326. https://doi.org/10.1105/tpc.107.056663

Song Q, Smith AD (2011) Identifying dispersed epigenomic domains from ChIP-Seq data. Bioinformatics 27:870–871

Souza DP, Andrade MO, Alvarez-Martinez CE, Arantes GM, Farah CS, Salinas RK (2011) A component of the Xanthomonadaceae type IV secretion system combines a VirB7 motif with a N0 domain found in outer membrane transport proteins. PLoS Pathol 7:e1002031

Soyer JL, Clairet C, Gay EJ, Lapalu N, Rouxel T, Stukenbrock EH, Fudal I (2021) Genome-wide mapping of histone modifications during axenic growth in two species of *Leptosphaeria maculans* showing contrasting genomic organization. Chromosome Res 29(2):219–236. https://doi.org/10.1007/s10577-021-09658-1

Spanu PD (2012) The genomics of obligate (and Non-obligate) biotrophs. Annu Rev Phytopathol 50:91–109. https://doi.org/10.1146/annurev-phyto-081211-173024

Spanu PD, Abbot JC, Amselem J, Burgis TA, Soanes DM, Stüber K, van Themaat EVL, Brown JKM, Butcher SA, Gurr SJ, Lebrun MH, Ridout CJ, Schulze-Lefert P, Talbot NJ, Ahmadinejad N, Ametz C, Barton GR, Benjdia M, Bidzinski P, Bindschedler LV, Both M, Brewer MT, Cadle-Davidson L, Cadle-Davidson MM, Collemare J, Cramer R, Frenkel O, Godfrey D, Harriman J, Hoede C, King BC, Klages S, Kleemann J, Knoll D, Koti PS, Kreplak J, López-Ruiz FJ, Lu X, Maekawa T, Mahanil S, Micali C, Milgroom MG, Montana G, Noir S, O'Connel RJ, Oberhaensli S, Parlange F, Pendersen C, Quesneville H, Reinhardt R, Rott M, Sacristán S, Schmidt SM, Schön M, Skaminoti P, Sommer H, Stephens A, Takahra H, Thordal-Christensen H, Vigouroux M, Weßling R, Wicker T, Panstruga R (2010) Genome expansion and gene loss in powdery mildew fungi reveal tradeoffs in extreme parasitism. Science 330: 1543–1546. https://doi.org/10.1126/science.1194573

Stavolone L, Alioto D, Ragozzino A, Laliberte J-F (1998) Variability among turnip mosaic potyvirus isolates. Phytopathology 88:1200–1204

Stobbs LW, Shattuck VI (1989) *Turnip mosaic virus* strains in southern Ontario, Canada. Plant Dis 73:208–212

Studholme DJ, Kemen E, MacLean D, Schornack S, Aritua V, Thwaites R, Grant M, Smith J, Jones JDG (2010) Genome-wide sequencing data reveals virulence factors implicated in banana *Xanthomonas* wilt. FEMS Microbiol Lett 310:182–192

Stukenbrock EH (2013) Evolution, selection and isolation: a genomic view of speciation in fungal plant pathogens. New Phytol 99:895–907. https://doi.org/10.1111/nph.12374

Suehiro N, Natsuaki T, Watanabe T, Okuda S (2004) An important determinant of the ability of Turnip mosaic virus to infect *Brassica* spp. and/or *Raphanus sativus* is in its P3 protein. J Gen Virol 85:2087–2098

Sundelin T (2008) Identification of expressed *Plasmodiophora brassicae* genes during plant infection and molecular diagnosis of plant pathogens. Ph D Thesis, University of Copenhagen

Sutherland JB (1992) Detoxification of polycyclic aromatic hydrocarbons by fungi. J Ind Microbiol 9:53–61. https://doi.org/10.1007/BF01576368

Swofford DL (1993) Phylogenetic analysis using parsimony (PAUP version 3.1.1). Illinois Natural History Surv, Champaign, IL

Syme RA, Tan KC, Hane JK, Dodhia K, Stoll T, Hastie M, Furuki E, Ellwood SR, Williams AH, Tan YF, Testa AC, Gorman JJ, Oliver RP (2016) Comprehensive annotation of the *Parastagonospora nodorum* reference genome using next-generation genomics, transcriptomics and proteogenomics. PLoS One 11:e0147221

Takahara H, Dolf A, Endl E, O'Connell R (2009) Flow cytometric purification of *Colletotrichum higginsianum* biotrophic hyphae from *Arabidopsis* leaves for stage-specific transcriptome analysis. Plant J 59:672–683. https://doi.org/10.1111/j.1365-313X.2009.03896.x

Takamatsu S, Kano Y (2001) PCR primers useful for nucleotide sequencing of rDNA of the powdery mildew fungi. Mycoscience 42:135–139

Tamura K, Dudley J, Nei M, Kumar S (2007) MEGA4: molecular evolutionary genetics analysis (MEGA) software version 4.0. Mol Biol Evol 24:1596–1599

Tan Z, Gibbs AJ, Tomitaka Y, Sanchez F, Ponz F, Ohshima K (2005) Mutations in Turnip mosaic virus genomes that have adapted to *Raphanus sativus*. J Gen Virol 86:501–510

Tang JL, Liu YN, Barber CE, Dow JM, Wootton JC, Daniels MJ (1991) Genetic and molecular analysis of a cluster of *rpf* genes involved in positive regulation of synthesis of extracellular enzymes and polysaccharide in *Xanthomonas campestris* pathovar *campestris*. Mol Gen Genet 226:409–417

Tanifuji G, Archibald JM, Hashimoto T (2016) Comparative genomics of mitochondria in chlorarachniophyte algae: endosymbiotic gene transfer and organellar genome dynamics. Sci Rep 6:21016. https://doi.org/10.1038/srep21016

Tarailo-Graovac M, Chen N (2009) Using repeat masker to identify repetitive elements in genomic sequences. Curr Prot. Bioinformatics 5:4.10.1–4.10.14

Tatsuzawa F, Saito N, Toki K, Shinoda K, Honda T (2012) Flower colors and their anthocyanins in *Matthiola incana* cultivars (Brassicaceae). J Jpn Soc Hortic Sci 81:91–100

Terby J (1924) Les divisions sporogoniques du *Plasmodiophora brassicae* Woron. Bull Acad R Sci Belgique Cl Sci 5:519–537

Testa AC, Oliver RP, Hane JK (2016) Occulter Cut: a comprehensive survey of AT-rich regions in fungal genomes. Genome Biol Evol 8:2044–2064

Thieme F, Koebnik R, Bekel T, Berger C, Boch J, Buttner D, Caldana C, Gaigalat L, Goesmann A, Kay S, Kirchner O, Lanz C, Linke B, McHardy AC, Meyer F, Mittenhuber G, Nies DH, Niesbach-Klosgen U, Patschkowski T, Ruckert C, Rupp O, Schneiker S, Schuster SC, Vorholter FJ, Weber E, Puhler A, Bonas U, Bartels D, Kaiser O (2005) Insights into genome plasticity and pathogenicity of the plant pathogenic bacterium *Xanthomonas campestris* pv. *vesicatoria* revealed by the complete genome sequence. J Bacteriol 187:7254–7266

Thines M (2014) Phylogeny and evolution of plant pathogenic oomycetes—a global overview. Eur J Plant Pathol 138:431–447. https://doi.org/10.1007/s10658-013-0366-5

Thines M, Kamoun S (2010) Oomycete-plant co-evolution: recent advances and future prospects. Curr Opin Plant Biol 13:427–433. https://doi.org/10.1016/j.pbi.2010.04.001

Thines M, Choi YJ, Kemen E, Ploch S, Holub EB, Shin HD, Jones JDG (2009a) A new species of *Albugo* parasitic to *Arabidopsis thaliana* reveals new evolutionary patterns in white blister rusts (Albuginaceae). Persoonia 22:123–128

Thines M, Voglmayr H, Goker M (2009b) Taxonomy and phylogeny of the downy mildews (Peronosporaceae). In: Lamour K, Kamoun S (eds) Oomycete genetics and genomics. John Wiley & Sons, London, pp 47–75

Thomma BPHJ, Eggermont K, Penninckx IAMA, Mauch-Mani B, Vogelsang R, Cammue BPA, Broekaert WF (1998) Separate jasmonate-dependent and salicylate-dependent defense-response pathways in *Arabidopsis* are essential for resistance to distinct microbial pathogens. Proc Natl Acad Sci U S A 95:5107–15111

Thomma BPHJ, Seidl MF, Shi-Kunne X, Cook DE, Bolton MD, van Kan JAL, Faino L (2016) Mind the gap; seven reasons to close fragmented genome assemblies. Fungal Genet Biol 90:24–30

Tomimura K, Gibbs AJ, Jenner CE, Walsh JA, Ohshima K (2003) The phylogeny of *Turnip mosaic virus*; comparisons of 38 genomic sequences reveal a Eurasian origin and a recent 'emergence' in east Asia. Mol Ecol 12:2099–2111

Tomimura K, Spak J, Katis N, Jenner CE, Walsh JA, Gibbs AJ, Ohshima K (2004) Comparison of the genetic structure of populations of Turnip mosaic virus in West and East Eurasia. Virology 330:408–423

Tommerup IC, Ingram DS (1971) The life-cycle of *Plasmodiophora brassicae* Woron. in *Brassica* tissue cultures and intact roots. New Phytol 70:327–332

Tor M, Holub EB, Brose E, Musker R, Gunn N (1994) Map positions of three loci in *Arabidopsis thaliana* associated with isolate-specific recognition of *Peronospora parasitica* (downy mildew). Mol Plant-Microbe Interact 7:214–222. https://doi.org/10.1094/MPMI-7-0214

Torres PS, Malamud F, Rigano LA, Russo DM, Marano MR, Castagnaro AP, Zorreguieta A, Bouarab K, Dow JM, Vojnov AA (2007) Controlled synthesis of the DSF cell–cell signal is required for biofilm formation and virulence in *Xanthomonas campestris*. Environ Microbiol 9: 2101–2109

Tyler BM, Tripathy S, Zhang XM, Dehal P, Jiang RH, Aerts A, Arredondo FD, Baxter L, Bensasson D, Beynon JL, Chapman J, Damasceno CM, Dorrance AE, Dou D, Dickerman

AW, Dubchak IL, Garbelotto M, Gijzen M, Gordon SG, Govers F, Grunwald NJ, Huang W, Ivors KL, Jones RW, Kamoun S, Krampis K, Lamour KH, Lee MK, McDonald WH, Medina M, Meijer HJ, Nordberg EK, Maclean DJ, Ospina-Giraldo MD, Morris PF, Phuntumart V, Putnam NH, Rash S, Rose JK, Sakihama Y, Salamov AA, Savidor A, Scheuring CF, Smith BM, Sobral BW, Terry A, Torto-Alalibo TA, Win J, Xu Z, Zhang H, Grigoriev IV, Rokhsar DS, Boore JL (2006) *Phytophthora* genome sequences uncover evolutionary origins and mechanisms of pathogenesis. Science 313:1261–1266. https://doi.org/10.1126/Science.1128796

Umate P, Tuteja N, Tuteja R (2011) Genome-wide comprehensive analysis of human helicases. Commun Integr Biol 4:118–137. https://doi.org/10.4161/cib.4.1.13844

Urcuqui-Inchima S, Haenni A-L, Bernardi F (2001) Potyvirus proteins: a wealth of functions. Virus Res 74:157–175

Van de Wouw AP, Thomas VL, Cozijnsen AJ, Marcroft SJ, Salisbury PA, Howlett BJ (2008) Identification of *Leptosphaeria biglobosa* 'canadensis' on *Brassica juncea* stubble from northern New South Wales, Australia. Aust Plant Dis Notes 3:124–128

Van de Wouw AP, Cozijnsen AJ, Hane JK, Brunner PC, McDonald BA, Oliver RP, Howlett BJ (2010) Evolution of linked avirulence effectors in *Leptosphaeria maculans* is affected by genomic environment and exposure to resistance genes in host plants. PLoS Pathol 6: e1001180. https://doi.org/10.1371/journal.ppat.1001180

Van Wees SCM, Chang HS, Zhu T, Glazebrook J (2003) Characterization of the early response of *Arabidopsis* to *Alternaria brassicicola* infection using expression profiling. Plant Physiol 132: 606–617

Veluchamy S, Rollins JA (2008) A CRY-DASH-type photolyase/cryptochrome from *Sclerotinia sclerotiorum* mediates minor UV-A-specific effects on development. Fungal Genet Biol 45: 1265–1276. https://doi.org/10.1016/j.fgb.2008.06.004

Veluchamy S, Williams B, Kim K, Dickman MB (2012) Physiological and molecular plant pathology the CuZn superoxide dismutase from *Sclerotinia sclerotiorum* is involved with oxidative stress tolerance, virulence, and oxalate production. Physiol Mol Plant Pathol 78:14–23. https://doi.org/10.1016/j.pmpp.2011.12.005

Venn L (1979) The genetic control of sexual compatibility in *Leptosphaeria maculans*. Australas Plant Pathol 8:5–6. https://doi.org/10.1071/APP9790005

Vicente JG, Holub EB (2013) *Xanthomonas campestris* pv. *campestris* (cause of black rot of crucifers) in the genomic era is still a worldwide threat to Brassica crops. Mol. Plant Pathol 14(1):2–18

Vicente JG, Conway J, Roberts SJ, Taylor JD (2001) Identification and origin of *Xanthomonas campestris* pv. *campestris* races and related pathovars. Phytopathology 91:492–499

Vincenot L, Balesdent MH, Li H, Barbetti MJ, Sivasithamparam K, Gout L, Rouxel T (2008) Occurrence of a new subclade of *Leptosphaeria biglobosa* in Western Australia. Phytopathology 98:321–329

Visser RGF, Bachem CWB, de Boer JM, Bryan GJ, Chakrabati SK, Feingold S, Gromadka R, van Ham RCHJ, Huang S, Jacobs JME, Kuznetsov B, de Melo PE, Milbourne D, Orjeda G, Sagredo B, Tang X (2009) Sequencing the potato genome: outline and first results to come from the elucidation of the sequence of the world's third most important food crop. Am J Potato Res 86:417

Voglmayr H (2003) Phylogenetic relationships of *Peronospora* and related genera based on nuclear ribosomal ITS sequences. Mycol Res 107:1132–1142

Voglmayr H, Greilhuber J (1998) Genome size determination in peronosporales (Oomycota) by Feulgen image analysis. Fungal Genet Biol 25:181–195. https://doi.org/10.1006/fgbi.1998.1097

Voigt K, Cozijnsen AJ, Kroymann J, Pöggeler S, Howlett BJ (2005) Phylogenetic relationships between members of the crucifer pathogenic *Leptosphaeria maculans* species complex as shown by mating type (MAT1-2), actin, and β-tubulin sequences. Mol Phylogenet Evol 37: 541–557

Vojnov AA, Slater H, Daniels MJ, Dow JM (2001) Expression of the gumoperon directing xanthan biosynthesis in *Xanthomonas campestris* and its regulation *in planta*. Mol Plant-Microbe Interact 14:768–774

Vorhoelter FJ, Schneiker S, Goesmann A, Krause L, Bekel T, Kaiser O, Linke B, Patschkowski T, Rueckert C, Schmid J, Sidhu VK, Sieber V, Tauch A, Watt SA, Weisshaar B, Becker A, Niehaus K, Puehler A (2008) The genome of *Xanthomonas campestris* pv. *campestris* B100 and its use for the reconstruction of metabolic pathways involved in xanthan biosynthesis. J Biotechnol 134:33–45

Vorholter FJ, Niehaus K, Puhler A (2001) Lipopolysaccharide biosynthesis in *Xanthomonas campestris* pv. *campestris*: a cluster of 15 genes is involved in the biosynthesis of the LPS O-antigen and the LPS core. Mol Gen Genomics 266:79–95

Walker J, Smith AM (1972) *Leptosphaeria narmari* and *L. korrae* spp. nov., two long-spored pathogens of grasses in Australia. Trans Br Mycol Soc 58:459–466

Wallenhammar AC, Arwidsson O (2001) Detection of *Plasmodiophora brassicae* by PCR in naturally infested soils. Eur J Plant Pathol 107:313–321

Walsh JA (1989) Genetic control of immunity to Turnip mosaic virus in winter oilseed rape (*Brassica napus* ssp. *oleifera*) and the effect of foreign isolates of the virus. Ann Appl Biol 115:89–99

Walsh JA, Jenner CE (2002) Turnip mosaic virus and the quest for durable resistance. Mol Plant Pathol 3:289–300

Walsh JA, Tomlinson JA (1985) Viruses infecting winter oilseed rape (*Brassica napus* ssp. *oleifera*). Ann Appl Biol 107(3):485–495

Wang LF, Tang XY, He CZ (2007) The bifunctional effector *AvrXccC* of *Xanthomonas campestris* pv. *campestris* requires plasma membrane-anchoring for host recognition. Mol Plant Pathol 8: 491–501

Wang L, Liu Y, Liu J, Zhang Y, Zhang X, Pan H (2016) The *Sclerotinia sclerotiorum* FoxE2 gene is required for apothecial development. Mycology 106:484–490

Wei K, Tang DJ, He YQ, Feng JX, Jiang BL, Lu GT, Chen BS, Tang JL (2007) hpaR, a putative marR family transcriptional regulator, is positively controlled by HrpG and HrpX and involved in the pathogenesis, hypersensitive response, and extracellular protease production of *Xanthomonas campestris* pathovar *campestris*. J Bacteriol 189:2055–2062

Wei L, Jian H, Lu K, Filardo F, Yin N, Liu L, Qu C, Li W, Du H, Li J (2016) Genome-wide association analysis and differential expression analysis of resistance to *Sclerotinia* stem rot in *Brassica napus*. Plant Biotechnol J 14(6):1368–1380

Weld RJ, Eady CC, Ridgway HJ (2005) *Agrobacterium* mediated transformation of *Sclerotinia sclerotiorum*. J Microbiol Methods 65(1):202–207. https://doi.org/10.1016/jmimet.2005.07.010

Wengelnik K, Rossier O, Bonas U (1999) Mutations in the regulatory gene *hrpG* of *Xanthomonas campestris* pv. *vesicatoria* result in constitutive expression of all *hrp* genes. J Bacteriol 181(21): 6828–6831

Weßling R, Schmidt SM, Micali CO, Knaust F, Reinhardt R, Neumann U, van Themaat EVL, Panstruga R (2012) Transcriptome analysis of enriched *Golovinomyces orontii* haustoria by deep 454 pyrosequencing. Fungal Genet Biol 49:470–482

Weßling R, Epple P, Altmann S, He Y, Yang L, Henz SR, McDonald N, Wiley K, Bader KC, Gläßer C, Mukhtar MS, Haigis S, Ghamsari L, Stephens AE, Ecker JR, Vidal M, Jones JDG, Mayer KFX, van Themaat EVL, Weigel D, Schulze-Lefert P, Dangl JL, Panstruga R, Braun P (2014) Convergent targeting of a common host protein-network by pathogen effectors from three kingdoms of life. Cell Host Microbe 16(3):364–375. https://doi.org/10.1016/j.chom.2014.08.004

White TJ, Bruns T, Lee S, Taylor JW (1990) Amplication and direct sequencing of fungal ribosomal RNA genes for phylogenetics. In: Innis MA, Gelfand DH, Sninsky JJ, White TJ (eds) PCR protocols. A guide to methods and application. Academic Press, London, pp 315–322

White FF, Potnis N, Jones JB, Koebnik R (2009) The type III effectors of *Xanthomonas*. Mol Plant Pathol 10:749–766

Wicker T, Oberhaensli S, Parlange F, Buchmann JP, Shatalina M, Roffler S, Ben-David R, Dolezel J, Simkova H, Schulze-Lefert P, Spanu PD, Bruggmann R, Amselem J, Quesneville H, van Themaat EVL, Paape T, Shimizu KK, Keller B (2013) The wheat powdery mildew genome shows the unique evolution of an obligate biotroph. Nat Genet 45:1092–1096

Wilson TJG, Bertrand N, Tang JL, Feng JX, Pan MQ, Barber CE, Dow JM, Daniels MJ (1998) The rpfA gene of *Xanthomonas campestris* pathovar *campestris*, which is involved in the regulation of pathogenicity factor production, encodes an aconitase. Mol Microbiol 28:961–970

Wiltshire SP (1947) Species of *Alternaria* on Brassicae. Mycol Pap 20:1–15

Wolters PJ, Faino L, van den Bosch TBM, Evenhuis B, Visser RGF, Seidl MF, Vleeshouwers VGAA (2018) Gapless genome assembly of the potato and tomato early blight pathogen *Alternaria solani*. Mol Plant-Microbe Interact 31:692–694

Woudenberg JHC, Groenewald JZ, Binder M, Crous PW (2013) *Alternaria* redefined. Stud Mycol 75:171–212

Woudenberg JH, Seidl MF, Groenewald JZ, de Vries M, Stielow JB, Thomma BP, Crous PW (2015) *Alternaria* section *Alternaria*: species, formae speciales or pathotypes? Stud Mycol 82: 1–21

Wu Y (2012) Unwinding and rewinding: double faces of helicase? J Nucl Acids 2012:140601. https://doi.org/10.1155/2012/140601

Wu J, Cai G, Tu J, Li L, Liu S, Luo X, Zhou L, Fan C, Zhou Y (2013) Identification of QTLs for resistance to *Sclerotinia* stem rot and *BnaC. IGMT5*, a as a candidate gene of the major resistant QTL SRC6 in *Brassica napus*. PLoS One 8:e67740. https://doi.org/10.1371/journal.pone.0067740

Wu J, Zhao Q, Liu S, Shahid M, Lan L, Cai G, Zhang C, Fan C, Wang Y, Zhou Y (2016) Genome-wide association study identifies new loci for resistance to Sclerotinia stem rot in *Brassica napus*. Front Plant Sci 7:1418

Xia S, Xu Y, Hoy R, Zhang J, Qin L, Li X (2020) The notorious soil borne pathogenic fungus *Sclerotinia sclerotiorum*: an update on genes studied with mutant analysis. Pathogens 9(1):27. https://doi.org/10.3390/pathogens9010027

Xiao X, Xie J, Cheng J, Li G, Yi X, Jiang D, Fu Y (2014) Novel secretory protein Ss-Caf1 of the plant-pathogenic fungus *Sclerotinia sclerotiorum* is required for host penetration and normal sclerotial development. Mol Plant-Microbe Interact 27:40–55. https://doi.org/10.1094/MPMI-05-13-0145-R

Xu L, Chen W (2013) Random T-DNA mutagenesis identifies a Cu/Zn superoxide dismutase gene as a virulence factor of *Sclerotinia sclerotiorum*. Mol Plant-Microbe Interact 26:431–441. https://doi.org/10.1094/MPMI-07-12-0177-R

Xu RQ, Li XZ, Wei HY, Jiang B, Li K, He YQ, Feng JX, Tang JL (2006) Regulation of eight avr genes by hrpG and hrpX in *Xanthomonas campestris* pv. *campestris* and their role in pathogenicity. Prog Nat Sci 16:1288–1294

Xu RQ, Blanvillain S, Feng JX, Jiang BL, Li XZ, Wei HY, Kroj T, Lauber E, Roby D, Chen B, He YQ, Lu GT, Tang DJ, Vasse J, Arlat M, Tang JL (2008) AvrAC(Xcc8004), a type III effector with a leucine-rich repeat domain from *Xanthomonas campestris* pathovar *campestris* confers avirulence in vascular tissues of *Arabidopsis thaliana* ecotype Col-0. J Bacteriol 190:343–355

Xu L, Xiang M, White D, Chen W (2015) pH dependency of sclerotial development and pathogenicity revealed by using genetically defined oxalate-minus mutants of *Sclerotinia sclerotiorum*. Environ Microbiol 17:2896–2909. https://doi.org/10.1111/1462-2920.12818

Xu T, Li J, Yu B, Liu L, Zhang X, Liu J, Montes HM (2018) Transcription factor SsSte12 was involved in mycelium growth and development in *Sclerotinia sclerotiorum*. Front Microbiol 9: 1–13. https://doi.org/10.3389/fmicb.2018.02476

Yajima W, Liang Y, Kav NNV, Sclerotinia A, Bary D (2009) Gene disruption of an arabinofuranosidase/β-xylosidase precursor decreases *Sclerotinia sclerotiorum* virulence on canola tissue. Mol Plant-Microbe Interact 22:783–789. https://doi.org/10.1094/MPMI-22-7-0783

Yan Y, Yuan Q, Tang J, Huang J, Hsiang T, Wei Y, Zheng L (2018) *Colletotrichum higginsianum* as a model for understanding host–pathogen interactions: a review. Int J Mol Sci 19(7):2142. https://doi.org/10.3390/ijms19072142

Yang G, Tang L, Gong Y, Xie J, Fu Y, Jiang D, Li G, Collinge DB, Chen W, Cheng J (2018) A cerato-platanin protein SsCP1 targets plant PR1 and contributes to virulence of *Sclerotinia sclerotiorum*. New Phytol 217:739–755. https://doi.org/10.1111/nph.14842

Yano S, Tanaka S, Ito S, Kameya-Iwaki M (1997) Variations of random amplified polymorphic DNA (RAPD) patterns among field populations of *Plasmodiophora brassicae*. Ann Phytopathol Soc Jpn 63:179–182

Yarden O, Veluchamy S, Dickman MB, Kabbage M (2014) Physiological and molecular plant pathology *Sclerotinia sclerotiorum* catalase SCAT1 affects oxidative stress tolerance, regulates ergosterol levels and controls pathogenic development. Physiol Mol Plant Pathol 85:34–41

Yasaka R, Fukagawa H, Ikematsu M, Soda H, Korkmaz S, Golnaraghi A, Katis N, Ho SYW, Gibbs AJ, Ohshima K (2017) The timescale of emergence and spread of *Turnip mosaic* potyvirus. Sci Rep 7:4240

Yerkes WD, Shaw CG (1959) Taxonomy of the *Peronospora* species on Cruciferae and Chenopodiaceae. Phytopathology 49:499–507

Yin X, Yi B, Chen W, Zhang W, Tu J, Fernando WGD, Fu T (2010) Mapping of QTLs detected in a *Brassica napus* DH population for resistance to *Sclerotinia sclerotiorum* in multiple environments. Euphytica 173:25–35. https://doi.org/10.1007/s10681-009-0095-1

Yoshii M, Yoshioka N, Ishikawa M, Naito S (1998) Isolation of an *Arabidopsis thaliana* mutant in which accumulation of *Cucumber mosaic virus* coat protein is delayed. Plant J 13:211–219

Yoshimi A, Imanishi J, Gafur A, Tanaka C, Tsuda M (2003) Characterization and genetic analysis of laboratory mutants of *Cochliobolus heterostrophus* resistant to dicarboximide and phenylpyrrole fungicides. J Gen Plant Pathol 69:101–108

You MP, Akhatar J, Mittal M, Barbetti MJ, Maina S, Banga SS (2021) Comparative analysis of draft genome assemblies developed from whole genome sequences of two *Hyaloperonospora brassicae* isolate samples differing in field virulence on *Brassica napus*. Biotech Rep 31: e00653. https://doi.org/10.1016/j.btre.2021.e00653

Yu SC, Zhang FL, Zhao XY, Yu YJ, Zhang DS (2011) Sequence–characterized amplified region and simple sequence repeat markers for identifying the major quantitative trait locus responsible for seedling resistance to downy mildew in Chinese cabbage (*Brassica rapa* ssp. *pekinensis*). Plant Breed 130:580–583. https://doi.org/10.1111/j.1439-0523.2011.01874.x

Yu F, Lydiate DJ, Gugel RK, Sharpe AG, Rimmer SR (2012) Introgression of *Brassica rapa* subsp. *sylvestris* blackleg resistance into *Brassica napus*. Mol Breed 30:1495–1506. https://doi.org/10.1007/s11032-012-9735-6

Yu Y, Xiao J, Yang Y, Bi C, Qing L, Tan W (2015) Physiological and molecular plant pathology *Ss-Bi1* encodes a putative BAX Inhibitor-1 protein that is required for full virulence of *Sclerotinia sclerotiorum*. Physiol Mol Plant Pathol 90:115–122. https://doi.org/10.1016/j.pmpp.2015.04.005

Yu F, Zhang X, Huang Z, Chu M, Song T, Falk KC, Deora A, Chen Q, Zhang Y, McGregor L, Gossen BD, McDonald MR, Peng G (2016) Identification of genome–wide variants and discovery of variants associated with *Brassica rapa* clubroot resistance gene Rcr1 through bulked segregant RNA sequencing. PLoS One 11:e0153218. https://doi.org/10.1371/journal.pone.0153218

Yu Y, Xiao J, Zhu W, Yang Y, Mei J, Bi C, Qian WEI, Qing L, Tan W (2017) Ss-Rhs1, a secretory Rhs repeat-containing protein, is required for the virulence of *Sclerotinia sclerotiorum*. Mol Plant Pathol 18:1052–1061. https://doi.org/10.1111/mpp.12459

Yu Y, Du J, Wang Y, Zhang M, Huang Z, Cai J, Fang A, Yang Y, Qing L, Bi C, Cheng J (2019) Survival factor 1 contributes to the oxidative stress response and is required for full virulence of *Sclerotinia sclerotiorum*. Mol Plant Pathol 20:895–906. https://doi.org/10.1111/mpp.12801

Yu H, Ayhan DH, Diener AC, Ma LJ (2020) Genome sequence of *Fusarium oxysporum* f. sp. *matthiolae*, a Brassicaceae pathogen. Mol Plant-Microbe Interact 33(4):569–572

Yun MH, Torres PS, El Oirdi M, Rigano LA, Gonzalez-Lamothe R, Marano MR, Castagnaro AP, Dankert MA, Bouarab K, Vojnov A (2006) Xanthan induces plant susceptibility by suppressing callose deposition. Plant Physiol 141:178–187

Zampounis A, Pigne S, Dallery JF, Wittenberg AH, Zhou S, Schwartz DC, Thon MR, O'Connell RJ (2016) Genome sequence and annotation of *Colletotrichum higginsianum*, a causal agent of crucifer anthracnose disease. Genome Announc 4:e00821–e00816. https://doi.org/10.1128/genomeA.00821-16

Zander M, Patel DA, Van de Wouw A, Lai K, Lorenc MT, Campbell E, Hayward A, Edwards D, Raman H, Batley J (2013) Identifying genetic diversity of avirulence genes in *Leptosphaeria maculans* using whole genome sequencing. Funct Integr Genomics 13:295–308. https://doi.org/10.1007/s10142-013-0324-5

Zang N, Tang DJ, Wei ML, He YQ, Chen BS, Feng JX, Xu J, Gan YQ, Jiang BL, Tang JL (2007) Requirement of a mip-like gene for virulence in the phytopathogenic bacterium *Xanthomonas campestris* pv. *campestris*. Mol Plant-Microbe Interact 20:21–30

Zeiner CA, Purvine SO, Zink EM, Paša-Tolić L, Chaput DL, Haridas S, Wu S, LaButti K, Grigoriev IV, Henrissat B, Santelli CM, Hansel CM (2016) Comparative analysis of secretome profiles of manganese(II)-oxidizing ascomycete fungi. PLoS ONE 1:e0157844

Zhang Y, Lamm R, Pillonel C, Lam S, Xu JR (2002) Osmoregulation and fungicide resistance: the *Neurospora crassa* os-2 gene encodes a HOG1 mitogen-activated protein kinase homologue. Appl Environ Microbiol 68:532–538

Zhang X, Liu B, Zou F, Shen D, Yin Z, Wang R, He F, Wang Y, Tyler BM, Fan W, Qian W, Dou D (2019) Whole genome re-sequencing reveals natural variation and adaptive evolution of *Phytophthora soja*e. Front Microbiol 10:2792

Zhao JW, Meng JL (2003) Genetic analysis of loci associated with partial resistance to Sclerotinia sclerotiorum in rapeseed (*Brassica napus* L.). Theor Appl Genet 106:759–764

Zhao J, Udall JA, Quijada PA, Grau CR, Meng J, Osborn TC (2006) Quantitative trait loci for resistance to *Sclerotinia sclerotiorum* and its association with a homeologous non-reciprocal transposition in *Brassica napus* L. Theor Appl Genet 112:509–516

Zheng LY, Guo XS, He B, Sun LJ, Peng Y, Dong SS, Liu TF, Jiang S, Ramachandran S, Liu CM, Jing HC (2011) Genome-wide patterns of genetic variation in sweet and grain sorghum (*Sorghum bicolor*). Genome Biol 12:R114. https://doi.org/10.1186/gb-2011-12-11-r114

Zhu W, Wei W, Fu Y, Cheng J, Xie J, Li G, Yi X (2013) A secretory protein of necrotrophic fungus *Sclerotinia sclerotiorum* that suppresses host resistance. PLoS One 8:e53901. https://doi.org/10.1371/journal.pone.0053901

Zhu GL, Yu G, Zhang YH, Rollins JA, Li J, Pan H (2019) The formaldehyde dehydrogenase SsFdh1 is regulated by and functionally cooperates with the GATA transcription. MSystems 4:1–19. https://doi.org/10.1128/mSystems.00397-19

Zou Z, Zhang X, Parks P, du Toit LJ, van de Wouw AP, Fernando WGD (2019) A new subclade of *Leptosphaeria biglobosa* identified from *Brassica rapa*. Int J Mol Sci 20:1668. https://doi.org/10.3390/ijms20071668

3 Genomics of Host–Pathogen Interaction

Abstract

Genomics of host–pathogen interactions have greatly advanced in perception and understanding of molecular and genetical mechanisms of host–pathogen recognition system and pathogenesis. Next-generation sequencing and omics technologies have enabled the sequencing of entire host and pathogen genome or transcriptome more efficiently for better comprehension of host-pathosystem. Pathogenomics have revealed how pathogens adopt a specific host, subvert innate immune responses and change host range, and develop new virulent races/strains/pathotypes. In *Brassica–Albugo* interactions, the host–*R* genes control pathogen infection and development in compatible (susceptible) and incompatible (resistant/tolerant) cultivars. Differentially expressed proteins regulate defense reactions in the resistant varieties. The boost in cysteine and GSH levels is crucial for incompatible interaction between *Albugo candida* and *Brassica juncea*. In the *A. candida–B. juncea* pathosystem, the timing of the induction of PR-5 in R and S varieties is crucial for mounting of an effective defense to the pathogen. During infection and pathogenesis of *Albugo* in *Brassica* genotypes differing in level of resistance (R, PR, MS, S), *R* genes regulate functions and molecular mechanisms. In *Brassica–Alternaria* interactions species-specific genes of *Alternaria* with putative roles in virulence have been identified. Transcriptomic analysis has revealed six genes strongly associated with *A. brassicicola* pathogenesis. *AbPf2* gene activates pathogenicity in *A. brassicicola*. *AbNPS2* gene plays an important role in disease development and virulence. Among several genes identified as pathogenicity factors, *Amr1* gene is important for success of *A. brassicicola* as competitive saprophyte and plant parasite. Microarray analysis of *Brassica* infected with *Alternaria* at 48 hpi revealed differential expression of 87 genes mainly involved in cell response signaling, cell wall degradation, protein

G. Singh Saharan et al., *Genomics of Crucifer's Host-Pathosystem*, https://doi.org/10.1007/978-981-19-3812-2_3

degradation, auxins-induced metabolism, being downregulated. The change in host cell ultrastructure and transcriptome reprogramming has been observed in *Alternaria*-infected *Brassica.* Defense-related genes are activated also in *Brassica–Alternaria* susceptible interactions. Mitogen-activated protein (MAP) kinase and CAMP/PKA signaling pathways are involved in fungal growth, conidial formation, stress tolerance, and pathogenesis in *Arabidopsis* by *Colletotrichum higginsianum.* During pathogenesis, host peroxisomes, microtubules, and Golgi are novel targets for fungal effector genes. Fus3/Ksi1-related MAPK chMK1 plays a significant role in cell wall integrity. In *Brassica–Erysiphe* interactions, three closely related genes, *MLO2, MLO6,* and *MLO12*, play a role in powdery mildew pathogenesis through suppression of basal defense responses. Powdery mildew effector genes promote pathogenesis through host transcriptional regulation. The WRKY 18/40 regulate induced expression of genes whose products are required for pathogen growth and reproduction. In response to powdery mildew infection, the expression of genes is regulated for different biological functions by transcriptional genes. The powdery mildew pathogenesis and disease development is governed by the genes *AtLFG1* and *AtLFG2* in *A. thaliana*. Several genes viz., *RSP1, RSP2, DMR1, LRR-RLK, DMR6,* and *DLO1* govern pathogenesis of *Brassica* by *Hyaloperonospora* enabling susceptibility process like attraction and attachment of pathogen to host cells, accommodation of specialized infection and feeding structures inside host cells, and nutrient production and transport from host to pathogen. Haustorium formation is initiated when pathogen secrete hydrolyses to locally degrade the host cell wall and establish an entry point. Haustorial penetration is accompanied by reorganization of GFP-tagged cytoskeleton and organelles in *Arabidopsis* mesophyll cells.

In *Brassica–Leptosphaeria* pathosystem, *BnMPK9–AvrLm1* interaction enhances cell death caused by *BnMPk9.* The virulence mechanisms are the transition period from biotrophy to necrotrophy to facilitate the acquisition of host nutrients by the pathogen. The *AvrLM1* and *AvrLM4-7* contribute to the aggressiveness of *L. maculans* analysis during leaf infection. Genome and transcriptome analysis during *Brassica* pathogenesis by *Leptosphaeria* has revealed several genes for 20 major functional categories. The GFP is expressed by the Cht promoter in stem lesions of *Brassica* during infection of *Leptosphaeria.* The gene *Pro1* is involved in clubroot pathogenesis by stimulating resting spore germination through its proteolytic activity. The genes *PbPP2A* and *PbHMG* are expressed in the early phases of infection in *B. rapa* roots. The BjN1t1 expression in *B. juncea* during infection promotes over production of auxins to form galls in infected roots. The role of different genes has been revealed through analysis of *P. brassicae* genome and life stage specific transcriptomics. Several effectors (genes) involved in the infection process (life cycle) of *P. brassicae* and subsequent clubroot development in its host have been identified. *Plasmodiophora* pathogenesis is influenced by lipids in resting spores, glucosinolates content of the host and host resistance level. Gene expression levels of primary and secondary zoospores govern different mechanisms utilized by the pathogen in causing primary and secondary infections of *Plasmodiophora*. Differential expressions of up- and downregulated genes and functional groups have been determined after 4, 7, and 10 DAI during life cycle stages of *Plasmodiophora*. In *Brassica* during *Plasmodiophora* infections, there is differential expression of several genes in host shoots, and roots regulating different biological, biochemical, and hormonal responses. In *Arabidopsis* roots infected with *Plasmodiophora*, hypoxia and SWEET genes are expressed to form galls. Host species-specific genes are expressed during *Brassica–Plasmodiophora* interactions. The genes *Pb-YPT*, *Pb-Brep9*, and *Pb-PSA* regulate *P. brassicae* infection. During clubroot development, there is expression of MiRNA and immunophilins. *Plasmodiophora* manipulates host defense through methyl transferase by inactivating SA to Me SA. The GPCR signal transduction pathways play important roles in the growth, development, and pathogenicity of *P. brassicae* to crucifers. During *Sclerotinia* pathogenesis of *Brassica,* four endo-polygalacturonase genes (*sspg1d, sspg3, sspg5,* and *sspg 6*) and two exo-PG genes (*ssxpg1* and *ssxpg2*) are expressed. Molecular analysis of *Sclerotinia sclerotiorum* has revealed its biology, mycelial growth, virulence, pathogenesis, and signal events for sclerotia formation. Genes like *Mats*, *FoxE2, CAMP*, and Nsd1 control apothecial and ascospore development. *Sclerotinia* infection and pathogenesis is regulated by toxins, CWDEs, ROS, OA, Ssvp1, V263, Svf1, PKs13, and Nacα. During the course of *Brassica* infections by *Sclerotinia* transcriptomic analysis revealed genes for functional groups like hydrolytic enzymes, CAZymes, secondary metabolites, detoxification, signaling, development, secreted effectors, oxalic acid, and reactive oxygen species production. The genes *SsSvf1* is involved in coping with ROS for pathogenesis of *Sclerotinia*. There is differential expression of genes during *Sclerotinia* interaction with R and S genotypes of *Brassica*. Integrated mRNA, sRNA, and degradome sequencing revealed complex responses to *Sclerotinia* infection in *Brassica*. The gene SSITL is an effector which suppresses resistance in the host at early stage of *Sclerotinia* infection. Oxaloacetate acetyl-hydrolase gene is involved in *Sclerotinia* pathogenesis. Three *Sclerotinia* genes (*SsCTR1*, *SsCCS*, and *SsATX1*) are upregulated during infection of *B. oleracea*. Several spliceosome genes are differentially altered during *Sclerotinia* infections. P3 proteins are important in the infection cycle of *TuMV* and involved in virus replication, accumulation, symptomatology, resistance breeding, and cell to cell movement. The *BnSGS3* gene in *Brassica* has differentiated effects on the pathogenesis of viruses. Nitric oxide signaling and calcium protein kinases can modulate NADPH oxidase activity in modulating *A. thaliana–TuMV* pathosystem. Differential expression of proteins has been identified which are activated in response to *Verticillium longisporum* in the leaf apoplast and xylem sap of *Brassica* in relation to disease development, photosynthetic electron transport, gas exchange and nutrient status. The secretion system of *Xanthomonas* genome regulates pathogenesis of black rot in crucifers. The features and functions of Type III effector genes of *Xanthomonas* in pathogenesis of crucifers have been characterized. The effectors/genes of *Xanthomonas* governing virulence and pathogenicity have been identified. Key pathogenicity determinants and virulence genes of *Xanthomonas* have been characterized and exported by Type ii and Type iii secretion system (T3S). The hormonal signaling in the pathways of *Xanthomonas* infection and pathogenesis of *Brassica* has been revealed.

The use of evolutionary genomics on the host–pathogen interaction and its impact can clear road map to combat the continuous threat of pathogens in a changing world. New tools will facilitate to resolve challenges of genomics of host with multiple pathogens interactions. The whole genome analysis approaches offer the potential to reveal new unbiased insight into the genetic basis of host-pathosystem of crucifers'.

Keywords

Genomics of host–pathogen interaction · *Albugo* pathosystem · *R* genes function · Molecular mechanisms of interaction · *Alternaria* pathosystem · Identification of species-specific genes · Transcription factors · *AbPf2* gene activation · Microarray analysis · Transcriptome reprogramming · Defense-related genes · *Colletotrichum* pathosystem · Mitogen-activated protein (MAP) kinase · cAMP/PKA signaling · Identification of effectors/genes · *Erysiphe* pathosystem · Genomics of powdery mildew · MLO proteins · Transcriptional programming · Lifeguard protein (LFG) · Gene expression levels · *Hyaloperonospora* pathosystem · Accommodation of infection · *Leptosphaeria* pathosystem · Molecular mechanisms · Characterization of genome and transcriptomes · *Plasmodiophora* pathosystem · Gene functions–characterization · Effector proteases · Serine protease pro 1 · Endogenous glucosinolate · Host gene expression · Differential expression of genes · Identification of effectors genes · Expression of hypoxia genes · SWEET genes · Host-specific genes · miRNAs · Immunophilins · Transcriptome analysis · G-protein-coupled receptors (GPCR) signal transduction · *Sclerotinia* pathosystem · Expression of genes · Enzymes · Signaling events · Identification of *Arabidopsis* NLR genes · Differential expression of multiple metabolic enzymes · Differential gene expression · Integrated mRNA · sRNA · Degradome sequencing · Secretory protein · Oxaloacetate acetylhydrolase gene · Effector like proteins · Role of copper · Genome-wide alternative splicing · *Turnip Mosaic Virus* pathosystem · Determinant genes · *BnSGS3* gene · Identification of modulating genes · *Verticillium* pathosystem · Differential expression of proteins · *Xanthomonas* pathosystem · Secretion system regulation · Type III effector genes · *Xanthomonas* gene · Hormonal signaling pathways · Evolutionary genomics · Challenges of genomics · Advantages of gene fragments

3.1 Introduction

Due to their vulnerability to a wide range of pathogens, the genomics of crucifers' host-pathogen interaction is extremely complex and intriguing, including pathogens that are necrotrophs (*Alternaria, Leptosphaeria*) as well as biotrophs (*Albugo, Hyaloperonospora*) and hemibiotrophs (*Sclerotinia, Colletotrichum*). Some pathogens are strictly foliar (*Alternaria*, *Colletotrichum*) to root invaders (*Fusarium, Verticillium*) and ectoparasites (*Erysiphe*) to endoparasites (*Plasmodiophora*). The mode of host entry is through natural openings (stomata, hydathodes), injury and direct through appressorial formation and enzymatic action after softening of hard tissues. Mode of acquisition of nutrition is from absorption to holophytic. Mixed infection and co-infection of pathogens have synergistic effects on host response. Host range is very wide to narrow and exhibiting host specificity. The effect on the host tissues is from specific (foliage, stem) to systemic and vascular system. Each of these factors has great bearing on the host–pathogen recognition process during pre- and post-host–pathogen interaction. Immediately after establishment of recognition process, the pathogens respond through initiation and expression of virulence genes to suppress immune system of the host for successful pathogenesis. The expression level of virulence genes differs in host–pathogen interaction to manipulate host defense in favor of pathogens to feed them for infection, development, and survival to continue the cycle.

The evaluation of published literature on host–pathogen systems reveals the expected recent increase in whole genome data sets for a broad range of host–pathogen studies, and also what is missing when addressing complex systems on ecological and spatiotemporal scales. This highlights a gap that future efforts might be specifically focused on in order to be able to project and test whether the identified underlying genetics of host–pathogen interactions scale-up from simple systems, such as a laboratory study, to ecological and spatiotemporal complex situations in the wild. The quantity and complexity of the sequence data generated in many recent and ongoing studies of host–pathogen interactions presents a unique opportunity for further investigations addressing as yet unexplored aspects of the data. Perhaps most critically, however, it is suspected that this massive volume of complex data poses an increasing challenge for comparisons across studies. The lack of comprehensive cross-taxon comparative databases of host–pathogen interactions likely impedes the synthesis of individual host–pathogen studies and translation of new knowledge into solutions for real world problems. In the real-world, (a) pathogens attack hosts in the context of changing host environments, (b) these environments are increasingly impacted by anthropogenic forces such as climate change, and (c) are usually characterized by diverse communities of pathogenic and nonpathogenic organisms. Studies of host–pathogen interactions suggest that these complexities are rarely captured in a single study. Thus, it will be essential to conduct comparative studies and perform meta-analyses of existing data across systems in order to achieve a comprehensive synthesis of how genomics can address host–pathogen interactions at different scales.

3.2 *Brassica–Albugo* Pathosystem

3.2.1 Introduction

In all species of *Albugo*, oospores are the primary source of inoculum (Butler 1918; Butler and Jones 1961; Chupp 1925; Heald 1926; Verma et al. 1975; Verma and Petrie 1980; Walker 1957; Saharan and Verma 1992; Saharan et al. 2014). In perennial hosts like horse radish, the mycelium persists in the crowns, and occasionally in lateral roots (Kadow and Anderson 1940). Secondary infection and spread of disease during the growing season is by means of sporangia, which are readily carried short distances by splashed water droplets or, to a certain degree, by air currents. Moisture on the host surface is essential both for sporangial germination and infection by zoospores. The most likely primary infection sites are the emerging cotyledons (Verma et al. 1975). Zoospores derived from germinating oospores are also capable of causing infection (Verma et al. 1975), but no evidence exists of infection arising directly by germ tubes produced from germinating oospores (Verma and Petrie 1975; Verma 2012).

On germination, each sporangium and oospore releases about 4–12 and 40–60 biflagellate zoospores, respectively (Verma and Petrie 1975). Raabe and Pound (1952) reported a persistent terminal bead on one flagellum of *A. occidentalis* zoospores. In *A. tragopogonis,* Whipps and Cooke (1978) observed the appearance and subsequent disappearance of flagellar beads during zoospore differentiation, and the absorption of beads by encysting zoospores. After swimming for a time, a zoospore encysts and forms a germ tube, which enters the host through a stoma (Verma et al. 1975). Encystment of zoospores, penetration of germ tube through a stoma, and formation of a haustorium within 8 h after inoculation, were similar in both resistant and susceptible hosts (Verma et al. 1975). In the former, mycelial growth of the pathogen ceases in the substomatal chamber, and within 72 h after inoculation, a marked encapsulation forms around each single haustorium (Plate 3.1, Verma et al. 1975). In the congenial or susceptible host, however, the mycelium advances intercellularly with the production of several haustoria per infected cell (Liu and Rimmer 1990; Liu et al. 1989; Napper 1933; Verma et al. 1975; Walker 1957). According to Liu and Rimmer (1990), the host genotype, inoculum concentration, and incubation temperature, as well as the interactions among these factors had significant effects on infection. The day/night temperatures of 22/17 °C were more favorable to fungal growth than 15/10 °C. Infection levels increased with the concentration of inoculum on each line/cultivar at both the two-leaf and five-leaf growth stages of *B. napus* and *B. rapa* (Saharan et al. 2014).

3.2.2 *R* Genes Function During *Albugo* Pathogenesis

During their study of histopathological compatibility and incompatibility between *Brassica rapa* and *Albugo candida* race 7, Liu et al. (1989) observed the formation of haustoriums within palisade mesophyll cells adjacent to substomatal chambers 8 hours following inoculation. In the resistant cultivars, most primary hyphae

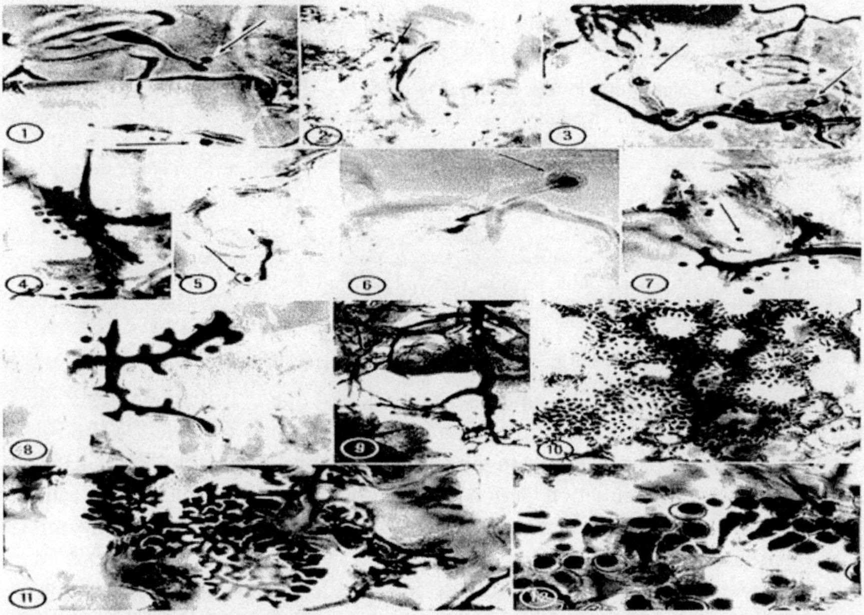

Plate 3.1 Development of *Albugo candida* in the cotyledons of several *Brassica* spp. with increasing time after inoculation (**1**) *Brassica juncea* 18 h after inoculation × 100. Note penetration of stomata by germ tube and formation of first haustorium (arrowed) (**2**) *Brassica rapa* 48 h × 530 haustorium arrowed (**3**) *Brassica juncea* 46 h × 770. Haustoria arrowed (**4**) *Brassica rapa* 9 days × 640. Note several haustoria in one cell. (**5**). *Brassica napus* 4–5 days × 530. Encapsulation arrowed (**6**) *Brassica napus* 4–5 days × 820. Encapsulation arrowed (**7**) *Brassica rapa* 9 days × 640. Note encapsulation around one haustorium (arrowed) (**8**) *Brassica rapa* 3 days × 650. Note typical branching of coenocytic mycelium (**9**) *Brassica rapa* 6 days × 300. Note varying thickness of individual hyphae (**10**) *Brassica hirta* 9 days × 130. Note massive amount of intercellular fungal thallus and production of zoosporangia from club-shaped sporangiophores (**11**) *Brassica campestris* 9 days × 800. Note characteristically branched mycelium from which sporangiophores develop (**12**) *Brassica rapa* 9 days × 900. Note club-shaped sporangiophores and production of chains of zoosporangia (Verma et al. 1975)

produced only one haustorium, necrosis of the invaded host cells occurred within 12 h, and mycelial growth ceased 72 h after inoculation (Liu et al. 1989). The death of host cells was largely restricted to the penetration site; the adjacent nonpenetrated cells remained apparently unaffected. In the susceptible hosts, necrosis of infected cells occurred only infrequently, and hyphal growth continued unabated, resulting in mycelial ramifications into the mesophyll. Numerous haustoria per infected-cell were produced. The presence of as many as 14 haustoria in a single cell in "green island" tissue of artificially infected *B. juncea* cotyledons have been reported (Plate 3.1; Verma et al. 1975). Lakra and Saharan (1988a, b) found that in host varieties RH-30, RLM 1357, Kranti, and RC 781, the pathogen colonizes within 76.8, 74.4, 79.2, and 204.0 h of inoculation, and with an incubation period of 98.4, 91.2, 124.8,

Table 3.1 Colonization and incubation period of *Albugo candida* in resistant and susceptible varieties of Indian mustard (Lakra and Saharan 1988b)

Varieties	Colonization in host tissues (h)	SD	Incubation period (h)	SD
RH 30 (S)	76.8	4.6	94.4	4.8
RLM 1357 (S)	74.4	3.8	91.2	3.6
Kranti (S)	79.2	5.8	124.8	5.8
RC 781 (R)	204.0	5.7	283.2	6.5
RH 8541 (R)	–	–	–	–

R Resistant to white rust, *S* susceptible to white rust, *SD* standard deviation, – no infection

and 283.2 h, respectively (Table 3.1). In RC 781, exceptionally long periods of colonization and expression of symptoms were required due to its inherent resistant character. The pathogen could not establish in the host tissues of variety RH 8541 due to its high degree of resistance. In the susceptible cultivars RH 30, RLM 1357, and Kranti, a shorter incubation period was required for both colonization and expression of symptoms. The decision between compatible (susceptible) and incompatible (resistant) reactions of *Albugo–Brassica* system is made within 80 h of inoculation (Verma et al. 1975). Colonization in the leaves of all susceptible cultivars occurred within 80 h of inoculation compared to 204 h in the resistant cultivar-RC 781. Verma et al. (1975) found that up to the formation of the first haustorium, there is no difference in the infection process in resistant and susceptible hosts of *B. juncea, B. rapa, B. nigra,* and *B. napus.* However, Napper (1933) found that *A. candida* entered through stomata of a resistant host as readily as those of a susceptible host. The delay in colonization and incubation period of resistant host formed a practical base to identify host resistance under both field and laboratory conditions. Besides, age of plant, age of leaf involved, prevailing environmental conditions, and nutrients applied may also influence the incubation periods. Coffey (1975) observed an incubation period of 192 h in cabbage cotyledons under favorable conditions of high humidity and 15 °C. Under field conditions, longer incubation periods in highly tolerant and resistance cultivars (Lakra and Saharan 1988b), will help to reduce buildup of primary inoculum (oospores) in the soil.

Progression of *A. candida* infection was studied in cotyledons of susceptible (*B. rapa, B. juncea)*, moderately resistant (*B. hirta*), and immune (*B. napus*) hosts (Verma et al. 1975). Cotyledons of all four *Brassica* species were inoculated with zoospores of *A. candida* produced from germinating oospores or zoosporangia. At different times after inoculation, whole cotyledons were fixed in 95% ethanol-acetic acid (v/v) solution, cleaned in 70% lactic acid for 3–4 days, and stained with cotton blue in lactophenol. The preparation was examined under the compound microscope. Generally, the sequence of events from zoospore encystment to formation of the first haustorium was the same in all hosts, although under field conditions, *B. hirta* is moderately resistant and *B. napus* is essentially "immune." In *B. juncea,* the first haustorium was observed 16–18 h after inoculation, while in *B. rapa, B. hirta,* and *B. napus*, the first haustorium was observed about 48 h after inoculation (Plate 3.1; Verma et al. 1975). In the susceptible hosts, after the formation of the first haustorium, the hyphae grew rapidly and produced variable number of haustoria in each cell. The profusely branched, coenocytic mycelium appeared to fill all available intercellular spaces, and in 5–6 days after inoculation, the club-shaped zoosporangia develop from a dense layer of mycelium.

In the immune host, usually only one haustorium was formed, after which the hyphae ceased to elongate. At about 72 h after inoculation, a fairly thick, densely stained encapsulation was usually detected around each haustorium and later only "ghost" outline of hyphae and haustorium were observed. Encapsulations were not observed around haustoria of susceptible hosts. From these observations (Verma et al. 1975), it seems very probable that zoospores derived from germinating oospores constitute the primary inoculum for infection of cotyledons of susceptible *Brassica* species. No evidence of direct infection by the simple or branched germ tubes from germinating oospores was seen (Verma and Petrie 1975). The establishment and maintenance of compatible relationship between *A. candida* and its hosts hinges on the successful formation of the first haustorium. A similar sequence of events in both the susceptible and immune hosts up to this point suggests that there appears to be no morphological barrier to zoospore encystment, germination and subsequent penetration through stomata. In the incompatible combination, however, it is not clear whether the parasite fails to produce a functional haustorium, or whether a viable haustorium is formed within the host cell and is subsequently killed by the host's defense mechanism. The fairly dense, thick encapsulation observed around haustorium of immune host tissue suggests that the later may be the case. In any event, it does seem that the decision between compatibility and incompatibility is made within 48 h after inoculation (Verma et al. 1975).

3.2.3 Molecular Mechanisms of Host–Pathogen Interaction

To gain insight into the molecular mechanisms underlying activation of complex plant defense responses that occur in resistant and susceptible varieties of *B. juncea*, challenged with *A. candida*, a time-course protein profiling experiment followed by transcript studies was conducted (Kaur et al. 2011). The latter studies investigated whether the observed proteome changes were substantiated by changes in mRNA levels. Results revealed a clear differentiation between resistant and susceptible interactions and led to the identification of six differentially regulated proteins potentially involved in defense reactions in the resistant variety. A seventh differentially expressed protein is potentially involved in suppressing the host defense system, as it was detected only in the susceptible variety. Plant defense responses often begin with gene-for-gene recognition of the pathogen. The production of virulence effectors by a pathogen leads to its recognition by a host plant (Jones and Dangl 2006). Recognition results in the rapid activation of defense responses which are usually accompanied by an oxidative burst, that is, the rapid production of reactive oxygen species (ROS). ROS production is also required for the hypersensitive response (HR), a type of programmed cell death thought to limit the access of the pathogen to water and nutrients (Glazebrook 2005). The plastid localized RCCR protein was found to be downregulated in the pathogen-inoculated compared with the mock-treated resistant cultivar. The protein was only detected from 8 hpi onwards in the mock-treated seedlings and, to a lesser extent, the pathogen inoculated resistant seedlings. The RCCR transcript levels, however, showed a transient upregulation in both resistant and susceptible varieties under both treatments at 2–8 hpi suggesting a diurnal control of gene expression. This different trend in the 2-D gel-based proteome analysis compared with the observed changes in transcript abundance recorded by qRT-PCR may be due to post-transcriptional or post-translational modifications which may have led to the altered steady-state protein amounts in the two cultivars.

Pheophorbide (Pheide) an oxygenase (PaO) and RCCR catalyze the key reaction of chlorophyll catabolism, the porphyrin macrocycle cleavage of Pheide to a primary fluorescent catabolite (pFCC) (Wuthrich et al. 2000). Disruption of RCCR activity is likely to result in the accumulation of RCC and Pheide which are phototoxic and cause cell death by the light-dependent production of free oxygen radicals (Mach et al. 2001). Alternatively, accumulation of RCC itself may cause a specific signal that triggers cell death (Mach et al. 2001). Cell death and chlorophyll breakdown have been related to plant senescence and disease progression (Rodoni et al. 1997; Wuthrich et al. 2000; Mach et al. 2001; Pruzinska et al. 2007).

The *Arabidopsis* ACD2 gene encodes RCCR and was found to suppress the spread of disease symptoms caused by *Pseudomonas syringae* (Mach et al. 2001). Inoculated plants that lack RCCR/ACD2 activity show spreading cell death beyond the initial containment zone around the initial lesion. On the other hand, plants expressing high levels of the RCCR/ACD2 protein showed higher tolerance to infection with a virulent *Pseudomonas syringae* strain. This suggests that an increase in RCCR/ACD2 protein amount may alter the flux of chlorophyll catabolites that normally accumulate during disease and trigger cell death (Mach et al. 2001). The inability to detect RCCR protein at any stage of the infection cycle in the susceptible variety might point toward the inability of these seedlings to initiate a successful HR during the early stages of pathogen attack.

The production of ROS presents a challenge to a plant because of the potential of these compounds to damage cellular components. The plant must therefore maintain a balanced system that produces ROS for defense at the same time as it produces antioxidants to protect against ROS-mediated oxidative damage. One of the critical metabolites for maintaining redox balance in the face of oxidative stress is glutathione, which is a ubiquitous supplier of reducing power for cellular processes (Noctor and Foyer 1998). Glutathione (GSH) is a conjugate of glutamate, cysteine, and glycine. The last two steps in cysteine biosynthesis are catalyzed by a bi-enzyme complex of serine acetyl-transferase (SAT) and cysteine synthase, also called O-acetylserine (thiol) lyase (OAS-TL). Kaur et al. (2011) found that a *B. juncea* homologue of the plastid-localized isoform AtOAS-TL B was present in the resistant variety, but not detectable in the susceptible variety. The plastid localized isoform of OAS-TL B contributes up to 65% of the total OAS-TL enzyme activity in wild-type *Arabidopsis* plants (Heeg et al. 2008).

Cysteine is incorporated into GSH by a two-step enzymatic reaction catalyzed by *c*-glutamyl cysteine synthetase, an enzyme localized exclusively to the plastid, and GSH synthetase, an enzyme that shows dual targeting to both the cytosol and plastids (Wachter et al. 2005). The GSH biosynthesis is strongly regulated by cysteine availability (Strohm et al. 1995). Overexpression of a plastidic OAS-TL isoform or an inactive SAT isoform, caused moderate to very high increases of cysteine and GSH content in tobacco, respectively (Noji et al. 2001; Wirtz and Hell 2007). The pool size and redox status of GSH are important for determining a plant's ability to control ROS production and combat microbial pathogens (Noctor and Foyer 1998).

In *Arabidopsis*, both primary and secondary sulfur metabolism is induced upon pathogen attack at transcriptional, translational, and metabolite levels when it is grown under sufficient sulfate supply, while sulfate-limited conditions compromise basal resistance to necrotrophic fungi (Kruse et al. 2007). An increase in GSH content was also reported to occur in leaves attacked by virulent biotrophic pathogens (El-Zahaby et al. 1995; Fodor et al. 1997; Vanacker et al. 1998, 1999). This suggests that a boost in cysteine and GSH levels is important for the incompatible interaction between *A. candida* and *B. juncea* (Kaur et al. 2011).

One of the processes consuming GSH could involve the *B. juncea* GSTF9 homologue. This GST is encoded by one of 13 phi-class GST genes in *Arabidopsis*. In *B. juncea*, GSTF9 was detected only in the resistant variety and increased in abundance as the infection progressed. The plant-specific phi-class GSTs are related to fungal and bacterial proteins and little is known about their substrate specificities (Dixon et al. 2010). The *B. juncea* peptides identified closely match a jasmonic-acid inducible *B. juncea* GST, with close homology to the cytosolic dimeric GSTF9 protein from *Arabidopsis*. Abundance of GSTF6, F9, U5, and U13 transcripts were found to be methyl-jasmonate responsive in *Arabidopsis* (Jost et al. 2005). A quadruple knock-out line of *Arabidopsis* lacking GSTF6, F7, F9, and F10 showed no impairment in growth or development even under numerous stress conditions, but the subtle metabolic changes observed suggest an impaired response to oxidative stress conditions (Sappl et al. 2009). The endogenous products of oxidative damage, including membrane lipid peroxides and products of oxidative DNA degradation, are highly cytotoxic. The GSTs conjugate GSH with such endogenously produced electrophiles, resulting in their detoxification (Pickett and Lu 1989; Dudler et al. 1991; Bartling et al. 1993; Berhane et al. 1994). Some GSTs also function as glutathione peroxidises to detoxify oxidative products directly (Williamson and Beverley 1987; Bartling et al. 1993) and there are many reports that suggest a central role for GSH in plant defense activation (Marrs 1996; Dixon et al. 2010). Kaur et al. (2011) also detected two plastid localized isoforms of SOD in the resistant *B. juncea* variety that were absent from the susceptible variety. One of these SOD isoforms, the FSD1 homologue, showed an early and sustained increase in abundance only in the resistant *B. juncea* variety on both transcript and protein level. Interestingly, the other isoform, the CSD2 homologue, was induced to a similar extent on transcript level in both varieties under pathogen attack, but protein accumulation was only detectable in the resistant cultivar upon pathogen challenge. This again suggests strong post-transcriptional regulation of this particular isoform that is resulting in different outcomes in the resistant and susceptible varieties. The SODs are known to play an important role in protecting cells against the toxic effects of ROS produced in various subcellular compartments (Fridovich 1986; Halliwell and Gutteridge 1989). An oxidative burst is key feature underlying successful pathogen recognition at the site of the initial infection (Alvarez et al. 1998; Van Breusegem and Dat 2006). Rapid induction of SOD and accumulation of H_2O_2 are characteristic early features of the HR following perception of avirulence signals from the pathogen (Lamb and Dixon 1997).

The ROS-detoxifying enzymes such as SOD catalyze the dismutation of ROS such as superoxide, O_2 to hydrogen peroxide, H_2O_2. The H_2O_2 itself can trigger cell death, but at the same time functions as a messenger in cellular communication demanding a finely tuned adjustment of its cellular concentration (Foyer and Noctor 2000). The SODs therefore play an important role in ameliorating stress during conditions such as pathogen infection where overproduction of ROS is induced (De Gara et al. 2003). Various ROS species have similar deleterious effects on plant cells but they each activate different signaling pathways (Laloi et al. 2007). Moreover, the cellular compartment in which ROS are generated determines the nature of this signal (Miller et al. 2007). Kaur et al. (2011) could find two plastid-localized SODs responding differently to *A. candida*. The CSD2 protein seems to accumulate transiently in the resistant variety upon pathogen inoculation, while the FSD1 protein is very strongly induced and keeps increasing throughout. There was a transient increase in both FSD1 and CSD2 transcript abundances upon pathogen inoculation in the susceptible variety but these are not sustained at the protein level.

A cyclophilin was found to be necessary for host–pathogen recognition in *Arabidopsis* (Coaker et al. 2005). The PPIase subfamily of cyclophilins is known to have chaperone function assisting in protein folding and the assembly of large protein complexes (Kromina et al. 2008). One of the seven cytosolic cyclophilin isoforms of *Arabidopsis*, ROC1, is the closest homologue of human cyclophilin A and yeast CPR1, and is required for activation of the *Pseudomonas syringae* effector protein AvrRpt2 (Coaker et al. 2005). However, many of the 29 cyclophilins in *Arabidopsis* are targeted to the plastid. While most of these reside in the thylakoid lumen, one, CYP20-3/ROC4, is found in the chloroplast stroma. However, CYP20-3 protein was detected only in the susceptible *B. juncea* variety, and this was confirmed by a very low CYP20-3 transcript abundance in the resistant variety. This suggests that the *B. juncea* CYP20-3 protein might play an important role in the compatible interaction between *A. candida* and *B. juncea*. But its subcellular localization makes it less likely that it is a direct target for pathogen-derived effector proteins (Kaur et al. 2011). Earlier studies suggest that its activity might be sensitive to changes in redox balance and that it is induced by light. It was demonstrated in *Arabidopsis* that CYP20-3 may play a role in the regulation of cysteine biosynthesis in the chloroplast stroma through its interaction with the plastid isoform SAT1. This interaction seems to be required for optimal enzyme activity under stress conditions, most likely by stabilizing the assembly of one SAT trimer and three OAS-TL dimers into the cysteine synthase complex. It was shown that cyp20-3 mutants do not show an increase in thiol content in plants subjected to salt stress or high light conditions (Dominguez-Solis et al. 2008). It is recommended that ROS elicit the transcriptional activation of SAT1 as well as the oxidation of CYP20-3 in a peroxiredoxin-dependent manner. Thioredoxin would then be needed to regenerate active CYP20-3 (Motohashi et al. 2003). It is interesting to note that the *B. juncea* CYP20-3 protein could only be detected in leaves of the susceptible, but not the resistant cultivar. This could indicate that oxidized CYP20-3 accumulates in this cultivar due to a redox imbalance in the chloroplast rendering the cysteine synthase complex inactive, but this need to be investigated further (Kaur et al. 2011).

Resistance to pathogens is associated with the activation of signaling pathways that lead to the expression of certain pathogenesis-related (PR) proteins (Selitrennikoff 2001). While the precise mechanism of action of the PR-5 protein is not yet completely understood (Richardson et al. 1987; Wang et al. 1996; Coca et al. 2000; Ibeas et al. 2000), it is most likely involved in the degradation of the pathogen cell wall. Several PR-5 proteins have been reported to cause cell permeability changes in fungal cells with intact cell walls (Roberts and Selitrennikoff 1990). The maize PR-5 protein zeamatin inhibits growth of suspension cultures of *Candida albicans*, *Neurospora crassa*, and *Trichoderma reesei* by causing rapid cell lysis, even at low temperatures (Roberts and Selitrennikoff 1990). PR-5 proteins may have a direct role in cell wall degradation as they bind (1, 3)-b-glucans and have in vitro (1, 3)-b-glucanase activity (Grenier et al. 1993; Trudel et al. 1998). The induction of the PR-5 protein following inoculation with *A. candida* in the resistant *B. juncea* variety suggests an antipathogenic role. This role is further indicated by its absence from the proteome profiles of the susceptible variety and by the lower transcript abundance in the susceptible variety compared with the resistant variety. The delayed induction of PR-5 transcript accumulation in the susceptible line compared with the resistant line suggests that PR-5 needs to be present in the early stages of infection to exert its antipathogenic effect. Thus, in the context of disease progression in the *A. candida–B. juncea* pathosystem, the timing of the induction of PR-5 in resistant and susceptible varieties appears to be crucial for mounting an effective defense response to this pathogen.

All but two of the candidate proteins identified by Kaur et al. (2011) reside in the chloroplast stroma and are directly or indirectly involved in redox homeostasis. This is quite remarkable given that chloroplasts are increasingly being regarded as sensors of environmental changes and mediators of plant stress responses (Fernandez and Strand 2008). Defense against pathogens might therefore require the activation of photo inhibitory ROS production combined with the balancing power of the reactive oxygen regulatory network of the chloroplast (Mittler et al. 2004; Dietz et al. 2006). This may be important for generating an oxidative burst that limits the uncontrolled spread of cell death once a HR has been triggered by the presence of a pathogen (Belhaj et al. 2009). This is the first time that the host–pathogen interaction of the *B. juncea–A. candida* pathosystem has been dissected on a molecular level to differentiate between compatible and incompatible interactions. These results demonstrate that the timing of the expression of defense-related genes plays a crucial role during pathogenesis and incompatible interactions and that the redox balance within

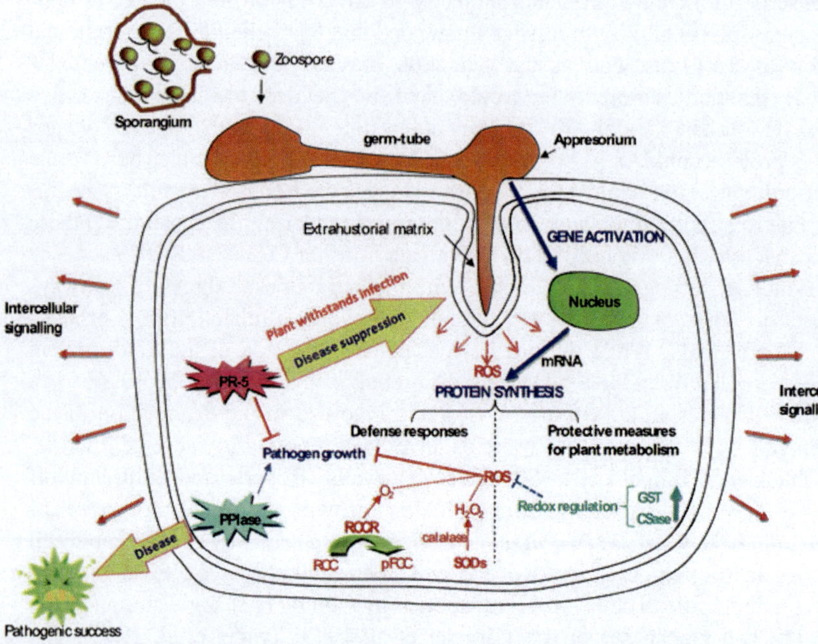

Fig. 3.1 Major elicitor-induced changes during the interaction of *Brassica juncea* and *Albugo candida* (Kaur et al. 2011)

the chloroplast may be of crucial importance for mounting a successful defense response. These findings also indicate that synergistic and conserved strategies are utilized by the resistant host to fight off the *A. candida* attack. Determining the changes in protein expression patterns elicited by *A. candida* in a resistant versus a susceptible variety of *B. juncea* has not only led to the likely identification of key molecular components of this specific host–pathogen interaction, but now opens the way for developing genetic markers that could be used to screen for resistance within *B. juncea* germplasm collections. Such studies may also open novel avenues for engineering durable resistance to this pathogen. Kaur et al. (2011) identified steps in the metabolic processes and defense-related proteins that are required for mounting a successful defense response in *B. juncea* against *A. candida*. Looking at functional relationships between different genes-of-interest using transgenic overexpression and RNAi lines will probably be the next step in the characterization of this pathosystem (Fig. 3.1).

3.2.4 Regulation of Host–Pathogen Interaction by *R* Genes

The function and molecular mechanisms of *R* genes during infection and pathogenesis of *A. candida* has been assessed using genotypes differing in level of *R* genes (R, PR) through histopathology. Histopathology with light and electron microscopes has revealed detailed fine structures during the process of host–pathogen interaction. Cotyledons of *B. napus, B. rapa,* and *B. juncea* inoculated with zoospores of race 7v were observed under light microscope at 1, 3, and 7 days after inoculation (Bansal et al. 2005). One day after inoculation, zoospore germination was generally high and exhibited a similar behavior on all three host genotypes (Plate 3.2). In most cases, germ tubes penetrated stomata after forming appressoria, but in a few cases, the germ tubes penetrated directly without forming appressoria. Occasionally, more than one germ tubes penetrated each stoma. In general, the frequency of zoospore germ tube penetration without appressoria was lower in the resistant (R) and partial resistant (PR) than in the moderately susceptible (MS) and susceptible (S) genotypes.

A hypersensitive response was also observed in the resistant host–pathogen interaction (Plate 3.2). Haustorium formation was observed 1 day after inoculation

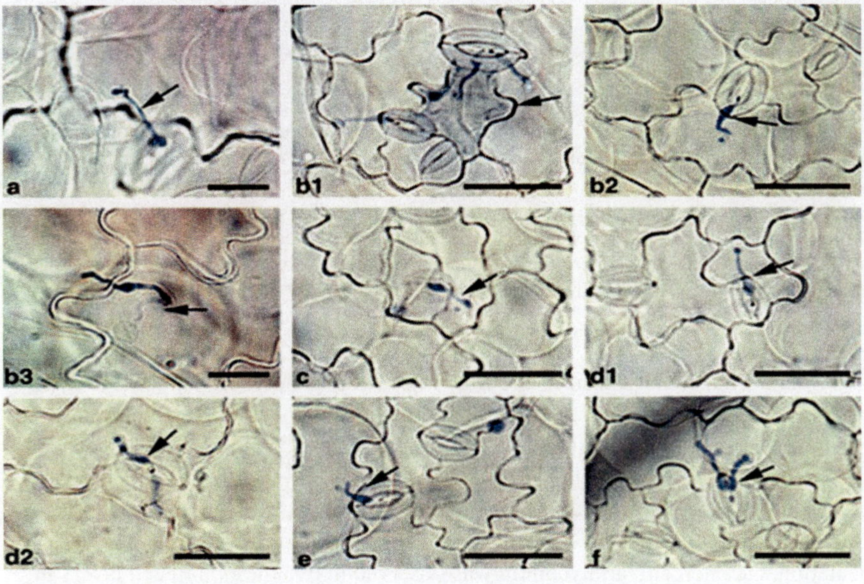

Plate 3.2 Histology of host–pathogen interaction of *Albugo candida* race 7v on *Brassica napus*, *B. rapa,* and *B. juncea* after 1 day of inoculation. (**a**) Spore germination on the stomatal opening of the partially resistant genotype *B. napus* A00-63N; (**b1**) Spore germination and hypersensitive response of the host epidermal cell of resistant *B. napus* A00-65N genotype; spore germination on the stomatal opening (**b2**) and on the epidermal cell (**b3**); (**c**) Spore germination and penetration through the epidermal cell of moderately susceptible *B. napus* 88-1409K; (**d1**, **d2**) through the stomatal opening of susceptible *B. rapa* "Torch." (**e**, **f**) Spore germination on the stomatal opening of resistant *B. napus* "Westar," and *B. juncea* "commercial brown." In each figure, an arrow points to the feature indicated. (**a**, **b3**) Bar = 0.02 mm; (**b1**, **b2**, **c–f**) Bar = 0.05 mm (Bansal et al. 2005)

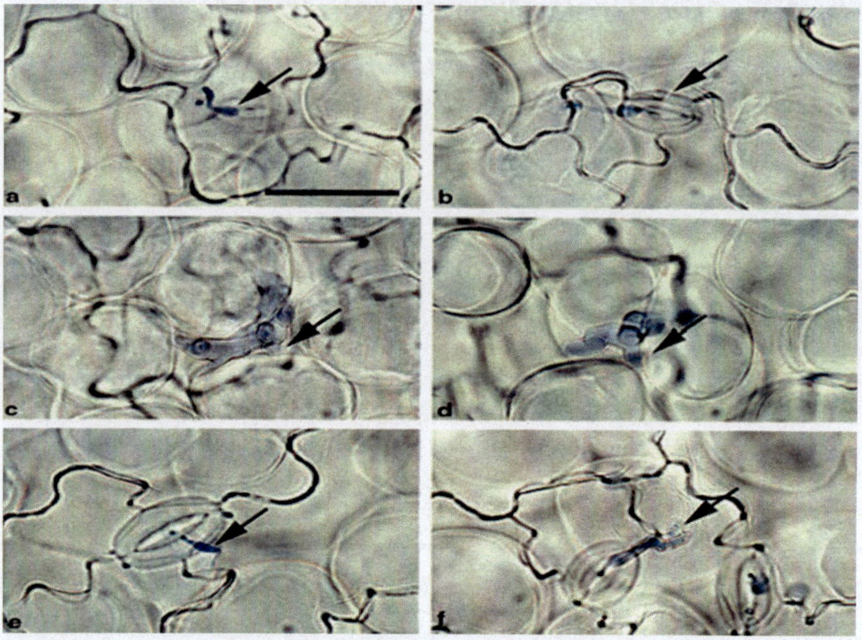

Plate 3.3 Histology of host–pathogen interaction of *Albugo candida* race 7v on *Brassica napus, B. rapa,* and *B. juncea* 3 days after inoculation. (**a**) Spore germination on the stomatal opening of the partially resistant genotype *B. napus* A00-63N, (**b**) and in resistant *B. napus* A00-65N genotype. (**c**) Sub-epidermal fungal mycelium in the mesophyll cell layer of moderately susceptible *B. napus* 88-1409K and, (**d**) in the intercellular spaces of mesophyll tissue of susceptible *B. rapa* "Torch." (**e**) Spore germination on the stomatal opening of resistant *B. napus* "Westar," and (**f**) Spore germination with some mycelial growth close to the stomatal opening of resistant *B. juncea* "commercial brown." In each figure, an arrow points to the feature indicated, Bar = 0.05 mm (Bansal et al. 2005)

in the S, and occasionally, also in the MS, PR, and R genotypes. Three days after inoculation, most histological features of the pathogen (Plate 3.3) were similar to those formed 1 day after inoculation (Plate 3.3), in the susceptible control "Torch," while mycelium was present in the deeper mesophyll layers of the cotyledonary leaves (Plate 3.2), in the R, PR, and MS genotypes, most of the fungal growth was restricted only to the first layer of mesophyll cells (Plate 3.3). Seven days after inoculation, the intercellular mycelium was observed in two mesophyll cell layers in the PR and MS genotypes, but in three mesophyll cell layers in the S genotype. In "*R*" and PR genotypes, guard cells of stomata through which penetration occurs, appeared shrunken and necrotic, and this hypersensitive response was more frequent in these than in the MS genotypes. There was no hypersensitive response observed in the guard cells of the S genotype, and the hyphal growth was more profuse and penetrated more deeply than in any of the other genotypes. Seven days after inoculation, generally, the histology of host–pathogen interactions remained essentially the same as in 3 days after inoculation (Bansal et al. 2005). Mesophyll tissue was colonized occasionally, but only up to one cell layer. Hypersensitive response and shrinking of guard cells were also noticed (Bansal et al. 2005). In general, greater numbers of haustoria were observed in the S than in the MS, PR, and R genotypes. Localized cell necrosis (hypersensitive reaction) of cells in the area of infection was observed in the R, PR and MS genotypes, and the extent of necrosis among the genotypes was MS < PR < R. Pustule development appeared to be inversely correlated with the level of necrosis. The pustules were few, diminutive and poorly sporulating in the PR host, 7 days after inoculation. In contrast, the S hosts produced copious quantities of pustules with high levels of sporulation. In the R genotypes, no pustules were observed in this investigation except in "Commercial Brown" where a small number of seedlings developed a few small pustules.

The histological observations clearly indicated that fungal growth in the PR genotype was more restricted than in the MS and S genotypes. The differences in the extent of mycelial growth and host invasion correlated well with the final phenotypic expression of the disease symptoms (Bansal et al. 2005). Pidskalny and Rimmer (1985) have reported that races 7 and 2 from *B. rapa* and *B. juncea*, respectively, are not cross pathogenic. But, mycelial growth sometimes developed in intercellular spaces of the first mesophyll cell layer of the highly resistant genotype "Commercial Brown," and very small pinhead-size pustules were occasionally noticed on some of these plants (<5%). This may reflect partial resistance in this genotype (Bansal et al. 2005). Transverse section of the hypertrophied inflorescence axis revealed a single layered epidermis with idioblasts at some places, 4–5 layered cortexes having parenchymatous and chlorenchymatous cells, and conjoint collateral vascular bundles with 2–5 layered cambium and parenchymatous pith.

In hypertrophied axis, epidermal cells were elongated because of the pressure exerted by the developing fungal oospores within the host cells. The cortical cells were penetrated by fungal haustoria. The vascular cylinder was stretched due to the development of the antheridia and oogonia in cortex and pith. The xylem elements increased in length and became thin-walled. There was poor lignification of the xylem elements after invasion of *A. cruciferarum.* Thus, the hypertrophied tissue represents an example of hypoplasty. The malformation is mainly due to hypertrophy and hyperplasia of the infected cortical and pith cells. The differentiation of tissues in the infected inflorescence axis appeared of low order, but cell enlargement and cell division were very conspicuous (Maheshwari et al. 1985a, b).

Histopathology of compatibility and incompatibility between oilseed rape and *A. candida* has been studied by Liu et al. (1989). Germination of encysted zoospores occurred 2–3 h after inoculation, infection was initiated with germ-tubes penetrating through stomata, and haustorium formation was first observed 8 h after inoculation in the palisade mesophyll cells adjacent to the substomatal chambers. Only after the establishment of the first haustorium did compatible and incompatible interactions begin to differentiate. In the resistant cultivar, most primary hyphae produced only single haustorium. Necrosis of the invaded host cell was first observed 12 h after inoculation followed by cessation of fungal growth. The death of host cells was largely restricted to the penetration site; the adjacent nonpenetrated cells remained apparently unaffected. In the susceptible hosts, necrosis of infected cells occurred only infrequently, and hyphal growth continued unabated, resulting in mycelial ramification into the mesophyll. Numerous haustoria per invaded host cell were produced (Liu et al. 1989). Histological studies showed that the earliest event distinguishing a compatible from an incompatible interaction occurred after formation of the first haustorium, and that resistance was not manifested until the host mesophyll cells had come into contact with the first haustorium.

Transmission electron microscopy (TEM) was used to examine details of the host–pathogen interface in *A. thaliana* cotyledons infected with *A. candida* (Soylu 2004; Soylu et al. 2003). After successful entry through stomatal pores, the pathogen developed a substomatal vesicle and subsequently produced intercellular hyphae. Observations revealed that coenocytic intercellular hyphae ramified and spread intercellularly throughout the host tissues forming several haustoria in host mesophyll cells. Intracellular haustoria were spherical and 4.5 μm in diameter. Each haustorium was connected to intercellular hyphae by a narrow, slender haustorium neck. The cytoplasm of the haustorium included the organelles characteristic of the pathogen. No obvious response was observed in host cells following formation of haustoria. Most of the mesophyll cells contained normal haustoria and the host cytoplasm displayed a high degree of structural integrity. Absence of host cell wall alteration and cell death in penetrated host cells suggested that the pathogen exerts considerable control over basic cellular processes and in this respect, response to this biotrophic. Oomycete differs considerably from responses to other necrotrophic pathogens. Modification of the host plasma membrane (PM) along the cell wall, and around the haustoria, was detected by applying the periodic acid-chromic acid-phosphotungstic (PACP) staining technique. After staining with PACP, the host PM adjacent to the host cell wall and the distal region of the haustorial neck was found to be intensely electron dense. By contrast, the extra haustorial membrane, where the host PM surrounded the haustorium, was consistently very lightly stained (Soylu et al. 2003; Soylu 2004).

3.3 *Brassica–Alternaria* Pathosystem

3.3.1 Introduction

The *Alternaria* belonging to the class of Dothideomycetes contains many important plant pathogens. Diseases in the Brassicaceae family caused by *Alternaria* spp. result in significant yield losses. *Alternaria* spp. has a wide host range within the Brassicaceae, infecting both the vegetable as well as the oilseed crops. Some of the most damaging species include *Alternaria brassicae, A. brassicicola, A. alternata, A. raphani, A. japonicus*, and *A. tenuissima.* The *A. brassicae* preferentially infects the oleiferous *Brassicas* while the others are more devastating on the vegetable *Brassicas*. The *A. brassicae* is particularly more damaging in the hilly regions of the Indian subcontinent, where conducive climatic conditions allow it to profusely reproduce and cause infections on almost all parts of the plant. Extensive screening for resistance to *A. brassicae* in the cultivated *Brassica* germplasms has not revealed any source of resistance.

The factors that contribute to the pathogenicity of *A. brassicae* are relatively unknown. Pathogenicity of many *Alternaria* spp. has been mainly attributed to the secretion of host-specific toxins (HSTs). The HSTs induce pathogenesis on a rather narrow species range and are mostly indispensable for pathogenicity. At least 12 *A. alternata* pathotypes have been reported to produce HSTs and thereby cause disease on different species. Many of the HST producing genes/gene clusters has been found on super numerary chromosomes or dispensable chromosomes. The *A. brassicae* has been reported to produce low molecular weight cyclic depsipeptides named destruxins. Destruxin B is known to be a major phytotoxin and is reported to be a probable HST of *A. brassicae*. Additionally, a proteinaceous HST (ABR-toxin), was isolated from the spore germination fluid of *A. brassicae* but was only partially characterized (Saharan et al. 2016).

3.3.2 Identification of Species-Specific Genes of *Alternaria* with Putative Roles in Virulence

Differences in gene content and diversity within genes contribute to adaptation, growth, and pathogenicity. In order to catalogue the differences in the gene content within the *Alternaria* genus and the Dothideomycetes, Rajarammohan et al. (2019) carried out an orthology analysis on the combined set of 3, 60, 216 proteins from 30 different species (including 16 *Alternaria* species) belonging to Dothideomycetes out of which 3, 45, 321 proteins could be assigned to at least one of the orthogroups. They identified 460 *A. brassicae* specific genes were present in *A. brassicae* but absent in all other *Alternaria* species. These species-specific genes included 35 secreted protein coding genes out of which 11 were predicted to be effectors. Additionally, 20 of these species-specific genes were present on the DC. A large number of these proteins belonged to the category of uncharacterized proteins with no known function. In order to test whether these species-specific genes are the result of adaptive evolution taking place in the repeat-rich regions of the genome, a permutation test was carried out to compare the overlap of repeat-rich regions and transposable elements with a random gene set against the overlap of these species-specific genes. It was found that these species-specific genes overlapped significantly with repeat-rich regions (*P*-value: 9.99e − 05; *Z*-score: −4.825) and transposable elements (*P*-value: 0.0460; *Z*-score: 2.539) in the genome (Rajarammohan et al. 2019).

3.3.3 Transcription Factors Associated with Pathogenesis

A total of 138 *Alternaria brassicicola* transcription factor genes with zinc finger domains, and other regulatory genes analyzed to date, only six genes including *AbSte12*^{C}, *AbPac*^{C}, *AbVf19*^{C}, *Amr1*, *AbVf8*, and *AbPro1*^{C} are strongly associated

Table 3.2 Summary of *Alternaria brassicicola* transcription factor domains based on Pfam scans (Cho et al. 2012)

Description	Gene	KO 1[a]	KO 2[b]	Pathogenesis-associated genes
Zn finger domain	221	4	117	*AbVf8*, *AbVf19*[C], *AbSte12*[C], *AbPac*[C], *AbPro1*[C], *Amr1*
Fungal-specific TF	72	1	0	
Ph d finger	1	–	–	
Helix turn-helix	29	1	–	
Regulatory receiver	23	5	–	
Leucine zipper	21	2	–	
High mobility box group	15	3	–	
Helix–loop-helix	13	5	–	
DNA binding myb	10	–	–	
Jumonji	7	–	–	
Homeobox	4	–	–	
β-Regulatory protein	3	–	–	
T-Regulatory	2	–	–	
Total	421	21	117	

[a] Genes of previously screened targeted deletion mutants (Cho et al. 2009)
[b] Genes of mutants created and screened (Cho et al. 2012)
[C] Genes whose functions were previously characterized (Cho et al. 2009; Srivastava et al. 2012)

with pathogenesis (Table 3.2). The *AbSte12* is a homolog of the *Ste12* gene with mutants that are either nonpathogenic or reduced in virulence in several fungi (Wong Sak Hoi and Dumas 2010). The *Ste12* is regulated by a MAP kinase, *Fus3/Kss1*. Mutants of either homologous gene in *A. brassicicola* shows defective conidial development, and are nonpathogenic (Cho et al. 2007, 2009). The *AbPac*[C] gene has a single copy in the genome sequence of *A. brassicicola* and shows a high sequence similarity ($E = 0.0$) with the *Pac*[C] gene in other fungi. The *AbVf19* regulates suite of genes that are important for decomposing and utilizing plant materials during the late stage of plant infection. Regardless of their function, however, all mutants studied to dates except *Amr* 1 are nonpathogenic or shows a reduction in virulence (Cho et al. 2012).

3.3.4 The *AbPf2* Gene Activates Pathogenicity in *Alternaria brassicicola*

A conserved mitogen-activated protein kinase encoded by *Amk1*, and its downstream transcription factor encoded by *AbSte12*, are essential for pathogenesis (Cho et al. 2007, 2009). However, mutation of either *Amk1*, *AbSte12* or their homologs in other fungi caused defects in other cellular processes in addition to the loss of pathogenicity (You and Chung 2007; Zhao et al. 2007; Wong Sak Hoi and Dumas 2010). Most transcription factors associated with pathogenesis in parasitic fungi are also linked to additional cellular processes (Joubert et al. 2011; Son et al. 2011; Wang et al. 2011; Chen et al. 2012). Because of the multiple functions of transcription factors, it is difficult to identify the pathogenicity genes regulated by them. Cho et al. (2013) identified a transcription factor gene that was essential for initiating pathogenesis in *A. brassicicola*. Targeted mutants of the gene were nonpathogenic and grew on the surface of host tissue without causing disease symptoms.

The *Alternaria brassicicola* is a successful saprophyte and necrotrophic plant pathogen. To identify molecular determinants of pathogenicity, Cho et al. (2013) created nonpathogenic mutants of a transcription factor-encoding gene, *AbPf2*. The frequency and timing of germination and appressorium formation on host plants were similar between the nonpathogenic Δ abpf2 mutants and wild-type *A. brassicicola*. The mutants were also similar in vitro to wild-type *A. brassicicola* in terms of vegetative growth, conidium production, and responses to a phytoalexin, reactive oxygen species and osmolites. The hyphae of the mutants grew slowly but did not cause disease symptoms on the surface of host plants. Transcripts of the *AbPf2* gene increased exponentially soon after wild-type conidia contacted their host plants. A small amount of AbPf2 protein, as monitored using GFP fusions, was present in young, mature conidia. The protein level decreased during saprophytic growth, but increased and was located primarily in fungal nuclei during pathogenesis. Levels of the proteins and transcripts sharply decreased following colonization of host tissues beyond the initial infection site. When expression of the transcription factor was induced in the wild-type during early pathogenesis, 106 fungal genes were also induced in the wild-type but not in the *Δ abpf2* mutants. Notably, 33 of the 106 genes encoded secreted proteins, including eight putative effector proteins. Plants inoculated with *Δ abpf2* mutants expressed higher levels of genes associated with photosynthesis, the pentose phosphate pathway and primary metabolism, but lower levels of defense-related genes. The AbPf2is an important regulator of pathogenesis, but does not affect other cellular processes in *A. brassicicola* (Cho et al. 2013).

3.3.5 Identification, Cloning, and Sequencing of Virulence Genes

In many plant pathosystems, fungal secondary metabolites derived from nonribosomal peptide synthetases (*NPSs*) are phytotoxic virulence factors or are antibiotics thought to be important for niche competition with other microorganisms. However, many of the functions of *NPS* genes and their products are largely unknown. Kim et al. (2007) investigated the function of one of the *A. brassicicola NPS* genes, *AbNPS2*. The predicted amino acid sequence of *AbNPS2* shows high sequence similarity with *Alternaria brassicae, AbrePsy1*, *Cochliobolus heterostrophus, NPS4*, and a *Stagonospora nodorum NPS*. The *AbNPS2* open reading frame was predicted to be 22 kb in length, and encodes a large protein (7195 amino acids) showing typical *NPS* modular organization. Gene expression analysis of *AbNPS2* in wild-type fungus indicated that it is expressed almost exclusively in conidia and conidiophores, broadly in the reproductive developmental phase. The *AbNPS2* gene disruption mutants show abnormal spore cell wall morphology, and a decreased hydrophobicity phenotype. Conidium of *abnps2* mutants displays an aberrantly inflated cell wall, and an increase in lipid bodies compared with wild-type. Further, phenotypic analyses of *abnps2* mutants showed decreased spore germination rates both in vitro and in vivo, and a marked reduction in sporulation in vivo compared with wild-type fungus. Moreover, virulence tests on *Brassicas* with *abnps2* mutants reveals a significant reduction in lesion size compared with wild-type but only when aged spores are used. Collectively, these results indicate that *AbNPS2* plays an important role in disease development and virulence (Kim et al. 2007; Saharan et al. 2016).

3.3.6 Identification of Pathogenicity Factors/Genes

The functional redundancy of lipases with regards to *Alternaria* pathogenicity has been revealed by many workers (Yao and Koller 1994, 1995; Berto et al. 1999; Cho et al. 2006). Interestingly, one of the factors responsible for the pathogenicity has been predicted to be secondary metabolite production. A nonribosomal peptide synthase gene (*NPS6*) in *Cochliobolus heterostrophus* and *A. brassicicola* is found to direct the biosynthesis of a siderophores metabolite important for oxidative stress tolerance, and pathogenicity (Oide et al. 2006). The secondary metabolite corresponding to or synthesized via *AbNPS2* has to be characterized. There is a need to further characterize secondary metabolite biosynthetic genes and their role in pathogenicity and fungal development.

Another important area of investigation in the *Alternaria*–Brassicaceae pathosystem is the fungal signal transduction. The disruption of the *Fus3*/*Kss1MAP* kinase homolog (*Amk1*) in *A. brassicicola* resulted in a complete loss of pathogenicity as in other fungi has been observed (Cho et al. 2006, 2007). Interestingly, in the latter study, it was shown that addition of long polypeptide nutrients partially restored pathogenicity to the mutants. In addition, two novel virulence factors by Cho et al. (2009) were predicted to encode a transcription factor (*AbPro1*), and a two component histidine kinase gene (*AbNIK1*). Both of these kinases are pathogenicity factors in phytopathogenic fungi. The *Slt2* was found to be associated with cell wall integrity and *HOG* with oxidative stress tolerance (Xu 2000). The identification of virulence factors is the disruption of *Aso-1*, a gene required for hyphal fusion (anastomosis) which is also found to be required for pathogenicity in *Alternaria* species (Craven et al. 2008). Eventually over a hundred genes have been functionally analyzed through various techniques like gene knockout, and overexpression experiments making *A. brassicicola,* the species of choice for functional genomics research to define conserved virulence mechanisms for this important genus of fungi (Oide et al. 2006; Cho et al. 2006, 2007, 2009; Kim et al. 2007). With the objective of identification of *A. brassicicola*, an attempt was made to examine the role of cutinase genes in *A. brassicicola* pathogenesis (Yao and Koller 1994, 1995). In these studies, biolistic transformation was used to disrupt the *CUTAB1* gene. It affected the saprophytic growth since cutin was no longer able to be utilized as a sole carbon source, but this disruption had no significant effect on *A. brassicicola* pathogenicity. An extracellular lipase was found to be produced by *A. brassicicola* in vitro (Berto et al. 1999). The antilipase antibodies were found to significantly decrease the ability of *A. brassicicola* to cause disease on cauliflower leaves. However, disruption of four predicted *A. brassicicola* lipase genes expressed during plant infection did not result in reduced virulence on cabbage (Cho et al. 2006).

To study regulatory mechanisms of pathogenesis, Cho et al. (2012) mined 421 genes in silico encoding putative transcription factors in a machine annotated, draft genome sequence of *A. brassicicola*. Targeted gene disruption mutants for 117 of the transcription factor genes were produced and screened. Three of these genes are associated with pathogenesis. Disruption mutants of one gene (*AbPacC*) are nonpathogenic and another gene (*AbVf8*) causes lesions less than half the diameter of wild type lesions. Unexpectedly, mutants of the third gene, *Amr1* causes lesions with two-fold larger diameter than the wild type and complementation mutants. The *Amr1* is a homolog of *Cmr1*, a transcription factor that regulates melanin biosynthesis in several fungi. Cho et al. (2012) created gene deletion mutants of Δ*amr1* and characterized their phenotypes. The Δ*amr1* mutants used pectin as carbon source more efficiently than the wild type, were melanin-deficient, and more sensitive to UV light and glucanase digestion. The *AMR1* protein is localized in the nuclei of hyphae and in highly melanized conidia during the late stage of plant pathogenesis. RNA *seq* analysis reveals that three genes in the melanin biosynthesis pathway, along with the deleted *Amr1* gene are expressed at low levels in the mutants. In contrast, many hydrolytic enzymes coding genes are expressed at higher levels in the mutants than in the wild type during pathogenesis. A gene important for survival in nature negatively affects virulence, probably by less efficient use of plant cell wall materials. The functions of the *Amr1* gene are important to be the success of *A. brassicicola* as competitive saprophyte and plant parasite (Saharan et al. 2016).

3.3.7 Microarray Analysis of *Brassica* Infected with *Alternaria*

The transcriptional profiling of *B. olerace*a leaves infected with *A. brassicicola* using 29 k *Arabidopsis* Oligonucleotide Microarrays was performed. Out of 29,100 microarray oligonucleotide probes, 8654 could be reliably mapped to *B. oleracea* nucleotide sequences, which represented 8014 unique genes. The expression of these genes in infected plants at 12, 24, and 48 hpi was compared to control plants (0 hpi). Significant changes in gene expression (FDR <0.1) were observed as early as 12 hpi (27 genes) and they intensified while the infection progressed, with 189 genes regulated at 24 hpi and 410 genes at 48 hpi. Overall, 487 distinct genes were significantly affected in infected *B. oleracea* leaves.

Based on the results of the statistical analysis, the Gene Ontology (GO) and MapMan analyses were performed to annotate the significantly up- and downregulated genes, and to determine the most affected biological processes and activation of individual genes during *B. oleracea* infection with *A. brassicicola*. Biotic stress response-related processes, such as the cell wall macromolecule

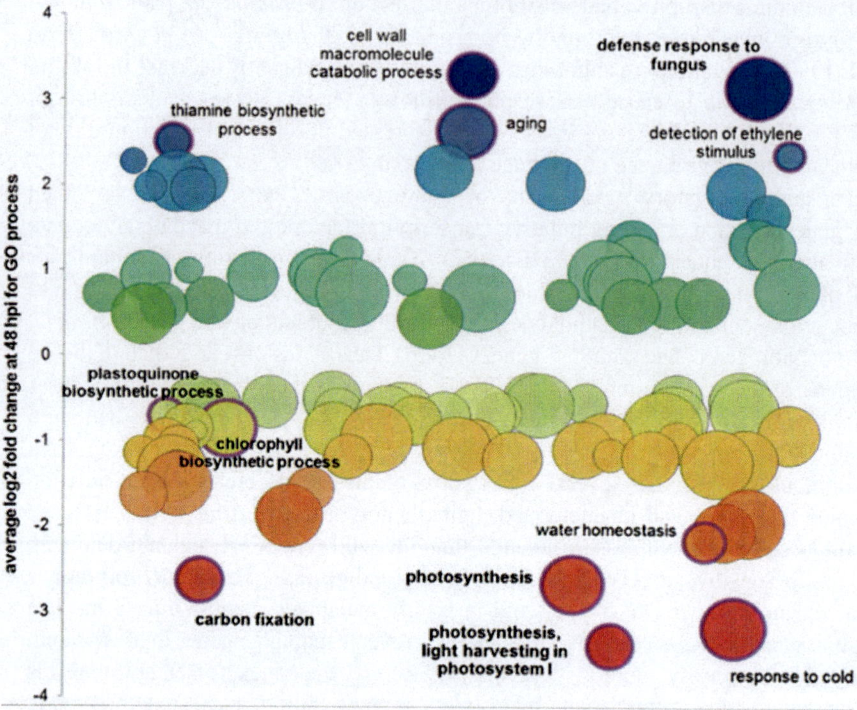

Fig. 3.2 Gene ontology of biological processes in *Brassica oleracea* leaves during *Alternaria brassicicola* infection (Macioszek et al. 2020)

catabolic process and defense response to fungus, were among the most extensively upregulated GO terms (Fig. 3.2). They were correlated to the microscopic results obtained in this study to uncover the differential responses of host cells to the infection within the inoculation site (Fig. 3.2, Plate 3.4). The upregulation of host genes involved in aging, detection of ethylene stimulus and thiamine biosynthesis indicated general reprogramming of the host transcriptome in response to the infection. Analysis of stress-related genes at 48 hpi revealed differential expression of 167 genes, with 87 of them, mainly involved in cell response signaling, cell wall degradation, protein degradation, auxin-induced metabolism, being downregulated. However, transcriptional activation of 80 stress-related genes, among others LRR-RLK7 (AT1G09970) encoding leucine-rich repeat transmembrane protein receptor-like kinase 7, WRKY33 (AT2G38470) encoding transcription factor, PDF1.2 (AT5G44420) encoding jasmonate-responsive plant defensin or FAH1 (AT4G36220) encoding ferulate-5-hydroxylase 1 and CCoAMT (AT1G67980) encoding caffeoyl-CoA 3-*O*-methyl transferase involved in lignin biosynthesis, even at 48 hpi, indicated that, despite the successful colonization of the susceptible host by *A. brassicicola*, some of the host cells battled with the fungus and basal defense signaling was still active.

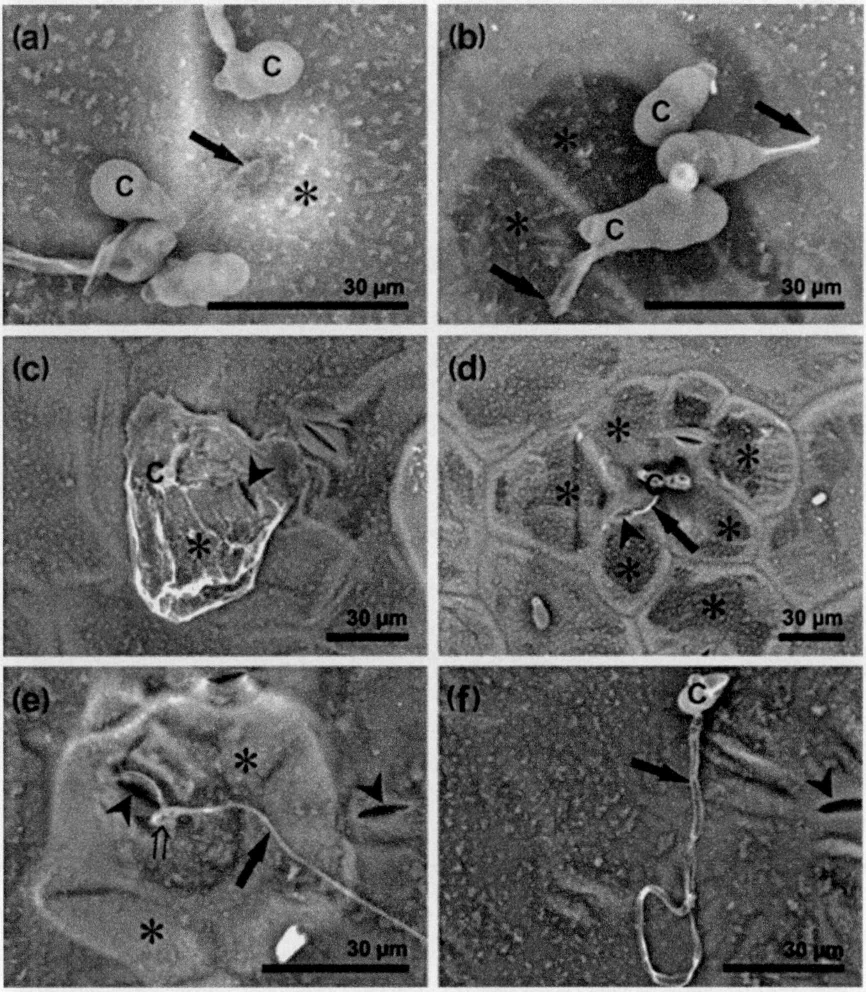

Plate 3.4 The *Alternaria brassicicola* penetration modes and different host cell reactions. The images were capture using SEM at 24 hpi. (**a**) a bright "halo" (asterisk) around the appressorium (arrow) penetrating the host cell; (**b**) direct penetration (arrows) caused a collapse of epidermal cells (asterisks) underneath germinating conidia; (**c**) dissolved wax layer (asterisk) around germinating conidium (arrow head points to stoma); (**d**) germination tube (arrow) entering the leaf via stoma (arrowhead) surrounded by collapsed epidermal cells (asterisks); (**e**) germination tube (arrow) forming an appressorium (double tail arrow) over stoma (arrowhead). Neighboring epidermal cells (asterisks) have fortified cell walls; (**f**) symptom less growth of hypha (arrow) on leaf surface (arrow head points to stoma). *C* conidium (Macioszek et al. 2020)

The photosynthesis-related processes were strongly downregulated (Fig. 3.2) in agreement with the observed severe changes in the ultrastructure of chloroplasts during infection (Plate 3.5). Interestingly, gradual down-regulation of photosynthesis-related gene expression was observed during the entire course of

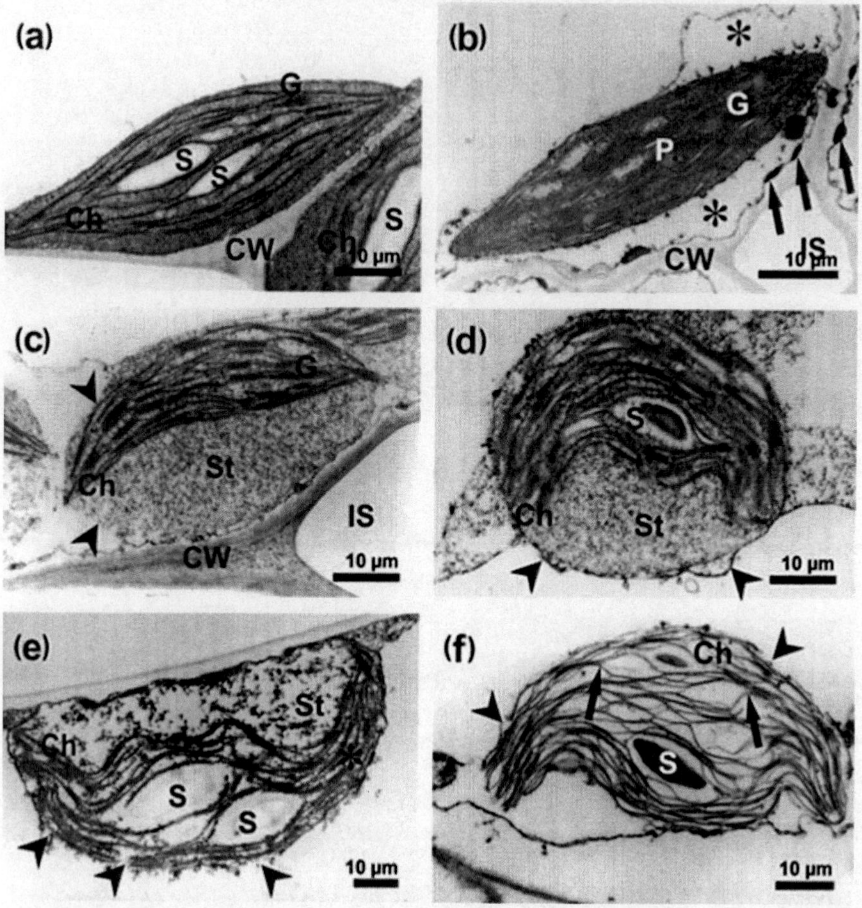

Plate 3.5 Disintegration of chloroplasts in *Brassica oleracea* leaf mesophyll cells infected with *Alternaria brassicicola* at 48 hpi. The images were captured using TEM. (**a**) lenticular-shaped chloroplast with typical thylakoid and grana structure, and small starch grains in uninfected leaf; (**b**) fragment of mesophyll cell with degraded cytoplasm (asterisks), osmiophilic granules in plasma lemma (arrows) and atypical well-preserved chloroplast with plastoglobuli; (**c**) round-shaped chloroplast with a still preserved thylakoid and grana system, disintegrated chloroplast envelope (arrow heads) and swollen electron-translucent stroma; (**d**) round-shaped chloroplast with cup-shaped thylakoid system, reduced number of grana, locally swollen envelope (arrow heads) and electron-translucent stroma; (**e**) strongly disintegrated chloroplast with flocculent remnants of stroma, locally disintegrated envelope (arrow heads) and collapsed thylakoid system (asterisks) devoid of grana; (**f**) remnants of chloroplast with destroyed envelope (arrow heads) and thylakoids with widened lumens (arrows). Grana and stroma are missing. *Ch* chloroplast, *CW* cell wall, *G* granum, *P* plastoglobule, *S* starch grain, *St* stroma (Macioszek et al. 2020)

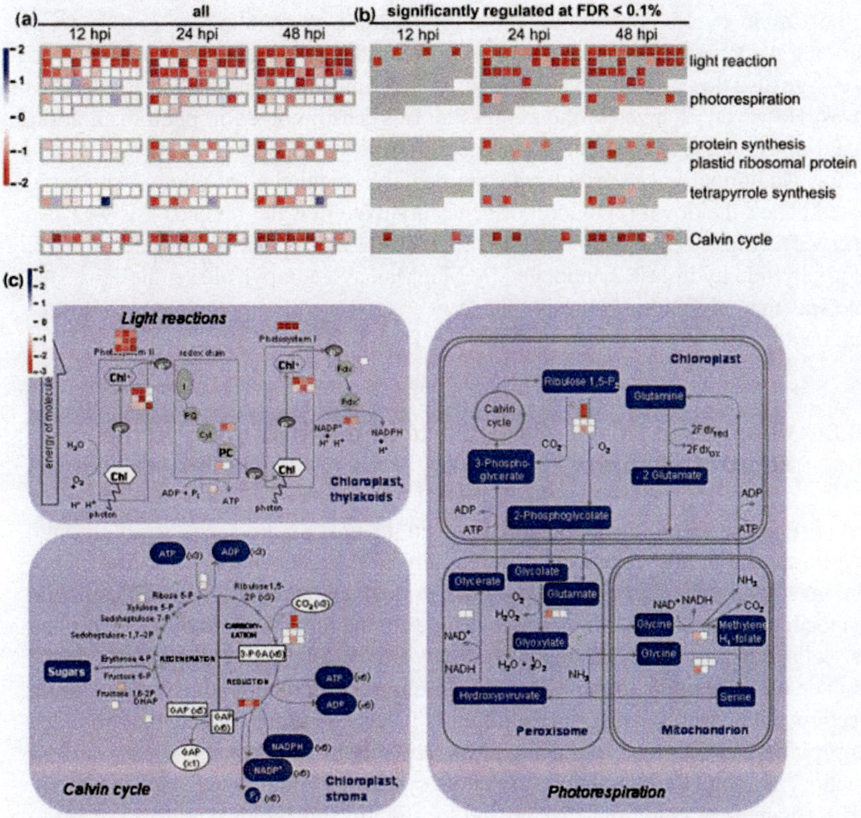

Fig. 3.3 Microarray-based analysis of expression changes of photosynthesis-related *Brassica oleracea* genes during the course of *Alternaria brassicicola* infection. (**a**) MapMan analysis of all mapped photosynthesis related genes (104 genes); (**b**) MapMan analysis of photosynthesis-related genes significantly regulated at FDR <0.1; (**c**) MapMan diagram of significant expression changes of photosynthesis (Macioszek et al. 2020)

B. oleracea–A. brassicicola interaction investigated in this study. The majority of the 104 genes that were mapped to photosynthesis-related processes by the MapMan analysis, were negatively regulated at one or more time points (Fig. 3.3) and the number of significantly downregulated genes (at FDR <0.1) increased from 6 at 12 hpi, through 32 at 24 hpi, up to 44 genes at 48 hpi (Fig. 3.3). All crucial photosynthesis-related processes, such as light reactions, Calvin cycle, photorespiration, plastid ribosomal protein synthesis, and tetrapyrrole synthesis, were affected (Fig. 3.3). The distinctive genes in the photosynthesis light reaction category were LHCb and LHCa, encoding chlorophyll a/b-binding proteins, which are the part of photosystem II and photosystem I light harvesting complexes, respectively, and thus are important components of the chloroplast membrane system. These genes were downregulated as early as 12 hpi, with an increasing effect at subsequent time points.

In the Calvin cycle, constituting the dark reaction of photosynthesis, the genes encoding the RuBisCO small subunit and glycer-aldehyde 3-phosphate dehydrogenase were also downregulated at all investigated time points in infected *B. oleracea* leaves. However, changes in the expression of genes involved in photorespiration, plastid ribosomal protein synthesis and tetrapyrrole synthesis started later, at 24 and 48 hpi. Additionally, consistent with the negative regulation of photo-respiration-related genes, the down-regulation of genes involved in water homeostasis was also observed (Fig. 3.2). The AtCLH2 (AT5g43860) encoding chlorophyll-chlorophyllido hydrolase 2 annotated by MapMan as a tetrapyrrole synthesis-related gene was the only photosynthesis-related gene consistently upregulated at all three examined time points of infection (Macioszek et al. 2020).

3.3.8 Changes in Host Cell Ultrastructure and Transcriptome Reprogramming

Alternaria brassicicola invaded and colonized the host tissues successfully through necrotic spots, but host cells were differentially affected depending on their distance from invading hyphae, as evidenced by the gradual degradation of organelles within the inoculation site. Similar individual changes in the ultrastructure of infected plant host cells, such as cell lysis, disintegration of the nuclei and chloroplasts, or the presence of osmiophilic granules have been observed in other pathosystems during infection with viruses, bacteria, and fungi. It has to be mentioned that rounded chloroplasts were also observed in *B. juncea* cells in response to *A. brassicicola* infection, although the cup-shaped thylakoid system was not found (Macioszek et al. 2020). Changes of chloroplast shape and location of thylakoids could be related to cell lysis and, as a consequence, changes of osmotic pressure of stroma could have occurred. However, the underlying processes behind chloroplast degradation during infection are complex and require further extensive investigation. In the case of infected *B. oleracea* mesophyll cells, changes of organelle ultrastructure could be a result of the action of toxins and secondary metabolites secreted by *A. brassicicola* and general reprogramming of the host transcriptome and metabolome in response to the infection.

Alternaria species secrete various nonhosts (nHSTs) and host-selective toxins (HSTs) which are responsible for the degradation of different organelles within an infected cell in over 200 plant species.

The *Alternaria alternata* pathotypes produce different HSTs depending on the plant species, such as the AM-toxin that degrades the plasma membrane and chloroplasts in apple; the ACR-toxin that causes damage of the mitochondrial in lemon; the AK-toxin, AF-toxin and ACT-toxin that target the plasma membrane in susceptible cultivars of Japanese pear, strawberry and tangerine, respectively (Chen et al. 2019a). A possible effect of identified *A. brassicicola* secreted compounds during infection on plant cell structures or their influence on the host transcriptome has been described by Macioszek et al. (2020). It has to be emphasized that visualized differential *B. oleracea* cell responses to *A. brassicicola* infection, from defensive to highly susceptible, were confirmed by microarray results, and were also concordant with the previous studies on plant gene expression during infection of susceptible *B. oleracea* and *B. napus* with *A. brassicicola* (Schenk et al. 2008). The most upregulated categories revealed in the analysis of Gene Ontology of Biological Processes were aging, the detection of ethylene stimulus, thiamine biosynthetic processes, cell wall macromolecule catabolic processes and defense response to fungus. Additionally, analysis of stress-related genes revealed upregulation of the individual genes associated with hormone metabolism (brassinosteroids, ethylene and salicylic acid biosynthesis), peroxidases, glutathione S-transferases and secondary metabolism connected to flavonoid biosynthesis. Many of these genes have been previously described as important factors in general plant cell response to biotic stress and identified in transcriptome profile analyses of other susceptible hosts infected with *Alternaria* species. Some of these genes undoubtedly are characteristic for the *B. oleracea* response to infection by *A. brassicicola*, able it confirmation of their roles requires further investigations based on comparative transcriptomics and proteomic and metabolomic approaches. It has to be emphasized that the up- and downregulation of the host genes in response to *A. brassicicola* revealed in microarray analysis is concordant with a general scheme of transcriptomics response of a host to necrotrophic pathogens.

Perception of necrotrophic fungi activates hormone-dependent and hormone-independent signaling pathways. In resistant interactions, a hormone-dependent response to necrotrophs is mainly based on induction of genes associated with JA and ET-dependent signaling pathways to restrict pathogen spread. In susceptible interactions, genes involved in abscisic acid (ABA), brassinosteroid (BR) and auxin-dependent signaling are also often induced. Regardless of the mode of interaction, hormone-independent signaling is mediated by induction of genes involved in biosynthesis of phytoalexins and pathogenesis-related proteins (PR). If a pathogen is unable to detoxify host-specific phytoalexins, they often restrict its growth as in the case of *Arabidopsis* and *A. brassicicola* interaction. In addition, PR proteins are plant species-specific and play important roles in a pathogen recognition processes by a host as well as a host defense response (Pandey et al. 2016).

3.3.9 Defense-Related Genes are Activated in Susceptible Interaction

The upregulation of several stress-related genes involved in the defense response to fungus in the examined susceptible cultivar of *B. oleracea* even at 48 hpi, when necrotic lesions were fully developed, indicates that some of the host cells still attempted to combat the invasion of the fungus. The fungal hyphae spreading beyond inoculation site, even at a later stage of infection, apparently induced host defense response through PTI (pattern-triggered immunity). Probably, the perception of DAMP (damage-associated molecular pattern, e.g., products of the *A. brassicicola* cell wall degrading enzymes or plant secreted peptides) or PAMP (pathogen-associated molecular pattern, e.g., chitin) through a PRR (pattern recognition receptor) triggered the PTI-associated signaling cascade (albeit transcription of MEK1 was negatively regulated) and activated WRKY33 in infected *B. oleracea* cells, similarly to the situation observed in *Arabidopsis* during fungal infection. Interestingly, the only upregulated gene encoding PRR in microarray analysis was RLK7. The RLK7 belongs to the category XI of RLKs and acts as a receptor for PIP1 secreted peptide. Elevated expression of PIP1 has been described in *Arabidopsis* guard cells during *Pseudomonas*-induced PTI, and thus the peptide plays a role in stomatal immunity. Moreover, RLK7 also contributes to *Arabidopsis* resistance to *B. cinerea* and *Pythium irregulare*. However, LYK5, the other gene encoding RLK, which is an important receptor engaged in chitin perception in *Arabidopsis*, was downregulated. The upregulated genes involved in defense signaling such as WRKY33 and PDF1.2 have been identified in *A. thaliana* signaling pathways and are required for signaling resistance to necrotrophic fungi *A. brassicicola* and *B. cinerea*. In *Arabidopsis*, WRKY33 transcription factor is responsible for the activation of camalexin biosynthesis, which efficiently inhibits development of necrotrophic fungi. Possibly, WRKY33 activates de novo biosynthesis of *B. oleracea* fungus-induced phytoalexins, such as brassinin, which can be metabolized by *A. brassicicola*, and thus the fungus suppresses the first line of host defense. Subsequently, JA-dependent signaling, which is characteristic for host cells infected by fungal necrotrophs, activates transcription of JA-responsive defense genes encoding PR (pathogenesis-related) proteins such as PDF 1.2 involved in resistance against *B. cinerea* and *A. brassicicola*. However, the down-regulation of LOX2 chloroplast lipooxygenase required for JA accumulation may be also responsible for the impaired defense of the host and, in turn, salicylic acid-dependent genes become activated, which promote host cell death. Moreover, local fortification of the host cell walls microscopically observed within inoculation site in a study was reflected in the upregulation of several stress-responsive cell walls and secondary metabolism-related genes, such as FAH1 involved in lignin biosynthesis and an important component of *A. thaliana* resistance mechanism to *B. cinerea* and CCoAMT, which is involved in the biosynthesis of cell wall-bound phenolics and lignin. The strengthening of host cell walls through lignin biosynthesis and its deposition at a pathogen entry site also constitutes the first line of the host defense reaction in order to slow down or restrict pathogen development, especially as lignin is not metabolized by most pathogens. Thus, even successful invasion of host tissues within inoculation sites does not signify that every single host cell immediately surrenders to a fungal invader. However, downregulation of numerous stress-related genes found in the study indicates that *A. brassicicola* effectively overcomes the host arsenal of defenses. Although, it has been reported that *A. brassicicola* secretes phytotoxin brassicicolin A, histone deacetylase inhibitor depudicin, siderophore N, N-dimethylcoproge, a proteinaceous host-specific toxin–AB-toxin and many low molecular weight secondary metabolites e.g., brassicenes A to F, there is not yet any known putative *A. brassicicola* effector(s) targeting also still unknown *Brassica* receptor(s) and triggering ETS (effector-triggered susceptibility) or ETI (effector-triggered immunity), albeit many research groups all over the world work on *A. brassicicola* pathogenicity factors (Cho 2015; Macioszek et al. 2020).

3.4 *Brassica–Arabidopsis–Colletotrichum* Pathosystem

3.4.1 Introduction

The *Colletotrichum higginsianum* causing anthracnose disease of crucifers is emerging as an attractive model for studying both fungal pathogenicity and plant immune responses due to the availability of genomic and transcriptomic data bases (O'Connell et al. 2012; Gan et al. 2013), and the ease of genetic manipulation of both the host plant and the pathogen (O'Connell et al. 2004; Zhang et al. 2006; Huser et al. 2009; Narusaka et al. 2009). By recognizing host physical and chemical cues, *C. higginsianum* conidia differentiate melanized appressorium, an infection structure, at the tips of conidial germ tubes. Appressorium formation is required for successful infection since the fungus penetrates the cuticle and plant cell wall by utilization of enormous turgor pressure in melanized appressoria for further invasive growth (O'Connell et al. 2012). Thus, inhibition of melanized appressorium formation will facilitate the efficient control of anthracnose disease (Wei et al. 2016).

3.4.2 Mitogen-Activated Protein (MAP) Kinase and cAMP/PKA Signaling Pathway for Pathogenesis

Adhesion to the plant surface is the first step in initiation of the infection process in many plant pathogenic fungi. Following adhesion, physical signals such as those involving tissue hardness and hydrophobicity or chemical signals (cutin monomers and leaf waxes) induce germination and appressorial formation in several plant pathogenic fungi. In eukaryotic cells, the transduction of a variety of extracellular signals and the regulation of different developmental processes are regulated by mitogen-activated protein (MAP) kinase pathways and cAMP/PKA signaling pathways. Thus, some intensive studies on these two pathways have been carried out based on the infection process of *C. higginsianum*. The ChSte7, encoding a MAPKK orthologue gene Ste7 in yeast, was highly expressed in vegetative and invasive growth stages in *C. higginsianum.* Deletion of ChSte7 resulted insignificant reduction in vegetative growth, inability to form appressoria and also reduced invasive growth inside host plant tissues, which was similar for *M. oryzae, B. cinerea,* and *Ustilago maydis*.

A Fus3/Kss1-related MAPK gene in *C. higginsianum*, ChMK1, was also reported to play an important role in cell wall integrity, colony melanization, and pathogenicity on *A. thaliana*. The ChRgf, encoding a Ras guanine-nucleotide exchange factor protein, might be a control element of MAPK pathway (REF), and its deletion resulted in some phenotypes similar to those involving deletion of the two MAPK pathway genes: defects in vegetative growth, altered hyphal morphology, reduced conidiation, poor surface attachment, and low germination on hydrophobic surfaces (Gu et al. 2017; Wei et al. 2016). These results indicate that the MAPK pathway is involved in a critical conserved role to control the pathogenicity and growth of *C. higginsianum* through the extracellular signal transmission compared to other phytopathogenic fungi. Moreover, another extracellular signal transmission pathway, cAMP/PKA signaling pathway, was also studied in *C. higginsianum*. The PKA catalytic subunits ChPKA1 and adenylate cyclase ChAC deletion mutants were significantly reduced in hyphal growth rate, tolerance to cell wall inhibitors and conidiation, but had an increased tolerance to elevated temperatures and exogenous H_2O_2. In contrast, the ChPKA2 mutant had no detectable alteration of phenotypes, suggesting that ChPKA1 contributes mainly to PKA activities in *C. higginsianum*. These findings suggest that the cAMP/PKA signaling pathway also contributes to growth, conidial formation, stress tolerance, and pathogenesis in *C. higginsianum* (Yan et al. 2018).

The MAPK and cAMP/PKA signaling pathways are well known in regulation of appressorial morphogenesis and plant infection in *M. oryzae, C. truncatum* and some other phytopathogenic fungi, but investigation of these two signaling pathways in *C. higginsianum* still helped to provide insights into the mechanism of the *C. higginsianum*–cruciferous crop interaction, and to facilitate investigation of efficient management of anthracnose disease. Functional comparisons of MAPK and cAMP/PKA signaling pathways in *C. higginsianum* with other phytopathogenic fungi may provide a deeper understanding of pathogenic mechanisms of this fungus. After adhesion to plant surfaces, the infection strategy of *C. higginsianum* includes two phases: an initial biotrophic phase and a subsequent necrotrophic growth phase. For biotrophic growth and transition to necrotrophic growth, fine-tuned regulation of cell walls developmental processes are essential. For fungi, management of these processes involves many pathways, including kinases and the co-activator Mob-family proteins. These pathways have been called morphogenesis-related and the septum initiation network in *Saccharomyces pombe*, or regulation of Ace2p activity and morphogenesis and the mitotic exit network both in *S. cerevisiae*. In this latter fungus, Cbk1 is the terminal kinase in the RAM pathway, and it is classified in the kinase subfamily of nuclear Dbf2-related or large tumor suppressor. The function of Mob-family proteins which are essential for the activation of NDR kinases was also studied in *C. higginsianum*. The results showed that the *C. higginsianum* genome encodes three members of the Mob1/phocein protein family. The ChMob1 is required for conidiation, cytokinesis, and plant infection. The ChMob2 binds to the conserved NDR/LATS kinase ChCbk1, and is involved in virulence on *A. thaliana* and is required for both conidiation and formation of functional appressoria. ChMob3 knockout mutants have no obvious phenotype in vegetative cells or during infection. Moreover, Mob2 and Cbk1 co-localize to the cytoplasm and are excluded from nuclei in conidia and during appressorial formation in vitro. Mutants in the two potential Mob2/Cbk1 complex targets ChSSD1 and ChACE2 genes show defects in pathogenicity (Schmidpeter et al. 2017; Yan et al. 2018).

3.4.3 Identification of Effectors/Genes during Pathogenesis

In the case of plant pathogenic hemibiotrophs, colonization and the initial biotrophic interaction with host cells is facilitated by pathogen-encoded small, secreted proteins termed effectors. Biotrophy-specific hyphal cells play important roles in transporting effectors into the host cells and in obtaining nutrition from the host. Depending on the fungal species, the interfacial matrix is either continuous with the host plant apoplast such as for Puccinales or separated into an interfacial apoplastic compartment like that in *M. oryzae*. The interfacial membrane among a variety of pathosystems ranges from undifferentiated plant plasma membranes to highly specialized membranes with complicated elaborations and unique components. Vesicles are abundant in both host and pathogen cytoplasm near the interface, implying that both are involved in active production of secreted compounds into and across interfacial zones. Shimada et al. (2006) found some evidence for localized specialization of the interfacial membrane around BH of *C. higginsianum*. Biotrophic infection by *C. higginsianum* has differences from other biotrophic fungus, and the process of interfacial membrane exchange with the fungus deserves further study (Yan et al. 2018).

In *Colletotrichum higginsianum*, inventories of putative effectors have been predicted from the annotated genomes of *C. higginsianum* revealing 18 genes, of which six were not predicted to be secreted, and two were chitinases, leaving 10 putative secreted LysM effectors. The LysM domain comprises 40–60 amino acid residues and mediates binding to chitin and peptidoglycans. Chitin is a microbe-associated molecular pattern (MAMP) that can be detected by plant pattern recognition receptors (PRRs) to activate a variety of MAMP-triggered immune responses. To avoid recognition by host receptors, several ascomycetes are known to produce effector proteins which either block the activity of host plant chitinases or compete with host plant receptors which bind chitin fragments. The function of two effectors, ChELP1 and ChELP2, homologs of LysM proteins were characterized in *C. higginsianum*. ChELP2 has been found to be located on the surface of bulbous biotrophic hyphae at the interface with living plant cells, but it has not been discovered in necrotrophic hyphae. In previous experiments, recombinant ChELP1 and ChELP2 were found to bind chitin oligomers in vitro with high specificity and high affinity. Both proteins suppressed chitin-triggered activation of two immunity-related plant mitogen-activated protein kinases in *Arabidopsis*. These results suggested a double role for these LysM proteins as effectors for suppressing chitin-triggered immunity and as proteins essential for appressorial development and function (Takahara et al. 2016).

The role of secreted effector proteins during infection by hemibiotrophic plant pathogens is poorly understood. Based on deep transcriptome sequencing and computational mining of Expressed Sequence Tags from precise infection stages, a large of planta-expressed effector candidates were found in *C. higginsianum*. Most biotrophy-associated ChEC genes were dramatically upregulated exclusively in planta and distinct sets of effectors are deployed in successive waves by particular fungal cell types. With fluorescent protein tagging and transmission electron microscopy-immunogold labeling, early expressed effector proteins are observed to be focally secreted from appressorial penetration pores before host invasion. In addition, later-expressed effectors accumulate in structures formed at the interface between primary hyphae and living host cells, implicating these hyphae in effector delivery. Furthermore, the coordinated expression and secretion of antagonistic biotrophy effectors and toxin effectors contribute to fungal virulence and the regulation of hemibiotrophy in *C. higginsianum.* These findings indicate new functions for fungal infection structures that have not been reported previously, specifically the localized release of effector proteins at the interface between fungal pathogen and plant host, and associated with the penetration pore. This provides the basis to model the switch to necrotrophy from biotrophy by this fungus. Future research should attempt to decipher the nature of the plant signals inducing effector gene expression and the way that they are sensed by the pathogen. A total of 61 putative effector proteins were separately cloned into a plant expression vector providing an N-terminal GFP tag, and the tagged proteins were transiently expressed directly in side plant cells using ATMT. Among them, subcellular localization of 16 candidate effectors was verified, nine were imported into plant nuclei, three were imported into the matrix of peroxisomes, three decorated cortical microtubule arrays, and one was associated with Golgi stacks. These findings revealed that plant peroxisomes, microtubules and Golgi are novel targets for fungal effectors (Robin et al. 2018; Yan et al. 2018).

3.4.4 Role of MAPK Gene in Development of Anthracnose Disease

The *Colletotrichum higginsianum* is an economically important pathogen that causes anthracnose disease in a wide range of cruciferous crops. To facilitate the efficient control of anthracnose disease, it will be important to understand the mechanism by which the cruciferous crops and *C. higginsianum* interact. A key step in understanding this interaction is characterizing the mitogen-activated protein kinases (MAPK) signaling pathway of *C. higginsianum.* The MAPK plays important roles in diverse physiological processes of multiple pathogens. Wei et al. (2016) analyzed, a Fus3/Kss1-related MAPK gene, ChMK1, from *C. higginsianum* was analyzed. The result showed that the Fus3/Kss1-related MAPK ChMK1 plays a significant role in cell wall integrity. Targeted deletion of ChMK1 resulted in a hypersensitivity to cell wall inhibitors, reduced conidiation and albinistic colonies. Further, the deletion mutant was also unable to form melanized appressorium, a specialized infection structure that is necessary for successful infection. Therefore, the deletion mutant loses pathogenicity on *A. thaliana* leaves, demonstrating that ChMK1 plays an essential role in the early infection step. In addition, the ChMK1 deletion mutant showed an attenuated growth rate that is different from that of its homolog in *C. lagenarium*, indicating the diverse roles that Fus3/Kss1-related MAPKs plays in phytopathogenic fungi. Furthermore, the expression level of three melanin synthesis associated genes were clearly decreased in the albinistic ChMK1 mutant compared to that of the wild type strain, suggesting that ChMK1 is also required for colony melanization in *C. higginsianum* (Wei et al. 2016).

3.4.5 Molecular Events during Host–Pathogen Interaction

Plants usually defend against microbial pathogens by activating both localized and systemic resistance responses. These responses include hypersensitive response, cell-wall fortification, and synthesis of phytoalexins and production of other antimicrobial secondary metabolites or pathogenesis-related proteins (PR proteins). Signaling molecules implicated in these inducible defense systems include salicylic acid (SA), jasmonic acid (JA), ethylene (ET), abscisic acid (ABA), auxin, gibberellins (GAs), cytokinins (CKs), brassino steroids (BRs), and reactive oxygen species (ROS). These phytohormones can induce defense responses individually, and also interact synergistically or antagonistically to further orchestrate downstream signaling. A large number of fungal and oomycete pathogens have been reported to infect the model plant *A. thaliana*, either naturally or in the laboratory. As a typical hemibiotrophic fungus, *C. higginsianum* develops a series of specialized infection structures. In particular, the intracellular BH of *C. higginsianum* are equivalent to haustoria, and hence this pathosystem can provide insights into the molecular basis of biotrophy in obligatory parasitic organisms, such as rusts, powdery mildews, and downy mildews, all of which are not readily culturable or genetically manipulated (Shigenaga and Argueso 2016; Yan et al. 2018).

3.5 *Brassica–Erysiphe* Pathosystem

3.5.1 Introduction

Powdery mildew fungi of crucifers (*Erysiphe cruciferarum, E. cichoracearum, E. orontii, Oidium neolycopersici*) are ectoparasites on their respective host plant species. Primary infection takes place through ascospores or conidia after landing on the host surface through germ tube which develops an appressorium at the tip, a thickened infection structure. After cell wall penetration, specialized hyphae feeding structure called haustorium is formed which gets necessary nutrients from the host tissues for the growth, and development of the pathogen to produce mycelium, conidiophores, and conidia to continue its asexual state. Secondary spread of the disease is through air borne conidia causing infection, and pathogenesis under congenial environmental conditions. At the maturity stage of the crop, sexual state of the pathogen is visible in the form of small, pinhead, dark brown, spherical chasmothecia which over summer or over winter to serve as a source of primary inoculum for the next season. Chasmothecia imbibe to burst, and release asci, and ascospores which germinate after landing on host surface, and cause primary infection. The pathogenesis of powdery mildew fungi is achieved through the suppression of the hosts preformed, and inducible penetration, resistance mechanisms as well as maintenance of the compatible interaction after host cell penetration. After host cell penetration, the haustorial complex is established, which consists of the fungal haustorium, an extrahaustorial modified plant plasma membrane, and the haustorial matrix. The haustorial complex is the site of the most intimate contact of the pathogen with its host cell, and functions in nutrient uptake, and provides a platform for the delivery of pathogens effector genes for pathogenesis. The expression of hundreds of host genes is altered within 8 hpi indicating that signaling between the pathogen, and individual host cells already occurs a few hours postcontact of the powdery mildew conidia or ascospores with the leaf surface of about the stage of penetration, but certainly prior to establishment of the extra-haustorial matrix. For the successful pathogenesis of crucifer's powdery mildew, there is great role of genetic factors (genes), MLO proteins, and host transcriptional changes. Powdery mildew of crucifers is activated by SA signaling pathways at the early stage of infection, and by the JA/ET signaling pathways at the later stage of pathogenesis. Infection is highly activated by chitinase in the genotypes. Adjustment of the plant host metabolism to support the growth of the powdery mildew pathogen is consistent with an increased ploidy level of the mesophyll cells underlying infected epidermal cells. Powdery mildew disease cycle operates at both asexual and sexual stages depending on its host plant species infected under prevailing environmental conditions (Saharan et al. 2019).

3.5.2 Genomics of Powdery Mildew Pathogenesis

The biotrophic lifestyle of powdery mildew fungi dictates that they infect plants, and establish colonies in a "quiet" manner, so as not to arouse the defense mechanisms of the host (Panstruga 2003). They likely also usurp host infrastructure for transport of nutrients, and building blocks, and coerce the host cell into accommodating their impressive haustorial feeding structures (Plate 3.6, Mendgen and Hahn 2002;

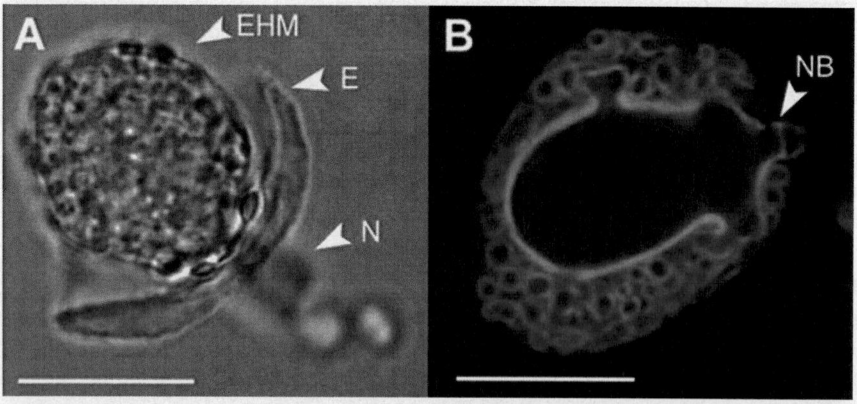

Plate 3.6 Haustorial complexes of *Golovinomyces orontii*. Phase contrast (**a**) and epifluorescence (**b**) micrographs of *G. orontii* haustoria isolated from *Arabidopsis* leaves. Notice the highly convoluted, and complex folding of the haustorial cell surface providing a large area for nutrient uptake from, and effector delivery into the host. Haustoria were labeled with wheat germ agglutinin-FITC. *EHM* extra haustorial membrane, *E* encasement, *N* haustorial neck, *NB* neck band. Scale bars = 20 μm (Micali et al. 2008)

Schulze-Lefert and Panstruga 2003). Specific host genes, and/or proteins termed compatibility or susceptibility factors are believed to be essential for successful pathogenesis by a given microbe, and a lack of these factors is predicted to result in resistance to an otherwise virulent pathogen (Vogel and Somerville 2000; Panstruga 2003). To isolate compatibility factors, two independent genetic screens were performed in the late 1990s, and identified many new components that paint an ever more complex picture of *Arabidopsis* cellular mechanisms of defense. A hall mark of constitutive defense is the transcriptional activation of PR genes such as PR1 (Loake and Grant 2007). In an attempt to identify novel elements of resistance, three powdery mildew, (*Golovinomyces cichoracearum*) resistant mutants that do not constitutively express PR1 were isolated in the lab of R. Innes, and named enhanced disease resistance (*edr*) 1, *edr2,* and *edr3* (Frye and Innes 1998; Frye et al. 2001; Tang and Innes 2002; Tang et al. 2005, 2006).

All three mutants display late-onset of resistance (5–8 days post-inoculation) that is characterized by accelerated cell death in mesophyll cells leading to large necrotic lesions, and limited or absent conidiation of the pathogen. Resistance in all cases is SA-dependent, and JA-independent. *EDR1* encodes a CTR1-like protein kinase, and negative regulator of disease resistance (Frye et al. 2001). Presence of *EDR1* limits the transcriptional amplification of *RPW8*.1, and reduces RPW8-conditioned host cell death in response to pathogen attack (Xiao et al. 2005; Fig. 3.4). *EDR2* encodes a mitochondrial protein (Fig. 3.4) with a pleckstrin homology domain, and a steroidogenic acute regulatory protein-related lipid transfer (START) motif that are each indicative of a role in lipid binding, and signaling. Both *EDR1* and *EDR2* function in a common pathway since double mutants display resistance phenotypes identical to the respective single mutant plants (Tang et al. 2005). In addition, both mutants display enhanced senescence in response to ethylene. In contrast, *EDR3* appears to function in an independent pathway from *EDR1,* and *EDR2,* and the corresponding mutant does not display an early senescence phenotype.

Like *EDR2*, *EDR3* localizes at least in part to mitochondria, and encodes a dynamin-like protein with an N-terminal GTPase domain, and a C-terminal GTPase effector region (Tang et al. 2006; Fig. 3.4). In mammals, such proteins are responsible for membrane tabulation, and vesicle pinching, and regulate mitochondrial dynamics associated with apoptosis (programmed cell death; Tang et al. 2006). The characterization of *edr* mutants suggests a link between SA-mediated resistance, mitochondrial function, and programmed cell death, and further stresses the broad parallels that exist between animal, and plant immune responses (Ausubel 2005). Indeed, a central feature of defense responses in mammalian systems is the ability to undergo apoptosis in response to pathogen (or nonself) detection (Hiscott et al. 2006). Often the apoptotic program is either initiated or amplified in mitochondria, and components of the mitochondrial apoptosome are absolutely required for completion of programmed cell death (Keeble and Gilmore 2007).

In an independent screen performed by Vogel and Somerville (2000) and Vogel et al. (2002) to recover loss-of-susceptibility to *G. cichoracearum* mutants in *Arabidopsis*, the powdery mildew resistant (pmr) mutants pmrl to *pmr6* were isolated, and to date four of them have been cloned (Vogel and Somerville 2000;

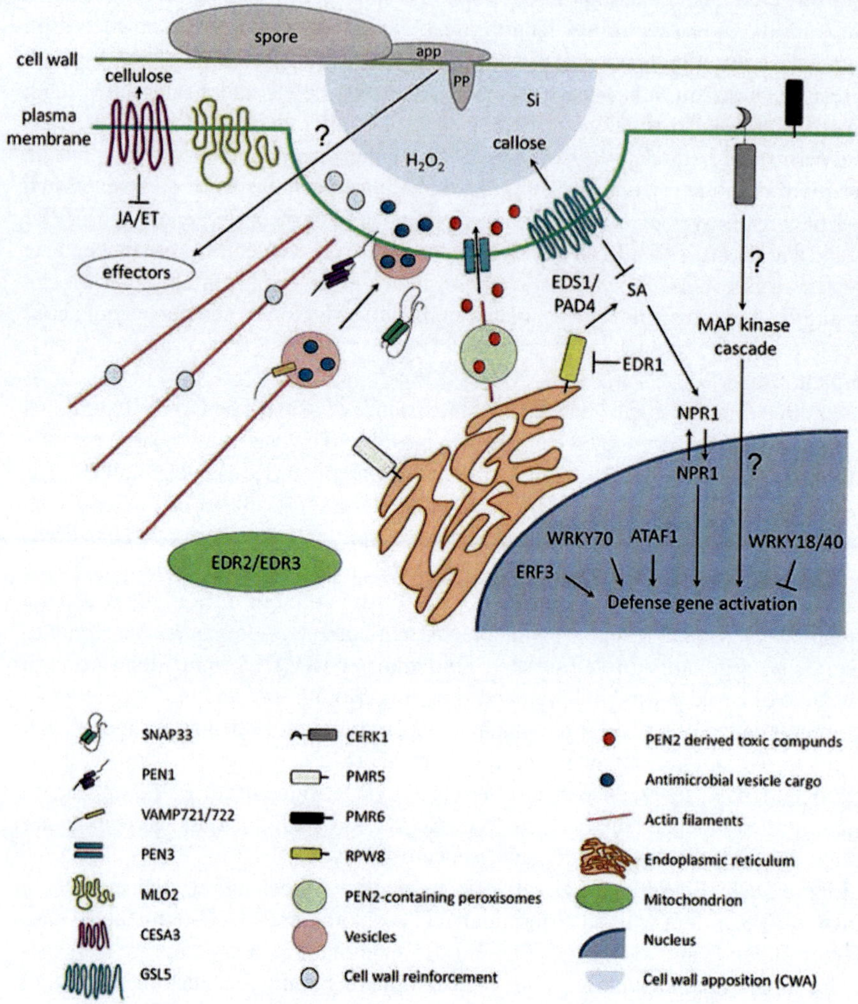

Fig. 3.4 Schematic diagram illustrating genetically anchored components in *Arabidopsis* powdery mildew susceptibility/resistance. The figure depicts a section of a host cell attacked by a powdery mildew germ tube. Components coded by shape and color are explained in the legend below the scheme. *App* appressorium, *pp* penetration peg, *Si* silicon. Question marks indicate presumed links/activities (Micali et al. 2008)

Vogel et al. 2002, 2004; Jacobs et al. 2003; Nishimura et al. 2003; Consonni et al. 2006; Plate 3.7). The *PMR2* is allelic to MLO2, PMR4 (synonyms CALS12, and GSL5) encodes a wound, and pathogen associated callose synthase, PMR5 belongs to a large family of plant-specific genes of unknown function, and PMR6 encodes a glycosyl-phosphatidyl-inositol (GPI)-anchored pectate-lyase-like protein (Vogel et al. 2002, 2004; Jacobs et al. 2003; Nishimura et al. 2003; Consonni et al. 2006).

Plate 3.7 Macroscopic infection phenotypes of susceptible and resistant *Arabidopsis* lines. Rosette leaves of 5–6-week-old *A. thaliana* ecotypes Col-0 (**a**), Do-0 (**b**), and Sorbo (**c**) as well as the powdery mildew resistant mutant pmr6-3 (**d**) at 13 days post-inoculation with *Golovinomyces orontii*. Completion of the asexual powdery mildew life cycle is evidenced by the occurrence of abundant sporulation (white powder) on inoculated rosette leaves of the susceptible accession, Col-0. Younger leaves without disease symptoms emerged after inoculation with fungal conidiospores. Note the difference in appearance of infected leaves of resistant accessions Do-0 and Sorbo. Resistance in both accessions is assumed to be governed by RPW8 (Gollner et al. 2008)

In pmr4, pmr5, and pmr6 mutant's resistance is entirely established postpenetration, and results in reduced colony size, and conidiophore formation in interactions with both *G. cichoracearum* and *G. orontii*. The *PMR* genes cloned to date are involved in different cellular activities, underscoring the diversity of plant biological processes that are required for successful powdery mildew pathogenesis (Micali et al. 2008).

The *PMR4* controls callose synthesis at cell wall appositions (papillae) that form beneath infection, and wound sites, and that are believed to provide a physical barrier to fungal penetration (Aist and Bushnell 1991; Plate 3.7). It is, therefore came as a surprise that a loss-of-function in a callose synthase renders a plant more resistant to powdery mildew infection (Jacobs et al. 2003; Nishimura et al. 2003). A comparative analysis of genes that are differentially expressed in wild-type, and pmr4 plants upon infection with *G. cichoracearum* revealed that resistance to powdery mildews is likely the result of constitutive activation of the SA signaling pathway that is further enhanced upon pathogen attack (Nishimura et al. 2003).

Mutations in genes encoding components of the salicylic acid signaling pathway indeed abolished pmr4-based resistance without restoring callose deposition at papillae. Callose and/or callose-synthase-based signaling were shown to activate basal defense responses that are effective against a variety of pathogens including *G. orontii,* and *H. parasitica* (Jacobs et al. 2003; Nishimura et al. 2003). In this sense, *PMR4* is likely less a fungal-specific compatibility molecule, and more a general basal defense switch located at the cell wall that is effective against a broad range of pathogens. Two additional pmr mutants stress the potential importance of cell wall integrity surveillance in resistance mechanisms. The *PMR6 encodes* a putative pectate lyase with proposed pectin-degrading activity localized at the cell wall. The pmr6 mutant displays an increased pectin, and uronic acid content in the cell wall, and exhibits a stunted phenotype (Vogel et al. 2002).

In pmr5, similar changes in cell wall components have been reported. The *PMR5* codes for a member of a large family of plant-specific proteins of unknown function that is likely targeted to the endoplasmic reticulum/secretory pathway (Plate 3.7). Although the exact mechanism leading to powdery mildew resistance is unknown, it is independent of SA, ET, and JA signaling, since mutations in any of the known defense pathways have no effect on pmr5- or pmr6-based resistance. However, pmr5/pmr6 double mutants show further increased resistance compared to the respective single mutants, indicating that the two genes likely control parallel, and independent defense responses. Furthermore, the fact that pmr5, and pmr6 are resistant to *G. cichoracearum,* and *G. orontii*, but fully susceptible to unrelated pathogens such as virulent strains of either *P. syringae* or *H. parasitica*, suggests that these two proteins, in contrast to the *PMR4* callose synthase gene, may in fact be true powdery mildew specific compatibility factors (Vogel and Somerville 2000; Vogel et al. 2002, 2004).

3.5.3 The MLO Proteins (Genes) in Powdery Mildew Pathogenesis

Isolated in the same genetic screen as genes *PMR4*, *PMR5,* and *PMR6*, *PMR2* encoding *MLO2* (*MILDEW LOCUS O*) has an essential role in the establishment of compatibility with the powdery mildew species *G. cichoracearum,* and *G. orontii*. Together with the phylogenetically closely related paralogs *MLO6,* and *MLO12,* which act in concert with *MLO2* in partial functional redundancy (Consonni et al. 2006), *MLO2* controls entry of powdery mildew fungi in *Arabidopsis* epidermal cells (Plate 3.8). In this respect, *MLO2*, *MLO6,* and *MLO12,* and the barley homologue, and first member of this protein family to be identified—*Mlo* (Panstruga 2005), represent the prototype for molecules that mediate compatibility between powdery mildews, and their plant hosts. Loss of particular MLO protein(s) in barley, tomato, and *Arabidopsis* render these plant hosts immune to powdery mildew infection (Bai et al. 2008). The conserved function of MLO proteins in plant defense predates the divergence of dicots, and monocots, and underscores the importance of this protein family in the establishment of compatible plant-powdery mildew interactions (Consonni et al. 2006). The MLO proteins are integral membrane proteins with seven trans-membrane domains (Plate 3.8) that have so far only been identified in plants (Devoto et al. 2003). In *Arabidopsis*, they are encoded by a family of 15 genes, and appear to have diverse functions in addition to plant defense.

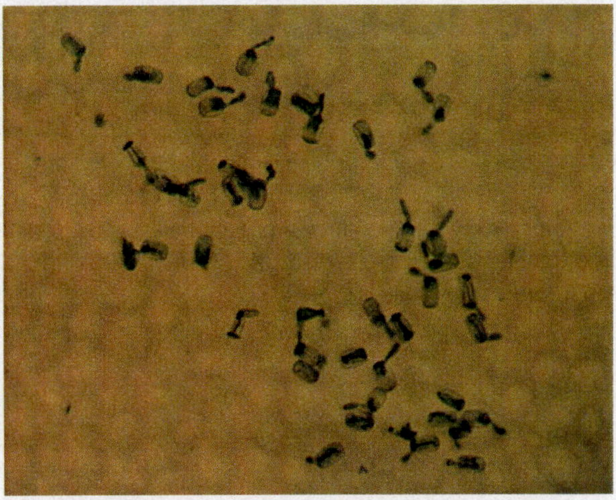

Plate 3.8 Resistance in *Arabidopsis* mlo2 mlo6 mlo12 triple mutants is characterized by complete failure of successful host cell invasion. The micrograph shows attempted penetration of numerous *Golovinomyces orontii* sporelings at 48 h post-inoculation on a highly resistant mlo2 mlo6 mlo12 triple mutant in the genetic background of otherwise susceptible Col-0 (Consonni et al. 2006). Note the aborted fungal entry evidenced by a lack of secondary hyphae compared to the situation in a Col-0 wild type plant (Plate 4.4a). Scale bar = 100 μm (Micali et al. 2008)

However, only the three closely related members *MLO2*, *MLO6,* and *MLO12* seem to play a role in powdery mildew pathogenesis. Consistent with previous findings in barley (Jarosch et al. 1999; Kumar et al. 2001), *mlo2mlo6* double, and *mlo2mlo6mlo12* triple mutants show enhanced susceptibility to the necrotrophic pathogens *Alternaria alternata,* and *A. brassicicola,* and the hemi-biotrophic oomycete *Phytophthora infestans*, indicating that MLO proteins modulate the infection process of pathogens with diverse life styles (Consonni et al. 2006).

Barley MLO, and MLO2 from *Arabidopsis* each interact via a conserved peptide domain in their cytoplasmic carboxyl terminus with the Ca^{2+} sensor calmodulin (Kim et al. 2002; Bhat et al. 2005), and are thought to regulate defense responses against powdery mildews via PEN1 (ROR2 in barley)-dependent mechanisms at the cell periphery (Collins et al. 2003; Schulze-Lefert 2004; Panstruga 2005). Similar to nonhost resistance, *mlo*-mediated resistance in *A. thaliana* does not depend on the JA/ET or SA signaling pathways (Collins et al. 2003; Lipka et al. 2005; Stein et al. 2006; Consonni et al. 2006). The physical proximity of MLO to the PEN1-dependent nonhost defense machinery at the sites of fungal attack (Bhat et al. 2005), the recently observed transcriptional co-expression of these genes in *Arabidopsis* (Micali et al. 2008), and the genetic, and cytological similarities between mlo-based, and nonhost resistance (Humphry et al. 2006) point to a tight mechanistic link between MLO, and the basal defense apparatus. The fact that presence of MLO is absolutely required by powdery mildew fungi to successfully infect host plants suggests that these pathogens possibly exploit MLO function(s) to suppress basal defense responses (Panstruga 2005) during the process of pathogenesis.

3.5.4 Host Transcriptional Regulation of Powdery Mildew Pathogenesis

Besides defense-related transcriptional changes induced by microbe recognition, host gene expression are potentially impacted by the action of powdery mildew effectors to promote fungal pathogenesis. Consequently, adaptation of the host metabolism to the presence of the pathogen has been reported (Wildermuth 2010). Laser micro-dissection-assisted site-specific profiling of transcript abundance during late *Go* infection stages (5 dpi) suggests a suppression of photosynthesis, and points to a carbon source-to-sink transition in powdery mildew-infected cells (Fig. 3.5; Chandran et al. 2009, 2010). The powdery mildew-triggered induction of genes associated with sugar metabolism and hexose transporters associated with sink organs further reinforces this idea. The respective proteins might contribute to the availability of carbohydrates at infection sites, and consistently their elevated expression levels may reflect an increased demand for hexoses by the fungus (Fabro et al. 2008; Chandran et al. 2009, 2010). Increased transcript abundance of genes related to respiration, including glycolysis, the tricarboxylic acid cycle, and the mitochondrial electron transport chain further strengthen the notion of an adaptation to the elevated energy consumption by the infected tissue (Fabro et al. 2008; Chandran et al. 2010).

Adjustment of the plant host metabolism to support the growth of the biotrophic pathogen is consistent with an increased ploidy level of the mesophyll cells underlying infected epidermal cells at later stages of compatible interactions (Fig. 3.5; Chandran et al. 2010, 2013). This correlates with the accumulation of *PLANT UBX DOMAIN-CONTAINING PROTEIN 2* (*PUX2*: At2g01650) transcripts after *Go* infection (Chandran et al. 2009). Strikingly, the onset of *PUX2* induction at 5 dpi overlaps with the occurrence of endo-reduplication in mesophyll cells, and corresponds with fungal growth, and reproduction (Chandran et al. 2013). The resulting polyploidy might compensate for the increased metabolic activity resulting from the nutritional demands of the fungus. This is supported by decreased spore formation coinciding with reduced basal ploidy in *pux2*, and thus identifies endo-reduplication as a potential determinant of susceptibility to powdery mildew (Chandran et al. 2010, 2013). The presence of UBX (ubiquitin regulatory X) and PUB (peptide: N-glycanase/UBA or UBX-containing proteins) domains in PUX2 suggests that, like other proteins with similar domain structures, it might act as a regulatory co-factor of cell division control protein 48 (CDC48: At3g09840). Indeed, this AAA-ATPase (ATPase associated with diverse cellular activities) interacts with PUX2 in vitro (Rancour et al. 2004). As CDC48 complexes contribute to cell cycle progression, its interaction with PUX2 might regulate cell ploidy (Rancour et al. 2004; Madsen et al. 2009; Yamanaka et al. 2012; Gallois et al. 2013).

The MYB3R4 (At5g11510), a cell cycle control-associated MYB3R TF activating G2/M progression, is locally induced 5 dpi with *Go*. As genome duplication is a controlled process that occurs during mitosis, it is conceivable that MYB3R4 is required for powdery mildew-induced polyploidy, which is supported by the phenotype of the *myb3r4* mutant (Haga et al. 2007; Chandran et al. 2010,

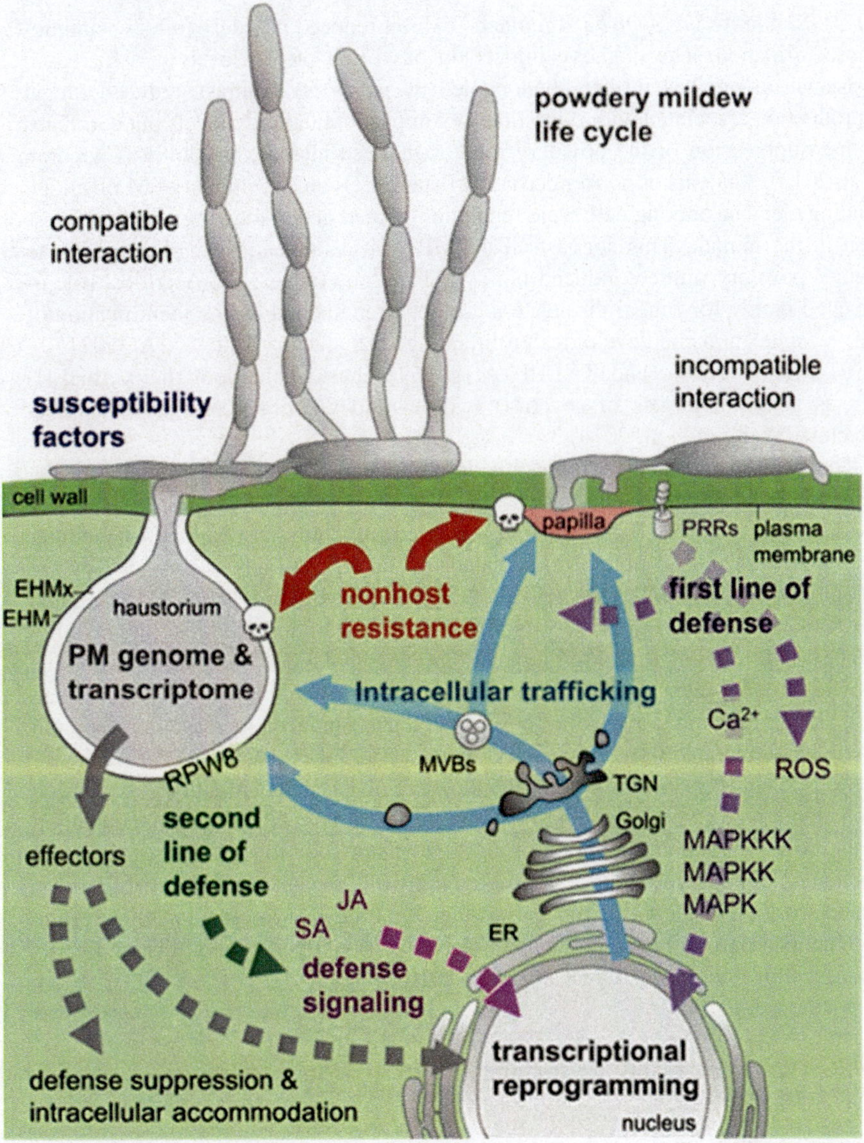

Fig. 3.5 Proposed model of endo-reduplication in powdery mildew pathogenesis. The scheme depicts regulators, and mechanisms involved in the control of endo-reduplication, and consequences of powdery mildew-induced mesophyll polyploidy. A powdery mildew (gray)-colonized leaf epidermal cell and an underlying mesophyll cell are shown. Gray arrows indicate the proposed translocation of components between cells. Solid lines indicate proven regulatory impacts, and dashed lines indicate speculative regulatory impacts. *M* mitosis (Kuhn et al. 2016)

2013). Similar to *pux2*, *myb3r4* mutants' exhibit reduced powdery mildew conidiophore formation. The negative impact on basal cell ploidy levels and increased resistance of *pux2* is further phenocopied by *pmr6*. By contrast, reduced fungal reproduction associated with *pmr5* does not impact basal ploidy levels but correlates with a suppression of the powdery mildew-induced increase in ploidy (Chandran et al. 2013). Analysis of *pmr5* microarray data reveals an enrichment of MYB3R TF binding elements among cell cycle regulation-related genes showing altered expression in the mutant. This suggests that PMR5 acts upstream of a MYB3R TF to control powdery mildew-induced ploidy (Chandran et al. 2013). A critical role of elevated ploidy for fungal virulence is further strengthened by the identification of the teosinte branched/cycloidea/proliferating cell factor TCP13 (At3g02150), TCP14 (At3g47620), and TCP15 (At1g69690) basic helix–loop–helix (bHLH) TFs as common targets of several powdery mildews, oomycetes, and bacterial effectors (Weßling et al. 2014).

The TCP14 and TCP15 repress endo-reduplication by directly regulating the expression of cell-cycle genes (Peng et al. 2015). Mutation of TCP13, TCP14, and to a lesser extent of TCP15, results in increased susceptibility toward *Go* (Weßling et al. 2014). A link to ubiquitin-mediated regulation of ploidy is provided by the ubiquitin receptors DA1 (At1g19270; "Da" is Chinese for "large"), DA-related (DAR) 1, and DAR2 (At2g39830), which interact with and modulate the stability of TCP14/15 to regulate endo-reduplication (Peng et al. 2015). Remarkably, DEL1, known to repress genes required for the onset of endo-reduplication (Vlieghe et al. 2005; Lammens et al. 2008), does not impact the *Go*-induced increase in mesophyll ploidy when mutated or overexpressed (Chandran et al. 2014). Instead, microarray analyses of *del1* plants revealed an induction of basal defense gene expression compared to wild type. The identification of effector targets involved in adaptation of the host metabolism (Weßling et al. 2014) marks one of the first steps toward elucidation of pathogen-induced reprogramming of the plant transcriptome. Further, characterization of the mechanisms by which the fungus enforces adjustment of the plant cellular program to promote its intracellular accommodation will be an important aspect of future research (Kuhn et al. 2016).

3.5.5 Transcriptional Programming of Powdery Mildew Pathogenesis

Transcriptional programming during powdery mildew pathogenesis was analyzed by Pandey et al. (2010). The transcriptional reprogramming induced by *G. orontii* during early events of *Arabidopsis* infection, and the role of two closely related WRKY transcription factors, 18, and 40, was observed. It was observed that WRKY18/40 negatively affects pre-invasion defenses in *Arabidopsis,* and deduced a subset of genes that appear to be under WRKY18/40 control. The direct, in vivo physical interaction of WRKY40 to W box containing promoter regions of the key regulatory genes EDS1, and JAZ8, and of the AP2/ERF transcription factor gene, RRTF1 was demonstrated.

Previous transcriptomic studies mainly focused on later stages of powdery mildew infection i.e. haustoria formation (18–24 hpi; (Fabro et al. 2008), or pathogen growth, and reproduction (1–7 days post-inoculation (Chandran et al. 2009). Pandey et al. (2010) found that expression of hundreds of host genes is altered within 8 hpi indicating that signaling between the pathogen, and individual host cells already occurs a few hours postcontact of the spores with the leaf surface at about the stage of penetration, but certainly prior to establishment of the extra-haustorial matrix. It is important to note that *G. orontii* invades only epidermal cells (Lipka et al. 2008). Hence, local gene expression differences within such cells may actually be higher than the values observed using whole leaf samples. Interestingly, nine out of the 24 genes most highly, and differentially expressed (>3-fold) in uninfected wrky18 wrky40 plants encode transcriptional modulators. This implies that WRKY18/40 is central negative regulators controlling distinct downstream transcriptional programs. The role of phyto-hormones and distinct signaling pathways during compatible interactions of *Arabidopsis,* and powdery mildews is not firmly established (Chandran et al. 2009; Consonni et al. 2006; Fabro et al. 2008). Although, WRKY18/40 expression was observed to be dependent on the SA biosynthetic gene isochorismate synthases1 (ICS1) at later stages of *G. orontii* infection (1–7 dpi), induction of these genes at the early infection stage (6 hpi) was ICS1 independent (Chandran et al. 2009). Compared with WT, wrky18 wrky40 plants constitutively express 2.4–8-fold higher levels of five co-regulated JAZ family members, namely JAZ1, JAZ5, JAZ7, JAZ8, and JAZ10, and their promoters contain several W box elements. JAZ proteins are key repressors of JA signaling (Chico et al. 2008). Although Pandey et al. (2010) could not detect in vivo associations of WRKY40 at JAZ7, and JAZ10 promoter sites nor to LOX2, and AOS, but observed specific interaction with a region of the JAZ8 promoter containing two perfect W boxes. Now, only one other transcription factor, MYC2, has been identified as a direct regulator of JAZ promoter activity. However, expression of JAZ7 and 8 does not require MYC2 (Chung et al. 2009). It implies that WRKY40 affects JA signaling by directly controlling the expression of a subset of these negative regulators. Lack of mutants for JAZ7, JAZ8, and JAZ10 currently hampers functional testing of their role in plant–pathogen interaction. Also, the contribution of WRKY18 in this process remains to be elucidated, although it exerts a positive regulatory function, along with WRKY40, in JA signaling (Wang et al. 2008).

The role of RRTF1 in plant defense remains to be rigorously tested, but recent reports indicate its involvement in regulating redox homeostasis during stress, and expression of RRTF1 is dependent on COI1, a central regulator of JA signaling (Khandelwal et al. 2008; Wang et al. 2008). Moreover, RRTF1 is part of a core redox signaling sub-network that also includes EDS1, and WRKY33 (Khandelwal et al. 2008). Possibly, all three of these genes are under WRKY18/40 control. WRKY factors form a functionally interconnected transcriptional network (Eulgem and Somssich 2007), and WRKY18/40 may thus indirectly modulate gene expression by directly controlling other WRKY activities. Indeed, loss-of-WRKY18/WRKY40 function appears to limit the pathogen-induced accumulation of WRKY11, and

WRKY33 transcripts. The PAMP induced expression of WRKY33 was shown to depend on W box promoter elements that are bound by WRKY factors in planta (Lippok et al. 2007). Compared with WT, wrky18 wrky40 mutants displayed rapid transient induction of CYP71A13, and an overall increased accumulation of PAD3 transcripts. WRKY33 binding to the PAD3 promoter in vivo has been demonstrated, but no binding was detected within the CYP71A13 promoter (Qiu et al. 2008). It suggests that CYP71A13 may represent a direct target of WRKY18/ WRKY 40 while PAD3 is an indirect target that is directly controlled by WRKY33 (Pandey et al. 2010).

In general, expression of stress response genes is transient, and their transcription relies on a combination of distinct sets of co-activators, and negative regulators (Lopez-Maury et al. 2008). It remains to be elucidated how WRKY18/40 negatively regulate transcriptional outputs. In the case of RRTF1, and JAZ8, one can envision a direct repressor function on basal gene expression. However, CYP71A13, EDS1, and PAD4 regulation may be different. Expression of these genes in wrky18 wrky40 is not significantly elevated in the absence of pathogen infection. Rather, loss of WRKY18/WRKY40 function results in a pathogen-dependent but exaggerated increase in their transcript levels. A simplistic model would be that other WRKY factors (i.e., WRKY60) could substitute to form weaker but sufficient repressor complexes at these promoters. Replacement of WRKY factors at specific W box sites has been demonstrated for two parsley promoters (Turck et al. 2004). Activation of these genes would require an activator complex that is dependent of pathogen-triggered signaling. Elevated transcript levels observed in the mutant may be the consequence of a higher association of the activator(s) to the transcription complex due to failure of the weaker corepressers to counter this effect. This could imply that binding sites for default repressors and sites for signal-induced activators overlap, a situation often found in inducible eukaryotic promoters (Affolter et al. 2008). Naturally, alternative scenarios are conceivable that may also involve altered regulation at the post-transcriptional level.

The consequences of transcriptional reprogramming should be reflected in qualitative/quantitative altered host metabolite levels resulting in resistance of wrky18 wrky40 plants toward *G. orontii*. Although wrky18 wrky40 mutants accumulated high camalexin levels, and were resistant to this powdery mildew lends support to the importance of this compound in determining the outcome in this interaction. Proteases and endo-membrane components involved in protein quality control, and secretion have been associated with hormone responsive signal transduction events, and disease resistance (Beers et al. 2004; Wang et al. 2005). Several protease genes including *At3g49340* were more strongly upregulated in challenged wrky18 wrky40 mutants than in WT plants. Genes encoding components of the endo-membrane system, and endoplasmic reticulum (ER) response (e.g., CYP71A12, CYP81F2), wall-associated kinase (At2g34500), BiP3 (At1g09080), and type-1 phospho-diesterase (At4g29700)) were also differentially regulated. CYP81F2 encodes a P450 mono-oxygenase essential for pathogen induced accumulation of antifungal compounds (Bednarek et al. 2009). The BiP3 has been associated with the ER response, and its upregulation in wrky18 wrky40-resistant plants intuitively suggests that in susceptible WT plants, *G. orontii* may manipulate part of the host ER response system by impinging on WRKY18/40 functions. Defense activation is often preceded by mitogen activated protein kinase (MAPK) signaling (Colcombet and Hirt 2008). Uninfected wrky18wrky40 plants constitutively expressed five-fold elevated transcript levels of a mitogen-activated triple kinase gene MAPKKK19 (*At5g67080*).

Although wrky18 wrky40 lines overexpressing WRKY40 showed a high degree of susceptibility, penetration efficiency did not reach WT levels. Because WRKY18/40/60 can form homo- and heterodimers (Xu et al. 2006), appropriate WRKY18/40 hetero-dimerization may be partly required for maximal transcriptional reprogramming. Alternatively, or additionally, WRKY18, and 40 may target different subsets of genes. Due to extreme difficulties in generating plants expressing a WRKY40 transgene under its native promoter, interpretation of results relies on ectopic expression of this gene. Despite this caveat, these lines show no obvious phenotypic differences to WT plants, and do not constitutively express defense genes. Moreover, as illustrated by ChIP, overexpressed of WRKY40 shows binding selectivity toward some promoters and even at such sites only to a subset of the overall number of W box elements present.

During the early course of *G. orontii* infection, the JA signaling pathway is required for plant susceptibility, whereas the EDS1 signaling pathway is needed to maintain resistance (Pandey et al. 2010). WRKY18/40 negatively modulates the expression of positive regulators of defense such as EDS1, and the camalexin biosynthetic genes, but positively enhance JA-signaling by partly suppressing the expression of genes encoding negative regulators of this pathway. This enables the pathogen to overcome early host defenses, and ultimately leads to susceptibility. Alternatively, WRKY18/40 regulates induced expression of genes whose products are required for pathogen growth, and reproduction. However, very few such host susceptibility factors enhancing powdery mildew pathogenesis have been identified (Eichmann and Hückelhoven 2008). Moreover, despite the fact that constitutive activation of JA and SA signaling can result in enhanced resistance toward powdery mildew infection, comprehensive studies employing a range of SA, JA, and ET signaling mutants failed to reveal a clear-cut role of these pathways in *G. orontii* resistance. Thus, a model is required in which *G. orontii*, in part via the regulatory action of WRKY18/40, temporally alters the balance between SA and JA signaling at early stages of infections, keeping host defenses suppressed, and thereby allowing the pathogen to gain access to host nutrients. Future investigations will aim at functionally dissecting the WRKY40/18-dependent pathways, and the role of key WRKY40 downstream targets in modulating host plant defenses. Such studies will include extensive evaluation of phytohormone dynamics (especially of JA and SA), analysis of selected mutants, and the generation of transgenic lines modified in the expression of multiple key components identified (Pandey et al. 2010).

3.5.5.1 Transcriptional (Genes) Regulation and Expression in Response to Powdery Mildew Infection

The broad range of transcriptional changes that occur in *Arabidopsis* during powdery mildew infection (Zimmerli et al. 2004; Fabro et al. 2008) reflect both defense responses mounted by the plant as well as possible manipulation of the host by the fungal pathogen. With the sequencing of the *Arabidopsis* genome and availability of novel technologies for transcript profiling (oligonucleotide-based microarray chips) it has been possible to peek into the gene expression changes that occur during the infection process. Several dedicated websites such as Genevestigator (https://www.genevestigator.ethz.ch/) and ATGen Express (http://www.arabidopsis.org/info/expression/ATGenExpress.jsp) provide publicly available collections of microarray data on *Arabidopsis* challenged with a variety of pathogens including powdery mildews, which represent a largely unexplored treasure of information. However, the spatial resolution of microarray data is limiting, since most microarray experiments are performed on whole leaf samples, and as such only provide a gross overview about changes in transcript abundance in pathogen-challenged leaf tissue. Since powdery mildews exclusively colonize the epidermal layer of the leaf, many tissue-specific transcriptional changes are likely to be masked by proportionately larger amounts of mesophyll transcripts. Even within the epidermis, attacked cells, and successfully colonized cells each differ in their transcript profile from nonattacked cells (Gjetting et al. 2004, 2007). Owing to these limitations, expression levels of numerous genes in infected host cells might be considerably underestimated. Employment of techniques that allow transcript profiling of powdery mildew-challenged *Arabidopsis* leaves at cellular resolution, e.g., based on RNA extraction via micro capillaries or by laser micro-dissection (Kehr 2003), is thus urgently required to refine the currently existing data sets (Micali et al. 2008).

Earlier transcriptional changes in *Arabidopsis* compared the response observed in host (*G. cichoracearum*), and nonhost interactions (Bgh; Zimmerli et al. 2004) at 8, 18, and 24 h post-inoculation. Many of the differentially regulated genes displayed a similar response in both host, and nonhost combinations. However, in the nonhost interaction, the response was more rapid, and of greater amplitude than in the compatible interaction, supporting the hypothesis that common defense machinery is activated upon attack by adapted, and nonadapted pathogens, and reinforcing the notion that adapted pathogens are able to partially suppress basal defense in their host species. Several genes that exhibit lower transcript abundance after pathogen challenge, a majority encode proteins that are involved in photosynthesis, metabolism, transport, and transcription/ translation. Consistent with these functions, half of the gene products are either known or predicted to be chloroplast residents. This observation corroborates further studies indicating that *Arabidopsis* suffers up to 30% reduction in the gain of dry weight following Bgh infections (Zimmerli et al. 2004). This decrease is more modest in compatible infections with *G. cichoracearum*. An inverse relationship between defense responses, and plant growth has been also observed in interactions between cereals, and powdery mildews, and under scores the metabolic costs associated with activation of defense pathways (Swarbrick et al. 2006; Wright et al. 1995a, b). Intriguingly, a recent microarray analysis of the compatible interaction between *Arabidopsis* and *G. cichoracearum* revealed that host photosynthesis is likely stimulated by the fungus, through the induction of genes involved in chlorophyll binding, and chlorophyll an oxygenase (Fabro et al. 2008). Though, it is conceivable that an increase in photosynthetic rate is required, as the metabolic demands of the infected epidermal tissue increase, these findings are somewhat contradictory to those obtained in the study of Zimmerli et al. (2004). Manipulation of the host metabolism by the fungal parasite thus likely involves up- as well as downregulation of photosynthetic genes.

However, many gene transcripts related to defense responses accumulate upon both host, and nonhost powdery mildew attack (transcripts encoding glycosyl hydrolases and β-1, 3-glucanase). Consistently, many of the respective genes contain in their up-stream promoter region the cis-regulatory element OCS, which is known to be responsive to auxin, SA, and hydrogen peroxide. Transcripts specifically accumulating during nonhost interactions encode defensins, and derive from genes that are regulated by JA, and ethylene. SA signaling therefore does not appear to play an important role in nonhost defense against Bgh. Alternatively, SA-controlled genes may only become activated once the penetration defenses have failed. However, analysis of the transcriptional response of *Arabidopsis* leaves to chitin (the main fungal cell wall polysaccharide), and chitin derivatives identified several genes involved in defense responses (Ramonell et al. 2002). Genes encoding plant-derived chitinases exhibit higher transcript accumulation upon pathogen challenge compared to control conditions, and chitin oligomers obtained as fungal cell wall degradation products constitute potent signaling molecules responsible for induced resistance to *G. cichoracearum*. This resistance response appears to be activated through signaling routes that are distinct from the SA, JA, and ET pathways (Ramonell et al. 2002). Genes encoding PR5, two WRKY transcription factors, and several disease resistance proteins belonging to the TIR–NB–LRR family were induced upon chitin treatment (Ramonell et al. 2002). Genes with reduced transcript abundance included ETR2, encoding a transmembrane protein kinase with a LRR domain, the defensin PDF2.5, RAR1 (encoding a protein with two zinc-finger-like domains that is required for the accumulation of many R proteins), and several other genes coding for zinc-finger proteins. To functionally validate the microarray experiments, T-DNA insertion mutants within seven of the differentially expressed genes were tested for susceptibility to *G. cichoracearum*. Three of the seven mutants, encoding a putative RING zinc-finger, and two putative disease resistance proteins, displayed enhanced susceptibility to the pathogen, further reinforcing the role of chitin sensing, and signaling in plant defense responses (Ramonell et al. 2002). The critical role played by chitin perception in plant defense responses was recently strengthened by the cloning of the presumptive *Arabidopsis* chitin (co-)receptor LysM Receptor-Like Kinase1 gene (LysM RLK1 or CERK1) by two independent groups (Miya et al. 2007; Wan et al. 2008). The gene codes for a plasma membrane-resident protein with a single trans-membrane domain, three extracellular LysM domains and a cytoplasmic serine/threonine kinase domain (Plate 4.1; Chap. 4). Mutations in this gene abolish virtually all changes in gene transcription normally observed upon chitin perception by the plant, without affecting other elicitor perception such as flagellin sensing, suggesting that the protein is part of the chitin receptor complex (Wan et al. 2008). It is conceivable that in analogy to the FLS2 receptor kinase, which is responsible for flagellin-triggered signaling (Asai et al. 2002), LysM RLK1/CERK feeds into MAP kinase signaling, and defense gene activation (Plate 4.1; Chap. 4). Interestingly, although loss of LysM RLK1 caused enhanced susceptibility to the host fungus *G. cichoracearum* at the postpenetration stage (Wan et al. 2008), it did not affect the plant response to the adapted hemibiotrophic fungus *C. higginsianum,* and resulted in only a marginal increase in susceptibility to the nonadapted necrotrophic fungus *A. brassicicola*. These observations suggest that pathogen perception by the plant is microbe-specific, and implies the existence of more elicitor-specific receptors yet to be discovered (Micali et al. 2008).

In addition to genes involved in defense responses, genes that are induced because they are likely required for pathogen growth have also been identified (Fotopoulos et al. 2003). The observation that *Arabidopsis* leaves infected with *G. cichoracearum* have an enhanced sugar uptake compared to noninfected leaves prompted an investigation of sugar transporter expression upon powdery mildew challenge. *G. cichoracearum* induces expression of STP4, encoding a glucose transporter as well as β-fruct1, coding for a cell wall invertase, in several cell types of the leaf (most notably the epidermis, and vascular bundle) irrespective of whether they are penetrated by the fungus or not. This transcriptional induction is sustained for up to 6 days post-inoculation, and is not reproducible in a nonhost interaction with Bgh. A similar role for sugar transporters in fungal development was also found in the interaction between the rust fungus *Uromyces fabae,* and broad bean (*Vicia faba*). However, in this case, the hexose transporter is of fungal origin, and resides in the haustorial membrane (Voegele et al. 2001), suggesting that exchange of nutrients at the plant/biotroph interface requires transport facilitators on both sides. It is not surprising that sugars, and therefore sugar biosynthesis and transport, would be important for fungal growth, and development, and enhanced by successful fungal pathogens. The mechanisms leading to this genetic and physiological manipulation remain an enigma. However, new efforts into powdery mildew studies promise to lift the veil on the mystery (Micali et al. 2008).

3.5.6 Genes Governing Powdery Mildew Pathogenesis

Plant disease occurs, if a virulent pathogen encounters a susceptible host plant. Since plants are considered basically resistant to most pathogenic microbes, this requires evolutionary adaptation of the pathogen to its host (Jones and Dangl 2006). Host adaptation involves pathogen effector-mediated suppression of defense mechanisms initiated by the plant, e.g., after recognition of pathogen-associated molecular patterns (PAMPs) (da Cunha et al. 2006; Koeck et al. 2011). Biotrophic pathogens often show a highly specific adaptation to their host plant. *Formae speciales* of the cereal powdery mildew fungus *Blumeria graminis* are able to colonize only one specific cereal species, while all others remain nonhosts. This may be explained by the inability of nonadapted powdery mildew fungi to suppress PAMP-triggered immunity, possibly due to an inadequate effector repertoire or timing of delivery or differences in host effector targets for susceptibility (Niks and Marcel 2009). For successful pathogenesis, powdery mildew fungi need an efficient suppression of the plant's preformed and inducible penetration resistance mechanisms as well as maintenance of the compatible interaction after host cell penetration, which may include suppression of programmed cell death (PCD; Huckelhoven and Panstruga 2011). After host cell penetration, the haustorial complex is established, which consists of the fungal haustorium, an extra-haustorial modified plant plasma membrane, and the haustorial matrix in between (Koh et al. 2005; Micali et al. 2011). The haustorial complex represents the site of the most intimate contact of the fungus with its host cell, and likely functions in nutrient uptake, but may also provide a platform for the delivery of fungal effectors (Micali et al. 2011). Components of defense or susceptibility to powdery mildew fungi are conserved among monocotyledonous and dicotyledonous plants. In *Arabidopsis,* and barley (nonhost) penetration resistance to nonadapted powdery mildew fungi involves the functionally homologous soluble *N*-ethyl-maleimide-sensitive factor attachment (SNARE) protein PEN1/ROR2, which form complexes with VAMP721/722, and SNAP33/SNAP 34 (Collins et al. 2003; Douchkov et al. 2005; Kwon et al. 2008). Conversely, one considers the plasma membrane-resident mildew locus O (MLO) protein as a major susceptibility factor for powdery mildew fungi in monocot, and dicot plants. Lack of functional MLO protein(s) results in complete race nonspecific penetration resistance to the barley powdery mildew fungus *B. graminis* f.sp. *hordei* (*Bgh*) in barley (Buschges et al. 1997), and abolishment of host cell penetration, and conidiophore production of *Golovinomyces orontii* or *G. cichoracearum* on *Arabidopsis mlo2/6/12* triple mutants (Consonni et al. 2006). The conserved cell death suppressor protein Bax inhibitor-1 (BI-1) is also required for full powdery mildew susceptibility in barley. Transient or stable overexpression of *BI-1* support penetration of *Bgh* into barley epidermal cells, while transient or stable knock down of *BI-1* expression restrict it (Huckelhoven et al. 2003; Eichmann et al. 2006, 2010; Babaeizad et al. 2009). BI-1 proteins are endoplasmic reticulum-resident proteins, which are suppressors of cell death responses to biotic, and abiotic stress stimuli in plants, and animals (Huckelhoven 2004; Watanabe and Lam 2009; Henke et al. 2011; Ishikawa et al. 2011; Robinson et al. 2011). As well as having six or seven predicted transmembrane domains, BI-1 amino acid sequences harbor the "BI-1-like protein family motif" (NCBI conserved domain database motif cd06181) or "inhibitor of apoptosis-promoting Bax1-related motif" (InterPro IPR006214). This motif spans the transmembrane domains of the proteins, hence represents the overall domain architecture of these proteins and is of unknown function (Huckelhoven 2004). In mammals, various transmembrane BI-1-like protein family motif-containing proteins have been identified, which do not exhibit obvious sequence similarity to BI-1 proteins. Among them are the glutamate-binding protein (GBP, also known as glutamate-binding subunit of a NMDA receptor, GRINA), responsive to centrifugal force, and shear, Golgi antiapoptotic protein (GAAP), and FAS apoptotic inhibitory molecule 2, which is also known as life guard (LFG) (Reimers et al. 2006, 2008; Hu et al. 2009). Comprehensive structure prediction and phylogenetic analyses showed that all of these proteins belong to the evolutionarily conserved LFG protein family in eukaryotes, named after the mammalian member with the best-annotated function (Hu et al. 2009). Some members of the mammalian LFG protein family (human LFG and GAAPs), possess an antiapoptotic activity, which they have in common with BI-1 proteins (Gubser et al. 2007; Reimers et al. 2008; Hu et al. 2009). LFG proteins are conserved in plants (Reimers et al. 2006, 2008; Hu et al. 2009), but plant LFG protein functions are largely uncharacterized. The barley (*Hordeum vulgare* L.), and *Arabidopsis thaliana* genomes harbor at least five *LFG* genes (*HvLFGa–HvLFGe,* and *AtLFG1–AtLFG5*). At least *HvLFGa*, *AtLFG1*, and *AtLFG2* are involved in susceptibility to adapted powdery mildew fungi, further supporting functional similarity between BI-1, and LFG proteins.

Pathogens manipulate eukaryotic cells during invasion, and target plant proteins to achieve host susceptibility. Bax inhibitor-1 (BI-1) is an endoplasmic reticulum-resident cell death suppressor in plants and animals, and is required for full susceptibility of barley to the barley powdery mildew fungus *Blumeria graminis* f.sp. *hordei*. Lifeguard (LFG) proteins resemble BI-1 proteins in terms of predicted membrane topology, and cell-death-inhibiting function in metazoans, but display clear sequence-specific distinctions. The barley (*Hordeum vulgare* L.), and *Arabidopsis thaliana* genomes harbor five *LFG* genes, *HvLFGa–HvLFGe,* and *AtLFG1–AtLFG5*, whose functions are largely uncharacterized. As observed for *HvBI-1*, single-cell overexpression of *HvLFGa* supports penetration success of *B. graminis* f. sp. *hordei* into barley epidermal cells, while transient-induced gene silencing restricts it. In penetrated barley epidermal cells, a green fluorescent protein-tagged HvLFGa protein accumulates at the site of fungal entry, around fungal haustoria, and in endosomal or vacuolar membranes. It suggests a role of LFG proteins in plant–powdery mildew interactions in both monocot, and dicot plants, because stable overexpression or knockdown of *AtLFG1* or *AtLFG2* also support or delay development of the powdery mildew fungus *E. cruciferarum* on the respective *Arabidopsis* mutants (Fig. 3.6). Weis et al. (2013) have identified new modulators of plant–powdery mildew interactions, which support functional similarities between BI-1, and LFG proteins beyond cell death regulation.

3.5.7 Function of Lifeguard Protein (LFG) in Powdery Mildew Pathogenesis

In order to examine the function of AtLFG proteins in powdery mildew susceptibility, powdery mildew disease development on *Arabidopsis LFG* T-DNA insertion mutants was analyzed. Two T-DNA insertion mutants were available for *AtLFG1* (SALK-147263C and SALK-111590), and one for *AtLFG2* (SALK-052507C). The integration and positions of T-DNA insertions were confirmed by PCR. Both *AtLFG1* mutants carry their T-DNA insertion at different positions in the promoter region. The T-DNA position in the *AtLFG2* mutant is in the 3′ UTR. Expression data retrieved from Genevestigator (https://www.genevestigator.com/gv/, Hruz et al.

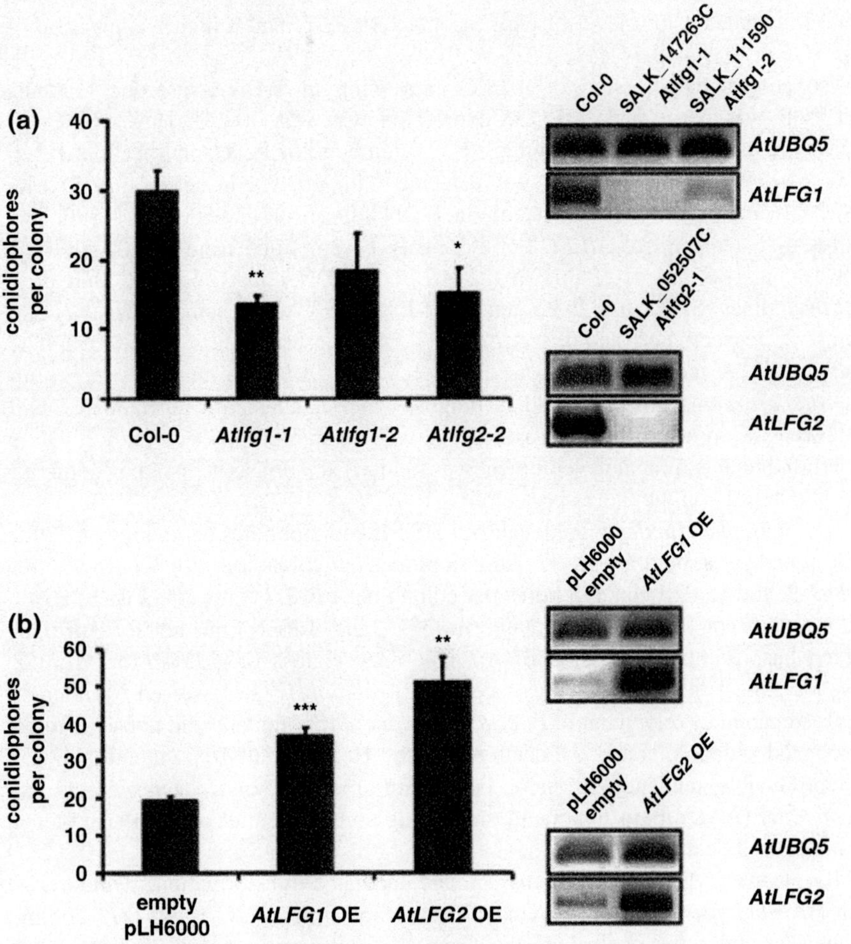

Fig. 3.6 Impact of *AtLFG1* or *AtLFG2* over- or under expression on the outcome of the *Arabidopsis*–powdery mildew interaction. (**a**) Development of *Erysiphe cruciferarum* on 5-week-old Col-0, and *Atlfg* T-DNA insertion mutants at 5 dpi. Conidiophores per colony were counted on five individual plants per mutant, respectively; 50 colonies per line were evaluated. Expression of the full length target genes was analyzed by RT-PCR. Amplification of a *UBIQUITIN 5* fragment indicated similar quantity of template cDNA. The upper panel shows *AtLFG1* expression in wild-type Col-0, *Atlfg1-1*, and *Atlfg1-2* plants; the lower panel shows *AtLFG2* expression in wild-type Col-0, and in *Atlfg2-1*. (**b**) Development of *E. cruciferarum* on 5-week-old *AtLFG* overexpression mutants, and empty pLH6000 vector control plants at 5 dpi. Conidiophores per colony were counted on one leaf of five individual plants per mutant, respectively. *AtLFG1* or *2* expressions in plants transformed with the empty vector pLH6000, and *AtLFG1* or *2* overexpression mutants was examined by RT-PCR. Amplification of a *UBQ5* fragment indicated similar quantity of template cDNAs. The upper panel shows *AtLFG1* expression in empty pLH6000 vector, and *Atlfg1* overexpression plants; the lower panel *AtLFG2* expression in empty pLH6000 vector, and in *AtLFG2* overexpression plants. Values are mean, and standard errors; *, **, and *** indicate significance at $P < 0.05$, 0.01, and 0.001, respectively (Weis et al. 2013)

2008) indicated a low level of expression of *AtLFG1,* and a higher expression of *AtLFG2* in *Arabidopsis* leaves.

To confirm suppression of *AtLFG* expression in T-DNA insertion mutants, RT-PCR was performed. *AtLFG1* expression was very low in leaves of Col-0 wild-type plants as indicated by high cycle numbers for PCR amplification. However, *AtLFG1* expression was not detectable in T-DNA insertion line SALK-147263C and only weakly detectable in T-DNA insertion line SALK-111590. The full-length transcript of *AtLFG2* could easily be amplified from Col-0 wild-type leaves, but was not detectable in the *AtLFG2* T-DNA insertion line. Hereafter, *AtLFG1* lines SALK-147263C and SALK-111590 were named *Atlfg1-1* and *Atlfg1-2*, respectively, and the *AtLFG2* T-DNA insertion line SALK-052507C *Atlfg2-1.* The inoculated *Atlfg1-1*, *Atlfg1-2,* and *Atlfg2-1* mutants were with *E. cruciferarum*, a powdery mildew fungus, which is adapted to *Arabidopsis*, and analyzed the infection by macroscopic as well as microscopy. On *Atlfg* T-DNA insertion mutants, fungal development was reduced at 5 dpi, as assessed by counting the number of conidiophores per colony.

The *Erysiphe cruciferarum* developed 30 ± 3 conidiophores per colony on Col-0 wild-type leaves, but only 14 ± 1 conidiophores per colony on *Atlfg1-1*, 18 ± 5 on *Atlfg1-2*, and 15 ± 4 conidiophores per colony on *Atlfg2-1*. This effect was statistically significant for the null mutants *Atlfg1-1,* and *Atlfg2-1* but not for *Atlfg1-2*, which has residual expression of *AtLFG1*. Hence, loss of *AtLFG1* or *AtLFG2* delayed fungal development on susceptible *Arabidopsis*. As assessed by microscopy, impaired development of *E. cruciferarum* on *Atlfg* mutants did not seem to be associated with accelerated cell death responses. However, all *Atlfg* mutants showed a reproducible tendency to more cell death upon *Alternaria alternate* f. sp. *lycopersici* (AAL) toxin treatment, suggesting some potential of AtLFG proteins to inhibit cell death.

By means of *Agrobacterium tumefaciens*-mediated transformation, stable transgenic *Arabidopsis* plants were created, that express *AtLFG1* or *AtLFG2* coding sequences under the control of the CaMV *35S* promoter. Control plants were transformed with the empty vector pLH6000. Presence of the integrated T-DNA was confirmed by PCR. Transgenic plants showed elevated levels of *AtLFG1* or *AtLFG2* transcripts, respectively, when compared to empty vector control plants as analyzed by RT-PCR. On *AtLFG1* as well as on *AtLFG2* overexpression mutants, *E. cruciferarum* showed an accelerated development when compared to the empty vector control plants. The number of conidiophores per colony increased from 19 ± 1 on empty vector control plants to 37 ± 2 on *AtLFG1,* and to 51 ± 7 on *AtLFG2* overexpressing plants. Another two biological replications gave similar results. Together, data indicate a function of AtLFG1 and AtLFG2 in supporting development of *E. cruciferarum* in a compatible interaction with *Arabidopsis* (Weis et al. 2013).

3.5.8 Gene Expression Levels of Healthy and Powdery Mildew-Infected Plants

The disease is associated with the expression of disease resistance genes in plant. To determine the ability of *B. napus* and *R. alboglabra* to resist *E. cruciferarum* infection during different time periods (1, 2, 4, 6, 8, and 10 dpi), Alkooranee et al. (2015) investigated the expression profile of defense-related genes in plants by qRT-PCR. Six primers for each genotype were used, including PR-1, PDF1.2, PR-2, PR-3, CHI620, and CHI570 genes. The housekeeping gene, GAPDH, was used as reference gene. For both genotypes, the PR-1, and PR-2 genes were used as markers for the SA signal pathway, whereas the PDF1.2 genes was used as a marker for the JA/ET signal pathway, as well as the putative marker genes PR-3, CHI620, and CHI570. The results indicated that most of the genes were amplified (90.0%) after *E. cruciferarum* infection. The gene expression increased over time in both genotypes, which was higher in plants inoculated with the fungus pathogen compared with the healthy plants. The gene expression levels increased in *R. alboglabra* infected compared with noninfected, and with *B. napus*, it showed increased gene expression levels of PR1, and PR2 which were upregulated by 218.86-fold, and 43.08-fold at 1 dpi, respectively, compared with an average of 2-fold increased expression in noninfected (healthy) plants (Fig. 3.7a). The expression levels of the PR-1, and PR-2 genes increased in *B. napus* infected were upregulated by 69.05-fold at 8 dpi, and 6.14-fold at 6 dpi, respectively, compared with an average of 1.2-fold increased expression in noninfected plants (Fig. 3.7b). The PR-3 and PDF1.2 genes were used as markers for the JA/ET signal pathway. The expression of the PR-3 gene increased, and peaked 10.10-fold (1.04- to 10.50-fold) at 10 dpi in infected *B. napus* were upregulated by 67.77-fold (from 0.117- to 7.93-fold) at 8 dpi in infected *R. alboglabra*, respectively, compared with the noninfected ones (Fig. 3.7c). The expression levels of the PDF.1.2 gene showed significant differences between infected, and noninfected plants were upregulated by 8.151-fold (from 1.85- to 15.07-fold) at 8 dpi in *B. napus* infected (Fig. 3.7d). Other defense-related genes were investigated for the comparison between the signal transduction pathways in both genotypes. Thus, qRT-PCR was used to analyze the effects of different times, and infections on the expression of the following putative marker genes: CHI620 and CHI570 (chitinase), which are antipathogenic enzymes. The expression levels of CHI620 gene increased clearly with time, and reached its peak at 1 dpi were upregulated by 58.62-fold (from 1.64- to 96.15-fold) in *B. napus*, and reached 37.29-fold (from 1.21- to 45.13-fold) at 2 dpi in *R. alboglabra* infected compared with the noninfected (healthy) ones (Fig. 3.7e). The expression levels of the CHI570 gene increased clearly with time, and reached its peak at 1 dpi in *B. napus,* and in *R. alboglabra* than in noninfected (healthy), which were upregulated by 40.86-fold (from 1.64- to 66.73-fold) and 59.87-fold (from 1.34- to 80.23-fold), respectively (Fig. 3.7f).

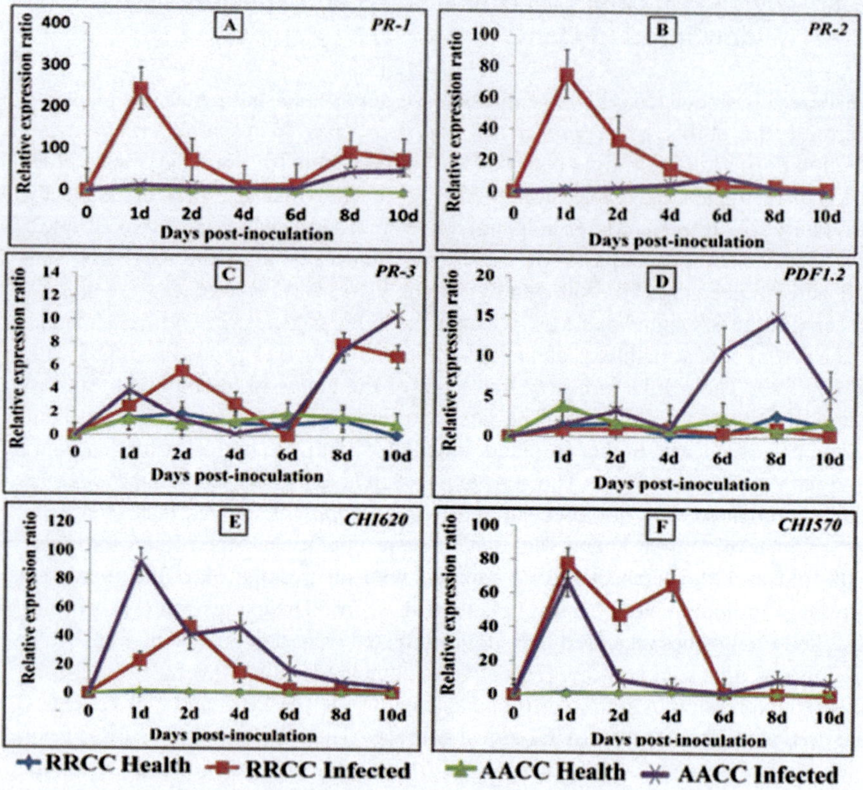

Fig. 3.7 Expression of defense-related genes in three potted for each time of *Brassica napus*, and *R. alboglabra* genotypes of 6-weeks-old inoculated by pressing diseased leaves by powdery mildew onto leaves. Leaves were collected 1, 2, 4, 6, 8, and 10 days' postinfection. Total RNA was extracted, and cDNA was synthesized. Expression levels of the PR-1, PR-2, PDF1.2 (glucanase; BGL2), PR-3 (basic chitinase), CHI620 and CHI570 (chitinase) genes were monitored by RT q-PCR. The expression levels of genes were compared with the expression level of GAPDH (Alkooranee et al. 2015)

3.6 *Brassica–Hyaloperonospora* Pathosystem

3.6.1 Introduction

In crucifers, the downy mildew infection may be either general or local. In the former case, all or most of the leaves, and inflorescence, may bear conidiophores. Although some parts (especially the stem) may show no external injury, microscopic examination shows the mycelia are in the tissues. Generalized infection is restricted to young tissues, and seedlings may show completely infected leaves. Localized infection also occurs in young tissues, especially those still in active division. In the hypertrophy caused by *Albugo*, the cells of the epidermis and cortex are dividing and may readily give entrance to *Hyaloperonospora*. Young inflorescence may wholly or partly be infected, while normal tissues of older stems and leaves below the initial site of infection may remain free.

3.6.2 Genes Governing Pathogenesis

Much non-immunity-related processes are thought to contribute to plant disease susceptibility. Plants often inadvertently facilitate pathogens in establishing disease. Three different levels of susceptibility-enabling processes can be distinguished: (1) attraction and attachment of pathogens to host cells, (2) accommodation of specialized infection and feeding structures inside plant cells, and (3) nutrient production and transport from host to pathogen (Lapin and Van den Ackerveken 2013). One can imagine that mutation of host genes that affect any of these processes leads to plants with reduced disease susceptibility. However, such mutations could also affect basal plant processes that affect the overall physiology of the plant. In genetic terms, susceptibility (*S*) genes can be defined as genes that are involved in disease susceptibility and contribute positively to the infection process and pathogenesis.

In the *Arabidopsis,* several genes for susceptibility to oomycetes have been identified. Several of these are involved in host amino acid metabolism, e.g., the *Aspartate Kinase2/RAR1 Suppressor 1 (AK2/RSP1), Dihydrodipicolinate Synthase2/RAR1 Suppressor 2 (DHDPS2/RSP2)* and *Downy Mildew Resistant 1 (DMR1)*. The *rsp1* and *rsp2* mutants accumulate methionine, threonine, and isoleucine, which are linked to increased resistance to *Hpa* (Stuttmann et al. 2011). Similarly, the *Arabidopsis-dmr1* mutant that is mutated in Homoserine Kinase (HSK) has increased levels of homoserine, which is a common precursor of methionine, threonine, and isoleucine. L-homoserine application was sufficient to make *Arabidopsis* resistant to *Hpa* (van Damme et al. 2009). The resistance in *rsp1*, *rsp2,* and *dmr1* mutants is independent of salicylic acid (SA)-induced defense (van Damme et al. 2009; Stuttmann et al. 2011).

A very different group of susceptibility genes are those that encode negative regulators of immunity. When impaired, for instance by mutation, these regulators are no longer able to suppress certain immune responses and therefore show an enhanced disease resistance. Two examples from *Arabidopsis* are the IMPAIRED OOMYCETE SUSCEPTIBILITY1 (IOS1) and the Downy Mildew Resistant 6 (DMR6) proteins. IOS1 is a leucine-rich repeat receptor-like kinase (LRR-RLK) and is required for full susceptibility to *H. parasitica*. In addition, the *ios1* mutant is also resistant to the oomycete *H. parasitica* (Hok et al. 2011, 2014). The loss of susceptibility of the *ios1* mutant seems linked to ABA hypersensitivity and independent of plant defense mechanisms. Impairment of negative regulation of ABA signaling in *ios1* is suggested to lead to impaired susceptibility to *H. parasitica* (Hok et al. 2014). Also, *Arabidopsis dmr6* mutants are resistant to *H. parasitica,* the oomycete *Phytophthora capsici*, and other biotrophic pathogens (van Damme et al.

2009; Zeilmaker et al. 2015). DMR6 and its close paralog DMR6-LIKE OXYGENASE 1 (DLO1/S3H) are needed for hydroxylation of SA to 2, 5- and 2, 3-dihydroxybenzoic acid, respectively. Overexpression of *DMR6* and *DLO1* leads to depletion of SA; this renders the plants highly susceptible to *H. parasitica* (Zeilmaker et al. 2015).

When analyzing mutants with lesions in genes encoding negative regulators of immunity is whether they truly have a suppressive role in defense. This is particularly important for mutants that show autoimmune phenotypes, also referred to as lesion mimic mutants (Rodriguez et al. 2016). There are several examples of "negative regulator" mutants that can be suppressed by mutations in NLR genes encoding for cytoplasmic immune receptors. For instance, the *Arabidopsis mpk4* mutant exhibits constitutive activation of defense and was classified as a negative regulatory mutant. However, it appeared that the *mpk4* mutation required the presence of the NLR SUMM2 (Zhang et al. 2012). The idea is that the MPK4 protein, which is a target of the *Pseudomonas* effector HopAI1, is guarded by the NLR SUMM2. In the *mpk4* mutant, SUMM2 would be constitutively triggered resulting in an autoimmune phenotype (Herlihy et al. 2019).

3.6.3 Accommodation of Infection Structures during Pathogenesis

Pathogens manipulate the host for accommodation of infection structures, particularly the haustorium which penetrates the plant cell wall to create intimate associations with individual host cells (Lapin and Van den Ackerveken 2013). The haustorium is a specialized feeding structure that is formed by biotrophic and hemibiotrophic oomycetes during plant colonization (Spencer-Phillips 1997; Szabo and Bushnell 2001; Wang et al. 2017). Effectors are secreted from haustoria to the exterior and interior of host cells (Wang et al. 2017). Haustorium formation is initiated when oomycetes secrete hydrolases to locally degrade the plant cell wall and establish an entry point. Microscopy of infected *Arabidopsis* leaves has illustrated the ultrastructure of oomycete haustoria (Mims et al. 2004; Soylu et al. 2004; Lu et al. 2012). Following penetration, haustoria grow into the host cell by invagination of the plant cell membrane. The pathogen haustorial membrane is distinct from the host extrahaustorial membrane and the two membranes are separated by a glycan-rich extrahaustorial matrix. The haustorium itself contains numerous large organelles including vacuoles, mitochondria, and lipid bodies (Plate 3.9a). These structures suggest the haustoria maintain an active metabolism and reinforce the hypothesis that it is the primary site of pathogen feeding (Mims et al. 2004). The extra-haustorial membrane displays a highly convoluted topology, perhaps resulting from vesicle traffic to and from the haustoria.

The signaling and trafficking differences that constitute host resistance versus successful haustorial formation are not well understood. It is clear that many responses originate from the plant, but the pathogen likely co-opts some of these functions for its own purposes. The extrahaustorial membrane is synthesized de novo and lacks many integral proteins typically found in the plant PM (Lu et al.

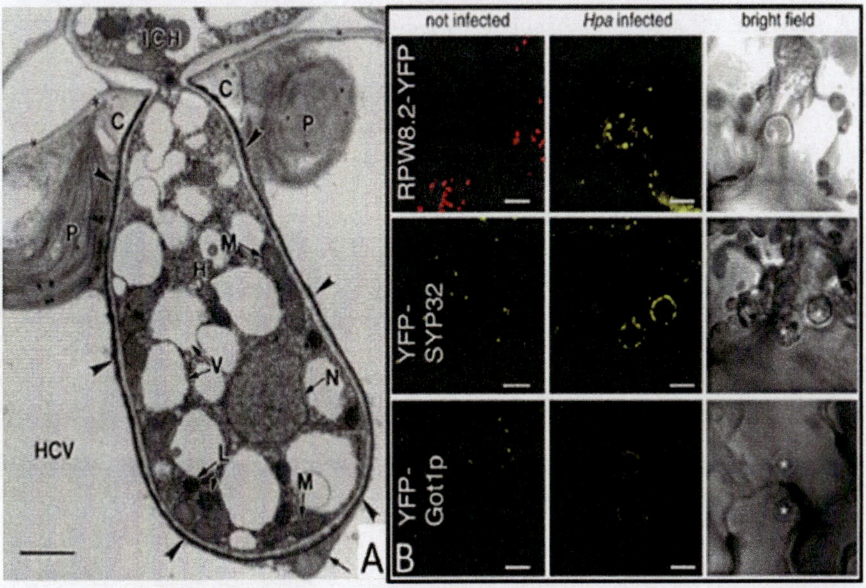

Plate 3.9 Manipulation of host cell structure and metabolism. (**a**) Transmission electron micrograph of fixed *Hpa* haustoria in leaf cells of *At*. Near-median longitudinal section of a haustorium (H) showing its continuity with an intercellular hypha (ICH). Collar of host cell wall material (C). Host cell structures: the cell wall (asterisks), plastids (P), the central vacuole (HCV), and cytoplasm (arrow). Pathogen structures: nucleus (N), mitochondria (M), lipid bodies (L), vacuoles (V), and extrahaustorial matrix (arrow heads). Scale bar = 1 μm. Image reproduced from (Mims et al. 2004). (**b**) Secretory vesicles differentially localize around *Hpa*. Confocal micrographs of *At* transgenic lines expressing the indicated fluorophore fusions show cross-sections of noninfected and *Hpa*-infected leaves at 3 dpi. *Hpa* haustoria are shown in bright field images indicated by asterisks. RPW8.2–YFP, YFP–SYP32 and YFP–Got1p are detected in vesicles around *Hpa* haustoria. Red signals seen in uninfected RPW8.2–YFP leaves are chlorophyll auto fluorescence. Bar = 10 μm. (Image reproduced from Lu et al. (2012))

2012). Some plant defensive proteins localize to the extrahaustorial membrane, including immune receptors such as FLS2 and trafficking proteins such as PENETRATION1 (PEN1, Plate 3.9b). Studies of early *H. parasitica* infection reveals haustorial penetration is accompanied by reorganization of GFP-tagged cytoskeleton and organelles in *Arabidopsis* mesophyll cells (Takemoto et al. 2003; Mims et al. 2004; Lu et al. 2012). This reorganization is observed in compatible and incompatible interactions. Some of these responses are stimulated by plant immune signaling as well as mechanical disturbance (Hardham et al. 2008). However, a successful infection would likely benefit from plant organelles close to haustorium for access to plant cell trafficking and metabolism. Host cell rearrangements are accompanied by accumulation of secretary vesicles, flush with lipids for the new membrane (Takemoto et al. 2003). Other cargo contained in these vesicles remains to be identified. The purpose of these vesicles as defense mechanism or transporter of nutrients to the pathogen is unclear.

A signature of successful *Arabidopsis* resistance to *H. parasitica* and *Albugo* invasion is robust production of cell wall appositions, or encapsulation of haustorium in callose (Donofrio and Delaney 2001). These callose-containing structures are hypothesized to provide physical protection against pathogen feeding structures (Allen and Friend 1983). *Arabidopsis* PLASMODESMATA-LOCATED PROTEIN 1 moves to haustorial sites and stimulates callose production to impede haustorial entry. Suppression of immunity likely includes reduction in *PDLP1* expression and activity, as the gene is not expressed in fully colonized cells (Caillaud et al. 2014). Plants deficient in salicylic acid signaling fail to produce robust callose depositions and often produce callose bands at the neck of haustorium (Donofrio and Delaney 2001). While abundant callose appears to stymie haustorium formation, callose neck bands at the site of cell wall penetration of *H. parasitica* and *Albugo* haustorium may promote haustorial accommodation (Herlihy et al. 2019).

3.7 *Brassica–Leptosphaeria* Pathosystem

3.7.1 Introduction

The *Leptosphaeria maculans*, the fungal agent of blackleg disease (phoma stem canker), causes major yield loss on *Brassica napus* (canola/rapeseed) crops worldwide (West et al. 2001). During infection it remains extracellular and exhibits a range of modes of infection from biotrophy to necrotrophy on *Brassica* hosts. Genome-wide transcriptomic analyses in the *B. napus–L. maculans* pathosystem revealed that the majority of known and predicted effectors had no expression during *in vitro* culture but were highly upregulated during infection, supporting their roles as virulence factors (Haddadi et al. 2016). Resistance against *L. maculans* at the cotyledon stage is race specific. So far, 19 race-specific resistance (R) genes have been reported from *Brassica* species (Larkan et al. 2016), but only two R genes, *LepR3* and *Rlm2*, have been cloned, both encoding membrane-localized receptor-like proteins (RLPs) (Larkan et al. 2013, 2015). The LepR3 perceives *L. maculans* AvrLm1 and triggers *Brassica* defense, leading to hypersensitive response (HR) at the site of infection (Larkan et al. 2013). Transient expression of both LepR3 and AvrLm1 induces HR in *Nicotiana benthamiana* leaves, and SOBIR1 and BAK1 receptor like kinases (RLK), two components of LepR3 complex, are required for the perception of AvrLm1 (Ma and Borhan 2015). Now, up to today seven Avr genes (*AvrLm1, 2, 3, 4–7, 5–9, 6,* and *11*) have been cloned from *L. maculans*, all encoding cysteine-rich proteins, except for AvrLm1, which contains only one cysteine residue (Gout et al. 2006). Cysteine enrichment is a feature found in most of the effectors of apoplastic fungi such as *L. maculans* and is believed to protect the effector protein, by formation of disulfide bonds, from plant proteases released into the apoplastic space during pathogen invasion (van den Burg et al. 2003). Lack of cysteine enrichment in the AvrLm1 protein could indicate that AvrLm1 is translocated inside the plant host cells. Given that LepR3 is a cell surface receptor, recognition of AvrLm1 by LepR3 likely occurs in the apoplast (Gout et al. 2006;

Larkan et al. 2013; Ma and Borhan 2015). However, it is still possible that AvrLm1 is translocated inside the plant cells to modulate host immunity. Owing to technical challenges, host cellular location for a vast majority of effector proteins of fungal and oomycete plant pathogens has not been determined and localization of pathogen effectors in the host plant is often inferred from their structure and their plant target proteins.

Although, AvrLm1 was cloned a decade ago, the host target and molecular mechanism underlying AvrLm1 virulence have remained unknown. Ma et al. (2018) reported the identification of MPK 9 as a novel target of *L. maculans* effector AvrLm1. Stable expression of BnMPK9 in *B. napus* enhances the growth of *L. maculans* by suppressing the salicylic acid (SA) pathway, suggesting that BnMPK9 negatively regulates plant defense. Results indicate that AvrLm1 as a virulence factor functions by stabilizing BnMPK9 and inducing its phosphorylation, leading to enhanced cell death and plant susceptibility to the hemi-biotrophic pathogen *L. maculans*.

3.7.2 Molecular Mechanisms of Pathogenesis

The *Brassica napus* resistance at the seedling stage to *L. maculans* is conferred by race-specific resistance genes. Despite significant advances in defining the genetics of the *B. napus–L. maculans* interaction including isolation of two *R* genes and several *Avr* genes (Raman et al. 2013), up to now limited information about the molecular interaction and function of R/Avr proteins for the *B. napus–L. maculans* pathosystem has been generated. The virulence functions of a *L. maculans* effector by identifying its target in *B. napus,* Ma et al. (2018) identified BnMPK9 as the host target of AvrLm1 and determined their interaction both *in vitro* and in *planta*. The BnMPK9–AvrLm1 interaction enhances cell death caused by BnMPK9, and mutation within kinase domain of BnMPK9 abolishes cell death. The AvrLm1 stabilizes the accumulation and facilitates the phosphorylation of BnMPK9, leading in the development of enhanced cell death in tobacco. Furthermore, overexpression of BnMPK9 perturbs hormone signaling pathways and enhances the growth of *L. maculans* in *B. napus*. Together, these results uncover a novel role for BnMPK9 in plant defense and its modulation by a fungal effector, causing enhanced plant susceptibility.

The *Leptosphaeria maculans* is an intercellular pathogen, and not surprisingly, the LepR3 protein, the *B. napus* receptor for *AvrLm1*, is a membrane-bound RLP. To gain insight into the localization of AvrLm1, where tracked the AvrLm1-mCherry fusion protein expressed in *L. maculans* during infection of *B. napus* cotyledon tissues. There is no deletion of mCherry fluorescent signals inside the host plant cells, but frequently observed accumulation of mCherry signals at the focal points of contact between the fungal hyphae and plant host mesophyll cells. In addition, proof of function of AvrLm1 signal peptide in yeast and enrichment of AvrLm1 protein in the apoplastic fluid of *B. napus* tissue infected with *L. maculans* provided further evidence that *L. maculans* secretes AvrLm1 into the host apoplast. The AvrLm1 encodes a protein with one cysteine residue, which is atypical for the commonly reported apoplastic fungal effector proteins (Fudal et al. 2007), indicating that AvrLm1 may enter into plant cell to avoid the harsh environment of the apoplast. Therefore, lack of discernible red-fluorescent signals inside the host cell may be due to the rapid diffusion of AvrLm1-mCherry protein, reducing fluorescent signals to a level below detection limits. Similar technical limitations have been reported for the translocation studies of fluorescently labeled effectors from other filamentous pathogens such as the RXLR effectors from oomycetes or effectors from *C. higginsianum* (Jing et al. 2016; Kleemann et al. 2012). Alternative localization techniques reported, such as the split fluorescent GFP system, may overcome these technical limitations (Henry et al. 2017; Park et al. 2017).

The physical interaction between AvrLm1 and BnMPK9 results in increased accumulation and enhanced phosphorylation of BnMPK9. So far, there has been no report of a fungal effector directly targeting the MAPK pathway. The only example from filamentous pathogens is the *Phytophthora infestans* RXLR effector PexRD2, which has been reported to interact with host MAPKKKε and suppress MAPKKKε-dependent cell death to perturb plant immune signaling (King et al. 2014). However, accumulating evidences from studies on a variety of bacterial pathogens support that manipulation of components of MAPK kinase signaling cascade is an important virulence strategy of bacterial pathogens (Bigeard et al. 2015; Feng and Zhou 2012). Two known MAPK modules, MKK4/MKK5-MPK3/MPK6 and MEKK1-MKK1/MKK2-MPK4 involved in plant defense signaling, are targeted by different bacterial effectors either to inactivate or enhance their kinase activities. For example, HopAI1 inactivates AtMPK3, AtMPK4, and AtMPK6 by irreversibly removing the phosphate group from the threonine residue of the MAPK activation loop (Zhang et al. 2007). A *P. syringae* effector *HopF2* target AtMKK5, and the *P. syringae* effector AvrB interacts with AtMPK4 and induces the phosphorylation and activation of MPK4 (Cui et al. 2010; Wang et al. 2010). These results provide that a fungal pathogen also targets MAPK modules to manipulate plant defense response. The RLK proteins, SOBIR1 and BAK1, are components of the LepR3 receptor complex and are required for the recognition of AvrLm1 (Ma and Borhan 2015), indicating the importance of MAPK signaling cascade downstream of the LepR3 receptor complex.

Previous studies showed that AtMPK9, together with AtMPK12, functions in ABA- and methyl jasmonate-induced stomatal closure in *Arabidopsis* (Lee et al. 2016; Jammes et al. 2009; Khokon et al. 2015) and they act redundantly. However, Ma et al. (2018) showed that AvrLm1 does not interact with MPK12, proving the specificity of AvrLm1-MPK9 interaction and indicating that in the Brassicaceae, MPK12 is dispensable for the virulence function of AvrLm1. Furthermore, transient overexpression of BnMPK9 in *N. benthamiana* resulted in cell death. However, Ma et al. (2018) did not detect any visible cell death phenotype when BnMPK12 was overexpressed in tobacco. These observations indicated that MPK9's function in inducing cell death and mediating *L. maculans* virulence in *B. napus* is independent of MPK12. The importance of MPKs in regulating plant immune response is well documented (Pitzschke et al. 2009; Popescu et al. 2009). Overexpression of the

N. benthamiana homologs of AtMPK3 and AtMPK6 and *Arabidopsis* MKK4, MKK5, MKK7, and MKK9 cause cell death in tobacco (Popescu et al. 2009; Ren et al. 2006; Yang et al. 2001). Also, overexpression of *B. napus* MAPKKK4 causes upregulation of genes related to reactive oxygen species accumulation, cell death, and defense response (Li et al. 2015a). Plant cell death is an effective mechanism of containing the growth of biotrophic pathogens, whereas induction of cell death is a virulence mechanism employed by hemibiotrophic and necrotrophic fungi to more readily absorb nutrients from the host plant (Ding et al. 2011; Greenberg and Yao 2004). Interestingly, previous study of the *B. napus* transcriptome in response to *L. maculans* infection revealed that the peak expression of AvrLm1 and NLP (necrosis- and ethylene-inducing protein) coincide with the transition from biotrophy to necrotrophy of *L. maculans* during *B. napus* cotyledon infection (Haddadi et al. 2016). Taken together, heightened expression of AvrLm1 by *L. maculans* leading to enhanced induction of cell death through the interaction of AvrLm1 with BnMPK9 could be perceived as a virulence mechanism supporting the transition from biotrophy to necrotrophy to facilitate the acquisition of host nutrients by *L. maculans*.

In the BnMPK9 transgenic *B. napus* plants, transcription of PR-1 and WYKR70 is significantly reduced but expression levels of LOX3 and AOS are elevated, indicating suppression of the SA-dependent pathway and activation of the JA-dependent pathway. However, expression of *LOX2* gene in AvrLm1-overexpressing *Arabidopsis* is downregulated. This apparent discrepancy in LOX gene expression could be attributed to the suggested presence of a cytoplasm- and a chloroplast-localized pathway for JA biosynthesis. LOX2 is reported to be targeted to the chloroplast (He et al. 2002). Although, LOX3 is possibly targeted to the chloroplast, its expression pattern is not necessarily similar to that of LOX2. It was reported that LOX3 was upregulated during the leaf senescence, but expression of LOX2 was significantly downregulated (He et al. 2002). The importance of SA and JA pathways during the *L. maculans–B. napus* interaction has also been previously reported (Becker et al. 2017; Haddadi et al. 2016). In *Brassica napus* cotyledon tissues infected with *L. maculans*, SA-related genes were induced at the initial phases of infection (2–4 dpi), but as the infection progressed and at the initiation of necrotrophic stage, genes related to the JA pathway were upregulated (Haddadi et al. 2016). Thus, Ma et al. (2018) hypothesize that AvrLm1 as a virulence factor facilitates the phosphorylation of BnMPK9, leading to the suppression of SA-dependent pathway and supporting the colonization of *L. maculans* at the initial infection stage. The AvrLm4-7, another *L. maculans* effector, has been reported to act as a virulence factor by suppressing the host SA signaling pathway (Nováková et al. 2016). The AvrLm1 and AvrLm4-7 also contribute to the aggressiveness of *L. maculans* during leaf infection (Huang et al. 2006, 2009). An effector from a fungal plant pathogen targets the MPK pathway to suppress plant host defense. AvrLm1 supports the *L. maculans* infection by stabilizing the accumulation and inducing the phosphorylation of BnMPK9.

3.7.3 Characterization of Genome and Transcriptomes During Interaction Between *Brassica* and *Leptosphaeria* Species

As more fungal genome sequences become available, it is apparent that their complement of genes and transcriptomes reflects fungal life styles. Life styles of plant pathogenic fungi are often classified into three broad categories: biotrophy, where the pathogen feeds from live host cells, necrotrophy, where the host cells are killed ahead of colonization, and hemi-biotrophy, where the pathogen feeds from living cells before switching to a necrotrophic style of growth. These designations are imprecise and as the mechanisms of pathogenicity in a range of fungi are elucidated, life style boundaries become more blurred (Oliver and Solomon 2010). *Leptosphaeria maculans* "brassicae" has numerous "gene for gene" interactions with *B. napus* where by avirulence allele in the fungus renders it unable to attack cultivars with the corresponding resistance gene. This "gene for gene" resistance is expressed in seedlings; five avirulence genes have been cloned so far - *AvrLm1, AvrLm4-7, AvrLm6, AvrLm1,* and *AvrLmJ1*. Only one resistance gene, *LepR3*, has been cloned from *B. napus*.

During infection fungi derive nutrition from the host plant, often by enzymatic degradation of proteins and carbohydrates. These latter enzymes are classed as Carbohydrate-Active enzymes (CAZys) and they often have well characterized domains. As well as providing nutrition, CAZy activity releases cell wall products that can act as DAMPs (Damage Associated Molecular Patterns) that activate the host immune system. The biotrophic plant pathogens including *Hyaloperonospora arabidopsis* have fewer CAZy genes than necrotrophic plant pathogens do, perhaps because biotrophs do not need to digest plant cell walls for nutrition and they must evade the host immune system. Genomic sequences are now available for many Dothideomycetes and an extensive comparative analysis of 18 of them, including *L. maculans* "*brassicae*" has been published (Ohm et al. 2012). These include the *Brassica*-infecting pathogen, *Alternaria brassicicola*, as well as three wheat-infecting pathogens. The *L. maculans* "*brassicae*" genome is compartmentalized into AT-rich -that are gene-poor comprising up to 35% of the genome, and gene-rich regions that have a high GC content, while the genome of *L. biglobosa* "canadensis" is 30 Mb and lacks AT-rich regions. A limited amount of oligo-array transcriptome data has been produced for *L. maculans* "*brassicae*" during *in vitro* culture and infection of *B. napus*, but transcriptome data have not been reported for *L. biglobosa* or any closely related *Brassica* pathogen. Patterns of global gene expression can be generated by RNAseq, a technique that enables analysis of dual transcriptomes; for instance, during a plant pathogen interaction. Furthermore, RNAseqcan be exploited to analyze nonmodel organisms for which genomic resources are not well developed. Few dual transcriptomes for plants and pathogenic fungi have been reported. Lowe et al. (2014) described the genomes and transcriptomes of *L. biglobosa* "canadensis" and *L. maculans* "*brassicae*" and canola (*Brassica napus*), during infection and *in vitro*. The aim was to characterize genes that each pathogen uses to evade detection by the host or to derive nutrition from the host, *viz.*, the carbohydrate-active enzymes and peptidases. The genes upregulated by the host during infection by each pathogen has also been examined.

The *Leptosphaeria maculans* "*brassicae*" is a damaging fungal pathogen of canola (*Brassica napus*), causing lesions on cotyledons and leaves, and cankers on the lower stem. A related species, *L. biglobosa* "canadensis," colonizes cotyledons but causes few stem canker. Lowe et al. (2014) described the complement of genes encoding carbohydrate-active enzymes (CAZys) and peptidases of these fungi, as well as of four related plant pathogens. They also reported dual-organism RNA-seq transcriptomes of these two *Leptosphaeria* species and *B. napus* during infection. During the first 7 days of infection *L. biglobosa* "canadensis," a necrotroph, expressed more cell wall degrading genes than *L. maculans* "*brassicae,*" a hemibiotroph. *L. maculans* "*brassicae*" expressed many genes in the Carbohydrate Binding Module class of CAZy, particularly CBM50 genes, with potential roles in the evasion of basal innate immunity in the host plant. At this time, three avirulence genes were amongst the top 20 most highly upregulated *L. maculans* "*brassicae*" genes *in planta*. The two fungi had a similar number of peptidase genes, and trypsin was transcribed at high levels by both fungi early in infection. *L. biglobosa* "canadensis" infection activated the jasmonic acid and salicylic acid defense pathways in *B. napus*, consistent with defense against necrotrophs. *L. maculans* "*brassicae*" triggered a high level of expression of isochorismate synthase 1, a reporter for salicylic acid signaling.

The *Leptosphaeria biglobosa* "canadensis" infection triggered coordinated shut down of photosynthesis genes, and a concomitant increase in transcription of cell wall remodeling genes of the host plant. Expression of particular classes of CAZy genes and the triggering of host defense and particular metabolic pathways are consistent with the necrotrophic life style of *L. biglobosa* "canadensis," and the hemibiotrophic life style of *L. maculans* "brassicae." Top 20 functional categories regulated during infection by *Leptosphaeria* are presented in Table 3.3 (Lowe et al. 2014).

3.7.4 Green Fluorescent Protein as a Reporter in the *Brassica–Leptosphaeria* Interaction

Green fluorescent protein has been used as a reporter in the *Brassica–Leptosphaeria* interaction causing blackleg disease of oilseed *Brassica* spp. including *B. napus* L. (canola) and *B. juncea* (Indian mustard). Little is known about genes involved in pathogenicity of this fungus. Northern analysis and reverse transcription-polymerase chain reaction (RT-PCR) have been used to detect gene expression during infection, but these techniques are limited by the amount of fungal biomass present in the tissue. Therefore, a visual marker of gene expression in a single fungal spore or hypha would be useful in detecting gene activity early in the infection process. A reporter molecule would also be useful for observing disease progression in general. Previous studies have relied upon time-consuming staining and fixing techniques that do not readily allow observation of live infected tissues. So far, the only reporter

Table 3.3 Top 20 functional categories that are significantly regulated in *Brassica napus* in response to infection by *Leptosphaeria maculans* "*brassicae*" compared to infection by *Leptosphaeria biglobosa* "canadensis" (Lowe et al. 2014)

MapMan function category	MapMan function description	Number of genes	*P*-value
1	Photosynthesis	313	0.00E + 00
1.1	Photosynthesis, light reactions	210	0.00E + 00
1.1.1	Photosynthesis, light reactions, photosystem II	105	0.00E + 00
26	Miscellaneous	365	4.42E − 25
10	Cell wall	116	4.31E − 12
1.1.2	Photosynthesis light reaction photosystem I	44	5.06E − 11
1.1.1.2	Photosynthesis light reaction. photosystem II, PSII polypeptide subunits	63	5.06E − 11
1.3	Photosynthesis Calvin cycle	76	2.19E − 10
1.1.1.1	PS light reactions photosystem II, LHC-II	42	3.08E − 10
1.1.2.2	PS light reactions photosystem I, PSI polypeptide subunits	39	3.56E − 10
20.1	Stress, biotic	123	1.98E − 09
10.6	Cell wall, degradation	28	3.88E − 09
26.12	Misc peroxidases	26	1.07E − 08
26.8	Misc nitrilases, nitrile lyases, berberine bridge enzymes, reticuline oxidases, troponine reductases	42	3.72E − 08
26.10	Misc cytochrome P450	46	1.93E − 07
26.3	Misc gluco-, galacto- and mannosidases	42	1.35E − 06
10.6.3	Cell wall degradation, pectate lyases and polygalacturonases	17	1.74E − 05
35	Not assigned	1969	6.69E − 05
35.2	Not assigned, unknown	1969	6.69E − 05
17	Hormone metabolism	107	8.21E − 05
2.2.1	Major CHO metabolism, degradation sucrose	22	6.47E − 04
20.1.7.6	Stress biotic, PR-proteins, proteinase inhibitors	12	6.63E − 04
1.2	Photosynthesis, photorespiration	27	7.32E − 04
17.5	Hormone metabolism, ethylene	30	7.77E − 04
34	Transport	208	8.58E − 04
33	Development	107	1.62E − 03
26.9	Misc glutathione-S-transferases	32	1.92E − 03
10.8	Cell wall pectin esterases	18	1.92E − 03
17.5.1	Hormone metabolism, ethylene synthesis-degradation	21	1.92E − 03
20.1.7.6.1	Stress biotic, PR-proteins, proteinase inhibitors, trypsin inhibitor	11	1.92E − 03
1.1.3	Photosynthesis, light reactions, cytochrome b6/f	8	2.25E − 03
10.8.1	Cell wall, pectin esterases, PME	13	2.35E − 03
31.4	Cell vesicle transport	32	2.41E − 03
1.3.6	Photosynthesis, Calvin cycle, aldolase	13	2.49E − 03

(continued)

Table 3.3 (continued)

MapMan function category	MapMan function description	Number of genes	*P*-value
1.1.4	Photosynthesis, light reaction, ATP synthase	15	2.55E – 03
29.5.3	Protein degradation, cysteine protease	18	3.11E – 03
29.5.9	Protein degradation, AAA type	15	3.41E – 03
2.2.1.3	Major CHO metabolism, degradation, sucrose invertases	13	4.18E – 03
13.2.3	Amino acid metabolism, degradation, aspartate family	11	5.62E – 03
1.1.5	Photosynthesis, light reaction, other electron carrier (oxidation/ reduction)	18	5.77E – 03

Brassica napus genes expressed during infection at 7 dpi were assigned to functional categories and overall pathway regulation was examined. The top 20 categories determined by MapMan analysis are listed. *PS* photosynthesis, *LHC* light harvesting complex, *CHO* carbohydrate, *PR* pathogen response, *PME* pectin methyl esterase, *ATP* adenosine triphosphate. *AAA type* ATPases Associated with diverse cellular Activities. *P* values indicate the likelihood of the observed pathway regulation being due to chance

gene that has been used in *L. maculans* is β-D-glucuronidase (GUS) from *Escherichia coli*, which relies upon treating leaves with a substrate for the reporter enzyme. Green Fluorescent Protein (GFP) from the jelly fish *Aequorea victoria* is particularly useful as a reporter molecule in living tissues, as it does not require a substrate and is visualized under UV light without further treatment of the tissue.

Various forms of GFP have been used as a reporter in many organisms. However, despite its advantages, GFP has been used in only a few plant–pathogenic fungi. Sexton and Howlett (2001) have transformed *L. maculans* with a gene encoding GFP under the control of two different promoters: firstly, a constitutive promoter from *Aspergillus nidulans*, and secondly a promoter from a regulated *L. maculans* gene, cyanide hydratase (*Cht*). Cyanide hydratase catalyzes the detoxification of cyanide, and transcription of the gene is induced by adding cyanide or propionitrile to liquid cultures. *Cht* messenger RNA is detectable by RT-PCR in lesions on infected cotyledons of *B. napus* and *B. juncea*. The application of GFP in *L. maculans* for following disease progress and for gene expression during infection has been studied. *Leptosphaeria maculans* was transformed with a gene encoding the reporter molecule Green Fluorescent Protein (GFP). The *Aspergillus nidulans gpdA* promoter was fused to the GFP gene to enable constitutive expression, and this was used to follow disease progression. Also, the GFP gene was fused to the *L. maculans* promoter for cyanide hydratase (*Cht*), a fungal enzyme that catalyzes the detoxification of cyanide. Northern analyses and fluorescence image analyses of *L. maculans* transformed with this construct confirmed that GFP transcription occurred under similar conditions to that of *Cht*. GFP was expressed from the *Cht* promoter in cotyledon lesions on *B. napus* and *B. juncea*, and the high level of fluorescence in stem lesions indicated that *Cht* is also transcribed during infection of the stem (Sexton and Howlett 2001).

3.8 *Brassica–Plasmodiophora* Pathosystem

3.8.1 Introduction

The main characteristics of the *Plasmodiophora brassicae* genome are its small size of 24.2–25.41 Mb with high gene density and compactness. The genome contains genes potentially involved in host hormones manipulation. The enzymes (CAZymes) involved in synthesis, metabolism and transport of carbohydrates are present. Mitochondrial genome is variable (34–180 kbp) in size because of variation in the intergenic regions and intron content. There is higher G+C content in mitochondrial sequence than in nuclear genome. Molecular mechanisms of pathogenesis revealed expression of *Y10* gene during plasmodial stages of *P. brassicae*. A trehalose 6-phosphate synthase gene and *PbSTK* L1 genes are expressed during resting spore's formation.

A serine protease (*PRO1*) gene is expressed during host infection as well as in dormant and germinating resting spores. More than 58 genes are upregulated in the secondary zoospores relative to 55 genes downregulated in primary zoospores. The genes *PbPP2A* and *PbHMG* are expressed at the early phases of infection. The gene *BjNIT1* expression in *B. juncea* promotes over production of auxins to form root galls. The main plant hormones pathways modified are cytokinin biosynthesis, auxin homeostasis and salicylic acid, and jasmonic acid metabolism. The methyl transferase gene of *P. brassicae* acts by methylation of salicylic acid, benzoic acid, and anthranilic acids to suppress salicylic acid induced defense in the host plant. Bioinformatics have been used to identify putative effectors/genes of *P. brassicae* for pathogenesis. Transcriptomics analysis has revealed 162 candidates' enzymes/genes involved in lipid metabolism differentially expressed in the *P. brassicae* life cycle. The rate of infection and pathogen development is most rapid in susceptible host in comparison to resistant. The transcriptomic and proteomic approaches have elucidated mechanisms of *P. brassicae* infection and coordination in host reactions. Functional genomics approaches like expressed sequence tags (EST's), serial analysis of gene expression (SAGE), massively parallel signature sequencing (MPSS), next-generation sequencing (NGS), RNA sequencing (RNA-seq), microarray analysis, and construction of suppression subtractive hybridization (SSHcDNA) liberty have been used to achieve better insight into the molecular mechanisms of host parasite interaction and expression of functional genes. Microarray chips, two dimensional electrophoresis and high throughout sequencing techniques like metabotyping, laser micro-dissection, transcriptional profiling, and mi RNA sequencing have been exploited to observe expression of genes regulating metabolic pathways, and signal transduction at early events during pathogenesis. Differential expression of genes in host shoot and root has been observed during *P. brassicae* infection. The members of TF families' viz., MYB, bHLH, and AP2/EREBP are highly up- or downregulated in both shoot and root. However, both shoot and root also responded to pathogen in a number of common ways such as induction of JA, ABA, and shikimate pathways metabolism. The hypoxia gene *PDC1* and *PDC2* are induced during secondary infection in cortical cells of root galls. Several BrSWEET genes of clade 1 and clad III are upregulated on infection suggest close relationship between sugar translocation and *P. brassicae* growth. RNA's like arth-mi R160, arth-miR164 and arth-miR167 are involved in pathogenesis and target auxin-response-factor (ARF) in auxin homeostasis at clubroot development. Immunopropilin gene (*PbCYP3*) is differentially regulated in *P. brassicae* life stages at germinating, and maturing spores and plasmodial formation in infected *Brassica* hosts. Hydrolases and peptidases are the most common differentially regulated secreted proteins in *P. brassicae* infected *B. napus*. The *Pb3-BSMT* gene (benzoic acid/salicylic methyl-transferases) inactivates host defense by converting salicylic acid to methyl salicylate (Saharan et al. 2021).

3.8.2 Molecular Biology of *Plasmodiophora brassicae* Pathogenesis

A search of the National Center for Biotechnology Information (NCBI) database for *P. brassicae*-originated accessions yielded 180 nucleotide sequences, 138 expressed sequence tags, and 53 proteins. Of these sequences, only a small number have a potential importance in pathogenicity, largely based on sequence similarity with genes from other species. Most of these sequences are derived from cDNA libraries constructed by suppression subtractive hybridization (SSH). Among them, there are 76 *P. brassicae* genes isolated from a SSH library constructed between RNA from *P. brassicae*-infected and uninfected *Arabidopsis* tissue (Bulman et al. 2006). Many of the sequences are predicted to contain signal peptides for translocation of the encoded proteins. Using *Brassica rapa* L. clubroot galls, Sundelin et al. (2011) isolated about 140 clones. Half of these clones, which represented 24 unisequences (10 of which are newly characterized), originated from *P. brassicae*. Through the use of next-generation sequencing (NGS) technologies, Burki et al. (2010) generated 3167 *P. brassicae* specific contigs from infected *B. rapa*. These sequences, however, are very short, and their functions or relevance to clubroot development is not known. Feng et al. (2010) constructed a SSH library at the early stages of infection (7 days after inoculation) of canola roots by *P. brassicae*. The library consisted of 797 clones that represented 439 unigenes. Thirty-two of these genes are demonstrated to be of a *P. brassicae* origin. Differential expression of a subset of these genes in the resting spores, and at 7 days after inoculation was observed by quantitative PCR (qPCR), suggesting that some of these genes may play a role during pathogenesis. In addition to these libraries, a few other genes have been identified from *P. brassicae,* and postulated to be related to its pathogenicity. Ito et al. (1997) identified a gene (Y10) whose expression is exclusively correlated with the vegetative plasmodial stage of *P. brassicae*. Brodmann et al. (2002) found a trehalose-6-phosphate synthase gene with expression correlated with an accumulation of trehalose in resting spores. Ando et al. (2006) demonstrated that the expression of PbSTKL1 increases strongly beginning 30 days after inoculation and is coincident with resting spore formation. Two cDNAs, PbBrip9, and PbCC249, are found to be strongly expressed at stages of disease development corresponding to the occurrence of sporulating plasmodia (Siemens et al. 2009b). All these genes may have a potential role in clubroot pathogenesis.

Until now, only one *P. brassicae* gene, a serine protease (PRO1), has been functionally studied, and proven experimentally to be important for pathogenesis (Feng et al. 2010). Southern analysis and specific PCR amplification indicated that PRO1 is a single-copy gene present in a broad range of *P. brassicae* pathotypes. Northern analysis revealed expression of PRO1 during plant infection, as well as in dormant, and germinating resting spores. The expression pattern assessed by Northern analysis was further confirmed by qPCR in a later study (Feng et al. 2013a, b). The optimal temperature of Pro1 is 25 °C, and it exhibits maximum activity at pH 6.2–6.4, and minimal activity at pH 6.8–7.6. These values coincide with the temperature and pH conditions favorable for *P. brassicae* resting spore germination in the field. When Pro1 is used to treat canola root exudates, it enhances the stimulatory effect of the root exudates on *P. brassicae* resting spore germination, indicating that Pro1 plays a role during clubroot pathogenesis by stimulating resting spore germination through its proteolytic activity (Feng et al. 2014). Bulman et al. (2007) identified DNA sequences spanning or flanking 24 *P. brassicae* genes, eventually sequencing close to 44 kb of genomic DNA. Some general characters of *P. brassicae* genes emerged from this preliminary genome survey: (1) *P. brassicae* genes are rich in spliceosomal introns with consensus splice sites, and branch-point sequences similar to those found in other eukaryotes, (2) as a consequence of being intron-rich, splicing is an important feature of gene expression, (3) *Plasmodiophora brassicae* transcription is likely to begin from initiator elements rather than TATA-box containing promoters, and (4) intergenic distances are short, ranging from 44 to 470 bp. Feng et al. (2013a, b) reported a gene expression study in which all database available, *P. brassicae* genes (118) were investigated by dot-blot hybridization and qPCR. The expression of these genes in primary and secondary zoospores was compared. Both dot blot and qPCR identified up- and downregulated genes, and the correlation between these two techniques was confirmed. Quantitative PCR indicated that 58 genes are upregulated in the secondary zoospores relative to the primary zoospores, whereas 55 are downregulated.

3.8.3 Gene Functions at the Host–Pathogen Interaction

Genomic functions of *P. brassicae* are hampered by both the small number of *P. brassicae* genes isolated, and the limited homology of many *Plasmodiophora* sequences with known genes. The approximately 100 partial cDNA fragments or full-length clones of *P. brassicae* have been isolated up to now. About half of these *P. brassicae* genes, a biochemical function are strongly indicated by DNA or amino acid similarities with characterized genes of other organisms, but until now, no function for a *P. brassicae* gene has been proven experimentally. The remaining DNA fragments from *P. brassicae* reveal little or no similarity to sequence data in the databases, reflecting the paucity of genomic data available for protists from the Cercozoa (Keeling et al. 2005). Nucleotide or amino-acid-sequence similarity with known genes from other organisms can give a hint as to gene function, but the presence of a protein domain in a sequence might also be misleading.

The small roots of *A. thaliana* facilitate the observation of clubroot symptoms in this host plant, revealing a reasonably reproducible pattern of disease development under controlled environmental conditions (Kobelt 2000; Kobelt et al. 2000; Siemens et al. 2002, 2006). Correlating transcription with distinct stages of symptom development caused by the pathogen will be an opportunity to define landmarks, and milestones of disease development with a possibility for gaining insights into the function of *P. brassicae* genes. A prominent actin cytoskeleton has been detected by immune-histochemistry over that part of the life cycle after secondary infection (Kobelt 2000). The PbActin and PbGST products appeared strongly expressed, and well detectable over the whole infection cycle within *A. thaliana* roots, and can therefore be used as indicators for an actively growing pathogen (Siemens et al. 2009a, b).

Two unidentified single-copy genes, *PbBrip9*, and *PbCC249*, were strongly expressed at disease stages corresponding with the occurrence of sporulating plasmodia (Siemens et al. 2009b), whereas PbSTKL1 transcription coincided with resting spore formation (Ando et al. 2006). The expression of gene fragment Y10 was exclusively correlated with the vegetative plasmodial stage (Ito et al. 1999), and the transcription of PbTPS correlated with an accumulation of trehalose (Brodmann et al. 2002). Finally, preliminary evidence indicates that the genes PbPP2A (protein phosphatase 2A), and PbHMG (DNA binding protein with high-mobility group box) are preferentially expressed in the early phases of infection in *B. rapa* roots (Sundelin 2008). The gene *Bj N1T1* expression in *B. juncea* during infection promotes over production of auxins to form galls in infected roots (Liu et al. 2018b). Altogether, the very first molecular landmarks of clubroot disease development in *Arabidopsis* are established. The whole genome transcription data from the host plant *A. thaliana* has revealed new insights into clubroot development (Siemens et al. 2006; Jubault et al. 2008; Ludwig-Muller et al. 2009). Combining *Arabidopsis* microarray data with expression analyses of *P. brassicae* genes will provide a strong new approach to clubroot research in the future.

Plasmodiophora brassicae probably secretes proteins into host cells to alter their structure, as with other biotrophic pathogens. Predicted proteins with signal peptides from *P. brassicae* are thus of much interest (Sundelin 2008). The *P. brassicae* sequences with signal peptides known to date, two matched with functionally characterized genes with roles in pathogenicity, namely, PbPPI1, which contains an FKBP family peptidyl prolyl cis-trans isomerase domain, and PbPDA1, which contains a "Nod B-type" chitooligo-saccharide deacetylase domain (Bulman et al. 2006). Surface-exposed or secreted PPIase domain-containing proteins have been found in several intracellular pathogens (Engleberg et al. 1989; Moro et al. 1995). The chit oligosaccharide deacetylase family includes members known to deacetylate substrates, including xylan, and chitin, which might play roles in cell wall penetration during myxamoeboid movement of secondary plasmodia. The intracellular pathogen *P. brassicae* lives in close contact with the host, therefore, using a plant-derived RNAi strategy might work when an important parasitism gene of *P. brassicae* is identified (Siemens et al. 2009a, b).

3.8.4 The *Plasmodiophora brassicae* Genome and Life Stage-Specific Transcriptomics

The *Plasmodiophora brassicae* genomes are available, including one reference genome from a European isolate, and additional re-sequencing of five Canadian, and one Chinese isolates. The pathotype classification of these isolates is difficult to compare because these have been characterized using different differential hosts sets. The main characteristics of the *P. brassicae* genome were described, such as its small size (24.2–25.5 Mb), and high gene density. Thirteen chitin synthases were described, suggesting an important role for this gene family in resting spore formation. Secreted proteins, described to play a role in plant pathogenic organisms in suppressing the plant defense responses, and modifying the host metabolism, were previously predicted in silico in *P. brassicae*, but for most of them, no putative function could be assigned. The *P. brassicae* genome also contains genes potentially involved in host hormone manipulation, such as the auxin-responsive Gretchen Hagen 3, isopentenyl-transferases, methyl-transferase, and cytokinin oxidase. The clubroot genome also appears to be lacking several metabolic pathways, a characteristic of the eukaryotic biotrophic plant pathogens. The missing genes encode proteins involved in sulfur, and nitrogen uptake, as well as the arginine, lysine, thiamine, and fatty acid biosynthesis pathways. In addition, only a few carbohydrate active enzymes (CAZymes), involved in the synthesis, metabolism, and transport of carbohydrates, have been observed (Bi et al. 2016; Rolfe et al. 2016; Spanu et al. 2010).

The current life stage-specific transcriptomics data related to the molecular events in the life cycle of *P. brassicae* was summarized by Schwelm et al. (2016). During spore germination, and at the primary zoospore stage, enzymes involved in chitinous cell wall digestion are highly expressed. Several metabolism pathways such as the starch, citrate cycle, pentose phosphate pathway, as well as pyruvate, and trehalose metabolism were also very active. Although the molecular mechanisms involved in primary root hair infection are unknown, the secondary phase has been better described. During re-infection by the secondary zoospores in root hairs or the root cortex, genes involved in basal, and lipid metabolism were highly expressed. The plant clubroot symptoms (hyperplasia and hypertrophy of the infected tissues leading to the gall phenotype) were found to result from the deregulation of the host plant primary, and secondary metabolism as well as from the modification of plant hormone homeostasis by the pathogen. The main plant hormone pathways modified were cytokinin biosynthesis, auxin homeostasis, and salicylic acid, and jasmonic acid metabolism. However, the mechanisms by which the pathogen may drive plant hormone homeostasis are still not completely worked out. At the end of the *P. brassicae* life cycle, secondary plasmodia transform into resting spores. In these structures, trehalose could play an important role for their long-term survival capacity. The description of the current transcriptomics data in *P. brassicae* shows that some important steps in *P. brassicae* developmental stages are still not well-characterized at the molecular level (Siemens et al. 2006; Wagner et al. 2012; Ludwig-Muller et al. 2009, 2015).

The genome of eH, a second European isolate was presented by Daval et al. (2018). Refinement of the annotation of the eH genome led to the identification of the mitochondrial genome sequence, which was found to be bigger than that of *Spongospora subterranea*, another plant parasitic Plasmodiophorid phylogenetically related to *P. brassicae*. New pathways were also predicted, such as those for the synthesis of spermidine, a polyamine upregulated in clubbed regions of roots. A *P. brassicae* pathway genome database was created to facilitate the functional study of metabolic pathways in transcriptomics approaches. These available tools can help in understanding the regulation of *P. brassicae* metabolism during infection, and in response to diverse constraints.

3.8.5 Functional Characterization of Genes During Pathogenesis

In order to manipulate plant defenses and enable pathogen colonization, many eukaryotic biotrophic plant pathogens have evolved advanced strategies to deliver effector proteins into the host cell during infection. Successful infection relies primarily on the success of the release of the effectors, which in many cases are responsible for the suppression of plant immunity. The initial recognition of conserved microbial features, known as pathogen-associated molecular patterns (PAMPs), leads to PAMP-triggered immunity (PTI) in the host. PAMP-triggered immunity is a first level of immune response that can be overcome by effector proteins produced by adapted pathogens. Resistance(R) to adapted pathogens is achieved through specific recognition of effectors, also known as avirulence proteins, by corresponding R proteins produced by the plant host. This effector-R protein recognition constitutes the second level of immune response, effector triggered immunity (ETI). *P. brassicae* is a well-adapted pathogen of *Brassica* hosts; though indirect evidence suggests lack of either response, both PTI and ETI have not been well-characterized in the host- *P. brassicae* pathosystem (Dodds and Rathjen 2010; Jones and Dangl 2006; Siemens et al. 2011).

Understanding and identifying the proteins that comprise the secretome of *P. brassicae* is an important step toward identifying the complementary R-proteins in the plant host. Putative effectors from biotrophic plant pathogens such as oomycetes and fungi are emerging from the sequencing and assembly of their genomes or transcriptomes, followed by comparative analysis with candidate effector genes. This strategy has been followed by Schwelm et al. (2015) and Rolfe et al. (2016) to identify putative effectors released by *P. brassicae* during the infection process. However, the obligate parasitic nature of *P. brassicae* has made it impossible to obtain a complete genome, with six draft genomes barely annotated for five Canadian path types and one European pathotype. Transcriptomes of *P. brassicae* infecting *B. napus* and *Arabidopsis thaliana* in the Canadian study, and infecting

B. rapa, *B. napus*, and *B. oleracea* var. *capitata* in the European study, have identified some candidate effectors. However, while the information obtained in both studies is a good start, the effectors specifically responsible for *P. brassicae* infection and subsequent disease progression have still to be identified (Perez-Lopez et al. 2018). Therefore, the overexpression of a benzoic acid/salicylic acid methyl transferase-encoding gene [PBRA_T000444 in *P. brassicae* European pathotype 3 (Pbe3), PbPT3Sc00026_A_1.308_1 in *P. brassicae* pathotype 3 (Pb3) draft genomes; Genbank accession number AFK13134] during the second week of infection, with expression peaking 3 and 4 weeks after infection was analyzed. Salicylic acid (SA) is essential for the activation of plant defense. In the plant cell, regulation of active SA is managed through the maintenance of different inactive forms of SA, such as methyl salicylate (MeSA). The methyl transferase identified in *P. brassicae* has been characterized in detail by Ludwig-Muller et al. (2015), identifying its role in the methylation of salicylic, benzoic and anthranilic acids, thereby contributing to the suppression of the salicylic acid-induced defense in a plant host.

3.8.6 Effector Proteases

The basic criteria to identify candidate secreted effector proteins are: proteins with a signal peptide (within the initial 60 amino acids at the N-terminus), no trans-membrane domains, small size between 300 and 450 amino acids, and mostly species-specific. These parameters have been used in the identification of putative *P. brassicae* effectors (Table 3.4). In addition, several other characteristics, motifs, and domains have been associated with effector proteins and have been used to improve identification and functional characterization of *P. brassicae* effectors/ genes as follows:

Table 3.4 Steps to identify putative effectors within the secretome of *Plasmodiophora brassicae* in the European strain Pbe3 and the Canadian strain Pb3 (Perez-Lopez et al. 2018)

	Europe strain (Pbe3)	Canadian strain (Pb3)
Signal peptide/No transmembrane domain[a]	533[b]	590[b]
D score >0.7	NA	431
Size (Small secreted proteins)[c]	416	221
Overexpression[d]	300	NA
Plant-free library[e]	92	NA

NA Not applied

[a] In the Canadian strain, other subcellular localization signals were also used to remove putative proteins from further analysis

[b] Putative secreted proteins

[c] For the European strain the cutoff was <450 aa, while for the Canadian protein proteins <300 aa were selected

[d] At least 10 expected fragments per kilobase of transcript per million fragments (FPKM)

[e] FPKM $\log_2$ fold change >5 in plant-free library

3.8.6.1 A-Cysteine-Rich Proteins

Cysteine rich small proteins have been identified as effectors in several plant pathogens, especially fungi, such as *Cladosporium fulvum* (syn. *Passalora fulva*), an asexual extracellular fungal pathogen of tomato. In *C. fulvum*, cys-rich effectors can inhibit and protect against plant hydrolytic enzymes, such as proteases, glucanases, and chitinases. Cys-rich small-secreted proteins have also been identified as major effectors in the obligate biotrophic pathogens *Melampsora larici-populina*, and the Asian soybean rust fungus *Phakopsora pachyrhizi*, where one of the cys-rich small proteins identified as an effector has been shown to suppress plant immunity. However, its role in *P. brassicae* is still not known.

3.8.6.2 The Rxlr Motifs

The motif RxLR, arginine-any amino acid-leucine arginine, has been identified in the N-terminal of some oomycete and fungal effectors. Although, the function of the RxLR motif in effector proteins remains unclear, it has been shown to be necessary for translocation into the host cell and to elicit immune responses in plant cells. Curiously, despite the divergence between *P. brassicae* and oomycetes, RxLR motifs have been reported in effectors in both the *P. brassicae* Canadian strain Pb3 and the European strain Pbe3, but the IDs of these RxLR protein-encoding genes have yet to be determined. Many putative effectors containing the RxLR motif also contain the second conserved motif, DEER (aspartate-glutamate-glutamate arginine), located toward the C-terminus (Yaeno et al. 2011).

3.8.6.3 Chitin-Binding Domains

Chitin, a recognized microbial PAMP, is a major structural component of fungal cell walls. Some fungal effectors have been shown to contain chitin binding domains that are able to protect the pathogen against plant chitinases. These effectors can also act as scavengers of chitin fragments released by the pathogen during infection, thereby, avoiding a PAMP-triggered immunity response by the host plant. The resting spores of *P. brassicae*, formed at the end of the pathogen lifecycle, contain chitin in their cell walls. The presence of the carbohydrate/chitin-binding (CBM18) domain, enriched in the plasmodiophorid secretome, suggests that these putative effectors might be involved in the formation and possibly the germination of resting spores. A blastp (https://blast.ncbi.nlm.nih.gov) search with a chitin-binding domain protein, identified in the genome of Canadian strain Pb3 (PbPT3Sc00048_S_5.266_1), showed identity with a *Fusarium fujikuroi* chitinase (Genbank accession number CCT72994). In thePbe3 genome, chitin recognition proteins, like the protein encoded by the PBRA_002543 genes have also been identified (Schwelm et al. 2015).

3.8.6.4 Protease/Protease Inhibitors

Sequence identity between plant pathogen effectors and other protein sequences is often low such that the assessment of functions based on putative orthology alone has been limited. In many cases, the three-dimensional structure of the protein, the disulfide bond pattern, and the cysteine spacing have been used to identify protease and/or protease inhibitors as putative effectors in obligate biotrophic soil-born plant pathogens. These effectors target host proteins/proteases during infection, thereby, manipulating the host response to infection. During resting spore formation, Pbe3 overexpresses releases Kazal-like (PBRA_001430), and papain protease inhibitors. The Kazal family of serine protease inhibitors, characterized by the presence of ten cysteine residues including the characteristic CX7CX6YX3CX2–3C signature, has been reported as effectors in fungi and oomycetes.

A putative serine protease (Gene Bank accession number AM411657) was identified among the *P. brassicae* genes that were expressed during infection. This protease carried a predicted signal peptide sequence but lacked homologs in other plant pathogens. Further, studies identified the serine protease as Pro1, a member of the S28 family of proteases that, due to its proteolytic activity, may play a role during clubroot pathogenesis by stimulating resting spore germination. Curiously, none of these studies suggested that this protein was an effector, probably because its suggested role occurs outside of the plant cell (Feng et al. 2010).

3.8.6.5 Nuclear Localization Domains

Differing from many fungi and oomycetes, *P. brassicae* is an intracellular pathogen. Another criterion used to identify putative effectors from obligate intracellular pathogens has been the presence of nuclear localization domains, which allows effectors to directly modulate plant gene expression. The effectors capable of migrating to the plant cell nucleus have only been described in bacteria for many years, but more recently these motifs together with nuclear localization of effectors has been described in fungi and nematodes. In bioinformatics pipelines designed to identify putative effectors, the inclusion of steps to remove proteins containing sub-cellular localization signals will remove these effectors, although researchers could analyze these amino acid sequences directly using the online tool Target P 1.1 Server (http://www.cbs.dtu.dk/services/TargetP). To date, none of the studies on the secretome and putative effectors of *P. brassicae* have detected or identified secreted proteins with such domains and none of the putative secreted proteins reported for *P. brassicae* have been identified as containing nuclear localization domains (Schwelm et al. 2015).

3.8.6.6 The Pexel Motif

The *Plasmodiophora brassicae* is evolutionarily closer to the malaria parasite, *Plasmodium* spp., than to oomycetes or fungal pathogens. While there are many differences between the immune system of animals and plants, they both share the common characteristic of being targeted by pathogen effectors. The discovery of the Pexel motif (Plasmodium export element) was a ground breaking finding that contributed to the understanding of the infection process of Plasmodium. Pexel is a pentameric motif present in the N-terminal portion of all the proteins translocated through the parasitophorous vacuole membrane. It is comprised of a positively charged, hydrophilic amino acid in position one (Arg or Lys), a hydrophobic amino acid in position three (Leu or Ile), and another less conserved amino acid in position 5 (predominantly Asp, Glu, or Gln), with noncharged amino acids in positions two and four (Ser, Thr, Cys, Met, Asn, or Gln). The N-terminal domain of an effector protein from the soybean cyst nematode *Heterodera glycines*, containing unique sequence similarity to domains of an effector of *Plasmodium* spp., indicates that the use of analogous effectors by highly diverse parasites of plants and animals occurs and is worth exploring (Marti et al. 2004).

3.8.6.7 Plant Pathogenic Plasmodiophorids

The *Plasmodiophora brassicae* is not the only plant pathogenic protists, nor is it the only plasmodiophorid affecting economically important crops: the group includes *Spongospora subterranea*, the causal agent of powdery scab on potato tubers, and *Polymyxa* spp., *Polymyxa graminis* and *Polymyxa betae*, which affect graminaceous plants and Chenopodiaceae plants, respectively. Genomic sequences, although limited, are available from *S. subterranea* and comprehensive *S. subterranea* transcriptomic datasets are available from root galls. These data suggest intron-rich genes and an enrichment of chitin-related enzymes in the *S. subterranean* transcriptome. Transposable elements are more expressed in *S. subterranea* than in *P. brassicae*, but evidence for recombination in *S. subterranea* is limited and there is little understanding of sexual recombination in phytomyxids. A study of secreted proteins in *S. subterranea* has been carried out, identifying the benzoic acid/salicylic acid methyl transferase-encoding gene overexpression (Schwelm et al. 2015, 2018).

3.8.6.8 Role of Effectors (Genes) in Clubroot Life Cycle

Clubroot has three main stages to its life cycle, (1) survival in the soil and germination of resting spores, (2) root hair or epidermal cell infection, and (3) cortical infection, although the pathogen has also been observed in phloem (Kageyama and Asano 2009). The serine protease, Pro1, is thought to be involved in step 1, the germination of resting spores (Feng et al. 2010). It is from step 2 to 5 of the life cycle, it is expected that other proteases and protease inhibitors, such as cys-rich proteases and methyl transferases will be produced to suppress plant immunity, thereby preventing the host plant from mounting responses such as programmed cell death and increasing the probability of success of the infection process (Siemens et al. 2011). During the formation of primary and secondary plasmodia, infected root tissues develop into swollen galls; it is expected that *P. brassicae* will secret an array of effector proteins triggering growth, expansion and differentiations of infected host cells. During zoospore maturation, as expected, *P. brassicae* effectors with chitin-binding domains will express to remove chitin fragments that otherwise could trigger PAMP–associated immunity during late stages of the *P. brassicae* life cycle (Fig. 3.8). Based on previous studies, might be also expected overexpression of effectors implies on manipulating plant defense such as PbBSMT (Malinowski et al. 2012; Ludwig-Muller et al. 2015) and effectors disturbing plant meristematic activity during the formation of secondary zoospore and secondary plasmodia. While there are still steps in the life cycle of *P. brassicae* that remain to be clarified, the study of effectors and their roles in life cycle completion and clubroot disease progression is an arena waiting to be explored (Perez-Lopez et al. 2018).

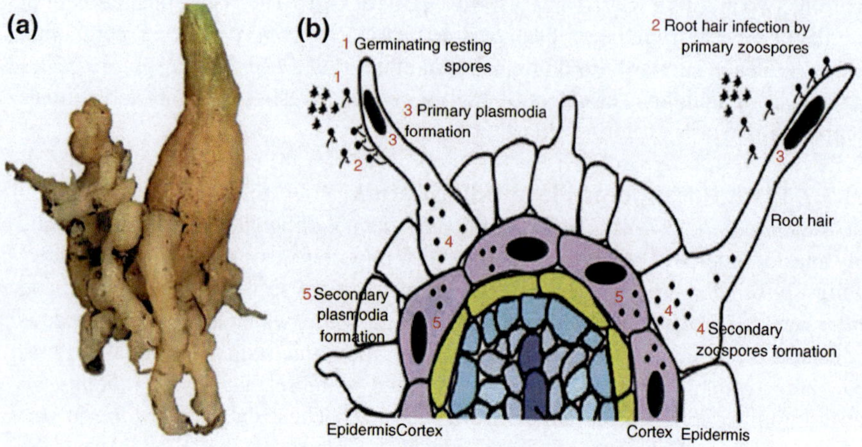

Fig. 3.8 *Brassica* plant root affected by *Plasmodiophora brassicae*. (**a**) Canola root with typical galls after 1 month of inoculation with *P. brassicae* resting spores. (**b**) Life cycle of *P. brassicae* showing the steps involved in infection through to the production of secondary plasmodia in the host plant cortical cells. Scheme based on that of Kageyama and Asano (2009), representing spindle-shaped resting spores, biflagellate primary zoospores, zoospores, and primary and secondary plasmodia (oval black figure in root hairs and cortical cells, respectively). Further steps in *P. brassicae*'s life cycle, such as the formation of resting spores in cortical cells and its ejection to the soil, are not shown in this scheme (Perez-Lopez et al. 2018)

3.8.6.9 A Coherent Pipeline to Identify Effectors in Future

Computational prediction is an excellent starting point to screen for putative effectors, to identify functional domains, and to help us understand the evolution, distribution, and characteristics of effectors. Utilizing the information available, Perez-Lopez et al. (2018) have designed a coherent pipeline aimed at identifying putative effectors involved in the infection process by *P. brassicae,* and subsequent clubroot disease progression in its plant host (Fig. 3.9). The pipeline makes use of tools to identify the origin of reads, and identify motifs, and functions of the predicted secreted proteins. The bioinformatics tools referred to in the pipeline, together with their descriptions and websites, are listed in Table 3.5. This pipeline is only a suggestion, although key parameters should be set in order to run it properly.

Bioinformatics analysis is the route to the identification of candidate effector proteins, but laboratory confirmation of function will always be required. Validation of the RNA-seq data through qPCR or ddPCR, identification of the sub-cellular localization of the candidate effector proteins through transient expression or stable transformation and identification of the plant proteins interacting with the pathogen effectors are some of the logical next steps toward the identification of *P. brassicae* effector proteins (Bulman et al. 2018; Irani et al. 2018).

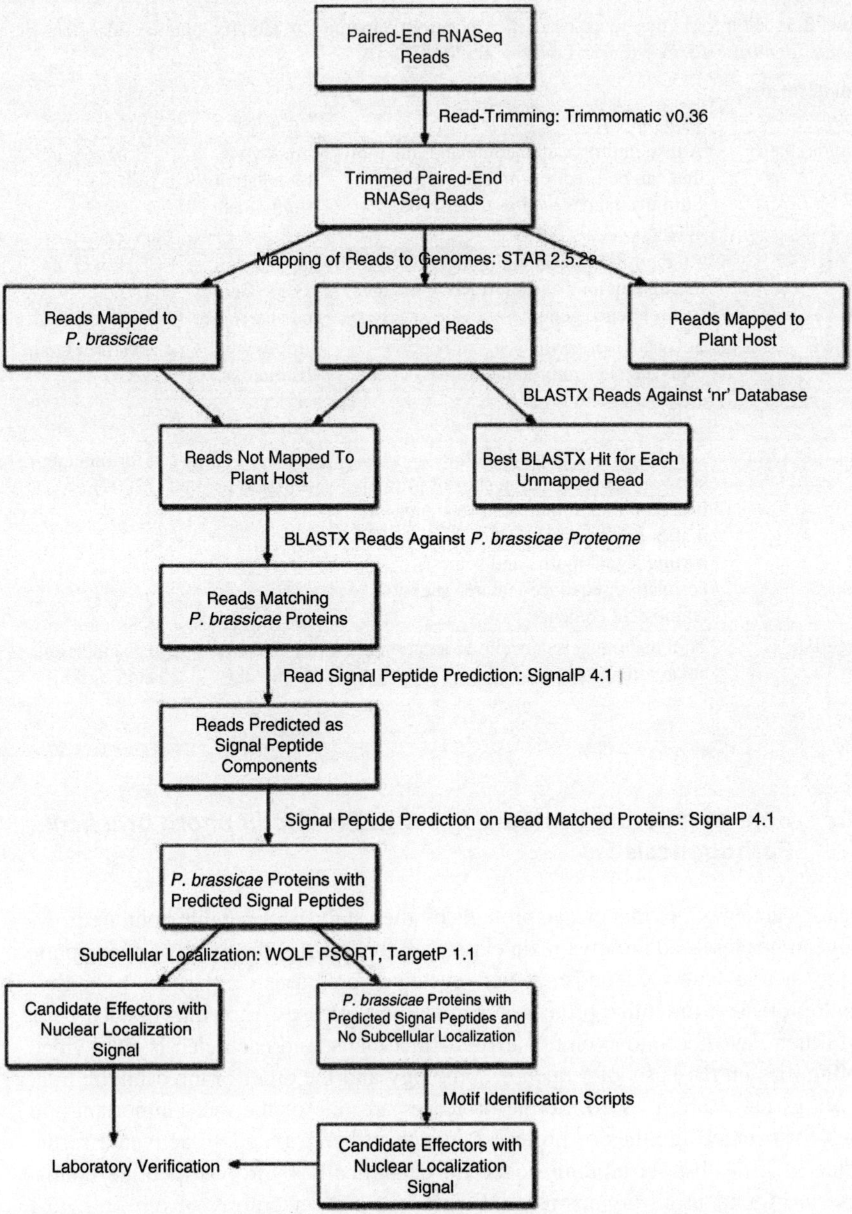

Fig. 3.9 Coherent pipeline to identify putative effector proteins of *Plasmodiophora brassicae*. The pipeline assumes: (1) Researchers are starting with RNA-Seq reads from a host plant infected with *P. brassicae*; (2) The draft genomes available for *P. brassicae*, other plasmodiophorids, oomycetes, and other plant pathogens are used; (3) Motifs mentioned in this chapter or structural similarities with previously described effector proteins were identified (Perez-Lopez et al. 2018)

Table 3.5 Bioinformatics tools used in a coherent pipeline to identify putative effectors of *Plasmodiophora brassicae* (Perez-Lopez et al. 2018)

Bioinformatics tool	Description	Website	References
Trimmomatic	A fast, multithreaded command line tool that can be used to trim and crop Illumina data as well as to remove adapters	http://www.usadellab.org/trimmomatic	Bolger et al. (2014)
STAR	An algorithm for aligning high-throughput long and short RNA-seq data to a reference genome	http://code.google.Com/p/rna-star	Dobin et al. (2013)
SignalP	Software that predicts signal peptide cleavage sites in proteins from eukaryotes and prokaryotes	http://www.cbs.dtu.dk/services/SignalP	Petersen et al. (2011)
WoLF PSORT	An algorithm that predicts the sub-cellular localization sites of proteins based on their amino acid sequences. It makes predictions based on both known sorting signal motifs and some correlative sequence features such as amino acid content	www.wolfpsort.org	Horton et al. (2007)
TargetP	Tool that predicts subcellular location of eukaryotic proteins	http://www.cbs.dtu.dk/services/TargetP	Emanuelsson et al. (2000)

3.8.7 Role of a Serine Protease Pro 1 in *Plasmodiophora brassicae* Pathogenesis

Proteases are enzymes that cleave proteins by the catalysis of peptide bond hydrolysis. It can be classified into five main classes on the basis of the reactive nucleophile in the active site of the enzyme: cysteine proteases, aspartate proteases, metalloproteases, threonine proteases, and serine proteases. Enzymes in each class are further classified into evolutionarily distinct clans, and each clan is subdivided into families on the basis of sequence homology and the order of the catalytic triad (Rawlings and Barrett 1993). Serine proteases are one of the most important and widely distributed families of proteases, which utilize a uniquely activated serine residue in the substrate binding pocket to catalytically hydrolyze peptide bonds. These are found in all organisms, and carry out a diverse array of physiological functions, such as digestion, defense, and the production of active peptides (Yousef et al. 2003). The pathogenicity function of serine protease has been well studied in protozoans that are pathogenic to humans. Serine proteases have also been reported to be important for pathogenicity in plant pathogens. A serine protease and a metalloprotease from *Xanthomonas campestris* pv. *campestris* play important roles for pathogenesis in black rot on turnip leaves (Dow et al. 1990).

A serine protease gene (*PRO1*) was cloned from *P. brassicae* and its molecular characteristics were investigated by Feng et al. (2010). Southern analysis and specific polymerase chain reaction (PCR) amplification indicated that *PRO1* is a single-copy gene present in a broad range of *P. brassicae* pathotypes. Northern analysis revealed that the expression of *PRO1* was induced during plant infection, and that the quantity of transcript fluctuated according to the stage of pathogenesis. Amino acid sequence analysis suggested that the encoded protein (Pro1) belongs to the S28 family of proteases, with a predicted signal peptide, and a theoretical molecular mass of 49.4 kDa. The open reading frame (ORF) of *PRO1* was transferred into *Pichia pastoris,* and Pro1 was heterologously produced. Pro1 showed proteolytic activity on skimmed milk and *N*-succinyl-Ala-Ala-Phe-7-amido-4-methylcoumarin, and the activity could be inhibited by serine protease inhibitors with the chelating agent ethylenediaminetetraacetic acid. The optimal temperature of Pro1 was 25 °C, and it exhibited high activity at pH 6.0–6.4. These values coincide with the temperature and pH conditions favorable for *P. brassicae* resting spore germination in the field. When Pro1 was used to treat *Brassica* root exudates, it enhanced the stimulating effect of the root exudates on *P. brassicae* resting spore germination, indicating that Pro1 may play a role during clubroot pathogenesis by stimulating resting spore germination through its proteolytic activity (Feng et al. 2010).

3.8.8 Role of Lipids in *Plasmodiophora brassicae* Pathogenesis

In pathogen dormancy, carbohydrate and lipid composition of spores play a determinant role. Plasmodiophorid spores contained an abundant amount of lipid bodies. The most abundant fatty acid in resting spores of *P. brassicae* is Arachidonic acid (ARA 20:4). Transcriptional data suggest that fatty acids are synthesized in the plasmodia and degraded in resting spore. Lipid droplet is a cellular organelle that stores neutral lipids triacylglycerol (TAG) and cholesterol esters as a source of energy, and carbon. They serve as the energy reservoir for the entire organism. Lipid droplets have been shown to be physiological important organelles in lipid metabolism rather than just storage organelles. Numerous proteins have been shown by immune detection or proteomic approaches to be associated with lipid droplets and to change under different physiological conditions. Lipid droplets biogenesis and function have been studied in many organisms from archae, bacteria, plant to mammalian cells. Lipid with lipid droplets and their associated proteins will facilitate to improve the understanding of their functional roles for *P. brassicae*. In addition to comprising cellular membranes, fatty acids are stored within cells as energy-rich TAGs in lipid droplets. In response to metabolic signals, mobilized fatty acids and other precursors derived from stored neutral lipids are used for a wide variety of functions within lipoproteins. Fatty acids are essential metabolites used for energy production via β-oxidation. Because an excess of free fatty acids in the cytoplasm is toxic to cells, cells esterify fatty acids into neutral lipids, and package them into lipid droplets. Bi et al. (2016) found that lipid droplets are present in the resting spores, and secondary plasmodia of *P. brassicae* by transmission electron microscopy (TEM). Nile red staining showed that different zoosporic stages of *P. brassicae* contained lipid droplets. Two hundred and ninety-five lipid droplet-associated proteins were identified by proteomic study, and seven types of fatty acid compositions were chemically characterized using GC-MS in lipid droplets. Genome integrated transcriptome, proteome of its lipid synthesis, storage, and metabolism are very essential aspect in *P. brassicae* life cycle.

Lipid droplets are observed in the resting spores of *P. brassicae*. Lipid droplet-associated proteins were identified, and categorized into nine groups. Transcriptome analysis of these proteins during three different zoosporic stages revealed differences in gene expression pattern. GO enrichment analysis revealed that these proteins associated with lipid droplets were mainly linked to biosynthesis and metabolism. GC-MS analysis revealed that lipid droplets contain seven types of free fatty acids: saturated fatty acids C16:0 and C18:0 and unsaturated fatty acids C18:1Δ9, C18:1Δ11, C18:2, C20:4, and C20:5. *P. brassicae* accumulated a large amount of triacylglycerols (TAGs). Bi et al. (2016) systematically analyzed the putative proteins involved in TAG biosynthesis, and its metabolic pathway. KEGG pathway analysis defined 3390 genes, including 167 genes involved in lipid metabolism. Transcriptome analysis revealed that 162 candidate enzymes involved in lipid metabolism were differential expressed.

3.8.9 Relationship of Endogenous Glucosinolate Content to *Plasmodiophora brassicae* Infection

The infection by *P. brassicae* may occur in non-*Brassica* species, leading to the potential formation of resting spores. Structures resembling *P. brassicae* are found in the root cortex of *Tropaeolum majus* (*Nasturtium*), *Carica papaya* (Pawpaw), *Reseda alba* (White mignonette or White upright mignonette), and *Beta vulgaris* (Beet) as demonstrated by scanning electron microscopy (SEM). Inoculation of *Brassica rapa* roots with spores extracted from either *T. majus* or *B. vulgaris* roots which had been previously inoculated with *P. brassicae* led to development of clubroot in the roots of *B. rapa*. It is also observed that the development of the symptom might be correlated with glucosinolate content, although other host factors are implicated in the *B. vulgaris* interaction with *P. brassicae*. In the glucosinolate-containing non-*Brassicas*, *T. majus,* and *C. papaya*, the concentrations of benzyl-glucosinolate increased markedly in roots inoculated with *P. brassicae*, compared with the controls. There were also increases in concentrations of benzyl glucosinolate in leaves of *T. majus* after *P. brassicae* infection. However, in *R. alba* roots, the total glucosinolate content decreased after inoculation with *P. brassicae* compared with the controls. High root concentrations of 2-OH-2-phenylethylglucosinolate (glucobarbarin) compared with low root indole glucosinolates in this species might limit *P. brassicae* infection and development. The importance of cultivation of non-*Brassica* species on fields infested with *P. brassicae* needs careful selection of nonhost species (Ludwig-Müller et al. 1999).

3.8.10 Effect of Host Resistance on *Plasmodiophora brassicae* Pathogenesis

Infection of root hairs by *P. brassicae* occurs in a broad range of plants, not only in susceptible host cultivars but also in resistant cultivars (Diederichsen et al. 2009) and even nonhost plants (Feng et al. 2012). Root hairs serve as a niche for pathogen increase prior to invasion of the root cortex, but also provide an opportunity for close interaction between host and pathogen, which facilitates the evolution of mechanisms by which *P. brassicae* can break down basal resistance to cortical infection (Feng et al. 2013a). In susceptible plants, both root hair infection and clubroot severity increase with higher inoculum density (Hwang et al. 2011b, c). Root hair infection occurs more frequently with a compatible isolate (susceptible reaction) compared with an incompatible isolate (resistant reaction) and, subsequently, fewer cells within the root become infected when the host is resistant (Tanaka et al. 2006). The compatible isolate forms secondary plasmodia with many nuclei and eventually resting spores are formed in the host root tissue. In contrast, plasmodia formed by the incompatible isolate remain immature with only a small number of nuclei and do not produce resting spores (Hwang et al. 2011b; Deora et al. 2013). It suggests that resistance in host species is associated with a small amount of suppression of infection during primary infection, but a much greater suppression of secondary infection and subsequent plasmodial development in the root cortex. One important difference between the resistant and susceptible host reactions is that in a resistant host, the secondary thickening of the cell walls in the xylem are not degraded and there is fewer cell wall breakages (Donald et al. 2008). The influence of cultivar resistance on root hair infection was studied across a range of inoculum densities in canola in one resistant and one susceptible cultivar (Hwang et al. 2011b). Subsequently it was observed that root hair infection in two resistant and three susceptible canola cultivars (Hwang et al. 2011a). Infection and clubroot severity were higher and plant height was reduced in the susceptible cultivars relative to the resistant cultivars. Root hair infection and the amount of *P. brassicae* DNA (assessed using qPCR analysis) increased in both resistant and susceptible cultivars over time, but the increases were greater in the susceptible cultivars. Also, there was a strong linear relationship between root infection and the amount of DNA of *P. brassicae* in the root hairs. The slow increase in the amount of pathogen DNA and root hair infection observed in the resistant cultivars, and the sharp increases observed between 4 and 10 days after inoculation in the susceptible cultivars, indicate that secondary infection and pathogen development are occurring quickly in the susceptible cultivars. However, sharp declines in detectable levels of pathogen DNA in the seedling roots occurred during the transition from zoospore development in the root hairs to secondary infection (Hwang et al. 2011a). These levels quickly rebounded as the pathogen spread into the root cortex after secondary infection (Gludovacz 2013).

The rate of infection and pathogen development is most rapid in susceptible canola lines (Deora et al. 2012, 2013). However, resistant plants showed reduced growth and delayed development when inoculated with resting spores of an avirulent pathotype (Hwang et al. 2011a; Deora et al. 2013), which indicates that resistance is an active process. In at-least one vegetable *Brassica* line, microscopic analysis has revealed substantial levels of pathogen infection in inoculated roots of symptomless plants (Gludovacz 2013; Gludovacz et al. 2013). The mechanism for this response has not yet been determined, but it is possible that it involves genes for tolerance to *P. brassicae.* Successful management of *P. brassicae* through the deployment of genetically resistant cultivars will require very careful control of susceptible weeds and canola volunteers to maximize the impact of resistant cultivars on reducing resting spore levels in the soil and to prevent the build-up of virulent populations (Hwang et al. 2014).

3.8.11 Molecular Mechanisms of Pathogenesis

The information on the genomic makeup of *P. brassicae* and the molecular basis of host resistance/susceptibility is very limited, largely because of the nature of this organism as an obligate parasite. Graf et al. (2004) isolated a few partial cDNAs that had no significant homology to Gen Bank sequences. Suppression subtractive hybridization (SSH) has been conducted between RNA from *P. brassicae*-infected and uninfected *Arabidopsis* tissue, and a total of 76 new *P. brassicae* gene sequences have been identified (Bulman et al. 2006). Many of the sequences are predicted to contain signal peptides for translocation of the encoded proteins. Again using SSH, but starting with *B. rapa* rather than *Arabidopsis* clubroot galls, Sundelin et al. (2011) isolated about 140 clones. Half of these clones, which represented 24 unisequences, originated from *P. brassicae*. Out of these, 10 unisequences are newly characterized *P. brassicae* genes. By using next-generation sequencing (NGS) technologies, Burki et al. (2010) generated 27,333 contigs assembled from 454 reads from *P. brassicae* infected *B. rapa*. Among these contigs, 3167 are from *P. brassicae*. A few genes have been postulated to be related to the pathogenicity of *P. brassicae* (Ito et al. 1999; Brodmann et al. 2002; Ando et al. 2006; Siemens et al. 2009b), mainly based on the observation that the expression of the gene is correlated with certain stage of disease development. One gene, a serine protease (*PRO1*), has been heterologously expressed and proven experimentally to be important for resting spore germination (Feng et al. 2010).

The transcriptomic and proteomic approaches have elucidated the mechanisms of *P. brassicae* infection and coordination in plant reactions. Using a full genome Affymetrix chip, Siemens et al. (2006) found that more than 1000 *Arabidopsis* genes are differentially expressed in *P. brassicae*-infected roots at either 10 or 23 days after inoculation versus noninfected roots. Two proteomic studies, both using two-dimensional gel electrophoresis and quantitative image analysis in conjunction with mass spectrometry-based protein identification, illustrated changes in protein abundance at 4 days after inoculation (dai) in *Arabidopsis* (Devos et al. 2006), and at 0.5, 1, 2, and 3 dai in *B. napus* (Cao et al. 2008). The observed differential expression on mRNA or protein levels at the different stages of pathogenesis suggested that the corresponding genes might be involved in the plant response to the clubroot pathogen. Functional genomic approaches provide powerful tools for identifying expressed genes. Among these techniques, expressed sequence tags (ESTs, Adams et al. 1991), serial analysis of gene expression (SAGE, Velculescu et al. 1995), and massively parallel signature sequencing (MPSS, Brenner et al. 2000) have been successfully employed. Due to its relative simplicity and inexpensive nature, single pass EST sequencing has been the most widely used method to characterize genes associated with cellular development, with biotic and abiotic stress in plant research.

Although many EST clones lack the complete sequences of mRNAs, they can be very helpful to investigate the biological function of particular genes, especially in poorly annotated genomes. Additionally, it is possible to obtain full-length clones starting from ESTs. EST resources are also important for comparative transcriptomics because gene expression profiles can be obtained by analyzing ESTs generated under varying conditions. The whole transcriptome sequencing using next-generation sequencing (NGS) technologies or RNA sequencing (RNA-seq) has been started to be utilized to reveal the complex landscape of the transcriptome from various organisms at an unprecedented level of sensitivity and accuracy (Ozsolak and Milos 2011). Compared with traditional EST sequencing and other high-throughput technologies such as microarrays, NGS and RNA-seq offer a near-complete snapshot of a transcriptome with a higher dynamic range of expression level. However, common NGS platforms including Illumina, SOLiD, and 454 suffer from the disadvantage that the sequence reads are often very short (35–500 bp). As a consequence, high technologies on transcriptome assembly are required after NGS.

Despite the advantages of NGS and RNA-seq, EST cloning incorporated with SSH (Diatchenko et al. 1996) has been used to identify genes involved in host–pathogen interactions. This method reduces, by subtractive hybridization, the cloning of abundantly expressed housekeeping genes or genes commonly expressed in both control and treated plants, thereby normalizing expressed cDNA profiles during library construction. Moreover, it significantly enhances, by suppression PCR, the chances of cloning differentially expressed genes. This is particularly important because many pathogenesis-related genes are expressed at low levels, and can be limited to a particular tissue or cell type. This technique has been used to isolate plant and pathogen genes that are expressed during infection (Huang et al. 2007). Both of the two cDNA libraries of *P. brassicae*-infected plants, one using *Arabidopsis* (Bulman et al. 2006) and the other *B. rapa* (Sundelin et al. 2011), were constructed by using the SSH technique.

The construction of a SSH cDNA library followed by EST analysis was used to achieve a better insight into the genes expressed during the compatible interaction between a susceptible *Brassica* cultivar and *P. brassicae*. The infection of plants by pathogenic and the subsequent establishment of disease involves substantial changes in the biochemistry and physiology of both partners. Analysis of genes that are expressed during these interactions represents a powerful strategy to obtain insights into the molecular events underlying these changes. Feng et al. (2013b) have used suppression subtractive hybridization (SSH) and expressed sequence tag (EST) analysis to investigate gene expression during the early stages of colonization of *Brassica* roots by *P. brassicae*. A cDNA library was constructed by SSH which consisted of 797 clones that represented 439 unigenes. Thirty-two of these genes were demonstrated to be of *P. brassicae* origin. The remaining 407 genes, which were of a *Brassica* origin, were subjected to gene ontology and in silico analyses. Real-time PCR analysis of ten *P. brassicae* and seven *Brassica* genes indicated that seven of the former and five of the latter were upregulated at 7 days after infection, suggesting the importance of these genes in pathogenesis or host resistance.

Molecular mechanisms of *P. brassicae* infection during primary and secondary zoospores cannot be differentiated morphologically. When inoculated onto healthy *Brassica* roots, secondary zoospores cause primary infections similar to those caused by primary zoospores (Feng and Zhou 2012; Feng et al. 2012). In contrast, when inoculated onto *Brassica* roots that are already under primary infection, primary zoospores can cause secondary infections without going through a primary infection phase (Feng et al. 2013b). Thus, from a developmental biology standpoint, neither morphogenesis nor functional differentiation can clearly distinguish between these two kinds of zoospores. The primary and secondary zoospores maybe essentially same or they undergo a divergence. No matter the state of divergence between the spore types, the "arms race" with the host would induce changes in the secondary zoospores. These changes are most likely at the gene expression level, and can be both genetic and epigenetic in nature. Knowledge of the molecular differences between primary and secondary zoospores may be useful for rational clubroot resistance breeding efforts as well as for the development of pesticides effective against *P. brassicae*. If these two kinds of zoospores are the same with respect to gene expression, a gene or quantitative trait loci (QTL) for resistance against either primary or secondary infection will be also effective against the other. Thus, the selection for resistance and subsequent genetic studies can reliably be conducted at the primary infection stage, which would save time and resources. This scenario is also applicable to fungicide development. Despite this importance, data on gene expression in these two kinds of zoospores is limited with information on the genetic makeup of *P. brassicae*, with only a small number of expressed sequence tags (ESTs) and protein sequences available in GenBank. Similarly, very few *P. brassicae* genes have been functionally characterized (Feng et al. 2010). Most of the database-available genes are derived from cDNA libraries that were constructed using the suppression subtractive hybridization techniques (Bulman et al. 2006; Feng and Zhou 2012; Sundelin et al. 2011), which are enriched in genes upregulated during pathogenesis. Thus, these genes are all good candidates for gene expression studies during the different stages of pathogenesis. All database-available of *P. brassicae* genes were analyzed in silico. To compare the gene regulation profiles of the primary and secondary zoospores, these genes were subjected to both dot blot and real time PCR analyses to understand clubroot pathogenesis and provide gene candidates for gene functional studies, which will in turn facilitate breeding for resistance and the identification of potential targets for the development of novel pesticides.

To get information on the molecular basis of pathogenicity of the clubroot pathogen *P. brassicae* specific primers were designed and used to amplify genomic fragments of *P. brassicae* genes that represent all database-available proteins and ESTs. The PCR products were blotted on membranes and hybridized with digoxigenin-labeled double-stranded cDNA, derived from either primary or secondary zoospores of *P. brassicae*. The same primers were also used in real-time PCR against the single-stranded cDNA synthesized from the two types of zoospores. Both dot blot and real-time PCR identified up- and downregulated genes provide the correlation between these two techniques. Real-time PCR indicated that 58 genes were upregulated in the secondary zoospores relative to the primary zoospores, whereas 55 were downregulated. These data suggest that different mechanisms are utilized by the pathogen in causing primary and secondary infections. The expression patterns of genes with known or putative functions suggest the relative importance of these genes during pathogenesis. In contrast, highly expressed or regulated genes with unknown function can be further studied in identifying pathogenicity factors (Feng et al. 2013b).

3.8.12 Host Gene Expression During the Primary Phase of *Plasmodiophora brassicae* Interaction

Wide-scale gene expression analysis offers an opportunity to explore the molecular basis of plant–pathogen interactions, particularly with respect to mechanisms of resistance and basal defense, as well as host versus nonhost resistance, and biotrophy versus necrotrophy (Oliver and Ipcho 2004; Anderson et al. 2005; Glazebrook 2005; Wise et al. 2007). Microarray technology has enabled the identification of defense-related genes for many plant–pathogen interactions as well as the mechanisms by which plants develop resistance.

The infection of *P. brassicae* occurs in two distinct phases: the primary phase, which occurs predominantly in root hairs, and the secondary phase, which occurs predominantly in the root cortex (Ingram and Tommerup 1972). It is the secondary phase that produces the characteristic disease symptoms of root tissue hypertrophy and hyperplasia, or "clubbing." Broadly speaking, in resistant hosts, the primary phase of the lifecycle can still be observed, but not in the secondary phase (Dekhuijzen 1979; Takahashi et al. 2006; Donald et al. 2008). In a compatible (susceptible) interaction with *Arabidopsis thaliana* (L.) Heynh., the pathogen can proceed through its entire lifecycle from 3 to 28 days (Siemens et al. 2002). *P. brassicae* attachment and penetration of the root hair can occur from 4 days after inoculation onwards (Agarwal et al. 2009). This occurs after a resting spore releases a single motile zoospore that attaches to the wall of the root hair and encysts by means of a tubular structure (the rohr and the stachel), injecting its protoplast into the host cytoplasm (Aist and Williams 1971). Between 5 and 12 days after inoculation, plasmodia and then zoosporangia (producing secondary zoospores) develop within the root hairs with little visual sign of damage to the host cells. Between 10 and 13 days after inoculation, *P. brassicae* undergoes a transition from the primary phase to the secondary phase of its lifecycle as observed by Devos et al. (2005) and Siemens et al. (2006) depending on the isolate used (single spore or field population). Within 28 days after inoculation, the secondary plasmodia multiply into resting spores within the cortical cells proliferating into galls in *A. thaliana* (Agarwal et al. 2009). Siemens et al. (2006) have reported the use of microarray analysis to examine gene expression following inoculation of a susceptible *A. thaliana* ecotype Col-0 with *P. brassicae* at 10 and 23 days after inoculation (DAI) using a single spore isolate. These two time points were within the so-called secondary infection stage, which takes place in the root cortex, the 10-day time point being described as the early secondary plasmodial stage and the 23-day time point as the late secondary stage, by which time the disease has progressed substantially in the roots to develop galls. Their study found wide-scale changes in the expression of genes in several important functional groups including those involved in growth, sugar phosphate metabolism, defense, and plant hormone synthesis. The two time points analyzed, however, were relatively late in the interaction for studying the potentially important events that take place during initial pathogen invasion. Proteomic-based approaches have also been reported (Devos et al. 2006; Cao et al. 2008). Devos et al. (2006) examined the susceptible response in *A. thaliana* roots at 4 days after inoculation (DAI) with a field population of *P. brassicae*. Most of the differentially regulated proteins (12%) were involved in plant defense, cell differentiation, hormone metabolism and detoxification. Cao et al. (2008) examined the susceptible response in canola (*B. napus*) roots 12, 24, 48, and 72 h after inoculation using the field population of *P. brassicae* collected from central Alberta, Canada. Only 20 protein spots in total were differentially regulated including proteins involved in lignin biosynthesis, cytokinin metabolism, glycolysis, intracellular calcium homeostasis, and the detoxification of reactive oxygen species (ROS). Further, a review of the interactions of *P. brassicae* with its hosts (Ludwig-Müller and Schuller 2008), emphasized that although the components of early signaling in the interaction between *A. thaliana* and leaf pathogens are well understood, there are relatively few studies with root pathogens and indeed none conducted with *P. brassicae*. A comparative study between resistance and susceptibility is not possible for the interaction of *P. brassicae* with *A. thaliana*, as almost all the *A. thaliana* ecotypes tested so far are susceptible to *P. brassicae* (Siemens et al. 2002; Agarwal et al. 2009). Also, the enormous variation in the virulence of *P. brassicae* populations reported worldwide (Jones et al. 1982) often makes it difficult to determine the host's genomic and metabolomic responses.

Microarray analysis was used to investigate changes in host gene expression during the primary stages of the interaction between the susceptible plant *Arabidopsis thaliana* (L.) Heynh ecotype Col-0 and the biotrophic pathogen *Plasmodiophora brassicae* Woronin. Analyses were conducted at 4, 7, and 10 days after inoculation (DAI) and revealed significant induction or suppression of a relatively low number of genes in a range of functional categories. At 4 DAI, there was induced expression of several genes known to be critical for pathogen recognition and signal transduction in other resistant host–pathogen interactions. As the pathogen further colonized root tissue and progressed through the primary plasmodium stage for production of zoosporangia at 7 and 10 DAI, respectively, fewer genes showed changes in expression. The microarray results were validated by examining a subset of induced genes at 4 DAI by quantitative real-time reverse transcriptase PCR (RT-qPCR) analysis all of which correlated positively with the microarray data. The two *A. thaliana* mutants *jar1* and *coil* tested were found to be susceptible to *P. brassicae*. The involvement of defense-related hormones in the interaction was further investigated (Agarwal et al. 2011) and the findings indicate that addition of salicylic acid can suppress clubroot disease in *A. thaliana* plants (Fig. 3.10).

UP	DOWN
Signaling pathway/transcription and transduction Leucine-rich repeat protein CDPK protein TIR-NBS-LRR protein MAPKK protein Serine/threonine protein kinase WRKY transcription factor (*WRKY18*) **Cell wall modification** Expansin protein Xyloglucan:xyloglucosyl transferase Pectinesterase protein **Phenylproponoid pathway (lignin biosynthesis)** 4-coumarate-CoA ligase (*4CL*) **Jasmonic acid biosynthesis** Lipoxygenase (*LOX4*) **Lipid metabolism** Phytoalexin-deficient 4 protein *PAD4*)	**Oxidative burst/stress** Peroxidase 72 (*PER72*) Glutathione S-transferase Heat shock protein 101 (*HSP101*) NADP oxidoreductase **Cell wall modification** Hydroxyproline-rich glycoprotein protein Arabinogalactan protein **Phenylproponoid pathway (lignin biosynthesis** Cinnamoyl-CoA reductase (*CCR1*) 4-coumarate-CoA ligase (*4CL1*) Cinnamyl-alcohol dehydrogenase (*CAD5*) **Cytokinin biosynthesis** UDP-glucoronosyl/UDP-glucosyl transferase protein **Salicylic acid biosynthesis** Phenylalanine ammonia-lyase (*PAL4*) **Ethylene biosynthesis** 1-aminocyclopropane-1-carboxylate synthase 2 (*ACS2*) Ethylene receptor (*ETR2*) **Antimicrobial proteins** Chitinase Class IV chitinase
Signaling pathway/transcription and transduction Protein kinase **Oxidative burst/stress** NADPH oxidase **Cell wall modification** Pectinesterase Endo-1-4 beta glucanase	
Signaling pathway/transcription and transduction Zinc finger protein Leucine-rich repeat protein Serine/threonine protein kinase Protein phosphatase 2C Rapid alkalinization factor protein **Oxidative burst/stress** Glutathione peroxidase **Cytokinin biosynthesis** UDP-glucoronosyl/UDP-glucosyl transferase protein **Ethylene biosynthesis** 1-aminocyclopropane-1-carboxylate oxidase (*ACC oxidase*)	**Oxidative burst/stress** L-ascorbate peroxidase

Fig. 3.10 Left panel: the lifecycle stages of *Plasmodiophora brassicae* (primary root hair phase) at 4, 7, and 10 days after inoculation (DAI). Right panel: summary table of up- and downregulated genes and functional groups at 4, 7, and 10 DAI (Agarwal et al. 2011)

3.8.13 Differential Expression of Genes During Infection of *Plasmodiophora brassicae*

Microarray chips, two-dimensional electrophoresis and high throughput sequencing technology have been used to study gene expression in the host plant in response to *P. brassicae* infection. Siemens et al. (2002, 2006, 2009b) explored the gene expression of *A. thaliana* Col-0 inoculated with *P. brassicae* after 10 and 23 days using a microarray chip. They found that more than 1000 host genes were differentially expressed in the infected roots as compared to the control roots. Starch, sulfur, and secondary metabolism, auxin, and cytokinin synthesis, and the expression of transport-related genes changed significantly, where as that of genes associated with defense and lignin synthesis did not. In addition, some flavonoid genes were also differentially expressed (Siemens et al. 2006). Devos et al. (2006) and Cao et al. (2008) discovered protein changes in *Arabidopsis* following inoculation with *P. brassicae* using two-dimensional electrophoresis. Devos et al. (2006) observed 35 upregulated and 11 downregulated proteins 4 days after *P. brassicae* inoculation, which were mainly associated with defense, cell metabolism, cell differentiation, and active oxygen activity. Cao et al. (2008) observed changes in the expression of 20 proteins (with 13 spots increasing and seven spots decreasing) 12, 24, and 48 h post-inoculation, including lignin synthesis, cytokinin synthesis, calcium steady-state, glycolysis, and oxygen activity. Agarwal et al. (2011) detected 147, 27, and 37 differentially expressed genes (DEGs) in *A. thaliana* after 4, 7, and 10 days, respectively, using a microarray chip. They further deduced that changes observed at 4 days postinfection (dpi) were mainly related to host and pathogen recognition and signal transduction. Using transcriptome analysis, Chen et al. (2015a) observed major changes between sensitive and resistant varieties of *B. rapa* 0–96 h postinfection in metabolism, transport, and signal transduction. Nowadays, new high-throughput technologies are gradually being used to study the interaction between *P. brassicae* and the host, such as metabo-typing, laser micro-dissection coupled to transcriptional profiling coupled and miRNA sequencing (Wagner et al. 2012; Schuller et al. 2014; Verma et al. 2014). The *P. brassicae* genome was sequenced, which will be very convenient for studies on the interaction between host plants and *P. brassicae* in the future (Schwelm et al. 2015; Rolfe et al. 2016). In order to clarify the early events taking place between the pathogen and the host, Zhao et al. (2017) analyzed differentially expressed genes and pathways in *A. thaliana* 24 and 48 h following infection with *P. brassicae*, using transcriptome analysis. In addition, the molecular mechanism of *A. thaliana's* response to *P. brassicae* in the early infection phase was described. Zhao et al. (2017) demonstrated changes on gene expression and pathways in *A. thaliana* during early infection with *P. brassicae* using transcriptome analysis, and identified 1903 and 1359 DEGs at 24 and 48 h postinoculation (hpi), respectively. Flavonoids and the lignin synthesis pathways were enhanced, glucosinolates, terpenoids, and proanthocyanidins accumulated and many hormonal- and receptor-kinase related genes were expressed, caused by *P. brassicae* infection during its early phase (Fig. 3.11). Therefore, the early interaction between *A. thaliana* and *P. brassicae* plays an important role in the entire infection process.

A

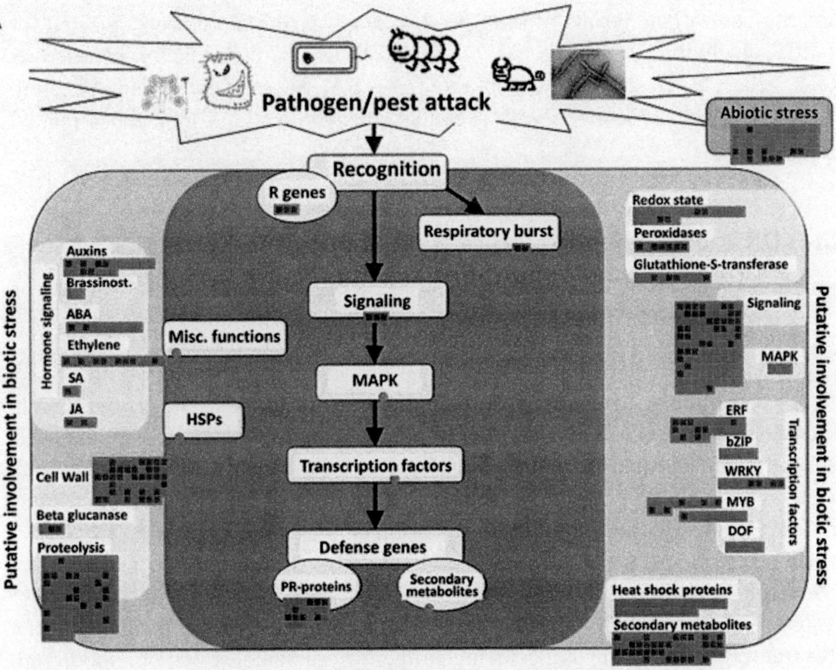

B

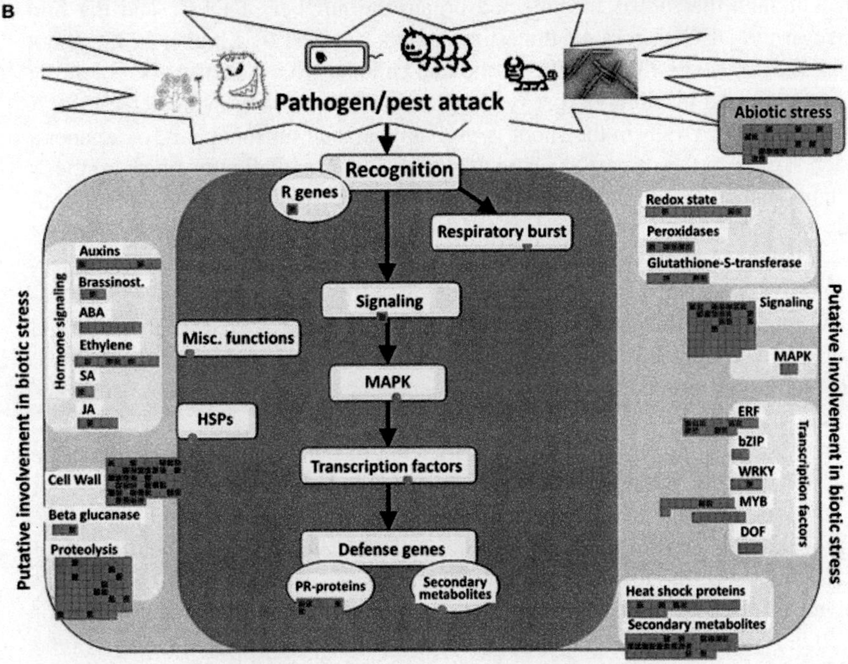

Fig. 3.11 Biotic stress pathway analyses of DEGs in *Arabidopsis thaliana* during early infection by *Plasmodiophora brassicae.* Analysis of the biotic stress pathway of DEGs was performed using MAPMAN software. Red boxes mean upregulated genes and green mean downregulated genes. (**a**) 24 h after inoculation; (**b**) 48 h after inoculation. The pathway frames are from the MAPMAN software database (Zhao et al. 2017)

In order to confirm the results of the RNA-seq analysis, 40 DEGs were selected for qRT-PCR, including 10 PR genes, three MYB transcription factor genes, one WRKY gene, 11 genes involved in the IAA, SA, JA, and cytokinin signaling pathways, 11 genes associated with secondary metabolism and four genes related to the cell wall.

3.8.14 Differential Expression of Host Shoots and Roots Genes During *Plasmodiophora brassicae* Infection

The *Plasmodiophora brassicae* has evolved several strategies to avoid host recognition and subsequent cell death to establish an intracellular parasitism within living host cells, enabling it to extract/absorb nutrients from its host and to direct cell proliferation of host root tissues in order to produce a nutrient-sink gall habitat. The limited resistant germplasm in the Brassicaceae are available, so exploring and understanding the genomics of disease progression in this host plant will contribute to the development of a clubroot resistance breeding strategy. Irani et al. (2018) provided a broad-spectrum view of the *Arabidopsis* transcriptomic response to *P. brassicae* inoculation, in above- and below-ground tissues of the host plant. They observed more DEGs at later stages of the disease establishment, with the largest transcriptomic changes occurring in the 24 dpi samples of both shoot and root. Although the root is the infected organ, the number of DEGs and the fold change in the DEGs across all three time points sampled was higher in the shoot (Fig. 3.12a–e), suggesting that metabolic and cellular processes of the shoot are in fact more affected by clubroot disease progression than those of the infected root. A higher number of DEGs in the shoot is probably a result, in some part, of a general stress response to the disease, as shown by the down-regulation of photosynthesis and light reaction genes, whereas, gene expression in diseased roots would predominantly be a specific response to the pathogen attack. A similar downregulation of genes involved in photosynthesis, together with an upregulation of pathogen protection and oxidative stress genes, has been reported in shoots in response to inoculation of roots with rhizobacterium. Although spatially separated, the above and below-ground tissues of plants do interact with each other and results suggest that future studies are needed to address this interaction in response to root pathogens (Rojas et al. 2014).

The identification and functional classification of highly DEGs showed that the expression of many of these genes was co-regulated or oppositely regulated in shoot and root. In the shoot, as expected, DEGs involved in photosynthesis were downregulated after root inoculation with *P. brassicae* and subsequent clubroot progression. Such a down-regulation has previously been reported for genes involved in photosynthesis, biosynthesis of chlorophyll and the small subunit of rubisco in response to pathogen attacks. Furthermore, several genes involved in the degradation of starch and triacylglycerols (lipid storage molecules) and the Calvin cycle were also downregulated in the shoot. Following decreased water and nutrient uptake by the host plant, the general downregulation of these genes in the shoot of

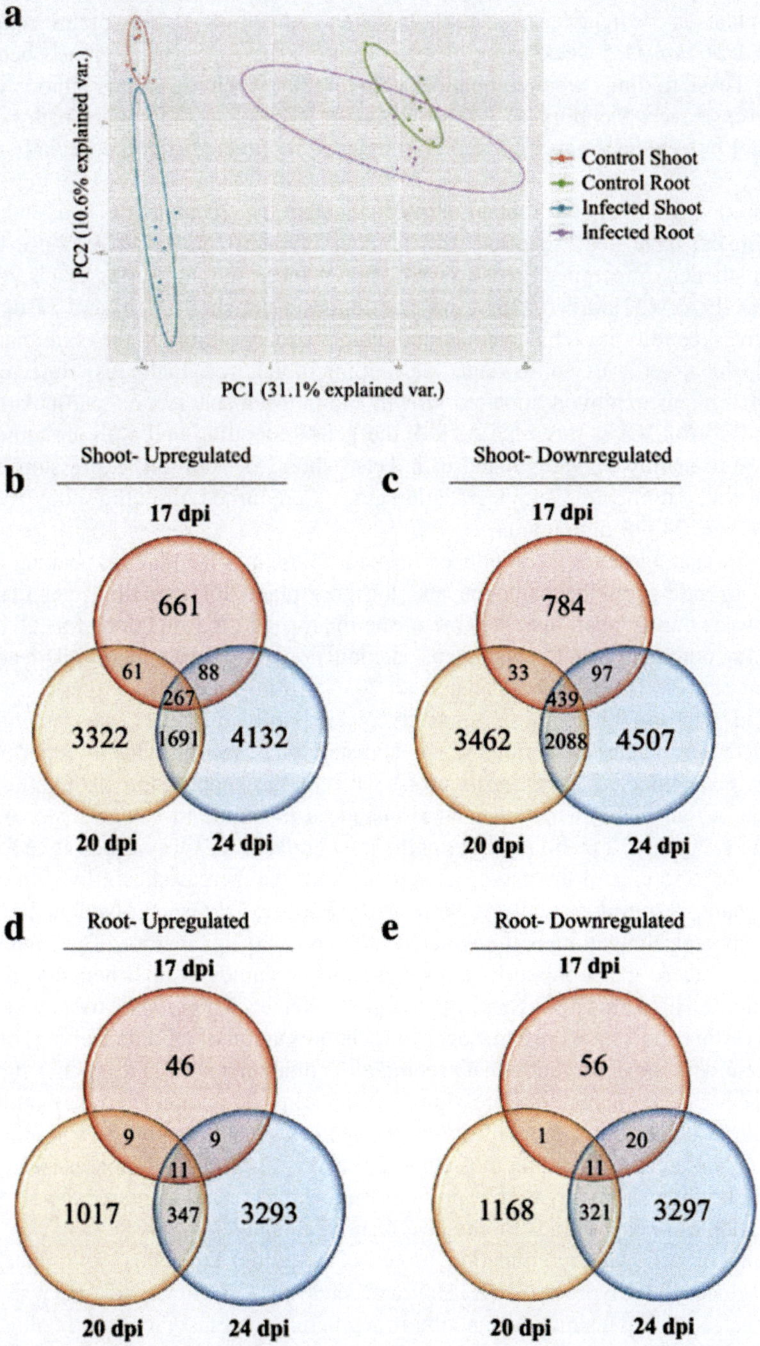

Fig. 3.12 (**a**) PCA plot displaying biological variation between infected and control samples in shoot and root. (**b–e**) Venn diagrams showing the total number of significantly DEGs (*P*-value ≥0.05) at 17, 20, and 24 dpi in infected shoot and root compared to mock-infected control samples. The overlapping regions correspond to the number of DEGs present at more than one-time point. (**b**) Upregulated DEGs in shoot. (**c**) Downregulated DEGs in shoot. (**d**) Upregulated DEGs in root. (**e**) Downregulated DEGs in root (Irani et al. 2018)

infected plants is predictable. A decrease in starch and soluble carbohydrates in the leaves of *Arabidopsis*, 5 weeks after infection with *P. brassicae*, has previously been reported. These findings were compatible with a reduced rate of photosynthesis in *Arabidopsis* leaves. Contrary to these decreases, tryptophan (shikimate pathway metabolite) biosynthesis genes were highly induced in infected shoot, especially at 24 dpi.

Increased levels of tryptophan-derived metabolites (camalexin and indol glucosinolates) have been reported in a number of plant species in response to various pathogen infections/disease progressions. Increased transcript levels for CHALCONE SYNTHASE (CHS) were also found in the shoot at 20 and 24 dpi. Under stress conditions, chalcone compounds produced through the shikimate pathway which, serve as antioxidants and antimicrobials (phytoalexins). Infected *Arabidopsis* plants exhibited impaired growth and induced senescence, particularly at 20, and 24 dpi. Over this time period, the genes encoding cell wall structural proteins were mainly downregulated in infected shoot. In addition, expression of many cell wall modification and degradation genes were turned off in the shoot over the 17, 20, and 24 dpi time points.

The ET signaling genes were induced in shoot following inoculation, peaking at 24 dpi. Depending on the pathogen and the host plant, ET signaling can play different roles during plant disease. One of the main suggested functions for ET in response to non-necrotrophic pathogens is induction of programmed cell death (Consonni et al. 2010; Sanchez-Vallet et al. 2010). In infected root, expression of starch and sucrose biosynthesis and glycolysis pathway genes was mainly upregulated. An increase in various sugars and starch has been previously reported in *P. brassicae* infected *Arabidopsis* roots, whereas the suppression of invertase expression in transgenic plants resulted in enhanced tolerance to *P. brassicae*. An increased availability of carbon sources at the infection sites is a way to resource for the metabolic activities of the pathogen as well as for the increased cell division of the host plant, resulting in gall sat these sites. Further, DEGs involved in lipid biosynthesis and elongation is upregulated at 17 and 20 dpi in root. The higher abundance of these genes might be a response for the pathogen lipid demand. An upregulation of lipid transfer genes was found in infected *Arabidopsis* roots when galls and resting spores are present, suggesting the formation of a lipid reserve. This is consistent with the observation that accumulation of abundant lipid droplets in the pathogen cytoplasm. Increased expression of phospholipases in infected root, mainly at 24 dpi, may be attributed to their roles in the synthesis of JA, oxylipins, and other important plant defense signaling molecules. However, since phospholipases hydrolyze phospholipids, an increased expression may indicate an increased level of catabolization of phospholipids by the host plant (Canonne et al. 2011).

A group of cell wall modification proteins, including expansins, xyloglucan endotransglucosylase/hydrolases (XTHs) and endo-β-1, 4-glucanases (EGases) facilitate cell wall expansion and function in regulating the plasticity and rheology of the cell wall. The upregulation of cell wall modification genes in the root is likely linked to cell wall loosening and cell expansion in infected root. However, several DEGs encoding cell wall degradation enzymes were strongly downregulated in infected root at 24 dpi. The *P. brassicae*-induces regulation of cell wall modification proteins involved in an alteration of cell wall structure during infection. Moreover, in root, an upregulation of *SAUR* genes that are an early auxin-responsive gene family is consistent with the suggestion that IAA induces cell division and elongation during gall formation in *P. brassicae*-infected *Arabidopsis*. An involvement of auxin in the host—*P. brassicae* interaction has been observed by Ludwig-Muller et al. (2015).

In both shoot, and root at 24 dpi, the expression of several genes in major branches of the shikimate pathway is upregulated. At 17 dpi, genes involved in the biosynthesis of sinapate esters through the phenylpropanoid branch of the shikimate pathway were upregulated in both tissues. Both a higher level of transcription of shikimate pathway genes, together with an increased production of the pathway's metabolites in plants, has been reported in response to pathogenic attacks. In *P. brassicae*-infected roots, an accumulation of phenylpropanoid pathway compounds or a higher expression of related genes has been reported. The shikimate pathway metabolites have important roles in both root and shoot responses in infected host plants (Irani et al. 2018).

Transcriptome levels for genes in the JA biosynthesis pathway are elevated at all three time points in shoot and at 24 dpi in root. However, several genes involved in the JA response pathway are strongly downregulated in root at 24 dpi, perhaps by the pathogen, indicating that the transcript regulation of JA response genes might be manipulated by the pathogen. *P. brassicae* is able to manipulate a number of hormone pathways and can alter hormone homeostasis in the host plant to its benefit. On the other hand, JA response genes may be downregulated once the pathogen is established and having successfully disarmed the plant immune response by 24 dpi. An enhanced JA level was reported, in Chinese cabbage roots, 3–5 weeks after infection with *P. brassicae*. High levels of jasmonate-related compounds were also reported in *Arabidopsis* ecotype Col-0 roots, 21 days after inoculation with *P. brassicae*. In addition, during secondary infection by *P. brassicae,* the JA pathway is induced in susceptible *Arabidopsis* Col-0 but not in the partially resistant ecotype, Bur-0. The JA pathway is involved in clubroot disease during *P. brassicae* secondary infection (Rolfe et al. 2016; Lemarie et al. 2015).

A number of DEGs in the ABA response pathway are upregulated in both shoot and root at 24 dpi. The level of ABA has been reported to increase in *P. brassicae*-infected *B. rapa* plants during gall formation. Moreover, upregulation of ABA response and signaling genes during late time points after *P. brassicae* infection has been reported. Upregulation of ABA response genes might be a response to abiotic stress conditions in infected tissues that are subsequently under water and nutrient stress. However, it has also been proposed that ABA plays a role in the induction of plant defense response genes and JA-induced defense genes (Adie et al. 2007).

Members of the TF families, MYB, bHLH, and AP2/EREBP, were highly up- or downregulated in both shoot and root (Table 3.6). In *Arabidopsis*, the functions of MYBTFs are very diverse, including anthocyanin, and flavonol biosynthesis, terminal differentiation and cell cycle regulation. Similarly, members of the bHLH family

Table 3.6 DEGs belonging to TF families with $\log_2$ fold change $\geq \pm 2$ at three tested time points (Irani et al. 2018)

TF family	Total number	Common expression	Opposite expression	Change	Shoot	Root
AP2/ EREBP	17	0	0	Up	5	4
				Down	7	1
Aux/IAA	2	0	0	Up	0	0
				Down	2	0
B3 (DBP)	6	1 (Up) At1g49475	0	Up	1	2
				Down	3	0
bHLH	36	1 (Up) At4g29930	0	Up	9	7
				Down	17	3
bZIP	8	0	0	Up	4	1
				Down	2	1
C2C2-CO-like	8	1 (Up) At3g21890	1 (Up in shoot, Down in root)	Up	4	1
			At3g21150	Down	2	1
C2H2	9	1 (Up) At3g53600	0	Up	2	1
				Down	6	0
C3H	5	2 (Down) At5g44260	0	Up	1	0
		At2g25900		Down	2	2
CCAAT	8	2 (Up) At3g05690	0	Up	2	5
		At5g06510		Down	1	0
G2-like	4	1 (Down) At4g37180	0	Up	0	0
				Down	2	2
HB	11	0	0	Up	0	4
				Down	6	1
HSF	4	0	0	Up	3	1
				Down	0	0
LOB	8	1 (Down) At4g37540	0	Up	2	1
				Down	4	1
MYB	39	1 (Up) At5g56840	0	Up	20	9
				Down	7	3
Orphans	4	0	0	Up	2	0
				Down	2	0
Putative	8	0	0	Up	2	1
				Down	3	2
Unclassified	21	2 (Down) At1g32540	1 (Up in shoot, Down in root)	Up	3	4
		At4g01335	At2g05160	Down	11	3
WRKY	6	0	0	Up	6	0
				Down	0	0

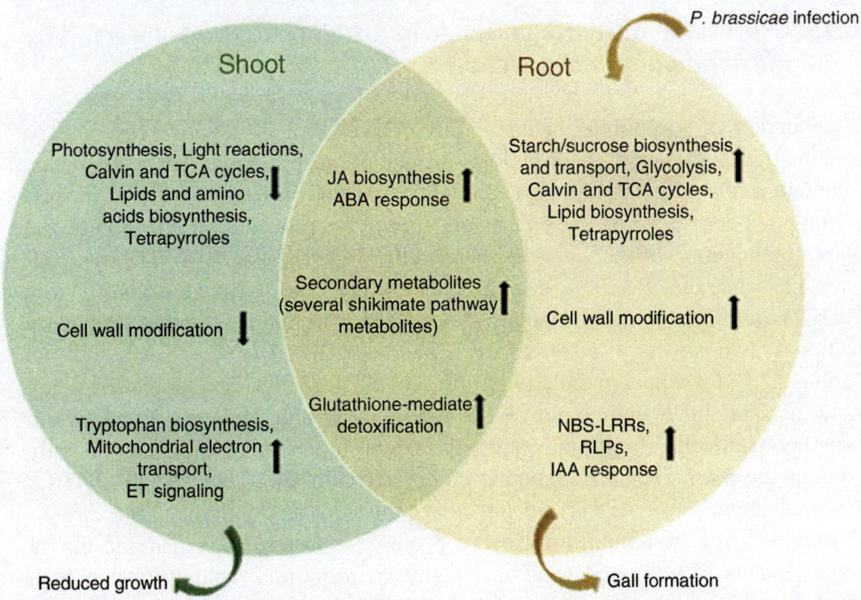

Fig. 3.13 Simplified model to describe the major transcriptional regulation in shoot, and root of *Plasmodiophora brassicae*-infected *Arabidopsis*. The over lapping region corresponds to common DEGs in shoot, and root. The upward black arrows show upregulated DEGs and the downward arrows show. Downregulated DEGs. *ABA* abscisic acid, *ET* ethylene, *IAA* indole acetic acid (auxin), *JA* jasmonic acid, *RLPs* receptor-like proteins (Irani et al. 2018)

play various roles in the regulation of ABA and gibberellins signaling pathways, stomata development and root development in *Arabidopsis*. The comparison of expression profiles of TFs, in shoot and root tissues, highlights the major differences in response of each tissue to the infection. As shown in Table 3.6, only a few TFs are commonly up- or downregulated in both infected shoot and root (Feller et al. 2011). *Arabidopsis* has a complex transcriptomic response to *P. brassicae* infection, and in general, the plant response to pathogen infection of the root is different between shoot and root. Irani et al. (2018) proposed a simplified model of the major molecular changes in shoot and root in response to *P. brassicae* infection and clubroot disease progression (Fig. 3.13). In the root, there is a general induction of genes related to cell wall modification, starch/sucrose biosynthesis and transport to provide energy. In contrast, in the shoot, and possibly as a result of the impairment of nutrient and water uptake, a downregulation of photosynthesis, chlorophyll synthesis, starch/sucrose biosynthesis and an inhibition of synthesis of proteins and enzymes involved in cell wall formation and its modification. As a result, the leaves show symptoms of stress such as reduced growth and necrosis. However, the shoot and root also responded to the pathogen in a number of common ways, such as an induction of JA biosynthesis, ABA response and the production of shikimate pathway metabolites.

3.8.15 Expression of Hypoxia Genes in *Arabidopsis* Roots Infected by *Plasmodiophora brassicae*

The induction of alcohol fermentation in roots is a plant adaptive response to flooding stress and oxygen deprivation. Available transcriptomic data suggest that fermentation-related genes are also frequently induced in roots infected with gall forming pathogens. Hypoxia responses in *Arabidopsis* roots during infection by the clubroot pathogen *P. brassicae* were assessed by Gravot et al. (2016). The hypoxia-related gene markers Pyruvate decarboxylase 1 (PDC1), Pyruvate decarboxylase 2 (PDC2) and Alcohol dehydrogenase 1 (ADH1) were induced during secondary infection by two isolates of *P. brassicae*, eH and e2. The *PDC2* was highly induced as soon as 7 days post-inoculation (dpi), i.e., before the development of gall symptoms, and GUS staining revealed that ADH1 induction was localized in infected cortical cells of root galls at 21 dpi. Clubroot symptoms were significantly milder in the *pdc1* and *pdc2* mutants compared with Col-0, but a null T-DNA insertional mutation of ADH1 did not affect clubroot susceptibility. The Arg/N-end rule pathway of ubiquitin-mediated proteolysis controls oxygen sensing in plants. Mutants of components of this pathway, *ate1, ate2,* and *prt6*, that both exhibit constitutive hypoxia responses, showed enhanced clubroot symptoms. In contrast, gall development was reduced in quintuple and sextuple mutants where the activity of all oxygen-sensing Group VII Ethylene Response Factor transcription factors (ERFVIIs) is absent (erfVII and *prt6* erfVII). The induction of PDC1 and PDC2 during the secondary infection of roots by *P. brassicae* contributes positively to clubroot development, and which is controlled by oxygen-sensing through ERFVIIs. The absence of any major role of ADH1 in symptom development may also suggest that PDC activity could contribute to the formation of galls through the activation of a PDH bypass.

3.8.16 Expression of SWEET Genes During *Plasmodiophora brassicae* Infection

After infection of the plant roots, *P. brassicae* colonization leads to swelling roots and gall formation, eventually inhibiting roots to uptake nutrients and water from the soil. Sugars are the main carbon source for the living organisms including pathogens and their host plants. In plants, sugars are synthesized in the photosynthetic organs and allocated into the sink organs via long-distance transport. Sugars produced in the living plants not only play key roles for plant growth and development, but also are transferred or secreted into the plant surface. Sugars secreted by plants could be used as carbon supply for pathogen growth (Patrick 1989; Chen et al. 2010). Especially, biotrophic microbes like *P. brassicae* obtain nutrients from host cells by extracting sugars (Manck-Götzenberger and Requena 2016; Roman and Rathjen 2017). In *P. brassicae*, it has been reported that 2% sucrose solution can promote germination of resting spores (Guo 2001).

In addition, accumulation of soluble sugars (hexoses and sucrose) are found in the galls of *P. brassicae*-infected *Arabidopsis thaliana* (Evans and Scholes 1995; Brodmann et al. 2002) and *Brassica* plants (Williams et al. 1968; Keen and Williams 1969), indicating that the galls might act as a newly established carbon sink during clubroot development. Sugar transporters, including well-known gene families such as MSTs (monosaccharide transporters) (Toyofuku et al. 2000), SUTs (sucrose transporters) (Reinders et al. 2008, 2011), and SWEETs (sugar will be eventually exported transporters) (Chen et al. 2012) are responsible to translocate mono-saccharides or di-saccharides.

Transcriptome studies revealed that several genes involved in sugar transport and metabolism strongly upregulated during gall formation in clubroot-infected *Arabidopsis* (Siemens et al. 2006; Malinowski et al. 2012). A transcriptome profiling performed on both shoot and root tissues of *A. thaliana* challenged by *P. brassicae* also pointed that genes involved in sucrose and starch biosynthesis are differently expressed in shoot and root (Irani et al. 2018). These transcriptomic and metabolic experiments provided insights that sugar accumulation is important for gall formation, probably depending on some sugar transport genes. Moreover, during the process of plant–pathogen co-evolution, the pathogens have evolved a mechanism to compete for sugars with infected host cells to sustain their life cycle (Doidy et al. 2012; Lemoine et al. 2013). Competition for sugars at the plant–pathogen interface is likely controlled by the sugar transporters (Toyofuku et al. 2000; Doidy et al. 2012), thus regulation of those transporters would give a new insight into disease control.

The SWEETs are a class of proteins responsible for transporting sugar across cell membranes, involving in diverse physiological processes of various plant species, such as in reproductive development (Ge et al. 2000; Yang et al. 2006a; Guan et al. 2008), leaf senescence (Quirino et al. 1999; Seo et al. 2011), and plant responses to abiotic and biotic stresses in *Arabidopsis* and rice (Chu et al. 2006; Yang et al. 2006a, b; Antony et al. 2010; Chen et al. 2010; Liu et al. 2011). The SWEETs genes are evolutionally conserved in both prokaryotes and eukaryotes (Gamas et al. 1996; Artero et al. 1998; Chong et al. 2014; Jian et al. 2016), but plant species possess more SWEETs than that in animals and prokaryotes (Yuan and Wang 2013). Phylogenetic analysis of plant SWEET genes revealed that they can be grouped into four clades as firstly defined in *A. thaliana* (Chen et al. 2010). The SWEETs grouped into clades I and II have priority to transport hexoses, while clade III and IV proteins are preferentially transport sucrose and fructose, respectively (Chen et al. 2010, 2012; Lin et al. 2014). Characterization of SWEET proteins indicated that they contain seven transmembrane (TM) helices carrying two MtN3/saliva domains in plants and animals, while only one MtN3/saliva domain with three TM helices exists in SWEET proteins of prokaryotes (Wang et al. 2014). The SWEETs localized on the plasma membrane to import or export sugars (Chen et al. 2012). However, SWEET is a distinct transporter family which mediates energy-independent influx and efflux of sugars compared to MSTs and SUTs, both of which are energy-dependent (Reinders et al. 2008, 2011; Yuan and Wang 2013). Some SWEETs harboring two MtN3/saliva domains functioned as low-affinity glucose and sucrose transporters, such as glucose transportersAtSWEET4, AtSWEET5, AtSWEET7, and

AtSWEET13, and sucrose transporters AtSWEET11 and AtSWEET12 in *Arabidopsis* as well as OsSWEET11 and OsSWEET14 in rice (Chen et al. 2010, 2012).

The SWEETs are also crucial to regulate carbon transport in parasitism and pathogens interaction. Some members of the SWEET genes have been demonstrated as the targets of extracellular pathogens. During the host–microbe interaction, the pathogen infection regulates expression patterns of the SWEET genes to facilitate the pathogen obtaining the sugar. Indeed, induction of SWEET genes upon pathogen infection has also been reported in *Arabidopsis*, rice, and other plant species. In *Arabidopsis*, at least 9 SWEET members (AtSWEET 2, 4, 7, 8, 10, 11, 12, 15, and 17) were differentially regulated following infection of different types of pathogens (Ferrari et al. 1994; Chen et al. 2010). In the *Arabidopsis–P. brassicae* interaction, *P. brassicae* infection strongly induced the expression of AtSWEET15 during the gall formation (Siemens et al. 2006). The role of SWEETs during *P. brassicae*-induced gall formation in Chinese cabbage was examined for sugar contents in roots and leaves of the Chinese cabbage clubroot resistant (CR) and susceptible (CS) near isogenic lines (NILs) after infection of *P. brassicae*. In addition, the response of *B. rapa* SWEET (hereafter BrSWEET) sugar transporters to *P. brassicae* infection as well as clubroot disease development in *Arabidopsis* SWEET11 mutant was examined. The pathogen lives inside roots, and hijacks nutrients from the host plants. It is suggested that clubroot galls created an additional nutrient sink in infected roots. Sugar contents in leaves and roots before and after *P. brassicae* infection using a pair of Chinese cabbage near-isogenic lines (NILs), carrying either a clubroot resistant (CR) or susceptible (CS) allele at the CRb locus were assessed. *P. brassicae* infection caused significant increase of glucose and fructose contents in the root of CS-NIL compared to CR-NIL, suggesting that sugar translocation and *P. brassicae* growth are closely related. Among 32 *B. rapa* SWEET homologs, several BrSWEETs belonging to Clade I and III were significantly upregulated, especially in CS-NIL upon *P. brassicae* infection. Moreover, *Arabidopsis* sweet 11 mutant exhibited slower gall formation compared to the wild-type plants. *P. brassicae* infection probably triggers active sugar translocation between the sugar producing tissues and the clubbed tissues, and the SWEET family genes are involved in this process (Li et al. 2018).

3.8.17 Expression of Host-Specific Genes

3.8.17.1 *Brassica rapa*

Orphan genes are often associated with biotic and abiotic stress, and they usually function in stress-response signal transduction pathways. Using RNA sequencing, transcriptome profiling was co-paired between clubroot-resistant and clubroot-susceptible *B. rapa* genotypes after infection with *P. brassicae* during early infection (Chen et al. 2015a, b). Jiang et al. (2018) reassembled respective RNA sequencing data to analyze the BSGs expression patterns, and the results indicated that 180 unique BSGs were supported by RNA-Seq data (FPKM >2 for least one

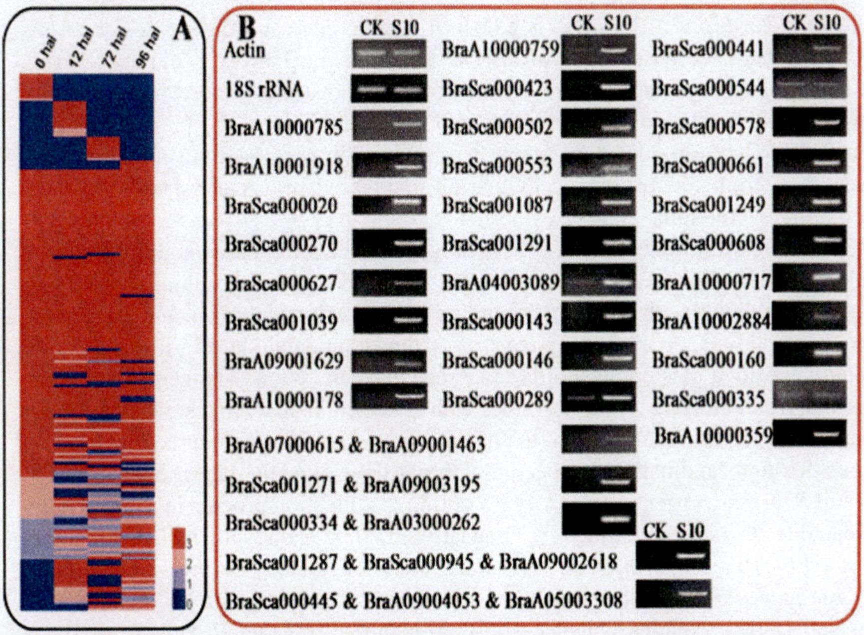

Fig. 3.14 Expression patterns of BSGs in *Brassica rapa* after *Plasmodiophora brassicae* infection. (**a**) Heat map of BSGs after *P. brassicae* treatment of Chinese cabbage (FPKM >2). Zero, 12, 72, and 96 hai (hai: hours after inoculation) represents clubroot-susceptible Chinese cabbage (BJN3-2) that was infected by *P. brassicae*. Gene expressions were analyzed by semi-qPCR. *B. rapa* 18S rRNA and Actin genes were used as the internal control. (**b**) *P. brassicae* treatment (Jiang et al. 2018)

sampling stage (Fig. 3.14a, b). Further, 24 BSGs exhibited high expression abundance at least in one sampling stage (FPKM >50). Seventy-eight BSGs displayed constitutive expression at every point (FPKM >2), and 32 BSGs showed period-specific expression (FPKM >2). Deora et al. (2012) indicated that *P. brassicae* is blocked at middle or late stages of infection in resistant *Brassica* crops. However, BSGs may play more important roles in middle or late periods [4–20 days after inoculation (dai)]. To confirm this hypothesis, roots from *B. rapa* plants were collected at 10 dai with *P. brassicae* (referred to CK and S10), and 52 successfully amplified BSGs in Chiifu were analyzed using semi-qPCR. As expected, 78.85% (41 genes) of analyzed genes responded to *P. brassicae* treatments, and 39 out of 41 BSGs were upregulated (Fig. 3.14b). However, only 8 out of these 52 BSGs have transcript support (FPKM >2) (Fig. 3.14a). The results strongly suggest that BSGs play important roles in *P. brassicae* stress responses (Jiang et al. 2018).

3.8.17.2 *Arabidopsis thaliana*

The crucifer *Arabidopsis thaliana* has been used to characterize the expression of several *Plasmodiophora brassicae* genes during pathogenesis (Siemens et al.

2009a, b). *Arabidopsis thaliana* is a host of *P. brassicae* and the small size of the roots facilitates the observation of disease development. However, clubroot development in *Brassica*, takes a long time and the roots show more homogeneous developmental stages of plasmodia in comparison to *A. thaliana* (Gustafsson and Fält 1986). Therefore, the short life span of this small weedy species might prevent researchers from identifying genes that are expressed in a stage-specific manner during the interaction.

The expression of several *P. brassicae* genes over the time course of the host–pathogen interaction in *Brassica rapa* were analyzed by semi quantitative RT-PCR in order to find molecular markers for the actively growing pathogen, and for distinct developmental stages of *P. brassicae*, early developmental stages in root hairs or young vegetative plasmodia shortly after secondary infection. The gene expression was analyzed in three *B. rapa* host lines with a virulent isolate, to test whether semi-quantitative RT-PCR can be used to characterize host–pathogen interactions in crop plants. To find landmarks for specific clubroot development stages, expression of several *P. brassicae* genes was analyzed during the infection process in a universally susceptible *Brassica rapa* line (cv. Granaat, ECD 05). The results revealed that the genes *Pb-YPT, Pb-Brip9*, and *Pb-PSA* were regulated during infection of *P. brassicae*, *Pb-Brip9*, and *Pb-PSA* were expressed in early stages of infection. Wu et al. (2012) provides evidence that semi-quantitative RT-PCR could be used for discriminating developmental stages of *P. brassicae* in *B. rapa*. Stage-specific gene expression of *P. brassicae* should be studied in *B. rapa* roots due to longer disease development in comparison to *Arabidopsis thaliana* (Wu et al. 2012).

3.8.18 Expression of miRNAs in Clubroot Pathogenesis

MicroRNAs are a highly conserved class of small noncoding RNAs that regulate gene expression by post-transcriptional repression. Emerging evidence indicates that hosts' endogenous small RNAs represent an important mechanism of control in plant immune responses and hormone signaling at the time of stress. RNAs like ath-miR 160 and ath-miR167 are involved in pathogenesis and target the auxin-response-factor (ARF). Another microRNA, ath-miR 164, has been implicated in auxin homeostasis and lateral root development, which may have a bearing on clubroot development. Therefore, it is conceivable that miRNAs may be involved in mediating important plant processes following infection with *P. brassicae*, which results in the development of root galls, leading to stunting and decreased productivity. Now, microarrays of miRNAs have become available for high-throughput global analysis of miRNAs expression patterns in response to pathogens. This valuable tool has been further exploited in a number of plants to study the miRNA: mRNA interactions in tomato, rice, and *Arabidopsis* which may lead to the identification of miRNAs that are involved in plant responses to specific stresses, including *P. brassicae* challenge (Li et al. 2010; Mallory et al. 2005; Guo et al. 2005; Liu et al. 2008). Clubroot disease of Brassicaceae family has potential to substantially limit *Brassica* production. A miRNA microarray-based approach was used to identify miRNA expression and regulation during the disease initiation and progression (Verma et al. 2014). Results revealed the useful basis toward understanding the dynamics of miRNAs related to *P. brassicae* infection of *Brassica*. The differential expression analysis of miRNA during pathogenesis of *P. brassicae* was carried out at two developmentally distinct time points. The expression of various miRNAs was observed to be modulated during the process of clubroot initiation and development. Many of these miRNAs were involved in the regulation of gene activity by targeting, among others, TFs, ARFs, MYB, MADS box, AP2, and ERFs. Many of these are known to be involved in stress-responses and developmental events related to pathogenesis. A diverse array of miRNAs and their target genes are modulated during the pathogenesis of *Brassica* by *P. brassicae*. It enhances the understanding related to the biological processes involved in *P. brassicae-Brassica* interactions and may lead to unique ways to generate disease resistant phenotypes (Fig. 3.15).

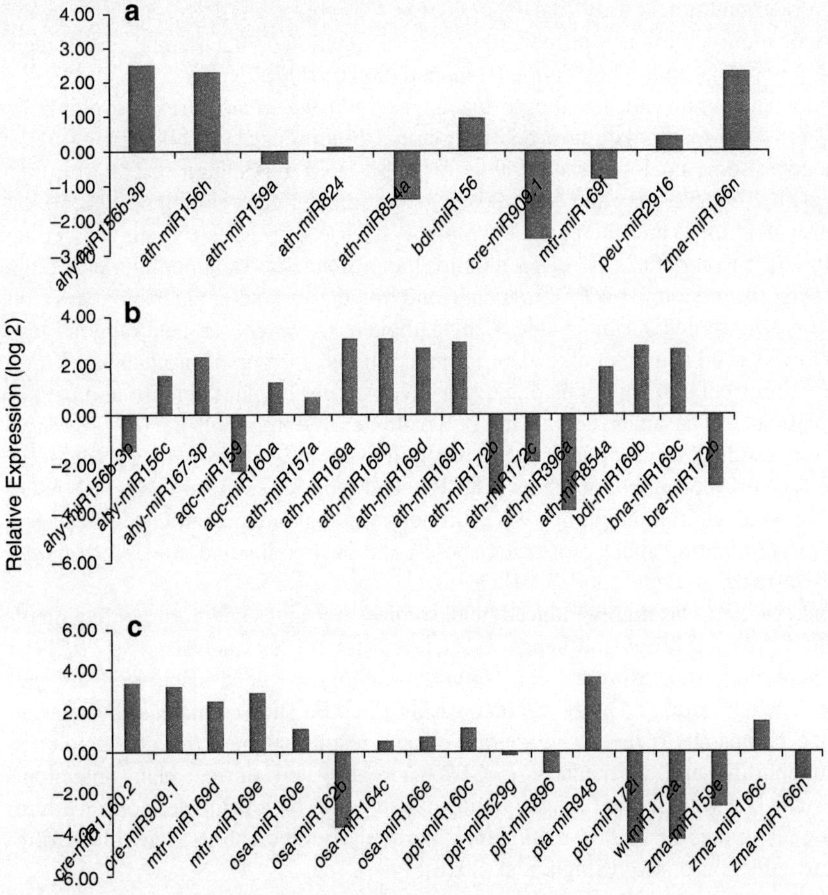

Fig. 3.15 miRNA-microarray expression of *Plasmodiophora brassicae* responsive miRNAs exhibiting differential expression at 10- (**a**) and 20-dpi (**b**, **c**) following pathogen infection (Verma et al. 2014)

3.8.19 Expression of Immunophilins During *Plasmodiophora brassicae* Infection

Immunophilins are ubiquitous proteins with properties that allow them to regulate protein structure, activity, and stability (Wang and Heitman 2005; Hanes 2015). They operate either by peptidyl-propyl isomerization of selected targets, as chaperons or by binding of small ligands. The immunophilins comprise of three structurally unrelated sub-families: the cyclophilins (CYPs), the FK506-binding proteins (FKBPs), and the parvulin-like proteins (Hanes 2015). Besides modulating the formation of cis-trans isomers of proline (Galat 2003), immunophilins have two important amino acid properties: (a) prolyl-isomerase activity, which catalyzes the rotation of the X-Pro peptide bonds from the cis to trans configuration, a rate-limiting step in protein folding (Wang and Heitman 2005), and (b) affinity to bind to immune suppressive drugs. The stramenopile human parasite *Blastocystis* sp. lives under anaerobic conditions partly like *P. brassicae* (Gravot et al. 2016), and secretes a range of immunophilins with potential effector functions, which could lead to cell death (apoptosis) in the host tissue (Denoeud et al. 2011).

Another animal parasite, the protozoan *Toxoplasma gondii* uses a cyclophilin (TgCYP18) to manipulate host cell responses (Ibrahim et al. 2009). Among plant pathogens, some cyclophilins are able to inhibit RNA replication of plant viruses (Lin et al. 2012; Kovalev and Nagy 2013), while some effector proteins interact with plant cyclophilins, and thereby induce plant defense responses (Domingues et al. 2010, 2012; Kong et al. 2015). To monitor the presence of immunophilin encoding genes in the genome of *P. brassicae*, and to analyze potential candidates for function, the molecular amenable plant pathogen *M. oryzae* as heterologous test system was used (Singh et al. 2018). Immunophilins, a group of proteins known to have various cellular functions, including involvement in plant defense and pathogen virulence was analyzed in clubroot pathogen. Searches in the *P. brassicae* genome resulted in 20 putative immunophilins comprising of 11 cyclophilins (CYPs), 7 FK506-binding proteins (FKBPs), and 2 parvulin-like proteins. RNAseq data showed that immunophilins were differentially regulated in enriched life stages such as germinating spores, maturing spores, and plasmodia, and infected *Brassica* hosts (*B. rapa, B. napus,* and *B. oleracea*).

The *PbCYP3* was highly induced in all studied life stages and during infection of all three *Brassica* hosts, and hence was selected for further analysis. The *PbCYP3* was heterologously expressed in *Magnaporthe oryzae* gene-inactivated Δ*Cyp1* strain. The new strain Δ*Cyp1*+ overexpressing *PbCYP3* showed increased virulence on rice compared to the Δ*Cyp1* strain. These results suggest that the predicted immunophilins and particularly *PbCYP3* are activated during plant infection. *M. oryzae* is a well-studied fungal pathogen and could be a valuable tool for future functional studies of *P. brassicae* genes, particularly elucidating their role during various infection phases (Singh et al. 2018).

3.8.20 Analysis of Host Response Through Transcriptome Data

Two RNASeq experiments were performed to gain insight into the genes involved in host penetration, infection and gall formation. RNA extracted from the roots of the clubroot-susceptible *B. napus* line DH12075sampled at weekly intervals (from 1 to 6 weeks) after inoculation with Pb3 spores was sequenced. Sequences generated for RNA prepared from the roots of DH12075 plants mock inoculated with water at the same time points served as controls. In addition, Rolfe et al. (2016) sequenced RNA from control and infected roots or hypocotyls of *A. thaliana* plants at the onset (16 DPI) and at late stages (26 DPI) of gall formation with three independent biological replicates. This second experiment used a population of *P. brassicae* described by Malinowski et al. (2012). Analysis of differential expression in infected *B. napus* found significant changes in the expression of 617 *Pb3* genes during the 6-week course of infection (genes in the *B. napus* data were considered to be differentially regulated if they showed a twofold or greater change inexpression in at least two consecutive samples and an adjusted P value of ≤ 0.05). Within the differentially expressed genes there were genes involved in carbohydrate processing, amino-sugar and nucleotide sugar metabolism, transporters, cytoskeleton, motor proteins and flagella-associated proteins. Of the differentially regulated transcripts, 41 were genes encoding for predicted secreted proteins with 14 of them being SSPs. Hydrolases and peptidases were the most common differentially regulated secreted proteins. The infected *A. thaliana* samples, ~25 million paired-end reads were obtained from each sample. For the 16 DPI samples, ~13% (3.5 million) of the paired reads matched the *P. brassicae* genome in both hypocotyl and root samples. For the 26 DPI samples, 48% of the paired reads (12 million) from hypocotyl tissue and somewhat less (34%, 9 million) from root tissue matched the *P. brassicae* genome. The root and hypocotyl samples at each time point were very similar to each other. Only 53 and 32 genes showed differential expression (twofold or greater with corrected probability of $P \leq 0.05$) between the 16 and 26 DPI hypocotyl and root tissue, respectively. In contrast, gene expression differed markedly between the 16 and 26 DPI samples within each tissue (1661 genes for hypocotyl tissue and 957 genes for root tissue, of which 763 were in common). Genes that were differentially up- or downregulated ($\geq$2-fold, $P \leq 0.05$ after correction for false discovery) between 16 and 26 DPI samples in both tissue types were identified using DESeq2 and GO terms analyzed using BLAST2GO (Rolfe et al. 2016).

Several GO terms associated with cytoskeletal and membrane activities (cytoskeletal protein binding, plasma membrane, ion transport, cytoskeleton, cell differentiation, protein transport) were downregulated at 26 DPI compared with 16 DPI, indicative of a shift in metabolic processes associated with the change from plasmodial growth form to spore formation. Many genes associated with general cellular metabolism (carboxylic, lipid and organo-nitrogen metabolic processes, catalytic activity) showed the opposite pattern, with enrichment in the later stages of gall development and spore formation (Rolfe et al. 2016).

3.8.20.1 Transcriptome Analysis of *Brassica–Plasmodiophora* Interaction at Early Infection Stage

To explore the resistant molecular mechanism of the interaction between *Plasmodiophora brassicae* and *Brassica napus* during early infection, gene differential expression analysis of the clubroot-resistant rapeseed line ZHE -226 (R) and clubroot-susceptible line10159 (S) were performed by RNA seq. A total of 809 and 1082 genes were commonly up- and downregulated at 0, 12, 48, 72 h (hpi) time points in R line. Pattern recognition receptors (PRRs), chitinase, R gene, WRKY transcription factor, salicylic acid (SA) and jasmonic acid (JA) were involved in the early response in clubroot infection in lines of R and S, but showed different expression patterns. Compared to S line, most of the PRRs genes were downregulated, *R* genes were upregulated and SA genes were downregulated in earlier stage but upregulated at 72 hpi in R line. Chitinase-, WRKY-, and JA-related genes were both induced in R and S lines. These results showed that PTI mediated by PRRs was significantly induced in S line, but played a small role in R line. ETI-mediated by R gene and resistance signal pathways mediated by SA played great role in clubroot resistance. It provides a better understanding of the resistance genes of *B. napus* in response to *P. brassicae* and assistance for clubroot resistant breeding (Luo et al. 2019).

3.8.21 Manipulation of Host Defense/Resistance

A number of lines of evidence support the view that *Plasmodiophora brassicae*, in common with many other biotrophic pathogens, manipulates the defense responses of its hosts. Analysis of differentially expressed putative Pb3 effectors within the first 6 weeks after post-inoculation has provided evidence for the manipulation of plant defense, by the induction of protease inhibitors and possibly suppression of salicylic acid (SA)-which induced defense. Pair wise comparisons of *B. napus* genes differentially expressed in response to Pb3 infection using the application HORMONOMETER showed that salicylic acid mediated responses at the earliest stages of infection was negatively correlated with the *A. thaliana* gene expression profile in response to SA. In addition, the expression of a *P. brassicae* predicted secreted protein (PbPT3Sc00026_A_1.308_1) identified as a benzoic acid/salicylic acid methyl transferase (Pb3-BSMT) (NBCI accession number AFK13134) was extremely low in *B. napus*-infected roots during the first week after inoculation, but increased in 2 weeks and peaked at weeks 3 and 4. Likewise, expression was higher in infected *A. thaliana* tissue at 16 DPI than at 26 DPI. A similar expression pattern for this gene was reported by Ludwig-Muller et al. (2015). Rolfe et al. (2016) confirmed the methyl transferase activity of this enzyme and suggested a possible role for this enzyme in inactivating plant defense by converting SA (Salicylic Acid) to MeSA (Methyl Salicylate), which acts as a mobile signal in systemic acquired resistance (SAR) in plants.

3.8.22 Role of G-Protein-Coupled Receptors (GPCR) Signal Transduction Pathways in *Plasmodiophora brassicae* Pathogenesis

The comparative genomics analysis demonstrated the presence of a set of significantly expanded and species-specific genes within the *Plasmodiophora brassicae* genome. GO enrichment analysis in this set of genes revealed that several functional groups, including G-protein-coupled receptors (GPCR) signaling pathways, were significantly enriched in the *P. brassicae*. Results have shown that GPCRs, a large family of extremely versatile cell-surface receptors, are able to sense signaling molecules (such as biogenic amines, purines, odorants, ions, and pheromones) outside the cells, then activate signal transduction pathways inside the cells and, ultimately, elicit cellular responses (Prabhu et al. 2007). The GPCRs play a central role in eukaryotic signal transduction and are involved in the regulation of many cell functions, including cell proliferation, survival, and motility (Dorsam and Gutkind 2007; Krishnan et al. 2014). The transcriptome analysis revealed that these GPCR pathway-related genes are highly expressed during the infection process, including the germinating resting spore stage and the cortical infection stage. The treatment with a GPCR inhibitor relieves, the symptoms of clubroot disease, suggesting that a significantly enriched GPCR signaling pathway plays an important role in plant–*P. brassicae* interaction. Inhibitor treatment suppressed both the root hair and cortical infection, indicating that the GPCR signaling pathway of *P. brassicae* is involved in the perception and recognition of external signals, including chemical molecules (from host root exudate or soil environment) and signal stimulation from the intracellular space of the host. As a soil-borne obligate parasite, *P. brassicae* survives in soil in the form of resting spores. Precise sensing of signal molecules (such as calcium ions or other signals from the host plant) is important for resting spores to exit dormancy and begin germination, as initiation of the initial step of the life cycle must take place under suitable conditions. During the root hair infection stage and the cortical infection stage, *P. brassicae* acts as an intracellular parasite that takes up nutrients from the host and is able to sense the intracellular environment of the host cells (Kageyama and Asano 2009; Perez-Lopez et al. 2018). For *P. brassicae*, a close interaction with its host may be the driving force behind the evolution of enriched GPCR signaling pathways, as these are key molecular mechanisms for adapting to the intracellular environment.

The ability to sense, and transduce extracellular signals into intracellular responses using a sophisticated molecular signaling system is crucial for living organisms. G-protein-coupled receptors (GPCRs), characterized by a seven transmembrane domain structure, are involved in a wide variety of physiological processes. Signals from a diverse range of extracellular signaling molecules, including hormones, neuro-transmitters, ions, photons, and odorants, can be relayed by GPCRs to intracellular downstream signaling pathways to ultimately elicit cellular responses (Hilger et al. 2018). Based on their sequence homology and functional similarity, the super family of GPCRs can be generally divided into six classes (A, B, C, D, E, and F) (Qiu et al. 2009; Alexander et al. 2017). The signal transduction of

GPCRs is dependent upon receptor-mediated activation of heterotrimeric G proteins consisting of three subunits, Gα, Gβ, and Gγ (Kamp et al. 2016). As one of the main signaling pathways in eukaryotes, the signaling system of GPCRs is involved in numerous processes including developmental control, cell growth, migration, density sensing, and neuro-transmission (Rosenbaum et al. 2009; de Mendoza et al. 2014). As important central sensors that activate eukaryotic signaling, GPCRs are the primary targets of human drugs (van den Hoogen et al. 2018). Despite this, the specific roles of these receptors and their downstream signaling pathways in protists remain poorly understood. Bi et al. (2019) conducted a comparative genomics analysis to determine the role of the GPCR signaling pathway in *P. brassicae* pathogenesis of crucifers. It was observed that *P. brassicae* is an important biotrophic eukaryotic plant pathogen, and a member of the rhizarian protists. This biotrophic pathogen causes clubroot in cruciferous plants via novel intracellular mechanisms that are markedly different from those of other biotrophic organisms. To date, genomes from six single spore isolates of *P. brassicae* have been sequenced. An accurate description of the evolutionary status of biotrophic protists, however, remains lacking. Bi et al. (2019) determined the draft genome of the *P. brassicae* ZJ-1 strain. A total of 10,951 protein-coding genes were identified from a 24.1 Mb genome sequence. After a comparative genomics approach to prove the Rhizaria super group is an independent branch in the eukaryotic evolutionary tree. It was found that the GPCR signaling pathway, the versatile signal transduction to multiple intracellular signaling cascades in response to extracellular signals in eukaryotes, is significantly enriched in *P. brassicae*-expanded and *P. brassicae*-specific gene sets. Additionally, treatment with a GPCR inhibitor relieved the symptoms of clubroot and significantly suppressed the development of plasmodia. These findings suggest that GPCR signal transduction pathways play important roles in the growth, development, and pathogenicity of *P. brassicae* to the crucifers at large.

3.9 *Brassica–Sclerotinia* Pathosystem

3.9.1 Introduction

The *Sclerotinia sclerotiorum* (Lib.) de Bary is a soil borne pathogenic fungus, which was first described as *Peziza sclerotiorum* by Libert. Whetzel (1945) established *S. sclerotiorum* (Lib.) de Bary as the major species of *Sclerotinia*, as it is the best-known and most economically important pathogen of the genus. It was proposed as the representative species for the genus *Sclerotinia.* The family Sclerotiniaceae also includes two other very closely related species, *S. minor* and *S. trifoliorum.* All three species produce vegetative resting structures, termed sclerotia, and mushroom-like reproductive structures, named apothecia, during their life cycles. However, S. minor forms small sesame-sized sclerotia, whereas *S. sclerotiorum* and *S. trifoliorum* can form large sclerotia as big as peas. They all lack an obvious conidial stage. *S. sclerotiorum* has the broadest host range, which includes both herbaceous and succulent plants, and even some woody ornamentals and monocots.

The *Sclerotinia sclerotiorum* is a highly damaging pathogen with diverse infection modes and a double feeding lifestyle of both biotroph and necrotroph. It attacks host plants either by means of ascospores that can be discharged forcibly upwards from apothecia into the air, or by mycelium arising from infected tissue or from germinated sclerotia. When ascospores land on susceptible host tissue, they can germinate under favorable conditions and start a new cycle of infection. To date, the micro-conidia of *S. sclerotiorum* have been observed amongst the mycelium and on the surfaces of sclerotia. However, no evidence of their functionality has been identified. Under moist and cool conditions, this fungus rapidly grows inside the infected host tissues and develops symptoms of browning, water-soaking, and a white, cotton-like mycelium, which leads to necrosis, stunting, premature ripening, and wilting of the host. Therefore, the diseases that it causes have been given names, including stem rot, drop, crown rot, cottony rot, watery soft rot, blossom blight, and Sclerotinia blight. Upon killing the host, the fungus can saprophytically grow on the dead plant tissue. Sclerotia are later abundantly formed on the host surface and cavities, in plant debris, and in soil, where they are able to remain dormant for up to 10 years. *S. sclerotiorum* has an extremely broad host range, which consists of more than 600 plant species, including almost all dicotyledonous and some monocotyledonous plants (Saharan and Mehta 2008; Liang and Rollins 2018; Xia et al. 2020).

In oilseed rape, infection by *Sclerotinia* is usually via ascospore bearing petals. At petal fall ascospore infected petals stick to leaves initiating infections that can develop into stem rot lesions (McCartney et al. 2001a, b). Ascospores discharged from the apothecia at the base of the plants in soil constitute an important primary source of infection. Mycelium in soil or mycelium arising from the sclerotia is less important initial source of infection because of the low competitive saprophytic ability of the fungus (Newton and Sequeira 1972). The ascospores can germinate in the presence of a thin film of water, in less than 24 h at 5–30 °C, with an optimum range being 5–10 °C. On germinating, the ascospores give rise to infection hyphae and initial penetration of the host tissue takes place directly by mechanical pressure through the cuticle or the infection hypha may penetrate already wounded or injured tissue also. After entrance of the fungus into the host, the mycelium ramifies intercellularly or intracellularly causing colonization of tissue primarily because of enzymatic dissolution of the cell wall in advance and cells die some distance ahead of the invading hyphae. Pectolytic enzymes are responsible for tissue maceration indirectly damaging the cell membrane, which results in subsequent death of cells (Morrall et al. 1972). Rai and Dhawan (1976) and Dhawan and Srivastava (1987) reported production of polymethyl galacturonase transeliminase (PGTE) and cellulase (CX) enzymes by *S. sclerotiorum* infecting rapeseed-mustard plants. According to them, virulence of different isolates appears to be associated with the activity of PMG and CX enzymes. The role of protease activity in infection of plants of *B. juncea* is also demonstrated (Dhawan 1980). It appears that invasion of tissues of *B. juncea* is also related to the infection process, mediated by production of a toxin which is identified as oxalic acid (Rai and Dhawan 1976), which is formed in infected *B. juncea* plants. The toxic substance is thermostable, translocatable and treatment of the host plant with culture filtrate results in infection.

3.9.2 Role of Enzymes in Pathogenesis

Pathogenesis of *S. sclerotiorum* is complex and not well understood (Dickman and Mitra 1992). The fungus produces a wide array of degradative, lytic enzymes (e.g., endo and exopectinases, cellulases, hemicellulases, proteases) which are believed to facilitate colonization and host cell wall degradation (Marciano et al. 1983; Riou et al. 1991). Infection of canola pollen by *S. sclerotiorum* occurs by direct hyphal penetration of cell walls without the formation of appressoria or infection cushions. Hyphal penetration is more commonly observed through the germinative pores than other parts of pollen walls. Plasmolysis and disintegration of the pollen cytoplasm occurs as a result of hyphal ramification within the infected pollen grains (Huang et al. 1998). The cell walls degrading enzymes capable of destroying cellular components and production of oxalic acid are associated with disease development. de Bary (1887) stated "The power of infecting is shown by the power of penetrating the membranes (cuticle) which are evidently dissolved at the point of penetration. Hence, it is very probable that this power depends on the presence of a substance which can dissolve a membrane, ferment (enzyme) in fact." In addition, Prior and Owen (1964) mentioned possible enzymatic action on the cuticle in association with infection cushions on clover leaves. However, Purdy (1958) and Boyle (1921) were unable to discern any alteration of host tissue due to the diffusion of substances before penetration. Until the possibility of enzymatic action on host tissue prior to penetration is demonstrated experimentally, ingress into the host must be considered a mechanical process as based on histological evidence.

Pectolytic enzymes are always associated with diseases caused by *Sclerotinia* (Barkai-Golan 1974; Calonge et al. 1969; Echandi and Walker 1957; Hancock 1966; Held 1955; Lumsden 1976, 1979; Maxwell and Lumsden 1970; Morrall et al. 1972; Newton 1972; Riou et al. 1991; Sharma and Sharma 1984; van den Berg and Yang 1969). Qualitative and quantitative information is rather limited, but these enzymes have been associated with quantitative decreases in the pectic substance content of diseased tissue (Hancock 1966) and with histochemical and structural changes in host cell middle lamellae (Calonge et al. 1969; Lumsden 1976). In addition, pectolytic enzyme activity has been localized in infected tissue.

3.9.3 Expression of Genes During Pathogenesis

To identify genes involved in fungal development and pathogenesis, Li et al. (2004a, b) generated 2232 expressed sequence tags (ESTs) from two cDNA libraries constructed using either mycelia grown in pectin medium or tissues from infected *Brassica napus* stems. A total of 774 individual fungal genes are identified of which 39 are representing only among the infected plant EST collection. Annotation of

534 unigenes is possible following the categories applied to *Saccharomyces cerevisiae* and the Universal Gene Ontology scheme. cDNAs are identified that encodes potential pathogenicity factors including four endopolygalacturonases, two exopolygalacturonases and several metabolite transporters. The potential role of these genes, as well as those encoding signal transduction factors, in the infection process is very vital.

Five major and several minor PG isoenzymes have been identified in a *Sclerotinia sclerotiorum* isolate from *Brassica napus* by iso-electric focusing and pectin gel overlays. Using a combination of degenerate PCR and expressed sequence tags (ESTs) four endo-polygalacturonase (PG) genes, designated as *sspg1d, sspg3, sspg5*, and *sspg6* and two exo-PG genes, *ssxpg1* and *ssxpg2* are identified. *SSPG1d* is a member of the PG gene family previously described by Fraissinet-Tachet et al. (1995). The mature *SSPG1d* is a neutral PG, whereas fully processed *SSPG3, SSPG5*, and *SSPG6* are acidic enzymes. Under saprophytic growth conditions, *sspg1d, sspg3, sspg5,* and *ssxpg1* expression is induced by pectin and galacturonic acid and subject to catabolite repression by glucose. Conditions could not be identified under which *sspg6* or *ssxpg2* are expressed well. Transfer of mycelia from liquid media to solid substrates induces expression of sspg1d suggesting that it may also be regulated by thigmotropic interactions. Under pathogenic conditions, *sspg1d*is highly expressed during infection. The *sspg3* is also expressed during infection, albeit at lower levels than sspg1d, whereas *sspg5, sspg6,* and *ssxpg1* are expressed only weakly (Li et al. 2004a, b).

3.9.4 Molecular Biology of *Sclerotinia sclerotiorum* for Regulation of Growth, Virulence, and Pathogenesis

The gene/mutant/protein nomenclature adopted is most commonly used by *Sclerotinia* researchers, as there are many discrepancies in *S. sclerotiorum* literature. As an, Abc1 protein is encoded by wild-type gene that is italicized as *Abc1*. Mutant is denoted as abc1. Exceptions are specifically explained. Another consideration to keep in mind is the presence of huge variations in the field isolates and local hosts used in these studies (Table 2.11; see Chap. 2, Sect. 2.10.6.2), which likely contribute to some of the discrepancies that were observed in the phenotypes of mutants of the same genes (Xia et al. 2020).

3.9.4.1 Regulation of Mycelial Growth and Virulence

During the last two decades, the availability of whole genome sequences and improved transformation as well as knockdown/knockout methods have greatly facilitated the molecular study of *S. sclerotiorum*. Most of the genes studied so far that affect growth are relatively conserved genes that also affect differentiation, such as sclerotial and apothecial development (Fig. 2.32, See Chap. 2, Sect. 2.10.6.2), affecting hyphal growth rate and virulence, but not differentiation, during development, as mycelial growth precedes and processes of virulence and differentiation.

The mutants here with major growth defects often exhibit virulence attenuation, although there are certainly exceptions such as mat1-2-1, which exhibits slower growth, aberrant apothecial morphogenesis, but normal virulence (Doughan et al. 2016). The RNAi knockdown of highly conserved transcription factors Ams2 or Mads lead to reduced mycelial growth rate and virulence, which suggested that they are early growth regulators. Ams2 is a GATA-box domain transcription factor. Ams2 is required for proper expression of histone and cell cycle-related genes, similar as in *Schizosaccharomyces pombe*. Mads is orthologous to yeast Mcm1 with aMADS-box domain, which is essential for viability, cell cycle, mating, mini-chromosome maintenance, recombination, and stress tolerance. Although mutants of these early growth factors severely affect growth, they can still form sclerotia, which indicated that the later stages of fungal differentiation pathways, such as sclerotial formation and apothecial development, might be less dependent on these early growth factors (Qu et al. 2014; Liu et al. 2018a).

The BAX inhibitor-1 (Bi1) is a highly conserved protein that is found in both eukaryotes and prokaryotes, exhibiting apoptosis-inhibiting activity. RNAi knock-down mutants in *S. sclerotiorum* showed more aerial hyphal growth and reduced virulence. The mutants also exhibit reduced tolerance to chemical stress, as its gene expression is induced by various stresses. Bi1 is likely a stress specific regulator dedicated to both biotic and abiotic stress responses, due to its minor roles in growth and development. Unlike the previously described genes, *PemG1* serves an opposite function in regards to growth and virulence. The RNAi mutants of *PemG1* exhibit faster mycelial growth and enhanced virulence, indicating that it is a negative regulator in these processes. Consistently, more infection cushions, more CWDE activities, and less susceptibility to salt stress were observed. However, the exact mechanism on how this fungal specific protein negatively regulates growth and virulence is unclear (Xia et al. 2020).

3.9.4.2 Signaling Events for Sclerotial Formation

Sclerotia of *Sclerotinia sclerotiorum* initiate with small primordia and rapidly develop into white compact hyphal masses. They mature with dehydration and pigmentation once they cease growth. The melanized outer surface of the sclerotium is mostly responsible for resisting adverse conditions. As expected for broad regulators, these mutants show not only defective growth and sclerotial malformation, but also almost always loss of virulence and other pleiotropic defects (Table 2.11, Chap. 2, Sect. 2.10.6.2). The melanization of sclerotia is an important process for the long-term survival of *Sclerotinia* species in nature. The deletion of melanin biosynthesis genes scytalone dehydratase (*Scd1*) or trihydroxynaphthalene reductase (*Thr1*) resulted in slower mycelial growth, reduced number, size, and pigmentation of sclerotia, which suggested that melanization, is critical for sclerotial maturation and perhaps feedback promotion of growth. However, unlike in other pathogenic fungi, the virulence of these knockout strains is not affected, suggesting that melanin does not contribute to pathogenesis in *Sclerotinia* as it does for other fungi, like *Magnaporthe* species (Liang et al. 2017).

Knocking out some other conserved genes in *S. sclerotiorum* can also lead to diverse defects. For example, the peroxisomal carnitine acetyl transferase (CAT) encoding gene in *S. sclerotiorum* is Pth2. The CAT activity is required for Acetyl-CoA transport, which is needed for fatty acid metabolism. Thepth2 knockout mutants show aberrant growth, nonsclerotial formation, and reduced oxalic acid (OA) levels and virulence (Liberti et al. 2013). In another instance, the silencing of integrin-like gene *Itl* leads to altered colony morphology, smaller and irregular shaped sclerotia, and hypovirulence. Although *Cvnh* homologs seem to only exist in *Sclerotinia* and *Botrytis*, the CVNH domain is a conserved domain. The silencing of this gene lead to reduced growth rate, nonsclerotial formation, and lower virulence (Lyu et al. 2015). Phosphorylation and de-phosphorylation events are critical for almost all signal transductions. Therefore, it is not surprising that a number of kinases and phosphatases are involved in both growth and sclerotial formation. Mutation of two-component histidine kinase Shk1 leads to slower and altered hyphal growth and failed sclerotial formation. Although the mutant is also sensitive to osmotic stress and shows heightened resistance to fungicides, it exhibits normal virulence. Shk1 likely works upstream of the MAPK cascade to control these processes. Smk1 and Smk3 are two MAPK kinases that have been studied in *S. sclerotiorum* so far. Mutants of these two kinases show similar growth defects and nonsclerotial formation, suggesting that they may function downstream of Shk1 (Bashi et al. 2016; Chen et al. 2004). cAMP could act upstream of the MAPKs, as exogenous application of cAMP inhibits Smk1 expression and MAPK phosphorylation. Type 2A phosphoprotein phosphatase Pph1 possibly functions downstream of the MAPK. The mutants of pph1 show arrested growth. Consistently, PP2A R2 B regulatory subunit rgb1 mutants show aberrant sclerotial formation and failed penetration during infection (Erental et al. 2007).

Reactive oxygen species (ROS) play diverse roles in development and virulence in fungi. This is supported by the mutant analysis of type I catalase Scat1. The deletion of Scat1 leads to slower radial growth, higher number of small sclerotia that are not properly melanized, and hypovirulence. Calcium is another common factor that is involved in signaling. The study of Cna1 and Caf1 supports its contribution in *S. sclerotiorum* biology. Calcineurin Ser/Thr protein phosphatase Cna1 is conserved in eukaryotes, whose activity is dependent on calcium and calmodulin. Antisense cna1 mutants show altered growth, sclerotial development, and hypovirulence. Caf1 is a secreted protein with putative Ca^{2+}-binding EF-hand motif, which seems to be conserved in fungi. The caf1 T-DNA mutants exhibit slightly slower growth, deformed sclerotia, and hypovirulence (Xiao et al. 2014).

Conserved transcription factors are another large group of genes that are responsible for signaling. Sfh1 is a GATA transcription factor, but with an SNF5 domain. It is a member of the house keeping RSC (Remodels Structure of Chromatin) complex, which is an ATP-dependent chromatin remodeler that is essential for cell cycle. Its diverse functions explain its involvement in growth, sclerotial development, and virulence. Transcription factor Ste12 likely acts downstream of the MAP kinase cascades. In ste12 RNAi lines, slow growth, smaller sclerotia, and fewer appressoria, leading to hypovirulence, were observed. Forkhead (FKH) box (FOX) family transcription factors are important regulators of primary metabolism, cell cycle, and morphogenesis in animals and fungi. The fkh1RNAi mutants show slower growth, nonsclerotial formation, hypovirulence, and defective stress responses, suggesting that it is a transcriptional regulator that is shared by multiple pathways (Fan et al. 2017).

The CWDEs have been speculated to play key roles in fungal pathogenesis. In support of this, the deletion of Xyl1 that encoded an endo-β-1, 4-xylanase leads to a loss of pathogenicity. Interesting, growth retardation, less sclerotial formation, and dense hyphal branching were also observed in the mutant, suggestive of a role of xylanase in development as well (Yu et al. 2016).

Opsins are conserved light-sensitive proteins involved in circadian rhythms. The sop1 RNAi lines show reduced growth, nonsclerotial formation, hypovirulence, and heightened sensitivities to osmotic and fungicide stresses. These observations suggest that light may have diverse roles in various processes in *Sclerotinia* through opsins. Rearrangement hotspot (Rhs) repeat-containing proteins are widely distributed in bacteria and eukaryotes. The *Rhs1* is a secreted protein and its orthologs seem to only be present in *Sclerotinia* and *Botrytis* species. The *Rhs1* RNAi mutants show slightly slower colony growth, fewer and larger sclerotial formation, and hypovirulence. These data suggest that Rhs1is likely a key virulence factor for *Sclerotinia* and it also slightly contributes to growth and sclerotial development. Cerato-platanins are fungal secreted elicitors that can trigger plant immunity and cell death. Sm1 can cause host hypersensitive response. Interestingly, besides virulence, RNAi mutants of Sm1 also exhibit slower growth, impaired sclerotial development, and heightened sensitivities to different stresses, suggesting a broader role of the elicitor. However, these defects were not detected in a separate study in deletion mutant *cp1* mutating the same gene. Future careful transgenic complementation experiments are needed to solve these discrepancies (Yang et al. 2018).

The existence of mutants that have normal mycelial growth, but do not form sclerotia, suggest that later stages of sclerotial development are separated from growth. Although most of these mutants still show attenuated virulence, some exhibit normal pathogenicity, such as *sl2* and *ggt1* (7-glutamyl transpeptidase). The RNAi knockdown of Sl2, GAPDH (*Gpd*), or *Hex1* lead to altered sclerotial formation or melanization, similar to its interacting proteins glyceraldehyde-3-phosphatedehydrogenase (GAPDH) and Hex. However, the virulence of *sl2* is not affected, suggesting that sclerotial formation and virulence can be separately controlled by distinct downstream factors. As a fungal-specific secreted protein localizing to fungal cell wall, how Sl2 affects sclerotial development is unclear. Ggt is a highly conserved enzyme that contributes to redox homeostasis. The deletion mutants of *ggt1* show much higher GSH + GSSG and hydrogen peroxide accumulation, and separation of the cortex layer of mature sclerotia from the medulla. In addition, the mutants are defective in compound appressoria production, only affecting the penetration stage of infection. These mutant phenotypes suggest that GSH, GSSG, or GSH/GSSG ratio contribute to the regulation of sclerotia maturation and appressoria development (Li et al. 2012b).

The rest of the genes studied so far that affect sclerotial formation, but not growth, all show attenuated virulence. They likely act upstream of *Sl2* and *Ggt*. ROS increase was observed in sclerotial initials and infection cushions in *S. sclerotiorum*, likely due to the combined activities of NADPH oxidases Nox1 and Nox2. Similar to exogenous application of NADPH inhibitor or ROS scavenger, the RNAi mutants of either Nox1 or Nox2 exhibit failed sclerotial development, indicating that ROS is required for sclerotial initiation. Interestingly, reduced OA production and virulence was only observed in nox1, but not nox2, which suggests that these two genes are not fully redundant, likely due to their different expression patterns. Consistently, when the thioredoxin reductase (*Trr1*) gene that is partly responsible for ROS detoxification is silenced, less and larger sclerotia are formed. However, virulence is attenuated in *Trr1* silenced strains, suggesting that an optimum ROS is needed for virulence. Low pH was known to promote sclerotial proliferation. The pH-sensitive transcription factor Pac1 is responsible for ambient pH-induced gene expression. Consistently, the deletion of Pac1 caused aberrant sclerotial formation and reduced OA levels, thus leading to hypovirulence. As higher levels of OA reduce pH and may affect ROS homeostasis, proper balance among ROS, OA, and pH is likely paramount for the biology of this fungal pathogen (Rollins et al. 2003; Xia et al. 2020).

3.9.4.3 Control of Apothecial and Ascospore Development

The apothecium consisting of a stalk (stipe) and a saucer-shaped disc is characteristic of the *Sclerotinia* genus. Asci develop on the upper surface of apothecia and when they mature, the ascospores are dispersed. As mating-type (Mat) genes have generally been established as key regulators of sexual development of ascomycetes, it is not surprising that they affect apothecial development in the homothallic *S. sclerotiorum*. Similar to other fungi, the *S. sclerotiorum Mat* locus contains a cluster of four *Mat* genes encoding transcription factors, with *MAT1-1s* encoding an alpha-domain and *Mat1-2s* encoding high mobility group (HMG) proteins. By deletion knockout analysis, each of the four Mat genes was analyzed with regards to their contributions to regular mycelial growth, virulence, sclerotial formation, and apothecial development. While these Mats contribute little to vegetative growth, development, and pathogenesis, *Mat1-1-5, Mat1-1-1*, and *Mat1-2-1* are essential for early, and *Mat1-2-4* is critical for late sexual stages, including apothecial initiation and ascospore development. As transcription factors, it is also not surprising that these *Mat* gene mutants exhibit altered expression of the *Mat* locus and putative pheromone and pheromone receptor genes, explaining their potential regulatory mechanism and specific phenotypes (Doughan et al. 2016).

Similar to the *Mats*, another gene that specifically contributes to apothecial development is *FoxE2*. FoxE2 is a member of the forkhead box (FOX) transcription factor (TF) family. FOX proteins play diverse roles in morphogenesis, development, pathogenicity, and stress responses in animals and fungi. Knockout analysis of *FoxE2* in *S. sclerotiorum* showed that it does not affect mycelial growth, sclerotial formation, or virulence. The striking mutant phenotypic similarity between the *Mats* and *FoxE2* brings the question on the relationship among these transcription factors. It could be possible that a transcriptional network that includes these transcription factors needs to be orchestrated for proper sexual stage development in *S. sclerotiorum* (Wang et al. 2016). As a UV-A photoreceptor, Cry1 was suspected to be involved in apothecial development, as its expression was induced by UV during this specific developmental phase. However, no obvious defects in apothecial development were detected in the cry1 deletion mutant, suggesting that it is not a major regulator, or there exists redundancy masking its effects on photomorphogenesis (Veluchamy and Rollins 2008).

All other mutants with defects in apothecial development show pleiotropic defects, including sac1 (*Adenylate cyclase*) and *nsd1* (GATA-type IVb zinc-finger transcription factor), both encoding conserved proteins. The cAMP is a well-known molecule with diverse roles for many biological processes in plant fungal pathogens. Although protein kinase A catalytic subunit gene (*Pka1*) was hypothesized to mediate cAMP signaling, the *pka1* single mutant exhibits nondetectable defects, likely due to genetic redundancy of *Pka2*. In comparison, with dramatic cAMP level reduction, thesac1 mutant displays slower growth, abolished pathogenicity, and failed apothecial development. In other fungi, Nsd1 serves as transcription factor for development and environmental responses. In nsd1 knockout mutants, altered hyphal growth, smaller sclerotia, failed appressorium formation leading to failed pathogenicity, and failed apothecial initiation were observed. These indicate that Nsd1 is likely a general transcription factor that regulates many processes in *S. sclerotiorum*, or a specific regulator that regulates a general downstream factor that is shared by many pathways. A NO homeostasis regulating glutathione-dependent formaldehyde dehydrogenases (Fdh1) encoding gene was revealed to be one of the targets of Nsd1. The deletion mutants of *fdh1* exhibit slower mycelial growth, smaller sclerotia, and defective compound appressoria, leading to failed penetration into host tissue, being suggestive of the keys roles NO plays in these processes. The high similarity between the fdh1 and nsd1 mutant phenotypes indicates that *Fdh1* could be the main target of Nsd1 during growth and development (Ii and Rollins 2007; Zhu et al. 2019; Pan and Rollins 2017).

3.9.4.4 Regulation of Pathogenesis

Following a short biotrophic phase after infection, *S. sclerotiorum* causes tissue maceration and necrosis leading to rapid cell death and host cell wall degradation during fungus colonization. Toxins and CWDEs are believed to play critical roles in promoting these processes. The study of axp mutants deleting a secreted arabinofuranosidase/β-xylosidase precursor reflects the importance of CWDEs in the virulence of *S. sclerotiorum*. Attenuated virulence was observed in canola. Similarly, the critical contribution of ROS in virulence is corroborated in sod1 mutants affecting Cu/Zn superoxide dismutase. Oxalic acid (OA) is another key factor that has long been hypothesized to contribute to the pathogenicity of this pathogen.

During infection, *S. sclerotiorum* produces large amounts of OA. Earlier physiological and pharmacological studies have highlighted its importance. In addition to virulence, UV mutants where the exact mutations were not identified could not produce OA and failed to form sclerotia, also suggesting a link between OA and development. It was not until the availability of genetically defined OA deficient mutants that the contribution of OA toward *Sclerotinia* biology was partly settled. In *odc2* RNAi strain where the gene encoding oxalate decarboxylase in charge of catabolism of OA is silenced, less and nonfunctional appressoria were formed, leading to hypovirulence. These data support the involvement of developmentally regulated OA accumulation in penetration-dependent infection. In generated *oah* (oxaloacetate acetyl hydrolase) mutants through the deletion or CRISPR, where OA biosynthesis is blocked, failed appressorium formation, hypovirulence, and delayed sclerotial formation was observed. However, the deletion or T-DNA mutants of *oah* in a separate study show pH-dependent virulence alteration and normal growth and development. These discrepancies can likely be resolved through more careful examination of other mutants that are deficient in OA biosynthesis (Xia et al. 2020). Many pathogen effectors are secreted into the host to dampen host immunity and/or promote virulence. The effector biology has been focusing on the study of secreted proteins, as these effector candidates can be relatively easily identified through secretome aaA and RNA-seq analysis. Many secreted proteins in *S. sclerotiorum* have diverse roles in both development and virulence (Table 2.11, Chap. 2, Sect. 2.10.6.2). Examples of effectors only affecting virulence in *S. sclerotiorum* include Ssvp1, V263, and Qdo (quercetin dioxygenase). Less conserved Ssvp1 and V263are both small secreted proteins that contribute to virulence. On the other hand, more conserved Qdo catalyzes the cleavage of the flavonol carbon skeleton, revealing the critical contributions of degrading host antimicrobial flavonol in fungal virulence. Besides secreted proteins, there are also nonsecreted regulators that are solely dedicated to pathogen penetration and virulence regulation. The examples in *S. sclerotiorum* include Svf1 (Survival factor 1), Pks13 (polyketide synthase 13), and Nac α (nascent-polypeptide-associated complex alpha subunit). In yeast, Svf1 promotes survival under oxidative stress conditions. Therefore, it likely assists in coping with ROS during infection in *Sclerotinia*. Pks13 is likely involved in melanin biosynthesis in appressoria, as its CRISPR mutant exhibits albino compound appressoria without affecting virulence. Nac α is a highly conserved protein with transcriptional co-activator activity. Interesting, Nac α negatively impacts the expression of polygalacturonase-encoding genes, which likely contributes to the enhanced virulence phenotypes of the mutant (Li et al. 2015a, b; Yu et al. 2019; Chen et al. 2019b; Xia et al. 2020).

3.9.4.5 Changes in *Sclerotinia sclerotiorum* Transcriptome During Infection of *Brassica napus*

The RNA-Seq analysis was performed with the events occurring through the early (1 h) to the middle (48 h) stages of infection. Transcript analysis revealed the temporal pattern and amplitude of the deployment of genes associated with aspects of pathogenicity or virulence during the course of *S. sclerotiorum* infection on *Brassica napus*. These genes were categorized into eight functional groups: hydrolytic enzymes, secondary metabolites, detoxification, signaling, development, secreted effectors, oxalic acid, and reactive oxygen species production. The induction patterns of nearly all of these genes agreed with their predicted functions. Principal component analysis delineated gene expression patterns that signified transitions between pathogenic phases, namely host penetration, ramification and necrotic stages, and provided evidence for the occurrence of a brief biotrophic phase soon after host penetration. *S. sclerotiorum* deploys an array of factors and complex strategies to facilitate host colonization and mitigate host defenses. A broad overview of the sequential expression of virulence/pathogenicity-associated genes during infection of *B. napus* by *S. sclerotiorum* provides information for further characterization of genes involved in the *S. sclerotiorum*-host plant interactions (Seifbarghi et al. 2017).

3.9.5 Identification of an *Arabidopsis* NLR Genes Conferring Susceptibility in *Brassica* to *Sclerotinia*

Resistance (R) genes mediating complete disease resistance belong to the nucleotide-binding site leucine-rich repeat (NLR) family, genes underlying QDR discovered to date span a broad range of molecular functions (Poland et al. 2011; Roux et al. 2014; Corwin and Kliebenstein 2017). Molecular function of genes associated with QDR include for instance transporters (Krattinger et al. 2009), kinases (Fu et al. 2009; Huard Chauveau et al. 2013), peptidases (Poland et al. 2011; Badet et al. 2017), or actin-related proteins (Moscou et al. 2011). NLRs (Debieu et al. 2015; Lee et al. 2016) and signaling components typically associated with R-mediated resistance (Iakovidis et al. 2016) can also mediate QDR, illustrating the tight integration of QDR and R-mediated immunity components.

Because of their complex interplay and their small phenotypic effect, the functional characterization of QDR genes is challenging. With large collections of mutant lines available, studies in the model plant *A. thaliana* enable the rapid functional characterization of individual candidate QDR genes (Huard Chauveau et al. 2013; Corwin et al. 2016; Rajarammohan et al. 2018). Fully quantitative readouts are often required to reveal the contribution of individual *A. thaliana* genes to QDR against *S. sclerotiorum*, such as ethylene and ROS detection (Perchepied et al. 2010; Zhang et al. 2013), lesion area measurements, and estimation of fungal biomass (Zhang et al. 2013; Badet et al. 2017). Quantitative image analysis can be used as a proxy to evaluate disease severity (Baranowski et al. 2015; Mutka et al. 2016; Karisto et al. 2017; Badet et al. 2017) and the impact of stress on plant fitness (Chen et al. 2014; Nelson et al. 2018; Czedik Eysenberg et al. 2018). Increasing the accuracy and robustness of such quantitative phenotyping often involves increasing the number of measurements. This allowed for instance to reveal the role of individual effectors from the bacterial pathogen *Xanthomonas axonopodis* pv. *manihotis* in virulence (Mutka et al. 2016). High-throughput image-based plant phenotyping proved valuable in crop breeding and for research purposes (Araus and Cairns 2014; Fahlgren et al. 2015; Coppens et al. 2017). There is therefore a need to improve ability to generate large datasets of quantitative plant disease measurements at low cost, low footprint, and reduced human intervention (Czedik Eysenberg et al. 2018). Because it is generally nondestructive, image-based disease measurement also gives access to the dynamics of disease progression. This enabled for instance to study the spatial and temporal distribution of pathogens in plants tissues (Mutka et al. 2016) and distinguish between infected and noninfected plants long before qualitative symptoms are visible (Czedik Eysenberg et al. 2018).

In spite of remarkable progress in methods and technology, the fundamental advances in plant pathology enabled by automated plant phenotyping are still relatively limited. A method for the image-based phenotyping of plant disease symptoms combining high throughput, high resolution, low foot print, and low cost is currently lacking. This design is crucial to identify novel QDR genes and advances our conceptual understanding of this complex trait. Barbacci et al. (2020) presented the Navautron system, an integrated hardware and software solution capable of automatically calculating disease lesion area over time on detached leaves inoculated by pathogens causing necrotic lesions. The Navautron system is able to capture 800 *A. thaliana* leaves per m^2. Its automatic leaf and lesion tracking features allow discriminating disease symptom dynamics with differences in lesion doubling time, as small as 12 min, with nearly no human intervention.

The speed of lesion growth was highly dependent on the host plant genotype and a good indicator of QDR level. It was hypothesized that large effect polymorphisms common to the two most resistant *A. thaliana* accessions, but absent from the four other accessions, would point toward disease susceptibility genes. Highly divergent alleles of LAZ 5 in the resistant accessions Rubezhnoe and Lip-0, were identified while identical alleles existed in the other *A. thaliana* accessions analyzed. As expected, two LAZ 5-deficient mutant lines in theCol-0 genetic background showed enhanced QDR to *S. sclerotiorum*, whereas plants mutated in the closely related CSA 1 gene responded like the wild-type. Considering that the ectopic activation of LAZ 5 triggers cell death (Palma et al. 2010). Results suggest that the broad host range necrotrophic fungus *S. sclerotiorum* exploits this *R* gene induced plant cell death to its benefit (Barbacci et al. 2020).

Plant genes conferring complete resistance against *S. sclerotiorum* have not been reported. Instead, plant populations challenged by *S. sclerotiorum* exhibit a continuum of partial resistance designated as quantitative disease resistance (QDR). Because of their complex interplay and their small phenotypic effect, the functional characterization of QDR genes remains limited. How broad host range necrotrophic fungi manipulate plant programmed cell death is for instance largely unknown. Barbacci et al. (2020) designed a time-resolved automated disease phenotyping pipeline enabling high-throughput disease lesion measurement with high resolution, low footprint at low cost. It could accurately recover contrasted disease responses in several pathosystems using this system. Phenotyping pipeline was used to assess the kinetics of disease symptoms caused by seven *S. sclerotiorum* isolates on six *A. thaliana* natural accessions with unprecedented resolution. Large effect polymorphisms common to the most resistant *A. thaliana* accessions identified highly divergent alleles of the nucleotide-binding site leucine-rich repeat gene LAZ 5 in the resistant accessions Rubezhnoe and Lip-0. The impaired LAZ 5 expression in LAZ 5.1 mutant lines and in *A. thaliana* Rub natural accession correlate with enhanced QDR to *S. sclerotiorum*. These findings illustrate the value of time-resolved image-based phenotyping for unraveling the genetic bases of complex traits such as QDR. The results suggest that *S. sclerotiorum* manipulates plant sphingolipid pathways guarded by LAZ 5 to trigger programmed cell death and cause disease (Barbacci et al. 2020).

3.9.6 Survival Factor 1 Contributes in Pathogenesis of *Sclerotinia sclerotiorum*

The survival factor 1 (*SVF1*) gene was first identified in *Saccharomyces cerevisiae* in a screen for mutations that could be functionally complemented by exogenous expression of the human antiapoptotic gene *Bcl*-x_L (Brace et al. 2005; Vander Heiden et al. 2002). However, *SVF1* and Bcl-x_L have distinct roles in regulating cell survival (Brace et al. 2005). The *S. cerevisiae* cells lacking Svf1 protein showed hypersensitivity to direct chemical precursors of ROS, suggesting that *SVF1* is necessary for survivalist possible under oxidative stress (Brace et al. 2005). Deeper research has shown that Svf1-mediated cell survival under conditions of oxidative stress by affecting the sphingolipid metabolism in *S. cerevisiae* (Brace et al. 2007). A gene in *S. sclerotiorum* (SS1G_01919) named *SsSvf1* (*Sclerotinia sclerotiorum* survival factor 1) was predicted to encode a yeast Svf1 homologous protein. The function of *SsSvf1* was determined via a reverse genetic approach, and its role in oxidative stress response and pathogenicity was investigated. The research may help clarify the function of Svf1 in fungal plant pathogens and the pathogenicity of *S. sclerotiorum* in more detail (Yu et al. 2019).

Reactive oxygen species (ROS) modulations are critical for the pathogenic development of *S. sclerotiorum*. The fungus applies enzymatic and nonenzymatic antioxidants to cope with the oxidative stress during the infection processes. Survival factor 1 was identified and characterized to promote survival under conditions of oxidative stress in *Saccharomyces cerevisiae*. It was found that a gene named *SsSvf1* was predicted to encode a survival factor 1 homologue in *S. sclerotiorum*. The *SsSvf1* transcripts showed high expression levels in hyphae under oxidative stress. Silencing of *SsSvf1* resulted in increased sensitivity to oxidative stress in culture and increased levels of intracellular ROS. Transcripts of *SsSvf1* showed a dramatic increase during the initial stage of infection and the gene-silenced strains displayed reduced virulence on oilseed rape and *Arabidopsis thaliana*. Inhibition of plant ROS production partially restores virulence of *SsSvf1* gene-silenced strains. *SsSvf1* gene-silenced strains exhibited normal oxalate production, but were impaired in compound appressorium formation and cell wall integrity. The results suggest that *SsSvf1* is involved in coping with ROS during fungal-host interactions and plays a crucial role in the pathogenicity of *S. sclerotiorum* (Yu et al. 2019).

3.9.7 Differential Expression of Multiple Metabolic Enzymes on Different Hosts During *Sclerotinia* Pathogenesis

Necrotrophic phyto-pathogens are thought to use a variety of pathogenicity factors to facilitate infection of their plant hosts. A variety of carbohydrate-active enzymes

(CAZymes) are secreted by necrotrophs such as *S. sclerotiorum*, including xylanases, pectinases, and cellulases. CAZymes that degrade plant cell walls are known more specifically as cell wall-degrading enzymes (CWDEs), and may facilitate the penetration of plant tissues by fungal appressoria. In *S. sclerotiorum*, a suite of candidate CAZyme-encoding genes have been predicted and some have been characterized. These include a variety of enzymes associated with the degradation of cell wall components such as cellulose, hemicellulose, and pectins. Many necrotrophs are also known to secrete broad-spectrum phytotoxic secondary metabolites to induce host tissue necrosis. Fungal necrotrophs also have means of mitigating and manipulating host defense responses. Phytotoxic secondary metabolites produced by plant hosts during infection, such as phytoalexins, are metabolized by *S. sclerotiorum* and other necrotrophs to limit their effects. Alternatively, plant-derived phytotoxins may be exported from fungal cells by transporter-based efflux. However, fungal strategies for managing the plant immune response may be considerably more sophisticated than detoxification. The infection process of *S. sclerotiorum* is thought to exhibit an early biotrophic phase, during which the pathogen secretes oxalic acid (OA) as a "compatibility factor" to subdue host defenses. Due to the presence of an early biotrophic phase, it has been suggested that *S. sclerotiorum* may be better described as a hemibiotroph than a necrotroph. This shift in paradigm, from the prototypical necrotroph infection strategy to one with an early biotrophic phase, is aligned with a broader concept that has emerged of a full continuum of trophic strategies among phytopathogens, in which necrotrophs may have more in common with biotrophs than once thought (Liang and Rollins 2018).

The limited available evidence suggests that the infection strategy of broad host range necrotrophs may involve a combination of broad-spectrum and host-specific facets. The proteinaceous effector SsSSVP1, which is produced by *S. sclerotiorum*, is a notable example of a broad-spectrum pathogenicity factor. Infection assays demonstrate that mutant SsSSVP1-silenced *S. sclerotiorum* strains are significantly less virulent than the control in both *B. napus* and *Arabidopsis thaliana*, indicating the effector has a significant interspecific role in the infection process. The metabolite oxalic acid also appears to be employed by *S. sclerotiorum* as a broad-spectrum pathogenicity factor. Infection assays of OA-deficient *S. sclerotiorum* strains on a range of hosts reduced virulence, demonstrating that OA plays an important role in the infection strategy of *S. sclerotiorum* across a wide range of host species. These broad-spectrum pathogenicity factors may contribute to the ability of *S. sclerotiorum* to infect a wide range of plant hosts. By comparison with broad-spectrum pathogenicity determinants, host-specific infection mechanisms are relatively poorly understood in host-generalist necrotrophs. Gene expression studies suggest that certain pathogenicity factors may be targeted to specific hosts through differential gene expression, though the interactions of these differentially expressed genes have not been characterized. Blanco-Ulate et al. (2014) examined differential expression of CAZyme-associated genes in *B. cinerea* on a selection of plant host tissues consisting of lettuce leaves, tomato fruit, and grape berries. Pectin-modifying enzymes were more highly expressed on pectin-rich tomato fruit and grape berries than on lettuce leaves, which contain lower levels of pectin. This was thought to suggest a role for differential expression of CAZyme-associated genes in the adaptation of *B. cinerea* to different hosts, possibly contributing to its broad host range.

Host-specific differential gene expression has also been observed in *S. sclerotiorum.* Guyon et al. (2014) demonstrated distinct differences in the expression patterns of a selection of 16 effector candidates between the hosts *S. lycospersicum, Nicotiana benthamiana* and resistant and susceptible accessions of *A. thaliana*. More broadly, variations in gene expression over time have also been examined in *S. sclerotiorum* using RNA-seq. However, host-specific differential expression across the entire *S. sclerotiorum* genome has not yet been characterized (Allan et al. 2019). A RNA-sequencing analysis of host-specific differential gene expression in *S. sclerotiorum* during the infection of *Lupinus angustifolius* and *B. napus*, which are the most widely grown broadleaf crops in the West Australian wheat belt have been described. To identify genes expressed in a host-dependent fashion by *S. sclerotiorum* across the entire genome, in contrast with earlier studies that focused on specific groups of putative pathogenicity factors. It was hypothesized that host-specific differential gene expression in *S. sclerotiorum* would be observed primarily among putative pathogenicity-related genes, and particularly CAZymes. However, these genes were found to make up only a small portion of all host-specific differentially expressed genes. The diverse set of genes differentially expressed by *S. sclerotiorum in planta* suggests a possible role of differential expression as a means of adaptation to plant defense responses (Allan et al. 2019).

The molecular processes that underpin broad host range necrotrophy are not fully understood. RNA sequencing was used to assess whether *S. sclerotiorum* genes are differentially expressed in response to infection of the two different host crops canola (*Brassica napus*) and lupin (*Lupinus angustifolius*). A total of 10,864 of the 11,130 genes in the *S. sclerotiorum* genome were expressed. Of these, 628 were upregulated *in planta* relative to in vitro on at-least one host, suggesting involvement in the broader infection process. Among these genes were predicted carbohydrate-active enzymes (CAZymes) and secondary metabolites. A considerably smaller group of 53 genes were differentially expressed between the two plant hosts. Of these host-specific genes, only six were either CAZymes, secondary metabolites or putative effectors. The remaining genes represented a diverse range of functional categories, including several associated with the metabolism and efflux of xenobiotic compounds, such as cytochrome P450s, metal-beta-lactamases, tannases, and major facilitator superfamily transporters. These results suggest that *S. sclerotiorum* may regulate the expression of detoxification-related genes in response to phytotoxins produced by the different host species. It is the comparative whole transcriptome analysis of *S. sclerotiorum* during infection of different hosts (Allan et al. 2019).

3.9.7.1 Differential Gene Expression in *Brassica* Genotypes in Response to *Sclerotinia* Infection

Gene expression changes associated with *S. sclerotiorum* infection were assessed in a partially resistant and a susceptible genotype of oilseed *Brassica napus* using a whole genome microarray from *Arabidopsis.* A total of 686 and 1547 genes were

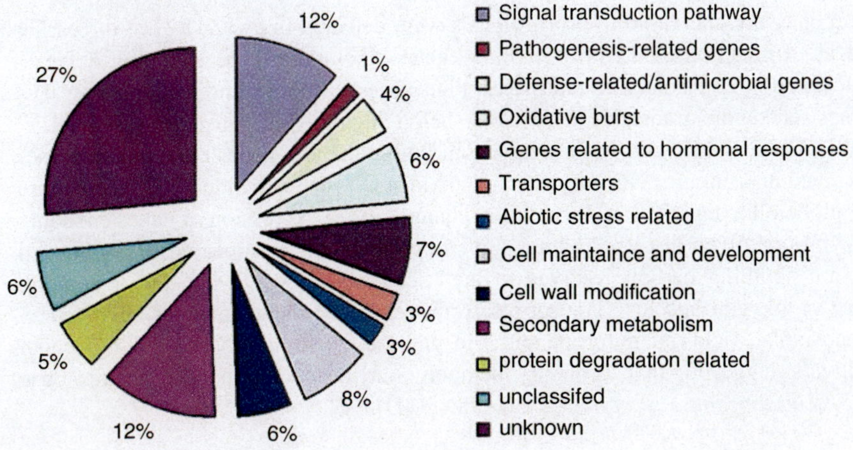

Fig. 3.16 Functional classification of genes which displayed significantly differential expression in RV289 at 24 h post-inoculation with *Sclerotinia sclerotiorum* (Zhao et al. 2007)

found to be differentially expressed after infection in the resistant and susceptible genotypes, respectively. The number of differentially expressed genes increased over infection time with the majority being upregulated in both genotypes. The putative functions of the differentially expressed genes included pathogenesis related (PR) proteins, proteins involved in the oxidative burst, protein kinase, molecule transporters, cell maintenance and development, abiotic stress, as well as proteins with unknown functions (Fig. 3.16). The gene regulation patterns indicated that a large part of the defense response exhibited as a temporal and quantitative difference between the two genotypes. Genes associated with jasmonic acid (JA) and ethylene signal transduction pathways were induced, but no salicylic acid (SA) responsive genes were identified. Candidate defense genes were identified by integration of the early response genes in the partially resistant line with previously mapped quantitative trait loci (QTL). Expression levels of these genes were verified by Northern blot analyses. These results indicate that genes encoding various proteins involved in diverse roles, particularly WRKY transcription factors and plant cell wall-related proteins may play an important role in the defense response to *S. sclerotiorum* disease (Zhao et al. 2007).

3.9.7.2 Differential Gene Expression in *Brassica napus* Cultivars Infected with *Sclerotinia sclerotiorum*

The fungal pathogen *Sclerotinia sclerotiorum* infects a broad range of dicotyledonous plant species and causes stem rot in *Brassica napus*. To elucidate the mechanisms underlying the defense response, the patterns of gene expression in the partially resistant *B. napus* cultivar ZhongYou 821 (ZY821) and the susceptible cultivar Westar were studied using a *B. napus* oligonucleotide microarray. Although maximum differential gene expression was observed at 48 h post-inoculation (hpi) in both cultivars, increased transcript levels were detected in cv. ZY821 at the earlier stages of infection (6–12 hpi) for many genes, including those encoding defense-associated proteins, such as chitinases, glucanases, osmotins and lectins, as well as genes encoding transcription factors belonging to the zinc finger, WRKY, APETALA2 (AP2), and MYB classes. In both cultivars, genes encoding enzymes involved in jasmonic acid, ethylene and auxin synthesis were induced, as were those for gibberellin degradation. In addition, changes in the expression of genes encoding enzymes involved in carbohydrate and energy metabolism appeared to be directed toward shuttling carbon reserves to the tricarboxylic acid cycle and generating reactive oxygen species. Transcripts from genes encoding enzymes involved in glucosinolate and phenylpropanoid biosynthesis were highly elevated in both cultivars, suggesting that secondary metabolites are also components of the response to *S. sclerotiorum* in *B. napus* (Zhao et al. 2009).

3.9.8 Integrated mRNA, sRNA, and Degradome Sequencing Reveal Complex Responses to *Sclerotinia* Infection in *Brassica*

Transcriptome studies have shown that *B. napus* genes involved in the response to pathogen invasion e.g., genes encoding receptor-like kinases, proteins harboring nucleotide binding site leucine-rich repeats, proteins that function in mitogen-activated protein kinase cascades, G-proteins, calcium binding proteins, hormones, TFs, pathogenesis-related (PR) proteins, proteins involved in phenylpropanoid and glucosinolate metabolism, and transport proteins) were significantly altered following infection (Li et al. 2012a, b). MicroRNAs (miRNAs) are endogenous, noncoding RNAs that play crucial roles in various biological processes at the transcriptional and post-transcriptional level by negatively regulating gene expression. The miRNAs act as key components in gene regulatory pathways associated with many processes, such as germination, development, organ maturation, signal transduction, and stress responses.

Moreover, several studies have shown that miRNAs function in responses to pathogen invasion. miR393 contributed to the defense response in *Arabidopsis thaliana* against *Pseudomonas syringae* pv. *tomato* (Pst) by suppressing target genes, including auxin receptors, transport inhibitor response 1 (TIR1), auxin signaling F-box protein 2 (AFB2), and AFB3, which negatively regulate auxin signaling. Furthermore, miR393 acts as a key regulator in the glucosinolate pathway, which is involved in plant responses to pathogens. In *Brassica rapa*, miR158 and miR1885 were induced after infection by *Turnip mosaic virus* (*TuMV*) and suppressed disease resistance protein genes-Nucleotide-binding site leucine-rich repeat. MiR396 from *A. thaliana* functions as a positive regulator during cyst nematode infection. In *Nicotiana tabacum* (tobacco), miR6019 and miR6020 guided to incise N genes, which increases resistance to the *Tobacco mosaic virus* (TMV) (Li et al. 2012a, b). Several studies have been conducted to identify miRNAs in oilseed rape. Xie et al. (2007) first identified 21 potential *B. napus* miRNAs using computational methods. Later, next-generation sequencing technologies were used to identify miRNAs in different tissues or in response to different stresses. There are only 92 mature miRNA sequences of *B. napus* in the miR Base database (version: 21.0). Compared with the number of miRNAs that have been identified in other species, such as *A. thaliana*, *Arabidopsis lyrata*, and *Oryza sativa*. It is likely that there are many *B. napus* miRNAs that have not yet been discovered.

The molecular mechanisms underlying the interaction between *S. sclerotiorum* and *B. napus* are only partially understood, due to the lack of resistant lines. Jian et al. (2016) identified five relatively resistant lines in a previous study. To investigate the response of oilseed rape to *S. sclerotiorum* infection at the mRNA and miRNA levels, they performed RNA-Seq, sRNA-Seq, and degradome deep sequencing on both mock-inoculated and *S. sclerotiorum* inoculated stems at 48 h post-inoculation (hpi). The identified differentially expressed genes were associated with protein kinases, signal transduction (CDPKs, G proteins, and MAPKs), TFs, hormones, PR proteins, secondary metabolism, and transport proteins. In sRNA-Seq analysis, 10 known and 41 novel miRNAs were differentially expressed at 48 hpi with *S. sclerotiorum*. Furthermore, 80 cleavage sites with 64 miRNAs were predicted by degradome sequencing. The data provide a broad view of gene expression changes after *S. sclerotiorum* infection in oilseed rape (Jian et al. 2016).

The molecular mechanisms underlying the response of oilseed rape to *S. sclerotiorum* infection at the transcriptional and post-transcriptional levels are poorly understood. Jian et al. (2018) used an integrated omics approach (transcriptome, sRNAome, and degradome sequencing) on the Illumina platform to compare the RNA expression and post-transcriptional profiles of oilseed rape plants inoculated or not with *S. sclerotiorum*. In total, 7065 differentially expressed genes (DEGs) compared with the mock-inoculated control at 48 h post-inoculation was identified. These DEGs were associated with protein kinases, signal transduction, transcription factors, hormones, pathogenesis related proteins, secondary metabolism, and transport. In the sRNA-Seq analysis, 77 known and 176 novel miRNAs were identified; however, only 10 known and 41 novel miRNAs were differentially expressed between the samples inoculated or not with *S. sclerotiorum*. Degradome sequencing predicted 80 cleavage sites with 64 miRNAs. Integrated mRNA, sRNA, and degradome sequencing analysis reveal oilseed rape complex responses to *S. sclerotiorum* infection. This provides a view of miRNA and mRNA expression profiles in oilseed rape following *S. sclerotiorum* infection (Jian et al. 2016).

3.9.9 The Interplay of Apoptosis and Autophagy in the Pathogenesis of *Sclerotinia sclerotiorum*

Programmed cell death is characterized by a cascade of tightly controlled events that culminate in the orchestrated death of the cell. In multicellular organism's autophagy and apoptosis are recognized as two principal means by which these genetically determined cell deaths occur. During plant–microbe interactions, cell death programs can mediate both resistant and susceptible events. Via oxalic acid (OA), the necrotrophic phytopathogen *Sclerotinia sclerotiorum* hijacks host pathways and induces cell death in host plant tissue resulting in hallmark apoptotic features in a time and dose dependent manner. OA-deficient mutants are nonpathogenic and trigger a restricted cell death phenotype in the host that unexpectedly exhibits markers associated with the plant hypersensitive response including callose deposition and a pronounced oxidative burst, suggesting that the plant can recognize and in this case respond, defensively. Using a combination of electron and fluorescence microscopy, chemical effectors and reverse genetics, Kabbage et al. (2013) showed that this restricted cell death is autophagic. Inhibition of autophagy rescued the nonpathogenic mutant phenotype. These findings indicate that autophagy is a defense response in this necrotrophic fungus/plant interaction and suggest a novel function associated with OA; namely, the suppression of autophagy. These data suggest that not all cell deaths are equivalent, and though programmed cell death occurs in both situations, the outcome is predicated on who is in control of the cell death machinery. Based on data, it is not cell death per se that dictates the outcome of certain plant–microbe interactions, but the manner by which cell death occurs that is crucial (Kabbage et al. 2013).

3.9.10 A Secretory Protein of *Sclerotinia sclerotiorum* Suppresses Host Resistance in Early Stage of Pathogenesis

The *Sclerotinia sclerotiorum integrin-like* (SSITL) (SS1G_14133) of *Sclerotinia sclerotiorum* encodes a protein with 302 amino acid residues including a signal peptide, its secretion property was confirmed with immune localization and immune-fluorescence techniques. SSITL was classified in the integrin alpha N-terminal domain superfamily, and its 3D structure is similar to those of human integrin a4-subunit and a fungal integrin-like protein. When *S. sclerotiorum* was inoculated to its host, high expression of *SSITL* was detected during the initial stages of infection (1.5–3.0 hpi). Targeted silencing of *SSITL* resulted in a significant reduction in virulence; on the other hand, inoculation of *SSITL* silenced transformant A10 initiated strong and rapid defense response in *Arabidopsis*, the highest expressions of defense genes *PDF1*.2 and *PR*-1 appeared at 3 hpi which was 9 h earlier than that time when plants were inoculated with the wild-type strain of *S. sclerotiorum*. Systemic resistance induced by A10 was detected by analysis of the expression of *PDF1*.2 and *PR*-1, and confirmed following inoculation with *Botrytis cinerea*. A10 induced much larger lesions on *Arabidopsis* mutant *ein2* and *jar1*, and slightly larger lesions on mutant *pad4* and *NahG* in comparison with the wild-type plants. Furthermore, both transient and constitutive expression of *SSITL* in *Arabidopsis* suppressed the expression of *PDF1*.2 and led to be more susceptible to A10 and the wild-type strain of *S. sclerotiorum* and *B. cinerea*. The results suggested that *SSITL* is an effector possibly and plays significant role in the suppression of jasmonic/ethylene (JA/ET) signal pathway mediated resistance at the early stage of infection (Zhu et al. 2013).

3.9.11 Mechanisms of Oxaloacetate Acetyl Hydrolase Gene of *Sclerotinia* in Pathogenesis

During pathogenesis, *S. sclerotiorum* accumulates mill molar levels of oxalic acid (OA) which has been implicated as being critical to its broad host range necrotrophic pathogenicity (Andrew et al. 2012). The most compelling evidence for the importance of OA in pathogenesis comes from ultraviolet (UV)-induced oxalate-deficient mutants ("A" mutants), reported to be nonpathogenic on *Phaseolus vulgaris* (Godoy et al. 1990) and *Arabidopsis thaliana* (Dickman and Mitra 1992). These mutants have since been shown to accumulate low levels of OA and to produce limited lesions on various hosts (Cessna et al. 2000; Kim et al. 2008; Williams et al. 2011). In addition to pathogenicity, OA accumulation may also affect the growth and developmental processes of *S. sclerotiorum*. The "A" mutants exhibit reduced growth and aberrant development, including fluffy hyphae and abolished sclerotium formation, whereas, in the wild-type (WT), low ambient pH promotes both sclerotium morphogenesis and vegetative growth (Rollins 2003; Rollins and Dickman 2001). These effects of OA may be derived from its effect on ambient pH, as well as its metabolic connections with primary metabolism. The oxalic acid is a strong dicarboxylic organic acid synthesized by a broad range of pathogenic and nonpathogenic organisms, including bacteria, fungi, plants, and some mammals. In filamentous fungi, OA biogenesis has been demonstrated to occur through the hydrolysis of oxaloacetate, the oxidation of glyoxylate or the oxidation of glycolaldehyde (Dutton and Evans 1996). Previous biochemical studies have indicated that the conversion of oxalo acetate to OA and carbon dioxide, catalyzed by oxalo acetate acetyl hydrolase (OAH, EC 3.7.1.1), is the primary activity associated with OA biogenesis in Ascomycota fungi, including *S. sclerotiorum* (Kubicek et al. 1988; Maxwell 1973; Ruijter et al. 1999). In *S. sclerotiorum*, neutral pH strongly induces OA accumulation and OAH activity (Maxwell 1973; Rollins and Dickman 2001). This neutral pH induction is dependent on the zinc finger ambient pH-dependent transcriptional regulator Ss-Pac1 (Rollins 2003; Rollins and Dickman 2001). Diverse carbon sources can support OA biogenesis in *S. sclerotiorum* with intermediates derived from the tricarboxylic acid (TCA) cycle or the glyoxylate cycle depending on the carbon growth substrate (Liberti et al. 2013).

To determine the role of OAH in *S. sclerotiorum,* OA metabolism and to evaluate its requirement for pathogenesis, Liang et al. (2015) isolated and performed gene deletion analysis of a putative OAH-encoding gene *Ss-oah1*. It was demonstrated that *Ss-oah1* is required for OA accumulation, and that it affects pH-responsive growth, morphogenesis, and virulence of *S. sclerotiorum* on a variety of plant hosts. The observation that the *Ss-oah1* knockout mutants (Δ*ss-oah1* mutants) retain the ability to form primary lesions suggests that factors other than OA contribute significantly toward the establishment of initial host compatibility, i.e. primary lesions. Histological features of infection suggest that *S. sclerotiorum* can initially colonize its hosts to a limited degree without OA accumulation and that the infection and colonization process involves a dynamic regulation of OA accumulation.

The oxaloacetate acetyl hydrolase (OAH, EC 3.7.1.1)-encoding gene *Ss-oah1* was cloned and functionally characterized from *Sclerotinia sclerotiorum. Ss-oah1* transcript accumulation mirrored oxalic acid (OA) accumulation with neutral pH induction is dependent on the pH-responsive transcriptional regulator Ss-Pac1. Unlike previously characterized ultra violet (UV)-induced oxalate-deficient mutants ("A" mutants) which retain the capacity to accumulate OA, gene deletion Δ*ss-oah1* mutants did not accumulate OA in culture or during plant infection. This defect in OA accumulation was fully restored on reintroduction of the wild-type (WT) *Ss-oah1* gene. The Δ*ss-oah1* mutants were also deficient in compound appressorium and sclerotium development and exhibited a severe radial growth defect on medium buffered at neutral pH. On a variety of plant hosts, the *Δss-oah1* mutants established very restricted lesions in which the infectious hyphae gradually lost viability. Cytological comparisons of WT and Δ*ss-oah1* infections revealed low and no OA accumulation, respectively, in subcuticular hyphae. Both WT and mutant hyphae exhibited a transient association with viable host epidermal cells at the infection front. The data establish a critical requirement for OAH activity in *S. sclerotiorum* OA biogenesis and pathogenesis, but also suggest that factors independent of OA contribute to the establishment of primary lesions (Liang et al. 2015).

3.9.12 Functional Characterization of *BNaNPR1* Gene Family During *Sclerotinia* Pathogenesis

Nonexpressor of PR1 (*NPR1*) is an important SA perception gene encoding a transcription co-factor protein (Cao et al. 1994). A large number of studies on *A. thaliana* have revealed that *NPR1* plays an important role in regulating defense response. The mutations in *AtNPR1* block induction of multiple defense genes by SA (Cao et al. 1994; Roetschi et al. 2001), suggesting that *AtNPR1* is a key regulator for SA-mediated defense response (Mou et al. 2003). The *NPR1* also inhibits the defense response mediated by jasmonic acid (JA), another important signal molecular, and thus regulates cross-talk between SA and JA (Spoel et al. 2003).

In plant species, two to six *NPR1*-like genes have been found. In *Arabidopsis*, *AtNPR1* has five paralogs, including *AtNPR2*, *AtNPR3*, *AtNPR4*, *AtNPR5* (*AtBOP1*, BLADE ON PETIOLE 1), and *AtNPR6* (*BOP2*). *AtNPR2* may also have a role in the perception of SA (Canet et al. 2010). However, the gene was also thought to possibly be a simple redundant or nonfunctional paralog of *AtNPR1* (Backer et al. 2019). *AtNPR3* and *AtNPR4* (*AtNPR3/4*) were showed to negatively regulate SAR (Liu et al. 2005; Zhang et al. 2006). The *AtBOP1*, as well as *AtBOP2*, is involved in the development of lateral organs (Hepworth et al. 2005). This family of proteins contains several domains: an N-terminal BTB/POZ (broad complex, tramtrack, and bric-a-brac/poxvirus and zinc finger) domain, a central ankyrin repeat domain, and a C-terminal transactivation domain (Spoel et al. 2009). Some *NPR1-like* genes have been identified in several plant species. However, to date, a systematic analysis of this gene family in *B. napus* has yet not been reported, although the genome of the crop has been released in 2014 (Chalhoub et al. 2014).

In the *NPR1*-like gene family, the most extensively studied member is *AtNPR1* due to its importance in plant immunity. Mutants of the protein encoded by *AtNPR1* are more susceptible to virulent pathogens and display compromised expression of *PR1* when compared to plants expressing wild type *NPR1* (Cao et al. 1994; Glazebrook et al. 1996). In the case of the interaction of *Arabidopsis* with *S. sclerotiorum*, information on *AtNPR1* involvement is argued. Guo and Stotz (2007) showed that the *Arabidopsis NPR1* loss-of-function mutant (*npr1*) was hyper susceptible to *S. sclerotiorum*, and the induction of *PR1* was reduced in this mutant (Guo and Stotz 2007). However, this hyper susceptibility for the *npr1* mutant was not observed in another experiment in which *npr1* did not show any significant difference in resistance to *S. sclerotiorum* compared with the wild type (Perchepied et al. 2010). The role of *AtNPR1* in resistance to *S. sclerotiorum* is contradictory. The roles *B. napus NPR1* (*BnaNPR1*), the important regulator of defense responses, plays in resistance of the crop to *S. sclerotiorum* because determining its roles in this resistance could provide insights into the interaction of *B. napus* with *S. sclerotiorum.* Wang et al. (2020) performed a genome-wide survey for *NPR1*-like genes in *B. napus*. Then analyses, including phylogenetic analysis, gene structure, conserved motifs, gene chromosome location, and expression profiling, were performed on these genes. Further, a loss-of-function analysis of *BnaNPR1* on resistance to *S. sclerotiorum* was performed using RNA interference (RNAi) (Wang et al. 2020).

The *BnaNPR1*-like gene family was identified in *B. napus,* and it was revealed that repression of BnaNPR1 significantly reduces resistance to *S. sclerotiorum,* intensifies ROS accumulation, and changes the expression of genes associated with SA and JA/ET signaling in response to pathogen. The nonexpressor of pathogenesis-related genes 1 (*NPR1*) and related *NPR1-like* genes play an important role in regulating plant defense. A total of 19 *BnaNPR1-like* genes were identified in the *B. napus* genome, and then named according to their respective best match in *Arabidopsis thaliana* (*At*), which led to the determination of *B. napus* homologs of every *AtNPR1-like* gene. Analysis of important protein domains and functional motifs indicated the conservation and variation among these homologs. Phylogenetic analysis of these BnaNPR1-like proteins and their *Arabidopsis* homologs revealed six distinct subclades, consequently indicating that their name classification totally conformed to their phylogenetic relationships. Further, *B. napus* transcriptomic data showed that the expression of three *BnaNPR1s* was significantly downregulated in response to infection with *Sclerotinia sclerotiorum*, the most important pathogen of this crop, whereas *BnaNPR2/3/4/5/6s* did not show the expression differences in general. Wang et al. (2020) generated *B. napus BnaNPR1*-RNAi lines to interpret the effect of the downregulated expression of *BnaNPR1s* on resistance to *S. sclerotiorum*. The results showed that *BnaNPR1*-RNAi significantly decreased this resistance. It was revealed that *BnaNPR1*-RNAi intensified ROS production and changed defense responses in the interaction of plants with this pathogen. These results indicated that *S. sclerotiorum* might use *BnaNPR1* to regulate specific physiological processes of *B. napus*, such as ROS production and SA defense response, for the infection (Wang et al. 2020).

3.9.13 The Effector-Like Proteins Regulate *Sclerotinia* Pathogenesis

In *Sclerotinia sclerotiorum*, thus so far, there have been several proteins described with effector-like properties and/or effector-like activities *in planta*. These proteins include a polygalacturonase enzyme, sspg1d, that interacts with a *B. napus* C2 domain containing protein (Wang et al. 2009); an integrin-like protein, SSITL, that suppresses defense responses in *A. thaliana* (although, in addition to the activity of SSITL *in planta*, deletion mutants for the cognate gene exhibited abnormal hyphal tip branching, slower growth in vitro and abnormally small sclerotia) (Zhu et al. 2013); two necrosis and ethylene inducing proteins, SsNep1 and SsNep2, that cause necrosis when heterologously expressed in *Nicotiana benthamiana* (Dallal Bashi et al. 2010); a cutinase, SsCut, that causes cell death in a range of plant species, including the nonhost, wheat (*Triticum aestivum*) (Zhang et al. 2014); a polygalacturonase protein that induces calcium signaling and host cell death in soybean (Zuppini et al. 2005); a hypothetical secreted protein that, when disrupted, attenuates virulence in *S. sclerotiorum* (Liang et al. 2013); a small, cysteine-rich secreted protein with a cyanoviron-N homology (CVNH) domain, which attenuates virulence when deleted (Lyu et al. 2015); and, a secreted protein with no known functional domains, SsSSVP1, seemingly restricted to *S. sclerotiorum* and the closely related species *B. cinerea*, which causes necrosis in *N. benthamiana*. This protein was shown to interact with *N. benthamiana* QCR8, a subunit of the cytochromeb–c1 complex of the mitochondrial respiratory chain. It is thought that this activity is what caused necrosis in host tissue (Lyu et al. 2016).

The starting point for investigations into the evolution and molecular activity of fungal effector genes often begins with analysis of the fungal genome. Numerous studies, including two studies on *S. sclerotiorum*, have attempted to identify effector-like or secreted proteins potentially involved in infection (Amaral et al. 2012; Nemri et al. 2014; Guyon et al. 2014; Heard et al. 2015). These studies have used various criteria to differentiate effector-like from non-effector-like genes. Primarily, the focus has been on small, cysteine-rich proteins with a predicted secretion signal, as many experimentally characterized effectors exhibit these properties (Sperschneider et al. 2015). Recently, a machine learning approach based on numerous criteria derived from properties of known effectors was applied to effector prediction. This approach was shown to be more sensitive and specific than the traditional, and to some degree arbitrary, approach of filtering through manually selected criteria (Sperschneider et al. 2016; Derbyshire et al. 2017).

3.9.14 Role of Copper in *Brassica–Sclerotinia* Interaction

Copper is an essential nutrient for microbial pathogens and serves as an important cofactor for enzymes of scavenging ROS. Ding et al. (2020) simultaneously analyzed the transcriptome changes of *Brassica oleracea* and *S. sclerotiorum* during inoculation. The results showed that *S. sclerotiorum* induces the genes involved in copper ion import and "copper ion transport" during infection to promote the ROS detoxification and fungal virulence. While the resistance host maintains the genes involved in "copper ion homeostasis," which promote the resistance to *S. sclerotiorum* and antioxidant activity. Furthermore, copper concentration is higher in the necrotic area than the margin and healthy area, which may indicate that copper flow from healthy area into the necrotic area during infection. It highlights subtler and complex role of copper during the interaction of *S. sclerotiorum* and its host plants.

Necrotrophic plant pathogen induces host reactive oxygen species (ROS) production, which leads to necrosis in the host, allowing the pathogen to absorb nutrients from the dead tissues. *Sclerotinia sclerotiorum* is a typical necrotrophic pathogen that causes *Sclerotinia* stem rot in more than 400 species, resulting in serious economic losses. Three *S. sclerotiorum* genes involved in copper ion import/transport, *SsCTR1*, *SsCCS* and *SsATX1*, were significantly upregulated during infection of *Brassica oleracea*. Function analysis revealed that these genes involved in fungal ROS detoxification and virulence. On the host side, four genes putatively involved in copper ion homeostasis, *BolCCS*, *BolCCH*, *BolMT2A*, and *BolDRT112*, were significantly downregulated in susceptible *B. oleracea*, but stably expressed in resistant *B. oleracea* during infection. Their homologs were found to promote resistance to *S. sclerotiorum* and increase antioxidant activity in *Arabidopsis thaliana*. Further, copper concentration analysis indicated that copper flow from healthy area into the necrotic area during infection. A model (Fig. 3.17) was proposed that *S. sclerotiorum* utilizes host copper to detoxify ROS in its cells, whereas the resistant hosts may restrict the supply of essential copper nutrients to *S. sclerotiorum* by maintaining copper ion homeostasis during infection (Ding et al. 2020).

3.9.15 Genome-Wide Alternative Splicing Profile of *Sclerotinia* During the Colonization of Diverse Host Families

The *Sclerotinia sclerotiorum* is a notorious generalist plant pathogen that threatens more than 600 host plants, including wild and cultivated species. The molecular bases underlying the broad compatibility of *S. sclerotiorum* with its hosts was elucidated. Alternate splicing (AS) is a common regulated cellular process that increases cell protein and RNA diversity. Ibrahim et al. (2021) annotated spliceosome genes in the genome of *S. sclerotiorum* and characterized their expression in vitro and during the colonization of six host species. Several spliceosome genes were differentially expressed in planta, suggesting that AS was altered during infection. Using stringent parameters, 1487 *S. sclerotiorum* genes differentially expressed in planta and exhibiting alternative transcripts were identified. The most common AS events during the colonization of all plants were retained introns and the alternative 3′ receiver site. Ibrahim et al. (2021) identified *S. sclerotiorum* genes expressed *in planta* for which (a) the relative accumulation of alternative transcripts

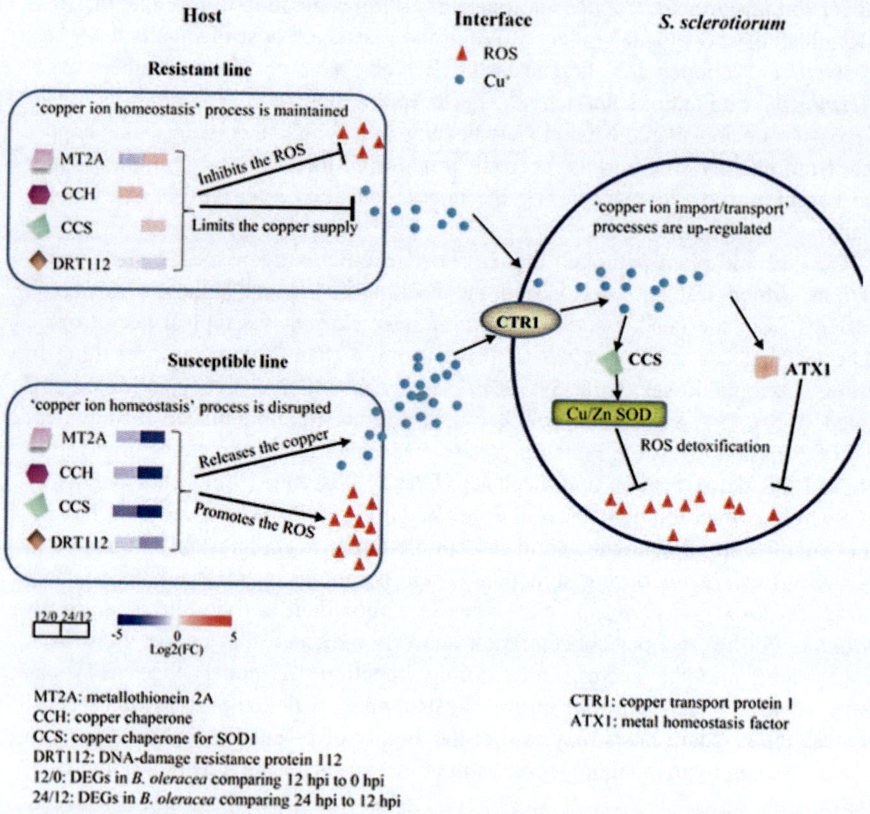

Fig. 3.17 Model depicting the battle for copper acquisition between host and *Sclerotinia sclerotiorum* during infection. *S. sclerotiorum* promotes its copper uptake systems to acquire host-derived copper, and activates its ROS scavenging system for survival and virulence, while a resistant host induces "nutritional immunity" that restricts the supply of essential copper nutrients to *S. sclerotiorum* by maintaining "copper ion homeostasis." 12/0: gene expression comparison of 12 hpi to 0 hpi in host, 24/12: gene expression comparison of 24 hpi to 12 hpi in host (Ding et al. 2020)

varies according to the host being colonized and (b) alternative transcripts harbor distinct protein domains. This notably included 42 genes encoding predicted secreted proteins showing high-confidence AS events. This indicated that AS events are taking place in the plant pathogenic fungus *S. sclerotiorum* during the colonization of host plants and could generate functional diversity in the repertoire of proteins secreted by *S. sclerotiorum* during infection (Figs. 3.18 and 3.19).

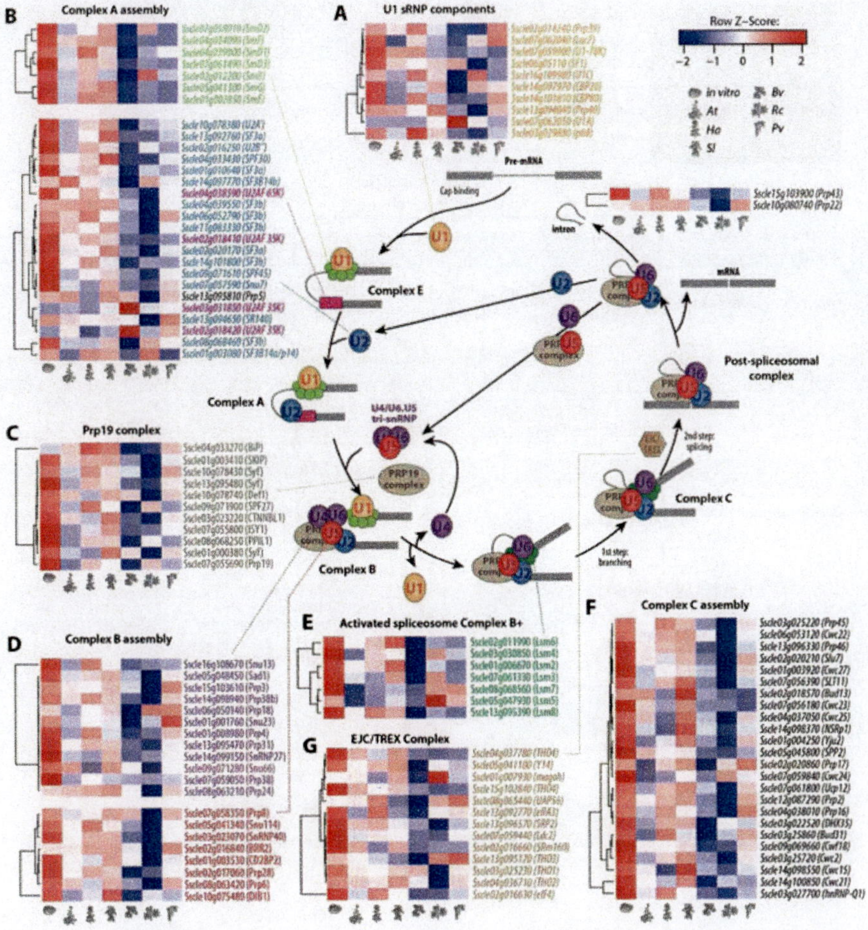

Fig. 3.18 Identification of *Sclerotinia sclerotiorum* spliceosome components and their transcriptional regulation during the infection of plants from six botanical families. Diagrammatic representation of mRNA splicing process featuring *S. sclerotiorum* genes involved in each step. A hypothetical pre-mRNA molecule is depicted with two exons shown as dark gray boxes and an intron shown as dark gray line. Circles, rounded rectangles, and hexagons show protein complexes. The relative gene expression for 116 spliceosome genes at the edge of *S. sclerotiorum* mycelium during infection of plants from six species and in vitro is shown as heatmaps. Pre-mRNA splicing involves multiple spliceosomal complexes. First, complex E is established by binding of U1 snRNP (**a**) to small nuclear ribonucleo-protein-associated proteins (Sm), U2 associated factors (U2AF), and splicing factors (SF), leading to the recruitment of U2 and the formation of complex A (**b**). The PRP19C/Prp19 complex/NTC/19 complex (**c**) stabilizes the U4/U5/U6 tri-snRNP spliceosomal complex leading to complex B assembly (**d**). The U1/U4 snRNPs are released to form the activated spliceosome complex B+ (**e**) triggering branching, intron excision, conformational re-arrangements into complex C (**f**) and ligation of the proximal and distal exons. EJC/TREX is recruited to spliced mRNAs to mediate export to the cytoplasm (**g**). *At*, *Arabidopsis thaliana*; *Bv*, *Beta vulgaris*; *Ha*, *Helianthus annuus*; *Pv*, *Phaseolus vulgaris*; *Rc*, *Ricinus communis*; *Sl*, *Solanum lycospersicum* (Ibrahim et al. 2021)

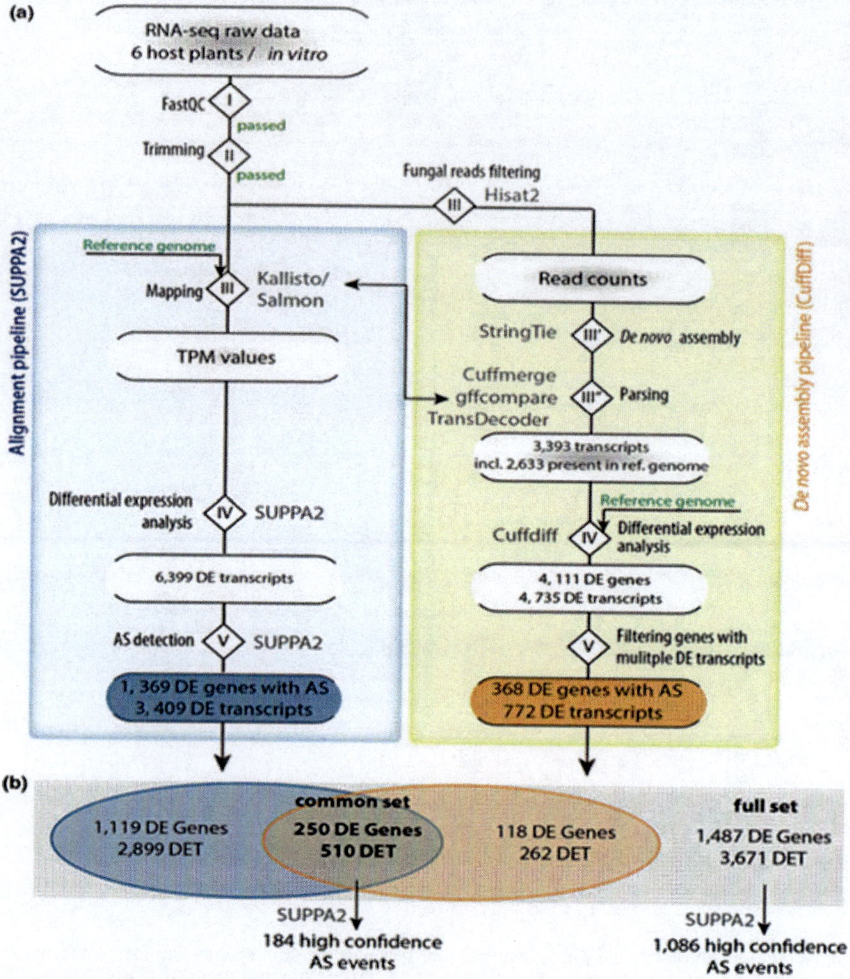

Fig. 3.19 Pipeline for genome-wide detection of alternative splicing (AS) in *Sclerotinia sclerotiorum*. (**a**) Raw RNA-seq data was first inspected with FastQC (I) and quality-trimmed using Trimmomatic (II). Applied two pipelines for detection and analysis of novel transcripts, the de novo assembly pipeline (yellow box) and the transcriptome alignment pipeline (blue box). For detection of novel transcripts, Mapped reads with HISAT (III) to the *S. sclerotiorum* reference genome; this data was used in a modified String Tie de novo assembly (III′). Using Cuffmerge, gffcompare, and Transdecoder, identified novel transcripts compared with the reference gene annotation (III″) and generated a new reference annotation. Using the new annotation and the reference genome, performed differential expression analysis (IV) and filtered the differentially expressed (DE) genes for those encoding at least two DE transcripts (DET; V). In the transcriptome alignment pipeline, mapping was done with Kallisto or Salmon (III), DE analysis and AS detection with SUPPA2 (IV and V). (**b**) A Venn diagram summarizing the results from DE analysis in (**a**) for both pipelines; numbers are given only for genes encoding multiple transcripts. *AS* alternative splicing, *DE* differentially expressed, *DET* differentially expressed transcript, *TPM* transcripts per million (Ibrahim et al. 2021)

3.10 *Brassica–Turnip Mosaic Virus* Pathosystem

3.10.1 Introduction

The *Turnip mosaic virus* (*TuMV*) is a member of the genus *Potyvirus* (Berger et al. 2000) and infects cruciferous plants throughout the world (Walsh and Jenner 2002). Cruciferous plants include a large variety of economically important crops belonging to different genera (Gomez-Campo 1999). The *TuMV* has adapted to such diversified cruciferous crops, and the many *TuMV* isolates in the world have been classified into several strains or pathotypes (Fujisawa 1990; Green and Deng 1985; Jenner and Walsh 1996; Provvidenti 1980; Stavolone et al. 1998; Stobbs and Shattuck 1989). Systemic infection of plants occurs when a virus establishes genome amplification, then movement cell-to-cell and over long distances via the plasmodesmata and phloem cells (Carrington et al. 1996; Lucas and Gilbertson 1994). The factors that influence virus host range, symptomatology and/or pathogenicity have been studied using various virus mutants as well as recombinants constructed from closely related viruses (Rao 1999). In many cases, these factors involve one or more proteins involved in virus replication and/or transport, including coat protein (CP), movement protein and/or other proteins conferring the function necessary for virus movement. The genus *Potyvirus* contains more than 200 members or possible members and belongs to the largest plant virus family, Potyviridae (Berger et al. 2000). Potyviruses have flexuous filamentous particles that contain an approximately 10 kb positive-sense single-stranded RNA that is covalently linked to a virus genome-linked protein (VPg) at the 59 end and polyadenylated at the 39 end (Revers et al. 1999). The RNA has a single open reading frame that is translated into a large polyprotein, which is proteolytically cleaved into mature proteins by three virus-encoded proteinases (Riechmann et al. 1992; Urcuqui-Inchima et al. 2001).

3.10.2 Molecular Mechanism of *Turnip Mosaic Virus* Pathogenesis

The potyvirus P3 protein is proteolytically cleaved from polyprotein by HC-Pro and NIa-Pro, resulting in either the P3 protein or its precursor P3-6K1 (Riechmann et al. 1992). Among potyviruses, there is relatively little similarity in P3proteins compared to other proteins (Urcuqui-Inchima et al. 2001). The P3 proteins are thought to be involved in virus replication (Merits et al. 1999), accumulation (Klein et al. 1994), symptomatology (Chu et al. 1997; Saenz et al. 2000), resistance breaking (Hjulsager et al. 2002; Jenner et al. 2002, 2003; Johansen et al. 2001) and cell-to-cell movement (Dallot et al. 2001; Johansen et al. 2001).

Many *TuMV* isolates have been collected around the world and grouped into four lineages by sequencing and phylogenetic analysis of sequences encoding P1 and CP (Ohshima et al. 2002), CP (Sanchez et al. 2003), or the entire polyprotein (Tomimura et al. 2003). These classifications were based on host range, but the factors affecting the virus host range in cruciferous plants are not clear. Suehiro et al. (2004) used two Japanese *TuMV* isolates, Tu-2R1 and Tu-3 to assess their virulence. Tu-2R1 can systemically infect not only Japanese radish (*Raphanus sativus* L.) but also *Brassica* spp., although Tu-2R1 induces very mild symptoms in cabbage (*Brassica oleracea* L.). On the other hand, Tu-3 infects *Brassica* spp. but not radish. To identify the genetic determinant that plays a role in systemic infection and symptomatology of *TuMV* in cabbage and radish, they constructed two infectious full-length cDNA clones and a series of chimeric viruses between them. Examination of the pathogenicity of these chimeric viruses in cabbage and radish showed that the determinant defining the differential infection phenotype is the P3 protein.

The *Turnip mosaic virus* (*TuMV*, genus *Potyvirus*, family Potyviridae) infects mainly cruciferous plants. Isolates Tu-3 and Tu-2R1 of *TuMV* exhibited different infection phenotypes in cabbage (*Brassica oleracea* L.) and Japanese radish (*Raphanus sativus* L.). Infectious full-length cDNA clones, pTuC and pTuR1, were constructed from isolates Tu-3 and Tu-2R1, respectively. Progeny virus derived from infections with pTuC induced systemic chlorotic and ringspot symptoms in infected cabbage, but no systemic infection in radish. Virus derived from plants infected with pTuR1 induced a mild chlorotic mottle in cabbage and infected radish systemically to induce mosaic symptoms. By exchanging genome fragments between the two virus isolates, theP3-coding region was shown to be responsible for systemic infection by *TuMV* and the symptoms it induces in cabbage and radish. Moreover, exchanges of smaller parts of the P3 region resulted in recombinants that induced complex infection phenotypes, especially the combination of pTuC-derived N-terminal sequence and pTuR1-derived C-terminal sequence. Analysis by tissue immuno-blotting of the inoculated leaves showed that the distributions of P3-chimeric viruses differed from those of the parents, and that the origin of the P3 components affected not only virus accumulation, but also long-distance movement. These results suggest that the P3 protein is an important factor in the infection cycle of *TuMV* and in determining the host range of this and perhaps other potyviruses (Suehiro et al. 2004).

3.10.3 The *Turnip Mosaic Virus* Pathogenicity-Determinant Genes

The *Brassica napus* differential line 165 is resistant to infection by *Turnip mosaic virus* (*TuMV*) isolates belonging to pathotypes 1 and 3. Nucleotide sequences of resistance-breaking mutants of pathotype 1(UK 1), pathotype 3 (CHN 12), and wild-type isolates have been determined. When the mutations identified were introduced into an infectious clone of UK 1, a single mutation in the viral P3 protein induced a hypersensitive (necrotic) response in inoculated leaves of line 165 plants. Full systemic non-necrotic infection was only possible when another mutation (in the cylindrical inclusion protein) was introduced. Tests on segregating populations derived from line 165 indicated that the two viral genes were pathogenicity determinants for two different resistance genes in line 165. One gene responsible for an extreme form of resistance (no symptoms seen) was peristatic to a second responsible for the hypersensitive reaction. These results help to explain the relative stability of the resistance in line 165 and to further define the genetic basis of the *TuMV* pathotyping system (Jenner et al. 2002).

3.10.4 The *BnSGS3* Gene in *Brassica* has Differential Effects on the Pathogenesis of Viruses

The RNA silencing, including post-transcriptional gene silencing (PTGS), is an important mechanism of antiviral defense in plants. In the mechanism and relevant antiviral, plant genes are reported to search for new methods and resources in disease resistance breeding. Suppressor of gene silencing 3 (SGS3) has antiviral activity and specifically exists in plant functioning as a RNA-binding protein. It has been showed in *Arabidopsis* that SGS3 is required for PTGS and natural virus resistance. In the cytoplasm of plant cells, SGS3 stabilizes and recruits the viral single-stranded RNAs (ssRNAs) transcripts to RNA-dependent RNA polymerase 6 (RDR6) and RDR6 converts ssRNAs into double-stranded RNAs (dsRNAs) in SGS3/RDR6 bodies. Next, Dicer-like 4 (DCL4) digests the dsRNAs into 21–24 nt primary small interfering RNAs (siRNAs) assisted by double-stranded-RNA-binding protein 4 (DRB4) and part of the dsRNAs are processed into secondary siRNAs by DCL2. These two classes of siRNAs are protected from degradation aided by HUA ENHANCER 1 (HEN1) and then are combined onto the Argonaute 1/2 (AGO1/AGO2) protein and RNA-induced silencing complexes (RISCs) to silence the complementary viral RNAs. Furthermore, SGS3 can also function in controlling other kinds of RNA silencing such as those induced by sense transgenes (S-PTGS) or DNA virus-induced gene silencing (VIGS) (Mourrain et al. 2000).

The *Cucumber mosaic Virus (CMV)*, *Turnip mosaic virus* (*TuMV*) and *Turnip vein-cleaning virus* (*TVCV*) have different degrees of inhibitory effects on TGS. The *CMV* can only partially inhibit PTGS and produces only a little amount of *CMV* virus RNA in *Arabidopsis* L1 plants, which is significantly lower than that in sgs3 mutants. On the contrary, *TuMV* and *TVCV* completely inhibit PTGS. As a result, these two viruses accumulated to equally high levels in the sgs3 mutants and L1 plants. These results indicate that the PTGS-mediated resistance efficiency on virus infection is related to the inhibitory capability of virus to PTGS (Fukunaga and Doudna 2009). The SGS3 is an essential component of PTGS and closely associated with plant virus resistance. So far, however, the function of SGS3 in oilseed rape has not been reported. Little is known about the expression of SGS3 in oilseed rape or the interaction between major viruses of oilseed rape and SGS3. Chen et al. (2015a, b) investigated the expression pattern of the suppressor of gene silencing 3 in *Brassica napus* (designated *BnSGS3*, accession number KP292910) in various oilseed rape tissues. Furthermore, they generated transgenic oilseed rape plants with overexpressed (*BnSGS3*-Ov) and silenced (*BnSGS3*-Si) levels of *BnSGS3*. Using these transgenic plants upon inoculation with three different viruses Oilseed rape mosaic virus (*ORMV*, *TuMV* and *CMV*), results revealed that *CMV* accumulation was negatively correlated with expression levels of *BnSGS3*, whereas accumulations of both *TuMV* and *ORMV* were positively correlated with expression levels of

BnSGS3, suggesting that *BnSGS3* selectively inhibits *CMV* accumulation but promotes *TuMV* and *ORMV* accumulation.

Virus diseases greatly affect oilseed rape (*Brassica napus*) production and productivity. Investigating antiviral genes may lead to the development of disease-resistant varieties of oilseed rape. Chen et al. (2015a, b) examined the effects of the *suppressor of gene silencing 3* in *Brassica napus* (*BnSGS3*, a putative antiviral gene) with different genus viruses by constructing *BnSGS3*-overexpressing (*BnSGS3*-Ov) and *BnSGS3*-silenced (*BnSGS3*-Si) oilseed rape (cv. Zhongshuang No. 6) plants. These three viruses are *Oilseed rape mosaic virus* (*ORMV*), *Turnip mosaic virus* (*TuMV*) and *Cucumber mosaic virus* (*CMV*). The native *BnSGS3* expressed in all examined tissues with the highest expression in siliques. All three viruses induced *BnSGS3* expression, but *ORMV* induced a dramatic increase in the *BnSGS3*-Ov plants, followed by *TuMV* and *CMV*. Upon inoculation with three different viruses, transcript abundance of *BnSGS3* gene follows: *BnSGS3*-Ov > nontransgenic plants >*BnSGS3*-Si. The accumulation quantities of *ORMV* and *TuMV* exhibited a similar trend. However, *CMV* accumulation showed an opposite trend where virus accumulations were negatively correlated with *BnSGS3* expression. The results suggest that *BnSGS3* selectively inhibits *CMV* accumulation but promotes *ORMV* and *TuMV* accumulation. The *BnSGS3* should be used in different ways (up- and down-regulation) for breeding virus-resistant oilseed rape varieties (Chen et al. 2015a, b).

3.10.5 Identification of Pathogenesis Modulating Genes in *Arabidopsis–Turnip Mosaic Virus* Interaction

The ultrastructural effects of *TuMV* infection on *A. thaliana* Col-0 and NADPH oxidase rbohD and rbohF transposon mutants were compared, and performed *in situ* localization and quantification of virus particles, H_2O_2, and PR1 proteins. The *TuMV* infection in rbohD led to a systemic development of virus infection. The virus concentration and localization significantly increased between 3 and 7 days after inoculation in rbohD, even more dynamically than in Col-0. In this mutant, in situ localization of H_2O_2 corresponds with virus particles and was observed in vacuoles and between the cell wall and plasmalemma. Moreover, quantification of H_2O_2 showed that rbohD-inoculated plants displayed a considerable reduction in ROS accumulation compared to Col-0. ROS derived from RbohD plays a role in the resistance response to *TuMV* in this pathosystem. Opposite to rbohD, a significant decrease in virus infection cycle was observed in rbohF and rbohD/rbohF compared to Col-0 orrbohD. Interestingly, rbohD/rbohF response was accompanied by local necrosis, typical for HR-like reactions with the characteristic rebuilding of cell walls, phenolic-like compounds deposited inside xylem elements, together with the rare occurrence of virus particles and complete lack of virus inclusions. Noticeably, in situ localization of H_2O_2 in these more susceptible genotypes revealed that the deposits of hydrogen peroxide were differentially linked to the cell wall, near plasmo desmata and in the area around necrotizing cells. This pattern of H_2O_2 deposition evidenced a possible longer distance transport of these reactive species in these genotypes that could contribute to the enhanced resistance observed. Deposition of PR1 protein was activated in all *A. thaliana* variants compared to mock-inoculated plants, but the more susceptible rbohD plants displayed a higher level of PR1, whereas rbohF and rbohD/F displayed a lower accumulation. The qualitative and quantitative analyses indicated that PR1 location strictly parallels the virus alterations, although it is not clear that the PR1 induction acts as an antiviral factor. The important role of ROS produced by NADPH oxidases in regulating the *Arabidopsis* response to *TuMV* and a differential role of RbohD and RbohF in this interaction were revealed. Further molecular and cellular studies are needed to elucidate the possible role of other signaling molecules (e.g., nitric oxide metabolism and calcium signature distribution) that may potentially serve as active components in the *A. thaliana–TuMV* pathosystem. It has been reported that nitricoxide signaling and calcium protein kinases (among other kinases) can modulate NADPH oxidase activity; these analyses may help provide a better understanding of the mechanisms modulating *A. thaliana–TuMV* interactions and pathogenesis (Otulak-Kozieł et al. 2020).

3.11 *Brassica–Verticillium* Pathosystem

3.11.1 Introduction

The *Verticillium longisporum* is a soil-borne fungus which causes vascular disease primarily in Brassicaceae (Karapapa et al. 1997; Zeise and von Tiedemann 2002; Fradin and Thomma 2006). Triggered by plant root exudates, fungal hyphae develop from microsclerotia, penetrate the root epidermis, grow through the root cortex and enter the xylem elements from where *V. longisporum* spreads throughout the plant (Eynck et al. 2007). The life cycle is completed after formation of microsclerotia on senescent tissue (Wilhelm 1955). *Verticillium longisporum* differs in molecular and morphological traits from its close relatives *V. dahliae* and *V. albo-atrum* (Karapapa et al. 1997; Zeise and von Tiedemann 2001): The microsclerotia of *V. longisporum* are elongated in contrast to the spherical morphology of those of *V. dahliae* and its conidia are nearly twice as long as those from other *Verticillium* species. The DNA content was estimated to be nearly 1.8 times that of *V. dahliae*, suggesting that *V. longisporum* is a "nearly diploid" hybrid (Inderbitzin et al. 2011; Tran et al. 2013). Whereas *V. dahliae* and *V. albo-atrum* are able to infect a broad variety of crop species, *V. longisporum* primarily colonizes members of the family of the Brassicaceae (Zeise and von Tiedemann 2002). In contrast to *V. dahliae* and many other vascular pathogens, *V. longisporum* (isolate Vl43) does not induce wilting of the host plant (Reusche et al. 2012), which may result from only partial colonization of the vessel elements (Eynck et al. 2007).

Innovations in genomic techniques have the potential to bring a synthesis to the study of host–pathogen interactions across systems and environmental conditions. Several recent trends in this perspective for genomic studies of host–pathogen systems viz., (1) evolutionary genomics approaches have allowed the field to move from a candidate gene approach to investigations at the scale of whole genomes; (2) the use of genomics for the detection and surveillance of host–pathogen systems; (3) the challenges of the integrating natural history and ecological metadata and genomic data across systems and time scales due to infrastructural challenges of database integration and transparency; and (4) the impact of anthropogenic stressors on host–pathogen systems that have consequences for global health. Any single host–pathogen study is constrained by limited resources or genomic tractability, the geographical and evolutionary time scales involved as well as environmental complexities. Accordingly, transparent and open science will help to achieve a comprehensive understanding of host–pathogen interactions in general. This will contribute to the integration of findings across the different scales. A large repertoire of comparable and inter-communicative studies will facilitate a more generalizable understanding of the causes and consequences of host–pathogen interactions and a clearer road map to combating the continuous threat of pathogens in a changing world (Näpflin et al. 2019).

3.11.2 Differential Expression of Proteins During *Verticillium* Pathogenesis of *Brassica*

The proteomic approaches have been introduced to obtain a comprehensive overview of proteins responsive in plants to pathogen attack. The apoplast, which includes the xylem sap, has received specific attention because it is the first compartment where recognition and defense may take place. When plants or cell cultures were challenged with pathogens or elicitors pronounced changes in the extracellular proteome were found such as accumulation of novel protease inhibitors, stimulation of defense proteins, modification in cell proteins, etc. Activation of apoplastic defenses has not been characterized in this species, although knowledge on these interactions may help to device strategies for improving plant resistance. Floerl et al. (2008) identified proteins activated in response to *V. longisporum* in the leaf apoplast and xylem sap of *Brassica napus* and to investigate whether changes in the composition of extracellular fluid were efficient in attenuating fungal proliferation. They have characterized these responses in relation to the development of disease symptoms, photosynthetic electron transport, gas exchange and nutrient status.

The *Verticillium longisporum* is one of the most important pathogens of *Brassicaceae* that remains strictly in the xylem during most stages of its development. It has been suggested that disease symptoms are associated with clogging of xylem vessels. Floerl et al. (2008) investigated extracellular defense reactions induced by *V. longisporum* in the xylem sap and leaf apoplast of *Brassica napus* var. *napus* in relation to the development of disease symptoms, photosynthesis and nutrient status. *V. longisporum* (strain VL43) did not overcome the hypocotyl barrier until 3 weeks after infection although the plants showed massive stunting of the stem and mild leaf chlorosis. During this initial infection phase photosynthetic carbon assimilation, transpiration rate and nutrient elements in leaves were not affected in

VL43-infected compared to noninfected plants. Proteome analysis of the leaf apoplast revealed 170 spots after 2-D-protein separation, of which 12 were significantly enhanced in response to VL43-infection. LS-MS/MS analysis and data base searches revealed matches of VL43-responsive proteins to an endo chitinase, a peroxidase, a PR-4 protein and a β-1,3-glucanase. In xylem sap, three upregulated proteins were found of which two were identified as PR-4 and β-1, 3-glucanase. Xylem sap of infected plants inhibited the growth of *V. longisporum*. The *V. longisporum* infection did not result in drought stress or nutrient limitations. Stunting and mild chlorosis were, therefore, not consequences of insufficient water and nutrient supply due to VL43-caused xylem obstruction. A distinct array of extracellular PR-proteins was activated that might have limited *Verticillium* spreading above the hypocotyl. In silico analysis suggested that ethylene was involved in upregulating VL43-responsive proteins (Floerl et al. 2008; Tables 3.7 and 3.8).

3.12 *Brassica–Xanthomonas* Pathosystem

3.12.1 Introduction

The *Xanthomonas* is a group of plant-associated bacteria, well known for causing several diseases in crop plants, such as cotton, rice, and cruciferous vegetables. The *Xanthomonas* has an extensive host range documented, including mono- and dicotyledonous plants. A well-known species of this genus is *X. campestris,* previously subdivided into 6 pathovars. However, a recent study, based on pathogenicity assays, reclassified *X. campestris* into three pathovars (pv): *X. campestris* pv. *campestris*, *X. campestris* pv. *Raphani,* and *X. campestris* pv. *incanae.* The strains that did not induce disease symptoms in *Brassicaceae* plants were denominated *X. campestris*. *X. campestris* pv. *campestris* (*Xcc*) is the causing agent of black rot disease in cruciferous crops, which causes severe yield and economic losses. Several races of this pathovar have been reported, showing the complexity of *Xcc*. The molecular processes involved in the ability of *Xcc* to infect the plant have been explored by different "omic" techniques, which have contributed to a better understanding of the plant-pathogen interactions (Alkhateeb et al. 2017; Santos et al. 2017).

3.12.2 Secretion System of *Xanthomonas* Genome Regulates Pathogenicity

Genome wide comparisons have revealed three secretion systems (types II, III and IV) in National Collection of Plant Pathogenic Bacteria, Sand Hutton, UK with regard to the *Xcc* that have been associated with pathogenic bacteria of plants or animals. The type II secretion system (T2SS) mediates the transport of proteins into the extracellular space. This system can secrete plant cell wall-degrading enzymes, including cellulose, polygalacturonase, xylanase and protease. Two classes of T2SS

Table 3.7 Differentially expressed proteins in apoplastic washing fluids of leaves of oilseed rape (*Brassica napus* var. *napus*) after infection with *Verticillium longisporum* VL43 (Floerl et al. 2008)

Spot no	Similarity (organism) peptide	NCBI (ACC)	Protein score (threshold)	MW	Target peptide	Factor	*P*
21	**Basic glucanase (Bj)**	gi|86371164	223 (56)	30722	S (RC 1)	3.9	0.037
	LYDPNQEVLSALR	(ABC94638)					
	YISVGNEVQPSDPTSR						
	GISGFPPSSGTFTPEFR						
	β-1,3-Glucanase (Ath)	gi|6735303	91 (56)	37632	S (RC 1)		
	LASSQAEADTWVR	(At3g57240)					
	GISGFPPSSGTFTPEF						
22	**Basic glucanase (Bj)**	gi|86371164	405 (56)	30722	S (RC 1)	3.1	0.001
	IASSQAEADTWVR	(ABC94638)					
	LYDPNQEVLSALR						
	YISVGNEVQPSDPTSR						
	GISGFPPSSGTFTPEFR						
	NLFHALVDTVYAALEK						
	GSNIDLLLDVPNPDLQR						
	LDYALFTSPSTVVNDGSNAYR						
	β-1,3-Glucanase (Ath)	gi|6735303	180 (56)	37632	S (RC 1)		
	LASSQAEADTWVR	(At3g57240)					
	GISGFPPSSGTFTPEFR						
	TYVNNLIQTVK						
23	**β-1,3-Glucanase (Ath)**	gi|6735303	222 (56)	37632	S (RC 1)	4.8	0.049
	LASSQAEADTWVR	(At3g57240)					
	GISGFPPSSGTFTPEFR						
	TYVNNLIQTVK						
	GSNIELVLDVPNPDLQR						
30	**Not identified**					5.8	0.007

32	**Not identified**					8.3	0.000
34	**Basic endochitinase (CHB4) (Bn)**	gil584929	95 (56)	29684	S (RC 1)	7.7	0.009
	DSFINAANTFPNFANSVTR	(Q06209)					
	AINGMECNGGNSGAVNAR						
36	**Not identified**					4.8	0.014
51	**Basic endochitinase (CHB4) (Bn)**	gil584929	159 (44)	29684	S (RC 1)	4.2	0.020
	DSFINAANTFPNFANSVTR	(Q06209)					
	AINGMECNGGNSGAVNAR						
	DYCGQLGVDPGPNLS						
52	**Not identified**					3.5	0.045
53	**Not identified**					4.4	0.005
54	**Korean horseradish peroxidase (Rs)**	gil1518388	94 (44)	33993	S (RC 3)	2.4	0.040
	MGDISPLTGSSGEIR	(X91172)					
	Peroxidase (Ath)	gil15239075	94 (44)	34650	S (RC 1)		
	MGDISPLTGSSGEIR	(At5g05340)					
55	**Not identified**					3.6	0.005

Data indicate means of 5 gels per treatment. NCBI and accession number (ACC) of the proteins were identified by comparing the peptide sequence with NCBI databank. Protein score, predicted molecular weight (MW) and predicted target peptide with reliability class (RC) are indicated. Factor indicates mean intensity of spots in apoplastic washing fluid of VL43-infected plants/mean intensity of spot in apoplastic washing fluid of control plants. *P*-values were calculated using Protein Weaver software. *Ath*: *Arabidopsis thaliana*, *Bj*: *Brassica juncea*, *Bn*: *Brassica napus*, *Rs*: *Raphanus sativus*

Table 3.8 Differentially regulated proteins in xylem sap of oilseed rape (*Brassica napus* var. *napus*) after infection with *Verticillium longisporum* VL43 (Floerl et al. 2008)

Spot no	Similarity (organism) peptide	NCBI (ACC)	Protein score (threshold)	MW	Target peptide
XVL 1	**Basic glucanase (Bj)**	gi| 118763538	510 (44)	30722	S (RC 1)
	IASSQAEADTWVR	(ABC94638)			
	LYDPNQEVLSALR				
	YISVGNEVQPSDPTSR				
	GISGFPPSSGTFTPEFR				
	NLFHALVDTVYAALEK				
	GSNIDLLLDVPNPDLQR				
	LDYALFTSPSTVVNDGSNAYR				
	β-1,3-Glucanase (Ath)	gi|6735303(	222 (45)	37632	S (RC 1)
	LASSQAEADTWVR	At3g57240)			
	GISGFPPSSGTFTPEFR				
	TYVNNLIQTVK				
	GSNIELVLDVPNPDLQR				
XVL 2	**Not identified**				
XVL 3	**PR4-Typ protein (Br)**	gi|73671284	162 (44)	15950	S (RC 2)
	VTNTGTQAQATVR	(AF528181)			
	QIDTDGQGYAR				
	ATYHFYNPAQNGWDLYR				

NCBI and accession number of the proteins identified by comparing the peptide sequence with NCBI databank are indicated. Protein score, predicted molecular weight (MW) and predicted target peptide with reliability class (RC) are indicated. *Ath*: *Arabidopsis thaliana*, *Bj*: *Brassica juncea*, *Br*: *Brassica rapa*

have been identified in *Xanthomonas*: *xps* and *xcs*. The *xps* cluster is present in all *Xanthomonas* genomes currently sequenced and in other genera (*Xylella*), whereas the *xcs* genes are only present in some of the *Xanthomonas* genomes, including *Xcc* and *X. campestris* pv. *raphani* (Lu et al. 2008). Qian et al. (2005) obtained six pathogenicity-deficient mutants (*xpsD*, *xpsE*, *xpsF*, *xpsK*, *xpsL,* and *xpsM*) related to the *xps* system, whereas no pathogenicity deficient mutants were found on the 12 annotated genes related to the *xcs* system; therefore, the *xps* genes are related to pathogenicity, whereas the *xcs* genes may have other roles not essential for pathogenicity or may be nonfunctional.

The mutation of DsbB, a protein involved in disulfide bond formation in the periplasm of *Xcc*, reduced virulence, cell mobility and bacterial growth *in planta* and resulted in ineffective T2SS and T3SS (Jiang et al. 2008b). The type IV secretion system (T4SS) is important for the release of macromolecules, and this system is used by Gram-negative bacteria to translocate protein and DNA substrates across the cell envelope into target cells (Souza et al. 2011). However, the role of T4SS in the pathogenicity of *Xanthomonas* is uncertain. The T4SS carried in the race 9 isolate 8004 is encoded mainly by genes in the *virD4* and the *virB* cluster that consists of nine open reading frames (ORFs). Qian et al. (2005) identified a mutated gene encoding the channel-forming protein VirB8, and concluded that the T4SS contributes to the virulence of the pathogen. However, He et al. (2007) have shown that deletion of *virD4* and the *virB* cluster of strain 8004 does not affect the virulence of the pathogen. The T3SS is generally thought to be essential for pathogenicity and for the initiation of disease in susceptible host plants, and can be involved in the elicitation of cell death and other defense responses in resistant plants.

The T3SS is encoded by the *hrp* cluster of genes and translocates effector proteins directly into the plant cells; it is also possible that effectors can exit the bacterial phytopathogens via the Hrp pathways (Lindgren 1997). Individual genes have been named *hrp*, *hrp*-conserved (*hrc*) or *hrp* associated (*hpa*). Most *hrp* gene sequences are only found in *Xanthomonas* and related genera, whereas *hrc* genes are conserved in many plant and animal pathogens. In *Xcc*, the *hrp* cluster consists of 26 genes extending from *hpa2* to *hrpF* (da Silva et al. 2002). The *Xcc* genomes do not contain *hpa3* and the N-terminal region of *xopF1*, although these genes are present in other *Xanthomonas* species, including *X. campestris* pv. *raphani* (Lu et al. 2008). Four mutants of genes encoding the T3SS machinery (*hrcJ*, *hrcU*, *hrcV*, *hrpE*) and one encoding a regulatory protein (*hrpG*) were nonpathogenic (Qian et al. 2005). Mutations in the regulator *hrpG* of strain 8004 can make all the *hrp* genes express constitutively and enhance the intensity of the HR reaction in pepper (Jiang et al. 2006). Wei et al. (2007) showed that *hpaR*, a putative *marR* family transcriptional regulator, is essential for the pathogenicity of strain 8004 on cabbage, is required for the HR response on the nonhost pepper and is under the positive control of the two key *hrp* gene regulators HrpG and HrpX. The GntR family is a frequent group of helix–turn–helix transcriptional regulators in bacteria. One of the six GntR regulators, HpaR1, positively and negatively affects the expression of HR and pathogenicity *hrp* genes that encode the type III secretion via hrpG (An et al. 2011).

Several genes that are regulated in a Hrp X-dependent manner possess a consensus nucleotide sequence TTCGB–N15–TTCGB (B can be base C, G or T), which has been termed the plant-inducible promoter box, or PIP box. The detection of PIP boxes provides a refined tool for the identification of candidate genes in the hrpX regulon, as well as effector protein genes of the type III pathway. The 17 putative PIP box sequences have been described in the promoter regions of *Xcc* ATCC33913 (da Silva et al. 2002) and 56 genes have been predicted to have PIP boxes in strain 8004 (Jiang et al. 2009).

3.12.3 Functional Characterization of Type III Effector Genes of *Xanthomonas* in Pathogenesis

There has been considerable research effort to identify and understand type III effector function in *Xanthomonas* species. White et al. (2009) observed type III effectors from *Xanthomonas* and classified them into 39 different groups (called *Xop*) reflecting sequence similarity. From these, 21 appear to be present in the *Xcc* strain ATCC33913. Nine effectors, considered to be core effectors, are present in almost all *Xanthomonas* isolates, with the exception of *X. albilineans,* and in some cases, *X. campestris* pv. *raphani*; other effectors are found in a more limited number of species/isolates (Bogdanove et al. 2011; Ryan et al. 2011). From the extensive list of genes encoding putative T3SS proteins in *Xcc* and *X. campestris* pv. *raphani* (Table 3.9), effort has begun to identify potential effector proteins (which are important for pathogenicity in a susceptible host) and avirulence determinants (which elicit defense in a resistant host). Effector proteins are generally thought to be involved in the suppression of the host defense system or in the alteration of host metabolism to favor nutrient uptake of the pathogen. The deletion of individual effector genes can often result in little or no reduction in pathogenicity because of the presence of alternative functional (redundant) effector proteins (Cunnac et al. 2009). Some T3SS effectors may activate defenses (rapid host cell death) when infiltrated into plant tissue, and thus indicate a potential role as avirulence (Avr) proteins in a resistant host (Flor 1971; Keen 1990).

It has been shown that the mutagenesis of eight candidate genes from strain 8004 had no effect on pathogenicity in Chinese radish and cabbage (Xu et al. 2006). However, He et al. (2007) showed that mutants of *avrBs1* lost the ability to elicit an HR in the pepper line ECW10R that contains the corresponding R gene *Bs1*. The AvrBs1 effector is responsible for the elicitation of HR on pepper dependent on T3SS (Xu et al. 2006). However, Castaneda et al. (2005) did not detect an HR variation between the *avrBs1* mutants and the wild-type 528T. He et al. (2007) suggested that it is possible that the function of *avrBs1* is redundantly encoded in the *Xcc* type strain and that the expression is different in the two strains. Ignatov et al. (2002) reported that an *avrBs2* homologue, designated *avrRxc1/3*, determined the avirulence of *Xcc* 512/2 (a natural mutant strain of a race 3 isolate) on *Brassica* plants with the B genome, including *B. juncea* cv. Florida Mustard and *B. carinata* line PI 199947, and on some *B. oleracea* cultivars (including the lines SR1/3, Badger Inbred-16/2 and cv. Miracle F1). This gene was also responsible for a mesophyllic HR on leaves of Florida Mustard and nonhost pepper plants carrying *Bs2*. Xu et al. (2006) showed that *avrBs2* is required for full virulence of strain 8004 on Chinese radish and cabbage. In contrast, Castaneda et al. (2005) showed that the deletion of the entire putative *avrBs2* homologue in the type strain had no effect on the interaction with Florida Mustard and pepper. The strains employed by Ignatov et al. (2002) are probably derived from the type strain used by da Silva et al. (2002) and Castaneda et al. (2005); therefore, it is difficult to reconcile these results and the role of the *avrBs2* homologue on pathogenicity/race specificity still needs to be clarified.

Table 3.9 Occurrence, features, and function of putative and proven effector/avirulence genes of *Xanthomonas campestris* pv. *campestris* (Vicente and Holub 2013)

Effector class	Synonym	Gene ID in ATCC33913	Gene ID in 8004	Gene ID in B100	Gene ID in *Xcr* 756C	PIP box	Features and function on pathogenicity
Core effectors[a] AvrBs2		XCC0052	XC 0052	xcc-b100 0057	No	Yes	Glycerophosphoryl diester phosphodiesterase. Avirulence in *Brassica juncea*, *B. carinata* and *B. oleracea* and HR on pepper (Ignatov et al. 2002) or no effect (Castaneda et al. 2005)
XopK		XCC2899	XC 1210	xcc-b100 1254	No		Unknown
XopL	XopLR	XCC4186	XC 4273	xcc-b100 4400	No	Yes	Leucine-rich protein
							Reduced virulence on radish (Jiang et al. 2009)
XopN		XCC0231	XC 0241	xcc-b100 0253	No		ARM/HEAT repeat
							Reduced virulence on radish (Jiang et al. 2008a)
XopP		XCC1247	XC 2994	xcc-b100 3057	XCA_1500	Yes	Unknown
							Reduced virulence on radish (Jiang et al. 2009)
XopQ		XCC1072	XC 3177	xcc-b100 3274	No	Yes	Putative inosine–uridine nucleotide N-ribohydrolase
							Reduced virulence on radish (Jiang et al. 2009)
XopR		XCC0258	XC 0268	xcc-b100 0280	XCA_4254		Unknown
XopX		XCC0529	XC 0541	xcc-b100 0558	No		Methionine-rich protein
		XCC0530	XC 0542	and 0559			
XopZ		XCC1975	XC 2210	xcc-b100 2274	No		Unknown

(continued)

Table 3.9 (continued)

Effector class	Synonym	Gene ID in ATCC33913	Gene ID in 8004	Gene ID in B100	Gene ID in *Xcr* 756C	PIP box	Features and function on pathogenicity
Other effectors[b]							
AvrBs1		XCC2100	XC 2081	xcc-b100 2396	No	No	HR on pepper (He et al. 2007)
XopD		XCC2896	XC 1213	xcc-b100 1256	No		SUMO cysteine protease (C48 family), EAR motif, DNA binding, nuclear localization (Canonne et al. 2012)
XopE	AvrXccE1	XCC1629	XC 2602	No	No	Yes	Putative transglutaminase
							Avirulence on *B. rapa* (He et al. 2007)
XopF		XCC1218	XC 3024	xcc-b100 3087	XCA_1470		Unknown
XopG		XCC3258	XC 0967	xcc-b100 2655	No		M27 family peptidase (Clostridium toxin)
XopH	AvrBs1.1	XCC2099	XC 2082	xcc-b100 2395	No	No	Putative tyrosine phosphatase
XopJ	AvrXccB	XCC3731	XC 3802	No	No	Yes	Putative cysteine protease (C55 family) or serine/threonine acetyltransferase. Ubiquitin-like protease
							No effect on pathogenicity (Jiang et al. 2009)
XopAC	AvrAC	XCC2565	XC 1553	xcc-b100 1596	XCA_2914	Yes	Leucine-rich protein
							Avirulence recognized in vascular tissues of *Arabidopsis* (Xu et al. 2008)
XopAD		No	No	No	XCA_1464		SKWP repeat protein
XopAG		XCC3600	XC 0563	xcc-b100 0580	No		Unknown
XopAH	AvrXccC	XCC2109	XC 2004	xcc-b100 2071	No	Yes	Dual function. Avirulence in *B. juncea* and *B. rapa*; virulence in *B. oleracea* (Castaneda et al. 2005; He et al. 2007; Wang et al. 2007)

XopAL1	AvrPhpE	XCC1246	XC 2995	xcc-b100 3058	XCA_1499	Yes	Reduced virulence on radish (Jiang et al. 2009)
XopAL2		Downstream XCC3574	Downstream XC_3916	xcc-b100 0616	No		Unknown
XopAM	XopR1	XCC1089	XC 3160	xcc-b100 3256	No	Yes	Reduced virulence on radish (Jiang et al. 2009)
XopAT		No	No	No	XCA_1464a	Yes	Unknown. No similarity with other known proteins (Bogdanove et al. 2011)
XopA		XCC1240	XC 3002	xcc-b100 3065	XCA_1492		Harpin, may not be a t3e
HpaA		XCC1224	XC 3018	xcc-b100 3081	XCA_1476		T3 secretion control protein, may not be a t3e
HrpW		XCC1219	XC 3023	xcc-b100 3086	XCA_1471		Pectate lyase, may not be a t3e
AvrXccA1	AvrXca	XCC4229	XC 4318	xcc-b100 4450	XCA_4581	No	May not be a t3e. Virulence of *Xcr* in *Arabidopsis* (Parker et al. 1993)
AvrXccA2		XCC2396	XC 1716	xcc-b100 1770	XCA_2696	No	Unknown, may not be a t3e

Adapted from da Silva et al. (2002), White et al. (2009), Jiang et al. (2009) and Ryan et al. (2011)
t3e Type III effector
[a] Nine core effectors are found in most *Xanthomonas* spp., with the exception of *X. albilineans* and, in some cases, *X. campestris* pv. *raphani* (*Xcr*)
[b] Other effectors are found in a more limited number of *Xanthomonas* spp. isolates

Genes from the *AvrBs3* family have been found in a number of *Xanthomonas* species and pathovars, and in *Ralstonia solanacearum* (White et al. 2009). This large family of closely related T3 effectors constitutes the TAL effector family. All members of the family have a common structure with tandem repeats of a sequence of amino acids, typically of 34 amino acids in the central part of the proteins, nuclear localization signals (NLSs) and an acidic activation domain (AD); the avirulence and virulence specificities of an AvrBs3 member are dependent on the number and order of the repeats and the NLSs and AD (Boch and Bonas 2010; Gurlebeck et al. 2006).

Interestingly, the three *Xcc* strains sequenced do not contain genes from the *avrBs3/pthA* family, whereas three members of this family have been identified in an isolate named as *X. campestris* pv. *armoraciae*. These three genes were designated *hax2*, *hax3,* and *hax4* (homologue of *avrBs3* in *Xanthomonas*) (Kay et al. 2005). The three Hax proteins are translocated via the T3SS. The Hax3 and Hax4 proteins have the typical structure of the AvrBs3-like effectors, with 34 amino acid repeats in the central part, but Hax2 has 35 amino acid repeats that contain an additional proline residue. The three *hax* genes have an additive effect on disease symptoms in radish, with hax2 having the strongest influence, and two *hax* genes (*hax3* and *hax4*) have a *Bs4*-dependent avirulence activity in tomato (Kay et al. 2005).

Transconjugants of Chinese isolates containing *avrXccE1* (*xopE*) became avirulent on Chinese cabbage (*B. rapa*) cv. Zhongbai-83 (He et al. 2007). In contrast, Castaneda et al. (2005) did not notice any effect on pathogenicity on the plants tested, but it is possible that the lines tested by these authors did not possess the *R* gene responsive to *avrXccE1*. Jiang et al. (2008a) showed that XopN from strain 8004 (designated XopXccN) is required for full virulence. A mutant of the strain 8004 with an insertion in *xopXccN* was significantly weaker than strain 8004 when inoculated on Chinese radish (*Raphanus sativus* var. *radiculus*) cv. Manshenhong. The expression of *xopXccN* is regulated by the key *hrp* regulators HrpG and HrpX. The protein sequence of XopD has been identified in strain B100 (Canonne et al. 2012). This type III effector targets the *Arabidopsis* transcription factor MYB30, possibly as part of a virulence strategy that allows *Xcc* to suppress the plant cell responses to infection (Canonne et al. 2011).

Mutagenesis of the eight candidate effector genes from type strain ATCC33913 (syn. 528T) had no effect on pathogenicity when the mutants were used to inoculate six crucifer species, and none of the mutated genes singly or in any combination affected the nonhost HR elicited by the type strain in pepper. However, insertion or deletion mutants in a locus of a gene designated *avrXccFM* (XopAH) became virulent on Florida Mustard, and therefore changed the race specificity of the isolate (Castaneda et al. 2005). The T3SS effector AvrXccC of strain 8004, which belongs to the AvrB effector family of *Xanthomonas*, has been shown to have a dual effect: this effector is required for full bacterial virulence in a susceptible *B. oleracea* cv. Jingfeng 1 and for avirulence in a *B. napiformis* L.H. Bailey [syn. *B. juncea* (L.) Czern. var. *napiformis* (Pailleux and Bois) Kitam.] accession (Wang et al. 2007). The *avrXcc C* gene was expressed in the race 2 strain HRI3849A (which causes symptoms on the mustard accession used by these authors) and the resulting strain was avirulent. He et al. (2007) also showed that an *avrXccC* mutant of 8004 became pathogenic on mustard (*B. juncea* var. *megarrhiza* Tsen et Lee) cv. Guangtou and Chinese cabbage (*B. rapa* ssp. *pekinensis*) cv. Zhongbai-83. The *avrXccC* gene of strain 8004, which confers avirulence on some mustard cultivars, is identical to *avrXccFM* of strain ATCC33913 (syn. 528T), which confers avirulence on Florida Mustard (He et al. 2007). Wang et al. (2007) showed that AvrXcc C is anchored to the plant plasma membrane and its avirulence function for host recognition depends on its location.

The expression of *avrXccC* is *hrpG/hrpX* dependent. The *Xcc* also contains genes encoding leucine-rich repeat (LRR) proteins. These motifs are commonly involved in protein–protein interactions and are found in several classes of plant disease resistance genes. Xu et al. (2008) showed that one of these genes, *xopAC*, named *avrACXcc8004*, encodes a protein containing LRRs; this gene is a type III effector that appears to be restricted to strains of *X. campestris*. The *A. thaliana* accession Col-0 was resistant to the wild-type 8004 strain and to an *avrACXcc8004* mutant when the leaf mesophyll was infiltrated with bacterial suspensions, but the Col-0 ecotype became susceptible to the mutant when the bacterium was introduced into the vascular system by piercing the central vein of the leaves, indicating that the product of *avrACXcc8004* is recognized in vascular tissues and this gene might be related to the ability to colonize the xylem of a host. Some *avr*/effector genes have also been identified in closely related pathovars, including isolates identified as *X. campestris* pv. *armoraciae* and *raphani* (possibly synonymous). Parker et al. (1993) cloned an avirulence gene of an isolate identified as *X. campestris* pv. *raphani* (isolate 1067). This isolate was avirulent in all the *A. thaliana* accessions tested. When the *avrXca* gene was transferred to *Xcc* strain 8004, it strongly reduced symptom development and bacterial growth in Columbia plants, but did not affect virulence to *Brassica* plants. The *avrXca* gene encodes a protein of 67 kDa that has no homology with known sequences and confers avirulence in a number of *A. thaliana* accessions, except one ("Kas-1"). The avirulence phenotype is not *Hrp* dependent and the interaction with *A. thaliana* did not lead to a characteristic HR, indicating that this may not be a type III effector. Corbett et al. (2005) have characterized a virulence factor, designated Svx, of *E. carotovora* ssp. a*troseptica* which shows homology to AvrXca. This protein is secreted by the type II secretion apparatus and the transcription of the *svx* gene is regulated by quorum sensing; the function of this protein is unknown, but it may play a role in the pathogenicity of *E. carotovora* pv. *atroseptica.* So far, there is some evidence that 12 of the putative effector genes from *Xcc* are phenotypically functional in a very restricted number of isolates (Gabriel 1999).

3.12.4 Virulence and Pathogenicity Effectors/Genes of *Xanthomonas*

The surface structure and appendages of phytopathogenic bacteria are important for the attachment, colonization, and infection of the host. Type IV pili may contribute to bacterial pathogenesis by affecting surface motility, micro colony and biofilm formation, adhesion, immune evasion, and cell signaling (Craig et al. 2004). The comparison of genomes from isolates ATCC33913 and 8004 showed that at least 26 genes related to the pili assembly are highly conserved, and mutations in two of the assembly genes (*pilB* and *pilC*) have reduced virulence (Qian et al. 2005).

Many phytopathogenic bacteria produce a large number of factors that might be essential or contribute to cause disease. The bacteria from the genus *Xanthomonas* typically produce yellow, membrane-bound pigments, called xanthomonadins. These pigments have a role in the maintenance of the ecological fitness of the bacteria, protecting the cells against photo oxidative stress. Xanthomonadin biosynthesis is encoded by the *pig* cluster of genes. Poplawsky and Chun (1998) have shown that the mutation of *pigB* results in a reduction in epiphytic survival and decrease in infection via the hydathodes. The biosynthesis of xanthomonadin is regulated by 3-hydroxybenzoic acid, a diffusible factor (DF) that is associated with a range of biological functions; DF-deficient mutants are nonpigmented, impaired in survival ability and less virulent (He et al. 2011). The *Xcc* produces a range of extracellular enzymes (including proteases, pectinases, and endoglucanase). The extracellular enzymes are capable of degrading the plant cell components and may be required to overcome plant defense responses, to allow bacteria to move into uncolonized plant tissues and to mobilize plant polymers for nutritional purposes (Torres et al. 2007). In *Xcc*, the synthesis of extracellular degrading enzymes and exopolysaccharides is regulated by products of the *rpf* genes (rpfABFCHGDIE) (da Silva et al. 2002). The *rpf* gene cluster of *Xcc* regulates the genes involved in the synthesis of extracellular hydrolytic enzymes and extracellular polysaccharides (EPSs) (including xanthan) and plays a role in motility, toxin production, oxidative stress resistance, aerobic respiration, and biofilm formation (Dow et al. 2003; Wilson et al. 1998). A diffusible signal factor (DSF) responsible for cell–cell signaling is involved in the transcriptional control of these genes (Barber et al. 1997). Some genes under the control of the rpf/DSF system are required for the first stages of endophytic colonization (Gudesblat et al. 2009). Genes that encode proteins involved in the synthesis or degradation of cyclic-di-guanosine-3′,5-′-monophosphate (cyclic-di-GMP), a second messenger involved in the regulation of cellular functions, have been shown to play significant roles in virulence to plants. Complex regulatory networks possibly allow the bacteria to adapt to different environments and to modulate virulence factors (Ryan et al. 2007). The expression levels of proteases and endoglucanases are reduced when the *rpfI* gene is inactivated in *Xcc*, suggesting that *rpfI* may have a function in the extensive tissue degeneration that is characteristic of black rot (Dow et al. 2000). RpfF is involved in the synthesis of DSF, and a two component system comprising RpfG and the complex sensor kinase RpfC is involved in sensing and responding to DSF (Ryan et al. 2010). Mutants deficient in RpfG, a phosphodiesterase that degrades cyclic-di-GMP, or the elevation of cyclic-di-GMP via the overexpression of a GGDEF domain protein WspR, caused the aggregation of cells, reduction in mobility and decrease in production of virulence factors, extracellular enzymes and exopolysaccharides (Hsiao et al. 2011). Mutations of *rpfF* or *rpfC* also lead to a reduction in the synthesis of virulence factors. The cell–cell interactions are highly regulated and dynamic; the interactions involving HD-GYP and GGDEF domain proteins are dependent on DSF signaling influence the localized expression of cyclic-di-GMP and mediate virulence (Ryan et al. 2007).

The Mips (macrophage infectivity potentiators) genes encode proteins reported as virulence factors in human pathogens. Zang et al. (2007) showed that a *mip*-like gene of *Xcc* is involved in pathogenicity through an effect on the production of exopolysaccharides and on the activity of extracellular proteases. EPSs play an important role in the pathogenicity and virulence of many bacteria, both in terms of direct interactions with host cells and in conferring protection against hostile environments (Coplin and Cook 1990). The most important EPS secreted by *Xcc* is xanthan gum, an EPS that has many industrial applications. Xanthan is a complex EPS with a cellulosic backbone and trisaccharide side chains of two mannoses and one glucuronate residues that are attached to every second glucose in the backbone (Becker et al. 1998; Jansson et al. 1975; Vorhoelter et al. 2008; Yun et al. 2006). Yun et al. (2006) have shown that a xanthan minus mutant strain and a mutant strain that produces truncated xanthan fail to cause disease in *A. thaliana* and *Nicotiana benthamiana*, and induce the deposition of callose in these plants; pretreatment of the plant with xanthan restores the pathogenicity of both strains. The results indicate that xanthan induces susceptibility to *Xcc* by suppressing callose deposition. The bacterial cyclic b-(1,2)-glucan also has an effect on plant defense suppression and callose deposition, but, unlike xanthan, which acts locally, cyclic glucan acts systemically as part of a counter-defensive strategy that might facilitate the spread of the pathogen in the plants (Rigano et al. 2007).

The Xanthan has a broad range of applications in food and nonfood products from the oil, pharmaceutical, cosmetic, paper, paint, and textile industries, and is mainly used as a thickening, stabilizing, gelling, and emulsifying agent (Becker et al. 1998). The biosynthesis of xanthan production and the genes that encode for the enzymes involved have been studied (Ielpi et al. 1993; Becker et al. 1998). Xanthan synthesis is encoded by twelve *gum* genes (*gumBCDEFGHIJKLM*), which are located in a single gene cluster of 12 kb that is mainly expressed as an operon from a promoter upstream of the first gene, *gumB* (Vojnov et al. 2001). The genes required for the synthesis of xanthan and its nucleotide sugar precursors are highly conserved among the three sequenced strains (ATCC33913, 8004 and B100), but are significantly different when compared with the other sequenced *Xanthomonas*; a model for the biosynthesis of xanthan based on the genome annotation has been proposed (Vorhoelter et al. 2008). The disruption of the *rmlA*, *xanA*, and *gumK* genes, which are responsible for the biosynthesis of intermediates for xanthan production, can lead to a loss of pathogenicity (Qian et al. 2005).

The lipopolysaccharide (LPS) is an essential component of the outer membrane of Gram-negative bacteria, and is an important virulence factor in *Xanthomonas* (Dow et al. 1995; He et al. 2007; Patil et al. 2007). Over 20 genes for LPS synthesis have been characterized in *Xcc*, including *xanAB*, *rmlABCB*, *rfaXY*, *lpsIJ* and 15 genes that constitute the *wxc* cluster (Vorholter et al. 2001). The *wxc* gene cluster is involved in the synthesis of the LPS O-antigen, the most variable part of LPS. The *wxc* genes are highly divergent between *Xcc* strains, indicating that different isolates might produce varied LPSs (He et al. 2007). The LPS gene cluster of *Xcc* B100 is significantly different from those of 8004 and ATCC33913, but is almost identical to the LPS cluster of *X. campestris* pv. *raphani* 756C. This could indicate that the LPS cluster of B100 was acquired recently via horizontal gene transfer from *X. campestris* pv. *raphani* 756C or, more likely, that the *X. campestris* pv. *raphani* strain diverged recently from an *Xcc* with a B100-type LPS locus (Patil et al. 2007). Qian et al. (2005) showed that the disruption of a number of *wxc* genes resulted in a significant reduction in virulence (Vicente and Holub 2013).

3.12.5 Characterization of *Xanthomonas* Gene for Virulence

Extensive molecular genetics of *Xanthomonas campestris* has identified key pathogenicity determinants such as extracellular polysaccharides, lipo-polysaccharides, DSF-dependent quorum sensing, extracellular enzymes (proteases, cellulases) exported by the type II secretion systems or proteins secreted by the type III secretion (T3S) system. Type III-secreted proteins (T3SP) include type III effector (T3E) proteins which are injected inside the plant cells where many of them interfere with cell physiology and plant immunity. Because the Hrp (hypersensitive response and pathogenicity) T3S system is essential for *Xanthomonas* virulence, it is assumed that the type III effectome is also globally essential for any interaction with the plant. Yet, individual effectors usually have limited or nonsignificant contribution to pathogenicity when studied alone, probably reflecting functional redundancy and/or additivity between effectors. *hrp* genes are not expressed in *Xanthomonas* cultivated in rich media. Yet, *hrp* gene expression can be induced in *Xanthomonas* in specific minimal media (XVM2, MME, MMX) that were thought to mimic in planta conditions. Two master regulators of the *hrp* systems which are both required for virulence and *hrp* gene expression in minimal medium have been identified: *HrpX* is an AraC-type transcriptional activator inducing the expression of all *hrp* operons but hrpA upon binding to the plant-inducible promoter (PIP) box (TTCGB-N15-TTCGB; B represents C, G, or T) in the promoter region. Expression of most T3E genes is under the control of *hrpX*. HrpG, an OmpR-type transcriptional regulator, and its putative cognate sensor kinase HpaS control the expression of *hrpX*, the *hrpA* operon and other genes. Interestingly, several point mutations in *hrp*G (hrpG*) can render its activity constitutive in the absence of inducing condition and result in an increased aggressiveness on plants (Koebnik et al. 2006; Li et al. 2014; Wengelnik et al. 1999).

Four complete genome sequences are available in the *X. campestris* species (pathovar *campestris*: 8004, ATCC33913, B100; pathovar *raphani*: 756C) and six draft genomes (pathovar *campestris*: Xca5, JX, B-1459, CN14, CN15 and CN16). These strains have in common a ca. 5-Mb circular chromosome with approx. 65% G + C content. The presence of plasmids has only been reported in strains B-1459, CN14, CN15 and CN16. No genomic information is available yet on the pathovars incanae and the nonpathogenic group of *X. campestris*. Comparative genomics has proven to be an important tool to mine for pathogenicity determinants, host specificity factors and in particular for T3E. Due to the lack of conserved secretion/translocation signals in T3SP, those can only be predicted by indirect means: in silico proteomes can be studied by homology searches with known T3E sequences, by searching for eukaryotic signatures (nuclear localization signals, myristoylation/palmitoylation signals, F-box motifs,. . .) and/or predicted genes can be scrutinized for the presence of a PIP box. Indeed, co-expression of T3SP with the Hrp T3S system is the rule and has proven to be a powerful tool for the identification of novel T3E in *Xanthomonas*, *Ralstonia solanacearum* and *Pseudomonas syringae*. The analysis of *Xanthomonas transcriptomes* in hrp-inducing conditions or in de-regulated mutants expressing constitutively the hrp regulon has been a major source for the discovery of novel effectors (Roux et al. 2015).

The genome sequencing of four *X. campestris* strains of the pathovars *incanae*, *raphani* and NP and their comparison are available. This approach identified a very limited core type III effectome for this species of *Xanthomonas*. A transcriptomic analysis of the hrpG regulon has also been performed in a *X. campestris* pv. *raphani* strain CFBP 5828R and provides important information for the structural annotation of the genes and precious hints for the identification of candidate type III effector proteins as well as genes which could contribute to pathogenicity of *X. campestris*. The observed genomic coverage of the species by sequencing and annotating the genomes of strains from pathovar *incanae* (CFBP 1606R and CFBP 2527R), pathovar *raphani* (CFBP 5828R) and a pathovar formerly named *barbareae* (CFBP 5825R) was characterized. While comparative analyses identified a large core ORFeome at the species level, the core type III effectome was limited to only three putative type III effectors (XopP, XopF1 and XopAL1). In *Xanthomonas*, these effector proteins are injected inside the plant cells by the type III secretion system and contribute collectively to virulence. A deep and strand-specific RNA sequencing strategy was adopted in order to experimentally refine genome annotation for strain CFBP 5828R. This approach also allowed the experimental definition of novel ORFs and noncoding RNA transcripts. Using a constitutively active allele of hrpG, a master regulator of the type III secretion system, a HrpG-dependent regulon of 141 genes co-regulated with the type III secretion system was identified. Importantly, all these genes but seven are positively regulated by HrpG and 56 of those encode components of the Hrp type III secretion system and putative effector proteins. This dataset is an important resource to mine for novel type III effector proteins as well as for bacterial genes which could contribute to pathogenicity of *X. campestris* (Roux et al. 2015).

3.12.6 Hormonal Signaling in the Pathways of Biotrophy-Necrotrophy Transition of *Xanthomonas* Infection in *Brassica*

The *Xanthomonas campestris* pv. *campestris* (*Xcc*) is a hemibiotrophic bacterial pathogen that causes black rot disease in brassicaceous crops and, thus, causes huge economic losses on an annual basis. Infection by hemibiotrophic pathogens includes sequential biotrophic and necrotrophic phases. During the initial biotrophic phase, which is generally asymptomatic, the pathogen obtains nutrients from its host's living tissue, and during the following necrotrophic phase, which is relatively destructive, the pathogen damages or destroys host cells for nutrients. The infection of plants with *Xcc* is also associated with the formation of V-shaped necrotic lesions on leaf margins and the induction of oxidative stress, including the accumulation of reactive oxygen species (ROS). Phytohormones including salicylic acid (SA) and jasmonic acid (JA) play leading roles in the activation of basal defenses against plant pathogens. The SA signaling mediates the resistance of plants against biotrophic pathogens, and JA-signaling regulates the resistance of plants against necrotrophs. Moreover, SA and JA signaling are mutually antagonistic, and the stress-responsive hormone abscisic acid (ABA) antagonizes JA signaling, thereby inducing susceptibility to necrotrophs. However, the plant growth hormone cytokinin (CK) synergistically interacts with SA to activate defense responses against biotrophic plant pathogens. Studies have investigated variations in endogenous hormone levels in various *B. napus* cultivars, as well as the effects of signaling and hormone regulation on phenylpropanoid metabolism during the necrotrophic phase of *Xcc* infection. Higher levels of SA and ABA, which are associated with higher SA/JA and ABA/JA ratios, were reported to regulate the development of disease symptoms during the necrotrophic phase of Xcc infection by enhanced the glutathione (GSH)-redox balance, and JA-mediated increase in phenolic accumulation was associated with Xcc resistance (Vicente and Holub 2013; Islam et al. 2021).

Hemi-biotrophic pathogens manipulate endogenous hormone signaling in order to disrupt the basal defense responses of their hosts, i.e., to induce disease susceptibility. However, little is known about *Xcc*-induced alterations in the phytohormonal signaling of infected plants during the pathogen's transition between biotrophy and necrotrophy. Investigating the alteration of hormone levels during different phases of *Xcc* infection may reveal new information about the complex hormonal changes that occur during biphasic plant–microbe interactions. Therefore, Islam et al. (2021) monitored levels of endogenous hormones during both the biotrophic and necrotrophic phases of *Xcc* infection in *B. napus* and characterized physiological association of hormonal status, with ROS accumulation and GSH-redox homeostasis and infection symptoms.

Plant pathogens are typically classified as biotrophic, necrotrophic, or hemibiotrophic, depending on their lifestyle. Biotrophic pathogens maintain a feeding relationship with living host cells, rather than killing them, whereas necrotrophs actively kill the host tissues to acquire nutrients from dead or dying cells, and hemibiotrophs exhibit both strategies of infection, generally shifting from an early biotrophic phase to a subsequent necrotrophic phase. In responses to these different classes of pathogens, plant shows differences in immune responses because of their modes of nutrient uptake. Islam et al. (2021) identified two distinct phases in the infection of *B. napus* cv. Mosa by *Xcc*, namely an early asymptomatic biotrophic phase that could be observed at 2 DPI and a subsequent necrotrophic phase that occurred at 7–24 DPI (Fig. 3.20a). In addition, the length of the V-shaped lesions increased with the progression of infection (Fig. 3.20b). Reactive oxygen species

Fig. 3.20 Effects of *Xanthomonas campestris* pv. *campestris* (*Xcc*) inoculation on the disease symptoms of *Brassica napus* leaves. (**a**) V-shaped lesion development, (**b**) lesion length, (**c**) H_2O_2 content, and (**d**) *NADPH oxidase* expression. Data are presented as means ± SE ($n = 3$). Bars labeled with different letters are significantly different at $P < 0.05$ according to Duncan's multiple range test. Asterisks indicate significant differences between treatment groups: *$P < 0.05$; **$P < 0.01$; ***$P < 0.001$ (Islam et al. 2021)

(ROS) have a contrasting role in plant defense that depends on the pathogen's lifestyle. For example, the accumulation of some ROS including H_2O_2 effectively prevents the growth of biotrophic pathogens but also serves as a virulence factor for necrotrophic pathogens. Interestingly, biotrophic-phase hemibiotrophs suppress oxidative bursts, whereas necrotrophic-state pathogens promote ROS-induced cell death and tissue necrosis. The expression of *NADPH oxidase*, which is involved in H_2O_2 production, was downregulated during the biotrophic phase but upregulated during the necrotrophic phase (Fig. 3.20d), as was H_2O_2 accumulation (Fig. 3.20c). Greater H_2O_2 production has been reported to occur during the necrotrophic phase of infection caused by hemibiotrophs (e.g., *Septoria tritici* and *Verticillium dahlia*. Susceptibility to the hemibiotrophic pathogen *Fusarium oxysporum* has been correlated with the upregulation of *NADPH oxidase* during a later phase of infection. It suggested that Xcc initially suppresses the *NADPH oxidase*-mediated production of H_2O_2, which could contribute to basal defenses against biotrophic pathogens but later utilizes H_2O_2, during the necrotrophic phase, to facilitate H_2O_2-induced tissue necrosis.

The invasion of plants by pathogens induces complex defense responses, with phytohormones playing a central role. However, plant defenses against hemibiotrophic pathogens depend on the parallel and orderly expression of SA and JA signaling. Higher SA/JA ratios were previously correlated with black rot disease susceptibility in *B. napus* at a later phase of infection, and RNA sequencing previously revealed that SA and JA signaling were activated in *Brassica* during the biotrophic and necrotrophic phases of infection caused by *Leptosphaeria*,

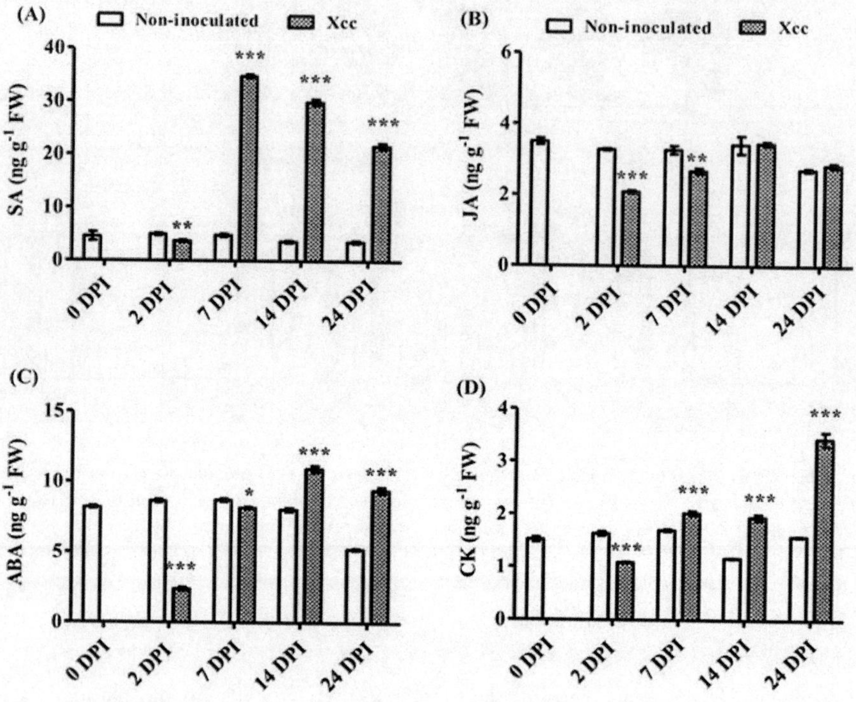

Fig. 3.21 Changes of the phytohormone levels of *Brassica napus* leaves after *Xanthomonas campestris* pv. *campestris* (*Xcc*) inoculation. (**a**) Salicylic acid (SA), (**b**) jasmonic acid (JA), (**c**) abscisic acid (ABA), and (**d**) cytokinin (CK). Data are presented as means ± SE ($n = 3$). Asterisks indicate significant differences between treatment groups: *$P < 0.05$; **$P < 0.01$; ***$P < 0.001$ (Islam et al. 2021)

respectively. Interestingly, it has been reported that necrotrophic pathogens manipulate SA synthesis and signaling to promote disease development through the inhibition of JA signaling. Indeed, the delayed induction of SA accumulation was correlated with the increased susceptibility to the hemibiotroph *F. graminearum*. In the present study, SA level was significantly increased to greater extent (9.5-fold) at the beginning of the necrotrophic phase of *Xcc*-infection (at 7 DPI) than at the asymptomatic phase (at 2 DPI), as was the expression levels of SA synthesis-related genes (*EDS1* and *ICS1*) and SA signaling-related (*NPR1* and *PR1*; Figs. 3.21a and 3.22a–d). Therefore, *Xcc*-induced changes in SA level and related gene expression levels are crucial during the transition between the biotrophic and necrotrophic phases of *Xcc* infection. In addition, JA levels were restricted during both phases of Xcc infection (Fig. 3.21b), and expression of the JA-signaling gene *PDF1.2* was downregulated by Xcc infection, regardless of infection phase (Fig. 3.22e). The JA-mediated resistance was reported to occur during the necrotrophic phase of *Xcc* infection. The role of SA and JA antagonism during the evolution of plant-pathogen interactions has been well documented. The *Xcc* infection induced significant

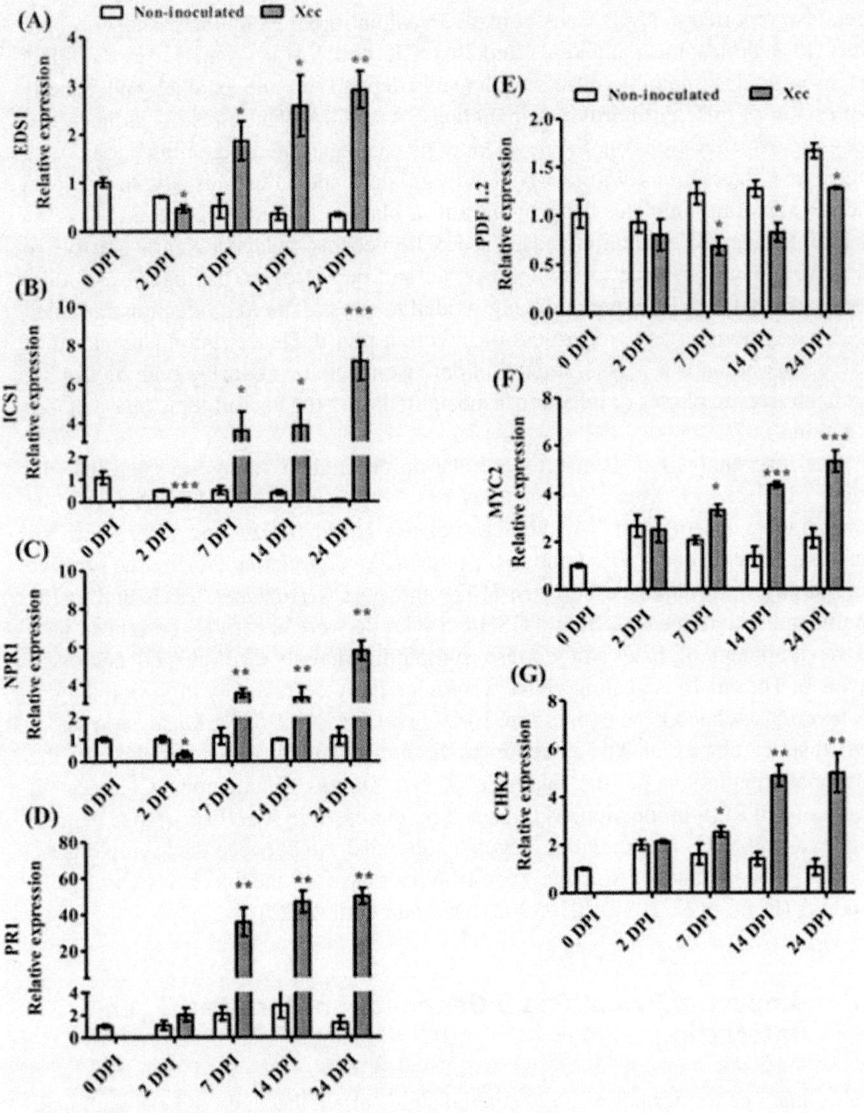

Fig. 3.22 Relative expression levels of phytohormonal synthesis- and signaling-related genes in the noninoculated and *Xanthomonas campestris* pv. *campestris* (*Xcc*)—inoculated leaves. (**a**) *EDS1*, (**b**) *ICS1*, (**c**) *NPR1*, (**d**) *PR1*, (**e**) *PDF1.2*, (**f**) *MYC2*, and (**g**) *CHK2*. Data are presented as means ± SE ($n = 3$). Asterisks indicate significant differences between treatment groups: *$P < 0.05$; **$P < 0.01$; ***$P < 0.001$ (Islam et al. 2021)

changes in ABA and CK levels. More specifically, both ABA and CK increased gradually as the infection progressed (Fig. 3.21c, d), as did the expression levels of corresponding signaling genes (*MYC2* and *CHK2*, respectively; Fig. 3.22f, g). It has been also reported that ABA antagonizes JA signaling, thereby increasing susceptibility to necrotrophic pathogens, and that CK and SA act synergistically. Plant defenses against hemibiotrophic pathogens depend on the parallel and orderly expression of different hormones including SA and JA. These results indicate that *Xcc*-induced alterations in SA level and its synthesis and signaling along with synergistic interactions with ABA, CK regulates, and antagonism with JA-level and it's signaling regulates the progression of black rot disease symptoms.

Maintenance of glutathione-based redox homeostasis plays a significant role in the developing resistance to hemibiotrophic and necrotrophic pathogens including *Xcc* by detoxifying ROS and reducing oxidative stress. The *Xcc* infection disrupted GSH-redox homeostasis, regardless of infection phase. Hence, the maintenance of GSH-redox homeostasis is an important defense mechanism during both biotrophic and necrotrophic phases of infection, especially during the necrotrophic phase of *Xcc* infection.

Heat map analysis of Pearson's correlation coefficient values revealed that the length of black rot disease lesions was positively correlated with SA level, the expression levels of both SA synthesis-related genes (*EDS1* and *ICS1*) and SA signaling-related genes (*NPR1* and *PR1*), and CK level. Relatively increased positive correlations were observed between H_2O_2 and *NADPH oxidase* levels and lesion length, and both reduced GSH and GSH/GSSG ratio were negatively correlated with the development of black rot disease symptoms (length of V-shaped necrosis). Levels of JA and JA-signaling genes were negatively correlated with lesion length, SA level, SA-related gene expression, H_2O_2 level, and *NADPH oxidase* expression. Two distinct phases of *Xcc* infection *in B. napus* and provides insight into the hormonal regulation of the biphasic *Xcc-B. napus* interaction. Changes in SA-mediated ROS production, as indicated by increases in *NADPH oxidase* expression level and H_2O_2 accumulation, regulate the transition between the asymptomatic biotrophic phase and destructive necrotrophic phase of the black rot disease in *B. napus* (Islam et al. 2017, 2019, 2021; Mamun et al. 2020).

3.13 Impact of Evolutionary Genomics on Host–Pathogen Interaction

Evolutionary genomics has recently entered a new era in the study of host–pathogen interactions. A variety of novel genomic techniques has transformed the identification, detection and classification of both hosts and pathogens, allowing a greater resolution that helps decipher their underlying dynamics and provides novel insights into their environmental context. Nevertheless, many challenges to a general understanding of host–pathogen interactions remain, in particular in the synthesis and integration of concepts and findings across a variety of systems and different spatiotemporal and ecological scales. The majority of studies used whole genome resolution to address their research objectives across a broad range of ecological scales, especially when focusing on the pathogen side of the interaction. Nevertheless, genomic studies conducted in a complex spatiotemporal context are currently rare. Because, the processes of host–pathogen interactions can be understood at multiple scales, from molecular-, cellular-, and physiological-scales to the levels of populations and ecosystems. A major obstacle for synthesis across diverse host–pathogen systems is that data are collected on widely diverging scales with different degrees of resolution. This disparity not only hampers effective infrastructural organization of the data but also data granularity and accessibility. Comprehensive metadata deposited in association with genomic data in easily accessible databases will allow greater inference across systems in the future, especially when combined with open data standards and practices. The standardization and comparability of such data will facilitate early detection of emerging infectious diseases as well as studies of the impact of anthropogenic stressors, such as climate change, on disease dynamics in humans and wildlife. Innovations in genomic techniques have the potential to bring a synthesis to the study of host–pathogen interactions across systems and environmental conditions. Several recent trends in this perspective for genomic studies of host–pathogen systems viz., (1) evolutionary genomics approaches have allowed the field to move from a candidate gene approach to investigations at the scale of whole genomes; (2) the use of genomics for the detection and surveillance of host–pathogen systems; (3) the challenges of the integrating natural history and ecological metadata and genomic data across systems and timescales due to infrastructural challenges of database integration and transparency; and (4) the impact of anthropogenic stressors on host–pathogen systems that have consequences for global health. Any single host–pathogen study is constrained by limited resources or genomic tractability, the geographical and evolutionary time scales involved as well as environmental complexities. Accordingly, transparent and open science will help to achieve a comprehensive understanding of host–pathogen interactions in general. This will contribute to the integration of findings across the different scales. A large repertoire of comparable and inter-communicative studies will facilitate a more generalizable understanding of the causes and consequences of host–pathogen interactions and a clearer roadmap to combating the continuous threat of pathogens in a changing world (Näpflin et al. 2019).

3.14 Challenges of Genomics of Host–Pathogens Interaction

Pathogens are widely agreed to be among the strongest agents of natural selection in nature, and their influence on the genomes of host species is often readily detectable (Quach et al. 2016; Shultz and Sackton 2019). With the advent of rapid DNA sequencing technologies, genetic studies of host–pathogen interactions have moved from single gene perspectives to genome-wide approaches interrogating whole genomes of hosts and/or pathogens. At the same time, these studies have begun to tackle an increasingly diverse array of systems in both the field and laboratory, and have expanded from analysis of single pathogens to multiple pathogens under a variety of conditions. Environmental factors and gene-by-environment interactions, such as those beginning to be studied in microbiome research (Libertucci and Young 2019), are increasingly appreciated as important in modulating the severity and fitness consequences of infections (Villarino et al. 2016; Chomwong et al. 2018). As genomic approaches become increasingly accessible and affordable, it is becoming clear that the limiting factor in host–pathogen research is often not the technical aspects of sequencing pathogens or host genomes, but rather the ecological, immunological and epigeneticcon text in which genomic data are embedded (Kratochwil and Meyer 2015). To mentionone example, post-translational modifications of proteins in the mucus are known to play critical roles in pathogen defense in addition to genetic factors (Linden et al. 2008a, b).

Host–pathogen studies encompass an extraordinary variety of temporal and spatial scales, including wide ranges of ecological settings and pathogen complexities such as experimental versus field studies or single versus multiple pathogens as well as genomic complexities, ranging from candidate gene studies to whole genome scans (Fig. 3.23). Any single study can be classified according to these scales, with concomitant benefits and deficiencies in capturing the details of host–pathogen interactions in the real world. For example, studies aiming to link the evolution of host and pathogen genomes and to detect genomic signatures of host–pathogen interactions have arisen from searches for associations with single host candidate genes, such as genes of the major histocompatibility complex (MHC) (Meyer-Lucht and Sommer 2005; Savage and Zamudio 2011), to genome-wide scans for associations with resistance or susceptibility (Fumagalli et al. 2011; Bartha et al. 2013). Now the genomic insights into host–pathogen interactions that have stem from field studies investigating temporal and spatial patterns (Penczykowski et al. 2015; Bourgeois et al. 2017); to experimentally evolving populations or ancient DNA studies encompassing hundreds or thousands of host generations (Bos et al. 2011; Cairns et al. 2017; Tso et al. 2018; Spyrou et al. 2019); to phylogenetic and comparative studies spanning tens of millions of years (Enard et al. 2016; Koonin et al. 2017; Shultz and Sackton 2019). This variety makes it challenging to draw broad generalizations linking processes on different scales and, to date, few syntheses have attempted to bridge the many temporal and spatial scales on which host–pathogen studies take place. This represents an important deficit, as generalizability is key to identifying fundamental principles in the field of host–pathogen interactions and enables accurate predictions to be made about host–pathogen interactions in new study systems.

To address the complexities and commonalities of diverse studies of host–pathogen interactions through the lens of evolutionary genomics, Näpflin et al. (2019) emphasized the wide range of approaches used recently and focus primarily on evolutionary responses of hosts to pathogens (Fig. 3.23). They documented the diversity of recent studies of host–pathogen interactions through a comprehensive analysis of the recent literature on the subject. This survey documents the sheer diversity of temporal and spatial scales on which host–pathogen studies are conducted, but also reveals that the heterogeneity of results across studies, from laboratory to field to experimental settings, poses a challenge for synthesis. The survey identified gaps in emphasis on research on host–pathogen interactions, but also revealed opportunities for discovering common principles and methodologies that are likely to drive the research field forward. The review on major themes in the

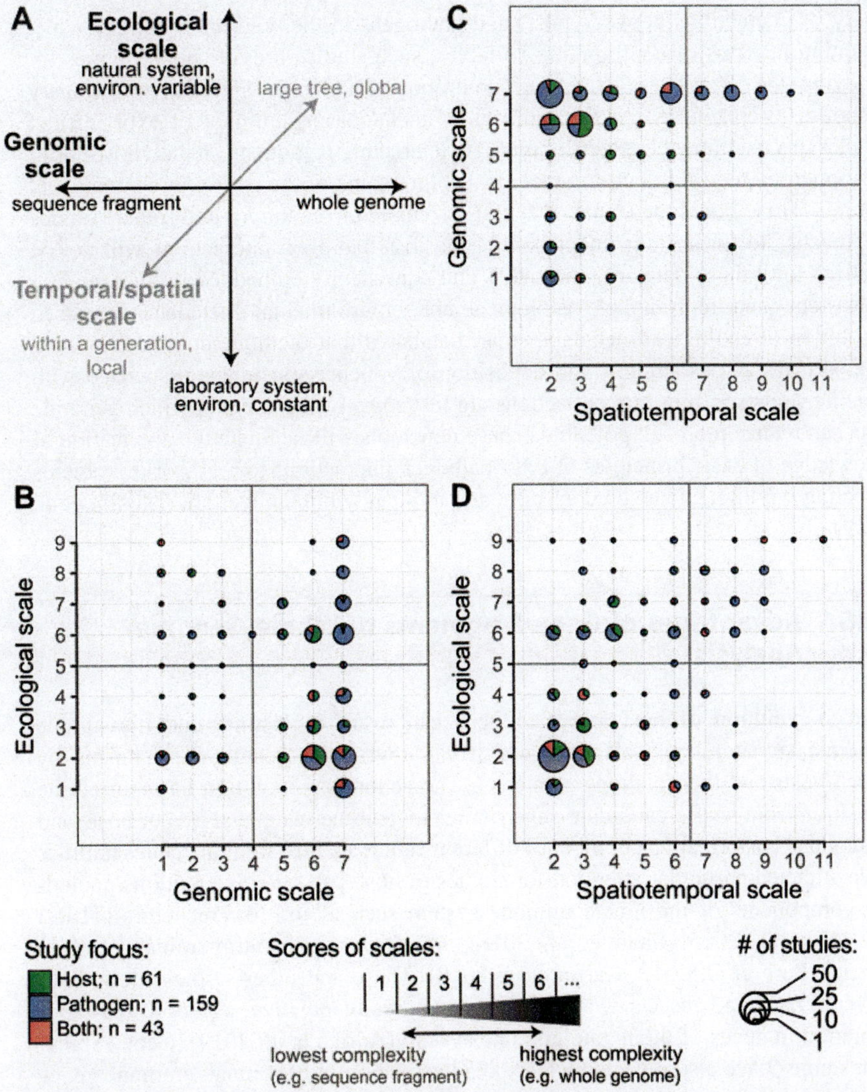

Fig. 3.23 The diversity of recent studies of host–pathogen interactions. (**a**) Each of three scales of complexity genomic, ecological and spatiotemporal is represented as an axis in this illustration. A study of host–pathogen interaction is placed into this three-dimensional space based on the level of genetic, ecological, and spatiotemporal detail that is being studied. (**b–d**) Pie charts summarize the results of the scores for the level of genetic, ecological, and spatiotemporal complexity investigated in host–pathogen studies published between 2014 and 2018. (**b**) The complexity of the ecological and genomic settings across studies are not correlated (Spearman's D 0:02, *P*-value adjusted D 1.00; (**c**) nor are the genomic and spatio-temporal scale (D 0:16, *P*-value adj. D 0.13. (**d**) In contrast, the ecological scale positively correlates with the score of spatio-temporal scale across studies (D0:50, *P*-value adj. D 0.00) (Näpflin et al. 2019)

study of interactions between hosts and pathogens in the wild are required. While daunting in terms of confounding variables, such studies provide opportunities for studying the synergistic effects of anthropogenic change and the evolutionary response to epizootics. At the same time, an increasing number of experimental studies that examine the effects of multiple interacting pathogens on their hosts, or of host microbiome on infection outcome, capture some of the reality of epizootics in nature. They concluded that the full promise of genomic and other -omics approaches to further our understanding of host–pathogen interactions will not be realized until these data are thoroughly and consistently embedded in high quality, consistent, and reproducible ecological and environmental metadata. Increased resolution of ecological metadata, smart databases that facilitate collaboration and comparisons across studies, and deposition of voucher specimens associated with specific studies in museum collections are just some of the ways in which genomic data can realize their full potential. These new tools will facilitate the application of knowledge of basic principles of host–pathogen interactions to real world problems affecting wildlife, endangered species, and ultimately human welfare (Näpflin et al. 2019).

3.15 Advantages of Gene Fragments to Whole-Genome Analysis

The co-evolution of host and pathogens can result in distinct and measurable genomic signatures of selection, which can reveal the genetic mechanisms by which hosts and their pathogens interact. The genomics revolution has spurred the transition from single-candidate gene studies to genome wide analyses of hosts and pathogens. Historically, a number of different candidate host immune genes families have attracted particular attention for studies of host–pathogen interactions, including components of the innate immune system such as toll-like receptors (TLRs) (Zhang et al. 2015; Shan et al. 2018), interferons and antimicrobial peptides (Franzenburg et al. 2013; Carlin et al. 2018). These and other studies show, both in vertebrates and insects, wide spread signatures of positive selection and rapid evolution in genes of the innate immune system (Adrian et al. 2019). Gene expression studies have also revealed wide spread activation of host innate immune genes upon natural or experimental infection with pathogens, such as *Pseudomonas* and Daphnia (Kumar et al. 2018). As such, these studies have contributed much to our general understanding of the host's responses to pathogen exposure and common pathways to resistance evolution over time.

The candidate gene family that has attracted the most attention in ecological and evolutionary host–pathogen studies is the major histocompatibility complex (MHC) (Spurgin and Richardson 2010). MHC genes encode cell surface molecules that play a central role in pathogen recognition as part of the adaptive immune response. T-cells act to destroy infected cells both directly, as cytotoxic T-cells, and indirectly, as T-helper cells which activate other immune cells, but they can only determine what is self or foreign from peptides presented by MHC molecules. The number of

MHC gene copies carried by individuals varies widely between, and even within, species (O'Connor et al. 2019). Additionally, the allelic diversity recorded within and between gene copies makes the MHC genes the most polymorphic loci to date (Reche and Reinherz 2003; Robinson et al. 2015). This exceptional polymorphism is believed to be primarily maintained by selection from a wide range of pathogens (Prugnolle et al. 2005; Qutob et al. 2011). Overall, it is clear that MHC genes play a pivotal role in the fight against pathogens and numerous studies have established associations between MHC genotypes and infections with a particular pathogen (Bolnick and Stutz 2017; O'Connor et al. 2019).

Although the candidate gene approach has been the standard method for studying immune genes in the context of host–pathogen interactions, the rapidly decreasing costs of high throughput sequencing are making whole-genome approaches much more feasible. Whole-genome population genetic and comparative genomic studies allow unbiased detection of regions of the genome that are evolving non-neutrally across a variety of time scales. When combined with functional annotations or association studies, such comprehensive genome-wide surveys permit incisive tests of the effects of pathogens on host genomes that are simply not possible from candidate gene studies. Furthermore, whole genome studies are not restricted by a priori predictions of which genes are important in responding to pathogen challenges. Thus, whole-genome approaches offer the potential to reveal new unbiased insights into the genetic basis of host–pathogen interactions, e.g. (Enard et al. 2016; Shultz and Sackton 2019). Since multiple genes are most often involved in a host's response to a particular pathogen, whole-genome approaches also have the potential to reveal these under studied polygenic responses (Daub et al. 2013). However, a major caveat associated with the whole-genome approach is that genomic regions of high repeat content, or highly duplicated genes, often do not assemble well or at all, whether considering host or pathogen. This is particularly problematic in the case of de novo genome assemblies in nonmodel species (Peona et al. 2018). Genome assembly problems may be a significant disadvantage for host–pathogen studies given that some key genes which play a role in innate and adaptive immunity are not only highly duplicated but also to some degree physically linked in the genome, such as the beta-defensin and MHC genes (Kaufman et al. 1999; Hellgren and Ekblom 2010; Balakrishnan et al. 2010). Improved sequence and scaffolding techniques are being developed to remedy problems of assembling such regions (Dilthey et al. 2015) and some may be possible to overcome with long-read sequencing methods (Peona et al. 2018). Thus, while the whole-genome approach may become a gold standard for many host–pathogen studies in the future, we currently see a continued need for sequencing methods that target focal genes or a reduced representation of the whole genome, in particular in studies of nonmodel species (Näpflin et al. 2019).

References

Adams MD, Kelley JM, Gocayne JD, Dubnick M, Polymeropoulos MH, Xiao H (1991) Links complementary DNA sequencing: expressed sequence tags and human genome project. Science 252:1651–1656

Adie B, Chico JM, Rubio-Somoza I, Solano R (2007) Modulation of plant defenses by ethylene. J Plant Growth Regul 26:160–177

Adrian J, Bonsignore P, Hammer S, Frickey T, Hauck CR (2019) Adaptation to host specific bacterial pathogens drives rapid evolution of a human innate immune receptor. Curr Biol 29: 616–630. https://doi.org/10.1016/j.cub.2019.01.058

Affolter M, Pyrowolakis G, Weiss A, Basler K (2008) Signal-induced repression: the exception or the rule in developmental signaling? Dev Cell 15(1):11–22. https://doi.org/10.1016/j.devcel.2008.06.006

Agarwal A, Kaul V, Faggian R, Cahill DM (2009) Development and use of a model system to monitor clubroot disease progression with an Australian field population of *Plasmodiophora brassicae*. Australas Plant Pathol 38:120–127

Agarwal A, Kaul V, Faggian R, Rookes JE, Ludwig-Müller JE, Cahill DM (2011) Analysis of global host gene expression during the primary phase of the *Arabidopsis thaliana–Plasmodiophora brassicae* interaction. Func Plant Biol 38:462–478

Aist JR, Bushnell WR (1991) Invasion of plants by powdery mildew fungi, and cellular mechanisms of resistance. In: Cole GT, Hoch HC (eds) The fungal spore and disease initiation in plants and animals, vol 6, 1st edn. Plenum Press, New York, pp 321–345

Aist JR, Williams PH (1971) The cytology and kinetics of cabbage root hair penetration by *Plasmodiophora brassicae*. Can J Bot 49(11):2023–2034. https://doi.org/10.1139/b71-284

Alexander SP, Christopoulos A, Davenport AP, Kelly E, Marrion NV, Peters JA, Faccenda E, Harding SD, Pawson AJ, Sharman JL, Southan C, Davies JA, CGTP Collaborators (2017) The concise guide to PHARMACOLOGY 2017/18: G protein-coupled receptors. Br J Pharmacol 174(Suppl 1):S17–S129. https://doi.org/10.1111/bph.13878

Alkhateeb RS, Rückert C, Rupp O, Pucker B, Hublik G, Wibberg D, Niehaus K, Pühler A, Vorhölter FJ (2017) Refined annotation of the complete genome of the phytopathogenic and xanthan producing *Xanthomonas campestris* pv. *campestris* strain B100 based on RNA sequence data. J Biotechnol 253:55–61. https://doi.org/10.1016/j.jbiotec.2017.05.009

Alkooranee JT, Yin Y, Aledan TR, Jiang Y, Lu G, Wu J, Li M (2015) Systemic resistance to powdery mildew in *Brassica napus* (AACC) and *Raphanus alboglabra* (RRCC) by *Trichoderma harzianum* TH12. PLoS One 10(11):e0142177. https://doi.org/10.1371/journal.pone.0142177

Allan J, Regmi R, Denton-Giles M, Kamphuis LG, Derbyshire MC (2019) The host generalist phytopathogenic fungus *Sclerotinia sclerotiorum* differentially expresses multiple metabolic enzymes on two different plant hosts. Sci Reptr 9:19966. https://doi.org/10.1038/s41598-019-56396-w

Allen FHE, Friend J (1983) Resistance of potato tubers to infection by *Phytophthora infestans* a structural study of haustorial encasement. Physiol Plant Pathol 1983:285–IN284

Alvarez ME, Pennel RI, Meijer PJ, Ishikawa A, Dixon RA, Lamb C (1998) Reactive oxygen intermediates mediate a systemic signal network in the establishment of plant immunity. Cell 92:773–784

Amaral AMD, Antoniw J, Rudd JJ, Hammond-Kosack KE (2012) Defining the predicted protein secretome of the fungal wheat leaf pathogen *Mycosphaerella graminicola*. PLoS One 7:e49904

An SQ, Lu GT, Su HZ, Li RF, He YQ, Jiang BL, Tang DJ, Tang JL (2011) Systematic mutagenesis of all predicted *GntR* genes in *Xanthomonas campestris* pv. *campestris* reveals a gntR family transcriptional regulator controlling hyper-sensitive response and virulence. Mol Plant-Microbe Interact 24:1027–1039

Anderson JP, Thatcher LF, Singh KB (2005) Plant defense responses: conservation between models and crops. Func Plant Biol 32:21–34. https://doi.org/10.1071/FP04136

Ando S, Yamada T, Asano T, Kamachi S, Tsushima S, Hagio T, Tabei Y (2006) Molecular cloning of *PbSTKLl* gene from *Plasmodiophora brassicae* expressed during club root development. J Phytopathol 154:185–189

Andrew M, Barua R, Short SM, Kohn LM (2012) Evidence for a common toolbox based on necrotrophy in a fungal lineage spanning necrotrophs, biotrophs, endophytes, host generalists and specialists. PLoS One 7:e29943

Antony G, Zhou J, Huang S, Li T, Liu B, White F, Yang B (2010) Rice xa13 recessive resistance to bacterial blight is defeated by induction of the disease susceptibility gene Os-11N3. Plant Cell 22:3864–3876. https://doi.org/10.1105/tpc.110.078964

Araus JL, Cairns JE (2014) Field high-throughput phenotyping: the new crop breeding frontier. Trends Plant Sci 19:52–61

Artero RD, Terol-Alcayde J, Paricio N, Ring J, Bargues M, Torres A et al (1998) Saliva, a new drosophila gene expressed in the embryonic salivary glands with homologues in plants and vertebrates. Mech Dev 75:159–162. https://doi.org/10.1016/S0925-4773(98)00087-2

Asai T, Tena G, Plotnikova J, Willmann MR, Chiu WL, Gomez-Gomez L, Boller T, Ausubel FM, Sheen J (2002) MAP kinase signalling cascade in *Arabidopsis* innate immunity. Nature 4156(1): 977–983

Ausubel FM (2005) Are innate immune signaling pathways in plants and animals conserved. Nat Immunol 66(1):973–979

Babaeizad V, Imani J, Kogel KH, Eichmann R, Hückelhoven R (2009) Over-expression of the cell death regulator BAX inhibitor-1 in barley confers reduced or enhanced susceptibility to distinct fungal pathogens. Theor Appl Genet 118:455–463

Backer R, Naidoo S, van den Berg N (2019) The NON-EXPRESSOR OF PATHOGENESIS-RELATED GENES 1 (NPR1) and related family: mechanistic insights in plant disease resistance. Front Plant Sci 10:102

Badet T, Voisin D, Mbengue M, Barascud M, Sucher J, Sadon P, Balagué C, Roby D, Raffaele S (2017) Parallel evolution of the POQR prolyl oligo peptidase gene conferring plant quantitative disease resistance. PLoS Genet 13:e1007143

Bai YL, Pavan S, Zheng Z, Zappel NF, Reinstädler A, Lotti C, De Giovanni C, Ricciardi L, Lindhout P, Visser R, Theres K, Panstruga R (2008) Naturally occurring broad-spectrum powdery mildew resistance in a Central American tomato accession is caused by loss of *Mlo* function. Mol Plant-Microbe Interact 216(1):30–39

Balakrishnan CN, Ekblom R, Völker M, Westerdahl H, Godinez R, Kotkiewicz H, Burt DW, Graves T, Griffin DK, Warren WC, Edwards SV (2010) Gene duplication and fragmentation in the zebra finch major histocompatibility complex. BMC Biol 8:29. https://doi.org/10.1186/1741-7007-8-29

Bansal VK, Tewari JP, Stringam GR, Thiagarajah MR (2005) Histological and inheritance studies of partial resistance in the *Brassica napus–Albugo candida* host–pathogen interaction. Plant Breed 124:27–32

Baranowski P, Jedryczka M, Mazurek W, Babula Skowronska D, Siedliska A, Kaczmarek J (2015) Hyperspectral and thermal imaging of oilseed rape (*Brassica napus*) response to fungal species of the genus *Alternaria*. PLoS One 10:e0122913

Barbacci A, Navaud O, Mbengue M, Barascud M, Godiard L, Khafif M, Lacaze A, Raffaele S (2020) Rapid identification of an *Arabidopsis* NLR gene as a candidate conferring susceptibility to *Sclerotinia sclerotiorum* using time-resolved automated phenotyping. Plant J 103(2): 903–917. https://doi.org/10.1111/tpj.14747

Barber CE, Tang JL, Feng JX, Pan MQ, Wilson TJG, Slater H, Dow JM, Williams P, Daniels MJ (1997) A novel regulatory system required for pathogenicity of *Xanthomonas campestris* is mediated by a small diffusible signalmolecule. Mol Microbiol 24:555–566

Barkai-Golan R (1974) Production of cellulose and poly-galacturonase by *Sclerotinia minor*. Mycopathol Mycol Appl 54:297–302

Bartha I, Carlson JM, Brumme CJ, McLaren PJ, Brumme ZL, John M, Haas DW, Martinez-Picado J, Dalmau J, López-Galíndez C, Casado C, Rauch A, Günthard HF, Bernasconi E,

Vernazza P, Klimkait T, Yerly S, O'Brien SJ, Listgarten J, Pfeifer N, Lippert C, Fusi N, Kutalik Z, Allen TM, Müller V, Harrigan PR, Heckerman D, Telenti A, Fellay J (2013) A genome-to-genome analysis of associations between human genetic variation, HIV-1 sequence diversity, and viral control. eLife 2:e1001123. https://doi.org/10.7554/eLife.01123

Bartling D, Radzio R, Steiner U, Weiler EW (1993) A glutathione *S*-transferase with glutathione peroxidase activity from *Arabidopsis thaliana*: molecular cloning and functional characterization. Eur J Biochem 216:579–586

Bashi ZD, Gyawali S, Bekkaoui D, Coutu C, Lee L, Poon J, Rimmer SR, Khachatourians GG, Hegedus DD (2016) The *Sclerotinia sclerotiorum Slt2* mitogen-activated protein kinase ortholog, *SMK3*, is required for infection initiation but not lesion expansion. Can J Microbiol 850:836–850. https://doi.org/10.1139/cjm-2016-0091

Becker A, Katzen F, Puhler A, Ielpi L (1998) Xanthan gum biosynthesis and application: a biochemical/genetic perspective. Appl Microbiol Biotechnol 50:145–152

Becker MG, Zhang X, Walker PL, Wan JC, Millar JL, Khan D, Granger MJ, Cavers JD, Chan AC, Fernando DWG (2017) Transcriptome analysis of the *Brassica napus–Leptosphaeria maculans* pathosystem identifies receptor, signaling and structural genes underlying plant resistance. Plant J 90:573–586

Bednarek P, Pislewska-Bednarek M, Svatos A, Schneider B, Doubsky J, Mansurova M, Humphry M, Consonni C, Panstruga R, Sanchez-Vallet A, Molina A, Schulze-Lefert P (2009) A glucosinolate metabolism pathway in living plant cells mediates broad-spectrum antifungal defense. Science 323(5910):101–106. https://doi.org/10.1126/science.1163732

Beers EP, Jones AM, Dickerman AW (2004) The S8 serine, C1A cysteine and A1 aspartic protease families in *Arabidopsis*. Phytochemistry 65:43–58

Belhaj K, Lin BQ, Mauch F (2009) The chloroplast protein *RPH1* plays a role in the immune response of *Arabidopsis* to *Phytophthora brassicae*. Plant J 58:287–298

Berger PH, Barnett OW, Brunt AA et al (2000) Family potyviridae. In: van Regenmortel MHV, Fauquet CM, Bishop DHL et al (eds) Virus taxonomy: seventh report of the international committee on taxonomy of viruses. Academic Press, San Diego, pp 703–724

Berhane K, Widersten M, Engstrom A, Kozarich J, Mannervik B (1994) Detoxication of base propenals and other α, β-unsaturated aldehyde products of radical reactions and lipid peroxidation by human glutathione S-transferases. Proc Natl Acad Sci U S A 91:1480–1484

Berto P, Commenil P, Belingheri L, Dehorter B (1999) Occurrence of a lipase in spores of *Alternaria brassicicola* with a crucial role in the infection of cauliflower leaves. FEMS Microbiol Lett 180:183–189

Bhat RA, Miklis M, Schmelzer E, Schulze-Lefert P, Panstruga R (2005) Recruitment and interaction dynamics of plant penetration resistance components in a plasma membrane microdomain. Proc Natl Acad Sci U S A 102:3135–3140

Bi K, He Z, Gao Z, Zhao Y, Fu Y, Cheng J, Xie J, Jiang D, Chen T (2016) Integrated omics study of lipid droplets from *Plasmodiophora brassicae*. Sci Rep 6:36965

Bi K, Chen T, He Z, Gao Z, Zhao Y, Liu H, Fu Y, Xie J, Cheng J, Jiang D (2019) Comparative genomics reveals the unique evolutionary status of *Plasmodiophora brassicae* and the essential role of GPCR signaling pathways. Phytopathol Res 1:12. https://doi.org/10.1186/s42483-019-0018-6

Bigeard J, Colcombet J, Hirt H (2015) Signaling mechanisms in pattern-triggered immunity (PTI). Mol Plant 8:521–539

Blanco-Ulate B, Morales-Cruz A, Amrine KC, Labavitch JM, Powell AL, Cantu D (2014) Genome-wide transcriptional profiling of *Botrytis cinerea* genes targeting plant cell walls during infections of different hosts. Front Plant Sci 5:435. https://doi.org/10.3389/fpls.2014.00435

Boch J, Bonas U (2010) *Xanthomonas AvrBs3* family-type III effectors: discovery and function. Annu Rev Phytopathol 48:419–436

Bogdanove AJ, Koebnik R, Lu H, Furutani A, Angiuoli SV, Patil PB, Van Sluys MA, Ryan RP, Meyer DF, Han SW, Aparna G, Rajaram M, Delcher AL, Phillippy AM, Puiu D, Schatz MC, Shumway M, Sommer DD, Trapnell C, Benahmed F, Dimitrov G, Madupu R, Radune D,

Sullivan S, Jha G, Ishihara H, Lee SW, Pandey A, Sharma V, Sriariyanun M, Szurek B, Vera-Cruz CM, Dorman KS, Ronald PC, Verdier V, Dow JM, Sonti RV, Tsuge S, Brendel VP, Rabinowicz PD, Leach JE, White FF, Salzberg SL (2011) Two new complete genome sequences offer insight into host and tissue specificity of plant pathogenic *Xanthomonas* spp. J Bacteriol 193:5450–5464

Bolger AM, Lohse M, Usadel B (2014) Trimmomatic: a flexible trimmer for illumina sequence data. Bioinformatics 30:2114–2120

Bolnick DI, Stutz WE (2017) Frequency dependence limits divergent evolution by favouring rare immigrants over residents. Nature 546(7657):285

Bos KI, Schuenemann VJ, Golding GB, Burbano HA, Waglechner N, Coombes BK, McPhee JB, DeWitte SN, Meyer M, Schmedes S, Wood J, Earn DJD, Herring DA, Bauer P, Poinar HN, Krause J (2011) A draft genome of *Yersinia pestis* from victims of the Black Death. Nature 478: 506–510. https://doi.org/10.1038/nature10549

Bourgeois Y, Roulin AC, Müller K, Ebert D (2017) Parasitism drives host genome evolution: insights from the *Pasteuria ramosa*-Daphnia magna system. Evolution 71:1106–1113. https://doi.org/10.1111/evo.13209

Boyle C (1921) Studies in the physiology of parasitism. VI. Infection by *Sclerotinia libertiana*. Ann Bot 35:337–347

Brace JL, Vander Weele DJ, Rudin CM (2005) *Svf1* inhibits reactive oxygen species generation and promotes survival under conditions of oxidative stress in *Saccharomyces cerevisiae*. Yeast 22: 641–652

Brace JL, Lester RL, Dickson RC, Rudin CM (2007) SVF1 regulates cell survival by affecting sphingolipid metabolism in *Saccharomyces cerevisiae*. Genetics 175(1):65–76. https://doi.org/10.1534/genetics.106.064527

Brenner S, Johnson M, Bridgham J, Golda G, Lloyd DH, Johnson D (2000) Gene expression analysis by massively parallel signature sequencing (MPSS) on microbead arrays. Nat Biotechnol 18:630–634

Brodmann D, Schuller A, Ludwig-Muller J, Aeschbacher RA, Wiemken A, Boller T, Wingler A (2002) Induction of trehalase in *Arabidopsis* plants infected with the trehalose-producing pathogen *Plasmodiophora brassicae*. Mol Plant Microbe Interact 15(7):693–700

Bulman S, Siemens J, Ridgway HJ, Eady C, Conner AJ (2006) Identification of genes from the obligate intracellular plant pathogen, *Plasmodiophora brassicae*. FEMS Microbiol Lett 264: 198–204

Bulman S, Ridgway HJ, Eady C, Conner AJ (2007) Intron-rich gene structure in the intracellular plant parasite *Plasmodiophora brassicae*. Protist 158:423–433

Bulman S, Richter F, Marschollek S, Benade F, Julke S, Ludwig-Muller J (2018) Arabidopsis thaliana expressing PbBSMT, a gene encoding a SABATH-type methyltransferase from the plant pathogenic protist *Plasmodiophora brassicae*, show leaf chlorosis and altered host susceptibility. Plant Biol 21(Suppl 1):120–130. https://doi.org/10.1111/plb.12728

Burki F, Kudryavtsev A, Matz MV, Aglyamova GV, Bulman S, Fiers M, Keeling PJ, Pawlowski J (2010) Evolution of *Rhizaria*: new insights from phylogenomic analysis of uncultivated protists. BMC Evol Biol 10:377

Buschges R, Hollricher K, Panstruga R, Simons G, Wolter M, Frijters A, van Daelen R, van der Lee T, Diergaarde P, Groenendijk J, Töpsch S, Vos P, Salamini F, Schulze-Lefert P (1997) The barley *Mlo* gene: a novel control element of plant pathogen resistance. Cell 88:695–705

Butler EJ (1918) White rust (*Cystopus candidus*) (Pers. Lev.). In: Butler EJ (ed) Fungi and disease in plants. Thacker Spink, Calcutta, pp 291–297

Butler EJ, Jones SG (1961) Plant pathology. Macmillan, London

Caillaud M-C, Wirthmueller L, Sklenar J, Findlay K, Piquerez SJ, Jones AM, Robatzek S, Jones JD, Faulkner C (2014) The plasmodesmal protein PDLP1 localises to haustoria-associated membranes during downy mildew infection and regulates callose deposition. PLoS Pathog 2014:e1004496

Cairns J, Frickel J, Jalasvuori M, Hiltunen T, Becks L (2017) Genomic evolution of bacterial populations under co-selection by antibiotics and phage. Mol Ecol 26:1848–1859. https://doi.org/10.1111/mec.13950

Calonge FD, Fielding AH, Byrde RJW, Akinrefon DA (1969) Changes in ultra-structure following fungal invasion and the possible relevance of extra cellular enzymes. J Exp Bot 20:350–357

Canet JV, Dobón A, Roig A, Tornero P (2010) Structure-function analysis of npr1 alleles in *Arabidopsis* reveals a role for its paralogs in the perception of salicylic acid. Plant Cell Environ 33:1911–1922

Canonne J, Marino D, Jauneau A, Pouzet C, Briere C, Roby D, Rivas S (2011) The Xanthomonas type III effector *XopD* targets the *Arabidopsis* transcription factor *MYB30* to suppress plant defense. Plant Cell 23:3498–3511

Canonne J, Pichereaux C, Marino D, Roby D, Rossignol M, Rivas S (2012) Identification of the protein sequence of the type III effector XopD from the B100 strain of *Xanthomonas campestris* pv *campestris*. Plant Signal Behav 7:184–187

Cao H, Bowling SA, Gordon AS, Dong XN (1994) Characterization of an *Arabidopsis* mutant that is nonresponsive to inducers of systemic acquired resistance. Plant Cell 6:1583–1592

Cao T, Srivastava S, Rahman MH, Kav NNV, Hotte N, Deyholos MK, Strelkov SE (2008) Proteome-level changes in the roots of *Brassica napus* as a result of *Plasmodiophora brassicae* infection. Plant Sci 174:97–115

Carlin AF, Vizcarra EA, Branche E, Viramontes KM, Suarez-Amaran L, Ley K, Heinz S, Benner C, Shresta S, Glass CK (2018) Deconvolution of pro- and antiviral genomic responses in Zika virus-infected and by stander macrophages. Proc Natl Acad Sci U S A 115:E9172–E9181. https://doi.org/10.1073/pnas.1807690115

Carrington JC, Kasschau KD, Mahajan SK, Schaad MC (1996) Cell-to-cell and long-distance transport of viruses in plants. Plant Cell 8:1669–1681

Castaneda A, Reddy JD, El-Yacoubi B, Gabriel DW (2005) Mutagenesis of all eight *avr* genes in *Xanthomonas campestris* pv. *campestris* had no detected effect on pathogenicity, but one *avr* gene affected race specificity. Mol Plant-Microbe Interact 18:1306–1317

Cessna SG, Sears VE, Dickman MB, Low PS (2000) Oxalic acid, a pathogenicity factor for *Sclerotinia sclerotiorum*, suppresses the oxidative burst of the host plant. Plant Cell 12:2192–2199

Chalhoub B, Denoeud F, Liu S, Parkin IA, Tang H, Wang X, Chiquet J, Belcram H, Tong C, Samans B, Corréa M, Da Silva C, Just J, Falentin C, Koh CS, Le Clainche I, Bernard M, Bento P, Noel B, Labadie K, Alberti A, Charles M, Arnaud D, Guo H, Daviaud C, Alamery S, Jabbari K, Zhao M, Edger PP, Chelaifa H, Tack D, Lassalle G, Mestiri I, Schnel N, Le Paslier MC, Fan G, Renault V, Bayer PE, Golicz AA, Manoli S, Lee TH, Thi VH, Chalabi S, Hu Q, Fan C, Tollenaere R, Lu Y, Battail C, Shen J, Sidebottom CH, Wang X, Canaguier A, Chauveau A, Bérard A, Deniot G, Guan M, Liu Z, Sun F, Lim YP, Lyons E, Town CD, Bancroft I, Wang X, Meng J, Ma J, Pires JC, King GJ, Brunel D, Delourme R, Renard M, Aury JM, Adams KL, Batley J, Snowdon RJ, Tost J, Edwards D, Zhou Y, Hua W, Sharpe AG, Paterson AH, Guan C, Wincker P (2014) Early allopolyploid evolution in the post-Neolithic *Brassica napus* oilseed genome. Science 345(6199):950–953

Chandran D, Tai YC, Hather G, Dewdney J, Denoux C, Burgess DG, Ausubel FM, Speed TP, Wildermuth MC (2009) Temporal global expression data reveal known and novel salicylate-impacted processes and regulators mediating powdery mildew growth and reproduction on *Arabidopsis*. Plant Physiol 49:1435–1451

Chandran D, Inada N, Hather G, Kleindt CK, Wildermuth MC (2010) Laser micro-dissection of *Arabidopsis* cells at the powdery mildew infection site reveals site-specific processes and regulators. Proc Natl Acad Sci U S A 107:460–465

Chandran D, Rickert J, Cherk C, Dotson BR, Wildermuth MC (2013) Host cell ploidy underlying the fungal feeding site is a determinant of powdery mildew growth and reproduction. Mol Plant Microbe Interact 26:537–545

Chandran D, Rickert J, Huang Y, Steinwand MA, Marr SK, Wildermuth MC (2014) A typical E2F transcriptional repressor DEL1 acts at the intersection of plant growth and immunity by controlling the hormone salicylic acid. Cell Host Microbe 15:506–513

Chen C, Harel A, Gorovoits R, Yarden O, Dickman MB (2004) MAPK regulation of sclerotial development in *Sclerotinia sclerotiorum* is linked with pH and CAMP sensing. Mol Plant Microbe Interact 17:404–413. https://doi.org/10.1094/MPMI.2004.17.4.404

Chen LQ, Hou BH, Lalonde S, Takanaga H, Hartung ML, Qu XQ, Guo WJ, Kim JG, Underwood W, Chaudhuri B, Chermak D, Antony G, White FF, Somerville SC, Mudgett MB, Frommer WB (2010) Sugar transporters for intercellular exchange and nutrition of pathogens. Nature 468:527–532. https://doi.org/10.1038/nature09606

Chen LQ, Qu XQ, Hou BH, Sosso D, Osorio S, Fernie AR, Frommer WB (2012) Sucrose efflux mediated by SWEET proteins as a key step for phloem transport. Science 335:207–211. https://doi.org/10.1126/science.1213351

Chen D, Neumann K, Friedel S, Kilian B, Chen M, Altmann T, Klukas C (2014) Dissecting the phenotypic components of crop plant growth and drought responses based on high-throughput image analysis. Plant Cell Online 26:4636–4655

Chen J, Pang W, Chen B, Zhang C, Piao Z (2015a) Transcriptome analysis of *Brassica rapa* near-isogenic lines carrying clubroot-resistant and -susceptible alleles in response to *Plasmodiophora brassicae* during early infection. Front Plant Sci 6:1183

Chen Q, Wang J, Hou M, Liu S, Huang J, Cai L (2015b) BnSGS3 has differential effects on the accumulation of *CMV*, *ORMV* and *TuMV* in oilseed rape. Viruses 7(8):4169–4185. https://doi.org/10.3390/v7082815

Chen F, Ma R, Chen XL (2019a) Advances of metabolomics in fungal pathogen–plant interactions. Metabolites 9:169. https://doi.org/10.3390/metabo9080169

Chen J, Ullah C, Reichelt M, Gershenzon J, Hammerbacher A (2019b) *Sclerotinia sclerotiorum* circumvents flavonoid defenses by catabolizing flavonol glycosides and aglycones. Plant Physiol 180:1975–1987

Chico JM, Chini A, Fonseca S, Solano R (2008) JAZ repressors set the rhythm in jasmonate signaling. Curr Opin Plant Biol 11:486–494

Cho Y (2015) How the necrotrophic fungus *Alternaria brassicicola* kills plant cells remains an enigma. Eukaryot Cell 14:335–344. https://doi.org/10.1128/EC.00226-14

Cho Y, Davis JW, Kim KH, Wang J, Sun QH, Cramer RA, Lawrence CB (2006) A high throughput targeted gene disruption method for *Alternaria brassicicola* functional genomics using linear minimal element (LME) constructs. Mol Plant-Microbe Interact 19:7–15

Cho Y, Cramer RA, Kim KH, Davis J, Mitchell TK, Figuli P, Pryor BM, Lemasters E, Lawrence CB (2007) The Fus3/Kss1 MAP kinase homolog Amk1 regulates the expression of genes encoding hydrolytic enzymes in *Alternaria brassicicola*. Fungal Genet Biol 44:543–553

Cho Y, Kim K-H, La Rota M, Scott D, Santopietro G, Callihan M, Mitchell TK, Lawrence CB (2009) Identification of virulence factors by high throughput targeted gene deletion of regulatory genes in *Alternaria brassicicola*. Mol Microbiol 72:1316–1333

Cho Y, Srivastava A, Ohm RA, Lawrence CB, Wang KH, Grigoriev IV, Marahatta SP (2012) Transcription factor *Amr1* induces melanin biosynthesis and suppresses virulence in *Alternaria brassicicola*. PLoS Pathog 8(10):e1002974

Cho Y, Ohm BA, Grigoriev IV, Srivastava A (2013) Fungal-specific transcription factor *AbPf2* activates pathogenicity in *Alternaria brassicicola*. Plant J 75:498–514

Chomwong S, Charoensapsri W, Amparyup P, Tassanakajon A (2018) Two host gut-derived lactic acid bacteria activate the proPO system and increase resistance to an AHPND-causing strain of *Vibrio parahaemolyticus* in the shrimp *Litopenaeus vannamei*. Dev Comp Immun 89:54–65. https://doi.org/10.1016/j.dci.2018.08.002

Chong J, Piron MC, Meyer S, Merdinoglu D, Bertsch C, Mestre P (2014) The SWEET family of sugar transporters in grapevine: Vv SWEET 4 is involved in the interaction with *Botrytis cinerea*. J Exp Bot 65:6589–6601. https://doi.org/10.1093/jxb/eru375

Chu M, Lopez-Moya JJ, Llave-Correas C, Pirone TP (1997) Two separate regions in the genome of the tobacco etch virus contain determinants of the wilting response of Tabasco pepper. Mol Plant Microbe Interact 10:472–480

Chu Z, Yuan M, Yao L, Ge X, Yuan B, Xu C, Li X, Fu B, Li Z, Bennetzen JL, Zhang Q, Wang S (2006) Promoter mutations of an essential gene for pollen development result in disease resistance in rice. Genes Dev 20:1250–1255. https://doi.org/10.1101/gad.1416306

Chung HS, Niu Y, Browse J, Howe GA (2009) Top hits in contemporary JAZ: an update on jasmonate signaling. Phytochemistry 70:1547–1559

Chupp C (1925) Manual of vegetable garden diseases. Macmillan, New York

Coaker G, Falick A, Staskawicz B (2005) Activation of a phytopathogenic bacterial effector protein by a eukaryotic cyclophilin. Science 308:548–550

Coca MA, Damsz B, Yun DJ, Hasegawa PM, Bressan RA, Narasimhan M (2000) Heterotrimeric G-proteins of a filamentous fungus regulates cell wall composition and susceptibility to a plant PR-5 protein. Plant J 22:61–69

Coffey MD (1975) Ultra-structural features of the haustorial apparatus of the white blister fungus *A. candida*. Can J Bot 53:1285–1299

Colcombet J, Hirt H (2008) *Arabidopsis* MAPKs: a complex signalling network involved in multiple biological processes. Biochem J 413:217–226

Collins NC, Thordal-Christensen H, Lipka V, Bau S, Kombrink E, Qiu JL, Hückelhoven R, Stein M, Freialdenhoven A, Somerville SC, Schulze-Lefert P (2003) SNARE-protein-mediated disease resistance at the plant cell wall. Nature 425:973–977

Consonni C, Humphry ME, Hartmann HA, Livaja M, Durner J, Westphal L, Vogel J, Lipka V, Kemmerling B, Schulze-Lefert P, Somerville SC, Panstruga R (2006) Conserved requirement for a plant host cell protein in powdery mildew pathogenesis. Nat Genet 386(1):716–720

Consonni C, Bednarek P, Humphry M, Francocci F, Ferrari S, Harzen A, Ver Loren van Themaat E, Panstruga R (2010) Tryptophan-derived metabolites are required for antifungal defense in the Arabidopsis mlo2 mutant. Plant Physiol 152:1544–1561. https://doi.org/10.1104/pp.109.147660

Coplin DL, Cook D (1990) Molecular genetics of extracellular polysaccharide biosynthesis in vascular phytopathogenic bacteria. Mol Plant-Microbe Interact 3:271–279

Coppens F, Wuyts N, Inzé D, Dhondt S (2017) Unlocking the potential of plant phenotyping data through integration and data-driven approaches. Curr Opin Syst Biol 4:58–63

Corbett M, Virtue S, Bell K, Birch P, Burr T, Hyman L, Lilley K, Poock S, Toth I, Salmond G (2005) Identification of a new quorum-sensing-controlled virulence factor in *Erwinia carotovora* subsp. *atroseptica* secreted via the type II targeting pathway. Mol Plant-Microbe Interact 18:334–342

Corwin JA, Kliebenstein DJ (2017) Quantitative Resistance: more than just perception of a pathogen. Plant Cell 29:655–665

Corwin JA, Copeland D, Feusier J, Subedy A, Eshbaugh R, Palmer C, Maloof J, Kliebenstein DJ (2016) The quantitative basis of the Arabidopsis innate immune system to endemic pathogens depends on pathogen genetics. PLoS Genet 12:1–29

Craig L, Pique ME, Tainer JA (2004) Type IV pilus structure and bacterial pathogenicity. Nat Rev Microbiol 2:363–378

Craven KD, Velez H, Cho Y, Lawrence CB, Mitchell TK (2008) Anastomosis is required for virulence of the fungal necrotroph, *Alternaria brassicicola*. Eukaryot Cell 7:675–683

Cui H, Wang Y, Xue L, Chu J, Yan C, Fu J, Chen M, Innes RW, Zhou JM (2010) *Pseudomonas syringae* effector protein AvrB perturbs *Arabidopsis* hormone signaling by activating MAP kinase 4. Cell Host Microbe 7:164–175

Cunnac S, Lindeberg M, Collmer A (2009) *Pseudomonas syringae* type III secretion system effectors: repertoires in search of functions. Curr Opin Microbiol 12:53–60

Czedik Eysenberg A, Seitner S, Guldener U, Koemeda S, Jez J, Colombini M, Djamei A (2018) The 'PhenoBox', a flexible, automated, open-source plant phenotyping solution. New Phytol 219:808–823

Da Cunha L, McFall AJ, Mackey D (2006) Innate immunity in plants: a continuum of layered defenses. Microbes Infect 8:1372–1381

da Silva ACR, Ferro JA, Reinach FC, Farah CS, Furlan LR, Quaggio RB, Monteiro-Vitorello CB, Van Sluys MA, Almeida NF, Alves LMC, do Amaral AM, Bertolini MC, Camargo LEA, Camarotte G, Cannavan F, Cardozo J, Chambergo F, Clapina LP, Cicarelli RMB, Coutinho LL, Cursino-Santos JR, El-Dorry H, Faria JB, Ferreira AJS, Ferreira RCC, Ferro MIT, Formighieri EF, Franco MC, Greggio CC, Gruber A, Katsuyama AM, Kishi LT, Leite RP, Lemos EGM, Lemos MVF, Locali EC, Machado MA, Madeira A, Martinez-Rossi NM, Martins EC, Meidanis J, Menck CFM, Miyaki CY, Moon DH, Moreira LM, Novo MTM, Okura VK, Oliveira MC, Oliveira VR, Pereira HA, Rossi A, Sena JAD, Silva C, de Souza RF, Spinola LAF, Takita MA, Tamura RE, Teixeira EC, Tezza RID, dos Santos MT, Truffi D, Tsai SM, White FF, Setubal JC, Kitajima JP (2002) Comparison of the genomes of two *Xanthomonas* pathogens with differing host specificities. Nature 417:459–463

Dallal Bashi Z, Hegedus DD, Buchwaldt L, Rimmer SR, Borhan MH (2010) Expression and regulation of *Sclerotinia sclerotiorum* necrosis and ethylene-inducing peptides (NEPs). Mol Plant Pathol 11:43–53

Dallot S, Quiot-Douine L, Saenz P, Cervera MT, Garcia J-A, Quiot J-B (2001) Identification of Plum pox virus determinants implicated in specific interactions with different *Prunus* spp. Phytopathology 91:159–164

Daub JT, Hofer T, Cutivet E, Dupanloup I, Quintana-Murci L, Robinson-Rechavi M, Excoffier L (2013) Evidence for polygenic adaptation to pathogens in the human genome. Mol Biol Evol 30: 1544–1558. https://doi.org/10.1093/molbev/mst080

Daval S, Belcour A, Gazengel K, Legrand L, Gouzy J, Cottret L, Lebreton L, Aigu Y, Mougel C, Manzanares-Dauleux Maria J (2018) Computational analysis of the *Plasmodiophora brassicae* genome: mitochondrial sequence description and metabolic pathway database design. BMA Genomics 111(6):1629–1640. https://doi.org/10.1016/j.ygeno.2018.11.013

de Bary A (1887) Comparative morphology and biology of fungi, mycetozoa and bacteria. London, Clerenden, p 525

De Gara L, de Pinto MC, Tommasi F (2003) The antioxidant systems *vis-a-vis* reactive oxygen species during plant–pathogen interaction. Plant Physiol Biochem 41:863–870

de Mendoza A, Sebé-Pedrós A, Ruiz-Trillo I (2014) The evolution of the GPCR signaling system in eukaryotes: modularity, conservation, and the transition to metazoan multi-cellularity. Genome Biol Evol 6(3):606–619

Debieu M, Huard-Chauveau C, Genissel A, Roux F, Roby D, Huard-Chauveau C, Genissel A, Roux F, Roby D (2015) Quantitative disease resistance to the bacterial pathogen *Xanthomonas campestris* involves an *Arabidopsis* immune receptor pair and a gene of unknown function. Mol Plant Pathol 17:510–520

Dekhuijzen HM (1979) Electron microscopic studies on the root hairs and cortex of a susceptible and a resistant variety of *Brassica campestris* infected with *Plasmodiophora brassicae*. Netherlands J Plant Pathol 85:1–17. https://doi.org/10.1007/BF01976714

Denoeud F, Roussel M, Noel B, Wawrzyniak I, Da Silva C, Diogon M, Viscogliosi E, Brochier-Armanet C, Couloux A, Poulain J, Segurens B, Anthouard V, Texier C, Blot N, Poirier P, Ng GC, Tan KSW, Artiguenave F, Jaillon O, Aury J-M, Delbac F, Wincker P, Vivarès CP, El Alaoui H (2011) Genome sequence of the stramenopile Blastocystis, a human anaerobic parasite. Genome Biol 12:R29

Deora A, Gossen BD, McDonald MR (2012) Infection and development of *Plasmodiophora brassicae* in resistant and susceptible canola cultivars. Can J Plant Pathol 34:239–247

Deora A, Gossen BD, Mcdonald MR (2013) Cytology of infection, development and expression of resistance to *Plasmodiophora brassicae* in canola. Ann Appl Biol 163:56–71. https://doi.org/10.1111/aab.12033

Derbyshire M, Denton-Giles M, Hegedus D, Seifbarghy S, Rollins J, van Kan J, Seidl MF, Faino L, Mbengue M, Navaud O, Raffaele S, Hammond-Kosack K, Heard S, Oliver R (2017) The complete genome sequence of the phytopathogenic fungus Sclerotinia sclerotiorum reveals insights into the genome architecture of broad host range pathogens. Genome Biol Evol 9(3): 593–618. https://doi.org/10.1093/gbe/evx030

Devos S, Vissenberg K, Verbelen JP, Prinsen E (2005) Infection of Chinese cabbage by *Plasmodiophora brassicae* leads to a stimulation of plant growth: impacts on cell wall metabolism and hormone balance. New Phytol 166:241–250. https://doi.org/10.1111/j.1469-8137.2004.01304.x

Devos S, Laukens K, Deckers P, Straeten DVD, Beeckman T, Inze D, Onckelen HV, Witters P, Prinsen E (2006) A hormone and proteome approach to picturing the initial metabolic events during *Plasmodiophora brassicae* infection on *Arabidopsis*. Mol Plant-Microbe Interact 19: 1431–1443. https://doi.org/10.1094/MPMI-19-1431

Devoto A, Piffanelli P, Elliott C, Simmons C, Taramino G, Goh CS, Cohen FE, Emerson BC, Schulze-Lefert P, Panstruga R (2003) Molecular phylogeny and evolution of the plant-specific seven-transmembrane MLO family. J Mol Evol 566(1):77–88

Dhawan S (1980) Production of protease by *Sclerotinia sclerotiorum* cause of white rot of *Brassica juncea*. Curr Sci 49:361–362

Dhawan S, Srivastava SK (1987) Trans-eliminative (lytic) pectic enzymes in *Sclerotinia* infection of *Brassica juncea*. Indian J Mycol Plant Pathol 17:325–327

Diatchenko L, Lau YF, Campbell AP, Chenchik A, Moqadam F, Huang B, Lukyanov S, Lukyanov K, Gurskaya N, Sverdlov ED, Siebert PD (1996) Suppression subtractive hybridization: a method for generating differentially regulated or tissue-specific cDNA probes and libraries. Proc Natl Acad Sci U S A 93:6025–6030

Dickman MB, Mitra A (1992) *Arabidopsis thaliana* as a model for studying *Sclerotinia sclerotiorum* pathogenesis. Physiol Mol Plant Pathol 41:255–263

Diederichsen E, Frauen M, Linders EG, Hatakeyama K, Hirai M (2009) Status and perspectives of clubroot resistance breeding in crucifer crops. J Plant Growth Regul 28:265–281

Dietz KJ, Jacob S, Oelze ML, Laxa M, Tognetti V, de Miranda SMN, Baier M, Finkemeier I (2006) The function of peroxiredoxins in plant organelle redox metabolism. J Exp Bot 57:1697–1709

Dilthey A, Cox C, Iqbal Z, Nelson MR, McVean G (2015) Improved genome inference in the MHC using a population reference graph. Nat Genet 47:682–688. https://doi.org/10.1038/ng.3257

Ding L, Xu H, Yi H, Yang L, Kong Z, Zhang L, Xue S, Jia H, Ma Z (2011) Resistance to hemi-biotrophic F. graminearum infection is associated with coordinated and ordered expression of diverse defense signaling pathways. PLoS One 6(e19008)

Ding Y, Mei J, Chai Y, Yang W, Mao Y, Yan B, Yu Y, Disi JO, Rana K, Li J, Qian W (2020) *Sclerotinia sclerotiorum* utilizes host-derived copper for ROS detoxification and infection. PLoS Pathol 16(10):e1008919. https://doi.org/10.1371/journal.ppat.1008919

Dixon DP, Skipsey M, Edwards R (2010) Roles for glutathione transferases in plant secondary metabolism. Phytochemistry 71:338–350

Dobin A, Davis CA, Schlesinger F, Drenkow J, Zaleski C, Jha S, Batut P, Chaisson M, Gingeras TR (2013) STAR: ultrafast universal RNA-seq aligner. Bioinformatics 29:15–21

Dodds PN, Rathjen JP (2010) Plant immunity: towards an integrated view of plant–pathogen interactions. Nat Rev Genet 11:539–548

Doidy J, Grace E, Kühn C, Simonplas F, Casieri L, Wipf D (2012) Sugar transporters in plants and in their interactions with fungi. Trends Plant Sci 17:413–422. https://doi.org/10.1016/j.tplants.2012.03.009

Domingues MN, De Souza TA, Cernadas RA, de Oliveira MLP, Docena C, Farah CS, Benedetti CE (2010) The *Xanthomonas citri* effector protein *PthA* interacts with citrus proteins involved in nuclear transport, protein folding and ubiquitination associated with DNA repair. Mol Plant Pathol 11:663–675

Domingues MN, de Campos BM, de Oliveira ML, de Mello UQ, Benedetti CE (2012) TAL effectors target the C-terminal domain of RNA polymerase II (CTD) by inhibiting the prolyl-isomerase activity of a CTD-associated cyclophilin. PLoS One 7:e41553

Dominguez-Solis JR, Zengyong H, Amparo L, Julie T, Buchanan BB, Sheng L (2008) A cyclophilin links redox and light signals to cysteine biosynthesis and stress responses in chloroplasts. Proc Natl Acad Sci U S A 105:16386–16391

Donald EC, Jaudzens G, Porter IJ (2008) Pathology of cortical invasion by *Plasmodiophora brassicae* in clubroot resistant and susceptible *Brassica oleracea* hosts. Plant Pathol 33:585–589

Donofrio NM, Delaney TP (2001) Abnormal callose response phenotype and hyper susceptibility to *Peronospora parasit*ica in defense-compromised *Arabidopsis nim1-1* and salicylate hydroxylase - expressing plants. Mol Plant-Microbe Interact 2001:439–450

Dorsam RT, Gutkind JS (2007) G-protein-coupled receptors and cancer. Nat Rev Cancer 7(2): 79–94

Douchkov D, Nowara D, Zierold U, Schweizer P (2005) A high-throughput gene-silencing system for the functional assessment of defense-related genes in barley epidermal cells. Mol Plant Microbe Interact 18:755–761

Doughan B, Rollins JA, Avery SV (2016) Characterization of *MAT* gene functions in the life cycle of *Sclerotinia sclerotiorum* reveals a lineage-specific *MAT* gene functioning in apothecium morphogenesis. Fungal Biol 120:1105–1117. https://doi.org/10.1016/j.funbio.2016.06.007

Dow JM, Clarke BR, Milligan DE, Tang JL, Daniels MJ (1990) Extracellular proteases from *Xanthomonas campestris* pv. *campestris*, the black rot pathogen. Appl Environ Microbiol 56: 2994–2998

Dow JM, Osbourn AE, Wilson TJG, Daniels MJ (1995) A locus determining pathogenicity of *Xanthomonas campestris* is involved in lipopolysaccharide biosynthesis. Mol Plant Microbe Interact 8:768–777

Dow JM, Feng JX, Barber CE, Tang JL, Daniels MJ (2000) Novel genes involved in the regulation of pathogenicity factor production within the *rpf* gene cluster of *Xanthomonas campestris*. Microbiology 146:885–891

Dow JM, Crossman L, Findlay K, He YQ, Feng JX, Tang JL (2003) Biofilm dispersal in *Xanthomonas campestris* is controlled by cell–cell signaling and is required for full virulence to plants. Proc Natl Acad Sci U S A 100:10995–11000

Dudler R, Hertig C, Rebman G, Bull J, Mauch F (1991) A pathogen-induced wheat gene encodes a protein homologous to glutathione S-transferases. Mol Plant Microbe Interact 4:4–18

Dutton MV, Evans CS (1996) Oxalate production by fungi: its role in pathogenicity and ecology in the soil environment. Can J Microbiol 42:881–895

Echandi E, Walker JC (1957) Pectolytic enzymes produced by *Sclerotinia sclerotiorum*. Phytopathology 47:303–306

Eichmann R, Hückelhoven R (2008) Accommodation of powdery mildew fungi in intact plant cells. J Plant Physiol 165:5–18

Eichmann R, Dechert C, Kogel KH, Hückelhoven R (2006) Transient over-expression of barley BAX inhibitor-1 weakens oxidative defence and *MLA12*-mediated resistance to *Blumeria graminis* f.sp. *hordei*. Mol Plant Pathol 7:543–552

Eichmann R, Bischof M, Weis C, Shaw J, Lacomme C, Schweizer P, Duchkov D, Hensel G, Kumlehn J, Hückelhoven R (2010) BAX INHIBITOR-1 is required for full susceptibility of barley to powdery mildew. Mol Plant Microbe Interact 23:1217–1227

El-Zahaby HM, Gullner G, Kiraly Z (1995) Effects of powdery mildew infection of barley on the ascorbate-glutathione cycle and other antioxidants in different host-pathogen interactions. Phytopathology 85:1225–1230

Emanuelsson O, Nielsen H, Brunak S, von Heijne G (2000) Predicting subcellular localization of proteins based on their N-terminal amino acid sequence. J Mol Biol 300:1005–1016. https://doi.org/10.1006/jmbi.2000.3903

Enard D, Le C, Gwennap C, Petrov DA (2016) Viruses are a dominant driver of protein adaptation in mammals. Elife 5:e12469. https://doi.org/10.7554/eLife.12469

Engleberg NC, Carter C, Weber DR, Cianciotto NP, Eisenstein BI (1989) DNA sequence of *MIP*, a Legionella pneumophila gene associated with macrophage infectivity. Infect Immun 57:1263–1270

Erental A, Harel A, Yarden O (2007) Type 2A phosphoprotein phosphatase is required for asexual development and pathogenesis of *Sclerotinia sclerotiorum*. Mol Plant Microbe Interact 20:944–954. https://doi.org/10.1094/MPMI-20-8-0944

Eulgem T, Somssich IE (2007) Networks of WRKY transcription factors in defense signaling. Curr Opin Plant Biol 10:366–371

Evans JL, Scholes JD (1995) How does clubroot alter the regulation of carbon metabolism in its host? Aspects Appl Biol 42:125–132

Eynck C, Koopmann B, Grunewaldt-Stoecker G, Karlovsky P, von Tiedemann A (2007) Differential interactions of *Verticillium longisporum* and *V. dahliae* with *Brassica napus* detected with molecular and histological techniques. Eur J Plant Pathol 118:259–274

Fabro G, Di Rienzo JA, Voigt CA, Sevchenko T, Dehesh K, Somerville S, Alvarez ME (2008) Genome-wide expression profiling *Arabidopsis* at the stage of *Golovinomyces cichoracearum* haustorium formation. Plant Physiol 1466(1):1421–1439

Fahlgren N, Feldman M, Gehan MA, Wilson MS, Shyu C, Bryant DW, Hill ST, McEntee CJ, Warnasooriya SN, Kumar I, Ficor T, Turnipseed S, Gilbert KB, Brutnell TP, Carrington JC, Mockler TC, Baxter I (2015) A versatile phenotyping system and analytics platform reveals diverse temporal responses to water availability in Setaria. Mol Plant 8:1520–1535

Fan H, Yu G, Liu Y, Zhang X, Liu J, Zhang Y, Rollins JA, Sun F, Pan H (2017) An atypical fork head-containing transcription factor SsFKH1 is involved in sclerotial formation and is essential for pathogenicity in *Sclerotinia sclerotiorum*. Mol Plant Pathol 18:963–975. https://doi.org/10.1111/mpp.12453

Feller A, Machemer K, Braun EL, Grotewold E (2011) Evolutionary and comparative analysis of MYB and bHLH plant transcription factors. Plant J 66:94–116. https://doi.org/10.1111/j.1365-313X.2010.04459.x

Feng F, Zhou JM (2012) Plant–bacterial pathogen interactions mediated by type III effectors. Curr Opin Plant 15:469–476

Feng J, Hwang R, Hwang SF, Strelkov SE, Gossen BD, Zhou QX, Peng G (2010) Molecular characterization of a serine protease *Pro1* from *Plasmodiophora brassicae* that stimulates resting spore germination. Mol Plant Pathol 11(4):503–512

Feng J, Xiao Q, Hwang SF, Strelkov SE, Gossen BD (2012) Infection of canola by secondary zoospores of *Plasmodiophora brassicae* produced on a nonhost. Eur J Plant Pathol 132:309–315

Feng J, Hwang SF, Strelkov SE (2013a) Assessment of gene expression profiles in primary and secondary zoospores of *Plasmodiophora brassicae* by dot blot and real-time PCR. Microbiol Res 168:518–524

Feng J, Hwang SF, Strelkov SE (2013b) Studies into primary and secondary infection processes by *Plasmodiophora brassicae* on canola. Plant Pathol 62:177–183. https://doi.org/10.1111/j.1365-3059.2012.02612.x

Feng BZ, Zhu XP, Fu L, Lv RF, Storey D, Tooley P, Zhang XG (2014) Characterization of necrosis-inducing NLP proteins in *Phytophthora capsici*. BMC Plant Biol 14:126. https://doi.org/10.1186/1471-2229-14-126

Fernandez AP, Strand A (2008) Retrograde signaling and plant stress: plastid signals initiate cellular stress responses. Curr Opin Plant Biol 11:509–513

Ferrari S, Galletti R, Denoux C, Lorenzo GD, Ausubel FM, Dewdney J (1994) Resistance to *Botrytis cinerea* induced in *Arabidopsis* by elicitors is independent of salicylic acid, ethylene, or jasmonate signaling but requires Phytoalexin Deficent3. Plant Physiol 144:367–379. https://doi.org/10.1104/pp.107.095596

Floerl S, Druebert C, Majcherczyk A, Karlovsky P, Kües U, Polle A (2008) Defence reactions in the apoplastic proteome of oilseed rape (*Brassica napus* var. *napus*) attenuate *Verticillium longisporum* growth but not disease symptoms. BMC Plant Biol 8:129

Flor HH (1971) Current status of gene-for-gene concept. Annu Rev Phytopathol 9:275–296

Fodor J, Gullner G, Adam AL, Barna B, Komives T, Kiraly Z (1997) Local and systemic responses of antioxidants to tobacco mosaic virus infection and to salicylic acid in tobacco. Plant Physiol 114:1443–1451

Fotopoulos V, Gilbert MJ, Pittman JK, Marvier AC, Buchanan AJ, Sauer N, Hall JL, Williams LE (2003) The monosaccharide transporter gene, AtSTP4, and the cell-wall invertase, *Atβfruct1*, are induced in *Arabidopsis* during infection with the fungal biotroph *Erysiphe cichoracearum*. Plant Physiol 1326(1):821–829

Foyer CH, Noctor G (2000) Oxygen processing in photosynthesis: regulation and signaling. New Phytol 146:359–389. https://doi.org/10.1046/j.1469-8137.2000.00667.x

Fradin EF, Thomma BPHJ (2006) Physiology and molecular aspects of *Verticillium* wilt diseases caused by *V. dahliae* and *V. albo-atrum*. Mol Plant Pathol 7:71–86

Fraissinet-Tachet L, Reymond-Cotton P, Fevre M (1995) Characterization of a multi-gene family encoding an endopolygalacturonase in *Sclerotinia sclerotiorum*. Curr Microbiol 29:96–100

Franzenburg S, Walter J, Kuenzel S, Wang J, Baines JF, Bosch TCG, Fraune S (2013) Distinct antimicrobial peptide expression determines host species-specific bacterial associations. Proc Natl Acad Sci U S A 110:E3730–E3738. https://doi.org/10.1073/pnas.1304960110

Fridovich I (1986) Superoxide dismutases. Adv Enzy Related Areas Mol Biol 58:61–97

Frye CA, Innes RW (1998) An *Arabidopsis* mutant with enhanced resistance to powdery mildew. Plant Cell 106(1):947–956

Frye CA, Tang DZ, Innes RW (2001) Negative regulation of defense responses in plants by a conserved MAPKK kinase. Proc Natl Acad Sci U S A 986(1):373–378

Fu D, Uauy C, Distelfeld A, Blechl A, Epstein L, Chen X, Sela H, Fahima T, Dubcovsky J (2009) A kinase START gene confers temperature dependent resistance to wheat stripe rust. Science 323: 1357–1360

Fudal I, Ross S, Gout L, Blaise F, Kuhn ML, Eckert MR, Cattolico L, Bernard-Samain S, Balesdent MH, Rouxel T (2007) Heterochromatin-like regions as ecological niches for avirulence genes in the *Leptosphaeria maculans* genome: map-based cloning of *AvrLm6*. Mol Plant Microbe Interact 20:459–470

Fujisawa I (1990) *Turnip mosaic virus* strains in cruciferous crops in Japan. Jpn Agric Res Q 23: 289–293

Fukunaga R, Doudna JA (2009) dsRNA with 5′ overhangs contributes to endogenous and antiviral RNA silencing pathways in plants. EMBO J 28:545–555

Fumagalli M, Sironi M, Pozzoli U, Ferrer-Admettla A, Pattini L, Nielsen R (2011) Signatures of environmental genetic adaptation pinpoint pathogens as the main selective pressure through human evolution. PLoS Genet 7(11):1371. https://doi.org/10.1371/journal.pgen.1002355

Gabriel DW (1999) Why do pathogens carry avirulence genes? Physiol Mol Plant Pathol 55:205–214

Galat A (2003) Peptidylprolyl cis/trans isomerases (immunophilins): biological diversity, targets, functions. Curr Top Med Chem 3:1315–1347

Gallois JL, Drouaud J, Lecureuil A, Guyon-Debast A, Bonhomme S, Guerche P (2013) Functional characterization of the plant ubiquitin regulatory X (UBX) domain-containing protein AtPUX7 in *Arabidopsis thaliana*. Gene 526:299–308

Gamas P, Niebel FC, Lescure N, Cullimore J (1996) Use of a subtractive hybridization approach to identify new *Medicago truncatula* genes induced during root nodule development. Mol Plant Microbe Interact 9:233–242. https://doi.org/10.1094/MPMI-9-0233

Gan P, Ikeda K, Irieda H, Narusaka M, O'Connell RJ, Narusaka Y, Takano Y, Kubo Y, Shirasu K (2013) Comparative genomic and transcriptomic analyses reveal the hemibiotrophic stage shift of *Colletotrichum* fungi. New Phytol 197(4):1236–1249. https://doi.org/10.1111/nph.12085

Ge YX, Angenent GC, Wittich PE, Peters J, Franken J, Busscher M et al (2000) *Nec1*, a novel gene, highly expressed in nectary tissue of petunia hybrida. Plant J 24:725–734. https://doi.org/10.1046/j.1365-313x.2000.00926.x

Gjetting T, Carver TL, Skøt L, Lyngkjaer MF (2004) Differential gene expression in individual papilla-resistant and powdery mildew-infected barley epidermal cells. Mol Plant Microbe Interact 17(7):729–738

Gjetting T, Hagedorn PH, Schweizer P, Thordal-Christensen H, Carver TL, Lyngkjaer MF (2007) Single-cell transcript profiling of barley attacked by the powdery mildew fungus. Mol Plant Microbe Interact 20(3):235–246

Glazebrook J (2005) Contrasting mechanisms of defense against biotrophic and necrotrophic pathogens. Annu Rev Phytopathol 436(1):205–227

Glazebrook J, Rogers EE, Ausubel FM (1996) Isolation of *Arabidopsis* mutants with enhanced disease susceptibility by direct screening. Genetics 143:973–982

Gludovacz TV (2013) Clubroot in canola and cabbage in relation to soil temperature, plant growth and host resistance. M Sc Thesis, The University of Guelph, Ontario, Canada, p 238

Gludovacz TV, Gossen BD, McDonald MR (2013) Evidence for tolerance to *Plasmodiophora brassicae* in a cabbage cultivar. Can J Plant Pathol 35:111

Godoy G, Steadman JR, Dickman MB, Dam R (1990) Use of mutants to demonstrate the role of oxalic acid in pathogenicity of *Sclerotinia sclerotiorum* on *Phaseolus vulgaris*. Physiol Mol Plant Pathol 37:179–191

Gollner K, Schweizer P, Bai Y, Panstruga R (2008) Natural genetic resources of *Arabidopsis thaliana* reveal a high prevalence and unexpected phenotypic plasticity of *RPW8*-mediated powdery mildew resistance. New Phytol 177(3):725–742

Gomez-Campo C (1999) Taxonomy. In: Gomez-Campo C (ed) Biology of *Brassica* coenospecies. Elsevier Science, Amsterdam, pp 3–32

Gout L, Fudal I, Kuhn ML, Blaise F, Eckert M, Cattolico L, Balesdent MH, Rouxel T (2006) Lost in the middle of nowhere: the AvrLm1 avirulence gene of the Dothideomycete *Leptosphaeria maculans*. Mol Microbiol 60:67–80

Graf F, Fahling M, Siemens J (2004) Chromosome polymorphism of the obligate biotrophic parasite *Plasmodiophora brassicae*. J Phytopathol 152:86–91

Gravot A, Gautier R, Lime T, Lemarié S, Jubault M, Lariagon C, Lemoine J, Vicente J, Robert Seilaniantz A, Holdsworth MJ, Manzanares-Dauleux MJ (2016) Hypoxia response in *Arabidopsis* roots infected by *Plasmodiophora brassicae* supports the development of clubroot. BMC Plant Biol 16:251

Green SK, Deng TC (1985) Turnip mosaic virus strains in cruciferous hosts in Taiwan. Plant Dis 69:28–31

Greenberg JT, Yao N (2004) The role and regulation of programmed cell death in plant–pathogen interactions. Cell Microbiol 6:201–211

Grenier JC, Potvin Asselin A (1993) Barley pathogenesis-related proteins with fungal cell wall lytic activity inhibits the growth of yeasts. Plant Physiol 103:1277–1283

Gu Q, Chen M, Huang J, Wei Y, Hsiang T, Zheng L (2017) Multifaceted roles of the Ras guanine-nucleotide exchange factor ChRgf in development, pathogenesis, and stress responses of *Colletotrichum higginsianum*. Phytopathology 107:433–443

Guan YF, Huang XY, Zhu J, Gao JF, Zhang HX, Yang ZN (2008) RUPTURED POLLEN GRAIN1, a member of the MtN3/saliva gene family, is crucial for exine pattern formation and cell integrity of microspore in *Arabidopsis*. Plant Physiol 147:852–863. https://doi.org/10.1104/pp.108.118026

Gubser C, Bergamaschi D, Hollinshead M, Lu X, van Kuppeveld FJ, Smith GL (2007) A new inhibitor of apoptosis from vaccinia virus and eukaryotes. PLoS Pathog 3:e17

Gudesblat GE, Torres PS, Vojnov AA (2009) *Xanthomonas campestris* overcomes *Arabidopsis* stomatal innate immunity through a DSF cell-to-cell signal-regulated virulence factor. Plant Physiol 149:1017–1027

Guo XH (2001) Biological characteristics and pathogenesis of *Plasmodiophora brassicae*. Southwest Agriculture University, Chongqing

Guo X, Stotz HU (2007) Defense against *Sclerotinia sclerotiorum* in *Arabidopsis* is dependent on jasmonic acid, salicylic acid, and ethylene signaling. Mol Plant Microbe Interact 20(11): 1384–1395. https://doi.org/10.1094/MPMI-20-11-1384

Guo HS, Xie Q, Fei JF, Chua NH (2005) MicroRNA directs mRNA cleavage of the transcription factor NAC1 to down-regulate auxin signals for *Arabidopsis* lateral root development. Plant Cell 17:1376–1386

Gurlebeck D, Thieme F, Bonas U (2006) Type III effector proteins from the plant pathogen *Xanthomonas* and their role in the interaction with the host plant. J Plant Physiol 163:233–255

Gustafsson M, Fält AS (1986) Genetic studies on resistance to clubroot in *Brassica napus*. Ann Appl Biol 108:409–415

Guyon K, Balague C, Roby D, Raffaele S (2014) Secretome analysis reveals effector candidates associated with broad host range necrotrophy in the fungal plant pathogen *Sclerotinia sclerotiorum*. BMC Genomics 15:336. https://doi.org/10.1186/1471-2164-15-336

Haddadi P, Ma L, Wang H, Borhan MH (2016) Genome-wide transcriptomic analyses provide insights into the lifestyle transition and effector repertoire of *Leptosphaeria maculans* during the colonization of *Brassica napus* seedlings. Mol Plant Pathol 8:1196–1210

Haga N, Kato K, Murase M, Araki S, Kubo M, Demura T, Suzuki K, Müller I, Voß U, Jürgens G, Ito M (2007) R1R2R3-Myb proteins positively regulate cytokinesis through activation of KNOLLE transcription in *Arabidopsis thaliana*. Development 134:1101–1110

Halliwell B, Gutteridge JMC (1989) Free radicals in biology and medicine. Clarendon Press, Oxford

Hancock JG (1966) Degradation of pectic substances associated with pathogenesis by *Sclerotinia sclerotiorum* in sunflower and tomato stems. Phytopathology 56:975–979

Hanes SD (2015) Prolylisomerases in gene transcription. Biochem Biophys Acta 1850:2017–2034

Hardham AR, Takemoto D, White RG (2008) Rapid and dynamic sub-cellular reorganization following mechanical stimulation of *Arabidopsis* epidermal cells mimics responses to fungal and oomycete attack. BMC Plant Biol 8:63

He Y, Fukushige H, Hildebrand DF, Gan S (2002) Evidence supporting a role of jasmonic acid in *Arabidopsis* leaf senescence. Plant Physiol 128:876–884

He YQ, Zhang L, Jiang BL, Zhang ZC, Xu RQ, Tang DJ, Qin J, Jiang W, Zhang X, Liao J, Cao JR, Zhang SS, Wei ML, Liang XX, Lu GT, Feng JX, Chen B, Cheng J, Tang JL (2007) Comparative and functional genomics reveals genetic diversity and determinants of host specificity among reference strains and a large collection of Chinese isolates of the phytopathogen *Xanthomonas campestris* pv. *campestris*. Genome Biol 8:R218

He YW, Wu J, Zhou L, Yang F, He YO, Jiang BL, Bai LQ, Xu YQ, Deng ZX, Tang JL, Zhang LH (2011) *Xanthomonas campestris* diffusible factor is 3-hydroxybenzoic acid and is associated with Xanthomonas in biosynthesis, cell viability, antioxidant activity, and systemic invasion. Mol Plant Microbe Interact 24:948–957

Heald FD (1926) Diseases due to downy mildew and allies. In: Heald FD (ed) Manual of plant diseases. McGraw-Hill, New York, pp 390–436

Heard S, Brown NA, Hammond-Kosack K (2015) An interspecies comparative analysis of the predicted secretomes of the necrotrophic plant pathogens *Sclerotinia sclerotiorum* and *Botrytis cinerea*. PLoS One 10(6):e0130534. https://doi.org/10.1371/journal.pone.0130534

Heeg C, Kruse C, Jost R, Gutensohn M, Ruppert T, Wirtz M, Hell R (2008) Analysis of the *Arabidopsis* *O*-acetylserine (thiol) lyase gene family demonstrates compartment-specific differences in the regulation of cysteine synthesis. Plant Cell 20:168–185

Held YM (1955) Physiological differences between a normal and a degenerate strain of *Sclerotinia trifoliorum*. Phytopathology 45:39–42

Hellgren O, Ekblom R (2010) Evolution of a cluster of innate immune genes (beta-defensins) along the ancestral lines of chicken and zebra finch. Immun Res 6. https://doi.org/10.1186/1745-7580-6-3

Henke N, Lisak DA, Schneider L, Habicht J, Pergande M, Methner A (2011) The ancient cell death suppressor BAX inhibitor-1. Cell Calcium 50:251–260

Henry E, Toruño TY, Jauneau A, Deslandes L, Coaker GL (2017) Direct and indirect visualization of bacterial effector delivery into diverse plant cell types during infection. Plant Cell 29:1555–1570

Hepworth SR, Zhang Y, McKim S, Li X, Haughn GW (2005) BLADEON-PETIOLE-dependent signaling controls leaf and foral patterning in *Arabidopsis*. Plant Cell 17:1434–1448

Herlihy J, Ludwig NR, van den Ackerveken G, McDowell JM (2019) Oomycetes used in *Arabidopsis* research. Arabidopsis Book 17:1–26

Hilger D, Masureel M, Kobilka BK (2018) Structure and dynamics of GPCR signaling complexes. Nat Struct Mol Biol 25(1):4–12

Hiscott J, Rongtuan L, Peyman N, Paz S (2006) Master CARD: a priceless link to innate immunity. Trends Mol Med 126(1):53–56

Hjulsager CK, Lund OS, Johansen IE (2002) A new pathotype of Pea seed borne mosaic virus explained by properties of the *P3-6K1*- and viral genome-linked protein (VPg)-coding regions. Mol Plant Microbe Interact 15:169–171

Hok S, Danchin EG, Allasia V, Panabieres F, Attard A, Keller H (2011) An *Arabidopsis* (malectin-like) leucine-rich repeat receptor-like kinase contributes to downy mildew disease. Plant Cell Environ 34(11):1944–1957

Hok S, Allasia V, Andrio E, Naessens E, Ribes E, Panabieres F, Attard A, Ris N, Clement M, Barlet X, Marco Y, Grill E, Eichmann R, Weis C, Huckelhoven R, Ammon A, Ludwig-Muller J, Voll LM, Keller H (2014) The receptor kinase IMPAIRED OOMYCETE SUSCEPTIBILITY1 attenuates abscisic acid responses in *Arabidopsis*. Plant Physiol 166:1506–1518

Horton P, Park K-J, Obayashi T, Fujita N, Harada H, Adams-Collier CJ, Nakai K (2007) WoLF PSORT: protein localization predictor. Nucleic Acids Res 35:W585–W587

Hruz T, Laule O, Szabo G, Wessendorp F, Bleuler S, Oertle L, Widmayer P, Gruissem W, Zimmermann P (2008) Genevestigator v3: a reference expression database for the meta-analysis of transcriptomes. Adv Bioinform 420747

Hsiao YM, Liu YF, Fang MC, Song WL (2011) *XCC2731*, a GGDEF domain protein in *Xanthomonas campestris*, is involved in bacterial attachment and is positively regulated by Clp. Microbiol Res 166:548–565

Hu L, Smith TF, Goldberger G (2009) *LFG*: a candidate apoptosis regulatory gene family. Apoptosis 14:1255–1265

Huang HC, Chang C, Kozub GC (1998) Effect of temperature during sclerotial formation, sclerotial dryness and relative humidity on myceliogenic germination of sclerotia of *Sclerotinia sclerotiorum*. Can J Bot 76:494–499

Huang Y-J, Li Z-Q, Evans N, Rouxel T, Fitt BDL, Balesdent M-H (2006) Fitness cost associated with loss of the *AvrLm4* avirulence function in *Leptosphaeria maculans* (phoma stem canker of oilseed rape). Eur J Plant Pathol 114:77–89

Huang X, Li Y, Niu Q, Zhang K (2007) Suppression subtractive hybridization (SSH) and its modifications on microbiological research. Appl Microbiol Biotechnol 76:753–760

Huang Y-J, Balesdent M-H, Li Z-Q, Evans N, Rouxel T, Fitt BDL (2009) Fitness cost of virulence differs between the *AvrLm1* and *AvrLm4* loci in *Leptosphaeria maculans* (phoma stem canker of oilseed rape). Eur J Plant Pathol 126:279

Huard Chauveau C, Perchepied L, Debieu M, Rivas S, Kroj T, Kars I, Bergelson J, Roux F, Roby D (2013) An atypical kinase under balancing selection confers broad-spectrum disease resistance in *Arabidopsis*. PLoS Genet 9:e1003766

Huckelhoven R (2004) BAX inhibitor-1, an ancient cell death suppressor in animals and plants with prokaryotic relatives. Apoptosis 9:299–307

Huckelhoven R, Panstruga R (2011) Cell biology of the plant–powdery mildew interaction. Curr Opin Plant Biol 14:738–746

Huckelhoven R, Dechert C, Kogel KH (2003) Overexpression of barley BAX inhibitor 1 induces breakdown of *mlo*-mediated penetration resistance to *Blumeria graminis*. Proc Natl Acad Sci U S A 100:5555–5560

Humphry M, Consonni C, Panstruga R (2006) *mlo*-based powdery-mildew immunity: silver bullet or simply non-host resistance Mol. Plant Pathol 76(1):605–610

Huser A, Takahara H, Schmalenbach W, O'Connell R (2009) Discovery of pathogenicity genes in the crucifer anthracnose fungus *Colletotrichum higginsianum*, using random insertional mutagenesis. Mol Plant Microbe Interact 22:143–156. https://doi.org/10.1094/MPMI-22-2-0143

Hwang SF, Ahmed HU, Strelkov SE, Gossen BD, Turnbull GD, Peng G, Howard RJ (2011a) Seedling age and inoculum density affect clubroot severity and seed yield in canola. Can J Plant Sci 91:183–190

Hwang SF, Ahmed HU, Zhu Q, Strelkovb SE, Gossenc BD, Peng G, Turnbulla GD (2011b) Influence of cultivar response and inoculum density on root hair infection by *Plasmodiophora brassicae*. Plant Pathol 60:820–829

Hwang SF, Strelkov SE, Feng J, Gossen BD, Howard RJ (2011c) *Plasmodiophora brassicae*: an emerging pathogen of the Canadian canola (*Brassica napus*) crop. Mol Plant Pathol 12:105–113

Hwang SF, Howard RJ, Strelkov SE, Gossen BD, Peng G (2014) Management of clubroot (*Plasmodiophora brassicae*) on canola (*Brassica napus*) in western Canada. Can J Plant Pathol 36:49–65. https://doi.org/10.1080/07060661.2013.863806

Iakovidis M, Teixeira PJPL, Exposito-Alonso M, Cowper M, Leerink SVB, Law TF, Liu Q, Vu MC, Dang TM, Corwin JA, Weigel D, Dangl JL, Grant SR (2016) Effector triggered immune response in *Arabidopsis thaliana* is a quantitative trait. Genetics 204(1):337–353. https://doi.org/10.1534/genetics.116.190678

Ibeas JI, Lee H, Damsz B, Prasad DT, Pardo JM, Hasegawa PM, Bressan RA, Narasimhan ML (2000) Fungal cell wall phosphomannans facilitate the toxic activity of a plant PR-5 protein. Plant J 23:375–383

Ibrahim HM, Bannai H, Xuan X, Nishikawa Y (2009) *Toxoplasma gondii* cyclophilin 18-mediated production of nitric oxide induces bradyzoite conversion in a CCR5-dependent manner. Infect Immun 77:3686–3695

Ibrahim HMM, Kusch S, Didelon M, Raffaele S (2021) Genome-wide alternative splicing profiling in the fungal plant pathogen *Sclerotinia sclerotiorum* during the colonization of diverse host families. Mol Plant Pathol 22(2):31–47

Ielpi L, Couso RO, Dankert MA (1993) Sequential assembly and polymerization of the polyprenol-linked pentasaccharide repeating unit of the xanthan polysaccharide in *Xanthomonas campestris*. J Bacteriol 175:2490–2500

Ignatov AN, Monakhos GF, Djalilov FS, Pozmogova GV (2002) Avirulence gene from *Xanthomonas campestris* pv. *campestris* homologous to the *avrBs2* locus is recognized in race-specific reaction by two different resistance genes in Brassicas. Russ J Genet 38:1404–1410

Ii WMJ, Rollins JVA (2007) Deletion of the adenylate cyclase (Sac1) gene affects multiple developmental pathways and pathogenicity in Sclerotinia sclerotiorum. Fungal Genet Biol 44: 521–530

Inderbitzin P, Davis RM, Bostock RM, Subbarao KV (2011) The ascomycete *Verticillium longisporum* is a hybrid and a plant pathogen with an expanded host range. PLoS One 6(3): e18260. https://doi.org/10.1371/journal.pone.0018260

Ingram DS, Tommerup IC (1972) The life history of *Plasmodiophora brassicae* Woron. Proc R Soc Lond 180:103–112

Irani S, Trost B, Waldner M, Nayidu N, Tu J, Kusalik AJ, Todd CD, Wei Y, Bonham-Smith PC (2018) Transcriptome analysis of response to *Plasmodiophora brassicae* infection in the *Arabidopsis* shoot and root. BMC Genomics 19:23. https://doi.org/10.1186/s12864-017-4426-7

Ishikawa T, Watanabe N, Nagano M, Kawai-Yamada M, Lam E (2011) Bax inhibitor-1: a highly conserved endoplasmic reticulum-resident cell death suppressor. Cell Death Differ 18:1271–1278

Islam MT, Lee BR, Park SH, La VH, Bae DW, Kim T-H (2017) Cultivar variation in hormonal balance is a significant determinant of disease susceptibility to *Xanthomonas campestris pv. campestris* in *Brassica napus*. Front. Plant Sci 8:2121

Islam MT, Lee BR, Park SH, La VH, Jung WJ, Bae DW (2019) Hormonal regulations in soluble and cell-wall bound phenolic accumulation in two cultivars of *Brassica napus* contrasting susceptibility to *Xanthomonas campestris* pv. *campestris*. Plant Sci 285:132–140

Islam MT, Mamun MA, Lee BR, La VH, Jung WJ, Bae DW, Kim TH (2021) Role of salicylic acid signaling in the biotrophy-necrotrophy transition of *Xanthomonas campestris* pv. *campestris* infection in *Brassica napus*. Physiol Mol. Plant Pathol 113:101578. https://doi.org/10.1016/j.pmpp.2020.101578

Ito S, Yano S, Kishi F, Kameya-Iwaki M, Maehara T, Tanaka S (1997) Cloning of a single-copy DNA sequence unique to *Plasmodiophora brassicae*. Physiol Mol Plant Pathol 50:289–300

Ito S, Ichinose H, Yanagi C, Tanaka S, Kameya-Iwaki M, Kishi F (1999) Identification of an in planta-induced mRNA of *Plasmodiophora brassicae*. J Phytopathol 147:79–82

Jacobs AK, Lipka V, Burton RA, Panstruga R, Strizhov N, Schulze-Lefert P, Fincher GB (2003) An *Arabidopsis* callose synthase, GSL5, is required for wound and papillary callose formation. Plant Cell 156(1):2503–2513

Jammes F, Song C, Shin D, Munemasa S, Takeda K, Gu D, Cho D, Lee S, Giordo R, Sritubtim S (2009) MAP kinases MPK9 and MPK12 are preferentially expressed in guard cells and positively regulate ROS-mediated ABA signaling. Proc Natl Acad Sci U S A 106:20520–20525

Jansson PE, Kenne L, Lindberg B (1975) Structure of extracellular polysaccharide from *Xanthomonas campestris*. Carbohydr Res 45:275–282

Jarosch B, Kogel KH, Schaffrath U (1999) The ambivalence of the barley *Mlo* locus: Mutations conferring resistance against powdery mildew (*Blumeria graminis* f. sp. *hordei*) enhance susceptibility to the rice blast fungus *Magnaporthe grisea*. Mol Plant Microbe Interact 126(1):508–514

Jenner CE, Walsh JA (1996) Pathotypic variation in *turnip mosaic virus* with special reference to European isolates. Plant Pathol 45:848–856

Jenner CE, Tomimura K, Ohshima K, Hughes SL, Walsh JA (2002) Mutations in *Turnip mosaic virus* P3 and cylindrical inclusion proteins are separately required to overcome two *Brassica napus* resistance genes. Virology 300:50–59

Jenner CE, Wang X, Tomimura K, Ohshima K, Ponz F, Walsh JA (2003) The dual role of the potyvirus P3 protein of *Turnip mosaic virus* as a symptom and avirulence determinant in *Brassicas*. Mol Plant Microbe Interact 16:777–784

Jian H, Ma J, Wei L, Liu P, Zhang A, Yang B, Li J, Xu X, Liu L (2018) Integrated mRNA, sRNA, and degradome sequencing reveal oilseed rape complex responses to *Sclerotinia sclerotiorum* (Lib.) infection. Sci Rep 8(1):10987. https://doi.org/10.1038/s41598-018-29365-y

Jian H, Lu K, Yang B, Wang T, Zhang L, Zhang A, Wang J, Liu L, Qu C, Li J (2016) Genome-wide analysis and expression profiling of the SUC and SWEET gene families of sucrose transporters in oilseed rape (*Brassica napus* L). Front Plant Sci 7:1464. https://doi.org/10.3389/fpls.2016.01464

Jiang B, Xu RQ, Li XZ, Wei HY, Bai F, Hu X, He YQ, Tang JL (2006) Construction and characterization of a *hrpG* mutant rendering constitutive expression of *hrp* genes in *Xanthomonas campestris* pv. *campestris*. Prog Nat Sci 16:480–485

Jiang BL, He YQ, Cen WJ, Wei HY, Jiang GF, Jiang W, Hang XH, Feng JX, Lu GT, Tang DH, Tang JL (2008a) The type III secretion effector xopXccN of *Xanthomonas campestris* pv. *campestris* is required for full virulence. Res Microbiol 159:216–220

Jiang BL, Liu J, Chen LF, Ge YY, Hang XH, He YQ, Tang DJ, Lu GT, Tang JL (2008b) DsbB is required for the pathogenesis process of *Xanthomonas campestris* pv. *campestris*. Mol Plant Microbe Interact 21:1036–1045

Jiang W, Jiang BL, Xu RQ, Huang JD, Wei HY, Jiang GF, Cen WJ, Liu J, Ge YY, Li GH, Su LL, Hang XH, Tang DJ, Lu GT, Feng JX, He YQ, Tang JL (2009) Identification of six type III effector genes with the PIP box in *Xanthomonas campestris* pv. *campestris* and five of them contribute individually to full pathogenicity. Mol Plant Microbe Interact 22:1401–1411

Jing M, Guo B, Li H, Yang B, Wang H, Kong G, Zhao Y, Xu H, Wang Y, Ye W (2016) A *Phytophthora sojae* effector suppresses endoplasmic reticulum stress-mediated immunity by stabilizing plant binding immunoglobulin proteins. Nat Commun 7:11685

Johansen IE, Lund OS, Hjulsager CK, Laursen J (2001) Recessive resistance in *Pisum sativum* and potyvirus pathotype resolved in a gene-for-cistron correspondence between host and virus. J Virol 75:6609–6614

Jones JDG, Dangl JL (2006) The plant immune system. Nature 2006:323–329

Jones DR, Ingram DS, Dixon GR (1982) Characterization of isolates derived from single resting spores of *Plasmodiophora brassicae* and studies of their interaction. Plant Pathol 31:239–246. https://doi.org/10.1111/j.1365-3059.1982.tb01274.x

Jost R, Altschmied L, Bloem E, Bogs J, Gershenzon J, Hähnel U, Hänsch R, Hartmann T, Kopriva S, Kruse C (2005) Expression profiling of metabolic genes in response to methyl jasmonate reveals regulation of genes of primary and secondary sulfur-related pathways in *Arabidopsis thaliana*. Photosynth Res 86:491–508

Joubert A, Simoneau P, Campion C, Bataille-Simoneau N, Iacomi-Vasilescu B, Poupard P, Francois JM, Georgeault S, Sellier E, Guillemette T (2011) Impact of the unfolded protein response on the pathogenicity of the necrotrophic fungus *Alternaria brassicicola*. Mol Microbiol 79:1305–1324

Jubault M, Hamon C, Gravot A, Lariagon C, Delourme R, Bouchereau A, Manzanares-Dauleux MJ (2008) Differential regulation of root arginine catabolism and polyamine metabolism in clubroot-susceptible and partially resistant *Arabidopsis* genotypes. Plant Physiol 146:2008–2019. https://doi.org/10.1104/pp.108.117432

Kabbage M, Williams B, Dickman MB (2013) Cell death control: the interplay of apoptosis and autophagy in the pathogenicity of *Sclerotinia sclerotiorum*. PLoS Pathog 9:e1003287

Kadow KJ, Anderson HW (1940) A study of horse radish diseases and their control. Univ Illinois Agril Expt Sta Bull 469:531–543

Kageyama K, Asano T (2009) Life cycle of *Plasmodiophora brassicae*. J Plant Growth Regul 28(3):203–211

Kamp ME, Liu Y, Kortholt A (2016) Function and regulation of heterotrimeric G proteins during chemotaxis. Intl J Mol Sci 17(1):90. https://doi.org/10.3390/ijms17010090

Karapapa VK, Bainbridge BW, Heale JB (1997) Morphological and molecular characterization of *Verticillium longisporum* comb. nov., pathogenic to oilseed rape. Mycol Res 101:1281–1294

Karisto P, Hund A, Yu K, Anderegg J, Walter A, Mascher F, McDonald BA, Mikaberidze A (2017) Ranking quantitative resistance to *Septoria tritici* blotch in elite wheat cultivars using automated image analysis. Phytopathology 108:568–581

Kaufman J, Milne S, Göbel TW, Walker BA, Jacob JP, Auffray C, Zoorob R, Beck S (1999) The chicken B locus is a minimal essential major histocompatibility complex. Nature 401:923–925. https://doi.org/10.1038/44856

Kaur P, Ricarda Jost R, Sivasithamparam K, Barbetti MJ (2011) Proteome analysis of the *Albugo candida–Brassica juncea* pathosystem reveals that the timing of the expression of defence-related genes is a crucial determinant of pathogenesis. J Exp Bot 62(3):1285–1298. https://doi.org/10.1093/jxb/erq365

Kay S, Boch J, Bonas U (2005) Characterization of *AvrBs3*-like effectors from a Brassicaceae pathogen reveals virulence and avirulence activities and a protein with a novel repeat architecture. Mol Plant Microbe Interact 18:838–848

Keeble JA, Gilmore AP (2007) Apoptosis commitment-translating survival signals into decisions on mitochondria. Cell Res 176(1):976–984

Keeling PJ, Burger G, Durnford DG, Lang BF, Lee RW, Pearlman RE, Roger AJ, Gray MW (2005) The tree of eukaryotes. Trends Ecol Evol 20:670–676

Keen NT (1990) Gene-for-gene complementarity in plant–pathogen interactions. Annu Rev Genet 24:447–463

Keen NT, Williams PH (1969) Translocation of sugars into infected cabbage tissues during clubroot development. Plant Physiol 44:748–754. https://doi.org/10.1104/pp.44.5.748

Kehr J (2003) Single cell technology. Curr Opin Plant Biol 66(1):617–621

Khandelwal A, Elvitigala T, Ghosh B, Quatrano RS (2008) *Arabidopsis* transcriptome reveals control circuits regulating redox homeostasis and the role of an AP2 transcription factor. Plant Physiol 148:2050–2058

Khokon MAR, Salam MA, Jammes F, Ye W, Hossain MA, Uraji M, Nakamura Y, Mori IC, Kwak JM, Murata Y (2015) Two guard cell mitogen-activated protein kinases, *MPK9* and *MPK12*, function in methyl jasmonate-induced stomatal closure in *Arabidopsis thaliana*. Plant Biol (Stuttg.) 17:946–952

Kim MC, Panstruga R, Elliott C, Müller J, Devoto A, Yoon HW, Park HC, Cho MJ, Schulze-Lefert P (2002) Calmodulin interacts with MLO protein to regulate defence against mildew in barley. Nature 4166(1):447–451

Kim TG, Kim HM, Lee HJ, Shin YJ, Kwon TH, Lee NJ, Jang YS, Yang MS (2007) Reduced protease activity in transformed rice cell suspension cultures expressing a proteinase inhibition. Prot Expres Purif 53:270–274

Kim KS, Min JY, Dickman MB (2008) Oxalic acid is an elicitor of plant programmed cell death during *Sclerotinia sclerotiorum* disease development. Mol Plant Microbe Interact 21:605–612

King SRF, McLellan H, Boevink PC, Armstrong MR, Bukharova T, Sukarta O, Win J, Kamoun S, Birch PRJ, Banfield MJ (2014) *Phytophthora infestans* RXLR effector PexRD2 interacts with host MAPKKKε to suppress plant immune signaling. Plant Cell 26:1345–1359

Kleemann J, Rincon-Rivera LJ, Takahara H, Neumann U, van Themaat EVL, van der Does HC, Hacquard S, Stüber K, Will I, Schmalenbach W, Schmelzer E, O'Connell RJ (2012) Sequential delivery of host-induced virulence effectors by appressoria and intracellular hyphae of the phytopathogen *Colletotrichum higginsianum*. PLoS Pathog 8:e1002643

Klein PG, Klein RR, Rodrıguez-Cerezo E, Hunt AG, Shaw JG (1994) Mutational analysis of the tobacco vein mottling virus genome. Virology 204:759–769

Kobelt P (2000) Die Verbreitung von sekundären Plasmodien von *Plasmodiophora brassicae* (Wor.) im Wurzelgewebe von *Arabidopsis thaliana* nach immunhistologischer Markierung des plasmodialen Zytoskeletts. Freie Universitat Berlin, dissertation, Berlin, Germany

Kobelt P, Siemens J, Sacristan MD (2000) Histological characterization of the incompatible interaction between *Arabidopsis thaliana* and the obligate biotrophic pathogen *Plasmodiophora brassicae*. Mycol Res 104:220–225. https://doi.org/10.1017/S0953756299001781

Koebnik R, Kruger A, Thieme F, Urban A, Bonas U (2006) Specific binding of the *Xanthomonas campestris* pv. *vesicatoria AraC*-type transcriptional activator HrpX to plant-inducible promoter boxes. J Bacteriol 188(21):7652–7660. https://doi.org/10.1128/JB.00795-06

Koeck M, Hardham AR, Dodds PN (2011) The role of effectors of biotrophic and hemibiotrophic fungi in infection. Cell Microbiol 13(12):1849–1857

Koh S, Andre A, Edwards H, Ehrhardt D, Somerville S (2005) *Arabidopsis thaliana* subcellular responses to compatible *Erysiphe cichoracearum* infections. Plant J 446(1):516–529

Kong G, Zhao Y, Jing M, Huang J, Yang J, Xia Y, Kong L, Ye W, Xiong Q, Qiao Y, Dong S, Ma W, Wang Y (2015) The activation of *Phytophthora* effector *Avr3b* by plant cyclophilin is required for the nudix hydrolase activity of *Avr3b*. PLoS Pathog 11:e1005139

Koonin EV, Makarova KS, Wolf YI (2017) Evolutionary genomics of defense systems in archaea and bacteria. Annu Rev Microbiol 71:233–261. https://doi.org/10.1146/annurev-micro-090816-093830

Kovalev N, Nagy PD (2013) Cyclophilin A binds to the viral RNA and replication proteins, resulting in inhibition of tombus viral replicase assembly. J Virol 87:13330–13342

Kratochwil CF, Meyer A (2015) Closing the genotype–phenotype gap: emerging technologies for evolutionary genetics in ecological model vertebrate systems. Bioessays 37:213–226. https://doi.org/10.1002/bies.201400142

Krattinger SG, Lagudah ES, Spielmeyer W, Singh RP, Huerta-Espino J, McFadden H, Bossolini E, Selter LL, Keller B (2009) A putative ABC transporter confers durable resistance to multiple fungal pathogens in wheat. Science 323:1360–1363

Krishnan A, Dnyansagar R, Almen MS, Williams MJ, Fredriksson R, Manoj N, Schioth HB (2014) The GPCR repertoire in the demosponge *Amphimedon queenslandica*: insights into the GPCR system at the early divergence of animals. BMC Evol Biol 14(1):270. https://doi.org/10.1186/s12862-014-0270-4

Kromina KA, Ignatov AN, Abdeeva IA, Adhikari TB (2008) Role of peptidyl-prolyl *cis/tra*ns isomerases in pathological processes. Memory Cell Biol 2:195–202

Kruse C, Jost R, Lipschis M, Kopp B, Hartmann M, Hell R (2007) Sulfur-enhanced defence: effects of sulfur metabolism, nitrogen supply, and pathogen lifestyle. Plant Biol 9:608–619

Kubicek CP, Schreferl-Kunar G, Wohrer W, Rohr M (1988) Evidence for a cytoplasmic pathway of oxalate biosynthesis in *Aspergillus niger*. Appl Environ Microbiol 54:633–637

Kuhn H, Kwaaitaal M, Kusch S, Acevedo-Garcia J, Wu Hanstruga R (2016) Biotrophy at its best: novel findings and unsolved mysteries of the *Arabidopsis*-powdery mildew pathosystem. Arabidopsis Book 30(14):e0184

Kumar J, Hückelhoven R, Beckhove U, Nagarajan S, Kogel KH (2001) A compromised *Mlo* pathway affects the response of barley to the necrotrophic fungus *Bipolaris sorokiniana* (teleomorph: *Cochliobolus sativus*) and its toxins. Phytopathology 91(2):127–133

Kumar SS, Tandberg JI, Penesyan A, Elbourne LDH, Suarez-Bosche N, Don E, Skadberg E, Fenaroli F, Cole N, Winther-Larsen HC, Paulsen IT (2018) Dual transcriptomics of host-pathogen interaction of cystic fibrosis isolate *Pseudomonas aeruginosa* PASS1 with Zebrafish. Front Cell Infect Microbiol 8:406. https://doi.org/10.3389/fcimb.2018.00406

Kwon C, Neu C, Pajonk S, Yun HS, Lipka U, Humphry M, Bau S, Straus M, Kwaaitaal M, Rampelt H, El Kasmi F, Jürgens G, Parker J, Panstruga R, Lipka V, Schulze-Lefert P (2008) Co-option of a default secretory pathway for plant immune responses. Nature 451:835–840

Lakra BS, Saharan GS (1988a) Morphological and pathological variations in *Albugo candida* associated with *Brassica* species. Indian J Mycol Plant Pathol 18:149–156

Lakra BS, Saharan GS (1988b) Influence of host resistance on colonization and incubation period of *A. candida* in mustard. Cruciferae NewsLett 13:108–109

Laloi C, Stachowiak M, Pers-Kamczyc E, Warzych E, Murgia I, Apel K (2007) Cross-talk between singlet oxygen- and hydrogen peroxide-dependent signaling of stress responses in *Arabidopsis thaliana*. Proc Natl Acad Sci U S A 104:672–677

Lamb C, Dixon R (1997) The oxidative burst in plant disease resistance. Annu Rev Plant Physiol Plant Mol Biol 48:251–275

Lammens T, Boudolf V, Kheibarshekan L, Zalmas LP, Gaamouche T, Maes S, Vanstraelen M, Kondorosi E, La Thangue NB, Govaerts W, Inzé D, De Veylder L (2008) Atypical E2F activity restrains APC/CCCS52A2 function obligatory for endocycle onset. Proc Natl Acad Sci U S A 105:14721–14726

Lapin D, Van den Ackerveken G (2013) Susceptibility to plant disease: more than a failure of host immunity. Trends Plant Sci 18:546–554

Larkan NJ, Lydiate DJ, Parkin IAP, Nelson MN, Epp DJ, Cowling WA, Rimmer SR, Borhan MH (2013) The *Brassica napus* blackleg resistance gene *LepR3* encodes a receptor-like protein triggered by the *Leptosphaeria maculans* effector *AVRLM1*. New Phytol 197:595–605. https://doi.org/10.1111/nph.12043

Larkan NJ, Ma L, Borhan MH (2015) The *Brassica napus* receptor-like protein RLM2 is encoded by a second allele of the LepR3/Rlm2 blackleg resistance locus. Plant Biotechnol J 13:983–992

Larkan NJ, Yu F, Lydiate DJ, Rimmer SR, Borhan MH (2016) Single R gene introgression lines for accurate dissection of the *Brassica* - *Leptosphaeria* pathosystem. Front Plant Sci 7:1771

Lee TG, Diers BW, Hudson ME (2016) An efficient method for measuring copy number variation applied to improvement of nematode resistance in soybean. Plant J 88:143–153

Lemarie S, Robert-Seilaniantz A, Lariagon C, Lemoine J, Marnet N, Jubault M, Manzanares-Dauleux MJ, Gravot A (2015) Both the jasmonic acid and the salicylic acid pathways contribute to resistance to the biotrophic clubroot agent *Plasmodiophora brassicae* in *Arabidopsis*. Plant Cell Physiol 56(11):2158–2168

Lemoine R, Camera SL, Atanassova R, Dedaldechamp F, Allario T, Pourtau N, Bonnemain JL, Laloi M, Coutos-Thévenot P, Maurousset L, Faucher M, Girousse C, Lemonnier P, Parrilla J, Durand M (2013) Source-to-sink transport of sugar and regulation by environmental factors. Front Plant Sci 4:272. https://doi.org/10.3389/fpls.2013.00272

Li RG, Rimmer R, Buchwaldt L, Sharpe AG, Senguin-Swartz G, Hegedus DD (2004a) Interaction of *Sclerotinia sclerotiorum* with *Brassica napus* cloning and characterization of endo and exo-polygalacturonases expressed during saprophytic and parasitic modes. Fungal Genet Biol 41:754–765

Li RG, Rimmer R, Buchwaldt L, Sharpe A, Seguin-Swart G, Coutu C, Hegedus D (2004b) Interaction of *Sclerotinia sclerotiorum* with a resistant *Brassica napus* cultivar: Expressed sequence tag analysis identifies genes associated with fungal pathogenesis. Fungal Genet Biol 41(8):735–753. https://doi.org/10.1016/j.fgb.2004.03.001

Li Y, Zhang Q, Zhang J, Wu L, Qi Y, Zhou JM (2010) Identification of micro RNAs involved in pathogen-associated molecular pattern-triggered plant innate immunity. Plant Physiol 152: 2222–2231

Li F, Pignatta D, Bendix C, Brunkard JO, Cohn MM, Tung J, Sun H, Kumar P, Baker B (2012a) MicroRNA regulation of plant innate immune receptors. Proc Natl Acad Sci U S A 109(5): 1790–1795. https://doi.org/10.1073/pnas.1118282109

Li M, Liang X, Rollins JA (2012b) *Sclerotinia sclerotiorum* γ-glutamyl transpeptidase (Ss-Ggt1) is required for regulating glutathione accumulation and development of sclerotia and compound appressoria. Mol Plant Microbe Interact 25:412–420

Li RF, Lu GT, Li L, Su HZ, Feng GF, Chen Y, He YQ, Jiang BL, Tang DJ, Tang JL (2014) Identification of a putative cognate sensor kinase for the two-component response regulator HrpG, a key regulator controlling the expression of the *hrp* genes in *Xanthomonas campestris* pv. *campestris*. Environ Microbiol 16(7):2053–2071

Li L, Ye C, Zhao R, Li X, Liu WZ, Wu F, Yan J, Jiang YQ, Yang B (2015a) Mitogen-activated protein kinase kinase kinase (MAPKKK) 4 from rapeseed (*Brassica napus* L.) is a novel member inducing ROS accumulation and cell death. Biochem Biophys Res Commun 467: 792–797

Li XL, Guo M, Xu DF, Chen FX, Zhang HJ, Pan Y, Li MM, Gao Z (2015b) The nascent-polypeptide associated complex alpha subunit regulates the polygalacturonases expression negatively and influences the pathogenicity of *Sclerotinia sclerotiorum*. Mycologia 107: 1130–1137

Li H, Li X, Xuan Y, Jiang J, Wei Y, Piao Z (2018) Genome wide identification and expression profiling of *SWEET* genes family reveals its role during *Plasmodiophora brassicae*-induced formation of clubroot in *Brassica rapa*. Front Plant Sci 9:207. https://doi.org/10.3389/fpls.2018.00207

Liang X, Rollins JA (2018) Mechanisms of broad host range necrotrophic pathogenesis in *Sclerotinia sclerotiorum*. Phytopathology 108:1128–1140. https://doi.org/10.1094/PHYTO-06-18-0197-RVW

Liang Y, Yajima W, Davis MR, Kav NNV, Strelkov SE (2013) Disruption of a gene encoding a hypothetical secreted protein from *Sclerotinia sclerotiorum* reduces its virulence on canola (*Brassica napus*). Can J Plant Pathol 35:46–55

Liang X, Liberti D, Li M, Kim YT, Hutchens A, Wilson R, Rollins JA (2015) Oxaloacetate acetylhydrolase gene mutants of *Sclerotinia sclerotiorum* do not accumulate oxalic acid, but do produce limited lesions on host plants. Mol Plant Pathol 16(6):559–571. https://doi.org/10.1111/mpp.12211

Liang YUE, Xiong WEI, Steinkellner S, Feng JIE (2017) Deficiency of the melanin biosynthesis genes *SCD1* and *THR1* affects sclerotial development and vegetative growth, but not pathogenicity, in *Sclerotinia sclerotiorum*. Mol Plant Pathol 19:1–10. https://doi.org/10.1111/mpp.12627

Liberti D, Rollins JA, Dobinson KF (2013) Peroxysomal carnitine acetyle transferase influences host colonization capacity in *Sclerotinia sclerotiorum*. Mol Plant Microbe Interact 26:768–780

Libertucci J, Young VB (2019) The role of the microbiota in infectious diseases. Nat Microbiol 4: 35–45. https://doi.org/10.1038/s41564-018-0278-4

Lin JY, Mendu V, Pogany J, Qin J, Nagy PD (2012) The TPR domain in the host Cyp40-like cyclophilin binds to the viral replication protein and inhibits the assembly of the tombus viral replicase. PLoS Pathog 8:e1002491

Lin IW, Sosso D, Chen LQ, Gase K, Kim SG, Kessler D, Klinkenberg PM, Gorder MK, Hou BH, Qu XQ, Carter CJ, Baldwin IT, Frommer WB (2014) Nectar secretion requires sucrose phosphate synthases and the sugar transporter *SWEET9*. Nature 508:546–549. https://doi.org/10.1038/nature13082

Linden SK, Sutton P, Karlsson NG, Korolik V, McGuckin MA (2008a) Mucins in the mucosal barrier to infection. Mucosal Immun 1:183–197. https://doi.org/10.1038/mi.2008.5

Linden S, Mahdavi J, Semino-Mora C, Olsen C, Carlstedt I, Borén T, Dubois A (2008b) Role of ABO secretor status in mucosal innate immunity and *Helicobacter pylori* infection. PLoS Pathog 4:e2. https://doi.org/10.1371/journal.ppat.0040002

Lindgren PB (1997) The role of *hrp* genes during plant–bacterial interactions. Annu Rev Phytopathol 35:129–152

Lipka V, Dittgen J, Bednarek P, Bhat R, Wiermer M, Stein M, Landtag J, Brandt W, Rosahl S, Scheel D, Llorente F, Molina A, Parker J, Somerville S, Lefert PS (2005) Pre- and post-invasion defenses both contribute to nonhost resistance in *Arabidopsis*. Science 310:1180–1183

Lipka U, Fuchs R, Lipka V (2008) *Arabidopsis* non-host resistance to powdery mildews. Curr Opin Plant Biol 11:404–411

Lippok B, Birkenbihl RP, Rivory G, Brümmer J, Schmelzer E, Logemann E, Somssich IE (2007) Expression of AtWRKY33 encoding a pathogen-/PAMP-responsive WRKY transcription factor is regulated by a composite DNA motif containing W box elements. Mol Plant Microbe Interact 20:420–429

Liu JQ, Rimmer SR (1990) Effect of host .genotype, inoculum concentration, and incubation temperature on white rust development in oilseed rape. Can J Plant Pathol 12:389–392

Liu JQ, Rimmer SR, Scarth R (1989) Histopathology of compatibility and incompatibility between oilseed rape and *Albugo candida*. Plant Pathol 38:176–182

Liu G, Holub EB, Alonso JM, Ecker JR, Fobert PR (2005) An *Arabidopsis* NPR1-like gene, NPR4, is required for disease resistance. Plant J Cell Mol Biol 41:304–318

Liu HH, Tian X, Li YJ, Wu CA, Zheng CC (2008) Microarray-based analysis of stress-regulated microRNAs in *Arabidopsis thaliana*. RNA 14:836–843

Liu Q, Yuan M, Zhou Y, Li X, Xiao J, Wang S (2011) A paralog of the MtN3/saliva family recessively confers race-specific resistance to *Xanthomonas oryzae* in rice. Plant Cell Environ 34:1958–1969. https://doi.org/10.1111/j.1365-3040.2011.02391.x

Liu L, Wang Q, Zhang X, Liu J, Zhang Y, Pan H (2018a) Ssams2, a gene encoding GATA transcription factor, is required for appressoria formation and chromosome segregation in *Sclerotinia sclerotiorum*. Front Microbiol 9:3031. https://doi.org/10.3389/fmicb.2018.03031

Liu Y, Xu A, Liang F, Yao X, Wang Y, Liu X, Zhang Y, Dalelhan J, Zhang B, Qin M, Huang Z, Shaolin L (2018b) Screening of clubroot-resistant varieties and transfer of clubroot resistance genes to *Brassica napus* using distant hybridization. Breed Sci 68:258–267

Loake G, Grant M (2007) Salicylic acid in plant defense: the players and protagonists. Curr Opin Plant Biol 10:466–472

Lopez-Maury L, Marguerat S, Bahler J (2008) Tuning gene expression to changing environments: from rapid responses to evolutionary adaptation. Nat Rev Genet 9:583–593

Lowe RGT, Cassin A, Grandaubert J, Clark BL, Van de Wouw AP, Rouxel T, Howlett BJ (2014) Genomes and transcriptomes of partners in plant-fungal-interactions between canola (*Brassica napus*) and two *Leptosphaeria* species. PLoS One 9(7):e103098. https://doi.org/10.1371/journal.pone.0103098

Lu H, Patil P, Van Sluys MA, White FF, Ryan RP, Dow JM, Rabinowicz P, Salzberg SL, Leach JE, Sonti R, Brendel V, Bogdanove AJ (2008) Acquisition and evolution of plant pathogenesis-associated gene clusters and candidate determinants of tissue-specificity in *Xanthomonas*. PLoS One 3:e3828

Lu YJ, Schornack S, Spallek T, Geldner N, Chory J, Schellmann S, Schumacher K, Kamoun S, Robatzek S (2012) Patterns of plant subcellular responses to successful oomycete infections reveal differences in host cell reprogramming and endocytic trafficking. Cell Microbiol 14:682–697

Lucas WJ, Gilbertson RL (1994) Plasmodesmata in relation to viral movement within leaf tissues. Annu Rev Phytopathol 32:387–411

Ludwig-Müller J, Schuller A (2008) What can we learn from clubroots: alterations in host roots and hormone homeostasis caused by *Plasmodiophora brassicae*. Eur J Plant Pathol 121:291–302. https://doi.org/10.1007/s10658-007-9237-2

Ludwig-Müller J, Bennett R, Kiddle G, Ihmig S, Ruppel M, Hilgenberg W (1999) The host range of *Plasmodiophora brassicae* and its relationship to endogenous glucosinolate content. New Phytol 141:443–458. https://doi.org/10.1046/j.1469-8137.1999.00368.x

Ludwig-Muller J, Julke S, Bierfreund NM, Decker EL, Reski R (2009) Moss (*Physcomitrella patens*) GH3 proteins act in auxin homeostasis. New Phytol 181:323–338

Ludwig-Muller J, Julken S, Geib K, Richter F, Mithofer A, Sola I, Rusak G, Keenan S, Bulman S (2015) A novel methyltransferase from the intracellular pathogen *Plasmodiophora brassicae* methylates salicylic acid. Mol Plant Pathol 16(4):349–364

Lumsden RD (1976) Pectolytic enzymes of *Sclerotinia sclerotiorum* and their localization in infected bean. Can J Bot 54:2630–2641

Lumsden RD (1979) Histology and physiology of pathogenesis in plant disease caused by *Sclerotinia* species. Phytopathology 69:890–896

Luo Y, Wang Y, Zu F, Fu M, Zhao K, Zhang Y, Wang J, Tian Z, Chen W, Li J, Yuan X, Li G (2019) Transcriptome analysis of *Brassica napus-Plasmodiophora brassicae* interaction during early infection. Chin J Oil Crop Sci 41(3):421–434. https://doi.org/10.7505/j.issn.1007-9084.2019.03.015

Lyu X, Shen C, Fu Y, Xie J, Jiang D, Li G, Cheng J (2015) Comparative genomic and transcriptional analyses of the carbohydrate-active enzymes and secretomes of phytopathogenic fungi reveal their significant roles during infection and development. Sci Rep 5:15565. https://doi.org/10.1038/srep15565

Lyu X, Shen C, Fu Y, Xie J, Jiang D, Li G, Cheng J (2016) A small secreted virulence-related protein is essential for the necrotrophic interactions of *Sclerotinia sclerotiorum* with its host plants. PLoS Pathog 12(2):e1005435. https://doi.org/10.1371/journal.ppat.1005435

Ma L, Borhan H (2015) The receptor-like kinase SOBIR1 interacts with *Brassica napus LepR3* and is required for *Leptosphaeria maculans AvrLm1*-triggered immunity. Front Plant Sci 6:933

Ma L, Djavaheri M, Wang H, Larkan NJ, Haddadi P, Beynon E, Gropp G, Borhan MH (2018) *Leptosphaeria maculans* effector protein avrlm1 modulates plant immunity by enhancing MAP kinase 9 phosphorylation. iScience 3:177–191. https://doi.org/10.1016/j.isci.2018.04.015

Mach JM, Castillo AR, Hoogstraten R, Greenberg JT (2001) The *Arabidopsis*-accelerated cell death gene *ACD2* encodes red chlorophyll catabolite reductase and suppresses the spread of disease symptoms. Proc Natl Acad Sci U S A 98:771–776

Macioszek VK, Gapińska M, Zmienko A, Sobczak M, Skoczowski A, Oliwa J, Kononowicz AK (2020) Complexity of *Brassica oleracea–Alternaria brassicicola* susceptible interaction reveals down regulation of photosynthesis at ultra structural, transcriptional, and physiological levels. Cells 9(10):2329. https://doi.org/10.3390/cells9102329

Madsen L, Seeger M, Semple CA, Hartmann-Petersen R (2009) New ATPase regulators—p97 goes to the PUB. Int J Biochem Cell Biol 41:2380–2388

Maheshwari DK, Chaturvedi SN, Yadav BS (1985a) Histochemical studies on hypertrophied inflorescence axis of *Brassica juncea* due to *Albugo candida*. Indian Phytopathol 38:263–266

Maheshwari DK, Chaturvedi SN, Yadav BS (1985b) Structure and development of galls induced by *Albugo* in inflorescence axis of *Brassica juncea*. Indian Phytopathol 38:546–548

Malinowski R, Smith JA, Fleming AJ, Scholes JD, Rolfe SA (2012) Gall formation in clubroot-infected *Arabidopsis* results from an increase in existing meristematic activities of the host but is not essential for the completion of the pathogen life cycle. Plant J 71(2):226–238. https://doi.org/10.1111/j.1365-313X.2012.04983.x

Mallory AC, Bartel DP, Bartel B (2005) MicroRNA-directed regulation of *Arabidopsis* AUXIN RESPONSE FACTOR17 is essential for proper development and modulates expression of early auxin response genes. Plant Cell 17:1360–1375

Mamun MA, Islam MT, Lee BR, La VH, Bae DW, Kim TH (2020) Genotypic variation in resistance gene-mediated calcium signaling and hormonal signaling involved in effector-triggered immunity or disease susceptibility in the *Xanthomonas campestris* pv. *campestris–Brassica napus* pathosystem. Plants (Basel) 9(3):303. https://doi.org/10.3390/plants9030303

Manck-Götzenberger J, Requena N (2016) Arbuscular mycorrhiza symbiosis induces a major transcriptional reprogramming of the potato SWEET sugar transporter family. Front Plant Sci 7:487. https://doi.org/10.3389/fpls.2016.00487

Marciano P, Di Lenna P, Magro P (1983) Oxalic acid, cell wall-degrading enzymes and pH in pathogenesis and their significance in the virulence of two *Sclerotinia sclerotiorum* isolates on sunflower. Physiol Mol Plant Pathol 22:339–345

Marrs KA (1996) The functions and regulation of glutathione *S*-transferases in plants. Annu Rev Plant Biol 47:127–158

Marti M, Good RT, Rug M, Knuepfer E, Cowman AF (2004) Targeting malaria virulence and remodeling proteins to the host erythrocyte. Science 306:1930–1933

Maxwell DP (1973) Oxalate formation in *Whetzelinia sclerotiorum* by oxaloacetate acetyl hydrolase. Physiol Plant Pathol 3:279–288

Maxwell DP, Lumsden RD (1970) Oxalic acid production by *Sclerotinia sclerotiorum* in infected bean and in culture. Phytopathology 60:1395–1398

McCartney A, Heran A, Li Q (2001a) Infection of oilseed rape (*Brassica napus*) by petals containing ascospores of *Sclerotinia sclerotiorum*. In: Young CS, Hughes KJD (eds) Proc *Sclerotinia* 2001, the XI International *Sclerotinia* workshop, York, 8–12 July 2001, Central Science Laboratory, York, England, pp 1987–1090

McCartney A, Heran A, Li Q, Freeman J (2001b) Petal fall, petal retention and petal duration in oilseed rape crops. In: Young CS, Hughes KJD (eds) Proc *Sclerotinia* 2001, the XI International *Sclerotinia* workshop, York, 8–12 July 2001, Central Science Laboratory, York, England, pp 185–186

Mendgen K, Hahn M (2002) Plant infection and the establishment of fungal biotrophy. Trends Plant Sci 76(1):352–357

Merits A, Guo DY, Jarvekulg L, Saarma M (1999) Biochemical and genetic evidence for interactions between potato A potyvirus-encoded proteins P1 and P3 and proteins of the putative replication complex. Virology 263:15–22

Meyer-Lucht Y, Sommer S (2005) MHC diversity and the association to nematode parasitism in the yellow-necked mouse (*Apodemus flavicollis*). Mol Ecol 14:2233–2243. https://doi.org/10.1111/j.1365-294X.2005.02557.x

Micali C, Göllner K, Humphry M, Consonni C, Panstruga R (2008) The powdery mildew disease of Arabidopsis: a paradigm for the interaction between plants and biotrophic fungi. Arabidopsis Book 6:e0115. https://doi.org/10.1199/tab.0115

Micali CO, Neumann U, Grunewald D, Panstruga R, O'Connell R (2011) Biogenesis of a specialized plant-fungal interface during host cell internalization of *Golovinomyces orontii* haustoria. Cell Microbiol 13:210–226

Miller G, Suzuki N, Rizhsky L, Hegie A, Koussevitzky S, Mittler R (2007) Double mutants deficient in cytosolic and thylakoid ascorbate peroxidase reveal a complex mode of interaction between reactive oxygen species, plant development, and response to abiotic stresses. Plant Physiol 144:1777–1785

Mims CW, Richardson EA, Holt Iii BF, Dangl JL (2004) Ultrastructure of the host–pathogen interface in *Arabidopsis thaliana* leaves infected by the downy mildew *Hyaloperonospora parasitica*. Can J Bot 82(7):1001–1008

Mittler R, Vanderauwera S, Gollery M, Van Breusegem F (2004) Reactive oxygen gene network of plants. Trends Plant Sci 9:490–498

Miya A, Albert P, Shinya T, Desaki Y, Ichimura K, Shirasu K, Narusaka Y, Kawakami N, Kaku H, Shibuya N (2007) *CERK1*, a LysM receptor kinase, is essential for chitin elicitor signaling in Arabidopsis. Proc Natl Acad Sci U S A 1046(1):19613–19618

Moro A, Ruiz-Cabello F, Fernandez-Cano A, Stock RP, Gonzalez A (1995) Secretion by *Trypanosoma cruzi* of a peptidyl-prolyl cis-transisomerase involved in cell infection. EMBO J 14:2483–2490

Morrall RAA, Duczek IJ, Sheard JW (1972) Variation and correlation within and between morphology, pathogenicity and pectolytic enzyme activity in *Sclerotinia* from Saskatchewan. Can J Bot 50:767–786

Moscou MJ, Lauter N, Steffenson B, Wise RP (2011) Quantitative and qualitative stem rust resistance factors in barley are associated with transcriptional suppression of defense regulons. PLoS Genet 7:e1002208

Motohashi K, Koyama F, Nakanishi Y, Ueoka-Nakanishi H, Hisabori T (2003) Chloroplast cyclophilin is a target protein of thioredoxin. J Biol Chem 278:31848–31852

Mou Z, Fan W, Dong X (2003) Inducers of plant systemic acquired resistance regulates *NPR1* function through redox changes. Cell 113:1–10

Mourrain P, Béclin C, Elmayan T, Feuerbach F, Godon C, Morel JB, Jouette D, Lacombe AM, Nikic S, Picault N, Rémoué K, Sanial M, Vo TA, Vaucheret H (2000) *Arabidopsis SGS2* and *SGS3* genes are required for post transcriptional gene silencing and natural virus resistance. Cell 26:533–542. https://doi.org/10.1016/S0092-8674(00)80863-6

Mutka AM, Fentress SJ, Sher JW, Berry JC, Pretz C, Nusinow DA, Bart R (2016) Quantitative, image-based phenotyping methods provide insight into spatial and temporal dimensions of plant disease. Plant Physiol 172:00984

Näpflin K, O'Connor EA, Becks L, Bensch S, Ellis VA, Hafer-Hahmann N, Harding KC, Lindén SK, Olsen MT, Roved J, Sackton TB, Shultz AJ, Venkatakrishnan V, Videvall E, Westerdahl H, Winternitz JC, Edwards SV (2019) Genomics of host-pathogen interactions: challenges and opportunities across ecological and spatiotemporal scales. Peer J 7:e8013. https://doi.org/10.7717/peerj.8013

Napper ME (1933) Observations on spore germination and specialization of parasitism in *Cystopus candidus*. J Pomol Horti Sci 11:81–100

Narusaka M, Shirasu K, Noutoshi Y, Kubo Y, Shiraishi T, Iwabuchi M et al (2009) *RRS1* and *RPS4* provide a dual resistance-gene system against fungal and bacterial pathogens. Plant J 60:218–226. https://doi.org/10.1111/j.1365-313X.2009.03949.x

Nelson JM, Hauser DA, Hinson R, Shaw AJ (2018) A novel experimental system using the liverwort *Marchantia polymorpha* and its fungal endophytes reveals diverse and context-dependent effects. New Phytol 218:1217–1232

Nemri A, Saunders DGO, Anderson C, Upadhyaya NM, Win J, Lawrence GJ, Jones DA, Kamoun S, Ellis JG, Dodds PN (2014) The genome sequence and effector complement of the flax rust pathogen *Melampsora lini*. Front Plant, Sci 5:98

Newton HK (1972) *Sclerotinia sclerotiorum* an incitant of lettuce drop: sources of inoculum, host resistance and mechanism of pathogenesis. PhD thesis, Murofolm 72-V.292, Univ Murofilm Inc., Ann Arbor, MI, 135 p

Newton HK, Sequeira L (1972) Ascospores as the primary infective propagules of *Sclerotinia sclerotiorum* in Wisconsin. Plant Dis Rep 56:798–802

Niks RE, Marcel TC (2009) Nonhost and basal resistance: how to explain specificity? New Phytol 182:817–828

Nishimura MT, Stein M, Hou BH, Vogel JP, Edwards H, Somerville SC (2003) Loss of a callose synthase results in salicylic acid-dependent disease resistance. Science 301(5635):969–972

Noctor G, Foyer CH (1998) Ascorbate and glutathione: keeping active oxygen under control. Annu Rev Plant Physiol Plant Mol Biol 49:249–279

Noji M, Saito M, Nakamura M, Aono M, Saji H, Saito K (2001) Cysteine synthase overexpression in tobacco confers tolerance to sulfur-containing environment pollutants. Plant Physiol 126: 973–980

Nováková M, Šašek V, Trdá L, Krutinová H, Mongin T, Valentová O, Balesdent M-H, Rouxel T, Burketová L (2016) *Leptosphaeria maculans* effector *AvrLm4-7* affects salicylic acid (SA) and ethylene (ET) signalling and hydrogen peroxide (H2O2) accumulation in *Brassica napus*. Mol Plant Pathol 17:818–831

O'Connell RJ, Herbert C, Sreenivasaprasad S, Khatib M, Esquerré-Tugayé M-T, Dumas B (2004) A novel *Arabidopsis-Colletotrichum* pathosystem for the molecular dissection of plant-fungal interactions. Mol Plant Microbe Interact 17:272–282

O'Connell RJ, Thon MR, Hacquard S, Amyotte SG, Kleemann J, Torres MF, Damm U, Buiate EA, Epstein L, Alkan N, Altmüller J, Alvarado-Balderrama L, Bauser CA, Becker C, Birren BW, Chen Z, Choi J, Crouch JA, Duvick JP, Farman MA, Gan P, Heiman D, Henrissat B, Howard RJ, Kabbage M, Koch C, Kracher B, Kubo Y, Law AD, Lebrun MH, Lee YH, Miyara I, Moore N, Neumann U, Nordström K, Panaccione DG, Panstruga R, Place M, Proctor RH, Prusky D, Rech G, Reinhardt R, Rollins JA, Rounsley S, Schardl CL, Schwartz DC, Shenoy N, Shirasu K, Sikhakolli UR, Stüber K, Sukno SA, Sweigard JA, Takano Y, Takahara H, Trail F, van der Does HC, Voll LM, Will I, Young S, Zeng Q, Zhang J, Zhou S, Dickman MB, Schulze-Lefert P, Loren V, van Themaat E, Ma LJ, Vaillancourt LJ (2012) Lifestyle transitions in plant pathogenic *Colletotrichum* fungi deciphered by genome and transcriptome analyses. Nat Genet 44(9):1060–1065. https://doi.org/10.1038/ng.2372

O'Connor EA, Westerdahl H, Burri R, Edwards SV (2019) Avian MHC evolution in the era of genomics: phase 1.0. Cells 8:1152. https://doi.org/10.3390/cells8101152

Ohm RA, Feau N, Henrissat B, Schoch CL, Horwitz BA, Barry KW, Condon BJ, Copeland AC, Dhillon B, Glaser F, Hesse CN, Kosti I, LaButti K, Lindquist EA, Lucas S, Salamov AA, Bradshaw RE, Ciuffetti L, Hamelin RC, Kema GHJ, Lawrence C, Scott JA, Spatafora JW, Turgeon BG, de Wit PJGM, Zhong S, Goodwin SB, Grigoriev IV (2012) Diverse lifestyles and strategies of plant pathogenesis encoded in the genomes of eighteen *Dothideomycetes* fungi. PLoS Pathog 8(12):e1003037. https://doi.org/10.1371/journal.ppat.1003037

Ohshima K, Yamaguchi Y, Hirota R, Hamamoto T, Tomimura K, Tan Z, Sano T, Azuhata F, Walsh JA, Fletcher J, Chen J, Gera A, Gibbs A (2002) Molecular evolution of Turnip mosaic virus; evidence of host adaptation, genetic recombination and geographical spread. J Gen Virol 83: 1511–1521

Oide S, Moeder W, Krasnoff S, Gibson D, Haas H, Yoshioka K, Turgeon BG (2006) *NPS6*, encoding a non-ribosomal peptide synthetase involved in siderophore-mediated iron metabolism, is a conserved virulence determinant of plant pathogenic ascomycetes. Plant Cell 18:2836–2853

Oliver RP, Ipcho SV (2004) *Arabidopsis* pathology breathes new life into the necrotrophs-vs.-biotrophs classification of fungal pathogens. Mol Plant Pathol 5:347–352

Oliver RP, Solomon PS (2010) New developments in pathogenicity and virulence of necrotrophs. Curr Opin Plant Biol 13:415–419

Otulak-Kozieł K, Kozieł E, Bujarski JJ, Frankowska-Łukawska J, Torres MA (2020) Respiratory burst oxidase homologs *RBOHD* and *RBOHF* as key modulating components of response in turnip mosaic virus—*Arabidopsis thaliana* (L) Heyhn system. Int J Mol Sci 21(22):8510. https://doi.org/10.3390/ijms21228510

Ozsolak F, Milos PM (2011) RNA sequencing: advances, challenges and opportunities. Nat Rev Genet 12:87–98

Palma K, Thorgrimsen S, Malinovsky FG, Fiil BK, Nielsen HB, Brodersen P, Hofius D, Petersen M, Mundy J (2010) Autoimmunity in *Arabidopsis acd11* is mediated by epigenetic regulation of an immune receptor. PLoS Pathog 6:e1001137

Pan H, Rollins JA (2017) The GATA-Type IVb zinc-finger transcription factor SsNsd1 regulates asexual-Sexual development and appressoria formation in *Sclerotinia sclerotiorum*. Mol Plant Pathol 9:1679–1689

Pandey SP, Roccaro M, Schön M, Logemann E, Somssich IE (2010) Transcriptional reprogramming regulated by WRKY18 and WRKY40 facilitates powdery mildew infection of *Arabidopsis*. Plant J 64:912–923

Pandey D, Rajendran SRCK, Gaur M, Sajeesh PK, Kumar A (2016) Plant defense signaling and responses against necrotrophic fungal pathogens. J Plant Growth Regul 35:1159–1174. https://doi.org/10.1007/s00344-016-9600-7

Panstruga R (2003) Establishing compatibility between plants and obligate biotrophic pathogens. Curr Opin Plant Biol 66(1):320–326

Panstruga R (2005) Serpentine plant MLO proteins as entry portals for powdery mildew fungi. Biochem Soc Trans 336(1):389–392

Park E, Lee H-Y, Woo J, Choi D, Dinesh-Kumar SP (2017) Spatiotemporal monitoring of *Pseudomonas syringae* effectors via type III secretion using split fluorescent protein fragments. Plant Cell 29:1571–1584

Parker JE, Barber CE, Fan MJ, Daniels MJ (1993) Interaction of *Xanthomonas campestris* with *Arabidopsis thaliana*: characterization of a gene from *X. c.* pv. *raphani* that confers avirulence to most *A. thaliana* accessions. Mol Plant Microbe Interact 6:216–224

Patil PB, Bogdanove AJ, Sonti RV (2007) The role of horizontal transfer in the evolution of a highly variable lipopolysaccharide biosynthesis locus in xanthomonads that infect rice, citrus and crucifers. BMC Evol Biol 7:243

Patrick JW (1989) Solute Efflux from the host at plant-microorganism interfaces. Funct Plant Biol 16:53–67. https://doi.org/10.1071/PP9890053

Peng Y, Chen L, Lu Y, Wu Y, Dumenil J, Zhu Z, Bevan MW, Li Y (2015) The ubiquitin receptors DA1, DAR1, and DAR2 redundantly regulate endo-reduplication by modulating the stability of TCP14/15 in *Arabidopsis*. Plant Cell 27:649–662

Penczykowski RM, Laine A-L, Koskella B (2015) Understanding the ecology and evolution of host-parasite interactions across scales. Evol Appl 9:37–52. https://doi.org/10.1111/eva.12294

Peona V, Weissensteiner MH, Suh A (2018) How complete are "complete" genome assemblies?—an avian perspective. Mol Ecol Resour 18:1188–1195. https://doi.org/10.1111/1755-0998.12933

Perchepied L, Balague C, Riou C, Claudel-Renard C, Rivière N, Grezes-Besset B, Roby D (2010) Nitric oxide participates in the complex interplay of defense-related signaling pathways controlling disease resistance to *Sclerotinia sclerotiorum* in *Arabidopsis thaliana*. Mol Plant Microbe Interact 23:846–860

Perez-Lopez E, Waldner M, Hossain M, Kusalik AJ, Wei Y, Bonham-Smith PC, Todd CD (2018) Identification of *Plasmodiophora brassicae* effectors-a challenging goal. Virulence 9(1): 1344–1353

Petersen TN, Brunak S, von Heijne G, Nielsen H (2011) Signal P 4.0: discriminating signal peptides from transmembrane regions. Nat Methods 8:785–786

Pickett CB, Lu AYH (1989) Glutathione *S*-transferases: gene structure, regulation, and biological function. Annu Rev Biochem 58:743–764

Pidskalny RS, Rimmer SR (1985) Virulence of *Albugo candida* from turnip rape (*Brassica campestris*) and mustard (*Brassica juncea)* on various crucifers. Can J Plant Pathol 7:283–286

Pitzschke A, Schikora A, Hirt H (2009) MAPK cascade signalling networks in plant defence. Curr Opin Plant Biol 12:421–426

Poland JA, Bradbury PJ, Buckler ES, Nelson RJ (2011) Genome-wide nested association mapping of quantitative resistance to northern leaf blight in maize. Proc Natl Acad Sci U S A 108:6893–6898

Popescu S, Popescu G, Bachan S, Zhang Z, Gerstein M, Snyder M, Dinesh-Kumar S (2009) MAPK target networks in *Arabidopsis thaliana* revealed using functional protein microarrays. Genes Dev 23:80–92

Poplawsky AR, Chun W (1998) *Xanthomonas campestris* pv. *campestris* requires a functional *pigB* for epiphytic survival and host infection. Mol Plant Microbe Interact 11:466–475

Prabhu Y, Müller R, Anjard C, Noegel AA (2007) *GrlJ*, a dictyostelium GABAB-like receptor with roles in post-aggregation development. BMC Dev Biol 7:44. https://doi.org/10.1186/1471-213X-7-44

Prior GD, Owen JH (1964) Pathological anatomy of *Sclerotinia trifoliorum* on clover and alfalfa. Phytopathology 54:784–787

Provvidenti R (1980) Evaluation of Chinese cabbage cultivars from Japan and the People's Republic of China for resistance to *Turnip mosaic virus* and *Cauliflower mosaic virus*. J Am Soc Hort Sci 105:571–573

Prugnolle F, Manica A, Charpentier M, Guégan JF, Guernier V, Balloux F (2005) Pathogen-driven selection and worldwide HLA Class I diversity. Curr Biol 15:1022–1027. https://doi.org/10.1016/j.cub.2005.04.050

Pruzinska A, Anders I, Aubry S, Schenk N, Tapernoux-Luthi E, Muller T, Krautler B, Hortensteiner S (2007) *In vivo* participation of red chlorophyll catabolite reductase in chlorophyll breakdown. Plant Cell 19:369–387

Purdy LH (1958) Some factors affecting penetration and infection by *Sclerotinia sclerotiorum*. Phytopathology 48:605–609

Qian W, Jia YT, Ren SX, He YQ, Feng JX, Lu LF, Sun QH, Ying G, Tang DJ, Tang H, Wu W, Hao P, Wang LF, Jiang BL, Zeng SY, Gu WY, Lu G, Rong L, Tian YC, Yao ZJ, Fu G, Chen BS, Fang RX, Qiang BQ, Chen Z, Zhao GP, Tang JL, He CZ (2005) Comparative and functional genomic analyses of the pathogenicity of phytopathogen *Xanthomonas campestris* pv. *campestris*. Genome Res 15:757–767

Qiu JL, Fiil BK, Petersen K, Nielsen HB, Botanga CJ, Thorgrimsen S, Palma K, Suarez-Rodriguez MC, Sandbech-Clausen S, Lichota J, Brodersen P, Grasser KD, Mattsson O, Glazebrook J, Mundy J, Petersen M (2008) *Arabidopsis* MAP kinase 4 regulates gene expression through transcription factor release in the nucleus. EMBO J 27:2214–2221

Qiu JD, Huang JH, Liang RP, Lu XQ (2009) Prediction of G-protein-coupled receptor classes based on the concept of Chou's pseudo amino acid composition: an approach from discrete wavelet transform. Anal Biochem 390(1):68–73

Qu X, Yu B, Liu J, Zhang X, Li G, Zhang D, Li L, Wang X, Wang L, Chen J, Mu W, Pan H, Zhang Y (2014) MADS-box transcription factor SsMADS is involved in regulating growth and virulence in *Sclerotinia sclerotiorum*. Int J Mol Sci 15:8049–8062. https://doi.org/10.3390/ijms15058049

Quach H, Rotival M, Pothlichet J, Loh Y-HE, Dannemann M, Zidane N, Laval G, Patin E, Harmant C, Lopez M, Deschamps M, Naffakh N, Duffy D, Coen A, Leroux-Roels G, Clément F, Boland A, Deleuze J-F, Kelso J, Albert ML, Quintana-Murci L (2016) Genetic adaptation and neandertal admixture shaped the immune system of human populations. Cell 167:643–656. https://doi.org/10.1016/j.cell.2016.09.024

Quirino BF, Normanly J, Amasino RM (1999) Diverse range of gene activity during *Arabidopsis thaliana* leaf senescence includes pathogen-independent induction of defense-related genes. Plant Mol Biol 40:267–278. https://doi.org/10.1023/A:1006199932265

Qutob N, Balloux F, Raj T, Liu H, Marionde Procé S, Trowsdale J, Manica A (2011) Signatures of historical demography and pathogen richness on MHC class I genes. Immunogenetics 64:165–175. https://doi.org/10.1007/s00251-011-0576-y

Raabe RD, Pound GS (1952) Morphology and pathogenicity of *A. occidentalis*, the incitant of white rust of spinach. Phytopathology 42:473 (Abstr.)

Rai JN, Dhawan S (1976) Production of polymethyl galacturonase and cellulose and its relationship with virulence in isolates of *Sclerotinia sclerotiorum* (Lib.) de Bary. Indian J Exp Biol 14:197

Rajarammohan S, Pradhan AK, Pental D, Kaur J (2018) Genome-wide association mapping in *Arabidopsis* identifies novel genes underlying quantitative disease resistance to *Alternaria brassicae*. Mol Plant Pathol 19:1719–1732

Rajarammohan S, Paritosh K, Pental D, Kaur J (2019) Comparative genomics of *Alternaria* species provides insights into the pathogenic lifestyle of *Alternaria brassicae*—a pathogen of the Brassicaceae family. BMC Genomics 20:1036. https://doi.org/10.1186/s12864-019-6414-6

Raman H, Raman R, Larkan N (2013) Genetic dissection of blackleg resistance loci in rapeseed (*Brassica napus* L.). In: Andersen SB (ed) Plant breeding from laboratories to fields. Tech Ch. 04

Ramonell KM, Zhang B, Ewing RM, Chen Y, Xu D, Stacey G, Somerville S (2002) Microarray analysis of chitin elicitation in *Arabidopsis thaliana*. Mol Plant Pathol 36(1):301–311

Rancour DM, Park S, Knight SD, Bednarek SY (2004) Plant UBX domain-containing protein 1, *PUX1*, regulates the oligomeric structure and activity of *Arabidopsis* CDC48. J Biol Chem 279:54264–54274

Rao ALN (1999) Molecular basis of symptomatology. In: Mundahar CL (ed) Molecular biology of plant viruses. Kluwer Academic, Boston, MA, pp 201–210

Rawlings ND, Barrett AJ (1993) Evolutionary families of peptidases. Biochem J 290:205–218

Reche PA, Reinherz EL (2003) Sequence variability analysis of human class I and class II MHC molecules: functional and structural correlates of amino acid polymorphisms. J Mol Biol 331: 623–641. https://doi.org/10.1016/s0022-2836(03)00750-2

Reimers K, Choi CY, Mau-Thek E, Vogt PM (2006) Sequence analysis shows that lifeguard belongs to a new evolutionarily conserved cytoprotective family. Int J Mol Med 18:729–734

Reimers K, Choi CY, Bucan V, Vogt PM (2008) The Bax Inhibitor-1 (BI-1) family in apoptosis and tumorigenesis. Curr Mol Med 8:148–156

Reinders A, Sivitz AB, Starker CG, Gantt J, Stephen W, John M (2008) Functional analysis of *LjSUT4*, a vacuolar sucrose transporter from *Lotus japonicus*. Plant Mol Biol 68:289–299. https://doi.org/10.1007/s11103-008-9370-0

Reinders A, Sivitz AB, Ward JM (2011) Evolution of plant sucrose uptake transporters. Front Plant Sci 3:22. https://doi.org/10.3389/fpls.2012.00022

Ren D, Yang KY, Li GJ, Liu Y, Zhang S (2006) Activation of Ntf4, a tobacco mitogen-activated protein kinase, during plant defense response and its involvement in hypersensitive response-like cell death. Plant Physiol 141:1482–1493

Reusche M, Thole K, Janz D, Truskina J, Rindfleisch S, Drübert C, Polle A, Lipka V, Teichmann T (2012) Verticillium infection triggers VASCULAR-RELATED NAC DOMAIN7-dependent de novo xylem formation and enhances drought tolerance in *Arabidopsis*. Plant Cell 24:3823–3837. https://doi.org/10.1105/tpc.112.103374

Revers F, Le Gall O, Candresse T, Maule AJ (1999) New advances in understanding the molecular biology of plant/potyvirus interactions. Mol Plant Microbe Interact 12:367–376

Richardson M, Valdes-Rodriquez S, Blanco-Labra A (1987) A possible function for thaumatin and a TMV-induced protein suggested by homology to a maize inhibitor. Nature 327:432–434

Riechmann JL, Laín S, Garcia JA (1992) Highlights and prospects of potyvirus molecular biology. J Gen Virol 73:1–16

Rigano LA, Payette C, Brouillard G, Marano MR, Abramowicz L, Torres PS, Yun M, Castagnaro AP, El Oirdi M, Dufour V, Malamud F, Dow JM, Bouarab K, Vojnov A (2007) Bacterial cyclic beta-(1,2)-glucanacts in systemic suppression of plant immune responses. Plant Cell 19:2077–2089

Riou C, Freyssinet G, Fevre M (1991) Production of cell wall-degrading enzymes by the phytopathogenic fungus *Sclerotinia sclerotiorum*. Appl Environ Microbiol 57:1478–1484

Roberts W, Selitrennikoff CP (1990) Zeamatin, an antifungal protein from maize with membrane-permeabilizing activity. J Gen Microbiol 136:1771–1778

Robin GP, Kleemann J, Neumann U, Cabre L, Dallery JF, Lapalu N, O'Connell RJ (2018) Subcellular localization screening of *Colletotrichum higginsianum* effector candidates identifies fungal proteins targeted to plant peroxisomes, golgi bodies, and microtubules. Front Plant Sci 9: 562. https://doi.org/10.3389/fpls.2018.00562

Robinson KS, Clements A, Williams AC, Berger CN, Frankel G (2011) Bax inhibitor 1 in apoptosis and disease. Oncogene 30:2391–2400

Robinson J, Halliwell JA, Hayhurst JD, Flicek P, Parham P, Marsh SGE (2015) The IPD and IMGT/HLA database: allele variant databases. Nucleic Acids Res 43:D423–D431. https://doi.org/10.1093/nar/gku1161

Rodoni S, Vicentini F, Schellenberg M, Matile P, Hortensteiner S (1997) Partial purification and characterization of red chlorophyll catabolite reductase, a stroma protein involved in chlorophyll breakdown. Plant Physiol 115:677–682

Rodriguez E, El Ghoul H, Mundy J, Petersen M (2016) Making sense of plant auto-immunity and 'negative regulators'. FEBS J 283:1385–1391

Roetschi A, Si-Ammour A, Belbahri L, Mauch F, Mauch-Mani B (2001) Characterization of an *Arabidopsis-Phytophthora* pathosystem: resistance requires a functional PAD2 gene and is independent of salicylic acid, ethylene and jasmonic acid signalling. Plant J Cell Mol Biol 28: 293–305

Rojas CM, Senthil-Kumar M, Tzin V, Mysore KS (2014) Regulation of primary plant metabolism during plant-pathogen interactions and its contribution to plant defense. Front Plant Sci 5:17. https://doi.org/10.3389/fpls.2014.00017

Rolfe SA, Strelkov SE, Links MG, Clarke WE, Robinson SJ, Djavaheri M, Malinowski R, Haddadi P, Kagale S, Parkin IAP, Taheri A, Borhan MH (2016) The compact genome of the plant pathogen *Plasmodiophora brassicae* is adapted to intracellular interactions with host *Brassica* spp. BMC Genomics 17:272. https://doi.org/10.1186/s12864-016-2597-2

Rollins JA (2003) The *Sclerotinia sclerotiorum* pac1 gene is required for sclerotial development and virulence. Mol Plant Microbe Interact 16:785–795

Rollins JA, Dickman MB (2001) pH signaling in *Sclerotinia sclerotiorum* identification of a pac C/RIM 1 homolog. Appl Environ Microbiol 67:75–81

Rollins JA, Pathology P, Hall F (2003) The *Sclerotinia sclerotiorum* pac1 gene is required for sclerotial development and virulence. Mol Plant Microbe Interact 16:785–795

Roman RV, Rathjen JP (2017) Apoplastic sugar extraction and quantification from wheat leaves infected with biotrophic fungi. Methods Mol Biol 1659:125–134. https://doi.org/10.1007/978-1-4939-7249-4-11

Rosenbaum DM, Rasmussen SG, Kobilka BK (2009) The structure and function of G-protein-coupled receptors. Nature 459(7245):356–363

Roux F, Voisin D, Badet T, Balagué C, Barlet X, Huard-Chauveau C, Roby D, Raffaele S (2014) Resistance to phytopathogens e tutti quanti: placing plant quantitative disease resistance on the map. Mol Plant Pathol 15:427–432

Roux B, Bolot S, Guy E, Denancé N, Lautier M, Jardinaud MF, Fischer-Le Saux M, Portier P, Jacques MA, Gagnevin L, Pruvost O, Lauber E, Arlat M, Carrère S, Koebnik R, Noël LD (2015) Genomics and transcriptomics of *Xanthomonas campestris* species challenge the concept of core type III effectome. BMC Genomics 16:975. https://doi.org/10.1186/s12864-015-2190-0

Ruijter GJG, Van De Vondervoort PJI, Visser J (1999) Oxalic acid production by *Aspergillus niger*: an oxalate-non-producing mutant produces citric acid at pH 5 and in the presence of manganese. Microbiology 145:2569–2576

Ryan RP, Fouhy Y, Lucey JF, Jiang BL, He YQ, Feng JX, Tang JL, Dow JM (2007) Cyclic di-GMP signaling in the virulence and environmental adaptation of *Xanthomonas campestris*. Mol Microbiol 63:429–442

Ryan RP, McCarthy Y, Andrade M, Farah CS, Armitage JP, Dow JM (2010) Cell–cell signal-dependent dynamic interactions between HD-GYP and GGDEF domain proteins mediate virulence in *Xanthomonas campestris*. Proc Natl Acad Sci U S A 107:5989–5994

Ryan RP, Vorholter FJ, Potnis N, Jones JB, Van Sluys MA, Bogdanove AJ, Dow JM (2011) Pathogenomics of *Xanthomonas*: understanding bacterium–plant interactions. Nat Rev Microbiol 9:344–355

Saenz P, Cervera MT, Dallot S, Quiot L, Quiot J-B, Riechmann JL, Garcıa JA (2000) Identification of a pathogenicity determinant of Plum pox virus in the sequence encoding the C-terminal region of protein P3+6K1. J Gen Virol 81:557–566

Saharan GS, Mehta M (2008) *Sclerotinia* diseases of crop plants: biology, ecology and disease management. Springer, Dordrecht, p 485

Saharan G S, Verma P R (1992) White rusts: a review of economically important species. International Development Research Centre (IDRC), Ottawa, IDRC-MR315e: IV+65p

Saharan GS, Verma PR, Meena PD, Kumar A (2014) White rust of crucifers: biology, ecology and management. Springer, Dordrecht, p 244

Saharan GS, Naresh M, Meena PD (2016) Alternaria diseases of crucifers: biology, ecology and disease management. Springer, Singapore, p 297

Saharan GS, Mehta Naresh K, Meena PD (2019) Powdery mildew disease of crucifers: biology, ecology and disease management. Springer Verlag, Singapore, p 362

Saharan GS, Naresh M, Meena PD (2021) Clubroot disease of crucifers: biology, ecology and disease management. Springer, Singapore, p 757. ISBN 978-981-16-2132-1

Sanchez F, Wang X, Jenner CE, Walsh JA, Ponz F (2003) Strains of *Turnip mosaic* potyvirus as defined by the molecular analysis of the coat protein gene of the virus. Virus Res 94:33–43

Sanchez-Vallet A, Ramos B, Bednarek P, Lopez G, Pislewska-Bednarek M, Schulze-Lefert P, Molina A (2010) Tryptophan-derived secondary metabolites in *Arabidopsis thaliana* confers non-host resistance to necrotrophic *Plectosphaerella cucumerina* fungi. Plant J 63:115–127

Santos C, Maximiano MR, Ribeiro DG, Oliveira-Neto OB, Murad AM, Franco OL, Mehta A (2017) Differential accumulation of *Xanthomonas campestris* pv. *campestris* proteins during the interaction with the host plant: contributions of an *in vivo* system. Proteomics 17(12):1700086

Sappl PG, Carroll AJ, Clifton R, Lister R, Whelan J, Millar AH, Singh KB (2009) The *Arabidopsis* glutathione transferase gene family displays complex stress regulation and co-silencing multiple genes results in altered metabolic sensitivity to oxidative stress. Plant J 58:53–68

Savage AE, Zamudio KR (2011) MHC genotypes associate with resistance to a frog-killing fungus. Proc Natl Acad Sci U S A 108:16705–16710. https://doi.org/10.1073/pnas.1106893108

Schenk PM, Thomas-Hall SR, Nguyen AV, Manners JM, Kazan K, Spangenberg G (2008) Identification of plant defence genes in canola using *Arabidopsis* cDNA microarrays. Plant Biol 10:539–547. https://doi.org/10.1111/j.1438-8677.2008.00056.x

Schmidpeter J, Dahl M, Hofmann J, Kock C (2017) *ChMob2* binds to *ChCbk1* and promotes virulence and conidiation of the fungal pathogen *Colletotrichum higginsianum*. BMC Microbiol 17:22. https://doi.org/10.1186/s12866-017-0932-7

Schuller A, Kehr J, Ludwig-Muller J (2014) Laser microdissection coupled to transcriptional profiling of *Arabidopsis* roots inoculated by *Plasmodiophora brassicae* indicates a role for brassinosteroids in clubroot formation. Plant Cell Physiol 55:392–411

Schulze-Lefert P (2004) Knocking on heaven's wall: pathogenesis of and resistance to biotrophic fungi at the cell wall. Curr Opin Plant Biol 76(1):377–383

Schulze-Lefert P, Panstruga R (2003) Establishment of biotrophy by parasitic fungi and reprogramming of host cells for disease resistance. Annu Rev Phytopathol 41:641–467

Schwelm A, Fogelqvist J, Knaust A, Jülke S, Lilja T, Bonilla-Rosso G, Karlsson M, Shevchenko A, Dhandapani V, Choi SR, Kim HG, Park JY, Lim YP, Ludwig-Müller J, Dixelius C (2015) The *Plasmodiophora brassicae* genome reveals insights in its life cycle and ancestry of chitin synthases. Sci Rep 5(11153). https://doi.org/10.1038/srep11153

Schwelm A, Berney C, Dixelius C, Bass D, Neuhauser S (2016) The large subunit rDNA sequence of *Plasmodiophora brassicae* does not contain intra-species polymorphism. Protist 167:544–554

Schwelm A, Badstober J, Bulman S, Desoignies N, Etemadi M, Falloon RE, Gachon CMM, Legreve A, Lukes J, Merz U, Nenarokova A, Strittmatter M, Sullivan BK, Neuhauser S (2018) Not in your usual Top 10: protists that infect plants and algae. Mol Plant Pathol 19(4): 1029–1044. https://doi.org/10.1111/mpp.12580

Seifbarghi S, Borhan MH, Wei Y, Coutu C, Robinson SJ, Hegedus DD (2017) Changes in the *Sclerotinia sclerotiorum* transcriptome during infection of *Brassica napus*. BMC Genomics 18: 266

Selitrennikoff CP (2001) Antifungal proteins. Appl Environ Microbiol 67:2883–2894

Seo PJ, Park JM, Kang SK, Kim SG, Park CM (2011) An *Arabidopsis* senescence-associated SAG29 regulates cell viability under high salinity. Planta 233:189–200. https://doi.org/10.1007/s00425-010-1293-8

Sexton AC, Howlett BJ (2001) Green fluorescent protein as a reporter in the *Brassica–Leptosphaeria maculans* interaction. Physiol Mol Plant Pathol 58(1):13–21

Shan S, Liu D, Liu R, Zhu Y, Li T, Zhang F, An L, Yang G, Li H (2018) Non-mammalian Toll-like receptor 18 (*Tlr18*) recognizes bacterial pathogens in common carp (*Cyprinus carpio L.*): indications for a role of participation in the NF-κB signaling pathway. Fish Shellfish Immun 72:187–198. https://doi.org/10.1016/j.fsi.2017.09.081

Sharma RC, Sharma SL (1984) Studies on pectolytic enzymes associated with stalk rot of cauliflower caused by *Sclerotinia sclerotiorum* (Lib.) de Bary. Indian J Plant Pathol 2:69–72

Shigenaga AM, Argueso CT (2016) No hormone to rule them all: interactions of plant hormones during the responses of plants to pathogens. Semin Cell Dev Biol 56:174–189. https://doi.org/10.1016/j.semcdb.2016.06.005

Shimada C, Lipka V, O'Connell R, Okuno T, Schulze-Lefert P, Takano Y (2006) Nonhost resistance in *Arabidopsis-Colletotrichum* interactions acts at the cell periphery and requires actin filament function. Mol Plant Microbe Interact 19:270–279

Shultz AJ, Sackton TB (2019) Immune genes are hot spots of shared positive selection across birds and mammals. Elife 8:e41815. https://doi.org/10.7554/eLife.41815

Siemens J, Nagel M, Ludwig-Muller J, Sacristan MD (2002) The interaction of *Plasmodiophora brassicae* and *Arabidopsis thaliana*: parameters for disease quantification and screening of mutant lines. J Phytopathol 150:592–605. https://doi.org/10.1046/j.1439-0434.2002.00818.x

Siemens J, Keller I, Sarx J, Kunz S, Schuller A, Nagel W, Schmulling T, Parniske M, Ludwig-Muller J (2006) Transcriptome analysis of *Arabidopsis* clubroots indicate a key role for cytokinins in disease development. Mol Plant Microbe Interact 19:480–494. https://doi.org/10.1094/MPMI-19-0480

Siemens J, Bulman S, Rehn F, Sundelin T (2009a) Molecular biology of *Plasmodiophora brassicae*. J Plant Growth Regul 28(3):245–251. https://doi.org/10.1007/s00344-009-9091-x

Siemens J, Graf H, Bulman S, In O, Ludwig-Müller J (2009b) Monitoring expression of selected *Plasmodiophora brassicae* genes during clubroot development in *Arabidopsis thaliana*. Plant Pathol 58:130–136

Siemens J, González MC, Wolf S, Hofmann C, Greiner S, Du Y, Rausch T, Roitsch T, Ludwig-Müller J (2011) Extracellular invertase is involved in the regulation of clubroot disease in *Arabidopsis thaliana*. Mol Plant Pathol 12(3):247–262. https://doi.org/10.1111/j.1364-3703.2010.00667.x

Singh AK, Ganapathy Subramanian B, Sarkar S, Singh A (2018) Deep learning for plant stress phenotyping: trends and future perspectives. Trends Plant Sci 23:883–898

Son H, Seo YS, Min K, Park AR, Lee J, Jin JM, Lin Y, Cao P, Hong SY, Kim EK, Lee SH, Cho A, Lee S, Kim MG, Kim Y, Kim JE, Kim JC, Choi GJ, Yun SH, Lim JY, Kim M, Lee YH, Choi YD, Lee YW (2011) A phenome-based functional analysis of transcription factors in the cereal head blight fungus, *Fusarium graminearum*. PLoS Pathog 7:e1002310

Souza DP, Andrade MO, Alvarez-Martinez CE, Arantes GM, Farah CS, Salinas RK (2011) A component of the Xanthomonadaceae type IV secretion system combines a VirB7 motif with a N0 domain found in outer membrane transport proteins. PLoS Pathol 7:e1002031

Soylu S (2004) Ultra-structural characterization of the host–pathogen interface in white blister-infected *Arabidopsis* leaves. Mycopathol 158:457–464

Soylu S, Keshavarzi M, Brown I, Mansfield JW (2003) Ultra-structural characterization of interactions between *Arabidopsis thaliana* and *Albugo candida*. Physiol Mol Plant Pathol 63: 201–211

Soylu EM, Soylu S, Mansfield JW (2004) Ultrastructural characterisation of pathogen development and host responses during compatible and incompatible interactions between *Arabidopsis thaliana* and *Peronospora parasitica*. Physiol Mol Plant Pathol 65(2):67–78

Spanu PD, Abbot JC, Amselem J, Burgis TA, Soanes DM, Stüber K, van Themaat EVL, Brown JKM, Butcher SA, Gurr SJ, Lebrun MH, Ridout CJ, Schulze-Lefert P, Talbot NJ, Ahmadinejad N, Ametz C, Barton GR, Benjdia M, Bidzinski P, Bindschedler LV, Both M, Brewer MT, Cadle-Davidson L, Cadle-Davidson MM, Collemare J, Cramer R, Frenkel O, Godfrey D, Harriman J, Hoede C, King BC, Klages S, Kleemann J, Knoll D, Koti PS, Kreplak J, López-Ruiz FJ, Lu X, Maekawa T, Mahanil S, Micali C, Milgroom MG, Montana G, Noir S, O'Connel RJ, Oberhaensli S, Parlange F, Pendersen C, Quesneville H, Reinhardt R, Rott M, Sacristán S, Schmidt SM, Schön M, Skaminoti P, Sommer H, Stephens A, Takahra H, Thordal-Christensen H, Vigouroux M, Weßling R, Wicker T, Panstruga R (2010) Genome expansion and gene loss in powdery mildew fungi reveal trade offs in extreme parasitism. Science 330: 1543–1546. https://doi.org/10.1126/science.1194573

Spencer-Phillips P (1997) Function of fungal haustoria in epiphytic and endophytic infections. Adv Bot Res 24:309–333

Sperschneider J, Dodds PN, Gardiner DM, Manners JM, Singh KB, Taylor JM (2015) Advances and challenges in computational prediction of effectors from plant pathogenic fungi. PLoS Pathog 11(5):e1004806. https://doi.org/10.1371/journal.ppat.1004806

Sperschneider J, Gardiner DM, Dodds PN, Tini F, Covarelli L, Singh KB, Manners JM, Taylor JM (2016) EffectorP: predicting fungal effector proteins from secretomes using machine learning. New Phytol 210(2):743–761. https://doi.org/10.1111/nph.13794

Spoel SH, Koornneef A, Claessens SM, Korzelius JP, Van Pelt JA, Mueller MJ, Buchala AJ, Metraux JP, Brown R, Kazan K, Van Loon LC, Dong X, Pieterse CM (2003) NPR1 modulates cross-talk between salicylate- and jasmonate-dependent defense pathways through a novel function in the cytosol. Plant Cell 15:760–770

Spoel SH, Mou Z, Tada Y, Spivey NW, Genschik P, Dong X (2009) Proteasome-mediated turnover of the transcription co activator *NPR1* plays dual roles in regulating plant immunity. Cell 137: 860–872. https://doi.org/10.1016/j.cell.2009.03.038

Spurgin LG, Richardson DS (2010) How pathogens drive genetic diversity: MHC, mechanisms and misunderstandings. Proc R Soc Lond B Biol Sci 277:979–988. https://doi.org/10.1098/rspb.2009.2084

Spyrou MA, Bos KI, Herbig A, Krause J (2019) Ancient pathogen genomics as an emerging tool for infectious disease research. Nat Rev Genet 13:323–340. https://doi.org/10.1038/s41576-019-0119-1

Srivastava A, Ohm RA, Oxiles L, Brooks F, Lawrence CB, Grigoriev IV, Cho Y (2012) A zinc-finger-family transcription factor, *AbVf19*, is required for the induction of a gene subset important for virulence in *Alternaria brassicicola*. Mol Plant Microbe Interact 25:443–452

Stavolone L, Alioto D, Ragozzino A, Laliberte J-F (1998) Variability among *turnip mosaic* potyvirus isolates. Phytopathology 88:1200–1204

Stein M, Dittgen J, Sánchez-Rodríguez C, Hou BH, Molina A, Schulze-Lefert P, Lipka V, Somerville S (2006) *Arabidopsis PEN3/PDR8*, an ATP binding cassette transporter, contributes to non-host resistance to in appropriate pathogens that enter by direct penetration. Plant Cell 186(1):731–746

Stobbs LW, Shattuck VI (1989) *Turnip mosaic virus* strains in southern Ontario, Canada. Plant Dis 73:208–212

Strohm M, Jouanin L, Kunert KJ, Pruvost C, Polle A, Foyer CH, Rennenberg H (1995) Regulation of glutathione synthesis in leaves of transgenic poplar (*Populus tremula* × *P. alba*) over expressing glutathione synthetase. Plant J 7:141–145

Stuttmann J, Hubberten H-M, Rietz S, Kaur J, Muskett P, Guerois R, Bednarek P, Hoefgen R, Parker JE (2011) Perturbation of *Arabidopsis* amino acid metabolism causes incompatibility with the adapted biotrophic pathogen *Hyaloperonospora arabidopsidis*. Plant Cell 23(7): 2788–2803

Suehiro N, Natsuaki T, Watanabe T, Okuda S (2004) An important determinant of the ability of Turnip mosaic virus to infect *Brassica* spp. and/or Raphanus sativus is in its P3 protein. J Gen Virol 85:2087–2098

Sundelin T (2008) Identification of expressed *Plasmodiophora brassicae* genes during plant infection and molecular diagnosis of plant pathogens. Ph D thesis, University of Copenhagen, Copenhagen

Sundelin T, Jensen DF, Lubeck M (2011) Identification of expressed genes during infection of Chinese cabbage (*Brassica rapa* subsp. *pekinensis*) by *Plasmodiophora brassicae*. J Eukaryot Microbiol 58:310–314

Swarbrick PJ, Schulze-Lefert P, Scholes JD (2006) Metabolic consequences of susceptibility and resistance (race-specific and broad-spectrum) in barley leaves challenged with powdery mildew. Plant Cell Environ 296(1):1061–1076

Szabo LJ, Bushnell WR (2001) Hidden robbers: the role of fungal haustoria in parasitism of plants. Proc Natl Acad Sci U S A 98(14):7654–7655

Takahara H, Hacquard S, Kombrink A, Hughes HB, Halder V, Robin GP, Hiruma K, Neumann U, Shinya T, Kombrink E, Shibuya N, Thomma BPHJ, O'Connell RJ (2016) *Colletotrichum higginsianum* extracellular *LysM* proteins play dual roles in appressorial function and suppression of chitin-triggered plant immunity. New Phytol 211:1323–1337. https://doi.org/10.1111/nph.13994

Takahashi H, Ishikawa T, Kaido M, Takita K, Hayakawa T, Okazaki K, Itoh K, Mitsui T, Hori H (2006) *Plasmodiophora brassicae*-induced cell death and medium alkalization in clubroot-resistant cultured roots of *Brassica rapa*. J Phytopathol 154:156–162. https://doi.org/10.1111/j.1439-0434.2006.01076.x

Takemoto D, Jones DA, Hardham AR (2003) GFP-tagging of cell components reveals the dynamics of sub-cellular re-organization in response to infection of *Arabidopsis* by oomycete pathogens. Plant J 33(4):775–792

Tanaka S, Mido H, Ito S (2006) Colonization by two isolates of *Plasmodiophora brassicae* with differing pathogenicity on a clubroot-resistant cultivar of Chinese cabbage (*Brassica rapa* L. subsp. *pekinensis*). J Gen Plant Pathol 72:205–209

Tang D, Innes RW (2002) Over expression of a kinase-deficient form of the *EDR1* gene enhances powdery mildew resistance and ethylene-induced senescence in *Arabidopsis*. Plant J 32(6): 975–983

Tang D, Ade J, Frye CA, Innes RW (2005) Regulation of plant defense responses in *Arabidopsis* by *EDR2*, a PH and START domain-containing protein. Plant J 446(1):245–257

Tang D, Ade J, Frye CA, Innes RW (2006) A mutation in the GTP hydrolysis site of *Arabidopsis* dynamin-related protein 1E confers enhanced cell death in response to powdery mildew infection. Plant J 476(1):75–84

Tomimura K, Gibbs AJ, Jenner CE, Walsh JA, Ohshima K (2003) The phylogeny of *Turnip mosaic virus*; comparisons of 38 genomic sequences reveal a Eurasian origin and a recent 'emergence' in east Asia. Mol Ecol 12:2099–2111

Torres PS, Malamud F, Rigano LA, Russo DM, Marano MR, Castagnaro AP, Zorreguieta A, Bouarab K, Dow JM, Vojnov AA (2007) Controlled synthesis of the DSF cell–cell signal is required for biofilm formation and virulence in *Xanthomonas campestris*. Environ Microbiol 9: 2101–2109

Toyofuku K, Kasahara M, Yamaguchi J (2000) Characterization and expression of monosaccharide transporters (OsMSTs) in rice. Plant Cell Physiol 41:940–947. https://doi.org/10.1093/pcp/pcd016

Tran V, Braus-Stromeyer S, Timpner C, Braus G (2013) Molecular diagnosis to discriminate pathogen and apathogen species of the hybrid Verticillium longisporum on the oilseed crop Brassica napus. Appl Microbiol Biotechnol 97:4467–4483

Trudel J, Grenier J, Potvin C, Asselin A (1998) Several thaumatin like proteins bind to β-1,3-glucans. Plant Physiol 118:1431–1438

Tso GHW, Reales-Calderon JA, Tan ASM, Sem X, Le GTT, Tan TG, Lai GC, Srinivasan KG, Yurieva M, Liao W, Poidinger M, Zolezzi F, Rancati G, Pavelka N (2018) Experimental evolution of a fungal pathogen into a gut symbiont. Science 362:589–595. https://doi.org/10.1126/science.aat0537

Turck F, Zhou A, Somssich IE (2004) Stimulus-dependent, promoter specific binding of transcription factor *WRKY1* to its native promoter and the defense-related gene *PcPR1-1* in parsley. Plant Cell 16:2573–2585

Urcuqui-Inchima S, Haenni AL, Bernardi F (2001) Potyvirus proteins: a wealth of functions. Virus Res 74(1-2):157–175. https://doi.org/10.1016/s0168-1702(01)00220-9

Van Breusegem F, Dat JF (2006) Reactive oxygen species in plant cell death. Plant Physiol 141: 384–390

van Damme M, Zeilmaker T, Elberse J, Andel A, de Sain-van der Velden M, van den Ackerveken G (2009) Downy mildew resistance in *Arabidopsis* by mutation of *HOMOSERINE KINASE*. Plant Cell 21(7):2179–2189

van den Berg L, Yang SM (1969) Effect of relative humidity on production of extra-cellular pectolytic enzymes by *Botrytis cinerea* and *Sclerotinia sclerotiorum*. Can J Bot 47:1007–1010

van den Burg HA, Westerink N, Francoijs KJ, Roth R, Woestenenk E, Boeren S, de Wit PJGM, Joosten MHAJ, Vervoort J (2003) Natural disulfide bond-disrupted mutants of AVR4 of the tomato pathogen *Cladosporium fulvum* are sensitive to proteolysis, circumvent Cf-4-mediated resistance, but retain their chitin binding ability. J Biol Chem 278:27340–27346

van den Hoogen DJ, Meijer HJG, Seidl MF, Govers F (2018) The ancient link between G-protein-coupled receptors and C-terminal phospholipid kinase domains. MBio 9(1):e02119–7. https://doi.org/10.1128/mBio.02119-17

Vanacker H, Foyer CH, Carver TLW (1998) Changes in apoplastic antioxidants induced by powdery mildew attack in oat genotypes with race non-specific resistance. Planta 208:444–452

Vanacker H, Carver TLW, Foyer CH (1999) Pathogen-induced changes in the antioxidant status of the apoplast in barley leaves. Plant Physiol 117:103–1114

Vander Heiden MG, Choy JS, Vander Weele DJ, Brace JL, Harris MH, Bauer DE, Prange B, Kron SJ, Thompson CB, Rudin CM (2002) Bcl-x(L) complements *Saccharomyces cerevisiae* genes that facilitate the switch from glycolytic to oxidative metabolism. J Biol Chem 277:44870–44876

Velculescu VE, Zhang L, Vogelstein B, Kinzler KW (1995) Serial analysis of gene expression. Science 270(5235):484–487

Veluchamy S, Rollins JA (2008) A CRY-DASH-type photolyase/cryptochrome from *Sclerotinia sclerotiorum* mediates minor UV-A-specific effects on development. Fungal Genet Biol 45: 1265–1276. https://doi.org/10.1016/j.fgb.2008.06.004

Verma PR (2012) White rust of crucifers: an overview of research progress. J Oilseed Brassica 3(2): 78–87

Verma PR, Petrie GA (1975) Germination of oospores of *Albugo candida*. Can J Bot 53:836–842

Verma PR, Petrie GA (1980) Effect of seed infestation and flower bud inoculation on systemic infection of turnip rape by *Albugo candida*. Can J Plant Sci 60:267–271

Verma PR, Harding H, Petrie GA, Williams PH (1975) Infection and temporal development of mycelium of *Albugo candida* in cotyledons of four *Brassica* spp. Can J Bot 53:1016–1020

Verma SS, Rahman MH, Deyholos MK, Basu U, Kav NNV (2014) Differential expression of miRNAs in *Brassica napus* root following infection with *Plasmodiophora brassicae*. PLoS One 9(1):e86648. https://doi.org/10.1371/journal.pone.0086648

Vicente JG, Holub EB (2013) *Xanthomonas campestris* pv. *campestris* (cause of black rot of crucifers) in the genomic era is still a worldwide threat to *Brassica* crops. Mol Plant Pathol 14(1):2–18

Villarino NF, LeCleir GR, Denny JE, Dearth SP, Harding CL, Sloan SS, Gribble JL, Campagna SR, Wilhelm SW, Schmidt NW (2016) Composition of the gut microbiota modulates the severity of malaria. Proc Natl Acad Sci U S A 113(8):2235–2240. https://doi.org/10.1073/pnas.1504887113

Vlieghe K, Boudolf V, Beemster GTS, Maes S, Magyar Z, Atanassova A, de Almeida Engler J, De Groodt R, Inzé D, De Veylder L (2005) The DP-E2F-like gene *DEL1* controls the endocycle in *Arabidopsis thaliana*. Curr Biol 15:59–63

Voegele RT, Struck C, Hahn M, Mendgen K (2001) The role of haustoria in sugar supply during infection of broad bean by the rust fungus *Uromyces fabae*. Proc Natl Acad Sci U S A 986(1): 8133–8138

Vogel J, Somerville S (2000) Isolation and characterization of powdery mildew resistant *Arabidopsis* mutants. Proc Natl Acad Sci U S A 97:1897–1902

Vogel JP, Raab TK, Schiff C, Somerville SC (2002) PMR6, a pectate lyase-like gene required for powdery mildew susceptibility in *Arabidopsis*. Plant Cell 146(1):2095–2106

Vogel JP, Raab TK, Somerville CR, Somerville SC (2004) Mutations in *PMR5* result in powdery mildew resistance and altered cell wall composition. Plant J 406(1):968–978

Vojnov AA, Slater H, Daniels MJ, Dow JM (2001) Expression of the gumoperon directing xanthan biosynthesis in *Xanthomonas campestris* and its regulation in planta. Mol Plant Microbe Interact 14:768–774

Vorhoelter FJ, Schneiker S, Goesmann A, Krause L, Bekel T, Kaiser O, Linke B, Patschkowski T, Rueckert C, Schmid J, Sidhu VK, Sieber V, Tauch A, Watt SA, Weisshaar B, Becker A, Niehaus K, Puehler A (2008) The genome of *Xanthomonas campestris* pv. *campestris* B 100 and its use for the reconstruction of metabolic pathways involved in xanthan biosynthesis. J Biotechnol 134:33–45

Vorholter FJ, Niehaus K, Puhler A (2001) Lipopolysaccharide biosynthesis in *Xanthomonas campestris* pv. *campestris*: a cluster of 15 genes is involved in the biosynthesis of the LPS O-antigen and the LPS core. Mol Genet Genomics 266:79–95

Wachter A, Wolf S, Steininger H, Bogs J, Rausch T (2005) Differential targeting of GSH1 and GSH2 is achieved by multiple transcription initiation: implications for the compartmentation of glutathione biosynthesis in the Brassicaceae. Plant J 41:15–30

Wagner G, Charton S, Lariagon C, Laperche A, Lugan R, Hopkins J, Frendo P, Bouchereau A, Delourme R, Gravot A, Manzanares-Dauleux MJ (2012) Metabotyping: a new approach to investigate rapeseed (*Brassica napus* L.) genetic diversity in the metabolic response to clubroot infection. Mol Plant Microbe Interact 25:1478–1491

Walker JC (1957) Plant pathology. McGraw-Hill, New York, pp 214–219

Walsh JA, Jenner CE (2002) *Turnip mosaic virus* and the quest for durable resistance. Mol Plant Pathol 3:289–300

Wan JR, Zhang XC, Neece D, Ramonell KM, Clough S, Kim SY, Stacey MG, Stacey G (2008) A LysM receptor-like kinase plays a critical role in chitin signaling and fungal resistance in *Arabidopsis*. Plant Cell 206(1):471–481

Wang P, Heitman J (2005) The cyclophilins. Genom Biol 6:226

Wang X, Zafian P, Choudhary M, Lawton M (1996) The PR5K receptor protein kinase from *Arabidopsis thaliana* is structurally related to a family of plant defence proteins. Proc Natl Acad Sci U S A 93:2598–2602

Wang D, Weaver ND, Kesarwani M, Dong X (2005) Induction of protein secretory pathway is required for systemic acquired resistance. Science 308:1036–1040

Wang W, Devoto A, Turner JG, Xiao S (2007) Expression of the membrane-associated resistance protein RPW8 enhances basal defense against biotrophic pathogens. Mol Plant Microbe Interact 206(1):966–976

Wang Z, Cao G, Wang X, Miao J, Liu X, Chen Z, Qu LJ, Gu H (2008) Identification and characterization of COI1-dependent transcription factor genes involved in JA-mediated response to wounding in *Arabidopsis* plants. Plant Cell Reptr 27:125–135

Wang Z, Mao H, Dong C, Ji R, Cai L, Fu H, Liu S (2009) Overexpression of *Brassica napus* MPK4 enhances resistance to *Sclerotinia sclerotiorum* in oilseed rape. Mol Plant Microbe Interact 22: 235–244

Wang Y, Li J, Hou S, Wang X, Li Y, Ren D, Chen S, Tang X, Zhou JM (2010) A *Pseudomonas syringae* ADP-ribosyltransferase inhibits *Arabidopsis* mitogen-activated protein kinase kinases. Plant Cell 22:2033–2044

Wang Y, Liu W, Hou Z, Wang C, Zhou X, Jonkers W, Ding S, Kistler HC, Xu JR (2011) A novel transcriptional factor important for pathogenesis and ascosporo-genesis in *Fusarium graminearum*. Mol Plant Microbe Interact 24:118–128

Wang J, Yan C, Li Y, Hirata K, Yamamoto M, Yan N, Hu Q (2014) Crystal structure of a bacterial homologue of SWEET transporters. Cell Res 24:1486–1489. https://doi.org/10.1038/cr.2014.144

Wang L, Liu Y, Liu J, Zhang Y, Zhang X, Pan H (2016) The *Sclerotinia sclerotiorum* FoxE2 gene is required for apothecial development. Mycology 106:484–490

Wang S, Boevink PC, Welsh L, Zhang R, Whisson SC, Birch PRJ (2017) Delivery of cytoplasmic and apoplastic effectors from *Phytophthora infestans* haustoria by distinct secretion pathways. New Phytol 216:205–215

Wang Z, Ma LY, Li X, Zhao FY, Sarwar R, Cao J, Li YL, Ding LN, Zhu KM, Yang YH, Tan XL (2020) Genome-wide identification of the NPR1-like gene family in *Brassica napus* and functional characterization of BnaNPR1 in resistance to *Sclerotinia sclerotiorum*. Plant Cell Rep 39(6):709–722. https://doi.org/10.1007/s00299-020-02525-z

Watanabe N, Lam E (2009) Bax inhibitor-1, a conserved cell death suppressor, is a key molecular switch downstream from a variety of biotic and abiotic stress signals in plants. Int J Mol Sci 10:3149–3167

Wei K, Tang DJ, He YQ, Feng JX, Jiang BL, Lu GT, Chen BS, Tang JL (2007) *hpaR*, a putative *marR* family transcriptional regulator, is positively controlled by HrpG and HrpX and involved in the pathogenesis, hypersensitive response, and extracellular protease production of *Xanthomonas campestris* pathovar *campestris*. J Bacteriol 189:2055–2062

Wei L, Jian H, Lu K, Filardo F, Yin N, Liu L, Qu C, Li W, Du H, Li J (2016) Genome-wide association analysis and differential expression analysis of resistance to Sclerotinia stem rot in *Brassica napus*. Plant Biotechnol J 14(6):1368–1380

Weis C, Huckelhoven R, Eichmann R (2013) LIFEGUARD proteins support plant colonization by biotrophic powdery mildew fungi. J Exp Bot 64(12):3855–3867. https://doi.org/10.1093/jxb/ert217

Wengelnik K, Rossier O, Bonas U (1999) Mutations in the regulatory gene *hrpG* of *Xanthomonas campestris* pv. *vesicatoria* result in constitutive expression of all *hrp* genes. J Bacteriol 181(21):6828–6831

Weßling R, Epple P, Altmann S, He Y, Yang L, Henz SR, McDonald N, Wiley K, Bader KC, Glaßer C, Mukhtar MS, Haigis S, Ghamsari L, Stephens AE, Ecker JR, Vidal M, Jones JDG, Mayer KFX, van Themaat EVL, Weigel D, Schulze-Lefert P, Dangl JL, Panstruga R, Braun P (2014) Convergent targeting of a common host protein-network by pathogen effectors from three kingdoms of life. Cell Host Microbe 16(3):364–375. https://doi.org/10.1016/j.chom.2014.08.004

West JS, Kharbanda PD, Barbetti MJ, Fitt BDL (2001) Epidemiology and management of *Leptosphaeria maculans* (phoma stem canker) on oilseed rape in Australia, Canada and Europe. Plant Pathol 50:10–27

Whetzel HH (1945) A synopsis of the genera and species of the *Sclerotiniaceae*, a family of stromatic in-operculate discomycetes. Mycologia 37:648–714. https://doi.org/10.1080/00275514.1945.12024025

Whipps JM, Cooke RC (1978) Occurrence of flagellar beads on zoospores of *Albugo tragopogonis*. Trans Br Mycol Soc 71:141–142

White FF, Potnis N, Jones JB, Koebnik R (2009) The type III effectors of Xanthomonas. Mol Plant Pathol 10:749–766

Wildermuth MC (2010) Modulation of host nuclear ploidy: a common plant biotroph mechanism. Curr Opin Plant Biol 13:449–458

Wilhelm S (1955) Longevity of the *Verticillium* wilt fungus in the laboratory and in the field. Phytopathology 45:180–181

Williams PH, Keen NT, Strandberg JO, McNabola SS (1968) Metabolic synthesis and degradation during clubroot development in cabbage hypocotyls. Phytopathology 58:921–928

Williams B, Kabbage M, Kim HJ, Britt R, Dickman MB (2011) Tipping the balance: *Sclerotinia sclerotiorum* secreted oxalic acid suppresses host defenses by manipulating the host redox environment. PLoS Pathog 7:e1002107

Williamson G, Beverley MC (1987) The purification of acidic glutathione S-transferases from pea seeds and their activity with lipid peroxidation products. Biochem Soc Trans 15:1103–1104

Wilson TJG, Bertrand N, Tang JL, Feng JX, Pan MQ, Barber CE, Dow JM, Daniels MJ (1998) The rpfA gene of *Xanthomonas campestris* pathovar *campestris*, which is involved in the regulation of pathogenicity factor production, encodes an aconitase. Mol Microbiol 28:961–970

Wirtz M, Hell R (2007) Dominant-negative modification reveals the regulatory function of the multimeric cysteine synthase protein complex in transgenic tobacco. Plant Cell 19:625–639

Wise RP, Moscou MJ, Bogdanove AJ, Whitham SA (2007) Transcript profiling in host–pathogen interactions. Annu Rev Phytopathol 45:329–369. https://doi.org/10.1146/annurev.phyto.45.011107.143944

Wong Sak Hoi J, Dumas B (2010) *Ste12* and *Ste12*-like proteins, fungal transcription factors regulating development and pathogenicity. Eukaryot Cell 9:480–485

Wright DP, Baldwin BC, Shephard MC, Scholes JD (1995a) Source-sink relationships in wheat leaves infected with powdery mildew. I. Alterations in carbohydrate metabolism. Physiol Mol Plant Pathol 47(1):237–253

Wright DP, Baldwin BC, Shephard MC, Scholes JD (1995b) Source-sink relationships in wheat leaves infected with powdery mildew. II. Changes in the regulation of the Calvin cycle. Physiol Mol Plant Pathol 47(1):255–267

Wu L, Siemens J, Li S, Ludwig-Müller J, Gong Y, Zhong L, He J (2012) Estimating *Plasmodiophora brassicae* gene expression in lines of *B. rapa* by RT-PCR. Sci Hortic 133(1):1–5. https://doi.org/10.1016/j.scienta.2011.09.036

Wuthrich KL, Bovet L, Hunziker PE, Donnison IS, Hortensteiner S (2000) Molecular cloning, functional expression and characterization of RCC reductase involved in chlorophyll catabolism. Plant J 21:189–198

Xia S, Xu Y, Hoy R, Zhang J, Qin L, Li X (2020) The notorious soil borne pathogenic fungus *Sclerotinia sclerotiorum*: an update on genes studied with mutant analysis. Pathogens 9(1):27. https://doi.org/10.3390/pathogens9010027

Xiao SY, Calis O, Patrick E, Zhang GG, Charoenwattana P, Muskett P, Parker JE, Turner JG (2005) The atypical resistance gene, *RPW8*, recruits' components of basal defence for powdery mildew resistance in *Arabidopsis*. Plant J 426(1):95–110

Xiao X, Xie J, Cheng J, Li G, Yi X, Jiang D, Fu Y (2014) Novel secretory protein Ss-Caf1 of the plant-pathogenic fungus Sclerotinia sclerotiorum is required for host penetration and normal sclerotial development. Mol Plant Microbe Interact 27:40–55

Xie FL, Huang SQ, Guo K, Xiang AL, Zhu YY, Nie L, Yang ZM (2007) Computational identification of novel microRNAs and targets in *Brassica napus*. FEBS Lett 581(7):1464–1474. https://doi.org/10.1016/j.febslet.2007.02.074

Xu JR (2000) MAP kinases in fungal pathogens. Fungal Genet Biol 31:137–152

Xu RQ, Li XZ, Wei HY, Jiang B, Li K, He YQ, Feng JX, Tang JL (2006) Regulation of eight avr genes by hrpG and hrpX in *Xanthomonas campestris* pv. *campestris* and their role in pathogenicity. Prog Nat Sci 16:1288–1294

Xu RQ, Blanvillain S, Feng JX, Jiang BL, Li XZ, Wei HY, Kroj T, Lauber E, Roby D, Chen B, He YQ, Lu GT, Tang DJ, Vasse J, Arlat M, Tang JL (2008) AvrAC (*Xcc* 8004), a type III effector with a leucine-rich repeat domain from *Xanthomonas campestris* pathovar campestris confers avirulence in vascular tissues of *Arabidopsis thaliana* ecotype Col-0. J Bacteriol 190:343–355

Yaeno T, Li H, Chaparro-Garcia A, Schornack S, Koshiba S, Watanabe S, Kigawa T, Kamoun S, Shirasu K (2011) Phosphatidylinositol monophosphate-binding interface in the oomycete RXLR effector AVR3a is required for its stability in host cells to modulate plant immunity. Proc Natl Acad Sci U S A 108(35):14682–14687. https://doi.org/10.1073/pnas.1106002108

Yamanaka K, Sasagawa Y, Ogura T (2012) Recent advances in p97/VCP/Cdc48 cellular functions. Biochem Biophys Act 1823:130–137

Yan Y, Yuan Q, Tang J, Huang J, Hsiang T, Wei Y, Zheng L (2018) *Colletotrichum higginsianum* as a model for understanding host–pathogen interactions: a review. Int J Mol Sci 19(7):2142. https://doi.org/10.3390/ijms19072142

Yang KY, Liu Y, Zhang S (2001) Activation of a mitogen-activated protein kinase pathway is involved in disease resistance in tobacco. Proc Natl Acad Sci U S A 98:741–746

Yang B, Sugio A, White FF (2006a) Os8N3 is a host disease-susceptibility gene for bacterial blight of rice. Proc Natl Acad Sci U S A 103:10503–10508. https://doi.org/10.1073/pnas.0604088103

Yang TJ, Kim JS, Kwon SJ, Lim KB, Choi BS, Kim JA, Jin M, Park JY, Lim MH, Kim HI, Lim YP, Kang JJ, Hong JH, Kim CB, Bhak J, Bancroft I, Park BS (2006b) Sequence-level analysis of the diploidization process in the triplicated FLOWERING LOCUS C region of *Brassica rapa*. Plant Cell 18(6):1339–1347. https://doi.org/10.1105/tpc.105.040535

Yang G, Tang L, Gong Y, Xie J, Fu Y, Jiang D, Li G, Collinge DB, Chen W, Cheng J (2018) A cerato-platanin protein SsCP1 targets plant PR1 and contributes to virulence of *Sclerotinia sclerotiorum*. New Phytol 217:739–755

Yao C, Koller W (1994) Diversity of cutinases from plant pathogenic fungi: cloning and characterization of a cutinase gene from Alternaria brassicicola. Physiol Mol Plant Pathol 44:81–92

Yao CL, Koller W (1995) Diversity of cutinases from plant pathogenic fungi: different cutinases are expressed during saprophytic and pathogenic stages of *Alternaria brassicicola*. Mol Plant Microbe Interact 8:122–130

You BJ, Chung KR (2007) Phenotypic characterization of mutants of the citrus pathogen *Colletotrichum acutatum* defective in a PacC-mediated pH regulatory pathway. FEMS Microbiol Lett 277:107–114

Yousef GM, Kopolovic AD, Elliott MB, Diamandis EP (2003) Genomic overview of serine proteases. Biochem Biophys Res Commun 305:28–36

Yu F, Zhang X, Huang Z, Chu M, Song T, Falk KC, Deora A, Chen Q, Zhang Y, McGregor L, Gossen BD, McDonald MR, Peng G (2016) Identification of genome-wide variants and discovery of variants associated with *Brassica rapa* clubroot resistance gene *Rcr1* through bulked segregant RNA sequencing. PLoS One 11:e0153218. https://doi.org/10.1371/journal.pone.0153218

Yu Y, Du J, Wang Y, Zhang M, Huang Z, Cai J, Fang A, Yang Y, Qing L, Bi C, Cheng J (2019) Survival factor 1 contributes to the oxidative stress response and is required for full virulence of *Sclerotinia sclerotiorum*. Mol Plant Pathol 20(7):895–906. https://doi.org/10.1111/mpp.12801

Yuan M, Wang S (2013) Rice *MtN3*/saliva/SWEET family genes and their homologues in cellular organisms. Mol Plant 6:665–674. https://doi.org/10.1093/mp/sst035

Yun MH, Torres PS, El Oirdi M, Rigano LA, Gonzalez-Lamothe R, Marano MR, Castagnaro AP, Dankert MA, Bouarab K, Vojnov A (2006) Xanthan induces plant susceptibility by suppressing callose deposition. Plant Physiol 141:178–187

Zang N, Tang DJ, Wei ML, He YQ, Chen BS, Feng JX, Xu J, Gan YQ, Jiang BL, Tang JL (2007) Requirement of a *mip*-like gene for virulence in the phytopathogenic bacterium *Xanthomonas campestris* pv. *campestris*. Mol Plant Microbe Interact 20:21–30

Zeilmaker T, Ludwig NR, Elberse J, Seidl MF, Berke L, Van Doorn A, Schuurink RC, Snel B, Van den Ackerveken G (2015) DOWNY MILDEW RESISTANT 6 and DMR6-LIKE OXYGENASE 1 are partially redundant but distinct suppressors of immunity in *Arabidopsis*. Plant J 81: 210–222

Zeise K, von Tiedemann A (2001) Morphological and physiological differentiation among vegetative compatibility groups of *Verticillium dahliae* in relation to *V. longisporum*. J Phytopathol 149:469–475

Zeise K, von Tiedemann A (2002) Host specialization among vegetative compatibility groups of *Verticillium dahliae* in relation to *Verticillium longisporum*. J Phytopathol 150:112–119

Zhang K, McKinlay C, Hocart CH, Djordjevic MA (2006) The *Medicago truncatula* small protein proteome and peptidome. J Proteome Res 5:3355–3367

Zhang J, Shao F, Li Y, Cui H, Chen L, Li H, Zou Y, Long C, Lan L, Chai J (2007) A Pseudomonas syringae effector inactivates MAPKs to suppress PAMP-induced immunity in plants. Cell Host Microbe 1:175–185

Zhang Z, Wu Y, Gao M, Zhang J, Kong Q, Liu Y, Ba H, Zhou J, Zhang Y (2012) Disruption of PAMP-induced MAP kinase cascade by a *Pseudomonas syringae* effector activates plant immunity mediated by the NB-LRR protein SUMM2. Cell Host Microbe 11:253–263

Zhang W, Fraiture M, Kolb D, Loffelhardt B, Desaki Y, Boutrot FG, Tor M, Zipfel C, Gust A, Brunner F (2013) *Arabidopsis* receptor-like protein 30 and receptor-like kinase suppressor of BIR1-1/EVERSHED mediate innate immunity to necrotrophic fungi. Plant Cell 25:4227–4241

Zhang H, Wu Q, Cao S, Zhao T, Chen L, Zhuang P, Zhou X, Gao Z (2014) A novel protein elicitor (SsCut) from *Sclerotinia sclerotiorum* induces multiple defense responses in plants. Plant Mol Biol 86(4–5):495–511. https://doi.org/10.1007/s11103-014-0244-3

Zhang Y, Lun C-Y, Tsui S (2015) Metagenomics: a new way to illustrate the crosstalk between infectious diseases and host microbiome. Int J Mol Sci 16:26263–26279. https://doi.org/10.3390/ijms161125957

Zhao X, Mehrabi R, Xu JR (2007) Mitogen-activated protein kinase pathways and fungal pathogenesis. Eukaryot Cell 6:1701–1714

Zhao J, Buchwaldt L, Rimmer SR, Sharpe A, McGregor L, Bekkaoui D, Hegedus D (2009) Patterns of differential gene expression in *Brassica napus* cultivars infected with *Sclerotinia sclerotiorum*. Mol Plant Pathol 10(5):635–649. https://doi.org/10.1111/j.1364-3703.2009.00558.x

Zhao Y, Bi K, Gao Z, Chen T, Liu H, Xie J, Cheng J, Fu Y, Jiang D (2017) Transcriptome analysis of *Arabidopsis thaliana* in response to *Plasmodiophora brassicae* during early infection. Front Microbiol 8:673

Zhu W, Wei W, Fu Y, Cheng J, Xie J, Li G, Yi X, Kang Z, Dickman MB, Jiang D (2013) A secretory protein of necrotrophic fungus *Sclerotinia sclerotiorum* that suppresses host resistance. PLoS One 8:e53901

Zhu GL, Yu G, Zhang YH, Rollins JA, Li J, Pan H (2019) The formaldehyde dehydrogenase SsFdh1 is regulated by and functionally cooperates with the GATA transcription. MSystems 4:1–19

Zimmerli L, Stein M, Lipka V, Schulze-Lefert P, Somerville S (2004) Host and non-host pathogens elicit different jasmonate/ethylene responses in *Arabidopsis*. Plant J 406(1):633–646

Zuppini A, Navazio L, Sella L, Castiglioni C, Favaron F, Mariani P (2005) An endopolygalacturonase from *Sclerotinia sclerotiorum* induces calcium-mediated signaling and programmed cell death in soybean cells. Mol Plant Microbe Interact 18:849–855

Molecular Detection and Identification of Pathogens, Pathotypes, and Genes

4

Abstract

The accurate, rapid detection and identification of the pathogens is essential for effective management of crucifers' diseases. The traditional approaches are time consuming, expansive, laborious, and less in accuracy in comparison to molecular approaches like bar-coding markers, qRT-PCR, cDNA sequencing, pathogen enrichment sequencing (Pen seq), and use of sequence repositories. Pen Seq approach has been used to detect *Albugo* from infected leaves and to identify variations in pathogen genotypes. The distinct lineages that had diverged from each other were also identified. Many putative recombinant regions were identified between lineages. Pen Seq can identify *Albugo* race structure with phylogenetic lineage. *Albugo candida* ITS1 DNA has been detected in symptomatic and asymptomatic host tissues. Molecular markers like RAPD and ESTs have been used for identification, detection, and analysis of genetic variability in *Alternaria* species infecting crucifers. In the genome of *Colletotrichum*, two effectors (genes), ChELP1 and ChELP2, homologs of LysM have been characterized. Molecular quantification of *Erysiphe* infection and development in host tissues have been assessed using qPCR assessing host cell entry, area covered by hyphae, conidia, and conidiophores. The expression of genes has been detected in response to powdery mildew infection with a broad range of transcriptional changes conferring susceptibility, resistance, and bimolecular regulations. The phylogenetic relationship of powdery mildew of crucifers with other powdery mildew species and isolates variations have been determined through sequence analysis of ITSrDNA and sequence comparison of DNA encoding 5.85 rDNA, ITS1, and ITS2. DNA (RAPD) finger print approach has been used to distinguish isolates of *H. parasitica* infecting different *Brassica* species. The combination of LMD and RNA sequencing of *Brassica* improves

G. Singh Saharan et al., *Genomics of Crucifer's Host-Pathosystem*,
https://doi.org/10.1007/978-981-19-3812-2_4

Leptosphaeria detection at the early stage of infection. Sequencing of gene fragments of β-tublin, actin, and ITS region of rDNA from *L. biglobosa* isolates from different countries could show closer taxonomic similarities between different clades. Genomic prediction tools are valuable approaches in *Brassica* to assess *Leptosphaeria* species. *Plasmodiophora* virulence genes expressed during early stages of colonization of *Brassica* roots have been identified using SSH and EST analysis. AqPCR-based *Plasmodiophora* detection and spread in the soils has been developed for country-wide assessment. ITS sequencing and phylogenetic analysis of *Pseudocercosporella* isolates collected from *Brassica* species revealed a high degree of genetic similarity. Real-time PCR protocols have been developed to detect and discriminate anastomosis groups (AGs) of *R. solani* using ribosomal internal transcribed spacer (ITS) regions or β-Tubulin sequences of pathogen from field soils. Phenotypic pipeline has been developed to assess kinetics of disease symptoms caused by *Sclerotinia* in *A. thaliana* conferring susceptibility by NLR genes. A simple, rapid, sensitive, and cost effective mRT-PCR protocol has been developed for detection and differentiation of *Brassica* viruses. Molecular approaches like DAS-ELISA, Rt-PCR, and ELISA can detect, identify, and confirm phylogenetic relationship of *Brassica* viruses and their strains. Simplex and multiplex PCR assays based on DNA sequence from rITS region and coding and noncoding regions of actin, elongation factors 1-α, glyceraldehydes-3-phosphate dehydrogenase, and tryptophan synthase genes are valuable approaches for identification, detection, and phylogenetic analysis of *Verticillium* species. *Verticillium longisporum* isolates have been characterized on the basis of a group-1 Intron in SSU-rRNA gene of pathogen. Isolates from two different species, i.e., *V. dahliae* and *V. albo-atrum* have been characterized based on AFLP analysis of rRNA and DNA from mitochondria. Several virulence genes of *Xanthomonas* have been identified (pig, pigB, DF, EPSs, DSF, cyclic-di-GMP, rpfl, RpfF, RpfG, WspR, xantham, LPs) which regulate production of enzymes for infection and development of black rot in *Brassica* by multiplication and movement of bacteria in host tissues. Several novel genes for resistance to *Xanthomonas* in *Brassica* have been identified by SSH-PCR-based techniques. Molecular identification of different pathotypes, pathogens, elicitors/effectors have been revealed using *R* genes in the form of isogenic lines of *Brassica*. The accuracy of qRT-PCR analysis data expression of target genes is affected by factors such as RNA quality, purity, and PCR amplification efficiency, technical, and biological variations. To control these variables and avoid bias in qRT-PCR, selection of appropriate RGs is crucial under multiple stresses. The interplay between multiple pathogens and associated microbiomes requires disentangling of hidden histories in genes and genomics of both hosts and pathogens. *R* genes homologous DNA fragments in *A. thaliana* have been identified to use them for cloning and in breeding for resistance against major pathogens of crucifers. The molecular approaches are valuable tools and their efficiency and accuracy depends on proper sampling and cares in using them for correct diagnosis.

Keywords

Molecular detection · Identification · Pathogens · Pathotypes · Genes · *Albugo* pathogen · Sequence detection · PenSeq investigation · PenSeq identification with phylogenetic lineage · PCR-based assay · *Alternaria* pathogen · Diagnostic biomarker for the identification · Identification of pathotypes · *Colletotrichum* pathogen · Identification of effectors/genes · *Erysiphe* pathogen · Molecular quantification · Detection of genes expression · Phylogenetics · *Hyaloperonospora* pathogen · Molecular identification · *Leptosphaeria* pathogen · Combined LMD and RNA sequencing · Molecular identification · Genomic identification of *R* genes · *Plasmodiophora* pathogen · Identification of virulence gene · qPCR-based detection · Molecular identification · *Pseudocercosporella* pathogen · Phylogenetic analysis · *Rhizoctonia* pathogen · Molecular detection · *Sclerotinia* pathogen · Identification of an *Arabidopsis* NLR gene · *Turnip Mosaic Virus* pathogen · Detection · Differentiation by multiplex reverse transcription polymerase chain reaction · Identification of *Turnip yellows virus by* DAS-ELISA kits · Molecular detection · Identification · Biological and serological detection · *Verticillium* pathogen · Identification · Differentiation · Molecular characterization · *Xanthomonas* pathogen · Identification of virulence effectors/gene · Novel genes identification · Molecular identification of pathotype · qRT-PCR analysis of *Brassica* genes · Disentangling histories in genes and genomes · Identification of *R* genes homologous

4.1 Introduction

Development of methods for the rapid identification of new and already known pathogens and races is imperative for sustainable disease management. With the advent of several molecular and DNA sequencing technologies, researchers can now generate high-resolution genotypic data from any organism. However, the quality of the data depends on the quality and purity of the DNA sample, which means that the pathogen under study needs to be in excess compared to the host plant or cultivated axenically, which is not possible for the biotroph like *A. candida*. The field pathogenomics of wheat yellow rust has been investigated using cDNA sequencing from infected leaves in the field, resulting in ~37% of reads being mapped to the pathogen reference genome. In addition, complexity reduction methods have been developed for sequencing of specific subfractions of plant genomes (MutChromSeq and exome capture in wheat). Jouet et al. (2018) reported a complexity reduction method (Pathogen Enrichment Sequencing or "PenSeq") that enables capture and sequencing of plant-associated microbial sequences directly from field samples. It can explore race diversity in *A. candida* and the impact of *A. candida* on the host leaf microbiome.

To control disease outbreaks in crops, it is imperative that pathogen and pathogen population variability in the field can be quickly identified. This can be achieved by having a full set of genetic markers for different pathogens and pathotypes, not only contained within the local fields but universally. The initiatives for *P. brassicae* showed that *Cr811* was a good candidate gene to differentiate pathotype 5 from the rest of the pathotypes (Feng et al. 2016). Molecular characterization of pathogen effectors is gaining momentum, with better genome assemblies and higher gene prediction accuracies. This apparent have better insights into the structures and functions of effector proteins, which allows more efficient screening and identification of *R* genes (Gibriel et al. 2016; Neik et al. 2017).

The accurate and rapid identification of the organism(s) that causes plant disease is essential for effective disease control. It enables more informed decisions to be made about cultivar choice and how and when chemicals can be used most effectively to control disease epidemics. Pathogen detection and identification is important for preventing the spread of disease by screening seeds and other planting material and is needed to implement quarantine regulations. It is also important more generally in many areas of plant pathology research, for example in: studies of disease epidemiology; studies of yield loss disease relationships; and in designing new strategies for disease control. Pathogen or disease diagnosis is therefore, fundamental to virtually all aspects of plant pathology.

Traditional approaches to disease diagnosis generally first involve the interpretation of visual symptoms. This may be followed by laboratory identification, using for example selective media and microscopy, to confirm the diagnosis. In some cases, these methods are still the cheapest, simplest, and most appropriate. Conventional methods do, however, have a number of drawbacks, which has prompted the search for alternative diagnostic techniques. Traditional methods generally require skilled and specialized microbiological expertise, which often takes many years to acquire. There is a need to use more generic techniques that can be taught quickly and easily to relatively unskilled staff. Methods that involve culturing can often take days or weeks to complete and this is not acceptable when rapid, high throughput diagnosis is required. The results are not always conclusive, e.g., where similar symptoms can be caused by different pathogens or physiological conditions. Closely related organisms may be difficult to discriminate on the basis of morphological characters alone. It may also be necessary to discriminate between populations of the same pathogen that have specific properties, fungicide resistance, toxin production, or differences in virulence. Traditional methods may not be sensitive enough (e.g., where the detection of presymptomatic infection is needed) and as such much effort has been devoted to the development of novel methods for detecting and identifying plant pathogens over the last decade. The greatest effort has been on the development of diagnostics based on antibodies and nucleic acid technologies (Duncan and Torrance 1992; Schots et al. 1994; Ward 1994; Martin et al. 2000; McCartney et al. 2003). The details of protocols, methods, procedures, etc. have been covered in the chapter on techniques of crucifer's host–pathosystem (Table 4.1).

Table 4.1 Identification of host genes which response to pathogen at early host–pathogen interaction stages

Host pant	Gene	Pathogen	Disease	Reference
Brassica napus	*Rlm2, Rlm3, LepR1, LepR2, Rlm9, Cf9, Avr9*	*Leptosphaeria maculans*	Stem Canker (Blackleg)	Haddadi et al. (2019), Larkan et al. (2019), Hennin et al. (2001)
B. napus	*BnMPK4, NPR1, BnaMPK3*	*Sclerotinia sclerotiorum*	Sclerotinia rot	Wang et al. (2009, 2019, 2020)
B. napus	*Ac2V1*	*Albugo candida*	White rust	Somers et al. (2002)
Arabidopsis	*Bgh*	*Erysiphe cichoracearum*	Powdery mildew	Zimmerli et al. (2004)
Arabidopsis thaliana	*TIR-NB-LRR*	*Albugo candida*	White rust	Borhan et al. (2010)
B. juncea	*MAPK3*	*Alternaria brassicae*	Alternaria blight	Taj et al. (2011)
B. juncea	*BjuA046215, BjuWRR1*	*Albugo candida*	White rust	Bhayana et al. (2020), Arora et al. (2019)
Chinese rape variety Zhongyou 821	*MSI-99m*	*Sclerotinia sclerotiorum*	Sclerotinia rot	Sang et al. (2013)
Cruciferous crops	*ChMK1*	*Colletotrichum higginsianum*	Anthracnose disease	Wei et al. (2016)

Gene: the gene of host plant involved in the early interaction significantly

Compared to plants and animals, pathogen has simple body plans with often morphologically and ecologically obscure structures. This poses challenges for accurate and precise identifications. The identification of pathogen encouraging the approach of integrative (polyphasic) taxonomy for species delimitation, i.e., the combination of genealogy (phylogeny), phenotype (including autocology), and reproductive biology (when feasible) allows objective evaluation of diagnostic characters, either phenotypic or molecular or both. Verification of identifications is crucial but often neglected. Because of clade-specific evolutionary histories, there is currently no single tool for the identification of pathogen, although DNA barcoding using the internal transcribed spacer (ITS) remains a first diagnosis, particularly in meta-barcoding studies. Secondary DNA barcodes are increasingly implemented for groups where ITS does not provide sufficient precision. Issues of pair wise sequence similarity-based identifications and OTU clustering, and multiple sequences alignment-based phylogenetic approaches with subsequent verification are recommended as more accurate alternatives. In meta-barcoding approaches, the trade-off between speed and accuracy and precision of molecular identifications must be carefully considered. Intragenomic variation of the ITS and other barcoding markers should be properly documented, as phylotype diversity is not necessarily a proxy of species richness. Important strategies to improve molecular identification of pathogen are: (1) broadly document intraspecific and intragenomic variation of barcoding markers; (2) substantially expand sequence repositories, focusing on under sampled clades and missing taxa; (3) improve curation of sequence labels in primary repositories and substantially increase the number of sequences based on verified material; and (4) link sequence data to digital information of voucher specimens including imagery. In parallel, technological improvements to genome sequencing offer promising alternatives to DNA barcoding in the future. Despite the prevalence of DNA-based pathogens taxonomy, phenotype-based approaches remain an important strategy to catalog the global diversity of microbes and establish initial species hypotheses (Lucking et al. 2020).

4.2 Use of qRT-PCR Normalization in Crucifers Under Various Stress Conditions

During cultivation of crucifer's crops, they are exposed to various adverse environmental conditions, including abiotic and biotic stresses, such as drought, high-salinity, and pathogens, which result in production losses. In recent years, more and more researchers have focused on gene expression studies of crucifer's crops under various exogenous stimulations to understand its defense mechanisms. Quantitative real-time reverse transcription polymerase chain reaction (qRT-PCR), a powerful and efficient technique, has become the first choice in quantitative gene expression studies (Fang et al. 2004; Guenin et al. 2009). The qRT-PCR is a highly sensitive technique; however, the accuracy of its results can be affected by certain uncertainties, such as RNA quantity, purity, reverse-transcription efficiency, and PCR amplification (Zhu et al. 2013). To control these variables, the careful selection of reference genes as internal controls for normalizing gene expression is essential (Guenin et al. 2009; Kundu et al. 2013). Ideal reference genes should be stably expressed regardless of the experimental conditions (Goossens et al. 2005; Schmittgen and Zakrajsek 2000). Traditional reference genes are mostly "housekeeping" genes that are thought to have stable expression patterns, such as 18S ribosomal RNA, actin, polyubiquitin, glyceraldehyde-3-phosphate dehydrogenase, and elongation factor 1-α (Wang et al. 2013). However, to date, more and more studies have suggested that the expressions of these genes change significantly under different experimental treatments (Kim et al. 2003; Nicot et al. 2005; Stürzenbaum and Kille 2001). Therefore, for accurate normalization in gene expression studies, it is necessary to evaluate the expression stability of reference genes and to select the most suitable reference genes under different experimental conditions before real-time PCR amplification.

Several statistical algorithms have been developed to calculate gene expression stability and to determine the appropriate reference genes and their numbers required for normalization under specific experimental conditions (Kundu et al. 2013). In terms of geNorm algorithm, the M value of each gene, which is a stability measure, is calculated on the basis of the pair wise variation between it and all other control genes. The genes with the lowest M values have the most stable expression, and geNorm determines these genes by increasingly eliminating less stable genes from the candidate reference genes pool (Vandesompele et al. 2002). NormFinder, another algorithm, employs a statistical and mathematical model that not only estimates the overall expression variation of a candidate gene, but also considers the variation between the chosen subgroups (Huis et al. 2010). To date, a number of studies focusing on reference gene evaluation and validation have been undertaken in various plants, including *Arabidopsis* (Czechowski et al. 2005). The stability of nine potential reference genes for normalisation was investigated by Wang et al. (2014) using a sizable collection of biological samples that represented *B. napus* under various stress situations. Statistical algorithms implemented in geNorm and NormFinder were used. The results identified the best-performing reference genes in this model, and will benefit future gene expression studies in *B. napus* and other species of the genus *Brassica* (Wang et al. 2014).

Data normalization is essential for reliable output of quantitative real-time reverse-transcription polymerase chain reaction (qRT-PCR) assays, as the unsuitable choice of reference gene(s), whose expression might be influenced by exogenous treatments in plant tissues, could cause misinterpretation of results. Wang et al. (2014) investigated the expression variations of nine candidate reference genes in 40 samples of *B. napus* leaves subjected to various exogenous treatments. Parallel analyses by geNorm and NormFinder revealed that optimal reference genes differed across the different sets of samples. The best-ranked reference genes were *PP2A* and *TIP41* for salt stress, *TIP41* and *ACT7* for heavy metal (Cr6+) stress, *PP2A* and *UBC21* for drought stress, *Fbox* and *SAND* for cold stress, *F-box* and *ZNF* for salicylic acid stress, *TIP41*, *ACT7*, and *PP2A* for methyl jasmonate stress, *TIP41* and *ACT7* for abscisic acid stress, and *TIP41*, *UBC21*, and *PP2A* for *Sclerotinia sclerotiorum* stress. Two newly employed reference genes, *TIP41* and *PP2A*, showed better performances, suggesting their suitability in multiple conditions. To further validate the suitability of the reference genes, the expression patterns of *BnWRKY40* and *BnMKS1* were studied in parallel (Wang et al. 2014).

4.3 *Brassica–Albugo* Pathogen

4.3.1 Sequence Capture Detection of *Albugo* Sequences from Infected Leaves

Although some targets were not sequenced in some samples due to presence/absence polymorphisms between the strains used in the bait design and those in the samples, almost all targets were captured and sequenced from control organisms when present (min breadth of coverage = 94.14%, max = 100%; (Fig. 4.1a, b)). Read depth was variable between samples and organisms but in all cases sufficient for downstream analyses (mean read depth (±SD) = 1003 × (±696), min = 111). Additionally, targets were not sequenced where control organisms were absent (mean breadth of coverage (±SD) = 5.87% (±10.7), mean read depth (±SD) = 7 × (±10)). In all samples, however, a small proportion of reads were wrongly assigned to control organisms likely due to the presence of closely related species. Read misalignment

(a)

Breadth of coverage (>10x)

	#18	#19	#20	#21	#22	#114
A. candida	(–) 9.71	(Ac2v) 97.81	(AcBoT) 97.27	(AcEx1) 95.7	(AcEx1) 95.82	(–) 3.32
P. infestans	(–) 18.58	(–) 3.91	(–) 4.07	(88069^{td}) 98.01	(88069^{td}) 95.98	(–) 2.37
P. syringae	(–) 71.76	(–) 25.64	(–) 27.68	(–) 31.56	(DC3000) 94.14	(–) 27.63
A. laibachii	(Nc14) 99.99	(–) 36.12	(–) 39.7	(–) 30.07	(–) 34.13	(–) 1.91
H. arabidopsidis	(Emoy2) 99.99	(–) 4.02	(–) 6.25	(–) 11.41	(–) 8.78	(–) 0
E. cruciferarum	(–) 0	(+) 100	(–) 0	(–) 0	(–) 0	(–) 0

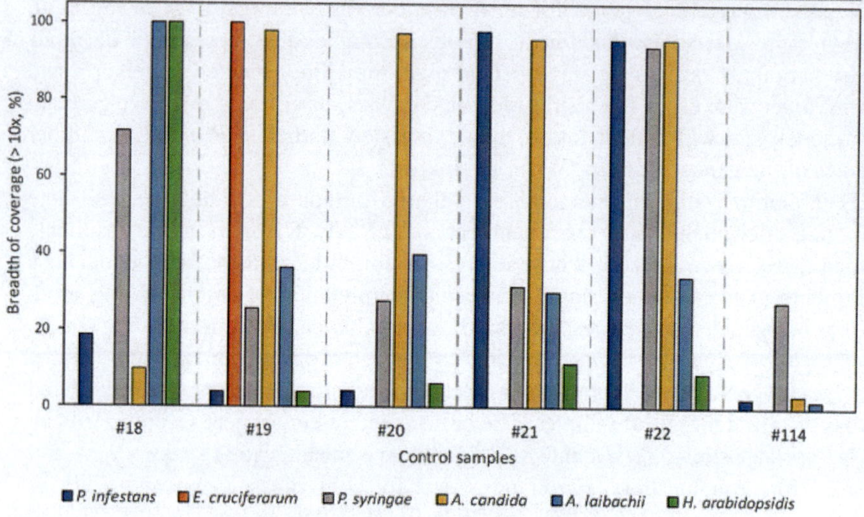

(b)

Average read depth at the ITS sequence

	#1	#97	#14	#17	#20	#38	#40
C. bursa-pastoris	314.761	452.931	7881.57	96.3702	113.133	15.4119	159.043
B. oleraceo	148.956	672.881	58.4938	538.238	7160.4	50.7887	15.3563
A. deltoidea	450.221	7870.25	267.536	358.035	613.802	82.4221	30.866
A. thaliana	140.998	516.282	116.121	227.395	169.324	14.1513	2606.42
C. hirsuta	18.2448	29.4233	14.376	64.5421	11.1981	20.9688	10.7515
A. montanum	10.3069	9.77704	31.2527	25.3104	10.3863	8.72038	18.7208
R. sativus	975.12	1428.56	305.174	2085.43	3388.02	7086.55	13.6535
S. officinale	7823.24	868.898	287.496	551.517	393.064	34.6176	14.6324
A. petiolata	3519.84	125.484	13.6838	27.0877	37.9307	9.53448	4
S. alba	709.304	87.0788	414.925	7853.12	2315.26	1783.36	114.293

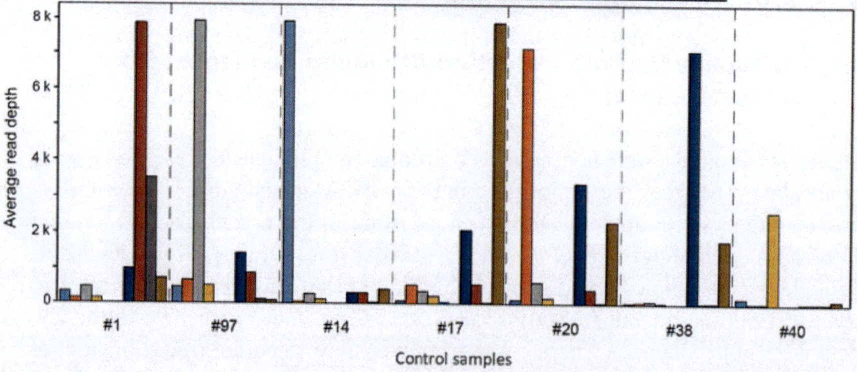

Fig. 4.1 (**a**) Reconstruction experiment for the detection of control organisms, using samples #18–22 and #114. (Upper panel) Description of the organisms present in control samples (#18–22) as well as the negative control (#114). The name of the strain used in the experiment is provided when the species is present in the sample and a minus sign indicates that the species is absent. In both cases, the breath of coverage at targeted loci is provided (proportion of targeted bases covered

was also observed at the ITS sequences of the hosts which were all fully covered by reads due to high homology between species. The highest read depth at the ITS was used to identify the host species. After verification that targeted loci were captured and sequenced to reasonable depth and coverage, the relationship between input DNA and output reads was examined using samples #56–59. Because read depth is highly correlated with the total number of reads generated per sample (Pearson $R^2 = 0.880$, $P < 0.001$). Linear regression analyses revealed a positive association between the proportion of reads mapping to control organisms and the amount of DNA that was used during library preparation for both *A. candida* and *P. syringae* ($R^2 = 0.88$ and 0.78). Due to the small sample size, *P* values were not significant for individual sets of samples ($P = 0.06$ and 0.12, respectively). However, combining both data sets renders the correlation significant ($R^2 = 0.83$, $P < 0.01$). Therefore, although additional samples are needed to generate more robust statistics, Pathogen Enrichment Sequencing (PenSeq) may provide an indication of the relative abundance of pathogen DNA across samples, as well as within samples, providing that the number of targeted base pairs per pathogen is taken into account. Jouet et al. (2018) investigated whether PenSeq could be used to investigate both the species diversity and genetic diversity of organisms within field samples.

4.3.2 The PenSeq Investigation of DNA Sequence Variation in Pathogen Genotypes from Field

A sequence capture-based protocol (PenSeq) that enables cost-effective assessment of pathogen genetic diversity infield samples have been developed by Jouet et al. (2018). The designed baits are based on specific sequences from ~50 microbial species. They were able to recover all targets, at high read depth, from control pathogens when present in reconstruction experiments. In addition, it showed the more abundant the microbe in the sample, the higher the proportion of reads observed.

The PenSeq thus reveals not only presence–absence differences but also relative abundance of microbes from field samples. The increased abundance and diversity of bacteria and fungi in the phyllospheres of *Albugo candida*-infected and noninfected plants supported the hypothesis that *A. candida* suppresses the host's immune response, influencing microbial communities. The most striking examples are the Rickettsiales (Proteobacteria) and Chytridiomycetes that were detected in all

Fig. 4.1 (continued) by at least 10 reads). (Lower panel) Histogram of the breadth of coverage at targeted loci for each control organism and sample. (**b**) Reconstruction experiment for the identification of the host species, using samples #1, 97, 14, 17, 20, 38, and 40. (Upper panel) The average read depth is provided for all internal transcribed spacer (ITS) sequences targeted in this study. The host corresponding to each control sample is highlighted in a color matching the histogram below. (Lower panel) Histogram of the average read depth at the ITS sequence of Brassicaceae in each sample (Jouet et al. 2018)

82 *A. candida*-infected plants but on none of the 12 noninfected plants. A caveat in this association analysis is that many variables, other than infection by *A. candida*, may play a role in the presence or absence of microbes. Therefore, it is advised to conduct microbiome-focused studies to confirm and quantitatively assess the impact of *A. candida* infection on the various associated microbes.

Jouet et al. (2018) sequenced a total of ~660 kb of the *A. candida* genome from 91 isolates (187 loci including putative effectors). The data was also used to identify distinct lineages that had diverged from each other by 0.29–1.15%. This could not be achieved using a limited number of loci (Saharan et al. 2014) or would have been very expensive and laborious using whole-genome sequencing of purified isolates. These lineages appeared to be mostly specialized on one host species and to be consistent with what has been defined as a pathotype or race in *A. candida* (Biga 1955; Pound and Williams 1963; Verma 2012). Conceivably, some *A. candida* lineages can infect closely related host species that were not sampled in this study, as observed for lineage I (Race 4) infecting *C. bursa-pastoris*, *A. thaliana,* and *Eutrema japonica* and lineage XII infecting *Lunaria annua* and *Lunaria rediviva*.

With PenSeq, it is possible to sequence ~115 samples on three HiSeq lanes, by multiplexing up to 47 samples in a lane. It can investigate not only the genetic diversity in *A. candida*, but also the presence and abundance of other microbes from samples collected in the field. This method enables inspection of many genetic marker loci and is a good alternative to expensive whole-genome sequencing of microbiota and to PCR-based methods such as internal transcriber spacing (ITS) and 16Ssequencing. Potential applications would be to study the coevolution of host resistance genes and pathogen effectors in natural populations or of microorganisms within specific environments. Sequence capture has been used to sequence viral genomes from clinical samples (VirCapSeq; Briese et al. 2015) and in a companion paper; PenSeq was used to investigate putative effector sequences in *P. infestans* and *Phytophthora capsici* (Thilliez et al. 2018). Although field pathogenomics can also be very helpful for assigning races to lineages (Hubbard et al. 2015), sequence capture enables the sequences under investigation to be predefined.

4.3.3 Identification of Recombination Break Points of Different Races

Hybridization between *A. candida* lineages may have enabled adaptation to the numerous hosts on which the pathogen has been reported (Adhikari et al. 2003; McMullan et al. 2015). Genetic exchange between lineages may occur if they share a common host (Pound and Williams 1963; Saharan et al. 2014) or if the immune system of the host is compromised by a compatible isolate (Cooper et al. 2008; Belhaj et al. 2017; Prince et al. 2017). By sampling more isolates from different host species, we set out to better understand the role of hybridization in *A. candida* evolution. Many putative recombinant regions were identified between lineages within the ~400 kb contig by Jouet et al. (2018). Some of these regions are quite polymorphic and are probably due to incomplete lineage sorting or represent trans-species polymorphism. However, other regions are (nearly) identical between races over long stretches of sequence, suggesting that they arose by recent hybridization. In addition, although the existence of tetraploid races of *A. candida* is unproven, the high levels of heterozygosity in triploid lineages (*B. oleracea* and *R. sativus*; lineages V and VII, Races 9 and 1) suggest that it was originated from intercrosses between diploids and tetraploids rather than from the absence of reduction division in either male or female germ cells. This idea is reinforced by the mid-branch placement of triploids in the network shown above (lineages V and VII (Races 9 and 1) mid-branch of lineages IV and VI (Races 7 and 1)). However, no mixed infections were detected which suggests that this scenario is rare. Conceivably, genetic exchanges between lineages that are adapted to diverged hosts (each with their unique sets of resistance genes) would most likely be maladaptive, in comparison with combinations of alleles which have been selected over extended evolutionary timescales (Brasier 2001).

Nonetheless, evidence for the emergence of novel pathogen races or species through hybridization continues to accumulate, particularly for crop species (Stukenbrock and McDonald 2008; Gladieux et al. 2011; Leroy et al. 2016; Stukenbrock 2016). *Phytophthora alni* on *Alnus* spp. arose via sexual reproduction between *Phytophthora cambivora* and a species related to *Phytophthora fragariae*, both nonpathogenic to *Alnus* spp. (Brasier et al. 2004). Likewise, *Blumeria graminis* f. sp. *triticale*, a powdery mildew found on triticale and wheat, is a hybrid between *B. graminis* f. sp. *tritici* pathogenic to wheat and *B. graminis* f. sp. *secalis* to rye (Menardo et al. 2015).

4.3.4 The PenSeq Identification of Race Structure with Phylogenetic Lineage

PenSeq data were used to investigate the genetic diversity of *A. candida* natural populations of 85 isolates analyzed that had high read depth at the ~400 kb contig (mean read depth (±SD) 479 × (±529); min = 18, max = 2336; Fig. 4.2). Initially, 32 neutrally evolving loci were used to build a phylogeny and measure genetic diversity (Fig. 4.3a). Two additional phylogenetic trees were built using the ~400 kb contig and seven conserved loci (mitochondrial and nuclear housekeeping genes), respectively (Fig. 4.3b). Discrimination of *A. candida* isolates was not possible due to low genetic variation within conserved loci. However, phylogenies built with the neutral loci and the 400 kb contig identified 17 *A. candida lineages* (I–XVII in Fig. 4.3a). These lineages consist of isolates collected on the same host or closely related hosts, irrespective of country of origin (mean within lineages *P* distance = 0–0.058%). Conversely, isolates collected on different hosts formed distinct lineages (mean between lineages *P* distance = 0.36–1.40%), with the exception of isolates collected on radish (*R. sativus*). In accordance with this observation, phylogenetic clustering mostly agreed with physiological race nomenclature with groups X and V corresponding to Races 5 and 9 in both trees, respectively. However, what would be considered as Races 1 (lineages VI and

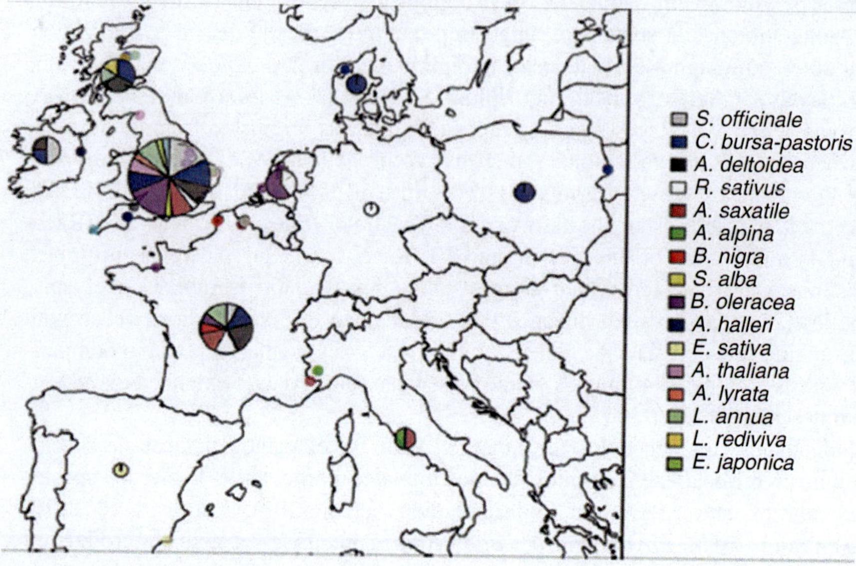

Fig. 4.2 Distribution of the 91 *A. candida* isolates collected during this work. Colors represent the hosts from which the isolates were collected. Dots indicate the positions of isolates as given by their GPS coordinates. When the exact location is not known, the isolate is shown at the center of the country from which it was collected. Pie charts summarize data from each country and their sizes are scaled to the number of samples collected. Samples from *Brassica juncea* (India), *Brassica carinata*, *Camelina sativa,* and *Brassica rapa* (Canada) were provided by collaborators and are not represented (Jouet et al. 2018)

VII), 2 (XIII) and 4 (I) consist of genetically more diverged isolates. In particular, isolates collected on radish (*R. sativus*) are so diverged that isolate #5 clusters distinctly from other radish isolates in both phylogenetic trees.

A phylogenetic tree was also constructed using presence/absence polymorphisms of putative CHxC and related effectors of *A. candida*. Isolates collected on the same or closely related hosts share similar effectors candidate repertoires. Thus, as with the two previous trees, phylogenetic clustering based on these sequences mainly agrees with physiological race designation, including Races 1 (lineages VI and VII) and 4 (I). However, Races 2 and 11 were not discriminated, as in the tree built using the neutral loci (lineages XIII and XIV). Genetic diversity at putative effectors was higher compared to other loci, emphasizing the important role that these proteins may have in host adaptation. The relatedness of some lineages to others is similar in the phylogenies in Fig. 4.3 with example, the *B. juncea* virulent races, including the reference genome Race 2 (isolate Ac2v), clustering distantly from most others in both data sets. However, this is not always the case and in Fig. 4.3a (neutral loci), isolates infecting *Brassica oleracea* (lineage V, Race 9) appear closely related to Capsella, *Camelina,* and *Arabidopsis*-infecting races (I, II, and III), whereas they appear more distantly related in the tree created using the 400 kb contig (Fig. 4.3b). This observation may be explained either by incomplete lineage sorting (the random

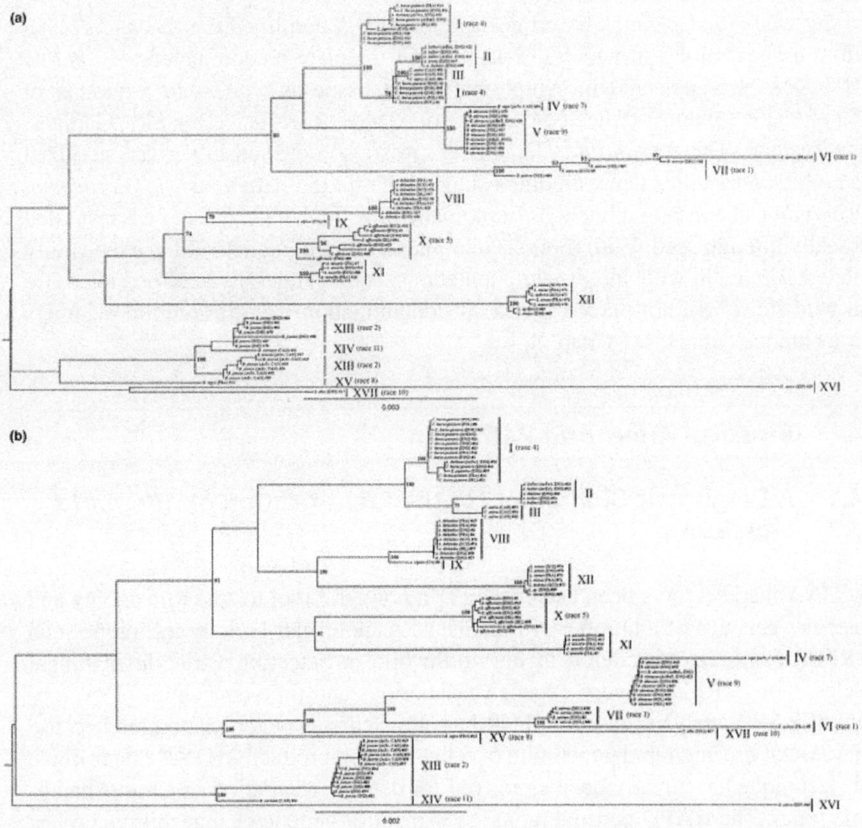

Fig. 4.3 Maximum-likelihood tree based on (**a**) the 32 concatenated neutrally evolving loci (24,249 bp) and (**b**) the ~400 kb contig of *A. candida* (398,508 bp) including 85 *A. candida* isolates collected from 20 host species and showing the 17 phylogenetic clusters. The tree was built with RAXML v.7.7.3 using gamma distributed rate variation among sites and the GTR model of DNA evolution. Bootstrap >70 are shown (100 replicates). The tree was viewed in FIGTREE v.1.3.1 and rooted at midpoint. The scale represents the number of substitutions per site (Jouet et al. 2018)

sorting of ancestral alleles into the descendant host-specific races) or hybridization by secondary contact of isolates from two host-specific races (Jouet et al. 2018).

4.3.5 A PCR-Based Assay for Detection of *Albugo*

The development and testing of a PCR-based assay for the detection of the internal transcribed spacer region 1 (ITS 1) of the ribosomal DNA (rDNA) of *A. candida* was described by Armstrong (2007). Similar PCR-based approaches have become widely used to detect *Albugo* spp. and other fungi within plant tissue. Primers were designed based on differences in the ITS 1 sequences between *A. candida* and *Lepidium oleraceum*. The primers preferentially amplified the *A. candida* ITS 1 from tissue samples in which the plant ITS 1 template predominated. *A. candida* ITS 1 DNA was detected in symptomatic host tissue as well as in a number of asymptomatic plants collected from Auckland Regional Botanic Gardens, New Zealand. The *A. candida* ITS 1 DNA was also detected in surface-sterilized seed of *L. oleraceum*. These findings suggest that in the native host *L. oleraceum*, *A. candida* can exists as a latent asymptomatic infection and can also be transmitted vertically through seed. With more development, this technique will have the ability to detect *Albugo* in wild and *ex situ* populations, and could assist in efforts to ensure that wild stocks are not placed at risk by contamination from restoration plantings (for technique details, see Chap. 8).

4.4 *Brassica–Alternaria* Pathogen

4.4.1 A Diagnostic Biomarker for the Identification of *Alternaria brassicae*

The DNA markers have been established as a powerful tool to study taxonomy and molecular genetics of a number of organisms. A molecular biology technique such as RFLP has been considered to be worthwhile in detecting variation in fungal species, though RFLP is very expensive, laborious and time consuming. Development of RAPD analysis, based on PCR has attracted considerable attention because of its ease of use and rapid generation of reliable and reproducible DNA fingerprints. This technique has already been employed for detecting variation in various pathogenic fungi. The RAPD permits rapid assessment of genetic changeability and has been brought in practice to study inter- and intraspecific changeability among the isolates of numerous fungal species. Reports have been documented concerning the genetic distinction intra- and inter-Alternaria spp., through RAPD molecular markers (Sharma and Tewari 1998). The genetic linking among the isolates of *A. brassicae* has been effectively analyzed by random primers to create reproducible polymorphisms. All amplified products revealed banding designs that directed the presence of genetic multiplicity. The RAPD has been employed earlier to course inter- and intraspecific genetic distinction in 14 isolates of *Alternaria brassicae*, which were gathered from different geographical regions of the world. Further, the data obtained by Singh (2016) showed genetic divergence among the 21 isolates of *A. brassicae* isolated from mustard. The genetic polymorphism within an *Alternaria* species has been positively studied employing RAPD molecular markers by many researchers. Genetic polymorphism among *A. brassicae* isolates from different geographical regions around the world was studied and low intraregional variation among Indian and Canadian isolates of *A. brassicae* with 75% resemblance among them was documented. This genetic changeability within an *Alternaria* species might be due to the existence of heterokaryosis, mutation, somatic hybridization, host selection, extensive dispersal, or of a cryptic sexual phase.

Genetic changeability was observed among 21 isolates of *A. brassicae* isolated from different cultivars of mustard growing in different parts of Uttar Pradesh, India. This could be the possible reason behind extreme and different disease reactions of genotypes at maximum locations. Correspondingly, RAPD has been employed earlier to investigate inter- and intraspecific variation in 32 *Alternaria brassicae* species, which were isolated from cauliflower (*Brassica oleracea* L. var., *botrytis*) and mustard [*Brassica juncea* (L.) Czern & Coss.] growing areas in India (Sharma et al. 2013).

Expressed Sequence Tags (ESTs) are the source of microsatellites, which can be employed to develop molecular markers for studies on the population genetics of *A. brassicicola* and closely associated *A. brassicae*, accountable for leaf spot of rapeseed-mustard. Usually, success rate of EST-SSR primers (percentage of SSR primers producing discrete amplification products) were observed to range from 50% to 100% between species within plants' genera and cross-species transmission of microsatellites within genera in fungi was more than 30%. Forty-five percent amplification of EST-derived microsatellites loci in *A. brassicicola* population was obtained. In cross-species amplification assay, 45% microsatellites loci of *A. dauci* showed amplification signal for *A. bataticola, A. solani,* and *A. zinnia*. This recommends that microsatellites developed from EST sequences are extremely transferable to other associated species. The more closely connected the organisms, the higher the rate of transferability because of more closely related species sharing more homology in microsatellite loci. Another probability for high rate of success in the amplification of EST-SSRs may be the consequence of numerous factors, such as the sequences from which the primers were derived, the adequate criteria used for primer design, and use of the species of same genus for the design and amplification of the primer set. The presence of high percentage of monomorphic SSR loci was observed in *A. brassicicola* population. This may be due to environmental and survival-based adaptations of fungus, which vary from habitat to small niche. There could be more explanations for the presence of monomorphic microsatellite in *A. brassicicola*: (1) it causes leaf spot on mustard and cauliflower, both have the same cropping period and environmental conditions and appears very late in season with low intensity, probably leading to less variability in population; (2) the number of isolates assessed was small and the genetic analysis was of low resolution; (3) there were repeated bottlenecks and/or single founder population/pathogen mating systems, causing in homogenizing effect; (4) microsatellite recognized may be located in the conserved region of the *A. brassicicola* genome; and (5) the incidence of *A. brassicicola* is less common over *A. brassicae* in India. All these arguments are reasonable, as evidenced by highly clonal nature with no known teleomorph of *A. brassicicola*. Five EST-SSR markers established from *A. brassicicola*, amplified *A. brassicae* isolates, and showed very low levels of polymorphism. Analysis of limited number of *A. brassicae* isolates might be one of the probable explanations for low level of transferability of microsatellites. Additionally, the wide distribution of *A. brassicae* in India over all mustard growing areas may lead to wide changeability in the population that may be possible reason for low level of transferability of microsatellites from *A. brassicicola*.

Nevertheless, there might be a high probability of transferability of *A. brassicicola* derived markers to *A. brassicae*, which necessities further investigation and statistical validation using large set of EST-derived SSRs. The results also established wide species transferability of developed EST primers and demonstrated that they may represent a set of well-conserved loci across the species. This may be consequent to the transfer of lineage specific genomic regions in fungi. It has been noticed from the study that the distribution of microsatellite in the *A. brassicicola* genome is not random. Tri-nucleotide repeats (AAC, GAA, and TAC) have been found to be common feature in EST derived microsatellites. High occurrence of these repeats in coding regions could be as a result of mutation and selection procedure for specific amino acids. The abundance of tri-nucleotide repeats EST-SSR is likely due to suppression of other kind of repeats in the coding region that reduces the frame-shift mutations in the coding regions. Additionally, there is a possibility that these tri-nucleotides in the coding region are translated into amino acid repeats (histidine, glutamic acid, threonine, serine, arginine, asparagines, glycines, etc.), which possibly contribute to the biological function of protein. Specific and speedy detection is very significant to monitor and quantify the occurrence of plant pathogens for efficient management of plant diseases. Presently, most of the assays used in *A. brassicicola* disease diagnostic rely on visual assessment of the symptoms, lesion diameter measurement or spore counting.

This monomorphic microsatellite marker was highly specific for *A. brassicicola* because no amplification signal was observed from other closely connected *Alternaria* species. No false positive or false negative results were observed with the assays revealed the stringency of specific primer set ABS28F and ABS28R, thus allowing specific and sensitive detection of *A. brassicicola* from cultures and plant samples. The assay was sensitive, reliably detecting 0.01 ng of genomic DNA per PCR of the target fungi. Consequently, the marker allows assessing fungal development from the initial stages of infection and eventually provides a fast and practical alternate to currently designated markers to differentiate and diagnose *A. brassicicola* from synchronously occurring fungus, *A. brassicae* and other fungi connected with rapeseed-mustard (Mishra et al. 2017).

4.4.2 Identification of Pathotypes Using Molecular Techniques

4.4.2.1 Genetic Variability

Genetic variability in nucleotide sequence of ITS region of four *Alternaria* species (*A. brassicae*, *A. brassicicola*, *A. raphani*, and *A. alternata*) infecting crucifers has not been recorded so far (Jasalavich et al. 1995). Cluster analysis of pathogenic variability data reveals a close relationship between Nazirhat (SS 04), Jaipur (SS 07), Sachha Khera (SS-10), and Samalakha (SS-11) isolates. Use of 100 Random Amplified Polymorphic DNA decamer primers indicates genetic variability among thirteen *A. brassicae* isolates. Almost all the isolates show relationship according to their geographical origin except Sachha Khera (SS-10) and Hatikhuti (SS-05) isolates. Pantnagar (SS-09) isolate was found closely related to Sachha Khera

(SS-10) isolate. No variability could be located among the *A. brassicae* isolates by Internally Transcribed Spacer—Amplified Fragment Length Polymorphism Molecular marker. Hence, pathogenic variability does exist among the isolates with variability also at genomic level but not in the highly conserved region of the genome of the pathogenic isolates of *A. brassicae* (Goyal et al. 2013). However, Internal Transcript Spacer analysis done by Sharma et al. (2013) shows that all isolates are 90–100% similar to each other, indicating genetic similarity among different *A. brassicae* isolates that vary pathogenically. However, analysis of 26 RAPD primers revealed a high level of genetic variability among 10 isolates of *A. brassicae* from different cvs. of *B. juncea* (Pramila et al. 2014).

RAPD analysis is more easy, efficient, fast and reproducible than RFLP analysis in the detection of intra-specific variation in three *Alternaria* species i.e., *A. brassicae*, *A. brassicicola* and *A. raphani* pathogenic to crucifers. Polymorphism within an Alternaria species by RAPD molecules marker has been described by many workers (Sharma and Tewari 1995, 1998; Kumar et al. 2008). Sharma and Tewari (1995) observed polymorphism among *A. brassicae* isolates from different geographical regions of the world. However, in 1998, they found low intraregional variations among Indian and Canadian isolates of *A. brassicae* with 75% similarity among them. However, RAPD analysis of *A. brassicae* isolates from different geographical regions of India by above hundred primers suggested a high degree of polymorphism among them. The dendrograms from both pathogenic and molecular analysis seems to indicate that the Pant Nagar (SS-09) and Hatikhuti (SS-05) isolates are quite different from the others and the two dendrograms follows the same trend (Goyal et al. 2013). BLAST analysis of the ITS of 32 *A. brassicae* isolates conducted by Sharma et al. (2013) showed high similarity among the isolates with *A. brassicae* of the NCBI database.

4.4.2.2 Proteome Analysis

Two isolates of *A. brassicae* with significant differences in virulence have been characterized at the proteome level. The morphological observations indicated the Ontario isolate to be more virulent by virtue of increased disease severity score as compared to the UAMH7476 isolate. This was further confirmed through histological observations that showed extensive colonization of the host tissue by the highly virulent isolate. Mycelial protein profiles of two differentially virulent *A. brassicae* isolates were compared using two dimensional gel electrophoresis (2DE) and mass spectrometry (MS) in order to identify proteins that may be responsible for the differences. Several differences in the mycelial proteomes of the two isolates were recorded. The proteins that were significantly abundant in the more virulent isolate included a protein with conserved actin related protein 2/3 domain, enolase, malate dehydrogenase and serine protease. The differential protein expression pattern can be exploited to identify putative virulence and pathogenicity factors in *A. brassicae* (Sharma et al. 2013).

4.4.2.3 Biochemical Variations

Biochemical constituents of *A. brassicae* isolates differ significantly (Khurana et al. 2005). The isolate BWL contains maximum ortho-dihydric phenols (O.D. phenols) and KTL-II has the least (Table 4.2). Isolate REW contains the highest amount of total phenols. The variation in amount of total phenols as well as ortho-dihydric phenols indicates the existence of variation in isolates. The isolate JD-II contains the maximum amount of reducing and nonreducing sugars while the isolate RTK has the minimum amount. The amount of reducing sugars present in different isolates also varies significantly in different isolates. The isolates BWL, CHR-II, CHR-I and RTK have significantly less amount of reducing sugars though least is in case of isolate BWL (0.74 mg). Variations in amount of sugars in isolates can be categorized into three groups as maximum, moderate, and minimum levels.

Vishwanath and Kolte (1997) had reported that *A. brassicae* isolate containing high amount of carbohydrate is an indicator of virulent nature of pathogen. The estimation of total RNA content in different isolates reveals significant differences among the isolates. Isolate KTL-II has the maximum amount of RNA while minimum is recorded in isolate SPT. On the basis of RNA contents, isolates can also be grouped in to three categories. The protein content in different isolates of the pathogen shows significant differences. Isolate CHR-I and JD-I contains the maximum amount of proteins. Isolate KTL-I have the least amount of proteins. Similarly, there are significant differences among the isolates in their free amino acids content. Isolate JD-I has the highest amount of free amino acids. Isolate SPT shows minimum amount of free amino acids. *Alternaria brassicae* isolates which have more amount of proteins contain moderate amount of free amino acids and RNA contents while isolates which has high amount of RNA also contained moderate amount of proteins and free amino acids. On the basis of similarity in biochemical composition, all isolates were grouped into three categories, maximum, moderate, and minimum. Isolates BHI, JHR, KTL-I, SRS, CHR-I, and SPT were placed in the first group containing maximum amount of bio-chemicals. Isolates BWL, RTK, CHR-II, HSR, KTL-II and REW were in the second group being moderate amount while isolates JD-I and JD-II were in the third group being minimum. The study reveals the considerable variability in the isolates of *A. brassicae* on the basis of biochemical contents.

4.5 *Brassica–Colletotrichum* Pathogen

4.5.1 Identification of Effectors/Genes of *Colletotrichum* for Virulence

In the case of plant pathogenic hemibiotrophs, colonization and the initial biotrophic interaction with host cells is facilitated by pathogen-encoded small, secreted proteins termed effectors/genes. Biotrophy-specific hyphal cells play important roles in transporting effectors into the host cells and in obtaining nutrition from the host. Depending on the fungal species, the interfacial matrix is either continuous with the

Table 4.2 Biochemical constituents (mg/g) * in isolates of *Alternaria brassicae* (Khurana et al. 2005)

		Biochemical constituents						
Isolates	Ortho-dihydric phenols	Total phenols	Nonreducing sugar	Reducing sugars	Total sugars	RNA	Proteins	Free amino acids
BHI	1.6	9.5	2.5	2.0	4.5	6.8	66.8	49.2
BWL	3.5	10.4	3.7	0.7	4.5	9.4	85.7	56.5
CHR-I	2.0	5.5	2.8	1.0	3.9	6.4	103.0	37.5
CHR-II	2.2	4.9	1.7	1.0	2.7	8.8	74.2	45.7
HSR	1.7	10.5	6.2	3.5	9.7	8.6	59.9	57.7
JD-I	2.1	6.1	6.1	1.4	7.5	9.6	100.4	63.9
JD-II	2.8	13.3	10.5	5.3	15.8	11.8	76.9	38.9
JHR	1.5	9.6	4.4	1.8	6.1	5.8	67.5	40.9
KTL-I	3.4	13.4	4.6	1.5	6.1	7.5	49.2	51.8
KTL-II	1.3	12.0	4.0	1.5	5.5	16.4	81.9	52.8
REW	2.2	13.5	2.2	1.6	3.8	6.9	85.2	50.2
RTK	1.4	9.7	1.5	1.1	2.6	11.6	89.9	41.3
SRS	2.2	7.9	4.8	3.2	8.0	6.5	68.7	40.3
SPT	2.4	8.2	2.8	2.2	5.0	4.5	70.4	25.6
C.D. (5%)	0.13	0.33	0.49	0.16	0.28	0.72	3.63	8.11
SEM±	0.06	0.15	0.23	0.07	0.13	0.33	1.69	3.78

*Based on dry mycelial weight in 1000 mg

host plant apoplast such as for Pucciniales or separated into an interfacial apoplastic compartment like that in *M. oryzae*. The interfacial membrane among a variety of pathosystems ranges from undifferentiated plant plasma membranes to highly specialized membranes with complicated elaborations and unique components. Vesicles are abundant in both host and pathogen cytoplasm near the interface, implying that both are involved in active production of secreted compounds into and across interfacial zones. Shimada et al. (2006) found some evidence for localized specialization of the interfacial membrane around BH of *C. higginsianum*. Biotrophic infection by *C. higginsianum* has differences from other biotrophic fungus, and the process of interfacial membrane exchange with the fungus deserves further study (Yan et al. 2018).

In *C. higginsianum*, inventories of putative effectors have been predicted from the annotated genomes of *C. higginsianum* revealing 18 genes, of which six were not predicted to be secreted, and two were chitinases, leaving 10 putative secreted LysM effectors/genes. The LysM domain comprises 40–60 amino acid residues and mediates binding to chitin and peptidoglycans. Chitin is a microbe-associated molecular pattern (MAMP) that can be detected by plant pattern recognition receptors (PRRs) to activate a variety of MAMP-triggered immune responses. To avoid recognition by host receptors, several ascomycetes are known to produce effector proteins which either block the activity of host plant chitinases or compete with host plant receptors which bind chitin fragments. The function of two effectors, ChELP1 and ChELP2, homologs of LysM proteins were characterized in *C. higginsianum*. The ChELP2 has been found to be located on the surface of bulbous biotrophic hyphae at the interface with living plant cells, but it has not been discovered in necrotrophic hyphae. In previous experiments, recombinant ChELP1 and ChELP2 were found to bind chitin oligomers in vitro with high specificity and high affinity. Both proteins suppressed chitin-triggered activation of two immunity-related plant mitogen-activated protein kinases in *Arabidopsis*. These results suggested a double role for these LysM proteins as effectors for suppressing chitin-triggered immunity and as proteins essential for appressorial development and function (Takahara et al. 2016). Sixty-one putative effector proteins were separately cloned into a plant expression vector providing an N-terminal GFP tag, and the tagged proteins were transiently expressed directly inside plant cells using ATMT. Among them, subcellular localization of 16 candidate effectors was verified, nine were imported into plant nuclei, three were imported into the matrix of peroxisomes, three decorated cortical microtubule arrays, and one was associated with Golgi stacks. These findings revealed that plant peroxisomes, microtubules, and Golgi are novel targets for fungal effectors (Robin et al. 2018).

4.6 *Brassica–Erysiphe* Pathogen

4.6.1 Molecular Quantification of Powdery Mildew

Two protocols for the quantification of the powdery mildew disease on *Arabidopsis* with reproducible results over a broad range of infection, phenotypes have been developed (Weßling and Panstruga 2012) (Fig. 4.4). The techniques are also useful for the quantification of subtler differences, although such instances may require higher number of experimental replications. For the qPCR-based method, efficient and specific primer combinations for the amplification of both host and pathogen DNA have been used. The quantified amount of *G. orontii* relative to *Arabidopsis* genomic DNA correlated well with the amount of fungal biomass. Fungal development was reflected accurately across both time, and different genotypes, and could also be determined upon the occurrence of localized powdery mildew-induced host cell death. 5 dpi was determined to be the optimal time point for comparative analysis among genotypes, but the method can also be used to resolve kinetics of powdery mildew pathogenesis (Plate 4.1).

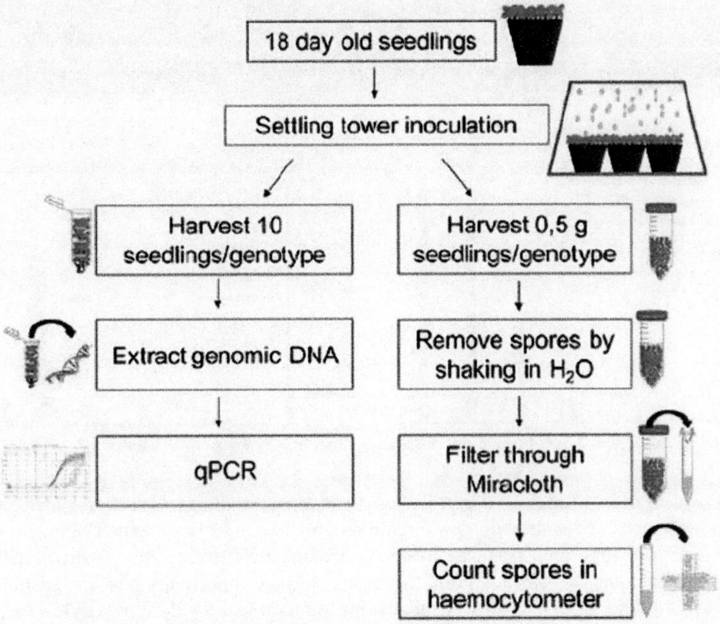

Fig. 4.4 Schematic overview of developed methods. Simplified view of the workflow of the qPCR- and spore count-based powdery mildew quantification procedures (Weßling and Panstruga 2012)

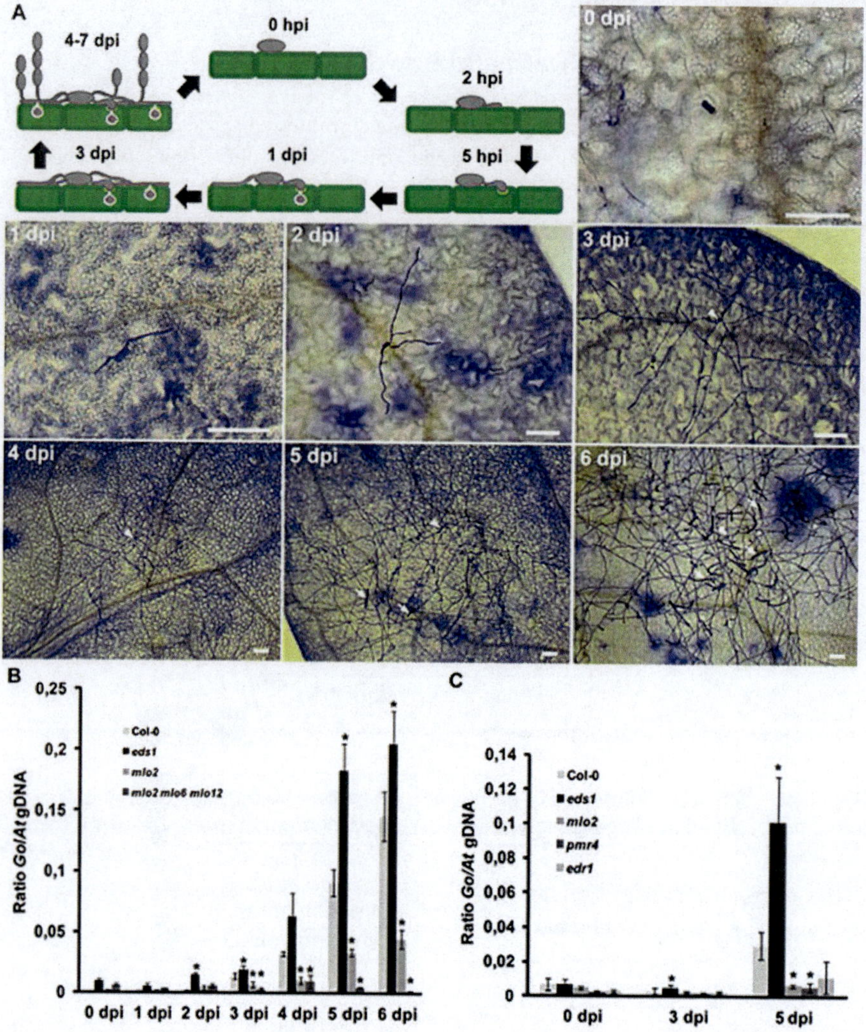

Plate 4.1 Powdery mildew disease progression on *Arabidopsis* seedlings. (**a**) Schematic overview and microscopic images of powdery mildew disease progression on Col-0 seedlings. Samples were harvested at indicated time points, and stained with Coomassie Brilliant Blue. Arrows indicate conidiospores chains, and arrowheads point to the initial spore. hpi, hours post inoculation (**b**) qPCR analysis of a time series of powdery mildew infection on Col-0 wild type, eds1, mlo2, and mlo2, mlo6 mlo2 seedlings. Ratios of *G. orontii* to *Arabidopsis* gDNA were determined by qPCR with primers R189/R192, and R193/R194, respectively. Bars represent the mean ± standard deviation of three technical replicates from a DNA sample of ten pooled seedlings grown in five different pots (two seedlings/pot used). (**c**) qPCR analysis of powdery mildew infection on *Arabidopsis* mutants that show powdery mildew-induced cell death. Representative time points of infection on Col-0 wild type, eds1, mlo2, pmr4, and edr1 seedlings were used. Ratios of *Golovinomyces orontii* to *Arabidopsis* gDNA were determined by qPCR with primers R189/R192, and R193/R194, respectively. Bars represent the mean ± standard deviation of three DNA samples (each derived from ten pooled seedlings grown in five different pots) with three technical replicates each. Asterisks indicate statistically significant differences to Col-0 in two-tailed Student's *t*-test ($P < 0.05$). Schematic overview in (**a**) is courtesy of Justine Lorek. Scale bars in

A spore count assay can also be useful for powdery mildew research. This assay has a wide dynamic range and accurately and reproducibly determined differences in conidiation by the pathogen (Plate 4.2). Both assays yielded comparable results for the susceptibility of the genotypes tested, and can easily be adapted to other *Arabidopsis*-infecting powdery mildew species, and other plant- powdery mildew pathosystems. Both methods require similar amounts of time for data generation. The qPCR method is intrinsically more cost-intensive but allows simultaneous DNA extraction, and PCR analysis of many samples. Spore counts are more cost-effective, and owing to the lower number of practical steps involved, offer less possibilities for experimental error (Fig. 4.4). Currently, powdery mildew infection is either assessed semi-quantitatively by coarse categorization and thus macroscopic symptoms or quantitatively by microscopic penetration rate counts at early, and/or conidiophores counts at late stages (Table 4.3). The latter methods require staining, mounting, and microscopy, and are therefore time consuming, and labor-intensive. Both assays introduced here are quantitative, and are applied at late stages of infection when differences between genotypes are more pronounced. They do not require staining, are easy to perform, and can generate quantitative data for larger numbers of genotypes (Table 4.3). In addition, the spore count assay is based on many seedlings rather than few mature plants, which offers the advantage of an averaging effect, lowering the possibility of experimental outliers. These assays can overcome limitations of currently used methods of powdery mildew quantification. They might be particularly suitable for the assessment of larger numbers of different mutant lines e.g. in the course of follow-up analyses of genes identified by omics approaches.

4.6.1.1 Plant Material and Inoculation Methods

This technique has been developed using the *Arabidopsis thaliana* Col-0 genotype, and the *edr1*-1, *eds1*-2, *pmr4*-1, *mlo2*-6 single, and *mlo2*-5 *mlo6*-2 *mlo12*-1 triple mutants in the Col-0 genetic background. Approximately 100 seeds were sown per pot of soil substrate, and five pots were used per genotype. After stratification for 2 days at 4 °C in darkness, plants were grown for 18 days at a day/night cycle of 10/14 h in a light chamber with 22°/20 °C day/night temperature, and a relative humidity of 60%. The *G. orontii* isolate MPIPZ was propagated on 4 week old *eds1*-2 plants, and conidia were used at 14–21 dpi. Inoculations were performed in a simple 80 cm high cardboard settling tower whose opening was covered with 80 μm nylon mesh (Reuber et al. 1998). The tower contained up to nine pots per inoculation. Three consecutive rounds of infection have been used. In each round, pots of all genotypes were included. A fine paint brush was used to harvest conidia from four heavily infected leaves, and to separate the conidia by brushing them through the

Plate 4.1 (continued) (**a**) are 100 μm. Data shown are representative of three independent experiments (Weßling and Panstruga 2012)

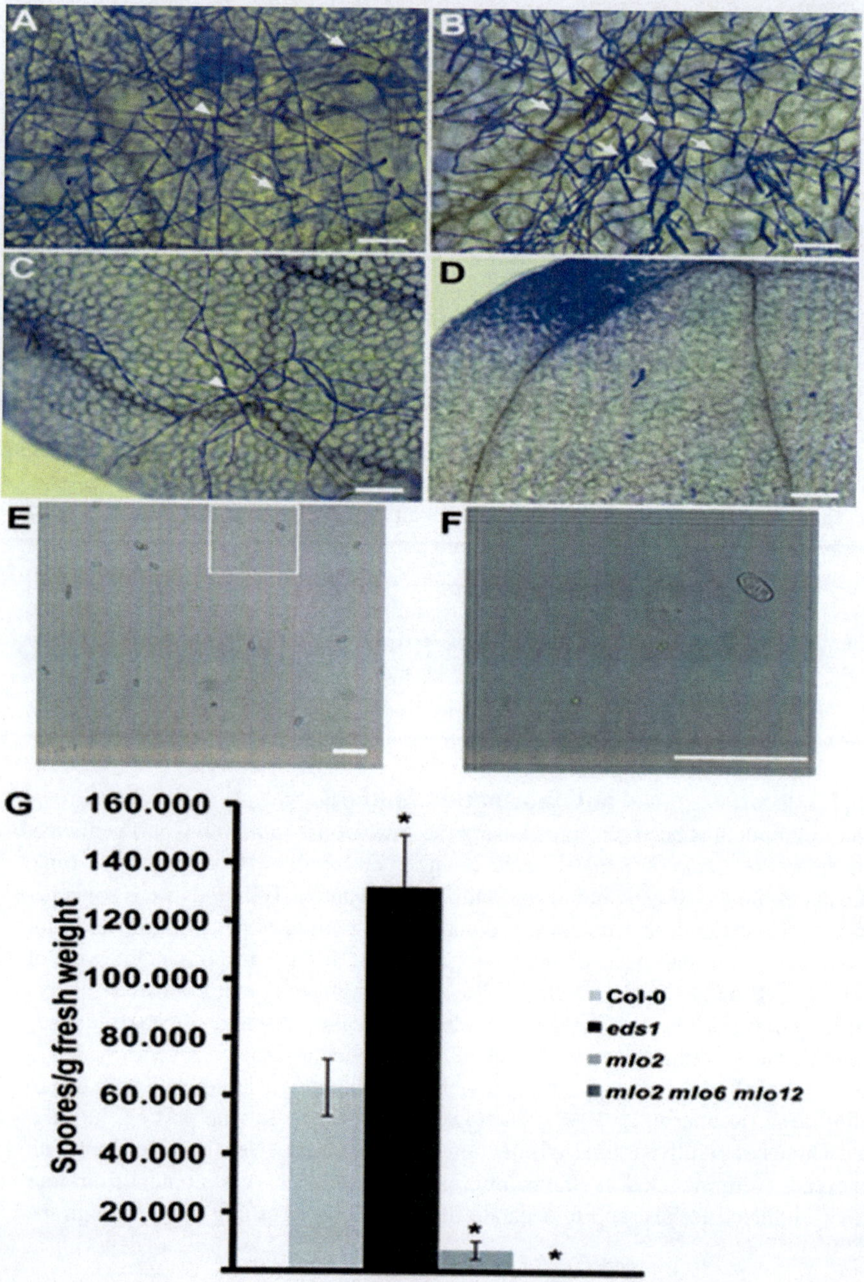

Plate 4.2 Analysis of powdery mildew infection by spore counts. *Golovinomyces orontii* infected leaves were harvested at 5 dpi from Col-0 wild type (**a**), eds1 (**b**), mlo2 (**c**), and mlo2 mlo6 mlo2 (**d**) seedlings, and stained with Coomassie Brilliant Blue. Arrows indicate conidiospores chains, and arrowheads point to the initial spore. (**e, f**) Bright field image of isolated spores in the hemocytometer. (**f**) is a close-up of the indicated area in (**e**). (**g**) Spore counts of indicated genotypes at 6 dpi normalized to seedling fresh weight. Bars represent the mean ± standard deviation of three samples (500 mg of seedlings each) from one experiment counting eight fields/sample. Asterisks indicate

nylon mesh. Inoculation density was approximately 750 spores/cm^2. Newly inoculated seedlings were then returned to the growth chamber.

4.6.1.2 Genomic DNA Extraction

Ten seedlings per genotype were harvested across pots, and frozen in liquid nitrogen. Genomic DNA was extracted essentially as previously described (Brouwer et al. 2003). Approximately 151 mm, and 100 mg 0.2 mm diameter glass beads were added and the frozen material was disrupted in a MM400 mixer mill for 2 × 1 min at 30 Hz. Subsequently, 300 μL lysis buffer (2.5 M LiCl, 50 mM Tris–HCl, 62.5 mM Na2-ethylenediamine tetra-acetic acid (EDTA), and 4.0% Triton X100, pH 8.0), and an equal volume of phenol: chloroform: isoamyl alcohol (25:24:1 v/v) were added, and samples were homogenized for 30 s at 30 Hz in the mixer mill. After centrifugation (5 min, 16,000 g), the supernatant was recovered, and the genomic DNA was precipitated by the addition of two volumes of 100% ethanol, incubation for 15 min at −20 °C, and another round of centrifugation. The DNA pellet was washed with 70% ethanol, air dried, and re-suspended in Millipore water. DNA quality and concentration were inspected on a Nanodrop system.

4.6.1.3 Quantitative Real-Time PCR

For qPCR, 15 μL samples were prepared using the Brilliant Sybr Green QPCR Reagent Kit according to the manufacturer's protocol. TheTaq DNA polymerase provided was replaced by another Taq polymerase, and the corresponding standard buffer. A final primer concentration of 0.4 μM, and three technical replicates per sample were used. QPCR was carried out according to the following protocol: denaturation at 95 °C for 3 min, 40 repeats of 95 °C for 20 s, 61 °C for 20 s, and 72 °C for 15 s. A melting curve analysis was conducted from 55 to 95 °C in 0.5 °C steps, and 10 s dwell time to confirm the amplification of single amplicons. Additionally, amplicon size, and identity were confirmed on a 2% agarose gel, and by DNA sequencing, respectively. The ratio of *G. orontii* to *Arabidopsis* genomic DNA was calculated using the ΔΔCt method (Pfaffi 2001).

4.6.1.4 Spore Counts

At 6 dpi, three samples of approximately 500 mg of seedlings were harvested per genotype. Five ml water was added, and spores liberated by vortexing for 30 s at maximum speed. The spore solution was filtered through Miracloth to remove large debris, and spores were four-fold concentrated by centrifugation (5 min, 4000 g). For each sample, spores were counted in eight 1 mm^2 fields of a Neubauer-improved hemocytometer, and results were averaged. Finally, spore counts were normalized to the initial weight of seedlings (Weßling and Panstruga 2012).

Plate 4.2 (continued) statistically significant differences to Col-0 in two-tailed Student's *t*-test ($P < 0.05$). Scale bars in (**a–f**) are 100 μm. Data shown are representative of three independent experiments (Weßling and Panstruga 2012)

Table 4.3 Comparison of methods to assess powdery mildew infection (Weßling and Panstruga 2012)

	Macroscopic categorization	Host cell entry counts	Conidiophores counts	Hyphal area quantification	qPCR	Conidia counts
Assay type	Semi-quantitative	Quantitative	Quantitative	Quantitative	Quantitative	Quantitative
Staining required	No	Yes	Yes	Yes	No	No
Microscopy required	No	Yes (hundreds of interaction sites)	Yes (multiple colonies)	Yes (multiple colonies)	No	Yes
Stage of pathogenesis scored	Late (conidiation)	Early (host cell entry)	Late (conidiation)	Early to late (host cell entry to conidiation)	Middle to late (hyphal expansion to conidiation)	Late (conidiation)
Importance of equal inoculation density	High (no normalization)	Low (internal normalization)	Low (internal normalization)	Low (internal normalization)	Low (internal normalization)	Medium (averaging effect of scoring multiple plants at once)
Suitable for high through put analysis	Yes	No	No	No	Yes	Yes

4.6.2 Detection of Genes Expression in Response to Powdery Mildew Infection

The broad range of transcriptional changes that occur in *Arabidopsis* during powdery mildew infection (Zimmerli et al. 2004; Fabro et al. 2008) reflect both defense responses mounted by the plant as well as possible manipulation of the host by the fungal pathogen. With the sequencing of the *Arabidopsis* genome, and availability of novel technologies for transcript profiling (oligonucleotide-based microarray chips) it has been possible to peek into the gene expression changes that occur during the infection process. Several dedicated websites such as Gene vestigator https://www.genevestigator.ethz.ch/ and ATGen Express (http://www.arabidopsis.org/info/expression/ATGenExpress.jsp) provide publicly available collections of microarray data on *Arabidopsis* challenged with a variety of pathogens including powdery mildews, which represent a largely unexplored treasure of information. However, the spatial resolution of microarray data is limiting, since most microarray experiments are performed on whole leaf samples, and as such only provide a gross overview about changes in transcript abundance in pathogen-challenged leaf tissue. Since, powdery mildews exclusively colonize the epidermal layer of the leaf, many tissue-specific transcriptional changes are likely to be masked by proportionately larger amounts of mesophyll transcripts. Even within the epidermis, attacked cells, and successfully colonized cells each differ in their transcript profile from nonattacked cells (Gjetting et al. 2004, 2007).

Owing to these limitations, expression levels of numerous genes in infected host cells might be considerably under-estimated. Employment of techniques that allow transcript profiling of powdery mildew-challenged *Arabidopsis* leaves at cellular resolution, e.g. based on RNA extraction via micro capillaries or by laser microdissection (Kehr 2003), is thus urgently required to refine the currently existing data sets (Micali et al. 2008).

Earlier transcriptional changes in *Arabidopsis* compared the response observed in host (*Golovinomyces cichoracearum*), and nonhost interactions (Bgh; Zimmerli et al. 2004) at 8, 18, and 24 h postinoculation. Many of the differentially regulated genes displayed a similar response in both host, and nonhost combinations. However, in the nonhost interaction, the response was more rapid, and of greater amplitude than in the compatible interaction, supporting the hypothesis that common defense machinery is activated upon attack by adapted, and nonadapted pathogens, and reinforcing the notion that adapted pathogens are able to partially suppress basal defense in their host species. Several genes that exhibit lower transcript abundance after pathogen challenge, a majority encode proteins that are involved in photosynthesis, metabolism, transport, and transcription/translation. Consistent with these functions, half of the gene products are either known or predicted to be chloroplast residents. This observation corroborates further studies indicating that *Arabidopsis* suffers up to 30% reduction in the gain of dry weight following *Bgh* infections (Zimmerli et al. 2004). This decrease is more modest in compatible infections with *G. cichoracearum*. An inverse relationship between defense responses, and plant growth has been also observed in interactions between cereals, and powdery mildews, and underscores the metabolic costs associated with activation of defense pathways (Swarbrick et al. 2006; Wright et al. 1995a, b). Intriguingly, a recent microarray analysis of the compatible interaction between *Arabidopsis,* and *G. cichoracearum*, revealed that host photosynthesis is likely stimulated by the fungus, through the induction of genes involved in chlorophyll binding, and chlorophyll a oxygenase (Fabro et al. 2008). Though it is conceivable that an increase in photosynthetic rate is required, as the metabolic demands of the infected epidermal tissue increase, these findings are somewhat contradictory to those obtained in the study of Zimmerli et al. (2004). Manipulation of the host metabolism by the fungal parasite thus likely involves up as well as down regulation of photosynthetic genes. However, many gene transcripts related to defense responses accumulate upon both host, and nonhost powdery mildew attack (transcripts encoding glycosyl hydrolases and β-1, 3-glucanase). Consistently, many of the respective genes contain in their up-stream promoter region the cis-regulatory element OCS, which is known to be responsive to auxin, SA, and hydrogen peroxide. Transcripts specifically accumulating during nonhost interactions encode defensins, and derive from genes that are regulated by JA, and ethylene. SA signaling therefore does not appear to play an important role in nonhost defense against Bgh. Alternatively, SA-controlled genes may only become activated once the penetration defenses have failed. However, analysis of the transcriptional response of *Arabidopsis* leaves to chitin (the main fungal cell wall polysaccharide), and chitin derivatives identified several genes involved in defense responses (Ramonell et al. 2002). Genes encoding plant-derived chitinases exhibit higher transcript accumulation upon pathogen challenge compared to control conditions, and chitin oligomers obtained as fungal cell wall degradation products constitute potent signaling molecules responsible for induced resistance to *G. cichoracearum.* This resistance response appears to be activated through signaling routes that are distinct from the SA, JA, and ET pathways (Ramonell et al. 2002). Genes encoding PR5, two WRKY transcription factors, and several disease resistance proteins belonging to the TIR-NB-LRR family were induced upon chitin treatment (Ramonell et al. 2002). Genes with reduced transcript abundance included ETR2, encoding a trans-membrane protein kinase with a LRR domain, the defensin PDF2.5, RAR1 (encoding a protein with two zinc-finger-like domains that is required for the accumulation of many R proteins), and several other genes coding for zinc-finger proteins. To functionally validate the microarray experiments, T-DNA insertion mutants within seven of the differentially expressed genes were tested for susceptibility to *G. cichoracearum.* Three of the seven mutants, encoding a putative RING zinc-finger, and two putative disease resistance proteins, displayed enhanced susceptibility to the pathogen, further reinforcing the role of chitin sensing, and signaling in plant defense responses (Ramonell et al. 2002). The critical role played by chitin perception in plant defense responses was strengthened by the cloning of the presumptive *Arabidopsis* chitin (co-) receptor LysM Receptor-Like Kinase1 gene (*LysM RLK1* or *CERK1*) by two independent groups (Miya et al. 2007; Wan et al. 2008). The gene codes for a plasma membrane-resident protein with a single transmembrane domain, three extracellular LysM domains and a cytoplasmic serine/threonine kinase domain (Plate 4.3). Mutations in this gene abolish virtually

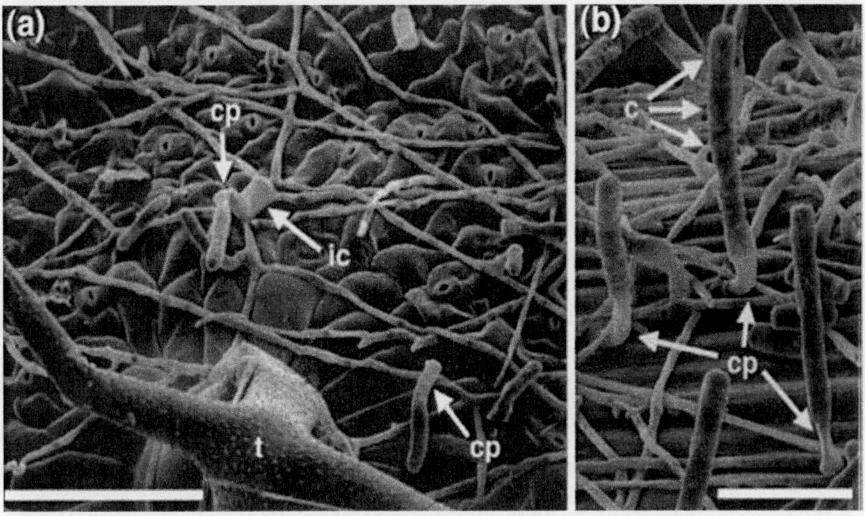

Plate 4.3 Scanning electron micrographs of *Erysiphe orontii*. (**a**) A 5-day-old colony of *E. orontii* on a Columbia leaf. *ic* initial conidium, *cp* conidiophore, *t* trichome. Scale bar = 100 μm. (**b**) Mature conidiophores on a pad4 stem. *cp* conidiophore, *c* conidium. Scale = 50 μm (Reuber et al. 1998)

all changes in gene transcription normally observed upon chitin perception by the plant, without affecting other elicitor perception such as flagellin sensing, suggesting that the protein is part of the chitin receptor complex (Wan et al. 2008). It is conceivable that in analogy to the FLS2 receptor kinase, which is responsible for flagellin-triggered signaling (Asai et al. 2002), LysM RLK1/CERK feeds into MAP kinase signaling, and defense gene activation (Plate 4.1). Interestingly, although loss of LysM RLK1 caused enhanced susceptibility to the host fungus *G. cichoracearum* at the postpenetration stage (Wan et al. 2008), it did not affect the plant response to the adapted hemibiotrophic fungus *Colletotrichum higginsianum,* and resulted in only a marginal increase in susceptibility to the nonadapted necrotrophic fungus *Alternaria brassicicola*. These observations suggest that pathogen perception by the plant is microbe-specific, and implies the existence of more elicitor-specific receptors yet to be discovered (Micali et al. 2008).

In addition to genes involved in defense responses, genes that are induced because they are likely required for pathogen growth have also been identified (Fotopoulos et al. 2003). The observation that *Arabidopsis* leaves infected with *G. cichoracearum* have an enhanced sugar uptake compared to noninfected leaves prompted an investigation of sugar transporter expression upon powdery mildew challenge. The *G. cichoracearum* induces expression of STP4, encoding a glucose transporter as well as βfruct1, coding for a cell wall invertase, in several cell types of the leaf (most notably the epidermis, and vascular bundle) irrespective of whether they are penetrated by the fungus or not. This transcriptional induction is sustained for up to 6 days postinoculation, and is not reproducible in a nonhost interaction with

Bgh. A similar role for sugar transporters in fungal development was also found in the interaction between the rust fungus *Uromyces fabae,* and broad bean (*Vicia faba*). However, in this case, the hexose transporter is of fungal origin, and resides in the haustorial membrane (Voegele et al. 2001), suggesting that exchange of nutrients at the plant/biotroph interface requires transport facilitators on both sides. It is not surprising that sugars, and therefore sugar biosynthesis and transport, would be important for fungal growth, and development, and enhanced by successful fungal pathogens. The mechanisms leading to this genetic, and physiological manipulation remain an enigma. However, new efforts into powdery mildew studies promise to lift the veil on the mystery (Micali et al. 2008).

4.6.3 Phylogenetics of *Erysiphe*

Phylogenetics relationship of powdery mildew of crucifers with other powdery mildew species has been determined through sequence analysis of the ITS rDNA (Choi et al. 2009), and sequence comparison of DNA encoding the 5.8S rRNA, ITS1, and ITS2 (Adam et al. 1999).

4.6.3.1 Sequence Analysis of the ITS rDNA

The rDNA gene sequences covering both highly conserved ribosomal sub-units, and variable internal transcribed spacer (ITS) regions are widely used to evaluate the polygenic relationships among related fungal species, and races within species (Saenz and Taylor 1999; Takamatsu and Kano 2001). The PCR amplification, and sequencing of the ITS rDNA was performed by Choi et al. (2009) using the following primer set: ITS 1: 5′TCCGTAGGT-GAACCTGCGG-3′, and ITS4: 5-′-TCCTCCGCTTATTGATATGGC-3′ (White et al. 1990). The resulting sequence of the complete ITS rDNA of *Erysiphe* sp. KUS-F23994 (accession no. FJ548627) has been deposited in the Gen Bank of National Centre for Biotechnology Information (NCBI, http://www.ncbi.nl.nih.gov/). The nearly complete ITS rDNA of *Erysiphe* sp. KUS-F23994 was aligned with other *Erysiphe* spp. nucleotide sequences from the NCBI Gen Bank database.

Molecular phylogenetic reconstructions of these ITS rDNA sequences were done using MEGA4, version 4.0 (Tamura et al. 2007), for neighbor-joining analysis using Tajima-Nei distances. To test the reproducibility of results, 1000 bootstrap replications (Felsenstein 1985) were performed by the parameters in default values. As shown in ITS based phylogenetic tree (Fig. 4.5), the *Erysiphe* sp. KUS-F23994 formed a well-supported group (a highly boot strap value of 97%) with three sequences of powdery mildew fungus *E. cruciferarum* which is able to infect *Arabidopsis thaliana, Brassica campestris,* and *B. rapa*. However, sequence divergences between the *Erysiphe* sp. KUS-23994, and other *Erysiphe* species, except *E. cruciferarum,* was more than 3%. Taken together, the polygenetic analysis revealed that the *Erysiphe* sp. KUS-F23994 is indeed identical to *E. cruciferarum* (Choi et al. 2009).

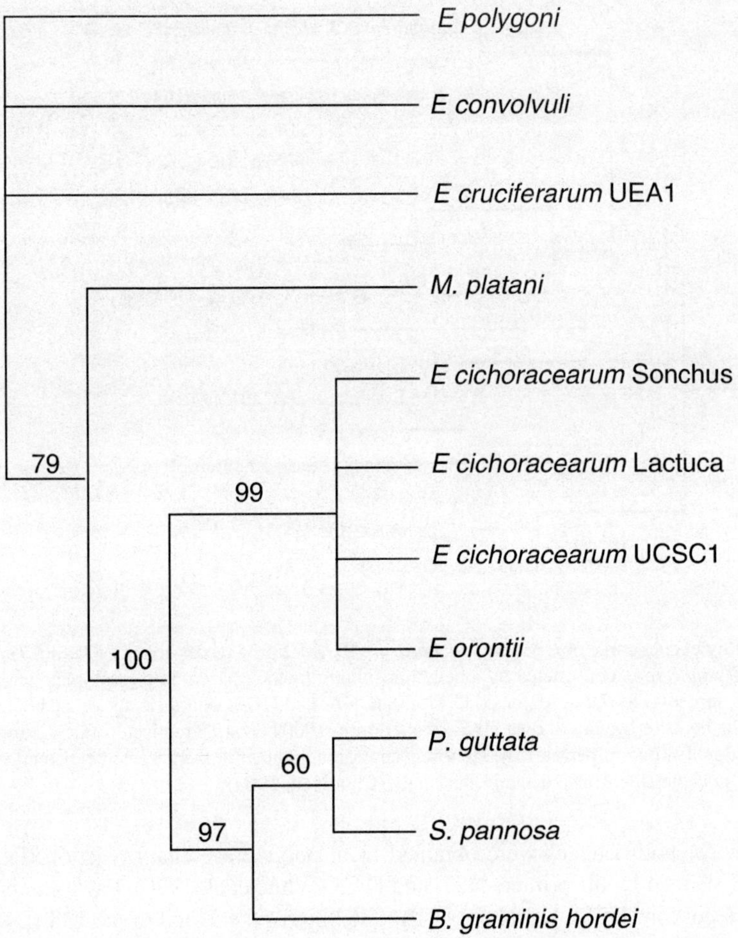

Fig. 4.5 Phylogenetic analysis of UEA1, UCSC1, and nine other powdery mildew fungi based on the sequences (Saenz and Taylor 1999). Strongly supported (95%) branches are shown in bold. Numbers shown are frequencies of branches in phylogenies estimated from bootstrap sampled data bases (Felsenstein 1985). Only values >60% are reported. *M. platani*: *Microsphaera platani*, *P. guttata*: *Phyllactinia guttata*, *S. pannosa*: *Sphaerotheca pannosa*, *B. graminis hordei*: *Blumeria graminis* f. sp. *hordei* (Adam et al. 1999)

4.6.3.2 Sequence Comparison of DNA Encoding the 5.8S rRNA, ITS1, and ITS2 Regions

To distinguish the powdery mildew isolates of *E. cruciferarum* (UEA1), and *E. cichoracearum* (UCSC1), sequence comparison of DNA was performed by Adam et al. (1999): The rDNA gene cluster contains both highly conserved ribosomal subunits, and variable internal transcribed spacer (ITS) regions that can be compared to evaluate the phylogenetic relationships among related fungal species, and races within species (White et al. 1990; O'Donnell 1992). The nucleotide

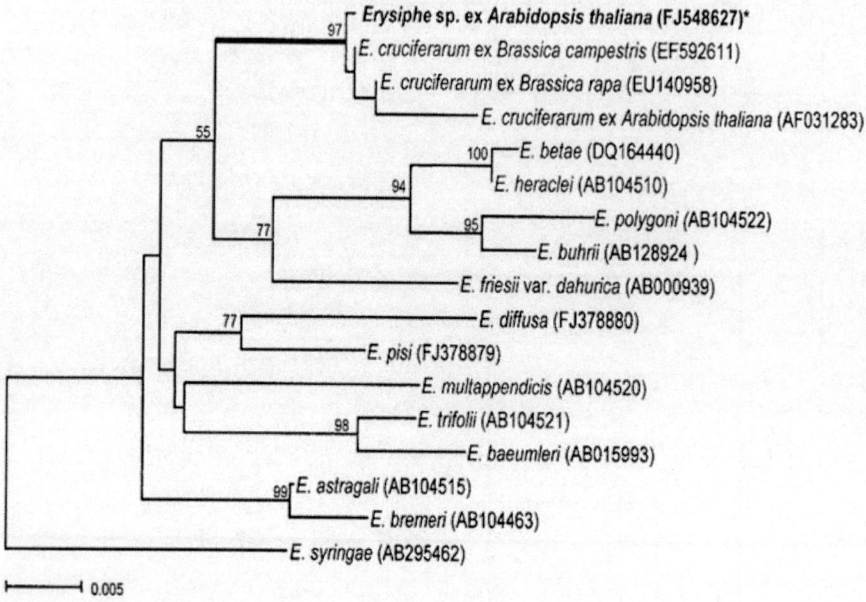

Fig. 4.6 Phylogenetic tree of *Erysiphe cruciferarum* KUS-F23994 (accession no. FJ548627), and other 16 *Erysiphe* spp., constructed by a neighbor-joining method based on the complete internal transcribed spacer (ITS) rDNA regions (ITS1, 5.8S rDNA, and ITS2). Numbers above the branches represent the bootstrap values of over 50% obtained from 1000 bootstrap replicates. Bar = Number of nucleotide substitutions per site. The GenBank accession numbers are represented in parentheses. Asterisk (*) denotes the isolate using in this study (Choi et al. 2009)

sequences for both isolates were obtained from polymerase chain reaction (PCR) fragments defined by the primers NS7, and ITS4 (White et al. 1990). Each sequence included a portion (UCSC1, 384 bp; UEA1, 363 bp) of the 3′ end of the 18S rDNA, the complete 5.8S rDNA subunit (154 bp), and the two flanking internal transcribed spacers, ITS1, and ITS2, as well as a 57- to 60-bp portion of the nuclear large 28S rDNA. ITS1 varied in size from 223 to 187 bp between UEA1, and UCSC1, respectively. The ITS2 region was also smaller in UCSC1 (165 bp) than in UEA1 (184 bp). As expected, the 3′ end of 18S, the 5.8S, and 5′ end of 28S rDNA sequences displayed a high degree of conservation between the two *Erysiphe* isolates, with 4, 1, and 2 bp changes in the respective segments. The occurrence of base pair changes in the conserved rDNA subunit genes in addition to a number of base pair changes in ITS1, and ITS2 confirmed, at the molecular level, that both isolates were from distinct phylogenetic groups (Saenz et al. 1994; Hirata and Takamatsu 1996; Saenz and Taylor 1999). The sequences were aligned relative to each other, matching sequences of *B. graminis* f. sp. *hordei,* and eight other powdery mildews for a phylogenetic reconstruction (Saenz and Taylor 1999) (Fig. 4.6). The tree is a strict consensus of the eight most parsimonious trees, which were found by a heuristic search employing the random stepwise addition option of PAUP (Swofford 1993). Branch support was determined by 1000 boot strapped data sets, which are displayed on the tree as numbers above the branches. The tree was 262 steps in length; the consistency index was 0.828, the retention index was 0.763, and the re-scaled consistency index was 0.632 (Farris 1989). This analysis showed that the UCSC1, and the UEA1 isolates are distinct *Erysiphe* spp. The UCSC1 isolate grouped with other *E. cichoracearum* isolates, and *E. orontii,* while the UEA1 isolate grouped with *E. polygoni,* and *E. convolvuli* (Adam et al. 1999).

4.7 *Brassica–Hyaloperonospora* Pathogens

4.7.1 Molecular Identification of *Hyaloperonospora parasitica* Virulence Using DNA Fingerprinting

Different pathotypes can be distinguished on the basis of host specificity (Sherriff and Lucas 1990), but such tests are time consuming, and may not reveal the full extent of variations present. Other phenotypic markers such as sexual compatibility type (Sherriff and Lucas 1989a, b) or fungicide sensitivity (Crute et al. 1985) can be assessed, but they provide limited information for epidemiological studies or genetic analysis. There is need, therefore, for alternative molecular markers (Michelmore and Hulbert 1987) to further define variations in the pathogen.

Randomly amplified polymorphic DNAs (RAPD) have been proposed as genetic markers that overcome many of the technical limitations of restriction fragment length polymorphism (RFLP) analysis (Williams et al. 1990; Welsh and McClelland 1990). RAPD scan can be used in the constructions of linkage maps (Williams et al. 1990; Reiter et al. 1992), in the identification of strains, and varieties by genomic fingerprinting (Welsh and McClelland 1990; Goodwin and Annis 1991; Hu and Quiros 1991; Schafer and Wostemeyer 1992; Kresovich et al. 1992; Klein-Lankhorst et al. 1991; Koller et al. 1993; Stiles et al. 1993) and following the cloning of amplified fragments may also serve as conventional RFLP probes. In the study carried out by Tham et al. (1994), the potential use of RAPDs as a source of genetic markers in *H. parasitica* was evaluated. Several isolates from different host species were compared to determine whether reproducible banding patterns correlated with host specificity. As *H. parasitica* can only be cultured on living plant tissues, particular attention was paid to possible artifacts arising through contamination of sample DNA from plant or other microbial sources. Random Amplified Polymorphic DNA (RAPD) fingerprints were generated from a range of *H. parasitica* isolates from different *Brassica* species (Table 4.4). Reaction conditions, in particular DNA template, primer and Mg^{2+} concentrations were optimized to ensure that amplifications were reproducible. Possible artifacts' arising through host plant DNA was assessed by including such DNA in control reactions. Confirmation that diagnostic RAPD bands were generated from fungal DNA was also obtained by Southern hybridization of a RAPD band to genomic fungal DNA. By screening 20 decamer primers, 2 were found to detect sufficient genetic variation to allow complete differentiation between pathotypes. These results illustrate the potential value of RAPDs for detecting polymorphisms between isolates of a nonculturable

Table 4.4 Isolates used for RAPD analysis (Tham et al. 1994)

Isolate code	Source host	Year of isolation	Geographical origin	Mating type[a]	Response to metalaxyl[b]
P001	*B. napus*	1982	Leicestershire	Homothallic	S
P1072	*B. napus*	1992	Hertfordshire	Homothallic	nt
P1118	*B. napus*	1993	Leicestershire	Homothallic	nt
P1119	*B. napus*	1993	Leicestershire	Homothallic	nt
P1130	*B. napus*	1993	Nottinghamshire	Homothallic	nt
P1121	*B. napus*	1993	Essex	Homothallic	nt
P1122	*B. napus*	1993	Hertfordshire	Homothallic	nt
P003[c]	*B. napus*	1983	–	P1	S
P004[c]	*B. napus*	1983	–	P1	S
P033[c]	*B. napus*	1983	–	P2	S
P1100[c]	*B. napus*	1992	–	P2	nt
P1105[c]	*B. napus*	1992	–	P1	nt
P005	*B. oleracea*	1977	Tyne and wear	P1	S
P006	*B. oleracea*	1983	Lincolnshire	P2	R
P1091	*B. oleracea*	1991	Lincolnshire	nt	nt
P1092	*B. oleracea*	1991	Lincolnshire	nt	nt

[a] Mating defined as in Sherriff and Lucas (1989a, b)
[b] *S* Sensitive, *R* resistant, *nt* not tested
[c] Single spore oospores progeny from self's of isolate P001, P003, and P033 segregated for virulence to the oilseed rape (*B. napus* var. *oleifera*) cv. Cresor

plant pathogenic fungus. Rehmany et al. (2000) compared downy mildew isolates from *A. thaliana,* and *B. oleracea* using AFLP, and ITS analysis dividing the isolates into five groups that correlated with taxonomic species, and in most of the cases with host origin.

4.7.2 Identification of Downy Mildew *R* Genes

Based on next-generation sequencing technology, specific-locus amplified fragment sequencing (SLAF-seq) was developed for mapping traits of interest, with the advantages of being a technique of high-throughput, high accuracy, low cost, and short time (Wang et al. 2012; Chen et al. 2013; Sun et al. 2013; Zhang et al. 2013). An improved SLAF-seq strategy was utilized to construct a high-density bin-map, providing abundant developed single-nucleotide polymorphisms (SNPs) markers which could be used to locate QTLs for resistance to downy mildew (Yu et al. 2016). Genome wide re-sequencing was used to obtain the sequences of different individuals based on reference genomic sequences. On account of the sequence variance analysis of different individuals, plenty of high density and high quality SNPs and InDel (insertion/deletion) markers were found which would be efficient tools to assist in QTL fine mapping.

The candidate downy mildew resistance genes encoding LRR-containing proteins in *Br-DM04*, within a 160-kb region have been identified. Using a transient expression system, over expression of *BrRLP48* could enhance resistance to downy mildew in a susceptible line. Otherwise, the leaves infiltrated with RNAi construct of *BrRLP48* could reduce the disease resistance of resistant parent MM. Subsequently, through sequence and expression analysis after downy mildew infection, it was found that *BrRLP48* was significantly induced in resistant line probably due to the presence of the putative SA and JA responsive elements in the promoter. Finally, the content of SA in the resistant line MM was dramatically elevated after inoculation, with only the expression of *BrRLP48* significantly induced under SA treatment, correlating with an increased expression following pathogen invasion. These results demonstrated the involvement of *R* genes in *Brassica* downy mildew resistance and shed light on the molecular mechanism for marker-assisted breeding of disease-resistant varieties in *B. rapa*.

A new QTL *Br-DM04* on Chromosome A04 for downy mildew resistance in DH and F_2 populations derived from the cross between the Chinese cabbage line BY (*B. rapa* ssp. *pekinensis*) and a European turnip line MM (*B. rapa L.* ssp. *rapifera*) was identified. The QTLs, on Chromosome A01 and A08, for downy mildew in *Brassica rapa* reported previously were identified from different resistant sources (Yu et al. 2009, 2016; Kim et al. 2011).

The minor QTL *hBrDM4* identified from resistant line T12-19, between Marker399582 and Marker 417609 (speculated physical position is between A04.2001507 and A04.3056256) on Chromosome A04, were active at the heading stage (Yu et al. 2016). The resistant QTL *Br-DM04* in this work was identified from a European turnip line MM. Although *hBrDM4* and *Br-DM04* (between A04.4285711 and A04.7136911) were both on A04, the physical locations of them implied that they are two independent QTLs. A new putative resistance gene (*R* gene), *BrRLP48*, encoding a RLP family protein was considered to confer the disease-susceptible line BY resistance toward downy mildew, in a transient expression system using *Agrobacterium* infiltration. *BrRLP47*, the expression of which was increased to about 27-fold following Agroinfiltration, could also confer resistance in BY toward downy mildew. However, the expression of *BrRLP47* was upregulated only 2.5 times at 2 h, with slight down-regulation at other time points after disease infection in the resistant line MM. Therefore, *BrRLP47* could not be the crucial gene conferring resistance in MM on account of not reaching functional expression levels in vivo after pathogen invasion.

Whole genome-wide gene expression profiles responding to downy mildew were analyzed in Chinese cabbage by high throughput RNA sequencing. A large number of DEGs were identified in both the resistant and susceptible lines. These genes were significantly enriched in plant–pathogen interaction pathway, plant hormone signal transduction pathway, biosynthesis of secondary metabolites pathway, and etc. Many genes, including defense response genes, transcriptional factor, and cell wall related genes, were only differentially expressed in the resistant line, or with greater change of expression in the resistant line. These genes were significantly enriched in plant–pathogen interaction pathway, protein processing in endoplasmic reticulum pathway, and metabolism pathways, which are important in regulating species specific resistance to downy mildew disease in Chinese cabbage (Li et al. 2018).

4.8 *Brassica–Leptosphaeria* Pathogen

4.8.1 Combined LMD and RNA Sequencing Improves Pathogen Detection at the Earliest Stages of Infection

To capture gene activity at the host–pathogen interface, Becker et al. (2019) used LMD-RNA-seq, targeting the first 100 μm of healthy plant tissue adjacent to the inoculation site, equal to approximately two plant cell layers (Fig. 4.7a, b). Following histological processing, cellular structure was well preserved in cross sections (Fig. 4.7b). To capture early signaling events during *L. maculans*' biotrophic phase, focused was on analysis at 1 and 3 dpi before visible lesion spread. Following read quality control and alignments to the *B. napus* genome, hierarchical clustering was performed to identify relationships between genotypes and treatments (Fig 4.7c). Samples clustered based on treatment and time point, apart from inoculated susceptible plants at 3 dpi that clustered with mock controls indicated a lack of a measurable defense response at this early stage in the susceptible line. Next, step detection levels of *L. maculans* transcripts were within each treatment (Fig. 4.7d). In all mock controls, few reads aligned to the *L. maculans* genome (<0.2%). At 1 dpi, pathogen alignment rates were similar in inoculated susceptible (5.7%) and resistant lines (5.2%). At 3 dpi, alignment rates resistant plants may be effectively stopping the pathogen by 3 dpi. Additionally, comparison was on resistant plants may be effectively stopping the pathogen by 3 dpi. Additionally, comparison was on 14.6% and 3.6% in susceptible and resistant lines, respectively (Fig. 4.7d), suggesting that alignment rates to those of RNA-seq reads derived from hand-dissected inoculation sites at 3 dpi (Fig. 4.7d). Data showed a marked increase in pathogen detection through the application of LMD, with an average 37-fold higher alignment rate in LMD-collected samples.

4.8.2 Molecular Identification of *Leptosphaeria* Subclades

Leptosphaeria maculans may be "younger" than *L. biglobosa*, in evolutionary terms. Since *L. biglobosa* is genetically more diverse than *L. maculans* (Gall et al. 1995; Purwantara et al. 2000; Mendes-Pereira et al. 2003; Voigt et al. 2005). Mendes-Pereira et al. (2003) studied the phylogeny of the *L. maculans–L. biglobosa* species complex and other related *Leptosphaeria* species, using parsimony analysis of the sequences of the entire ribosomal internal transcribed spacer (ITS) region, including the 5.8S rDNA. *L. biglobosa* isolates were clustered into five subclades, i.e., *L. biglobosa* "brassicae" (from various *Brassica* species, mostly in Europe), *L. biglobosa* "canadensis" (mostly found in central Canada), *L. biglobosa* "thlaspii" (from *Thlaspi arvense*), *L. biglobosa* "erysimii" (from *Erysimum* sp.), and

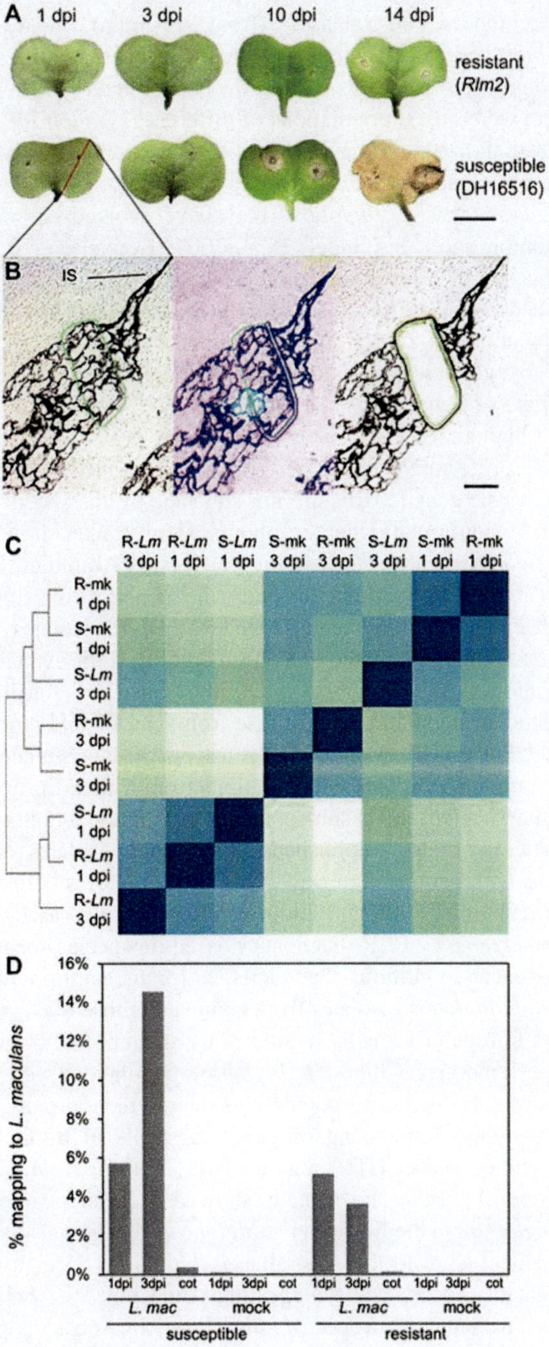

Fig. 4.7 Phenotypic response of resistant and susceptible hosts and preliminary data analysis. (**a**) Disease progression in resistant Topas *Rlm2* (top) and susceptible Topas DH16516. Lesion spread is observed in susceptible hosts by 10 days' postinoculation (dpi). Scale = 1 cm. (**b**) Cross section of histologically processed cotyledon showing excision of plant tissue by laser micro-dissection. Collected fragments were used to generate sequencing libraries. (**c**) Dissimilarity matrix, showing relationships between treatments and genotypes; *R* resistant, *S* susceptible, *Lm* *L. maculans*, *mk* mock. (**d**) Average mapping rates of all treatments to the *L. maculans* genome. Data show highest mapping rates in susceptible hosts at 3 dpi, and lowest in hand-dissection cotyledons (cot) and mock controls (Becker et al. 2019)

L. biglobosa "australensis" (from Australia). In Australia, a further subclade, *L. biglobosa* "occiaustralensis," has now been identified (Vincenot et al. 2008) and *L. biglobosa* "canadensis" has now been found (Van de Wouw et al. 2008). Whereas there were mixed populations of different *L. biglobosa* subclades in some countries (Australia), populations in Europe were exclusively *L. biglobosa "brassicae"* and those in Canada were exclusively *L. biglobosa* "canadensis." From an evolutionary point of view, *L. biglobosa* may have evolved earlier than *L. maculans* from a common ancestor (Gudelj et al. 2004; Rouxel and Balesdent 2005). Whilst two Chinese isolates from Guizhou province were classified as *L. biglobosa "brassicae"* (Mendes-Pereira et al. 2003), it is not clear how representative they are of populations of *L. biglobosa* in China. Furthermore, it is unclear how genetically diverse Chinese populations are in comparison to *L. biglobosa* populations from Europe or Canada.

Although, Hao et al. (2014) and Zhang et al. (2014) did surveys of *Leptosphaeria* species infecting Chinese *B. napus* and confirmed the presence of only *L. biglobosa*, neither the genetic structure nor subclade of this species was determined. Liu et al. (2014) confirmed whether phoma stem canker in China is caused solely by *L. biglobosa* and used banding patterns of Amplified Fragment Length Polymorphism (AFLP) markers to study the genetic diversity of Chinese *L. biglobosa* isolates and their relatedness to *L. biglobosa* isolates from different parts of the world. While this method has been employed for evaluation of geographical differentiation of *Leptosphaeria* spp. within Australia (Barrins et al. 2004) and between Australian populations and those from Europe and North America (Pongam et al. 1999; Purwantara et al. 2000), the use of AFLP was made for investigation of genetic diversity in *L. biglobosa* populations from China, Europe and Canada. The identity of the subclade of a sub-set of these Chinese Phoma stem canker *L. biglobosa* isolates was also established. Sixty fungal isolates obtained from oilseed rape stems with phoma stem canker disease symptoms collected from four provinces in China in1999, 2005 and 2006 were all identified as *Leptosphaeria biglobosa*, not *L. maculans*, by PCR diagnostics based on species-specific primers. There were no differences in cultural characteristics (pigmentation and in vitro growth) between these *L. biglobosa* isolates from China and those of 37 proven *L. biglobosa* isolates from Europe or Canada. In studies using amplified fragment length polymorphism (AFLP) markers, Chinese *L. biglobosa* populations were genetically more similar to European *L. biglobosa* populations than to the more diverse Canadian *L. biglobosa* populations. Sequencing of gene fragments of β-tubulin, actin and the internal transcribed spacer (ITS) region of rDNA from *L. biglobosa* isolates from China, Europe, Australia and Canada showed a closer taxonomic similarity of Chinese *L. biglobosa* to the European *L. biglobosa* "brassicae" than to Canadian *L. biglobosa* "canadensis" or to the Australian *L. biglobosa* "occiaustralensis" or "australensis" subclades. These results suggest that the Chinese *L. biglobosa* population is in the same subclade as European *L. biglobosa* "brassicae" populations (Liu et al. 2014).

4.8.3 Genomic Prediction to Identify Blackleg *R* Genes of *Brassica*

Most resistant cultivars contain major genes conferring resistance to specific blackleg pathotypes, which is qualitative resistance. While very effective initially, the pathogen has proven adept at overcoming the different major gene resistances in areas where high disease pressure was created by the intense cultivation of varieties with the same major gene (*Rlm1* in France, (Rouxel et al. 2003), and *LepR3* in Australia, (Sprague et al. 2006). In contrast, quantitative resistance, which is due to many minor genes, offers incremental resistance improvements. Initial blackleg leaf lesions may appear similar to those on susceptible varieties, but later in the season no or less severe stem cankers are observed on varieties with improved quantitative resistance (Delourme et al. 2014). Phenotypic and pedigree selection can potentially exploit both types of resistance to breed varieties with durable field resistance to *L. maculans* (Brun et al. 2010; Delourme et al. 2014). Phenotyping for both qualitative and quantitative resistance can be done in field disease nurseries against a full fungal population (Kaur et al. 2009; Light et al. 2011) as well as in controlled environment conditions against specific single spore fungal isolates (Rouxel et al. 2003; Marcroft et al. 2012a; Brun et al. 2010).

In the last 20 years, genomic data has been incorporated into breeding and quantitative trait loci (QTL) have been mapped that confer mainly qualitative resistance using a variety of statistical designs, such as F2-crosses (Ansan-Melayah et al. 1998), doubled haploid populations (Delourme et al. 2014), backcrosses (Long et al. 2011) and Winter and Spring diversity panels (Uzunova et al. 1995; Kaur et al. 2009; Jestin et al. 2011; Raman et al. 2012a; Delourme et al. 2014; Van de Wouw et al. 2016a, b). These efforts have identified a growing list of *L. maculans* resistance genes (e.g., *LepR1, LepR2, LepR3, LepR4, BLMR2, Rlm1, Rlm2, Rlm3, Rlm4, Rlm5, Rlm6, Rlm7, Rlm8, Rlm9, Rlm10, Rlm11, Rlm12,* and *LmFr1*) in *Brassica* species (Yu et al. 2005; Delourme et al. 2006; Yu et al. 2008; Marcroft et al. 2012b; Raman et al. 2012b; Balesdent et al. 2013; Larkan et al. 2014; Fomeju et al. 2014; Raman et al. 2016). Consequently, irrespective of some limitations, marker assisted selection has been used for mapping and validating resistance QTL to *L. maculans* using progenies/populations derived from breeding programs. However, studies in canola and other species show that genetic variance explained by known QTL tends to be low (Yang et al. 2010; Huo et al. 2016). Many plant genetic improvement schemes are transitioning from phenotype and pedigree based selection to genetic marker based genomic selection. Genomic selection uses a reference population that is genotyped and phenotyped to predict the genetic value for individuals that are only genotyped using genome-wide markers (Meuwissen et al. 2001). Several studies argue that genomic selection provides better accuracies than might be achieved on the basis of pedigree information alone (Crossa et al. 2010; Jia and Jannink 2012; Perez Enciso et al. 2015; Tsai et al. 2015; Vela Avitua et al. 2015). This paradigm shift is significantly improving genetic gain by reducing the breeding cycle, increasing selection accuracy in un-phenotyped germplasm, and increasing selection intensity. Furthermore, studies in a wide variety of plant species confirm the realized and potential benefits of using genomic selection (Heffner et al. 2011;

He et al. 2016; Bentley et al. 2014; Grenier et al. 2015). Various genomic prediction models have been proposed such as Genomic Best Linear Unbiased Prediction (GBLUP) (Nejati Javaremi et al. 1997; Habier et al. 2007; VanRaden 2008), Ridge Regression Best Linear Unbiased Prediction (RR-BLUP), Bayes A and B (Meuwissen et al. 2001), Bayes R (Erbe et al. 2012). Bayes R has been extended to incorporate prior information obtained from variant, gene, or regulatory regions (Bayes RC) (MacLeod et al. 2016). Genomic selection in canola was first highlighted by Wurschum et al. (2014), who reported prediction accuracies for agronomic traits using RR-BLUP and Bayes B in a lower diversity set of 391 doubled haploid lines genotyped with 251 SNPs. A more diverse set of 475 spring canola lines was used by Jan et al. (2016) to examine the performance of testcross combinations using 24,403 SNPs markers based on the RR-BLUP model. Using three different genomic selection methods (GBLUP, Bayes R, and Bayes RC), Fikere et al. (2018) predicted *R* genes in *B. napus* in a diverse set of 532 Winter and Spring canola lines genotyped for 98,054 SNPs and phenotyped for adult plant survival and stem internal infection in replicated field disease nurseries at two locations.

Genomic prediction is becoming a popular plant breeding method to predict the genetic merit of lines. While some genomic prediction results have been reported in canola, none have been evaluated for blackleg disease. The genomic prediction for seedling emergence, survival rate, and internal infection, using 532 Spring and Winter canola lines has been made. These lines were phenotyped in two replicated blackleg disease nurseries grown at Wickliffe and Green Lake, Victoria, Australia. A transcriptome genotyping-by-sequencing approach revealed 98,054 single nucleotide polymorphisms (SNPs) after quality control. Various genomic prediction scenarios based on Genomic Best Linear Unbiased Prediction (GBLUP), BayesR and BayesRC, which can make use of prior quantitative trait loci information, via cross-validation. Clustering based on genomic relationships showed that winter and spring lines were genetically distinct, indicating limited gene flow between sets. Genetic correlations within traits between Spring and Winter lines ranged from 0.68 to 0.90 (mean = 0.76). Based on GBLUP in the whole population, moderate to high genomic prediction accuracies were achieved within environments (0.35–0.74) and were reduced across environments (0.28–0.58). Prediction accuracy within the Spring set ranged from 0.30 to 0.69, and from 0.19 to 0.71 within the Winter set. The BayesR model resulted in slightly lower accuracy to GBLUP. The proportion of genetic variance explained by known blackleg quantitative trait loci (QTL) was <30%, indicating that there is a large reservoir of genetic variation in blackleg traits that remains to be discovered, but can be captured with genomic prediction. However, providing prior information of known QTL in the BayesRC method resulted in increased prediction accuracy for survival and internal infection, particularly with spring lines. Overall, these promising results indicate that genomic prediction will be a valuable tool to make use of all genetic variation to improve blackleg resistance in canola (Fikere et al. 2018).

4.9 *Brassica–Plasmodiophora* Pathogen

4.9.1 Identification of Virulence Genes of *Plasmodiophora*

The information on the genomic makeup of *P. brassicae* and the molecular basis of host resistance/susceptibility is very limited, largely because of the nature of this organism as an obligate parasite. Graf et al. (2004) isolated a few partial cDNAs that had no significant homology to Gen Bank sequences. Suppression subtractive hybridization (SSH) has been conducted between RNA from *P. brassicae*-infected and uninfected *Arabidopsis* tissue, and a total of 76 new *P. brassicae* gene sequences have been identified (Bulman et al. 2006). Many of the sequences are predicted to contain signal peptides for translocation of the encoded proteins. Again using SSH, but starting with *B. rapa* rather than *Arabidopsis* clubroot galls, Sundelin et al. (2011) isolated about 140 clones. Half of these clones, which represented 24 unisequences, originated from *P. brassicae*. Out of these, 10 unisequences are newly characterized *P. brassicae* genes. By using next generation sequencing (NGS) technologies, Burki et al. (2010) generated 27,333 contigs assembled from 454 reads from *P. brassicae* infected *B. rapa*. Among these contigs, 3167 are from *P. brassicae*. A few genes have been postulated to be related to the pathogenicity of *P. brassicae* (Ito et al. 1999; Brodmann et al. 2002; Ando et al. 2006; Siemens et al. 2009), mainly based on the observation that the expression of the gene is correlated with certain stage of disease development. One gene, a serine protease (*PRO1*), has been heterologously expressed and proven experimentally to be important for resting spore germination (Feng et al. 2010).

The transcriptomic and proteomic approaches have elucidated the mechanisms of *P. brassicae* infection and coordination in plant reactions. Using a full genome Affymetrix chip, Siemens et al. (2006) found that more than 1000 *Arabidopsis* genes are differentially expressed in *P. brassicae* infected roots at either 10 or 23 days after inoculation versus noninfected roots. Two proteomic studies, both using two-dimensional gel electrophoresis and quantitative image analysis in conjunction with mass spectrometry-based protein identification, illustrated changes in protein abundance at 4 days after inoculation (dai) in *Arabidopsis* (Devos et al. 2006), and at 0.5, 1, 2, and 3 dai in *B. napus* (Cao et al. 2008). The observed differential expression on mRNA or protein levels at the different stages of pathogenesis suggested that the corresponding genes might be involved in the plant response to the clubroot pathogen. Functional genomics approaches provide powerful tools for identifying expressed genes. Among these techniques, expressed sequence tags (ESTs, Adams et al. 1991), serial analysis of gene expression (SAGE, Velculescu et al. 1995), and massively parallel signature sequencing (MPSS, Brenner et al. 2000) have been successfully employed. Due to its relative simplicity and inexpensive nature, single pass EST sequencing has been the most widely used method to characterize genes associated with cellular development, with biotic and abiotic stress in plant research. Although many EST clones lack the complete sequences of mRNAs, they can be very helpful to investigate the biological function of particular genes, especially in poorly annotated genomes. Additionally, it is possible to obtain full-length clones starting from ESTs. EST resources are also important for comparative transcriptomics because gene expression profiles can be obtained by analyzing ESTs generated under varying conditions.

The whole transcriptome sequencing using next generation sequencing (NGS) technologies or RNA sequencing (RNA-seq) has been started to be utilized to reveal the complex landscape of the transcriptome from various organisms at an unprecedented level of sensitivity and accuracy (Ozsolak and Milos 2011). Compared with traditional EST sequencing and other high-throughput technologies such as microarrays, NGS and RNA-seq offer a near-complete snapshot of a transcriptome with a higher dynamic range of expression level. However, common NGS platforms including Illumina, SOLiD, and 454 suffer from the disadvantage that the sequence reads are often very short (35–500 bp). As a consequence, high technologies on transcriptome assembly are required after NGS. Despite the advantages of NGS and RNA-seq, EST cloning incorporated with SSH (Diatchenko et al. 1996) has been used to identify genes involved in host–pathogen interactions. This method reduces, by subtractive hybridization, the cloning of abundantly expressed housekeeping genes or genes commonly expressed in both control and treated plants, thereby normalizing expressed cDNA profiles during library construction. Moreover, it significantly enhances, by suppression PCR, the chances of cloning differentially expressed genes. This is particularly important because many pathogenesis-related genes are expressed at low levels, and can be limited to a particular tissue or cell type. This technique has been used to isolate plant and pathogen genes that are expressed during infection (Huang et al. 2007). Both of the two cDNA libraries of *P. brassicae*-infected plants, one using *Arabidopsis* (Bulman et al. 2006) and the other *B. rapa* (Sundelin et al. 2011), were constructed by using the SSH technique.

The construction of a SSH cDNA library followed by EST analysis was used to achieve a better insight into the genes expressed during the compatible interaction between a susceptible *Brassica* cultivar and *P. brassicae*. The infection of plants by pathogenic and the subsequent establishment of disease involves substantial changes in the biochemistry and physiology of both partners. Analysis of genes that are expressed during these interactions represents a powerful strategy to obtain insights into the molecular events underlying these changes. Feng et al. (2012a) have used suppression subtractive hybridization (SSH) and expressed sequence tag (EST) analysis to investigate gene expression during the early stages of colonization of *Brassica* roots by *P. brassicae*. A cDNA library was constructed by SSH which consisted of 797 clones that represented 439 unigenes. Thirty-two of these genes were demonstrated to be of *P. brassicae* origin. The remaining 407 genes, which were of a *Brassica* origin, were subjected to gene ontology and in silico analyses. Real-time PCR analysis of ten *P. brassicae* and seven *Brassica* genes indicated that seven of the former and five of the latter were upregulated at 7 days after infection, suggesting the importance of these genes in pathogenesis or host resistance.

Molecular mechanisms of *P. brassicae* infection during primary and secondary zoospores cannot be differentiated morphologically. When inoculated onto healthy *Brassica* roots, secondary zoospores cause primary infections similar to those caused by primary zoospores (Feng et al. 2012b). In contrast, when inoculated onto

Brassica roots that are already under primary infection, primary zoospores can cause secondary infections without going through a primary infection phase (Feng et al. 2013). Thus, from a developmental biology standpoint, neither morphogenesis nor functional differentiation can clearly distinguish between these two kinds of zoospores. The primary and secondary zoospores may be essentially same or they undergo a divergence. No matter the state of divergence between the spore types, the "arms race" with the host would induce changes in the secondary zoospores. These changes are most likely at the gene expression level, and can be both genetic and epigenetic in nature. Knowledge of the molecular differences between primary and secondary zoospores may be useful for rational clubroot resistance breeding efforts as well as for the development of pesticides effective against *P. brassicae*. If these two kinds of zoospores are the same with respect to gene expression, a gene or quantitative trait loci (QTL) for resistance against either primary or secondary infection will be also effective against the other. Thus, the selection for resistance and subsequent genetic studies can reliably be conducted at the primary infection stage, which would save time and resources. This scenario is also applicable to fungicide development. Despite this importance, data on gene expression in these two kinds of zoospores is limited with information on the genetic makeup of *P. brassicae*, with only a small number of expressed sequence tags (ESTs) and protein sequences available in GenBank. Similarly, very few *P. brassicae* genes have been functionally characterized (Feng et al. 2010). Most of the database-available genes are derived from cDNA libraries that were constructed using the suppression subtractive hybridization techniques (Bulman et al. 2006; Feng et al. 2012b; Sundelin et al. 2011), which are enriched in genes up-regulated during pathogenesis. Thus, these genes are all good candidates for gene expression studies during the different stages of pathogenesis. All database-available of *P. brassicae* genes were analyzed in silico. To compare the gene regulation profiles of the primary and secondary zoospores, these genes were subjected to both dot blot and real time PCR analyses to understand clubroot pathogenesis and provide gene candidates for gene functional studies, which will in turn facilitate breeding for resistance and the identification of potential targets for the development of novel pesticides.

To get information on the molecular basis of pathogenicity of the clubroot pathogen *Plasmodiophora brassicae*-specific primers were designed and used to amplify genomic fragments of *P. brassicae* genes that represent all database-available proteins and ESTs. The PCR products were blotted on membranes and hybridized with digoxigenin-labeled double-stranded cDNA, derived from either primary or secondary zoospores of *P. brassicae*. The same primers were also used in real-time PCR against the single-stranded cDNA synthesized from the two types of zoospores. Both dot blot and real-time PCR identified up- and downregulated genes provide the correlation between these two techniques. Real-time PCR indicated that 58 genes were upregulated in the secondary zoospores relative to the primary zoospores, whereas 55 were downregulated. These data suggest that different mechanisms are utilized by the pathogen in causing primary and secondary infections. The expression patterns of genes with known or putative functions suggest the relative importance of these genes during pathogenesis. In contrast, highly expressed or regulated genes with unknown function can be further studied in identifying pathogenicity factors (Feng et al. 2013).

4.9.2 The qPCR-Based Detection of *Plasmodiophora* Spread in Soil

Clubroot is a damaging disease of oilseed rape and vegetable *Brassicas* worldwide, caused by the soil-borne protist *Plasmodiophora brassicae* Wor. Due to the long life of resting spores, the assessment of the pathogen abundance in agricultural fields can serve as a guideline for disease control at the country-wide level or the regional scale. Bienkowska et al. (2020) collected 431 soil samples from fields cultivated with Brassicaceae crops throughout 16 provinces of Poland. The samples were subjected to qPCR based analysis of *P. brassicae* DNA concentration. Results reveal that the spore loads and gene copies g^{-1} soil were calculated and used to produce an assessment of the current clubroot risk potential at a country-wide and regional scale. The country-wide map, showing the spread of the pathogen in agricultural soils, was made using ArcGis software package implementing the interpolation with the Inverse Distance Weight method. The calculation of gene copies specific to *P. brassicae* helped to formulate the recommendations for farmers in respect to the cultivation guidelines. It showed a high risk of yield losses in defined regions of north, south-west and central Poland and an urgent need to undertake intensive preventative measures. The details of the protocol have been covered in Chap. 8 at Sect. 8.8 (Bienkowska et al. 2020).

4.9.3 Identification of Race/Pathotype-Specific Molecular Markers

The details for this protocol have been covered in Chap. 5 at Sect. 5.8.8.

4.9.4 Molecular Identification of *Plasmodiophora* Pathotype Effector/Gene

Molecular characterization of pathogen effectors is gaining momentum. The better genome assemblies and higher gene prediction accuracy will allow it to gain better insight into the structures and functions of effectors/proteins, allowing for more efficient identification of *R* genes. The initiatives of *P. brassicae* showed that *Cr811* gene is good candidate's gene to differentiate pathotypes 5 from the rest of the pathotypes.

4.10 *Brassica–Pseudocercosporella* Pathogen

4.10.1 Phylogenetic Analysis of *Pseudocercosporella* Isolates for Virulence

Pseudocercosporella capsellae has both sexual and asexual life cycles. When present, the teleomorph stage (*Mycosphaerella capsellae*), produced at the end of the crop season, and is primarily responsible for survival of the pathogen over the period between crops and for production of primary inoculums for the next season (Inman et al. 1991). Variation of pathogen populations can relate to differences in pathogenicity and/or virulence of isolates of the same pathogen (McDonald and Linde 2002), and such differences can be attributed to differences in genetic makeup of the fungal isolates (Kuninaga et al. 1997). Fifty-four single-conidial isolates collected from *Brassica juncea* (Indian mustard), *B. napus* (oilseed rape), *B. rapa* (turnip), and *Raphanus raphanistrum* (wild radish) across Western Australia were investigated for differences in pathogenicity and virulence using cotyledon screening tests, genetic differences using internal transcribed spacer (ITS) sequencing and phylogenetic analysis, and growth rates on potato dextrose, V8 juice and malt extract agars. All isolates from the four crucifer hosts were pathogenic on the three test species: *B. juncea*, *B. napus* and *R. raphanistrum*, but showed differences in levels of virulence. Overall, isolates from *B. juncea*, *B. napus* and *B. rapa* showed greatest virulence on *B. juncea*, least on *R. raphanistrum* and intermediate virulence on *B. napus*. Isolates from *R. raphanistrum* showed greatest virulence on *B. juncea*, least on *B. napus* and intermediate virulence on *R. raphanistrum*. Growth and production of a purple-pink pigment indicative of cercosporin was greatest on malt extract agar and cercosporin production on V8 juice agar was positively correlated with virulence of isolates on *B. juncea* and *B. napus*. ITS sequencing and phylogenetic analysis showed that isolates collected from *B. napus*, *B. juncea* and *B. rapa*, in general and with few exceptions, had a high degree of genetic similarity. In contrast, isolates from *R. raphanistrum* were clearly differentiated from isolate groups collected from *Brassica* hosts. *Pseudocercosporella capsellae* reference isolates from other countries generally grouped into a single separate cluster, highlighting the genetic distinctiveness of Western Australian isolates (Fig. 4.8; Gunasinghe et al. 2016).

4.11 *Brassica–Rhizoctonia* Pathogen

4.11.1 Molecular Detection of *Rhizoctonia solani* Virulence Distribution in Soil

Rhizoctonia solani is the anamorph of *Thanatephorus cucumeris,* a ubiquitous plant pathogen and a capable saprophyte, being able to compete effectively both above and below the soil surface (Menzies 1970). *Rhizoctonia solani* has long been recognized as a species complex, which can be subdivided into an astomosis groups

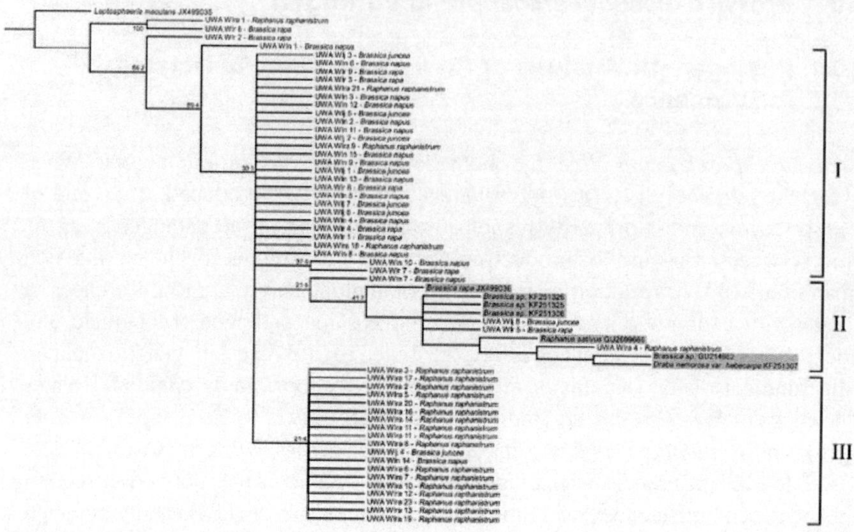

Fig. 4.8 Neighbor joining tree showing the phylogenetic relationships among 54 isolates of *Pseudocercosporella capsellae* generated from this study and seven reference isolates of *P. capsellae* from GenBank (shaded), based on their internal transcribed spacer sequences as generated by GENEIOUS with Tamura–Nei genetic distance model. Boot strap values of 1500 replications are shown for the major nodes. The host of origin of each isolate is given after each isolate name (Gunasinghe et al. 2016)

(AGs) based on an assessment of hyphal fusion with known tester isolates. The AGs can be further subdivided into intraspecific groups (ISGs) (Ogoshi and Ui 1978) or subgroups (Sneh 1996) based on cultural morphology and physiological characteristics. There are 13 recognized AGs with 14 recognized subgroups (Carling et al. 1994, 1999, 2002; Sneh 1996). Members from each AG may vary in morphology, pathogenicity and even in their susceptibility to different fungicides (Kataria et al. 1991; Fiddaman and Rossall 1995; Campion et al. 2003).

Therefore, it is necessary to know which AGs are causing disease to help target effective control strategies. However, AG determination using classical methods is time consuming and requires taxonomic expertise. The development of serological and molecular based assays has significantly advanced the detection of the fungus. Methods such as restriction fragment length polymorphism (RFLP) (Vilgalys and Gonzalez 1990), randomly amplified polymorphic DNA (RAPD) (Duncan et al. 1993) and amplified fragment length polymorphism (AFLP) (Julian et al. 1999) permit accurate and rapid AG and sub-group identification leading to an improved understanding of population structure. Universally primed PCR (UP-PCR) allows rapid identification of isolates representing 11 AGs in a cross-hybridization assay (Lubeck and Poulsen 2001). However, these methods require pure fungal cultures and are not suitable for testing plant material, soil or growing media directly. The design of primers for use in PCR (Liu et al. 1995) that recognize unique DNA sequences can result in direct detection of very low levels of target DNA in plant material and soil. Such PCR methods have been used to detect AG-1-IB (Grosch et al. 2007), AG-2 and subgroups (Salazar et al. 2000), AG-3 in potatoes (Bounou et al. 1999), AG-4 and AG-8 in wheat (Brisbane et al. 1995) and AG-8 in soil (Whisson et al. 1995).

Lees et al. (2002) reported a real-time assay designed to specifically detect AG-3 isolates in soil and potato tubers, though the method proved time consuming in a routine setting. Volossiouk et al. (1995) presented a rapid and simple method for DNA extraction from soil but the method is best suited to small soil samples (<1 g). Okubara et al. (2008) reported assays for several AG groups (AG-2-1, AG-8, AG-10, and AG-I) of *Rhizoctonia* spp. As well as *R. oryzae* from root and soil matrices, and Sayler and Yang (2007) reported real-time PCR assays for AG-1-IA in plant material. Finally, Okubara et al. (2007) also reported methods for the improved extraction of Rhizoctonia from cereal roots and soil, but from very small (0Æ2 g) soil sample sizes. For research into the epidemiology of *R. solani*, the ideal method would be rapid, quantitative, inexpensive and able to process larger (>1 g) soil samples.

A UK survey demonstrated AG-2-1 to be the most frequently isolated pathogenic group of *R. solani* on *B. oleracea* (cauliflower, calabrese, Brussels sprouts, green and savoy cabbage) crops (Budge et al. 2009), and an observation also confirmed such studies on cauliflower in Belgium (Pannecouque et al. 2008). Budge et al. (2009) developed real-time PCR protocols to detect pathogenic AGs, suitable for the extraction of *R. solani* DNA from soil and compost, and to investigate the spatial distribution of *R. solani* AG-2-1 infield soil before and after the planting of a susceptible *B. oleracea* crop. Real-time PCR protocols were developed to detect and discriminate 11 anastomosis groups (AGs) of *R. solani* using ribosomal internal transcribed spacer (ITS) regions (AG-1-IA, AG-1-IC, AG-2-1, AG-2-2, AG-4HGI +II, AG-4HGIII, AG-8) or β-tubulin (AG-3, AG-4HGII, AG-5, and AG-9) sequences. All real-time assays were target group specific, exceptAG-2-2, which showed a weak cross-reaction with AG-2tabac. In addition, methods were developed for the high throughput extraction of DNA from soil and compost samples. The DNA extraction method was used with the AG-2-1 assay and shown to be quantitative with a detection threshold of 10^{-7} g of *R. solani* per g of soil. A similar DNA extraction efficiency was observed for samples from three contrasting soil types. The developed methods were then used to investigate the spatial distribution of *R. solani* AG-2-1 in field soils. Soil from shallow depths of a field planted with *Brassica oleracea* tested positive for *R. solani* AG-2-1 more frequently than soil collected from greater depths. Quantification of *R. solani* inoculums infield samples proved challenging due to low level of inoculum in naturally occurring soils. The protocol has great uses of real-time PCR and DNA extraction to investigate the epidemiology of *R. solani* (Budge et al. 2009).

4.12 *Brassica–Sclerotinia* Pathogens

4.12.1 Rapid Identification of an *Arabidopsis* NLR Gene Conferring Susceptibility to *Sclerotinia sclerotiorum*

Resistance (R) genes mediating complete disease resistance all belong to the nucleotide-binding site leucine-rich repeat (NLR) family, genes underlying QDR discovered to date span a broad range of molecular functions (Poland et al. 2011; Roux et al. 2014; Corwin and Kliebenstein 2017). Molecular function of genes associated with QDR include for instance transporters (Krattinger et al. 2009), kinases (Fu et al. 2009; Huard Chauveau et al. 2013), peptidases (Poland et al. 2011; Badet et al. 2017), or actin-related proteins (Moscou et al. 2011). NLRs (Debieu et al. 2015; Lee et al. 2016) and signaling components typically associated with R-mediated resistance (Iakovidis et al. 2016) can also mediate QDR, illustrating the tight integration of QDR and R-mediated immunity components.

Because of their complex interplay and their small phenotypic effects, the functional characterization of QDR genes is challenging. With large collections of mutant lines available, studies in the model plant *A. thaliana* enable the rapid functional characterization of individual candidate QDR genes (Huard Chauveau et al. 2013; Corwin et al. 2016; Rajarammohan et al. 2018). Fully quantitative readouts are often required to reveal the contribution of individual A. thaliana genes to QDR against *S. sclerotiorum*, such as ethylene and ROS detection (Perchepied et al. 2010; Zhang et al. 2013), lesion area measurements, and estimation of fungal biomass (Zhang et al. 2013; Badet et al. 2017). Quantitative image analysis can be used as a proxy to evaluate disease severity (Baranowski et al. 2015; Mutka et al. 2016; Karisto et al. 2017; Badet et al. 2017) and the impact of stress on plant fitness (Chen et al. 2014; Nelson et al. 2018; Czedik Eysenberg et al. 2018). Increasing the accuracy and robustness of such quantitative phenotyping often involves increasing number of measurements. This allowed for instance to reveal the role of individual effectors from the bacterial pathogen *Xanthomonas axonopodis* pv. *manihotis* in virulence (Mutka et al. 2016). High-throughput image based plant phenotyping proved valuable in crop breeding and for research purposes (Araus and Cairns 2014; Fahlgren et al. 2015; Coppens et al. 2017). There is therefore a need to improve ability to generate large datasets of quantitative plant disease measurements at low cost, low footprint, and reduced human intervention (Czedik Eysenberg et al. 2018). Because it is generally nondestructive, image-based disease measurement also gives access to the dynamics of disease progression. This enabled for instance to study the spatial and temporal distribution of pathogens in plants tissues (Mutka et al. 2016) and distinguish between infected and noninfected plants long before qualitative symptoms are visible (Czedik Eysenberg et al. 2018).

In spite of remarkable progress in methods and technology in recent years, fundamental advances in plant pathology enabled by automated plant phenotyping are still relatively limited. A method for the image-based phenotyping of plant disease symptoms combining high throughput, high resolution, low footprint, and low cost is currently lacking. This design is crucial to identify novel QDR genes and advance our conceptual understanding of this complex trait. Barbacci et al. (2020) presented the Navautron system, an integrated hardware and software solution capable of automatically calculating disease lesion area over time on detached leaves inoculated by pathogens causing necrotic lesions. The Navautron system is able to capture 800 *A. thaliana* leaves per m^2. Its automatic leaf and lesion tracking features allow discriminating disease symptom dynamics with differences in lesion doubling time, as small as 12 min, with nearly no human intervention.

The speed of lesion growth was highly dependent on the host plant genotype and a good indicator of QDR level. It was hypothesized that large effect polymorphisms common to the two most resistant *A. thaliana* accessions, but absent from the four other accessions, would point towards disease susceptibility genes. Highly divergent alleles of LAZ5 in the resistant accessions Rubezhnoe and Lip-0, were identified while identical alleles existed in the other *A. thaliana* accessions analyzed. As expected, two LAZ5-deficient mutant lines in theCol-0 genetic background showed enhanced QDR to *S. sclerotiorum*, whereas plants mutated in the closely related CSA1 gene responded like the wild-type. Considering that the ectopic activation of LAZ5 triggers cell death (Palma et al. 2010). Results suggest that the broad host range necrotrophic fungus S. sclerotiorum exploits this *R* gene induced plant cell death to its benefit (Barbacci et al. 2020).

Plant genes conferring complete resistance against *S. sclerotiorum* have not been reported. Instead, plant populations challenged by *S. sclerotiorum* exhibit a continuum of partial resistance designated as quantitative disease resistance (QDR). Because of their complex interplay and their small phenotypic effect, the functional characterization of QDR genes remains limited. How broad host range necrotrophic fungi manipulate plant programmed cell death is for instance largely unknown. Barbacci et al. (2020) designed a time-resolved automated disease phenotyping pipeline enabling high-throughput disease lesion measurement with high resolution, low footprint at low cost. It could accurately recover contrasted disease responses in several pathosystems using this system. Phenotyping pipeline was used to assess the kinetics of disease symptoms caused by seven *S. sclerotiorum* isolates on six *A. thaliana* natural accessions with unprecedented resolution. Large effect polymorphisms common to the most resistant *A. thaliana* accessions identified highly divergent alleles of the nucleotide-binding site leucine-rich repeat gene LAZ5 in the resistant accessions Rubezhnoe and Lip-0. The impaired LAZ5 expression in laz5.1 mutant lines and in *A. thaliana* Rub natural accession correlate with enhanced QDR to *S. sclerotiorum*. These findings illustrate the value of time-resolved image-based phenotyping for unraveling the genetic bases of complex traits such as QDR. The results suggest that *S. sclerotiorum* manipulates plant sphingolipid pathways guarded by LAZ5 to trigger programmed cell death and cause disease (Barbacci et al. 2020).

4.13 *Brassica–Turnip Mosaic Virus* Pathogen

4.13.1 Detection and Differentiation of *Brassica* Virus by Multiplex Reverse Transcription Polymerase Chain Reaction

A multiplex reverse transcription-polymerase chain reaction (mRT-PCR) was developed for simultaneous detection and differentiation of the three genotypes of *BrYV*. The three genotypes of *BrYV* and *Tunip yellows virus* (*TuYV*) could be differentiated simultaneously using six optimized specific oligonucleotide primers, including one universal primer for detecting *BrYV*, three BrYV genotype-specific primers, and a pair of primers for specific detection of TuYV. Primers were designed from conserved regions of each virus and their specificity was confirmed by sequencing PCR products. The mRT-PCR products were 278 bp for *BrYV*-A, 674 bp for *BrYV*-B, 505 bp for *BrYV*-C, and 205 bp for *TuYV*. Amplification of three target genotypes was optimized by increasing the PCR annealing temperatures to 62 °C. One to three fragments specific for the virus genotypes were simultaneously amplified from infected samples and identified by their specific molecular sizes in agarose gel electrophoresis. No specific products could be amplified from cDNAs of other viruses which could infect crucifer crops. Detection limits of the plasmids for multiplex PCR were 100 fg for *BrYV*-A and *BrYV*-B, 10 pg for *BrYV*-C, and 1 pg for *TuYV*, respectively. The mRT-PCR was applied successfully for detection of three *BrYV* genotypes from field samples collected in China. The simple, rapid, sensitive, and cost-effective mRT-PCR was developed successfully for detection and differentiation of the three genotypes of *BrYV* (Zhang et al. 2016).

4.13.2 Detection and Identification of *Turnip Yellows Virus* in *Brassica* by DAS-ELISA Kits

A total of 86 oilseed rape samples from six crops in different localities were collected during 2014 and analyzed for the presence of *Turnip yellows virus* (*TuYV*), *Cauliflower mosaic virus* (*CaMV*) and *Turnip mosaic virus* (*TuMV*) using commercial double-antibody sandwich (DAS)-ELISA kits. *TuYV* was serologically detected in 60 collected samples (69.77%), and none of the samples tested were positive for *CaMV* and *TuMV*. Six selected *TuYV* isolates were successfully transmitted by *Myzus persicae* to three test plants, confirming the infectious nature of the disease. In the selected ELISA-positive samples, the presence of *TuYV* was further confirmed by RT-PCR and sequencing. A comparison of the obtained sequence with those available in GenBank confirmed the presence of *TuYV* in oilseed rape samples. An analysis of P0 gene sequence data for a subset of these isolates showed they clustered with the known *TuYV* and were distinct from *Beet western yellows virus* (*BWYV*) isolates (Milosevic et al. 2016).

4.13.3 Identification of *Turnip Mosaic Virus* From *Brassica napus* by DAS-ELISA

During March and April of 2011, 436 samples showing viral disease symptoms were collected from canola fields in the Khorasan Razavi province. The samples were tested by double-antibody sandwich (DAS)-enzyme linked immune sorbent assay (ELISA) for the presence of *Turnip mosaic virus* (*TuMV*). Among the 436 samples, 117 samples were found to be infected with *TuMV*. One of the infected samples from Govareshk region (*TuMV*-IRN GSK) was selected for biological purification. Total RNA of this isolate was extracted and reverse transcriptase (RT)-PCR was performed with specific primers according to the coat protein gene. PCR products (986 bp) was first purified and then directly sequenced. Phylogenetic analyses based on ClustalW multiple alignments with previously reported 33 isolates indicated 88–98% similarity in nucleotide and 94–99% in amino acid levels among isolates. *TuMV*-IRN GSK represented the highest identity to another Iranian isolate (IRN TRa6). Phylogenetic tree clustered all sequences into four groups and IRN GSK fell into the basal-B group. Nucleotide and amino acid distances between IRN GSK and other isolates in the basal-B group showed that this isolate was closely related to another Iranian isolate IRN TRa6, and distinct from other isolates in the basal-B group. These results indicate that *TuMV* is a common pathogen of canola crops in the Khorasan Razavi province (Sabokkhiz et al. 2012).

4.13.4 Molecular Detection, Identification, and Phylogenetic Analysis of *Turnip Yellows Virus* (*TuYV*)

The results of serological analyses of *TuYV* presence in oilseed rape in Serbia was further confirmed by the molecular RT-PCR method using the specific primersTuYVorf0F and TuYVorf0R, which amplify a fragment of the *TuYV* P0 gene. These primers successfully detected the presence of *TuYV* in all tested samples and amplified cDNA fragments of predicted size. One clear band of 780 bp was visible in all oilseed rape plants assayed, while no amplification products were observed in the healthy controls. After purification, the RT-PCR product derived from the isolate 119-*TuYV* was directly sequenced in both directions using the same primer pair as in RT-PCR, and deposited in GenBank (GenBank Accession No. KU351664). Sequence analysis of the P0 gene, conducted with MEGA5 software, revealed 99.7% nt identity (100% aa identity) between the two Serbian *TuYV* isolates from oilseed rape. One Serbian isolate (119-*TuYV*) showed the highest nucleotide identity of 99% (100% amino acid identity) with *TuYVGB* isolates from England (AF168608).

A maximum parsimony tree (Fig. 4.9), reconstructed using partial sequences of the P0 gene isolates, revealed that the *TuYV* isolate characterized in this study and the previously identified Serbian *TuYV* isolate (Milosevic et al. 2015), as well as the selected sequences of 11 characterized *TuYV, BWYV, BChV*, and *CABYV* isolates retrieved from GenBank database clustered into four groups depending on virus

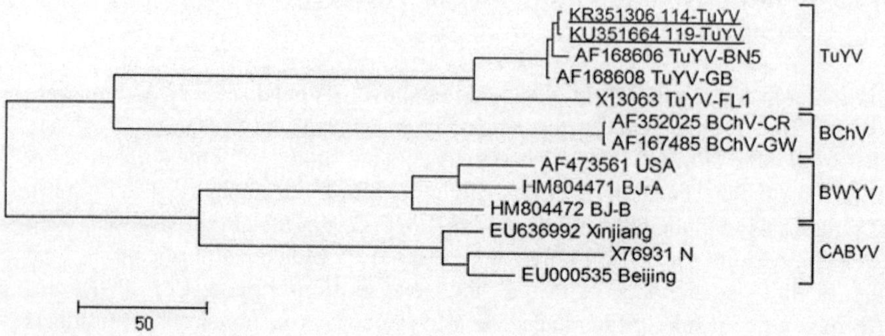

Fig. 4.9 Maximum parsimony tree based on partial sequences of the P0 gene of *Turnip yellows virus* (*TuYV*), *Beet western yellows virus* (*BWYV*), *Beet chlorosis virus* (*BChV*), and *Cucurbit aphid-borne yellows virus* (*CABYV*) isolates. The phylogram was generated with MEGA5 using boot strap analysis with 1000 replicates and boot strap values (>50%) are shown next to relevant branches. The *Turnip yellows virus* isolates from oilseed rape from Serbia are underlined (Milosevic et al. 2016)

species, with high boot strap values (100%). Genetic diversity among the four molecular groups of isolates ranged from 0.516 ± 0.038 to 0.583 ± 0.212, while diversity within each group was: 0.068 ± 0.008 (*TuYV*), 0.119 ± 0.013 (*BWYV*), 0.002 ± 0.002 (*BChV*), and 0.119 ± 0.013 (*CABYV*). An analysis of P0 gene sequence data for a subset of these isolates showed that they clustered with known *TuYV* and were distinct from *BWYV* isolates (Milosevic et al. 2016).

4.13.5 The *Turnip Mosaic Virus* Detection, Preservation, and Identification by ELISA

Initially, fresh field specimens were tested for viruses by enzyme-linked immune sorbent assays (ELISAs), using purified anti-virus immune globulins, alkaline phosphatase conjugates, and ELISA protocols from Agdia, Inc. Infection of fresh leaves of *B. napus* differential lines inoculated with *TuMV* isolates was identified by ELISA with *TuMV* monoclonal antibody EMA67 (Jenner et al. 1999). However, the majority of testing of field specimens and isolates was done on stored petiole and stem blots on nitro cellulose membranes (Schwinghamer et al. 2010) by tissue blot immune assays (TBIAs). These TBIAs were indirect procedures that involved incubation with antigen-specific primary antibodies followed by alkaline-phosphatase conjugated secondary antibodies (anti-rabbit or anti-mouse), and finally chromogenic substrate (nitroblue tetrazolium plus 5-bromo-4-chloro-3-indolyl phosphate) to detect virus in blots or tissue prints on nitrocellulose membranes. Primary antibodies for *TuMV* TBIA and potyvirus TBIA were, respectively, *TuMV* polyclonal and general potyvirus monoclonal PTY 1, both from ELISA kits supplied by Bio-Rad Phyto-Diagnostics, INRA (Institut National de la Recherche Agronomique, France). Primary antibodies for *CaMV* and Pea seed-borne mosaic virus (*PSbMV*)

TBIAs were polyclonals from DSMZ (Deutsche Sammlung von Mikroorganismen und Zellkulturen, Germany).

The TBIA procedure of Lin et al. (1990) and Hsu and Lawson (1991) was modified as follows: Tris-buffered saline (TBS) (0.02 M Tris–HCl, pH 7.5, 0.15 M NaCl) was used rather than phosphate buffered saline as basic buffer for antibody dilution and membrane washing (Hammond and Jordan 1990); 1 μg/mL polyvinyl alcohol (Sigma MW 30,000–70,000) was used as preincubation blocking agent (Makkouk and Comeau 1994); 2% polyvinyl pyrrolidone (Sigma PVP-40) and 0.2% egg albumin were added to basic buffer for dilution and incubation of monoclonal antibodies and alkaline phosphatase conjugated antibodies. For TBIAs that used polyclonals as primary antibodies, including *TuMV*, Cauliflower mosaic virus (*CaMV*), and other TBIAs, binding to host antigens was eliminated by dilution and preincubation at 37 °C for 2 h in strained leaf extract (1 g leaf per 50 mL TBS) of uninfected analogous host, usually wild radish, before adding the antibody mixture to membranes. Sensitivity of Potyvirus TBIA was maximized by heating membranes at 60 °C in 0.05 M Tris pH 9.5 (Jordan and Hammond 1991) for 5 min to expose antigenic determinants before blocking and incubating (Schwinghamer et al. 2014).

4.13.6 The *Turnip Mosaic Virus* Biological and Serological Detection

In order to confirm the presence/absence of the most commonly occurring oilseed rape viruses, comprehensive characterization of herbaceous plant responses was performed. Mechanically inoculated *TuMV* assay hosts *N. tabacum* and *Chenopodium amaranticolor*, showed chlorotic local lesions up to 10-days post inoculation (dpi). Chlorotic lesions regularly turned necrotic, but no systemic symptoms appeared. On the other hand, the local lesions that appeared in *C. quinoa* and *B. rapa* 10 dpi preceded systemic symptoms that were intensely expressed as vein clearing, mosaic, and distortion and stunting in *B. rapa* 24 dpi. The latter host was also suitable for the maintenance of the virus. *N. benthamiana*, *N. glutinosa*, and *N. megalosiphon* reacted with mild local chlorotic spots and systemic chlorosis of the veins with mild leaf crinkling, but *N. clevelandii* showed no signs of infection. *Pisum sativum* reacted strongly to the virus. The plants wilted and succumbed to the severe systemic necroses within 21 dpi, whereas *Phaseolus vulgaris*, known to be the nonhost for *TuMV* (Plant Viruses Online, http://pvo.bio-mirror.cn/descr855.htm), showed no sign of infection. Symptoms indicating the presence of other viruses (*CMV, TYMV,* and *RaMV*) commonly found in rapeseed in single or mixed infections (Rimmer et al. 2007) were not observed within the course of these experiments. The double-radial immune diffusion test with polyclonal antisera confirmed the *TuMV* presence, the *RaMV* and *TYMV* absence in the field oilseed rape leaf and floral tissue, as well as in the experimentally infected *B. rapa* and *N. benthamiana* plants (Seruga Musić et al. 2014).

4.14 *Brassica–Verticillium* Pathogen

4.14.1 Identification and Differentiation of *Verticillium* Species by Simplex and Multiplex PCR Assays

The *Verticillium sensu stricto* is a small group of agriculturally important, plant-associated fungi that cause *Verticillium* wilt, a type of vascular wilt that causes significant economic losses of numerous crops and ornamentals in many parts of the world. Among the ten species currently recognized in *Verticillium sensu stricto*, *V. dahliae* is most widespread and most economically important, but *V. albo-atrum*, *V. alfalfa*, *V. longisporum*, *V. nonalfalfae*, *V. tricorpus*, and *V. zaregamsianum* also cause significant losses, *V. nubilum* causes disease in pathogenicity tests, and both *V. isaacii* and *V. klebahnii* have been recovered from lettuce and artichoke, respectively. One of the characteristic features of *Verticillium* species is the formation of resting structures. The resting structures of *V. dahliae* consist of clusters of thick-walled cells called microsclerotia, which remain viable in the soil for at least 14 years. Because as few as two microsclerotia per gram of soil result in plant infection and yield losses, knowledge about the abundance of microsclerotia and other resting structures in the soil is an important factor to consider for disease management. *Verticillium* species also differ in host range and pathogenicity, thus, expedient detection, quantification and identification of *Verticillium* species has been the focus of extensive research efforts. These included the design of numerous PCR-based assays, targeting *V. albo-atrum*, *V. dahlia*, *V. longisporum*, and *V. tricorpus*, in a variety of substrates including alfalfa, oilseed rape, olive, pepper, potato, strawberry, soil, spinach seed, tomato, and herbaceous hosts in general (Steventon et al. 2002).

Five new *Verticillium* species were described, including *V. isaacii* and *V. klebahnii* that are morphologically indistinguishable from *V. tricorpus*, and *V. alfalfae* and *V. nonalfalfae* that resemble *V. albo-atrum* and cannot be differentiated based on morphology. Also, the relationship of *V. dahliae* to the diploid hybrid *V. longisporum* was clarified. It was found that *V. longisporum* consists of at least three groups that evolved independently by hybridization involving two unknown species and two lineages of *V. dahliae* (Fig. 4.10). The two unknown species have never been found except as parents of *V. longisporum*, and have informally been named Species A1 and species D1, and the two *V. dahliae* lineages are referred to as *V. dahliae* lineages D2and D3. The three groups or lineages of *V. longisporum* evolved by independent hybridization of species A1 with species D1, *V. dahlia* lineage D2 and *V. dahliae* lineage D3, respectively, and accordingly, are referred to as *V. longisporum* lineage A1/D1, *V. longisporum* lineage A1/D2 and *V. longisporum* lineage A1/D3, respectively.

The DNA sequence data generated in conjunction with phylogenetic and taxonomic studies of *Verticillium*, and designed PCR assays for the identification of *Verticillium* species and *V. longisporum* lineages will be useful for diagnostic slabs and research applications (Inderbitzin et al. 2013). Accurate species identification is essential for effective plant disease management, but is challenging in fungi

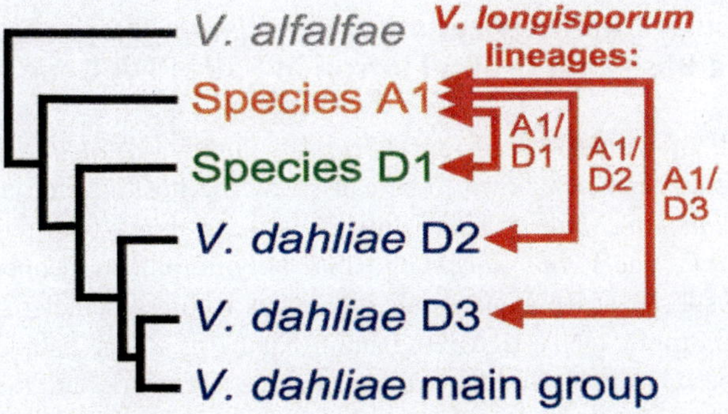

Fig. 4.10 Evolutionary history of *Verticillium longisporum* illustrated by a cartoon phylogenetic tree based on Inderbitzin et al. (2011). *Verticillium longisporum* evolved at least three different times by hybridization of species A1with species D1, *V. dahliae* lineage D2 and *V. dahliae* lineage D3, resulting in *V. longisporum* lineages A1/D1, A1/D2 and A1/D3, respectively. *Verticillium dahliae* isolates are in blue, species D1 in green, species A1 in orange and *V. alfalfae* in gray. Red arrows indicate parents of *V. longisporum*, connecting lines represent the three *V. longisporum* lineages. The *Verticillium dahliae* lineage D2 is marked as "*V. dahliae* D2" and comprises both *V. dahliae* isolates and D2- alleles of *V. longisporum* lineage A1/D2. The *Verticillium dahliae* lineage D3 ("*V. dahliae* D3") comprises only D3-alleles of *V. longisporum* lineage A1/D3. Most of the *V. dahliae* isolates belonged to the clade marked "*V. dahliae* main group" (Inderbitzin et al. 2013)

including *Verticillium sensu stricto* (Ascomycota, Sordariomycetes, Plectosphaerellaceae), a small genus of ten species that includes important plant pathogens. Inderbitzin et al. (2013) used fifteen PCR assays for the identification of all recognized *Verticillium* species and the three lineages of the diploid hybrid *V. longisporum*. The assays were based on DNA sequence data from the ribosomal internal transcribed spacer region, and coding and noncoding regions of actin, elongation factor 1-alpha, glyceraldehyde-3-phosphate dehydrogenase and tryptophan synthase genes. The eleven single target (simplex) PCR assays resulted in amplicons of diagnostic size for *V. alfalfae*, *V. albo-atrum*, *V. dahliae* including *V. longisporum* lineage A1/D3, *V. isaacii, V. klebahnii, V. nonalfalfae, V. nubilum, V. tricorpus, V. zaregamsianum*, and species A1 and species D1, the two undescribed ancestors of *V. longisporum*. The four multiple target (multiplex) PCR assays simultaneously differentiated the species or lineages within the following four groups: *Verticillium albo-atrum, V. alfalfae* and *V. nonalfalfae*; *Verticillium dahlia,* and *V. longisporum* lineages A1/D1, A1/D2 and A1/D3; *Verticillium dahliae* including *V. longisporum* lineage A1/D3, *V. isaacii, V. klebahnii,* and *V. tricorpus*; *Verticillium isaacii, V. klebahnii,* and *V. tricorpus*. Since *V. dahliae* is a parent of two of the three lineages of the diploid hybrid *V. longisporum*, no simplex PCR assay is able to differentiate *V. dahliae* from all *V. longisporum* lineages. PCR assays were tested with fungal DNA extracts from pure cultures, and were not evaluated for detection and quantification of *Verticillium* species from plant or soil samples. The DNA sequence alignments are provided and can be used for the design of additional primers.

4.14.2 Molecular Characterization of *Verticillium longisporum* on the Basis of a Group-I Intron in SSU-rRNA Gene

Verticillium wilt of winter oilseed rape (*Brassica napus* ssp. *oleifera*) and other cruciferous species has been shown to be caused by the host-adapted near-diploid species *Verticillium longisporum*. This pathogen has been shown to differ from other isolates of *V. dahlia* and *V. albo-atrum* by RFLP polymorphisms, sequence data of the ITS and the subrepeat RNA intergenic regions as well as characteristic random amplified polymorphic DNA (RAPD) banding patterns. This pathogen occurs widely in Europe (Germany, Sweden, Poland, France, Denmark, and the Ukraine), Asia, and Japan, and has been suggested to be also present in the UK and USA. A particular isolate responsible for the wilt in *Brussels* sprout in Evesham in the 1950s was suggested to be the first evidence of the presence of *V. longisporum* in the UK on the basis of some important characteristics. Additionally, *Verticillium* wilt in cauliflower has been attributed to a distinct strain of *Verticillium*, from polymorphisms in the intergenic spacer region of the nuclear rDNA and distinct RAPD banding patterns. This distinctive strain, with relatively large conidia and high nuclear DNA content, was suspected of being diploid and similar to *V. longisporum* and is considered the first evidence of the presence of *V. longisporum* in North America (Karapapa et al. 1997).

Group-I introns are present in many eukaryotes, prokaryotes, and viruses, and many of them in fungi are located in the nuclear and/or mitochondrial SSU-rRNA genes. They catalyze their own cleavage and have been characterized by a distinct common RNA secondary structure and a common splicing pathway. These introns with considerable differences in size and sequence are inserted in at least 17 positions in highly conserved regions of primary sequences. In particular, the distribution of group-1 introns is generally extremely irregular, with introns being optional, present in some isolates and absent from others. Their insertion position in the rRNA gene complex is considered to be directly correlated with the degree to which they are related to each other, and group-I introns with similar insertion sites across distantly related organisms are thought to have a common ancestry. Interestingly enough, almost all fungi containing group-I introns in their nuclear rRNA gene complex are parasitic or symbiotic.

Verticillium species are classified as Deuteromycetes, which is a subdivision comprising a diverse collection of genera related only by their apparent inability to reproduce sexually, and most of them are recognized to be anamorphs of Ascomycetes. Most species of *Verticillium* do not have proven teleomorphs. Links with sexual taxa have been reported for *V. tenerum*, *V. cinnabarinum*, *V. lateritium* (perfect stage *Nectria inventa*) and for *V. olivacearum* (perfect stage *N. cosmariospora*) placed in Hypocreales, and *Torrubiella/Cordyceps* is the teleomorph stage of *V. lecanii* placed in the Clavicipitaceae, but none for any of the asexually reproducing plant pathogenic species. Karapapa and Typas (2001) amplified the SSU of the rRNA gene complex and compared the amplicons from 36 isolates previously characterized as *V. longisporum*, including two isolates from cauliflower (USA), as well as from 47 isolates of 18 phytopathogenic and nonphytopathogenic species of *Verticillium*, and two isolates of *Cordyceps*. The SSU-rRNA of the former species was 839 bp larger that the ca. 1.65 kb typical for all the other *Verticillium* species tested, owing to the presence of an intervening sequence, characterized as group-I intron by the conserved core elements and secondary structure.

Verticillium wilt of oilseed rape is caused by the host-adapted pathogen *Verticillium longisporum* comb. nov. With one set of nuclear SSU-rRNA gene primers, a PCR amplification product of ca. 2.5 kb was generated from all isolates of *V. longisporum* tested (36 from Europe, Japan, and USA), with the exception of two recombinant isolates. On the contrary, all the other phytopathogenic and nonphytopathogenic species of *Verticillium* tested (18 species, 46 isolates), with the exception of one isolate of *V. lecanii* and two of *Cordyceps* sp., generated a product of ca. 1.65 kb. Sequence analysis of the SSU-rRNA gene of two typical isolates of *V. longisporum* (wild radish, Japan, and oilseed rape, Germany) revealed that this dimorphism was due to the presence of an identical 839-bp intron located in a highly conserved insertion position (nt 1165 of *Saccharomyces cerevisiae*). The intron sequence was classified as group-I intron on the basis of conserved sequence and secondary structural elements. Primers designed from the 839-bp intron sequence amplified only the *V. longisporum*. Phylogenetic analysis based on SSU-rDNA sequences showed that *V. longisporum* was closely related to the genera of other filamentous Ascomycetes with fruiting bodies (Karapapa and Typas 2001).

4.14.3 Molecular Characterization of *Verticillium* Isolates

The majority of isolates of the two main plant pathogenic species of *Verticillium* are not host adapted but are able to infect a wide range of hosts. Pathotypes do occur with perhaps the clearest example being *V. albo-atrum* isolates from Lucerne (alfalfa), which are molecularly and pathologically distinct from isolates from all other hosts. Infection of cruciferous hosts once seemed unusual with only isolated occurrences reported from Brussels sprouts and horseradish. The infections of crucifers by microsclerotial isolates have been found much more frequently, e.g., in oilseed rape in Europe, cauliflower in California, horseradish in Illinois, and various crucifers in Japan. Isolates of *V. dahliae* and *V. albo-atrum* from most hosts are short spored and have a low nuclear DNA content, but the majority of isolates from cruciferous hosts is long spored and has a high nuclear DNA content.

All crucifer isolates produce microsclerotia and have usually been classified as *V. dahliae,* often with the sub-specific addition of "var. *longisporum*," the name C. Stark gave to the first isolate from horse radish. On the basis of molecular and other studies, Karapapa et al. (1997) suggested that long-spored isolates should be classified as a separate species, *V. longisporum.* This new taxon was based mainly on European oilseed rape isolates and was not entirely satisfactory in that it excluded some long-spored isolates, including the original *V. dahliae* var. *longisporum,* which were described with the undefined term "recombinant." The new taxon also left unclear the status of short-spored crucifer isolates. It was further suggested by

Karapapa et al. (1997) that *V. longisporum* was an interspecific hybrid between a Lucerne isolate of *V. albo-atrum* and *V. dahliae*, and that genetically the new species was "near-diploid."

Diverse isolates of the soil borne wilt fungi *Verticillium dahliae* and *V. albo-atrum* were analyzed to understand the nature and origins infecting cruciferous hosts. All isolates from cruciferous crops produced microsclerotia, and the majority produced long conidia with a high nuclear DNA content; these isolates were divided into two groups by amplified fragment length polymorphism (AFLP) analysis. One group could be subdivided by other criteria such as rRNA sequences and mitochondrial DNA restriction fragment length polymorphism (RFLP) analysis. Two crucifer isolates were short spored and had a low nuclear DNA content. The results are consistent with the crucifer isolates being interspecific hybrids. The long-spored isolates are best regarded as amphihaploids (or allodiploids) with the AFLP groups probably each representing separate interspecific hybridization events. The short-spored crucifer isolates appear to be derived from interspecific hybrids and are here called "secondary haploids." Molecular evidence suggests that one parent in the crosses was similar to *V. dahliae.* The other parent of the amphihaploids seems to have been more similar to *V. albo-atrum* than to *V. dahliae*, but was distinct from all isolates of either species so far known (Collins et al. 2003).

4.15 *Brassica–Xanthomonas* Pathogen

4.15.1 Identification of Virulence Effectors/Genes of *Xanthomonas*

Many phytopathogenic bacteria produce a large number of factors that might be essential or contribute to cause disease. The bacteria from the genus *Xanthomonas* typically produce yellow, membrane-bound pigments, called xanthomonadins. These pigments have a role in the maintenance of the ecological fitness of the bacteria, protecting the cells against photooxidative stress. Xanthomonadin biosynthesis is encoded by the *pig* cluster of genes. Poplawsky and Chun (1998) have shown that the mutation of *pigB* results in a reduction in epiphytic survival and decrease in infection via the hydathodes. The biosynthesis of xanthomonadinis regulated by 3-hydroxybenzoic acid, a diffusible factor (DF) that is associated with a range of biological functions; DF-deficient mutants are nonpigmented, impaired in survival ability and less virulent (He et al. 2011).

The *Xcc* produces a range of extracellular enzymes (including proteases, pectinases and endoglucanase). The extracellular enzymes are capable of degrading the plant cell components and may be required to overcome plant defense responses, to allow bacteria to move into uncolonized plant tissues and to mobilize plant polymers for nutritional purposes (Torres et al. 2007). In *Xcc*, the synthesis of extracellular degrading enzymes and exopolysaccharides is regulated by products of the *rpf* genes (rpfABFCHGDIE) (da Silva et al. 2002). The *rpf* gene cluster of *Xcc* regulates the genes involved in the synthesis of extracellular hydrolytic enzymes and extracellular polysaccharides (EPSs)(including xanthan) and plays a role in motility, toxin production, oxidative stress resistance, aerobic respiration, and biofilm formation (Dow et al. 2003; Wilson et al. 1998). A diffusible signal factor (DSF) responsible for cell–cell signaling is involved in the transcriptional control of these genes (Barber et al. 1997). Some genes under the control of the rpf/DSF system are required for the first stages of endophytic colonization (Gudesblat et al. 2009). Genes that encode proteins involved in the synthesis or degradation of cyclic-di-guanosine-3′,5′-monophosphate (cyclic-di-GMP), a second messenger involved in the regulation of cellular functions, have been shown to play significant roles in virulence to plants. Complex regulatory networks possibly allow the bacteria to adapt to different environments and to modulate virulence factors (Ryan et al. 2007). The expression levels of proteases and endoglucanases are reduced when the *rpfI* gene is inactivated in *Xcc*, suggesting that *rpfI* may have a function in the extensive tissue degeneration that is characteristic of black rot (Dow et al. 2000). RpfF is involved in the synthesis of DSF, and a two component system comprising RpfG and the complex sensor kinase RpfC is involved in sensing and responding to DSF (Ryan et al. 2010). Mutants deficient in RpfG, a phosphodiesterase that degrades cyclic-di-GMP, or the elevation of cyclic-di-GMP via the over expression of a GGDEF domain protein WspR, caused the aggregation of cells, reduction in mobility and decrease in production of virulence factors, extracellular enzymes and exopolysaccharides (Hsiao et al. 2011). Mutations of *rpfF* or *rpfC* also lead to a reduction in the synthesis of virulence factors. The cell–cell interactions are highly regulated and dynamic; the interactions involving HD-GYP and GGDEF domain proteins are dependent on DSF signaling influence the localized expression of cyclic-di-GMP and mediate virulence (Ryan et al. 2007). The Mips (macrophage infectivity potentiators) genes encode proteins reported as virulence factors in human pathogens. Zang et al. (2007) showed that a *mip*-like gene of *Xcc* is involved in pathogenicity through an effect on the production of exopolysaccharides and on the activity of extracellular proteases. EPSs play an important role in the pathogenicity and virulence of many bacteria, both in terms of direct interactions with host cells and in conferring protection against hostile environments (Coplin and Cook 1990). The most important EPS secreted by *Xcc* is xanthan gum, an EPS that has many industrial applications. Xanthan is a complex EPS with a cellulosic backbone and trisaccharide side chains of two mannose and one glucuronate residues that are attached to every second glucose in the backbone (Becker et al. 1998; Jansson et al. 1975; Vorhoelter et al. 2008; Yun et al. 2006). Yun et al. (2006) have shown that a xanthan minus mutant strain and a mutant strain that produces truncated xanthan fail to cause disease in *A. thaliana* and *Nicotiana benthamiana*, and induce the deposition of callose in these plants, pretreatment of the plant with xanthan restores the pathogenicity of both strains. The results indicate that xanthan induces susceptibility to *Xcc* by suppressing callose deposition. The bacterial cyclic b-(1,2)-glucan also has an effect on plant defense suppression and callose deposition, but, unlike xanthan, which acts locally, cyclic glucan acts systemically as part of a counter-defensive strategy that might facilitate the spread of the pathogen in the plants (Rigano et al. 2007).

Xanthan has a broad range of applications in food and nonfood products from the oil, pharmaceutical, cosmetic, paper, paint and textile industries, and is mainly used as a thickening, stabilizing, gelling and emulsifying agent (Becker et al. 1998). The biosynthesis of xanthan production and the genes that encode for the enzymes involved have been studied (Ielpi et al. 1993; Becker et al. 1998). Xanthan synthesis is encoded by twelve *gum* genes (*gumBCDEFGHIJKLM*), which are located in a single gene cluster of 12 kb that is mainly expressed as an operon from a promoter upstream of the first gene, *gumB* (Vojnov et al. 2001).The genes required for the synthesis of xanthan and its nucleotide sugar precursors are highly conserved among the three sequenced strains (ATCC33913, 8004 and B100), but are significantly different when compared with the other sequenced *Xanthomonas*; a model for the biosynthesis of xanthan based on the genome annotation has been proposed (Vorhoelter et al. 2008). The disruption of the *rmlA*, *xanA,* and *gumK* genes, which are responsible for the biosynthesis of intermediates for xanthan production, can lead to a loss of pathogenicity (Qian et al. 2005). The lip polysaccharide (LPS) is an essential component of the outer membrane of Gram-negative bacteria, and is an important virulence factor in *Xanthomonas* (Dow et al. 1995; He et al. 2007; Patil et al. 2007). Over 20 genes for LPS synthesis have been characterized in *Xcc*, including *xanAB*, *rmlABCB*, *rfaXY*, *lpsIJ* and 15 genes that constitute the *wxc* cluster (Vorholter et al. 2001). The *wxc* gene cluster is involved in the synthesis of the LPSO-antigen, the most variable part of LPS. The *wxc* genes are highly divergent between *Xcc* strains, indicating that different isolates might produce varied LPSs (He et al. 2007). The LPS gene cluster of *Xcc* B100 is significantly different from those of 8004 and ATCC33913, but is almost identical to the LPS cluster of *X. campestris* pv. *raphani* 756C. This could indicate that the LPS cluster of B100 was acquired recently via horizontal gene transfer from *X. campestris* pv. *raphani* 756C or, more likely, that the *X. campestris* pv. *raphani* strain diverged recently from an *Xcc* with a B100-type LPS locus (Patil et al. 2007). Qian et al. (2005) showed that the disruption of a number of *wxc* genes resulted in a significant reduction in virulence (Vicente and Holub 2013).

4.15.2 Novel Genes Identification Through Molecular Approaches

Resistance (*R* gene) genes identification is the first step towards the resistant hybrid/variety development through appropriate breeding programs. Studying gene regulations in response to pathogens causing disease might give indications about the relevant defense genes. Suppression Subtractive Hybridization (SSH) is a powerful PCR based technique to identify and isolate differentially expressed genes such as in the response to change in environmental conditions or plant pathogen interaction (Diatchenko et al. 1996; Jin et al. 2010; Luo et al. 2010; Jiang et al. 2011; Guo et al. 2013; Roohie and Umesha 2015). The application of the SSH approach for the identification of genes associated with black rot resistance in Brassicaceae crops has been documented (Jiang et al. 2011; Roohie and Umesha 2015). Jiang and associates (Jiang et al. 2011) employed SSH combined with RT-PCR to identify the black rot resistance genes in cauliflower (*B. oleracea* var. *botrytis*). They constructed a SSH cDNA library using resistant line C712 as tester and its susceptible near-isogenic line C731 as driver. Roohie and Umesha (2015) employed the SSH approach for the identification of resistance genes and to decipher the resistance mechanism in cabbage after infection with Xcc. In the resistant cultivar of cabbage "Pusa Mukta" they constructed a cDNA library enriched in differentially expressed transcripts. Out of 150 unigenes studied that were classified into five functional categories, 35% of the total unigenes accounted for the defense-related unigenes. Defense-specific representation of the genes was confirmed through semi quantitative RT-PCR and their elevated expression in the resistant cultivar were validated by qPCR. The resistant cultivar evoked a strong hypersensitive response upon attack by the Xcc pathogen (Singh et al. 2018).

4.16 Molecular Identification of Pathotype and Pathogen Elicitors/Effectors

To control disease outbreaks in crops, it is imperative that pathogen population variability in the field can be quickly identified. This can be achieved by having a full set of genetic markers for different pathogen pathotypes, not only contained within the local fields but universally. For example, initiatives for *P. brassicae* showed that *Cr811* was a good candidate gene to differentiate pathotype 5 from the rest of the pathotypes (Feng et al. 2016).

Molecular characterization of pathogen effectors is gaining momentum. With better genome assemblies and higher gene prediction accuracies, we will soon have better insights into the structures and functions of effector proteins, which allow more efficient screening and identification of *R* genes (Gibriel et al. 2016). Elicitor candidates for *L. maculans* (Wilson and Howlett 2005; Novakova et al. 2016), and effector candidates for *S. sclerotiorum* (Guyon et al. 2014), are both responsible for inducing HR. RNAi studies established that the transcription factor LmStuA in *L. maculans* regulates expression of effector genes (Soyer et al. 2015). Genomic studies researching *L. maculans* effectors *AvrLmJ1* (Van de Wouw et al. 2014), *AvrLm1* (Gout et al. 2006), *AvrLm2* (Ghanbarnia et al. 2015), *AvrLm3* (Plissonneau et al. 2016), *AvrLm4*-7 (Parlange et al. 2009), *AvrLm1*-2-6 (Fudal et al. 2007), and *AvrLm11* (Balesdent et al. 2013) have provided new understandings on how they are strategically located within regions that are rich in transposable elements, such that they can be lost or inactivated rapidly during sexual reproduction (Rouxel et al. 2011; Howlett et al. 2015).This fungal genomic structure is unique to *L. maculans* and often the effectors display little or no sequence similarity to each other (Van De Wouw et al. 2010; Rouxel et al. 2011). Specific recognition between AvrLm4-7 and Rlm4 and Rlm7 genes and how the R and Avr effector proteins interact structurally has been examined using crystallography (Blondeau et al. 2015). There should be a focus on studies that provide additional molecular details at the gene level which would further our understanding of the factors leading to virulence and susceptibility associated with *R* gene mediated resistance (Neik et al. 2017).

4.17 The qRT-PCR Analysis of *Brassica* Reference Genes Under Various Stresses

Quantification of mRNA transcript levels provides important insights into the intricate metabolic pathways and signaling networks underlying the plant's physiological response to various abiotic and biotic stresses. The qRT-PCR is a sensitive, accurate and cost effective method to evaluate the expression of the target gene in different tissues, organs and treatments. However, the accuracy of the expression data is affected by many factors, such as RNA quality, purity, and PCR amplification efficiency, technical and true biological variations. To control these variables and avoid bias in qRT-PCR, selecting stably expressed RGs under different experimental conditions are crucial for normalization. Selecting an inappropriate RG leads to misinterpretation of the expression levels of the target gene. In theory, RGs are the genes which maintain a constant expression level in all cell types and under every experimental condition. However, studies have reported that based on the experimental conditions and species, the expression level of the traditionally used RG is variable. The housekeeping genes are required to maintain basic cellular functions such as components of the cytoskeleton, vesicular transport, cell cycle, glycolytic pathway, protein synthesis, and protein degradation. Traditionally, actin (*ACT*), elongation factor 1-alpha (*EF1A*), polyubiquitin (*UBQ*) and tubulin alpha (*TUA*) are used as RGs in expression analysis. With the development of microarray and Affymatrix Gene Chip many novel genes have been identified as suitable RGs and were found to exhibit stable expression in many species like *Arabidopsis thaliana*, *B. juncea, Brassica napus*, and other crucifers. Moreover, there is evidence to suggest that the expression of HKGs is differentially regulated under variable experimental conditions and species. Therefore, selecting stably expressed RG is crucial for accurate normalization of gene expression data.

Several methods have been developed to simplify the selection of appropriate RGs. geNorm gives expression stability value (M) of each RG under a given experimental condition. M_{value} is the average pair wise variation in the expression of particular RG with all the other RG analyzed in the experiment. Gene exhibiting lower value has more stable expression and higher value depicts less stable expression of RG in a given set of samples. geNorm also determines the minimum number of RGs required for accurate normalization by calculating normalization factors (NFs) with a cutoff value for $V_n/n + 1$ as 0.15, below this value no additional RG is required for normalization. Norm Finder evaluates expression stability of RGs with a different mathematical model wherein inter- and intra-group variation in the expression of candidate RGs is taken into consideration. This algorithm assigns an expression stability value to the RGs. A lower value is the determinant of higher expression stability and lower inter and intragroup variation. In the Best Keeper tool, the original C_t values are used for identification of the most stable RG under a given experimental condition.

It is a known fact that phytohormones like salicylic acid (SA), jasmonic acid (JA) and abscisic acid (ABA) and their cross-talk play a central role in determining the plant's response to a particular pathogen. To understand the mechanism of pathogen response operative in resistant/susceptible species, differential expression of target genes observed in *Brassica* in response to various challenges is an adequate approach. Therefore, it is important to identify suitable RGs which express stably in *Brassica* in order to perform an accurate comparative gene expression analysis in these species. Dixit et al. (2019) compared the performance of 7 candidate RGs across 8 sample subsets of *B. juncea* and *C. sativa* subjected to various treatments. Statistical algorithm geNorm, Norm Finder and Best Keeper we reutilized for the analysis, to identify stable RGs in both the species under a given experimental condition as well as to recognize those RGs which can be used for efficient comparison of target gene expression among *B. juncea* and *C. sativa*.

Quantitative real-time PCR (qRT-PCR) is an efficient method to estimate the gene expression levels but the accuracy of its result largely depends on the stability of the reference gene. Many studies have reported considerable variation in the expression of reference genes (RGs) in different tissue and conditions. Therefore, screening for appropriate RGs with stable expression is crucial for functional analysis of the target gene. Two closely related crucifers *Brassica juncea* (cultivated) and *Camelina sativa* (wild) respond differently towards various abiotic and biotic stress where *C. sativa* exhibits higher tolerance to various stress. Comparative gene expression analysis of the target genes between two such species is the key approach to understand the mechanism of a plant's response to stress. However, using an unsuitable RG can lead to mis-interpretation of expression levels of the target gene in such studies. In this investigation, the stability of seven candidate RGs including traditional housekeeping genes (HKGs) and novel candidate RGs were identified across diverse sample sets of *B. juncea* and *C. sativa* representing- hormone treated, wounded, *Alternaria brassicae* inoculated and combination treated samples (exogenous hormone treatment followed by *A. brassicae* inoculation). The stable RGs were identified in both the species and the most suitable RGs to perform an unbiased comparative gene expression analysis between *B. juncea* and *C. sativa*. Results revealed that *TIPS41* and *PP2A* were identified as the overall best performing RGs in both the species. However, the most suitable RG for each sample subset representing different condition must be individually selected. In Hormone treated and wounded samples*TIPS41* expressed stably in both the species and in *A. brassicae* inoculated and combination treatment performance of *PP2A* was the best. The identified and validated stable reference gene in *C. sativa* for accurate normalization of gene expression data will be useful resource for future (Dixit et al. 2019).

4.18 Disentangling Hidden Histories in Genes and Genomes of Hosts and Pathogens

The pathogenic lifestyle is ubiquitous across the host life, and pathogens are estimated to represent a substantial proportion of the diversity and biomass of many ecosystems (Poulin 2014; Padra et al. 2018). PCR-based technologies and the advent of high throughput sequencing, along with the associated reduction in sequencing costs, have facilitated the description of novel pathogens (Titcomb et al.

2019), with particular success in viral pathogen discovery (Chiu 2013; Datta et al. 2015). Furthermore, the application of metagenomic approaches has highlighted complex host–pathogen interactions and implicated host- and pathogen-associated microbial communities in successful pathogen infection and disease development (Sekirov et al. 2010; Kamada et al. 2013). This suggests that a pathogen rarely occurs alone, and instead may commonly be a member of a larger community (Gregory et al. 2019). Hence, understanding the interplay between multiple pathogens and associated microbiomes requires disentangling several levels of complexity. It is also crucial to gain an understanding of the fitness effects of each putative pathogen on its host, because the magnitude of the fitness cost (i.e., virulence) of a pathogen during infection determines its place on the mutualist–pathogen continuum. In principal, this requires demonstration of a fitness cost to the host, yet demonstrating fitness effects of many putative pathogens in nature is challenging and often requires datasets that are much larger than those obtained in a typical field study. Importantly, the fact that measuring the fitness consequences of infections in nature is challenging does not imply that pathogens are insignificant selective agents in the evolution of host genomes. In fact, pathogens are widely presumed to be among the strongest selective agents (Fumagalli et al. 2011; Pittman et al. 2016). Mutations conferring moderate or large benefits of resistance to hosts can become readily fixed by selection and are detectable through genome scans (Nielsen 2005; Vitti et al. 2013). Thus, comparative and population genetic studies of host genomes present compelling approaches for studying the presumed impact of pathogens (Fig. 4.11).

Genetic variation is typically studied at different levels, such as across species (Fig. 4.11a), across populations (Fig. 4.11b), within populations (Fig. 4.11c) or

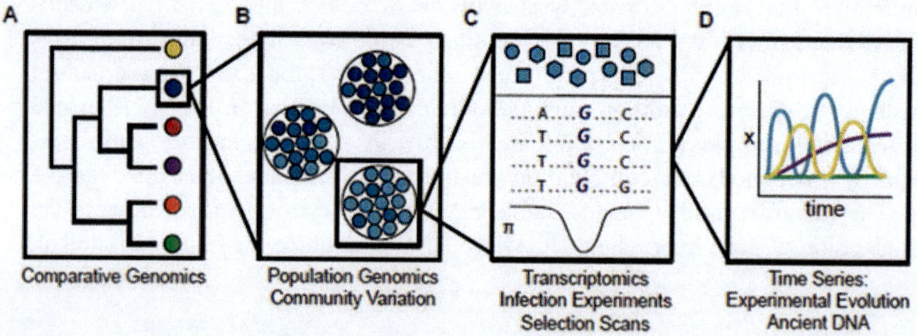

Fig. 4.11 Schematic illustration how genetic variation varies (**a**) across species, (**b**) across populations, (**c**) within a population, and (**d**) on an ecological time scale. (**a**) Comparative genomics across species can be used to identify genomic loci consistently under positive selection in particular lineages or all lineages. (**b**) Across populations, population genomic variation in different geographic populations can be correlated with pathogen communities. (**c**) Within a single population, phenotypic variation among individuals can be linked to pathogen variation or differentially expressed genes with transcriptomics. Genome scans may also identify regions of the genome under selection. (**d**) Finally, time series either derived through experimental evolution or studies of ancient DNA or diachronic samples can be used to identify the dynamics of a phenotype or allele frequency through time (Napflin et al. 2019)

through time (Fig. 4.11d) to disentangle the underlying genetics of host–pathogen interactions. For this purpose, two main approaches are typically employed. On the one hand, the underlying genetic architecture can be inferred using genotype–phenotype association studies. The statistical association between genomic loci and host–pathogen phenotypes is interpreted as evidence for the underlying genetics of a given phenotype (Hirschhorn et al. 2002). On the other hand, instead of determining fitness costs of pathogens in single experiments or surveys, biologists have turned to signals of natural selection over evolutionary time as recorded in host genomes (Sabeti et al. 2006). While these genome scans typically cannot directly test the causal selective agent, they do provide insight into the possible biological processes that are adapting most rapidly in host genomes (Biswas and Akey 2006). Indeed, analysis of signatures of selection in host genomes identified pathogens as the most likely drivers of the observed patterns in a number of studies.

Despite the success of genome-wide associations and selection scans to find genomic evidence of pathogen pressures on hosts, simultaneous genomic investigation of the co-evolutionary dynamics between host and pathogen within a single system remain rare. Indeed, today, few systems have the genomic resources available to truly interrogate ongoing genomic changes between pathogen and host in parallel. Additionally, emerging studies of experimental evolution in the field or laboratory, or multigene rational sampling of natural populations of hosts and pathogens, have successfully identified novel adaptive alleles in both hosts (Scanlan et al. 2015; Cairns et al. 2017) and pathogens (Pal et al. 2007). The reciprocal nature of the interactions between host and pathogens over time naturally lead researchers to ask whether host and pathogen species co-diversify over evolutionary time and to what extent genomics can inform the underlying processes. Indeed, attempts to detect co-speciation among hosts and pathogens date back to the beginning of the twentieth century. However, inferring co-evolutionary history through comparisons of host and pathogen phylogenies is challenging. For example, such comparisons can mistake a host shift followed by co-diversification as co-speciation (de Vienne et al. 2007). The former mechanism is more consistent with empirical data that suggests that the level of co-evolution necessary to drive co-speciation of host and pathogen is rarely encountered in nature (de Vienne et al. 2013). As models of molecular adaptation and gene tree evolution improve, we may be able to identify phylogenetic congruence at the gene-tree level or signatures of selection that co-vary among hosts and parasites with more confidence. In turn, we might be better able to interpret results in the light of co-diversification versus co-speciation of studies (Napflin et al. 2019).

4.19 Identification of *R* Genes Homologous DNA Fragments in *Arabidopsis thaliana*

Eight different bona-fide resistant gene like (RGL) PCR fragments have been obtained from *A. thaliana* accessions Col and *Ler*, using degenerate PCR primers based on the NBS-LRR type of previously cloned disease resistance genes

(Staskawicz et al. 1995). These fragments mapped to eight different loci on the *A. thaliana* genome. Six of these loci are closely linked to previously characterized disease resistance loci. Two fragments are actually derived from a functional *R* gene or from a member of an *R* gene family. The DNA sequences of all RGL fragments showed similarity to the NBS-LRR type of *R* genes, which demonstrates that the primers used can amplify a range of RGL sequences. These primers were designed with a combination of inosines or multiple nucleotides at the third codon position, by which it is theoretically possible to amplify all genes coding for the consensus amino acid sequences. However, not all NBS-LRR type of *R* genes present in Col or *Ler* was represented in the cloned RGL fragments. RGL fragments belonging to the *RPS2* or *RPM1* genes could not be cloned. An examination of the sequence of these *R* genes at the RG primer sites showed that especially the RG2 primer did not sufficiently match with the template (for both *RPM1* and *RPS2*, 23% mismatch). Slight alterations at the 3¢ end of the primers, selecting for different amino acid variants, will probably lead to the amplification of many other RGL fragments.

With the RG set of primers, a new type of RGL sequences represented by the L4, C4, and L3 fragments were observed. These are about 300 bp longer than what is expected based on the sequence of the *R* genes used to design the RG1 and RG2 primers. Comparison of the DNA sequences of these RGL fragments showed the presence of unique sequences in addition to homologous sequences, which are flanked by *A. thaliana* consensus sequences for exon-intron splice junctions (Brown 1996). This suggests the presence of introns in the NBS region of the corresponding RGL genes, which is supported by the presence of a continuous open reading frame after artificial splicing, and the continuous alignment with the derived amino acid sequences of other RGL fragments and of cloned gene products. The next step following the isolation of RGL sequences was to establish a correlation between an RGL gene sequence and disease resistance. Accessions *Col* and *Ler* have been used for the RGL sequence PCRs, so the cloned fragments can only be part of resistance genes known to be active in these two accessions. Kunkel (1996) and Holub (1997) summarized the map positions of many disease resistance loci determined so far. Not all disease resistance loci have been characterized as yet, and only a portion of the *R* genes residing at these loci will belong to the NBS-LRR gene class. Nevertheless, genetic linkage between an RGL fragment and a disease resistance locus has been found for nearly all isolated RGL fragments. The best correlation between RGL fragment and *R* gene was found for the C3 fragment, which turned out to be part of the *RPS5* gene against *Pseudomonas syringae*. The other RGL fragments, the presence of an *R* gene locus and a co-segregating RGL locus is often accession dependent. For instance, the *L4/C4* locus and the linked *RPP7*locus on chromosome 1 have been found in both *Col* and *Ler* (Crute et al. 1994), as well as the closely linked *L3A* and *EST177N18* loci. Also the *L2/C2* locus and the linked *RPP4*locus on chromosome 4 are present in both Col and *Ler* (Tor et al. 1994). Of the three segregating *L3* loci on chromosome5, the *L3C* locus has an extra *Ler*-specific gene copy, which correlates with the linked *RPP8* locus from *Ler* (Holub et al. 1994) and the *Ler*-specific *H. parasitica* recognitions against isolates Waco8, Waco11, and Boma1. This region of chromosome5 contains more disease resistance loci, which are not specifically found in *Ler*, such as the *RPS4* locus against *Pseudomonas syringae* pv. *pisi* (in accession Ws-3) (Hinsch and Staskawicz 1996), the *HRT* locus against turnip crinkle virus (not in Col or Ler) (Dempsey et al. 1997), the *TTR1* locus against tobacco ring spot virus (in Col) (Lee et al. 1996), and the *RAC2* locus against *A. candida* (in Ksk-2).

With a correlation established between the chromosomal positions of *R* gene loci and RGL loci, supportive evidence that RGL fragments are part of resistance genes was found in their genetic organization. Often, *R* genes have been found as multi-copy, clustered sequences such as the *PTO, FEN,* and *PRF* complex or the *Cf-9/Cf-4* and *Cf-2/Cf-5* clusters in tomato (Dixon et al. 1996; Jones et al. 1994), or as dispersed multicopy loci such as the *L6* and *M* loci in flax (Lawrence et al. 1995). In contrast, the LRR receptor protein kinase genes such as *ERECTA, RLK1, RLK4, RLK5,* and *PR5K* are single or low-copy genes (Torii et al. 1996; Walker 1993; Wang et al. 1996). Nearly all RGL fragments detected several hybridizing bands upon DNA blot hybridization and they often revealed a RFLP between accessions Col and Ler. An exception to the rule is the C1 fragment, which was present as a single-copy sequence in Col and not at all in Ler. A similar case was observed for the *RPM1* gene (Grant et al. 1995), which is present as a single-copy gene in resistant and absent in susceptible accessions. Especially intriguing is the genome organization of the gene family to which fragments L3, C4, and L4 belong. This family accommodates clustered and dispersed family members that appear to be correlated with the presence of *RPP* loci on two chromosomes. The DNA sequences and DNA blot hybridization patterns of the L3, L4, C4, L3.1, and CK1 clones and the EST clones with RGL DNA sequence homology showed that all these fragments are part of genes coding for both NBS and LRR domains. The PCR fragments L3.1 and CK1 were produced after annealing of the L3, RG1, and STK PCR primer sat unexpected positions. These annealing sites for the L3 andRG1 primers may be remnants of previous gene-shuffling events that are likely to have arisen during the generation of these multi-copy RGL genes.

Some of these genes are expressed, as demonstrated by the presence of the co-segregating EST sequences. For EST 108E20, this co-segregation was confirmed by partial DNA sequence analysis of BAC clone F24J19 containing part of the gene cluster at the *L4/C4* locus. Further analysis of the RGL copies present on this BAC clone may reveal clues about the organization and origin of this gene cluster and of the expression pattern of the individual genes in relation to disease resistance. The analysis of RGL sequences from the well-studied plant species *A. thaliana* may provide an insight into the evolution of gene families coding for NBS-LRR type of proteins toward a function in disease resistance, and offers starting points for direct cloning of these disease resistance genes (Aarts et al. 1998).

4.20 Comparisons Between Different Diagnostic Approaches

Both immunological and nucleic acid techniques offer considerable advantages over traditional diagnostic methods. However, the choice between these newer methods will depend on several factors including the application involved, the skills of the staff, the costs, the facilities needed and how many samples are to be analyzed. A comparison of immunological and PCR-based techniques is given in Table 4.5, showing how well each is able to fulfill different criteria that might be needed for a particular diagnostic application. Sometimes, a combination of diagnostic techniques is the best approach, for example immune capture can be used to improve the sensitivity or specificity of PCR assays and overcome problems with inhibitors in the sample. Culturing for a short time can be combined with PCR-detection (BIO-PCR) to increase the quantity of pathogen present, and to ensure that only viable microorganisms are detected (Schaad et al. 1999).

The new molecular-based diagnostic techniques that have been developed can be more rapid, sensitive and accurate than traditional methods. However, there are still problems that need to be addressed before they become widely used. With antibody-based techniques it can still be difficult to generate sufficiently specific antibodies for the target organism, particularly for fungi. For nucleic acid based methods, the limited availability of target DNA sequence data bases can restrict our ability to design appropriate diagnostic probes or primers. Thus, more effort will be needed to generate more comprehensive DNA sequence data bases, particularly for ribosomal DNA. Assay formats using nucleic acids are generally relatively complex, and improvements to simplify them would improve their usefulness. There is also the need to develop more rapid protocols for purifying DNA for testing that are amenable to automation. At the moment the costs of some of the newer techniques are high, especially real-time PCR. However, costs are reducing with increased use, due to products becoming available from a wider choice of manufacturers, and due to improvements in technology.

Immunological and DNA-based diagnostic tools are playing an increasingly important role in plant pathological research and will continue to do so. The specificity and sensitivity of these methods will open new opportunities for research on the interactions between pathogens and their hosts. The ability to target specific pathogen genes could lead to better understanding of host–pathogen resistance or the dynamics of fungicide resistance development within fungal populations. Diagnostics that can discriminate pathogens in disease complexes will improve our understanding of these complexes, which may lead to more efficient control strategies. The application of novel diagnostic methods to inoculum detection will allow progress to be made in our understanding of the temporal and spatial dynamics of epidemic development, and open up new opportunities for disease forecasting and management. However, new management tools, such as efficient sampling strategies, will have to be developed to make the best use of these new diagnostic methods. The methods will also play an important role in restricting the spread of disease via infected seeds and other plant material (Ward et al. 2004).

Table 4.5 Comparison of immunological and PCR techniques for the diagnosis and detection of plant pathogens (Ward et al. 2004)

Process	Immunological techniques	PCR techniques
Ease of development	Time-consuming and costly to develop and identify antibodies with the required specificity	Rapid if sequence information (e.g., rDNA sequences or specific gene sequence) is available. Alternatively, techniques such as RAPDs can be used to quickly identify species-specific markers
Sample preparation	Minimal to complex depending on plant material and presence of substances which may inhibit antibody–antigen interactions	Minimal, crude DNA preparations can be used in many cases; however, DNA purification may be required
Procedure	Generally complex but some formats can be used by nonspecialists. Can be performed in 1 day and some immune-diagnostic kits can give answers in 15 min	Complex, requires expert staff and appropriate measures should be taken to avoid cross-contamination of samples. Can be performed in 1 day
Portability	Assays can be adapted for use in the field	With a few exceptions[a] assays can only be done in a laboratory
Specificity	Often good for viruses, but cross reactivity between fungal species is a common problem	Excellent, technique can be designed to be genus-, species-, or isolate-specific
Sensitivity	Varies depending upon amount of antigen produced by the pathogen. Generally less sensitive than nucleic acid-based techniques	Excellent, can detect a single copy of the target DNA sequence which corresponds to a single cell of the target organism
Quantification	Some formats (e.g., ELISA) are directly quantitative	Quantification is laborious in standard PCR. Simple and rapid quantification is possible using real-time PCR
Detection of multiple pathogens	Difficult	Relatively easy using multiplex PCR
Detection of viable organism stages	Usually detects only viable material and may be specific for particular morphological forms/of the life cycle	Generally, will detect both living and dead material[b] and all morphological forms
Costs	Some formats e.g. ELISA are very cost effective for high-throughput applications	Expensive, particularly real-time PCR, although costs are decreasing with increased adoption of the technologies

[a] See Schaad et al. (2002)
[b] Except reverse transcriptase PCR

References

Aarts N, Metz M, Holub E, Staskawicz BJ, Daniels MJ, Parker JE (1998) Different requirements for *EDS1* and *NDR1* by disease resistance genes define at least two R gene-mediated signaling pathways in *Arabidopsis*. Proc Natl Acad Sci U S A 95:10306–10311

Adam L, Ellwood S, Wilson I, Xiao S, Saenz G, Oliver R, Turner JG, Somerville S (1999) Comparison of *Erysiphe cichoracearum* and *E. cruciferarum* and a survey of *Arabidopsis thaliana* accessions for their resistance to these two powdery mildew pathogens. Mol Plant Microbe Interact 12:1031–1043

Adams MD, Kelley JM, Gocayne JD, Dubnick M, Polymeropoulos MH, Xiao H, Merril CR, Wu A, Olde B, Moreno RF, Kerlavage AR, McCombie WR, Venter JC (1991) Complementary DNA sequencing: expressed sequence tags and human genome project. Science 252:1651–1656. https://doi.org/10.1126/science.2047873

Adhikari TB, Liu JQ, Mathur S, Wu CX, Rimmer SR (2003) Genetic and molecular analyses in crosses of race 2 and race 7 of *A. candida*. Phytopathology 93:959–965

Ando S, Yamada T, Asano T, Kamachi S, Tsushima S, Hagio T, Tabei Y (2006) Molecular cloning of *PbSTKLl* gene from *Plasmodiophora brassicae* expressed during club root development. J Phytopathol 154:185–189

Ansan-Melayah D, Balesdent MH, Delourme R, Pilet ML, Tanguy X, Renard M, Roulx T (1998) Genes for race-specific resistance against blackleg disease in *Brassica napus* L. Plant Breed 117: 373–378. https://doi.org/10.1111/j.1439-05231998.tb01956.x

Araus JL, Cairns JE (2014) Field high throughput phenotyping: the new crop breeding frontier. Trends Plant Sci 19:52–61

Armstrong T (2007) Molecular detection and pathology of the oomycete *Albugo candida* (white rust) in threatened coastal cresses. DOC Research & Development Series 274, Science & Technical Publishing Department of Conservation, New Zealand, pp 5–18

Arora H, Padmaja KL, Paritosh K, Mukhi N, Tewari A, Mukhopadhyay A, Gupta V, Pradhan AK, Pental D (2019) *BjuWRR1*, a CC-NB-LRR gene identified in *Brassica juncea*, confers resistance to white rust caused by *Albugo candida*. Theor Appl Genet 132:2223–2236. https://doi.org/10.1007/s00122-019-03350-z

Asai T, Tena G, Plotnikova J, Willmann MR, Chiu WL, Gomez-Gomez L, Boller T, Ausubel FM, Sheen J (2002) MAP kinase signalling cascade in *Arabidopsis* innate immunity. Nature 4156(1): 977–983

Badet T, Voisin D, Mbengue M, Barascud M, Sucher J, Sadon P, Balagué C, Roby D, Raffaele S (2017) Parallel evolution of the POQR prolyl oligo peptidase gene conferring plant quantitative disease resistance. PLoS Genet 13(12):e1007143. https://doi.org/10.1371/journal.pgen.1007143

Balesdent MH, Fudal I, Ollivier B, Bally P, Grandaubert J, Eber F, Chèvre AM, Leflon M, Rouxel T (2013) The dispensable chromosome of *Leptosphaeria maculans* shelters an effector gene conferring avirulence towards *Brassica rapa*. New Phytol 198:887–898. https://doi.org/10.1111/nph.12178

Baranowski P, Jedryczka M, Mazurek W, Babula Skowronska D, Siedliska A, Kaczmarek J (2015) Hyperspectral and thermal imaging of oilseed rape (*Brassica napus*) response to fungal species of the genus *Alternaria*. PLoS One 10:e0122913

Barbacci A, Navaud O, Mbengue M, Barascud M, Godiard L, Khafif M, Lacaze A, Raffaele S (2020) Rapid identification of an *Arabidopsis* NLR gene as a candidate conferring susceptibility to *Sclerotinia sclerotiorum* using time-resolved automated phenotyping. Plant J 103(2): 903–917. https://doi.org/10.1111/tpj.14747

Barber CE, Tang JL, Feng JX, Pan MQ, Wilson TJG, Slater H, Dow JM, Williams P, Daniels MJ (1997) A novel regulatory system required for pathogenicity of *Xanthomonas campestris* is mediated by a small diffusible signal molecule. Mol Microbiol 24:555–566

Barrins JM, Ades PK, Salisbury PA, Howlett BJ (2004) Genetic diversity of Australian isolates of Leptosphaeria maculans, the fungus that causes blackleg of canola (*Brassica napus*). Australas Plant Pathol 33:529–536

Becker A, Katzen F, Puhler A, Ielpi L (1998) Xanthan gum biosynthesis and application: a biochemical/genetic perspective. Appl Microbiol Biotechnol 50:145–152

Becker MG, Haddadi P, Wan J, Adam L, Walker P, Larkan NJ, Daayf F, Borhan MH, Belmonte MF (2019) Transcriptome analysis of Rlm2-mediated host immunity in the *Brassica napus*–

Leptosphaeria maculans pathosystem. Mol Plant Microbe Interact 32(8):1001–1012. https://doi.org/10.1094/MPMI-01-19-0028-R

Belhaj K, Cano LM, Prince DC, Kemen A, Yoshida K, Dagdas YF, Etherington GJ, Schoonbeek HJ, van Esse HP, Jones JD, Kamoun S, Schornack S (2017) *Arabidopsis* late blight: infection of a non-host plant by *Albugo laibachii* enables full colonization by *Phytophthora infestans*. Cell Microbiol 19(1):e12628

Bentley A, Scutari M, Gosman N, Faure S, Bedford F, Howell P, Cockram J, Rose GA, Barber T, Irigoyen J, Horsnell R, Pumfrey C, Winnie E, Schacht J, Beauchêne K, Praud S, Greenland A, Balding D, Mackay IJ (2014) Applying association mapping and genomic selection to the dissection of key traits in elite European wheat. Theor Appl Genet 127(12):2619–2633. https://doi.org/10.1007/s00122.014.2403.y

Bhayana L, Paritosh K, Arora H, Yadava SK, Singh P, Nadan D, Mukhopadhyay A, Gupta V, Pradhan AK, Pental D (2020) A mapped locus on LG A6 of *Brassica juncea* line tumida conferring resistance to white rust contains a CNL type R gene. Front Plant Sci 10:1690. https://doi.org/10.3389/fpls.2019.01690

Bienkowska AC, Kaczmarek J, Schmidt KM, Nieróbca A, Czajka A, Jedryczka M (2020) Country-wide qPCR based assessment of *Plasmodiophora brassicae* spread in agricultural soils and recommendations for the cultivation of Brassicaceae crops in Poland. Pathogens 9:1070. https://doi.org/10.3390/pathogens9121070

Biga MLB (1955) Review of the species of the genus *Albugo* based on the morphology of the conidia. Sydowia 9:339–358

Biswas S, Akey JM (2006) Genomic insights into positive selection. Trends Genet 22:437–446. https://doi.org/10.1016/j.tig.2006.06.005

Blondeau K, Blaise F, Graille M, Kale SD, Linglin J, Ollivier B, Labarde A, Lazar N, Daverdin G, Balesdent MH, Choi DHY, Tyler BM, Rouxel T, van Tilbeurgh H, Fuda I (2015) Crystal structure of the effector *AvrLm4–7* of *Leptosphaeria maculans* reveals insights into its translocation into plant cells and recognition by resistance proteins. Plant J 83:610–624. https://doi.org/10.1111/tpj.12913

Borhan MH, Holub EB, Kindrachuk C, Omidi M, Bozorgmanesh Frad G, Rimmer SR (2010) *WRR4,* a broad spectrum *TIR NB LRR* gene from *Arabidopsis thaliana* that confers white rust resistance in transgenic oilseed brassica crops. Mol Plant Pathol 11(2):283–291. https://doi.org/10.1111/j.1364-3703.2009.00599.x

Bounou S, Jabaji-Hare SH, Hogue R, Charest PM (1999) Polymerase chain reaction-based assay for specific detection of *Rhizoctonia solani* AG-3 isolates. Mycol Res 103:1–8

Brasier CM (2001) Rapid evolution of introduced plant pathogens via interspecific hybridization. BioScience 51:123–133

Brasier CM, Kirk SA, Delcan J, Cooke DE, Jung T, Man in't Veld WA (2004) *Phytophthora alni* sp. nov. and its variants: designation of emerging heteroploid hybrid pathogens spreading on Alnus trees. Mycol Res 108:1172–1184

Brenner S, Johnson M, Bridgham J, Golda G, Lloyd DH, Johnson D, Luo S, McCurdy S, Foy M, Ewan M, Roth R, George D, Eletr S, Albrecht G, Vermaas E, Williams SR, Moon K, Burcham T, Pallas M, DuBridge RB, Kirchner J, Fearon K, Mao J, Corcoran K (2000) Gene expression analysis by massively parallel signature sequencing (MPSS) on microbead arrays. Nat Biotechnol 18(6):630–634. https://doi.org/10.1038/76469

Briese T, Kapoor A, Mishra N, Jain K, Kumar A, Jabado OJ, Lipkin WI (2015) Virome capture sequencing enables sensitive viral diagnosis and comprehensive virome analysis. mBio 6: e01491-01415

Brisbane PG, Neate SM, Pankhurst CE, Scott NS, Thomas MR (1995) Sequence-tagged site markers to identify *Rhizoctonia solani* AG4 or AG8 infecting wheat in South Australia. Phytopathology 85:1423–1427

Brodmann A, Schuller A, Ludwig Muller J, Aeschbacher RA, Wiemken A, Boller T, Wingler A (2002) Induction of trehalase in *Arabidopsis* plants infected with the trehalose producing pathogen *Plasmodiophora brassicae*. Mol Plant Microbe Interact 15(7):693–700

Brouwer M, Lievens B, Van Hemelrijck W, Van den Ackerveken G, Cammue BPA, Thomma BPHJ (2003) Quantification of disease progression of several microbial pathogens on *Arabidopsis thaliana* using real-time fluorescence PCR. FEMS Microbiol Lett 228:241–248

Brown JW (1996) *Arabidopsis* intron mutations and pre-mRNA splicing. Plant J 10:771–780

Brun H, Chèvre AM, Fitt BD, Powers S, Besnard AL, Ermel M, Huteau V, Marquer B, Eber F, Renard M, Andrivon D (2010) Quantitative resistance increases the durability of qualitative resistance to *Leptosphaeria maculans* in *Brassica napus*. New Phytol 185:285–299. https://doi.org/10.1111/j.1469-8137.2009.03049.x

Budge GE, Shaw MW, Colyer A, Pietravalle S, Boonham N (2009) Molecular tools to investigate *Rhizoctonia solani* distribution in soil. Plant Pathol 58(6):1071–1080. https://doi.org/10.1111/j.1365-3059.2009.02139.x

Bulman S, Siemens J, Ridgway HJ, Eady C, Conner AJ (2006) Identification of genes from the obligate intracellular plant pathogen, *Plasmodiophora brassicae*. FEMS Microbiol Lett 264:198–204

Burki F, Kudryavtsev A, Matz MV, Aglyamova GV, Bulman S, Fiers M, Keeling PJ, Pawlowski J (2010) Evolution of Rhizaria: new insights from phylogenomic analysis of uncultivated protists. BMC Evol Biol 10:377. https://doi.org/10.1186/1471-2148-10-377

Cairns J, Frickel J, Jalasvuori M, Hiltunen T, Becks L (2017) Genomic evolution of bacterial populations under co-selection by antibiotics and phage. Mol Ecol 26:1848–1859. https://doi.org/10.1111/mec.13950

Campion C, Chatot C, Perraton B, Andrivon D (2003) Anastomosis groups, pathogenicity and sensitivity to fungicides of *Rhizoctonia solani* isolates collected on potato crops in France. Eur J Plant Pathol 109:983–992

Cao T, Srivastava S, Rahman MH, Kav NNV, Hotte N, Deyholos MK, Strelkov SE (2008) Proteome level changes in the roots of *Brassica napus* as a result of *Plasmodiophora brassicae* infection. Plant Sci 174:97–115

Carling DE, Rothrock CS, MacNish GC, Sweetingham MW, Brainard KA, Winters SW (1994) Characterization of anastomosis group-11 (AG-11) of *Rhizoctonia solani*. Phytopathology 84:1387–1393

Carling DE, Pope EJ, Brainard KA, Carter DA (1999) Characterization of mycorrhizal isolates of *Rhizoctonia solani* from an orchid, including AG-12, a new anastomosis group. Phytopathology 89:942–946

Carling DE, Baird RE, Gitaitis RD, Brainard KA, Kuninaga S (2002) Characterization of AG-13, a newly reported anastomosis group of *Rhizoctonia solani*. Phytopathology 92:893–899

Chen SQ, Huang ZF, Dai Y, Qin SW, Gao YY, Zhang LL, Gao Y, Chen J (2013) The development of 7E chromosome-specific molecular markers for *Thinopyrum elongatum* based on SLAF-seq technology. PLoS One 8:e65122. https://doi.org/10.1371/journal.pone.0065122

Chen D, Neumann K, Friedel S, Kilian B, Chen M, Altmann T, Klukas C (2014) Dissecting the phenotypic components of crop plant growth and drought responses based on high-throughput image analysis. Plant Cell Online 26:4636–4655

Chiu CY (2013) Viral pathogen discovery. Curr Opin Microbiol 16:468–478. https://doi.org/10.1016/j.mib.2013.05.001

Choi HW, Choi Y-J, Kim DS, Hwang IS, Choi DS, Kim NH, Lee DH, Shin H-D, Nam J, Hwang BK (2009) First report of powdery mildew caused by *Erysiphe cruciferarum* on *Arabidopsis thaliana* in Korea. Plant Pathol J 25(1):86–90

Collins NC, Thordal-Christensen H, Lipka V, Bau S, Kombrink E, Qiu JL, Hückelhoven R, Stein M, Freialdenhoven A, Somerville SC, Schulze-Lefert P (2003) *SNARE*-protein-mediated disease resistance at the plant cell wall. Nature 425(1):973–977

Cooper AJ, Latunde-Dada AO, Woods-Tor A, Lynn J, Lucas JA, Crute IR, Holub EB (2008) Basic compatibility of *Albugo candida* in *Arabidopsis thaliana* and *Brassica juncea* causes broad-spectrum suppression of innate immunity. Mol Plant Microbe Interact 21:745–756. https://doi.org/10.1094/MPMI-21-6-0745

Coplin DL, Cook D (1990) Molecular genetics of extracellular polysaccharide biosynthesis in vascular phytopathogenic bacteria. Mol Plant Microbe Interact 3:271–279

Coppens F, Wuyts N, Inzé D, Dhondt S (2017) Unlocking the potential of plant phenotyping data through integration and data driven approaches. Curr Opin Syst Biol 4:58–63

Corwin JA, Kliebenstein DJ (2017) Quantitative resistance: more than just perception of a pathogen. Plant Cell 29:655–665

Corwin JA, Copeland D, Feusier J, Subedy A, Eshbaugh R, Palmer C, Maloof J, Kliebenstein DJ (2016) The quantitative basis of the *Arabidopsis* innate immune system to endemic pathogens depends on pathogen genetics. PLoS Genet 12:1–29

Crossa J, de Los Campos G, Pérez P, Gianola D, Burgueño J, Araus JL, Makumbi D, Singh RP, Dreisigacker S, Yan J, Arief V, Banziger M, Braun HJ (2010) Prediction of genetic values of quantitative traits in plant breeding using pedigree and molecular markers. Genetics 186(2): 713–724. https://doi.org/10.1534/genetics.110.118521

Crute IR, Norwood JM, Gordon DL (1985) Resistance to phenylamide fungicides in lettuce and *Brassica* downy mildew. In: Proc, The Mixture Centenary Meeting, British Crop Protection Council Monograph No. 31, Bordeaux, Croydon, Surrey, pp 311–314

Crute I, Beynon J, Dangl J, Holub E, Mauch-Mani B, Slusarenko A, Staskawicz B, Ausubel F (1994) Microbial pathogenesis of *Arabidopsis*. In: Meyerowitz EM, Somerville CR (eds) *Arabidopsis*. Cold Spring Harbor Laboratory, Cold Spring Harbor, NY, pp 705–747

Czechowski T, Stitt M, Altmann T, Udvardi MK, Scheible WR (2005) Genome-wide identification and testing of superior reference genes for transcript normalization in *Arabidopsis*. Plant Physiol 139:5–17. https://doi.org/10.1104/pp.105.063743

Czedik Eysenberg A, Seitner S, Güldener U, Koemeda S, Jez J, Colombini M, Djamei A (2018) The 'PhenoBox', a flexible, automated, open source plant phenotyping solution. New Phytol 219: 808–823

Da Silva AC, Ferro JA, Reinach FC, Farah CS, Furlan LR, Quaggio RB, Monteiro Vitorello CB, Van Sluys MA, Almeida NF, Alves LM, Do Amaral AM, Bertolini MC, Camargo LE, Camarotte G, Cannavan F, Cardozo J, Chambergo F, Ciapina LP, Cicarelli RM, Coutinho LL, Cursino Santos JR, El Dorry H, Faria JB, Ferreira AJ, Ferreira RC, Ferro MI, Formighieri EF, Franco MC, Greggio CC, Gruber A, Katsuyama AM, Kishi LT, Leite RP, Lemos EG, Lemos MV, Locali EC, Machado MA, Madeira AM, Martinez Rossi NM, Martins EC, Meidanis J, Menck CF, Miyaki CY, Moon DH, Moreira LM, Novo MT, Okura VK, Oliveira MC, Oliveira VR, Pereira HA, Rossi A, Sena JA, Silva C, De Souza RF, Spinola LA, Takita MA, Tamura RE, Teixeira EC, Tezza RI, Trindade dos Santos M, Truffi D, Tsai SM, White FF, Setubal JC, Kitajima JP (2002) Comparison of the genomes of two *Xanthomonas* pathogens with differing host specificities. Nature 417:459–463

Datta S, Budhauliya R, Das B, Chatterjee S, Vanlalhmuaka, Veer V (2015) Next-generation sequencing in clinical virology: discovery of new viruses. World J Virol 4:265–276. https://doi.org/10.5501/wjv.v4.i3.265

de Vienne DM, Giraud T, Shykoff JA (2007) When can host shifts produce congruent host and parasite phylogenies? A simulation approach. J Evol Biol 20:1428–1438

De Vienne DM, Refrégier G, López Villavicencio M, Tellier A, Hood ME, Giraud T (2013) Cospeciation vs host-shift speciation: methods for testing, evidence from natural associations and relation to coevolution. New Phytol 198:347–385

Debieu M, Huard Chauveau C, Genissel A, Roux F, Roby D, Huard Chauveau C, Genissel A, Roux F, Roby D (2015) Quantitative disease resistance to the bacterial pathogen *Xanthomonas campestris* involves an *Arabidopsis* immune receptor pair and a gene of unknown function. Mol Plant Pathol 17:510–520

Delourme R, Chèvre AM, Brun H, Rouxel T, Balesdent MH, Dias JS, Salisbury P, Renard M, Rimmer SR (2006) Major gene and polygenic resistance to *Leptosphaeria maculans* in oilseed rape (*Brassica napus*). Eur J Plant Pathol 114:41–52. https://doi.org/10.1007/s10658-005-2108-9

Delourme R, Bousset L, Ermel M, Duffe P, Besnard AL, Marquer B, Fudal I, Linglin J, Chadœuf J, Brun H (2014) Quantitative resistance affects the speed of frequency increase but not the diversity of the virulence alleles overcoming a major resistance gene to *Leptosphaeria maculans* in oilseed rape. Infect Genet Evol 27:490–499. https://doi.org/10.1016/j.meegid.2013.12.019

Dempsey DA, Pathirana MS, Wobbe KK, Klessig DF (1997) Identification of an *Arabidopsis* locus required for resistance to turnip crinkle virus. Plant J 11:301–311

Devos S, Laukens K, Deckers P, Van Der Straeten D, Beeckman T, Inzé D, Van Onckelen H, Witters E, Prinsen E (2006) A hormone and proteome approach to picturing the initial metabolic events during *Plasmodiophora brassicae* infection on *Arabidopsis*. Mol Plant Microbe Interact 19:1431–1443. https://doi.org/10.1094/MPMI-19-1431

Diatchenko L, Lau YFC, Campbell AP, Chenchik A, Moqadam F, Huang B, Lukyanov K, Gurskaya K, Sverdlov E, Siebert PD (1996) Suppression subtractive hybridization: a method for generating differentially regulated or tissue-specific cDNA probes and libraries. Proc Natl Acad Sci U S A 93:6025–6030

Dixit S, Jangid VK, Grover A (2019) Evaluation of suitable reference genes in *Brassica juncea* and its wild relative *Camelina sativa* for qRT-PCR analysis under various stress conditions. PLoS One 14(9):e0222530. https://doi.org/10.1371/journal.pone.0222530

Dixon MS, Jones DA, Keddie JS, Thomas CM, Harrison K, Jones JD (1996) The tomato *Cf-2* disease resistance locus comprises two functional genes encoding leucine-rich repeat proteins. Cell 84:451–459. https://doi.org/10.1016/S0092-8674(00)81290-8

Dow JM, Osbourn AE, Wilson TJG, Daniels MJ (1995) A locus determining pathogenicity of *Xanthomonas campestris* is involved in lipopolysaccharide biosynthesis. Mol Plant Microbe Interact 8:768–777

Dow JM, Feng JX, Barber CE, Tang JL, Daniels MJ (2000) Novel genes involved in the regulation of pathogenicity factor production within the rpf gene cluster of *Xanthomonas campestris*. Microbiology 146:885–891

Dow JM, Crossman L, Findlay K, He YQ, Feng JX, Tang JL (2003) Biofilm dispersal in *Xanthomonas campestris* is controlled by cell–cell signaling and is required for full virulence to plants. Proc Natl Acad Sci U S A 100:10995–11000

Duncan JM, Torrance L (1992) In: Duncan JM, Torrance L (eds) Techniques for the rapid detection of plant pathogens. Blackwell Scientific Pub, Oxford, pp 47–62

Duncan S, Barton JE, O'Brien PA (1993) Analysis of variation in isolates of *Rhizoctonia solani* by random amplified polymorphic DNA assay. Mycol Res 97:1075–1082

Erbe M, Hayes BJ, Matukumalli LK, Goswami S, Bowman PJ, Reich CM (2012) Improving accuracy of genomic predictions within and between dairy cattle breeds with imputed high-density single nucleotide polymorphism panels. J Dairy Sci 95(7):4114–4129. https://doi.org/10.3168/jds.20115019

Fabro G, Di Rienzo JA, Voigt CA, Sevchenko T, Dehesh K, Somerville S, Alvarez ME (2008) Genome-wide expression profiling *Arabidopsis* at the stage of *Golovinomyces cichoracearum* haustorium formation. Plant Physiol 1466(1):1421–1439

Fahlgren N, Gehan MA, Baxter I (2015) Lights, camera, action: high-throughput plant phenotyping is ready for a close-up. Curr Opin Plant Biol 24:93–99

Fang X, Willis RC, Hoang Q, Kelnar K, Xu W (2004) High-throughput sample preparation for gene expression profiling and *in vitro* target validation. JALA 9(3):140–145

Farris JJ (1989) The retention index and the rescaled consistency index. Cladistics 5:417–419

Felsenstein J (1985) Confidence limits on phylogenies: an approach using the bootstrap. Evolution 39:783–791

Feng J, Hwang R, Hwang SF, Strelkov SE, Gossen BD, Zhou Q, Peng G (2010) Molecular characterization of a serine protease Pro1 from *Plasmodiophora brassicae* that stimulates resting spore germination. Mol Plant Pathol 11:503–512

Feng J, Hwang SF, Strelkov SE (2012a) Studies into primary and secondary infection processes by *Plasmodiophora brassicae* on canola. Plant Pathol 62:177–183. https://doi.org/10.1111/j.1365-3059.2012.02612.x

Feng J, Xiao Q, Hwang SF, Strelkov SE, Gossen BD (2012b) Infection of canola by secondary zoospores of *Plasmodiophora brassicae* produced on a nonhost. Eur J Plant Pathol 132:309–315

Feng J, Hwang SF, Strelkov SE (2013) Assessment of gene expression profiles in primary and secondary zoospores of *Plasmodiophora brassicae* by dot blot and real-time PCR. Microbiol Res 168(8):518–524

Feng J, Jiang J, Feindel D, Strelkov SE, Hwang S-F (2016) The gene Cr811 is present exclusively in pathotype 5 and new emerged pathotypes of the clubroot pathogen *Plasmodiophora brassicae*. Eur J Plant Pathol 145:615–620. https://doi.org/10.1007/s10658-016-0903-0

Fiddaman PJ, Rossall S (1995) Selection of bacterial antagonists for the biological-control of *Rhizoctonia solani* in oilseed rape (*Brassica napus*). Plant Pathol 44:695–703

Fikere M, Barbulescu DM, Malmberg MM, Shi F, Koh JCO, Slater AT, MacLeod IM, Bowman PJ, Salisbury PA (2018) Genomic prediction using prior quantitative trait loci information reveals a large reservoir of underutilized blackleg resistance in diverse canola (*Brassica napus* L.) lines. Plant Genome 11:1–16. https://doi.org/10.3835/plantgenome2017.11.0100

Fomeju BF, Falentin C, Lassalle G, Manzanares-Dauleux MJ, Delourme R (2014) Homoeologous duplicated regions are involved in quantitative resistance of *Brassica napus* to stem canker. BMC Genomics 15:498. https://doi.org/10.1186/1471-2164-15-498

Fotopoulos V, Gilbert MJ, Pittman JK, Marvier AC, Buchanan AJ, Sauer N, Hall JL, Williams LE (2003) The monosaccharide transporter gene, *AtSTP4*, and the cell-wall invertase, *Atβfruct1*, are induced in *Arabidopsis* during infection with the fungal biotroph *Erysiphe cichoracearum*. Plant Physiol 132(2):821–829

Fu D, Uauy C, Distelfeld A, Blechi A, Epstein L, Che X, Sela H, Fahima T, Dubcovsky J (2009) A kinase-*START* gene confers temperature-dependent resistance to wheat stripe rust. Science 323: 1357–1360

Fudal I, Ross S, Gout L, Blaise F, Kuhn ML, Eckert MR, Cattolico L, Bernard Samain S, Balesdent MH, Rouxel T (2007) Heterochromatin-like regions as ecological niches for avirulence genes in the *Leptosphaeria maculans* genome: map-based cloning of *AvrLm6*. Mol Plant Microbe Interact 20:459–470

Fumagalli M, Sironi M, Pozzoli U, Ferrer-Admettla A, Pattini L, Nielsen R (2011) Signatures of environmental genetic adaptation pin point pathogens as the main selective pressure through human evolution. PLoS Genet 7(11):e1002355. https://doi.org/10.1371/journal.pgen.1002355

Gall C, Balesdent MH, Desthieux I, Robin P, Rouxel T (1995) Polymorphism of Tox0 *Leptosphaeria maculans* isolates as revealed by soluble protein and isozyme electrophoresis. Mycol Res 99:221–229

Ghanbarnia K, Fudal I, Larkan NJ, Links MG, Balesdent MH, Profotova B, Fernando DWG, Rouxel T, Borhan H (2015) Rapid identification of the *Leptosphaeria maculans* avirulence gene *AvrLm2* using an intraspecific comparative genomics approach. Mol Plant Pathol 16:699–709

Gibriel HA, Thomma BP, Seidl MF (2016) The age of effectors: genome-based discovery and applications. Phytopathology 106:1206–1212. https://doi.org/10.1094/PHYTO-02-16-0110-FI

Gjetting T, Carver TL, Skøt L, Lyngkjaer MF (2004) Differential gene expression in individual papilla-resistant and powdery mildew-infected barley epidermal cells. Mol Plant Microbe Interact 17(7):729–738

Gjetting T, Hagedorn PH, Schweizer P, Thordal-Christensen, Carver TL, Lyngkjaer MF (2007) Single-cell transcript profiling of barley attacked by the powdery mildew fungus. Mol Plant Microbe Interact 20:235–246

Gladieux P, Vercken E, Fontaine MC, Hood ME, Jonot O, Couloux A, Giraud T (2011) Maintenance of fungal pathogen species that are specialized to different hosts: allopatric divergence and introgression through secondary contact. Mol Biol Evol 28:459–471

Goodwin PH, Annis SL (1991) Rapid identification of genetic variation and pathotype of *Leptosphaeria maculans* by random amplified polymorphic DNA assay. Appl Environ Microbiol 57:2482–2486

Goossens K, Poucke MV, Soom AV, Vandesompele J, Zeveren AV, Peelman LJ (2005) Selection of reference genes for quantitative real-time PCR in bovine pre-implantation embryos. BMC Dev Biol 5:27. https://doi.org/10.1186/1471-213X-5-27

Gout L, Fudal I, Kuhn ML, Blaise F, Eckert M, Cattolico L, Balesdent MH, Rouxel T (2006) Lost in the middle of nowhere: The *AvrLm1* avirulence gene of the Dothideomycete *Leptosphaeria maculans*. Mol Microbiol 60:67–80

Goyal P, Chattopadhyay C, Mathur AP, Kumar A, Meena PD, Datta S (2013) Pathogenic and molecular variability among *Brassica* isolates of *Alternaria brassicae* from India. Ann Plant Prot Sci 21(2):349–359

Graf H, Fahling M, Siemens J (2004) Chromosome polymorphism of the obligate biotrophic parasite *Plasmodiophora brassicae*. J Phytopathol 152(2):86–91. https://doi.org/10.1046/j.1439-0434.2003.00805.x

Grant MR, Godiard L, Straube E, Ashfield T, Lewald J, Sattler A, Innes RW, Dangl JL (1995) Structure of the *Arabidopsis RPM1* gene enabling dual specificity disease resistance. Science 269:843–846

Gregory AC, Zayed AA, Conceição Neto N, Temperton B, Bolduc B, Alberti A, Ardyna M, Arkhipova K, Carmichael M, Cruaud C, Dimier C, Domínguez-Huerta G, Ferland J, Kandels S, Liu Y, Marec C, Pesant S, Picheral M, Pisarev S, Poulain J, Tremblay J-É, Vik D, Babin M, Bowler C, Culley AI, De Vargas C, Dutilh BE, Iudicone D, Karp-Boss L, Roux S, Wincker P, Sullivan MB, Acinas SG, Babin M, Bork P, Boss E, Bowler C, Cochrane G, De Vargas C, Follows M, Gorsky G, Grimsley N, Guidi L, Hingamp P, Iudicone D, Jaillon O, Kandels-Lewis S, Karp-Boss L, Karsenti E, Not F, Ogata H, Pesant S, Poulton N, Raes J, Sardet C, Speich S, Stemmann L, Sunagawa S (2019) Marine DNA viral macro- and micro diversity from pole to pole. Cell 177(5):1109–1123.e14. https://doi.org/10.1016/j.cell.2019.03.040

Grenier CCT, Ospina Y, Quintero C, Châtel MH, Tohme J, Courtois B, Ahmadi N (2015) Accuracy of genomic selection in a rice synthetic population developed for recurrent selection Breeding. PLoS One 10(8):e0136594. https://doi.org/10.1371/journal.pone.0136594

Grosch R, Schneider JH, Peth A et al (2007) Development of a specific PCR assay for the detection of *Rhizoctonia solani* AG 1-IB using SCAR primers. J Appl Microbiol 102:806–819

Gudelj I, Fitt BDL, van den Bosch F (2004) Evolution of sibling fungal pathogens in relation to host specialization. Phytopathology 94:789–795

Gudesblat GE, Torres PS, Vojnov AA (2009) *Xanthomonas campestris* overcomes *Arabidopsis* stomatal innate immunity through a DSF cell-to-cell signal-regulated virulence factor. Plant Physiol 149:1017–1027

Guenin S, Mauriat M, Pelloux J, Van Wuytswinkel O, Bellini C, Gutierrez L (2009) Normalization of qRT-PCR data: the necessity of adopting a systematic, experimental conditions-specific, validation of references. J Exp Bot 60(2):487–493. https://doi.org/10.1093/jxb/ern305

Gunasinghe N, You M, Clode P, Barbetti MJ (2016) Mechanisms of resistance in Brassica carinata, *B. napus* and *B. juncea* to *Pseudocercosporella capsellae*. Plant Pathol 65:888–900. https://doi.org/10.1111/ppa.12484

Guo WL, Chen R-G, Gong ZH, Yin YX, Li DW (2013) Suppression subtractive hybridization analysis of genes regulated by application of exogenous abscisic acid in pepper plant (*Capsicum annuum* L.) leaves under chilling stress. PLoS One 8:e66667. https://doi.org/10.1371/journal.pone.0066667

Guyon K, Balague C, Roby D, Raffaele S (2014) Secretome analysis reveals effector candidates associated with broad host range necrotrophy in the fungal plant pathogen *Sclerotinia sclerotiorum*. BMC Genomics 15:336. https://doi.org/10.1186/1471-2164-15-336

Habier D, Fernando RL, Dekkers JCM (2007) The impact of genetic relationship information on genome assisted breeding values. Genetics 177:2389–2397. https://doi.org/10.1534/genetics.107.081190

Haddadi P, Larkan NJ, Borhan MH (2019) Dissecting R gene and host genetic background effect on the *Brassica napus* defense response to *Leptosphaeria maculans*. Sci Rep 9:6947. https://doi.org/10.1038/s41598-019-43419-9

Hammond J, Jordan RL (1990) Dot blots (viruses) and colony screening. In: Hampton R, Ball E, De Boer S (eds) Serological methods for detection and identification of viral and bacterial plant pathogens. A laboratory manual. APS Press, St. Paul, Minnesota, pp 237–241

Hao LF, Song PL, Li ZQ, Huangpu HY, Li QS (2014) Genetic diversity of Phoma stem canker pathogen *Leptosphaeria biglobosa* by ISSR. Chin J Oil Crop Sci 36(36):98–105

He YQ, Zhang L, Jiang BL, Zhang ZC, Xu RQ, Tang DJ, Qin J, Jiang W, Zhang X, Liao J, Cao JR, Zhang SS, Wei ML, Liang XX, Lu GT, Feng JX, Chen B, Cheng J, Tang JL (2007) Comparative and functional genomics reveals genetic diversity and determinants of host specificity among reference strains and a large collection of Chinese isolates of the phytopathogen *Xanthomonas campestris* pv. *campestris*. Genome Biol 8:R218

He YW, Wu J, Zhou L, Yang F, He YO, Jiang BL, Bai LQ, Xu YQ, Deng ZX, Tang JL, Zhang LH (2011) *Xanthomonas campestris* diffusible factor is 3-hydroxybenzoic acid and is associated with xanthomonadin biosynthesis, cell viability, antioxidant activity, and systemic invasion. Mol Plant Microbe Interact 24:948–957

He YY, Chen Q, Wu CW, Yang M, Zhou E (2016) *Colletotrichum truncatum*, a new cause of anthracnose on Chinese flowering cabbage (*Brassica parachinensis*) in China. Trop Plant Pathol 41:183–192

Heffner EL, Jannink JL, Sorrells ME (2011) Genomic selection accuracy using multifamily prediction models in a wheat breeding program. Plant Genome 4:65–75

Hennin C, Hofte M, Diederichsen E (2001) Functional expression of Cf9 and Avr9 genes in *Brassica napus* induces enhanced resistance to *Leptosphaeria maculans*. Mol Plant Microbe Interact 14(9):1075–1085

Hinsch M, Staskawicz B (1996) Identification of a new Arabidopsis disease resistance locus, *RPS4*, and cloning of the corresponding avirulence gene, *avrRps4*, from *Pseudomonas syringae* pv. *pisi*. Mol Plant Microbe Interact 9:55–61

Hirata T, Takamatsu S (1996) Nucleotide sequence diversity of rDNA internal transcribed spacers extracted from conidia and cleistothecia of several powdery mildew fungi. Mycoscience 37: 283–288

Hirschhorn JN, Lohmueller K, Byrne E, Hirschhorn K (2002) A comprehensive review of genetic association studies. Genet Med 4:45–61. https://doi.org/10.1097/00125817-200203000-00002

Holub EB (1997) Organization of resistance genes in *Arabidopsis*. In: Crute IR, Holub EB, Burdon JJ (eds) The gene-for-gene relationship in plant-parasite interaction. CAB Int, Wallingford, pp 5–26

Holub E, Beynon JL, Crute IR (1994) Phenotypic and genotypic characterization of interaction between isolates of *Peronospora parasitica* and accessions of *Arabidopsis thaliana*. Mol Plant Microbe Interact 7:223–239

Howlett BJ, Lowe RGT, Marcroft SJ, van de Wouw AP (2015) Evolution of virulence in fungal plant pathogens: exploiting fungal genomics to control plant disease. Mycologia 107(3): 441–451. https://doi.org/10.3852/14-317

Hsiao YM, Liu YF, Fang MC, Song WL (2011) *XCC2731*, a GGDEF domain protein in *Xanthomonas campestris*, is involved in bacterial attachment and is positively regulated by Clp. Microbiol Res 166:548–565

Hsu HT, Lawson RH (1991) Direct tissue blotting for detection of tomato spotted wilt virus in Impatiens. Plant Dis 75:292–295

Hu J, Quiros CE (1991) Identification of broccoli and cauliflowers cultivars with RAPD markers. Plant Cell Rep 10:505–511

Huang DW, Sherman BT, Tan Q, Collins JR, Alvord WG, Roayaei J, Stephens R, Baseler MW, Lane HC, Lempicki RA (2007) The DAVID gene functional classification tool: a novel biological module-centric algorithm to functionally analyze large gene lists. Genome Biol 8: R183. https://doi.org/10.1186/gb-2007-8-9-r183

Huard Chauveau C, Perchepied L, Debieu M, Rivas S, Kroj T, Kars I, Bergelson J, Roux F, Roby D (2013) An atypical kinase under balancing selection confers broad spectrum disease resistance in *Arabidopsis*. PLoS Genet 9:e1003766

Hubbard A, Lewis CM, Yoshida K, Ramirez-Gonzalez RH, de Vallavieille-Pope C, Thomas J, Kamoun S, Bayles R, Uauy C, Saunders DGO (2015) Field pathogenomics reveals the emergence of a diverse wheat yellow rust population. Genome Biol 16:23

Huis R, Hawkins S, Neutelings G (2010) Selection of reference genes for quantitative gene expression normalization in flax (*Linum usitatissimum* L.). BMC Plant Biol 10:71. https://doi.org/10.1186/1471-2229-10-71

Huo D, Ning Q, Shen X, Liu L, Zhang Z (2016) QTL mapping of kernel number related traits and validation of one major QTL for ear length in maize. PLoS One 11(5):e0155506. https://doi.org/10.1371/journal.pone.0155506

Iakovidis M, Teixeira PJ, Exposito-Alonso M, Cowper MG, Law TF, Liu Q, Vu MC, Dang TM, Corwin JA, Weigel D, Dangl JL, Grant SR (2016) Effector-triggered immune response in *Arabidopsis thaliana* is a quantitative trait. Genetics 204(1):337–353. https://doi.org/10.1534/genetics.116.190678

Ielpi L, Couso RO, Dankert MA (1993) Sequential assembly and polymerization of the polyprenol-linked pentasaccharide repeating unit of the xanthan polysaccharide in *Xanthomonas campestris*. J Bacteriol 175:2490–2500

Inderbitzin P, Davis RM, Bostock RM, Subbarao KV (2011) The ascomycete *Verticillium longisporum* is a hybrid and a plant pathogen with an expanded host range. PLoS One 6:e18260

Inderbitzin P, Davis RM, Bostock RM, Subbarao KV (2013) Identification and differentiation of *Verticillium* species and *V. longisporum* lineages by simplex and multiplex PCR assays. PLoS One 8(6):e65990. https://doi.org/10.1371/journal.pone.0065990

Inman A, Sivanesan A, Fitt B, Evans R (1991) The biology of *Mycosphaerella capsellae* sp. nov., the teleomorph of *Pseudocercosporella capsellae*, cause of white leaf spot of oilseed rape. Mycol Res 95:1334–1342. https://doi.org/10.1016/S0953-7562(09)80586-8

Ito S, Ichinose H, Yanagi C, Tanaka S, Kameya Iwaki M, Kishi F (1999) Identification of an in planta induced mRNA of *Plasmodiophora brassicae*. J Phytopathol 147:79–82

Jan H, Abbadi A, Lucke S, Nichols R, Snowdon R (2016) Genomic prediction of testcross performance in canola (*Brassica napus*). PLoS One 11:e0147769. https://doi.org/10.1371/journal.pone.0147769

Jansson PE, Kenne L, Lindberg B (1975) Structure of extracellular polysaccharide from *Xanthomonas campestris*. Carbohydr Res 45:275–282

Jasalavich CA, Morales VM, Pelcher LE, Seguin-Swartz G (1995) Comparison of nuclear ribosomal DNA sequences from *Alternaria* species pathogenic to crucifers. Mycol Res 99:604–614

Jenner CE, Keane GJ, Jones JE, Walsh JA (1999) Serotypic variation in *Turnip mosaic virus*. Plant Pathol 48:101–108

Jestin C, Lodé M, Vallée P, Domin C, Falentin C, Horvais R, Coedel S, Manzanares-Dauleux MJ, Delourme R (2011) Association mapping of quantitative resistance for *Leptosphaeria maculans* in oilseed rape (*Brassica napus* L.). Mol Breed 27:271–287. https://doi.org/10.1007/s11032-010-9429-x

Jia Y, Jannink JL (2012) Multiple trait genomic selection methods increase genetic value prediction accuracy. Genetics 192(4):1513–1522. https://doi.org/10.1534/genetics.112.144246

Jiang H, Song W, Li A, Yang X, Sun D (2011) Identification of genes differentially expressed in cauliflower associated with resistance to *Xanthomonas campestris* pv. *campestris*. Mol Biol Rep 38:621–629

Jin H, Sun Y, Yang Q, Chao Y, Kang J, Jin H, Li Y, Margaret G (2010) Screening of genes induced by salt stress from Alfalfa. Mol Biol Rep 37:745–753

Jones DA, Thomas CM, Hammond-Kosack KE, Balint-Kurti PJ, Jones JD (1994) Isolation of the tomato Cf-9 gene for resistance to *Cladosporium fulvum* by transposon tagging. Science 266: 789–793. https://doi.org/10.1126/science.7973631

Jordan R, Hammond J (1991) Comparison and differentiation of potyvirus isolates and identification of strain-, virus-, subgroup-specific and potyvirus group-common epitopes using monoclonal antibodies. J Gen Virol 72(1):25–36. https://doi.org/10.1099/0022-1317-72-1-25

Jouet A, Saunders DGO, McMullan M, Ward B, Furzer O, Jupe F, Cevik V, Hein I, Thilliez GJA, Holub E, van Oosterhout C, Jones JDG (2018) *Albugo candida* race diversity, ploidy and host-associated microbes revealed using DNA sequence capture on diseased plants in the field. New Phytol 221:1529–1543

Julian MC, Acero J, Salazar O, Keijer J, Rubio V (1999) Mating type-correlated molecular markers and demonstration of heterokaryosis in the phytopathogenic fungus *Thanatephorus cucumeris* (*Rhizoctonia solani*) AG 1-IC by AFLP DNA fingerprinting analysis. J Biotechnol 67:49–56

Kamada N, Chen GY, Inohara N, Nunez G (2013) Control of pathogens and pathobionts by the gut microbiota. Nat Immun 14:685–690. https://doi.org/10.1038/ni.2608

Karapapa VK, Typas MA (2001) Molecular characterization of the host-adapted pathogen *Verticillium longisporum* on the basis of a group-I intron found in the nuclear SSU-rRNA Gene. Curr Microbiol 42:217–224. https://doi.org/10.1007/s002840010207

Karapapa VK, Bainbridge BW, Heale JB (1997) Morphological and molecular characterization of *Verticillium longisporum* comb. nov., pathogenic to oilseed rape. Mycol Res 101:1281–1294

Karisto P, Hund A, Yu K, Anderegg J, Walter A, Mascher F, McDonald BA, Mikaberidze A (2017) Ranking quantitative resistance to *Septoria tritici* blotch in elite wheat cultivars using automated image analysis. Phytopathology 108:568–581

Kataria HR, Verma PR, Gisi U (1991) Variability in the sensitivity of *Rhizoctonia solani* anastomosis groups to fungicides. J Phytopathol 133:121–133

Kaur R, Shivani Saxena B, Kanwar HS, Dohroo NP, Majeed S, Sharma DR (2009) Detecting RAPD markers associated with black rot resistance in cabbage (*Brassica oleracea* var. *capitata*). Fruit Veg Cereal Sci Biotech 13:12–15

Kehr J (2003) Single cell technology. Curr Opin Plant Biol 66(1):617–621

Khurana AK, Naresh M, Sangwan MS (2005) Variation in bio-chemical contents of *Alternaria brassicae* (Berk.) Sacc. isolates. J Mycol Plant Pathol 35:343–345

Kim BR, Nam HY, Kim SU, Kim SI, Chang YJ (2003) Normalization of revere transcription quantitative-PCR with housekeeping genes in rice. Biotechnol Lett 25:1869–1872. https://doi.org/10.1023/A:1026298032009

Kim S, Song YH, Lee JY, Choi SR, Dhandapani V, Jang CS, Lim YP, Han T (2011) Identification of the BrRHP1 locus that confers resistance to downy mildew in Chinese cabbage (*Brassica rapa* ssp. *pekinensis*) and development of linked molecular markers. Theor Appl Genet 123: 1183–1192. https://doi.org/10.1007/s00122-011-1658-9

Klein-Lankhorst RM, Rietveld P, Machiels B, Verkerk R, Weide R, Gebhardt C, Koornneef M, Zabel P (1991) RFLP markers linked to the root knot nematode resistance gene Mi in tomato. Theor Appl Genet 81:661–667

Koller B, Lehmann A, McDermott JM (1993) Identification of apple cultivars using RAPD markers. Theor Appl Genet 85:901–904

Krattinger SG, Lagudah ES, Spielmeyer W, Singh RP, Huerta Espino J, McFadden H, Bossolini E, Selter LL, Keller B (2009) A putative ABC transporter confers durable resistance to multiple fungal pathogens in wheat. Science 323:1360–1363

Kresovich S, Williams JGK, McFerson JR, Routman EJ, Schaal BA (1992) Characterization of genetic identities and relationships of *Brassica oleracea* L. via a random amplified polymorphic DNA assay. Theor Appl Genet 85:190–196

Kumar V, Haldar S, Pandey KK, Singh RP, Singh AK, Singh PC (2008) Cultural, morphological, pathogenic and molecular variability amongst tomato isolates of *Alternaria solani* in India. World J Microbiol Biotechnol 24:1003–1009

Kundu A, Patel A, Pal A (2013) Defining reference genes for qPCR normalization to study biotic and abiotic stress responses in *Vigna mungo*. Plant Cell Rep 32:1647–1658. https://doi.org/10.1007/s00299-013-1478-2

Kunkel BN (1996) A useful weed put to work: genetic analysis of disease resistance in *Arabidopsis thaliana*. Trends Genet 12:63–69

Kuninaga S, Natsuaki T, Takeuchi T, Yokosawa R (1997) Sequence variation of the rDNA ITS regions within and between anastomosis groups in *Rhizoctonia solani*. Curr Genet 32:237–243

Larkan NJ, Lydiate DJ, Yu F, Rimmer SR, Borhan MH (2014) Co-localization of the blackleg resistance genes *Rlm2* and *LepR3* on *Brassica napus* chromosome A10. BMC Plant Biol 14: 1595. https://doi.org/10.1186/s12870-014-0387-z

Larkan NJ, Parkin IAP, Bohran MH (2019) Genetic mapping and characterization of the novel blackleg resistance genes *LepR5* and *LepR6*. In: 15th International Rapeseed Congress, Berlin June 16–19th, 2019

Lawrence GJ, Finnegan EJ, Ayliffe MA, Ellis JG (1995) The L6 gene for flax rust resistance is related to the Arabidopsis bacterial resistance gene *RPS2* and the tobacco viral resistance gene N. Plant Cell 7:1195–1206

Lee JM, Hartman GL, Domier LL, Bent AF (1996) Identification and map location of TTR1, a single locus in *Arabidopsis thaliana* that confers tolerance to tobacco ring spot nepovirus. Mol Plant-Microbe Interact 9:729–735

Lee Y, Kim YJ, Kim MH, Kwak JM (2016) *MAPK* cascades in guard cell signal transduction. Front Plant Sci 7:80

Lees AK, Cullen DW, Sullivan L, Nicolson MJ (2002) Development of conventional and quantitative real-time PCR assays for the detection and identification of *Rhizoctonia solani* AG-3 in potato and soil. Plant Pathol 51:293–302

Leroy T, Caffier V, Celton JM, Anger N, Durel CE, Lemaire C, Le Cam B (2016) When virulence originates from non-agricultural hosts: evolutionary and epidemiological consequences of introgressions following secondary contacts in *Venturia inaequalis*. New Phytol 210:1443–1452

Li J, Ding Q, Wang F, Li H, Zhang Y, Liu L, Jiao Z, Gao JW (2018) Genome-wide gene expression profiles in response to downy mildew in Chinese cabbage (*Brassica rapa* L. ssp. *pekinensis*). Eur J Plant Pathol 151:861–873. https://doi.org/10.1007/s10658-018-1427-6

Light KA, Gororo NN, Salisbury PA (2011) Usefulness of winter canola (*Brassica napus*) race-specific resistance genes against blackleg (causal agent *Leptosphaeria maculans*) in southern Australian growing conditions. Crop Pasture Sci 62:162–168. https://doi.org/10.1071/CP10187

Lin NS, Hsu YH, Hsu HT (1990) Immunological detection of plant viruses and a mycoplasma like organism by direct tissue blotting on nitrocellulose membranes. Phytopathology 80:824–828

Liu ZL, Domier LL, Sinclair JB (1995) Polymorphism of genes-coding for nuclear 18s ribosomal-RNA indicates genetic distinctiveness of anastomosis group-10 from other groups in the *Rhizoctonia solani* species complex. Appl Environ Microbiol 61:2659–2664

Liu Z, Latunde-Dada AO, Hall AM, Fitt BDL (2014) Phoma stem canker disease on oilseed rape (*Brassica napus*) in China is caused by *Leptosphaeria biglobosa* 'brassicae". Eur J Plant Pathol 140(4):841–857

Long Y, Wang Z, Sun Z, Fernando DWG, Mcvetty PBE, Li G (2011) Identification of two blackleg resistance genes and fine mapping of one of these two genes in a *Brassica napus* canola cultivar 'Surpass 400'. Theor Appl Genet 122:1223–1231. https://doi.org/10.1007/s00122-010-1526-z

Lubeck M, Poulsen H (2001) UP-PCR cross blot hybridization as a tool for identification of anastomosis groups in the *Rhizoctonia solani* complex. FEMS Microbiol Lett 201:83–89

Lucking R, Aime MC, Robbertse B, Miller AN, Ariyawansa HA, Aoki T, Cardinali G, Crous PW, Druzhinina IS, Geiser DM, Hawksworth DL, Hyde KD, Irinyi L, Jeewon R, Johnston PR, Kirk PM, Malosso E, May TW, Meyer W, Opik M, Robert V, Stadler M, Thines M, Vu D, Yurkov AM, Zhang N, Schoch CL (2020) Unambiguous identification of fungi: where do we stand and how accurate and precise is fungal DNA barcoding? IMA Fungus 11:14. https://doi.org/10.1186/s43008-020-00033-z

Luo Y, Lv GL, Wu WT, Chen SN, Cheng ZQ (2010) Analysis of genome expression in the response of *Oryza granulata* to *Xanthomonas oryzae* pv. *oryzae*. Mol Biol Rep 37(2):875–892. https://doi.org/10.1007/s11033-009-9694-0

MacLeod IM, Bowman PJ, Vander Jagt CJ, Haile-Mariam M, Kemper KE, Chamberlain AJ, Schrooten C, Hayes BJ, Goddard ME (2016) Exploiting biological priors and sequence variants enhances QTL discovery and genomic prediction of complex traits. BMC Genomics 17:144. https://doi.org/10.1186/s12864-016-2443-6

Makkouk KM, Comeau A (1994) Evaluation of various methods for the detection of barley yellow dwarf luteovirus by the tissue-blot immunoassay and its use for BYDV detection in cereals inoculated at different growth stages. Eur J Plant Pathol 100(1):71–80. https://doi.org/10.1007/BF01871967

Marcroft SJ, Elliott VL, Cozijnsen AJ, Salisbury PA, Howlett BJ, Van De Wouw AP (2012a) Identifying resistance genes to *Leptosphaeria maculans* in Australian *Brassica napus* cultivars based on reactions to isolates with known avirulence genotypes. Crop Pasture Sci 63:338–350. https://doi.org/10.1071/CP11341

Marcroft SJ, Van De Wouw AP, Salisbury PA, Potter TD, Howlett BJ (2012b) Effect of rotation of canola (*Brassica napus*) cultivars with different complements of blackleg resistance genes on disease severity. Plant Pathol 61:934–944. https://doi.org/10.1111/j.1365-3059.2011.02580.x

Martin RR, James D, Lévesque CA (2000) Impacts of molecular diagnostic technologies on plant disease management. Annu Rev Phytopathol 38:207–239

McCartney HA, Foster SJ, Fraaije BA, Ward E (2003) Molecular diagnostics for fungal plant pathogens. Pest Manage Sci 59(2):129–142. https://doi.org/10.1002/ps.575

McDonald BA, Linde C (2002) The population genetics of plant pathogens and breeding strategies for durable resistance. Euphytica 124:163–180. https://doi.org/10.1023/A:1015678432355

McMullan M, Gardiner A, Bailey K, Kemen E, Ward BJ, Cevik V, Robert-Seilaniantz A, Schultz-Larsen T, Balmuth A, Holub E, van Oosterhout C, Jones JDG (2015) Evidence for suppression of immunity as a driver for genomic introgressions and host range expansion in races of *Albugo candida*, a generalist parasite. eLife 4:e04550. https://doi.org/10.7554/eLife.04550

Menardo F, Praz C, Wyder S, Bourras SA, McNally KE, Parlange F, Riba A, Roffler S, Schaefer L, Shimizu KK, Valenti L, Zbinden H, Wicker T, Keller B (2015) Hybridization of powdery mildew strains gives raise to pathogens on novel agricultural crop species. Nat Genet 48:1–24

Mendes-Pereira E, Balesdent MH, Brun H, Rouxel T (2003) Molecular phylogeny of the *Leptosphaeria maculans* L. *biglobosa* species complex. Mycol Res 107:1287–1304. https://doi.org/10.1017/S0953756203008554

Menzies JD (1970) The first century of *Rhizoctonia solani*. In: Parmeter JR (ed) *Rhizoctonia solani*: biology and pathology. University of California Press, Berkeley, pp 3–5

Meuwissen TE, Hayes B, Goddard M (2001) Prediction of total genetic value using genome wide dense marker maps. Genetics 157(4):1819–1829

Micali C, Gollner K, Humphry M, Consonni C, Panstruga R (2008) The powdery mildew disease of *Arabidopsis*: a paradigm for the interaction between plants and biotrophic fungi. Arabidopsis Book 6:e0115. https://doi.org/10.1199/tab.0115

Michelmore R, Hulbert SH (1987) Molecular markers for genetic analysis of phytopathogenic fungi. Annu Rev Phytopathol 25(1):383–404. https://doi.org/10.1146/annurev.py.25.090187.002123

Milosevic D, Marjanovic-Jeromela A, Ignjatov M, Jovicic D, Stankovic I, Bulajic A, Krstic B (2015) First report of *Turnip yellows virus* on oilseed rape in Serbia. Plant Dis 99(12):1869. https://doi.org/10.1094/PDIS-05-15-0503-PDN

Milosevic D, Ignjatov M, Nikolic Z, Stankovic I, Bulajic A, Marjanovic-Jeromela A, Krstic B (2016) The presence of *Turnip yellows virus* in oilseed rape (*Brassica napus* L.) in Serbia. Pestic Phytomed (Belgrade) 31(1–2):37–44. https://doi.org/10.2298/PIF1602037M

Mishra S, Singh R, Kumar S, Kumar S (2017) Morphological, biochemical and molecular characterization of *Alternaria brassicicola* and *Alternaria brassicae*: a comparative overview. Am J Biochem Mol Biol 7(2):53–64. https://doi.org/10.3923/ajbmb.2017.53.64

Miya A, Albert P, Shinya T, Desaki Y, Ichimura K, Shirasu K, Narusaka Y, Kawakami N, Kaku H, Shibuya N (2007) CERK1, a LysM receptor kinase, is essential for chitin elicitor signaling in *Arabidopsis*. Proc Natl Acad Sci U S A 1046(1):19613–19618

Moscou MJ, Lauter N, Steffenson B, Wise RP (2011) Quantitative and qualitative stem rust resistance factors in barley are associated with transcriptional suppression of defense regulons. PLoS Genet 7:e1002208

Mutka AM, Fentress SJ, Sher JW, Berry JC, Pretz C, Nusinow DA, Bart R (2016) Quantitative, image-based phenotyping methods provide insight into spatial and temporal dimensions of plant disease. Plant Physiol 172:00984

Napflin K, O'Connor EA, Becks L, Bensch S, Ellis VA, Hafer-Hahmann N, Harding KC, Lindén SK, Olsen MT, Roved J, Sackton TB, Shultz AJ, Venkatakrishnan V, Videvall E, Westerdahl H, Winternitz JC, Edwards SV (2019) Genomics of host-pathogen interactions: challenges and opportunities across ecological and spatio-temporal scales. Peer J 7:e8013. https://doi.org/10.7717/peerj.8013

Neik TX, Barbetti MJ, Batley J (2017) Current status and challenges in identifying disease resistance genes in *Brassica napus*. Front Plant Sci 8:1788. https://doi.org/10.3389/fpls.2017.01788

Nejati Javaremi A, Smith CW, Gibson JP (1997) Effect of total allelic relationship on accuracy of evaluation and response to selection. J Ani Sci 75(7):1738–1745. https://doi.org/10.2527/1997.7571738x

Nelson JM, Hauser DA, Hinson R, Shaw AJ (2018) A novel experimental system using the liverwort *Marchantia polymorpha* and its fungal endophytes reveals diverse and context-dependent effects. New Phytol 218:1217–1232

Nicot N, Hausman JF, Hoffmann L, Evers D (2005) Housekeeping gene selection for real-time RT-PCR normalization in potato during biotic and abiotic stress. J Expt Bot 56:2907–2914. https://doi.org/10.1093/jxb/eri285

Nielsen R (2005) Molecular signatures of natural selection. Annu Rev Genet 39:197–218. https://doi.org/10.1146/annurev.genet.39.073003.112420

Novakova M, Kim PD, Sasek V, Burketova L, Jindrichova B, Santrucek J, Valentova O (2016) Separation and identification of candidate protein elicitors from the cultivation medium of *Leptosphaeria maculans* inducing resistance in *Brassica napus*. Biotechnol Prog 32:918–928. https://doi.org/10.1002/btpr.2266

O'Donnell KO (1992) Ribosomal DNA internal transcribed spacers are highly divergent in the phytopathogenic ascomycete *Fusarium sambucinum* (*Gibberella pulicaris*). Curr Genet 22: 213–220

Ogoshi A, Ui T (1978) Specificity of vitamin requirement among anastomosis groups of *Rhizoctonia solani* Kuhn. Ann Phytopathol Soc Japan 45:47–53

Okubara PA, Li C, Schroeder KL, Schumacher RT, Lawrence NP (2007) Improved extraction of *Rhizoctonia* and *Pythium* DNA from wheat roots and soil samples using pressure cycling technology. Can J Plant Pathol 29:304–310

Okubara PA, Schroeder KL, Patilitz TC (2008) Identification and quantification of *Rhizoctonia solani* and *R. oryzae* using real-time polymerase chain reaction. Phytopathology 98:837–847

Ozsolak F, Milos PM (2011) Transcriptome profiling using single-molecule direct RNA sequencing. Methods Mol Biol 733:51–61. https://doi.org/10.1007/978-1-61779-089-8_4

Padra M, Adamczyk B, Benktander J, Flahou B, Skoog EC, Padra JT, Smet A, Jin C, Ducatelle R, Samuelsson T, Haesebrouck F, Karlsson NG, Teneberg S, Lindén SK (2018) *Helicobacter suis* binding to carbohydrates on human and porcine gastric mucins and glycolipids occurs via two modes. Virulence 9:898–918. https://doi.org/10.1080/21505594.2018.1460979

Pal C, Maciá MD, Oliver A, Schachar I, Buckling A (2007) Co-evolution with viruses drives the evolution of bacterial mutation rates. Nature 450:1079–1081. https://doi.org/10.1038/nature06350

Palma K, Thorgrimsen S, Malinovsky FG, Fiil BK, Nielsen HB, Brodersen P, Hofius D, Petersen M, Mundy J (2010) Auto-immunity in *Arabidopsis acd11* is mediated by epigenetic regulation of an immune receptor. PLoS Pathog 6:e1001137

Pannecouque J, Van Beneden S, Höfte M (2008) Characterization and pathogenicity of *Rhizoctonia* isolates associated with cauliflower in Belgium. Plant Pathol 57:737–746

Parlange F, Daverdin G, Fudal I, Kuhn ML, Balesdent MH, Blaise F, Grezes-Besset B, Rouxel T (2009) *Leptosphaeria maculans* avirulence gene AvrLm4-7 confers a dual recognition specificity by the *Rlm4* and *Rlm7* resistance genes of oilseed rape, and circumvents *Rlm4*-mediated recognition through a single amino acid change. Mol Microbiol 71:851–863. https://doi.org/10.1111/j.1365-2958.2008.06547.x

Patil PB, Bogdanove AJ, Sonti RV (2007) The role of horizontal transfer in the evolution of a highly variable lipopolysaccharide biosynthesis locus in *Xanthomonads* that infect rice, citrus and crucifers. BMC Evol Biol 7:243

Perchepied L, Balagué C, Riou C, Claudel Renard C, Rivière N, Grezes Besset B, Roby D (2010) Nitric oxide participates in the complex interplay of defense related signaling pathways controlling disease resistance to *Sclerotinia sclerotiorum* in *Arabidopsis thaliana*. Mol Plant Microbe Interact 23:846–860. https://doi.org/10.1094/MPMI-23-7-0846

Perez Enciso M, Rincon JC, Legarra A (2015) Sequence vs. chip assisted genomic selection: accurate biological information is advised. Genet Sel Evol 47:43. https://doi.org/10.1186/s1271101501175

Pfaffi MW (2001) A new mathematical model for relative quantification in real-time RT–PCR. Nucleic Acids Res 29:e45

Pittman KJ, Glover LC, Wang L, Ko DC (2016) The legacy of past pandemics: common human mutations that protect against infectious disease. PLoS Pathog 12:e1005680. https://doi.org/10.1371/journal.ppat.1005680

Plissonneau C, Daverdin G, Ollivier B, Blaise F, Degrave A, Fudal I, Rouxel T, Balesdent MH (2016) A game of hide and seek between avirulence genes *AvrLm4-7* and *AvrLm3* in *Leptosphaeria maculans*. New Phytol 209:1613–1624. https://doi.org/10.1111/nph.13736

Poland JA, Bradbury PJ, Buckler ES, Nelson RJ (2011) Genome wide nested association mapping of quantitative resistance to northern leaf blight in maize. Proc Natl Acad Sci U S A 108:6893–6898

Pongam P, Osborn TC, Williams PH (1999) Assessment of genetic variation among *Leptosphaeria maculans* isolates using pathogenicity data and AFLP analysis. Plant Dis 83:149–154

Poplawsky AR, Chun W (1998) *Xanthomonas campestris* pv. *campestris* requires a functional *pigB* for epiphytic survival and host infection. Mol Plant Microbe Interact 11:466–475

Poulin R (2014) Parasite biodiversity revisited: frontiers and constraints. Int J Parasitol 44:581–589. https://doi.org/10.1016/j.ijpara.2014.02.003

Pound GS, Williams PH (1963) Biological races of *Albugo candida*. Phytopathology 53:1146–1149

Pramila, Giri P, Tasleem M, Taj G, Mal R, Kumar A (2014) Morphological, cultural, pathogenic and molecular variability amongst Indian mustard isolates of *Alternaria brassicae* in Uttarakhand. Afr J Biotech 13:441–448

Prince DC, Rallapalli G, Xu D, Schoonbeek HJ, Cevik V, Asai S, Kemen E, Cruz-Mireles N, Kemen A, Belhaj K, Schornack S, Kamoun S, Holub EB, Halkier BA, Jones JD (2017) *Albugo*-imposed changes to tryptophan-derived antimicrobial metabolite biosynthesis may contribute to suppression of non-host resistance to *Phytophthora infestans* in *Arabidopsis thaliana*. BMC Biol 15:20

Purwantara A, Barrins JM, Cozijnsen AJ, Ades PK, Howle BJ (2000) Genetic diversity of isolates of the *Leptosphaeria maculans* species complex from Australia, Europe and North America using amplified fragment length polymorphism analysis. Mycol Res 104:772–781. https://doi.org/10.1017/S095375629900235X

Qian W, Jia Y, Ren SX, He YQ, Feng JX, Lu LF, Sun Q, Ying G, Tang DJ, Tang H, Wu W, Hao P, Wang L, Jiang BL, Zeng S, Gu WY, Lu G, Rong L, Tian Y, Yao Z, Fu G, Chen B, Fang R, Qiang B, Chen Z, Zhao GP, Tang JL, He C (2005) Comparative and functional genomic analyses of the pathogenicity of phytopathogen *Xanthomonas campestris* pv. *campestris*. Genome Res 15:757–767

Rajarammohan S, Pradhan AK, Pental D, Kaur J (2018) Genome wide association mapping in *Arabidopsis* identifies novel genes underlying quantitative disease resistance to *Alternaria brassicae*. Mol Plant Pathol 19:1719–1732

Raman R, Taylor B, Lindbeck K, Coombes N, Barbulescu D, Salisbury P, Raman H (2012a) Molecular mapping and validation of *Rlm1* genes for resistance to *Leptosphaeria maculans* in canola (*Brassica napus* L). Crop Pasture Sci 63:1007–1017. https://doi.org/10.1071/CP12255

Raman R, Taylor B, Marcroft S, Stiller J, Eckermann P, Coombes N, Rehman A, Lindbeck K, Luckett D, Wratten N, Batley J, Edwards D, Wang X, Raman H (2012b) Molecular mapping of qualitative and quantitative loci for resistance to *Leptosphaeria maculans* causing blackleg disease in canola (*Brassica napus* L). Theor Appl Genet 125:405–418. https://doi.org/10.1007/s00122-012-1842-6

Raman R, Diffey S, Carling J, Cowley RB, Kilian A, Luckett DJ, Raman H (2016) Quantitative genetic analysis of grain yield in an Australian *Brassica napus* doubled-haploid population. Crop Pasture Sci 67:298–307. https://doi.org/10.1071/CP15283

Ramonell KM, Zhang B, Ewing RM, Chen Y, Xu D, Stacey G, Somerville S (2002) Microarray analysis of chitin elicitation in *Arabidopsis thaliana*. Mol Plant Pathol 36(1):301–311

Rehmany AP, Lynn JR, Tor M, Holub EB, Beynon JL (2000) A comparison of *Peronospora parasitica* (Downy mildew) isolates from *Arabidopsis thaliana* and *Brassica oleracea* using amplified fragment length polymorphism and internal transcribed spacer 1 sequence analyses. Fungal Genet Biol 30(2):95–103. https://doi.org/10.1006/fgbi.2000.1216

Reiter RS, Williams JGK, Feldman KA, Rafalski A, Tingey SV, Scolnik PA (1992) Global and local genome mapping in *Arabidopsis thaliana* by using recombinant inbred lines and random amplied polymorphic DNAs. Proc Natl Acad Sci U S A 89:1477–1481

Reuber TL, Plotnikova JM, Dewdney J, Rogers EE, Wood W, Ausubel MF (1998) Correlation of defense gene induction defects with powdery mildew susceptibility in *Arabidopsis* enhanced disease susceptibility mutants. Plant J 16(4):473–485

Rigano LA, Payette C, Brouillard G, Marano MR, Abramowicz L, Torres PS, Yun M, Castagnaro AP, El Oirdi M, Dufour V, Malamud F, Dow JM, Bouarab K, Vojnov A (2007) Bacterial cyclic beta-(1,2)-glucanacts in systemic suppression of plant immune responses. Plant Cell 19:2077–2089

Rimmer SR, Shattuck VI, Buchwaldt L (2007) Compendium of *Brassica* diseases. APS Press, St. Paul, MN

Robin GP, Kleemann J, Neumann U, Cabre L, Dallery J-F, Lapalu N, O'Connell RJ (2018) Subcellular localization screening of *Colletotrichum higginsianum* effector candidates identifies fungal proteins targeted to plant peroxisomes, golgi bodies, and microtubules. Front Plant Sci 9: 562. https://doi.org/10.3389/fpls.2018.00562

Roohie RK, Umesha S (2015) Identification of genes associated with black rot resistance in cabbage through suppression subtractive hybridization. 3 Biotech 5:1089–1100

Roux F, Voisin D, Badet T, Balague C, Barlet X, Huard-Chauveau C, Roby D, Raffaele S (2014) Resistance to phytopathogens e tutti quanti: placing plant quantitative disease resistance on the map. Mol Plant Pathol 15(5):427–432. https://doi.org/10.1111/mpp.12138

Rouxel T, Balesdent MH (2005) The stem canker (blackleg) fungus, *Leptosphaeria maculans*, enters the genomic era. Mol Plant Pathol 6:225–241

Rouxel T, Penaud A, Pinochet X, Brun H, Gout L, Delourme R, Schmit J, Balesdent MH (2003) A 10-year survey of populations of *Leptosphaeria maculans* in France indicates a rapid adaptation towards the *Rlm1* resistance gene of oilseed rape. Eur J Plant Pathol 109:871–881. https://doi.org/10.1023/A:1026189225466

Rouxel T, Grandaubert J, Hane JK, Hoede C, Van de Wouw AP, Couloux A, Dominguez V, Anthouard V, Bally P, Bourras S, Cozijnsen AJ, Ciuffetti LM, Degrave A, Dilmaghani A, Duret L, Fudal I, Goodwin SB, Gout L, Glaser N, Linglin J, Kema GH, Lapalu N, Lawrence CB, May K, Meyer M, Ollivier B, Poulain J, Schoch CL, Simon A, Spatafora JW, Stachowiak A, Turgeon BG, Tyler BM, Vincent D, Weissenbach J, Amselem J, Quesneville H, Oliver RP, Wincker P, Balesdent MH, Howlett BJ (2011) Effector diversification within compartments of the *Leptosphaeria maculans* genome affected by repeat-induced point mutations. Nat Commun 2:202

Ryan RP, Fouhy Y, Lucey JF, Jiang BL, He YQ, Feng JX, Tang JL, Dow JM (2007) Cyclic di-GMP signaling in the virulence and environmental adaptation of *Xanthomonas campestris*. Mol Microbiol 63:429–442

Ryan RP, McCarthy Y, Andrade M, Farah CS, Armitage JP, Dow JM (2010) Cell–cell signal-dependent dynamic interactions between HD-GYP and GGDEF domain proteins mediate virulence in *Xanthomonas campestris*. Proc Natl Acad Sci U S A 107:5989–5994

Sabeti PC, Schaffner SF, Fry B, Lohmueller J, Varilly P, Shamovsky O, Palma A, Mikkelsen TS, Altshuler D, Lander ES (2006) Positive natural selection in the human lineage. Science 312: 1614–1620. https://doi.org/10.1126/science.1124309

Sabokkhiz MA, Jafarpour B, Shahriari Ahmadi F, Tarighi S (2012) Identification of *Turnip mosaic virus* isolated from canola in northeast area of Iran. Afr J Biotechnol 11(80):14553–14560. https://doi.org/10.5897/AJB12.2121

Saenz GS, Taylor JW, Gargas A (1994) 18S rRNA gene sequences and supraordinal classification of the Erysiphales. Mycologia 86:212–216

Saenz GS, Taylor JW (1999) Phylogeny of Erysiphales (powdery mildews) inferred from internal transcribed spacer (ITS) ribosomal DNA sequences. Can J Bot 77:150–168

Saharan GS, Verma PR, Meena PD, Kumar A (2014) White rust of crucifers: biology, ecology and management. Springer, Frankfurt, p 244. ISBN 978-81-322-1791-6

Salazar O, Julian MC, Rubio V (2000) Primers based on specific rDNA-ITS sequences for PCR detection of *Rhizoctonia solani*, R-solani AG 2 subgroups and ecological types, and binucleate *Rhizoctonia*. Mycol Res 104:281–285

Sang X, Jue D, Yang L, Bai X, Chen M, Yang Q (2013) Genetic transformation of *Brassica napus* with *MSI-99m* gene increases resistance in transgenic plants to *Sclerotinia sclerotiorum*. Mol Plant Breed 4:30. https://doi.org/10.5376/mpb.2013.04.0030

Sayler RJ, Yang YN (2007) Detection and quantification of *Rhizoctonia solani* AG-1 IA, the rice sheath blight pathogen, in rice using real-time PCR. Plant Dis 91:1663–1668

Scanlan PD, Hall AR, Blackshields G, Friman V-P, Davis MR, Goldberg JB, Buckling A (2015) Co-evolution with bacteriophages drives genome-wide host evolution and constrains the acquisition of abiotic-beneficial mutations. Mol Biol Evol 32:1425–1435. https://doi.org/10.1093/molbev/msv032

Schaad NW, Berthier-Schaad Y, Sechler A, Knorr D (1999) Detection of *Clavibacter michiganensis* subsp. *sepedonicus* in potato tubers by BIO-PCR and an automated real-time fluorescence detection system. Plant Dis 83:1095–1100

Schaad NW, Opgenorth, Gaush P (2002) Real-time polymerase chain reaction for 1 h on-site diagnosis of Pierce's disease of grape in early season asymptomatic vines. Phytopathology 92:721–728

Schafer C, Wostemeyer J (1992) Random primer dependent PCR differentiates aggressive from non-aggressive isolates of the oilseed rape pathogen *Phoma lingam* (*Leptosphaeria maculans*). J Phytopathol 136:124–136

Schmittgen TD, Zakrajsek BA (2000) Effect of experimental treatment on housekeeping gene expression: validation by real-time, quantitative RT-PCR. J Biochem Biophys Methods 46:69–81

Schots A, Dewey FM, Oliver R (1994) Modern assays for plant pathogenic fungi: identification, detection and quantification. CAB International, Wallingford

Schwinghamer MW, Thomas JE, Schilg MA, Parry JN, Dann EK, Moore KJ, Kumari SG (2010) Mastre viruses in chickpea (*Cicer arietinum*) and other dicotyledonous crops and weeds in Queensland and northern New South Wales, Australia. Australas Plant Pathol 39:551–561. https://doi.org/10.1071/AP10032

Schwinghamer MW, Schilg MA, Walsh JA, Bambach RW, Cossu RM, Bambridge JM, Hind-Lanoiselet TL, McCorkell BE, Cross P (2014) *Turnip mosaic virus*: potential for crop losses in the grain belt of New South Wales, Australia. Australas Plant Pathol 43:663–678. https://doi.org/10.1007/s13313-014-0304-9

Sekirov I, Russell SL, Antunes LCM, Finlay BB (2010) Gut microbiota in health and disease. Physiol Rev 90:859–904. https://doi.org/10.1152/physrev.00045.2009

Seruga Musić M, Duc Nguyen H, Černi S, Mamula Đ, Ohshima K, Škorić D (2014) Multi-locus sequence analysis of '*Candidatus Phytoplasma asteris*' strain and the genome analysis of *Turnip mosaic virus* co-infecting oilseed rape. J Appl Microbiol 117(8):774–785

Sharma TR, Tewari JP (1995) Detection of genetic variation in *Alternaria brassicae* by RAPD fingerprints. J Plant Biochem Biotech 4:105–107

Sharma TR, Tewari JP (1998) RAPD analysis of three *Alternaria* species pathogenic to crucifers. Mycol Res 102:807–814

Sharma P, Deep S, Sharma M, Bhati DS (2013) Genetic variation of *Alternaria brassicae* (Berk) Sacc causing dark leaf spot of cauliflower and mustard in India. J Gen Plant Pathol 79(1):41–45

Sherriff C, Lucas JA (1989a) Cytogenetic study of heterothallic and homothallic isolates of *Peronospora parasitica*. Mycol Res 92(3):302–310

Sherriff C, Lucas JA (1989b) Heterothallism and homothallism in *Peronospora parasitica*. Mycol Res 92:311–316

Sherriff C, Lucas JA (1990) The host range of isolates of downy mildew, *Peronospora parasitica* from *Brassica* crop species. Plant Pathol 39:77–91

Shimada C, Lipka V, O'Connell R, Okuno T, Schulze-Lefert P, Takano Y (2006) Nonhost resistance in *Arabidopsis-Colletotrichum* interactions acts at the cell periphery and requires actin filament function. Mol Plant Microbe Interact 19:270–279. https://doi.org/10.1094/MPMI-19-0270

Siemens J, Keller I, Sarx J, Kunz S, Schuller A, Nagel W, Schmülling T, Parniske M, Ludwig-Müller J (2006) Transcriptome analysis of Arabidopsis clubroots indicate a key role for cytokinins in disease development. Mol Plant Microbe Interact 19:480–494. https://doi.org/10.1094/MPMI-19-0480

Siemens J, Graf H, Bulman S, In O, Ludwig-Müller J (2009) Monitoring expression of selected *Plasmodiophora brassicae* genes during clubroot development in *Arabidopsis thaliana*. Plant Pathol 58:130–136

Singh R (2016) Molecular diversity and development of diagnostic biomarkers for *Alternaria brassicae*. Ph.D. Thesis, IFTM University, Moradabad, India

Singh S, Dey SS, Bhatia R, Batley J, Kumar R (2018) Molecular breeding for resistance to black rot [*Xanthomonas campestris* pv. *campestris* (Pammel) Dowson] in Brassicas: recent advances. Euphytica 214:196

Sneh B (1996) Anastomosis groups of multinucleate *Rhizoctonia* spp. In: Sneh B, Jabaji-Hare S, Neate S, Dijst G (eds) *Rhizoctonia* species: taxonomy, molecular biology, ecology, pathology and disease control. Kluwer Academic Publishers, Dordrecht, pp 67–75

Somers DJ, Rakow G, Rimmer SR (2002) *Brassica napus* DNA markers linked to white rust resistance in *Brassica juncea*. Theor Appl Genet 104:1121–1124

Soyer JL, Hamiot A, Ollivier B, Balesdent MH, Rouxel T, Fudal I (2015) The APSES transcription factor *LmStuA* is required for sporulation, pathogenic development and effector gene expression in *Leptosphaeria maculans*. Mol Plant Pathol 16:1000–1005. https://doi.org/10.1111/mpp.12249

Sprague SJ, Marcroft SJ, Hayden HL, Howlett BJ (2006) Major gene resistance to blackleg in *Brassica napus* overcome within three years of commercial production in South-eastern Australia. Plant Dis 90:190–198. https://doi.org/10.1094/PD-90-0190

Staskawicz BJ, Ausubel FM, Baker BJ, Ellis JG, Jones JDG (1995) Molecular genetics of plant disease resistance. Science 268:661–667

Steventon LA, Fableson J, Hu Q, Dixelius C (2002) Identification of the causal agent of *Verticillium* wilt of winter oilseed rape in Sweden *V. longisporum*. Mycol Res 106:570–557

Stiles JI, Lemme C, Sondur S, Morshidi MB, Manshardt R (1993) Using randomly amplified polymorphic DNA for evaluating genetic relationships among papaya cultivars. Theor Appl Genet 85:697–701

Stukenbrock EH (2016) The role of hybridization in the evolution and emergence of new fungal plant pathogens. Phytopathology 106:104–112

Stukenbrock EH, McDonald BA (2008) The origins of plant pathogens in agro-ecosystems. Annu Rev Phytopathol 46:75–100

Stürzenbaum SR, Kille P (2001) Control genes in quantitative molecular biological techniques: the variability of invariance. Comp Biochem Physiol B 130:281–289

Sun X, Liu D, Zhang X, Li W, Liu H, Hong W, Jiang C, Guan N, Ma C, Zeng H, Xu C, Song J, Huang L, Wang C, Shi J, Wang R, Zheng X, Lu C, Wang X, Zheng H (2013) SLAF-seq: an efficient method of large-scale *de novo* SNP discovery and genotyping using high-throughput sequencing. PLoS One 8(3):e58700

Sundelin T, Jensen DF, Lubeck M (2011) Identification of expressed genes during infection of Chinese cabbage (*Brassica rapa* subsp. *pekinensis*) by *Plasmodiophora brassicae*. J Eukaryot Microbiol 58:310–314

Swarbrick PJ, Schulze-Lefert P, Scholes JD (2006) Metabolic consequences of susceptibility and resistance (race-specific and broad-spectrum) in barley leaves challenged with powdery mildew. Plant Cell Environ 296(1):1061–1076

Swofford DL (1993) Phylogenetic analysis using parsimony (PAUP version 3.1.1). Illinois Natural History Survey, Champaign, IL

Taj G, Agarwal P, Grant M, Kumar A (2011) Co-expression and in-silico interaction studies for inter-linking the activation of MAPK3 and LOX genes during pathogenesis of Alternaria blight (*Alternaria brassicae*) in *Brassica juncea*. J Oilseed Brassica 2(1):13–20

Takahara H, Hacquard S, Kombrink A, Hughes HB, Halder V, Robin GP, Hiruma K, Neumann U, Shinya T, Kombrink E, Shibuya N, Thomma BPHJ, O'Connell RJ (2016) *Colletotrichum higginsianum* extracellular LysM proteins play dual roles in appressorial function and suppression of chitin-triggered plant immunity. New Phytol 211:1323–1337. https://doi.org/10.1111/nph.13994

Takamatsu S, Kano Y (2001) PCR primers useful for nucleotide sequencing of rDNA of the powdery mildew fungi. Mycoscience 42:135–139

Tamura K, Dudley J, Nei M, Kumar S (2007) MEGA4: Molecular Evolutionary Genetics Analysis (MEGA) software version 4.0. Mol Biol Evol 24(8):1596–1599. https://doi.org/10.1093/molbev/msm092

Tham FY, Lucas JA, Wilson ZA (1994) DNA fingerprinting of *Peronospora parasitica*, a biotrophic fungal pathogen of crucifers. Theor Appl Genet 88:490–496

Thilliez GJA, Armstrong MR, Lim TY, Baker K, Jouet A, Ward B, Van Oosterhout C, Jones JDG, Huitema E, Birch PRJ, Hein I (2018) Pathogen enrichment sequencing (PenSeq) enables population genomic studies in oomycetes. New Phytol 4:903

Titcomb GC, Jerde CL, Young HS (2019) High-throughput sequencing for understanding the ecology of emerging infectious diseases at the wildlife-human interface. Front Ecol Evol 7:126

Tor M, Holub EB, Brose E, Musker R, Gunn N (1994) Map positions of three loci in *Arabidopsis thaliana* associated with isolate-specific recognition of *Peronospora parasitica* (downy mildew). Mol Plant Microbe Interact 7:214–222. https://doi.org/10.1094/MPMI-7-0214

Torii KU, Mitsukawa N, Oosumi T, Matsuura Y, Yokoyama R, Whittier RF, Komeda Y (1996) The *Arabidopsis ERECTA* gene encodes a putative receptor protein kinase with extracellular leucine rich repeats. Plant Cell 8:735–746

Torres PS, Malamud F, Rigano LA, Russo DM, Marano MR, Castagnaro AP, Zorreguieta A, Bouarab K, Dow JM, Vojnov AA (2007) Controlled synthesis of the DSF cell–cell signal is required for biofilm formation and virulence in *Xanthomonas campestris*. Environ Microbiol 9:2101–2109

Tsai HY, Hamilton A, Tinch AE, Guy DR, Gharbi K, Stear MJ, Matika O, Bishop SC, Houston RD (2015) Genome wide association and genomic prediction for growth traits in juvenile farmed

Atlantic salmon using a high density SNP array. BMC Genomics 16:969. https://doi.org/10.1186/s1286401521179

Uzunova M, Ecke W, Weissleder K, Robbelen G (1995) Mapping the genome of rapeseed (*Brassica napus* L). I. Construction of an RFLP linkage map and localization of QTLs for seed glucosinolate content. Theor Appl Genet 90:194–204. https://doi.org/10.1007/BF00222202

Van de Wouw AP, Thomas VL, Cozijnsen AJ, Marcroft SJ, Salisbury PA, Howlett BJ (2008) Identification of *Leptosphaeria biglobosa* 'canadensis' on *Brassica juncea* stubble from northern New South Wales, Australia. Australas Plant Dis Notes 3:124–128

Van de Wouw AP, Cozijnsen AJ, Hane JK, Brunner PC, McDonald BA, Oliver RP, Howlett BJ (2010) Evolution of linked avirulence effectors in *Leptosphaeria maculans* is affected by genomic environment and exposure to resistance genes in host plants. PLoS Pathog 6:e1001180

Van de Wouw AP, Marcroft SJ, Ware A, Lindbeck K, Khangura R, Howlett BJ (2014) Breakdown of resistance to the fungal disease, blackleg, is averted in commercial canola (*Brassica napus*) crops in Australia. Field Crops Res 166:144–151

Van de Wouw AP, Idnurm A, Davidson JA, Sprague SJ, Khangura RK, Ware AH, Lindbeck KD, Marcroft SJ (2016a) Fungal diseases of canola in Australia: identification of trends, threats and potential therapies. Australas Plant Pathol 45:415–423. https://doi.org/10.1007/s13313-016-0428-1

Van de Wouw AP, Marcroft SJ, Howlett BJ (2016b) Blackleg disease of canola in Australia. Crop Pasture Sci 67:273–283. https://doi.org/10.1071/CP15221

Vandesompele J, De Preter K, Pattyn F, Poppe B, Van Roy N, De Paepe A, Speleman F (2002) Accurate normalization of real-time quantitative RT-PCR data by geometric averaging of multiple internal control genes. Genome Biol 3:1–12. https://doi.org/10.1186/gb-2002-3-7-research0034

VanRaden PM (2008) Efficient methods to compute genomic predictions. J Dairy Sci 91(11): 4414–4423. https://doi.org/10.3168/jds.2007-0980

Vela Avitua S, Meuwissen TH, Luan T, Odegard J (2015) Accuracy of genomic selection for a sib evaluated trait using identity by state and identity by descent relationships. Genet Sel Evol 47:9. https://doi.org/10.1186/s1271101400842

Velculescu VE, Zhang L, Vogelstein B, Kinzler KW (1995) Serial analysis of gene expression. Science 270(5235):484–487. https://doi.org/10.1126/science.270.5235.484

Verma PR (2012) White rust of crucifers: an overview of research progress. J Oilseed Brassica 3: 78–87

Vicente JG, Holub EB (2013) *Xanthomonas campestris* pv. *campestris* (cause of black rot of crucifers) in the genomic era is still a worldwide threat to *Brassica* crops. Mol Plant Pathol 14(1):2–18. https://doi.org/10.1111/j.1364-3703.2012.00833.x

Vilgalys R, Gonzalez D (1990) Ribosomal DNA restriction fragment length polymorphisms in *Rhizoctonia solani*. Phytopathology 80:151–158

Vincenot L, Balesdent MH, Li H, Barbetti MJ, Sivasithamparam K, Gout L, Rouxel T (2008) Occurrence of a new subclade of *Leptosphaeria biglobosa* in Western Australia. Phytopathology 98:321–329

Vishwanath K, Kolte SJ (1997) Variability in *Alternaria brassicae*: response to host genotypes, toxin production and fungicides. Indian Phytopathol 50:373–381

Vitti JJ, Grossman SR, Sabeti PC (2013) Detecting natural selection in genomic data. Annu Rev Genet 47:97–120. https://doi.org/10.1146/annurev-genet-111212-133526

Voegele RT, Struck C, Hahn M, Mendgen K (2001) The role of haustoria in sugar supply during infection of broad bean by the rust fungus *Uromyces fabae*. Proc Natl Acad Sci U S A 986(1): 8133–8138

Voigt K, Cozijnsen AJ, Kroymann J, Pöggeler S, Howlett BJ (2005) Phylogenetic relationships between members of the crucifer pathogenic *Leptosphaeria maculans* species complex as shown by mating type (MAT1-2), actin, and β-tubulin sequences. Mol Phylogenet Evol 37: 541–557. https://doi.org/10.1016/j.ympev.2005.07.006

Vojnov AA, Slater H, Daniels MJ, Dow JM (2001) Expression of the gumoperon directing xanthan biosynthesis in *Xanthomonas campestris* and its regulation in planta. Mol Plant Microbe Interact 14:768–774

Volossiouk T, Robb JE, Nazar RN (1995) Direct DNA extraction for PCR-mediated assays of soil organisms. Appl Environ Microbiol 61:3972–3976

Vorhoelter FJ, Schneiker S, Goesmann A, Krause L, Bekel T, Kaiser O, Linke B, Patschkowski T, Rueckert C, Schmid J, Sidhu VK, Sieber V, Tauch A, Watt SA, Weisshaar B, Becker A, Niehaus K, Puehler A (2008) The genome of *Xanthomonas campestris* pv. *campestris* B 100 and its use for the reconstruction of metabolic pathways involved in xanthan biosynthesis. J Biotechnol 134:33–45

Vorholter FJ, Niehaus K, Puhler A (2001) Lipopolysaccharide biosynthesis in *Xanthomonas campestris* pv. *campestris*: a cluster of 15 genes is involved in the biosynthesis of the LPS O-antigen and the LPS core. Mol Genet Genomics 266:79–95

Walker JC (1993) Receptor-like protein kinase genes of *Arabidopsis thaliana*. Plant J 3:451–456

Wan JR, Zhang XC, Neece D, Ramonell KM, Clough S, Kim SY, Stacey MG, Stacey G (2008) A *LysM* receptor-like kinase plays a critical role in chitin signaling and fungal resistance in *Arabidopsis*. Plant Cell 206(1):471–481

Wang GL, Song WY, Wu RL, Sideris S, Ronald PC (1996) The cloned gene, *Xa27,* confers resistance to multiple *Xanthomonas oryzae* pv. *oryzae* isolates in transgenic plants. Mol Plant Microbe Interact 9:850–855

Wang Z, Mao H, Dong C, Ji R, Cai L, Fu H, Liu S (2009) Overexpression of *Brassica napus MPK4* enhances resistance to *Sclerotinia sclerotiorum* in oilseed rape. Mol Plant Microbe Interact 22: 235–244

Wang Z, Tan X, Zhang Z, Gu S, Li G, Shi H (2012) Defense to *Sclerotinia sclerotiorum* in oilseed rape is associated with the sequential activations of salicylic acid signaling and jasmonic acid signaling. Plant Sci Int J Expt Plant Biol 184:75–82

Wang Q, Li J, Hu L, Zhang T, Zhang G, Lou Y (2013) *OsMPK3* positively regulates the JA signaling pathway and plant; resistance to a chewing herbivore in rice. Plant Cell Rep 32:1075–1084. https://doi.org/10.1007/s00299-013-1389-2

Wang Y, Cheng X, Shan Q, Zhang Y, Liu J, Gao C, Qiu JL (2014) Simultaneous editing of three homoeo alleles in hexaploid bread wheat confers heritable resistance to powdery mildew. Nat Biotechnol 32:947–951. https://doi.org/10.1038/nbt.2969

Wang Z, Bao L-L, Zhao F-Y, Tang M-Q, Chen T, Li Y, Wang B-X, Fu B, Fang H, Li G-Y, Cao J, Ding L-N, Zhu K-M, Liu S-Y, Tan X-L (2019) *BnaMPK3*is a key regulator of defense responses to the devastating plant pathogen *Sclerotinia sclerotiorum* in oilseed rape. Front Plant Sci 10:91. https://doi.org/10.3389/fpls.2019.00091

Wang Z, Ma LY, Li X, Zhao FY, Sarwar R, Cao J, Li YL, Ding LN, Zhu KM, Yang YH, Tan XL (2020) Genome-wide identification of the *NPR1*-like gene family in *Brassica napus* and functional characterization of *BnaNPR1* in resistance to *Sclerotinia sclerotiorum.* Plant Cell Rep 39(6):709–722. https://doi.org/10.1007/s00299-020-02525-z

Ward E (1994) Use of the polymerase chain reaction for identifying plant pathogens. In: Blakeman JP, Williamson B (eds) Ecology of plant pathogens. CAB International, Wallingford, pp 143–160

Ward E, Foster SJ, Fraaije BA, Mccartney HA (2004) Plant pathogen diagnostics: immunological and nucleic acid-based approaches. Ann Appl Biol 145:1–16. https://doi.org/10.3109/1040841X.2010.489892

Wei C, Chen J, Kuang H (2016) Dramatic number variation of R genes in solanaceae species accounted for by a few R gene subfamilies. PLoS One 11(2):e0148708. https://doi.org/10.1371/journal.pone.0148708

Welsh J, McClelland M (1990) Fingerprinting genomes using PCR with arbitrary primers. Nucleic Acids Res 18:7213–7218

Weßling R, Panstruga R (2012) Rapid quantification of plant-powdery mildew interactions by qPCR and conidiospore counts. Plant Methods 8:35

Whisson DL, Herdina, Francis L (1995) Detection of *Rhizoctonia solani* AG-8 in soil using a specific DNA-probe. Mycol Res 99:1299–1302

White TJ, Bruns T, Lee S, Taylor JW (1990) Amplication and direct sequencing of fungal ribosomal RNA genes for phylogenetics. In: Innis MA, Gelfand DH, Sninsky JJ, White TJ (eds) PCR protocols: a guide to methods and application. Academic Press, London, pp 315–322

Williams JGK, Kubelik AR, Livak KJ, Rafalski JA, Tingey SV (1990) DNA polymorphisms amplified by arbitrary primers are useful as genetic markers. Nucleic Acids Res 18:6531–6535

Wilson LM, Howlett BJ (2005) *Leptosphaeria maculans*, a fungal pathogen of *Brassica napus*, secretes a subtilisin-like serine protease. Eur J Plant Pathol 112:23–29. https://doi.org/10.1007/s10658-004-7088-7

Wilson TJG, Bertrand N, Tang JL, Feng JX, Pan MQ, Barber CE, Dow JM, Daniels MJ (1998) The *rpfA* gene of *Xanthomonas campestris* pathovar *campestris*, which is involved in the regulation of pathogenicity factor production, encodes an aconitase. Mol Microbiol 28:961–970

Wright DP, Baldwin BC, Shephard MC, Scholes JD (1995a) Source-sink relationships in wheat leaves infected with powdery mildew. I. Alterations in carbohydrate metabolism. Physiol Mol Plant Pathol 476(1):237–253

Wright DP, Baldwin BC, Shephard MC, Scholes JD (1995b) Source-sink relationships in wheat leaves infected with powdery mildew. II. Changes in the regulation of the Calvin cycle. Physiol Mol Plant Pathol 476(1):255–267

Wurschum T, Abel S, Zhao YS (2014) Potential of genomic selection in rapeseed (*Brassica napus* L.) breeding. Plant Breed 133:45–51. https://doi.org/10.1111/pbr.12137

Yan Y, Yuan Q, Tang J, Huang J, Hsiang T, Wei Y, Zheng L (2018) *Colletotrichum higginsianum* as a model for understanding host-pathogen interactions: a review. Int J Mol Sci 19(7):2142. https://doi.org/10.3390/ijms19072142

Yang M, Ding G, Shi L, Feng J, Xu F, Meng J (2010) Quantitative trait loci for root morphology in response to low phosphorus stress in *Brassica napus*. Theor Appl Genet 121(1):181–193

Yu F, Lydiate DJ, Rimmer SR (2005) Identification of two novel genes for blackleg resistance in *Brassica napus*. Theor Appl Genet 110:969–979. https://doi.org/10.1007/s00122-004-1919-y

Yu F, Lydiate DJ, Rimmer SR (2008) Identification and mapping of a third blackleg resistance locus in *Brassica napus* derived from *B. rapa* subsp. *sylvestris*. Genome 51:64–72. https://doi.org/10.1139/G07-103

Yu SC, Zhang FL, Yu RB, Zou YM, Qi JN, Zhao XY, Yu YJ, Zhang D, Li L (2009) Genetic mapping and localization of a major QTL for seedling resistance to downy mildew in Chinese cabbage (*Brassica rapa* ssp. *pekinensis*). Mol Breed 23:573–590

Yu F, Zhang X, Huang Z, Chu M, Song T, Falk KC, Deora A, Chen Q, Zhang Y, McGregor L, Gossen BD, McDonald MR, Peng G (2016) Identification of genome-wide variants and discovery of variants associated with *Brassica rapa* clubroot resistance gene *Rcr1* through bulked segregant RNA sequencing. PLoS One 11:e0153218. https://doi.org/10.1371/journal.pone.0153218

Yun MH, Torres PS, El Oirdi M, Rigano LA, Gonzalez-Lamothe R, Marano MR, Castagnaro AP, Dankert MA, Bouarab K, Vojnov A (2006) Xanthan induces plant susceptibility by suppressing callose deposition. Plant Physiol 141:178–187

Zang N, Tang DJ, Wei ML, He YQ, Chen BS, Feng JX, Xu J, Gan YQ, Jiang BL, Tang JL (2007) Requirement of a *mip*-like gene for virulence in the phytopathogenic bacterium *Xanthomonas campestris* pv. *campestris*. Mol Plant Microbe Interact 20:21–30

Zhang W, Fraiture M, Kolb D, Löffelhardt B, Desaki Y, BoutrotF FG, Tör M, Zipfel C, Gust AA, Brunner F (2013) *Arabidopsis* receptor like protein and receptor like kinase suppressor of BIR11/EVERSHED mediate innate immunity to necrotrophic fungi. Plant Cell 25:4227–4241

Zhang T, Zhao Z, Zhang C, Pang W, Choi SR, Lim YP, Piao Z (2014) Fine genetic and physical mapping of the *CRb* gene conferring resistance to clubroot disease in *Brassica rapa*. Mol Breed 34:1173–1183. https://doi.org/10.1007/s11032-014-0108-1

Zhang H, Feng J, Hwang SF, Strelkov SE, Falak I, Huang X, Sun R (2016) Mapping of clubroot (*Plasmodiophora brassicae*) resistance in canola (*Brassica napus*). Plant Pathol 65:435–440. https://doi.org/10.1111/ppa.12422

Zhu W, Wei W, Fu Y, Cheng J, Xie J, Li G, Yi X (2013) A secretory protein of necrotrophic fungus *Sclerotinia sclerotiorum* that suppresses host resistance. PLoS One 8:e53901. https://doi.org/10.1371/journal.pone.0053901

Zimmerli L, Stein M, Lipka V, Schulze-Lefert P, Somerville S (2004) Host and non-host pathogens elicit different jasmonate/ethylene responses in *Arabidopsis*. Plant J 406(1):633–646

Pathogenomics of Pathogenic Variability 5

Abstract

At different stages of host–pathogen interactions, natural conditions and modern agricultural settings foster the evolution of new and virulent pathogenic races by selection pressure on *R* genes in crucifer genotypes. Polyploidy in *Albugo candida* contributes to race diversification and evolution by selection pressure imposed from cultivated *Brassica* species. Molecular studies have revealed high degree of genetic diversity within *Albugo* pathogenic to *Brassica* species with wide host range and six new specialized species along with virulent races. ITS and COX2, molecular approaches of phylogenetic analysis and morphological variations of oospores have divided *A. candida* into three groups (1, 2, 3) infecting different cruciferous with different races. Several races/pathotypes of *A. candida* have been identified at global level infecting different crucifers. The virulence of *A. candida* races varies from very narrow to wide infecting different *Brassica* species and genotypes. Phylogenetic analysis and nucleotide sequences have distinguished races of *A. candida* infecting different crucifers in different countries of the world. *Albugo candida* isolates from different crucifers have been characterized based on virulence genes specificity to their hosts in Australia. A single dominant gene *AvrAC1* controls avirulence in race 2 of *A. candida* infecting *B. rapa* cv. Torch. The genetic variation among geographically distinct isolates of *A. brassicae* has been assessed using RAPD-PCR markers. The genetic structure of *A. brassicicola* population indicates the occurrence of sexual recombination's with a cryptic sexual stage. Mutation of *Amr1* gene in *A. brassicicola* causes increased virulence. Three pathotypes of *A. brassicae* viz., RM1, RM2, and V3 have been identified using eight commonly cultivated *Brassica* species. The isolates of *A. brassicae* collected from different geographical areas are different in conidial morphology, cultural characteristics, and symptom production on different hosts. *A brassicae* isolates from different locations show high degree of genetic similarity. There is differential protein expression by

G. Singh Saharan et al., *Genomics of Crucifer's Host-Pathosystem*,
https://doi.org/10.1007/978-981-19-3812-2_5

the isolates of *A. brassicae*. The isolates of *A. brassicae* show variability in biochemical contents, sensitivity to different fungicides and plant extracts. The variation in relative levels of virulence in the form of disease incidence, severity, and defoliation within and between *Alternaria* spp. causing leaf spot disease on rapeseed-mustard has been recorded. Several pathotypes of *Alternaria* species have been identified with different designations using differential determinant attributes. The function of effectors/genes, *ChELP1*, and *ChELP2* homologs of Lysm proteins has been characterized in *C. higginsianum.* In India, powdery mildew pathogen produces both asexual and sexual stages on *B. juncea*, so there is every possibility for *Erysiphe cruciferarum* to express pathogenic variability. The isolates of *Hyaloperonospora parasitica* collected from different countries, locations, and host species have been characterized in the form of pathotypes showing host specificity and differential virulence. The isolates of *H. parasitica* from different host species and genotypes have been molecularly identified as pathotypes with their virulence potential using DNA (RAPD) finger printing. The isolates of *H. parasitica* collected and characterized on different crucifers as race and pathotypes cannot be compared in the absence of use of standardized host differentials (isogenic lines) and uniform designation/nomenclature system at international level. Hybridization of *Hyaloperonospora* isolates has also been observed. The evolution of virulent races in *Leptosphaeria* has been observed in the areas where crops with major gene resistance coupled with genetic uniformity have been sown. The avirulence genes of *L. maculans* races have been identified from Europe. *Leptosphaeria maculans* field populations have high level of gene diversity and genotypic diversity in France. An avirulent mutated *AvrLm* gene of *L. maculans* has been identified to detect new *R* genes for breeding blackleg resistant cultivars of *Brassica*. The genome-wide DNA variants and SNP haplotypes of *L. maculans* pathotypes have been identified. Pathotypes of *Plasmodiophora brassicae* has been identified from various countries using different differential host sets and designating pathotypes in their own way without following any sound uniform system of nomenclature. The genomes of *P. brassicae* pathotypes Pb2, Pb5, and Pb8 have been sequenced to gain insight into genome variations and its correlation with host specificity. Proteomics of *P. brassicae* has revealed potential effectors/genes of pathotypes. The isolates of *P. brassicae* from different countries have been phylogenetically analyzed for similarities and differences of pathotypes. Polymorphism in *P. brassicae* in case of Korean isolates collected from field has been recorded. The *Cr811* gene in *P. brassicae* pathotypes P5 plays an important role as molecular markers for identification of race P5 and other new virulences. The different geographical isolates of *P. brassicae* have been identified by rDNA sequence polymorphism. Pathogenic and genetic variations in *P. brassicae* pathotypes infecting a weed *C. flexuosa* and crucifers have been analyzed using RAPDs in Japan. The virulence of *P. brassicae* changes during interaction with CR genes in the host cultivars. The changes in *P. brassicae* pathotypes structure under field conditions have been analyzed through whole genome DNA similarity sequences. The application of RAPD seq has revealed distinct *P. brassicae* populations under Canadian conditions. ITS sequencing and phylogenetic analysis of *P. capsellae* isolates from *Brassica* hosts clearly differentiated them in to virulence factors/ races. High level of genetic diversity and recombination's in *Sclerotinia* populations in tropical countries has been identified with the groups of isolates as MCGs and MLHs infecting crucifers. The aggressiveness of *Sclerotinia* isolates on *Brassica* has been measured using AUDPC as one of the parameters. The sequence variation of ITS region of *S. sclerotiorum* isolates revealed higher heterogenecity and genetic variability showing presence of clonal and sexual progenies of the pathogen in *B. juncea* growing areas of India. The genetic diversity is higher in *Sclerotinia* populations from Canola and Sunflower in the NCUS. Distribution of microsatellite haplotypes regions is often on multiple crops. The genetic diversity and populations structure of *S. sclerotiorum* has been analyzed with DNA (RAPD) sequences from different countries of the world. The virulence factors of *Sclerotinia* has been identified on the basis of pathogenicity as virulent pathotypes (aggressiveness), genomic factors (dsRNA) controlled Hypo virulence, MCGs, cloned variables, haplotypes, genetic diversity of isolates, and population biology studies with markers such as MCGs, DNA finger printing or micro satellites. The *TuMV* isolates collected from different hosts could not be characterized on *Brassica* species for virulence. The molecular characterization of isolates from different species of *Verticillium* indicated that isolates infecting crucifers are long spored (amphihaploids or allodiploids) because of *Verticillium* interspecific hybridization events. Black rot of crucifers' bacterium has been characterized into pathovars of *X. campestris* on different species of crucifers. The *X. campestris* isolates from oilseed rape are more genetically diverse with specialization to their nonhost than their *Brassica* hosts. The virulent effectors/genes of pathogens have different evolutionary mechanisms to counteract with the defense mechanisms of the host. The identification, classification, and utilization of crucifer's pathogen pathotypes from different countries is most challenging task since scientists have not followed standardized uniform system and procedures in selecting host differentials, designation and nomenclature of pathotypes, pure isolates (single spore), and scoring of infection phenotypes. The expression of high genetic diversity by crucifers' pathogens under natural as well as modern agriculture settings is another challenge of variation in pathogenic populations.

Keywords

Pathogenomics · Pathogenic variability · *Albugo* Pathogenomics · Polyploidy to *Albugo* race diversification · Phylogenetic relationship · Virulence spectrum of virulence genes · Inheritance of avirulence · *Alternaria* pathogenomics · Genetic variation · Mutation of *Amr1* gene · Evolution of virulence · Identification of pathological variations · *Colletotrichum* pathogenomics · *Erysiphe* pathogenomics · Identification of pathotypes · *Hyaloperonospora* pathogenomics · Identification of virulence effectors/genes · Heterothallism and homothallism · *Leptosphaeria* pathogenomics · Evolution of virulence · Avirulence alleles · Genetic variability · Characterization of avirulent mutated *AvrLm* gene · Identification of genome-wide DNA variants · SNP haplotype · *Plasmodiophora* pathogenomics · Differentials for identification of pathotypes · Genome comparison of pathotypes · Proteomics of pathotypes · Phylogenetic analysis of pathotypes · Molecular marker for identification · Identification of geographical isolates · Pathogenic and genetic variation · Whole genome DNA of pathotype · *Pseudocercosporella* pathogenomics · Identification of virulence factors/races · *Sclerotinia* pathogenomics · Genetic diversity · Structure · Random amplified polymorphic DNA markers · *Turnip mosaic virus* pathogenomics · Pathological and biological characteristics · *Verticillium* pathogenomics · *Xanthomonas* pathogenomics · Identification of pathotypes · Evolution of virulent effectors · Challenges of pathotype identification · Genetic variation

5.1 Introduction

Pathogenic variability in the pathogens of crucifers has been determined in the form of races, pathotypes (fungal pathogens), pathovar (bacterial pathogens), strains (viral pathogens), and genetical variants (All pathogens). The standard procedure to detect pathogenic variability is to use isogenic lines of the host differentials and record differential interactions of different isolates of the pathogen. It is based on gene for gene interaction of host genotypes with pathogen isolates. The pathogen avirulent gene recognizes *R* gene in the host. Based on this differential interaction of host–pathogen genes, a race or pathotype is designated to identify pathogenic variability. These races and pathotypes differ in their virulence and aggressiveness to cause severe epidemics in different crucifer crops depending on host resistance genes in a cultivar and favorable environmental conditions for the development of the pathogen. Under natural conditions and modern agricultural settings, there is evolution of pathogenic variability in the form of new and virulent races through selection pressure of *R* genes in the genotypes/varieties of crucifer's crops during different stages of host–pathogen interactions. The brighter side of pathogenic variability is that a race or pathotype is excellent tool to identify *R* genes loci and *R* genes alleles on a locus of *Brassica* genotypes. It facilitates development of multigene-resistant cultivars to counteract so-called Boom and Bust cycle of resistant cultivars on the evolution of new virulent races of the pathogens. The propensity of a pathogen to evolve virulence depends on features of its biology (mode of reproduction) and its genome (amount of repetitive DNA). Avirulence effector genes in the pathogen are key pathogenic genes that are complementary to *R* genes of the host.

The plant disease resistance genes (*R* genes) that have been molecularly characterized so far can be grouped into several classes based on similarities in the function or amino acid sequence of the proteins they encode. The class containing the majority of cloned *R* genes is characterized by the presence of an N-terminal nucleotide-binding site (NBS) and a C-terminal stretch of leucine-rich repeats

(LRRs). A number of genes belonging to this class have been cloned, such as the *RPS2* (Bent et al. 1994; Mindrinos et al. 1994) and *RPM1* (Grant et al. 1995) genes (against *Pseudomonas syringae*) and the *RPP5* gene (against *Hyaloperonospora parasitica*) (Parker et al. 1997) from *Arabidopsis thaliana*, the *N* gene from tobacco (against *Tobacco Mosaic Virus*) (Whitham et al. 1994), the *PRF* gene from tomato (against *Pseudomonas syringae*) (Salmeron et al. 1996), and the *L6* gene from flax (against *Melampsora lini*) (Lawrence et al. 1995). Although these *R* genes are found in diverse plant species and are active against a range of pathogens, the conserved NBS-LRR features suggest a common function in the defense response against pathogen attack, probably as part of the signal transduction pathway (Staskawicz et al. 1995).

The isolation and characterization of more of these genes is important because they may provide clues about the complex mechanisms of resistance, the interactions involved in pathogen recognition, and the evolution of the *R* genes. Furthermore, cloned genes can be transferred to other species (Rommens et al. 1995; Thilmony et al. 1995) to study the resistance mechanism in a completely different genetic background. It is known from hybridization experiments with the cloned NBS-LRR type of *R* genes that several related sequences can be found, both in the same species as well as in other species (Lawrence et al. 1995; Mindrinos et al. 1994; Parker et al. 1997; Salmeron et al. 1996; Whitham et al. 1994). Isolation of these sequences by polymerase chain reaction (PCR) may be attainable. Despite the general lack in DNA sequence conservation between *R* genes, there are a few conserved protein motifs present in the NBS. With degenerate primers based on these homologous regions, it is possible to amplify several resistance-gene-like (RGL) DNA fragments, as has been shown for soybean and potato (Kanazin et al. 1996; Leister et al. 1996; Yu et al. 1996). Some of these DNA fragments have been mapped in the vicinity of known disease resistance loci. A similar approach in *Arabidopsis thaliana* has been applied with a degenerate primer based on the conserved amino acid sequence of the so-called P-loop motif and another degenerate primer downstream based on the sequence of the conserved domain 5 (Aarts et al. 1998; Lawrence et al. 1995). The *Arabidopsis thaliana* is a good model species to test the general applicability of a PCR-based isolation method, as well as many of the plant–pathogen interactions in this species have been studied (Crute et al. 1994). The genetic positions of many resistance loci have been determined (Kunkel 1996; Holub 1997), and some resistance gene homologues are represented in the extensive EST database (Aarts et al. 1998).

Much of our current knowledge of plant infecting fungi is derived from narrow host range pathosystems, which often exhibit particular characteristics resulting from a pronounced evolutionary arms race between host and pathogen. This arms race often results in pathogens evolving virulence genes that are counteracted by corresponding resistance loci in the host, which are themselves counteracted by new or variant virulence genes in the pathogen, ad infinitum. This is termed a "gene-for-gene" interaction. Plant pathogenic fungi that co-evolve with their hosts in a gene-for-gene manner often exhibit compartmentalized genomes that contain alternating no repetitive, gene-rich regions and repetitive, gene sparse regions.

Virulence-associated genes are often clustered in repetitive, gene sparse regions, and it is thought that this clustering and compartmentalization allows for their rapid evolution through transposable element (TE)-mediated duplication and mutation, without affecting housekeeping genes. This theory is known as the "two-speed-genome" hypothesis (Faino et al. 2016). Enhanced mutation rates in repeat-rich regions in fungi may arise through a process termed repeat-induced point mutation (RIP). This process is a genomic defense against excessive TE activity, which could be deleterious if left uninhibited. It causes point mutations in duplicated sequences through 50-methylation of cytosine followed by deamination to thymine. As a result, RIP-affected genomic regions often exhibit pronounced A/T content concomitant with a decrease in G/C content (Smith et al. 2012). Virulence-associated proteins are often found to have been affected by this process, which may be important for their rapid adaptation (Rouxel et al. 2011). Many of the virulence-associated genes of plant pathogenic fungi studied to date encode small secreted proteins, whose sole purpose is to manipulate plant defenses to advance fungal infection (Lo Presti et al. 2015). These proteins, termed "effectors," have been discovered in multiple plant pathogenic fungi and exhibit numerous different functions depending on fungal lifestyle. For example, necrotrophic fungi, which require dead tissue on which to feed, often produce effectors that promote cell death, whereas biotrophic fungi, which require living tissue, produce effectors that prevent cell death (Lyu et al. 2016). Hemibiotrophic fungi, which require both living and dead tissue at different life cycle stages, may produce different effector sat different time points during infection (Rudd et al. 2015). Many effector genes studied are species or even isolate specific and interact directly or indirectly with the products generated by particular intraspecific ally polymorphic loci in the host, thus conforming with the gene-for-gene hypothesis (Bourras et al. 2015, 2016). However, several effectors are also conserved across various fungal species and have seemingly similar roles in pathogenesis on plants, interacting with loci conserved within a species and/or between species (Sanz-Martín et al. 2016; Redkar et al. 2015; Hemetsberger et al. 2015; Derbyshire et al. 2017).

5.2 *Brassica–Albugo* Pathogenomics

5.2.1 Polyploidy as a Contributor to *Albugo* Race Diversification and Evolution

Polyploidy in *A. candida* may have important impacts on both the occurrence of sexual reproduction within and between lineages and the adaptive potential of the lineages. First, polyploidy isolates may have reduced fertility or be strictly asexual (especially triploids; Comai 2005). In triploid isolates from *B. oleracea* (V, race 9) and *R. sativus* (VII, race 1), heterozygous sites are mostly shared between isolates, suggesting strict clonal reproduction. The hypothesis of asexuality in these lineages is reinforced by the lack of observable oospores from isolates collected on *B. oleracea* and propagated in the laboratory (*AcBoT* and *AcBoL*; McMullan et al.

2015). Second, polyploid isolates may not be able to hybridize with other lineages (Pannell et al. 2004; Köhler et al. 2010). This may, in the long term, lead to speciation of the polyploid lineages. Finally, polyploidy may allow for relaxed selection pressure and therefore increased mutation rates at the duplicated genome (s), which may lead to the acquisition of novel gene functions (Comai 2005; Madlung 2012). It may also increase vigor (increased biomass, growth rate) in comparison to diploid races. Interestingly, polyploidy races identified had previously been classified as *A. candida macrospora* due to large sporangia, compared to those observed in *A. candida microspora* (isolates collected on *C. bursa-pastoris* or *Sisymbrium* spp. (Biga 1955; Pound and Williams 1963). It would be interesting to test for polyploidy in other lineages classified as *macrospora* (other *Brassica* spp. or *Erucastrum* spp.) and for an increased sporangia/zoospore resistance or survival in the polyploids (Jouet et al. 2018).

In *Albugo candida*, polyploidy appears to have been selected for only in cultivated host species. This is interesting because polyploidy may allow for (and indeed impose a need for) rapid clonal expansion of an adapted race on the relatively uniform genotypes of cultivated host populations. Jouet et al. (2018) have observed this with crop infecting (mostly polyploid) *A. candida* races showing a relative excess of intermediate frequency variants, both at neutral as well as effector genes. Conceivably, this observation is caused by a genetic bottleneck resulting from resistance breaking by a small number of genotypes during the widespread cultivation of *Brassica* crops in Northern Europe over the last 2000-year (Maggioni et al. 2000). In addition, asexuality associated with polyploidy may be beneficial, as sexual reproduction would break up combinations of alleles that are adapted to the host genotypes. It also can result in a phenomenon known as "frozen heterozygosity" (Schwarz 2017), in which the polymorphism contained within genetic loci, becomes fixed in the asexual lineage, which enables into resist the eroding effects of genetic drift. By contrast, clonal reproduction may be less advantageous for the pathogen in a co-evolutionary arms race with host species that are genetically diverse, such as pathogens infecting wild populations of *C. bursa-pastoris* or *S. officinale* (Jouet et al. 2018).

5.2.2 Pathogenic Diversity in *Albugo*

Molecular studies (Choi et al. 2006; Voglmayr and Riethmüller 2006) have supported not only its wide host range, but also the high degree of genetic diversity within *Albugo* parasitic to Brassicaceae. Subsequently, six new specialized species have been introduced, *A. koreana* from *C. bursa-pastoris* (Choi et al. 2007), *Albugo laibachii* from *A. thaliana* (Thines et al. 2009), *A. voglmayrii* from *Draba nemorosa* (Choi et al. 2008), *A. hesleri*, *A. hohenheimia*, and *A. leimonios* from species of the genus *Cardamine* (Ploch et al. 2010). In addition, *A. lepidii* (Choi et al. 2007) and *A. resedae* (Choi et al. 2011a, b) were confirmed as species distinct from *A. candida*, while *A. chardonii* from Cleomaceae was revealed synonymous with *A. candida* (Choi et al. 2007, 2009). It is believed that more than a dozen distinct species so far regarded to be *A. candida* await discovery (Thines and Voglmayr 2009). On the other hand, the phylogenetic results also demonstrated that the host range of *A. candida* s.str. extends from Brassicaceae to Cleomaceae and Capparaceae (Choi et al. 2007, 2009), but possibly not to Resedaceae (Choi et al. 2011a, b). The three genera, *Alyssum*, *Barbarea*, and *Rorippa*, are widely distributed and of these, about 50 species have been recorded as common host plants of *A. candida* in monographic studies (Wilson 1907; Biga 1955; Kochman and Majewski 1970a, b; Constantinescu and Negrean 1983; Novotelnova and Pystina 1985; Ul'yanishchev et al. 1985; Vanev et al. 1993; Farr and Rossman 2010).

The genetic diversity within the *A. candida* complex from various host plants was investigated by sequence analysis of the internal transcribed spacer (ITS) region of rDNA, and the cytochrome-*c*-oxidase subunit II (COX2) region of mtDNA (Choi et al. 2006). The aligned nucleotide sequences of *A. candida* shared significantly high distances, up to 20.4% and 8.9%, in two genes. The phylogenetic trees showed two separate groups that corresponded to the host genera. Group-I included *A. candida* isolates infecting *Arabis*, *Autrieta*, *Berteroa*, *Biscutella*, *Brassica*, *Cardaminopsis*, *Diplotaxis*, *Eruca*, *Erysimum*, *Heliophila*, *Iberis*, *Lunaria*, *Raphanus*, *Sinapis*, *Sisymbrium*, and *Thlaspi*. Group-II contained all isolates from *Capsella*, *Descurainia*, *Diptychocarpus*, *Draba*, and *Lepidium* (Table 5.1). The genetic similarities between the two genes among isolates within Group-I were 99.0–100% and 99.6–100%, while those within Group-II were 90.4–100% and 91.1–100%, respectively, showing considerably lower values for Group-II than for Group-I.

The *Albugo candida* isolates from *C. bursa-pastoris* in Korea are clearly separated by sequence analysis for the two genes compared to those from Wales, England, and the USA. Based on the molecular data from the two genes, the existence of high degree of genetic diversity within *A. candida* complexes warrants their division into several distinct species (Choi et al. 2006). *Albugo thaliana*, *A. candida*, and *A. laibachii* may occupy the same host within the same environment, but are phylogenetically distinct as inferred from analysis of both mitochondrial and nuclear DNA sequences (Thines et al. 2009). The strains obtained from *B. juncea* and *R. raphanistrum* (wild radish) are different in their host range. The isolate from *B. juncea* showed significant levels of sporulation on both differential cultivars of *B. juncea*, Vulcan and Commercial Brown (used for differentiating

Table 5.1 Diversity within the *Albugo candida*-complex (Choi et al. 2006)

Group-I		Group-II
Arabis	*Iberis*	*Capsella*
Autrieta	*Lunaria*	*Descurainia*
Berteroa	*Raphanus*	*Diptychocarpus*
Biscutella	*Sinapis*	*Draba*
Brassica	*Sisymbrium*	*Lepidium*
Cardaminopsis	*Thalaspi*	
Diplotaxis	*Eruca*	
Erysimum	*Heliophila*	

pathotype 2A from 2V) confirming the presence of pathotype 2V in Western Australia. This same isolate was able to infect *B. napus* from China (FAN 189), and *B. tournefortii* (wild turnip), *B. nigra*, and *R. sativus*. This serves as a caution to breeders when sourcing resistance against *A. candida* from *B. napus* germplasm. The isolate from *R. raphanistrum*, tested against the same set of cruciferous host differentials, caused significant sporulation on *B. juncea* differential Commercial Brown, *B. napus* (FAN 189), *B. nigra* (90745), *R. raphanistrum*, and *R. sativus*. Therefore, the strain from *R. raphanistrum*, while being a direct threat to *B. juncea*, may be a hazard, not only to any *B. napus* germplasm developed from *B. napus* breeding lines from China, but also to *B. nigra* and *R. sativus*, should these species be utilized commercially in Australia (Kaur et al. 2008). Isolates from Alberta and British Columbia in Canada appear to be hybrid pathotypes combining virulence characteristics of pathotypes 2A or 2V with 7A or 7V, since these isolates were pathogenic on both *B. juncea* and *B. rapa* cultivars (Rimmer et al. 2000).

5.2.3 Phylogenetic Relationship Between *Albugo* Species

Many monographic studies have recognized *A. candida* as the sole species infecting *C. bursa-pastoris* (Saharan and Verma 1992). But Choi et al. (2007) have observed that *A. candida* on this host collected in Korea was differentiated from those in other geographical regions on molecular grounds. However, the study included only a small number of specimens from a limited geographical area. Thirty six *Albugo* collections on *Capsella* specimens comprising 8 from North and South America (Argentina, Canada, USA), 8 from Asia (India, Korea, Palestine), 18 from Europe (England, Finland, Germany, Hungary, Ireland, Latvia, Netherlands, Romania, Russia, Sweden, Switzerland), and 2 from Oceania (Australia), were used for phylogenetic and morphological analysis by Choi et al. (2007).

Several studies have already proven that the sequence analyses of the ITS region of nrDNA (Cooke et al. 2000; Constantinescu and Fatehi 2002; Choi et al. 2003; Voglmayr 2003; Goker et al. 2004; Thines and Spring 2005; Spring et al. 2006; Kaur et al. 2011a), and the mitochondrial COX2 gene, which encodes subunit II of the cytochrome c oxidase complex (Hudspeth et al. 2003; Martin 2000; Martin and Tooley 2003), are useful to resolve closely related species within the *Oomycota*. Molecular analysis based on LSU nrDNA was used to uncover the phylogenetic relationship among several species of the *Albuginaceae* (Voglmayr and Riethmüller 2006). In morphological analysis characteristics of the oospore wall, which were previously suggested as useful to identify and compare species of *Albugo* (Choi and Priest 1995; Voglmayr and Riethmüller 2006), were found to be more helpful in distinguishing *Albugo* specimens on *C. bursa-pastoris,* than characters of the sporangiophores and sporangia. The use of morphological and molecular tools to investigate and to correctly identified *Albugo* species infecting *C. bursa-pastoris* and other hosts throughout the world and to help solving the systematic problem of the *A. candida* complex by stabilizing the taxonomy of this pathogen by selecting an appropriate type specimen is suggested (Choi et al. 2007).

Phylogenetic maximum parsimony and Bayesian analyses of 60 collections belonging to 12 species of *Albugo* were performed using nuclear large subunit ribosomal DNA sequences containing the D1 and D2 regions. These data were supplemented with detailed light and scanning electron microscopically analysis of oospore morphology. Molecular data revealed two main clades: one containing the collections from hosts belonging to the *Caryophyllales* and *Asteraceae*, and the other containing the collections from hosts belonging to the *Brassicaceae* and *Convolvulaceae*. Separation into these two clades was also corroborated by oospore morphology. Whereas the *Albugo* collections from *Caryophyllales* did not form a monophyletic lineage, the collections originating from *Brassicaceae*, *Convolvulaceae,* and *Asteraceae* each formed highly supported monophyletic clades. According to DNA sequence data and oospore morphology, the host genus *Amaranthus* harbors two distinct species, *Albugo amaranthi* and *Albugo bliti*. The DNA sequence data further indicate that *A. candida* and *Albugo tragopogonis* each may consist of several distinct lineages (Voglmayr and Riethmüller 2006).

The sequence analysis of *A. candida s.l.*, based on the COX2, mtDNA, ITS rDNA, and 28S rDNA genes, was used to elucidate the phylogenetic structure within this complex. Although *A. candida* is commonly regarded as the sole species causing WR on various plants of the *Brassicaceae*, including *Capsella*, but *A. candida* from *Capsella* was clearly divided into two clades, and that several distinct lineages are present in *A. candida s.l.* The phylogenetic partition between Korean and other continental specimens indicated by two nuclear and one mitochondrial locus suggests that there is a reproductive barrier between these groups, meeting the criteria of GCPSR (genealogical concordance phylogenetic species recognition) (Taylor et al. 2000). Therefore, Korean specimens warrant species rank, based on morphological and molecular data. White rust pathogens from various brassicaceous plants, including *Capsella* of other continental regions, formed a well-supported group in analysis of both COX2, ITS and 28S rDNA. The majority specimens from the *Brassicaceae* are positioned within this large group (group 1). In a study by Voglmayr and Riethmüller (2006), two clades are apparent in *A. candida s.l.*, which are both parasitic to a variety of *Brassicaceae*, and are supported with high bootstrap values. To this dataset, *Albugo* specimens from *Capsella* in Korea were added, so several interesting points can be addressed. Usually, *Albugo* specimens from the same host genus are grouped closely together, with the exception of *Capsella* (Choi et al. 2006, 2007) and *Sisymbrium* (Voglmayr and Riethmüller 2006). In the corresponding phylogenetic reconstruction, *Albugo* from *Sisymbrium loeselii* is identical in sequence with *Albugo* from *Erysimum cheiranthoides*. This might possibly be explainable by a misidentification of the host, which is not highly unlikely, as *E. cheiranthoides* and *S. loeselii* are quite similar in appearance. In the dataset of Voglmayr and Riethmüller (2006), *Albugo* specimens from *Lunaria*, *Capsella*, *Raphanus*, and *Berteroa* are grouped closely together, which is in accordance with the study of Choi et al. (2006), who also found the Korean specimens from *Brassica* and *Raphanus*, which have been reported to host *A. macrospora*, nested in the large, homogenous group of *A. candida s.str.* It is notable that the specimen from *Eutrema* originating from Taiwan, included in the first study, is also placed within this group. From this host species *A. wasabiae* has been described from Japan, a species ignored by subsequent authors. The specimens from *Arabidopsis* are placed in the more heterogeneous, smaller group in Voglmayr and Riethmüller (2006) and in a similar distinct, heterogeneous group in Choi et al. (2006), in the later study; this clade also contained *Albugo* from *Lepidium*, and *Albugo* from the Korean specimens from *C. bursa-pastoris*, which form a homogenous group. In the reanalysis of the dataset of Voglmayr and Riethmüller (2006), to which two *Albugo* specimens from *Capsella* gave similar results, suggesting placement of the Korean specimens from *Capsella* in a third distinct clade (group 3), which is placed sister to the clade containing *A. lepidii* and several further distinct lineages. Based on the molecular and morphological data available, it is believed that this phylogenetic group is to be regarded as new species, *A. koreana*. It seems also likely that other lineages in *A. candida s.l.* are distinct species (Choi et al. 2007).

The taxonomy of species within the *Albuginaceae* has been based primarily upon differences in the host range (Kochman and Majewski 1970b; Vanev et al. 1993), particularly host families. The size of oospores, sporangiophores and sporangia of the *Albuginaceae* may vary according to host plant and climatic factors (Togashi et al. 1931; Makinen and Hietajarvi 1965; Lakra and Saharan 1988b; Jat 1999; Saharan 1995). However, ornamentation of the oospore wall remains the critically important morphological characteristic for distinguishing species of *Albugo* (Biga 1955; Choi and Priest 1995; Voglmayr and Riethmüller 2006). The most useful characteristic for distinguishing *A. koreana* from *A. candida s.str.* is the surface ornamentation of the oospores, which are verrucose or tuberculate without ridges in the first, and with confluent ridges in the later species. As described and illustrated by Wilson (1907), oospores in *A. candida s.l.* can be "verrucose, or with low blunt ridges which are often confluent and irregularly branched"; judging from his drawings, his statement refers to characters which can be found on the same oospore (Wilson 1907). Corresponding to results of the phylogenetic analysis, WR pathogens that infect *Eruca* and *Heliophila* also had the same oospore ornamentation as *A. candida s.str.* Besides, *A. candida s.str.* from *C. bursa-pastoris*, the fungus from *Armoracia rusticana* (Voglmayr and Riethmüller 2006), *R. sativus* (Wilson 1907), *B. rapa* (Tewari and Skoropad 1977), and *B. juncea* (Nath et al. 2000) have been previously observed to have identical oospore characteristics.

Therefore, oospores with low blunt ridges that are often confluent and irregularly branched might be seen as a morphological characteristic of *A. candida s.str.* infecting various brassicaceous plants, including *Capsella* from countries other than Korea, and oospores without blunt ridges as a characteristic of Korean specimens of *Albugo* from *Capsella*. *Albugo* materials from five genera, *Capsella*, *Draba*, *Eruca*, *Heliophila*, and *Lepidium*, from which oospores were successfully observed, were divided into two types based on wall ornamentation; *Albugo* specimens from *Draba* and *Lepidium* were exhibiting some similarity compared to *A. koreana*, while those from *Eruca* and *Heliophila* were identical to those of *A. candida s.str.* Therefore, the three molecular groups of *A. candida s.l.* found by Choi et al. (2006), and in the amended dataset of Voglmayr and Riethmüller (2006), as well as ITS and COX2 molecular phylogenetic reconstructions by Choi et al. (2007), may be morphologically also differentiated by oospore characteristics, with groups 2 and 3, which are grouped with high support in LSU and ITS trees, exhibiting similar oospore ornamentation. Group 1 most likely contains a single species, *A. candida s.str.*, group 2 comprises several species, among which are *A. lepidii* and several other lineages, which might also warrant species rank, and its sister group (group 3) contains *Albugo* specimens from *Capsella* from Korea, which are assigned to the new species *A. koreana* (Choi et al. 2007).

Molecular studies conducted by Choi et al. (2006, 2007) revealed that *Lepidium* WR is different from *A. candida s.str.* and *Albugo* from *Capsella* in Korea. This supports the conclusion of Rao (1979), who described *Albugo* specimens from *Lepidium* as *A. lepidii* A. N. S. Rao. The presence of several species on the same host genus, on the basis of molecular and morphological properties, are also shown by Voglmayr and Riethmüller (2006), who found that two distinct species were parasitic to *Amaranthus*, based on nrLSU-DNA sequences and oospore ornamentation. As this species is clearly embed within *Wilsoniana* Thines, both morphologically and based on molecular phylogeny, as also shown by Voglmayr and Riethmüller (2006), and a reinvestigation of the dataset of Choi et al. (2007), *A. amaranthi* (Schwein.) Kuntze is combined into that genus below, to avoid nomenclatural confusion. The *A. gomphrenae*, which is most likely also a member of that genus, as molecular data suggests (Voglmayr and Riethmüller 2006), is not transfer to *Wilsoniana* (Choi et al. 2007). *A. candida* races (9) reported from W. Australia is similar phylogenetically. However, the isolate obtained from *B. tournefortii* was phylogenetically most distinct (Petkowski et al. 2010; Kaur et al. 2011b). *Albugo* from *Alyssum montanum*, *Barbarea vulgaris*, and *Rorippa* species is place in three phylo-genetically distinct clades, but closer to *A. candida* s. str. (Choi et al. 2011a, b).

5.2.4 Identification of Pathogenic Variability in *Albugo candida*

Physiologic specialization as virulence factor has long been known in *A. candida*. Eberhardt (1904) recognized two specialized groupings of *Albugo*, one attacking *Capsella*, *Lepidium* and *Arabis*, and the other attacking *Brassica*, *Sinapis* and *Diplotaxis*. He was, however, hesitant to use the phrase "biological forms" for this fungus. Later, Melhus (1911) also suggested the existence of specialization in *A. candida*. Pape and Rabbas (1920) demonstrated that this fungus on *C. bursa-pastoris* should be considered a distinct form. Pfister (1927) found several biological forms in *A. tragopogonis*. Taubenhaus (1923) and Ciferri (1928) reported that distinct races occurred on *I. batatas* and on *Ipomoea* spp. in the USA and West Indies. Ciferri (1928) divided *A. ipomoeae-panduratae* into two biologic species. Savulescu and Rayss (1930) distinguished eight morphological forms within *A. candida*. Later, Savulescu (1946) made ten varieties of *A. candida* based on host specialization and morphology. Hiura (1930) distinguished three biologic forms of *A. candida*, the first on *R. sativus*, the second on *B. juncea*, and the third on *B. rapa var. Chinensis*. Napper (1933) described 20 races of *A. candida* in Britain.

Togashi et al. (1934) found that sporangia of *Albugo* from *Brassica* and *Raphanus* were 20 × 18 μM in size, while those from *Cardamine*, *Capsella*, *Draba*, and *Arabis* measured 15.5 × 14.5 μM; these were classified as macrospora and microspora, respectively. Results of this Japanese study suggested that five distinct biological forms of *Albugo* were present. Subsequently, Ito and Tokunaga (1935) elevated the forms with the larger spores to the rank of the species *A. macrospora* (Togashi) Ito. Biga (1955) recognized two morphological taxa: *A. candida macrospora* and *A. candida microspora*, as proposed by Togashi et al. (1934), but renamed them *A. candida microspora* and *A. candida*, respectively. Based on sporangial measurements from 63 species, Biga (1955) reported that *A. candida microspora* (15–17.5 μM-diam.) is restricted to *Armoracia*, *Brassica*, *Erucastrum*, *Raphanus*, and *Rapistrum*, whereas *A. candida* (12.5–15 μm-diam.) has a wide range of cruciferous hosts. In addition, he identified two intraspecific taxa of *A. ipomoeae-panduratae,* and five of *A. tragopogonis*. *Albugo tragopogonis* was considered by Savulescu (1946) to have two varieties differing in spore shape each of which included morphological forms differing in size. Biga (1955) distinguished five morphological varieties in this species. Endo and Linn (1960) reported one race of *Albugo* on *Armoracia rusticana*.

It is clear that each of the above authors was hesitant in describing specialized races of *A. candida*. Pound and Williams (1963) identified six races of *A. candida*; race one from *R. sativus* var. Early Scarlet Globe, race 2 from *B. juncea* var. Southern Giant Curled, race 3 from *A. rusticana* var. common, race 4 from *C. bursa-pastoris*, race five from *S. officinale*, and race 6 from *R. islandica*. Race 7 from *B. rapa* Turnip or Polish rapeseed, and race 8 from *B. nigra* were added by Verma et al. (1975) and Delwiche and Williams (1977), respectively. In Russia, Novotel'Nova (1968), while analyzing intraspecific taxa, established that the *A. candida* species consists of separate morphological specialized forms confined to a particular range of host plants, i.e. to plants of certain species or groups of genera and species. Within the morphological forms, races can be differentiated, while within heterogeneous populations both races and forms can be differentiated. Geographical and climatic conditions leave their distinguishing mark on the processes of form development so that populations of the fungus encountered by investigators in different countries were not identical. Novotel'nova and Minasyan (1970) and Burdyukova (1980) studied the biology of *A. candida* and *A. tragopogonis* in Russia, and conducted an in-depth study of the extent of specialization in *A. candida*. In India, Singh and Bhardwaj (1984) tested 12 *Brassica* species and identified nine races from four hosts, viz., *B. juncea*, *B. rapa* var. Toria, *B. rapa* var. Brown Sarson, and *B. rapa* var. *Pekinensis.* Lakra and Saharan (1988a, b) identified five races of *A. candida* based on their reaction on a set of 16 host differentials. They identified two distinct races from *B. juncea*, which are different from previous records. One (race 2), attacks *B. nigra*, *B. juncea,* and *B. rapa* var. Brown Sarson, and the other (race 3) infects only *B. juncea* and *B. rapa* var. Toria. According to Petrie (1988) isolates of *A. candida* from *R. sativus*, *B. juncea*, *C. bursa-pastoris* and *B. rapa* represented four distinct races in Saskatchewan, Canada. There nomenclature, however, is not mentioned.

Bhardwaj and Sud (1988) evaluated 26 cultivated and wild cruciferous hosts and identified nine new biological races from nine hosts, viz., *B. rapa* var. Brown Sarson (cv. BSH I), *B. rapa* var. Toria (cv. OK-I), *B. juncea* (cv. Varuna), *B. chinensis*, *B. rapa* var. *Pekinensis* (cv. Local), *B. rapa* (cv. PTWG), *R. sativus* (cv. Chin I 'Pink), *R. raphanistrum* (Wild Radish), and *Lepidium virginicum* (Wild). They reported that reaction of nine isolates of *A. candida* differed from each other on 26 differential hosts revealing thereby, that the monotypic pathogen, *A. candida* on crucifers existed in the form of different biological races designated as new biological races or forms 1–9. The concept of races in *A. candida*, as proposed by Pound and Williams (1963) was based on species relationships. Studies have however, clearly demonstrated that cultivars of *Brassica* crops must be included in a set of host differentials to distinguish isolates of the pathogen within a present accepted race (Burdyukova 1980; Pidskalny and Rimmer 1985). There is an urgent need to standardize host differentials keeping in mind the homogeneity and purity of species and varieties. Petrie (1988) using North American race 2 and 7 from *B. juncea* and *B. rapa*, respectively, have screened accessions of several *Brassica* species including *B. rapa* var. Yellow Sarson, *B. rapa* var. Brown Sarson, *B. rapa* var. Toria, and *B. juncea* from India. Both yellow and brown sarson were equally highly susceptible to both races, toria only to race 7, and *B. juncea* only to race 2. A detailed study is needed to determine whether the races of *A. candida* attacking *B. juncea* and several *B. rapa* crops in India are similar to race 2 and 7 from Canada and the USA. Kolte et al. (1991) reported that the white rust isolate obtained from *B. rapa* appeared to be distinct in pathogenicity from the one obtained from *B. juncea* under Indian conditions. Petrie (1994) in Saskatchewan and Alberta discovered new races 7v in 1988 and race 2v in 1989. Verma et al. (1999) reported two new races of *A. candida* in India *viz.*, race 12 from *B. juncea* and race 13 from *B. rapa* var. Toria using 14 (including six standard) crucifer host differentials.

Mathur et al. (1995) and Rimmer et al. (2000) collected isolates of *A. candida* from different geographic locations in western Canada and tested for virulence on a number of cultivars and accessions of *Brassica* species to determine variability and distribution of different races in the area. Most isolates were identified as race 7, which could be subdivided into 7a and 7v on the basis of their virulence on *B. rapa* cv. Reward. Isolates 28-7 and 29-1 were less virulent as they were avirulent on all the differentials, except the rapid cycling *B. rapa* CrGCI-I8. *B. napus* cv. Tower isolates, 11-6 and 41-4, which could infect cultivars of both *B. rapa* and *B. juncea*, appeared to be hybrids between race 2 and race 7. Wu et al. (1995) studied genetic variation among isolates of *A. candida* using randomly amplified polymorphic DNA (RAPD) with five selected random primers fingerprint patterns generated for each isolates. Most polymorphism was found between different races than among isolates within a single race. Most Canadian field isolates were grouped as race 7 and could be further subdivided into two groups (7a and 7v). Classification of *A. candida* isolates based on the results from the RAPD analysis was identical to the virulence classification on ten *Brassica* differentials. Four distinct and new pathotypes of *A. candida* viz, ACI4 from RL 1359, AC 15, and AC 16 from Kranti, and AC 17 from RH 30 cultivars of *B. juncea* have been identified on the basis of their differential interactions with 11 host differentials by Gupta and Saharan (2002). Jat (1999) identified 20 distinct pathotypes of *A. candida;* 17 from *B. juncea* (AC 18–AC 34), 2 from *B. rapa* var. Brown Sarson (AC 35–AC 36), and one from *B. niger* (AC 37). From Western Australia, Kaur et al. (2008) identified pathotype AC 2A from *B. juncea*, and pathotype AC 2v from *Raphanus raphanistrum*. The pathogenic variability recorded in *A. candida* in the form of races from all over the world are: 9 from Australia, 20 from Britain, 4 from Canada, 2 from Germany, 49 from India, 8 from Japan, 18 from Rumania, and 7 from USA. However, nomenclature of *A. candida* races came into practice after the use of host differentials to distinguish races by Pound and Williams (1963). Global virulence of *A. candida* based on primary host is documented in Table 5.2. In total 117 races of *A. candida* has been reported from different countries (Table 5.3). In *A. candida*, sexual reproduction in the form of oospores is very common and enormous, especially on *B. juncea.* In addition, other mechanisms of variability including recombination, mutation, and heterokaryosis are also in operation in nature. To get the true picture of *A. candida* races and virulence spectrum, there is an urgent need to standardize host differentials for each crucifer species in the form of isogenic lines at international level. Standard nomenclature of the races viz, Acjun 1,2,—for *B. juncea* isolates, AC rap 1,2—for *B. rapa* isolates, AC nig 1,2—for *B. nigra* isolates, and A Col 1,2,—for *B. oleracea* isolates, etc. may be a very useful beginning (Saharan 2010; Saharan et al. 2014).

5.2.5 Virulence Spectrum of Virulence Genes/Pathotypes/Races

As per the gene-for-gene hypothesis, interaction of *Albugo-crucifers* for compatibility and incompatibility phenotype determines number of virulence genes in the pathotype and resistance genes in the host genotype. The virulence of *A. candida* from turnip rape (*B. rapa* var. Turnip Rape) and mustard (*B. juncea*) on various crucifers in Canada has been determined (Pidskalny and Rimmer 1985). It has been observed that pathotypes of *A. candida* from *B. juncea* have wide range of virulence genes. Pathotypes AC 23, AC 24, and AC 17 infect only one, two and three differential hosts indicating a limited virulence potential. However, pathotypes of wider virulence viz., AC 29, AC 27, AC 30, AC 18, and AC 21 infected 21, 18, 16, 12 and 10 host differentials, respectively (Jat 1999; Gupta and Saharan 2002). The intensity of virulence by the pathotypes is expressed through the incubation (IP), and latent periods (LP) of pathotypes interacting with host genotype under the influence of prevailing environmental conditions. Large amount of variability has been discovered in the IP and LP of genotypes from 6 to 17 days and 11–13 days, respectively in the genotypes–pathotypes interactions (Tables 5.4 and 5.5).

Availability of virulence variability in pathotypes from *B. juncea* suggested the possibility of identification of more number of resistant genes in the genotypes including identification of loci and alleles. In the absence of isogenic lines, it is not clear whether the races with wider virulence attack the same genes in the host

Table 5.2 Global distribution of virulence genes of *Albugo candida* pathotypes/races (Saharan 2010; Saharan et al. 2014)

Designate pathotype	Country	International primary host	Reference
AC1	North America	*R. sativus*	Pound and Williams (1963)
AC2	North America	*B. juncea*	Pound and Williams (1963)
AC2V	North America	*B. napus*	Petrie (1994)
AC3	North America	*A. rusticana*	Pound and Williams (1963)
AC4	North America	*C. bursa-pastoris*	Pound and Williams (1963)
AC5	North America	*S. officinale*	Pound and Williams (1963)
AC6	North America	*R. islandica*	Pound and Williams (1963)
AC7	North America	*B. rapa*	Verma et al. (1975)
AC7V	North America	*B. rapa* cv. Reward	Petrie (1994)
AC8	North America	*B. nigra*	Delwiche and Williams (1977)
AC9	North America	*B. oleracea*	Williams (1985)
AC10	North America	*B. alba*	Williams (1985)
AC11	North America	*B. carinata*	Williams (1985)
AC12	India	*B. juncea*	Verma et al. (1999)
AC13	India	*B. rapa* var. Toria	Verma et al. (1999)
AC1 to 9	India	*Brassica* species	Singh and Bhardwaj (1984)
AC1 to 5	India	*Brassica* species	Lakra and Saharan (1988a, b)
AC14	India	*B. juncea* cv. RL 1359	Gupta and Saharan (2002)
AC15	India	*B. juncea* cv. Kranti	Gupta and Saharan (2002)
AC16	India	*B. juncea* cv. Kranti	Gupta and Saharan (2002)
AC17	India	*B. juncea* cv. RH 30	Gupta and Saharan (2002)
AC18 to 34	India	*B. juncea* cv. RH 30; EC 182925; DVS 7-3-1	Jat (1999)
AC35 and AC36	India	*B. rapa* var. Brown Sarson	Jat (1999)

(continued)

Table 5.2 (continued)

Designate pathotype	Country	International primary host	Reference
AC37	India	*B. nigra*	Jat (1999)
AC2A	Western Australia	*B. juncea* cv. Vulcan; Commercial Brown	Kaur et al. (2008)
AC2V	Western Australia	*R. raphanistrum*	Kaur et al. (2008)

Table 5.3 Global number of *Albugo candida* virulence genes/pathotypes/races (Saharan et al. 2014)

Country	No. of pathotypes	References
Australia	9	Kaur et al. (2008, 2011b)
Britain	20	Napper (1933)
Canada	2 + 2 = 4	Verma et al. (1975), Delwiche and Williams (1977), Petrie (1994)
Germany	2	Eberhardt (1904)
India	9 + 4 + 20 + 5 + 9 + 2 = 49	Bhardwaj and Sud (1988), Gupta and Saharan (2002), Lakra and Saharan (1988a), Jat (1999), Verma et al. (1999)
Japan	3 + 5 = 8	Hiura (1930), Togashi et al. (1934)
Romania	8 + 10 = 18	Savulescu and Rayss (1930), Savulescu (1946)
USA	1 + 6 = 7	Endo and Linn (1960), Pound and Williams (1963)

differentials, or genes for susceptibility are different, or situated on different loci or tightly linked (Saharan 2010).

5.2.6 Phylogenetic Relationship of Pathotypes/Races

The prevalence of nine races of *A. candida* in Western Australia (WA) was reported by Kaur et al. (2011b). From comparing the complete ITS rDNA nucleotide sequence of the nine *A. candida* isolates from WA with those of 38 previously collected Australian isolates of *A. candida*, and as well as of those earlier reported from Europe and Asia, it was possible to distinguish distinct groups within these isolates. Despite the small overall sequence divergence, the majority of these races have already been reported from Australia (Petkowski et al. 2010); races analyzed by Kaur et al. (2011b) were phylogenetically similar to those reported previously. However, the isolate obtained from *B. tournefortii* was phylogenetically different from the most distinct isolates analyzed. The phylogenetic relationships identified are expected to help overcome the technical difficulties usually encountered with host range studies, allowing a better understanding of host specialization. The use of a molecular systematic classification of different strains of *A. candida* in conjunction with the pathogenic behavior on a variety of cruciferous hosts, the *A. candida*

Table 5.4 Incubation and latent period (in days) of *Albugo candida* isolates on host differentials under controlled conditions (Gupta and Saharan 2002)

	Incubation and latent period (in days) of *A. candida* isolates obtained from hosts							
	RL 1359		Kranti				RH 30	
	AC-14		AC-15		AC-16		AC-17	
Host differentials	IP	LP	IP	LP	IP	LP	IP	LP
EC-129126-1	–	–	–	–	–	–	–	–
EC-322090	–	–	–	–	–	–	–	–
EC-322092	–	–	–	–	–	–	–	–
EC-322093	–	–	–	–	–	–	–	–
Varuna	9	13	7	12	8	13	7	11
EC-287711	–	–	–	–	–	–	–	–
ZEM-1	–	–	6	11	–	–	–	–
RC 781	–	–	6	11	7	11	–	–
RH 30	17	©	7	11	9	13	6	11
RH 8113	17	©	9	13	7	11	–	–
Rajat	11	14	6	11	–	–	7	13

IP = Incubation period; LP = Latent period; – = No incubation period; © = Hypersensitive reaction

Table 5.5 Virulence spectrum of *Albugo candida* races (Jat 1999; Gupta and Saharan 2002)

A. candida race	Virulence spectrum[a]
AC 14	4
AC 15	6
AC 17	3
AC 18	12
AC 21	10
AC 23	1
AC 24	2
AC 25	9
AC 26	7
AC 27	18
AC 29	21
AC 30	16

[a] Number of host differentials infected

isolates from different hosts in WA provides reliability and repeatability in defining races of *A. candida.* Defining the race status of *A. candida* in WA has provided a sound basis for the selection and deployment of new cultivars of both oilseed and horticultural *Brassica* spp. In particular, this new information is critical not only to the planned rapid expansion of the *B. juncea* industry in Australia, but also in increasing interest to develop alternative *Brassica* spp. for industrial oil production (Kaur et al. 2011b).

5.2.7 Phylogenetic Analyses and Nucleotide Sequence Identities

For the nine *A. candida* isolates from W.A., complete rDNA-ITS region nucleotide sequence analysis showed a range of 72.4–100% nucleotide identity to the isolates used for comparison with previous Australian collections of *A. candida* and also those previously reported in Europe and Asia. The W.A. isolates from *B. juncea* showed close identity to *B. juncea* isolates of *A. candida* from Korea. Isolates from *B. juncea* from other states of Australia were not included in the phylogenetic studies of Petkowski et al. (2010) in Victoria, Australia. The *B. tournefortii* isolate of *A. candida* from W.A. was the most distant from all others compared, forming a distinct clade on its own, with an identity range of 77.4–80.5% compared with the rest of the isolates. *A. candida* isolates from *R. raphanistrum* and *R. sativus* from W.A. were least similar to others, with a nucleotide identity similarity of only 72.4% (Kaur et al. 2011b).

5.2.8 Characterization of *Albugo candida* Virulence Genes on Crucifers

Nine isolates of *A. candida*, representing strains collected from *B. juncea*, *B. rapa*, *B. oleracea*, *B. tournefortii*, *Raphanus raphanistrum*, *R. sativa*, *Eruca vesicaria* subsp. *sativa*, *Capsella bursa-pastoris*, and *Sisymbrium irio*, from different locations in Western Australia (WA), were tested on cruciferous host differentials to characterize their pathogenicity. Pathogenicity tests with appropriate differentials demonstrated the presence in WA of a unique strain from *B. rapa* that did not show characteristics of either race 7A or 7V and clearly is a distinct new pathogenic strain within race 7. Different strains collected from WA differed in their host range, with the strains from *B. tournefortii* and *S. irio* being highly host specific, failing to be pathogenic on any other differentials. *B. tournefortii* was host to as train attacking *B. juncea* and *E. vesicaria* subsp. *sativa*. The strain from *R. raphanistrum* showed a relatively wide host range among the differentials tested. *B. tournefortii*, *C. bursa-pastoris*, *R. raphanistrum*, and *S. irio* are common weeds within grain belt and horticultural regions in Australia. The *B. oleracea* isolate (race 9) was pathogenic to *B. juncea* "Vulcan" whereas the isolate from *B. juncea* (race 2V) was not pathogenic on *B. oleracea*. Similarly, the strain from *C. bursa pastoris* (race 4) was pathogenic on *B. juncea* Vulcan but the *B. juncea* strain was not pathogenic on *C. bursa pastoris*. In contrast, the strain from *R. sativus* (race 1) was pathogenic on *B. juncea* and the *B. juncea* strain was pathogenic on *R. sativus*. Field isolates from *B. rapa*, *B. tournefortii*, *E. vesicaria* subsp. *sativa*, and *S. irio* were all nonpathogenic on *B. juncea*. Isolates from *B. juncea* and *R. raphanistrum* were pathogenic on *B. napus* (FAN 189). For the nine *A. candida* isolates from WA, complete rDNA internal transcribed spacer region nucleotide sequence analysis showed a nucleotide identity range of 72.4–100% in comparison with previous Australian collections of *A. candida* and those previously reported in Europe and Asia. The *B. tournefortii* isolate of *A. candida* from WA formed a distinct clade on its own, with an identity range of 77.4–80.5% compared with the other isolates. Isolates from *R. raphanistrum* and *R. sativus* from WA were least similar to the other isolates, with a nucleotide identity similarity of only 72.4%. Characterization of the races of *A. candida* in Western Australia adds to the current knowledge regarding the diversity of this pathogen, allows choice of *Brassica* spp. or cultivars with resistance to races across different regions, and highlights the particular cruciferous weeds involved in pathogen inoculum carryover between successive cruciferous crops, particularly *B. juncea* crops (Kaur et al. 2011b).

5.2.9 Inheritance of Avirulence in *Albugo candida*

The *Albugo candida* (Pers.) Kunze, a diploid, oomycete biotroph, causes white rust of cultivated *Brassica* spp. At least ten pathogenic races (pathotypes) of *A. candida* have been identified based on specificity to different species of crucifers (Delwiche and Williams 1977; Hill et al. 1988; Pound and Williams 1963; Rimmer et al. 2000). Among the races, race 2 (Ac2) causes severe symptoms on most cultivars of brown mustard (*Brassica juncea* [L.] Czern & Coss.), but also infects some genotypes of other *Brassica* spp., including oilseed turnip rape, *Brassica rapa* L. (Petrie 1988; Pidskalny and Rimmer 1985; Rimmer et al. 2000). Race 7 (Ac7) is largely restricted to *B. rapa*, but has been reported to cause disease on some cultivars of *B. napus* (Fan et al. 1983), and some genotypes of *B. juncea*. The genetic control of resistance to *A. candida* has been studied in several *Brassica* spp. For example, dominant alleles at three unlinked loci (*Ac7-1*, *Ac7-2*, and *Ac7-3*) conferred resistance in *B. napus* cv. Regent to race Ac7 of *A. candida* (Fan et al. 1983; Liu et al. 1996). Two loci also controlled resistance in *B. napes* to *A. candida* race Ac2 collected from *B. juncea* (Verma and Bhowmik 1989). The Chinese *B. napes* accession 2282-9, which is susceptible to Ac7, has one locus controlling resistance to an isolate of *A. candida* collected from *B. carinata* (Liu and Rimmer 1992). These studies indicated that only one allele for resistance was sufficient to condition an incompatible reaction in this path system. In addition, a single locus controlling resistance to Ac2 in *B. napus* and *B. rapa* was mapped using restriction fragment length polymorphism (RFLP) markers (Ferreira et al. 1995; Kole et al. 1996). A dominant allele at single locus or two tightly linked loci was reported to confer resistance to both races Ac2 and Ac7 of *A. candida* (Kole et al. 2002).

The information on the genetics of the pathogen is minimal and genetic studies were considered intractable due to the lack of a reliable method to germinate oospores. Liu and Rimmer (1993) developed an efficient method to germinate oospores of *A. candida* and isolated metalaxyl-insensitive strains derived from Ac2 and Ac7. These developments provided an opportunity to analyze genetically the inheritance of cultivar specificity and metalaxyl resistance and to conduct other genetic studies of *A. candida*. Among oomycete pathosystems, most studies have been focused on heterothallic species such as *Phytophthora infestans* (Shaw 1991), *P. parasitica* (Förster and Coffey 1990), and *Bremia lactucae* (Crute and Johnson 1976; Hulbert et al. 1988; Michelmore et al. 1984; Norwood et al. 1983). For homothallic oomycetes, where inbreeding is common, occasional out crossing has been reported (Francis and St Clair 1993; Whisson et al. 1994) and used to produce F_2 progeny in which both avirulence genes and molecular markers segregated in a Mendelian fashion (Tyler et al. 1995; Whisson et al. 1995). The use of molecular markers based on the polymerase chain reaction (PCR), such as random amplified polymorphic DNA (RAPD) (Williams et al. 1990), is advantageous because it requires no prior information about genomic sequences and is simple compared with other molecular marker methods. The RAPD markers are usually dominant because polymorphisms are detected as the presence or absence of bands. The RAPDs were used successfully as genetic markers in genetic studies of oomycetes (Francis and St Clair 1993; Tyler et al. 1995; Whisson et al. 1995). Amplified fragment length polymorphism (AFLP) markers also have been used for genetic analysis and genome mapping of both plants and pathogens. AFLP markers are allele specific and the technique allows the simultaneous analysis of a large number of DNA and two populations were produced to determine the genetic mechanisms controlling avirulence in *A. candida*. They were derived from crosses MiAc2-B1 × MsAc7-A1 (Cross-1) and MiAc2-B5 × MsAc7-A1 (Cross-2). Because *A. candida* is homothallic, these two progeny populations were expected to consist of both selfed and hybrid spores. The main objectives of this study were to (1) identify F_1 hybrids by phenotypic screen and confirm them with RAPD markers, (2) elucidate the genetic mechanism of avirulence of F_2 populations on *B. rapa* cv. Torch, and (3) analyze the segregation of AFLP markers in F_2 individuals derived from two F_1 hybrids (Cr2-5 and Cr2-7) of Cross-2 (Adhikari et al. 2003).

The in heritance of avirulence and polymorphic molecular markers in *Albugo candida*, was studied in crosses of race 2 (Ac2), using isolates MiAc2-B1 or MiAc2-B5 (metalaxyl in sensitive and virulent to *Brassica juncea* cv. Burgonde) with race 7 (Ac7), using isolate MsAc7-A1 (metalaxyl-sensitive and virulent to *B. rapa* cv. Torch). Hybrids were obtained via co-inoculation on to a common susceptible host. Putative F_1 progeny were selfed to produce F_2 progeny. The parents and F_1 progeny were examined for virulence on the differential cultivars *B. juncea* cv. Burgonde and *B. rapa* cv. Torch. Segregation of avirulence or virulence of F_2 populations was analyzed on cv. Torch. Putative F_1 hybrids were confirmed by random amplified polymorphic DNA markers specific for each parent. Avirulence or virulence of F_2 progeny to *B. rapa* cv. Torch suggested 3:1 in each of three populations, supporting the hypothesis of a single dominant avirulence gene. Amplified fragment length polymorphism markers also segregated in regular Mendelian fashion among F_2 progeny derived from two F_1 hybrids (Cr2-5 and Cr2-7) of Cross-2. This first putative avirulence gene in *A. candida* was designated *AvrAc1*. These results suggest that a single dominant gene controls avirulence in race Ac2 to *B. rapa* cv. Torch and provides further evidence for the gene-for-gene relationship in the *Albugo–Brassica* pathosystem (Adhikari et al. 2003).

5.3 *Brassica–Alternaria* Pathogenomics

5.3.1 Genetic Variation Among *Alternaria brassicae* Isolates

For genotypic identification of pathogen, PCR-based marker techniques were used extensively. DNA markers provide an opportunity for characterization of genotypes and measurement of genetic relationships. RAPD markers are very fast and simple to develop because of the arbitrary sequence of the primer. RAPD markers have been used for phylogenetic relationship among and inside species of genus *Brassica* and its connected genera. A fast assessment of genetic variability and study of intra-specific variability among different isolates of fungus can be easily done by using RAPD markers (Sharma et al. 2013a). Genetic variation in populations of *Alternaria* species has been characterized by RAPD.

The Rapid amplified polymorphic DNA (RAPD) markers was used to know the genetic relationships among the different isolates. Sixteen 10-mer primers of arbitrary nucleotide sequences were tested for amplification of *Alternaria brassicae* DNA by using PCR. Of these nine primers amplified the fungal genomic DNA from all the 20 isolates produced reproducible RAPD profiles. The UPGMA analysis showed that the isolates collected from different regions could be classified into four major group. A close relationship was revealed between two isolates based on the cluster analysis data. The results proved the existence of genetic variability among *Alternaria brassicae* isolates infecting oilseed *Brassica*, which is not based on their location and could be utilized for development of resistance against this pathogen. The genetic variation among geographically distinct isolates of *A. brassicae* has been assessed using RAPD-PCR markers (Chanyal et al. 2018).

5.3.2 Genetic Structure of *Alternaria* Populations Suggest the Occurrence of Sexual Recombination

The genetic basis of resistance in the host, virulence, and aggressiveness in the pathogen, and whether these are qualitatively or quantitatively based, provides a frame work for understanding the nature of the selective forces each player imposes on its partner and the likely rate of evolution any response (Burdon et al. 1996). Equally, the spatial and temporal scale of interactions, the extent of population isolation, and microclimatic differences in the suitability of the environment for disease development, are all likely to affect the epidemiology and genetic structure of pathogen populations (Burdon and Thrall 1999). Finally, knowledge of the structure of host and pathogen populations and their respective recombination systems provides insight into the relative importance of mutation, recombination, and gene flow. In agricultural path systems, various combinations of this information have promoted better understanding of the epidemiology of specific interactions and the capacity of the pathogen to evolve. In such systems, biology, space and time all can have marked effects on the level of recombination that occurs (Taylor et al. 1999). While high levels of pathogen genetic diversity are typically associated with substantial genetic recombination (although mutation and migration also play a role; Burdon and Silk 1997), however, the diversity observed is spatially partitioned depends on the size and connectedness of populations.

Little is known about the genetic structure of *A. brassicicola* populations (Bock et al. 2002). However, the occurrence of substantial genetic variability in a related species, *A. alternata* occurring on tomato crops in the USA (Morris et al. 2000) prompted those authors to speculate variously that diversity was maintained by mutation, uniform host selection, extensive dispersal or the existence of a cryptic sexual stage. Bock et al. (2005) showed genetic differences among small sample of isolates of *A. brassicicola* using AFLP markers (Bock et al. 2002). The extent of diversity in five populations of *A. brassicicola* occurring along an 11 km stretch of Australian coastline has been assessed to know the sources of that variation (Bock et al. 2005). Substantial polymorphism was detected between isolates from five populations of *Alternaria brassicicola* attacking *Cakile maritima* along the New South Wales coast of Australia, with a maximum of two genotypes being shared between population pairs. Ten pair-wise population comparisons, six had no pathogen genotypes in common; only one genotype occurred five times, and most (93%) were found only once. Although, an UPGMA based on Nei's measure of genetic distance separated the five populations, a cluster analysis using individual isolates failed to group them according to population, indicating significant gene flow. An analysis of molecular variance indicated ca 14% of the variation occurred between populations, representing moderate population differentiation over the spatial scale of the study. Tests of the relative contribution of clonality and sexual recombination indicated low, albeit significant levels of link age disequilibrium in all populations. The level of linkage disequilibrium, and the high genotype diversity, provides support for the contention that a hither to unidentified sexual stage might be a significant factor in the life-cycle of *A. brassicicola* (Bock et al. 2005).

5.3.3 Mutation of *Amr1* Gene Causes Increased Virulence

The *Amr1* gene knockout mutants, both gene disruption (*amr1*) and gene deletion (Δ*amr1* and Δ*amr1*; *amr1p-GEP*), are easily recognized by the lack of melanin in their conidia, orange colored colonies, and the secretion of a pink pigment. These phenotypes are similar to those found in mutants of its homologs in other fungi (Kihara et al. 2008). The Δ*amr1* mutants are more susceptible to UV light, and enzyme digestion. Unexpectedly, the *Amr1* gene knockout mutants are more virulent than the wild type, producing larger lesions, and require fewer conidia for successful infection. Mutation of *Amr1* homologs in the phytopathogenic fungi, *Magnaporthe grisea* and *Colletotrichum lagenarium,* produces melanin-deficient conidia. The increased virulence of the Δ*amr1* mutants is unexpected, and unique among melanin deficient mutants of pathogenic fungi.

5.3.4 The Cause of Increased Virulence in Δ*amr1* Mutants

The pink pigment secreted by Δ*amr1* mutants is neither toxic to host plants, nor beneficial for the wild type *A. brassicicola* to infect host plants. Furthermore, melanin deficient mutants of another gene, *abpks7*, show no changes in virulence. Neither the presence of pink pigment, nor the lack of melanin is associated with the increased virulence of the Δ*amr1* mutants. Additionally, there is no difference in germination or vegetative growth between the Δ*amr1* mutants, and the wild type in the presence of various chemicals. However, the mutants are more susceptible than wild type to UV light, and the cell wall degrading enzymes, glucanase. The mutant's ability to respond to stress is not associated with their increased virulence, since plant glucanases are known to be defense enzymes (Cho et al. 2012).

The ability of the mutants to outgrow the wild type when citrus pectin was the major carbon source suggested that the mutants were more efficient in utilizing the protein component of plant cell walls and middle lamellae. The expression profile comparisons identified 30 hydrolytic enzymes genes expressed more in the mutant than in the wild type *A. brassicicola* during pathogenesis. However, they included only two pectate lyase genes and one pectin esterase gene. Other genes up regulated in the mutant compared to the wild type included gene coding for two cutinases, and a lipolytic enzyme, and 20 other various glycoside hydrolases (GH). Interestingly, 6 of the 22 genes in the GH61-gene family are expressed at higher levels in the Δ*amr1* mutant. The GH61 family might have a role in degrading cellulose, lignocelluloses, chitin, or other polysaccharides (Karkehabadi et al. 2008). The GH 61 family may be involved in digesting pectin's side chains, which are structurally diverse in constituting sugars and glycosidic linkages. These may digest other cell wall components, such as cellulose and hemicelluloses along with other glycoside hydrolases. Induction of the *Amr1* gene in *A. brassicicola* did not occur until 64 h post inoculation (hpi). Suppression of the *Amr1* gene might be important during pathogenesis when energy is allocated for colonization of host tissues, and overcoming host defense mechanisms, rather than producing reproductive dispersal structures like conidia. In the wild type, the transcripts of hydrolytic enzyme genes were slightly increased compared to the actin transcripts, as previously reported (Karkehabadi et al. 2008). In addition, when conidiation was initiated, the induction of *Amr1* was accompanied by a sharp increase in transcripts of its downstream genes in the melanin synthesis pathway. There was also small increase in the transcripts of hydrolytic enzymes coding genes in the wild type. However, many of these genes mostly expressing glycoside hydrolases, a few cutinases, and pectate digestion enzymes were expressed at higher levels in the Δ*amr1-4* and Δ*amr1; Amr1p*-GFP mutants, did not express *Amr1*. This suggests that *Amr1* suppressed genes predicted to be involved in the digestion of lipids, and carbohydrates. This occurred during late pathogenesis (88 hpi) when they were still needed to some extent for either killing (Noda et al. 2010) or digesting host tissues. Whether the *Amr1* gene is involved in, the regulation of most hydrolytic enzymes sequentially induced in *A. brassicicola* during pathogenesis has yet to be investigated. Other genes encoding a necrosis inducing protein, two P450s, the aegerolysin peptidase homolog, and other hypothetical proteins might have contributed to the increased virulence of the *amr1* mutant, and warrant further investigation. Interestingly, 24 genes expressed at higher levels in Δ*amr1* were expressed at lower levels in the Δ*abvf19* deletion mutants that displayed reduction in virulence, suggesting opposite roles of the two-transcription factor gene in the regulation of pathogenesis-associated genes. This finding may aid in the selection of specific targets for future analysis of their functions in pathogenesis (Cho et al. 2012).

5.3.5 Evolution of Virulence in *Alternaria brassicicola*

The melanin deficient mutant phenotype, and expression pattern of the *Amr1*, *Brn*, *Brn2,* and *Scd1* genes in *A. brassicicola* suggest that the function of the *Amr1* gene is to regulate melanin synthesis during conidiogenesis. This function is highly conserved in other fungi. In addition to regulating the melanin biosynthesis pathway, *Amr1* negatively regulated virulence_ENREF_1. The evolution of virulence has been a subject of extensive theoretical analysis. The tradeoff hypothesis produced a number of models showing a contested relationship between virulence, and transmissibility. It challenges the hypothesis that strong virulence can seriously damage or kill host plants and threaten pathogen survival, so parasites will ultimately evolve to be less virulent. However, analysis of ecological, and epidemiological factors related to virulence has produced inconsistent results. Therefore, a modified trade off theory predicted that strong parasites became, moderately virulent to their hosts over time, but did not completely lose their virulence. Cho et al. (2012) identified a transcription factor suppressing virulence but enhancing the ability of a fungus to preserve itself in space, and time via melanin synthesis, primarily in conidia. This study provides an example of a transcription factor gene important for survival in nature that plays a critical role in negatively regulating virulence in the necrotrophic plant pathogens, *A. brassicicola.* The suppression of virulence observed is unique, and might have been acquired by this fungus, as this phenotype has not been observed to loss function of mutants *Amr1* homologs in other species of plant pathogenic fungi. An ancestral fungal strain of *A. brassicicola* without the ability to suppress virulence by the activity of *Amr1* could have been a more virulent pathogen. Negative regulation of virulence is rare but exists in other phytopathogenic fungi. Currently, *A. brassicicola* is a prolific saprophyte and necrotroph, efficiently colonizing susceptible, weakened or dead host plants. Its pathogenicity, however, is inhibited on more resistant host plants. Cho et al. (2012) speculated that the suppressive functions of *Amr1* contribute to the specialized adaptation of *A. brassicicola* as an efficient and successful facultative parasite (Saharan et al. 2016).

5.3.6 Identification of Pathological Variations/Pathotypes

Out of four species of *Alternaria* known to occur on crucifers, *Alternaria brassicae* (Berk.) Sacc. is more severe and variable (Verma and Saharan 1994). Preliminary reports on variability in *Alternaria* species were reported from Holland (Van Schreven 1953) and UK (Mridha 1983). Isolates of *A. brassicae* obtained from rapeseed (colza) show differences in cultural growth on cherry agar and were differ in their pathogenesis on seedlings of colza. The pathogenic variability in *A. brassicae* has been observed by various workers (Verma and Saharan 1994), however, information on existence of distinct pathotypes using standard host differentials is rather limited (Saharan 1992a, b). According to Mridha (1983), 13 isolates of *A. brassicae* tested on selected cultivars of winter rape differed in their virulence. Similarly, Kolte et al. (1989, 1991), and Awasthi and Kolte et al. (1989) also reported variability in *A. brassicae*.

None of these workers used different Brassica species to distinguish isolates of *A. brassicae* on the basis of their reaction on host differentials. Saharan and Kadian (1983) used eight commonly cultivated *Brassica* species to distinguish isolates of *A. brassicae* infecting *Brassica* species in India. In a cross infection study, they could distinguish between three isolates showing differential reactions. These isolates were distinguished as RM1, RM2, and V3. All these three races are found to be virulent on rapeseed and mustard group. Mehta et al. (2003) collected ten isolates from different agro climatic zones of India and inoculated on a set of 17 host differentials. These isolates behaved differentially i.e., isolate DLK appeared to be most virulent infecting 16 differentials followed by RSR-1 and GDP which caused infection on 15 host differentials but could not distinguish various isolates into pathotypes.

Eleven genotypes of *Brassica juncea* were selected as differentials to identify pathotypes of *Alternaria brassicae* causing blight disease of Indian mustard by Gupta et al. (2004). Four distinct pathotypes of *A. brassicae* viz., Bj-4 (BWL), Bj-5 (HSR), Bj-6 (RTK), and Bj-7 (REW) were identified. Pathotypes Bj-4 was most virulent infecting 11 host differentials and Bj-5 was least virulent which infected only six host differentials. Incubation and latent periods exhibited greater variability for host genotype × isolate interactions of *Alternaria brassicae* isolates. Minimum incubation period of 3 days was observed for pathotypes Bj-4 (BWL) and Bj-6 (RTK) on commonly cultivated cultivars like Varuna and RH-30. Vishwanath and Kolte (1997) recorded differential interaction between *Brassica* crop species and three different (A, C, and D) isolates of *A. brassicae* which shows significantly lower degree of average disease score against D isolate indicating that the D isolate is an avirulent pathotypes of *A. brassicae*. *Alternaria brassicae* isolate A shows significantly higher disease score than the C isolate on *B. napus* PPNS1, *B. juncea* cv. PR15, *B. campestris* var. toria cv. PT 303, PT 30, *B. campestris* var. yellow sarson cv. T-151, whereas the isolate C shows significantly higher degree of average disease score on *B. campestris* var. Yellow Sarson cv. PYST-6, *B. campestris* ssp. *rapifera* cv. Turnip red and *B. alba* in comparison to isolate A. *Alternaria brassicae* A isolate is a highly virulent pathotype and C isolate is a moderately virulent pathotype. The toxigenicity of three isolates of *A. brassicae* on leaves of various hosts shows isolate A causes more severe symptoms than the isolate C and D at both 1:10 and 1:100 dilutions. However, isolate C toxin produces significantly higher symptom severity score on *E. sativa*. Toxin from both A and C isolates give significantly minimum symptoms severity score on *E. sativa*, whereas toxin from D isolate give maximum symptom severity score on *E. sativa* but it does not produce symptoms on leaves of other host cultivars. Isolate A toxin give significantly less seed germination and minimum plumule and radical lengths as compared to isolate C and D at 1:10 and 1:100 dilutions. However, some differences among different genotypes are observed with respect to seed germination and seedling growth with respect to toxins produced by the three isolates.

Pathogenic diversity of fifteen isolates of *Alternaria brassicae* collected from different locations of Haryana (India) infecting rapeseed and mustard were tested on a set of 17 host differentials in green house conditions. All these isolates behave differently on host differentials. Isolates CHR-CHR-II, JND-II, JHR, SRS, and I are most virulent infecting all the host differentials followed by REW, RTK and SPT causing infection on 16 host differentials. Isolates BWL, BHI, and KTL-II infects all the differentials except *E. sativa* and *B. alba* whereas isolate HSR is least virulent as it produces symptoms only on 13 host differentials. Out of 17 host differentials, six differentials namely *B. juncea*, *B. carinata*, *B. nigra*, *B. oleracea* var. *botrytis*, *B. rapa,* and *B. alba* differentiated 15 isolate in to eight pathotypes/races. The eight pathotypes identified are CHR-I, CHR-II, JND-II, JHR, SRS fall in first group; CHR-III, JND-I, KTL-I in second group; BHI, KTL-II in third group and BWL, HSR, REW, RTK, and SPT separately as each group, respectively (Kumar et al. 2003).

On host differentials, each isolate collected from different locations of India behaves differentially. Isolate DLK is most virulent as it infects 16 host differentials; RSR-I and GDP on 15 host differentials; HSR-I, HSR-III and RSR-II on 14 host differentials and isolates HSR-II, GNR, LDH and KNR infects only 13 differentials. The comparative study reveals that all the differentials are susceptible to different isolates of *A. brassicae*. The genotype *Brassica alba* is infected by only four isolates whereas genotypes *B. oleracea* var. *botrytis* is infected by five isolates. The range of incubation period varies from 3 to 13 days in isolates but in general, majority took 3–5 days to cause the infection (Mehta et al. 2003). Sangwan and Mehta (2007) analyzed virulence pattern of 24 *A. brassicae* isolates on a set of 17 host differentials consisting of *Brassica* species. The isolates BTD, BBK, DSA, GNR, HSR, and PNT appeared to have wide virulence pattern as these isolates can infect all the 17 host differentials. Isolates BHP, BRT, GDP, HSRP, JPR, NGN, *B. alba,* and Midas-1 are able to infect 16 host differentials. Virulence of isolates *B. chin* and VRN is confined to 13 host differentials. The host differential *B. alba* variety "Local" is least susceptible as only 12 isolates are pathogenic on it. In terms of incubation period, majority of the isolates took 3–5 days to initiate the disease whereas, isolates ASM, BHP, FRD, JPR, and RSR took longer period to produce symptoms on cruciferous vegetables. Eight host differentials differentiated all the 24 isolates into 14 categories and each category can be designated as one pathotype/race. Meena et al. (2012)

measured aggressiveness of *A. brassicae* isolates on host differentials in the form of lesion size. The host differentials *B. alba*, *B. juncea* (PAB, EC-399299), *E. sativa*, *B. carinata,* and *B. napus* expressed minimum lesion size.

Pathogenic variability of purified 98 isolates collected from *Brassica* species were analyzed to identify virulent pathotype for screening and breeding resistant genotypes of oilseed *Brassica* by Singh et al. (2013). Eight types of variants were grouped based on resistant and susceptible reactions, incubation and latent period, lesion size and *Alternaria* blight severity. Isolates from Rewari and Fatehabad districts are able to infect all host differentials followed by Bhiwani district isolates on ten differentials and minimum numbers of host differentials (7) are infected by Rohtak district isolates. Minimum incubation (IP) and latent period (LP) of 4–5 days and 6–7 days respectively are required by Rewari district isolates and maximum IP and LP 6–8 days and 8–10 days respectively are observed from Rohtak district isolates. Average maximum size of spots formed by Rewari district isolates and minimum size of spot is also induced by Rohtak district isolates. Similarly, maximum Alternaria blight severity (24.6%) is expressed by Rewari district isolates (group-1) and least by Rohtak district isolates (group-6). *Brassica juncea* var. Varuna contracted maximum disease severity (24.6%) and minimum by *Brassica alba* (2.9%). Therefore, Rewari isolates are most virulent, while Rohtak district isolates are least virulent. Hence, Rewari district isolates can be used for screening of oilseed *Brassica* germplasm. *B. alba* can be harnessed as donor parent for breeding resistance/tolerant variety of oilseed *Brassica* against *Alternaria* blight.

5.3.6.1 Symptomatological Variations

The symptom variability exhibited on leaves of different host differentials of *B. juncea* due to *A. brassicae* isolates is in the form of medium size, circular, greyish brown spots, 6–8 mm in diameter, with three regular concentric raised rings of dark brown color. No yellow halo is produced (BWL). Another type of symptom produced by other isolate is in the form of large size, circular spots (8–10 mm), light green in center with concentric rings with dark yellow halo around the spot (RTK). Third isolate produced large size, irregular spot of 10 mm in diameter, dark brown in center and grayish around the margins with concentric rings without yellow halo (HSR). In fourth case no yellow halo is produced, medium size spots, light green in color with only one concentric ring and very light pale ring around the spots is observed (REW). The symptoms produced by four different isolates are stable and distinct on *Brassica juncea* host differentials. Each distinct pathotype produces its characteristic symptom on *B. juncea* host differentials. This shows that symptom variability is a function of specific pathotype rather than host differentials genetic variations (Gupta et al. 2004). These pathotypes are designated as Bj-4 (BWL), Bj-5 (HSR), Bj-6 (RTK) and Bj-7 (REW) in order of their discovery as suggested by Saharan and Kadian (1983). Kolte et al. (1991) identified three pathotypes A, C, and D on the basis of virulence and some spot characters viz., spot color, periphery color, presence or absence of concentric ring and central region of the spot. Goyal et al. (2013) reported pathogenic variability among *A. brassicae* isolates on host genotypes, on the basis of many qualitative characters such as spot color, periphery color, central point color, presence or absence of concentric rings, central point, yellow halo region and one quantitative character viz., percentage disease severity. Three characters i.e., central point color, presence or absence of central point and yellow halo region of the spot should be used to study variability among *A. brassicae* isolates. Pathogenic variability test conducted on 12 host differentials reveals that all the isolates collected from rapeseed-mustard are pathogenic or aggressive at different rates on all the host differentials and produces different types of spots on different hosts (Saharan et al. 2016).

5.3.6.2 Morphological and Cultural Variations

Various isolates were collected from different locations on the basis of morphology of *A. brassicae* spores indicating place of collection viz., Hisar (HSR-I)- *Brassica juncea* (var. RH-30); HSR-II- *B. campestris* var. *yellow sarson* (var. YSPb-24); HSR-III- *B. tournefortii* (var. Local); Sri Ganganagar (GNR)-*B. juncea* (var. Kranti); Ludhiana (LDH)-*B. juncea* (var. RH-30); Kanpur (KNR)-*B. juncea* (var. RH-30); Dhaula Kuan (DLK)-*B. juncea* (var. RH-30); Gurdaspur (GDP)-*B. juncea* (var. RH-30); R.S. Pura (RSR-I)-*B. juncea* (var. RH-30); (RSR-II)-*B. juncea* (var.RH-30), which were analyzed for their morphological and cultural variations by Mehta et al. (2003). The morphological characteristics of each isolate such as size i.e. length, breadth, number of septa, beak length, beak septa etc. was recorded from 15 days old culture. Isolates differed in their spore size. Based on spore length, isolates are categorized into four groups i.e. small (<100 μm); medium (101–150 μm); long (151–200 μm), and very long (>200 μm). The group includes I-GDP; group-2- HSR-I, HSR-III, GNR, KNR, RSR-I; group 3- LDH, DLK and group 4-HSR-II and RSR-II. The longest spore length is observed in case of HSR-II and RSR-II >200 μm and shortest in case of GDP (94.45 μm). The breadth fall in range of 13.5–36.0 μm with the maximum spore breadth in GNR and minimum in case of HSR-II.

The number of septa (horizontal) varies between 5 and 13, and maximum in case of HSR-II and minimum in case of GDP. The maximum beak length is observed in case of HSR-II, and minimum in case of GNR. The number of septa in beak varies from zero to six. Morphological variations i.e. conidial size in the isolates of *A. brassicae* is due to nutrition rather than a characteristic pathological variation (Saharan and Kadian 1983). However, glaring differences in conidial size are noticed among the isolates even when same medium is used for the growth of the isolates. It can be assumed that variation in the isolates may be inherent since isolates were collected from diverse agro climatic zones. Each isolate differed in their conidial size. Hence, these variations in the conidial size indicate the existence of variability in this pathogen. Conidial/spore measurement recorded on each isolate from Haryana (India) revealed that isolates differed in their conidial size. The average conidial length varied from 118.62 to 194.52 μm, being maximum of isolate RTK and minimum of isolate HSR. The range of conidial size varied from 81 to 300 μm. The average breadth of conidia also varied from 14 to 23 μm. The thickest conidium was of isolate JND-I and thinnest of isolate CHR-III. The septation both horizontal and vertical did not show any variation. However, horizontal septation varied from 3 to 12 and vertical from 0 to 6. Some variations in beak size were also recorded. The average beak length varied from 39.99 to 119.07 μm. The longest beak was of isolate RTK and smallest of JND-I. The average beak septation varied from 1.9 to 5.8. The variations in the conidial size indicated the existence of variability in pathogen from Haryana (India).

Variation in morphology and cultural characteristics among 13 representative Indian geographical isolates collections of *Alternaria brassicae* were analyzed by Goyal et al. (2011). All the isolates show high level of variability in vitro in respect of conidial length, width, beak length, and number of septa. Conidia of Nazirhat isolate (SS 04) are smallest with lowest number of septa. Substantial variation is found in mycelial growth, sporulation among these isolates in different nutrient media and artificial environmental conditions, viz., temperature, relative humidity, light and hydrogen ion concentration. Different temperature ranges (25–30 °C; 15–35 °C) are found optimum for different isolates for mycelial growth and sporulation, respectively. All the 13 isolates grow best at 100% relative humidity. However, they sporulate the most at different percentage relative humidity (40–100%). This reflects the adaptation of the respective isolates to the ambient conditions in the different cropping areas, where the disease occurs in varied proportions in different years, which may have also induced the available cultural variability. All the isolates do not grow and sporulate abundantly on the same nutrient medium. However, on an average Asthana and Hawker's medium is good for all the cultures. Variation in optimum pH and light condition for mycelial growth, sporulation is also observed. Cluster analysis for data on cultural variability among 13 *A. brassicae* isolates found a close relationship among isolates from Uttar Pradesh, Uttaranchal, and Haryana but distantly related to other (Goyal et al. 2011).

Morphological characteristic of different isolates of *Alternaria brassicae* reveals variation in growth, shape, and pigmentation of colony, conidial measurements and number of septa. Conidial length varies from 106.7 to 285.9 μm, width 33.5–57 μm, and beak length 41.4–180.0 μm. Number of horizontal septa varies from 3.2 to 8.0 and vertical 0.3–1.4. Different synthetic media shows profound variation in mycelial growth of *A. brassicae* isolates and poor sporulation indicates that fungus requires some organic sources of nutrition for better growth and sporulation. It seems that the degree of sporulation of *A. brassicae* isolates is a function of nutrition. Pathogens aggressiveness demonstrates the existence of considerable variation in tolerance to *Brassica* species to *A. brassicae* (Meena et al. 2012). Variation in morphology and cultural characteristics among 32 representative Indian geographical isolates of *Alternaria brassicae* from cauliflower and rapeseed-mustard were observed by Sharma et al. (2013a). All the isolates show high level of variability in vitro in respect of conidial length, width, and number of septa. Conidia of Uttar Pradesh isolate (CaAB U4) are smallest with lowest number of septa. Substantial variation is found in mycelial growth, and sporulation among these isolates in different nutrient media. All the isolates do not grow and sporulate abundantly on the same nutrient medium. However, Potato Dextrose agar, cauliflower (host) agar medium, and carrot Potato agar are good for all the cultures. Variations in mycelial growth, and sporulation are also observed. Cluster analysis of data on cultural variability among 32 *A. brassicae* isolates found a close relationship among isolates of both origins viz., from cauliflower and mustard. Isolates from Uttar Pradesh, Delhi, Haryana, and West Bengal are found to be similar to each other whereas the Rajasthan isolates along with Tamil Nadu and Kerala isolate are distantly related to others. All the isolates are pathogenic in nature but directly related to the cultural and morphological characteristics. These isolates are further molecularly characterized by using internal transcribed spacer region where all the isolates are found 56% similar to each other and 99% similar to the *A. brassicae* isolates present in NCBI database.

Alternaria brassicae colonies varied in their cultural behavior ranging from cottony, flurry to feathery, with smooth to rough margins. Color of the colonies ranged between white, off white to light brown. Colony growth varies from slow, medium to fast with fast growth in isolate KM and least in JD on the potato dextrose agar (PDA) medium. Significant morphological variations in conidial length, width and number of horizontal septa are observed in the isolates. Average conidial size ranged from 105 to 135 × 10–20 μm. Isolates exhibit variations in disease index, number, and size of the lesions. The dendogram analysis based on molecular (DNA, RAPD) basis reveals two groups at 14% similarity coefficient. Group I is composed of seven isolates namely VR, DV, P7, LM, P10, KR, and ND with 18% similarity (82% dissimilarity) while group II is composed of only three isolates namely JD, KA and AS with only 24% similarity (76% dissimilarity) (Pramila et al. 2014).

5.3.7 Identification of Pathotypes Using Molecular Approaches

5.3.7.1 Genetic Variability

Genetic variability in nucleotide sequence of ITS region of four *Alternaria* species (*A. brassicae*, *A. brassicicola*, *A. raphani*, and *A. alternata*) infecting crucifers has been recorded. Cluster analysis of pathogenic variability data reveals a close relationship between Nazirhat (SS 04), Jaipur (SS 07), Sachha Khera (SS-10), and Samalakha (SS-11) isolates. Use of 100 Random Amplified Polymorphic DNA decamer primers indicates genetic variability among 13 *A. brassicae* isolates. Almost all the isolates show relationship according to their geographical origin except Sachha Khera (SS-10) and Hatikhuti (SS-05) isolates. Pantnagar (SS-09) isolate was found closely related to Sachha Khera (SS-10) isolate. No variability could be located among the *A. brassicae* isolates by Internally Transcribed Spacer—Amplified Fragment Length Polymorphism Molecular marker. Hence, pathogenic variability do exist among the isolates with variability also at genomic level but not in the highly conserved region of the genome of the pathogenic isolates of *A. brassicae* (Goyal et al. 2013). However, Internal Transcript Spacer analysis done by Sharma et al. (2013a) shows that all isolates are 90–100% similar to each other, indicating genetic similarity among different *A. brassicae* isolates that vary pathogenically. However, analysis of 26 RAPD primers revealed a high level of genetic variability among ten isolates of *A. brassicae* from different cultivars of *B. juncea* (Pramila et al. 2014).

The RAPD analysis is more easy, efficient, fast, and reproducible than RFLP analysis in the detection of intra-specific variation in three *Alternaria* species, i.e., *A. brassicae*, *A. brassicicola,* and *A. raphani* pathogenic to crucifers. Many workers (Sharma and Tewari 1995, 1998; Kumar et al. 2008) have described polymorphism within an *Alternaria* species by RAPD molecules marker. Sharma and Tewari (1995) observed polymorphism among *A. brassicae* isolates from different geographical regions of the world. However, in 1998, they found low intraregional variations among Indian and Canadian isolates of *A. brassicae* with 75% similarity among them. However, RAPD analysis of *A. brassicae* isolates from different geographical regions of India by 100 primers suggested a high degree of polymorphism among them. The dendograms from both pathogenic and molecular analysis seems to indicate that the Pant Nagar (SS-09) and Hatikhuti (SS-05) isolates are quite different from the others and the two dendograms follows the same trend (Goyal et al. 2013). BLAST analysis of the ITS of 32 *A. brassicae* isolates conducted by Sharma et al. (2013a) showed high similarity among the isolates with *A. brassicae* of the NCBI database.

5.3.7.2 Proteome Analysis

Two isolates of *A. brassicae* with significant differences in virulence have been characterized at the proteome level. The morphological observations indicated the Ontario isolate to be more virulent by virtue of increased disease severity score as compared to the UAMH7476 isolate. This was further confirmed through histological observations that showed extensive colonization of the host tissue by the highly virulent isolate. Mycelial protein profiles of two differentially virulent *A. brassicae* isolates were compared using two dimensional gel electrophoresis (2DE) and mass spectrometry (MS) in order to identify proteins that may be responsible for the differences. Several differences in the mycelial proteomes of the two isolates were recorded. The proteins that were significantly abundant in the more virulent isolate included a protein with conserved actin related protein 2/3 domain, enolase, malate dehydrogenase and serine protease. The differential protein expression pattern can be exploited to identify putative virulence and pathogenicity factors in *A. brassicae* (Sharma et al. 2010).

5.3.8 Biochemical Variations

Biochemical constituents of *A. brassicae* isolates differ significantly (Khurana et al. 2005). The isolate BWL contains maximum ortho-dihydric phenols (OD phenols) and KTL-II has the least. Isolate REW contains the highest amount of total phenols. The variation in amount of total phenols as well as ortho-dihydric phenols indicates the existence of variation in isolates. The isolate JD-II contains the maximum amount of reducing and nonreducing sugars while the isolate RTK has the minimum amount. The amount of reducing sugars present in different isolates also varies significantly in different isolates. The isolates BWL, CHR-II, CHR-I and RTK have significantly less amount of reducing sugars though least is in case of isolate

BWL (0.74 mg). Variations in amount of sugars in isolates can be categorized into three groups as maximum, moderate, and minimum levels. Vishwanath and Kolte (1997) had reported that *A. brassicae* isolate containing high amount of carbohydrate is an indicator of virulent nature of pathogen. The estimation of total RNA content in different isolates reveals significant differences among the isolates. Isolate KTL-II has the maximum amount of RNA while minimum is recorded in isolate SPT. On the basis of RNA contents, isolates can also be grouped in to three categories. The protein content in different isolates of the pathogen shows significant differences. Isolate CHR-I and JD-I contains the maximum amount of proteins. Isolate KTL-I has the least amount of proteins. Similarly, there are significant differences among the isolates in their free amino acids content. Isolate JD-I have the highest amount of free amino acids. Isolate SPT shows minimum amount of free amino acids. *Alternaria brassicae* isolates which have more amount of proteins contain moderate amount of free amino acids and RNA contents while isolates which has high amount of RNA also contained moderate amount of proteins and free amino acids. On the basis of similarity in biochemical composition, all isolates were grouped into three categories, maximum, moderate and minimum. Isolates BHI, JHR, KTL-I, SRS, CHR-I, and SPT were placed in the first group containing maximum amount of bio-chemicals. Isolates BWL, RTK, CHR-II, HSR, KTL-II, and REW were in the second group being moderate amount while isolates JD-I and JD-II were in the third group being minimum. The study reveals the considerable variability in the isolates of *A. brassicae* on the basis of biochemical contents.

5.3.9 Variations in Fungicidal and Plant Extracts Sensitivity

The variation in field performance of fungicides in the control of *Alternaria* disease of crucifers may be due to response of pathotypes prevalent in a region. According to Vishwanath and Kolte (1997), isolate A of *A. brassicae* shows more tolerance to Ziram and Ridomil-72 at 50 and 100 ppm as compared to isolate C. Similarly, isolate A exhibits maximum tolerance followed by isolates C and D against mancozeb and iprodione. Efficacy of different fungicides and neem products were tested against fifteen isolates of *Alternaria brassicae* collected from different locations of Haryana (India). The fungicide Kitazin was highly effective against all the isolates in inhibiting the spore germination. It was followed by Dithane M-45 and Ridomil MZ-72. Similarly, among four neem products tested, Achook and Bio-neem were quite effective as compared to Furpume and Nimbicidine. Variations were also observed among isolates in their sensitivity against these fungicides. Isolates BHI, CHR-I, and CHR-III were sensitive to all the fungicides whereas JHR was sensitive only to Dithane M-45, Kitazin, and Bavistin. In case of neem products, HSR isolate was not sensitive to Achook whereas Bio-neem proved to be effective against CHR-I, CHR-III, HSR, KTL-I, KTL-II, REW, and SPT. All the isolates according to their sensitivity against different fungicides and neem products fall more or less in same group. Isolates were categorized into eight groups on the basis of their differential sensitivity to fungicides and neem formulations. These groups were designated into eight pathotypes/races present under Haryana, Indian conditions infecting rapeseed and mustard. Variations in the sensitivity of 14 isolates of *A. brassicae* to the plant extracts of Bougainvillea, garlic, Lawsonia, neem, mint, and Eucalyptus have been recorded (Kumar et al. 2003, 2004).

The response of the *A. brassicae* isolates collected from various parts of India were also evaluated against various fungicides and results revealed that among the ten fungicides used, Emisan-6 in general proved most effective inhibiting the spore germination followed by Ridomil MZ-72. The least inhibition was observed in case of Sulfex and Blitox. Among the isolates PNT, BHP, CAUL and *B. alba* were highly sensitive to fungicide whereas isolate FRD, B. chin, and ASM were least affected. The differential behavior of the isolates was quite evident in terms of their sensitivity even to most commonly used fungicide Dithane M-45 where the spore germination among isolates varied from 17.1% to 43.59% (Sangwan and Mehta 2006). Differential behavior of various isolates also indicated that isolates BHP, and *B alba* were more sensitive followed by CAUL, TRN, PNT and HSR. The isolates RC-781, FRD, B. chin, ASM, GRN, and GDP responded similarly.

5.3.10 Virulence Variability of *Alternaria* species on Crucifers

The variability in virulence across the *Alternaria* spp. population associated with Alternaria leaf spot on rapeseed determines incidence, severity, and consequent defoliation. The potential of *A. japonica* and *A. ethzedia* to be significant pathogens of rapeseed in Australia, in addition to *A. brassicae* and *A. alternata* has been demonstrated. All *Alternaria* spp. isolates tested were virulent and the level of virulence varied across isolates within and between *Alternaria* spp. in relation to the disease incidence, severity, and consequent defoliation. *A. japonica* was the most virulent among the species found on rapeseed in Australia, followed by *A. brassicae*, *A. ethzedia*, and *A. alternata*. The remaining *Alternaria* spp. expressed varying levels of virulence dependent upon the disease assessment parameter, i.e., disease incidence, severity, or defoliation. *Alternaria hordeicola* caused least disease and defoliation across isolates of the ten *Alternaria* spp. In general, *A. brassicae* has been categorized as the most virulent pathogen across Brassicaceae hosts (Nowicki et al. 2012; Saharan et al. 2016). Furthermore, the variation across isolates of some *Alternaria* spp. is well known. Goyal et al. (2013) noticed the pathogenic variability of 13 isolates of *A. brassicae* on 12 differential host *Brassicas*; tested isolates varied in aggressiveness across the differentials and produced different lesion symptoms on different hosts (Table 5.6). Furthermore, Siciliano et al. (2017) found three distinct virulences for *A. alternata* on cultivated rocket (high, moderate, and low virulence) and two distinct of virulences for *A. japonica* on wild rocket (high and moderate virulence). Finally, variability in virulence has also been demonstrated in *A. alternata*, *A. brassicicola*, and *A. raphani* on Brassicaceae in other studies (Atkinson 1950; Stoll 1952; Czyżewska 1971; Nowakowska et al. 2019) (Table 5.6).

The *Alternaria* spp. investigated, it is to noteworthy that *A. japonica* was the most virulent species and caused severe defoliation on rapeseed. This species has been

Table 5.6 Determinants of variability in *Alternaria* infecting crucifers (Saharan et al. 2016)

Determinant attributes	*Alternaria* species	Host	Pathotypes	Reference
Pathological	*A. alternata*	Crambe	A, B, C	Czyzewska (1969), Czyżewska (1971)
	A. brassicae	Crucifers	RM-1, RM-2, V-3	Saharan and Kadian (1983)
	A. brassicae	Rape	13	Mridha (1983)
	A. brassicicola	Cauliflower	3	Stoll (1952)
	A. raphani	Radish	Wild variants	Atkinson (1953)
	A. brassicae	*B. juncea*	Bj-4, Bj-5, Bj-6, Bj-7	Gupta et al. (2004)
	A. brassicae	*B. juncea*	DLK, RSR-1, GDP	Mehta et al. (2003)
	A. brassicae	*B. juncea*	A, C, D	Vishwanath and Kolte (1997)
	A. brassicae	*B. juncea*	8	Kumar et al. (2003)
	A. brassicae	*Brassica* spp.	10	Mehta et al. (2003)
	A. brassicae	*Brassica* spp.	14	Sangwan and Mehta (2007)
	A. brassicae	*Brassica* spp.	8	Singh et al. (2008)
Symptomatological	*A. brassicae*	*Brassica* spp.	Bj-4, Bj-5, Bj-6, Bj-7	Gupta et al. (2004)
	A. brassicae	*Brassica* spp.	A, C, D	Kolte et al. (1991)
	A. brassicae	*Brassica* spp.	12	Goyal et al. (2013)
Morphological Cultural and Nutritional	*A. brassicae*	*B. carinata*	A, B, C, D	Kolte et al. (1989, 1991)
	A. brassicae	*Brassica* spp.	4	Mehta et al. (2003)
	A. brassicae	*Brassica* spp.	4	Goyal et al. (2011)
	A. brassicae	*Brassica* spp.	5	Meena et al. (2012)
	A. brassicae	Cauliflower; Rapeseed-mustard	2	Sharma et al. (2013a)
	A. brassicae	*B. juncea*	2	Pramila et al. (2014)

(continued)

Table 5.6 (continued)

Determinant attributes	*Alternaria* species	Host	Pathotypes	Reference
	A. brassicae	Colza	–	Van Schreven (1953)
	A. brassicae	Rapeseed-Mustard	2	Mehta et al. (2005)
Biochemical	*A. brassicae*	Rapeseed-Mustard	3	Khurana et al. (2005)
	A. brassicae	Rapeseed-Mustard	3	Vishwanath and Kolte (1997)
Genetical	A. species (4)	Crucifers	Genetical similar in ITS region	Jasalavich et al. (1995); Goyal et al. (2013)
	A. brassicae	*B. juncea*	Vary pathogenically	Sharma et al. (2013a)
	A. brassicae	*B. juncea*	Isolates genetically variable	Pramila et al. (2014)
Molecular	*A. brassicae*	Crucifers	Polymorphism in isolate by RAPD analysis	Sharma and Tewari (1995, 1998)
	A. brassicicola	Crucifers	-do-	Goyal et al. (2013)
	A. raphani	Crucifers	-do-	Kumar et al. (2008)
Proteome level	*A. brassicae*	Crucifers	Variation in protein level of virulent and avirulent isolates	Sharma et al. (2010)
Thermo Sensitivity	*A. alternata*	Crambe	A, B, C	Czyzewska (1970)
	A. brassicae	*B. juncea*	3	Kumar et al. (2003)
Fungicidal sensitivity	*A. brassicae*	Crucifers	A, C, D	Vishwanath and Kolte (1997)
	A. brassicae	Crucifers	8	Kumar et al. (2003, 2004)

frequently reported as a pathogen on Brassicaceae in Australia on rapeseed and mustard (Al-lami et al. 2019); on Chinese cabbage, hoary stock (*Matthiola incana*), wild radish (Australian Plant Pest Database 2019) and is considered a significant pathogen in Italy on wild and cultivated rocket (Garibaldi et al. 2011; Gilardi et al. 2014; Siciliano et al. 2017), as well as on turnip (*B. rapa* subsp. *rapa*) in Spain (Bassimba et al. 2013). *Alternaria brassicae*, *A. alternata*, *A. brassicicola*, and *A. raphani* are also significant pathogens on Brassicaceae and are virulent across different plant stages resulting in significant yield loss, particularly at mature leaf, stem, and silique stage infections (Saharan et al. 2016). The severe disease caused by *A. alternata* and *A. brassicae* reconfirmed their importance as pathogens on Brassicaceae both in Australia (Sampson and Walker 1982; Shivas 1989; You et al. 2005; Van de Wouw et al. 2016; Al-lami et al. 2019; Australian Plant Pest Database 2019; Farr and Rossman 2010) and elsewhere (Farr and Rossman 2019).

The pathogen *A. ethzedia*, levels of disease incidence, severity, and consequent defoliation level of 84.1%, 89.5%, and 89.4%, respectively, were indicative of moderate to high virulence rapeseed. This pathogen has been reported in Australia by Al-lami et al. (2019), with >40 isolates collected from 2015 to 2016 surveys of rapeseed crops across southern Australia. This high lights the capability of *A. ethzedia* to be an important pathogen of rapeseed. As Farr and Rossman (2019) reported for *A. ethzedia* on rapeseed in Switzerland, it is known to have a wide host range, including Poaceae (Andersen et al. 2009; Farr and Rossman 2019). More than 50% of *A. conjuncta* and *A. infectoria* and >65.0% of *A. metachromatica* isolates caused moderately severe disease. This is in agreement with the investigation of Al-lami et al. (2019) where both *A. metachromatica* and *A. infectoria* were the predominant *Alternaria* spp. recorded in 2015–2016 rapeseed crop surveys, with 30 and 59 isolates of *A. infectoria* and 50 and 55 isolates of *A. metachromatica* recorded in 2015 and 2016, respectively. However, more than 70.0% of *A. hordeicola* isolates showed only low virulence and little defoliation on rapeseed, but this species was found to be relatively common on rapeseed in the Australian 2016 rapeseed survey (Al-lami et al. 2019). It is likely that *A. conjuncta*, *A. hordeicola*, *A. metachromatica*, and *A. infectoria* are pathogens with little host specialization. This is perhaps due to their saprophytic and/or opportunistic growth, as observed by Thomma (2003), who found *Alternaria* spp. in weakened plants and noted that alteration of *Alternaria* sp. behavior from saprophytic to parasitic may not be distinguished. Additionally, whereas Scheffer (1989) identified three types of *Alternaria* spp., i.e., saprophytic, specialized opportunistic, and virulent, and also highlighted them as opportunistic pathogens with a broad host range on older and enervated plants, but still with ability to initiate appressoria and penetrate living tissues.

In a different study, McRoberts and Lennard (1996) noted host specificity at pre-and post-penetration stages of *Alternaria* spp. development. They also distinguish had between *Alternaria* spp. associated with *B. napus* (*A. brassicae*, *A. brassicicola*, *A. raphani*) and those associated with *Solanum lycopersicum* (*A. solani*). This difference was highlighted even for nonspecialized species such as *A. alternate* and *A. infectoria*, where *A. alternata* showed a close association with *S. lycopersicum* but *A. infectoria* showed close association with *Papaver rhoeas*. More than 60% of *A. eureka* isolates showed moderate incidence and severity with accompanying defoliation. This pathogen is also probably a nonspecialized species as there is no concomitant host specialization known for this species (Farr and Rossman 2019). However, there are few reports of this pathogen, although it has been found on *Medicago rugosa* in Australia (Australian Plant Pest Database 2019), on *Dactylis glomerata* in Spain and *Malus domestica* in South Africa (Farr and

Rossman 2019). Likewise, for *A. tenuissima*, >50% of isolates showed moderate incidence and severity with accompanying defoliation. In Australia, this pathogen has been reported on *B. campestris* (Australian Plant Pest Database 2019) and *B. napus* (Al-lami et al. 2019); and on non-Brassicaceae hosts *Vaccinium corymbosum* (You et al. 2014) and *Malus* sp. (Harteveld et al. 2013). Elsewhere, *A. tenuissima* has been reported on various non-Brassicaceae hosts (Farr and Rossman 2019). The *A. tenuissima*, together with *A. alternata* are widespread pathogens without host specialization that are also known to be both saprophytic and pathogenic species (Farr and Rossman 2019). Smith et al. (1988) noted that *A. alternata* and *A. tenuissima* were common saprophytic species on various plant materials, including infected flowers, fruits, leaves, and seeds, and they noted the disease association between *A. alternata* (syn. *A. tenuis*) and *A. tenuissima* with *Pleospora* spp.

There is variation in relative levels of virulence, in terms of disease incidence, severity and defoliation, within and between *Alternaria* spp. associated with leaf spot on rapeseed. These variations in virulence explain the relative importance of different *Alternaria* spp. pathogen populations associated with rapeseed, not only in Australia but probably elsewhere. Four *Alternaria* spp.—*A. alternata*, *A. brassicae*, *A. ethzedia*, and *A. japonica*—were overall the most virulent species. In contrast, *A. conjuncta*, *A. eureka*, *A. infectoria*, *A. metachromatica*, and *A. tenuissima* expressed varying levels of virulence according to the disease parameter used: disease incidence, severity, or accompanying defoliation. The *A. hordeicola* was the least virulent species. It is evident that there is great fluidity within the *Alternaria* spp. population on rapeseed, not just in terms of variation in virulence, but also in that, the species population composition is strongly influenced by location and year (Al-lami et al. 2019). Across these, ten *Alternaria* spp., there is clearly need for further investigation into *Alternaria*–host specificities, between-species interactions, within-species pathotypes, and a need to understand the drivers that determine the changes in pathogen components within the *Alternaria* spp. population in relation to rapeseed (Al-lami et al. 2019).

5.3.11 Identification and Nomenclature of Pathotypes

Physiologic races or pathotypes of plant pathogens are identified on the basis of infection types produced by them on specific set of cultivars called "Differentials." The procedures and problems involved in the collection of diseased samples, isolation, and purification of cultures, maintenance of specific isolates, techniques of inoculation, and scoring of infection types have been described (Verma and Saharan 1994). In biotrophs, host–pathosystem (*Puccinia*–wheat, *Melampsora*–flax) norm and standards of selection of host differentials acceptable at international level has been followed, but it has not been met in the studies conducted in *Alternaria*–crucifers system. Selection of standard host differentials consists of a set of host varieties termed "differentials," supplemented differentials (additional host varieties), single gene lines, and near isogenic lines. Use of such a set of host differentials can clear the picture of presence, and identification of pathotypes in *Alternaria* spp. infecting crucifers.

Nomenclature of a race or pathotype has been done earlier as: (1) Arbitrary numbers: races are generally designated as number, or letters in an arbitrary manner, generally in order of their discovery e.g., cereal rusts; (2) Black's nomenclature: A race is designated on the basis of its virulence on a host resistance gene, e.g., *Phytophthora*-potato system, race R1, R2, Race R1, 4 etc.); (3) Virulence formulae: It is based on a race virulent, and avirulent of particular gene for resistance e.g., the formula 6,7,10/5,8,9a,11 for a race of *Puccinia* virulent on Sr 6, Sr7, Sr10, but avirulent on Sr5, Sr8, Sr9a, and Sr11; (4) Habgood nomenclature; and (5) Virulence analysis. However, out of these criteria of nomenclature, *Brassica* researchers have adopted first method in isolation (not taken in account earlier reports) giving their own arbitrary numbers not keeping parity with others and order of discovery. Even researchers had hesitation in designating of pathotypes of *Alternaria* species except Saharan and Kadian (1983), and Gupta et al. (2004). The procedure and method of *Alternaria* species pathotypes designation adopted by Gupta et al. (2004) seems to be logical since it is based on interaction of a pathogen isolates with one specific genotype of a host species. Apparently, it meets gene for gene hypothesis in the absence of standard monogenic/isogenic host differentials sets. The determinant attributes used by different workers for identification of pathotypes of *Alternaria* species infecting crucifers are given in Table 5.6. The utility and advantages of race/pathotype identification in *Alternaria* can boost *Brassica* production through: (1) Development of resistant cultivars; (2) Identification of new genes for resistance; (3) Development of multigene resistant cultivars, and (4) Identification of favorable gene combinations (Singh and Chand 1983; Saharan et al. 2016).

5.4 *Brassica–Colletotrichum* Pathogenomics

5.4.1 Expression of Virulence Effectors/Genes

Genome and transcriptome analyses of *C. higginsianum* infecting *A. thaliana* have shown that this fungus has many virulence factors. However, relatively few molecular determinants of virulence in *C. higginsianum* have been experimentally verified (Table 5.7). The function of two effectors (genes), ChELP1, and ChELP2 homologs of Lysm proteins were characterized in *C. higginsianum*. ChELp2 is located on the surface of bulbous biotrophic hyphae at the interface with living host cells. Co-ordination in expression and secretion of antagonistic bio trophy effectors/genes and toxin effectors contribute to fungal virulence and to the regulation of hemibiotrophy in *C. higginsianum*. The role of secreted effector proteins during infection by hemibiotrophic plant pathogens is poorly understood.

Based on deep transcriptome sequencing and computational mining of Expressed Sequence Tags from precise infection stages, a large of planta-expressed effector candidates were founding *C. higginsianum*. Most biotrophy-associated ChEC genes were dramatically upregulated exclusively in planta and distinct sets of effectors are

Table 5.7 Genes involved in virulence of *Colletotrichum higginsianum* (Yan et al. 2018)

Gene	ID	Description	Reference
path-19	CH063_11554	Putative Arg6 precursor	Takahara et al. (2012)
path-35	CH063_15109	Carbamoyl–phosphate synthetase	Takahara et al. (2012)
Ch-MEL1	unknown	Hypothetical protein	Liu et al. (2013)
ChPma2	CH063_09060	Plasma-membrane proton-efflux P-type ATPase	Korn et al. (2015)
ChELP1	CH063_13023	LysM effectors	Takahara et al. (2016)
ChELP2	CH063_04445	LysM effectors	Takahara et al. (2016)
ChSte7	CH063 02455	Serine/threonine protein kinases	Yuan et al. (2016)
ChRgf	CH063_04363	Ras guanine-nucleotide exchange factor	Gu et al. (2017)
ChMK1	CH063_08490	Fus3/Kss1-relatedMAPKgene	Wei et al. (2016)
ChMfs1	CH063_12120	Major facilitator super family (MFS) transporter	Liu et al. (2017)
ChMob2	CH063_12012	Mob1/phocein family protein	Schmidpeter et al. (2017)
ChCbk1	CH063_12968	NDR/LATS kinase	Schmidpeter et al. (2017)

deployed in successive waves by particular fungal cell types. With fluorescent protein, tagging and transmission electron microscopy-immunogold labeling, early expressed effector proteins are observed to be focally secreted from appressorial penetration pores before host invasion. In addition, later-expressed effectors accumulate in structures formed at the interface between primary hyphae and living host cells, implicating these hyphae in effector delivery. Furthermore, the coordinated expression and secretion of antagonistic biotrophy effectors and toxin effectors contribute to fungal virulence and the regulation of hemibiotrophy in *C. higginsianum*. These findings indicate new functions for fungal infection structures that have not been reported previously, specifically the localized release of effector proteins at the interface between fungal pathogen and plant host, and associated with the penetration pore. This provides the basis to model the switch to necrotrophy from biotrophy by this fungus (Yan et al. 2018). Future research should attempt to decipher the nature of the plant signals inducing effector gene expression and the way that they are sensed by the pathogen (Heath 2000).

5.5 *Brassica–Erysiphe* Pathogenomics

5.5.1 Identification of Pathotypes of *Erysiphe*

To determine whether a powdery mildew pathogen growing on one host species can cause infection on another host species, which is known to be infected by same morphological species, has been studied extensively. Neger (1902) was the first person to show that some of the morphological species of powdery mildews are broken into biologic forms, limited to definite host plants. Marchal (1902, 1903); Reed (1905, 1907, 1908, 1909, 1912), Salmon (1903a, b, 1904a, b, c, 1905a, b), Steiner (1908) and Voglino (1905) have very extensively studied regarding the occurrence of biological/physiologic specialization in different powdery mildews. The cereals powdery mildews, *Erysiphe graminis* has been exceptionally favorable host–pathosystem for studies on biologic specialization/pathogenic variability. The biologic forms are restricted to the species of a single genus. Conidia from wheat powdery mildew will not infect barley, oats, and rye. Similarly, conidia from barley powdery mildew will not infect wheat, oats, and rye. In other words, in every case, the powdery mildew on one cereal is unable to infect the species of other cereals. It is also known as pathogenic variability or host specificity of powdery mildew pathogens. In fact, all individuals produced because of sexual process are expected to be different from each other, and from their parents in a number of characteristics, although they retain most similarities with then, and belong to the same species. It is true in case of powdery mildews reproducing through sexual fruiting bodies (chasmothecia), and ascospores infecting crucifers. However, in asexually reproducing powdery mildew through conidia, frequency, and degree of variability among the progeny are reduced greatly but even then, certain individuals among the progeny will show different characteristics. The mechanisms of variability may be hybridization, mutation, and cytoplasmic inheritance. However, so far, pathogenic variability or host specificity has not been records in any of powdery mildew fungi infecting crucifers. However, variations in the development of *E. cruciferarum* on two cvs. *B. napus* has been observed by Munro and Lennard (1982). The genotypes of *A. thaliana* (Adam and Somerville 1996; Adam et al. 1999; Plotnikova et al. 1998; Göllner et al. 2008; Kuhn et al. 2016), *B. juncea*, *B. napus,* and *B. carinata* have been recorded to express infection phenotypes (Singh 1994; Uloth et al. 2017). Similarly, Xiao et al. (1997) observed differentiated interaction of *E. cruciferarum* isolates UEA1 on 50 accessions of *A. thaliana*. In India, powdery mildew pathogen of mustard completes its anamorph (asexual) and teleomorph (sexual) stages on *B. juncea* under field conditions (Saharan and Kaushik 1981), there is every possibility for evolution of pathogenic variability with more virulent strains of *E. cruciferarum* (Meena et al. 2018).

5.6 *Brassica–Hyaloperonospora* Pathogenomics

5.6.1 Identification of Virulence Effectors/Genes/Pathotypes

The identification of pathogenic variability seems to be initiated in 1944 with the observation of Wang (1944) who classified the reaction of the hosts into four categories: (1) susceptible with normal symptoms; (2) resistant, showing large necrotic spots; (3) para-immune, showing slightly visible necrotic dots; and (4) immune, with no visible symptoms. Three pathotypes of *H. parasitica* have been differentiated: *H. parasitica Brassicae* on *Brassica*, *H. parasitica Raphani* on *Raphanus,* and *H. parasitica Capsellae* on *Capsella*. The three pathotypes were not mutually compatible with each other's host. Six forms of *H. parasitica Brassicae* were differentiated by their reaction to *B. chinensis*, *B. oleracea*, *B. juncea,* and *B. napobrassicae*. Wang (1944) prepared a dichotomous key to physiological forms from China as follows:

A. *Capsella bursa-pastoris*, immune
B. *Raphanus sativus*, immune or para-immune variety Brassicae
C. *B. oleracea*, resistant or para-immune
D. *B. chinensis*, susceptible
E. *B. juncea* (Meitan Dav Yu Tsai), susceptible Ph. fm. 1.
EE. *B. juncea* (Meitan Dav Yu Tsai), resistant
F. *B. napobrassica*, immune Ph. fm. 2.
FF. *B. napobrassica*, resistant Ph. fm. 3.
DD. *B. chinensis*, resistant
E. *B. juncea* (Dav Ching Tsai), susceptible Ph. fm. 4
EE. *B. juncea* (Dav Ching Tsai), resistant Ph. fm. 5
CC. *B. oleracea*, susceptible Ph. fm. 6
BB. *R. sativus*, susceptible variety *Raphani*
AA. *C. bursa-pastoris*, susceptible variety *capsellae*

Felton and Walker (1946), and Natti (1958) differentiated the races of *H. parasitica* found on *R. sativus*, and *B. oleracea* on the basis of their host specificity. Morris and Knox-Davies (1980) also indicated distinct races of *H. parasitica* on *B. oleracea,* and *R. raphanistrum* based on host specificity. *H. parasitica* f. *brassicae* on cabbage, f. *rapae* on turnip, f. *rapiferae* on *B. rapa*, f. *rapifera*, f. *napi* on *rape*, f. *raphani* on radish, and f. *sinapsidis* on *Sinapis alba* have been distinguished as special forms of *H. parasitica* from Leningrad though all are similar morphologically (Dzhanuzakov 1963).

Three varieties of *H. parasitica* have been differentiated in 35 samples of downy mildew from *B. pekinensis,* and other crucifers, namely f. sp. *brassicae* on *Brassica*, f. sp. *raphani* on *Raphanus*, and f. sp. *capsellae* on *Capsella* (Chang et al. 1964). *Hyaloperonospora parasitica* f. sp. *brassicae* exists in at least three different subforms (*pekinensis*, *oleracea,* and *juncea*). Isolates from *B. pekinensis*, *B. chinensis,* and turnip were classified in the same group, and can attack all three hosts but did not infect *Capsella bursa-pastoris*, radish, cabbage, Chinese mustard, and *B. juncea* var. *multiceps*. *B. juncea* var. *megarrhiza* expressed various reactions to these isolates. Isolates from Chinese mustard, *B. juncea* var. *megarrhiza*, and *B. juncea* var. *multiceps* were limited to these hosts, except that those from Chinese mustard did not infect some vars. of *B. pekinensis*, *B. chinensis,* and turnip. Radish isolates are of two types, one infects only radish, the other vars. of *B. pekinensis*, cabbage, and turnips as well as radish. Isolates from cabbage, and *C. bursa-pastoris* are host specific. In Norway, cross-inoculation experiments with downy mildew from cabbage, turnip rape, and radish indicated the occurrence of different races on cabbage, and radish (Semb 1969).

According to Natti et al. (1967) the predominant physiologic race of *H. parasitica* pathogenic to broccoli, and other types of *B. oleracea* grown commercially in New York were race 1, and race 2. The later race was pathogenic to plants resistant to race 1. Dickinson and Greenhalgh (1977) observed a wide variation in the reaction of seedlings of different crucifers' species to isolates of *Hyaloperonospora* derived from *Brassica,* and *Raphanus* species. Masheva et al. (1996a) found that the pathogenicity of two isolates of *H. parasitica* (One from Plovdiv, and one from Gama Oriahovitsa, Bulgaria) to wards ten hybrids of cabbage, broccoli, and cauliflower differed in pathogenicity, and host specificity.

In India, *Hyaloperonospora parasitica* isolates from different hosts vary in host range. Isolates from *Brassica*, *Raphanus*, *Eruca,* and *Sisymbrium* are not cross-infective (Bains and Jhooty 1983). Mehta and Saharan (1994) tested the host range of 9 isolates of *H. parasitica* collected from the leaves, and stag heads of 6 host species on 17 host differentials (Tables 5.8 and 5.9). Isolates from *Brassica* oilseeds

Table 5.8 Host differentials of *Hyaloperonospora parasitica* (Mehta and Saharan 1994)

Common name	Botanical name	Cultivar
Indian Mustard (Raya)	*Brassica juncea*	RH-30
Toria	*Brassica campestris* var. Toria	TH-68
Yellow Sarson	*B. campestris* var. Yellow Sarson	YSPb-24
Brown Sarson	*B. campestris* var. Brown Sarson	BSH-1
Ethiopian mustard	*B. carinata*	HC-1
White mustard	*B. alba*	Local
Black mustard	*B. nigra*	Local
Chinese mustard	*B. chinensis*	Local
Chinese mustard	*B. pekinensis*	Local
Rapeseed	*B. napus*	GSL-1
Wild turnip	*B. tournefortii*	Local
Cabbage	*B. oleracea* var. *capitata*	Pride of India
Cauliflower	*B. oleracea* var. *botrytis*	Snowball-16
Turnip	*B. rapa*	White Purple Top
Knol khol	*B. caulorapa*	Early White Vienna
Taramira	*Eruca sativa*	Local
Radish	*Raphanus sativus*	HR-1

Table 5.9 Response of 17 *Brassica* species to nine isolates of *Hyaloperonospora parasitica* (Mehta and Saharan 1994)

	Sources of *H. parasitica* isolates/reactions								
	Toria	Yellow Sarson		Brown Sarson		Raya		Cauliflower	*B. nigra*
Differential hosts	Leaf inoculum	Leaf inoculum	Inflorescence inoculum	Leaf inoculum	Inflorescence inoculum	Leaf inoculum	Inflorescence inoculum	Leaf inoculum	Inflorescence inoculum
Brassica campestris var. Toria	+ (4)	+ (4)	+ (4)	+ (4)	+ (4)	+ (4)	+ (4)	+ (4)	+ (4)
B. campestris var. Yellow Sarson	+ (6)	+ (4)	+ (4)	+ (4)	+ (4)	+ (4)	+ (4)	+ (4)	+ (4)
B. campestris var. Brown Sarson	+ (4)	+ (4)	+ (4)	+ (4)	+ (4)	+ (4)	+ (4)	+ (4)	+ (4)
Brassica juncea	+ (4)	+ (4)	+ (4)	+ (4)	+ (4)	+ (4)	+ (4)	+ (4)	+ (4)
B. carinata	+ (6)	+ (6)	+ (4)	+ (6)	+ (4)	+ (4)	+ (4)	–	+ (4)
B. alba	–	–	–	–	–	–	–	–	–
B. nigra	+ (6)	+ (6)	+ (4)	+ (4)	+ (4)	+ (4)	+ (4)	–	+ (4)
B. chinensis	+ (6)	+ (6)	+ (4)	+ (4)	+ (4)	+ (4)	+ (4)	–	+ (4)
B. pekinensis	+ (6)	+ (6)	+ (4)	+ (4)	+ (4)	+ (4)	+ (4)	–	+ (4)
B. napus	+ (4)	+ (4)	+ (4)	+ (4)	+ (4)	+ (4)	+ (4)	–	+ (4)
B. tournefortii	+ (6)	+ (6)	+ (4)	+ (6)	+ (4)	+ (4)	+ (4)	+ (4)	+ (4)
Eruca sativa	+ (4)	+ (4)	+ (4)	+ (4)	+ (4)	+ (4)	+ (4)	+ (4)	+ (4)
Raphanus sativus	+ (4)	+ (4)	+ (6)	+ (4)	+ (4)	+ (4)	+ (4)	+ (4)	+ (4)
B. oleracea var. *capitata*	+ (4)	+ (4)	+ (4)	+ (4)	+ (4)	+ (4)	+ (4)	+ (4)	+ (4)
B. oleracea var. *botrytis*	+ (4)	+ (4)	+ (4)	+ (4)	+ (4)	+ (4)	+ (4)	+ (4)	+ (4)
B. rapa	+ (4)	+ (4)	+ (4)	+ (4)	+ (4)	+ (4)	+ (4)	+ (4)	+ (4)
B. caulorapa	+ (4)	+ (4)	+ (4)	+ (4)	+ (4)	+ (4)	+ (4)	+ (4)	+ (4)

() = Incubation period in days; + = Infection; – = No infection B = *Brassica*; 1 = Leaf inoculum; 2 = Inflorescence inoculum

Table 5.10 Conidial size of *Hyaloperonospora parasitica* isolates derived from 11 *Brassica* species (Mehta and Saharan 1994)

Sources of isolates	*H. parasitica* isolates[a]	Conidial dimensions (μm) Range Length	Range Width	Average Length	Average Width
Brassica campestris var. Toria	T_1	19.50–29.25	16.62–27.30	25.93	19.30
B. campestris var. Yellow Sarson	YS_1	21.45–29.25	19.50–24.37	25.35	21.74
B. campestris var. Yellow Sarson	YS_2	19.50–29.25	19.50–24.37	26.81	23.39
B. campestris var. Brown Sarson	BS_1	21.45–29.25	19.00–24.37	25.35	21.45
B. campestris var. Brown Sarson	BS_2	21.93–29.25	19.50–26.81	26.56	23.64
Brassica juncea	R_1	20.47–29.25	19.50–26.32	25.64	21.84
Brassica juncea	R_2	19.50–29.25	19.50–26.81	27.05	22.90
Brassica oleracea var. *botrytis*	C_1	19.50–29.25	17.55–24.37	23.30	20.76
B. tournefortii	BT_2	19.50–24.37	16.57–19.50	23.59	20.96
Raphanus sativus	RS_2	19.50–24.37	18.52–21.45	22.03	19.69
B. nigra	BN_2	21.93–29.25	19.50–26.81	26.81	22.90
LSD ($P < 0.05$)		–	–	2.43	1.78

1 = Leaves inoculum; 2 = Hypertrophied inflorescence
[a] Source of inoculum

infected all species, except *B. alba*, whereas isolates from cauliflower leaves did not infect *B. carinata*, *B. alba*, *B. nigra*, *B. chinensis*, *B. pekinensis*, and *B. napus* (Table 5.9). There was no significant difference among the conidial size of the isolates collected from leaves, and stag heads, but significant differences were observed among these groups (Table 5.10). The isolates were classified into two distinct pathotypes, one from Cauliflower, and other from oilseeds *Brassica*. There was no significant difference between the isolates in percentages of spore germination. Specific populations of *H. parasitica* differing in pathogenesis, and host specificity were reported from Bulgaria (Masheva et al. 1996a, b). The populations formed at a lower temperature were more aggressive on cabbage heads.

In the United Kingdom, differential host resistance in relation to pathogenic variation of isolates derived from the same host species were identified in *B. rapa* (Moss et al. 1991; Silue et al. 1996), *B. napus* (Nashaat and Rawlinson 1994), *B. juncea* (Nashaat and Awasthi 1995), and *B. oleracea* (Silue et al. 1996). Isolates from different *Brassica* species found to be most virulent on their species of origin, were never the less able to grow to less extent on other *Brassica* species (Sherriff and Lucas 1990). Nashaat and Awasthi (1995) identified five groups of *B. juncea* accessions with differential resistance to U.K. isolates Rl and P003 derived from oilseed rape (*B. napus* ssp. *oleifera*), and Indian isolates IP01, and IP02 derived from mustard (*B. juncea*). All *B. juncea* accessions were resistant to isolates from *B. napus*, but at the same time *B. napus* cv. *Arian* was resistant to isolates from *B. juncea*. Twenty-one differential responses to *H. parasitica* isolates from *B. oleracea,* and two from *B. rapa* were identified. Of the seven isolates tested, four were from crops of cauliflower in France, two from oilseed rape in the U.K., and one was from mustard in India. All *Raphanus sativus* accessions were resistant to all seven isolates (Silue et al. 1996). Differential resistance of rapid cycling, and commercial genotypes of *B. rapa* were also reported from the U.K. (Moss et al. 1988). A range of differential host responses was characterized by four homologous isolates. Pathogen isolates were also characterized in relation to their sexual compatibility type, and response to phenylamide fungicides (Moss et al. 1988). Eleven isolates of *H. parasitica* tested on rapid cycling populations of *B. rapa* (aa, CrGc-1-1), *B. nigra* (bb, CrGc-2-1), *B. oleracea* (cc, CrGc-3-1), *B. juncea* (aabb CrGc-4-1), *B. napus* (aacc CrGc-5-1), *B. carinata* (bbccCrGc-6-1), and *R. sativus* (rr CrGc-7-1) indicated specificity toward particular genotypes within each rapid cycling population (Hill et al. 1988).

Variation in the response of different host lines to *H. parasitica* has also been detected within wild crucifer species such as Shepherd's Purse (*Capsella bursa-pastoris*), and *Arabidopsis thaliana* (Lucas et al. 1994). Moss et al. (1994) attempted to cross fungal isolates originating from different *Brassica* spp. by co-inoculating host lines previously identified as susceptible to these isolates. A proportion of oospore progeny recovered from these crosses appeared to be hybrids, and had reduced virulence on both hosts of origin. The differential resistance to *H. parasitica* identified in *Brassica* spp. can be used for future studies of the genetics of the host–pathogen interaction, and for breeding for disease resistance. The pathogenicity of four isolates of *H. parasitica* obtained from Japanese radish, rape, broccoli, and Shepherd's purse (*Capsella bursa-pastoris*) were compared by observing infected cotyledons.

These isolates eventually sporulated on host plant cotyledons (compatible combinations), but formed only small necrotic lesions on nonhost cotyledons (incompatible combinations). Sheath encasement of haustoria, and host cell death frequently occurred in the incompatible combinations. The *C. bursa-pastoris* isolate could only infect the host plant with haustorium formation in nonhosts inhibited mainly by papillae. Cytochemical observation showed that these papillae contained callose and polyphenol compounds but not lignin (Yoshida and Ohguchi 1998). Silue et al. (1996) identified 21 differential responses to *H. parasitica* isolates from *B. oleracea,* and 2 from *B. rapa*. All *Raphanus sativus* accessions were resistant to all seven isolates. Accessions for which seedling populations exhibited a heterogeneous reaction to some isolates were classified in a separate category. However, Mehta and Saharan (1994) classified 9 isolates of *H. parasitica* (collected from *Brassica* species) into two distinct pathotypes, one from cauliflower, and the other from oil-yielding *Brassica* species on the basis of their reaction to a set of 17 host differentials.

Variation has been found among interaction phenotypes when a range of *A. thaliana* (*At*), and *Brassica* accessions are inoculated with different isolates of *H. parasitica* (*Hp*). In *Brassica*, genetic characterization of this variation was rudimentary, and only one allele at a single locus associated with isolate specific resistance to *H. parasitica* in *B. napus* has been defined. In contrast, and considering only those phenotypes that can be discriminated readily among segregating individuals, alleles controlling genotype specific variation for response to these pathogens have been identified at 21 loci (RPP, and RAC) in the at genome. Most of these loci now have been mapped with varying degrees of resolution, and occur in several linkage groups in four of the five at chromosomes (Leckie et al. 1996).

Eight *Brassica oleracea* lines were evaluated for their phenotypic reaction to the 13 European isolates of *H. brassicae* by Coelho et al. (2012). The host lines exhibiting a similar phenotypic pattern to different isolates were clustered into six classes (A–F) according to their response to the isolates. The control lines cv. Senna was highly susceptible with no visible host response to all of the pathogen isolates tested (designated class A) mean IP $\geq$ 5.9; showing a consistent susceptible response. The two class F lines (PCA 20.14, and EBH 527) were resistant or moderately resistant to all 13 isolates (mean IP $\leq$ 3.0). The remaining five lines exhibited resistance to at least one, but not all of the isolates including a single lines in class B (EBH508) which exhibited resistance to three isolates (*Hp006*, *Hp* 539, and *Hp* 704), and was highly susceptible to the remaining isolates; a class C line (EBH502) which was resistant at least to some degree to seven isolates; a class D line (EBH525) which was fully resistant to 11 isolates; and two class E lines (PC10, and PC11) which were resistant to all of the isolates except for high susceptibility to the single FP 06 isolates (from France).

Six different pathotypes of *H. brassicae* were identified by classifying the mean IP as either being compatible (mean IP $\geq$ 4.5), incompatible (mean IP < 2.5) or intermediate. Isolates showing the same virulence pattern on the differential set of host lines were grouped, and considered to belong to the same pathotype. Pathotypes 1, 4, and 6 were represented by a single isolate, *Hp006*, *Hp806*, and FP06, respectively. Other pathotypes were represented by more than one isolate including two examples of pathotype 2 (Hp539, and Hp704), five of pathotype 3 (*Hp* 530, *Hp533*, *Hp535*, *Hp717*, and *Hp710*) and three of pathotype 5 (*Hp541*, *Hp702*, and *Hp801*). The three referred cases with multiple isolates included examples from different countries. For example, pathotype 3 includes isolates from Italy, and two from diverse locations in Portugal (Odemira and Azores Island), and the UK (Coelho et al. 2012). Tham et al. (1994) used RAPD analysis to compare 16 isolates of *H. parasitica* from two different *Brassica* species, *B. napus,* and *B. oleracea*. Two out of 20 random primers screened gave reproducible band patterns capable of discriminating between the different host-adopted isolates.

Different pathotypes can be distinguished on the basis of host specificity (Sherriff and Lucas 1990), but such tests are time consuming, and may not reveal the full extent of variations present. Other phenotypic markers such as sexual compatibility type (Sherriff and Lucas 1989) or fungicide sensitivity (Crute et al. 1985) can be assessed, but they provide limited information for epidemiological studies or genetic analysis. There is need, therefore, for alternative molecular markers (Michelmore and Hulbert 1987) to further define variations in the pathogen.

Randomly amplified polymorphic DNAs (RAPD) have been proposed as genetic markers that overcome many of the technical limitations of restriction fragment length polymorphism (RFLP) analysis (Williams et al. 1990; Welsh and McClelland 1990). RAPD scan can be used in the constructions of linkage maps (Williams et al. 1990; Reiter et al. 1992), in the identification of strains, and varieties by genomic fingerprinting (Welsh and McClelland 1990; Goodwin and Annis 1991; Hu and Quiros 1991; Schafer and Wostemeyer 1992; Kresovich et al. 1992; Klein-Lankhorst et al. 1991; Koller et al. 1993; Stiles et al. 1993) and following the cloning of amplified fragments may also serve as conventional RFLP probes. In the study carried out by Tham et al. (1994), the potential use of RAPDs as a source of genetic markers in *H. parasitica* was evaluated. Several isolates from different host species were compared to determine whether reproducible banding patterns correlated with host specificity. As *H. parasitica* can only be cultured on living plant tissues, particular attention was paid to possible artifacts arising through contamination of sample DNA from plant or other microbial sources. Random Amplified Polymorphic DNA (RAPD) fingerprints were generated from a range of *H. parasitica* isolates from different *Brassica* species. Reaction conditions, in particular DNA template, primer and Mg^{2+} concentrations were optimized to ensure that amplifications were reproducible. Possible artifacts arising through host plant DNA were assessed by including such DNA in control reactions. Confirmation that diagnostic RAPD bands were generated from fungal DNA was also obtained by Southern hybridization of a RAPD band to genomic fungal DNA. By screening 20 decamer primers, two were found to detect sufficient genetic variation to allow complete differentiation between pathotypes. These results illustrate the potential value of RAPDs for detecting polymorphisms between isolates of a nonculturable plant pathogenic fungus. Rehmany et al. (2000) compared downy mildew isolates from *A. thaliana,* and *B. oleracea* using AFLP, and ITS analysis dividing the isolates into five groups that correlated with taxonomic species, and in most of the cases with host origin.

5.6.2 Identification of Host Differentials and Nomenclature of Pathotypes

Physiologic specialization/pathogenic variability have long been recorded in *H. parasitica* infecting cruciferous plants all over the world (Gardner 1920; Kobel 1921; Thung 1926; Gaumann 1926; Hiura and Kanegae 1934; Wang 1944; Saharan et al. 2017). However, nomenclature, and designation of *H. parasitica* pathotypes/races was initiated by Natti et al. (1967) based on differential interaction of isolates on cruciferous hosts viz., broccoli, and *B. oleracea*. Natti et al. (1967) identified, and designated race 1, and race 2 of *H. parasitica* showing differential reactions on these two hosts. Later, several reports have come from different countries on the host specificity of *H. parasitica* isolates collected from numerous cruciferous hosts (Felton and Walker 1946; Natti 1958; Dzhanuzakov 1963; Chang et al. 1964; Semb 1969; Dickinson and Greenhalgh 1977; Morris and Knox-Davies 1980; Hill et al. 1988; Sherriff and Lucas 1990; Moss et al. 1991; Mehta and Saharan 1994; Lucas et al. 1994; Tham et al. 1994; Silue et al. 1996; Masheva et al. 1996b; Yoshida and Ohguchi 1998; Saharan et al. 2017). The methodology, and host species/varieties/lines used in these studies were not comparable, and reproducible in the absence of nonmaintenance of isolates, and use of international standard set of host differential. It is difficult to compare the results, and determine whether distinct pathotypes are same or different. Mehta and Saharan (1994) designated distinct pathotypes, one from cauliflower, and other from oilseed *Brassica* on the basis of differential interaction of 9 isolates on a set of 17 host cruciferous host differential set (Table 5.11). Nashaat and Awasthi (1995) identified five groups of *B. juncea* accessions with differential resistance to UK isolates R1, and P003 derived from oilseed rape (*B. napus* ssp. *oleifera*), and Indian isolates IP01, and IP02 derived from mustard (*B. juncea*). Coelho et al. (2012) assembled set of genetically uniform lines of *B. oleracea* (doubled haploid or self-pollinated inbred lines) to use as host differentials set for the characterization of pathotypic variation of *H. parasitica* (*H. brassicae*) within a European sample of the pathogen collected from various crop types of *B. oleracea*. Six pathotypes (HP1 to HP6) were distinguished indicating a potential use of the host differentials for monitoring frequencies in *H. parasitica* populations (Table 5.11). A naming system for the isolates of *H. arabidopsidis* was introduced by Dangl et al. (1992), Holub et al. (1994), and Slusarenko and Schlaich (2003) on the basis of geographical location, and ecotypes infected (AHCO, ASWA, CALA, EDCO, EMWA etc.). However, still there is a need to standardize a set of host differentials (isogenic lines), and a system of nomenclature, and designation of pathotypes of *H. parasitica* at International level. It is essential to know whether pathotypes of *H. parasitica* reported so far from different countries are same or different (Table 5.11). It will be very helpful to identify resistant genes, and their combinations to breed resistant cvs. and their durability extent under different agro-ecological situations.

5.6.3 Heterothallism and Homothallism

Induction of the sexual process in several downy mildew species requires the presence of two strains of opposite mating type. Such heterothallic behavior has been reported for *H. parasitica* (De Bruyn 1937; McMeekin 1960; Kluczewski and Lucas 1983; Sherriff and Lucas 1989; Sequeira and Monteiro 1996). Homothallic forms of *H. parasitica* have also been observed (De Bruyn 1937; Sherriff and Lucas 1989; Sequeira and Monteiro 1996). No relationship has been observed between geographic origin, and mating types. Both heterothallic, and homothallic isolates have been isolated from the same fields (Sequeira and Monteiro 1996). The two forms of sexual reproduction are very important for the maintenance, and evolution of the fungal strains, and epidemiology of the disease. On *B. oleracea,* and

Table 5.11 Identification of pathotypes of *Hyaloperonospora parasitica, and Hyaloperonospora arabidopsidis* (Saharan et al. 2017)

Designate pathotype	Country	International primary host	References
Isolate 1	USA	Turnip	Gardner (1920)
Biological forms 2	Holland	Cabbage	Thung (1926)
Biological strains 3	Germany	Crucifers	Gaumann (1926)
Biological forms	Japan	Crucifers	Hiura and Kanegae (1934)
Pathotypes 3	China	*Brassica, Raphanus, Capsella*	Wang (1944)
Races	USA	*Raphanus, B. oleracea*	Felton and Walker (1946), Natti (1958)
Races	South Africa	*B. oleracea, Raphanus*	Morris and Knox-Davies (1980)
Special forms	Leningrad	Crucifers	Dzhanuzakov (1963)
Three forms	China	Crucifers	Chang et al. (1964)
Races	Norway	Cabbage, radish	Semb (1969)
Race 1 and 2	USA (New York)	Broccoli, *B. oleracea*	Natti et al. (1967)
Isolates	India	Crucifers	Bains and Jhooty (1983)
Pathotypes 2 (9 Isolates)	India	*Brassica*, Cauliflower	Mehta and Saharan (1994)
Isolates P007, P008, P013, P014	UK	*B. rapa (B. campestris)*	Moss et al. (1991), Silue et al. (1996)
Isolates	UK	*B. napus*	Nashaat and Rawlinson (1994)
IP01 and IP02	UK	*B. juncea*	Nashaat and Awasthi (1995)
R1 and P003	UK	*B. napus*	Nashaat and Awasthi (1995)
7 Isolates	UK, France, India	Cauliflower (4), Rape (2), Mustard (1)	Silue et al. (1996)
11 Isolates	USA	Crucifers rapid cycling population	Hill et al. (1988)
2 Isolates	Bulgaria	Plovdiv, Goma Oriahovitsa	Masheva et al. (1996a, b)
4 Isolates	Japan	Japanese radish, rape, broccoli, Shepherd's purse	Yoshida and Ohguchi (1998)
21 Isolates	UK	*Brassica oleracea*	Silue et al. (1996)
2 Isolates	UK	*B. rapa*	Silue et al. (1996)
6 Pathotypes HP1 to HP6	Portugal, UK	*B. oleracea*	Coelho et al. (2012)

(continued)

Table 5.11 (continued)

Designate pathotype	Country	International primary host	References
5 Groups of Isolates through DNA finger-printing (P001 to P005)	UK	*A. thaliana, oleracea*	Tham et al. (1994)
AHCO, ASWA, CALA 2, CALA, CAND 3, EDCO, EMCO, EMOY, EMOY 2, EMWA 1, GOCO, GOWA, HIKS1, HIKS, HIND, HIND 2, HIND 4, MADI, MAKS, MAKS9, NOCO, NOCO 2, NOKS 1, WAND, WELA, WELA1, WELA 3	Germany, U.K.	*Arabidopsis accessions* Col-0, Ler-0, Oy-0, Nd-1, Ws-0, Wei-0	Dangl et al. (1992), Holub et al. (1994), Slusarenko and Schlaich (2003), Schlaich and Slusarenko (2009)

B. campestris heterothallic isolates of two mating types designated as P1, and P2 have been identified.

Isolates from oil seed rape *B. napus* have been found to be uniformly homothallic, and remained self-fertile even after months of laboratory subculture (Sherriff and Lucas 1989). In a cytogenetic study of heterothallic, and homothallic isolates of *H. parasitica* at metaphase 1 of meiosis, a ring of four chromosomes is found (Sherriff and Lucas 1989). This ring is interpreted as a reciprocal translocation complex between chromosomes carrying the mating type alleles. In homothallic isolates a fifth chromosome is associated with the ring of four. The self-fertility of these isolates may therefore be due to the presence of a third mating type allele on the fifth chromosome, a condition known as secondary homothallism. The determination of sexual compatibility type (SCT) of an unknown isolate can be achieved by mixing conidia in a 1:1 ratio with isolates of known SCT, and inoculating to a common compatible host. In heterothallic isolates, oospores may form in combination with isolates of opposite SCT. Isoenzyme markers are particularly important in discriminating between self, and true hybrid progeny (Moss et al. 1988).

5.6.4 Hybridization of *Hyaloperonospora* Isolates

The production of viable oospore of *H. parasitica* under laboratory conditions, and described recovery of isolates (sexual progeny) from these oospore populations has been described (Moss et al. 1994). Oospores were produced when isolates of opposite sexual compatibility type, specialized to the same or different *Brassica* species, were grown together in seedling cotyledons of a host line capable of supporting growth of both isolates. Recovery of sexual progeny from oospore populations produced from two out of four pairing between isolates, specialized to the same host species (homologous pairings), and proved relatively easy. On the basis of their characterization with respect to virulence, response to phenylamide fungicides, sexual compatibility type, and isoenzyme polymorphisms, there was evidence that the sexual progeny from these homologous pairing could be of hybrid origin. For the first time in a member of the Peronosporaceae, it provided possibility to recover, and successfully characterize a new sexual progeny from pairing between isolates specialized to different host species (heterologous pairing). However, the majority of such isolates sporulated weakly, and as consequence, proved difficult to maintain, and were lost. Nevertheless, it is concluded that some evidence for the hybrid nature of progeny from heterologous pairing was obtained.

5.7 *Brassica–Leptosphaeria* Pathogenomics

5.7.1 Evolution of Virulence in *Leptosphaeria*

Major gene resistance coupled with genetic uniformity of crop plants provides strong selection pressure for pathogen populations to overcome disease resistance, for blackleg resistance of canola. Furthermore, the high density sowing of crops lets canopies overlap and enable ready transmission of inocula between plants, thus increasing disease pressure. In contrast, plants in natural environments generally have high genetic diversity and often are sparsely distributed without overlap of canopies. The accumulation of genomic sequence data from many fungi has revealed relationships between particular genome features and parasitic life strategy. For instance, several plant pathogenic fungi have significantly larger genomes than related saprophytes. This increased genome size generally is due to the presence of repetitive DNA that comprises inactivated or truncated transposable elements, which often house effectors, as is the case for *L. maculans*. Other plant pathogens with large numbers of repetitive DNA include *Mycosphaerella fijiensis*, *Cladosporium fulvum*, *Blumeria graminis*, *Phytophthora infestans*, and *Hyaloperonospora arabidopsidis* (Raffaele and Kamoun 2012). Evolution of virulence in fungal pathogens of crops can takes place over different time frames. The increase in frequency of virulence in populations of *L. maculans* on the Eyre Peninsula in the two resistance breakdowns described above took 3 years. Virulence can evolve in a single event when pathogens acquire virulence by horizontal gene transfer.

The fitness cost of mutation of avirulence genes of fungi can affect the effectiveness of recycling the corresponding resistance gene in crops (Zhan and McDonald 2013). Fitness costs are associated with mutation of *AvrLm1* and *AvrLm4* in *L. maculans* (Huang et al. 2010). This is consistent with the observations of much higher frequencies of *AvrLm1* in fungal populations in regions of Australia where cultivars with *Rlm1* resistance have not been sown, compared to areas where these cultivars have been extensively sown. A similar situation is seen for *AvrLm4* (Marcroft et al. 2012a). Molecular data support the roles of *AvrLm1* and AvrLm4 in virulence of *L. maculans*, particularly at early stages of infection where disease symptoms are not visually apparent. Global transcriptome RNA-seq data revealed that three avirulence genes, including *AvrLm1* and *AvrLm4*, are among the 20 highest upregulated genes 7 days after inoculation of canola cotyledons with a virulent isolate of *L. maculans* (Lowe et al. 2014). Transcript levels of these genes were much lower at 14 days post inoculation, when extensive necrosis was established. The role of avirulence genes as virulence factors in several other fungal-plant systems including *Cladosporium fulvum on* tomato has been reported (Mesarich et al. 2014). Genome structure of fungal pathogens can affect evolution of virulence as effectors including avirulence genes are often housed in dynamic genomic regions such as inactivated transposable elements where they can be readily mutated or deleted. The approach of monitoring virulence in fungal populations and then with drawing particular canola cultivars from sale on a regional level has increased the commercial life of cultivars in Australia, despite blackleg resistance being overcome in one geographic region. This approach of monitoring disease and selectively withdrawing cultivars that harbor genes at risk of being defeated should be applicable to manage diseases of other crops, particularly those where the pathogen, like *L. maculans*, has high evolutionary potential and where resistance genes have been identified in the crop plant (Howlett et al. 2015).

5.7.2 Avirulence Alleles of *Leptosphaeria maculans* in Europe

The *Leptosphaeria maculans* populations can change under selection, particularly when there is large-scale cropping of cultivars with resistance genes (Howlett 2004). In France, cultivars with the resistance gene *Rlm1* quickly became popular with growers and were grown over a large proportion of the oil seed rape area. This resulted in rapid adaptation of the *L. maculans* population and breakdown of *Rlm1* resistance (Ansan-Melayah et al. 1997; Rouxel et al. 2003). Similarly, in Australia, breakdown of a single dominant resistance gene derived from *Brassica rapa* subsp. *sylvestris* was observed after only 2–3 years of widespread commercial cultivation (Li et al. 2003; Sprague et al. 2006). In Europe, the life cycle of *L. maculans* begins each season when ascospores land on oilseed rape cotyledons and/or leaves in autumn. At present, both polygenic (quantitative trait loci based) resistance to *L. maculans* (Pilet et al. 1998) and at least nine specific resistance genes (*Rlm1–Rlm9*) are known in oilseed rape. *Rlm1-Rlm4*, *Rlm7* and *Rlm9* are from *Brassica napus* (Ansan-Melayah et al. 1998; Balesdent et al. 2001, 2002; Delourme et al. 2004), *Rlm5* and *Rlm6* from *B. juncea* (Chevre et al. 1997; Barret et al. 1998; Balesdent et al. 2002, 2005), and *Rlm8* from *B. rapa* (Balesdent et al. 2002). These resistance genes correspond to *L. maculans* avirulence genes *AvrLm1-AvrLm9*. Stachowiak et al. (2006) described the structure of races of field populations of *L. maculans* across the main oil seed rape growing regions of Europe, including the UK, Poland, Germany and Sweden, by comparison to that described in a large-scale survey in France (Balesdent et al. 2006).

The first large-scale Europe-wide survey of avirulence alleles and races of *L. maculans* was described by Stachowiak et al. (2006). Isolates were collected from the spring rape cultivar Drakkar, with no known genes for resistance against

L. maculans, at six experimental sites across the main oilseed rape growing regions of Europe, including the UK, Germany, Sweden, and Poland (Table 5.12). Additionally, in Poland, isolates were collected from cv. Darmor, which has resistance gene, *Rlm9*. In total, 603 isolates were collected during autumn in 2002 (287 isolates from Germany and the UK) and 2003 (316 isolates from Poland and Sweden). The identity of alleles at eight avirulence loci was determined for these isolates. No isolates had the virulence allele *avrLm6* and three virulence alleles (*avrLm2*, *avrLm3,* and *avrLm9*) were present in all isolates. The isolates were polymorphic for *AvrLm1*, *AvrLm4*, *AvrLm5,* and *AvrLm7* alleles, with virulence alleles at AvrLm1 and AvrLm4 loci and avirulence alleles at AvrLm7 and AvrLm5 loci predominant in populations. Virulent *avr Lm7* isolates were found at only one site in Sweden. Approximately 90% of all isolates belonged to one of two races (combinations of avirulence alleles), Av5-6-7 (77% of isolates) or Av6-7 (12%). Eight races were identified, with four races at frequencies less than 1% (Table 5.13). The study suggested that *Rlm6* and *Rlm7* are still effective sources of resistance against *L. maculans* in oilseed rape in Europe. The results are comparable to those of a similar survey done in France in autumn 2000 and 2001.

5.7.3 Genetic Variability in *Leptosphaeria maculans* from France

The knowledge of genetic variation in *L. maculans* populations and the role of the evolutionary processes such as mutation, migration, genetic drift, selection, and/or recombination in these rapid shifts in the field population is still very limited. Previous studies of genetic variation in *L. maculans* populations from Canada and Australia found high levels of genetic variation in field populations of *L. maculans*, which is consistent with the hypothesis that sexual reproduction is an important part of the life cycle of *L. maculans*. However, reports of the distribution of genetic variation among and within field populations have been inconsistent. There have been no studies of the genetic structure of *L. maculans* field populations in France and other countries of Western Europe. In these regions, previous work has focused mainly on field population race structure and has used virulence markers to assess genetic diversity in the fungus. Gout et al. (2006) assessed genetic variation in French populations of *L. maculans* with mini satellite loc, mating type, and virulence markers to test the hypotheses that the field populations were randomly mating, that populations from different fields were genetically separable, and that epidemics are initiated by ascospores.

Leptosphaeria maculans is the ubiquitous fungal pathogen of *Brassica* crops and causes the devastating stem canker disease of oilseed rape worldwide. Gout et al. (2006) used mini-satellite markers to determine the genetic structure of *L. maculans* in four field populations from France. Isolates were collected at three different spatial scales (leaf, 2-m^2 field plot, and field) enabling the evaluation of spatial distribution of the mating type alleles and of genetic variability within and among field populations. Within each field population, no gametic disequilibrium between the mini-satellite loci was detected and the mating type alleles were present at equal

Table 5.12 List of the oilseed rape cultivars/lines used in a differential set to differentiate isolates of *Leptosphaeria maculans* in seedling pathogenicity tests (Stachowiak et al. 2006)

	Frequency (%)						
Race[a]	Boxworth (UK)	Rothamsted (UK)	Teendorf (D)	Svalov (S)	Poznan (PL)	Pulawy (PL)	Mean
Av5-6-7-(8)	77.9	60.2	86.8	76.0	81.4	80.0	77.1
Av6-7-(8)	8.8	16.3	10.8	19.3	11.6	6.6	12.2
Av1-5-6-7-(8)	9.4	14.7	1.2	0.9	6.5	10.0	7.1
Av4-5-6-7-(8)	2.9	3.9	1.2	0.0	0.0	1.7	1.6
Av1-6-7-(8)	1.0	2.0	0.0	0.9	0.5	0.0	0.7
Av4-6-7-(8)	0.0	2.0	0.0	0.0	0.0	1.7	0.6
Av6-(8)	0.0	0.0	0.0	2.9	0.0	0.0	0.5
Av1-4-5-6-7-(8)	0.0	0.9	0.0	0.0	0.0	0.0	0.2
Number of isolates	103	100	84	109	147	60	100
Number of races	5	7	4	5	4	5	5
Margalef index	0.86	1.30	0.68	0.85	0.60	0.98	1.09
Simpson diversity index	0.38	0.60	0.24	0.39	0.32	0.35	0.39

[a] Race nomenclature according to Balesdent et al. (2005). The numbers indicate the *AvrLm* locus for which the isolate is avirulent. The number in parentheses indicates that the corresponding locus (*AvrLm8*) has not been characterized

Table 5.13 The frequency of *Leptosphaeria maculans* races on oilseed rape at different sites in the UK, Germany (D), Sweden (S), or Poland (PL) (Stachowiak et al. 2006)

Cultivar	Resistance gene
Columbus	*Rlm1* + *Rlm3*
Bristol	*Rlm2* + *Rlm9*
22-1-1	*Rlm3*
Jet Neuf	*Rlm4*
150-2-1	*Rlm5*
Darmor MX[a]	*Rlm6*
23-1-1	*Rlm7*
Goéland	*Rlm9*

[a] Not characterized at the *Rlm9* locus

frequencies. Both sexual and asexual reproduction occurs in the field, but the genetic structure of these populations is consistent with annual cycles of randomly mating sexual reproduction. All *L. maculans* field populations had a high level of gene diversity (*H*-0.68–0.75) and genotypic diversity. Within each field population, the number of genotypes often was very close to the number of isolates. Analysis of molecular variance indicated that >99.5% of the total genetic variability was distributed at a small spatial scale, i.e., within 2-m^2 field plots. Population differentiation among the four field populations was low ($GST < 0.02$), suggesting a high degree of gene exchange between these populations. The high gene flow evidenced here in French populations of *L. maculans* suggests a rapid countrywide diffusion of novel virulence alleles whenever novel resistance sources are used (Gout et al. 2006).

5.7.4 Characterization of Avirulent-Mutated *AvrLm* Gene of *Leptosphaeria*

The predominant tools employed to achieve site-directed double strand breaks are ZFNs (zinc-finger nucleases) and TALENS (transcription activator-like effector nucleases), both of which have provided invaluable assistance in targeted gene mutation in a diverse range of organisms. The clustered regularly interspaced short palindromic repeats (CRISPR)-Cas (CRISPR-associated) protein system has gained much attention and has been widely used for editing the genomes of animals, plants and fungi. In comparison to ZFNs and TALENS, CRISPR/Cas9 is easier to manipulate, as it employs a RNA-guide nuclease, Cas9, to induce double-strand breaks for mutation and can be recycled to edit different targets. CRISPR/Cas9 has been used for targeted mutagenes is in various plant species, including Arabidopsis (Jiang et al. 2013). Blackleg disease survey, revealed the dynamic virulence allele profiles of *L. maculans* isolates in Manitoba, Canada, it was evident that the fungal populations carry a diverse number of *Avr* genes.

A total of 180 *Avr* gene races were identified from 964 isolates, with three major races observed: *AvrLm*-2-4-5-6-7, *AvrLm2*-4-5-6-7-S, and *AvrLm*-1-4-5-6-7-11-(S). Very few isolates carried a single *Avr* gene conferring virulence to most commercial canola cultivars (Fernando et al. 2018). However, *L. maculans* isolate (DS103) with few *Avr* genes, termed *UMAvr7*, was identified and carried *avrLm1-2-3-4-9-11-LepR1-LepR2-S-AvrLm5-6-7*, which is virulent on *B. napus* genotypes with different resistance genes (except *Rlm7*). This isolate was considered the ideal candidate, not only for elucidating the role of *Avr* effectors in pathogenicity, but also with which to examine the role of *Aver*-R interaction in the *B. napus* host defense. For breeding programs, an ideal isolate that does not carry any of the identified *Aver* genes in *L. maculans*, and therefore, can be adopted for new resistance gene identification in *Brassica* genotypes. *Brassica* genotype showing incompatible reaction to this isolate should have a corresponding novel resistance *R* gene. The *AvrLm4*-7 gene has been cloned and determined by a single nucleotide mutation in coding region, where codon 358 shifted from "C" to "G" conferring recognition to *Rlm7* and *Rlm4/7* in *B. napus*, respectively (Parlange et al. 2009). To knock out *AvrLm7* gene for resistance screening in *Brassica* germplasm, Zou et al. (2020) generated a deletion in *AvrLm7* by employing CRISPR/Cas9. To validate this gene deletion, the pathogenicity of both UMAvr7 (wild type) and the umavr7 (CRISPR mutant) was analyzed on the corresponding *B. napus* cultivars 01-23-2-1 (with Rlm7) and Westar (no *R* gene), along with sequencing so that a suitable isolate (mutant) altering the pathogenicity on *B. napus Rlm7* genotype can be evaluated.

The clustered regularly interspaced short palindromic repeat (CRISPR)/CRISPR-associated (Cas) system to generate the mutant isolate umavr7 from a point mutation of the*AvrLm7* coding region in a *L. maculans* isolate (UMAvr7) has been employed by Zou et al. (2020). Reverse transcription PCR and transcript to me data confirmed that the *AvrLm7* gene was knocked out in the mutant isolate. Pathogenicity tests indicated that umavr7 could cause large lesions on one set of *Brassica* differential/ genotypes that express different resistance (*R*) genes. Comparative pathogenicity tests between UMAvr7 (wild type) and umavr7 on the corresponding *B. napus* genotype 01-23-2-1 (with *Rlm7*) showed that umavr7 is a mutant isolate, producing large gray/green lesions on cotyledons. The pathogenicity of the mutant isolate was shifted from avirulent to virulent on the *B. napus Rlm7* genotype. Therefore, this mutant is virulence on the identified resistant genes to blackleg disease in *B. napus* genotypes. Superoxide accumulated differently in cotyledons in response to infection with UMAvr7 and umavr7, especially in resistant *B. napus* genotype 01-23-2-1. Resistance/susceptibility was further evaluated on 123 *B. napus* genotypes with the mutant isolate, umavr7. Only six of the 123 genotypes showed resistance to umavr7. The identification of these six resistant *B. napus* genotypes will lead to further studies on the development of blackleg disease resistance through breeding and the identification of novel *R* genes (Zou et al. 2020).

5.7.5 Identification of Genome-Wide DNA Variants and SNP Haplotypes of *Leptosphaeria maculans* Pathotypes

Monitoring changes in race structure of *L. maculans* populations is important for selecting effective *R* genes for blackleg resistance breaking and management.

Toward this goal, a reliable approach for *L. maculans* race structure assessment becomes essential to gain under-standing of the frequency of avirulence genes and race groups of the pathogen, and provide guidance for deployment of resistant *Brassica* cultivars. A total of 162 representative isolates collected in western Canada were selected for genome resequencing with an Illumina platform. Assembly of the short reads against the reference genome of *L. maculans*' brassicae' isolate v23.1.3 led to the discovery of 21,016 DNA variants (SNPs and In Dels), 93% SNPs and 7% In Dels, with a transition/transversion (Ts/Tv) ratio of 3.1 genome wide. In Dels occurred mainly in GC-blocks and the Ts/Tv ratio of SNPs in AT-blocks was >2 times higher than that in GC-blocks. The number of variants was positively correlated with supercontig size, GC-block size and gene numbers. DNA variants in most avirulence genes were SNPs, except a deletion in *AvrLm1*. The number of SNPs varied from 1 to 2 in *AvrLm2*, *AvrLmJ1-5-9*, *AvrLm6*, *AvrLm10A*, *AvrLm10B*, and *AvrLm11*, 8 in *AvrLm3* and 38 in *AvrLm4-7*. It is the first report of triallelic SNPs in *AvrLm2*, and *AvrLm4-7*, and premature STOP codons in *AvrLm4-7*. Nine SNP haplotypes were identified in*AvrLm4-7*, however, only 2–3 haplotypes occurred in other avirulence genes, and in total 47 haplotype groups were identified from the isolates. The 47 SNP haplotype groups were translated into 44 protein haplotype groups and then isolates of *L. maculans* collected in western Canada were classified into ten races. There are at least three advantages with genotyping-by-haplotyping when compared to their methods of analysis for *L. maculans* Avr profile or race structure, including SNP genotyping. First, it considers all DNA variants in an *Avr* gene. Second, it can be readily translated into protein isoforms. Third, it is able to capture the emergence of new SNPs in any pathogen populations. Therefore, Chen et al. (2020) proposed genotyping-by-haplotyping as a new method for large-scale *L. maculans* Avr profile analysis (Fig. 5.1). All existing SNP haplotypes with connection with *Avr* gene functions can be categorized, indexed, and stored in a database for inquiries. Any new haplotype will be added to the database once its relationship with the avirulence/virulence has been determined through the conventional phenotyping process. The identification of genome-wide DNA variants and SNP haplotypes associated with avirulence genes of *Leptosphaeria maculans* were performed with 158 isolates selected from 1590 isolates originating from western Canada.

There were 21,016 polymorphic variants identified in the isolates. Forty-eight SNP haplotype groups were discovered and linked with different avirulence gene functionality. Being more informative and accurate than SNP genotyping, SNP haplotyping was hence proposed to be a more reliable and informative strategy for large-scale survey of *L. maculans* race structure (Chen et al. 2020). Several *AVR* genes of *L. maculans* have been molecularly characterized with genome size and features. Some of the *AVR* genes have been cloned. Pathotypes distribution in different countries has been assessed (Table 5.14) along with breakdown of resistance in Canada, France and Australia.

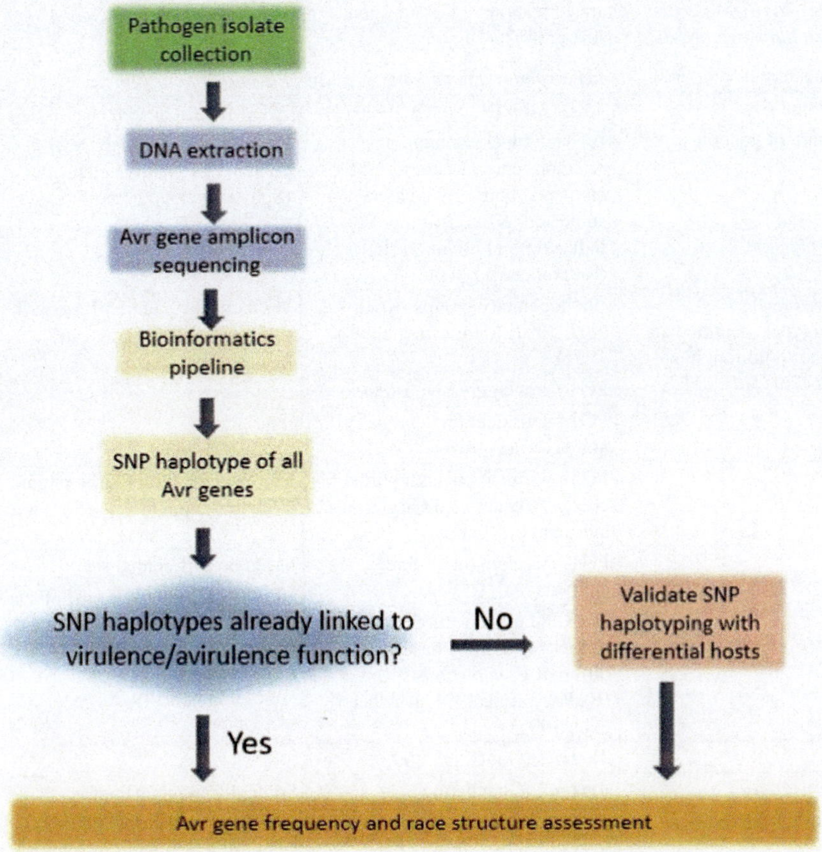

Fig. 5.1 Proposed approach for *Leptosphaeria maculans Avr* profiling using haplotyping (Chen et al. 2020)

5.8 *Brassica–Plasmodiophora* Pathogenomics

5.8.1 Differentials for Identification of Pathotypes

Researchers from most of the countries of the world have used differential sets of Williams (1966), Some et al. (1996), and European clubroot differential (ECD) set (Buczacki et al. 1975) to identify clubroot pathotypes occurrence, and distribution in their respective region/areas, and on cruciferous species grown for vegetable, and oilseed *Brassicas*. However, Strelkov et al. (2018) used a modified Canadian Clubroot Differential (CCD) set of 13 *Brassica* hosts which includes differential sets of Williams (1966), Some et al. (1996), selected eight hosts of ECD set (Buczacki et al. 1975), and three genotypes of *B. napus* (Mendel, Wester, 45 H29)

Table 5.14 Molecular characterization of *AVR* genes in *Leptosphaeria maculans and Plasmodiophora brassicae* (Neik et al. 2017)

Characters	*Leptosphaeria maculans*	*Plasmodiophora brassicae*
Genome size	45.12 Mbp (Rouxel et al. 2011)	25.5 Mbp (Schwelm et al. 2015)
Feature of genome	AT-rich blocks containing effector genes and transposable elements, both of which is subjected to frequent repeat-induced point mutation (RIP) (Rouxel et al. 2011)	Low number of genes, low content of repetitive elements (Schwelm et al. 2015)
Races/Isolates/ Pathotypes identified and cloned (bold font has been cloned)	[a]Pathogenicity groups (Koch et al. 1991; Mengistu et al. 1991)	Williams differential: Race 2, 3, 4, 5, 6, 7, 8 (Williams 1966)
	PG1—non aggressive isolate	
	PG2—avirulent on "Quinta" and "Glacier"	
	PG3—virulent on susceptible check "Westar" and Glacier but avirulent on "Quinta"	European Clubroot Differential: Coding system
	PG4—virulent on all three cultivars	based on differential set comprising five hosts each of *B. campestris*, *B. napus* and *B. oleracea* using 34 European *P. brassicae* collections (Buczacki et al. 1975)
	[a]This PG classification was further reclassified based on different *B. napus* hosts (Badawy et al. 1991; Kutcher et al. 1993)	
	AvrLm1 (Gout et al. 2006)	
	AvrLm2 (Ghanbarnia et al. 2015)	Somé differential (France population): Five pathotypes, P1, P2, P3, P4, and P5 (Some et al. 1996)
	AvrLm3 (Plissonneau et al. 2016)	
	AvrLm4-7 (Parlange et al. 2009)	
	AvrLm5 later known as ***Avr LmJ1*** (Plissonneau et al. 2017)	
	Avr Lm6 (Fudal et al. 2007)	
	AvrLm7 (Balesdent et al. 2002; Parlange et al. 2009)	
	AvrLm8 (Balesdent et al. 2002)	
	AvrLm9 (Balesdent et al. 2005)	
	AvrLm10 (Petit-Houdenot et al. 2016)	
	AvrLm11 (Balesdent et al. 2013)	
	AvrLmS (Van de Wouw et al. 2009)	
	AvrLepR1 (not genetically characterized)	

(continued)

Table 5.14 (continued)

Characters	*Leptosphaeria maculans*	*Plasmodiophora brassicae*
	AvrLepR2 (not genetically characterized)	
	AvrLepR3 (not genetically characterized)	
	AvrLepR4 (not genetically characterized)	
Affecting countries and prevalence of isolate pathotype/race found	Western Canada (2010–2011, Alberta, Saskatchewan, Manitoba): 674 isolates majority *AvrLm2*, *4*, *6*, *7*, (S) (Liban et al. 2016)	Canada (2005, Edmonton, Alberta): homogenous, ECD-/15/12 or Williams pathotype 3 or Somé P2 (Strelkov et al. 2007)
	Germany (2011–2013): out of 644 isolates, majority of the tested isolates (85%) were virulent toward *B. napus Rlm* genes (*Rlm1*, *3*, *4*, *9*). *Rlm7* remains effective. Hence, the prevalent isolates harbor: *AvrLm5*, *6*, *7*, *8* (Winter and Koopmann 2016)	Canada (2003, Alberta): ECD-16/15/12 and 16/15/0 or Williams pathotype 3 and 5 respectively; British Columbia ECD 16/2/12 or Williams pathotype 6;
	UK, Germany, Sweden, Poland (2002–2003)—603 isolates genotyped, majority *AvrLm5*, *6*, *7* (Stachowiak et al. 2006)	
	France (2000–2001)—1797 isolates genotyped, majority *AvrLm5*, *6*, *7*, *8* (Balesdent et al. 2006)	Ontario ECD 16/0/14 or Williams pathotype 6 (Strelkov et al. 2006)
	American continent (2004–2006): *AvrLm1*, *2*, *4*, *6*, *7* (Dilmaghani et al. 2009)	China: Williams race 4 (Chai et al. 2014)
	Australia (1987–2015): *AvrLm1*, *2*, *3*, *4*, *5*, *6*, *7*. Frequency of *AvrLm1*, *4* and *6* fluctuated across the years depending	Czech Republic: Williams 7 most prevalent others include 6, 4, 9, 10, 2, 3, and 1 (Říčarová et al. 2016)
	On which type of cultivar carrying specific *R* gene was commonly grown. Overall, *AvrLm4* was extremely low compared	
	To *AvrLm1* and *AvrLm6*. Frequency of *AvrLm3* was also low (Marcroft et al. 2012a, b; Van de Wouw et al. 2017)	
Resistance Breakdown	(within 5 years in France) (Rouxel et al. 2003; Sprague et al. 2006) *Rlm1* (within 3 years in Australia) (Brun et al. 2010;	"Mendel" succumbed from susceptible volunteers inGermany (Diederichsen et al. 2014)

(continued)

Table 5.14 (continued)

Characters	*Leptosphaeria maculans*	*Plasmodiophora brassicae*
	Sprague et al. 2006), first sign of breakdown was detected within LepR3 a year (Li et al. 2003) (France, 6 years after commercially released 2004–2010) (Mollier 2015). *Rlm7* (France) field trial experiment, breakdown after 3 years (Brun et al. 2010) *Rlm6* (Canada) (Zhang et al. 2016) *Rlm3*	

[a] This PG classification etc.

as shown in Table 5.15. Some other attempts have been made to develop a novel clubfoot differential set, which can be applicable widely with fair system to detect spectrum of virulence of *P. brassicae* on different species of crucifers with wider validity, and applicability. Sharma et al. (2013b) determined reaction of lines of the rapid cycling *Brassica* collection (RCBC), and lines of *Arabidopsis thaliana* to pathotypes 2, 3, 5, and 6 classified on the system of Williams. In Japan Kuginuki et al. (1999) proposed a new race differentiation system using two CR cultivars (CR Kukai 65, and CR Utage 70) of Chinese cabbage as hosts. Hatakayema et al. (2006) proposed a modified differential system using Chinese cabbage cv. Super CR Hiroki as an alternative for CR Utage 70. In Turkey, Apaydin et al. (2010) used 11 hosts of cruciferous species to identify pathotypes of *P. brassicae*. A new race designation system for clubroot with four clubroot races and differentials from several *B. oleracea* species was developed to enable a relevant and reproducible description of clubroot resistant *B oleracea* varieties (Table 5.16; Smilde et al. 2012; International Seed Federation 2019). However, so far no sound system for identification of pathotypes of *P. brassicae* has been developed at International level, which can detect virulence on different species of crucifers. There is a need to develop an ideal differential set with isogenic lines of crucifers for each species separately so that right spectrum of virulence on each species can be identified with number of pathotypes occurring in a country. Global distribution of *P. brassicae* is given in Table 5.17. It is clear from the Table 5.17 that similar virulence is distributed all over the world. It indicates limitation of differential sets used in each country (same number of differential) which could not detect any new pathotypes. The differential system used was Williams (1966), Some et al. (1996), ECD set (Buczacki et al. 1975), and Canadian clubroot differential set (Strelkov et al. 2018).

Addition of others more number of genotypes could have detected new or more number of pathotypes/virulence. The other limitation is in most of the cases instead of single spore (resting spore), populations of isolates have been used consisting of mixture of resting spores, which may be unable to detect new pathotype. Single resting spore may also consist of more than one race. No uniform system has been followed to designate or name in nomenclature of pathotypes. The pathotypes have been designated as serial numbers (1, 2, 3, ... *n*), as triplicates (16/2/15, 16/22/30,

Table 5.15 Host differentials for identification of *Plasmodiophora brassicae* pathotypes (Saharan et al. 2021)

Brassica species/varieties constituting differential set			
1	2	3	4
B. napus var. *napobrassicae*	*B. rapa* ssp. *rapifera*	*B. napus* var. *napus* Brutor	ECD 02 AAbbCC
Laurentian	Polish rape	Nevin	ECD 05 Granaat
B. napus var. *rapifera*	Line aaBBCC ECD 01	*B. napus* var. *rapifera*	ECD 06 Nevin
Wilhelmsburger	Line AAbbCC ECD 02	Wilhelmsburger	ECD 08 Giant rape selection
B. oleracea var. *capitata*	Line AABBcc ECD 03		ECD 09 New Zealand resistant rape
Badger Shipper	Line AABBCC ECD 04		ECD 10 Wilhelmsburger
Jersey Queen	*B. rapa* var. *pekinensis*		ECD 11 Badger Shipper
	Chinese cabbage cv. Granaat ECD 05		
	B. napus var. *napus*		ECD 13 Jersey Queen
	Fodder rape DC 101 Nevin ECD 06		Bruter
	DC119 Giant Rape ECD 07		Laurentian
	DC 128 selection from DC 119 ECD 08		Mendel
	New Zealand resistant rape ECD 09		Wester
	B. napus var. *rapifera* DC 130, Wilhelmsburger ECD 10		45H29
	B. oleracea var. *capitata*		
	Cabbage, Badger Shipper ECD 11		
	Bindasachsener ECD 12		
	Jersey Queen, ECD 13		
	Septa ECD 14		
	Verheul (Kale) ECD 15		

. . . *n*), and as letters (P1, P2, P5, . . . *n*); (A, B, C, . . . X . . . *n*). It does not indicate which pathotype has been detected by which host genotype. The nomenclature of pathotype should indicate the host species, which have been recognized by a specific pathotype. It is proposed that a uniform nomenclature should be followed which can indicate from which species it has been identified. The pathotypes may be designated as BoPb1, BoPb2—from *B. oleracea*, BrPb1; BrPb2—from *B. rapa*, BnPb1; BnPb2, —from *B. napus*, At Pb1; AtPb2 from *A. thaliana,* and so on.

In this scheme, first two letters indicate host species, and second two letters indicate pathotypes number. It is more benign and expedient for the nomenclature

Table 5.16 Identification of *Plasmodiophora brassicae* races using *Brassica oleracea* varieties (International Seed Federation 2019)

Differential hosts as *B. oleracea* varieties		Pathotypes/races			
		Pb: 0	Pb: 1	Pb: 2	Pb: 3
Bartolo[a]	White cabbage	S	S	S	S
Lodero[a]	Red Cabbage	HR	HR	HR	S
051632[a]	White cabbage	HR	S	S	HR
Kilaton[a]	White cabbage	HR	HR	S	HR
Monclano[a]	Broccoli	HR	HR	S	HR
Clapton[a]	Cauliflower	HR	HR	S	HR
Crispus[a]	Brussel Sprouts	HR	HR	S	HR

S susceptible, *HR* highly resistant
[a] Differential hosts and isolates that are used by seed sector

of pathotypes of *P. brassicae*. Although binary notation system of nomenclature used to designate ECD pathotypes allows users to deduce which differentials are resistant or susceptible to a given isolate of *P. brassicae* but it is viewed as cumbersome, and confusing. Recently, race, and isolate specific molecular markers have been developed through genome realignment, which enables detection of *P. brassicae* isolates (Jeong et al. 2018). However, genomic characterization of pathotypes has not been found to correlate with pathogenesis.

5.8.2 Genome Comparison of *Plasmodiophora brassicae* Pathotypes

Alignment of the Pb3 and Pb6 genomes revealed a close to perfect synteny with rare examples of inversion of small segments of the Pb6 genome. The majority of predicted Pb6 genes had an orthologue in the Pb3 genome except for 14 genes that were unique to Pb6, with six of these having no homologue in the NCBI nonredundant (nr) protein database and eight that matched hypothetical proteins (*E* value <0.001). To gain insight into genome variation and its correlation with host specificity, Rolfe et al. (2016) re-sequenced the genomes of Pb2, Pb5 and Pb8. A total of 173, 134 SNPs were discovered by alignment of the genome sequence of Pb6 and the re-sequenced pathotypes against the Pb3 reference. Paired-end Illumina reads generated for Pb3 were also aligned to the Pb3 genome. After filtering (on allele frequency, SNP positions with heterozygous genotypes and indels) 132, 429 SNPs remained and were clustered into 90 haplotype blocks. One haplotype block was found to result from a single SNP between the reference genome and Pb3 Illumina reads; investigation of this SNP indicated that it resulted from a sequencing error. Out of the remaining 89 haplotype blocks, 86.5% contained more than one SNP. Comparison of whole genome SNP profiles revealed that Pb2, 3, 5, 8 were highly similar and clustered as one group, while the SNP profile of Pb6 was more divergent. The haplotypes correlate with the host range and geographical distribution of these pathotypes, with Pb2, 3, 5, 8 being virulent on *B. napus* and present in the prairie

Table 5.17 Global distribution of *Plasmodiophora brassicae* pathotypes (Saharan et al. 2021)

Countries	Pathotypes	References
Australia	3, 6, 7	Williams (1966)
Canada	1, 2, 3, 5, 6, 8	Strelkov and Hwang (2014)
	16/0/14, 16/2/12	Strelkov et al. (2006), Strelkov et al. (2007)
	16/15/0, 16/15/12, –/15/12	
	P2, P3	Xue et al. (2008)
	A, B, C, D, E, F, G, H, I, J, K, L, M, N, O, P, X	Strelkov et al. (2018)
China	1, 2, 4, 7, 9, 10, 11, 13	Chai et al. (2014)
Czech Republic	1, 2, 4, 6, 7, 10	Říčarová et al. (2016)
	16/2/15, 16/2/14, 16/10/4, 16/14/13	
	16/14/15, 16/14/31, 16/18/15, 16/14/12	
	16/31/8, 16/15/15, 16/14/4, 16/14/14	
	16/15/31, 16/31/12, 16/22/31, 16/22/30	
	16/31/15, 16/3/15	
	P1, P2, P3, P4, P5	
Finland	1, 2, 3, 4, 6, 7	Linnasalmi and Toiviainen (1981)
	16/31/12, 16/3/31, 16/19/31, 16/18/30	
	16/31/31, 16/00/30, 16/00/31/, 16/02/31	
	16/03/31	
France	P1, P2, P3, P4, P5, P6, P7, P8	Manzanares-Dauleux et al. (2001)
GDR	1, 3, 4, 6, 7, 9	Williams and Seidel (1968)
Iceland	7	Linnasalmi and Weisaeth (1978)
India	Pathotype present	Bhattacharya et al. (2014)
Japan	1, 2, 3, 4, 5, 6, 7, 8, 9, 16, 20	Tanaka and Ito (2013)
	A, B, C, D	Osaki (2008), Osaki et al. (2008)
Korea	1, 2, 3, 4, 5, 9	Kim et al. (2016)
New Zealand	1, 2, 4, 5, 6, 7, A, B, C, D, E, F, G, H	Williams (1966), Lammerink (1965, 1986)
Norway	1, 3, 4, 5, 6, 7, 9	Linnasalmi and Weisaeth (1978)
Poland	2, 3, 4, 6, 7, 9, 10, 12, 16	Říčarová et al. (2016)
	17/31/31, 16/7/31, 16/31/8, 16/31/31	
	16/14/30, 16/14/31, 16/31/14, 16/6/31	
	16/14/15, 16/7/28, 16/31/29, 16/14/12	
	16/2/14, P1, P3	
Sweden	1, 2, 3, 4, 6, 7, 15	Jonsson (1981)
Turkey	1, 16/31/31	Apaydin et al. (2010)
UK	1	Williams (1966)
USA	6, 7, 16/02/31, 16/03/31	Williams and Walker (1963), Dobson et al. (1983)
USSR	1, 2, 6, 7	Krivchenko et al. (1982)

provinces and Pb6 failing to cause disease on *B. napus* and being prevalent in British Columbia and Eastern provinces of Canada. Without access to read data, analyses of the e3 genome could not be completed as with the other pathotypes. However, by direct sequence comparison 48,578 out of the identified 132, 429 SNPs were found to differentiate the genome of e3 reported by Schwelm et al. (2015) and Pb3. These SNPs were grouped into 58 haplotypes and showed that e3 clustered with Pb6.

5.8.3 Proteomics of *Plasmodiophora brassicae* Pathotypes for Potential Effectors (Genes)

To annotate and assign GO functions to the predicted genes, predicted Pb3 proteins were searched for homology to entries in the NCBI nr, GO and Interpret databases using Blast2Go-PRO (Conesa et al. 2005). Out of the 10,851 total predicted proteins for Pb3, 3460 did not have hits in the NCBI database (E value $<10^{-3}$). An OrthoMCL search of protein families identified 341 paralogous groups in Pb3 with the largest groups having up to 37 paralogues. A Blast search of the NCBI nr database identified matches for only 51 of the paralogous groups. A search for orthologous proteins identified *P. ramorum* as the organism with the highest number of orthologues matching Pb3 proteins (755 Pb3 proteins). Among these were genes involved in protein and nucleic acid binding, ATP and metal ion binding, oxidoreductases, kinases and ATPase activity. Rolfe et al. (2016) searched the genome of Pb3 for potential secreted proteins, which might act as effectors. The total number of predicted secreted proteins was 590, with 431 of these proteins having a probability score (D score) of greater than 0.7. The 221 proteins from this list were defined as small (less than 300 aa) secreted proteins (SSPs). Nearly 60% of the Pb3 SSPs (132 out of 221) had no homology with any protein in the NCBI nr database. Sixty-three of the 590 Pb3 SSPs did not have any homologue among the e3 predicted proteins (E value $<10^{-3}$). A blastn search confirmed the presence of all but one (PbPT3Sc00048_Am_3.106_1) of these genes in the e3 genome. Comparing Pb3 SSPs with those predicted for Pb6 revealed that 119 of the Pb3 SSPs did not have a homologue in the Pb6 genome. Further study is required to validate the uniqueness of these 119 SPs for Pb3 and to reveal their possible role in determining host specificity. Rolfe et al. (2016) searched the SSPs of Pb3 for the presence of RXLR motifs. RXLR motifs identified in the effectors of oomycete pathogens are believed to function in translocation of effector proteins into the plant host cells. RXLR type effectors are highly prevalent in *Phytophthora* and downy mildew species. A potential RXLR motif was found in only 13 of the 221 predicted SSPs indicating that this motif is not prevalent among the effectors of *P. brassicae*. The clubroot pathogen is evolutionary distinct from oomycetes and the lack or under representation of RXLR like motifs might be expected; however, this pathogen resides within the host cell and in this regard it resembles malaria parasites (plasmodium species). Malaria parasites are also evolutionary distinct from oomycetes but have adapted a similar mechanism for translocation of effector proteins

(Sperschneider et al. 2015). Therefore, it is possible that a similar mechanism for translocation of virulent proteins operates in *P. brassicae*.

5.8.4 Phylogenetic Analysis of Pathotypes

The *Plasmodiophora brassicae* isolates collected from Korea have been compared with the isolates from Japan, and Europe for their phylogenic relationship. Two separate clusters were formed in the phylogenetic tree based on the rDNA sequences that separated *P. brassicae* isolates from other *Cercozoa* species according to the Maximum Likelihood method (Fig. 5.2). Similar phylogenetic separation was also found in a circular cladogram constructed following the unweighted pair-group method with arithmetic mean (UPGMA) method, and Kimura 80 nucleotide distance measure method. The 11 Korean isolates, and three published sequences AB526843, KX011115, and KX011135 were in the same cluster (Fig. 5.2). However, all 11 Korean isolates formed a separate subcluster with the Japanese NGY isolate (Fig. 5.2). The Korean isolates having higher dG values between −131, and −163 kcal mol in their predicted secondary structures, due to sequence variations in region 2 (Geosan, Geumsan, Seosan, Haenam2, and Pyeongchang), further subclustered together. The other six Korean isolates with smaller than −168 kcal mol dG clustered together, indicating that the variation in region 2 accounted for the majority of variation in phylogenetic clustering (Fig. 5.2). The Japanese reference isolate had a closer sequence similarity with the Korean isolate Geosan that predicted the largest dG value (Fig. 5.2). In the neighbor-joining tree, two Korean isolates, Haenam1, and Hoengseong, were closely clustered with the European sequences KX011115, and KX011135 (Laila et al. 2017).

5.8.5 Polymorphisms in *Plasmodiophora brassicae*

A number of single-nucleotide polymorphisms (SNPs), oligonucleotide polymorphisms, and insertions-deletions (In Dels) among the 11 Korean field isolates were observed. Such variation was predominant in the large subunit. To develop specific markers based on single-nucleotide or oligonucleotide polymorphisms, ten probes were designed and tested through high-resolution melting. Four out of the ten probes resulted in significant melting temperature differences (Fig. 5.3); however, the remaining six probes showed no differences (Fig. 5.3). Probes 661, 856, and 903 clearly separated Yeoncheon, Haenam1, and Hoengseong (red peaks) from the other eight isolates. The melting temperature difference between the two groups for probes 661, 856, and 903 were 5 °C, 8 °C, and 7 °C, respectively (Fig. 5.3). Probes 661 and 856 separated the Korean isolates based on SNPs, whereas probe 903 separated the two groups based on an InDel of four nucleotides. Probe 629 had about a 6 °C melting temperature difference between peaks produced by Daejon, Haenam2, Seosan, Pyeongchang, Geumsan (red peaks, Fig. 5.3), and the other six isolates (Gangneung1, Yeoncheon, Gangneung2,

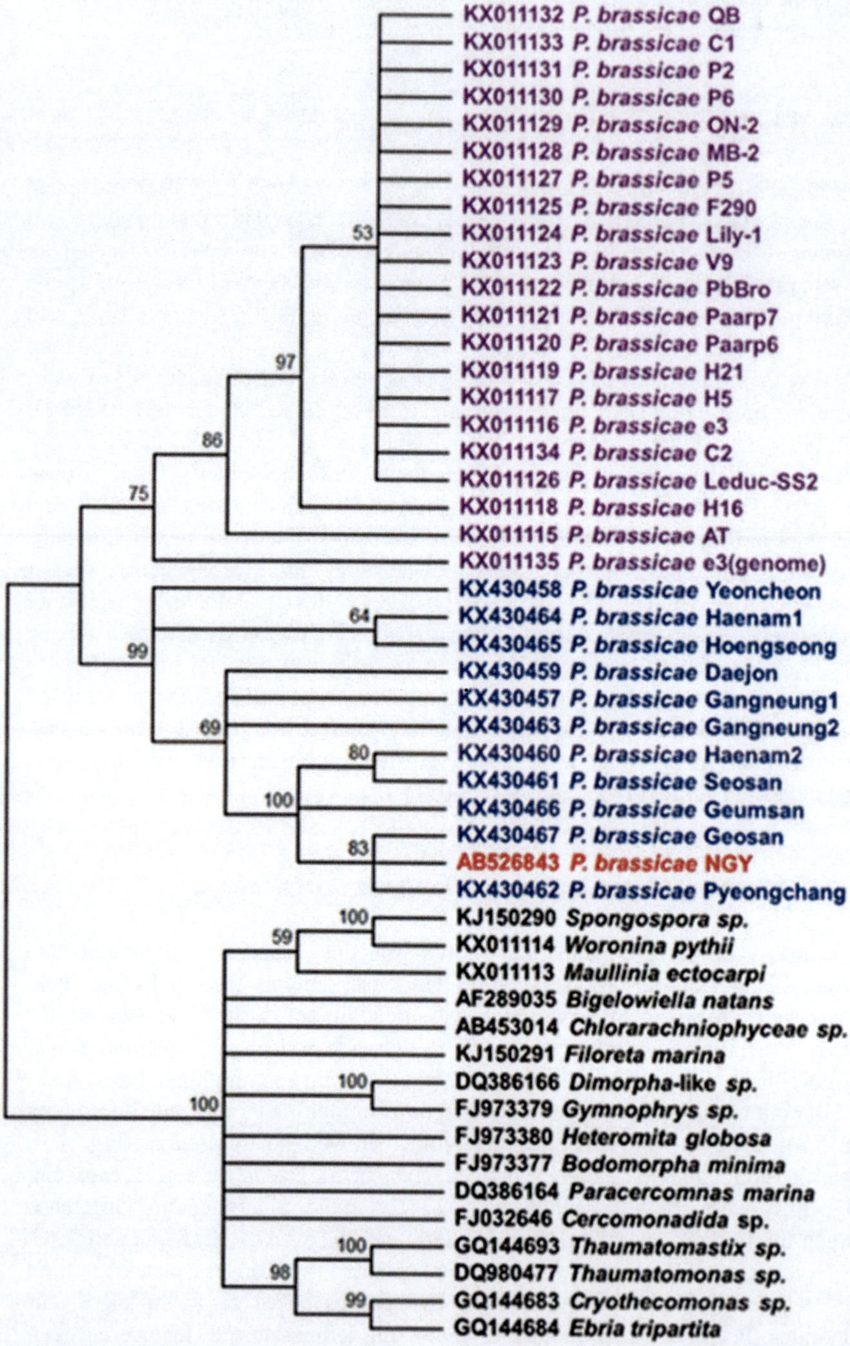

Fig. 5.2 Phylogenetic tree constructed following the Maximum Likelihood method, and Jukes-Cantor model in Mega 6.06 software between the 11 Korean field isolates, and the reference isolates of *P. brassicae*, based on variations in the nucleotide sequences of rDNA. Complete rDNA, if available or LSU sequences were used. Accessions with "KX4304" were obtained in this study.

Haenam1, Hoengseong, Goesan, blue peaks in Fig. 5.3) due to variation in three adjacent nucleotides (Laila et al. 2017).

5.8.6 Molecular Marker for Identification of *Plasmodiophora brassicae* Pathotypes

Breeding of clubroot-resistant plants has been hampered by the large variation of pathogenicity in *P. brassicae* and by the lack of an efficient means for detecting specific isolates. To improve the practicality of *P. brassicae* pathotype-identification, a molecular approach was developed. RAPD profiles of 37 single-spore-derived isolates belonging to seven different pathotypes were compared. A RAPD marker, OPL141200, was found in the molecular pattern of all the isolates belonging to one particular pathotype (P1), pathogenic on all differential hosts tested. The DNA band corresponding to this marker was cloned and sequenced. No significant homology to previously characterized nucleotide sequences was found. Primers were designed to specifically amplify the OPL141200 band. The SCAR marker was observed in all isolates belonging to pathotype P1 and was absent in isolates belonging to other pathotypes and in the different plant hosts analyzed. The SCAR marker was also generated from direct amplification of DNA from clubs (mixture of host and pathogen DNA) developed after infection by P1 isolates. This molecular marker may be a valuable tool for rapid and reliable identification of *P. brassicae* P1 isolates in areas where resistant varieties are cultivated (Manzanares-Dauleux et al. 2000).

To investigate genetic diversity, and geographic differentiation of *P. brassicae* populations from different regions, Strehlow et al. (2014) compared three regions of Germany that differ in oilseed rape cropping history, oilseed rape acreage, and incidence of clubroot. These regions were either spatially separated or separated by the former inner German border. *Plasmodiophora* isolates were collected from 59 fields (29, 17, and 13 fields per region, respectively), and 174 amplified fragment length polymorphism (AFLP) markers were analyzed. Every field isolate showed a unique genotype pattern; that is, no genotype was shared among the regions and different fields. The mean gene diversity was 0.27, suggesting that *P. brassicae* is a genetically diverse species. The comparison of indexes (gene diversity, genotypic diversity, and linkage disequilibrium) between the regions does not support the hypotheses that cropping history, oilseed rape acreage, and incidence of clubroot affect these estimates. Principal component analysis (PCA), fixation index (FST), and generalized linear model (GLM) were suitable to specify regional differences. PCA revealed two clusters of isolates based on the geographic origin of the isolates, and FST showed that these clusters were highly differentiated. Hypotheses about

Fig. 5.2 (continued) Purple, red, and blue colors indicate *P. brassicae* rDNA sequences obtained from Schwelm et al. (2016), Niwa et al. (2011) and this study. Black colors indicate rDNA sequences from other *Cercozoa* species (Laila et al. 2017)

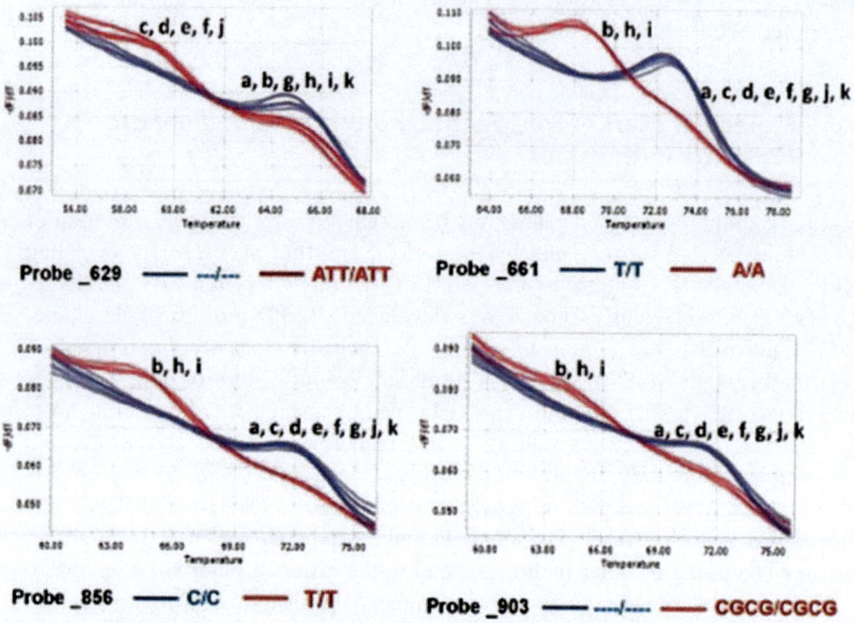

Fig. 5.3 Variation due to polymorphism between Korean *Plasmodiophora brassicae* isolates. The melting temperatures are shown for probes (single-nucleotide, and oligonucleotide polymorphism, and In Del) after high-resolution melting analysis in a Roche light cycler. a, Gangneung1; b, Yeoncheon; c, Daejon; d, Haenam2; e, Seosan; f, Pyeongchang; g, Gangneung2; h, Haenam1; i, Hoengseong; j, Geumsan, and k, Goesan (Laila et al. 2017)

association of genotypes with different spatial scales were tested with GLM: the region, reflecting the cropping history, and the individual field had a significant effect on the AFLP pattern. Strehlow et al. (2014) proposed that individual field isolates represent a discrete population and that geographic differentiation results from low levels of gene flow due to the limited dispersal of this soil borne pathogen and from localized selection pressure as unifying force on the genotypes. By using PCR with gene-specific primers, 117 nonhouse keeping *P. brassicae* genes were investigated for their presence/absence in isolates representing pathotypes 2, 3, 5, 6 or 8 (Williams' system), and in new strains of the pathogen recently identified from Alberta, Canada, which are highly virulent on clubroot resistant canola (*Brassica napus*) cultivars. One gene, named *Cr811*, was present in isolates of pathotype 5 and in a new strain of *P. brassicae* that is classified as pathotype 5 on the system of Williams, but causes severe clubroot on resistant canola. Three additional newly identified virulent populations, similar to pathotype 3, also were tested, and found to carry *Cr811*. In contrast, the *Cr811* gene was absent in all isolates representing the original pathotypes 2, 3, 6, and 8. Quantitative real-time PCR analysis indicated greater copy numbers of *Cr811* in the new virulent pathotype 3-like and pathotype 5-like populations versus the original pathotype 5, indicating a potential method to identify new pathotypes of *P. brassicae* before the breakdown of resistance can be detected in a canola crop. The association of *Cr811* with infection by *P. brassicae* of previously resistant canola cultivars, as revealed by qPCR analysis, suggests that *Cr811* plays an important role in the interaction between host resistance and *P. brassicae* pathogenicity (Feng et al. 2016).

5.8.7 Identification of Geographical Isolates of *Plasmodiophora brassicae* by Ribosomal DNA Sequence Polymorphisms

In plant disease diagnostics, the application of marker-based detection of isolates of a pathogen has become essential. Polymerase chain reaction (PCR)-marker-based detection is advantageous for accurate identification of nonculturable obligate parasites. PCR-based techniques for detection of *P. brassicae* have targeted both a single-copy sequence, and the multicopy ribosomal DNA. Ribosomal DNA (rDNA) is a nuclear DNA sequence that codes for ribosomal RNA (rRNA), which is necessary for the synthesis of proteins, cellular growth, and development of organisms. About 80% of the rRNA present in a growing cell takes part in modulating approximately half of all cellular translation. An rDNA repeat unit in eukaryotes generally consists of several coding regions including a smaller subunit, a larger subunit, a comparatively larger intergenic spacer (IGS), and two comparatively larger internal transcribed spacers, ITS1 and ITS2, on either side of a 5.8s subunit. Ribosomes translate mRNA molecules to produce proteins. Figure 5.4 reveals that the rDNA of *P. brassicae* consists of a small subunit (SSU), ITS1, 5.8s, ITS2, and a large subunit (LSU). Designing, testing, and confirming molecular markers that can quickly identify *P. brassicae* isolates collected from different geographical locations may improve efforts to halt the dissemination, and better control this pathogen. *P. brassicae* isolates with geographical variation (Swedish versus UK) had single-nucleotide polymorphisms (SNP) in the ITS, and 18s smaller subunits. Such variation can be exploited for accurate detection of a particular isolate with SNP markers. To design these markers, molecular characterization of the rDNA

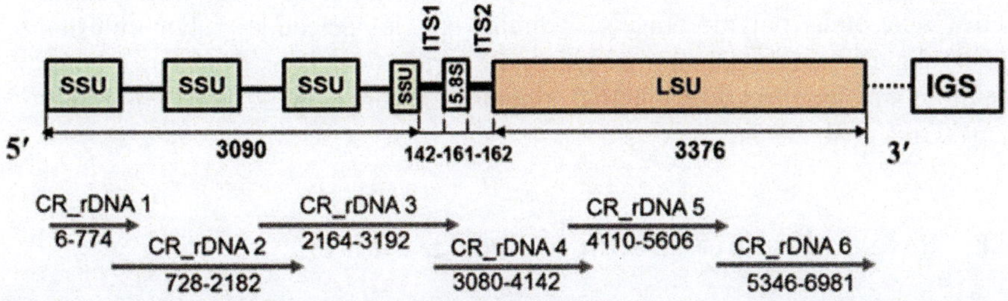

Fig. 5.4 Schematic structure of the ribosomal DNA (rDNA) of *Plasmodiophora brassicae*. SSU, small subunit; LSU, large subunit; ITS, internal transcribed spacer; IGS, intergenic spacer based on literature; CR DNA (from 1 to 6) indicates marker sequences developed for cloning the target regions. Numbers in the figure are lengths in base pairs (bp) of each subunit cloned, and sequenced. The dotted line in-between LSU, and IGS at the 3′ end indicates uncompleted sequences. Arrows from 5′ to 3′ direction indicate length of primers (Laila et al. 2017)

is essential. Niwa et al. (2011) published the complete rDNA structure of *P. brassicae* of Japanese origin (accession number AB526843; clone NGY; 9513 base pairs (bp)). They also sequenced partial rRNA sequences of the LSU from 25 clones from eight Japanese geographical isolates (field populations); these sequences were between 1131 and 1479 bp, and were quite diverse in nucleotide sequence compared to the NGY isolate. Schwelm et al. (2016) published nearly complete rDNA sequences of two isolates of *P. brassicae* of European origin, AT (KX011115, 7909 bp) and e3 (KX0111135, 7382 bp), which were nearly identical to each other at their 28s LSU. They also published partial sequences for 28s rRNA (1143 bp) from 19 other *P. brassicae* isolates originating from Germany (KX011116), Sweden (KX011117–KX011123), New Zealand (KX011124), Canada (KX011125–KX011132), and South Korea (KX011133–KX011134). These 19 partial rDNA sequences originating from five different countries were identical to AT and e3 isolates originating from Germany and Austria, respectively (Schwelm et al. 2016). In Korea, a number of geographical isolates of *P. brassicae* have been collected (Jo et al. 2011). Laila et al. (2017) determined the complete structure of the rDNA repeat unit of *P. brassicae* and identified the polymorphisms among Korean field populations of different geographic origin in order to develop molecular markers for characterization of the isolates. Understanding the *P. brassicae* genetics, and developing efficient molecular markers, is essential for effective detection of harmful races of this pathogen. Samples from 11 Korean field populations of *P. brassicae* (geographic isolates); collected from nine different locations in South Korea were used in this study. Genomic DNA was extracted from the clubroot-infected samples to sequence the ribosomal DNA. Primers and probes for *P. brassicae* were designed using a ribosomal DNA gene sequence from a Japanese strain available in GenBank (accession number AB526843; isolate NGY). The nuclear ribosomal DNA (rDNA) sequence of *P. brassicae*, comprising 6932 base pairs (bp), was cloned and sequenced and found to include the small subunits (SSUs) and a large subunit (LSU), internal transcribed spacers (ITS1 and ITS2), and a 5.8s. Sequence variation was observed in both the SSU and LSU (Fig. 5.4). Four markers showed useful differences in high-resolution melting (HRM) analysis to identify nucleotide polymorphisms including single-nucleotide polymorphisms (SNPs), oligonucleotide polymorphisms, and insertions/deletions (InDels). A combination of three markers was able to distinguish the geographical isolates into two groups (Laila et al. 2017).

5.8.8 Race and Isolate-Specific Molecular Marker

In Korea, there are at least five races that belong to four pathotype groups. A study conducted in Korea attempted to develop molecular markers based on ribosomal DNA polymorphism to detect *P. brassicae* isolates, but none of those markers was either race-specific or pathotype-specific. Jeong et al. (2018) developed race and isolate-specific markers by exploiting genomic sequence variations. A total of 119 markers were developed based on unique variation existing in genomic sequences of each of the races. Only 12 markers were able to detect *P. brassicae* strains of each isolate or race. Ycheon14 markers were specific to isolates of race 2, Yeoncheon, and Hoengseong. Ycheon9, and Ycheon10 markers were specific to Yeoncheon isolate (race2, pathotype 3), ZJ1-3, ZJ1-4, and ZJ1-5 markers were specific to Haenam2 (race 4) isolate, ZJ1-35, ZJ1-40, ZJ1-41, and ZJ1-49 markers were specific to Hoengseong isolate and ZJ1-56, and ZJ1-64 markers were specific to Pyeongchang isolate (race 4, pathotype 3). The PCR-based sequence characterized amplified region (SCAR) markers developed are able to detect five Korean isolates of *P. brassicae.* These markers can be utilized in identifying four Korean *P. brassicae* isolates from different regions.

Pathotype differentiation via virulence on sets of differential hosts is time consuming, labor intensive, and can be affected by environmental conditions. Further, due to the fact that different pathotypes may be present in single galls, hence complicated results may be obtained from such tests. Development of molecular markers specific to particular pathotypes would be a valuable tool to identify, and monitor *P. brassicae* populations. A random amplified polymorphic DNA (RAPD) derived sequenced characterized amplified region (SCAR) marker specific to P1 as defined on the differential of Some et al. (1996) has been developed, and proven to be useful for the identification of this pathotype from field isolates. The molecular determinants of a pathotype are most likely related to the presence/absence or differential expression of genes related to pathogenicity. In GenBank, there are approximately 120 nonhouse keeping genes of *P. brassicae* represented by expressed sequence tags (EST) or predicted proteins. 118 primers pairs specific to most of these genes were used by Zhang et al. (2015) to evaluate the presence of the corresponding genes in genomic DNA samples prepared from single spore isolates belonging to five pathotypes as classified on the differential set of Williams (1966). The members of these 118 genes may be specific to certain pathotypes. To identify gene(s) that is deferentially present in pathotypes, and to molecularly characterize the identified gene(s), and to test the usefulness of these gene(s), as markers to differentiate unknown pathotypes, such markers are very useful. One gene, designated *Cr811*, was identified to be exclusively present in P5. The in planta expression of this gene was studied using an infection time course on both a resistant, and a susceptible canola genotype. Primers specific to this gene were used to identify P5 from field isolates.

Isolates of the pathogen can be classified into pathotypes according to their pathogenicity on differential hosts. The presence or absence of all database available nonhouse keeping *P. brassicae* genes (118 in total) were assessed by polymerase chain reaction (PCR) analysis in isolates belonging to five *P. brassicae* pathotypes (2,3,5,6, and 8 according to Williams differential set). One gene designated *Cr811* was present exclusively in the isolate of pathotype 5. This was further confirmed by dot blot hybridization, and by PCR using alternate DNA preparations, and primers. Reverse transcription quantitative PCR analysis indicated that in planta, expression of *cr811* was upregulated during canola infection, especially in the stage of secondary plasmodia. Primers specific to *cr811* could distinguish a field isolate of *P. brassicae* belonging to pathotype 5 from two other field isolates representing pathotypes 3, and 8. These findings suggest that *cr811* is a gene that is potentially involved in clubroot pathogenesis, and that it might serve as a molecular marker for differentiation of pathotype 5 from other pathotypes (Zhang et al. 2015). Differential expression of genes between pathotypes 4 (P4) and pathotype 7 (P7) was assessed according to transcriptome data of molecular marker screening. Among the pathotypes (P2, P4, P7, P9, P10, and P11) tested, six genes were exclusive to P4, dividing the isolates into three types: *PBRA_003263* and *PBRA_003268* were present in all P4 isolates. *PBRA_000003/Novel 1512* were found in a type of P4 (P4-1) and Novel 137/PBRA_005772 were found in another P4 type, P4-2 Amplicons for all six genes were produced for only one isolate, which was named P4-3. The genes identified might serve as molecular markers for differentiation of P4 from other pathotypes and may distinguish different types of P4 (Zheng et al. 2018).

5.8.9 Pathogenic and Genetic Variation in *Plasmodiophora brassicae* Pathotypes

The molecular markers have been used to analyze the relationship between genetic variation and pathogenicity or host preferences in population and single-spore isolates of *P. brassicae* (Buhariwalla et al. 1995; Fahling et al. 2003; Ito et al. 1997, 1998; Manzanares-Dauleux et al. 2001; Möller and Harling 1996; Osaki et al. 2008; Yano et al. 1997). These findings are contributing to understanding the genetic background involved in the pathogenic variation of *P. brassicae* populations. In Japan, Yano et al. (1996, 1997) have shown that two populations (Williams' race 9) are clearly distinguishable from 14 other populations (Williams' races 1 and 4) by random amplified polymorphic DNAs (RAPDs), and the result corresponds well with results using isozyme patterns in the same populations. However, their methods could not distinguish between pathogenic and nonpathogenic populations on club-root resistance cultivars of Chinese cabbage bred in Japan. Thereafter, Osaki et al. (2008) analyzed the relationship between RAPDs, and pathogenic variation (pathotypes determined using clubroot resistance cultivars of Chinese cabbage, and Williams' races) among additional populations, including those from the weed, *C. flexuosa*, from Japan. The cluster analysis based on RAPD data has indicated that three populations (Williams' race 9) are clearly distinguishable from other populations (Williams' races 1 and 4), as described by Yano et al. (1997). It reveals that partial correspondence exists between pathotypes and genetic variations in populations as follows: all pathotype B, and D populations were distinguishable from all pathotype C populations except one (Fig. 5.5). However, their method could not distinguish pathotype B populations from those of pathotype D or pathotype A population from other locations.

Pathotype populations A are extremely heterogenic, and are presumed to have arisen through plural evolutionary processes (Osaki et al. 2008). Genetic variation among populations of *P. brassicae* from Japan is relatively high, but the correlation between pathogenic, and genetic variations in these populations is not very high (Yano et al. 1997; Osaki et al. 2008). Heterogeneity in each population may have

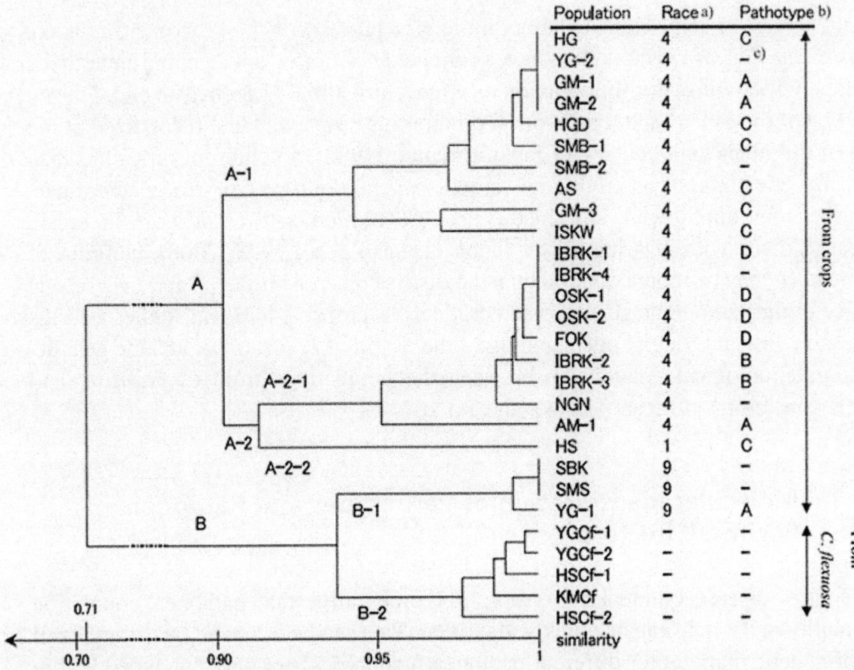

Fig. 5.5 Dendrogram based on RAPD data for 28 populations of *Plasmodiophora brassicae* from cruciferous crops, and the wild crucifer *Cardamine flexuosa*. The dendrogram was constructed by the unweighted paired group method with arithmetic averaging from the matrix of coefficients of simple matching similarity. (**a**) Williams' race, (**b**) pathotype determined on the basis of pathogenicity on two clubroot resistant cultivars of Chinese cabbage, CR Kukai 65, and CR Utage 70, pathotype A pathogenic on both cultivars, pathotype B pathogenic on CR Kukai 65 but nonpathogenic on CR Utage 70, pathotype C nonpathogenic on CR Kukai 65 but pathogenic on CR Utage 70, pathotype D nonpathogenic on both cultivars, c not tested (Tanaka and Ito 2013)

affected the results of the authors' analyses of pathogenicity, and DNA polymorphisms. Manzanares-Dauleux et al. (2001) used single-spore isolates for similar analyses, the correlation between pathogenic, and genetic variations was relatively low, even among these isolates. Specialization of races and pathotypes seems to reflect only a limited aspect within the high level of genetic diversity in this organism. In a dendrogram based on RAPD analysis, *P. brassicae* populations from *C. flexuosa* grouped into a single subcluster, and were located in a single cluster with all Williams' race 9 populations from crops (Osaki et al. 2008; Fig. 5.5). The results indicate that these two groups of populations are genetically closely related. The race 9 populations are minor within *P. brassicae* from Japan, and have unique pathogenicity (avirulent on many common cultivars of cabbage, and have relatively lower virulence than race 4 populations on common cultivars of Chinese cabbage, but are virulent on many clubroot resistance cultivars of Chinese cabbage) (Tanaka et al. 1991, 1997, 1998). The significance of the genetically close relationship between the two groups of populations from crops, and *C. flexuosa* is not clear; however, available results may point to an evolutionary process of the pathogen in Japan. *Cardamine flexuosa* is thought to have originated in Eurasia, and become naturalized in Japan following the introduction of rice cultivation (Takematsu and Ichizen 1993). On the other hand, clubroot on cruciferous crops was first recognized at the end of the nineteenth century in Japan (Ikegami 1992), so clubroot on *C. flexuosa* may precede clubroot on cruciferous crops. *Plasmodiophora brassicae*, an obligate parasite, may thus have a long history of co-evolution with *C. flexuosa* in Japan. Because of its wide distribution in Japan (Tanaka et al. 1993, 2006), clubroot of *C. flexuosa* may be more common than the disease on crops in Japan and may be one source of the pathogenic genotypes on cruciferous crops. Additional studies are thus necessary for a thorough understanding of the origin of *P. brassicae* and the genetic, and epidemiological relationship between the populations from *C. flexuosa*, and cruciferous crops in Japan (Tanaka and Ito 2013).

5.8.10 Molecular Identification of Pathotypes and Pathogen Elicitors/Effectors

To control disease outbreaks in crops, it is imperative that pathogen population variability in the field can be quickly identified. This can be achieved by having a full set of genetic markers for different pathogen pathotypes, not only contained within the local fields but universally. The initiatives for *P. brassicae* showed that *Cr811* was a good candidate gene to differentiate pathotype 5 from the rest of the pathotypes (Feng et al. 2016). Molecular characterization of pathogen effectors is gaining momentum. With better genome assemblies and higher gene prediction accuracies, it will soon have better insights into the structures and functions of effector proteins, which allow more efficient screening and identification of *R* genes (Gibriel et al. 2016). There should be a focus on studies that provide additional molecular details at the gene level which would further our understanding of the factors leading to virulence and susceptibility associated with *R* gene-mediated resistance (Table 5.14; Neik et al. 2017).

5.8.11 Role of Pathogenic Variability in the Genetic Complexity of Clubroot Resistance (CR) Genes

The virulence of pathogen and resistance of host plant also changes along with the continuous game between them. The virulence of *P. brassicae* with many races (pathotypes) has been identified. At present, most CR resources identified in *Brassica* were pathotype specific. However, the resistance to single race controlled by single gene was unstable, which was continuously eroded by the pathogen's adaptation. The above dynamics and instability made the genetic regulation of host plant became more complexity. Until now, there is no certain conclusion about genetic regulation mode of CR. Genetic analysis showed that both qualitative and quantitative manners existed in the important *Brassica* crops due to the different resistant sources. In *B. rapa*, CR was controlled in qualitative and quantitative manners both, whereas it was quantitative traits in *B. oleracea* (Piao et al. 2009; Lee et al. 2016). Genetic studies on CR of rapeseed were mostly based on the sources, of which the resistance was derived from *B. rapa* (Zhang et al. 2015). The inheritance of (CR) in rapeseed was governed by one or two single independent dominant genes (Piao et al. 2009). The differences of genetic regulation mode in CR among different species may be related to the various races. So enhancing the study to physiological race, including developing of new identification system, is the basis to perform intensive study.

5.8.12 Whole Genome DNA Similarity and Pathotype Population Structure of *Plasmodiophora brassicae*

Pathotypes of clubroot have been characterized based on their ability to cause disease on differential hosts. The most widely used differential sets include the Williams' differential set, the European Clubroot Differential set, the Somé set, and the Canadian Clubroot Differential set. Molecular markers have been identified to distinguish the predominant pathotypes (P11, P9, and P4) in China and pathotype 5X, one of the virulent new pathotypes from canola in Canada, but such markers are not yet in wide spread use and no markers exist for the other pathotypes (Sedaghatkish et al. 2019). Genome diversity in plant pathogen populations is the result of migration, random genetic drift, mutation, recombination, and natural selection. *Plasmodiophora brassicae* is a soil-borne pathogen, so migration and gene flow among regional populations almost certainly occurs most commonly as the result of human activities, although movement via wind and water also contribute to pathogen dispersal. The *P. brassicae* genome has been sequenced using different pathotypes (see Chap. 2 at Sect. 2.9). Although *P. brassicae* cannot be grown in pure culture, dual tissue cultures of *P. brassicae* with a host species have been used to culture this pathogen. Dual cultures of *P. brassicae* with callus of *Brassica* spp. have been used to observe the development of *P. brassicae* within host cells with minimal contamination by soil microbes and have been used to provide pure samples of DNA for genomic analysis of other plasmophoroids.

The genetic similarity of *P. brassicae* strains from across Canada and selected locations in China and the USA were assessed using whole-genome DNA sequences and compared with a published genome sequence from Europe (Sedaghatkish et al. 2019). The limited numbers of single spore isolates assessed were often nearly identical to the collections they were taken from A phylogenetic tree that consisted of five clades was developed based on SNP polymorphisms and validated using several approaches (Fig. 5.6). Strains of *P. brassicae* from vegetables across North America formed Clade 1. The Clade 2 was mainly from canola outside of the Canadian Prairies at sites where the change in pathotype had been detected. Three of five collections from China clustered together in Clade 3, which was substantially different from strains in North America and the single-spore isolate from Europe. Strains from canola on the Prairies mostly clustered into Clades 4 and 5. Clade

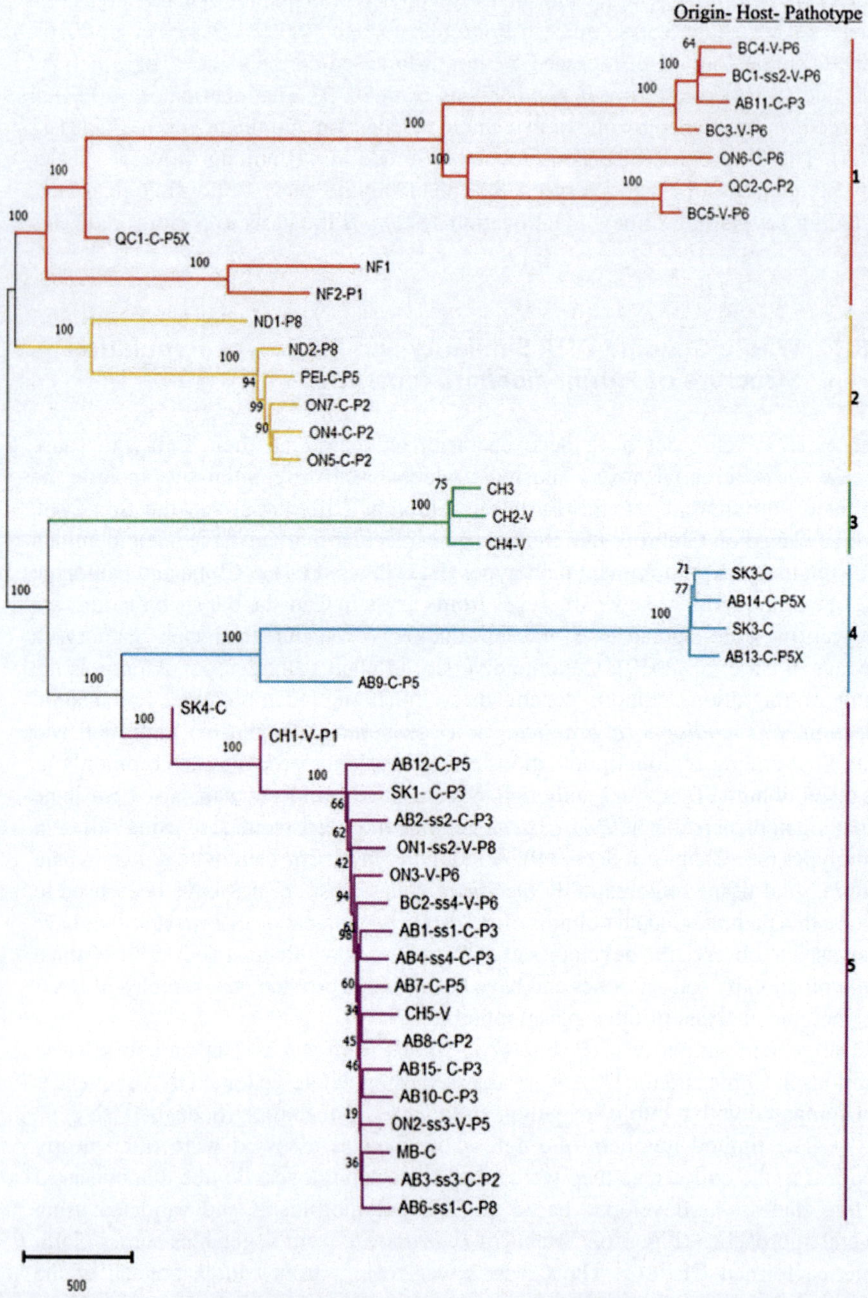

Fig. 5.6 SNP-based phylogenetic tree of *Plasmodiophora brassicae* based on whole-genome alignment of 43 *Plasmodiophora brassicae* strains from Canada, China, and the USA mapped against e3 from Europe. The scale represents the evolutionary distances. The percentage of 1000 bootstrap trials (1000 replicates) is shown above the branches. Strains details are summarized in their names: location by province (e.g., AB = Alberta), Williams' pathotype (e.g., P3) where

4 consisted of the new, virulent pathotypes from Alberta. Most of the strains from canola in Saskatchewan and Manitoba, which represent recent expansion of the geographic distribution of the pathogen, clustered together with the older collections from canola in Alberta and two collections from China in Clade 5. Clades 1 and 2 clustered separately from the initial strains from canola on the Canadian Prairies (Clade 5), but differed even more from the cluster of new, virulent pathotypes that are currently increasing in the Prairie region (Clade 4). A separate clade consisting of collections from China was expected, given the relatively long period of geographic separation and differential selection between the pathogen populations in North America and China, but the observation that 2 of 5 collections from China clustered with the original strains from canola in Alberta indicated that selection pressure may be driving both populations in a similar direction.

The collections AB13-C-P5X and AB14-C-P5X (new pathotype 5X from Alberta) were very different from the older collections from Alberta, with SNP disruption of function in thousands of genes. It is unlikely that this new, virulent pathotype could have accumulated thousands of mutations since the initial registration of a clubroot-resistant canola cultivar in 2009. However, the possibility of mixed populations of pathotypes being present within a club might have influenced the gene diversity among field collections. Similarly, genetic diversity was high between collections taken before and after a change of pathotype at sites in Quebec and Ontario. The high diversity between these collections indicated that the likelihood that the new pathotypes developed from the original pathotypes was very low. These results indicated that the "new" pathotypes were the result of selection of pre-existing genotypes present at low levels in the population (rather than from mutation to virulence) that increased rapidly in response to selection first on a new host and later on resistant cultivars. Population-structure analysis was also used to identify the migrants between two populations. Population structure analysis indicated that collections AB9-C-P5 from Alberta, NF2-P1 and NF1 from New found land and ND1-V-P8 from North Dakota were admixtures of two populations. Admixtures are generally the result of directional selection due to migration and isolation (Sung et al. 2012). The NF2-P1 and NF1 strains collected from Newfoundland were found in Clade 2 but were also occasionally assigned to Clades 1 or 5. The isolate Ab9-C-P5 was an admixture of Clades 4 and 5, while ND1-V-P8 from North Dakota was an admixture of Clades 2 and 4. Nucleotide diversity (π) depends on factors such as population size, mutation rate, and reproduction states (Sung et al. 2012). A previous study estimated nucleotide diversity per site of 0.32 for *P. brassicae* strains in Alberta (Holtz et al. 2018). The diversity was high between strains from the Prairies and the rest of Canada; for the whole genome, $\theta\pi = 0.0009$, and for average diversity per site, $\pi = 0.22$.

Fig. 5.6 (continued) available, and SS = single-spore isolate, V = vegetable host, and C = canola host (Sedaghatkish et al. 2019)

Sexual recombination is another important reason for genetic variation, which shapes the impact of selection in the genome. The timing and occurrence of sexual reproduction, if it occurs at all in the life cycle of *P. brassicae*, is not well understood. It has been suggested that secondary zoospores may fuse before the cortical infection phase, but no such fusion was observed in a study of secondary zoospores (Ludwig-Müller and Schuller 2008).

Sexual reproduction between different genotypes is clearly not an absolute requirement for infection by *P. brassicae*, since inoculation of a host with a single resting spore can result in infection and symptom development. No sexual recombination was detected in a study with repetitive molecular probes also. However, nonmeiotic recombination processes such as gene conversion and mitotic crossing-over have been reported in *P. brassicae* (Fahling et al. 2004).

To gain more knowledge about the occurrence of recombination in *P. brassicae*, linkage disequilibrium was calculated. Linkage disequilibrium generally represents a pronounced deviation from the random pattern associated with sexual reproduction. Linkage disequilibrium can be also caused by migration, genetic drift, selection, or other factors. Linkage disequilibrium between collections from the Prairies and the rest of Canada was assessed to better understand the genetic forces contributing to the development of new pathotypes. Disequilibrium was higher in strains from the Prairies relative to strains from the rest of Canada. This demonstrated that fewer genetic changes had occurred in Prairie populations compared with those in the other provinces in Canada. The rate that linkage disequilibrium decays with time over generations is influenced by the recombination frequency. In self-fertile plants, linkage disequilibrium decays more slowly and at a large genetic distance (up to 20 cM) than in out crossing species. In contrast, decay occurs faster (within 100–1500 bp) in a heterozygous background with several recombination events. One approach used to estimate linkage disequilibrium decay is to find the distance at which half of the maximum linkage disequilibrium has decayed.

The half decay of disequilibrium occurred at 29 kb when measured across all 43 *P. brassicae* strains. This value was in the range that has been reported for ten fungal species, in which the half decay value was 110 bp for species with a mixed reproductive mode and >100 kb for clonal species. Relatively rapid decay of linkage disequilibrium in *P. brassicae* populations indicated a high frequency of recombination in the population, which intern supports the hypothesis that *P. brassicae* is not a clonal species (Vos et al. 2017). Tajima's test of neutrality was performed to determine if balancing selection was apparent in the *P. brassicae* strains. When balancing selection is present, populations maintain variation, so alleles with low frequency or that less fit is not removed from the population. Balancing selection might help the population adapt to a sudden change, such as exposure to a new host. Tajima's D was 0.73 for 43 *P. brassicae* strains. The positive value of D indicated that balancing selection was present in these populations. This supported the observations of several previous studies, which reported a high percentage of diversity and the preservation of rare pathotypes in *P. brassicae* populations. The unique infection process of *P. brassicae*, which starts with thousands or millions of separate infections of root hairs from individual resting spores that, collectively result in production of one or a few large clubs, appears to be suited to result in balancing selection (Strehlow et al. 2014).

At present, no information is available on the mutation rate of *P. brassicae*. Random genetic drift is unlikely to result in rapid genome variation of *P. brassicae* because of the large populations of long-lived resting spores in soil. Only a limited number of studies of genetic diversity among regional populations of *P. brassicae* are available, generally based on partial genome sequences rather than the entire genome. Populations of *P. brassicae* were highly diverse among regions of Germany, and among fields and even years at a single site in north western France. In contrast, the diversity was so low among 21 isolates from Alberta the population was clonal. None of these studies showed a correlation among host source, virulence or DNA pattern of *P. brassicae*, which is consistent with the observation of Sedaghatkish et al. (2019).

In another study, a single-spore isolate from vegetables in British Columbia (ABOT-JE ss1) differed substantially from 20 field collections and single-spore isolates in Alberta. ABOT-JE ss1 was not available for the study, but the original field collection BC3-V-P6 (ABOT-JE-04-01) and two single-spore isolates from the same initial collection (BC1-ss2-V-P6 and BC2-ss4-VP6) were included. Field collection BC3-V-P6 and single-spore isolate BC1-ss2-V-P6 were very similar to e3 from Europe, which is in consistent with the previous study (Rolfe et al. 2016). Genetic similarity with e3 from Europe was high for Clades 1 and 4, and much lower for Clades 2 and 5. This same pattern was apparent in the lack of similarity between the pathotypes present before and after resistance breakdown, as illustrated both in the phylogenetic tree and heat map analysis of strains from the same site. Both of these broad patterns support the hypothesis that pathotype 3 had likely been present in the initial population(s) introduced to North America from Europe but was present only at very low levels on *Brassica* vegetables on the Prairies. Pathotype 3 became dominant on canola in Alberta only after repeated exposure (and selection for virulent pathotypes) on a new host, canola. Prior to the release of the first clubroot-resistant canola cultivar in 2009, all canola cultivars on the Prairies were susceptible to pathotype 3. Currently, selection pressure from wide spread use of clubroot-resistant cultivars has resulted in the increase of numerous new pathotypes capable of overcoming the genetic resistance available in the initial generation of clubroot-resistant canola cultivars. The diversity and rapid development of these new, virulent pathotypes further supports this conclusion. It also indicates that there may be many pathotypes still to be found among the quadrillions of resting spores currently present in the soil on the Prairies.

It was demonstrated that genetic diversity is high in *P. brassicae* populations in Canada. This high diversity may be largely the result of two conflicting forces: balancing selection in *P. brassicae*, which results in the maintenance of a wide variety of genetic lineages in the pathogen population, and selection pressure associated with differences in host and source of genetic resistance. Genetic diversity in *P. brassicae* was lower on the Prairies, where it encounters a narrow host range (primarily canola and susceptible weeds) relative to other production areas where the pathogen encounters a broad range of *Brassica* vegetables. It is not known when the pathogen was introduced to Alberta, but data from the study support the hypothesis that pathotypes virulent on canola likely took many years to build up to detectable levels because they occurred at low level in the initial populations introduced to the region on *Brassica* vegetables. In addition, new pathotypes appeared on the Prairies due to selection pressure from the wide spread use of resistant canola cultivars. Each clade in the phylogenetic tree contained multiple pathotypes and strains from various hosts, which indicated that pathotype and host were not the principal factors in grouping the strains. The dual tissue cultures provided an excellent source of DNA for sequencing this difficult and unique pathogen. Assessments of genetic diversity based on whole-genome sequences indicated that the "new" pathotypes on canola had likely been present in the pathogen population at very low levels, maintained through balancing selection, and became dominant due to selection associated with a host shift. The high linkage of loci in strains from the Prairies should make it possible to identify markers that correlate with pathotype (Sedaghatkish et al. 2019).

5.8.13 Genotyping of *Plasmodiophora brassicae* Reveals Distinct Population

The efficacy of the resistant cultivars as a clubroot management tool has not been as durable as first hoped. Severe clubroot symptoms were found on CR cultivars in some fields, with these symptoms shown to result from the emergence of new virulence phenotypes of *P. brassicae*. These new strains of the pathogen are highly virulent on nearly all CR canola cultivars tested, suggesting that the utility of genetic resistance as a management tool is at risk (Strelkov et al. 2016). Despite its importance, little is known about the population genetics of *P. brassicae*. Traditionally, the pathogen has been classified into pathotypes based on its virulence patterns on various host differential sets. In Canada, six pathotypes have been identified using the classification system of Williams, including pathotypes 1, 2, 3, 5, 6, and 8. All of these pathotypes, with the exception of pathotype 1, occur within the Canadian prairies. The recently identified *P. brassicae* field isolates capable of overcoming the resistance in CR canola cultivars are also classified as pathotype 5, but this designation does not reflect their increased virulence on these hosts. Infected host plants may contain multiple *P. brassicae* genotypes, collectively referred to as a field isolate. Single-spore isolates are sometimes obtained to produce genetically uniform samples, enabling more detailed analysis of the pathogen and avoiding the potential complications of working with heterogenetic field isolates.

Although molecular methods have been used with great success to study other plant pathogens, their successful application with *P. brassicae* has been difficult. Since the pathogen cannot be cultured, DNA extractions are contaminated by the host DNA which can interfere with traditional molecular marker techniques such as RAPD or AFLP analysis. Recent advances in genomics, however, have made examination of *P. brassicae* more feasible, and its genome was recently sequenced, as were the genomes of it shots *B. rapa* and *B. napus*.

The development of restriction site-associated sequencing (RAD seq) has facilitated the rapid and cost effective identification of large numbers of polymorphisms with in species. The RADseq is a form of reduced-representation library sequencing. It involves the digestion of DNA with one or more restriction enzymes followed by sequencing of the resulting fragments in a high through put DNA sequence. This method facilitates targeting of only a fraction of the genome, which allows for sequencing of a larger number of isolates and achieves a greater depth of coverage per locus for the same budget. Although, RADseq is employed frequently on species with large genomes, including agricultural crop species, this technique can also be employed on species with smaller genomes, typically by varying the restriction enzymes used in order to reduce the amount of genome reduction. Although contamination from host DNA is an issue with species such as *P. brassicae*, RADseq analysis does provide methods to mitigate that problem. RADseq data can be analyzed either de novo in the absence of a reference genome or by alignment to a reference genome. By aligning to the reference genome, contaminating DNA from other species should be removed from the analysis. Additionally, for host species, which are likely to contaminate DNA extractions, sequences aligning to the host genome can be, removed prior to RADseq data analysis, further reducing the chances that DNA that is not from the target species will affect the analysis. Holtz et al. (2018) performed a genome-wide examination of the diversity within *P. brassicae* using a RADseq approach. Specifically, the aims were to: (1) determine if there is any differentiation between *P. brassicae* populations from Alberta that are virulent on CR canola and other *P. brassicae* populations that are not virulent on CR canola, and (2) compare these pathogen populations with populations from other regions of Canada.

Restriction site-associated DNA sequencing (RADseq) was used to examine the genetic diversity within *P. brassicae* single-spore and field isolates collected from across Canada. The isolates included individuals that were either capable or incapable of causing disease on clubroot resistant canola cultivars. Over 8750 variants were identified through RADseq. Population analysis indicated that most isolates belonged to one of two distinct populations, corresponding with the ability of isolates to cause disease on resistant cultivars. Within each population, there were low levels of genetic diversity. One thousand and fifty of the genetic variants that distinguished the two populations were non synonymous, altering the coding sequences of genes. The application of RADseq revealed two distinct populations of *P. brassicae* in Canada, suggesting multiple introductions of the pathogen into the country. The genetic variation found here would be important for future research and monitoring of the pathogen (Holtz et al. 2018).

5.9 *Brassica–Pseudocercosporella* Pathogenomics

5.9.1 Identification of Virulence Factors/Races

White leaf spot (*Pseudocercosporella capsellae*) is an important disease across most oilseed *Brassica*-growing countries including France, UK, Canada, Germany and Australia (Barbetti and Sivasithamparam 1981; Howlett et al. 1999; Eshraghi et al. 2005; Gunasinghe et al. 2016). *Pseudocercosporella capsellae* has a wide host range infecting diverse wild and cultivated crucifers, including oilseed, forage and vegetable Brassicas. In Western Australia, *P. capsellae* was first recorded in 1942 on *Brassica napus* (Shivas 1989), subsequently on turnip rapeseed (*B. rapa*), Chinese cabbage (*B. rapa* var. *chinensis*), broccoli (*B. oleracea* var. *italica*), swede (*B. napus* var. *napobrassica*) (Shivas 1989), and more recently on *B. juncea* (Eshraghi et al. 2005). Some species of the genus *Cercospora* are known as producers of cercosporin, visible as a purple-pink pigment adjacent to the margins of colonies growing on artificial media (Petrie and Vanterpool 1978; Daub and Ehrenshaft 2000).

The *Pseudocercosporella capsellae* has both sexual and asexual life cycles. When present, the teleomorph stage (*Mycosphaerella capsellae*), produced at the end of the crop season, is primarily responsible for survival of the pathogen over the period between crops and for production of primary inoculum for the next season (Inman et al. 1991; Gunasinghe et al. 2016). Variation of pathogen populations can relate to differences in pathogenicity and/or virulence of isolates of the same pathogen (McDonald and Linde 2002), and such differences can be attributed to differences in genetic makeup of the fungal isolates (Kuninaga et al. 1997). Understanding this variation within populations of agronomically significant pathogens is an essential criterion toward developing and applying management strategies, and in understanding the pathogen's ability to respond to implemented management practices. For instance, pathogens with high variation are more likely to develop fungicide resistance (Attanayake et al. 2013). Furthermore, this information provides a strong foundation for plant breeding programs to generate sustainable genotypes with enduring disease resistance. If the variation of the pathogen is not considered, supposedly resistant cultivars may become susceptible when sown in locations where races that overcome this resistance develop and/or are already present (Ma et al. 2009). Differences observed across isolates of *P. capsellae* on potato dextrose agar (PDA), such as the production of diffusible purple-pink pigment and growth habits, suggested possible variation within pathogen populations. Furthermore, in previous studies by Gunasinghe et al. (2013) and Eshraghi et al. (2007), where plants were sprayed with a mixture of isolates of *P. capsellae*, different reactions were observed across diverse *Brassica* species, including the oilseed species *B. napus*, *B. carinata*, and *B. juncea*. Gunasinghe et al. (2016) identified phenotypic and genetic variability of the 54 pathogen isolates from Western Australia and elucidated possible relationships with virulence. For this purpose, cotyledon screening tests of *B. napus* (oilseed rape), *B. juncea* (Indian mustard), and *R. raphanistrum* (wild radish), under controlled environmental conditions, were employed for 52 isolates of *P. capsellae* to assess their pathogenicity and virulence differences. Fifty-four *P. capsellae* isolates were characterized using ITS sequencing and phylogenetic analysis. Growth rate and ability to produce the purple-pink pigment indicative of cercosporin for the same 54 isolates were evaluated on three different media: potato dextrose, V8 juice and malt extract agars.

Fifty-four single-conidial isolates collected from *Brassica juncea* (Indian mustard), *B. napus* (oilseed rape), *B. rapa* (turnip), and *Raphanus raphanistrum* (wild radish) across Western Australia were investigated for differences in pathogenicity and virulence using cotyledon screening tests, genetic differences using internal transcribed spacer (ITS) sequencing and phylogenetic analysis, and growth rates on potato dextrose, V8 juice and malt extract agars. All isolates from the four-crucifer hosts were pathogenic on the three test species: *B. juncea*, *B. napus,* and *R. raphanistrum*, but showed differences in levels of virulence. Overall, isolates from *B. juncea*, *B. napus* and *B. rapa* showed greatest virulence on *B. juncea*, least on *R. raphanistrum* and intermediate virulence on *B. napus*. Isolates from *R. raphanistrum* showed greatest virulence on *B. juncea*, least on *B. napus* and intermediate virulence on *R. raphanistrum*. Growth and production of a purple-pink pigment indicative of cercosporin was greatest on malt extract agar and cercosporin production on V8 juice agar was positively correlated with virulence of isolates on *B. juncea* and *B. napus*. ITS sequencing and phylogenetic analysis showed that isolates collected from *B. napus*, *B. juncea* and *B. rapa*, in general and with few exceptions, had a high degree of genetic similarity. In contrast, isolates from *R. raphanistrum* were clearly differentiated from isolate groups collected from *Brassica* hosts. *Pseudocercosporella capsellae* reference isolates from other countries generally grouped into a single separate cluster, highlighting the genetic distinctiveness of Western Australian isolates (Fig. 5.7; Gunasinghe et al. 2016).

5.10 *Brassica–Sclerotinia* Pathogenomics

5.10.1 Genetic Diversity of *Sclerotinia sclerotiorum* Isolates

Out of the 17 isolates, one *S. sclerotiorum* isolate tested for vegetative compatibility, 46 paired with at least one other isolate, while 18 isolates were incompatible with all other isolates. Seven of the isolates lost viability during testing. A total of 12 MCGs were identified. Clarkson et al. (2017) demonstrated that the 71 isolates were grouped into three IGS haplotype groups, namely groups 3, 5, and 7. Notably, isolates within each of the 11 designated MCGs also grouped based on IGS haplotype. Single isolates from each of the 11 MCGs that collectively included all threats haplotypes, were chosen for further in vitro and in planta analyses. A range of growth rates on PDA were observed for the 11 isolates, varying from 53.0 mm for isolate CU10.12 to 70.6 mm for isolate CU11.19. All isolates were able to acidify the PDA medium from pH 7 to pH 4 as demonstrated by a color change of pH-indicator bromophenol blue from purple to yellow. Population diversity and genetic structure of *S. sclerotiorum* have been well studied around the world describing clonality

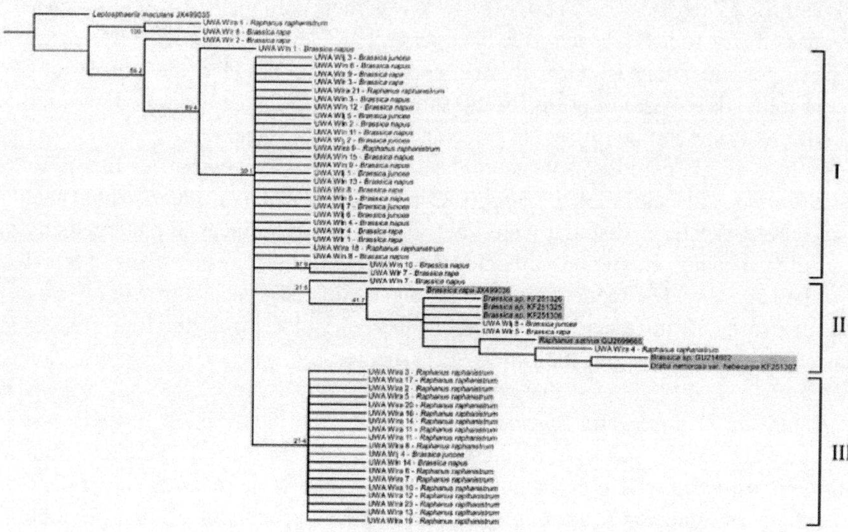

Fig. 5.7 Neighbor-joining tree showing the phylogenetic relationships among 54 isolates of *Pseudocercosporella capsellae* generated from this study and seven reference isolates of *P. capsellae* from GenBank (shaded), based on their internal transcribed spacer sequences as generated by GENEIOUS with Tamura–Nei genetic distance model. Bootstrap values of 1500 replications are shown for the major nodes. The host of origin of each isolate is given after each isolate name (Gunasinghe et al. 2016)

based on the self-fertility and predominant asexual reproduction by sclerotia. In contrast, recent studies conducted in Brazil, China, Iran, New Zealand, USA, and UK reported genetic recombination and mixed population structures of *S. sclerotiorum*. However, except for few studies conducted in tropical countries, almost all the studies have been conducted using pathogen samples from temperate and subtropical countries. Gomes et al. (2011) and Júnior et al. (2011) reported high genetic diversity and clonal as well as recombining populations of *S. sclerotiorum* in a tropical country, Brazil. Lehner et al. (2017) reported that there is no difference in genetic diversity in tropical and temperate conditions. Under conducive conditions and high level of pathogen population diversity, multiple infections per single host plant are often reported. When a host is infected by different species/genotypes of a pathogen, inhibition, co-survival or generation of new genotypes via recombination is possible ultimately compromising the host fitness.

Multiple genotypes of a pathogen species infecting a single host plant have been reported in several crops. Even though population structure and genetic diversity of *S. sclerotiorum* is well studied in temperate climatic conditions, only a few studies have been conducted in tropical countries. It is also not clear whether the populations are clonal or recombining in the tropics. Forty-seven isolates of *S. sclerotiorum* were collected from commercial cabbage (*Brassica oleracea* L.) fields in Nuwara Eliya district of Sri Lanka, where the disease has been previously reported. All the isolates were subjected to genetic diversity study using mycelial compatibility grouping and microsatellite markers. Fourteen mycelial compatibility groups (MCGs) and 23 multilocus haplotypes (MLHs) were recorded. Mean expected heterozygosity of the population was 0.56. MLHs were weakly correlated with MCGs. Population genetic structure analysis and principal coordinator identified three genetic clusters. Genetic recombination was inferred within each genetic cluster when isolates were subjected to clone correction. There was evidence of multiple infections on single plant as detected by the presence of more than one MCG on each cabbage plant. However, multiple infections did not increase the disease severity in detached leaf assay. Mahalingam et al. (2020) found high genetic diversity and recombination of *S. sclerotiorum* population in a tropical country, Sri Lanka (Table 5.18).

5.10.2 Aggressiveness of *Sclerotinia sclerotiorum* Isolates

The mean AUDPC values for each isolate × variety/accession combination in germplasm set 1 showed that the Australian *S. sclerotiorum* isolates were pathogenic on all nine Australian oilseed rape varieties and the Chinese accession, while mock inoculated plants did not show any symptoms. Significant differences in isolate aggressiveness were evident (Fig. 5.8). The five most aggressive isolates were CU8.24, CU10.20, CU10.17, CU8.20 and CU11.19, while CU10.12 and CU11.4 produced smaller lesions. Interestingly, these two isolates also had the slowest growth rate on PDA. Taken together, isolates CU8.24, CU10.17 and CU11.19 were highly aggressive on *B. napus*, demonstrated a rapid growth rate in vitro, and were genetically diverse. Therefore, these isolates were considered suitable for subsequent screening of *B. napus* accessions identified as partially resistant in a Canadian study (Gyawali et al. 2016).

5.10.3 Genetic Diversity in *Sclerotinia sclerotiorum* Population from *Brassica juncea*

The morphological and molecular characterization of pathogenic *S. sclerotiorum* geographical isolates from oilseed *Brassica* including *B. juncea* (Indian mustard) were investigated. The isolates of *S. sclerotiorum* originated from different agro-climatic conditions were analyzed for comparison similarity or differences between them as well as to examine the virulence of the pathogen specifically in *Brassica*. The collections of *S. sclerotiorum* isolates from symptomatic *Brassica* plants were analyzed for morphological features, and molecular characterization. The virulence evaluation test of 65 isolates on four *Brassica* cultivars has shown 5 of them as highly virulent, 46 as virulent and 14 as moderately virulent. Phylogenetic analysis encompassing all the morphological features, SSR polymorphism, and ITS sequencing has shown the existence of high genetic diversity among the isolates that categorized all the isolates in three evolutionary lineages in the derived dendrogram. Further, genetic variability analysis based on sequences variation in ITS region of all the isolates has shown the existence of either insertions or deletions of the

Table 5.18 Origin of *Sclerotinia sclerotiorum* isolates, their corresponding mycelial compatibility groups (MGCs) and multilocus haplotypes (MLHs) (Mahalingam et al. 2020)

Isolate name[a]	Location[a]	MCG	MLH	Isolate name[a]	Location[a]	MCG	MLH
SS5	Pattipola Field 2	1	1	SS11	Pattipola Field 4	3	2
SS26	Ambewela Field 1	1	1	SS15	Pattipola Field 5	3	2
SS35	Ambewela Field 3	1	1	SS19	Pattipola Field 6	3	2
SS36	Ambewela Field 4	1	1	SS31	Ambewela Field 1	3	2
A25	Seetha Eliya Field 5	1	1	A20	Seetha Eliya Field 3	3	2
A27	Seetha Eliya Field 6	1	1	SS18	Pattipola Field 6	3	15
G2	Meepilimana Filed 1	1	3	A15	Seetha Eliya Field 1	4	3
SS6	Pattipola Field 2	1	4	A26	Seetha Eliya Field 6	4	3
SS21	Pattipola Field 7	1	7	G1	Meepilimana Filed 1	4	3
A21	Seetha Eliya Field 3	1	7	G3	Meepilimana Filed 1	4	3
A16	Seetha Eliya Field 1	1	8	SS17	Pattipola Field 5	4	14
A22	Seetha Eliya Field 3	1	8	SS20	Pattipola Field 6	4	16
SS9	Pattipola Field 3	1	12	SS24	Pattipola Field 7	4	17
SS32	Ambewela Field 2	1	21	SS2	Pattipola Field 1	5	4
SS8	Pattipola Field 3	2	4	SS4	Pattipola Field 1	6	6
A23	Seetha Eliya Field 4	2	4	A18	Seetha Eliya Field 2	7	6
SS10	Pattipola Field 3	2	5	A19	Seetha Eliya Field 2	8	6
SS13	Pattipola Field 4	2	5	A24	Seetha Eliya Field 4	9	9
SS22	Pattipola Field 7	2	5	A17	Seetha Eliya Field 1	10	10
SS25	Pattipola Field 8	2	5	SS27	Ambewela Field 1	11	18
SS3	Pattipola Field 1	2	11	SS29	Ambewela Field 2	12	20
SS14	Pattipola Field 5	2	13	SS33	Ambewela Field 3	13	22

(continued)

Table 5.18 (continued)

Isolate name[a]	Location[a]	MCG	MLH	Isolate name[a]	Location[a]	MCG	MLH
SS28	Ambewela Field 1	2	19	SS34	Ambewela Field 3	14	23
SS7	Pattipola Field 2	3	2				

[a] Isolate names and locations are ordered according to mycelial compatibility groups (MCGs)

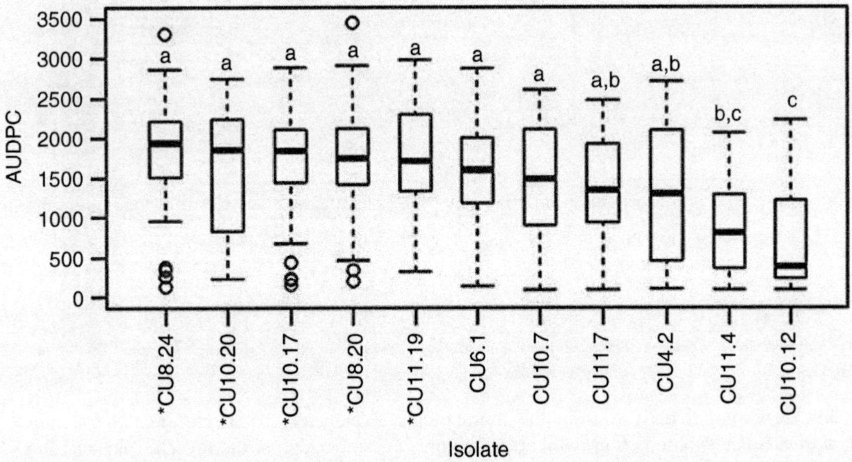

Fig. 5.8 Area under the disease progress curve (AUDPC) using stem lesion length measurements for 10 *Brassica napus* entries (germplasm set 1) inoculated with 11 Australian *Sclerotinia sclerotiorum* isolates. Each box plot displays the median and the lower and upper quartiles for each isolate. Letters above bars represent significant differences between entries as determined by ANOVA and Tukey's HSD post-hoc test ($p \leq 0.05$). Isolate names tagged with an asterisk were selected for screening of *B. napus* germplasm set 2 (Denton-Giles et al. 2018)

nucleotides in the ITS region which led to the interspecies variability and variation in a clade-specific manner. This analysis revealed the existence of higher heterogenecity and genetic variability in *S. sclerotiorum* isolates collection and indicates the presence of clonal and sexual progenies of the pathogen in the mustard growing regions of India surveyed (Figs. 5.9, 5.10, 5.11). The higher level of genetic variability and diversity among the *S. sclerotiorum* population needs robust screening approaches to identify the donor parent and utilize them in resistance breeding program to counter the menace of stem rot disease in *Brassica* (Sharma et al. 2018).

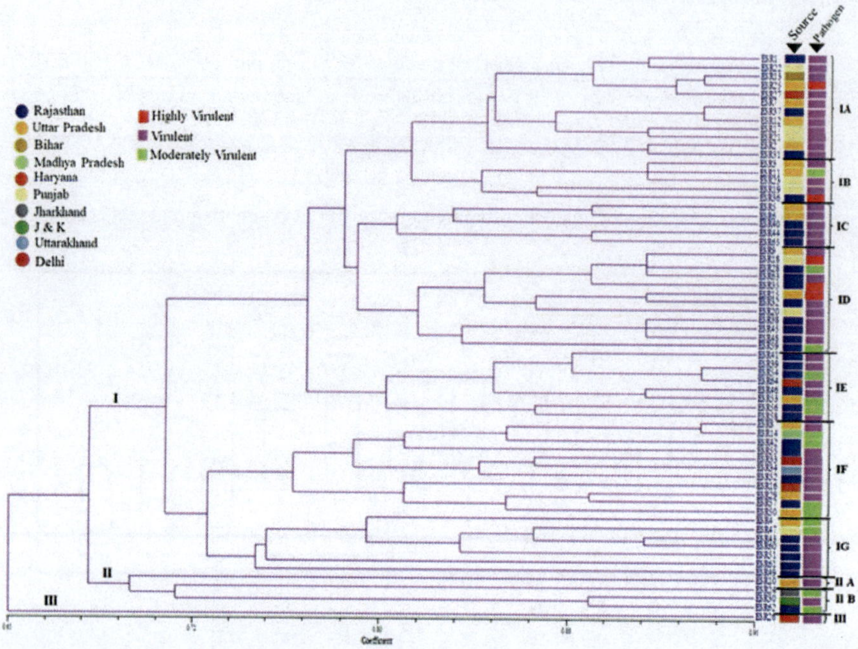

Fig. 5.9 Dendrogram derived from the morphological characters of 65 different *Sclerotinia sclerotiorum* isolates, based on growth, color, texture, shape and size of the mycelia and sclerotia grown on PDA medium and their respective pathogenicity over *Brassica* species (Sharma et al. 2018)

5.10.4 Genetic Diversity of *Sclerotinia sclerotiorum* in the North Central United States

The genetic diversity of 145 isolates of the pathogen from multiple hosts in the North Central region of USA was investigated. Mycelial compatibility groups (MCG) and micro satellite haplotypes were determined and analyzed for standard estimates of population genetic diversity and the importance of host and distance for genetic variation. MCG tests identified 49 different MCGs in the population and 52 unique micro satellite haplotypes.

There was an association between MCG and haplotypes for isolates belonging to the same MCG, which either shared identical haplotypes or differed at no more than 2 of the 12 polymorphic loci. For the majority of isolates, there was a one-to-one correspondence between MCG and haplotype. Eleven MCGs shared haplotypes. A single haplotype was found to be prevalent throughout the region. The majority of genetic variation in the isolate collection was found within rather than among host crops, suggesting little genetic divergence of *S. sclerotiorum* among hosts. There was only weak evidence of isolation by distance. Pair wise population comparisons among isolates from canola, dry bean, soybean, and sunflower suggested that gene flow between host-populations is more common for some crops than others.

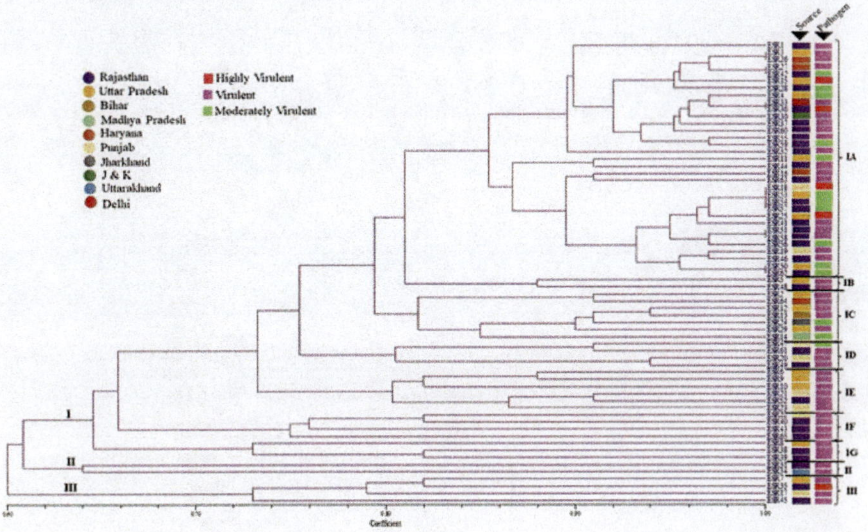

Fig. 5.10 Un-weighted pair group method with arithmetic mean (UPGMA) dendrogram showing the genetic relationships between the 65 isolates of *Sclerotinia sclerotiorum* assessed in this study, based on simple sequence repeat (SSR) markers (Jaccard's coefficient) and their respective pathogenicity over the *Brassica* species (Sharma et al. 2018)

Analysis of link age disequilibrium in the isolates from the four major crops indicated primarily clonal reproduction, but also evidence of genetic recombination for isolates from canola and sunflower. Accordingly, genetic diversity was highest for populations from canola and sunflower. Distribution of microsatellite haplotypes across the study region strongly suggest that specific haplotypes of *S. sclerotiorum* are often found on multiple crops, movement of individual haplotypes among crops is common and host identity is not a barrier to gene flow for *S. sclerotiorum* in the north central United States (Aldrich-Wolfe et al. 2015).

5.10.5 Genetic Diversity and Structure of *Sclerotinia sclerotiorum* Populations Detected by Random Amplified Polymorphic DNA Markers

The genetic diversity and genetic structure of a population of isolates of *Sclerotinia sclerotiorum* from different regions and host plants were investigated using the random-amplified polymorphic DNA (RAPD) method with 20 random decamer primer pairs in order to provide some information on the phylogenetic taxa and breeding for resistance to Sclerotinia stem rot. A minimum of three and a maximum of 15 unambiguously amplified bands were generated, furnishing a total of 170 bands ranging in size from 100 to 3200 bp, corresponding to an average of 8.5 bands per primer pair. One hundred and four of these 170 bands (61.2%) were

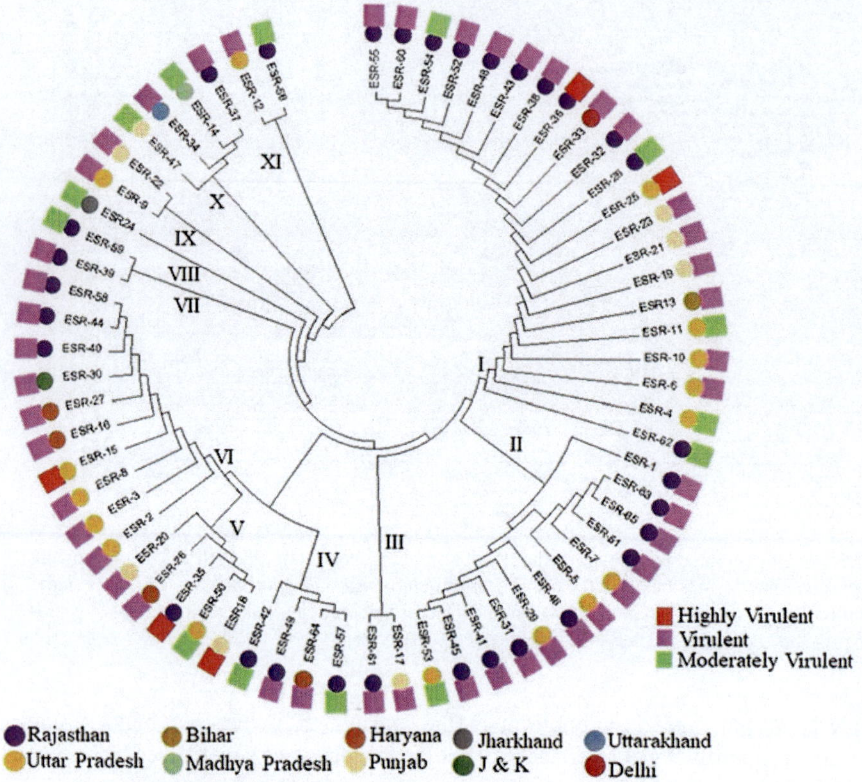

Fig. 5.11 Phylogenetic analysis of *Sclerotinia sclerotiorum* isolates based on rONA ITS sequences by Neighbor-Joining method and respective pathogenicity over *Brassica* species (Sharma et al. 2018)

polymorphic, the percentage of polymorphic bands for each primer pair ranging from 0.0% to 86.7%. The genetic relationships among the isolates, based on the results of RAPD analysis, were examined. The genetic similarity of all selected isolates was quite high. At the species level, the genetic diversity estimated by Nei's gene diversity (*h*) was 0.197 and Shannon's index of diversity (*I*) was 0.300. The unweighted pair-group mean analysis (UPGMA) cluster analysis showed that most isolates from the same regions were grouped in the same cluster or a close cluster. The population of isolates from Hefei (Anhui Province, China) was uniform and relatively distant to other populations. The Canadian population collected from carrot (*Daucus carota* var. *sativa* DC.) was relatively close to the Polish population collected from oilseed rape (*Brassica napus* L.) plants. There was no relationship between isolates from the same host plants. An analysis of molecular variance (AMOVA) revealed that the percentage of variance attributable to variation among and within populations was 50.62% and 49.38%, respectively. When accessions from China, Europe, and Canada were treated as three separate groups, the variance components among groups, among populations within groups, and within populations were—0.96%, 51.48%, and 49.47%, respectively. The genetic differentiations among and within populations were highly significant ($p < 0.001$). Similarly, the coefficient of gene differentiation (*Gst*) in total populations calculated by population genetic analysis was 0.2294, which indicated that the genetic variation among populations was 22.94%. The gene flow (*Nm*) was 1.68, which indicated that the gene permutation and interaction among populations was relatively high (Sun et al. 2005).

5.10.6 Identification of Virulence Factors

5.10.6.1 Pathogenicity

Considerable variation in morphology (Marukawa et al. 1975; Morrall et al. 1972; Price and Colhoun 1975) and pathogenicity (Marciano et al. 1983; Morrall et al. 1972; Price and Colhoun 1975) has been observed among isolates of *S. sclerotiorum.* Variation in pathogenicity have been associated with the production of pectolytic enzymes (Hancock 1966; Lumsden 1976, 1979), cellulase (Lumsden 1969), hemicellulase (Hancock 1967), phosphatidase (Lumsden 1970), and oxalic acid (Maxwell and Lumsden 1970; Marciano et al. 1983; Noyes and Hancock 1981). However, variation in these factors does not appear to be the primary determinant of pathogenicity as correlations have not always been apparent (Marciano et al. 1983; Morrall et al. 1972). In *Sclerotinia trifoliorum*, a degenerate nonvirulent isolate of the pathogen produces more protopectinase than a normal strain but only the normal isolate secretes a toxin and is considered virulent (Held 1955). Little information is available on additional mechanisms that affect virulence among isolates of *S. sclerotiorum.* Ghasolia and Shivpuri (2007) studied the morphological and pathogenic variability in 38 isolates of *Sclerotinia sclerotiorum* collected from different rapeseed-mustard fields in Rajasthan State (India). Studies revealed that there are nine groups of the isolates and among these, group G_6 was the most virulent causing maximum disease incidence (82.44%) followed by G_1 (77.50%) and G_8 (75.01%) while G_9 had minimum disease incidence (11.08%). Based on the degree of aggressiveness, nine groups were divided into two pathotypes i.e., pathotypes-I (G_2, G_5, G_9 as these caused less than 18% disease incidence and pathotypes II includes G_1, G_3, G_4, G_6, G_7 and G_8 caused more than 66% disease incidence.

5.10.6.2 Genomic Factors

Hypovirulence has been viewed as the reduced ability of a pathogen to infect, colonize, kill, and reproduce in susceptible host tissue. Hypovirulence has been associated with genetic or cytoplasmic determinants such as nuclear and extra nuclear genetic elements, heterokaryons, organelles, viruses, virus-like agents and plasmids (Elliston 1982). Several fungal plant pathogens have been reported to contain mycoviruses, virus-like particles, or double-stranded ribonucleic acids (ds RNA) that are associated with reduced pathogenicity and/or reduced growth rate and reproductive fitness (Buck 1986, 1987; Elliston 1982; Ghabrial 1980; Nuss and Koltin 1990; Van Alfen 1982). The detection of ds RNA in fungi is often considered indicative of the presence of mycoviruses, most of which contain ds RNA as genetic material (Lemke and Nash 1974; Ralph 1969). A slow growing isolate of *S. sclerotiorum* was recovered in culture that, in preliminary experiments was hypovirulent on bean and celery in comparison to typical, virulent isolates. The presence of mycoviruses, virus-like particles, or ds RNA in *Sclerotinia* spp. has not been reported but the widespread presence of these agents in fungi and their possible influence on virulence suggests that such parasites may contribute to variations in cultural morphology and virulence among isolates of *S. sclerotiorum* (Boland 1987, 1988). Boland (1992) compared five isolates of *S. sclerotiorum* for morphological differences in agar culture, virulence on celery petioles, and the presence of double stranded ribonucleic acids (ds RNA). Four isolates grow rapidly in culture, develop typical colony morphologies are virulent on celery petioles and do not contain ds RNA. The fifth isolate (isolate 91) grows slowly in culture, develops typical colony morphology, produces significantly smaller lesions on celery than virulent isolates and contains ds RNA. Isolate 91 is considered to be hypovirulent. Hypovirulence is transmitted from isolate 91 to virulent, recipient isolates by hyphal anastomosis. Recipient isolates grow slowly in culture, develop a typical colony morphologies are hypovirulent and contain ds RNA. Initiating cultures from hyphal tips and treating colonies with heat or cycloheximide does not eliminate ds RNA from isolate 91. Hypo-virulence in isolate 91 of *S. sclerotiorum* is associated with the presence of ds RNA. The hypo-virulent phenotype of isolate 91 includes symptoms of both reduced virulence and debilitation. Hypovirulence is transmissible in *S. minor* (Melzer and Boland 1996). The existence of dsRNA-free hypovirulence in S10 progenies, observed by Li-Guo et al. (2003) suggests that another hypovirulence mechanism may exist in *S. sclerotiorum.*

In a study (Kohn et al. 1990, 1991) conducted in Ontario Canada, each of two field populations of *S. sclerotiorum* on canola has been shown to be genetically heterogeneous, i.e., composed of several clones, each clone capable of increasing through asexual or homothallic sexual reproduction. These clones are identified by two independent approaches, determination of mycelial compatibility groups (MCGs) and assay of molecular markers. All three molecular markers demonstrating intraspecific variability shows most MCGs to be genetically uniform. One of the molecular markers is a dispersed repetitive DNA sequence, pLKUu 20, which when used as a cloned probe in southern hybridizations produces a unique, complex hybridization pattern, a DNA fingerprint, for each MCG. Monosporous isolates of field-collected apothecia show no evidence of recombination among the determinants of mycelial compatibility or DNA fingerprint. This homogeneity suggests that clones reproduce by inbred homothallic fruiting, in addition to asexual means. This study determines that mycelial compatibility and DNA fingerprinting with pLKUμ 20 are effective tools for identifying clones of *S. sclerotiorum.* Clonal variability within and among field populations of *S. sclerotiorum* isolates from canola petals in Western Canada has been determined (Kohli et al. 1992) by analysis of two independent criteria, mycelial compatibility and DNA fingerprinting. Strains are considered to belong to the same clone if they are mycelially compatible and also have identical DNA fingerprints. Thirty-nine clones have been identified among 66 strains from seven locations in Alberta, Saskatchewan, and Manitoba. The most widely distributed clone, accounting for 18% of the isolates has been recorded in all three provinces. In 33 out of 36 mycelial compatibility groups (MCGs), each MCG has a unique DNA fingerprint; each of the remaining three MCGs includes strains with one of two fingerprints and is interpreted as two clones. A comparison of strains from Western Canada with those from a previous study of two fields in Ontario shows that the one clone identified in both Ontario fields is also present in Manitoba and Saskatchewan. This study demonstrates that clones of *S. sclerotiorum* are distributed over long distances geographically and confirms the results of the Ontario study in demonstrating that field populations of *S. sclerotiorum* on canola are composed of more than one clone. Analysis of monosporous siblings from homothallic sexual reproduction in each of two clones shows no meiotic segregation for determinants of either mycelial compatibility or DNA fingerprints. Therefore, intact clonal genotypes can potentially be dispersed as ascospores.

According to Errampolli and Kohn (1996), Electrophoretic karyotypes (EKs) of 83 isolates are variable within agricultural and natural populations of *Sclerotinia sclerotiorum* as well as among *S. minor* and *S. trifoliorum.* Variation in EKs is not observed within six mitotic or three meiotic lineages of isolates. EKs of 8–10 chromosome-size DNAs are observed. Homologous and heterologous probes hybridize to four linkage groups. Molecular techniques have been used to characterize different field isolates of *S. sclerotiorum.* Chromosome DNA resolves by pulsed field gel electrophoresis (PFGE) revealed that *S. sclerotiorum* contains at least 16 chromosomes ranging from 1.5 to 4.0 Mb. The size of the haploid genome is estimated to be 43.5 Mb. Six field isolates with different levels of virulence on sunflower germlings or green beans are differentiated by random amplification of polymorphic DNA (RAPD), and analyzed by clamped homogenous electric field electrophoresis. This analysis reveals few chromosome-length polymorphisms among these strains. Chromosomal DNA hybridization indicates that the end polygalacturonase-encoding *pgl* gene is localized on the smallest chromosome of all the strains, whereas the ribosomal DNA mapped to different sized chromosomes. The less aggressive strain is characterized by the presence of a supernumerary small band, presumably consisting of ds RNA. In contrast to numerous other phytopathogenic fungi, this study reveals a strong karyotypic stability among the strains of *S. sclerotiorum, which* may be preserved by the sexual mode of reproduction of this species (Fraissinet-Tachet et al. 1996).

Sclerotinia sclerotiorum produces several polygalacturonases, which together with other pectinolytic enzymes are also involved in the degradation of pectin. A number of different genomic clones have been isolated by screening a genomic DNA library in phage EMBL 3. Southern-blot and restriction mapping indicates that seven genes constitute two subfamilies of a multigene family encoding endopolygalacturonase. Using pulsed field gel electrophoresis to separate *S. sclerotiorum* chromosomes each subfamily is found to hybridize to a different chromosome. A comparison of the nucleotide sequence for the coding region of three members of the gene family reveals surprisingly few base substitutions suggesting that this gene family arise from recent multiple duplication events (Fraissinet-Tachet et al. 1995).

Analysis of mitochondrial DNA (mtDNA) haplotypes of *S. sclerotiorum* points to a common origin of some genotypes from agricultural populations especially when compared with two wild populations, which are sharply distinguished from the agricultural sample and from each other. Five agricultural population samples from canola (Alberta, Canada, and Norway), cabbage (North Carolina, USA), sunflower (Manitoba, Canada, and Queensland, Australia) and two Norwegian populations from a wild plant, *Ranunculus ficaria* have been compared. Haplotypes are determined by southern hybridization of purified organelle DNA from *S. sclerotiorum* and *Neurospora crassa* to total genomic DNA of *S. sclerotiorum.* Each isolate has one haplotype. Haplotypes of *S. sclerotiorum* from *R. ficaria* are different between the two wild populations and from all haplotypes observed in the agricultural populations. Among the wild isolates, DNA fingerprint, mtDNA haplotype and location in the sampling transect are all associated. Among the agricultural isolates, four haplotypes have been observed in at least two agricultural populations and one haplotype has been observed in all agricultural populations. In the Canadian canola sample, some clones have one mtDNA haplotype, indicating association with DNA fingerprint, some clones have more than one haplotype and some groups of clones share haplotypes. Some of the haplotype diversity may be due to the presence of extra-chromosomal elements associated with the mitochondria of *S. sclerotiorum* (Kohli and Kohn 1995). Eighty-four isolates of *S. sclerotiorum* from four cabbage production fields in North Carolina and 16 isolates from an experimental cabbage field plot in Louisiana were DNA fingerprinted and tested for mycelial compatibility by Cubeta et al. (1997). In a comparison with 594 unique DNA fingerprints of *S. sclerotiorum* from Canadian canola, no fingerprints are shared among Canadian, North Carolina, and Louisiana populations. DNA fingerprints from the North Carolina sample are distinctive from those of the Canadian and Louisiana samples with significantly more hybridizing fragments in the 7.7–18 kb range. Forty-one mycelial compatibility groups (MCGs) and 50 unique DNA fingerprints are identified from the North Carolina sample. Three MCGs and three fingerprints are identified from the Louisiana sample. From the North Carolina sample, 32 MCGs are each associated with a unique fingerprint and results revealed that there are 11 clones (i.e., cases in which two or more isolates belong to the same MCG and share the same DNA fingerprint). Six clones sampled from two or more fields represent approximately 29% of the total sample (24 of 84 isolates) with six clones recovered from fields 75 km apart. There are ten cases in which one MCG is associated with more than one DNA fingerprint and two cases in which one DNA fingerprint is associated with more than one MCG. The small sample from Louisiana is strictly clonal. The North Carolina sample has a clonal component, but deviates from one to one, association of MCG with DNA fingerprint to an extent consistent with more recombination or transposition than the other two populations sampled.

5.10.6.3 Mycelial Compatibility

Population variability of *S. sclerotiorum* causing stem rot of soybean has been determined by mycelial compatibility grouping (MCG) and isolate aggressiveness comparisons (Kull et al. 2004). Within widely dispersed MCGs, isolate aggressiveness varies. Mycelial compatibility groups of 47 strains of *S. sclerotiorum* have been investigated from Hungary by Zandoki et al. (2005) along with great number of vegetative compatibility groups.

According to Maltby and Mihail (1997), populations of *Sclerotinia sclerotiorum* are often composed of multiple genotypes. In examining 35 naturally infected canola plants, 29 supported reproduction (i.e., sclerotium formation) by a single *S. sclerotiorum* genotype, as defined by the mycelial compatibility test. Only six plants supported reproduction by two genotypes. To test the hypothesis that infrequent multiple genotype infections are due to differences in virulence or competitive ability among isolates, four greenhouse experiments have been conducted in which four isolates, representing three genotypes are used in pair wise co-inoculations of canola. There are no differences among the isolates in four virulence parameters. Mean reduction in sclerotial mass produced by a co-inoculated isolate is calculated by comparison with the mean sclerotial mass of that isolate in the absence of competition and used as the measure of competition. In all experiments, at least half of the co inoculation treatments resulted in reduced fungal reproduction for one or both of the co-inoculated isolates, providing evidence of competitive differences. Generally, the magnitude of reproductive reduction is the same for each isolate in the pair. However, the magnitude is nonreciprocal when the more competitive isolate is given an advantage of early temporal arrival or spatial placement of inoculum at the lower position on the stem. Competitive differences among *S. sclerotiorum* isolates affecting reproduction represent one possible mechanism to explain temporal shifts in genotypic frequencies.

5.10.6.4 Genetic Diversity of Isolates

The heterokaryon formation and vegetative compatibility in *S. sclerotiorum* has been observed by Ford et al. (1995). When auxotrophic mutants derived from six prototrophic wild types are paired in combinations then some combinations of auxotrophs lead to the development of prototrophic strains, the prototrophic strains, and heterokaryons. Heterokaryotic formation is under the control of a regulatory system that results in incompatibility in some strains. Vegetative compatibility groups do not directly correspond to mycelial compatibility groups. Telomere sequence based for revealing genotypic differences among isolates of *S. sclerotiorum* have been suggested (Meinhardt et al. 2002). According to Steadman et al. (1998), Random Amplified Polymorphic DNA (RAPD) distinguishes three species of *Sclerotinia* but not pathogenic variability in *S. sclerotiorum* isolates from diverse host and geographic origin. Virulence as a trait may be environmentally plastic. Isolates from the same host differs in virulence.

The genetic diversity and genetic structure of a population of isolates of *Sclerotinia sclerotiorum* (Lib.) de Bary from different regions and host plants have been investigated using the random amplified polymorphic DNA (RAPD) method with 20 random decamer primer pairs in order to provide some information on the phylogenetic taxa and breeding for resistance to *Sclerotinia* stem rot. A minimum of three and a maximum of 15 unambiguously amplified bands are generated, furnishing a total of 170 bands ranging in size from 100 to 3200 bp, corresponding to an average of 8.5 bands per primer pair. One hundred and four of these 170 bands (61.2%) are polymorphic, the percentage of polymorphic bands for each primer pair ranging from 0.0% to 86.7%. The genetic relationships among the isolates, based on the results of RAPD analysis are examined. The genetic similarity of all selected isolates is quite high. At the species level, the genetic diversity estimated by Nei's gene diversity (*h*) is 0.197 and Shannon's index of diversity (*I*) is 0.300. The Unweighted Pair-Group Mean Analysis (UPGMA) cluster analysis shows that most isolates from the same regions are grouped in the same cluster or a close cluster. The population of isolates from Hefei (Anhui Province, China) is more uniform and relatively distant to other populations. The Canadian population collected from carrot (*Daucus carota* var. *sativa* DC.) is relatively close to the Polish population collected from oilseed rape (*Brassica napus* L.) plants. There is no relationship between isolates from the same host plants. An analysis of molecular variance (ANOVA) revealed that the percentage of variance attributable to variation among and within populations is 50.62% and 49.38% respectively. When accessions from China, Europe, and Canada are treated as three separate groups, the variance components among groups, among populations within groups, and within populations are 0.96%, 51.48%, and 49.47% respectively. The genetic differentiations among and within populations are highly significant. Similarly, the coefficient of gene differentiation (Gst) in total populations calculated by population genetic analysis is 0.2294, which indicates that the genetic variation among populations is 22.94%. The gene flow (Nm) is 1.68, which indicates that the gene permutation and interaction among populations is relatively high (Sun et al. 2005). According to Noonan et al. (1996), isolates of *S. minor* and *S. trifoliorum* are clearly separate from *S. sclerotiorum* using RAPD, but no differences are detected in r DNA. Genetic diversity exists in the isolates of *S. sclerotiorum* from rapeseed, soybean, and sunflower but there is no correlation between the genetic diversities and virulence differentiation (Li et al. 2005).

5.10.6.5 Population Biology

To achieve genetic diversity in *Sclerotinia* populations worldwide, there is need to use the same genetic markers, expand and cross reference data bases, use the same positive controls standards and voucher isolates and provide genotyped and tested isolates of epidemiological significance for research on pathogenesis, control and resistance (Kohn 2001). Population studies on *S. sclerotiorum* have revealed a predominantly clonal mode of reproduction (Cubeta et al. 1997; Kohli and Kohn 1998; Kohli et al. 1995) with some evidence of out crossing contributing to the population structure in a few regions (Atallah et al. 2004; Kohli and Kohn 1998; Sexton and Howlett 2004). In Australia and some temperate regions of North America, populations of *S. sclerotiorum* show some genetic diversity throughout crossing in addition to clonal reproduction (Atallah et al. 2004; Sexton and Howlett

2004) while in Canadian oilseed rape fields, there is no evidence of sexual recombination within a large population composed of a small number of clones, with a single clone repeatedly isolated across 2000 km over a 4-year period (Anderson and Kohn 1995; Kohli et al. 1992). The *S. sclerotiorum* has a haploid somatic phase where clonality is the result of both asexual reproductions by means of sclerotia and sexual reproduction by self-fertilization (Kohn 1995) with the expectation that intracolonic variation is due to mutation (Carbone and Kohn 2001; Carbone et al. 1999).

Individual isolates are classified into clonal lineages by the use of two or more independent markers such as mycelial compatibility groups (MCGs), DNA fingerprinting or microsatellites (Auclair et al. 2004; Carbone et al. 1999; Hambleton et al. 2002; Kohn et al. 1991; Sirjusingh and Kohn 2001). MCG testing is a phenotypic, macroscopic assay of the self/non-self-recognition system controlled by multiple loci common in fungi (Carbone et al. 1999). Mycelial incompatibility is a failure of different strains to fuse and form one cohesive colony and is characterized by the formation of dead cells and reduced growth between the two incompatible colonies (Kohn et al. 1991). DNA fingerprinting utilizes a probe for a multicopy transposon like element in Southern analyses (Kohli et al. 1992). Microsatellite loci have high mutation rates and are multi-allelic in nature, making them useful in phylogenetic inference (Sirjusingh and Kohn 2001). In general, these studies have shown an association between MCGs, DNA fingerprints and microsatellite markers (Carpenter et al. 1999; Hambleton et al. 2002; Sexton and Howlett 2004), but Atallah et al. (2004) recently reported no association between MCGs and microsatellite markers with isolates from potato.

MCGs or micro-satellite markers have not been associated with specific virulence characteristics or ecological adaptations of the pathogen indeed, a lack of variation in virulence among isolates from defined geographical areas has been reported in a number of studies on agricultural populations (Atallah et al. 2004; Auclair et al. 2004; Kull et al. 2004; Sexton and Howlett 2004). Differences in virulence may be detected when comparing isolates form widely separate geographical regions. There has been no conclusive evidence to suggest host specialization among isolates of *S. sclerotiorum* (Kull et al. 2004). However, comparison of *S. sclerotiorum* populations on cultivated oilseed rape and on the wild perennial host *Ranunculus ficaria* indicated major differences between agricultural and wild populations (Kohn 1995). DNA fingerprint diversity is high in agricultural populations but low in wild populations and there is no evidence of out crossing in agricultural populations even though recombination occurs in wild populations.

5.11 *Brassica–Turnip Mosaic Virus* Pathogenomics

5.11.1 Pathological and Biological Characteristics of *Turnip Mosaic Virus* Isolates

The 74 European, 65 East Asian, and 16 other *TuMV* isolates were examined by Nguyen et al. (2013). Most *Brassica* plants, but not *Raphanus sativus*, were systemically infected by most isolates. Thus, they were of the B-infecting host-type (pathotype), although they had minor differences in pathogenicity. Only six European isolates were among this group: Cal1, DEU 4, ITA 2, ITA 7, ITA 8, and PV0104. These had been collected in Italy and Germany and had been isolated from *Abutilon* sp., *Calendula officinalis*, *Cheiranthus cheiri*, *R. raphanistrum*, *R. sativus*, and *Lactuca sativa*. In contrast, approximately 90% of isolates from Europe were B host-type (Tomimura et al. 2003, 2004). Nineteen of 74 isolates collected in Europe were from non-Brassicas. Four isolates, which called as the OM isolates or *TuMVOM*, came from *Orchis* spp., OM-N was isolated from *Orchis militaris* and OM-A was isolated from OM-N by single-lesion isolations, whereas ORM and OS were isolated from *Orchis morio* and *Orchis simia* plants, respectively. The fact that OM isolates have only been found in orchids in a single glasshouse collection supports the conclusion that the OM isolates did not come from *Brassica* plants growing near the glasshouse, but came from one or more of the orchids in the collection, all of which were from Central or Mediterranean Europe.

Isolates OM-N and OS did not infect *Brassica* test plants despite repeated testing. By contrast, isolate ORM occasionally infected *Brassica rapa* and *B. juncea*, but was only detected by reverse transcription-polymerase chain reaction (RT-PCR) and double anti-body sandwich enzyme-linked immunosorbent assay (DAS-ELISA) in the upper two uninoculated leaves of the inoculated plants and not in leaves that emerged subsequently. Hence, as no systemic infection with clear symptoms was produced by OM isolates in standard *Brassica* test plants, it was clear that they were biologically distinct from *TuMV*-BIs. OM isolates were also inoculated to plants of *Camelina sativa* ("gold of pleasure") and *Eruca sativa* (rocket), both of which are West Eurasian *Brassicas* that have been grown, unselected, as crop plants for several thousand years. OM and ORM isolates infected both *C. sativa* and *E. sativa*, but the OS isolate only infected *C. sativa*. Isolates OS and ORM were transmitted by *Myzus persicae* (Sulzer) and *Aphis gossypii* (Glover) in a nonpersistent manner, whereas OM-N and OM-A isolates were not, even though they were tested several times using more than 100 apterous aphids for each test; *E. sativa* and *C. sativa* were used as source and test plants. The three OM isolates were also mechanically inoculated to, but failed to infect, leaves of *Narcissus tazetta* var. *chinensis* (narcissus), *Dioscorea japonica* (Japanese yam), and *Allium fistulosum* (green or Welsh onion, scallion), which are the original hosts of *NYSV*, *JYMV*, and *ScMV*, respectively, and form the sister group to all *TuMVs* (Nguyen et al. 2013).

5.12 *Brassica–Verticillium* Pathogenomics

5.12.1 Molecular Characterization of *Verticillium* Isolates from Crucifers

The majority of isolates of the two main plant pathogenic species of *Verticillium* are not host adapted but are able to infect a wide range of hosts. Pathotypes do occur with perhaps the clearest example being *V. albo-atrum* isolates from lucerne (alfalfa), which are molecularly and pathologically distinct from isolates from all other hosts. Infection of cruciferous hosts once seemed unusual with only isolated occurrences reported from Brussels sprouts and horseradish. The infections of crucifers by microsclerotial isolates have been found much more frequently, e.g., in oilseed rape in Europe, cauliflower in California, horseradish in Illinois, and various crucifers in Japan. Isolates of *V. dahliae* and *V. albo-atrum* from most hosts are short spored and have a low nuclear DNA content, but the majority of isolates from cruciferous hosts is long spored and has a high nuclear DNA content.

All crucifer isolates produce microsclerotia and have usually been classified as *V. dahliae*, often with the sub specific addition of "var. *longisporum*," the name C. Stark gave to the first isolate from horseradish. On the basis of molecular and other studies, Karapapa et al. (1997) suggested that long-spored isolates should be classified as a separate species, *V. longisporum.* This new taxon was based mainly on European oilseed rape isolates and was not entirely satisfactory in that it excluded some long-spored isolates, including the original *V. dahliae* var. *longisporum*, which were described with the undefined term "recombinant." The new taxon also left unclear the status of short-spored crucifer isolates. It was further suggested by Karapapa et al. (1997) that *V. longisporum* was an interspecific hybrid between a lucerne isolate of *V. albo-atrum* and *V. dahliae*, and that genetically the new species was "near-diploid."

Diverse isolates of the soil borne wilt fungi *Verticillium dahliae* and *V. albo-atrum* were analyzed to understand the nature and origins infecting cruciferous hosts. All isolates from cruciferous crops produced microsclerotia, and the majority produced long conidia with a high nuclear DNA content; these isolates were divided into two groups by amplified fragment length polymorphism (AFLP) analysis. One group could be subdivided by other criteria such as rRNA sequences and mitochondrial DNA restriction fragment length polymorphism (RFLP) analysis. Two crucifer isolates were short spored and had a low nuclear DNA content. The results are consistent with the crucifer isolates being interspecific hybrids. The long-spored isolates are best regarded as amphihaploids (or allodiploids) with the AFLP groups probably each representing separate interspecific hybridization events. The short-spored crucifer isolates appear to be derived from interspecific hybrids and are here called "secondary haploids." Molecular evidence suggests that one parent in the crosses was similar to *V. dahliae.* The other parent of the amphihaploids seems to have been more similar to *V. albo-atrum* than to *V. dahliae*, but was distinct from all isolates of either species so far known (Collins et al. 2003).

5.13 *Brassica–Xanthomonas* Pathogenomics

5.13.1 Identification of Pathotypes of *Xanthomonas*

The genus *Xanthomonas* includes economically important pathogenic bacteria that are generally associated with plants. The taxonomy of this genus was initially determined according to host preference (typically the host of origin) and, consequently, a large number of species and pathovars have been defined (Burkholder

1957). Morphological, physiological, and biochemical characters were subsequently used to classify the *Xanthomonas* isolates into eight phenotypic groups (Van Den Mooter and Swings 1990). The *Xanthomonas* species were later reclassified on the basis of DNA–DNA hybridization, leading to *X. campestris* being restricted to comprise only the vascular pathogen *X. campestris* pv. *campestris* (Pammel) Dowson (*Xcc*), which causes black rot of *Brassica* species, and additional pathovars that cause vascular or leaf spot diseases in cruciferous hosts, including *X. campestris* pv. *aberrans* (Knosel) Dye, *armoraciae* (McCullock) Dye, *barbareae* (Burkholder) Dye, *incanae* (Kendrick and Baker) Dye, and *raphani* (White) Dye (Vauterin et al. 1995).

There was different *Xcc* with regard to what constitutes different pathovars. Some authors, namely Alvarez et al. (1994), have considered that *X. campestris* pv. *raphani*, is a pathovars originally described by White (1930), which has a broad range of hosts within the Brassicaceae and Solanaceae, and *X. campestris* pv. *armoraciae*, described earlier by McCulloch (1929) as a leaf spot disease of horseradish, are synonymous. Other authors, such as Tamura et al. (1994) and Vicente et al. (2006), have considered them to be distinct pathovars with a different host range. Other *X. campestris* pathovars have received less attention. Some of these pathovars, such as *X. campestris* pv. *aberrans*, may not be distinct from *Xcc* (Fargier and Manceau 2007; Fargier et al. 2011; Vicente et al. 2001). Fargier and Manceau (2007) considered that the species can be restricted to three pathovars (*campestris*, *raphani*, and *incanae*), but some isolates from ornamental crucifers, which are currently identified as pv. *campestris* or *incanae*, may still belong to distinct pathovars (Vicente et al. 2006). Black rot was first described by Garman (1894) as a disease of cabbage in Kentucky, USA. He isolated two types of bacteria from diseased plants, but could not determine which type of bacterium was causing the disease. In Iowa, USA, Pammel (1895a, b) observed a similar disease in rutabaga and turnip, and showed that the disease was caused by a bacterium (*Bacillus campestris*) with yellow pigmented colonies in culture. Reports from Wisconsin also attributed the disease of turnips and cabbage to the yellow bacterium (Russell 1898; Smith 1898). Since then, the disease has been identified in all continents wherever Brassicaceae crops are grown (Bradbury 1986), and is considered to be the most important disease of vegetable Brassica crops worldwide (Williams 1980). *Brassica oleracea* (including cabbage, cauliflower, broccoli, Brussels sprouts, kale) is economically the most important host of *Xcc*. However, disease also occurs in other *Brassica* crops, radish, ornamental crucifers and related weed species (Bradbury 1986). Some accessions of *Arabidopsis thaliana*, the model plant for molecular plant research, are also susceptible when inoculated with *Xcc* (Vicente and Holub 2013).

5.13.2 Genetic Diversity in Virulence of *Xanthomonas* Isolates from *Brassica*

The 11 *Xcc* races (1–11) were determined on the basis of interactions with differential cultivars (Cruz et al. 2017). In view of the importance of identifying in this plant pathogenic bacterium at the species, pathovars or race level, many genotyping methods have lately been proposed that are superior to classical methods of identification. Investigators have developed *Xcc*-specific primer sets (Park et al. 2004; Berg et al. 2005; Singh et al. 2016) and race 3-specific molecular markers (Afrin et al. 2018) for fast molecular identification. The most commonly used techniques for accurate identification and typing of *Xanthomonas* isolates at species and subspecies level include DNA fingerprinting techniques based on the amplification of repetitive elements, such as REP (repetitive extragenic palindromic), ERIC (enterobacterial repetitive intergenic consensus), BOX (Rademaker et al. 1998; Singh et al. 2016), as well as multilocus sequence typing (MLST) and multilocus sequence analysis (MLSA) using two schemes, both using different housekeeping genes: *fusA* (elongation factor 4), *gap*-1 (glyceraldehyde-3-phosphate dehydrogenase A), *gltA* (citrate synthase), gyrB1 (DNA gyrase B), *lacF* (PTS system lactose-specific EIIA component) and *lepA* (elongation factor 4) (Almeida et al. 2010), or *rpoD* (RNA polymerase sigma factor RpoD), *dnaK* (chaperone protein DnaK), *fyuA* (TonB dependent receptor) and *gyrB2* (Young et al. 2008). A combination of these methods is the most effective way to identify bacteria and assess genetic diversity, although they all have advantages and limitations in terms of the process of application, prices of the reagents needed, reproducibility and levels of phylogenetic and taxonomic resolution (Rademaker et al. 1998). The presence of *Xcc* on cultivated *Brassicas* (broccoli, cabbage, cauliflower, collard greens, kale, kohlrabi and winter oilseed rape) was studied in detail in Serbia, to characterize *Xcc* isolates obtained from crucifers from different regions on the basis of their pathogenic and genetic features by Popovic et al. (2019).

The genetic diversity of 147 *Xanthomonas campestris* pv. *campestris* (*Xcc*) isolates obtained from six *Brassica oleracea* vegetable crops (broccoli, cabbage, cauliflower, collard greens, kale, kohlrabi) and the winter oilseed rape crop *Brassica napus*, collected from different regions of Serbia was assessed. The XCF/XCR pathovar-specific primer set was used for fast preliminary identification. In repetitive sequence-based PCR (BOX, ERIC and REP) of all isolates, a higher level of genetic diversity was found in winter oilseed rape isolates compared to isolates from the other hosts. ERIC and REP-PCR showed the highest heterogeneity, with 10 and 9 banding patterns, respectively. The REP-PCR results showed the highest correlation (70%) with those obtained with multilocus sequence analysis (MLSA), performed with 10 housekeeping genes (*fusA*, *gap-1*, *gltA*, *gyrB1*, *lacF*, *lepA*, *rpoD*, *dnaK*, *fyuA*, and *gyrB2*). Three distinct phylogenetic groups of winter oilseed rape isolates were detected using MLSA. Two genes, *gltA* and *rpoD*, showed the greatest ability to identify and discriminate winter oilseed rape Xcc isolates from isolates of the other six hosts. The *lepA* gene exhibited specific three-nucleotide changes in sequences of some of the isolates. Results of virulence testing of 18 representative isolates showed statistically significant host–pathogen specialization for *Xcc* isolates from winter oilseed rape, cauliflower, kale, and kohlrabi. The oilseed rape isolates are more genetically diverse and show greater specialization to their host in comparison to the rest of the tested isolates from other *Brassica* hosts (Popovic et al. 2019; Fig. 5.12).

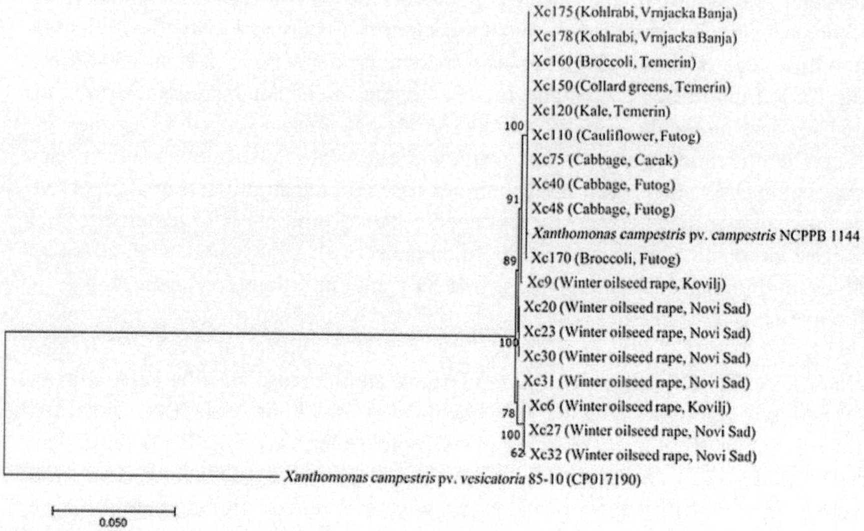

Fig. 5.12 Concatenated neighbor-joining phylogenetic tree based on *gltA*, *gyrB1*, *gap-1*, *lepA*, *lacF*, *fusA*, *rpoD*, *dnaK*, *gyrB2*, and *fyuA* partial sequences obtained from 18 representative Serbian *Xanthomonas campestris* pv. *campestris* isolates and reference strain NCPPB 1144 used for comparison. The tree was rooted with an out-group strain, *X. campestris* pv. *vesicatoria* (CP017190), from the NCBI database (Popovic et al. 2019)

5.14 Evolution of Virulent Effectors of Pathogens

The evolutionary mechanism was systematically mapped by physical interactions between proteins from the reference plant *Arabidopsis thaliana* and effector proteins from two pathogens: the Gram-negative bacterium *Pseudomonas syringae* (*Psy*) and the obligate biotrophic oomycete *Hyaloperonospora arabidopsidis* (*Hpa*). These two pathogens last shared a common ancestor over 1 billion years ago and use vastly different mechanisms to colonize plants. Despite independent evolution of virulence mechanisms, Mukhtar et al. (2012) hypothesized that these two pathogens would deploy effectors to manipulate a largely overlapping set of core cellular MTI machinery.

The analyses revealed that oomycete and bacterial effectors separated by ~1 billion years of evolution target an over lapping subset of plant proteins that include well-connected cellular hubs. The functional validation supports the notion that effectors are likely to converge on to inter-connected host machinery to suppress effective host defense and facilitate pathogen fitness. It was predicted that many of the 165 effector targets, defined will also be targets of additional, independently evolved effectors from other plant pathogens. It is anticipated that effectors that target highly connected cellular proteins fine tune cellular networks to increase pathogen fitness and that evolutionary forces integrate appropriate immune responses with those perturbations. As proposed in the Guard Hypothesis, the data are consistent with indirect connections between pathogen effectors and NB-LRR immune receptors, at least for the NB-LRR fragments represented in PPIN-1. The high degree of the effector targets argues against a "decoy" role for these proteins. While the concept of cellular decoys evolved to intercept pathogen effectors is attractive, and likely true in one case in the plant immune system, these are expected to have few, if any, additional cellular functions and as such would likely have very few other interaction partners in the protein interaction network.

Most of the 673 immune interactors have no previously described immune system function. The results bridge plant immunology, which predicted that effectors should target common proteins, and network science, which proposes that hubs should be targets for network manipulation. Derivation of general rules regarding the organization and function of host cellular machinery required for effective defense against microbial infection, and detailed mechanistic understanding of how pathogen effectors manipulate these machines to increase their fitness, will facilitate improvement of plant immune system function (Mukhtar et al. 2012).

5.15 Challenges of Pathotype Identification, Classification, and Utilization

Pathotype classification with the exception of the recent development of host differentials to delineate pathotypes of *Sclerotinia* Stem Rot (Ge et al. 2012), host differential sets to delineate pathotypes/races of the other pathogens have generally been established for some time (as early as the 1960s). Examples include clubroot (Williams 1966; Buczacki et al. 1975; Some et al. 1996), Blackleg (Mengistu et al. 1991; Purwantara et al. 2000; Balesdent et al. 2002, 2005; Van de Wouw et al. 2009) and downy mildew differentials (Natti 1958). The usefulness of pathotype classification became obscured when research groups used unknown, different, and/or universally unobtainable sources of host differentials that prevented application of a universal and consistent set of host differentials. The "loss" of the Downy Mildew differentials used by Natti (1958) well illustrates these challenges. Application of different host genotype differentials not commonly available across different locations and/or countries has meant findings in terms of definition of pathogen races/pathotypes is, at best of limited use and then only to that particular locality. Use of pathogen isolates without appropriate pathotype/race classification, can easily lead to misleading *R* gene identification when it comes to qualitative or quantitative resistance classifications. To overcome this, an additional differential host differential set was established for Clubroot in Japan using local Chinese cabbage cultivars and local field isolates (Kuginuki et al. 1999; Hatakayema et al. 2006) and similarly for downy mildew using European isolates from *B. oleracea* (Coelho et al. 2012). Furthermore, as new pathogen isolates carrying new virulence genes emerge as a normal consequence of pathogen evolutionary dynamics, host differential sets need to be updated and standardized to enable them to effectively delineate future new pathotypes/races.

Application of different scoring systems could also lead to complications in the identification of *R* genes. In downy mildew, at least three scoring systems have been applied (Dickson and Petzoldt 1993; Wang et al. 2000). It is also imperative that a robust race/pathotype differentiation system be utilized to separate different and group similar races/pathotypes. While for *L. maculans* and *P. brassicae* established race differentiation systems are robust, for *H. parasitica* and some other pathogens, this is not the case. This allows continuity with any previous studies that have different nomenclatures. Its suitability is illustrated by its recent successful application to define the first pathotypes within *S. sclerotiorum* populations on *B. napus* within Western Australia (Ge et al. 2012). Despite these differentials being made available worldwide, upon request, there has been little uptake outside of Australia, India, and Eastern Europe to date. The Goodwin et al. (1990) pathogen isolate differentiation system applied to *H. parasitica* isolates would also be highly beneficial for the same reasons (Neik et al. 2017).

5.16 Challenges of Genetic Variation in Pathogenic Populations

A high level of genetic variation in *S. sclerotiorum* populations was found in *B. napus* fields in Australia (Sexton et al. 2006) and Canada (Kohn et al. 1991). In particular, the Australian *S. sclerotiorum,* with different isolates/pathotypes, caused variable resistance outcomes ranging from intermediate resistance to a HR to complete resistance in different host *B. napus* genotypes (Garg et al. 2010a, b; Ge et al. 2015). In case of *L. maculans*, high genetic diversity has been found within populations. In France, evidence for this was found through an isolation by distance (IBD) study where there was high gene flow and high dispersal rates of spores among populations across large geographical distances (Travadon et al. 2011). These characteristics offer the pathogen a survival advantage to cope with the host resistance in the field, highlighting the role of co-evolutionary dynamics between pathogen and host in conferring resistance. Similarly, in Australia, *L. maculans* also displayed high genetic variation between geographic locations such that the isolates from Western Australia were shown to be genotypically different from those collected from the eastern states. Studies using SNPs (Zander et al. 2013; Patel et al. 2015), microsatellite and mini satellite markers (Hayden et al. 2007) indicated that some populations are panmictic. Gene loss in *L. maculans* has contributed to genetic variation between isolates (Golicz et al. 2015). The source of genetic diversity for *L. maculans* could also depend on the specific host morphotype from which the pathogen is adapted to. For instance *L. maculans* "*brassicae*" from cultivated *Brassica* and *L. maculans* "*lepidii*" from *Lepidium* sp. were identified in the International Blackleg of Crucifers Network (IBCN) collection as one of the seven subgroups of the species complex based on ITS-RFLPs (Mendes-Pereira et al. 2003). This species complex, differing in specificity and pathogenicity on *B. napus*, potentially complicates the cloning of *R* genes in *B. napus* (Voigt et al. 2005).

High genetic diversity was reported for *P. brassicae*, in Japan where the pathogen showed different pathogenicity levels on the cruciferous crops and cruciferous weeds (Jones et al. 1982; Tanaka and Ito 2013). However, pure pathotype/genotype isolates of *P. brassicae* can still be obtained from single root hairs, which minimizes the variability of pathogenicity, thus, a more reliable interpretation of the host–pathogen relationship can be established (Diederichsen et al. 2016). For *Hyaloperonospora parasitica*, a molecular study also revealed high genetic diversity within the species, probably due to the broad host range of this pathogen where it can infect, various *Brassica* and *Raphanus* species (Choi et al. 2003).

References

Aarts N, Metz M, Holub E, Staskawicz BJ, Daniels MJ, Parker JE (1998) Different requirements for *EDS1* and *NDR1* by disease resistance genes define at least two R gene-mediated signaling pathways in *Arabidopsis*. Proc Natl Acad Sci U S A 95:10306–10311

Adam L, Somerville SC (1996) Genetic characterization of five powdery mildew disease resistance loci in *Arabidopsis thaliana*. Plant J 9(3):341–356

Adam L, Ellwood S, Wilson I, Saenz G, Xiao S, Oliver RP, Turner JG, Somerville S (1999) Comparison of *Erysiphe cichoracearum* and *E. cruciferarum* and a survey of 360 *Arabidopsis thaliana* accessions for resistance to these two powdery mildew pathogens. Mol Plant Microbe Interact 12(12):1031–1043

Adhikari TB, Liu JQ, Mathur S, Wu CX, Rimmer SR (2003) Genetic and molecular analyses in crosses of race 2 and race 7 of *Albugo candida*. Phytopathology 93:959–965

Afrin KS, Rahim MA, Park JI, Natarajan S, Kim H-T, Nou IS (2018) Identification of NBS-encoding genes linked to black rot resistance in cabbage (*Brassica oleracea* var. *capitata*). Mol Biol Rep 45:773–785

Aldrich-Wolfe L, Travers S, Nelson BDJ (2015) Genetic variation of *Sclerotinia sclerotiorum* from multiple crops in the North Central United States. PLoS One 10(9):e0139188. https://doi.org/10.1371/journalpone0139188

Al-lami HFD, You MP, Barbetti MJ (2019) Incidence, pathogenicity and diversity of *Alternaria* spp. associated with Alternaria leaf spot of canola (*Brassica napus*) in Australia. Plant Pathol 68:492–503

Almeida NF, Yan S, Cai R, Clarke CR, Morris CE, Schaad NW, Schuenzel EL, Lacy GH, Sun X, Jones JB, Castillo JA, Bull CT, Leman S, Guttman DS, Setubal JC, Vinatzer BA (2010) PAMDB, a multilocus sequence typing and analysis database and website for plant-associated microbes. Phytopathology 100(3):208–215. https://doi.org/10.1094/PHYTO-100-3-0208

Alvarez AM, Benedict AA, Mizumoto CY, Hunter JE, Gabriel DW (1994) Serological, pathological and genetic diversity among strains of Xanthomonas campestris infecting crucifers. Phytopathology 84:1449–1457

Andersen B, Sørensen JL, Nielsen KF, van den Ende BG, de Hoog S (2009) A polyphasic approach to the taxonomy of the *Alternaria infectoria* species–group. Fungal Genet Biol 46:642–656

Anderson JB, Kohn LM (1995) Clonality in soil borne plant-pathogenic fungi. Annu Rev Phytopathol 33:369–391

Ansan-Melayah D, Balesdent HM, Delourme R, Pilet LM, Tanguy X, Renard M, Rouxel T (1997) Gene-for-gene interactions in the *Leptosphaeria maculans-Brassica napus* pathosystem. Acta Hortic 459:329–334

Ansan-Melayah D, Balesdent MH, Delourme R, Pilet ML, Tanguy X, Renard M, Roulx T (1998) Genes for race-specific resistance against blackleg disease in *Brassica napus* L. Plant Breed 117: 373–378. https://doi.org/10.1111/j.1439-05231998.tb01956.x

Apaydin A, Deligoz I, Kar H, Kibar B, Karaagac O (2010) An investigation on clubroot disease (*Plasmodiophora brassicae* Worn) races in the black sea region of Turkey. GOÜ Ziraat Fakültesi Dergisi 27(2):57–60

Atallah ZK, Larget B, Chen X, Johnson DA (2004) High genetic diversity, phenotypic uniformity, and evidence of out crossing in *Sclerotinia sclerotiorum* in the Columbia basin of Washington state. Phytopathology 94:737–742

Atkinson RG (1950) Studies on the parasitism and variation of *Alternaria raphani*. Can J Res 28: 288–317

Atkinson RG (1953) Survival and pathogenicity of *Alternaria raphani* after five years in dried soil cultures. Can J Bot 31:542–547

Attanayake RN, Carter PA, Jiang D, del Río-Mendoza L, Chen W (2013) *Sclerotinia sclerotiorum* populations infecting canola from China and the United States are genetically and phenotypically distinct. Phytopathology 103:750–761. https://doi.org/10.1094/PHYTO-07-12-0159-R

Auclair J, Boland GJ, Kohn LM, Rajcan I (2004) Genetic interactions between *Glycine max* and *Sclerotinia sclerotiorum* using a straw inoculation method. Plant Dis 88:891–895

Australian Plant Pest Database (2019). http://www.appd.ala.org.au. Accessed 12 Aug 2019

Badawy HMA, Hoppe HH, Koch E (1991) Differential reactions between the genus *Brassica* and aggressive single spore isolates of *Leptosphaeria maculans*. Phytopathology 131:109–119. https://doi.org/10.1111/j.1439-0434.1991.tb04737.x

Bains SS, Jhooty JS (1983) Host range and morphology of *Peronospora parasitica* from different sources. Indian J Mycol Plant Pathol 13:372–375

Balesdent MH, Attard A, Ansan-Melayah D, Delourme R, Renard M, Rouxel T (2001) Genetic control and host range of avirulence toward *Brassica napus* cultivars Quinta and Jet Neuf in *Leptosphaeria maculans*. Phytopathology 91:70–76. https://doi.org/10.1094/PHYTO.2001.91.1.70

Balesdent MH, Attard A, Kuhn ML, Rouxel T (2002) New avirulence genes in the phytopathogenic fungus *Leptosphaeria maculans*. Phytopathology 92:1122–1133. https://doi.org/10.1094/PHYTO.2002.92.10.1122

Balesdent MH, Barbetti M, Li H, Sivasithamparam K, Gout L, Rouxel T (2005) Analysis of *Leptosphaeria maculans* race structure in a worldwide collection of isolates. Phytopathology 95:1061–1071

Balesdent MH, Louvard K, Pinochet X, Rouxel T (2006) A large-scale survey of races of *Leptosphaeria maculans* occurring on oilseed rape in France. Eur J Plant Pathol 114:53–65

Balesdent MH, Fudal I, Ollivier B, Bally P, Grandaubert J, Eber F, Chèvre AM, Leflon M, Rouxel T (2013) The dispensable chromosome of *Leptosphaeria maculans* shelters an effector gene conferring avirulence towards *Brassica rapa*. New Phytol 198:887–898. https://doi.org/10.1111/nph.12178

Barbetti MJ, Sivasithamparam K (1981) *Pseudocercosporella capsellae* and *Myrothecium verrucaria* on rapeseed in Western Australia. Australas Plant Pathol 10:43–44. https://doi.org/10.1071/APP9810043

Barret P, Delourme R, Foisset N, Renard M (1998) Development of a SCAR (sequence characterised amplified region) marker for molecular tagging of the dwarf *BREIZH* (*Bzh*) gene in *Brassica napus* L. Theor Appl Genet 97:828–833. https://doi.org/10.1007/s001220050962

Bassimba DDM, Mira JL, Vicent A (2013) First report of *Alternaria japonica* causing black spot of turnip in Spain. Plant Dis 97:1505

Bent AF, Kunkel BN, Dahlbeck D, Brown KL, Schmidt R, Giraudat J, Leung J, Staskawicz BJ (1994) *RPS2* of *Arabidopsis thaliana*: a leucine-rich repeat class of plant disease resistance genes. Science 265:1856–1860

Berg T, Tesoriero L, Hailstones DL (2005) PCR-based detection of *Xanthomonas campestris* pathovars in *Brassica* seed. Plant Pathol 54:416–427. https://doi.org/10.1111/j.1365-3059.2005.01186.x

Bhardwaj CL, Sud AK (1988) A study on the variability of *Albugo candida* from Himachal Pradesh. J Mycol Plant Pathol 18:287–291

Bhattacharya I, Dutta S, Mondal S, Mondal B (2014) Clubroot disease on *Brassica* crops in India. Can J Plant Pathol 36:154–160. https://doi.org/10.1080/07060661.2013.875064

Biga MLB (1955) Review of the species of the genus *Albugo* based on the morphology of the conidia. Sydowia 9:339–358

Bock CH, Thrall PH, Brubaker CL, Burdon JJ (2002) Detection of genetic variation in *Alternaria brassicicola* using AFLP fingerprinting. Mycol Res 106:428–434

Bock CH, Thrall PH, Burdon JJ (2005) Genetic structure of populations of *Alternaria brassicicola* suggests the occurrence of sexual recombination. Mycol Res 109:227–236

Boland GJ (1987) Hypovirulence, debilitation and double-stranded RNA in *Sclerotinia sclerotiorum*. Phytopathology 77:1613

Boland GJ (1988) Hypovirulence in *Sclerotinia sclerotiorum* associated with double-stranded RNA. Plant Pathol 10:361

Boland GJ (1992) Hypovirulence and double-stranded RNA in *Sclerotinia sclerotiorum*. Can J Plant Pathol 14:10–17

Bourras S, McNally KE, Ben-David R, Parlange F, Roffler S, Praz CR, Oberhaensli S, Menardo F, Stirnweis D, Frenkel Z, Schaefer LK, Flückiger S, Treier G, Herren G, Korol AB, Wicker T, Keller B (2015) Multiple avirulence loci and allele-specific effector recognition control the *PM3* race-specific resistance of wheat to powdery mildew. Plant Cell 27(10):2991–3012. https://doi.org/10.1105/tpc.15.00171

Bourras S, McNally KE, Müller MC, Wicker T, Keller B (2016) Avirulence genes in cereal powdery mildews: the gene-for-gene hypothesis 2.0. Plant Biol Interact 7:241

Bradbury JF (1986) Guide to plant pathogenic bacteria. CABI, Slough

Brun H, Chèvre AM, Fitt BDL, Powers S, Besnard AL, Ermel M, Huteau V, Marquer B, Eber F, Renard M, Andrivon D (2010) Quantitative resistance increases the durability of qualitative resistance to *Leptosphaeria maculans* in *Brassica napus*. New Phytol 185:285–299. https://doi.org/10.1111/j.1469-8137.2009.03049.x

Buck KW (1986) Fungal virology-an overview. In: Buck KW (ed) Fungal virology. CRC Press, Boca Raton, FL. 84 p

Buck KW (1987) Viruses of plant pathogenic fungi. In: Day PR, Jellis GJ (eds) Genetics and plant pathogenesis. Blackwell Scientific, Oxford, pp 111–126

Buczacki ST, Toxopeus H, Mattusch P, Johnston TD, Dixon GR, Hobolth LA (1975) Study of physiologic specialization in *Plasmodiophora brassicae*: proposals for attempted rationalization through an international approach. Trans Br Mycol Soc 65:295–303. https://doi.org/10.1016/s0007-1536(75)80013-1

Buhariwalla S, Greaves S, Magrath R, Mithen R (1995) Development of specific PCR primers for the amplification of polymorphic DNA from the obligate root pathogen *Plasmodiophora brassicae*. Physiol Mol Plant Pathol 47:83–94

Burdon JJ, Silk J (1997) Sources and patterns of diversity in plant-pathogenic fungi. Phytopathology 87:664–669

Burdon JJ, Thrall PH (1999) Spatial and temporal patterns in coevolving plant and pathogen associations. Am Nat 153:S15–S33

Burdon JJ, Wennström A, Elmqvist T, Kirby GC (1996) The role of race specific resistance in natural plant populations. Oikos 76:411–416

Burdyukova LI (1980) Albuginacae fungi. Taxonomy, morphology, biology, and specialization. Ukrainskyi Botanichnyi Zhumal 37:65–74

Burkholder WH (1957) Genus II. *Xanthomonas* Dowson 1939. In: Breed RS, Murray EGD, Smith NR (eds) Bergey's manual of determinative bacteriology, 7th edn. Williams and Wilkins Co., Baltimore, MD, pp 152–183

Carbone I, Kohn LM (2001) Multi-locus nested haplo-type networks extended with DNA fingerprints show common origin and fine-scale, ongoing genetic divergence in a wild microbial meta-population. Mol Ecol 10:2409–2422. https://doi.org/10.1046/j.09621083.2001.01380.x

Carbone I, Anderson JB, Kohn LM (1999) Patterns of descent in clonal lineages and their multilocus fingerprints are resolved with combined gene genealogies. Evolution 53:11–21

Carpenter MA, Frampton C, Stewart A (1999) Genetic variation in New Zealand populations of the plant pathogen *Sclerotinia sclerotiorum*. N Z J Crop Hortic Sci 27:13–21

Chai AL, Xie XW, Shi YX, Li BJ (2014) Research status of clubroot (*Plasmodiophora brassicae*) on cruciferous crops in China. Can J Plant Pathol 36:142–153. https://doi.org/10.1080/07060661.2013.868829

Chang IH, Shin NL, Chiu WF (1964) A preliminary study on the physiological differentiation of the downy mildews (*Peronospora parasitica* (Pers.) Fr.) of Chinese cabbage and other cruciferous vegetables in the vicinity of Peking and Tientsin. Acta Phytopathol Sin 7:33–44

Chanyal S, Srivastava S, Meena PD, Taj G (2018) Comparative genetic variation among *Alternaria brassicae* isolates infecting oilseed Brassica in India. Int J Pure App Biosci 6(4):394–400. https://doi.org/10.18782/2320-7051.6570

Chen Q, Peng G, Kutcher R, Yu F (2020) Identification of genome-wide DNA variants and SNP haplotypes associated with avirulence genes of *Leptosphaeria maculans* in Western Canada. Res Square:1–17. https://doi.org/10.21203/rs.3.rs-24766/v1

Chevre AM, Barret P, Eber F, Dupuy P, Brun H, Tanguy X, Renard M (1997) Selection of stable *Brassica napus-B. juncea* recombinant lines resistant to blackleg (*Leptosphaeria maculans*). 1. Identification of molecular markers, chromosomal and genomic origin of the introgression. Theor Appl Genet 95:1104–1111

Cho Y, Srivastava A, Ohm RA, Lawrence CB, Wang KH, Grigoriev IV, Marahatta SP (2012) Transcription factor *Amr1* induces melanin biosynthesis and suppresses virulence in *Alternaria brassicicola*. PLoS Pathog 8:e1002974. https://doi.org/10.1371/journal.ppat.1002974

Choi D, Priest MJ (1995) A key to the genus *Albugo*. Mycotaxon 53:261–272

Choi YJ, Hong SB, Shin HD (2003) Diversity of the *Hyaloperonospora parasitica* complex from core Brassicaceous hosts based on ITS rDNA sequences. Mycol Res 107:1314–1322. https://doi.org/10.1017/S0953756203008578

Choi YJ, Hong SB, Shin HD (2006) Genetic diversity within the *Albugo candida* complex (Peronosporales, Oomycota) inferred from phylogenetic analysis of *ITS rDNA* and *COX2 mtDNA* sequences. Mol Phylogenet Evol 40:400–409

Choi YJ, Shin HD, Hong SB, Thines M (2007) Morphological and molecular discrimination among *Albugo candida* materials infecting *Capsella bursa-pastoris* world-wide. Fungal Divers 27:11–34

Choi YJ, Shin HD, Ploch S, Thines M (2008) Evidence for uncharted biodiversity in the *A. candida* complex, with the description of a new species. Mycol Res 112:1327–1334

Choi YJ, Shin HD, Hong SB, Thines M (2009) The host range of *Albugo candida* extends from Brassicaceae through Cleomaceae to Capparaceae. Mycol Prog 8:329–335

Choi YJ, Shin HD, Ploch S, Thines M (2011a) Three new phylogenetic lineages are the closest relatives of the widespread species *Albugo candida*. Fungal Biol 115:598–607

Choi YJ, Thines M, Shin HD (2011b) A new perspective on the evolution of white blister rust: *Albugo* s.str. (Albugonales; Oomycota) is not restricted to Brassicales but also present on Fabales. Org Divers Evol 11:193–199

Ciferri R (1928) Observations on the specialization of A. ipomoeae-panduratae (Schw.). SW Nuovo Giom Bot Ital 35:112–134

Clarkson JP, Warmington RJ, Walley PG, Denton-Giles M, Barbetti MJ, Brodal G, Nordskog B (2017) Population structure of *Sclerotinia subarctica* and *Sclerotinia sclerotiorum* in England, Scotland and Norway. Front Microbiol 8:490. https://doi.org/10.3389/fmicb.2017.00490

Coelho PS, Vicente JG, Monteiro AA, Holub EB (2012) Pathotypic diversity of *Hyaloperonospora brassicae* collected from *Brassica oleracea*. Eur J Plant Pathol 134:763–771. https://doi.org/10.1007/s10658-012-0052-z

Collins NC, Thordal-Christensen H, Lipka V, Bau S, Kombrink E, Qiu JL, Hückelhoven R, Stein M, Freialdenhoven A, Somerville SC, Schulze-Lefert P (2003) SNARE-protein-mediated disease resistance at the plant cell wall. Nature 4256(1):973–977

Comai L (2005) The advantages and disadvantages of being polyploid. Nat Rev Genet 6:836–846. https://doi.org/10.1038/nrg1711

Conesa A, Gotz S, Garcia-Gomez JM, Terol J, Talon M, Robles M (2005) Blast2GO: a universal tool for annotation, visualization and analysis in functional genomics research. Bioinformatics 21(18):3674–3676

Constantinescu O, Fatehi J (2002) Peronospora-like fungi (Chromista, Peronosporales) parasitic on Brassicaceae and related hosts. Nova Hedwigia 74:291–338

Constantinescu O, Negrean G (1983) Check-list of Romanian Peronosporales. Mycotaxon 16:537–556

Cooke DEL, Drenth A, Duncan JM, Wagels G, Brasier CM (2000) A molecular phylogeny of *Phytophthora* and related oomycetes. Fungal Genet Biol 30:17–32

Crute IR, Johnson AG (1976) The genetic relationship between races of *Bremia lactucae* and cultivars of *Lactuca sativa*. Ann Appl Biol 83:125–137

Crute IR, Norwood JM, Gordon DL (1985) Resistance to phenylamide fungicides in lettuce and *Brassica* downy mildew. In: Proc mixture centenary meeting, Br Crop Prot Council Monograph No. 31, Bordeaux, Croydon, Surrey, pp 311–314

Crute I, Beynon J, Dangl J, Holub E, Mauch-Mani B, Slusarenko A, Staskawicz B, Ausubel F (1994) Microbial pathogenesis of *Arabidopsis*. In: Meyerowitz EM, Somerville CR (eds) *Arabidopsis*. Cold Spring Harbor Laboratory, Cold Spring Harbor, NY, pp 705–747

Cruz J, Tenreiro R, Cruz L (2017) Assessment of diversity of *Xanthomonas campestris* pathovars affecting cruciferous plants in Portugal and disclosure of two novel *X. campestris* pv. *campestris* races. J Plant Pathol 99:403–414

Cubeta MA, Cody BR, Kohli Y, Kohn LM (1997) Clonality in *Sclerotinia sclerotiorum* on infected cabbage in Eastern North Carolina. Phytopathology 87:1000–1004

Czyzewska S (1969) Alternaria blight of *Crambe abyssinica*. Acta Mycol 5:175–211

Czyzewska S (1970) Wplyw temperatury na wzrost i zarodniowanie grzybow z rodzaju *Alternaria* wyizolowanych z modraka abisynskiego. Acta Mycol 6(2):261–276

Czyżewska S (1971) Pathogenicity of *Alternaria* species isolated from *Crambe abyssinica* Hochst. Acta Mycol 7:171–240

Dangl JL, Holub EB, Debener T, Lehnackers H, Ritter C, Crute IR (1992) Genetic definition of loci involved in *Arabidopsis*-pathogen interactions. In: Koncz C, Chua NH, Schell J (eds) Methods in *Arabidopsis* research. World Scientific Publishing Co, London, pp 393–418

Daub ME, Ehrenshaft M (2000) The photo activated *Cercospora* toxin cercosporin: contributions to plant disease and fundamental biology. Annu Rev Phytopathol 38:461–490

De Bruyn HLG (1937) Heterothallism in *Peronospora parasitica*. Genetica 19:553–558

Delourme R, Pilet-Nayel ML, Archipiano M, Horvais R, Tanguy X, Rouxel T, Brun H, Renard M, Balesdent MH (2004) A cluster of major specific resistance genes to *Leptosphaeria maculans* in *Brassica napus*. Phytopathology 94:578–583. https://doi.org/10.1094/PHYTO.2004.94.6.578

Delwiche PA, Williams PH (1977) Genetic studies in *Brassica nigra* (L.) Koch. Crucif Newsl 2:39

Denton-Giles M, Derbyshire MC, Khentry Y, Buchwaldt L, Kamphuis LG (2018) Partial stem resistance in *Brassica napus* to highly aggressive and genetically diverse *Sclerotinia sclerotiorum* isolates from Australia. Can J Plant Pathol 40(4):551–561. https://doi.org/10.1080/07060661.2018.1516699

Derbyshire M, Denton-Giles M, Hegedus D, Seifbarghy S, Rollins J, van Kan J, Seidl MF, Faino L, Mbengue M, Navaud O, Raffaele S, Hammond-Kosack K, Heard S, Oliver R (2017) The complete genome sequence of the phytopathogenic fungus *Sclerotinia sclerotiorum* reveals insights into the genome architecture of broad host range pathogens. Genome Biol Evol 9:593–618

Dickinson CH, Greenhalgh JR (1977) Host range and taxonomy of *Peronospora* on crucifers. Trans Br Mycol Soc 69:111–116. https://doi.org/10.1016/S0007-1536(77)80121-6

Dickson MH, Petzoldt R (1993) Plant age and isolate source affect expression of downy mildew resistance in Broccoli. Hortic Sci 28:730–731

Diederichsen E, Frauen M, Ludwig-Müller J (2014) Clubroot disease management challenges from a German perspective. Can J Plant Pathol 36:85–98. https://doi.org/10.1080/07060661.2013.861871

Diederichsen E, Wagenblatt B, Schallehn V (2016) Production of pure genotype isolates of Plasmodiophora brassicae Wor. -comparison of inoculations with root hairs containing single sporangiosori or with single resting spores. Eur J Plant Pathol 145:621–627

Dilmaghani A, Balesdent MH, Didier JP, Wu C, Davey J, Barbetti MJ, Li H, Moreno-Rico O, Phillips D, Despeghel JP, Vincenot L, Gout L, Rouxel T (2009) The *Leptosphaeria maculans* – *Leptosphaeria biglobosa* species complex in the American continent. Plant Pathol 58:1044–1058. https://doi.org/10.1111/j.1365-3059.2009.02149.x

Dobson RL, Robak J, Gabrielson RL (1983) Pathotypes of *Plasmodiophora brassicae* in Washington, Oregon and California. Plant Dis 67:269–271

Dzhanuzakov A (1963) Specialization and variability in some Peronosporaceous fungi. Bot Zh 47: 862–867

Eberhardt A (1904) Contribution attitude de *Cystopus candidus* Lev. Zentr Bakteriol Parasitenk 12: 235–249

Elliston JE (1982) Hypovirulence. Adv Plant Pathol 1:1–33

Endo RM, Linn MB (1960) The white rust disease of horseradish. IL Agric Exp Stn Bull 655:56

Errampolli D, Kohn LM (1996) Electrophoretic karyotypes of *Sclerotinia sclerotiorum*. Appl Environ Microbiol 62:4247–4251

Eshraghi L, You M, Barbetti M (2005) First report of white leaf spot caused by *Pseudocercosporella capsellae* on *Brassica juncea* in Australia. Plant Dis 89:1131. https://doi.org/10.1094/PD-89-1131B

Eshraghi L, Barbetti MJ, Li H, Danehloueipour N, Sivasithamparam K (2007) Resistance in oilseed rape (*Brassica napus*) and Indian mustard (*Brassica juncea*) to a mixture of *Pseudocercosporella capsellae* isolates from Western Australia. Field Crop Res 101:37–43

Fahling M, Graf H, Siemens J (2003) Pathotype separation of *Plasmodiophora brassicae* by the host plant. J Phytopathol 151(7–8):425–430. https://doi.org/10.1046/j.1439-0434.2003.00744.x

Fahling M, Graf H, Siemens J (2004) Characterization of a single-spore isolate population of *Plasmodiophora brassicae* resulting from a single club. J Phytopathol 152(7):438–444

Faino L, Seidi MF, Shi-Kunne X, Pauper M, van den Berg GCM, Wittenberg AHJ, Thomma BPHJ (2016) Transposons passively and actively contribute to evolution of the two-speed genome of a fungal pathogen. Genome Res 26:1091–1100

Fan Z, Rimmer SR, Stefansson BR (1983) Inheritance of resistance to *Albugo candida* in rape (*Brassica napus* L.). Can J Genet Cytol 25:420–424

Fargier E, Manceau C (2007) Pathogenicity assays restrict the species Xanthomonas campestris into three pathovars and reveal nine races within X. campestris pv. campestris. Plant Pathol 56:805–818

Fargier E, Saux MFL, Manceau C (2011) A multilocus sequence analysis of Xanthomonas campestris reveals a complex structure within crucifer-attacking pathovars of this species. Syst Appl Microbiol 34:156–165

Farr DF, Rossman AY (2010) Fungal databases, systematic mycology and microbiology laboratory. ARS, USDA, Washington DC. http://nt.ars-grin.gov/fungaldatabases/

Farr D, Rossman A (2019) Fungal databases. US National Fungus Collections. ARS, USDA, Washington DC. https://nt.ars-grin.gov/fungaldatabases. Accessed 12 Aug 2019

Felton MW, Walker JC (1946) Environmental factors affecting downy mildew of cabbage. J Agric Res 72:69–81

Feng J, Jiang J, Feindel D, Strelkov SE, Hwang S-F (2016) The gene Cr811 is present exclusively in pathotype 5 and new emerged pathotypes of the clubroot pathogen *Plasmodiophora brassicae*. Eur J Plant Pathol 145:615–620. https://doi.org/10.1007/s10658-016-0903-0

Fernando WGD, Zhang X, Selin C, Zou Z, Liban SH, McLaren DL, Kubinec A, Parks PS, Rashid MH, Padmathilake KRE, Rong K, Yang C, Gnanesh BN, Huang S (2018) A six-year investigation of the dynamics of Avirulence allele profiles, blackleg incidence, and mating type alleles of

Leptosphaeria maculans population associated with canola crops in Manitoba, Canada. Plant Dis 12:790–798. https://doi.org/10.1094/PDIS-05-17-0630-RE

Ferreira ME, Rimmer SR, Williams PH, Osborn TC (1995) Mapping loci controlling *Brassica napus* resistance to *Leptosphaeria maculans* under different screening conditions. Phytopathology 85:213–217

Ford EJ, Miller RV, Gray H, Sherwood JE (1995) Heterokaryon formation and vegetative compatibility in *Sclerotinia sclerotiorum*. Mycol Res 99:241–247

Förster H, Coffey MD (1990) Mating behavior of *Phytophthora parasitica*-evidence for sexual recombination in oospores using DNA restriction fragment length polymorphism as genetical markers. Exp Mycol 14:351–359

Fraissinet-Tachet L, Reymond-Cotton R, Fèvre M (1995) Characterization of a multigene family encoding an endopolygalacturonase in *Sclerotinia sclerotiorum*. Curr Microbiol 29:96–100

Fraissinet-Tachet L, Reymond-Cotton P, Fevre M (1996) Molecular karyotype of the phytopathogenic fungus *Sclerotinia sclerotiorum*. Curr Sci 29:496–501

Francis DM, St Clair DA (1993) Out-crossing in the homothallic oomycete, *Pythium ultimum*, detected with molecular markers. Curr Genet 24:100–106

Fudal I, Ross S, Gout L, Blaise F, Kuhn ML, Eckert MR, Cattolico L, Bernard-Samain S, Balesdent MH, Rouxel T (2007) Heterochromatin-like regions as ecological niches for avirulence genes in the *Leptosphaeria maculans* genome: map-based coning of *AvrLm6*. Mol Plant-Microbe Interact 20:459–470. https://doi.org/10.1094/MPMI-20-4-0459

Gardner MW (1920) *Peronospora* in turnip roots. Phytopathology 10:321–323

Garg H, Atri C, Sandhu PS, Kaur B, Renton M, Banga SK, Singh H, Singh C, Barbetti MJ, Banga SS (2010a) High level of resistance to *Sclerotinia sclerotiorum* in introgression lines derived from hybridization between wild crucifers and the crop *Brassica* species, *B. napus and B. juncea*. Field Crop Res 117:51–58. https://doi.org/10.1016/j.fcr.2010.01.013

Garg H, Kohn LM, Andrew M, Li H, Sivasithamparam K, Barbetti MJ (2010b) Pathogenicity of morphologically different isolates of *Sclerotinia sclerotiorum* with *Brassica napus* and *B. juncea* genotypes. Eur J Plant Pathol 126:305–315. https://doi.org/10.1007/s10658-009-9547-7

Garibaldi A, Gilardi G, Bertoldo C, Gullino ML (2011) First report of leaf spot of wild (*Diplotaxis tenuifolia*) and cultivated (*Eruca vesicaria*) rocket caused by *Alternaria japonica* in Italy. Plant Dis 95:1316

Garman H (1894) A bacterial disease of cabbage. Kentucky Agr Exp Stn Rep 3:43–46

Gaumann E (1926) On the specialization of downy mildew (*Peronospora brassicae* Gaum.) on cabbage and related species. Landwirtsch Jahrbuch Schweiz 40:463–468

Ge XT, Li YP, Wan ZJ, You MP, Finnegan PM, Banga SS, Sandhu PS, Garg H, Salisbury PA, Barbetti MJ (2012) Delineation of *Sclerotinia sclerotiorum* pathotypes using differential resistance responses on *Brassica napus* and *B. juncea* genotypes enables identification of resistance to prevailing pathotypes. Field Crop Res 127:248–258. https://doi.org/10.1016/j.fcr.2011.11.022

Ge XT, You MP, Barbetti MJ (2015) Virulence differences among Sclerotinia sclerotiorum isolates determines host cotyledon resistance responses in Brassicaceae genotypes. Eur J Plant Pathol 143:527–541. https://doi.org/10.1007/s10658-015-0696-6

Ghabrial SA (1980) Effects of fungal viruses on their hosts. Annu Rev Phytopathol 18:44–61

Ghanbarnia K, Fudal I, Larkan NJ, Links MG, Balesdent MH, Profotova B, Fernando WGD, Rouxel T, Borhan MH (2015) Rapid identification of the *Leptosphaeria maculans* avirulence gene AvrLm2 using an intraspecific comparative genomics approach. Mol Plant Pathol 16:699–709. https://doi.org/10.1111/mpp.12228

Ghasolia RP, Shivpuri A (2007) Morphological and pathogenic variability in rapeseed and mustard isolates of *Sclerotinia sclerotiorum*. Indian Phytopathol 60:76–81

Gibriel HA, Thomma BP, Seidl MF (2016) The age of effectors: genome-based discovery and applications. Phytopathology 106:1206–1212. https://doi.org/10.1094/PHYTO-02-16-0110-FI

Gilardi G, Demarchi S, Ortu G, Gullino ML, Garibaldi A (2014) Occurrence of *Alternaria japonica* on seeds of wild and cultivated rocket. J Phytopathol 163:419–422

Goker M, Riethmüller A, Voglmayr H, Weiss M, Oberwinkler F (2004) Phylogeny of *Hyaloperonospora* based on nuclear ribosomal internal transcribed spacer sequences. Mycol Prog 3:83–94

Golicz AA, Martinez PA, Zander M, Patel DA, Van De Wouw AP, Visendi P, Fitzgerald T, Edwards D, Batley J (2015) Gene loss in the fungal canola pathogen *Leptosphaeria maculans*. Funct Integr Genomics 15:189–196. https://doi.org/10.1007/s10142-014-0412-1

Göllner K, Schweizer P, Bai Y, Panstruga R (2008) Natural genetic resources of *Arabidopsis thaliana* reveal a high prevalence and unexpected phenotypic plasticity of *RPW8*-mediated powdery mildew resistance. New Phytol 1776(1):725–742

Gomes EV, Nascimento LBD, De Freitas MA, Nasser LCB, Petrofeza S (2011) Microsatellite markers reveal genetic variation within *Sclerotinia sclerotiorum* populations in irrigated dry bean crops in Brazil. J Phytopathol 159:94–99. https://doi.org/10.1111/j.1439-0434.2010.01724.x

Goodwin SB, Allard RW, Webster RK (1990) A nomenclature for Rhynchosporium secalis pathotypes. Phytopathology 80:1330–1336. https://doi.org/10.1094/Phyto-80-1330

Goodwin PH, Annis SL (1991) Rapid identification of genetic variation and pathotype of *Leptosphaeria maculans* by random amplified polymorphic DNA assay. Appl Environ Microbiol 57:2482–2486

Gout L, Fudal I, Kuhn ML, Blaise F, Eckert M, Cattolico L, Balesdent MH, Rouxel T (2006) Lost in the middle of nowhere: the *AvrLm1* avirulence gene of the Dothideomycete *Leptosphaeria maculans*. Mol Microbiol 60:67–80

Goyal P, Chahar M, Mathur AP, Kumar A, Chattopadhyay C (2011) Morphological and cultural variation in different oilseed Brassica isolates of *Alternaria brassicae* from different geographical regions of India. Indian J Agric Sci 81:1053–1058

Goyal P, Chattopadhyay C, Mathur AP, Kumar A, Meena PD, Datta S (2013) Pathogenic and molecular variability among *Brassica* isolates of *Alternaria brassicae* from India. Ann Plant Prot Sci 21(2):349–359

Grant MR, Godiard L, Straube E, Ashfield T, Lewald J, Sattler A, Innes RW, Dangl JL (1995) Structure of the *Arabidopsis RPM1* gene enabling dual specificity disease resistance. Science 269:843–846

Gu Q, Chen M, Huang J, Wei Y, Hsiang T, Zheng L (2017) Multifaced roles of the *Ras* Guanine-nucleotide exchange factor *ChRgf* in development, pathogenesis, and stress response of *Colletotrichum higginsianum*. Phytopathology 107:433–443. https://doi.org/10.1094/PHYTO-03-16-0137-R

Gunasinghe N, You MP, Banga SS, Barbetti MJ (2013) High level resistance to Pseudocercosporella capsellae offers new opportunities to deploy host resistance to effectively manage white leaf spot disease across major cruciferous crops. Eur J Plant Pathol 138:873–890

Gunasinghe N, You MP, Barbetti MJ (2016) Phenotypic and phylogenetic studies associated with the crucifer white leaf spot pathogen, *Pseudocercosporella capsellae*, in Western Australia. Plant Pathol 65:205–217. https://doi.org/10.1111/ppa.12402

Gupta K, Saharan GS (2002) Identification of pathotypes of *Albugo candida* with stable characteristic symptoms on Indian mustard. J Mycol Plant Pathol 32:46–51

Gupta K, Saharan GS, Mehta N, Sangwan MS (2004) Identification of pathotypes of *Alternaria brassicae* from Indian mustard [*Brassica juncea* (L) Czern & Coss.]. J Mycol Plant Pathol 34:15–19

Gyawali S, Harrington M, Durkin J, Horner K, Parkin IA, Hegedus DD, Bekkaoui D, Buchwaldt L (2016) Microsatellite markers used for genome-wide association mapping of partial resistance to *Sclerotinia sclerotiorum* in a world collection of *Brassica napus*. Mol Breed 36:72. https://doi.org/10.1007/s11032-016-0496-5

Hambleton S, Walker C, Kohn LM (2002) Clonal lineages of *Sclerotinia sclerotiorum* previously known from other crops predominate in 1999–2000 from Ontario and Quebec soybean. Can J Plant Pathol 24:309–315

Hancock JG (1966) Degradation of pectic substances associated with pathogenesis by *Sclerotinia sclerotiorum* in sunflower and tomato stems. Phytopathology 56:975–979

Hancock JG (1967) Hemicellulose degradation in sunflower hypocotyls infected with *Sclerotinia sclerotiorum*. Phytopathology 57:203–206

Harteveld DOC, Akinsanmi OA, Drenth A (2013) Multiple *Alternaria* species groups are associated with leaf blotch and fruit spot diseases of apple in Australia. Plant Pathol 62:289–297

Hatakayema K, Fujimura M, Ishida M, Suzuki T, Sato T (2006) Classification of pathogenicity of *Plasmodiophora brassicae* field isolates in Japan based on resistance of F1 cultivars of Chinese cabbage (*Brassica rapa* L.) to clubroot. Acta Hortic 706:323–328. https://doi.org/10.17660/ActaHortic.2006.706.38

Hayden HL, Cozijnsen AJ, Howlett BJ (2007) Microsatellite and mini-satellite analysis of *Leptosphaeria maculans* in Australia reveals regional genetic differentiation. Phytopathology 97:879–887. https://doi.org/10.1094/PHYTO-97-7-0879

Heath MC (2000) Hypersensitive response-related death. In: Lam E, Fukuda H, Greenberg J (eds) Programmed cell death in higher plants. Springer, Holland, pp 77–90

Held YM (1955) Physiological differences between a normal and a degenerate strain of *Sclerotinia trifoliorum*. Phytopathology 45:39–42

Hemetsberger C, Mueller AN, Matei A, Herrberger C, Hensel G, Kumlehn J, Mishra B, Sharma R, Thines M, Hückelhoven R, Doehlemann G (2015) The fungal core effector *Pep1* is conserved across smuts of dicots and monocots. New Phytol 206:1116–1126

Hill C, Crute I, Sherriff C, Williams PH (1988) Specificity of *Albugo candida* and *Peronospora parasitica* pathotypes toward rapid-cycling crucifers. Crucif Newsl 13:112–113

Hiura M (1930) Biologic forms of *A. candida* (Pers.) Kuntze on some cruciferous plants. Jpn J Bot 5:1–20

Hiura M, Kanegae H (1934) Studies on the downy mildews of cruciferous vegetables in Japan. Trans Sapporo Nat Hist Soc 13:125–133

Holtz MD, Hwang S-F, Strelkov SE (2018) Genotyping of *Plasmodiophora brassicae* reveals the presence of distinct populations. BMC Genomics 19(1):254

Holub EB (1997) Organization of resistance genes in *Arabidopsis*. In: Crute IR, Holub EB, Burdon JJ (eds) The gene-for-gene relationship in plant-parasite interaction. CAB Int, Wallingford, pp 5–26

Holub EB, Beynon JL, Crute IR (1994) Phenotypic and genotypic characterization of interactions between isolates of *Peronospora parasitica* and accessions of *Arabidopsis thaliana*. Mol Plant-Microbe Interact 7:223–239

Howlett BJ (2004) Current knowledge of the interaction between *Brassica napus* and *Leptosphaeria maculans*. Can J Plant Pathol 26(3):245–252. https://doi.org/10.1080/07060660409507141

Howlett B, Ballinger D, Barbetti MJ (1999) Diseases. In: Salisbury PA, Potter TD, Mcdonald G (eds) Canola in Australia. Organising Committee of the 10th International Rapeseed Congress, Canberra, Australia, pp 47–52

Howlett BJ, Lowe RGT, Marcroft SJ, van de Wouw AP (2015) Evolution of virulence in fungal plant pathogens: exploiting fungal genomics to control plant disease. Mycologia 107(3): 441–451. https://doi.org/10.3852/14-317

Hu J, Quiros CE (1991) Identification of broccoli and cauliflowers cultivars with RAPD markers. Plant Cell Rep 10:505–511

Huang YJ, Balesdent MH, Li Z-Q, Evans N, Rouxel T, Fitt BDL (2010) Fitness cost of virulence differs between the *AvrLm1* and *AvrLm4* loci in *Leptosphaeria maculans* (phoma stem canker of oilseed rape). Eur J Plant Pathol 126:279–291

Hudspeth DSS, Stenger D, Hudspeth MES (2003) A COX2 phylogenetic hypothesis of the downy mildews and white rusts. Fungal Divers 13:47–57

Hulbert SH, Illott TW, Legg EJ, Lincoln SE, Lander ES, Michelmore RW (1988) Genetic analysis of the fungus, *Bremia lactucae*, using restriction fragment length polymorphisms. Genetics 120: 947–958

Ikegami H (1992) Proliferation and pathogenicity of *Plasmodiophora brassicae* in infected callus tissue (studies on the clubroot of cruciferous plant X). Proc Kansai Plant Prot Soc 34:17–28

Inman A, Sivanesan A, Fitt B, Evans R (1991) The biology of *Mycosphaerella capsellae* sp. nov., the teleomorph of *Pseudocercosporella capsellae*, cause of white leaf spot of oilseed rape. Mycol Res 95:1334–1342. https://doi.org/10.1016/S0953-7562(09)80586-8

International Seed Federation (2019) Differential sets *Plasmodiophora brassicae* (Pb)-*Brassica oleracea*. Chem Reposoir 7:1260. 1–2

Ito S, Tokunaga Y (1935) Notae mycologicae Asiae orientalis. I. Trans Sapporo Nat Hist Soc 14:11–33

Ito S, Maehara T, Tanaka S, Kameya-Iwaki M, Yano S, Kishi F (1997) Cloning of a single-copy DNA sequence unique to *Plasmodiophora brassicae*. Physiol Mol Plant Pathol 50:289–300

Ito S, Nakamura T, Matsumoto T, Maehara T, Tanaka S, Kameya-Iwaki M, Kishi F (1998) Characterization of DNA sequences cloned from resting spores of *Plasmodiophora brassicae*. J Phytopathol 146(2–3):143–147. https://doi.org/10.1111/j.1439-0434.1998.tb04671.x

Jasalavich CA, Morales VM, Pelcher LE, Seguin-Swartz G (1995) Comparison of nuclear ribosomal DNA sequences from *Alternaria* species pathogenic to crucifers. Mycol Res 99:604–614

Jat RR (1999) Pathogenic variability and inheritance of resistance to *Albugo candida* in oilseed *Brassica*. PhD Thesis, CCSHAU Hisar, 129 p

Jeong JY, Robin AHK, Natarajan S, Laila R, Kim HT, Park J-I, Nou I-S (2018) Race- and isolate-specific molecular marker development through genome-realignment enables detection of Korean *Plasmodiophora brassicae* isolates, causal agents of clubroot disease. Plant Pathol J 34(6):506–513. https://doi.org/10.5423/PPJ.OA.12.2017.0266

Jiang W, Zhou H, Bi H, Fromm M, Yang B, Weeks DP (2013) Demonstration of CRISPR/Cas9/sgRNA-mediated targeted gene modification in *Arabidopsis*, tobacco, sorghum and rice. Nucleot Acid Res 41e:188. https://doi.org/10.1093/nar/gkt780

Jo SJ, Shim SA, Jang KS, Choi YH, Kim JC, Choi GJ (2011) Resistance of cultivars of Chinese cabbage to *Plasmodiophora brassicae* isolates of several races collected in Korea. Korean J Hortic Sci Technol 29:610–616

Jones DR, Ingram DS, Dixon GR (1982) Factors affecting tests for differential pathogenicity in populations of *Plasmodiophora brassicae*. Plant Pathol 31:229–238. https://doi.org/10.1111/j.1365-3059.1982.tb01273.x

Jonsson R (1981) Breeding for improved resistance to clubroot in oil rape. In: Proc *Brassica* Conf, pp 44–46

Jouet A, Saunders DGO, McMullan M, Ward B, Furzer O, Jupe F, Cevik V, Hein I, Thilliez GJA, Holub E, van Oosterhout C, Jones JDG (2018) *Albugo candida* race diversity, ploidy and host-associated microbes revealed using DNA sequence capture on diseased plants in the field. New Phytol 221:1529–1543

Júnior CL, Gomes E, Junior ML, Nasser L, Petrofeza S (2011) Genetic diversity and mycelial compatibility groups of the plant-pathogenic fungus *Sclerotinia sclerotiorum* in Brazil. Genet Mol Res 10:868–877. https://doi.org/10.4238/vol10-2gmr937

Kanazin V, Marek LF, Shoemaker RC (1996) Resistance gene analogs are conserved and clustered in soybean. Proc Natl Acad Sci U S A 93:11746–11750

Karapapa VK, Bainbridge BW, Heale JB (1997) Morphological and molecular characterization of *Verticillium longisporum* comb. nov., pathogenic to oilseed rape. Mycol Res 101:1281–1294

Karkehabadi S, Hansson H, Kim S, Piens K, Mitchinson C, Sandgren M (2008) The first structure of a glycoside hydrolase family member, *Cel61B* from *Hypocrea jecorina* at 1.6 Å resolution. J Mol Biol 383:144–154

Kaur P, Sivasithamparam K, Barbetti MJ (2008) Pathogenic behaviour of strains of *Albugo candida* from *Brassica juncea* (Indian mustard) and *Raphanus raphanistrum* (wild radish) in Western Australia. Aust J Plant Pathol 37:353–356

Kaur P, Jost R, Sivasithamparam K, Barbetti MJ (2011a) Proteome analysis of the *A. candida-B. juncea* pathosystem reveals that the timing of the expression of defence-related genes is a crucial determinant of pathogenesis. J Exp Bot 62:1285–1298. https://doi.org/10.1093/jxb/erq365

Kaur P, Sivasithamparam K, Barbetti MJ (2011b) Host range and phylogenetic relationships of *Albugo candida* from cruciferous hosts in Western Australia, with special reference to *Brassica juncea*. Plant Dis 95:712–718

Khurana AK, Mehta N, Sangwan MS (2005) Variation in biochemical contents of *Alternaria brassicae* (Berk) Sacc. isolates. Indian J Mycol Plant Pathol 35:343–345

Kihara J, Moriwaki A, Tanaka N, Tanaka C, Ueno M, Arase S (2008) Characterization of the BMR1gene encoding a transcription factor for melanin biosynthesis genes in the phytopathogenic fungus *Bipolaris oryzae*. FEMS Microbiol Lett 281:221–227

Kim H, Jo EJ, Choi YH, Jang KS, Choi GJ (2016) Pathotype classification of *Plasmodiophora brassicae* isolates using Clubroot-resistant cultivars of Chinese cabbage. Plant Pathol J 32:423. https://doi.org/10.5423/PPJ.OA.04.2016.0081

Klein-Lankhorst RM, Rietveld P, Machiels B, Verkerk R, Weide R, Gebhardt C, Koornneef M, Zabel P (1991) RFLP markers linked to the root knot nematode resistance gene *Miin* tomato. Theor Appl Genet 81:661–667

Kluczewski SM, Lucas JA (1983) Host infection and oospore formation by *Peronospora parasitica* in agricultural and horticultural *Brassica* species. Trans Br Mycol Soc 81:591–596

Kobel F (1921) The problem of host selection by parasitic fungi. Naturwissen Wochenschr 36:113–118

Koch E, Song K, Osborn TC, Williams PH (1991) Relationship between pathogenicity and phylogeny based on restriction fragment length polymorphism in *Leptosphaeria maculans*. Mol Plant-Microbe Interact 4:341–349. https://doi.org/10.1094/MPMI-4-341

Kochman J, Majewski T (1970a) Fungi (Mycota). Vol IV. Phycomycetes, peronosporales. Panstwowe Wydawnictwo Naukowe, Warszawa

Kochman J, Majewski T (1970b) Grzyby (Mycota). Glonowce (Phycomycetes). Wroslikowe (Pernosporales), vol 4. Panstwowe Wydawnictwo Naukowe, Warszawa, p 309

Köhler C, Mittelsten Scheid O, Erilova A (2010) The impact of the triploid block on the origin and evolution of polyploid plants. Trends Genet 26:142–148

Kohli Y, Kohn LM (1998) Random association among alleles in clonal populations of *Sclerotinia sclerotiorum*. Fungal Genet Biol 23:139–149

Kohli Y, Morrall RAA, Anderson JB, Kohn LM (1992) Local and trans-Canadian clonal distribution of *Sclerotinia sclerotiorum* on canola. Phytopathology 82:875–880

Kohli Y, Brunner LJ, Yoell H, Milgroom MG, Anderson JB, Morall RAA, Kohn LM (1995) Clonal dispersal and spatial mixing in populations of the plant-pathogenic fungus, *Sclerotinia sclerotiorum*. Mol Ecol 4:69–77

Kohli LM (1992) Developing new characters for fungal systematic: an experimental approach for determining the rank of *Sclerotinia*. Mycologia 84:139–153

Kohn LM (1995) The clonal dynamic in wild and agricultural plant pathogen populations. Can J Bot 73:1231–1240

Kohn LM (2001) Integrating our genotypic diversity data: towards a global picture of population subdivision and fine scale structure in Sclerotinia with implications for disease management. In: Young CS, Hughes KJD (eds) Proc of Sclerotinia 2001 of the XI International Sclerotinia Workshop, 8–12 July 2001. Central Science Laboratory, York, pp 17–19

Kohn LM, Carbone I, Anderson JB (1990) Mycelial interactions in *Sclerotinia sclerotiorum*. Exp Mycol 14:255–267

Kohn LM, Stasovski E, Carbone I, Royer J, Anderson JB (1991) Mycelial incompatibility and molecular markers identify genetic variability in field populations of *Sclerotinia sclerotiorum*. Phytopathology 81:480–485. https://doi.org/10.1094/Phyto-81-480

Kole C, Teutonico R, Mengistu A, Williams PH, Osborn TC (1996) Molecular mapping of a locus controlling resistance to *Albugo candida* in *Brassica rapa*. Phytopathology 86:367–369

Kole C, Williams PH, Rimmer SR, Osborn TC (2002) Linkage mapping of genes controlling resistance to white rust (*Albugo candida*) in *Brassica rapa* (syn. *campestris*) and comparative mapping to *Brassica napus* and *Arabidopsis thaliana*. Genome 45:22–27

Koller B, Lehmann A, McDermott JM, Gessler C (1993) Identification of apple cultivars using RAPD markers. Theor Appl Genet 85:901–904

Kolte SJ, Awasthi RP, Vishwanath K (1989) Disease problems in Brassicas and research activities at Pantnagar. In: IDRC (Canada) Oil Crops: Proc three meetings held at Pantnagar and Hyderabad, India, pp 43–48

Kolte SJ, Bardoloi DK, Awasthi RP (1991) The search for resistance to major diseases of rapeseed mustard in India. In: McGregor DI (ed) Proc 8th International Rapeseed Congress, 9–11 July 1991. Organizing Committee of 8th International Rapeseed Congress, Saskatoon, SK, pp 216–225

Korn M, Schmidpeter J, Dahl M, Müller S, Voll LM, Koch CA (2015) Genetic screen for pathogenicity genes in the hemibiotrophic fungus *Colletotrichum higginsianum* identifies the plasma membrane proton pump *Pma2* required for host penetration. PLoS One 10:e0125960. https://doi.org/10.1371/journal.pone.0125960

Kresovich S, Williams JGK, McFerson JR, Routman EJ, Schaal BA (1992) Characterization of genetic identities and relationships of *Brassica oleracea* L. via a random amplified polymorphic DNA assay. Theor Appl Genet 85:190–196

Krivchenko VI, Boos GV, Surmava ME (1982) Harakteristika genofonda kapusty po ustojcivosti k *Plasmodiophora brassicae* Woron. Tr Prikl Bot Genet Sel 72:113–120

Kuginuki Y, Yoshikawa H, Hirai M (1999) Variation in virulence of *Plasmodiophora brassicae* in Japan tested with clubroot resistant cultivars of Chinese cabbage (*Brassica rapa* L. spp. *pekinensis*). Eur J Plant Pathol 105:327–332

Kuhn H, Kwaaitaal M, Kusch S, Acevedo-Garcia J, Wu H, Panstruga R (2016) Biotrophy at its best: novel findings and unsolved mysteries of the *Arabidopsis*-powdery mildew pathosystem. Arabidopsis Book 14:e0184. https://doi.org/10.1199/tab.0184

Kull LS, Pedersen WL, Palmquist D, Hartman GL (2004) Mycelial compatibility grouping and aggressiveness of *Sclerotinia sclerotiorum*. Plant Dis 88:325–332

Kumar S, Sangwan MS, Mehta N, Kumar R (2003) Pathogenic diversity in isolates of *Alternaria brassicae* infecting rapeseed-mustard. J Mycol Plant Pathol 33:59–64

Kumar S, Sangwan MS, Mehta N, Kumar R (2004) Relative sensitivity of various isolates of *Alternaria brassicae* (Berk) Sacc to fungicides. J Mycol Plant Pathol 34:28–32

Kumar V, Haldar S, Pandey KK, Singh RP, Singh AK, Singh PC (2008) Cultural, morphological, pathogenic and molecular variability amongst tomato isolates of *Alternaria solani* in India. World J Microbiol Biotechnol 24:1003–1009

Kuninaga S, Natsuaki T, Takeuchi T, Yokosawa R (1997) Sequence variation of the rDNA ITS regions within and between anastomosis groups in *Rhizoctonia solani*. Curr Genet 32:237–243

Kunkel BN (1996) A useful weed put to work: genetic analysis of disease resistance in *Arabidopsis thaliana*. Trends Genet 12:63–69

Kutcher HR, Van Den Berg CGJ, Rimmer SR (1993) Variation in pathogenicity of *Leptosphaeria maculans* on *Brassica* spp. based on cotyledon and stem reactions. Can J Plant Pathol 15:253–258. https://doi.org/10.1080/07060669309501920

Laila R, Robin AH, Yang K, Choi GJ, Park JI, Nou IS (2017) Detection of ribosomal DNA sequence polymorphisms in the protist *Plasmodiophora brassicae* for the identification of geographical isolates. Int J Mol Sci 18:84. https://doi.org/10.3390/ijms18010084

Lakra BS, Saharan GS (1988a) Morphological and pathological variations in *Albugo candida* associated with *Brassica* species. Indian J Mycol Plant Pathol 18:149–156

Lakra BS, Saharan GS (1988b) Influence of host resistance on colonization and incubation period of *Albugo candida* in mustard. Crucif Newsl 13:108–109

Lammerink J (1965) Six pathogenic races of *Plasmodiophora brassicae* Worn. in New Zealand. N Z J Agric Res 8:156–164

Lammerink J (1986) Identification of an eighth race of *Plasmodiophora brassicae*, the cause of clubroot, in New Zealand. N Z J Agric Res 29:101–104

Lawrence GJ, Finnegan EJ, Ayliffe MA, Ellis JG (1995) The L6 gene for flax rust resistance is related to the *Arabidopsis* bacterial resistance gene *RPS2* and the tobacco viral resistance gene N. Plant Cell 7:1195–1206

Leckie D, Astley D, Crute IR, Ellis PR, Pink DAC, Boukema I, Monteiro AA, Dias S (1996) The location and exploitation of genes for pest and disease resistance in European gene bank collections of horticultural *Brassicas*. Acta Hortic 407:95–101

Lee J, Izzah NK, Choi B-S, Joh HJ, Lee S-C, Perumal S et al (2016) Genotyping-by-sequencing map permits identification of clubroot resistance QTLs and revision of the reference genome assembly in cabbage (*Brassica oleracea* L). DNA Res 23:29–41. https://doi.org/10.1093/dnares/dsv034

Lehner MS, Júnior TJDP, Del Ponte EM, Mizubuti ESG, Pethybridge SJ (2017) Independently founded populations of *Sclerotinia sclerotiorum* from a tropical and a temperate region have similar genetic structure. PLoS One 12:e0173915. https://doi.org/10.1371/journal.pone.0173915

Leister D, Ballvora A, Salamini F, Gebhardt C (1996) A PCR based approach for isolating pathogen resistance genes from potato with potential for wide application in plants. Nat Genet 14:421–429

Lemke PA, Nash CH (1974) Fungal viruses. Bact Rev 38:29–56

Li H, Sivasithamparam K, Barbetti MJ (2003) Breakdown of a *Brassica rapa* subsp. *sylvestris* single dominant blackleg resistance gene in *B. napus* rapeseed by *Leptosphaeria maculans* field isolates in Australia. Plant Dis 87:752. https://doi.org/10.1094/PDIS.2003.87.6.752A

Li YH, Wang H, Li JC, Wang D, Li DR (2005) Infection of *Sclerotinia sclerotiorum* to rapeseed, soybean and sunflower and its virulence differentiation. Acta Phytopathol Sin 35:486–492

Liban SH, Cross DJ, Kutcher HR, Peng G, Fernando WGD (2016) Race structure and frequency of avirulence genes in the western Canadian *Leptosphaeria maculans* pathogen population, the causal agent of blackleg in *Brassica* species. Plant Pathol 65:1161–1169. https://doi.org/10.1111/ppa.12489

Li-Guo Q, Huang HC, Laroche A, Acharya SN (2003) Occurrence and characterization of hypovirulence in the tan sclerotial isolate S10 of *Sclerotinia sclerotiorum*. Mycol Res 107:1350–1360

Linnasalmi A, Toiviainen A (1981) Races of *Plasmodiophora brassicae* Worn. in Finland. In: Proc Brassica Conf, p 24. Prelim Rep

Linnasalmi A, Weisaeth G (1978) Om klumprotraser I Trondelag *Plasmodiophora* rase 1, 7 og 9. Summary: races of clubroot in Trongdelag, Norway. Res Nor Agric 29:223–239

Liu Q, Rimmer SR (1992) Inheritance of resistance in *Brassica napus* to an Ethiopian isolate of *Albugo candida* from *Brassica carinata*. Can J Plant Pathol 14:116–120

Liu Q, Rimmer SR (1993) Production and germination of oospores of *Albugo candida*. Can J Plant Pathol 15:265–271

Liu JQ, Parks P, Rimmer SR (1996) Development of monogenic lines for resistance to *Albugo candida* from a Canadian *Brassica napus* cultivar. Phytopathology 86:1000–1004

Liu L, Zhao D, Zheng L, Hsiang T, Wei Y, Fu Y, Huang J (2013) Identification of virulence genes in the crucifer anthracnose fungus *Colletotrichum higginsianum* by insertional mutagenesis. Microb Pathog 64:6–17

Liu L, Yan Y, Huang J, Hsiang T, Wei Y, Li Y, Gao J, Zheng L (2017) A novel MFS transporter gene *ChMfs1* is important for hyphal morphology, conidiation, and pathogenicity in *Colletotrichum higginsianum*. Front Microbiol 8:1953. https://doi.org/10.3389/fmicb.2017.01953

Lo Presti L, Lanver D, Schweizer G, Tanaka S, Liang L, Tollot M, Zuccaro A, Reissmann S, Kahmann R (2015) Fungal effectors and plant susceptibility. Annu Rev Plant Biol 66:513–545. https://doi.org/10.1146/annurev-arplant-043014-114623

Lowe RG, Cassin A, Grandaubert J, Clark BL, Van De Wouw AP, Rouxel T, Howlett BJ (2014) Genomes and transcriptomes of partners in plant-fungal-interactions between canola (*Brassica napus*) and two *Leptosphaeria* species. PLoS One 9:e103098. https://doi.org/10.1371/journal.pone.0103098

Lucas JA, Hayter JBR, Crute IR (1994) The downy mildews: host specificity and pathogenesis. In: Kohmoto K (ed) Pathogenesis and host specificity in plant diseases, vol 2. Elsevier Science Ltd, Oxford

Ludwig-Müller J, Schuller A (2008) What can we learn from clubroots: alterations in host roots and hormone homeostasis caused by *Plasmodiophora brassicae*. Eur J Plant Pathol 121:291–302

Lumsden RD (1969) *Sclerotinia sclerotiorum* infection of bean and the production of cellulase. Phytopathology 59:653–657

Lumsden RD (1970) Phosphatidase of *Sclerotinia sclerotiorum* produced in culture and in infected bean. Phytopathology 60:1106–1110

Lumsden RD (1976) Pectolytic enzymes of *Sclerotinia sclerotiorum* and their localization in infected bean. Can J Bot 54:2630–2641

Lumsden RD (1979) Histology and physiology of pathogenesis in plant disease caused by *Sclerotinia* species. Phytopathology 69:890–896

Lyu X, Shen C, Fu Y, Xie J, Jiang D, Li G, Cheng J (2016) A small secreted virulence-related protein is essential for the necrotrophic interactions of *Sclerotinia sclerotiorum* with its host plants. PLoS Pathog 12:e1005435

Ma XL, Kong P, You MP, Li H, Sivasithamparam K, Barbetti MJ (2009) Molecular variation among isolates belonging to eight races of *Phytophthora clandestina*. Aust Plant Pathol 38:608–616

Madlung A (2012) Polyploidy and its effect on evolutionary success: old questions revisited with new tools. Heredity 110:99–104

Maggioni L, von Bothmer R, Poulsen G, Branca F (2000) Origin and domestication of cole crops (*Brassica oleracea* L): linguistic and literary considerations. Econ Bot 64:109–123

Mahalingam T, Chen W, Rajapakse CS, Somachandra KP, Attanayake RN (2020) Genetic diversity and recombination in the plant pathogen Sclerotinia sclerotiorum detected in Sri Lanka. Pathogens 9(4):306. https://doi.org/10.3390/pathogens9040306

Makinen Y, Hietajarvi L (1965) On Finnish micromycetes. 5. *Albugo candida* in Finland, with special reference to the variation in the size of the conidia. Ann Bot Fenn 2:33–46

Maltby AD, Mihail JD (1997) Competition among *Sclerotinia sclerotiorum* genotypes within canola stem. Can J Bot 75:462–468

Manzanares-Dauleux MJ, Delourme R, Baron F, Thomas G (2000) Mapping of one major gene and of QTLs involved in resistance to clubroot in *Brassica napus*. Theor Appl Genet 101(5): 885–891

Manzanares-Dauleux MJ, Divaret I, Baron F, Thomas G (2001) Assessment of biological and molecular variability between and within field isolates of *Plasmodiophora brassicae*. Plant Pathol 50:165–173. https://doi.org/10.1046/j.1365-3059.2001.00557.x

Marchal E (1902) De la specialization du parasitisme chez.-I: *Erysiphe graminis*. Comptes Rendus 135:210–212

Marchal E (1903) De la specialization de la parasitism chez.-I: *Erysiphe graminis*. Compt Rend Acad Sci 136:1280–1281

Marciano P, Di Lenna P, Magro P (1983) Oxalic acid, cell wall-degrading enzymes and pH in pathogenesis and their significance in the virulence of two *Sclerotinia sclerotiorum* isolates on sunflower. Physiol Plant Pathol 22:339–345

Marcroft SJ, Elliott VL, Cozijnsen AJ, Salisbury PA, Howlett BJ, Van De Wouw AP (2012a) Identifying resistance genes to *Leptosphaeria maculans* in Australian *Brassica napus* cultivars based on reactions to isolates with known avirulence genotypes. Crop Past Sci 63:338–350. https://doi.org/10.1071/CP11341

Marcroft SJ, Van De Wouw AP, Salisbury PA, Potter TD, Howlett BJ (2012b) Effect of rotation of canola (*Brassica napus*) cultivars with different complements of blackleg resistance genes on disease severity. Plant Pathol 61:934–944. https://doi.org/10.1111/j.1365-3059.2011.02580.x

Martin FN (2000) Phylogenetic relationships among some Pythium species inferred from sequence analysis of the mitochondrially encoded cytochrome oxidase II gene. Mycologia 92:711–727

Martin FN, Tooley PW (2003) Phylogenetic relationships among *Phytophthora* species inferred from sequence analysis of mitochondrially encoded cytochrome oxidase I and II genes. Mycologia 95:269–284

Marukawa S, Funakawa S, Satomura Y (1975) Some physical and chemical factors on formation of sclerotia in *Sclerotinia libertiana* Fuckel. Agric Biol Chem 39:463–468

Masheva S, Antonova G, Bahariev D (1996a) Seasonal variability in the *Peronospora parasitica* (Pers.) Fr. population. Crucif Newsl 18:120

Masheva S, Antonova G, Bahariev D (1996b) Pathogenicity of two isolates *Peronospora parasitica* with different district origin. Crucif Newsl 18:118

Mathur S, Wu C, Rimmer SR (1995) Virulence of isolates of *Albugo candida* from western Canada to *Brassica* species. In: Proc 9th Int Rapeseed Congr, Cambridge, UK, vol 2, pp 652–654

Maxwell DP, Lumsden RD (1970) Oxalic acid production by *Sclerotinia sclerotiorum* in infected bean and in culture. Phytopathology 60:1395–1398

McCulloch L (1929) A bacterial leaf spot of horse-radish caused by Bacterium campestris var. armoraciae, n. var. J Agric Res 38:269–287

Mcdonald BA, Linde C (2002) The population genetics of plant pathogens and breeding strategies for durable resistance. Euphytica 124:163–180

McMeekin D (1960) The role of the oospores of *Peronospora parasitica* from cabbage and radish. Phytopathology 50:93–97

McMullan M, Gardiner A, Bailey K, Kemen E, Ward BJ, Cevik V, Robert-Seilaniantz A, Schultz-Larsen T, Balmuth A, Holub EB, van Oosterhout C, Jones JDG (2015) Evidence for suppression of immunity as a driver for genomic introgressions and host range expansion in races of *Albugo candida*, a generalist parasite. elife 4:e0455

McRoberts N, Lennard JH (1996) Pathogen behaviour and plant cell reactions in interactions between *Alternaria* species and leaves of host and nonhost plants. Plant Pathol 45:742–752

Meena PD, Rani A, Meena R, Sharma P, Gupta R, Chowdappa P (2012) Aggressiveness, diversity and distribution of *Alternaria brassicae* isolates infecting oilseed *Brassica* in India. Afr J Microbiol Res 6:5249–5258

Meena PD, Mehta N, Rai PK, Saharan GS (2018) Geographical distribution of rapeseed- mustard powdery mildew disease in India. J Mycol Plant Pathol 48(3):284–302

Mehta N, Saharan GS (1994) Morphological and pathological variations in *Peronospora parasitica* infecting *Brassica* species. Indian Phytopathol 47:153–158

Mehta N, Sangwan MS, Srivastava MP (2003) Morphological and pathological variations in rapeseed-mustard isolates of *Alternaria brassicae*. Indian Phytopathol 56:188–190

Mehta N, Sangwan MS, Saharan GS (2005) Fungal diseases of rapeseed mustard. In: Saharan GS, Mehta N, Sangwan MS (eds) Diseases of oilseed crops. Indus Publishing Company, New Delhi, pp 1–86

Meinhardt LW, Wulff NA, Bellato CM, Tsai SM (2002) Telomere and micro satellite primers reveal diversity among *Sclerotinia sclerotiorum* isolates from Brazil. Fitopatol Bras 27:211–215

Melhus IE (1911) Experiments on spore germination and infection in certain species of Oomycetes. Wisconsin Agric Expt Stn Res Bull 15:25–91

Melzer MS, Boland GJ (1996) Transmissible hypo-virulence in *Sclerotinia minor*. Can J Plant Pathol 18:19–28

Mendes-Pereira E, Balesdent MH, Brun H, Rouxel T (2003) Molecular phylogeny of the *Leptosphaeria maculans* L *biglobosa* species complex. Mycol Res 107:1287–1304. https://doi.org/10.1017/S0953756203008554

Mengistu A, Rimmer SR, Koch E, Williams PH (1991) Pathogenicity grouping of isolates of *Leptosphaeria maculans* on *Brassica napus* cultivars and their disease reaction profiles on rapid-cycling Brassicas. Plant Dis 75:1279–1282. https://doi.org/10.1094/PD-75-1279

Mesarich CH, Griffiths SA, Van der Burgt A, Kmen B, Beenen HG, Etalo DW, Joosten MHA, de Wit PJGM (2014) Transcriptome sequencing uncovers the *Avr5* avirulence gene of the tomato leaf mold pathogen *Cladosporium fulvum*. Mol Plant-Microbe Interact 27:846–857. PMID: 24678832

Michelmore RW, Hulbert SH (1987) Molecular markers for genetic analysis of phytopathogenic fungi. Annu Rev Phytopathol 25:383–404

Michelmore RW, Norwood JM, Ingram DS, Crute IR, Nicholson P (1984) The inheritance of virulence in *Bremia lactucae* to match resistance factors 3, 4, 5, 6, 7, 8, 9, 10, and 11 in lettuce (*Lactuca sativa*). Plant Pathol 33:301–315

Mindrinos M, Katagiri F, Yu GL, Ausubel FM (1994) The *Arabidopsis thaliana* disease resistance gene *RPS2* encodes a protein containing a nucleotide-binding site and leucine-rich repeats. Cell 78:1089–1099

Möller M, Harling R (1996) Randomly amplified polymorphic DNA (RAPD) profiling of *Plasmodiophora brassicae*. Lett Appl Microbiol 22:70–75

Mollier P (2015) One avirulence precludes another (rapeseed). INRA News Portal

Morrall RAA, Duczek IJ, Sheard JW (1972) Variation and correlation within and between morphology, pathogenicity and pectolytic enzyme activity in *Sclerotinia* from Saskatchewan. Can J Bot 50:767–786

Morris MJ, Knox-Davies PS (1980) *Raphanus raphanistrum* as a weed host of pathogens of cultivated cruciferae in the Western Cape province of South Africa. Phytophylactica 12:53–55

Morris PF, Connolly MS, St Clair DA (2000) Genetic diversity of *Alternaria alternata* isolated from tomato in California assessed using RAPDs. Mycol Res 104:286–292

Moss NA, Crute IR, Lucas JA, Gordon PL (1988) Requirements for analysis of host species specificity in *Peronospora parasitica* (downy mildew). Crucif Newsl 13:114–116

Moss NA, Lucas JA, Crute IR (1991) Evidence for differential response to isolates of *Peronospora parasitica* (downy mildew) in *Brassica rapa*. Test Agro Chem cv. 12. Ann Appl Biol 118:96–97

Moss NA, Crute IR, Lucas JA (1994) Laboratory production of oospores of *Peronospora parasitica* (crucifer downy mildew) and the recovery and characterization of sexual progeny from crosses between isolates with different host specificity. Plant Pathol 43:713–725

Mridha MAU (1983) Virulence of different isolates of *Alternaria brassicae* on winter oilseed rape cultivars. In: Proc 6th international rapeseed conference, Paris, France, pp 1025–1029

Mukhtar MS, Carvunis AR, Dreze M, Epple P, Steinbrenner J, Moore J, Tasan M, Galli M, Hao T, Nishimura MT, Pevzner SJ, Donovan SE, Ghamsari L, Santhanam B, Romero V, Poulin MM, Gebreab F, Gutierrez BJ, Tam S, Monachello D, Boxem M, Harbort CJ, McDonald N, Gai L, Chen H, He Y, European Union Effectoromics Consortium, Vandenhaute J, Roth FP, Hill DE, Ecker JR, Vidal M, Beynon J, Braun P, Dangl JL (2012) Independently evolved virulence effectors converge onto hubs in a plant immune system network. Science 333:596–601. https://doi.org/10.1126/science.1203659

Munro JM, Lennard JH (1982) Variation in the development of *Erysiphe cruciferarum* Opiz. Ex.L Junnel on two cultivars of *Brassica napus*. Crucif Newsl 7:68–69

Napper ME (1933) Observations on spore germination and specialization of parasitism in *Cystopus candidus*. J Pomol Hortic Sci 11:81–100

Nashaat NI, Awasthi RP (1995) Evidence for differential resistance to *Peronospora parasitica* (downy mildew) in accessions of *Brassica juncea* (mustard) at the cotyledon stage. J Phytopathol 143:157–159

Nashaat NI, Rawlinson CJ (1994) The response of oilseed rape (*Brassica napus* ssp. *oleifera*) accessions with different glucosinolate and erucic acid contents to four isolates of *Peronospora parasitica* (downy mildew) and the identification of new sources of resistance. Plant Pathol 43: 278–285

Nath MD, Sharma SL, Kant U (2000) Growth of *Albugo candida* infected mustard callus in culture. Mycopathologia 152:147–153

Natti JJ (1958) Resistance of broccoli and other crucifers to downy mildew. Plant Dis Rep 42:656–662

Natti JJ, Dickson MH, Aktin JD (1967) Resistance of *Brasssica oleracea* cultivars to downy mildew. Phytopathology 57:144–147

Neger FW (1902) Beitrage zur Biologie der Erysipheen. Flora 90:221–272

Neik TX, Barbetti MJ, Batley J (2017) Current status and challenges in identifying disease resistance genes in *Brassica napus*. Front Plant Sci 8:1788. https://doi.org/10.3389/fpls.2017.01788

Nguyen HD, Tomitaka Y, Ho SYW, Duchêne S, Vetten H-J, Lesemann D, Walsh JA, Gibbs AJ, Ohshima K (2013) Turnip mosaic potyvirus probably first spread to Eurasian *Brassica* crops from wild orchids about 1000 years ago. PLoS One 8(2):e55336. https://doi.org/10.1371/journal.pone.0055336

Niwa R, Kawahara A, Murakami H, Tanaka S, Ezawa T (2011) Complete structure of nuclear rDNA of the obligate plant parasite *Plasmodiophora brassicae*: intraspecific polymorphisms in the exon and group I intron of the large subunit rDNA. Protist 162:423–434. https://doi.org/10.1016/j.protis.2011.02.005

Noda J, Brito N, Gonzalez C (2010) The *Botrytis cinerea* xylanase *Xyn11A* contributes to virulence with its necrotizing activity, not with its catalytic activity. BMC Plant Biol 10:38

Noonan MP, Glare TR, Harvey IC, Sands DC, Ocallaghan M (1996) Genetic comparison of *Sclerotinia sclerotiorum* isolates from New Zealand and USA. In: Proc Forty Ninth New Zealand Plant Prot Conference, New Zealand, 13–15 August, 1996, pp 126–131

Norwood JM, Michelmore RW, Crute IR, Ingram DS (1983) The inheritance of specific virulence in *Bremia lactucae* (downy mildew) to match specific resistance factors 1, 2, 4, 6 and 11 in *Lactuca sativa* (lettuce). Plant Pathol 32:177–186

Novotel'Nova NS (1968) Intraspecific taxa of C. candidus (Pers.) Lev. Novosti Sistematiki Nizsh Rast:88–96. (Russian)

Novotel'Nova NS, Minasyan MA (1970) Contribution to the biology of Cystopus candidus (Pers.) Lev. and C. tragopogonis (Pers.) Schroet. Trudy Vsesoyuznogo Nauchnoissledovaterskogo Instituta Zashchity rastenii 29:121–128. (Russian)

Novotelnova NS, Pystina KA (1985) Ordo Peronosporales (Fam. Pythiaceae, Phytophthoraceae, Peronosporeaceae, Cystopaceae). In: Gorlenko MV (ed) Cryptogamic flora of USSR, Fungi (3), vol 11. USSR: Nauka, Leningrad

Nowakowska M, Wrzesińska M, Kamiński P, Szczechura W, Lichocka M, Tartanus M, Kozik EU, Nowicki M (2019) *Alternaria brassicicola* – Brassicaceae pathosystem: insights into the infection process and resistance mechanisms under optimized artificial bio-assay. Eur J Plant Pathol 153:131–151. https://doi.org/10.1007/s10658-018-1548-y

Nowicki M, Nowakowska M, Niezgoda A, Kozik E (2012) Alternaria black spot of crucifers: symptoms, importance of disease, and perspectives of resistance breeding. Veg Crops Res Bull 76:5–19

Noyes RD, Hancock JG (1981) Role of oxalic acid in the Sclerotinia wilt of sunflower. Physiol Plant Pathol 8:123–132

Nuss DL, Koltin Y (1990) Significance of dsRNA genetic elements in plant pathogenic fungi. Annu Rev Phytopathol 28:37–58

Osaki K (2008) Studies on diversity of pathogenicity within *Plasmodiophora brassicae* (in Japanese with English summary). PhD dissertation, Tottori University

Osaki K, Fujiyama S, Nakayama A, Shimizu Y, Ito S, Tanaka S (2008) Relation between pathogenicity and genetic variation within *Plasmodiophora brassicae*. J Gen Plant Pathol 74: 281–288

Pammel LH (1895a) Bacteriosis of rutabaga (Bacillus campestris n sp.). Am Mon Microsc J 16:145–151

Pammel LH (1895b) Bacteriosis of rutabaga (Bacillus campestris n. sp.). Iowa State Coll Agric Exp Stn Bull 27:130–134

Pannell JR, Obbard DJ, Buggs RJA (2004) Polyploidy and the sexual system: what can we learn from Mercurialis annua? Biol J Linn Soc 82:547–560

Pape H, Rabbas P (1920) Inoculation tests with C. candidus Pers. Mitt Biol R-Anst L and U Forstw U 18:58–59. (German)

Park YJ, Lee BM, Ho-Hahn J, Lee GB, Park DS (2004) Sensitive and specific detection of *Xanthomonas campestris* pv. *campestris* by PCR using species-specific primers based on *hrpF* gene sequences. Microbiol Res 159:419–423

Parker JE, Coleman MJ, Szabo V, Frost LN, Schmidt R, van der Biezen E, Moores T, Dean C, Daniels MJ, Jones JDG (1997) The *Arabidopsis* downy mildew resistance gene *Rpp5* shares similarity to the Toll and Interleukin-1 receptors with N and L6. Plant Cell 9:879–894

Parlange F, Daverdin G, Fudal I, Kuhn ML, Balesdent MH, Blaise F, Grezes-Besset B, Rouxel T (2009) *Leptosphaeria maculans* avirulence gene *AvrLm4-7* confers a dual recognition specificity by the *Rlm4* and *Rlm7* resistance genes of oilseed rape, and circumvents *Rlm4*-mediated recognition through a single amino acid change. Mol Microbiol 71:851–863. https://doi.org/10.1111/j.1365-2958.2008.06547.x

Patel D, Zander M, Van De Wouw A, Mason A, Edwards D, Batley J (2015) Population diversity of *Leptosphaeria maculans* in Australia. Int J Biol 7:18–36. https://doi.org/10.5539/ijb.v7n3p18

Petit-Houdenot Y, Degrave A, Meyer M, Blaise F, Ollivier B, Rouxel T, Fudal I, Balesdent MH (2016) A two genes-for-one gene interaction between *Leptosphaeria maculans* and *Brassica napus*. In: Proc 11th meeting of phytopatics-mycology (Aussois)

Petkowski JE, Cunnington JH, Minchinton EJ, Cahill DM (2010) Molecular phylogenetic relationships between *Albugo candida* collections on the Brassicaceae in Australia. Plant Pathol 59:282–288

Petrie GA (1988) Races of *Albugo candida* (white rust and staghead) on cultivated cruciferae in Saskatchewan. Can J Plant Pathol 10:142–150

Petrie GA (1994) New races of *Albugo candida* (white rust) in Saskatchewan and Alberta. Can J Plant Pathol 16:251–252

Petrie GA, Vanterpool TC (1978) *Pseudocercosporella capsellae*, the cause of white leaf spot and grey stem of cruciferae in Western Canada'. Can Plant Dis Surv 58:69–72

Pfister R (1927) On the biology of Cystopus tragopogonis. Centralbl Mr Bakt Abt II 71:8–14:312–313

Piao ZY, Ramchiary N, Lim YP (2009) Genetics of clubroot resistance in *Brassica* species. J Plant Growth Regul 28:252–264. https://doi.org/10.1007/s00344-009-9093-8

Pidskalny RS, Rimmer SR (1985) Virulence of *Albugo candida* from turnip rape (*Brassica campestris*) and mustard (*Brassica juncea*) on various crucifers. Can J Plant Pathol 7:283–286

Pilet ML, Delourme R, Foisset N, Renard M (1998) Identification of loci contributing to quantitative field resistance to blackleg disease, causal agent *Leptosphaeria maculans* (Desm.) Ces. et de Not., in winter rapeseed (*Brassica napus* L.). Theor Appl Genet 96:23–30

Plissonneau C, Daverdin G, Ollivier B, Blaise F, Degrave A, Fudal I, Rouxel T, Balesdent MH (2016) A game of hide and seek between avirulence genes *AvrLm4*-7 and *AvrLm3* in *Leptosphaeria maculans*. New Phytol 209:1613–1624. https://doi.org/10.1111/nph.13736

Plissonneau C, Rouxel T, Chèvre AM, Van De Wouw AP, Balesdent MH (2017) One gene-one name: the *AvrLmJ1* avirulence gene of *Leptosphaeria maculans* is *AvrLm5*. Mol Plant Pathol 19(4):1012–1016. https://doi.org/10.1111/mpp.12574

Ploch S, Choi YJ, Rost C, Shin HD, Schilling E, Thines M (2010) Evolution of diversity in *Albugo* is driven by high host specificity and multiple speciation events on closely related Brassicaceae. Mol Phylogenet Evol 57:812–820

Plotnikova JM, Reuber TL, Ausubel FM, Pfister DH (1998) Powdery mildew pathogenesis of *Arabidopsis thaliana*. Mycologia 90(6):1009–1016

Popovic T, Mitrovic P, Jelusic A, Dimkic I, Marjanovic-Jeromela A, Nikolic I, Stankovic S (2019) Genetic diversity and virulence of *Xanthomonas campestris* pv. *campestris* isolates from *Brassica napus* and six *Brassica oleracea* crops in Serbia. Plant Pathol 68:1448–1457

Pound GS, Williams PH (1963) Biological races of *Albugo candida*. Phytopathology 53:1146–1149

Pramila PG, Tasleem M, Taj G, Mal R, Kumar A (2014) Morphological, cultural, pathogenic and molecular variability amongst Indian mustard isolates of *Alternaria brassicae* in Uttarakhand. Afr J Biotechnol 13(3):441–448

Price K, Colhoun J (1975) A study of variability of isolates of *Sclerotinia sclerotiorum* (Lib) de Bary from different hosts. Phytopathol Z 83:159–166

Purwantara A, Barrins JM, Cozijnsen AJ, Ades PK, Howle BJ (2000) Genetic diversity of isolates of the *Leptosphaeria maculans* species complex from Australia, Europe and North America using amplified fragment length polymorphism analysis. Mycol Res 104:772–781. https://doi.org/10.1017/S09537562-9900235X

Rademaker JLW, Louws FJ, De Bruijn FJ (1998) Characterization of the diversity of ecologically important microbes by rep-PCR genomic fingerprinting. Mol Microbial Ecol Manual 3:1–27

Raffaele S, Kamoun S (2012) Genome evolution in filamentous plant pathogens: why bigger can be better. Nat Rev Microbiol 10(6):417–430

Ralph RK (1969) Double-stranded RNA. Adv Virus Res 15:61–158

Rao ANS (1979) New species of white rust and sooty mold fungi. Indian J Mycol Plant Pathol 9: 283

Redkar A, Bonequi MV, Doehlemann G (2015) Conservation of the *Ustilago maydis* effector *See1* in related smuts. Plant Signal Behav 10:e1086855

Reed GM (1905) Infection experiments with *Erysiphe graminis* DC. Trans Wis Acad Sci Arts Lett 15:135–162

Reed GM (1907) Infection experiments with the mildew on cucurbits, *Erysiphe cichoracearum* DC. Trans Wis Acad Sci Arts Lett 17:527–547

Reed GM (1908) Infection experiments with Erysiphe cichoracearum DC. Bull Univ Wisc Sci Ser 3:337–416

Reed GM (1909) The mildews of the cereals. Bull Torrey Bot Club 36:353–388

Reed GM (1912) Infection experiments with the powdery mildew of wheat. Phytopathology 2:81–87

Rehmany AP, Lynn JR, Tor M, Holub EB, Beynon JL (2000) A comparison of *Peronospora parasitica* (Downy mildew) isolates from *Arabidopsis thaliana* and *Brassica oleracea* using amplified fragment length polymorphism and internal transcribed spacer 1 sequence analyses. Fungal Genet Biol 30(2):95–103. https://doi.org/10.1006/fgbi.2000.1216

Reiter RS, Williams JGK, Feldmann KA, Rafalski JA, Tingey SV, Scolnik PA (1992) Global and local genome mapping in *Arabidopsis thaliana* by using recombinant inbred lines and random amplified polymorphic DNAs. Proc Natl Acad Sci U S A 89:1477–1481

Říčarová V, Kazda J, Singh K, Ryšánek P (2016) Clubroot caused by *Plasmodiophora brassicae* Worn.: a review of emerging serious disease of oilseed rape in the Czech Republic. Plant Prot Sci 52:71–86. https://doi.org/10.17221/87/2015-PPS

Rimmer SR, Mathur S, Wu CR (2000) Virulence of isolates of *Albugo candida* from western Canada to *Brassica* species. Can J Plant Pathol 22:229–235

Rolfe SA, Strelkov SE, Links MG, Clarke WE, Robinson SJ, Djavaheri M, Malinowski R, Haddadi P, Kagale S, Parkin IAP, Taheri A, Borhan MH (2016) The compact genome of the plant pathogen *Plasmodiophora brassicae* is adapted to intracellular interactions with host *Brassica* spp. BMC Genomics 17:272. https://doi.org/10.1186/s12864-016-2597-2

Rommens CMT, Salmeron JM, Oldroyd GED, Staskawicz BJ (1995) Intergeneric transfer and functional expression of the tomato disease resistance gene *Pto*. Plant Cell 7:1537–1544

Rouxel T, Penaud A, Pinochet X, Brun H, Gout L, Delourme R et al (2003) A 10-year survey of populations of *Leptosphaeria maculans* in France indicates a rapid adaptation towards the *Rlm1* resistance gene of oilseed rape. Eur J Plant Pathol 109:871–881. https://doi.org/10.1023/A.1026189225466

Rouxel T, Grandaubert J, Hane JK, Hoede C, Van De Wouw AP, Couloux A, Dominguez V, Anthouard V, Bally P, Bourras S, Cozijnsen AJ, Ciuffetti LM, Degrave A, Dilmaghani A, Duret L, Fudal I, Goodwin SB, Gout L, Glaser N, Linglin J, Kema GHJ, Lapalu N, Lawrence CB, May K, Meyer M, Ollivier B, Poulain J, Schoch CL, Simon A, Spatafora JW, Anna

Stachowiak B, Turgeon G, Tyler BM, Vincent D, Weissenbach J, Amselem J, Quesneville H, Oliver RP, Wincker P, Balesdent M-H, Howlett BJ (2011) Effector diversification within compartments of the *Leptosphaeria maculans* genome affected by repeat-induced point mutations. Nat Commun 2:202. https://doi.org/10.1038/ncomms1189

Rudd JJ, Kanyuka K, Hassani-Pak K, Derbyshire M, Andongabo A, Devonshire J, Lysenko A, Saqi M, Desai NM, Powers SJ, Hooper J, Ambroso L, Bharti A, Farmer A, Hammond-Kosack KE, Dietrich RA, Courbot M (2015) Transcriptome and metabolite profiling of the infection cycle of *Zymoseptoria tritici* on wheat reveals a biphasic interaction with plant immunity involving differential pathogen chromosomal contributions and a variation on the hemibiotrophic lifestyle definition. Plant Physiol 167:1158–1185

Russell HL (1898) A bacterial rot of cabbage and allied plants. Wisc Agric Exp Stn Bull 65:1–39

Saharan GS (1992a) Disease resistance. In: Labana KS, Banga SS, Banga SK (eds) Breeding oilseed Brassicas, vol 12. Narosa Publ. House, New Delhi, pp 181–205

Saharan GS (1992b) Management of rapeseed and mustard diseases. In: Kumar D, Rai M (eds) Advances in oilseeds research. Scientific Publ, Jodhpur, pp 152–188

Saharan GS (1995) Nature and mechanism of resistance in rapeseed-mustard. In: Proc global conference on advances in research on plant diseases and their management. RCA, Udaipur, pp 35–36. (Abstr.)

Saharan GS (2010) Analysis of genetic diversity in *Albugo*-crucifer system. J Mycol Plant Pathol 40:1–13

Saharan GS, Kadian AK (1983) Physiologic specialization in *Alternaria brassicae*. Crucif Newsl 8: 32–33

Saharan GS, Kaushik JC (1981) Occurrence and epidemiology of powdery mildew of *Brassica*. Indian Phytopathol 35:17–21

Saharan GS, Verma PR (1992) White rusts: a review of economically important species. IDRC-MR315e, vol IV. International Development Research Centre, Ottawa, ON. 65 p

Saharan GS, Verma PR, Meena PD, Kumar A (2014) White rust of crucifers: biology, ecology and management. Springer Nature, Singapore

Saharan GS, Mehta N, Meena PD (2016) Alternaria diseases of crucifers: biology, ecology and disease management. Springer, Singapore. 244 p

Saharan GS, Mehta N, Meena PD (2017) Downy mildew disease of crucifers: biology, ecology and disease management. Springer Nature, Singapore. 357 p

Saharan GS, Mehta N, Meena PD (2021) Clubroot disease of crucifers: biology, ecology and disease management. Springer Nature, Singapore. 757 p

Salmeron JM, Oldroyd GED, Rommens CMT, Scofield SR, Kim H-S, Lavelle DT, Dahlbeck D, Staskawicz BJ (1996) Tomato *Prf* is a member of the leucine-rich repeat class of plant disease resistance genes and lies embedded within the Pto kinase gene cluster. Cell 86:123–133

Salmon ES (1903a) Infection powers of ascospores in Erysipheae. Aust J Bot 41(159):204–212

Salmon ES (1903b) On specialization of parasitism in the Erysiphaceae. Beih Bot Centralbl 14: 261–315

Salmon ES (1904a) On specialization of parasitism in the Erysiphaceae. New Phytol 3:109

Salmon ES (1904b) Mycological notes. Aust J Bot 42:182–186

Salmon ES (1904c) On *Erysiphe graminis* DC. and its adaptive parasitism within the genus *Bromus*. Ann Mycol 2:255–267

Salmon ES (1905a) Cultural experiments with an Oidium on *Euonymus japonicas* Linn. f. Ann Mycol 3:1–15

Salmon ES (1905b) On the variation shown by the conidial stage of *Phyllactinia corylea* (Pers.) Karst. Ann Mycol 3:493–505

Sampson PJ, Walker J (1982) An annotated list of plant diseases in Tasmania. Department of Agriculture Tasmania, Hobart, TAS

Sangwan MS, Mehta N (2006) Categorization of *Alternaria brassicae* isolates on the basis of differential sensitivity to fungicides. Plant Dis Res 21:114–117

Sangwan MS, Mehta N (2007) Pathogenic variability in isolates of *Alternaria brassicae* (Berk) Sacc. from different agro-climatic zones of India. Plant Dis Res 22:101–107

Sanz-Martín JM, Pacheco-Arjona JR, Bello-Rico V, Vargas WA, Monod M, Díaz-Mínguez JM, Thon MR, Sukno SA (2016) A highly conserved metalloprotease effector enhances virulence in the maize anthracnose fungus *Colletotrichum graminicola*. Mol Plant Pathol 17(7):1048–1062. https://doi.org/10.1111/mpp.12347

Savulescu O (1946) A study on the European species of the genus *Cystopus* Lev. with special reference to the species found in Rumania. Thesis, University of Bucarest, Rumania, p 213. (Abstract in Rev Appl Mycol, 27: 542, 1948)

Savulescu T, Rayss T (1930) Contribution to the knowledge of the Peronsoporaceae of Romania. Ann Mycol 28:297–320

Schafer C, Wostemeyer J (1992) Random primer dependent PCR differentiates aggressive from non-aggressive isolates of the oilseed rape pathogen *Phoma lingam* (*Leptosphaeria maculans*). J Phytopathol 136:124–136

Scheffer RP (1989) Ecological consequences of toxin production by *Cochliobolus* and related fungi. In: Graniti A, Durbin RD, Ballio A (eds) Phytotoxins and plant pathogenesis. Springer, Berlin, pp 285–300

Schlaich NL, Slusarenko A (2009) Downy mildew of *Arabidopsis* caused by *Hyaloperonospora arabidopsidis* (formerly *Hyaloperonospora parasitica*). In: Kurt L, Kamoun S (eds) Oomycete genetics and genomics: diversity, interactions and research tools. Wiley, Hoboken, NJ, pp 263–285. Chapter 13

Schmidpeter J, Dahl M, Hofmann J, Koch C (2017) *ChMob2* binds to *ChCbk1* and promotes virulence and conidiation of the fungal pathogen *Colletotrichum higginsianum*. BMC Microbiol 17:22. https://doi.org/10.1186/s12866-017-0932-7

Schwarz EM (2017) Evolution: a parthenogenetic nematode shows how animals become sexless. Curr Biol 27:R1064–R1066

Schwelm A, Fogelqvist J, Knaust A, Jülke S, Lilja T, Bonilla-Rosso G, Karlsson M, Shevchenko A, Dhandapani V, Choi SR, Kim HG (2015) The *Plasmodiophora brassicae* genome reveals insights in its life cycle and ancestry of chitin synthases. Sci Rep 5:11153. https://doi.org/10.1038/srep11153

Schwelm A, Berney C, Dixelius C, Bassc D, Neuhauser S (2016) The large subunit rDNA sequence of Plasmodiophora brassicae does not contain intra-species polymorphism. Protist 167:544–554. https://doi.org/10.1016/j.protis.2016.08.008

Sedaghatkish A, Gossen BD, Yu F, Torkamaneh D, McDonald MR (2019) Whole-genome DNA similarity and population structure of *Plasmodiophora brassicae* strains from Canada. BMC Genomics 20:744. https://doi.org/10.1186/s12864-019-6118-y

Semb L (1969) Cabbage downy mildew. Jord Avling 12:32–35

Sequeira P, Monteiro A (1996) Heterothallism and homothallism in Portuguese isolates of *Peronospora parasitica* (Pers. ex Fr.) Fr. Crucif Newsl 18:126–127

Sexton AC, Howlett BJ (2004) Microsatellite markers reveal genetic differentiation among populations of *Sclerotinia sclerotiorum* from Australian canola fields. Curr Genet 46:357–365

Sexton AC, Whitten AR, Howlett BJ (2006) Population structure of *Sclerotinia sclerotiorum* in an Australian canola field at flowering and stem-infection stages of the disease cycle. Genome 49: 1408–1415

Sharma TR, Tewari JP (1995) Detection of genetic variation in *Alternaria brassicae* by RAPD fingerprints. J Plant Biochem Biotechnol 4:105–107

Sharma TR, Tewari JP (1998) RAPD analysis of three *Alternaria* species pathogenic to crucifers. Mycol Res 102:807–814

Sharma S, Singh J, Munshi GD, Munshi SK (2010) Biochemical changes associated with application of biocontrol agents on Indian mustard leaves from plants infected with Alternaria Blight. Arch Phytopathol Plant Protect 43:315–323

Sharma M, Deep S, Bhati DS, Chowdappa P, Selvamani R, Sharma P (2013a) Morphological, cultural, pathogenic and molecular studies of *Alternaria brassicae* infecting cauliflower and mustard in India. Afr J Microbiol Res 7:3351–3363

Sharma K, Gossen BD, Greenshields D, Selvaraj G, Strelkov SE, McDonald MR (2013b) Reaction of lines of the rapid cycling *Brassica* collection and *Arabidopsis thaliana* to selected pathotypes of *Plasmodiophora brassicae*. Plant Dis 97(6):720–727. https://doi.org/10.1094/PDIS-08-12-0752-RE

Sharma P, Samkumar A, Rao M, Singh VV, Prasad L, Mishra DC, Bhattacharya R, Gupta NC (2018) Genetic diversity studies based on morphological variability, pathogenicity and molecular phylogeny of the *Sclerotinia sclerotiorum* population from Indian mustard (*Brassica juncea*). Front Microbiol 9:1169. https://doi.org/10.3389/fmicb.2018.01169

Shaw DS (1991) Genetics. In: Ingram DS, Williams PH (eds) *Phytophthora infestans*. Advances in plant pathology, vol 7. Academic Press, New York, NY, pp 131–170

Sherriff C, Lucas JA (1989) Heterothallism and homothallism in *Peronospora parasitica*. Mycol Res 92:311–316

Sherriff C, Lucas JA (1990) The host range of isolates of downy mildew, *Peronospora parasitica* from *Brassica* crop species. Plant Pathol 39:77–91

Shivas RG (1989) Fungal and bacterial diseases of plants in Western Australia. J R Soc WA 72:1–62

Siciliano I, Gilardi G, Ortu G, Gisi U, Gullino ML, Garibaldi A (2017) Identification and characterization of *Alternaria* species causing leaf spot on cabbage, cauliflower, wild and cultivated rocket by using molecular and morphological features and mycotoxin production. Eur J Plant Pathol 149:1–13

Silue D, Nashaat NI, Tirilly Y (1996) Differential responses of *Brassica oleracea* and *B. rapa* accessions to seven isolates of *Peronospora parasitica* at the cotyledon stage. Plant Dis 80:142–144

Singh K (1994) Studies on the ecofriendly management of powdery mildew (*Erysiphe cruciferarum* Opiz ex. Junell) of mustard [*Brassica juncea* (L) Czern & Coss]. MSc Thesis, Department of Plant Pathology, CCS HAU, Hisar, p 106, xvii

Singh BM, Bhardwaj CL (1984) Physiologic races of *Albugo candida* on crucifers in Himachal Pradesh. Indian J Mycol Plant Pathol 14:25. (Abstr)

Singh BM, Chand JN (1983) Identification of physiologic races-general consideration. In: Chand JN, Saharan GS (eds) Phytopathological techniques. HAU, Press, Hisar, pp 109–115

Singh S, Singh RP, Singh HK, Kumar K (2008) Screening of *Brassica* genotypes for quality traits and reaction to Alternaria blight and white rust. Crucif Newsl 27:35–36

Singh R, Singh D, Singh H (2013) Variability in *Alternaria brassicae* incitant of Alternaria blight of oilseed *Brassica*. Res Crops 14(4):1082–1088

Singh D, Rathaur PS, Vicente JG (2016) Characterization, genetic diversity and distribution of *Xanthomonas campestris* pv. *campestris* races causing black rot disease in cruciferous crops of India. Plant Pathol 65:1411–1418

Sirjusingh C, Kohn LM (2001) Characterization of microsatellites in the fungal plant pathogen *Sclerotinia sclerotiorum*. Mol Ecol Notes 1:267–269

Slusarenko AJ, Schlaich NL (2003) Downy mildew of *Arabidopsis thaliana* caused by *Hyaloperonospora parasitica* (formerly *Peronospora parasitica*). Mol Plant Pathol 4:159–170

Smilde WD, Linders EGA, Veenstra RM (2012) Characterization of clubroot resistance in *Brassica oleracea*. In: Program and abstract book 10th Conference of the European foundation of plant pathology, 1–5 October 2012. Poster abstract 46. http://www.efpp.net/ipm2/Program_and_abstract_book/posters.html

Smith EF (1898) The black rot of the cabbage. US Dep Agric Farm Bull 68:1–21

Smith IM, Dunez J, Lelliott RA, Phillips DH, Archer SA (1988) European handbook of plant diseases. Blackwell Scientific Publications, Oxford

Smith KM, Galazka JM, Phatale PA, Connolly LR, Freitag M (2012) Centromeres of filamentous fungi. Chromosom Res 20:635–656

Some A, Manzanares MJ, Laurens F, Baron F, Thomas G, Rouxel F (1996) Variation for virulence on Brassica napus L. amongst Plasmodiophora brassicae collections from France and derived single-spore isolates. Plant Pathol 45(3):432–439

Sperschneider J, Dodds PN, Gardiner DM, Manners JM, Singh KB, Taylor JM (2015) Advances and challenges in computational prediction of effectors from plant pathogenic fungi. PLoS Pathog 11(5):e1004806. https://doi.org/10.1371/journal.ppat.1004806

Sprague SJ, Marcroft SJ, Hayden HL, Howlett BJ (2006) Major gene resistance to blackleg in *Brassica napus* overcome within three years of commercial production in Southeastern Australia. Plant Dis 90:190–198. https://doi.org/10.1094/PD-90-0190

Spring O, Bachofer M, Thines M, Göker M, Oberwinkler F (2006) Intraspecific relationship of *Plasmopara halstedii* isolates differing in pathogenicity and geographic origin based on ITS sequence data. Eur J Plant Pathol 114:309–315

Stachowiak A, Olechnowicz J, Jedryczka M, Rouxel T, Balesdent MH, Happstadius I, Gladders P, Latunde-Dada A, Evans N (2006) Frequency of avirulence alleles in field populations of *Leptosphaeria maculans* in Europe. Eur J Plant Pathol 114:67–75. https://doi.org/10.1007/s10658-005-2931-z

Staskawicz BJ, Ausubel FM, Baker BJ, Ellis JG, Jones JDG (1995) Molecular genetics of plant disease resistance. Science 268:661–667

Steadman JR, Jung G, Adams MS, Powers K, Higgins B, Nelson BD, Gulya TJ (1998) Random amplified polymorphic DNA (RAPD) distinguishes three species of *Sclerotinia* but not pathogenic variability in *S. sclerotiorum* isolates from diverse host and geographic origin. In: Proc 1998 International Sclerotinia Workshop, Fargo, ND, 9–12 September, 1998, pp 10–13

Steiner JA (1908) Die Specialisation der Alchemillenbewohnenden *Sphaerotheca humuli* (DC.) Burr. Centrall blatt fur Bakt. Parasit Infext Krankheit Abstr II 21:677–736

Stiles JI, Lemme C, Sondur S, Morshidi MB, Manshardt R (1993) Using randomly amplified polymorphic DNA for evaluating genetic relationships among papaya cultivars. Theor Appl Genet 85:697–701

Stoll K (1952) The organ, injurious effect and control of *Brassica* blackening. Nachrichten Deutsch Pflanzensch 6:81–85

Strehlow B, De Mol F, Struck C (2014) History of oilseed rape cropping and geographic origin affect the genetic structure of *Plasmodiophora brassicae* populations. Phytopathology 104(5):532

Strelkov SE, Hwang SF (2014) Clubroot in the Canadian canola crop: 10 years into the outbreak. Can J Plant Pathol 36:27–36. https://doi.org/10.1080/07060661.2013.863807

Strelkov SE, Tewari JP, Smith-Degenhardt E (2006) Characterization of *Plasmodiophora brassicae* populations form Alberta, Canada. Can J Plant Pathol 28:467–474

Strelkov SE, Manolii VP, Cao TS, Xue SM, Hwang SF (2007) Pathotype classification of *Plasmodiophora brassicae* and its occurrence in *Brassica napus* in Alberta. Can J Phytopathol 155:706–712. https://doi.org/10.1111/j.1439-0434.2007.01303.x

Strelkov SE, Hwang SF, Manolii VP, Cao TS, Feindel D (2016) Emergence of new virulence phenotypes of *Plasmodiophora brassicae* on canola (*Brassica napus*) in Alberta, Canada. Eur J Plant Pathol 145:517–529. https://doi.org/10.1007/s10658-016-0888-8

Strelkov SE, Hwang SF, Manolii VP, Cao TS, Fredua-Agyeman R, Harding MW, Peng G, Gossen BD, McDonald MR, Feindel D (2018) Virulence and pathotype classification of *Plasmodiophora brassicae* populations collected from clubroot resistant canola (*Brassica napus*) in Canada. Can J Plant Pathol 40:284–298. https://doi.org/10.1080/07060661.2018.1459851

Sun JM, Irzykowski W, Jedryczka M, Xia HF (2005) Analysis of the genetic structure of *Sclerotinia sclerotiorum* (Lib.) de Bary populations from different regions and host plants by random amplified polymorphic DNA markers. J Integr Plant Biol 47:385–395

Sung W, Ackerman MS, Miller SF, Doak TG, Lynch M (2012) Drift-barrier hypothesis and mutation-rate evolution. PNAS 109(45):18488–18499

Takahara H, Huser A, O'Connell R (2012) Two arginine biosynthesis genes are essential for pathogenicity of *Colletotrichum higginsianum* on *Arabidopsis*. Mycology 3(2012):54–64

Takahara H, Hacquard S, Kombrink A, Hughes HB, Halder V, Robin GP, Hiruma K, Neumann U, Shinya T, Kombrink E, Shibuya N, Thomma BPHJ, O'Connell RJ (2016) *Colletotrichum higginsianum* extracellular LysM proteins play dual roles in appressorial function and suppression of chitin-triggered plant immunity. New Phytol 211:1323–1337. https://doi.org/10.1111/nph.13994

Takematsu T, Ichizen N (1993) Sekai no Zassou II. Zenkoku Noson Kyoiku Kyokai, Tokyo, pp 415–417

Tamura K, Takikawa Y, Tsuyumu S, Goto M (1994) Bacterial spot of crucifers caused by Xanthomonas campestris pv. raphani. Ann Phytopathol Soc Jpn 60:281–287

Tanaka S, Ito SI (2013) Pathogenic and genetic diversity in *Plasmodiophora brassicae* (clubroot) from Japan. J Gen Plant Pathol 79:297–306. https://doi.org/10.1007/s10327-013-0456-4

Tanaka S, Sakamoto Y, Kajima K, Fujieda K, Katumoto K, Nishi Y (1991) Pathogenicity of three isolates of clubroot fungus attacking clubroot-resistant cultivars of Chinese cabbage (in Japanese wih English summary). Bull Fac Agric Yamaguchi Univ 39:113–122

Tanaka S, Ito S, Kameya-Iwaki M, Katumoto K, Nishi Y (1993) Occurrence and distribution of clubroot disease on two cruciferous weeds, *Cardamine flexuosa* and *C. scutata*, in Japan. Trans Mycol Soc 34:381–388

Tanaka S, Yoshihara S, Ito S, Kameya-Iwaki M (1997) The influence of virulence of *Plasmodiophora brassicae* populations on epidemiology of Chinese cabbage clubroot and efficacy of fungicides (in Japanese with English summary). Ann Phytopathol Soc Jpn 63: 183–187

Tanaka S, Fujiyama S, Shigemori S, Nakayama A, Ito S, Kameya-Iwaki M (1998) Pathogenesis of isolates of *Plasmodiophora brassicae* from Japan (1) Race and pathogenesis in clubroot resistant cultivars (in Japanese with English summary). Kyushu Plant Prot Res 44:15–19

Tanaka S, Mizui Y, Terasaki H, Ito S (2006) Distribution of clubroot disease of a cruciferous weed, *Cardamine flexuosa*, in major isolated islands, Hokkaido and Okinawa in Japan. Mycoscience 47(2):72–77

Taubenhaus T (1923) The culture and diseases of the sweet potato. 120–123

Taylor JW, Jacobson DJ, Fisher MC (1999) The evolution of asexual fungi: reproduction, speciation and classification. Annu Rev Phytopathol 37:197–246

Taylor JW, Jacobson DJ, Kroken S, Kasuga T, Geiser DM, Hibbett DS, Fisher MC (2000) Phylogenetic species recognition and species concepts in fungi. Fungal Genet Biol 31:21–32

Tewari JP, Skoropad WP (1977) Ultrastructure of oospore development in *Albugo candida* on rapeseed. Can J Bot 55:2348–2357

Tham FY, Lucas JA, Wilson ZA (1994) DNA fingerprinting of *Peronospora parasitica*, a biotrophic fungal pathogen of crucifers. Theor Appl Genet 88:490–496

Thilmony RL, Chen Z, Bressan RA, Martin GB (1995) Expression of the tomato Pto gene in tobacco enhances resistance to *Pseudomonas syringae* pv *tabaci* expressing avr Pto. Plant Cell 7:1529–1536

Thines A, Spring O (2005) A revision of *Albugo* (Chromista, Peronosporomycetes). Mycotaxon 92: 443–458

Thines M, Voglmayr H (2009) An introduction to the white blister rusts (Albuginales). In: Lamour K, Sophien K (eds) Oomycete genetics and genomics: diversity, interactions and research tools. John Wiley & Sons, Inc, Hoboken, NJ, pp 77–92. Chapter 4

Thines M, Choi YJ, Kemen E, Ploch S, Holub EB, Shin HD, Jones JDG (2009) A new species of *Albugo* parasitic to *Arabidopsis thaliana* reveals new evolutionary patterns in white blister rusts (Albuginaceae). Persoonia 22:123–128

Thomma BP (2003) *Alternaria* spp.: from general saprophyte to specific parasite. Mol Plant Pathol 4:225–236

Thung TH (1926) *Peronospora parasitica* (Pers.) De Bary attacking cabbage heads. Phytopathology 16:365–366

Togashi K, Shibasaki Y, Sugana Y (1931) Morphological studies of white rust fungi in cruciferous plants. Jpn J Bot 5:82–83

Togashi G, Togashi K, Shibasaki Y (1934) Biometrical and biological studies of *Albugo candida* (Pers.) O. Kuntze in connection with its specialization, vol 18. Bull Impl College Agri Forestry, Morioka, p 88

Tomimura K, Gibbs AJ, Jenner CE, Walsh JA, Ohshima K (2003) The phylogeny of *Turnip mosaic virus*; comparisons of 38 genomic sequences reveal a Eurasian origin and a recent 'emergence' in east Asia. Mol Ecol 12:2099–2111

Tomimura K, Špak J, Katis N, Jenner CE, Walsh JA et al (2004) Comparisons of the genetic structure of populations of *Turnip mosaic virus* in West and East Eurasia. Virology 330:408–423

Travadon R, Sache I, Dutech C, Stachowiak A, Marquer B, Bousset L (2011) Absence of isolation by distance patterns at the regional scale in the fungal plant pathogen *Leptosphaeria maculans*. Fungal Biol 115:649–659. https://doi.org/10.1016/j.funbio.2011.03.009

Tyler BM, Forster H, Coffey MD (1995) Inheritance of avirulence factors and RFLP markers in outcrosses of the oomycete *Phytophthora sojae*. Mol Plant-Microbe Interact 8:515–523

Ul'yanishchev VI, Osipian LL, Kanchaveli LA, Akhundov TM (1985) Peronosporovye Griby (Peronosporaceous Fungi). In: Ul'yanishchev VI (ed) Opredelitel' Gribov Zakavkaz'ya. Erevan University, Erevan

Uloth MB, You MP, Barbetti MJ (2017) Plant age and ambient temperature: significant drivers for powdery mildew (*Erysiphe cruciferarum*) epidemics on oilseed rape (*Brassica napus*). Plant Pathol 67(2):445–456. https://doi.org/10.1111/ppa.12740

Van Alfen NK (1982) Biology and potential for disease control of hypovirulence of *Endothia parasitica*. Annu Rev Phytopathol 20:349–362

Van de Wouw AP, Marcroft SJ, Barbetti MJ, Hua L, Salisbury PA, Rouxel T, Howlett BJ, Balesdent MH (2009) Dual control of avirulence in *Leptosphaeria maculans* towards a *Brassica napus* cultivar with "sylvestris -derived" resistance suggests involvement of two resistance genes. Plant Pathol 58:305–313. https://doi.org/10.1111/j.1365-3059.2008.01982.x

Van de Wouw AP, Idnurm A, Davidson JA, Sprague SJ, Khangura RK, Ware AH et al (2016) Fungal diseases of canola in Australia: identification of trends, threats and potential therapies. Aust Plant Pathol 45:415–423

Van de Wouw AP, Howlett BJ, Idnurm A (2017) Changes in allele frequencies of avirulence genes in the blackleg fungus, *Leptosphaeria maculans*, over two decades in Australia. Crop Past Sci 69:CP16411. https://doi.org/10.1071/CP16411

Van Den Mooter M, Swings J (1990) Numerical analysis of phenotypic features of *Xanthomonas strains* and related strains and an improved taxonomy of the genus. Int J Syst Bacteriol 40:348–369

Van Schreven DA (1953) *Alternaria, Stemphylium* en *Botrytis* aantasting bij Koolzaad (*Brassica napus*). Tijdschr Plantenziekt 59:105–136

Vanev SG, Dimitrova EG, Ilieva EI (1993) Gybite v Bylgariya. 2 tom. Razred Peronosporales [Fungi Bulgaricae, vol. 2. Ordo Peronosporales]. Bulgarian Academy of Sciences, Sofia

Vauterin L, Hoste B, Kersters K, Swings J (1995) Re-classification of *Xanthomonas*. Int J Syst Bacteriol 45:472–489

Verma U, Bhowmik TP (1989) Inheritance of resistance to a Brassica juncea pathotype of Albugo candida in B. napus. Can J Plant Pathol 11:443–444

Verma PR, Saharan GS (1994) Monograph on *Alternaria* diseases of crucifers, Technical Bulletin 1994–6E. Saskatoon Research Centre, Agriculture and Agri-Food Canada, Saskatoon, SK. 162 p

Verma PR, Harding H, Petrie GA, Williams PH (1975) Infection and temporal development of mycelium of *Albugo candida* in cotyledons of four *Brassica* spp. Can J Bot 53:1016–1020

Verma PR, Saharan GS, Bartaria AM, Shivpuri A (1999) Biological races of *Albugo candida* on *Brassica juncea* and *Brassica rapa* var. Toria in India. J Mycol Plant Pathol 29:75–82

Vicente JG, Conway J, Roberts SJ, Taylor JD (2001) Identification and origin of Xanthomonas campestris pv. campestris races and related pathovars. Phytopathology 91:492–499

Vicente JG, Everett B, Roberts SJ (2006) Identification of isolates that cause a leaf spot disease of brassicas as Xanthomonas campestris pv. raphani and pathogenic and genetic comparison with related pathovars. Phytopathology 96:735–745

Vicente JG, Holub EB (2013) *Xanthomonas campestris* pv. *campestris* (cause of black rot of crucifers) in the genomic era is still a worldwide threat to Brassica crops. Mol Plant Pathol 14:2–18. https://doi.org/10.1111/j.1364-3703.2012.00833.x

Vishwanath K, Kolte SJ (1997) Variability in *Alternaria brassicae*: response to host genotypes, toxin production and fungicides. Indian Phytopathol 50:373–381

Voglino P (1905) Contribuzione allo studio della Phyllactinia corylea. Nuovo Giornale Bot Ital 12: 313–327

Voglmayr H (2003) Phylogenetic relationships of *Peronospora* and related genera based on nuclear ribosomal ITS sequences. Mycol Res 107:1132–1142

Voglmayr H, Riethmüller A (2006) Phylogenetic relationships of *Albugo* species (white blister rusts) based on LSU rDNA sequence and oospore data. Mycol Res 110:75–85

Voigt K, Cozijnsen AJ, Kroymann J, Pöggeler S, Howlett BJ (2005) Phylogenetic relationships between members of the crucifer pathogenic *Leptosphaeria maculans* species complex as shown by mating type (*MAT1-2*), actin, and β-tubulin sequences. Mol Phylogenet Evol 37: 541–557. https://doi.org/10.1016/j.ympev.2005.07.006

Vos P, Paulo M, Voorrips R, Visser R, Eck H, Eeuwijk F (2017) Evaluation of LD decay and various LD- decay estimators in simulated and SNP- array data of tetraploid potato. Theor Appl Genet 130(1):123–135. https://doi.org/10.1007/s00122-016-2798-8

Wang CM (1944) Physiological specialization in *Peronospora parasitica* and reaction of hosts. China J Sci Agric 1:249–257

Wang M, Farnham MW, Thomas CE (2000) Phenotypic variation for downy mildew resistance among inbred broccoli. Hortic Sci 35:925–929

Wei C, Chen J, Kuang H (2016) Dramatic number variation of R genes in solanaceae species accounted for by a few R gene subfamilies. PLoS One 11(2):e0148708. https://doi.org/10.1371/journal.pone.0148708

Welsh J, McClelland M (1990) Finger printing genomes using PCR with arbitrary primers. Nucleic Acids Res 18:7213–7218

Whisson SC, Drenth A, Maclean DJ, Irwin JAG (1994) Evidence for out crossing in *Phytophthora sojae* and linkage of a DNA marker to avirulence genes. Curr Genet 27:77–82

Whisson SC, Drenth A, Maclean DJ, Irwin JAG (1995) *Phytophthora sojae* avirulence genes, RAPD and RFLP markers used to construct a detailed genetic linkage map. Mol Plant-Microbe Interact 8:988–995

White HE (1930) Bacterial spot of radish and turnip. Phytopathology 20:653–662

Whitham S, Dinesh-Kumar SP, Choi D, Hehl R, Corr C, Baker B (1994) The product of the tobacco mosaic virus resistance gene N: similarity to toll and the interleukin-1 receptor. Cell 78:1101–1115

Williams PH (1966) A system for the determination of races of *Plasmodiophora brassicae* that infect cabbage and rutabaga. Phytopathology 56:624–626

Williams PH (1980) Black rot: a continuing threat to world crucifers. Plant Dis 64:736–742. https://doi.org/10.1094/PD-64-736

Williams PH (1985) Crucifer genetics cooperative (GrGc). Crucif Newsl 10:1–2

Williams PH, Seidel D (1968) Zum vorkommen von *Plasmodiophora brassicae*-rassen in der Deutschen Demokratischen Republik. Arch Pflanzensch 4:31–36

Williams PH, Walker JC (1963) Races of clubroot in North America. Plant Dis Rep 47:608–611

Williams JGK, Kubelik AR, Livak KJ, Rafalski JA, Tingey SV (1990) DNA polymorphisms amplified by arbitrary primers are useful as genetic markers. Nucleic Acids Res 18:6531–6535

Wilson GW (1907) Studies in North American Peronasporales. I. The genus *Albugo*. Bull Torrey Bot Club 34:61–84

Winter M, Koopmann B (2016) Race spectra of *Leptosphaeria maculans*, the causal agent of blackleg disease of oilseed rape, in different geographic regions in northern Germany. Eur J Plant Pathol 145:629–641. https://doi.org/10.1007/s10658-016-0932-8

Wu CR, Mathur S, Rimmer SR (1995) Differentiation of races and isolates of Albugo candida by random amplification of polymorphic DNA. In: Proc. 9th Intl Rapeseed Congr. GCIRC, Cambridge, pp 655–657

Xiao S, Ellwood S, Findlay K, Oliver RP, Turner JG (1997) Characterization of three loci controlling resistance of *Arabidopsis thaliana* accession Ms-0 to two powdery mildew diseases. Plant J 12(4):757–768

Xue SM, Cao TS, Howard RJ, Hwang SF, Strelkov SE (2008) Isolation and variation in virulence of single-spore isolates of *Plasmodiophora brassicae* from Canada. Plant Dis 92:456–462. https://doi.org/10.1094/pdis-92-3-0456

Yan Y, Yuan Q, Tang J, Huang J, Hsiang T, Wei Y, Zheng L (2018) *Colletotrichum higginsianum* as a model for understanding host-pathogen interactions: a review. Int J Mol Sci 19(7):2142. https://doi.org/10.3390/ijms19072142

Yano S, Tanaka S, Ito S, Kameya-Iwaki M (1996) Variations of isozyme and soluble proteins among field populations of *Plasmodiophora brassicae*. Ann Phytopathol Soc 62:365–371

Yano S, Tanaka S, Ito S, Kameya-Iwaki M (1997) Variations of random amplified polymorphic DNA (RAPD) patterns among field populations of *Plasmodiophora brassicae*. Ann Phytopathol Soc Jpn 63:179–182

Yoshida K, Ohguchi T (1998) Suppression of haustorium formation of *Peronospora parasitica* in heat-treated Japanese radish [*Raphanus sativus*] root tissues. Ann Phytopathol Soc Jpn 64:307–314

You MP, Simoneau P, Dongo A, Barbetti MJ, Li H, Sivasithamparam K (2005) First report of an Alternaria leaf spot caused by *Alternaria brassicae* on *Crambe abyssinicia* in Australia. Plant Dis 89:430

You MP, Lanoiselet V, Wang CP, Barbetti MJ (2014) First report of Alternaria leaf spot caused by Alternaria tenuissima on Blueberry (Vaccinium corymbosum) in Western Australia. Plant Dis 98(3):423. https://doi.org/10.1094/PDIS-07-13-0737-PDN

Young JM, Park DC, Shearman HM, Fargier E (2008) A multilocus sequence analysis of the genus *Xanthomonas*. Syst Appl Microbiol 31:366–377

Yu YG, Buss GR, Maroof MAS (1996) Isolation of a super family of candidate disease-resistance genes in soybean based on a conserved nucleotide-binding site. Proc Natl Acad Sci U S A 93: 11751–11756

Yuan Q, Chen M, Yan Y, Gu Q, Huang J, Zheng L (2016) *ChSte7* is required for vegetative growth and various plant infection processes in *Colletotrichum higginsianum*. Biomed Res Int 2016(5): 1–11

Zander M, Patel DA, Van De Wouw A, Lai K, Lorenc MT, Campbell E, Hayward A, Edwards D, Raman H, Batley J (2013) Identifying genetic diversity of avirulence genes in *Leptosphaeria maculans* using whole genome sequencing. Funct Integr Genomics 13:295–308. https://doi.org/10.1007/s10142-013-0324-5

Zandoki E, Szodi S, Turoczi G (2005) Mycelial compatibility of *Sclerotinia sclerotiorum* strains of different areas. Acta Phytopathol Entomol Hungarica 40:295–301

Zhan J, McDonald BA (2013) Experimental measures of pathogen competition and relative fitness. Annu Rev Phytopathol 51:131–153. https://doi.org/10.1146/annurev-phyto-082712-102302

Zhang H, Feng J, Zhang S, Zhang S, Li F, Strelkov SE, Sun R, Hwang SF (2015) Resistance to *Plasmodiophora brassicae* in *Brassica rapa* and *Brassica juncea* genotypes from China. Plant Dis 99:776–779. https://doi.org/10.1094/PDIS-08-14-0863-RE

Zhang X, Peng G, Kutcher HR, Balesdent MH, Delourme R, Fernando WGD (2016) Breakdown of Rlm3 resistance in the *Brassica napus–Leptosphaeria maculans* pathosystem in western Canada. Eur J Plant Pathol 145:659–674. https://doi.org/10.1007/s10658-015-0819-0

Zheng J, Wang X, Li Q, Yuan S, Wei S, Tian X, Huang Y, Wang W, Yang H (2018) Characterization of five molecular markers for pathotype identification of the clubroot pathogen *Plasmodiophora brassicae*. Phytopathology 103(3):495–503

Zou Z, Liu F, Selin C, Fernando WGD (2020) Generation and characterization of a virulent *Leptosphaeria maculans* isolate carrying a mutated *AvrLm7* gene using the CRISPR/Cas9 system. Front Microbiol 11:1969. https://doi.org/10.3389/fmicb.2020.01969

Check for updates

Biometabolomics of Crucifer's Host-Pathosystem

6

Abstract

During *Albugo–Brassica* interaction, host metabolism is activated to regulate photosynthesis, carbohydrate metabolism, and respiration. The accumulation of soluble carbohydrates increases invertase and triggers sugar signal transduction leading to photosynthetic gene expression. The induction of hyperplasia and hypertrophy of leaf, stem, and floral parts is regulated in the *Albugo*-invaded tissues due to production of IAA. The lipid profile in the form of fatty acid composition differs in *Albugo* species. The enzymes like Alfa amylase, invertase, peroxidase, IAA-oxidase, and ascorbic acid are differentially produced in healthy and *Albugo*-infected *B. juncea* leaves and inflorescence. NRPS-type SM gene clusters in *A. brassicae* produce toxins (Destruxin B, HC toxin, and siderophores) for virulence. An aspartate DtxS4 is required for the production of β-alanine from aspartic acid which is essential for production of the backbone structure of toxin Destruxin. The ROS production in *Brassica*-infected host is very complex. H_2O_2 and OH production is high and low, respectively, in moderately resistant host. Enzymes PPO, POX, and CAT increase in resistant host. Alternaria leaf infection in mustard influences production of carbohydrates (sugars), phenols, and chlorophyll contents. *Alternaria–Arabidopsis* pathosystem has revealed regulation of *COL1* gene linked to amount of jasmonic acid and its analogs for functional characterization. The ROS, cell death, and callose accumulation show a spatial and temporal correlation with host susceptibility. The *Amr1* gene synthesizes melanin in hyphal cells of *Alternaria* and reduces virulence due to lifestyle switch from pathogenesis to reproduction. *Alternaria* produces several metabolites/toxins which have effects on crucifers' physiology, biochemistry, and molecular process. Host-specific toxin has been used as tool to screen *Brassica* germ-plasm for sources of resistance to *Alternaria*. The *A. brassicae* toxin destruxin B plays role in signal transduction leading to programmed cell death, and suppression of defense system in *Brassica*. Several genes/proteins are differentially expressed in

G. Singh Saharan et al., *Genomics of Crucifer's Host-Pathosystem*,
https://doi.org/10.1007/978-981-19-3812-2_6

toxin-treated *Brassica* hosts for different molecular processes. The enzymes like cutinases and lipases contribute in the infection process of *Alternaria*. The *Brassica–Alternaria* interaction induces extreme alterations in host biochemistry to reduce photosynthetic area, defoliation, accelerated senescence, and low yield in susceptible genotypes. Arginine biosynthesis in conidia of *Colletotrichum* is essential for prepenetration process. Host metabolism supports the growth of *Erysiphe* with an increased level of ploidy of mesophyll cells in compatible interactions. *Hyaloperonospora* infection in *Brassica* host induces changes in metabolic processes like respiration, photosynthesis, nucleic acid, protein synthesis, phenol metabolism, endogenous growth substances, electrolyte leakage, and enzymes. The infection of *Leptosphaeria maculans* induces more than 20 genes in *B. napus* governing metabolic functional categories regulating pathogenesis. A virulence gene *AvrLm1* is secreted into the apoplast of *Brassica* during infection of *L. maculans* to modulate plant hormones signaling and metabolic activities. *Plasmodiophora brassicae* infection of *B. napus* increases aliphatic, indole, and aromatic GSLs. Aliphatic GSLs induced by JA in *B. napus* play a role during secondary infection stage of *Plasmodiophora* which is regulated by a gene *BnMYB* 28.1. The *P. brassicae* infection induces increase in sugar (glucose, fructose) translocation in roots of *Brassica*. Several Br Sweets belonging to clade I and III are upregulated in CS-NIL on *P. brassicae* infection. The metabolism of auxins, cytokinin, and brassinosteroids hormones is altered in *P. brassicae*-infected plants. The *PbBSMT* gene is expressed during complete infection process of *Brassica–Plasmodiophora* host-pathosystem and to minimize chitin-triggered defense responses of the host. The gall formation in the infected roots of *Brassica* is due to altered metabolisms, physiology, and modifications of the host cell walls induced by the *Plasmodiophora* through CWDEs genes. The genes like expansins, *XTHs, SHV3, SVL1, WAK*, and enzymes modification play a role in clubroot development during infection process. The genes encoding nutrient acquisition pathways have been identified during *Plasmodiophora* development in the host tissues. Polyamine biosynthesis (putrescine, spermidine, and spermine) is vital in *Brassica* for the survival and pathogenesis of *P. brassicae*. For the clubroot development, auxin signaling pathway is activated by TIR family and *ABP1* genes. Metabolites produced in *Sclerotinia* mycelium play a role in sclerotium formation and pathogenesis. The organic acids and enzymes are key metabolites of *Sclerotinia*. Enzymes like pectinolytic, proteases, cellulases, and gluco-amylases play significant role in pathogenesis by *Sclerotinia*. Several genes are differentially expressed by *Brassica–Sclerotinia* host-pathosystem for virulence at different stages of pathogenesis. *Sclerotinia* uses *Bna NPR1* gene to regulate specific physiological processes of *B. napus* to suppress defense response. The alteration of SA/JA ratio during infection of *Xanthomonas* determines a negative regulator of redox status and phenylpropanoid in the *Brassica–Xanthomonas* pathosystem.

Keywords

Biometabolomics · Host-pathosystem · *Albugo* host-pathosystem · Regulation of photosynthesis · Carbohydrate metabolism · Respiration · Induction of hypertrophy · Differential activity of enzymes · *Alternaria* host-pathosystem · Genomic regulation of secondary metabolites · Genetic control of Destruxin B · Induction of biomolecular changes · Transcriptional factor Amr1 regulate melanin biosynthesis · Production of metabolites · Effect of biomolecular alteration · Toxins on host biochemistry · *Colletotrichum* host-pathosystem · Biosynthesis of biomolecules · *Erysiphe* host-pathosystem · Host metabolism and ploidy of mesophyll cells · *Hyaloperonospora* host-pathosystem · Induction of biometabolites regulation · *Plasmodiophora* host-pathosystem · Aliphatic glucosinolates metabolism · Induction and translocation of sugars · Host hormone metabolism · Alterations of host cell wall rigidity · Genomics of nutrient acquisition · Induction of signaling molecules and genes · *Sclerotinia* host-pathosystem · Metabolites associated with pathogenesis · Cell wall degrading enzymes · *Bna NPR1* gene family · *Xanthomonas* host-pathosystem · Alteration in hormonal balance

6.1 Introduction

Biochemical studies on the growth and survival of a pathogen and of the changes it induces in its host can ultimately lead to a better understanding of epidemiology, disease development, and control. With a few exceptions, such studies on crucifers' host–pathogen interaction lag far behind those for diseases caused by other major groups of biotrophs. Ideal prerequisites for meaningful studies of the biochemistry of host–parasite interaction are (a) a clear understanding of the genetic control of virulence and avirulence in the parasite, and of susceptibility and resistance in the host; (b) precise histological and cytological descriptions of spore germination, infection, and the establishment and development of the infection; and (c) the availability of methods for growing the parasite alone and in combination with its host under controlled conditions. Unfortunately, these criteria have not been fully satisfied for some pathogens of crucifers.

6.2 *Brassica–Albugo* Host-Pathosystem

6.2.1 Regulation of Photosynthesis, Carbohydrate Metabolism, and Respiration

Increases in the activities of metabolites produced in the Kreb's cycle have been reported during host–pathogen interaction, but the relative contributions of this and the pentose phosphate pathways are not clear. A number of reports indicate that the respiration rates of tissues infected by members of the Albuginaceae also rise dramatically (Black et al.1968a, b; Williams and Pound 1964). However, the limited evidence so far suggests that such increases do not reflect any significant change in the pentose phosphate pathway. Radish cotyledons infected with *A. candida* increase respiration threefold, but the c1/c6 ratio, together with evidence from inhibition and feeding experiments, suggests that this was due to stimulation of existing pathways and did not involve the pentose phosphate path way (Black et al. 1968a, b; Williams and Pound 1964). Long and Cooke (1974) suggested that host–fungus movement of carbohydrates in *Albugo–Senecio squalidus* system is maintained by hydrolysis of host sucrose and uptake of hexoses, followed by accumulation of trehalose within the mycelium and spores. Trehalose was synthesized within pustules by the fungus, but no acyclic polyols were found. Accumulation of hexoses around pustules together with increased hydrolysis of exogenous sucrose by pustular material indicated increased invertase activity within infected tissues. Accumulation of dark-fixed carbon compounds in WR pustules of *Senecio squalidus* infected with *A. tragopogonis* has been reported (Thomton and Cooke 1970).

In the leaves of *C. maritima* infected with *A. candida,* there is reduction in ionic content (Na^+, K^+, Ca^{2+}), pigments like chlorophyll (chl_a, chl_b) and carotinoids, but increase in sucrose, reducing sugars and polysaccharides. In addition, increased amount of amino acids like glycine, alanine, methionine, and histidine has been observed by Aldesuquy and Baka (1992). Eight days after inoculation of leaves, white blisters became visible on the lower surface of the leaf, although no symptoms were apparent on the upper surface. After 14 days, the region of leaf invaded by fungal mycelium had become chlorotic (Chou et al. 2000). It has been hypothesized that an accumulation of soluble carbohydrates, following an increase in invertase activity, may trigger sugar signal transduction pathways leading to the repression of photosynthetic gene expression, and to the induction of defense proteins. This hypothesis was investigated by quantifying localized changes in carbohydrates and photosynthetic metabolism, and the expression of genes encoding photosynthetic and defense proteins. Quantitative imaging of chlorophyll fluorescence revealed that the rate of photosynthesis declined progressively in the invaded regions of the leaf. However, in noninfected regions of the infected leaf, the rate of photosynthesis was similar to that measured in the control leaf until later on during the infection cycle when it declined. Images of nonphotochemical fluorescence quenching (NPQ) suggested that the capacity of the Calvin cycle had been reduced in infected regions, and that there was a complex metabolic heterogeneity within the infected leaf (Chou et al. 2000).

The *Albugo candida* also caused localized changes in the carbohydrate metabolism of the leaf; soluble carbohydrates accumulated in the infected region whereas the amount of starch declined. The reverse was seen in no infected regions of the infected leaf; carbohydrates did not accumulate until later on during infection, and the amount of starch increased as the infection progressed. There was an increase in the activity of invertases which was confined to regions of the leaf invaded by the fungal mycelium. The increase in apoplastic invertase activity was of host origin, as mRNA levels of the ATb FRUCT1 gene (measured by semi-quantitative RT-PCR) increased 40-fold in the infected region. The increase in soluble invertase activity resulted from the appearance of a new isoform in the invaded region of the leaf. Current evidence suggests that this was of fungal origin. Northern blot analysis of cab and rbcS showed that photosynthetic gene expression was repressed in the infected leaf from 6 days after inoculation compared to noninoculated control leaves. In contrast, there was no detectable induction of defense proteins in the infected leaf (Chou et al. 2000).

The resistant and moderately resistant cultivars contained higher amounts of chlorophyll, sugars, and total phenols than the susceptible cultivar at all growth stages. Total proteins and free amino acids, however, were higher in the susceptible cultivar at all growth stages (Singh et al. 2011a). Higher starch contents were found in noninfected tissues which suggest that this could be due to the higher alpha amylase activity in diseased tissues (Debnath et al. 1998). Chlorophyll has a positive role in *A. candida* resistance in Indian mustard (Gupta et al. 1997a, b). Thaumatin-like protein (PR-5), a protein not previously associated with the resistance of *B. juncea* toward *A. candida,* was detected. One protein, peptidyl-prolyl cis/trans-isomerase (PPIase) isoform CYP20-3, was only detected in the susceptible variety and its amount increased abundantly in response to the pathogen. Peptidyl-prolyl cis/trans isomerases have recently been discovered to play an important role in pathogenesis by suppressing the host cell's immune response (Kaur et al. 2011).

1. The RNA content: In *Ipomoea*, there was the greater reduction of RNA content in WR infected tissues than in the healthy, adjacent tissues (Misra and Padhi 1981).
2. Photosynthesis: Using infrared CO_2 analysis, Black et al. (1968a, b) demonstrated that a decline in the photosynthetic rate in the *A. candida*-infected radish cotyledons was preceded with rise in respiration rate (Williams and Pound 1964). In another study, Harding et al. (1968) examined the pattern of pigment retention during green island development following infection of *B. juncea* cotyledons with *A. candida*, and also studied both the photosynthetic capacity and the ultrastructure of chloroplasts within the green island tissue. They found that labeled-glycine 2-^{14C} was incorporated into chlorophylls a and b in both infected and noninfected tissue. Both infected and noninfected tissue fixed $^{14}CO_2$ in the light, but 4 days after infection, green islands fixed five times more $^{14}CO_2$ in the light than did no infected tissue. Photosynthesis per mole of chlorophyll fell at the same rate in Green Island as in noninfected tissue. The maintenance of chlorophyll, and continued photosynthetic activity in green island tissue, was paralleled by delayed breakdown of chloroplasts. Electron microscopy showed that these retained their structural integrity longer in green islands than in noninfected tissue. The authors drew attention to the similarity in *B. juncea* between green islands and tissues treated with kinetin. Around the sites of infection, however, green islands of chlorophyll are retained, and within them, photosynthesis continues. Thus, the green islands are regions within which synthetic processes are maintained, and the movement of metabolites from host to parasite may continue even after the rest of the leaf has become nonfunctional. Enhanced fixation of CO_2 in the dark also occurs in green islands, but this appears to be due to the ability of the fungus itself to dark fix. The similarity between

green islands and the sites of exogenous application of cytokinins to plant tissues has been widely noted, although there is no clear evidence that these, or other plant growth substances, are actually responsible for the green island phenomenon.

6.2.2 Induction of Hypertrophy by Hormonal Regulation

Infection of host plants with *Albugo* causes hyperplasia and hypertrophy of leaf, stem, and floral parts. Kiermayer (1958) found that these symptoms are produced in plants infected with *A. candida* due to the production of indolacetic acid (IAA). Hirata (1954, 1956) found that infection with *A. candida* causes an initial increase in diffusible auxin in diseased stems and leaf sections, followed by a decrease before maximum development of the galls. The auxins in healthy and *Albugo*-infected inflorescences of *B. napus* have now been identified and estimated quantitatively by Srivastava et al. (1962). Malformed *B. napus* inflorescences produce IAA, IAN, accelerator L, and an ether-insoluble growth substance designated as Kumari et al. (1970) and Lal et al. (1980) studied the quantitative and qualitative changes in the amino acid contents of diseased (hypertrophied) and healthy tissues of mustard and radish, and suggested that the amino acid contents are greatly changed because of disruption in the metabolism of plant organs under the influence of a pathogen (Kumari et al. 1970). The infection causes the breakdown of plant proteins, releasing small quantities of tryptophan, which reacts with endogenic phenolic acid to produce IAA which is responsible for hypertrophied growth.

6.2.3 Differential Production of Lipid Profiles in *Albugo* species Sporangia

Fatty acids (FAs) play an important role as the major energy storage in oomycetes. Some of them may as well be an important factor for normal growth and development of these organisms (Herman and Herman 1985). The composition of fatty acid patterns in three WR species, *A. amaranthi, A. candida,* and *A. tragopogonis,* was determined by Spring et al. (2005). The fatty acid composition of these biotrophic WR species was identified by means of GC and GC/MS analyses. Besides the common saturated fatty acids of C14–C22 chain length, several unusual monounsaturated and polyunsaturated fatty acids were present in significant amounts in the lipid profiles of the three species; each taxon revealed its own characteristic pattern which was uniform among populations of different geographic origin, and independent from the host species. Fatty acids were obtained from minute amounts of sporangia collected from the surface of desiccated host plants. Comparison of herbarium specimens revealed a decrease of polyunsaturated compounds after storage for 30 months at room temperature, but still allowed unquestionable species differentiation.

6.2.4 Differential Activity of Enzymes in Infected and Healthy Host Tissues

The quantitative analysis of peroxidase, α-amylase, invertase, IAA-oxidase, ascorbic acid oxidase activity was observed in healthy and diseased *B. juncea* leaves and inflorescence. Maximum peroxidase activity observed in infected leaves and disease inflorescence as compared to healthy leaves and infected inflorescence; α-amylase was maximum in healthy leaves and minimum in infected inflorescence. Maximum invertase activity was found in infected leaves of *B. juncea* and minimum in healthy leaves of *B. juncea*. The IAA-oxidase activity was higher in infected leaves and inflorescence as compared to healthy leaves and inflorescence. The activity of ascorbic acid oxidase decreases with the increase in infection in plants (Singh et al. 2011b).

6.3 *Brassica–Alternaria* Host-Pathosystem

6.3.1 Genomic Regulation of *Alternaria Brassicae* Secondary Metabolites/Toxins

The genera of *Alternaria* and *Cochliobolus* are known to be the major producers of host-specific secondary metabolite toxins. *Alternaria* spp. especially are known for the production of chemically diverse secondary metabolites, which include the host-specific toxins (HSTs) and non-HSTs. These secondary metabolites are usually generated by nonribosomal peptide syntheses (NRPS) and polyketide synthases (PKS). Rajarammohan et al. (2019) identified five NRPS-type SM gene clusters, 12 PKS-type gene clusters, and seven terpene-like gene clusters in *A. brassicae*. Out of the five NRPS clusters, three clusters produced known secondary metabolites viz., Destruxin B, HC-toxin, and dimethyl-coprogen (siderophore) with known roles in virulence.

The gene clusters responsible for dimethyl coprogen (siderophore) production in *A. brassicae* consist of 22 genes, including the major biosynthetic genes, oxidoreductases, and siderophore transporters. Siderophores are iron-chelating compounds, used by fungi to acquire extracellular ferric iron and have been reported to be involved in fungal virulence. The identification of the gene cluster responsible for siderophore synthesis would enable the study of siderophores and their role in pathogenicity in *A. brassicae*. Additionally, a PKS-type cluster consisting of 12 genes, responsible for melanin production, was also identified. The melanin biosynthetic cluster has been described for *A. alternata* previously. Also, the transcription factor Amr1, which induces melanin production, has been characterized in *A. brassicicola* and is known to suppress virulence. However, the role of melanin in virulence is ambiguous and species-specific.

The plant pathogens belonging to the genus of *Alternaria* seem to have a dynamic capacity to acquire new secondary metabolite potential to colonize new ecological niches. The most parsimonious explanation for this dynamic acquisition of secondary metabolite potential is horizontal gene transfer within the genus of *Alternaria* and possibly with other genera. There is extensive evidence in the literature that much of the HSTs of *Alternaria* are carried on the dispensable chromosomes and exchange of these chromosomes can broaden the host specificity. The identified an NRPS cluster, possibly coding for HC-toxin is in one of the DCs (scaffold 18). The presence of HC-toxin gene cluster, avirulence determinant in *Cochliobolus carbonum*, in a DC of *A. brassicae,* points to the fact that interspecies horizontal gene transfer may be more common than expected. Apart from horizontal gene transfer, rapid duplication, divergence, and loss of the SM genes may also contribute to the pathogen evolving new metabolic capabilities. These processes of duplication and divergence may well be aided by the proximity of the secondary metabolite clusters to the repeat elements that makes them prone to RIP-mutations. A permutation test was used to compare the overlap of repeat-rich regions with a random gene set against the overlap of secondary metabolite cluster genes. The secondary metabolite clusters significantly overlapped repeat-rich regions as compared to the random gene set (*P*-value: 0.0017; *Z*-score: −2.7963). Also, these clusters overlapped significantly with transposable elements among the repeat-rich regions (*P*-value: 0.0087; *Z*-score: 2.9871). This shows that both the mechanisms described above for the acquisition of new secondary metabolite potential may be possible in the case of *A. brassicae*. Population-scale analyses at the species and genus level may throw light on the prevalence of these mechanisms within the *Alternaria* genus.

6.3.2 Genetic Control of Destruxin B Production by *Alternaria brassicae*

Destruxin B represents a class of cyclic depsipeptides that is known to be one of the key pathogenicity factors of *A. brassicae* and has been reported to be a host-specific toxin of *A. brassicae*. Destruxin B has not been reported to be produced by any of the other *Alternaria* species. The biosynthetic gene clusters are responsible for Destruxin B production in *A. brassicae*. The cluster consists of 10 genes, including the major biosynthetic enzyme encoded by an NRPS gene (DtxS1) and the rate-limiting enzyme, DtxS3 (aldo-keto reductase). Interestingly, synteny analysis of this cluster among the six *Alternaria* species showed that both these genes were not present in any of the other *Alternaria* spp., although the overall synteny of the cluster was maintained in all of these species. The absence of the key genes coding for the enzymes DtxS1 and DtxS3 in the Destrin B cluster in the other *Alternaria* species explains the absence of Destruxin B in those species. Wang et al. (2012) described the secondary metabolite biosynthetic cluster responsible for Destruxin production in *M. robertsii*. The Destruxin cluster in the *Metarhizium* species consists of two additional key genes viz., DtxS2 and DtxS4. DtxS2 is responsible for the conversion of Destruxin B into Destruxin A, C, D, and E. *Alternaria brassicae* is known to produce only Destruxin B and not the other forms of Destruxin, which might possibly explain the absence of DtxS2 in *A. brassicae*. The DtxS4 is an aspartate decarboxylase which is required for the production of β-alanine from aspartic acid and is essential for the production of the backbone structure of Destruxin. Moreover, deletion mutants of DtxS4 were able to produce Destruxin when they were supplemented with β-alanine in the medium. Even though DtxS4 is not present in the Destruxin cluster, *A. brassicae* has one functional copy of aspartate decarboxylase present elsewhere in the genome and therefore, can possibly ensure the supply of β-alanine for the synthesis of Destruxin B. The evolutionary history of Destruxin B production within the *Metarhizium* genus closely resembled that of *Alternaria*. The specialist pathogens such as *M. acridum* lack the two key enzymes and hence do not produce Destruxins. However, *M. robertsii*, a generalist with a wider host range, produces Destruxins. It was therefore hypothesized that Destruxins might be responsible for the establishment of the extended host range of *M. robertsii*. A similar hypothesis may also be true in the case of *A. brassicae*, which has a broad host range and can infect almost all the Brassicaceae (Rajarammohan et al. 2019).

6.3.3 Induction of Biomolecular Changes in *Brassica* by *Alternaria* Infection

Most of the 10 analyzed biochemical parameters (3 ROS species, 2 defensive substances, and 5 red-ox-enzymic activities) showed significant interaction between cultigen and time after infection with *A. brassicicola*. The sole exceptions were SOD and PGA, which lacked significant changes for neither of the factors nor their interaction. The dynamics of ROS production in response to *A. brassicicola* inoculation was complex. The O_2 showed decreasing trend for the (moderately) resistant cultigens and no major changes for the others. The H_2O_2 showed generally higher levels in the (moderately) resistant forms and a decreasing trend throughout. The OH showed a generally decreasing trend in time, with PGH05I building-up its amounts up to 2 dpi. The highest levels of lignins were recorded for *Camelina sativa*, but all cultigens showed their accumulation during infection. Free phenols showed no major differences throughout, although a noticeable peak manifested for PGH05I at 1 dpi. Among the tested red-ox enzymes, PPO showed clear increasing trend with the infection progress. The highest levels of POX activation were recorded for the resistance standard, *C. sativa*. Measurements of CAT showed its activation with time after infection, again with *C. sativa* showing the highest activation when compared with baseline (0 dpi) (Nowakowska et al. 2019).

The effect of *A. brassicicola* infection on the changes of nucleic acids of *B. oleracea* var. *botrytis* has been studied by Maitra and Samajpati (1985). An increase in total nucleic acids (97%), DNA (44%), and RNA (150%) in infected leaves is recorded. The increase in RNA content is due to an increase of each of the four nucleotides. However, guanylic acid increases significantly more than the other three nucleotides. The increase in G/C ratio is also due to an increase in guanylic acid. Atwal et al. (2004) have determined biochemical changes in relation to Alternaria leaf blight in Indian mustard. Healthy leaves of *B. juncea* possess significantly higher amount of total soluble sugars than those of different infected leaf parts at all growth stages. The depletion of sugars is highest in lesion areas of infected leaves (necrotic followed by chlorotic and green areas respectively) as compared to healthy leaves at the initial stages of infection. However, the accumulation of sugars is noticed in the lesion as well as lesion-free areas of the infected leaves with progress of infection. Sugar level is higher in chlorotic and green areas as compared to necrotic parts of the infected leaves. Both healthy and infected leaves have maximum sugar and starch content at 100 DAS, which declines thereafter. Higher levels of total sugars in chlorotic and green areas of infected leaves suggest that sugars are probably one of the factors responsible for resistance in *B. juncea*. They may contribute toward maintaining healthy flora of saprophytic microbes that help to inhibit pathogen growth (Juniper and Jeffree 1983). The lower accumulation in necrotic area can be due to utilization of sugars by the fungus for its own growth and establishment.

The chlorotic area of infected leaves possesses significantly higher levels of total and orthodihydroxy phenols, as compared to necrotic area, whereas no significant differences are observed in the green area of infected leaves as compared to healthy leaves. Total and orthodihydroxy phenols increase significantly with the increase in infection, and age of the plant, and are maximum at 120 DAS in all the cases. The presence of maximum levels of total and orthodihydroxy phenols in chlorotic area, the region next to the lesion area (necrotic), signifies their protective role to restrict the growth, spread of disease and invasion of the pathogen by formation of lignin and lignin like substances, which are more toxic to fungi. The content of flavonols is significantly lower in different infected parts of leaves as compared to the healthy leaves, and this decrease is more prominent in necrotic area followed by chlorotic and green, indicating less significant role of flavonols in the defense mechanism as compared to total and orthodihydroxy phenols. The chlorophyll is found to be completely absent in necrotic areas, and significantly low in chlorotic as compared to green areas of both infected and healthy leaves. Similar trend is observed in chlorophyll a and b content. However, the chlorophyll content of the green areas of the infected leaves is comparable to that of the healthy leaves at 80 DAS.

The biochemical analysis of leaves of different mustard varieties revealed that total phenol, *o*-dihydroxy phenol, total sugar, reducing sugar, chlorophyll content, and flavanol contents are more in resistant (*B. alba*) than in others. With progress of infection, total phenol, *o*-dihydroxy phenol, and protein content increased in all three genotypes, while the chlorophyll, total sugar, reducing sugar, and flavanol content decreased (Mathpal et al. 2011).

Green islands were observed around infected spots of *A. brassicicola,* and *A. brassicae* on mustard leaves by Mandahar and Suri (1983, 1987). These green islands have higher content of cytokinin-like substances, compared to the surrounding yellowed and healthy tissues. The presence of starch in green islands is correlated with the formation of metabolic sinks because of their higher content of cytokinin-like substances (Suri et al. 1983). Cytokinin-like substances appear to be actively involved in infection and pathogenesis of *A. brassicicola* (Suri and Mandahar 1984, 1985, 1986). Production of pectolytic and cellulolytic enzymes by *A. brassicae* causing leaf spot of *B. rapa* has been observed (Srivastava and

Srivastava 1982; Suri and Mandahar 1982). *Alternaria brassicae* produces invertase and amylase on cabbage decoctions (Weimer 1924), while infection in cauliflower leaves brings changes in polyphenol-oxidase and peroxidase (Maitra and Samajpati 1982). The presence of large number of enzymes viz., amylase, invertase, pectin methyl esterase, phosphorylase, aldolase, amidases, ribonuclease, alkaline phosphorylase, nucleophosphatases, catalase, dehydrogenases, deaminases, nucleodeaminases, glycerophosphatase, cellulose, and β-glucosidase in *A. alternata, A. brassicicola,* and *A. raphani* infecting crucifers has been demonstrated (Dasgupta and Verma 1961, 1962; Verma 1964, 1971).

6.3.4 *Alternaria–Arabidopsis* Pathosystem as Model to Predict Disease Pathophysiology

A model of JA signaling pathway with respect to plant-pathogen interaction has been developed to predict the pathophysiology of *Alternaria* blight. The integrated pathways help to reduce the effects of external perturbation to whole plant systems due to its robust properties. Each species in the pathway is considered as node and linked to other nodes at defined rate kinetics. The aim of metabolic modeling of Jasmonate signaling pathway is to provide a hypothesis based on computational simulation at various time extents and network analysis of *Arabidopsis–Alternaria* interaction for identification of key regulatory element present in the pathway. The integrated pathways will prove useful in understanding the pathophysiology of *Alternaria* blight disease in *Brassica*. It is established that the functional molecules of the living systems such as DNA, RNAs, receptor, enzymes, hormones, metabolites, etc., are necessary for the integration of network to modulate behavior of the metabolic pathways. Various in silico comprehensive maps have been constructed for dynamic analysis, pathophysiology and control mechanism as well as identification of biomarker for disease management.

The *Alternaria brassicicola–Arabidopsis* can be used as one of the excellent model system to understand phytophysiology of Alternaria blight in *Brassica*. *Alternaria brassicae*, being a hemibiotroph fungus, displays necrotrophic at some stages of pathogenesis. It has been reported that Jasmonic acid plays important role in mounting the defense response toward necrotrophic fungal pathogens. Deep insight into the JA-mediated signal transduction pathway involved during defense response will help in devising strategies for development of resistance against Alternaria blight provided that the key molecular components have been identified. The higher expression of defense-related genes during pathogenesis process caused by *A. brassicae* would have an immense potential to impart fungal resistance in the crop plants through their over expression.

Jasmonate signaling pathway is a well-characterized pathway in plant systems for defense responses particularly in case of necrotrophic pathogens but the knowledge about the key element of pathway during pathogenesis in term of quantity remain unclear. The developed model clearly demonstrated the fluctuations of key component involved in diseases resistance during simulation besides network analysis were also conducted to determine the value of topological parameter of the JA network for mapping of hubs based on out degree visualization. Degree is defined as a number of edges linked to a node, if it is a self-loop; it shall be counted as two edges of a single node. In directed node, the degree distribution is of two types, in and out degree distribution. The number of coming edges to anode is called in-degree distribution and number of out-going edges is called out degree distribution of a node. In case of biological network, the node and edge is defined in different ways; in metabolic network, the node is considered as metabolites and reactions are considered as edges; in protein–protein interaction network, protein is considered as node and edges are considered as bonding between one protein with other protein. Whereas in case of gene regulatory network, the nodes are considered as gene and edges are known for its regulatory effect.

The aim of modeling and simulation is to describe and understand the pathophysiology at molecular level where key regulatory element and its complex mechanisms that happen at several time scales were predicted. During simulation analysis at variable amount of jasmonic acid and its functional analogs, Methyl jasmonate (Me-JA); COI 1 is predicted as one of the potential molecular target because it controls the expression of various defense related genes during pathogenesis. It was also predicted that the up- and downregulation of COI1 during plant–pathogen interaction is totally linked to amount of jasmonic acid and its functional analogs. Pathak et al. (2017) have manually increased and decreased the amount of COI1 to determine the behavior of system during pathogenesis. The cloning and expression study conducted on *COI1 gene* of *Aquilaria sinensis* demonstrated that the Methyl jasmonate (MeJA), mechanical wounding, and heat stress could significantly induce its expression level. Network analysis has been publicized to be one of the powerful tools for decoding the key regulatory elements (hubs), which is responsible for controlling the complex biological machinery during diseases. Hub is a concept originated from the field of network science that refers to a node having number of links. In biology, a hub is metabolite, gene, or protein which is highly connected to the other nodes (i.e., metabolite, gene, or protein) in a network and that is known to regulate several other pathways associated with the main pathway. Recent advances in plant disease resistance research have provided exciting novel insights into decoding of defense signaling network. Diverse number of components plays essential roles in the regulation of whole network through cross-communication in antagonistic or synergistic manners. Biological properties of hubs in JA signaling pathway are very significant as these are closely related to the well-characterized components linked to the defense responses. Pathak et al. (2017) demonstrated the highly connected nodes (hubs) such as JAZ1, JAZ3, and MYC2 but it was predicted that the expression of these hubs is managed by COI1. These hub nodes are involved in the regulation of defense machinery through interaction of COI1 with jasmonate signal. COI1 is considered as one of the key components of JA pathway for that reason. Study suggested that COI1 has the structural traits for binding to jasmonoyl-isoleucine (JA-Ile/Coronatine) and determined as an intracellular signal with a hormone binding site. Therefore, it may serve as a potential molecular target for triggering de novo resistance in crop plant systems during plant–pathogen interaction via engineering of jasmonic acid pathway for sustainable production of jasmonic acid and its functional analogs during pathogenesis or providing exogenous compounds as defense induces that can mimic the expression of COI1.

It is evident that small molecule targeting COI1 leads to the production of appropriate defense whereas engineering of jasmonic acid pathway for proper production of jasmonic acid during diseases has a very complex and time taking process; therefore, it is concluded that there is a need of novel defense inducer molecule more efficient and stable as compared to jasmonic acid and its structural analogs to tackle today's problem of pathogen resistance by developing de novo resistance in crop plants for robust agricultural productivity and sustainability. Biologically active small molecules have been verified previously for crop plant protection by producing de novo resistance. This integrated approach shows that computational demonstration can construct and explain precisely the complete observable fact of JA signaling during pathogenesis at molecular level (Pathak et al. 2017).

Arabidopsis accession Gre-0 (Greenville, USA) supports substantial *A. brassicae* growth. *A. brassicae* infection processes and the induced host responses are analogous between the model host Gre-0 and its natural host Varuna. Unlike biotrophs, necrotrophs are believed to facilitate their entry into the host cells by degrading the host cell walls, thus relinquishing the need for developing specialized infection structures like appressoria. Although appressoria-like structures have been earlier reported for other *Alternaria* spp. e.g., *A. brassicicola*, *A. raphani*, *A. alternata*, these fungi habitually penetrate directly through cell junctions (McRoberts and Lennard 1996). The appearance of un-melanized ALS associated with host penetration events in a study was unforeseen since *A. brassicae* predominantly (90%) enters the host via stomata. Formation of similar un-melanized compound aspersoria for host penetration has been recently reported for several other necrotrophs.

The host responses to infection, like accumulation of ROS, cell death, and callose deposition, are reciprocally correlated with disease resistance in many biotrophic and hemi-biotrophic pathosystem (Koch and Slusarenko 1990; O'Connell et al. 2004; Soylu 2004). On the contrary, the successful necrotrophic infections are associated with extensive cell death and ROS production during the establishment and disease progression phase (Laluk and Mengiste 2010; Williams et al. 2011). Production of ROS and the concomitant cell death during the disease progression has been reported earlier in mustard and *Arabidopsis* (Col-0) upon infection with *A. brassicicola*—a closely related *Alternaria* sp. (Meur et al. 2015; Suudi et al. 2011). The restricted H_2O_2 accumulation and cell death during the early phase of infection and a more robust H_2O_2 generation accompanied by spreading cell death in the postestablishment phase was observed in both the compatible hosts and it positively correlated with the disease spread. These observations suggest that for a successful compatible interaction, a fine balance must be reached between the suppression of ROS and cell death at the initial stage of infection followed by extensive triggering of the same, postestablishment. Besides the host-induced ROS, *A. brassicae* also produces ROS at the hyphal tips during the compatible interaction. A close association of ROS generation and fungal aggressiveness has been explicitly recorded for various other necrotrophic fungi and bacteria. ROS-deficient mutants of *S. sclerotiorum* (*Ssnox1*), *B. cinerea* (*bcnox A* and *bcnox B*), and *A. alternata* (*AaNoxA*) have compromised virulence and have defective development (Kim et al. 2011; Segmuller et al. 2008; Takemoto et al. 2007).

It is generally believed that toxins secreted by necrotrophs induce host cell death ahead of extensive colonization which usually appears as a chlorotic halo surrounding the advancing necrotic zone (Glazebrook 2005). *Alternaria brassicae* is known to secrete Destruxin B both in vitro and *in planta* that has phytotoxic properties (Parada et al. 2007). Surprisingly, in *B. juncea* at the advancing infection front, intercellular fungal hyphal growth without any associated cell death was observed. No chlorotic zone surrounding the necrotic lesion was seen in these infected samples. Additionally, earlier observation that *A. brassicae* suppress plant ROS and cell death during penetration indicates that there might be a short biotrophic-like phase where the fungus suppresses plant defense responses before triggering massive cell death. Similar observations have been reported for the broad spectrum necrotroph *S. sclerotiorum* infecting tomato (Kabbage et al. 2013, 2015).

Callose deposition at the penetration site to stall the hyphal entry is a hallmark of plant response to infection especially against biotrophic fungi. Callose deposition does not appear to contribute to resistance as generalized callose deposits were observed at the sites of successful penetration and establishment in both the compatible hosts by Mandal et al. (2018). Rather, callose deposition appears to be a default plant response to infection. Nevertheless, the role of callose in basal resistance against *A. brassicae* as the callose synthase mutant gsl5/pmr4 showed enhanced susceptibility to the fungus. The extent of ROS, cell death, and callose accumulation show a spatial and temporal correlation with the developing symptom, thus, hinting at their putative role in susceptibility.

A comparative histopathological analysis for disease progression and plant responses to *A. brassicae* in *Arabidopsis* and *Brassica juncea* revealed significant similarities between the two compatible path systems. Interestingly, in both the compatible hosts, ROS accumulation, cell death, and callose deposition correlated with the development of the disease. The *Arabidopsis–Alternaria brassicae* can be an apt model pathosystem since it emulates the dynamics of the pathogen interaction with its natural host-*Brassicas*. The existing genetic diversity in *Arabidopsis* can be a starting point to screen for variation in responses to Alternaria leaf blight. Furthermore, several tools available for *Arabidopsis* can facilitate the dissection of genetic and molecular basis of resistance.

6.3.5 Transcriptional Factor *Amr1* Regulates Melanin Biosynthesis and Virulence in *Alternaria*

Melanin is a ubiquitous pigment that plays an important role in protecting fungi from the damaging effects of environmental stress. It accumulates in the cell walls of hyphae, and conidia during the late stationary phase of mycelial growth and also forms under stressful conditions, including ultraviolet irradiation, a hyperosmotic environment, nutrient deprivation, or an accumulation of toxic wastes during in vitro culture. Over expression of a *Cmr1* ortholog in *Bipolaris oryzae* (*Bmr1*) causes continuous induction of the three downstream genes in the melanin synthesis pathway and increases melanization of colonies (Kihara et al. 2008). The loss of function of *Cmr1*, its homologs, or their downstream genes in other fungi, results in melanin deficiency. Cho et al. (2012) identified a transcription factor *Amr1*, which regulates the melanin synthesis pathway in *A. brassicicola*. The melanin synthesis-associated structural genes *Bm1,* and *Pks* are located together in a 30 Kb region and their organization, and orientation is the same as in the closely related fungus *Cochliobolus heterostrophus*. Interestingly, melanin is not associated with virulence in *A. brassicicola* as has been reported in other fungi. However, there appears to be a linkage between melanin biosynthesis and reduction in virulence, perhaps due to a lifestyle switch from pathogenesis to reproduction (Cho et al. 2012).

6.3.6 Production of Metabolites by *Alternaria* Related to Pathogenesis

The *Alternaria brassicae* when grown in liquid still culture produces host-specific phytotoxins, which have been shown to be cyclo-depsipeptides. It also produces several other metabolites including the new drimane sesquiterpenes deoxyuvidin β-albrassitriol, and iso-albrassitriol, as well as *Brassica* diol, a C_{15}-prenylated pentaketide. These compounds show no phytotoxicity on canola (Ayer et al. 1987; Ayer and Pena-Rodriquez 1987). The earliest reports on phytotoxin production by *A. brassicae* or *A. raphani* are those of Degenhardt (1973, 1978) and Husain and Thakur (1966). Cytotoxicity tests with extracts of culture filtrates of *A. brassicae* and *A. brassicicola* show that both species produce compounds, which are toxic to human epithelial (HEP2) cells but there is no clear relationship between *Alternaria* infection of seed and its cytotoxicity. Metabolites produced by the two fungi are also phytotoxic, giving rise to necrosis of leaf discs, and inhibition of seedling root growth, but extracts of the less specialized parasite, *A. alternata,* causes more severe phytotoxic effects on crucifer test plants. Extracts of culture filtrates from *A. brassicae* and *A. brassicicola* are found to show nonspecific phytotoxicity (McKenzie et al. 1988).

Most fungi produce secondary metabolites such as mycotoxins or antibiotics. These compounds are synthesized from precursors derived from primary metabolism, and can be valuable as antifungal, antibacterial, or antitumor agents. *Alternaria brassicicola* is known to produce a number of interesting compounds including antitumoric depudecin (Matsumoto et al. 1992), antibiotic complex brassicicolin (Ciegler and Lindenfelser 1969), and phytotoxic brassicicenes (MacKinnon et al. 1999). Depudecin is C_{11} compound containing two epoxide groups and shows antitumor activity (Matsumoto et al. 1992). Phytotoxic brassicicenes, fusicoccane-like diterpenoids have been isolated from the culture filtrate of canola pathogen *A. brassicicola* (MacKinnon et al. 1999).

The fungal strain Sw-3 showing antimicrobial activity and isolated from soil of a Chinese cabbage patch in Korea was identified as *A. brassicicola* by its morphological characteristics, and confirmed by the analysis of the 18S gene and ITS region of rDNA. The fungus showed a similarity of 99% with *A. brassicicola* in the 18S rDNA sequence analysis. *Alternaria brassicicola* has been reported to produce an antitumor compound called depudecin. Strain SW-3 produces antimicrobial metabolites, in addition to depudecin, during sporulation under different growth conditions. The metabolite of the isolated fungus has strong antifungal activity against *Microsporum canis* and *Trichophyton rubrum,* and antibacterial activity against *Staphylococcus aureus* and *Pseudomonas aerogenes*. The amount and kind of metabolites produced by the isolates are affected by growth conditions such as nutrients and growth period (Jung et al. 2002). Metabolites and phytotoxins produced by *Alternaria* species infecting crucifers have been discussed in greater detail in our book "Alternaria Diseases of Crucifers: Biology, Ecology and Disease Management," Chapter 10 (Saharan et al. 2016).

The metabolites from *Alternaria* fungi can be grouped into several categories which include nitrogen-containing compounds, steroids, terpenoids, pyranones (pyrones), quinines, and phenolic. Several metabolites are unique to one *Alternaria* species, but most metabolites are produced by more than one species. The widest spread metabolite is alternariol which has been isolated from a few *Alternaria* species (Lou et al. 2013).

Alternaria brassicae when grown in liquid still culture produces host-specific phytotoxins cyclodepsipeptides, drimane sesquiterpenes deoxyuvidin B (4), alberassitriol (8), iso-albrassitriol (9), brassica dial (10), and a C_{15}-prenylated pentaketide. These compounds are not phytotoxic to canola (Ayer and Pena-Rodriquez 1987). *Alternaria brassicae* produces a host-specific, ABR toxin (protein) from germinating spores on host leaves (Otani et al. 1998; Parada et al. 2008). AB toxin from *A. brassicicola* germinating spores is induced by a host-derived oligosaccharide (Oka et al. 2005).

Metabolites from 76 mg fractionated cultures of *A. brassicicola* are phomapyrone F (4.1 mg), phomapyrone A (3.30 mg), phomapyrone G (5.1 mg), brassicicolin A (1.30 mg), brassicicene H (8, <1 mg), and 44 mg fractionated culture to yield brassicicene G (7.4 mg), brassicicene I (9.1 mg), and infectopyrone G (6.2 mg) (Pedras et al. 2009). The three new fusicoccane diterpenes, brassicenes G (7), H (8), and I (9) were named by analogy to previous designations (Mac Kinnon et al. 1999). From the least polar and nonphytotoxic fractions, phomapyrones A (3), F (4), G (5), and infectopyrone (6) were isolated and structures determined by comparison of their chromatographic and spectroscopic data with those of authentic samples isolated from *Leptosphaeria maculans* and from *L. biglobosa* (Pedras et al. 1994; Pedras and Biesenthal 2001; Pedras and Chumala 2005). From the most phytotoxic fractions F4 and F5, brassicicolin A (1) was obtained as a colorless and chromatographically homogeneous oil (HPLC R_t, 27.3 min); HR ESI-MS analysis indicated a molecular formula of $C_{32}H_{48}N_2O_{14}$, and the FT IR spectrum contained strong bands at 3400, 2145, and 1760 cm^{-1}, suggesting the presence of hydroxyl, is nitrile, and carbonyl groups, respectively.

The ^1H NMR spectrum was complex suggesting that the sample was actually a mixture of closely related compounds. Likewise, the ^{13}C NMR spectral data showed that each cluster of signals contained four resonances indicating the presence of four isomers. Comparison of these spectroscopic data and specific optical activity ($[\alpha]_D = +16$ (c 1.0, CH_2Cl_2); lit. $[\alpha]_D = 16.1$(c, 1.86 CH_2Cl_2), with those previously reported (Gloer et al. 1988), and along with chemical hydrolysis results indicated that this metabolite was brassicicolin A (1), a derivative of mannitol containing two acetyl, two 2-isocyano-3-methylbutanoyl, and two 2 hydroxy-3-methylbutanoyl substituent's. Fraction F5 was further separated to yield a chromatographically homogenous material ($C_{26}H_{41}NO_{13}$) shown to contain a mixture of nonsymmetrical mannitol derivates in which one of the two isocyano-3-methylbutanoyl group was missing. The ^{13}C NMR spectrum displayed clusters of four to six peaks, suggesting that the fraction contained a mixture of at-least four compounds. Given the complexity of this fraction, and the presence of inseparable epimers, no structures were assigned to each specific component.

Three related metabolites were isolated from fractions displaying low phytotoxicity. The ^1H NMR spectrum of each of these metabolites showed the presence of an isopropyl, a methoxy-methylene, and two methyl substituent's suggesting that these metabolites had fusicoccane structures, similar to those previously reported for brassicicenes A-F (Mac Kinnon et al. 1999). The most oxidized metabolite was obtained as a white solid whose molecular formula was determined to be $C_{21}H_{32}O_5$ (six double bond equivalent) based on NMR and HRE1-MS data. The ^1H NMR spectrum displayed four methyl doublets, a methoxyl group, and one olefinic proton.

Metabolites from leaves of *B. juncea* inoculated with *A. brassicicola* were detected by Pedras et al. (2009). No phytoalexins were detected 24 h postinoculation; four phytoalexins were detected 48 h postinoculation, and were quantified using calibration curves constructed with synthetic samples (Pedras et al. 2007). The phytoalexins spirobrassinin was produced in larger amounts than brassicic, cyclobrassinin, or rutalexin. As well, indolyl-3 acetonitrile was detected in higher concentration in leaves (0.8–4.0 nmol/g fresh weight) of infected plants than in control leaves (<0.1 nmol/g fresh weight). Furthermore, HPLC-MS analysis of leaf extracts did not detect any of the metabolites produced by *A. brassicicola* in liquid cultures.

Nine novel compounds closely related to ACTG-toxin, termed as tricycloalternarenes, were isolated from a strain of *A. alternata*, from *B. sinensis*, which was earlier used for the production of nonspecific phytotoxins 10 toxins (Nussbaum et al. 1999). Liebermann et al. (2000) also reported isolation of 11 new bicycloalternarenes as well as ACTG-toxins A and B. *Alternaria brassicae* (Berk.) Sacc. produces four cyclic depsipeptide phytotoxins including destruxin B, Homodestruxin, Destruxin B2, and desmethyl-destruxin B host-specific proteinaceous ABR-toxin (Ayer and Pena-Rodriquez 1987; Buchwaldt and Green 1992; Montemurro and Visconti 1992; Agarwal et al. 1994; Parada et al. 2008). In addition to proteinaceous AB-toxin (Otani et al. 1998) and Brassicillin A, a very selective phytotoxins metabolite in liquid cultures (Pedras et al. 2009), *A. brassicicola* also produces despeptides and fucicicin-like toxin compounds (Cooke et al. 1997; Mac Kinnon et al. 1999).

6.3.7 *Alternaria* Metabolites Influences Host Physiology, Biochemistry, and Molecular Process

The metabolites from plant pathogenic fungi are usually toxic to plants and are called phytotoxins. They are further divided into host-specific and host nonspecific toxins. The host-specific toxins (HSTs) are toxic only to host plants of the fungus that produces the toxin. Another definition seems to more acceptable that the host-specific toxins are toxic to plants that host the pathogen, but have lower phytotoxicity on nonhost plants. Most HSTs are considered to be the pathogenicity factors, which the pathogenic fungi producing them require to invade tissue and induce disease. All isolates of the pathogen that produce a HST are pathogenic to the specific host. All isolates that fail to produce HSTs lose their pathogenicity to the host plants. Plants that are susceptible to the pathogen are sensitive to the toxin. Such correlations between HST production and pathogenicity in the pathogens, and between toxin sensitivity and disease producing ability in plants, provide persuasive evidence that HSTs can be responsible for host-specific infection and disease development. Johnson et al. (2000, 2001) revealed that the genes involved in HST synthesis such as the cyclopeptide synthetase gene, whose product catalyzed AM toxin production in *A. alternata* apple pathotype, might reside on a conditionally dispensable (CD) chromosome. The loss of the CD chromosome led to loss of both toxin production and pathogenicity without affecting fungal growth. On the other hand, the exact roles of nonspecific toxins in pathogenesis are largely unknown, but some are thought to contribute to the features of virulence, such as the symptom development and in plants pathogen propagation. The virulence and host-specificity of these pathogens are based on production of the distinctive HSTs. Most of *Alternaria* HSTs are nitrogen containing metabolites causing black spot disease in *Brassicas*; Destruxin A, Destruxin B, Homodestruxin B, and Desmethyldestruxin B. Destruxins are group of HSTs produced both in vitro and *in planta* by *A. brassicae*, the causal agent of *Alternaria* black spot disease of rapeseed and canola. These cyclo-depsipeptides exhibit a wide variety of biological activities such as antitumor, antiviral, insecticidal, cytotoxic, immunosuppressant, and antiproliferative effect except their phytotoxicity. Interaction between *Alternaria* species and cruciferous plants were studied in detail by Pedras et al. (2009). Necrotrophic phytopathogens such as *A. alternata* and *A. brassicae* are known to synthesize phytotoxins that damage plant tissues and facilitate colonization, while in response to pathogens' attack crucifers biosynthesize phytoanticipins and phytoalexins. Phytoalexins are secondary metabolites produced de novo by plants in response to diverse forms of stress including microbial infection, UV irradiation, and heavy metal salts, whereas phytoanticipins are constitutive defenses whose concentrations can increase upon stress. To the detriment of cruciferous plants, the phytopathogens can overcome phytoanticipins and phytoalexins by producing detoxifying enzymes; phytoalexins brassinin is detoxified into 3-indolyl-methanamine and N″-acetyl-3-indolyl-methanamine by the pathogen *A. brassicae*. Very interestingly, cruciferous plants, (*B. napus* and *Sinapis alba*) can convert host-specific toxins destruxin B and homo-destruxin B into less phytotoxins hydro-destruxin B and hydroxyl-homo-destruxin B, respectively (Lou et al. 2013).

Host nonspecific *Alternaria* phytotoxins can affect many plants regardless of whether they are a host or nonhost of pathogen. Host nonspecific nitrogen-containing phytotoxins include tenuazonic acid, porritoxin, and 10 toxins. Tentoxin, a cyclic tetrapeptide from *A. alternata,* inhibits chloroplast development, which phenotypically manifests as chlorotic tissue. Tentoxin was suggested to exert its effect on chlorophyll accumulation through over energization of thylakoids (Lou et al. 2013). Destruxin B purified from culture filtrates (CFs) of *A. brassicae* induces chlorosis on host leaves at 50–100 μg mL^{-1}, and chlorosis or necrosis on nonhost leaves at 250–500 μg mL^{-1} (Bains and Tewari 1987). Destruxin B was detected in spore germination fluids (SGFs) on host and nonhost leaves, but not in sufficient amount to exert toxicity to host plants. When spores of nonpathogenic *A. alternata* are combined with destruxin B at 100 μg mL^{-1}, and inoculated on leaves, destruxin does not affect the infection behavior of the spores. Interestingly, SGF on host leaves allows nonpathogenic spores to colonize host leaves (Bains 1989). Moreover, a high molecular weight fraction (>5 kDa) without destruxin B obtained by ultrafiltration of SGF has host-specific toxic activity and infection inducing activity. According to Parada et al. (2007), destruxin B is not a HST and does not induce the accessibility of the host plant, which is essential for colonization of the pathogen. In addition, the results with SGF imply that a high molecular weight HST(s) is involved in the host–pathogen interaction.

According to Khandelwal et al. (2002), the protein like P^{53} is over expressed in toxin-treated and nutritionally depleted calli. Similar changes are seen in senescent damaged *Brassica* species indicating involvement of protein P^{53}-dependent pathways. The effects of destruxin B, a host-specific toxin of *A. brassicae* that causes black spot disease in oilseed Brassicas, were studied on in vitro pollen germination and pollen tube growth of *Brassica campestris* var. Brown Sarson, *B. juncea* cvs. Westar and Cresor, *B. nigra,* and *Sinapis alba*. Pollen grain of *B. nigra*, *B. juncea,* and *B. campestris* were the most sensitive and those of *S. alba* the least sensitive to the toxin. Effects of toxin on leaves of these species and the degree of sensitivity of leaves of different species are comparable to that of their pollen grains. The results on the responses of pollen grains as well as leaves to the toxin are in agreement with the degree of susceptibility/resistance of these species to *A. brassicae*, indicating that the genes imparting susceptibility/resistance are expressed in the pollen, a prerequisite for pollen selection. The toxin fed to the cut end of isolated inflorescence axis is readily taken up by the developing pollen, and results in the inhibition of germination of susceptible pollen (Hodgkin and Macdonald 1986; Shivanna and Sawhney 1993).

6.3.7.1 Effect of *Alternaria* Toxin on Host Physiology

The effect of phytotoxins on plants at the physiological level is characterized by the malfunctioning of many physiological processes including respiration, transpiration, photosynthesis, translocation, and growth and development. It is also characterized by the appearance of specific symptoms including wilting, growth suppression, chlorosis, necrosis, and spotting of aerial portions. *Alternaria* HSTs are classified into three groups in terms of primary site of action. AM toxin targets the chloroplast and plasma membrane, ACR (L) toxin targets mitochondria, and ACT, ACTG, AK, AF target the plasma membrane. Toxins produced by *A. brassicae* cause a brown necrotic spots on leaves and brown streaks on stem leading to yield losses. In general, respiration increases once parasitic relationship is established. *Alternaria* pathogens infect green aerial tissues and reduce photosynthetic activity, produce cytokinins, which lead to the formation of Green Island below the germinating conidia on senescing tissues, and cause deformation in chloroplast and mitochondria (Chaturvedi 1972; Zheng et al. 2006). In *B. juncea, A. brassicae,* HST Destruxin B, inhibits almost all-macromolecular biosynthesis, promotes ion leaching, and causes aberrations in mitochondria and chloroplast.

6.3.7.2 Effect of *Alternaria* Toxins on Host Biochemistry

Infection of plants by pathogenic fungi may trigger several biochemical defense responses including enzyme synthesis, cell wall deposition of lignin and suberin, and accumulation of specific metabolites (Daayf and Platt 2000; Abdel-Farid et al. 2009). The mechanisms of host selective pathogenesis are not well understood at the biochemical level, even in cases where the toxin site of action is known. The first physical barrier between plants and pathogen invasion is the cuticle (Schweizer et al. 1996; Fan and Koller 1998) and cell wall, which inhibit both initiation and spread of infection. One characteristic feature of many phytopathogenic organisms is their ability to produce an array of extracellular and highly stable enzymes capable of degrading the complex polysaccharides of the plant cell wall and membrane constituents. These enzymes usually are produced inductively in infected tissues; generally, they are extra cellular, and highly stable. Inoculation of *A. brassicae* on the leaves of *B. juncea* blight resistance cultivar RC-781 decreases the activities of cell wall degrading polygalacturonase and cellulose enzymes, but increases their activities in a susceptible cultivar Varuna up to 3 days of infection (Garg et al. 1999). Infection of *A. brassicicola* shows a differential and sequential induction of two classes of cutinolytic esterases; one class is expressed only in short duration contact (24 h) with intact cutin, but not induced by cutin monomer. The second class, however, is induced by cutin monomers only in prolonged exposure with intact cutin. This differential behavior indicates a sequential recognition of cutin as a barrier to be penetrated and utilized as a carbon source in saprophytic stages (Yao and Koller 1995; Fan and Koller 1998; Baker et al. 2005). The small (7–10 KD), lipid transfer proteins (LTPs) are expected to be involved in wax transport because of their increased expression during drought condition (Beisson et al. 2003; Cameron et al. 2006). During *A. brassicicola* infection, the *ltpgl* (GPI-anchored LTP) knock-out mutant showed increased susceptibility (Lee et al. 2009), but the genetically unidentified cutinase-deficient mutants were nonpathogenic (Tanabe et al. 1988). Results of a transcript profiling study of *hsr203J,* a known marker gene for HR, it was observed that the expression of *hsr*203J gene is more in *A. brassicae*-tolerant than in susceptible genotype of *B. juncea* (Mishra et al. 2011). In Chinese cabbage, *A. brassicae* infection significantly increases glucosinolates (aliphatic and indole) and anthocyanins, but decreases sucrose levels (Ludwig-Muller et al. 1999; Rosta's et al. 2002; Abdel-Farid et al. 2009).

Chitin, an important component of fungal cell wall, is absent in plants, but chitin degrading chitinase enzymes are present in abundant quality, which emphasizes that chitinase can directly affect the viability of fungal pathogens (Boller 1995; Stacey and Shibuya 1997; Shibuya and Minami 2001). This is proved by overexpression of chitinase in transgenic plants, which are generally more resistant to fungal pathogens. Further, it was observed that more chitinase is produced and accumulated at the site of fungal infection (Majeau et al. 1990; Roby et al. 1990; Wubben et al. 1992). Chhikara et al. (2012) reported that co-expression of chitinase and ribosomal—inactive protein in *B. juncea* provides more protection against *A. brassicae.*

The AAL-toxin, structurally related to sphinganine, a member of sphinganine-analogue mycotoxin (SAMS), is an inhibitor of sphinganine-N-acyltransferase (Gilchrist et al. 1992; Abbas et al. 1994) enzyme important in sphingolipid biosynthesis leading to accumulation of free sphingoid bases (Brandwagt et al. 2000; Spassieva et al. 2002; Gechev and Hille 2005). These long-chain sphingoid bases (LCBs) are determinant in the occurrence of programmed cell death (PCD) in susceptible plants (Shi et al. 2007; de Zelicourt et al. 2009). Sphinganine-analogue-mycotoxin (SAMS) inhibits ceramide biosynthesis. Biochemical and molecular data demonstrated that PCD triggered by AAL-toxin is also associated with H_2O_2 (Gechev et al. 2004) as *atr* mutant of *Arabidopsis* which shows enhanced tolerance to H_2O_2 and ROS-induced cell death (Gechev and Hille 2005; Gechev et al. 2008). Interaction with ROS regulates PCD (Gechev and Hille 2005). Brassinin hydrolase, a detoxifying enzyme (BHAb) of a crucifer phytoalexin brassinin, also plays an important role in development of Alternaria blight caused by *A. brassicicola* in *Brassica* (Pedras et al. 2009). Imazaki et al. (2010) reported that *A. alternata* pathotypes contain abundant peroxisomes, which are very important in both tissue colonization and pathogenesis. Other functions of peroxisomes include fatty acid metabolism, acetyl-CoA generation, secondary metabolism, cell wall biogenesis, and peroxisome homeostasis.

6.3.7.3 Effect of *Alternaria* Toxin on Molecular Process

The interaction between plants and pathogens at molecular level are specific, complex, and dynamic. Many responses are evoked in plants upon encountering pathogens but relatively very few have been studied in detail. Key events in plant–pathogen interaction include perception of pathogen on the plant or cell surface by receptors/sensors, and transduction of these perceptions through various transcription factors, and target genes which are involved in coordination of the appropriate responses (Hammond-Kosack and Jones 1996). Different types of both plant and pathogen genes have been shown to be involved in plant and necrotrophic fungal interaction. These responses range from genes that encode proteins like receptor/receptor kinase, cell wall degrading enzymes, toxins, and transporter proteins, to those involved in signal transduction cascades such as mitogen-activated protein (MAP) kinases, and various transcription factors like WRKY, NAC (Lawrence et al. 2008; Wang et al. 2009; Amselem et al. 2011). Ghose et al. (2008) also studied the differential profiling of selected defense-related genes induced on challenge by *A. brassicicola* in resistant white mustard, and their comparative expression pattern in susceptible Indian mustard. These genes have similarity with receptor-like-protein-kinase genes, genes involved with calcium-mediated signaling, salicylic acid-dependent genes, and other functional genes in *Arabidopsis.* After the attack, the fungal cell wall is hydrolyzed by an enzyme chitinase, and fragment of cell wall chitin is perceived by a receptor known as CERK 1 or Lys M RLK 1. A mutation in Lys M receptor-like protein (*Lys M RLK1*) gene blocks the induction of almost all chito-oligosaccharide-responsive genes and leads to increased susceptibility to fungal pathogens indicating that *LysM RLK1* is essential for chitin signaling in plants (Wan et al. 2008). Another receptor gene, PSK-receptor, PSK2 and PSKR I (Phytosulfokine Receptor), also induced in *Arabidopsis* leaves following *A. brassicicola* inoculation, also shows the role of PSK receptor in pathogenesis (Loivamaki et al. 2010).

Molecular mechanism of Alternaria blight in *Brassica* shows involvement of chlorotic and necrotic toxins and phytohormone. *Alternaria brassicae* produces a chlorotic toxin destruxin B that plays an important role in signal transduction leading to programmed cell death, and suppressing the defense system (Taj et al. 2004). Differential expression of cell cycle proteins in toxin-treated leaves and calli, and over expression of p53 suggest that the toxin-mediated perturbations in cell cycle eventually cause p53-induced-programmed cell death (PCD) (Khandelwal et al. 2002). Interestingly, *A. brassicae* pathotoxin-induced-cell-death pathway was antagonized by a phytohormone zeatin, in cell culture of *B. juncea* (Pandey et al. 2001). The antagonistic effect of these two structurally different entities strongly suggests the role of interactive signaling pathways in pathogenesis of Alternaria blight in *Brassica* species. Introgression of Osmotin, a known PR-5 protein, causes perturbation in resistance to biotic and abiotic stresses (Taj et al. 2004). The effect of toxin and zeatin treatments on the *B. juncea* leaf proteome was investigated by using two-dimensional electrophoresis and LCMS techniques; results showed that 20 proteins were uniquely expressed in toxin-treated leaf, while 27 proteins were expressed in together with toxin and zeatin. LCMS technique has also been used to identify a total of 15 proteins with differential expression in toxin treated leaves. The proteins identified in response to the toxin are glycosyl hydrolases, subtilisin like proteases, P-nitrophenyl phosphatase, malate dehydrogenase, heat shock, ribulose 1, 5 bis phosphate carboxylase, Cucumisin-like serine proteases, Globulin like protein, and ATP synthase. Sharma et al. (2007) also studied the proteome-level changes in *A. brassicae–B. napus* and suggested role of ROS-mediated-auxin signaling in the pathosystem.

Deciphering the MAP Kinase machinery components in *B. juncea,* RT-PCR amplification of all 20 known MAPK has been done. Among the MAP Kinases

8, 12 and 18 showed no expression, expression of MAP Kinase 3, 10, and 14 were validated with the Northern Blot, MAPK 3 important gene directly correlated with the transcription factors, and expresses in compatible interaction of *A. brassicae* and *B. juncea* (Taj et al. 2011). Expression of MAPKK 4, MAPKK 5, MAPKK 9, MAPKKK I, Wrky 22, and Wrky 29 has also been validated by real-time PCR. Expression analysis of MAP 2 K9 and MAP K-6 is also governed during pathogenesis of Alternaria blight in *Arabidopsis thaliana* where simultaneous increased levels up to middle stage of disease progression were observed (Kannan et al. 2012). Activation of more than one member of MAPK by *Alternaria* suggests that MAPK cascades act as points of convergence and divergence in signaling, and might play a pivotal role in deciding whether disease should progress, or defense pathways defeat the pathogen.

During the course of infection, various genes of pathogen are also expressed and play role in fungal pathogenesis; *AaFus3* gene, which encodes for *FUS3MAPK* in *A. alternata,* is required for conidial development and penetration of the fungus in plant (Lin et al. 2010). Another study shows that deletion of AbPro1 (transcription factor), and *AbNIK* 1 (two component histidine kinase) gene of *A. brassicicola* results in 70%, and complete loss of virulence, respectively (Cho et al. 2009). Cyanide hydratase, arsenic ATPase, format dehydrogenase is some other genes of *A. brassicicola,* which are being expressed and have a role in fungal pathogenesis (Cramer and Lawrance 2004). Unfolded protein response (UPR) pathway also regulates the fungal pathogenesis, and *AbHacA* gene encodes the major UPR transcriptional regulator; loss of UPR in mutants of *A. brassicicola* resulted in complete loss of virulence (Joubert et al. 2011).

6.3.7.4 Role of Metabolites in the Infection Process of *Alternaria*

Despite the taxonomic and pathogenic differences between *Alternaria* species, these cause similar infection patterns. Dormant spores have heavily melanized walls that under favorable conditions produce one or more germ tubes. Subsequently, the germ tubes penetrate stomata, cuticle or wounds with or without the formation of small appressoria. In less virulent species, wounds and stomata are targeted, while more virulent species can also penetrate directly (Rotem 1994): Enzymatic processes in *Alternaria* infections are essentially similar to those in other diseases. The cuticle, which consists of a combination of cutin (a hydroxyl fatty acid polyester), and waxes comprises the first line of defense to be overcome by directly penetrating fungal pathogens. In case of *A. brassicicola,* the differential expression of cutinase genes was monitored between saprophytic and pathogenic stages of the fungus (Yao and Koller 1995). Further, it was found that different cutinolytic enzymes are sequentially induced upon landing on, and penetration of the cabbage leaf (Fan and Koller 1998). Constitutively produced cutinases are expressed during the initial contact of *A. brassicicola* with the cuticle. After reaching subcuticular layer, different cutinases that are active during saprophytic growth are induced (Trail and Koller 1993; Yao and Koller 1994). These cutinases are inducible by cutin monomers. This implies a switch between the parasitic and the saprophytic stage of the fungal pathogen. So far, however, it has not been demonstrated that any extracellular hydrolase is crucially involved in fungal pathogenesis.

In addition to cutinases, lipases might also contribute to the establishment of infection. *A. brassicicola* was found to produce a lipase that acts as a virulence factor (Berto et al. 1997). Antilipase antibodies were found to have an inhibitory effect on the in vivo infection of cauliflower leaves, as symptom development was inhibited in a dose-dependent manner despite normal, germination of spores. However, the antibodies did not have any effect on fungal infection of dewaxed leaves, which suggests that this lipase has an early pathogenic activity during penetration (Berto et al. 1999). About one-third of the total cell wall components in dicotyledonous plants are pectic polysaccharides. These components can be hydrolyzed by fungal galacturonidases. On the other hand, the tangerine pathotype of *A. alternata* did not show this dependency, possibly because this particular pathogen largely depends on toxin production for the colonization of its host (Isshiki et al. 2001). For a specific *A. alternata* endoglucanase, it was demonstrated that its production is triggered by a pathogen-induced pH increase on the host (Eshel et al. 2002b). Correlation-studies between enzyme production and symptom development suggest that endoglucanases and also exoglucanases are involved in *A. alternata* pathogenicity (Eshel et al. 2000, 2002a, b).

6.3.7.5 Biomolecular Alteration During *Brassica–Alternaria* Interaction

Genetic control of rapeseed and mustard genotypes tolerant against *A. brassicicola* plays a critical role in determining the *Brassica* host-pathosystem interaction outcome. The assessment of *A. brassicicola* tolerance among rapeseed and mustard revealed that dark leaf spot disease induced an extreme alteration in plant biochemistry that cause reduction in photosynthetic area, defoliation, accelerated senescence, and ultimately poor yield in susceptible genotypes as compared to tolerant ones. Physio-biochemical defense response, as evidenced by tolerant genotypes via augmented activities of defense enzymes, is a vital sign of their role in *Brassica–Alternaria* interactions and its tolerance. These physiological, morpho-biochemical marker indices can be used as probes for rapid screening of germplasm. Although moderate tolerant germplasm EC250407 (mustard) and EC1494 (rapeseed) can also be utilized and can serves as tolerant material for black spot disease.

Several morpho-molecular parameters have been evaluated to test the performance of field mustard and rapeseed genotypes under artificial inoculation of the pathogen. During *Brassica–Alternaria* interaction, physio-biochemical defense response was witnessed in tolerant genotypes. Two tolerant genotypes (one for field mustard and one for rapeseed), i.e., EC 250407 and EC 1494 were identified. However, necrotic lesions were more prominent insusceptible genotypes with minimum chlorophyll (chlorophyll *a*, chlorophyll *b* and total chlorophyll) and carotenoids contents. Contrary to photosynthetic pigments, increase in total soluble protein (TSP) contents was observed with disease progression in susceptible genotypes. Tolerant genotypes of field mustard and rapeseed displayed remarkable increase in the activities of redox enzyme in infected leaves with least yield loss (6.47% and 5.74%) and disease severity index (DSI) of 2.9 and 2.1, respectively.

However, yield/plant showed close association with other morpho-yield parameters, photosynthetic pigments and redox enzymes (superoxide dismutase (SOD), catalase (CAT) and peroxidase (POD) activities except silique length and TSP. Based on the results of morpho-biochemical analyses, redox enzymes, and morphological parameters; their interplay has been proposed to determine the tolerance outcome of the *Brassica–A. brassicicola* interaction (Munir et al. 2020).

6.4 *Brassica–Colletotrichum* Host-Pathosystem

6.4.1 Biosynthesis of Biomolecules During Pathogenesis

In *Colletotrichum higginsianum*, two arginine auxotroph mutants showed reduced penetration and invasive growth ability, which was restored when L-arginine was supplied. Thus, arginine biosynthesis was shown to be dispensable for conidial germination and appressorial morphogenesis of *C. higginsianum*, suggesting that arginine reserves in conidia are sufficient for the completion of prepenetration development. However, arginine biosynthesis was critical for initial host penetration by appressoria and early biotrophic growth inside living host cells. Although not many genes related to nutritional requirements have been genetically analyzed in *C. higginsianum* so far, there are many such studies in other species of Colletotrichum. The GATA transcription factor, AreA, regulates the use of poor or complex nitrogen sources, and restricts their use when sufficient nitrogen sources are available within the organism. The AreA plays a critical role in fungal development, conidial production, and regulation of nitrogen metabolism and virulence in *C. gloeosporioides*. Kre5 and Kre6 are the key enzymes in β-1, 6-glucan synthesis and formation of branch points of the β-glucan network. In *C. graminicola*, RNAi-mediated reduction of KRE5 and KRE6 transcript abundance caused appressoria to burst and necrotrophic hyphae to swell, indicating that β-1, 6-glucosidic bonds are essential in these cells (Oliveira-Garcia and Deising 2016).

The homologous proteins in *C. higginsianum* should play important roles in fungal development or virulence. The ATP-binding cassette (ABC) and major facilitator super family (MFS) of transporters are two families that play important roles in transport processes. MFS transporters are usually demonstrated to be involved in multidrug resistance in fungi. MFS transporters are capable of transporting small molecules in response to ion gradients or function as drug: H+ antiporters in microorganisms. Mounting evidence indicates that MFS transporters may also indirectly control membrane potential by changing membrane lipid homeostasis and regulating internal pH and the stress response machinery and pathogenicity in fungi. Moreover, some MFS transporters also are involved in secretion of phytotoxins. A virulence-deficient mutant, Ch-1-T513, from a T-DNA insertion mutant library in *C. higginsianum* was found to have abnormal hyphae, which might be a key factor affecting virulence of the fungus. The study demonstrated that a MFS transporter named ChMfs1 is responsible for the mutant Ch-1-T513 phenotype, and ChMfs1 in *C. higginsianum* reported to be involved in pathogenicity and the production of intrahyphal hyphae (Liu et al. 2017).

The plasma membrane H + -ATPase is a proton pump that plays important energetic and regulatory roles in the physiology of plants and fungi controlling essential functions including nutrient uptake and intracellular pH regulation. In fungal cells, the activity of the proton pump is regulated by a large number of environmental factors at both transcriptional and post-translational levels. Structure and function of plasma membrane H + -ATPases have been extensively explored in fungi, revealing their role in vegetative growth, nutrient transport and pathogenicity. In *C. higginsianum*, a potential plasma membrane H + -ATPase Pma2 was frequently targeted in five independent insertion mutants from the T-DNA insertion mutant library. Chpma2 deletion mutants form fully melanized appressoria but entirely fail to penetrate the host tissue. Targeted gene knockout of another plasma-membrane pump gene, ChPMA1, gave a nonviable phenotype, indicating that ChPMA1 may be an essential gene and encode the major H + -transporting ATPase (Ghassemi et al. 2015; Yan et al. 2018).

6.5 *Brassica–Erysiphe* Host-Pathosystem

6.5.1 Host Metabolism and Increased Ploidy Level of Mesophyll Cells Regulates Powdery Mildew Infection

Besides defense-related transcriptional changes induced by microbe recognition, host gene expression are potentially impacted by the action of powdery mildew effectors to promote fungal pathogenesis. Consequently, adaptation of the host metabolism to the presence of the pathogen has been reported (Wildermuth 2010). Laser micro-dissection-assisted site-specific profiling of transcript abundance during late *Go* infection stages (5 dpi) suggests a suppression of photosynthesis and points to a carbon source-to-sink transition in powdery mildew-infected cells (Chandran et al. 2009, 2010). The powdery mildew-triggered induction of genes associated with sugar metabolism and hexose transporters associated with sink organs further reinforces this idea. The respective proteins might contribute to the availability of carbohydrates at infection sites, and consistently their elevated expression levels may reflect an increased demand for hexoses by the fungus (Fabro et al. 2008; Chandran et al. 2009, 2010). Increased transcript abundance of genes related to respiration, including glycolysis, the tricarboxylic acid cycle, and the mitochondrial electron transport chain further strengthen the notion of an adaptation to the elevated energy consumption by the infected tissue (Fabro et al. 2008; Chandran et al. 2010).

Adjustment of the plant host metabolism to support the growth of the biotrophic pathogen is consistent with an increased ploidy level of the mesophyll cells underlying infected epidermal cells at later stages of compatible interactions (Chandran et al. 2010, 2013). This correlates with the accumulation of *PLANT UBX DOMAIN-CONTAINING PROTEIN 2* (*PUX2*: At2g01650) transcripts after *Go* infection (Chandran et al. 2009). Strikingly, the onset of *PUX2* induction at 5 dpi overlaps with the occurrence of endo-reduplication in mesophyll cells, and corresponds with fungal growth, and reproduction (Chandran et al. 2013). The resulting polyploidy might compensate for the increased metabolic activity resulting from the nutritional demands of the fungus. This is supported by decreased spore formation coinciding with reduced basal ploidy in *pux2*, and thus identifies endo-reduplication as a potential determinant of susceptibility to powdery mildew (Chandran et al. 2010, 2013). The presence of UBX (ubiquitin regulatory X) and PUB (peptide: N-glycanase/UBA or UBX-containing proteins) domains in PUX2 suggests that, like other proteins with similar domain structures, it might act as a regulatory co-factor of cell division control protein 48 (CDC48: At3g09840). Indeed, this AAA-ATPase (ATPase associated with diverse cellular activities) interacts with PUX2 in vitro (Rancour et al. 2004). As CDC48 complexes contribute to cell cycle progression, its interaction with PUX2 might regulate cell ploidy (Rancour et al. 2004; Madsen et al. 2009; Yamanaka et al. 2012; Gallois et al. 2013). The MYB3R4 (At5g11510), a cell cycle control-associated MYB3R TF activating G2/M progression, is locally induced 5 dpi with *Go.* As genome duplication is a controlled process that occurs during mitosis, it is conceivable that MYB3R4 is required for powdery mildew-induced polyploidy, which is supported by the phenotype of the *myb3r4* mutant (Haga et al. 2007; Chandran et al. 2010, 2013). Similar to *pux2, myb3r4* mutant's exhibit reduced powdery mildew conidiophore formation.

The negative impact on basal cell ploidy levels, and increased resistance of *pux2* are further phenocopied by *pmr6*. By contrast, reduced fungal reproduction associated with *pmr5* does not impact basal ploidy levels but correlates with a suppression of the powdery mildew-induced increase in ploidy (Chandran et al. 2013). Analysis of *pmr5* microarray data reveals an enrichment of MYB3R TF binding elements among cell cycle regulation-related genes showing altered expression in the mutant. This suggests that PMR5 acts upstream of a MYB3R TF to control powdery mildew-induced ploidy (Chandran et al. 2013). A critical role of elevated ploidy for fungal virulence is further strengthened by the identification of the teosinte branched/cycloidea/proliferating cell factor TCP13 (At3g02150), TCP14 (At3g47620), and TCP15 (At1g69690) basic helix-loop-helix (bHLH) TFs as common targets of several powdery mildews, oomycetes, and bacterial effectors (Weßling et al. 2014). The TCP14, and TCP15 repress endo-reduplication by directly regulating the expression of cell-cycle genes (Peng et al. 2015). Mutation of TCP13, TCP14, and to a lesser extent of TCP15, results in increased susceptibility toward *Go* (Weßling et al. 2014). A link to ubiquitin-mediated regulation of ploidy is provided by the ubiquitin receptors DA1 (At1g19270; "Da" is Chinese for "large"), DA-related (DAR) 1, and DAR2 (At2g39830), which interact with and modulate the stability of TCP14/15 to regulate endo-reduplication (Peng et al. 2015). Remarkably, DEL1, known to repress genes required for the onset of endo-reduplication (Vlieghe et al. 2005; Lammens et al. 2008), does not impact the *Go*-induced increase in mesophyll ploidy when mutated or over-expressed (Chandran et al. 2014). Instead, microarray analyses of *del1* plants revealed an induction of basal defense gene expression compared to wild type. The identification of effector targets involved in adaptation of the host metabolism (Weßling et al. 2014) marks one of the first steps toward elucidation of pathogen-induced reprogramming of the plant transcriptome. Further, characterization of the mechanisms by which the fungus enforces adjustment of the plant cellular program to promote its intracellular accommodation will be an important aspect of future research (Kuhn et al. 2016).

6.5.2 Powdery Mildew Infection and Pathogenesis Regulates Transcriptional Events

Transcriptional programming during powdery mildew pathogenesis was analyzed by Pandey et al. (2010). The transcriptional reprogramming induced by *G. orontii* during early events of *Arabidopsis* infection, and the role of two closely related WRKY transcription factors, 18, and 40, was observed. It was observed that WRKY18/40 negatively affects preinvasion defenses in *Arabidopsis,* and deduced a subset of genes that appear to be under WRKY18/40 control. The direct, in vivo physical interaction of WRKY40 to W box containing promoter regions of the key regulatory genes EDS1, and JAZ8, and of the AP2/ERF transcription factor gene, RRTF1 was demonstrated.

Previous transcriptomic studies mainly focused on later stages of powdery mildew infection, i.e., haustoria formation (18–24 hpi, Fabro et al. 2008), or pathogen growth, and reproduction (1–7 days postinoculation, Chandran et al. 2009). Pandey et al. (2010) found that expression of hundreds of host genes is altered within 8 hpi indicating that signaling between the pathogen, and individual host cells already occurs a few hours post contact of the spores with the leaf surface at about the stage of penetration, but certainly prior to establishment of the extra-haustorial matrix. It is important to note that *G. orontii* invades only epidermal cells (Lipka et al. 2008). Hence, local gene expression differences within such cells may actually be higher than the values observed using whole leaf samples. Interestingly, nine out of the 24 genes most highly, and differentially expressed (>threefold) in uninfected wrky18 wrky40 plants encode transcriptional modulators. This implies that WRKY18/WRKY40 is central negative regulators controlling distinct downstream transcriptional programs. The role of phyto-hormones and distinct signaling pathways during compatible interactions of *Arabidopsis,* and powdery mildews is not firmly established (Chandran et al. 2009; Consonni et al. 2006; Fabro et al. 2008). Although WRKY18/WRKY40 expression was observed to be dependent on the SA biosynthetic gene iso-chorismate synthases1 (ICS1) at later stages of *G. orontii* infection (1–7 dpi), induction of these genes at the early infection stage (6 hpi) was ICS1 independent (Chandran et al. 2009). Compared with WT, wrky18 wrky40 plants constitutively express 2.4- to 8-fold higher levels of five co-regulated JAZ family members, namely JAZ1, JAZ5, JAZ7, JAZ8, and JAZ10, and their promoters contain several W box elements. The JAZ proteins are key repressors of JA signaling (Chico et al. 2008). Although, Pandey et al. (2010) could not detect in vivo associations of WRKY40 at JAZ7, and JAZ10 promoter sites nor to LOX2, and AOS, but observed specific interaction with a region of the JAZ8 promoter containing two perfect W boxes. Currently, only one other transcription factor, MYC2, has been identified as a direct regulator of JAZ promoter activity. However, expression of JAZ7 and 8 does not require MYC2 (Chung et al. 2009). It implies that WRKY40 affects JA signaling by directly controlling the expression of a subset of these negative regulators. Lack of mutants for JAZ7, −8 and − 10 currently hamper functional testing of their role in plant–pathogen interaction. Also, the contribution of WRKY18 in this process remains to be elucidated, although it exerts a positive regulatory function, along with WRKY40, in JA signaling (Wang et al. 2008).

The role of RRTF1 in plant defense remains to be rigorously tested, but recent reports indicate its involvement in regulating redox homeostasis during stress, and expression of RRTF1 is dependent on COI1, a central regulator of JA signaling (Khandelwal et al. 2008; Wang et al. 2008). Moreover, RRTF1 is part of a core redox signaling subnetwork that also includes EDS1, and WRKY33 (Khandelwal et al. 2008). Possibly, all three of these genes are under WRKY18/40 control. WRKY factors form a functionally interconnected transcriptional network (Eulgem and Somssich 2007), and WRKY18/40 may thus indirectly modulate gene expression by directly controlling other WRKY activities. Indeed, loss-of-WRKY18/40 function appears to limit the pathogen-induced accumulation of WRKY11, and WRKY33 transcripts. PAMP induced expression of WRKY33 was shown to depend on W box promoter elements that are bound by WRKY factors in planta (Lippok et al. 2007). Compared with WT, wrky18 wrky40 mutants displayed rapid transient induction of CYP71A13, and an overall increased accumulation of PAD3 transcripts. WRKY33 binding to the PAD3 promoter in vivo has been demonstrated, but no binding was detected within the CYP71A13 promoter (Qiu et al. 2008). It suggests that CYP71A13 may represent a direct target of WRKY18/40 while PAD3 is an indirect target that is directly controlled by WRKY33 (Pandey et al. 2010).

In general, expression of stress response genes is transient, and their transcription relies on a combination of distinct sets of co-activators, and negative regulators (Lopez-Maury et al. 2008). It remains to be elucidated how WRKY18/40 negatively regulate transcriptional outputs. In the case of RRTF1, and JAZ8, one can envision a direct repressor function on basal gene expression. However, CYP71A13, EDS1, and PAD4 regulation may be different. Expression of these genes in wrky18 wrky40 is not significantly elevated in the absence of pathogen infection. Rather, loss-of-WRKY18/40 function results in a pathogen-dependent but exaggerated increase in their transcript levels. A simplistic model would be that other WRKY factors (i.e., WRKY60) could substitute to form weaker but sufficient repressor complexes at these promoters. Replacement of WRKY factors at specific W box sites has been demonstrated for two parsley promoters (Turck et al. 2004). Activation of these genes would require an activator complex that is dependent of pathogen-triggered signaling. Elevated transcript levels observed in the mutant may be the consequence of a higher association of the activator(s) to the transcription complex due to failure of the weaker core pressers to counter this effect. This could imply that binding sites for default repressors, and sites for signal-induced activators overlap, a situation often found in inducible eukaryotic promoters (Affolter et al. 2008). Naturally, alternative scenarios are conceivable that may also involve altered regulation at the post-transcriptional level.

The consequences of transcriptional reprogramming should be reflected in qualitative/quantitative altered host metabolite levels resulting in resistance of wrky18 wrky40 plants toward *G. orontii.* Although wrky18 wrky40 mutants accumulated high camalexin levels, and were resistant to this powdery mildew lends support to the importance of this compound in determining the outcome in this interaction. Proteases and endo-membrane components involved in protein quality control, and secretion have been associated with hormone responsive signal transduction events, and disease resistance (Beers et al. 2004; Wang et al. 2005). Several protease genes including At3g49340 were more strongly upregulated in challenged wrky18 wrky40 mutants than in WT plants. Genes encoding components of the endo-membrane system, and endoplasmic reticulum (ER) response (e.g., CYP71A12, CYP81F2), wall-associated kinase (At2g34500), BiP3 (At1g09080), and type-1 phosphodiesterase (At4g29700) were also differentially regulated. CYP81F2 encodes a P450 mono-oxygenase essential for pathogen induced accumulation of antifungal compounds (Bednarek et al. 2009). The BiP3 has been associated with the ER response, and its upregulation in wrky18 wrky40-resistant plants intuitively suggests that in susceptible WT plants, *G. orontii* may manipulate part of the host ER response system by impinging on WRKY18/40 functions. Defense activation is often preceded by mitogen-activated protein kinase (MAPK) signaling (Colcombet and Hirt 2008). Uninfected wrky18 wrky40 plants constitutively expressed fivefold elevated transcript levels of a mitogen-activated triple kinase gene MAPKKK19 (At5g67080).

Although wrky18 wrky40 lines over-expressing WRKY40 showed a high degree of susceptibility, penetration efficiency did not reach WT levels. Because WRKY18/40/60 can form homo- and heterodimers (Xu et al. 2006), appropriate WRKY18/40 hetero-dimerization may be partly required for maximal transcriptional reprogramming. Alternatively, or additionally, WRKY18, and 40 may target different subsets of genes. Due to extreme difficulties in generating plants expressing a WRKY40 transgene under its native promoter, interpretation of results relies on ectopic expression of this gene. Despite this caveat, these lines show no obvious phenotypic differences to WT plants, and do not constitutively express defense genes. Moreover, as illustrated by ChIP, overexpressed WRKY40 shows binding selectivity toward some promoters and even at such sites only to a subset of the overall number of W box elements present.

During the early course of *G. orontii* infection, the JA signaling pathway is required for plant susceptibility, whereas the EDS1 signaling pathway is needed to maintain resistance (Pandey et al. 2010). The WRKY18/40 negatively modulates the expression of positive regulators of defense such as EDS1, and the camalexin biosynthetic genes, but positively enhances JA signaling by partly suppressing the expression of genes encoding negative regulators of this pathway. This enables the pathogen to overcome early host-defenses, and ultimately leads to susceptibility. Alternatively, WRKY18/40 regulates induced expression of genes whose products are required for pathogen growth, and reproduction. However, very few such host susceptibility factors enhancing powdery mildew pathogenesis have been identified (Eichmann and Huckelhoven 2008). Moreover, despite the fact that constitutive activation of JA, and SA signaling can result in enhanced resistance toward powdery mildew infection, comprehensive studies employing a range of SA, JA, and ET signaling mutants failed to reveal a clear-cut role of these pathways in *G. orontii* resistance. Thus, a model is required in which *G. orontii*, in part via the regulatory action of WRKY18/40, temporally alters the balance between SA, and JA signaling at early stages of infections, keeping host defenses suppressed, and thereby allowing the pathogen to gain access to host nutrients. Future investigations will aim at functionally dissecting the WRKY40/18-dependent pathways, and the role of key WRKY40 downstream targets in modulating host plant defenses. Such studies will include extensive evaluation of phytohormone dynamics (especially of JA, and SA), analysis of selected mutants, and the generation of transgenic lines modified in the expression of multiple key components identified (Pandey et al. 2010).

6.6 *Brassica–Hyaloperonospora* Host-Pathosystem

6.6.1 Induction of Biometabolites Regulates Pathogenesis

Many marked shifts were observed in the metabolic processes of plant tissues following infection by biotrophic parasites. These included changes in respiration, photosynthesis, nucleic acid, and protein synthesis, and phenol metabolism. There could also be changes in the translocation, and accumulation of nutrients, and in the levels of endogenous growth substances.

The respiration rate was raised sharply soon after *H. parasitica* infection of cabbage cotyledons, and reached a maximum, almost twice that of uninfected controls, at the time of the initiation of sporulation. The chlorophyll content of infected and noninfected cotyledons did not differ significantly at any time. The increased respiration rate of the infected tissues did not reflect any significant changes to the pentose phosphate pathway in this infection since no acyclic polyhydric alcohols were detected in soluble extracts of either infected leaves or fungal conidia (Thornton and Cooke 1974).

Pathogenesis in *Brassica–Hyaloperonospora* combinations was observed to be accompanied by large increases in electrolyte leakage, and increased activity of glucosidase, ribonuclease, and peroxidase (Kluczewski and Lucas 1982). The large increase in glucosidase was of pathogen origin while enhanced ribonuclease activity was due to a new postinflectional form of the enzyme. In vivo infected leaves of *B. juncea* produced cellulase, endo-PMG, and endo-PG (Singh et al. 1980).

6.7 *Brassica–Leptosphaeria* Host-Pathosystem

6.7.1 Expression of Metabolic Genes of *Brassica* During *Leptosphaeria* Pathogenesis

Fungal genome sequences become available now days; it is apparent that their complement of genes and transcriptomes reflects fungal lifestyles. Lifestyles of plant pathogenic fungi are often classified into three broad categories: biography, where the pathogen feeds from live host cells, necrotrophy, where the host cells are killed ahead of colonization, and hemi-biotrophy, where the pathogen feeds from living cells before switching to a necrotrophic style of growth. These designations are imprecise and as the mechanisms of pathogenicity in a range of fungi are elucidated, lifestyle boundaries become more blurred (Oliver and Solemon 2010). *Leptosphaeria maculans* "*brassicae*" has numerous "gene for gene" interactions with *B. napus* whereby an avirulence allele in the fungus renders it unable to attack cultivars with the corresponding resistance gene. This "gene for gene" resistance is expressed in seedlings; five avirulence genes have been cloned so far—*AvrLm1, AvrLm4-7, AvrLm6, AvrLm1,* and *AvrLmJ1*. Only one resistance gene, *LepR3*, has been cloned from *B. napus*.

During infection fungi derive nutrition from the host plant, often by enzymatic degradation of proteins and carbohydrates. These latter enzymes are classed as carbohydrate-active enzymes (CAZys) and they often have well characterized domains. As well as providing nutrition, CAZy activity releases cell wall products that can act as DAMPs (Damage Associated Molecular Patterns) that activate the host immune system. The biotrophic plant pathogens including *Hyaloperonospora arabidopsidis* have fewer CAZy genes than necrotrophic plant pathogens do, perhaps because biotrophs do not need to digest plant cell walls for nutrition and they must evade the host immune system. Genomic sequences are now available for many Dothideomycetes and an extensive comparative analysis of 18 of them, including *L. maculans* "brassicae" has been published (Ohm et al. 2012). These include the *Brassica*-infecting pathogen, *Alternaria brassicicola*, as well as three wheat-infecting pathogens. The *L. maculans* "*brassicae*" genome is compartmentalized into AT-rich that are gene-poor comprising up to 35% of the genome, and gene-rich regions that have a high GC content, while the genome of *L. biglobosa* "canadensis" is 30 Mb and lacks AT-rich regions. A limited amount of oligo-array transcriptome data have been produced for *L. maculans* "brassicae" during in vitro culture and infection of *B. napus*, but transcriptome data have not been reported for *L. biglobosa* or any closely related *Brassica* pathogen. Patterns of global gene expression can be generated by RNAseq, a technique that enables analysis of dual transcriptomes; for instance, during a plant–pathogen interaction. Furthermore, RNA seq can be exploited to analyze nonmodel organisms for which genomic resources are not well developed. Few dual transcriptomes for plants and pathogenic fungi have been reported. Lowe et al. (2014) described the genomes and transcriptomes of *L. biglobosa* "canadensis" and *L. maculans* "brassicae" and canola (*Brassica napus*), during infection and in vitro. The aim was to characterize genes that each pathogen uses to evade detection by the host, or to derive nutrition from the host, viz. the carbohydrate-active enzymes and peptidases. The genes upregulated by the host during infection by each pathogen has also been examined. Lowe et al. (2014) described the complement of genes encoding carbohydrate-active enzymes (CAZys) and peptidases of these fungi, as well as of four related plant pathogens. They also reported dual-organism RNA-seq transcriptomes of these two *Leptosphaeria* species and *B. napus* during infection. During the first 7 days of infection *L. biglobosa* "*canadensis*," a necrotroph, expressed more cell wall degrading genes than *L. maculans* "*brassicae*," a hemi-biotroph. *L. maculans* "*brassicae*" expressed many genes in the carbohydrate binding module class of CAZy, particularly CBM50 genes, with potential roles in the evasion of basal innate immunity in the host plant. At this time, three avirulence genes were among the top 20 most highly upregulated *L. maculans* "*brassicae*" genes *in planta*. The two fungi had a similar number of peptidase genes; antitrypsin was transcribed at high levels by both fungi early in infection. *L. biglobosa* "*canadensis*" infection activated the jasmonic acid and salicylic acid defense pathways in *B. napus*, consistent with defense against necrotrophs. *L. maculans* "*brassicae*" triggered a high level of expression of iso-chorismate synthase 1, a reporter for salicylic acid signaling. *L. biglobosa* "*canadensis*" infection triggered coordinated shutdown of photosynthesis genes, and a concomitant increase in transcription of cell wall remodeling genes of the host plant. Expression of particular classes of CAZy genes and the triggering of host defense and particular metabolic pathways are consistent with the necrotrophic lifestyle of *L. biglobosa* "*canadensis,*" and the hemibiotrophic life style of *L. maculans* "brassicae." Top 20 functional categories regulated during infection by *Leptosphaeria* are presented in Chap. 3, Table 3.3, Fig. 6.1 (Lowe et al. 2014).

6.7.2 Modulation of Host Metabolism by *Leptosphaeria* Effector Genes During Pathogenesis

A virulence factor modulates plant hormone signaling during *L. maculans* pathogenesis. Effectors contribute to fungal virulence by protection of fungal structures and modulation of host metabolism.

Comparison of near-isogenic *L. maculans* isolates with and without the effector *AvrLm1* suggested that isolates carrying *AvrLm1* were more aggressive on susceptible *B. napus* cultivars (Huang et al. 2009). The further examination of the virulence function of *AvrLm1*, the generated transgenic *Arabidopsis* Columbia-0 (Col-0) lines carrying AvrLm1 in which expression of the full-length *AvrLm1* gene was driven by the constitutive *CaMV* 35S promoter. A homozygous single-insertion line was generated (Col-AvrLm1), and expression of AvrLm1 was confirmed by quantitative reverse-transcriptase polymerase chain reaction (qRT-PCR). Expression of SA, jasmonic acid (JA), ethylene (ET), and ABA signaling response marker genes (PR1, LOX-2, PR4, and ABF2) in Col-AvrLm1 and Col-0 plants were compared. Expressions of PR1and LOX2 were significantly decreased in Col-AvrLm1, whereas expressions of ABF2 and PR4 were the same as in Col-0, indicating that

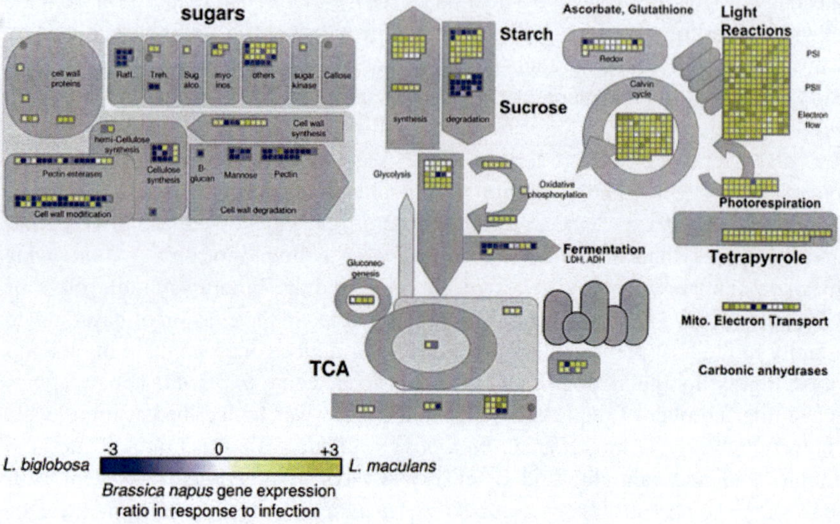

Fig. 6.1 Transcription of *Brassica napus* genes involved in metabolic processes including photosynthesis 7 days after inoculation with *Leptosphaeria maculans* "*brassicae*" (*Lmb*) or *L. biglobosa* "*canadensis*" (*Lbc*). RNA-seq gene expression values for a *B. napus* unigene set were used to calculate a ratio of expression values (log2) for *B. napus* genes after infection by *Lmb* or *Lbc*. Ratios were plotted on major metabolic pathways with Mapman software. A yellow square indicates a *B. napus* gene that is expressed more highly during *Lmb* infection, while a blue square indicates a *B. napus* gene with higher expression during *Lbc* infection. An expression ratio close to zero is shown with a white square and indicates equivalent expression during infection by either pathogen. Only genes with expression values greater than 10 FPKM were included. *LDH* lactate dehydrogenase, *ADH* alcohol dehydrogenases, *TCA* tricarboxylic acid cycle, *raff* raffinose, *Treh* trehalose, *PSI* photosystem one, *PSII* photosystem two, *ABA* abscisic acid, *ET* Ethylene, *SA* salicylic acid, *JA* jasmonic acid (Lowe et al. 2014)

AvrLm1 suppress SA and JA signaling but does not affect the ABA and ET pathways (Fig. 6.2). Col-AvrLm1 plants were challenged with the bacterial pathogen Pst DC3000.Col-0 was also inoculated with Pst-DC3000 as a control. Col-AvrLm1 showed enhanced disease symptoms (severe chlorosis) and higher bacterial growth compared with the Col-0 control. These findings indicate that AvrLm1 functions as a virulence factor by suppressing plant defense (Ma et al. 2018).

During *L. maculans* Infection, AvrLm1 is secreted into the apoplast and resides in the plant extracellular space during *Brassica* infection. To monitor AvrLm1 during *L. maculans* infection, AvrLm1 driven by its native promoter was tagged with the fluorescent protein mCherry and transferred to *L. maculans* isolate 3R11 (3R11: AvrLm1-mCherry). An additional transgenic isolate carrying the fluorescent gene YFP driven by a fungal constitutive promoter was generated to serve as a control. To ensure that addition of mCherry tag to AvrLm1 did not disrupt its Avr function (being recognized by LepR3) or the overall fitness of 3R11 isolate, the 3R11: AvrLm1-mCherry was inoculated on the susceptible *B. napus* cultivar Topas DH16516 and the Topas: LepR3 transgenic line NLA8-2 (Larkan et al. 2013). The

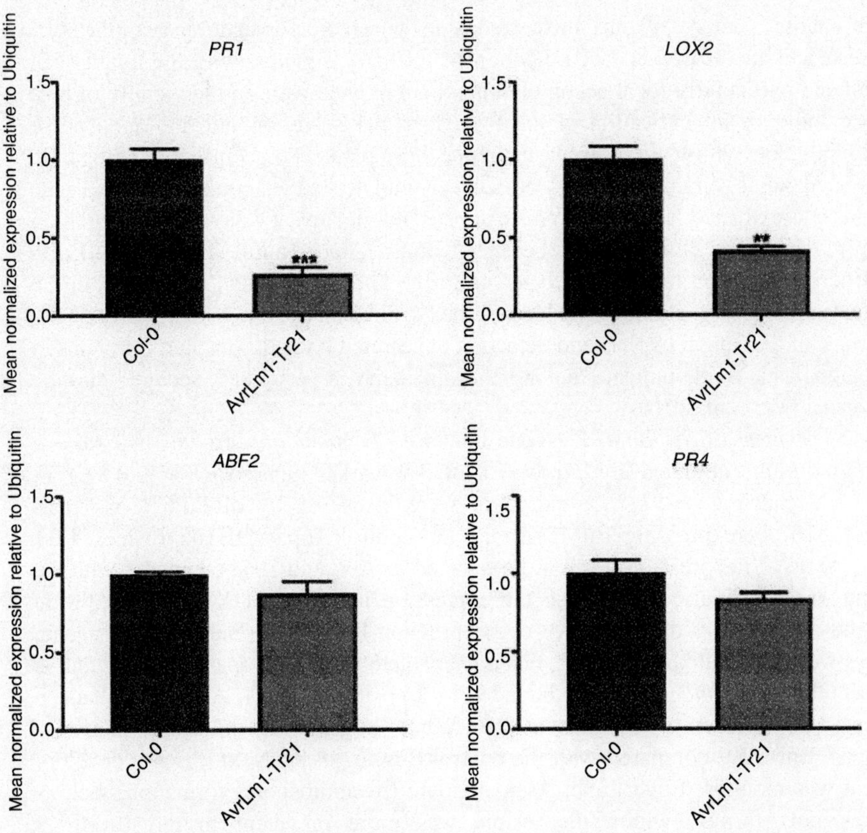

Fig. 6.2 Relative expression of salicylic acid (SA), ethylene (ET), jasmonic acid (JA), and abscisic acid (ABA) marker genes in *AvrLm1*-expressing *Arabidopsis* lines. Leaves of 4-week-old plants were collected for RNA extraction. Expressions of *Arabidopsis* defense marker genes including PR1, LOX2, ABF2, and PR4 were determined by quantitative reverse-transcriptase polymerase chain reaction and normalized to *Arabidopsis* ubiquity gene. Values represent means *G* standard error (SE) from three independent experiments. ** and ***represent the significant differences ($P < 0.01$, and $P < 0.001$, respectively, a two-tailed Student's *t* test) (Ma et al. 2018)

3R11:AvrLm1-mCherry was virulent on Topas DH16516. As anticipated, the transgenic strain 3R11:AvrLm1-mCherry was avirulent on the LepR3-transformed NLA8-2 line, confirming that addition of mCherry tag did not interfere with the function of AvrLm1. Subsequently, 3R11:AvrLm1-mCherry was inoculated on the host Topas DH16516 plants for the localization study.

Infected cotyledons of Topas DH16516 were examined by confocal microscopy, and red fluorescent signals of the tagged AvrLm1 were detected along the cell wall of the hyphae.

Intense focal fluorescent spots were observed where intercellular hyphae were in contact with the host mesophyll cells. In contrast, yellow fluorescent signals in the control 3R11-YFP were more uniformly observed throughout the fungal hyphae. It was unable to detect AvrLm1-mCherry inside the cell or apoplastic space of the host-infected tissue. However, accumulation of mCherry signals along the hyphal cell wall and particularly focal accumulation where hyphae were in the vicinity of host cells could be an indication of localized AvrLm1 secretion into the apoplast or accumulation within the cytoplasm of the host plant cells. Further investigation revealed that if AvrLm1-mCherry protein accumulates in the apoplast, the apoplastic fluids was extracted from *L. maculans*-infected cotyledons at 4 days' postinoculation (dpi). Apoplastic fluid was probed with anti-mCherry antibody, and AvrLm1-mCherry fusion protein was detected from the apoplastic fluids isolated from the infected plants. The studies revealed that the localization of AvrLm1-mCherry on the surface of the fungal hyphae and detection of secreted AvrLm1-mCherry proteins in the apoplastic fluids indicate that AvrLm1-mCherry is very likely secreted into the apoplast (Ma et al. 2018).

Overexpression of *BnMPK9* gene enhances *L. maculans* growth in *Brassica napus* during pathogenesis. Expression of SA marker genes *PR1* and *WRKY70*, and JA-related genes *LOX3* and *AOS*, were measured by qRT-PCR in Topas DH16516 overexpressing BnMPK9 and the control Topas DH16516 lines. PR1 and WRKY70 expression levels were reduced to 70% and 30% compared with the control, respectively. In contrast, the expression levels of LOX3 and AOS were slightly elevated in the BnMPK9 overexpression lines. Two BnMPK9 transgenic lines were inoculated with the *L. maculans* isolate v23.1.3. TopasDH16516 that is susceptible to *L. maculans* was included as a positive control. At 8 dpi enhanced disease development was evident as formation of larger lesion on the cotyledons of Topas: BnMPK9 compared with the control (Fig. 6.3a, b). Next, disease development was monitored over time. Asexual spore (pycnidiospore) formation, seen as black dots formed within the lesion, was more prevalent in the BnMPK9 overexpression lines at 10, 12, and 14 dpi (Fig. 6.3a). Furthermore, the amount of fungal DNA in the cotyledons of BnMPK9-overexpresing and wild-type Topas lines was quantified by quantitative PCR (qPCR) analysis of *L. maculans'* LmITS1 DNA at 8 dpi. Higher levels of *L. maculans* DNA in the Topas: BnMPK9 provided additional support that overexpression of BnMPK9 enhances the susceptibility of *B. napus* to *L. maculans* (Fig. 6.3c; Ma et al. 2018).

6.8 *Brassica–Plasmodiophora* Host-Pathosystem

6.8.1 Role of Aliphatic Glucosinolates Metabolism in Clubroot Pathogenesis

Due to the complicated composition of GSLs, their exact role in clubroot disease development remains unclear. The difference in the infection process between *Matthiola incana* L. (resistant) and *Brassica napus* L. (susceptible) was determined by Xu et al. (2018). Root hair infection was definitely observed in both resistant and susceptible hosts, but no infection was observed during the cortical infection stage in resistant roots; this finding was verified by molecular detection of *P. brassicae* via

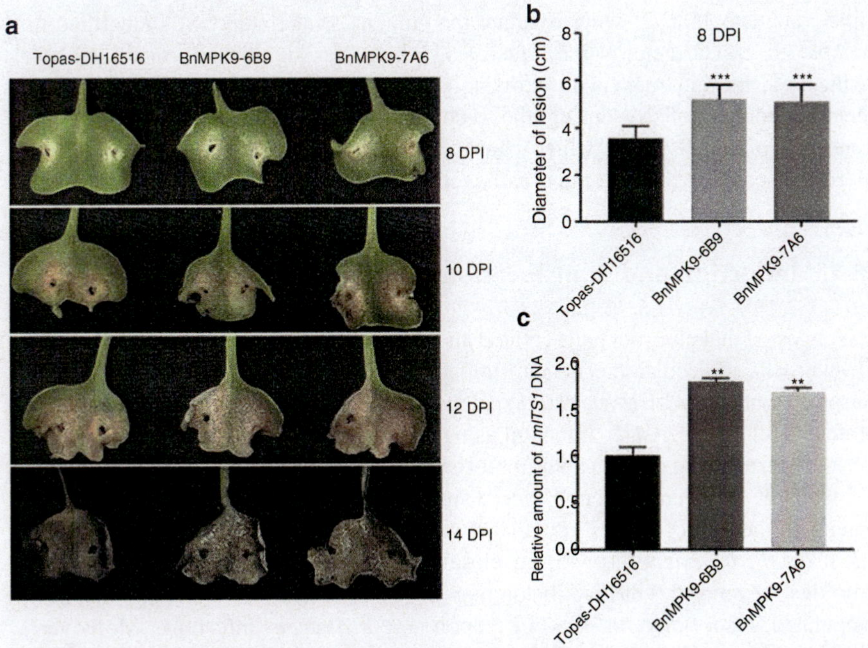

Fig. 6.3 Expression of BnMPK9 results in increased susceptibility to *Leptosphaeria maculans*. Cotyledons of 7-day-old Topas DH16516 seedlings and two BnMPK9-expressing lines (BnMPk9-6B9 and BnMPK9-7A6) were inoculated on wound sites using a spore suspension of *L. maculans* isolate v23.1.3. (**a**) Representative cotyledons at 8, 10, 12, and 14 dpi were photographed. (**b**) The size of the lesions was measured at 8 dpi and statistically analyzed. Values represent means G standard error (SE) ($n > 30$). Asterisks indicate statistically significant differences (***, $P < 0.001$, one-way analysis of variance [ANOVA]). Three independent biological experiments were conducted. (**c**) qRT-PCR analysis showing the fungal mass in *L. maculans*-infected cotyledons of *B. napus* at 8 dpi. Constitutively expressed *LmITS1* gene was used as a marker showing DNA level and pathogen mass. *LmITS1* gene expression levels were normalized to expression of *B. napus* actin. Values are means G SE of triplicate reactions of three independent biological samples. Significant differences are represented by ** (**, $P < 0.01$) (Ma et al. 2018)

PCR amplification at various times after inoculation. Based on the time course detection of the contents and compositions of GSLs after *P. brassicae* inoculation, susceptible roots exhibited increased accumulation of aliphatic, indolic, and aromatic GSLs in *B. napus*, but only aromatic GSLs were significantly increased in *M. incana*. Gluconapin, which was the main aliphatic GSL in *B. napus* and present only in *B. napus*, was significantly increased during the secondary infection stage. Quantification of the internal jasmonic acid (JA) concentration showed that both resistant and susceptible plants exhibited an enhanced level of JA, particularly in susceptible roots. The exogenous JA treatment induced aliphatic GSLs in *B. napus* and aromatic GSLs in *M. incana*. The jasmonic acid (JA)-induced aromatic GSLs may be involved in the defense against *P. brassicae*, whereas aliphatic GSLs induced by JA in *B. napus* likely play a role during the secondary infection stage.

Three candidate *MYB28* genes regulate the content of aliphatic GSLs identified in *B. napus*; one such gene was *BnMYB28.1*, which was significantly increased following both the treatment with exogenous JA and *P. brassicae* inoculation. The increased content of JA during the secondary infection stage may induce the expression of *BnMYB28.1*, which caused the accumulation of aliphatic GSLs in clubroot disease development and pathogenesis (Xu et al. 2018).

6.8.2 Induction and Translocation of Sugars in Clubroot Tissues

It is suggested that clubroot galls created an additional nutrient sink in infected roots. However, the molecular mechanism underlying *P. brassicae* infection and sugar transport is unclear. Li et al. (2018) analyzed sugar contents in leaves and roots before and after *P. brassicae* infection using a pair of Chinese cabbage near-isogenic lines (NILs), carrying either a clubroot resistant (CR) or susceptible (CS) allele at the *CRb* locus. *P. brassicae* infection caused significant increase of glucose and fructose contents in the root of CS-NIL compared to CR-NIL, suggesting that sugar translocation and *P. brassicae* growth are closely related. Among 32 *B. rapa SWEET* homologs, several *BrSWEETs* belonging to Clade I and III were significantly upregulated, especially in CS-NIL upon *P. brassicae* infection. Moreover, *Arabidopsis sweet11* mutant exhibited slower gall formation compared to the wild-type plants. It suggests that *P. brassicae* infection probably triggers active sugar translocation between the sugar producing tissues and the clubbed tissues, and the *SWEET* family genes are involved in this process (Li et al. 2018).

6.8.3 Role of Host Hormone Metabolism in Clubroot Development

The root cortex infection is accompanied with increased divisions and hypertrophy of the infected plant tissue leading to the gall phenotype (Ludwig-Muller et al. 2009). Therefore *P. brassicae* exploits existing meristematic activities of the host (Malinowski et al. 2012). *Arabidopsis* mutants impaired in ethylene synthesis develop larger clubs than wild type but other hormones are most likely more important (Knaust and Ludwig-Muller 2013). Auxin, cytokinin, and brassinosteroids are involved in controlling cell division and cell elongation in plants. The metabolism for all three hormones is altered in *P. brassicae*-infected plants (Siemens et al. 2006; Ludwig-Muller et al. 2009; Schuller et al. 2014). Cytokinin synthesis of *P. brassicae* plasmodia using adenine as a precursor has been demonstrated (Muller and Hilgenberg 1986). According to the genome data, there are two genes encoding for proteins with homologies to plant cytokinin biosynthesis genes (PbIPT1 and PbIPT2) (Schwelm et al. 2015). However, the plant proteins with highest homologies produce cis-zeatin and not trans-zeatin as described for *P. brassicae* (Muller and Hilgenberg 1986). The involvement of those two candidates' genes in cytokinin production remains to be shown. A sequence similar to a plant cytokinin oxidase/dehydrogenase has also been found in the genome of *P. brassicae* but a role for this cytokinin-degrading enzyme cannot be envisioned yet. Besides stimulating cell division and cell elongation in a healthy plant, auxins can also induce tumors after pathogen infections. Thus, the control of auxin levels significantly contributes to the defense network in plants (Ludwig-Muller 2015).

Several plant pathogens modify SA content in their hosts (Vlot et al. 2009). One *Arabidopsis* GH3 proteins also conjugates SA to various amino acids (Zhang et al. 2007). However, the PbGH3 showed no activity with SA as a substrate in vitro (Schwelm et al. 2015). In contrast, the secreted *P. brassicae* methyltransferase (PbBSMT) modifies SA to methylated SA (Ludwig-Muller et al. 2015). PbBSMT could be involved in the modulation of plant defense. Clubroot symptoms can be reduced by treatment with SA shortly after inoculation (Agarwal et al. 2011; Lovelock et al. 2013). The *PbBSMT* is already expressed 4 days after inoculation supporting an important role for the encoded protein from early on in the infection process. The *PbBSMT* gene is expressed during the complete infection process and one of the highest expressed genes in transcript me libraries of three *Brassica* hosts and during spore formation (Schwelm et al. 2015). A homologue was also detected in the transcriptome of *S. subterranea* suggesting that SA methylation might be a common strategy in the plasmodiophorids. In *P. brassicae,* the expression of PbBMST does also correlate with the synthesis of spore cell walls. As the cell walls contain chitin, PbBMST could play a role in minimizing the chitin triggered defense responses of the host (Schwelm et al. 2016).

6.8.4 *Plasmodiophora* Induces Alterations of Host Cell Wall Rigidity by Metabolic and Physiological Changes

The infection of *Plasmodiophora brassicae* results in molecular and physiological changes in the host root growth which results in gall formation. This is accompanied by altered metabolism, physiology, and modifications of the host cell walls (Dekhuijzen and Overeem 1971; Kavanagh and Williams 1981; Ludwig-Muller et al. 1993; Devos and Van-Oosten 2005; Ludwig-Muller et al. 2009; Malinowski et al. 2012). The growth of the host cells is resulting in surface increase; however, this is not automatically coupled to addition of new cell wall components (Cosgrove 2014). In clubroot-infected cells, the biosynthesis of wall components like cellulose, hemicelluloses, pectin (primary cell wall), and also lignin (secondary cell wall), which together form the load-bearing structures in the cell wall, is downregulated. Genes involved in lignin biosynthesis are also downregulated in infected *B. napes* (Cao et al. 2008; Ludwig-Muller et al. 2009). These results suggest an inhibition of cell wall component biosynthesis in clubroot-infected tissue. The observed downregulation of genes coding for biosynthesis of cellulose synthases (CesA), pectins (RHM, QUA), and hemicellulose (Csl genes) in infected plants supports this finding, which indicates that the cell walls of clubroot-infected cells are less rigid than their counterparts in uninfected plants. Also transcripts of genes coding for cell wall degrading enzymes (CWDEs) like pectinases, polygalacturonases (PGs), pectate lyases (PLs), pectin methylesterases (PMEs), and cellulases with hydrolase activity are upregulated (some strongly) in infected.

The *P. brassicae* is responsible for a decrease in root cell wall rigidity and stability in clubroots via downregulation of the biosynthesis of primary and secondary cell wall components like cellulose, pectin, callose, or lignin. On the other hand, enzymes which degrade these structural elements are upregulated in clubroots. This loss in stability does fit the biological development of *P. brassicae*, because the multinucleate plasmodia gradually fill the significantly enlarged host cells. By reducing the rigidity of the cell wall, *P. brassicae* can growth a larger size as the barrier posed by the cell wall gets more flexible. The processes seen at the transcriptomic level resemble the processes described during wall stress-related growth as processes that control rigidity, but also extensibility is differentially regulated—with the main aim to withstand increasing turgor pressure without rupture of the cell (Cosgrove 2016), while increased torpor-like pressure is built up by the growing parasite. That decreased rigidity of the cell wall has an important role for clubroot symptom development. This is supported by the finding that clubroot-tolerant plants show transcriptomic signatures of increased cell wall stability (Lahlali et al. 2017). The limited arsenal of CWDEs of *P. brassicae* has been notable when the genome of this parasite was analyzed (Schwelm et al. 2015; Rolfe et al. 2016). A lack of CWDEs is on the one hand a feature that is seen in many biotrophic plant parasites (Kamper et al. 2006). But on the other hand, the transcriptional changes observed in clubroot suggest that *P. brassicae* is able to compensate this by reprogramming the host metabolism. In addition to a reduced rigidity of the plant cell wall strength, depolymerized pectin can serve as carbon source for pathogen development (Jia et al. 2009) and increases the accessibility of other plant cell wall components like cellulose and hemi-celluloses to CWDEs, which can enable further and faster cell wall breakdown (Xue et al. 2018). This again would tie in very well with the biology of *P. brassicae*, which is obtaining nutrients from its host. So, an increased availability of cell wall degradation products (sugars) can provide energy for *P. brassicae*. In order to protect the plant cells from excessive degradation of wall components through pectinases like PMEs or PGs, the host is able to produce enzyme-inhibiting enzymes, called PMEIs (PME-inhibitor) and PGIPs (PG-inhibiting protein). These enzymes modulate PME sand PGs dynamically, and genes coding for these inhibitors were all downregulated in clubroot tissue. So, as expected from a biotroph, *P. brassicae* induces changes in its host via inhibiting a cell wall-related plant defense mechanism. Plant genes involved in the metabolism of sugars that are building blocks of cell wall components were upregulated in infected roots. Transcripts of genes coding for sucrose synthase (SuSy), which catalyzes the reaction from sucrose to UDP-glucose and fructose are one example. The UDP-glucose is needed as substrate for cellulose synthesis using CesA and for callose synthesis using CalS (Schneider et al. 2016). Genes coding for UDP-glycerinate decarboxylase (UXS), which produces UDP-xylose, a substrate for hemicelluloses and pectin, are upregulated in clubroot transcriptomes. Upregulated sugar metabolism could also be shown in several other studies of clubroot-infected *A. thaliana* (Siemens et al. 2006, 2011; Walerowski et al. 2018). It can be hypothesized that the increase in plant transcripts coding for CWDEs likely results in increased sugar availability via the degradation of cell wall components. On the other hand, the increase of transcripts belonging to sugar metabolic pathways likely results in an increased production of cell wall related sugars such as UDP-glucose or UDP-xylose. Overall it can be assumed that these sugars might be taken up by *P. brassicae* as energy source (Badstober et al. 2020).

Cell wall disruption can happen with the aid of CWDEs as described above, or via the enzymatic modification of wall rigidity and stability with expansins and xyloglucan endo-transglycosylases/hydrolases (XTHs). Expansins and XTHs are directly involved and responsible for such modifications. Transcripts of genes coding for expansins and for XTHs are significantly up-regulated (Siemens et al. 2006; Irani et al. 2018). XTH activity was also detected in *B. rapa* by Devos et al. (2005). The XTHs have two functions, either as xyloglucan endo-hydrolase (XEH), which cleaves xyloglucan chains, or as xyloglucan endo-transglucosylase (XET), which is able to re-anchor two xyloglucan microfibrils, which have ruptured as result of turgor driven cell wall disruption (Rose et al. 2002). The majority of *B. oleracea* genes that were upregulated in clubroots were annotated as XTHs possessing both functions (cleavage and re-anchoring). Notably, transcripts of genes that only feature the re-anchor activity (XET) were downregulated. In other words, when *P. brassicae* plasmodia are growing in the infected host cell, they might activate certain aspects of turgor driven cell growth in their host (up 385 regulation of expansins and XTHs with XEH and XET activity), but inhibit the plants ability tore-in force the cell wall (downregulation of XTHs with only XET activity). Other genes involved in cell expansion and alteration of wall rigidity with important roles in cell wall organization are genes coding for SHV3 (SHAVEN3), and SVL1 (SHV3-like). *Arabidopsis thaliana* SHV3 and SVL1 double mutants resulted in further defects like abnormal cell expansion, bloat guard cells, as well as ectopic lignin deposits. Also pectin modifications and altered cellulose contents were observed (Hayashi et al. 2008). Related genes were upregulated in clubroots (Badstober et al. 2020).

Wall-associated kinases (WAKs) belong to the receptor-like kinases with a cytoplasmic protein kinase domain. They are located in the plasma membrane and serve as pectin receptors (cell wall side), for longer pectins of native cell walls and short pectic fragments of oligogalacturonic acid (OGs) (Kohorn 2016). The WAKs do not just bind pectins; they also recognize pectin fragments, which can be produced by pectin-degrading plant pathogens. This gene family is also involved in triggering stress response cascades in the host plant (Sivaguru et al. 2003; Kohorn and Kohorn 2012). The WAK genes are also discussed to be involved in cell elongation, because WAK mutants with lower levels of WAK proteins led to loss of cell expansion (Lally et al. 2001), shorter roots and a reduction in vacuolar invertase what might control expansion indirectly through torpor pressure (Kohorn et al. 2006, 2009; Kohorn and Kohorn 2012). In clubroot-infected plants, all genes coding for WAKs were upregulated which fits very well to the above-discussed increase in cell expansion. Overall, stress response is downregulated in clubroots (Zhao et al. 2017), so an influence in stress response is not clear. All genes encoding WAKLs are downregulated in clubroots, what might influence resistance

410 negatively of host plants, because WAKLs are supposed to play a role in pathogen resistance (Verica and He 2002).

Among other physiological changes, clubroot infection leads to an altered biosynthesis of cell wall components. Genes coding for proteins involved in cellulose, hemicellulose, pectin, callose, and lignin biosynthesis like CesAs (cellulose synthase), Csls (cellulose synthase-like), and CalS/GSLs (callose synthase) are downregulated in infected plants. Also genes involved in pectin synthesis like Rhm and Qua1, or in the initial steps of lignin biosynthesis (PAL, C4H, and 4CL) are downregulated. Clubroot infection also leads to degradation and modification of existing wall components what results in decreased wall rigidity. Transcripts belonging to degrading/modifying enzymes, e.g., of pectins (PGs, PLs, PMEs, PAEs), cellulose through cellulases with hydrolase activity, or callose through degradation and deposition are upregulated in clubroots. Gene expressions for enzymes with degradation inhibiting activity are downregulated in infected tissue. Clubroot infection also leads to massive host cell expansion, as observed in transcriptomic data. Genes coding for XTHs and expansins, which allow cell wall loosening are upregulated. Also for WAK/L genes, which are involved in cell elongation and defense response, gene expression differed. Elongation processes are upregulated and defense response was downregulated in clubroot. Genes involved in plant defense mechanisms leading to reinforcement of cell walls (e.g., callose synthesis, lignification) are downregulated in clubroots, whereas deposition of callose is upregulated. All these processes concur with observations during clubroot development, and the typical clubroot phenotype is a result of efficient strategies of *P. brassicae* to influence host biology (Badstober et al. 2020).

6.8.5 Genomics of Nutrient Acquisition Pathways

The *P. brassicae* genome contains a complete molybdopterin biosynthetic pathway but the thiamine biosynthetic pathway is absent. Eleven putative thiamine transporters were identified within the Pb3 genome, consistent with the acquisition of this essential vitamin from its host. The genome encoded a sulfite oxidase but nitrate reductase was missing, with nitrogen acquired from the host in more complex forms. The shikimic acid pathway to chorismate was complete with incomplete evidence supporting the synthesis of Tyr and Phe. However, the branch via anthranilate to Trp was absent. The His and Val/Leu/Ile pathways were also absent. Incomplete pathways were present for Thr, Met, Arg, and Lys. A complete understanding of the biosynthetic capacity of *P. brassicae* will require further investigation, but it is clear that this obligate biotroph has lost the capacity to synthesize a number of amino acids and thus, must acquire them from the host. The predicted transcriptome contained 10 putative amino acid transporters. Inspection of the *P. brassicae* genome showed a complete glycolytic and TCA cycle, indicating a capacity for the utilization of carbohydrate from the host for energy production. However, unlike many biotrophic pathogens (including *P. brassicae, infestans*), there was no evidence for extracellular invertase genes that would degrade host apoplastic sucrose. Therefore, *P. brassicae* is likely to take up sugars directly. The genome contained 37 ABC transporters, some of which may be involved in nutrient acquisition. The PbPT3Sc00087_Am_0.68_1 showed homology (28%, *E*-value = 6e−11) to a SWEET sugar permease from *Ascaris suum*, PbPT3Sc00016_A_10.333_1 was predicted to encode a sucrose proton transporter, while PbPT3Sc00069_Am_1.146_1 and PbPT3Sc00030_A_2.294_1 encode putative monosaccharide transporters. The predicted sucrose transporters were expressed most strongly in developing galls (weeks 4–6 in *B. napus* and 26 DPI in *A. thaliana*) indicating that these genes may play a role in the development of sink strength during gall formation.

During the annotation process, in which predicted gene sequences were compared with known proteins in the KEGG, a series of matches were found against proteins with prokaryotic substrate binding domains. Inspection of these sequences found that these encoded a hitherto unknown class of eukaryotic transporters in which a prokaryotic substrate binding domain is fused to a 7 alpha-helical transmembrane domain. No such fusion proteins were found in any other eukaryotic genome suggesting that this class of transporter is unique to *P. brassicae*. The gene sequences are present within the e3 genome but either not annotated as genes or annotated as two separate genes. However, transcriptome analysis shows clearly that these are single coding regions in each case. Although characterization of the substrates transported by these proteins requires experimental confirmation, sequence homology suggests that PbPT3Sc00003_Am_0.106_1, PbPT3Sc00030_S_12. 269_1, PbPT3Sc00030_Am_12.97_1 and PbPT3Sc00018_Am_0.143_1 encode sugar (trehalose, maltose or oligosaccharide) transporters, while PbPT3Sc00024_A_2.266_1a, PbPT3Sc00024_ A_ 2.266_1b, PbPT3Sc00024_G_2.43_1 and PbPT3Sc00100_Sm_1.225_1 encode phosphate transporters. The majority of these transporters were expressed early in infection and during gall development, suggesting they play roles in nutrient acquisition at these key developmental stages (Rolfe et al. 2016). In common with many pathogens, *P. brassicae* has been shown to synthesize trehalose. The Pb3 genome contained 3 genes encoding trehalose phosphate synthase (TPS) (PbPT3Sc00016_A_8.299_1, PbPT3Sc00018_Sm_4.147_1, PbPT3Sc00059_A_1.289_1) and single genes encoding trehalose phosphate phosphatase (PbPT3Sc00030_A_6.312_1), and trehalase (PbPT3Sc00031_A_12.282_1) in agreement with the TPS gene PbPT3Sc00059_A_1.289_1 matched the fragment isolated by (Brodmann et al. 2002) and the corresponding complete gene reported elsewhere (Bulman et al. 2007).

6.8.6 Prediction of New Metabolic Pathways During *Plasmodiophora* Pathogenesis

New biochemical pathways and annotation data have been predicted by Daval et al. (2019) using *P. brassicae* crucifers host-pathosystem. A complete pathway for the spermidine biosynthesis was predicted in eH *P. brassicae*. Particularly, the three enzymes (EC 2.7.2.11, EC 1.2.1.41, and EC 2.6.1.13) required for conversion of glutamate to ornithine, were present in the eH genome although they were not found in the genomes described by Rolfe et al. (2016) and Schwelm et al. (2015). Ornithine is converted into putrescine, a precursor of the polyamine spermidine. The roles of polyamines synthesized by plants are well documented. They are involved in fundamental cellular processes, such as chromatin organization, cell proliferation, differentiation, and programmed cell death, but also in adaptive responses to abiotic, and biotic stresses. Polyamines have been found being accumulated in the galls caused by *P. brassicae*. In turnip (*Brassica rapa* var. *rapa*), the levels of putrescine, spermidine, and spermine were reported to be higher in the roots infected with clubroot than in the noninfected normal roots. In cabbage, broccoli, and Komatsuna clubroot galls, the levels of putrescine, spermidine, and spermine also increased after infection with *P. brassicae*. The high level of polyamines measured in the roots may be contributed by the host plant, as reported by Cao et al. (2008) who showed the upregulation of the *B. napus* polyamine biosynthetic enzyme spermidine synthase in response to *P. brassicae* infection. However, it could also be due to *P. brassicae,* since it is difficult to distinguish between metabolites of plant or pathogen origin. Polyamine biosynthesis could be vital to the survival and pathogenesis of *P. brassicae* as described in other pathogens, such as bacteria, fungi, and protists.

The chorismate/Shikimate pathway, which was predicted by Schwelm et al. (2015) but lacking the chorismate mutase (EC 5.4.99.5), Daval et al. (2019) found to be complete in the eH *P. brassicae* genome suggesting that this pathway is operational in *P. brassicae*. This pathway is known in plants, bacteria, fungi, and apicomplexan parasites, and is central for the biosynthesis of carbocyclic aromatic compounds. Chorismate is the common branch point for the production of these metabolites, and some intermediates in these pathways also have important physiological relevance. Salicylate, which is derived from chorismate, is involved in plant defense against *P. brassicae*. Anthranilate, another derivative of chorismate, is the precursor of the *Pseudomonas quinolone* signal, which regulates numerous virulence factors in *Pseudomonas aeruginosa*. Further, the chorismate mutase enzyme of this pathway, newly identified in the eH genome, was described as being involved in gall formation in *Meloidogyne javanica* and in *Ustilago maydis*. The potential role of this enzyme in *P. brassicae* pathogenicity would require functional studies.

Clubroot Cyc also predicted the L-Dopachrome biosynthesis pathway, which is involved in melanin synthesis. Melanin is a broad term for a group of natural pigments found in most organisms and localized in cell walls of different structures of fungi. They represent well-known virulence factors for several pathogenic fungi and bacteria. So Clubroot Cyc, a freely accessible tool that can be updated with newly published results, provides a way to easily visualize omics data and represent metabolic pathways potentially involved in particular experimental conditions of interest. Metabolic pathway prediction in *P. brassicae* is still, however, incomplete and poorly annotated because there is still not enough information in the reference databases. Further progress in metabolic network reconstruction and modeling, appropriate gap-filling methods are required, and software is being developed to identify and add lacking pathways, such as Meneco, a tool dedicated to the topological gap-filling of genome-scale draft metabolic network (Prigent et al. 2017; Daval et al. 2019).

6.8.7 Induction of Signaling Molecules and Genes During Gall Formation

The *Plasmodiophora brassicae* establishes an intricate interaction with their hosts during at least some parts of the infection process, because of their dependence on host carbon sources. They influence host physiology and alter host regulatory networks over a wide range of its genome. Especially the plant's hormonal balance is altered by this interaction. The changes in host hormone metabolism are connected to the intracellular life style of these protists (Ludwig-Muller et al. 2009). The infection process of plants by *P. brassicae* consists of two phases: the primary phase, which is restricted to root hairs and the secondary phase, which occurs in the cortex and stele of roots and hypocotyl and leads to abnormal development. Especially during this later phase, the host root responds to infection by increased cell division rates followed by hypertrophy of infected cells. These harbor first the plasmodia of *P. brassicae*, probably dividing together with the host cells. Later, the plasmodia grow and the host cell increases concomitantly in size. These enlarged host cells can reach at least 10 times the size of uninfected cells. While cell division has been attributed to the action of auxins, and cytokinins, cell enlargement has been so far linked exclusively to higher auxin concentrations, and synthesis. The induction of auxin in *Arabidopsis* correlates with an increase in seedling growth and Xyloglucan-Endo-Transferase/Hydrolase leading to cell elongation. In addition, a microarray revealed that genes involved in cell division, and expansion such as cell cycle genes, and expansins are upregulated, especially at the first analyzed time point of the disease (10 days after inoculation). Consequently, altering hormone concentrations has led to reduced clubroot symptoms. Auxin signaling pathway (s) contributes to clubroot formation (Siemens et al. 2006; Devos et al. 2005; Pasold et al. 2010; Neuhaus et al. 2000). Auxin signaling is regulated by two types of receptors: the nuclear-localized TIR/AFB family and the plasma membrane associated auxin binding protein 1 (ABP1) (Dharmasiri et al. 2005; Spartz and Gray 2008; Christian et al. 2006). The nuclear signaling pathway leads to the activation of the transcriptional response via auxin perception by the F-box protein TIR1 (transport inhibitor response 1). F-box proteins function as substrate recognition modules for the multisubunit complex of ubiquitin ligases (also called Skp1-Cullin1-F-box protein (SCF); SCFTIR). The F-box protein TIR1 is the receptor, which recruits, by binding to auxin, the protein target that is designated for degradation by ubiquitination. Polyubiquitinated proteins are then transferred into the 26S proteosome, and degraded.

Auxin-mediated transcriptional response is controlled by negative (Aux/IAA), and mainly positive (ARF) regulators. ARFs (auxin response factors) bind to the auxin responsive elements (AuxRe) in respective promoters of auxin-inducible genes for transcriptional activation, but some ARFs are also inhibitors (Guilfoyle and Hagen 2007). Both, Aux/IAAs, and ARFs are present as large families. The Aux/IAA proteins bind to the ARFs through homologous domains in both proteins and thereby repress auxin-regulated transcription (Reed 2001). The Aux/IAA proteins are short-lived, because they are degraded via the SCFTIR pathway and their degradation is promoted by auxin. Many mutations in *Aux/IAA* genes are stabilizing the resulting protein, because degradation domains are affected and thus constitute gain-of-function mutations (Reed 2001). Among the targets of ARFs are genes encoding proteins involved in the regulation of auxin homeostasis. The IAA amino acid conjugate synthetases (GH3) have been first recognized as auxin-induced genes. Later, it was discovered that it is a large gene family and that some proteins are involved in the conjugation of IAA to various amino acids. GH3 proteins are involved in various responses of plants to abiotic and biotic stresses. Especially the protein GH3.5 seems to play various roles in addition to the synthesis of IAA amino acid conjugates. It also conjugates the plant defense signal salicylic acid to amino acids and is involved in the synthesis of the *Arabidopsis* phytoalexin camalexin. One family member, GH3.11 (JAR1), catalyzes the formation of the isoleucine conjugate of jasmonic acid (JA). Contrary to the inactivation of IAA by conjugation, the isoleucine conjugate of JA is active, because it alone binds to the JA receptor COI1 (Hagen and Guilfoyle 2002; Park et al. 2007; Zhang et al. 2007).

The ABP1 has been shown to control events at the plasma membrane by the regulation of membrane fluxes leading in consequence to cell elongation. ABP1 has an ER retention sequence, but it is hypothesized that for action at the plasma membrane some protein escapes its intracellular location. In addition, there is no transmembrane sequence present in ABP1, which led to the speculation that a docking protein could be responsible for anchoring ABP1 in the membrane. ABP1 is involved in protoplast swelling, which is an indication for its role *in planta* in cell elongation. In addition, many growth processes seem to rely on ABP1 activity, for example an ABP1 knockout mutant is embryo lethal (Chen et al. 2001). Downstream of ABP1, the action of ATPases and ion channels was postulated. Both, H^+-ATPase and K^+-channels are needed for the auxin-mediated cell elongation response. K^+-channels necessary for the osmotic changes occurring during cell elongation can be induced by auxin in maize and *Arabidopsis* (Fuchs et al. 2006). In contrast, the TIR pathway can be responsible for the increased gene expression for these channels in response to auxin, linking the two pathways. Initial evidence for the involvement of the TIR pathway in the clubroot disease came from Alix et al. (2007) who reported the partial resistance of the *Arabidopsisaxr3-1* mutant. The *AXR3* gene encodes the transcriptional repressor IAA17, a member of the Aux/IAA family. The transcript is stabilized in the mutant, so that the transcriptional activation cannot take place. The contribution of both auxin signaling pathways to the clubroot symptom development and their roles during disease progression has been revealed by Jahn et al. (2013).

Two different classes of auxin receptors, the TIR family, and the auxin binding protein 1 (ABP1) in *Arabidopsis thaliana* are transcriptionally upregulated upon gall formation. Mutations in the *TIR* family resulted in more susceptible reactions to the root pathogen. As target genes for the different pathways have been investigated for the transcriptional regulation of selected transcriptional repressors (*Aux/IAA*), and transcription factors (*ARF*). As the TIR pathway controls auxin homeostasis via the upregulation of some auxin conjugate synthetases (GH3), the expression of selected *GH3* genes also show in most cases upregulation. A double *gh3* mutant showed also slightly higher susceptibility to *P. brassicae* infection, while all tested single mutants do not show any alteration in the clubroot phenotype. As targets for the ABP1-induced cell elongation the effect of potassium channel blockers on clubroot formation was investigated. Treatment with tetra-ethyl ammonium (TEA) resulted in less severe clubroot symptoms. This research provides evidence for the involvement of two auxin signaling pathways in *Arabidopsis* needed for the establishment of the root galls by *P. brassicae.*

6.9 *Brassica–Sclerotinia* Host-Pathosystem

6.9.1 Metabolites Associated with *Sclerotinia* Pathogenesis

The formation of sclerotia entails extensive cellular changes and the mobilization and deposition of many materials. The factors involved in the initiation phase have received little attention. As pointed out by Willetts (1978), vegetative hyphae usually grow away from one another. In the formation of initials, there must be an attraction so that hyphal fusions occur. Because initials usually form after growth has covered the substrate, their formation may be a response to changes in nutrient availability. The fact that several fungi, including *S. sclerotiorum,* form sclerotia when mycelial mats are transferred to a medium low in nutrients supports this hypothesis (Christias and Lockwood 1973). Initials may form only after certain metabolic products are produced in the mycelium or in the medium. Organic acids, long known to be produced by *Sclerotinia* spp. may play a role in sclerotium formation (Humpherson-Jones and Cook 1977; Wang and Le Tourneau 1971). While the amount and kinds of organic acids vary with the isolate and the medium, oxalic, fumaric, malic, and succinic acids usually are produced by *S. sclerotiorum* (Corsini and Le Tourneau 1973; Humpherson-Jones and Cook 1977). Other acids, e.g., citric, glycolic, glyoxylic, and a-ketoglutaric also have been detected (Colotello 1973; Humpherson-Jones and Cook 1977). Enzymes for the production of Krebs cycle acids and oxalic acid are present in *S. sclerotiorum* (Corsini and Le Tourneau 1973; Maxwell 1973).

When *S. sclerotiorum* is grown on a liquid medium, the pH of the culture filtrate drops to pH 3–4 and then tends to increase slightly (Wang and Le Tourneau 1971). The drop in pH occurs at a time of maximum acid production and when the specific activity of Krebs cycle enzymes is high. Numerous sclerotia form when the fungus grows on a medium which supports good growth and the initiation of sclerotia tends to coincide with the pH drop. When the fungus is grown on a medium which supports very little growth or sclerotium formation, the pH of the culture filtrate usually increases (Wang and Le Tourneau 1971). The addition of oxalic and Krebs cycle acids to the medium does not enhance sclerotium formation (Humpherson-Jones and Cooke 1977, Wang and LeTourneau 1971). Once initials have formed, materials must be supplied for the synthesis of reserves and other compounds that are present in the mature sclerotium. Developing sclerotia are covered with droplets that are continuous with the interior of the developing sclerotium (Colotelo 1974). These droplets are covered by a membrane and contain a variety of inorganic and organic materials, including several enzymes, which change in concentration as the sclerotium matures (Colotelo 1973). The precise role of the exudate is still open to interpretation. It may play an osmotic role in translocation or it may be involved in the loss of water from the sclerotium during maturation (Willetts 1978).

Although many enzymes have been detected in sclerotia, it is not yet clear how the activity of these enzymes is controlled or coordinated or which enzymes may play key roles in morphogenesis. From results of polyacrylamide–gel electrophoresis studies, Wong and Willetts (1974) suggested that suppression of glycolysis, the Krebs cycle, and stimulation of the pentose phosphate pathway are involved during the compaction and maturation of sclerotia. Various enzymes of phenol oxidation have been detected (Colotelo 1973; Jones 1970; Wong and Willetts 1974). Tyrosinase may be involved in the initiation of sclerotia and other phenol oxidases may be involved in the formation of the darkened rind (Wong and Willetts 1974).

6.9.2 Identification of Cell Wall Degrading Enzymes of *Sclerotinia*

Plant pathogenic fungi can facilitate colonization of their hosts by the production of a wide array of CWDEs, including pectinases, β-1, 3 glucanases, glycosidases, cellulases, xylanases, and cutinases (Annis and Goodwin 1997). Many of these CWDEs can occur in multiple isozymes that differ in isoelectric point, molecular weight, processing, or transcriptional regulation (Keon et al. 1987), giving great flexibility to the pathogen for penetration and colonization of the host and facilitating the creation of a pool of assimilable nutrients. In culture, expression of most CWDEs is tightly regulated at the transcription level by the availability of carbon and/or nitrogen sources (Alighisi and Favaron 1995). Such a carbon catabolite repression mechanism permits the fungus to adopt its metabolism to the availability of glucose or other carbon sources it can metabolize. In addition, ambient pH can also regulate enzyme levels at the transcription level (Cotton et al. 2003; Rollins and Dickman 2001).

During the interaction with its host, *S. sclerotiorum* secretes a full complement of CWDEs that can facilitate penetration, macerate tissues and degrade plant cell wall components (Hancock 1966; Lumsden 1969; Riou et al. 1991). Pectin is a major constituent of the plant cell wall and pectinases produced by *S. sclerotiorum* play a role in pectin degradation. Pectin hydrolysis weakens the cell wall to facilitate penetration and colonization of the host while also providing the fungus carbon sources for growth. *S. sclerotiorum* produces several forms of pectinolytic enzymes that are capable of killing plant cells and macerating plant tissues on their own (Alighisi and Favaron 1995), suggesting a role in pathogenicity. The polygalacturonases (PGs) are important pectinases that can degrade unesterified pectate polymers, the structural polysaccharides found in the middle lamella and the primary cell wall of higher plants. PG activity in *S. sclerotiorum* is induced by pectin or pectin monomers, such as galacturonic acid, but is repressed by the presence of simple sugars (Fraissinet-Tachet and Fevre 1996, Riou et al. 1992a, b), PGs have been implicated as virulence factors through targeted gene disruption in a few pathosystem (Garcia-Maceira et al. 2001; ten Have et al. 1998; Kars et al. 2005; Shieh et al. 1997; Wagner et al. 2000) but conclusive evidence for a universal role in virulence is lacking (Gao et al. 1996; Scott-Craig et al. 1990).

EndoPGs are endo-acting enzymes that catalyze the hydrolysis of homogalacturonan while exoPGs cleave monomeric or dimeric glycosyl groups from the pectic cell wall polysaccharides, resulting in substrate fragmentation and the release of potential nutrients (Kars and van Kan 2004). A number of both endo- and exo PGs have been cloned and characterized in *S. sclerotiorum.* Molecular genetics have shown that fungi can produce multiple forms (isozymes) of endoPGs through post-translational (glycosylation) or postsecretional (proteolysis) modifications, or alternatively are derived from a multigene family (Bussink et al. 1992; Caprari et al. 1993; Fraissinet-Tachet et al. 1995).

Sclerotinia sclerotiorum also secretes a number of molecular forms of pectinases that exhibit the same enzyme activity (Riou et al. 1991). Several *S. sclerotiorum* purified acidic endoPGs display differences in isoelectric point but have a similar molecular weight, suggesting that differential glycosylation (which will affect the molecular mass without modifying the net charge of proteins) may have little influence on PG multiplicity in *S. sclerotiorum* (Waksman et al. 1991). Fraissinet-Tachnet et al. (1995) demonstrated that the multiplicity of pectinolytic enzymes and poly-galacturonase isozymes of *S. sclerotiorum* are encoded by a multigene family comprising seven members and constituting two subfamilies. Although the advantages of multiple copies of pectinase genes have yet to be demonstrated experimentally for this fungus, one can speculate that multiple copies of functionally related genes confer flexibility and adaptability to pathogen with such a wide host range. Furthermore, analyses of *S. sclerotiorum* endoPGs have shown differential expression during pathogenesis (Cotton et al. 2002; Kasza et al. 2004; Li et al. 2004). In a recent study by Kasza et al. (2004), Northern blot analysis revealed endoPGs *pg1*, *pg2*, and *pg3* which are expressed maximally during the phase of colonization of healthy plant tissues at 36 h postinoculation (hpi), but are not detectable at 96 hpi. RT-PCR show *pg6* and *pg7* which are detected from 24 hpi until the end of the time course experiment at 96 hpi, but *pg5* transcripts are only detected between 48 and 72 hpi which correlates with the final phase of maceration. The depletion of inducers such as polygalacturonic acid during the course of infection, carbolite repression by accumulation of end products or an acidic ambient pH may create a sequence of environmental conditions that act to activate or repress transcription of individual endoPG gene (Kasza et al. 2004).

Plants are known to produce cell wall-associated glycol–proteins that are capable of inhibiting fungal endoPGs called (PGIPs) polygalacturonase-inhibiting proteins (De Lorenzo et al. 2001). Oligogalacturonides released from the plant cell wall by the enzymatic activity of endoPGs have been shown to acts as endogenous elicitors of the hypersensitive response (HR) (Davis et al. 1986). *Phaseolus vulgaris* PGIP was shown to prevent the HR induced by *S. sclerotiorum* endoPGs (Zuppini et al. 2005). Nonpectinolytic CWDEs have received less attention in *S. sclerotiorum.* However, enzymes such as proteases, cellulases, and glucoamylases have been characterized for this fungus (Lumsden 1969; Martel et al. 2002; Poussereau et al. 2001a, b; Riou et al. 1991). As up to 10% of the plant cell wall consists of proteins (Carpita and Gilbeaut 1993), proteases may be an important aspect of *S. sclerotiorum* pathogenesis.

6.9.3 Differential Expression of Genes by *Sclerotinia* for Virulence

As common with many broad-host range pathogens, the molecular aspects of *S. sclerotiorum* pathogenicity generally studied have concentrated on the roles of hydrolytic cell wall-degrading enzymes (CWDEs). *Sclerotinia sclerotiorum* is known to produce several pectinases, including both endo- (*Sspg1*, *Sspg3*, *Sspg5*, and *Sspg6*) and exo- (*Ssxpg1* and *Ssxpg2*) polygalacturonases. Apart from polygalacturonases, gene disruption mutants of an arabinofuranosidase/β-xylosidase precursor (*Ssaxp*) and an endo-β-1, 4-xylanase (*SsXyl1*) showed either reduction or loss of virulence, indicating their importance as virulence factors (Yang et al. 2016). Oxalic acid (OA) has been the most-studied *S. sclerotiorum* virulence factor so far. OA plays multiple roles in virulence of *S. sclerotiorum*, manipulating the host redox environment, inducing programmed cell death, detoxifying calcium, and mediating pH signaling. OA has long been considered an essential factor for pathogenicity of *S. sclerotiorum*. The studies using targeted mutants of the oxaloacetate acetyl hydrolase gene (*Ssoah1*), responsible for biogenesis and accumulation of OA, showed that OA is required for virulence, but not essential for pathogenicity on all hosts. The creation of an acidic pH environment during host colonization rather than OA production per se has been suggested to be primary requirement for the colonization stage of pathogenesis (Xu et al. 2015).

Several studies have reported the role of secreted effector genes in pathogenicity or virulence of this pathogen. A small cysteine rich secreted cyanovirin-N-homology domain protein encoding gene *SsCVNH*, and a gene encoding small secreted hypothetical protein Ssv263 were shown to be essential for full virulence of *S. sclerotiorum*. A compound appressorium formation-related gene1 *SsCaf1* and an Rhs repeat-containing protein encoding gene *Ss-Rhs1* were shown to be required for host penetration and initial hyphal infection. Secreted effectors including an integrin-like protein SsITL, and a chorismate mutase SsCM1 are known to suppress host resistance by interfering with jasmonic acid/ethylene signaling pathway and salicylic acid signaling pathways, respectively. Other secreted effectors, like the necrosis and ethylene-inducing peptides SsNep1 and SsNep2, a small secreted virulence related rote in SsSSVP1, and the cerato-platanin SsCP1, are known to induce host cell death and necrosis (Yang et al. 2018). Chittem et al. (2020) investigated the global transcriptional changes in *S. sclerotiorum* during infection of canola plants differing in their susceptibility to the pathogen. In contrast to the widely accepted necrotrophic nature of the pathogen, a two-phase infection model involving a brief biotrophic or basic compatibility phase characterized by host resistance suppression and subverting of host defenses followed by a necrotrophic phase was proposed based on cytological and molecular and genomic evidences. This brief initial phase appears to be partly facilitated by oxalic acid (OA) and by secreted effectors, which help *S. sclerotiorum* evade host recognition and suppress host defense signaling pathways. This transcriptome sequencing study to gain better understanding of the molecular mechanisms underlying the interaction of *S. sclerotiorum* and canola and it provided insights into the temporal aspects of important mechanisms/pathways employed by *S. sclerotiorum* for successful infection of canola.

The differential gene expression patterns of *S. sclerotiorum* during disease development on two canola lines differing in susceptibility to this pathogen were evaluated. Sequencing of the mRNA libraries derived from inoculated petioles and mycelium grown on liquid medium generated approximately 164 million Illumina reads, including 95 million 75-bp-single reads, and 69 million 50-bp-paired end reads. Overall, 36% of the quality filter-passed reads were mapped to the *S. sclerotiorum* reference genome. On the susceptible line, 1301 and 1214 *S. sclerotiorum* genes were differentially expressed at early (8–16 h postinoculation (hpi)) and late (24–48 hpi) infection stages, respectively, while on the resistant line, 1311 and 1335 genes were differentially expressed at these stages, respectively. Gene ontology (GO) categories associated with cell wall degradation, detoxification of host metabolites, peroxisome-related activities like fatty acid β-oxidation, glyoxylate cycle, oxido-reductase activity was significantly enriched in the upregulated gene sets on both susceptible and resistant lines. Quantitative RT-PCR of six selected DEGs further validated the RNA-seq differential gene expression analysis. The regulation of effector genes involved in host defense suppression or evasion during the early infection stage, and the expression of effectors involved in host cell death in the late stage of infection provide supporting evidence for a two-phase infection model involving a brief biotrophic phase during early stages of infection. The findings from this study emphasize the role of peroxisome-related pathways along with cell wall degradation and detoxification of host metabolites as the key mechanisms underlying pathogenesis of *S. sclerotiorum* on *B. napus* (Chittem et al. 2020).

6.9.4 Role of *BnaNPR1* Gene Family During *Sclerotinia* Pathogenesis

The *BnaNPR1*-like gene family was identified in *B. napus,* and it was revealed that repression of BnaNPR1 significantly reduces resistance to *S. sclerotiorum,* intensifies ROS accumulation, and changes the expression of genes associated with SA and JA/ET signaling in response to pathogen. The *nonexpresser of pathogenesis-related genes 1*(*NPR1*) and related *NPR1-like* genes play an important role in regulating plant defense. A total of 19 *BnaNPR1-like* genes were identified in the *B. napus* genome, and then named according to their respective best match in *Arabidopsis thaliana* (*At*), which led to the determination of *B. napus* homologs of every *AtNPR1-like* gene. Analysis of important protein domains and functional motifs indicated the conservation and variation among these homologs. Phylogenetic analysis of these BnaNPR1-like proteins and their *Arabidopsis* homologs revealed six distinct subclades, consequently indicating that their name classification totally conformed to their phylogenetic relationships. Further, *Brassica napus* transcriptomic data showed that the expression of three *BnaNPR1s* was significantly downregulated in response to infection with *Sclerotinia sclerotiorum*, the most important pathogen of this crop, whereas *BnaNPR2/BnaNPR3/BnaNPR4/BnaNPR5/BnaNPR6s* did not show the expression differences in general. Wang et al. (2020) generated *B. napus BnaNPR1*-RNAi lines to interpret the effect of the downregulated expression of *BnaNPR1s* on resistance to *S. sclerotiorum*. The results showed that *BnaNPR1*-RNAi significantly decreased this resistance. It was revealed that *BnaNPR1*-RNAi intensified ROS production and changed defense responses in the interaction of plants with this pathogen. These results indicated that *S. sclerotiorum* might use *BnaNPR1* to regulate specific physiological processes of *B. napus*, such as ROS production and SA defense response, for the infection (Wang et al. 2020).

6.10 *Brassica–Xanthomonas* Host-Pathosystem

6.10.1 Alteration of Hormonal Balance in *Brassica* Cultivars During *Xanthomonas* Pathogenesis

The role of cultivar variation in disease susceptibility and disease responses was elucidated in relation to hormonal status in the interaction of *Brassica napus* cultivars and *Xanthomonas campestris* pv. *campestris* (*Xcc*), the causal agent of black rot disease. Fully expanded leaves of six *B. napus* cultivars (cvs. Capitol, Youngsan, Saturnin, Colosse, Tamra, and Mosa) were inoculated with *Xcc*. At 14 days' postinoculation with *Xcc*, cultivar variation in susceptibility or resistance was interpreted with defense responses as estimated by redox status, defensive metabolites, and expression of phenyl propanoid synthesis-related genes in relation to endogenous hormonal status. Disease susceptibility of six cultivars was distinguished by necrotic lesions in the Xcc-inoculated leaves and characterized concurrently based on the higher increase in reactive oxygen species and lipid peroxidation. Among these cultivars, as the susceptibility was higher, the ratios of abscisic acid (ABA)/jasmonic acid (JA) and salicylic acid (SA)/JA tended to increase with enhanced expression of SA signaling regulatory gene *NPR1* and transcriptional factor *TGA1* and antagonistic suppression of JA-regulated gene *PDF* 1.2. In the resistant cultivar (cv. Capitol), accumulation of defensive metabolites with enhanced expression of genes involved in flavonoids (chalcone synthase), pro-anthocyanidins (anthocyanidin reductase), and hydroxycinnamic acids (ferulate-5-hydroxylase) biosynthesis and higher redox status were observed, whereas the opposite results were obtained for susceptible cultivars (cvs. Mosa and Tamra). These results clearly indicate that cultivar variation in susceptibility to infection by Xcc was determined by enhanced alteration of the SA/JA ratio, as a negative regulator of redox status and phenylpropanoid synthesis in the *Brassica napus*–Xcc pathosystem (Islam et al. 2017).

References

Abbas HK, Tanaka T, Duke SO, Porter JK, Wray EM, Hodges L, Sessions AE, Wang E, Merrill AH Jr, Riley RT (1994) Fumonisin-and AAL-toxin-induced disruption of sphingolipid metabolism with accumulation of free sphingoid bases. Plant Physiol 106:1085–1093

Abdel-Farid IB, Jahangir M, van den Hondel CAMJJ, Kim HK, Choi YH, Verpoorte R (2009) Fungal infection-induced metabolites in *Brassica rapa*. Plant Sci 176:608–615

Affolter M, Pyrowolakis G, Weiss A, Basler K (2008) Signal-induced repression: the exception or the rule in developmental signaling? Dev Cell 15(1):11–22. https://doi.org/10.1016/j.devcel.2008.06.006

Agarwal A, Garg GK, Singh US, Mishra DP (1994) Detection and role of chlorotic toxin and phytohormone in the pathogenesis of Alternaria blight in *Brassica napus*. Curr Sci 66:442–443

Agarwal A, Kaul V, Faggian R, Rookes JE, Ludwig-Müller J, Cahill DM (2011) Analysis of global host gene expression during the primary phase of the *Arabidopsis thaliana-Plasmodiophora brassicae* interaction. Funct Plant Biol 38:462–478

Aldesuquy HS, Baka ZAM (1992) Physiological and biochemical changes in host leaf tissues associated with the growth of two biotrophic fungi growing in Egypt. Phyton 32:129–142

Alighisi P, Favaron F (1995) Pectin degrading enzymes and plant parasite interactions. Eur J Plant Pathol 101:365–375

Alix K, Lariagon C, Delourme R, Manzanares-Dauleux MJ (2007) Exploiting natural genetic diversity and mutant resources of *Arabidopsis thaliana* to study the *Arabidopsis thaliana–Plasmodiophora brassicae* interaction. Plant Breed 126(2):218–221

Amselem J, Cuomo CA, van Kan JA, Viaud M, Benito EP, Couloux A, Coutinho PM, de Vries RP, Dyer PS, Fillinger S, Fournier E, Gout L, Hahn M, Kohn L, Lapalu N, Plummer KM, Pradier JM, Quévillon E, Sharon A, Simon A, ten Have A, Tudzynski B, Tudzynski P, Wincker P, Andrew M, Anthouard V, Beever RE, Beffa R, Benoit I, Bouzid O, Brault B, Chen Z, Choquer M, Collémare J, Cotton P, Danchin EG, Da Silva C, Gautier A, Giraud C, Giraud T, Gonzalez C, Grossetete S, Güldener U, Henrissat B, Howlett BJ, Kodira C, Kretschmer M, Lappartient A, Leroch M, Levis C, Mauceli E, Neuvéglise C, Oeser B, Pearson M, Poulain J, Poussereau N, Quesneville H, Rascle C, Schumacher J, Ségurens B, Sexton A, Silva E, Sirven C, Soanes DM, Talbot NJ, Templeton M, Yandava C, Yarden O, Zeng Q, Rollins JA, Lebrun MH, Dickman M (2011) Genomic analysis of the necrotrophic fungal pathogens *Sclerotinia sclerotiorum* and *Botrytis cinerea*. PLoS Genet 7:e1002230

Annis SL, Goodwin PH (1997) Recent advances in molecular genetics of plant cell wall degrading enzymes in plant pathogenic fungi. Eur J Plant Pathol 103:1–14

Atwal AK, Ramandeep Munshi SK, Mann APS (2004) Biochemical changes in relation to Alternaria leaf blight in Indian mustard. Plant Dis Res 19:57–65

Ayer WA, Pena-Rodriguez LM (1987) Metabolites produced by *Alternaria brassicae*, the black spot pathogen of Canola, part 2, Sesquiterpenoid metabolites. J Nat Prod 50:408–417

Ayer WA, Bains PS, Pena-Rodriguez LM, Tewari JP (1987) Production of a host specific phytotoxin by *Alternaria brassicae*. In: Proceedings of the 7th international rapeseed congress, Poznan, Poland (Abstr.), vol 257, pp 1256–1261

Badstöber J, Ciaghi S, Neuhauser S (2020) Dynamic cell wall modifications in *Brassicas* during clubroot disease. bioRxiv. https://doi.org/10.1101/2020.03.02.972901

Bains PS (1989) Purification, chemical characterization, host-specificity, bioassay, mode of action, and herbicidal use of the toxin produced by *Alternaria brassicae*. PhD thesis, University of Alberta, Edmonton, Alberta, Canada, 137 p

Bains PS, Tewari JP (1987) Purification, chemical characterization and host specificity of the toxin produced by *Alternaria brassicae*. Physiol Mol Plant Pathol 30:259–271

Bakar FDA, Murad AMA, Hamid AA, Zamrod Z, Mahadi NM, Sullivan P (2005) Induction and expression of cutinase activity during saprophytic growth of the fungal plant pathogen, *Glomerella cingulata*. Asia Pacific J Mol Biol Biotechnol 13:63–69

Bednarek P, Pislewska-Bednarek M, Svatos A, Schneider B, Doubsky J, Mansurova M, Humphry M, Consonni C, Panstruga R, Sanchez-Vallet A, Molina A, Schulze-Lefert P (2009) A glucosinolate metabolism pathway in living plant cells mediates broad-spectrum antifungal defense. Science 323:101–106. https://doi.org/10.1126/science.1163732

Beers EP, Jones AM, Dickerman AW (2004) The S8 serine, C1A cysteine and A1 aspartic protease families in *Arabidopsis*. Phytochemistry 65:43–58

Beisson F, Koo AJK, Ruuska S, Schwender J, Pollard M, Thelen JJ, Paddock T, Salas JJ, Savage L, Milcamps A, Mhaske VB, Cho Y, Ohlrogge JB (2003) *Arabidopsis* genes involved in acyl lipid metabolism. A 2003 census of the candidates, a study of the distribution of expressed sequence tags in organs, and a web-based database. Plant Physiol 132:681–697

Berto P, Belingheri L, Dehorter B (1997) Production and purification of a novel extracellular lipase from *Alternaria brassicicola*. Biotechnol Lett 19:533–536

Berto P, Commenil P, Belingheri L, Dehorter B (1999) Occurrence of a lipase in spores of *Alternaria brassicicola* with a crucial role in the infection of cauliflower leaves. FEMS Microbiol Lett 180:183–189

Black LL, Gorden DT, Williams PH (1968a) Carbon dioxide exchange by radish tissue infected with *Albugo candida* measured with an infrared CO2 analyzer. Phytopathology 58:173–178

Black LL, Williams PH, Pound GS (1968b) Anaerobic metabolism of *A. candida*—infected radish cotyledons. Phytopathology 58:672–675

Boller T (1995) Chemoperception of microbial signals in plant cells. Annu Rev Plant Physiol Plant Mol Biol 46:189–214

Brandwagt BF, Mesbah LA, Takken FLW, Laurent PL, Kneppers TJA, Hille J, Nijkamp HJJ (2000) A longevity assurance gene homolog of tomato mediates resistance to *Alternaria alternata* f. sp. *lycopersici* toxins and fumonisin B-1. Proc Natl Acad Sci 97:4961–4966

Brodmann A, Schuller A, Ludwig-Muller J, Aeschbacher RA, Wiemken A, Boller T, Wingler A (2002) Induction of trehalase in *Arabidopsis* plants infected with the trehalose producing pathogen *Plasmodiophora brassicae*. Mol Plant–Microbe Interact 15:693–700

Buchwaldt L, Green H (1992) Phytotoxicity of destruxin B and its possible role in the pathogenesis of *Alternaria brassicae*. Plant Pathol 41:55–63

Bulman S, Ridgway HJ, Eady C, Conner AJ (2007) Intron-rich gene structure in the intracellular plant parasite *Plasmodiophora brassicae*. Protist 158:423–433

Bussink H, Buxton F, Fraaye B, de Graaff L, Visser J (1992) The polygalacturonases of *Aspergillus niger* are encoded by a family of diverged genes. Eur J Biochem 208:83–90

Cameron KD, Teece MA, Smart LB (2006) Increased accumulation of cuticular wax and expression of lipid transfer protein in response to periodic drying events in leaves of tree tobacco. Plant Physiol 140:176–183

Cao T, Srivastava S, Rahman MH, Kav NNV, Hotte N, Deyholol MK, Strelkov SE (2008) Proteome-level changes in the roots of *Brassica napus* as a result of *Plasmodiophora brassicae* infection. Plant Sci 174:97–115

Caprari C, Bergmann C, Micheli Q, Salvi C, Alberheim P, Darvill A, Cervone F, De Lorenzo G (1993) *Fusarium moniliforme* secretes four endopolygalacturonases derived from a single gene product. Physiol Mol Plant Pathol 43:453–462

Carpita NC, Gilbeaut DM (1993) Structural models of primary cell walls in flowering plants: consistency of molecular structure with the physical properties of the wall during growth. Plant J 3:1–30

Chandran D, Tai YC, Hather G, Dewdney J, Denoux C, Burgess DG, Ausubel FM, Speed TP, Wildermuth MC (2009) Temporal global expression data reveal known and novel salicylate-impacted processes and regulators mediating powdery mildew growth and reproduction on *Arabidopsis*. Plant Physiol 49:1435–1451

Chandran D, Inada N, Hather G, Kleindt CK, Wildermuth MC (2010) Laser microdissection of *Arabidopsis* cells at the powdery mildew infection site reveals site-specific processes and regulators. Proc Natl Acad Sci 107:460–465. https://doi.org/10.1073/pnas.0912492107

Chandran D, Rickert J, Cherk C, Dotson BR, Wildermuth MC (2013) Host cell ploidy underlying the fungal feeding site is a determinant of powdery mildew growth and reproduction. Mol Plant-Microbe Interact 26:537–545

Chandran D, Rickert J, Huang Y, Steinwand MA, Marr SK, Wildermuth MC (2014) A typical E2F transcriptional repressor DEL1 acts at the intersection of plant growth and immunity by controlling the hormone salicylic acid. Cell Host Microbe 15:506–513

Chaturvedi C (1972) A new leaf spot of Popler incited by *Alternaria alternata*. (Fr.) Keissler. Indian Phytopathol 25:316–318

Chen JG, Ullah H, Young JC, Sussman MR, Jones AM (2001) *ABP1* is required for organized cell elongation and division in *Arabidopsis* embryogenesis. Genes Dev 15:902–911

Chhikara S, Chaudhury D, Dhankher OP, Jaiwal PK (2012) Combined expression of a barley class II chitinase and type I ribosome inactivating protein in transgenic *Brassica juncea* provides protection against *Alternaria brassicae*. Plant Cell Tiss Org Cult 08:83–89

Chico JM, Chini A, Fonseca S, Solano R (2008) JAZ repressors set the rhythm in jasmonate signaling. Curr Opin Plant Biol 11:486–494

Chittem K, Yajima WR, Goswami RS, del Rio MLE (2020) Transcriptome analysis of the plant pathogen *Sclerotinia sclerotiorum* interaction with resistant and susceptible canola (*Brassica napus*) lines. PLoS One 15(3):e0229844. https://doi.org/10.1371/journal.pone

Cho Y, Kim KH, La Rota M, Scott D, Santopietro G, Callihan M, Mitchell TK, Lawrence CB (2009) Identification of virulence factors by high throughput targeted gene deletion of regulatory genes in *Alternaria brassicicola*. Mol Microbiol 72:1316–1333

Cho Y, Srivastava A, Ohm RA, Lawrence CB, Wang KH, Grigoriev IV, Marahatta SP (2012) Transcription factor *Amr1* induces melanin biosynthesis and suppresses virulence in *Alternaria brassicicola*. PLoS Pathog 8:e1002974. https://doi.org/10.1371/journal.ppat.1002974

Chou HM, Bundock N, Rolfe SA, Scholes JD (2000) Infection of *Arabidopsis thaliana* leaves with *Albugo candida* (white blister rust) causes a reprogramming of host metabolism. Mol Plant Pathol 1:99–113

Christian M, Steffens B, Schenck D, Burmester S, Böttger M, Luthen H (2006) How does auxin enhance cell elongation? Roles of auxin binding proteins and potassium channels in growth control. Plant Biol 8:346–352

Christias C, Lockwood JL (1973) Conservation of mycelial constituents in four sclerotium forming fungi in nutrient deprived conditions. Phytopathology 63:602–605

Chung HS, Niu Y, Browse J, Howe GA (2009) Top hits in contemporary JAZ: an update on jasmonate signaling. Phytochemistry 70:1547–1559

Ciegler A, Lindenfelser LA (1969) An antibiotic complex from *Alternaria brassicicola*. Experientia 25:719–720

Colcombet J, Hirt H (2008) *Arabidopsis* MAPKs: a complex signalling network involved in multiple biological processes. Biochem J 413:217–226

Colotelo N (1973) Physiological and biochemical properties of the exudates associated with developing sclerotia of *Sclerotinia sclerotiorum* (Lib.) de Bary. Can J Microbiol 19:73–79

Colotelo N (1974) A scanning electron microscope study of developing sclerotia of *Sclerotinia sclerotiorum*. Can J Bot 52:1127–1130

Consonni C, Humphry ME, Hartmann HA, Livaja M, Durner J, Westphal L, Vogel J, Lipka V, Kemmerling B, Schulze-Lefert P, Somerville SC, Panstruga R (2006) Conserved requirement for a plant host cell protein in powdery mildew pathogenesis. Nat Genet 386(1):716–720

Cooke DEL, Jenkins PD, Lewis DM (1997) Production of phytotoxic spore germination liquids by *Alternaria brassicae* and *A. brassicicola* and their effect on species of the family Brassicaceae. Ann Appl Biol 131:413–426

Corsini DL, Le Tourneau D (1973) Organic acid metabolism in *Sclerotinia sclerotiorum*. Arch Microbiol 90:59–64
Cosgrove DJ (2014) Plant cell growth and elongation. eLS. https://doi.org/10.1002/9780470015902.a0001688
Cosgrove DJ (2016) Plant cell wall extensibility: connecting plant cell growth with cell wall structure, mechanics, and the action of wall-modifying enzymes. J Exp Bot 67:463–476
Cotton P, Rascale C, Fevre M (2002) Characterization of PG2, an early endoPG produced by *Sclerotinia sclerotiorum*, expressed in yeast. FEMS Microbiol Lett 213:239–244
Cotton P, Kasza Z, Bruel C, Rascle C, Fevre M (2003) Ambient pH controls the expression of endogalacturonase genes in the necrotrophic fungus *Sclerotinia sclerotiorum*. FEMS Microbiol Lett 277:163–169
Cramer R, Lawrence C (2004) Identification of *Alternaria brassicicola* genes expressed *in planta* during pathogenesis of *Arabidopsis thaliana*. Fungal Genet Biol 41:115–128
Daayf F, Platt HW (2000) Changes in metalaxyl resistance among glucose phosphate isomerase genotypes of *Phytophthora infestans* in Canada during 1997–1998. Am J Potato Res 77:311–318. Cross Ref Google scholar
Dasgupta SN, Verma JP (1961) Studies in the enzyme make-up of *Alternaria* I Qualitative demonstration of enzymes. Curr Sci 30:450–452
Dasgupta SN, Verma JP (1962) Studies in the enzyme make-up of *Alternaria* II. Quantitative distribution of enzymes. Proc Nat Inst Sci, India 28(B):201–205
Daval S, Belcour A, Gazengel K, Legrand L, Gouzy J, Cottret L, Lebreton L, Aigu Y, Mougel C, Manzanares-Dauleux MJ (2019) Computational analysis of the *Plasmodiophora brassicae* genome: mitochondrial sequence description and metabolic pathway database design. Genomics 111(6):1629–1640
Davis KR, Darvill AG, Albersheim P, Dell A (1986) Host pathogen interactions XXIX-Oligogalacturonides released from sodium polypectate by endopolygalacturonic acid lyase are elicitors of phytoalexin in soybean. Plant Physiol 80:568–577
De Lorenzo G, D'Ovidio R, Cervone F (2001) The role of polygalacturonase-inhibiting proteins (PGIPs) in defense against pathogenic fungi. Annu Rev Phytopathol 39:313–335
de Zelicourt A, Montiel G, Pouvreau JB, Thoiron S, Delgrange S, Simier P, Delavault P (2009) Susceptibility of *Phelipanche* and *Orobanche* species to AAL-toxin. Planta 230:1047–1055
Debnath M, Sharma SL, Kant U (1998) Changes in carbohydrate contents and hydrolysing enzymes in white rust of mustard [*Brassica juncea* (L.) Czern & Coss.] caused by *A. candida in vivo* and *in vitro*. J Phytol Res 11:81–82
Degenhardt KJ (1973) *Alternaria brassicae* and *A. raphani*: sporulation in culture, and their effects on yield and quality of rapeseed. MSc thesis, University of Alberta, Canada
Degenhardt KJ (1978) Alternaria black spot of rapeseed and mustard: phytotoxins and other aspects of the host–parasite interaction. PhD thesis, University of Saskatchewan, Saskatoon, Canada
Dekhuijzen HM, Overeem JC (1971) The role of cytokinins in clubroot formation. Physiol Plant Pathol 1:151–161
Devos M, Van-Oosten VR, Van-Poecke RMP, Van-Pelt JA, Pozo MJ, Mueller MJ, Buchala AJ, Métraux JP, Van Loon LC, Dicke M, Pieterse CMJ (2005) Signal signature and transcriptome changes of *Arabidopsis* during pathogen and insect attack. Mol Plant-Microbe Interact 18:923–937. https://doi.org/10.1094/MPMI-18-0923
Dharmasiri N, Dharmasiri S, Weijers D, Lechner E, Yamada M, Hobbie L, Ehrismann JS, Jurgens G, Estelle M (2005) Plant development is regulated by a family of auxin receptor F box proteins. Dev Cell 9:109–119
Eichmann R, Hückelhoven R (2008) Accommodation of powdery mildew fungi in intact plant cells. J Plant Physiol 165:5–18
Eshel D, Ben-Arie R, Dinoor A, Prusky D (2000) Resistance of gibberellin-treated persimmon fruit to *Alternaria alternata* arises from the reduced ability of the fungus to pro-duce endo-1,4-β-glucanase. Phytopathology 90:1256–1262

Eshel D, Lichter A, Dinoor A, Prusky D (2002a) Characterization of *Alternaria alternata* glucanase genes expressed during infection of resistant and susceptible persimmon fruits. Mol Plant Pathol 3:347–358

Eshel D, Miyara I, Ailing T, Dinoor A, Prusky D (2002b) pH regulates endoglucanase expression and virulence of *Alternaria alternata* in persimmon fruit. Mol Plant Microbe Interact 15:774–779

Eulgem T, Somssich IE (2007) Networks of WRKY transcription factors in defense signaling. Curr Opin Plant Biol 10:366–371

Fabro G, Di Rienzo JA, Voigt CA, Sevchenko T, Dehesh K, Somerville S, Alvarez ME (2008) Genome-wide expression profiling *Arabidopsis* at the stage of *Golovinomyces cichoracearum* haustorium formation. Plant Physiol 146(3):1421–1439

Fan CY, Köller W (1998) Diversity of cutinases from plant pathogenic fungi: differential and sequential expression of cutinolytic esterases by *Alternaria brassicicola*. FEMS Microbiol Lett 158:33–38

Fraissinet-Tachet L, Fevre M (1996) Regulation by galacturonic acid of pectinolytic enzyme production by *Sclerotinia sclerotiorum*. Curr Microbiol 33:49–53

Fraissinet-Tachet L, Reymond-Cotton R, Fevre M (1995) Characterization of a multigene family encoding an endopolygalacturonase in *Sclerotinia sclerotiorum*. Curr Genet 29:96–100

Fuchs LS, Fuchs D, Compton DL, Powell SR, Seethaler PM, Capizzi AM, Schatschneider C, Fletcher JM (2006) The cognitive correlates of third-grade skill in arithmetic, algorithmic computation, and arithmetic word problems. J Educ Psychol 98:29–43

Gallois JL, Drouaud J, Lecureuil A, Guyon-Debast A, Bonhomme S, Guerche P (2013) Functional characterization of the plant ubiquitin regulatory X (UBX) domain-containing protein AtPUX7 in *Arabidopsis thaliana*. Gene 526:299–308

Gao S, Choi GH, Shain L, Nuss DL (1996) Cloning and targeted disruption of enpg-1, encoding the major in vitro extracellular endopolygalacturonase of the chestnut blight fungus, *Cryphonectria parasitica*. Appl Environ Microbiol 62:1984–1990

Garcia-Maceira FI, Di Pietro A, Huertas-Gonzalez MD, Ruiz-Roldan MC, Roncero MIG (2001) Molecular characterization of an endopolygalacturonase from *Fusarium oxysporum* expressed during early stages of infection. Appl Environ Microbiol 67:2191–2196

Garg S, Dhawan K, Chawla HKL, Nainawatee HS (1999) *Alternaria brassicae* induced changes in tile activity of cell wall degrading enzymes in leaves of *Brassica juncea*. Biol Plants 30:387–392

Gechev TS, Hille J (2005) Hydrogen peroxide as a signal controlling plant programmed cell death. J Cell Biol 168:17–20

Gechev TS, Gadjev IZ, Hille J (2004) An extensive microarray analysis of AAL-toxin-induced cell death in *Arabidopsis thaliana* brings new insights into the complexity of programmed cell death in plants. Cell Mol Life Sci 61:1185–1197

Gechev T, Ferwerda M, Mehterov N, Laloi C, Qureshi MK, Hille J (2008) *Arabidopsis* AAL-toxin-resistant mutant *atr1* shows enhanced tolerance to programmed cell death induced by reactive oxygen species. Biochem Biophys Res Commun 375:639–644

Ghassemi S, Lichius A, Bidard F, Lemoine S, Rossignol MN, Herold S, Seiboth VS, Seiboth B, Espeso EA, Margeot A (2015) The β-importin *KAP8* (Pse1/Kap121) is required for nuclear import of the cellulase transcriptional regulator *XYR1*, asexual sporulation and stress resistance in *Trichoderma reesei*. Mol Microbiol 96:405–418. https://doi.org/10.1111/mmi.12944

Ghose K, Dey S, Barton H, Loake GJ, Basu D (2008) Differential profiling of selected defense-related genes induced on challenge with *Alternaria brassicicola* in resistant white mustard and their comparative expression pattern in susceptible Indian mustard. Mol Plant Pathol 9:763–775

Gilchrist DG, Ward B, Moussato V, Mirocha CJ (1992) Genetic and physiological response to fumonisins and AAL-toxin by intact tissue of a higher plant. Mycopathologia 117:57–64

Glazebrook J (2005) Contrasting mechanisms of defense against biotrophic and necrotrophic pathogens. Annu Rev Phytopathol 43:205–227. https://doi.org/10.1146/annurev.phyto.43.040204.135923

Gloer JB, Poch GK, Short DM, Mc Closkey DV (1988) Structure of Brassicicolin A: isocyanide antibiotic from the phyloplane fungus *Alternaria brassicicola*. J Org Chem 53:3758–3761
Guilfoyle TJ, Hagen G (2007) Auxin response factors. Curr Opin Plant Biol 10:453–460. https://doi.org/10.1016/j.pbi.2007.08.014
Gupta ML, Singh G, Raheja RK, Ahuja KL, Banga SK (1997a) Chlorophyll content in relation to white rust (*A. candida*) resistance in Indian mustard. Cruciferae Newslett 19:105–106
Gupta SK, Shyam KR, Dohroo NP (1997b) Sclerotinia wilt of french bean. Indian Phytopathol 50: 593
Haga N, Kato K, Murase M, Araki S, Kubo M, Demura T, Suzuki K, Muller I, Voß U, Jurgens G, Ito M (2007) *R1R2R3*-Myb proteins positively regulate cytokinesis through activation of *KNOLLE* transcription in *Arabidopsis thaliana*. Development 134:1101–1110
Hagen G, Guilfoyle T (2002) Auxin-responsive gene expression: genes, promoters and regulatory factors. Plant Mol Biol 49:373–385. https://doi.org/10.1023/A:1015207114117
Hammond-Kosack KE, Jones JDG (1996) Resistance gene-dependent plant defense responses. Plant Cell 8:1773–1791
Hancock JG (1966) Degradation of pectic substances associated with pathogenesis by *Sclerotinia sclerotiorum* in sunflower and tomato stems. Phytopathology 56:975–979
Harding H, Williams PH, McNabola SS (1968) Chlorophyll changes, photosynthesis, and ultrastructure of chloroplasts in *Albugo candida* induced "green islands" on detached *Brassica juncea* cotyledons. Can J Bot 46:1229–1234
Hayashi S, Ishii T, Matsunaga T, Tominaga R, Kuromori T, Wada T, Shinozaki K, Hirayama T (2008) The glycerophosphoryl diester phosphodiesterase-like proteins *SHV3* and its homologs play important roles in cell wall organization. Plant Cell Physiol 49:1522–1535
Herman RP, Herman CA (1985) Prostaglandins or prostaglandin like substances are implicated in normal growth and development in oomycetes. Prostaglandins 29:819–830
Hirata S (1954) Studies on the phytohormone in the malformed portion of the diseased plants. I. The relation between the growth rate and the amount of free auxin in the fungus galls and virus-infected plants. Ann Phytopathol Soc Jpn 19:33–38
Hirata S (1956) Studies on the phytohormone in the malformed portion of the diseased plants. II. On the reformation and the situation of free-auxin in the tissues of fungous galls. Ann Phytopathol Soc Jpn 19:185–190
Hodgkin T, MacDonald MV (1986) The effect of a phytotoxin from *Alternaria brassicicola* on *Brassica* pollen. New Phytol 104:631–636
Huang YJ, Balesdent MH, Li ZQ, Evans N, Rouxel T, Fitt BDL (2009) Fitness cost of virulence differs between the AvrLm1 and AvrLm4 loci in *Leptosphaeria maculans* (phoma stem canker of oilseed rape). Eur J Plant Pathol 126:279
Humpherson-Jones FM, Cooke RC (1977) Induction of sclerotium formation by acid stalling compounds in *Sclerotinia sclerotiorum* and *Sclerotium rolfsii*. Trans Br Mycol Soc 68:413–420
Husain A, Thakur RN (1966) Production of a toxin by *Alternaria brassicae* (Berk.) Sacc. *In vitro*. Labdev J Sci Technol 4:144–145
Imazaki A, Tanaka A, Harimoto Y, Yamamoto M, Akimitsu K, Park P, Tsuge T (2010) Contribution of peroxisomes to secondary metabolism and pathogenicity in the fungal plant pathogen *Alternaria alternata*. Eukaryot Cell 9:682–694
Irani S, Trost B, Waldner M, Nayidu N, Tu J, Kusalik AJ, Todd CD, Wei Y, Bonham-Smith PC (2018) Transcriptome analysis of response to *Plasmodiophora brassicae* infection in the *Arabidopsis* shoot and root. BMC Genomics 19:23
Islam MT, Lee BR, Park SH, La VH, Bae DW, Kim TH (2017) Cultivar variation in hormonal balance is a significant determinant of disease susceptibility to *Xanthomonas campestris* pv. *campestris* in *Brassica napus*. Front Plant Sci 8:2121. https://doi.org/10.3389/fpls.2017.02121
Isshiki T, Pearson B, Holbrook S, Doe CQ (2001) *Drosophila neuroblasts* sequentially express transcription factors which specify the temporal identity of their neuronal progeny. Cell 106: 511–521

Jahn L, Mucha S, Bergmann S, Horn C, Staswick P, Steffens B, Siemens J, Ludwig-Muller J (2013) The clubroot pathogen (*Plasmodiophora brassicae*) influences auxin signaling to regulate auxin homeostasis in *Arabidopsis*. Plant Theory 2:726–749

Jia YJ, Feng BZ, Sun WX, Zhang XG (2009) Polygalacturonase, pectate lyase and pectin methylesterase activity in pathogenic strains of *Phytophthora capsici* incubated under different conditions. J Phytopathol 157:585–591

Johnson RD, Johnson L, Itoh Y, Kodama M, Otani H, Kohmoto K (2000) Cloning and characterization of a cyclic peptide synthetase gene from *Alternaria alternata* apple pathotype whose product is involved in AM-toxin synthesis and pathogenicity. Mol Plant Microbe Interact 13: 742–753

Johnson LJ, Johnson RD, Akamatsu H, Salamiah A, Otani H, Kohmoto K, Kodama M (2001) Spontaneous loss of a conditionally dispensable chromosome from the *Alternaria alternata* apple pathotype leads to loss of toxin production and pathogenicity. Curr Genet 40:65–72

Jones D (1970) Ultrastructure and composition of the cell wall of *Sclerotinia sclerotiorum*. Trans Br Mycol Soc 54:351–360

Joubert A, Simoneau P, Campion C, Bataille-Simoneau N, Iacomi-Vasilescu B, Poupard P, Francois JM, Georgeault S, Sellier E, Guillemette T (2011) Impact of the unfolded protein response on the pathogenicity of the necrotrophic fungus *Alternaria brassicicola*. Mol Microbiol 79:1305–1324

Jung DS, Na YJ, Ryu KH (2002) Phylogenic analysis of *Alternaria brassicicola* producing bioactive metabolites. J Microbiol 40:289–294

Juniper BE, Jefree CE (1983) Plant surfaces. Edward Arnold, London, p 92

Kabbage M, Williams B, Dickman MB (2013) Cell death control: the interplay of apoptosis and autophagy in the pathogenicity of *Sclerotinia sclerotiorum*. PLoS Pathog 9:e1003287

Kabbage M, Yarden O, Dickman MB (2015) Pathogenic attributes of *Sclerotinia sclerotiorum*: switching from a biotrophic to necrotrophic life style. Plant Sci 233:53–60

Kämper J, Kahmann R, Bolker M, Ma LJ, Brefort T, Saville BJ, Banuett F, Kronstad JW, Gold SE, Muller O, Perlin MH, Wosten HA, De Vries R, Ruiz-Herrera J, Reynaga-Pena CG, Snetselaar K, Mccann M, Perez-Martin J, Feldbrugge M, Basse CW, Steinberg G, Ibeas JI, Holloman W, Guzman P, Farman M, Stajich JE, Sentandreu R, Gonzalez-Prieto JM, Kennell JC, Molina L, Schirawski J, Mendoza-Mendoza A, Greilinger D, Munch K, Rossel N, Scherer M, Vranes M, Ladendorf O, Vincon V, Fuchs U, Sandrock B, Meng S, Ho EC, Cahill MJ, Boyce KJ, Klose J, Klosterman SJ, Deelstra HJ, Ortiz-Castellanos L, Li W, Sanchez Alonso P, Schreier PH, HauserHahn I, Vaupel M, Koopmann E, Friedrich G, Voss H, Schluter T, Margolis J, Platt D, Swimmer C, Gnirke A, Chen F, Vysotskaia V, Mannhaupt G, Guldener U, Munsterkotte M, Haase D, Oesterheld M, Mewes HW, Mauceli EW, Decaprio D, Wade CM, Butler J, Young S, Jaffe DB, Calvo S, Nusbaum C, Galagan J, Birren BW (2006) Insights from the genome of the biotrophic fungal plant pathogen *Ustilago maydis*. Nature 444: 97–101

Kannan P, Pandey D, Gupta AK, Punetha H, Taj G, Kumar A (2012) Expression analysis of *MAP2K9* and *MAPK6* during pathogenesis of Alternaria blight in *Arabidopsis thaliana* ecotype Columbia. Mol Biol Rep 39:4439–4444. https://doi.org/10.1007/s11033-011-1232-1

Kars I, van Kan JAL (2004) Extracellular enzymes and metabolites involved in pathogenesis of *Botrytis*. In: Elad Y, Williamson B, Tudzynski P, Delen N (eds) *Botrytis*: biology, pathology and control. Kluwer, Dordrecht, pp 99–118

Kars I, Krooshof GH, Wagemakers L, Joosten R, Benen JAE, van Kan JAL (2005) Necrotizing activity of five *Botrytis cinerea* endopolygalacturonases produced in *Pichia pastoris*. Plant J 43: 213–225

Kasza Z, Vagvolgyi C, Fevre M, Cotton P (2004) Molecular characterization and *in planta* detection of *Sclerotinia sclerotiorum* endo-polygalacturonase genes. Curr Microbiol 48:208–213

Kaur P, Jost R, Sivasithamparam K, Barbetti MJ (2011) Proteome analysis of the *Albugo candida–Brassica juncea* pathosystem reveals that the timing of the expression of defence-related genes is a crucial determinant of pathogenesis. J Exp Bot 62(3):1285–1298. https://doi.org/10.1093/jxb/erq365

Kavanagh JA, Williams PH (1981) Indole auxins in *Plasmodiophora* infected cabbage roots and hypocotyls. Trans Br Mycol Soc 77:125–129

Keon JPR, Byrde RJW, Cooper RM (1987) Some aspects of fungal enzymes that degrade plant cell walls. In: Pegg GF, Ayres PG (eds) Fungal infection of plants. Cambridge University Press, Cambridge, pp 133–157

Khandelwal A, Kumar A, Banerjee M, Garg GK (2002) Effect of *Alternaria* pathotoxin(s) on expression of p53-like apoptotic protein in calli and leaves of *Brassica campestris*. Indian J Exp Biol 40:89–94

Khandelwal A, Elvitigala T, Ghosh B, Quatrano RS (2008) *Arabidopsis* transcriptome reveals control circuits regulating redox homeostasis and the role of an AP2 transcription factor. Plant Physiol 148:2050–2058

Kiermayer O (1958) Paper chromatographic studies of the growth substances of *Capsella bursa-pastoris* after infection by *Albugo candida* and *Peronospora parasitica*. Osterreichische botanische Zeitschrift 105:515–528

Kihara J, Moriwaki A, Tanaka N, Tanaka C, Ueno M, Arase S (2008) Characterization of the *BMR1* gene encoding a transcription factor for melanin biosynthesis genes in the phytopathogenic fungus *Bipolaris oryzae*. FEMS Microbiol Lett 281:221–227

Kim HJ, Chen C, Kabbage M, Dickman MB (2011) Identification and characterization of *Sclerotinia sclerotiorum* NADPH oxidases. Appl Environ Microbiol 77:7721–7729. https://doi.org/10.1128/AEM.05472-11

Kluczewski SM, Lucas JA (1982) Development and physiology of infection by the downy mildew fungus *Peronospora parasitica* (Pers. ex Fr.) Fr. in susceptible and resistant *Brassica* species. Plant Pathol 31:373–389

Knaust A, Ludwig-Müller J (2013) The ethylene signaling pathway is needed to restrict root gall growth in *Arabidopsis* after infection with the obligate biotrophic protist *Plasmodiophora brassicae*. J Plant Growth Regul 32:9–21. https://doi.org/10.1007/s00344-012-9271-y

Koch E, Slusarenko A (1990) *Arabidopsis* is susceptible to infection by a downy mildew fungus. Plant Cell 2:437–445

Kohorn BD (2016) Cell wall-associated kinases and pectin perception. J Exp Bot 67:489–494

Kohorn BD, Kohorn SL (2012) The cell wall-associated kinases, WAKs, as pectin receptors. Front Plant Sci 3:88

Kohorn BD, Kobayashi M, Johansen S, Riese J, Huang LF, Koch K, Fu S, Dotson A, Byers N (2006) An *Arabidopsis* cell wall-associated kinase required for invertase activity and cell growth. Plant J 46:307–316

Kohorn BD, Johansen S, Shishido A, Todorova T, Martinez R, Defeo E, Obregon P (2009) Pectin activation of MAP kinase and gene expression is WAK2 dependent. Plant J 60:974–982

Kuhn H, Kwaaitaal M, Kusch S, Acevedo-Garcia J, Wu H, Panstruga R (2016) Biotrophy at its best: novel findings and unsolved mysteries of the *Arabidopsis*-powdery mildew pathosystem. Arabidopsis Book 14:e0184. https://doi.org/10.1199/tab.0184

Kumari K, Varghese TM, Suryanarayana D (1970) Qualitative changes in the amino-acid contents of hypertrophied organs in mustard due to *Albugo candida*. Curr Sci 39:240–241

Lahlali R, Song T, Chu M, Yu F, Kumar S, Karunakaran C, Peng G (2017) Evaluating changes in cell-wall components associated with clubroot resistance using fourier transform infrared spectroscopy and RT-PCR. Int J Mol Sci 18(10):2058

Lal BB, Prasad M, Ram RP (1980) Amino acid constituents of inflorescence tissue of crucifers in health and disease, due to *Albugo candida* (Pers.) Kuntze. Zbl Bakt II Abt 135:240–245

Lally D, Ingmire P, Tong HY, He ZH (2001) Antisense expression of a cell wall-associated protein kinase, WAK4, inhibits cell elongation and alters morphology. Plant Cell 13(6):1317–1331

Laluk K, Mengiste T (2010) Necrotroph attacks on plants: Wanton destruction or covert extortion? Arabidopsis Book 8:e0136. https://doi.org/10.1199/tab.0136

Lammens T, Boudolf V, Kheibarshekan L, Zalmas LP, Gaamouche T, Maes S, Vanstraelen M, Kondorosi E, La Thangue NB, Govaerts W, Inze D, De Veylder L (2008) Atypical E2F activity restrains APC/CCCS52A2 function obligatory for endocycle onset. Proc Natl Acad Sci 105: 14721–14726

Larkan NJ, Lydiate DJ, Parkin IAP, Nelson MN, Epp DJ, Cowling WA, Rimmer SR, Borhan MH (2013) The *Brassica napus* blackleg resistance gene *LepR3* encodes a receptor-like protein triggered by the *Leptosphaeria maculans* effector *AVRLM1*. New Phytol 197:595–605. https://doi.org/10.1111/nph.12043

Lawrence CB, Mitchell TK, Craven KD, Cho YR, Cramer RA, Kim KH (2008) At death's door: Alternaria pathogenicity mechanisms. Plant Pathol J 24:101–111

Lee SB, Go YS, Bae HJ, Park JH, Cho SH, Cho HJ, Lee DS, Park OK, Hwang I, Suh MC (2009) Disruption of glycosylphosphatidylinositol-anchored lipid transfer protein gene altered cuticular lipid composition, increased plastoglobules, and enhanced susceptibility to infection by the fungal pathogen *Alternaria brassicicola*. Plant Physiol 150:42–54

Li RG, Rimmer R, Buchwaldt L, Sharpe AG, Seguin Swartz G, Coutu C, Hegedus DD (2004) Interaction of *Sclerotinia sclerotiorum* with a resistant *Brassica napus* cultivar: expressed sequence tag analysis identifies genes associated with fungal pathogenesis. Fungal Genet Biol 41:735–753

Li J, Zhang Y, Zhang Y, Yu PL, Pan H, Rollins JA (2018) Introduction of large sequence inserts by CRISPR-Cas9 to create pathogenicity mutants in the multinucleate filamentous pathogen *Sclerotinia sclerotiorum*. mBio 9(3):e00567–e00518. https://doi.org/10.1128/mBio.00567-18

Liebermann B, Nussbaum RP, Günther W (2000) Bicyclo-alternarenes produced by the phytopathogenic fungus *Alternaria alternata*. Phytochemistry 55:987–992

Lin CH, Yang SL, Wang NY, Chung KR (2010) The *FUS3 MAPK* signaling pathway of the citrus pathogen *Alternaria alternata* functions independently or cooperatively with the fungal redox-responsive AP1 regulator for diverse developmental, physiological and pathogenic processes. Fungal Genet Biol 47:381–391

Lipka U, Fuchs R, Lipka V (2008) *Arabidopsis* non-host resistance to powdery mildews. Curr Opin Plant Biol 11:404–411

Lippok B, Birkenbihl RP, Rivory G, Brummer J, Schmelzer E, Logemann E, Somssich IE (2007) Expression of *AtWRKY33* encoding a pathogen-/PAMP-responsive WRKY transcription factor is regulated by a composite DNA motif containing W box elements. Mol Plant-Microbe Interact 20:420–429

Liu Z, Jia Y, Ding Y, Shi Y, Li Z, Guo Y, Gong Z, Yang S (2017) Plasma membrane *CRPK1*-mediated phosphorylation of 14-3-3 proteins induces their nuclear import to fine-tune CBF signaling during cold response. Mol Cell 66:117–128. https://doi.org/10.1016/j.molcel.2017.02.016

Loivamaki M, Stuhrwohldt N, Deeken R, Steffens B, Roitsch T, Hedrich R, Sauter M (2010) A role for PSK signaling in wounding and microbial interactions in *Arabidopsis*. Physiol Plant 139(4): 348–357. https://doi.org/10.1111/j.1399-3054.2010.01371.x

Long DE, Cooke RC (1974) Carbohydrate composition and metabolism of *Senecio squalidus* leaves infected with *A. tragopogonis* (Pers.) SF Gray. New Phytol 73:889–899

Lopez-Maury L, Marguerat S, Bahler J (2008) Tuning gene expression to changing environments: from rapid responses to evolutionary adaptation. Nat Rev Genet 9:583–593

Lou J, Fu L, Peng Y, Zhou L (2013) Metabolites from *Alternaria* fungi and their bioactivities. Molecules 18:5891–5935

Lovelock DA, Donald CE, Conlan XA, Cahill DM (2013) Salicylic acid suppression of clubroot in broccoli (*Brassicae oleracea* var. *italica*) caused by the obligate biotroph *Plasmodiophora brassicae*. Australasian Plant Pathol 42:141–153

Lowe RGT, Cassin A, Grandaubert J, Clark BL, Van de Wouw AP, Rouxel T, Howlett BJ (2014) Genomes and transcriptomes of partners in plant-fungal- interactions between canola (*Brassica napus*) and two *Leptosphaeria* species. PLoS One 9(7):e103098. https://doi.org/10.1371/journal.pone.0103098

Ludwig-Muller J (2015) Bacteria and fungi controlling plant growth by manipulating auxin: balance between development and defense. J Plant Physiol 172:4–12. https://doi.org/10.1016/j.jplph.2014.01.002

LudwigMuller J, Bendel U, Thermann P, Ruppel M, Epstein E, Hilgenberg W (1993) Concentrations of indole-3-acetic acid in plants of tolerant and susceptible varieties of Chinese cabbage infected with *Plasmodiophora brassicae* Woron. New Phytol 125:763–769

Ludwig-Müller J, Pieper K, Ruppel M, Cohen JD, Epstein E, Kiddle G, Bennett R (1999) Indole glucosinolate and auxin biosynthesis in *Arabidopsis thaliana* L. glucosinolate mutants and the development of the clubroot disease. Planta 208:409–419

Ludwig-Muller J, Julke S, Bierfreund NM, Decker EL, Reski R (2009) Moss (*Physcomitrella patens*) GH3 proteins act in auxin homeostasis. New Phytol 181:323–338

Ludwig-Muller J, Julke S, GeiB K, Richter F, Mithofer A, Sola I, Rusak G, Keenan S, Bulman S (2015) A novel methyltransferase from the intracellular pathogen *Plasmodiophora brassicae* methylates salicylic acid. Mol Plant Pathol 16:349–364

Lumsden RD (1969) *Sclerotinia sclerotiorum* infection of bean and the production of cellulose. Phytopathology 59:653–657

Ma L, Djavaheri M, Wang H, Larkan NJ, Haddadi P, Beynon E, Gropp G, Borhan MH (2018) *Leptosphaeria maculans* effector protein *AvrLm1* modulates plant immunity by enhancing MAP Kinase 9 phosphorylation. iScience 3:177–191

MacKinnon SL, Keifer P, William AA (1999) Components from the phytotoxic extract of *Alternaria brassicicola*, a black spot pathogen of canola. Photochemistry 51:215–221

Madsen L, Seeger M, Semple CA, Hartmann-Petersen R (2009) New ATPase regulators—p97 goes to the PUB. Int J Biochem Cell Biol 41:2380–2388

Maitra S, Samajpati N (1982) Changes in polyphenoloxidase and peroxidase in the cauliflower (*Brassica oleracea* L.) leaves infected by *Alternaria brassicae* (Berk.) Sacc. In: Chattopadhyay SB, Samajpati N (eds) Advances in mycology and plant pathology. Oxford & IBH, New Delhi, pp 216–224

Maitra S, Samajpati N (1985) Nucleic acid content in leaf blight disease of *Brassica oleracea* var. Botrytis caused by *Alternaria brassicicola* (Schw.) Wiltshire. Phytopathol Z 114(1):69–72

Majeau N, Trudel J, Asselin A (1990) Diversity of cucumber chitinase isoforms and characterization of one seed basic chitinase with lysozyme activity. Plant Sci 68:9–16

Malinowski R, Smith JA, Fleming AJ, Scholes JD, Rolfe SA (2012) Gall formation in clubroot-infected *Arabidopsis* results from an increase in existing meristematic activities of the host but is not essential for the completion of the pathogen life cycle. Plant J 71:226–238

Mandahar CL, Suri RA (1983) Secretion of cytokinins *in vivo* and *in vitro* by *Alternaria brassicae* and their role in pathogenesis. Trop Plant Sci Res 1:285–288

Mandahar CL, Suri RA (1987) Cytokinin-like activity in green islands formed around infection spots of *Alternaria brassicicola* on mustard leaves in nature and their role in pathogenesis. Biol Plant 29:76–79

Mandal S, Rajarammohan S, Kaur J (2018) *Alternaria brassicae* interactions with the model Brassicaceae member *Arabidopsis thaliana* closely resembles those with mustard (*Brassica juncea*). Physiol Mol Biol Plants 24(1):51–59. https://doi.org/10.1007/s12298-017-0486-z

Martel MB, Herve Du-Penhoat C, Letoublon R, Fevre M (2002) Purification and characterization of a glucoamylases secreted by the plant pathogen *Sclerotinia sclerotiorum*. Can J Microbiol 48: 212–218

Mathpal P, Punetha H, Tewari AK, Agrawal S (2011) Biochemical defense mechanism in rapeseed-mustard genotypes against Alternaria blight disease. J Oilseed Brassicas 2:87–94

Matsumoto M, Matsutani S, Sugita K, Yoshida H, Hayashi F, Terui Y, Nakai H, Uotani N, Kawamura Y, Matsumoto K (1992) Depudecin: a novel compound inducing the flat phenotype of NIH3T3 cells doubly transformed by ras- and src-oncogene, produced by *Alternaria brassicicola*. J Antibiot 45:879–885

Maxwell DP (1973) Oxalate formation in *Whetzelinia sclerotiorum* by oxalo-acetate acetylhydrolase. Physiol Plant Pathol 3:279–288

McKenzie KJ, Robb J, Lennard JH (1988) Toxin production by *Alternaria* pathogens of oilseed rape (*Brassica napus*). Crop Res (Hort Res) 28:67–81

McRoberts N, Lennard JH (1996) Pathogen behaviour and plant cell reactions in interactions between *Alternaria* species and leaves of host and nonhost plants. Plant Pathol 45:742–752. https://doi.org/10.1046/j.1365-3059.1996.d01-4.x

Meur G, Shukla P, Dutta-Gupta A, Kirti PB (2015) Characterization of *Brassica juncea–Alternaria brassicicola* interaction and jasmonic acid carboxyl methyl transferase expression. Plant Gene 3:1–10. https://doi.org/10.1016/j.plgene.2015.06.001

Mishra MK, Suresh N, Bhat AM, Suryaprakash N, Kumar SS, Kumar A, Jayarama (2011) Genetic molecular analysis of *Coffea arabica* (Rubiaceae) hybrids using SRAP markers. Rev Biol Trop 59:607–617

Misra A, Padhi B (1981) Impact of brown rust and white rust on the RNA content of their host tissues. In: Bilgrami KS, Misra RS, Misra PC (eds) Advancing frontiers of mycology and plant pathology, pp 175–182

Montemurro N, Visconti A (1992) *Alternaria* metabolites—chemical and biological data. In: Chelkovski J, Visconti A (eds) *Alternaria* biology, plant diseases and metabolites. Elsevier, Amsterdam, pp 449–557

Muller P, Hilgenberg W (1986) Isomers of zeatin and zeatin riboside in clubroot tissue: evidence for trans-zeatin biosynthesis by *Plasmodiophora brassicae*. Physiologia Plant 66:245–250

Munir S, Shahzad AN, Qureshi MK (2020) Acuities into tolerance mechanisms via different bioassay during *Brassicaceae-Alternaria brassicicola* interaction and its impact on yield. PLoS One 15(12):e0242545. https://doi.org/10.1371/journal.pone.0242545

Neuhaus K, Grsic-Rausch S, Sauerteig S, Ludwig-Muller J (2000) *Arabidopsis* plants transformed with nitrilase 1 or 2 in antisense direction are delayed in clubroot development. J Plant Physiol 156(5/6):756–761

Nowakowska M, Wrzesińska M, Kamiński P, Szczechura W, Linhocka M, Tartanus M, Kozik EU, Nowicki M (2019) *Alternaria brassicicola*–Brassicaceae pathosystem: insights into the infection process and resistance mechanisms under optimized artificial bio-assay. Eur J Plant Pathol 153:131–151. https://doi.org/10.1007/s10658-018-1548-y

Nussbaum RI, Guenther W, Heinze S, Liebermann B (1999) New tricycloalternarenes produced by the phytopathogenic fungus *Alternaria alternata*. Phytochemistry 52:593–599

O'Connell R, Herbert C, Sreenivasaprasad S, Khatib M, Esquerré-Dugayé MT, Dumas B (2004) A novel *Arabidopsis-Colletotrichum* pathosystem for the molecular dissection of plant-fungal interactions. Mol Plant Microbe Interact 17:272–282

Ohm RA, Feau N, Henrissat B, Schoch CL, Horwitz BA, Barry KW, Condon BJ, Copeland AC, Dhillon B, Glaser F, Hesse CN, Kosti I, LaButti K, Lindquist EA, Lucas S, Salamov AA, Bradshaw RE, Ciuffetti L, Hamelin RC, Kema GH, Lawrence C, Scott JA, Spatafora JW, Turgeon BG, de Wit PJ, Zhong S, Goodwin SB, Grigoriev IV (2012) Diverse lifestyles and strategies of plant pathogenesis encoded in the genomes of eighteen Dothideomycetes fungi. PLoS Pathog 8(12):e1003037. https://doi.org/10.1371/journal.ppat.1003037

Oka K, Akamatsub H, Kodamab M, Nakajimab H, Kawadab T, Otani H (2005) Host-specific AB-toxin production by germinating spores of *Alternaria brassicicola* is induced by a host-derived oligosaccharide. Physiol Mol Plant Pathol 66:12–19

Oliveira-Garcia E, Deising HB (2016) Attenuation of PAMP-triggered immunity in maize requires down-regulation of the key β-1,6-glucan synthesis genes *KRE5* and *KRE6* in biotrophic hyphae of *Colletotrichum graminicola*. Plant J 87(4):355–375. https://doi.org/10.1111/tpj.13205

Oliver RP, Solomon PS (2010) New developments in pathogenicity and virulence of necrotrophs. Curr Opin Plant Biol 13:415–419

Otani H, Kohnobe A, Kodama M, Kohmoto K (1998) Production of a host-specific toxin by germinating spores of *Alternaria brassicicola*. Physiol Mol Plant Pathol 52:285–295

Pandey D, Kumar A, Garg GK (2001) Antagonism between *Alternaria brassicae* toxin and zeatin in cell culture of *Brassica juncea* (L) Czern & Coss. cv Divya. Physiol Mol Biol Plants 7:181–184

Pandey SP, Roccaro M, Schon M, Logemann E, Somssich IE (2010) Transcriptional reprogramming regulated by WRKY18 and WRKY40 facilitates powdery mildew infection of *Arabidopsis*. Plant J 64:912–923. https://doi.org/10.1111/j.1365-313X.2010.04387.x

Parada RY, Oka K, Yamagishi D, Kodama M, Otani H (2007) Destruxin B produced by *Alternaria brassicae* does not induce accessibility of host plants to fungal invasion. Physiol Mol Plant Pathol 71:48–54

Parada RY, Sakuno E, Mori N, Oka K, Egusa M, Kodama M, Otani H (2008) *Alternaria brassicae* produces a host-specific protein toxin from germinating spores on host leaves. Phytopathology 98:458–463

Park SW, Kaimoyo E, Kumar D, Mosher S, Klessig DF (2007) Methyl salicylate is a critical mobile signal for plant systemic acquired resistance. Science 318:113–116

Pasold S, Siegel I, Seidel C, Ludwig-Muller J (2010) Flavonoid accumulation in *Arabidopsis thaliana* root galls caused by the obligate biotrophic pathogen *Plasmodiophora brassicae*. Mol Plant Pathol 11:545–562

Pathak RK, Baunthiyal M, Shukla R, Pandey D, Taj G, Kumar A (2017) *In Silico* identification of mimicking molecules as defense inducers triggering jasmonic acid mediated immunity against Alternaria blight disease in *Brassica* species. Front Plant Sci 8:609. https://doi.org/10.3389/fpls.2017.00609

Pedras MSC, Biesenthal CJ (2001) Isolation, structure determination, and phytotoxicity of unusual dioxopiperazines from the phytopathogenic fungus *Phoma lingam*. Phytochemistry 58:905–909

Pedras MSC, Chumala PB (2005) Phomapyrones from blackleg causing phytopathogenic fungi: isolation, structure determination, biosyntheses and biological activity. Phytochemistry 66(1): 81–87. https://doi.org/10.1016/j.phytochem.2004.10.011

Pedras MSC, Morales VM, Taylor JL (1994) Phomapyrones: three metabolites from the blackleg fungus. Phytochemistry 36:1315–1318

Pedras MSC, Gadagi RS, Jha M, Sarma-Mamillapalle VK (2007) Detoxification of the phytoalexin brassinin by isolates of *Leptosphaeria maculans* pathogenic on brown mustard involves an inducible hydrolase. Phytochemistry 68:1572–1578

Pedras MSC, Chumala PB, Jin W, Islam MS, Hauck DW (2009) The phytopathogenic fungus *Alternaria brassicicola*: phytotoxin production and phytoalexin elicitation. Phytochemistry 70: 394–402

Peng G, Pageau D, Strelkov SE, Gossen BD, Hwang SF, Lahlali R (2015) A >2-year crop rotation reduces resting spores of Plasmodiophora brassicae in soil and the impact of clubroot on canola. Eur J Agron 70:78–84

Poussereau N, Creton S, Billon-Grand G, Rascle C, Fevre M (2001a) Regulation of acp1 encoding a non-aspatyl acid protease expressed during pathogenesis of *Sclerotinia sclerotiorum*. Microbiology 147:717–726

Poussereau N, Gente S, Rascle C, Billon-Grand G, Fevre M (2001b) aspS encoding an unusual aspartyl protease from *Sclerotinia sclerotiorum* is expressed during phytopathogenesis. FEMS Microbiol Lett 194:27–32

Prigent S, Frioux C, Dittami SM, Thiele S, Larhlimi A, Collet G, Gutknecht F, Got J, Eveillard D, Bourdon J, Plewniak F, Tonon T, Siegel A (2017) Meneco, a topology-based gap-filling tool applicable to degraded genome-wide metabolic networks. PLoS Comput Biol 13(1):e1005276. https://doi.org/10.1371/journal.pcbi.1005276

Qiu JL, Fiil BK, Petersen K, Nielsen HB, Botanga CJ, Thorgrimsen S, Palma K, Suarez-Rodriguez MC, Sandbech-Clausen S, Lichota J, Brodersen P, Grasser KD, Mattsson O, Glazebrook J, Mundy J, Petersen M (2008) *Arabidopsis* MAP kinase4 regulates gene expression through transcription factor release in the nucleus. EMBO J 27:2214–2221

Rajarammohan S, Pental D, Kaur J (2019) Near-complete genome assembly of *Alternaria brassicae*—a necrotrophic pathogen of *Brassica* crops. MPMI 32(8):928–930

Rancour DM, Park S, Knight SD, Bednarek SY (2004) Plant UBX domain-containing protein 1, PUX1, regulates the oligomeric structure and activity of *Arabidopsis* CDC48. J Biol Chem 279:54264–54274

Reed JW (2001) Roles and activities of Aux/IAA proteins in *Arabidopsis*. Trends Plant Sci 6:420–425

Riou C, Freyssinet G, Fevre M (1991) Production of cell wall-degrading enzymes by the phytopathogenic fungus *Sclerotinia sclerotiorum*. Appl Environ Microbiol 57:1478–1484

Riou C, Fraissinet-Tachet L, Freyssinet G, Fevre M (1992a) Secretion of pectic isoenzymes by *Sclerotinia sclerotiorum*. FEMS Microbiol Lett 91:231–238

Riou C, Fraissinet-Tachet L, Freyssinet G, Fevre M (1992b) Purification and characterization of extracellular pectinolytic enzymes produced by *Sclerotinia sclerotiorum*. Appl Environ Microbiol 58:578–583

Roby D, Broglie K, Cressman R, Biddle P, Chet I, Broglie R (1990) Activation of a bean chitinase promoter in transgenic tobacco plants by phytopathogenic fungi. Plant Cell 2:999–1008

Rolfe SA, Strelkov SE, Links MG, Clarke WE, Robinson SJ, Djavaheri M, Malinowski R, Haddadi P, Kagale S, Parkin IAP, Taheri A, Borhan MH (2016) The compact genome of the plant pathogen *Plasmodiophora brassicae* is adapted to intracellular interactions with host *Brassica* spp. BMC Genomics 17:272. https://doi.org/10.1186/s12864-016-2597-2

Rollins JA, Dickman MB (2001) pH signaling in *Sclerotinia sclerotiorum* identification of a pac C/RIM 1 homolog. Appl Environ Microbiol 67:75–81

Rose JK, Braam J, Fry SC, Nishitani K (2002) The XTH family of enzymes involved in xyloglucan endotransglucosylation and endohydrolysis: current perspectives and a new unifying nomenclature. Plant Cell Physiol 43:1421–1435

Rostás M, Bennett R, Hilker M (2002) Comparative physiological responses in Chinese cabbage induced by herbivory and fungal infection. J Chem Ecol 28:2449–2463

Rotem J (1994) The genus *Alternaria*: biology, epidemiology and pathogenicity. APS Press, St Paul, p 326

Schneider R, Hanak T, Persson S, Voigt CA (2016) Cellulose and callose synthesis and organization in focus, what's new? Curr Opin Plant Biol 34:9–16

Schuller A, Kehr J, Ludwig-Müller J (2014) Laser microdissection coupled to transcriptional profiling of *Arabidopsis* roots inoculated by *Plasmodiophora brassicae* indicates a role for brassinosteroids in clubroot formation. Plant Cell Physiol 55:392–411

Schweizer P, Jeanguenat A, Whitacre D, Métraux JP, Mosinger E (1996) Induction of resistance in barley against *Erysiphe graminis* f. sp. *hordei* by free cutin monomers. Physiol Mol Plant Pathol 49:103–120

Schwelm A, Fogelqvist J, Knaust A, Jülke S, Lilja T, Bonilla-Rosso G, Karlsson M, Shevchenko A, Dhandapani V, Choi SR, Kim HG, Park JY, Lim YP, Ludwig-Müller J, Dixelius C (2015) The *Plasmodiophora brassicae* genome reveals insights in its life cycle and ancestry of chitin synthases. Sci Rep 5:11153

Schwelm A, Dixelius C, Ludwig-Müller J (2016) New kid on the block—the clubroot pathogen genome moves the plasmodiophorids into the genomic era. Eur J Plant Pathol 145:531–542

Scott-Craig JS, Panaccione DG, Cervone F, Walton JD (1990) Endopolygalacturonase is not required for pathogenicity of *Cochliobolus carbonum* on maize. Plant Cell 2:1191–1200

Segmuller N, Kokkelink L, Giesbert S, Odinius D, van Kan J, Tudzynski P (2008) NADPH oxidases are involved in differentiation and pathogenicity in *Botrytis cinerea*. Mol Plant-Microbe Interact 21:808–819. https://doi.org/10.1094/MPMI-21-6-0808

Sharma N, Rahman MH, Strelkov SE, Thiagarajah M, Bansal VK, Kav NNV (2007) Proteome-level changes in two *Brassica napus* lines exhibiting differential responses to the fungal pathogen *Alternaria brassicae*. Plant Sci 172:95–110

Shi LH, Bielawski J, Mu JY, Dong HL, Teng C, Zhang J, Yang XH, Tomishige N, Hanada K, Hannun YA, Zuo JR (2007) Involvement of sphingoid bases in mediating reactive oxygen intermediate production and programmed cell death in *Arabidopsis*. Cell Res 17:1030–1040

Shibuya N, Minami E (2001) Oligosaccharide signaling for defence responses in plant. Physiol Mol Plant Pathol 59:223–233

Shieh MT, Brown RL, Whitehead MP, Cary JW, Cotty PJ, Cleveland TE, Dean RA (1997) Molecular genetic evidence for the involvement of a specific polygalacturonase P2c in the invasion and spread of *Aspergillus flavus* in cotton bolls. Appl Environ Microbiol 63:3548–3552

Shivanna KR, Sawhney VK (1993) Pollen selection for *Alternaria* resistance in oilseed *Brassicas*: responses of pollen grains and leaves to a toxin of *A. brassicae*. Theor Appl Genet 86:339–344

Siemens J, Keller I, Sarx J, Kunz S, Schuller A, Nagel W, Schmulling T, Parniske M, Ludwig-Muller J (2006) Transcriptome analysis of *Arabidopsis* clubroots indicate a key role for cytokinins in disease development. Mol Plant Microbe Interact 19:480–494. https://doi.org/10.1094/MPMI-19-0480

Siemens J, Gonzalez MC, Wolf S, Hofmann C, Greiner S, Du Y, Rausch T, Roitsch T, Ludwig-Muller J (2011) Extracellular invertase is involved in the regulation of clubroot disease in *Arabidopsis thaliana*. Mol Plant Pathol 12:247–262. https://doi.org/10.1111/j.1364-3703.2010.00667.x

Singh SB, Singh DV, Bais BS (1980) *In vivo* cellulase and pectinase production by *Albugo candida* and *Peronospora parasitica*. Indian Phytopathol 33:370–371

Singh Y, Rao DV, Batra A (2011a) Biochemical changes in *Brassica juncea* (L.) Czern. & Coss. infected with *Albugo candida* Kuntz. (Pers.). Int J Pharma Sci Rev 7:74–78

Singh Y, Rao DV, Batra A (2011b) Enzyme activity changes in *Brassica juncea* (L) Czern & Coss. in response to *Albugo candida* Kuntze (Pers). J Chem Pharma Res 3:18–24

Sivaguru M, Ezaki B, He ZH, Tong H, Osawa H, BaluskaF VD, Matsumoto H (2003) Aluminum-induced gene expression and protein localization of a cell wall-associated receptor kinase in *Arabidopsis*. Plant Physiol 132:2256–2266

Soylu EM, Soylu S, Mansfield JW (2004) Ultra-structural characterization of pathogen development and host responses during compatible and incompatible interactions between *Arabidopsis thaliana* and *Peronospora parasitica*. Physiol Mol Plant Pathol 65:67–78

Spartz AK, Gray WM (2008) Plant hormone receptors: new perceptions. Genes Dev 22:2139–2148

Spassieva SD, Markham JE, Hille J (2002) The plant disease resistance gene *Asc-1* prevents disruption of sphingolipid metabolism during AAL-toxin-induced programmed cell death. Plant J 32:561–572

Spring O, Haas K, Lamla I, Thurnhofer S, Vetter W (2005) The composition and taxonomic significance of fatty acid patterns in three white rust species: *Albugo amaranthi, A. candida* and *A. tragopogonis* (Peronosporales, Albuginaceae). Mycol Progr 4:179–184

Srivastava SK, Srivastava AK (1982) Polygalacturonase and cellulase production by *Alternaria brassicae* (Berk.) Sacc. var. Sarson New Bot 9:9–11

Srivastava BIS, Shaw M, Vanterpool TC (1962) Effect of *Albugo candida* (Pers. Ex Chev.) Kuntze. on growth substances in *Brassica napus* (L.). Can J Bot 40:53–59

Stacey G, Shibuya N (1997) Chitin recognition in rice and legumes. Plant and Soil 194:161–169

Suri RA, Mandahar CL (1982) Production of pectolytic enzymes by *Alternaria brassicae*. Indian J Mycol Plant Pathol 12:351–352

Suri RA, Mandahar CL (1984) Secretion of cytokinins *in vivo* and *in vitro* by *Alternaria brassicicola* and their role in pathogenesis. Experientia 40:462–463

Suri RA, Mandahar CL (1985) Involvement of cytokinin-like substances in the pathogenesis of *Alternaria brassicae* (Berk.) Sacc. Plant Sci 41:105–109

Suri RA, Mandahar CL (1986) Cytokinin-like substances and accumulation of labelled metabolites at infection sites of *Alternaria brassicicola*. Mycopathologia 94:153–156

Suri RA, Mandahar CL, Gill PS (1983) Study of secretion of cytokinins by *Alternaria brassicicola* and their possible role in pathogenesis in mustard. Indian J Plant Pathol 1:117–121

Suudi M, Kim MG, Park SR, Hwang DJ, Bae SC, Ahn IP (2011) *Arabidopsis* cell death in compatible and incompatible interactions with *Alternaria brassicicola*. Mol Cells 31:593–601. https://doi.org/10.1007/s10059-011-2203-z

Taj G, Kumar A, Bansal KC, Garg GK (2004) Introgression of osmotin gene for creation of resistance against Alternaria blight by perturbation of cell cycle machinery. Indian J Biotechnol 3:291–298

Taj G, Agarwal P, Grant M, Kumar A (2011) Co-expression and *in-silico* interaction studies for inter-linking the activation of *mapk3* and lox genes during pathogenesis of Alternaria blight (*Alternaria brassicae*) in *Brassica juncea*. J Oilseed Brassica 2:13–20

Takemoto D, Tanaka A, Scott B (2007) NADPH oxidases in fungi: diverse roles of reactive oxygen species in fungal cellular differentiation. Fungal Genet Biol 44:1065–1076. https://doi.org/10.1016/j.fgb.2007.04.011

Tanabe K, Nishimura S, Kohmoto K (1988) Pathogenicity of cutinase- and pectic enzymes deficient mutants of *Alternaria alternata* Japanese pear pathotype. Ann Phytopathol Soc Jpn 54:552–555

ten Have A, Mulder W, Visser J, van Kan JA (1998) The endopolygalacturonase gene Bcpg-1 is required for full virulence of *Botrytis cinerea*. Mol Plant-Microbe Interact 11:1009–1016

Thomton JH, Cooke RC (1970) Accumulation of dark-fixed carbon compounds in pustules of *Albugo tragopogonis*. Trans Br Mycol Soc 54:483–485

Thornton JD, Cooke RC (1974) Changes in respiration, chlorophyll content and soluble carbohydrates of detached cabbage cotyledons following infection with *Peronospora parasitica* (Pers. ex. Fr.) Fr. Physiol Plant Pathol 4:117–125

Trail F, Köller W (1993) Diversity of cutinases from plant pathogenic fungi: purification and characterization of two cutinases from *Alternaria brassicicola*. Physiol Mol Plant Pathol 42: 205–220

Turck F, Zhou A, Somssich IE (2004) Stimulus-dependent, promoter-specific binding of transcription factor *WRKY1* to its native promoter and the defense-related gene *PcPR1-1* in Parsley. Plant Cell 16:2573–2585

Verica JA, He ZH (2002) The cell wall-associated kinase (WAK) and WAK-like kinase gene family. Plant Physiol 129:455–459. https://doi.org/10.1104/pp.011028

Verma JP (1964) Studies in the enzyme make-up of *Alternaria*. V. Aldolase activity. Mycologia 56: 909–918

Verma JP (1971) Studies in the enzyme make-up of *Alternaria*. IX Nucleao-phosphatase activity. Indian Phytopathol 24:43–49

Vlieghe K, Boudolf V, Beemster GTS, Maes S, Magyar Z, Atanassova A, de Almeida EJ, De Groodt R, Inze D, De Veylder L (2005) The DP-E2F-like gene DEL1 controls the endocycle in *Arabidopsis thaliana*. Curr Biol 15:59–63

Vlot AC, D'MA D, Klessig DF (2009) Salicylic acid, a multifaceted hormone to combat disease. Annu Rev Phytopathol 47:177–206

Wagner F, Kusserow H, Schafer W (2000) Cloning and targeted disruption of two polygalacturonase genes in *Penicillium olsonii*. FEMS Microbiol Lett 186:293–299

Waksman G, Keon JP, Turner G (1991) Purification and characterization of two endopolygalacturonases from *Sclerotinia sclerotiorum*. Biochim Biophys Acta 1073:43–48

Walerowski P, Gündel A, Yahaya N, Truman W, Sobczak M, Olszak M, Rolfe S, Borisjuk L, Malinowski R (2018) Clubroot disease stimulates early steps of phloem differentiation and recruits SWEET sucrose transporters within developing galls. Plant Cell 30:3058–3073

Wan JR, Zhang XC, Neece D, Ramonell KM, Clough S, Kim SY, Stacey MG, Stacey G (2008) A *LysM* receptor-like kinase plays a critical role in chitin signaling and fungal resistance in Arabidopsis. Plant Cell 206(1):471–481

Wang SC, LeTourneau D (1971) Carbon sources, growth, sclerotium formation and carbohydrate composition of *Sclerotinia sclerotiorum*. Arch Mikrobiol 80:219–233

Wang D, Weaver ND, Kesarwani M, Dong X (2005) Induction of protein secretory pathway is required for systemic acquired resistance. Science 308:1036–1040

Wang Z, Cao G, Wang X, Miao J, Liu X, Chen Z, Qu L-J, Gu H (2008) Identification and characterization of COI1-dependent transcription factor genes involved in JA-mediated response to wounding in *Arabidopsis* plants. Plant Cell Rep 27:125–135

Wang L, Tsuda K, Sato M, Cohen JD, Katagiri F, Glazebrook J (2009) *Arabidopsis* CaM binding protein CBP60g contributes to MAMP-induced SA accumulation and is involved in disease resistance against *Pseudomonas syringae*. PLoS Pathol 5:e1000301

Wang Z, Tan X, Zhang Z, Gu S, Li G, Shi H (2012) Defense to *Sclerotinia sclerotiorum* in oilseed rape is associated with the sequential activations of salicylic acid signaling and jasmonic acid signaling. Plant Sci 184:75–82. https://doi.org/10.1016/j.plantsci.2011.12.013

Wang Z, Ma LY, Li X, Zhao FY, Sarwar R, Cao J, Li YL, Ding LN, Zhu KM, Yang YH, Tan XL (2020) Genome-wide identification of the NPR1-like gene family in *Brassica napus* and functional characterization of BnaNPR1 in resistance to *Sclerotinia sclerotiorum*. Plant Cell Rep 39(6):709–722. https://doi.org/10.1007/s00299-020-02525-z

Weimer JL (1924) Alternaria leaf spot and brown rot of cauliflower. J Agric Res 29:421–441

Weßling R, Epple P, Altmann S, He Y, Yang L, Henz SR, McDonald N, Wiley K, Bader KC, Gläßer C, Mukhtar MS, Haigis S, Ghamsari L, Stephens AE, Ecker JR, Vidal M, Jones JDG, Mayer KFX, van Themaat EVL, Weigel D, Schulze-Lefert P, Dangl JL, Panstruga R, Braun P (2014) Convergent targeting of a common host protein-network by pathogen effectors from three kingdoms of life. Cell Host Microbe 16(3):364–375. https://doi.org/10.1016/j.chom.2014.08.004

Wildermuth MC (2010) Modulation of host nuclear ploidy: a common plant biotroph mechanism. Curr Opin Plant Biol 13:449–458

Willetts HJ (1978) Sclerotium formation. In: Smith JE, Berry DR (eds) The filamentous fungi, vol 3. Wiley, New York, pp 197–203

Williams PH, Pound GS (1964) Metabolic studies on the host-parasite complex of *A. candida* on radish. Phytopathology 54:446–451

Williams B, Kabbage M, Kim HJ, Britt R, Dickman MB (2011) Tipping the balance: *Sclerotinia sclerotiorum* secreted oxalic acid suppresses host defenses by manipulating the host redox environment. PLoS Pathog 7:e1002107. https://doi.org/10.1371/journal.ppat.1002107

Wong JAL, Willetts HJ (1974) Polyacrylamide gel electrophoresis of enzymes during morphogenesis of sclerotia of *Sclerotinia sclerotiorum*. J Gen Microbiol 81:101–109

Wubben JP, Joosten MHAJ, Van Kan JAL, De Wit PJGM (1992) Subcellular localization of plant chitinases and 1,3-beta-glucanases in *Cladosporium fulvum* (syn. *Fulvia fulva*) infected tomato leaves. Physiol Mol Plant Pathol 41:23–32

Xu RQ, Li XZ, Wei HY, Jiang B, Li K, He YQ, Feng JX, Tang JL (2006) Regulation of eight avr genes by *hrpG* and *hrpX* in *Xanthomonas campestris* pv. *campestris* and their role in pathogenicity. Prog Nat Sci 16:1288–1294

Xu L, Xiang M, White D, Chen W (2015) pH dependency of sclerotial development and pathogenicity revealed by using genetically defined oxalate-minus mutants of *Sclerotinia sclerotiorum*. Environ Microbiol 17:2896–2909. https://doi.org/10.1111/1462-2920.12818

Xu T, Li J, Yu B, Liu L, Zhang X, Liu J, Montes HM (2018) Transcription factor *SsSte12* was involved in mycelium growth and development in *Sclerotinia sclerotiorum*. Front Microbiol 9: 1–13. https://doi.org/10.3389/fmicb.2018.02476

Xue CY, Zhou RJ, Li YJ, Xiao D, Fu JF (2018) Cell-wall-degrading enzymes produced in vitro and in vivo by *Rhizoctonia solani*, the causative fungus of peanut sheath blight. Peer J 6:e5580

Yamanaka K, Sasagawa Y, Ogura T (2012) Recent advances in p97/VCP/Cdc48 cellular functions. Biochem Biophys Act 1823:130–137

Yan DRL, Biederman JA, Barron-Gafford G, Yang X, Moore DJP, Dannenberg MP, Wang X, Knowles J F, Yang J, Smith W K (2018) Quantifying plant physiological response to water stress with high-frequency, near-surface observations of chlorophyll fluorescence and photochemical reflectivity. Abstr B31N-2694 presented at 2018 AGU Fall Meeting, Washington, DC, 10–14 Dec 2018

Yang J, Liu D, Wang X, Ji C, Cheng F, Liu B, Hu Z, Chen S, Pental D, Ju Y, Yao P, Li X, Xie K, Zhang J, Wang J, Liu F, Ma W, Shopan J, Zheng H, Mackenzie SA, Zhang M (2016) The genome sequence of allopolyploid *Brassica juncea* and analysis of differential homoeolog gene expression influencing selection. Nat Genet 48:1225–1232. https://doi.org/10.1038/ng.3657

Yang K, Nath UK, Biswas MK, Kayum MA, Yi G, Lee J, Yang TJ, Nou IS (2018) Whole-genome sequencing of *Brassica oleracea* var. *capitata* reveals new diversity of the mitogenome. PLoS One 13:e01194356

Yao C, Koller W (1994) Diversity of cutinases from plant pathogenic fungi: cloning and characterization of a cutinase gene from *Alternaria brassicicola*. Physiol Mol Plant Pathol 44:81–92

Yao CL, Koller W (1995) Diversity of cutinases from plant pathogenic fungi: different cutinases are expressed during saprophytic and pathogenic stages of *Alternaria brassicicola*. Mol Plant Microbe Interact 8:122–130

Zhang Z, LiQ LZ, Staswick PE, Wang M, Zhu Y, He Z (2007) Dual regulation role of GH3.5 in salicylic acid and auxin signaling during *Arabidopsis-Pseudomonas syringae* interaction. Plant Physiol 145:450–464

Zhao Y, Bi K, Gao Z, Chen T, Liu H, Xie J, Cheng J, Fu Y, Jiang D (2017) Transcriptome analysis of *Arabidopsis thaliana* in response to *Plasmodiophora brassicae* during early infection. Front Microbiol 8:673

Zheng ZY, Abu Qamar S, Chen ZX, Mengiste T (2006) Arabidopsis WRKY33 transcription factor is required for resistance to necrotrophic fungal pathogens. Plant J 48:592–605

Zuppini A, Navazio L, Sella L, Castiglioni C, Favaron F, Mariani P (2005) An endo-polygalacturonase from *Sclerotinia sclerotiorum* induces calcium-mediated signaling and programmed cell death in soybean cells. Mol Plant-Microbe Interact 18:849–855

7 Genomics of Crucifer's Host-Pathosystem at a Glance

Abstract

Crucifer's crops are very important crop commodity with very significant contribution in the world's need of human and animals yielding edible and industrial oil along with vegetables and forage crops of economic, trade, and food security. These groups of crops are challenged by numerous pathogens threatening their production at global level. Some of them have been exploited to reveal genomics of host–pathogen interaction to comprehend the molecular mechanisms of infection and pathogenesis for better management of diseases they cause and influence yield and quality losses. The genome of six *Brassica* crops has been sequenced after the sequencing of *Arabidopsis thaliana* genome. The genome size of *A. thaliana* is smallest with 125 Mb and *B. napus* genome is largest with 925 Mb. To get deeper insight into the molecular and biological functions of host–pathogen interaction, genome of major pathogens of crucifers has been sequenced and analyzed. Among pathogens, *P. brassicae* genome is compact and smallest with 24.2–25.5 Mb and *G. orontii* genome causing powdery mildew is largest with 160 Mb in size. The genome analysis of pathogens has facilitated phylogenetic, host specificity, pathogenicity factors/genes, and molecular events during crucifer's host–pathogen interactions. The perception and understanding of molecular and genetical mechanisms of host–pathogen recognition and events of pathogenesis are regulated by genomics modulation of interacting host and pathogen. Pathogenomics have revealed host specificity, change of host range, and evolution of pathogenic variability. The *R* genes regulate functions, and molecular mechanisms of host–pathogen interactions. The timing of the induction of genes in *R*, and *S* varieties is crucial for mounting effective host defense to the pathogens. Transcriptomic analysis has revealed genes strongly associated with pathogenesis. The genes involved in cell response signaling, cell wall degradation, protein degradation, enzymes production, host transcriptional, and hormonal regulation are differentially expressed. The virulence mechanisms are transition

G. Singh Saharan et al., *Genomics of Crucifer's Host-Pathosystem*,
https://doi.org/10.1007/978-981-19-3812-2_7

period from biotrophy to necrotrophy to facilitate the acquisition of host nutrients by the pathogen. Differential expression of up and down regulated genes, and functional groups has been identified during host–pathogen interaction. The whole genome analysis of host and pathogen offers potential unbiased insight into the molecular mechanisms of host-pathosystem in crucifers.

For rapid, accuracy, less laborious, and less expensive way of detection and identification of crucifer's pathogens, pathotypes, and effectors genes, molecular protocols have been used at field, and controlled laboratory conditions. The distinct lineages that had diverged from each other have also been detected. Molecular markers (RAPD, ESTs) have been used for detection and identification of genetical, and pathological variation and clades of pathogens. The molecular approaches have been allowed detection, identification, and quantification of pathogens host cell entry, area covered by pathogens in host tissues, seed, and soil. The RT-PCR protocols have been developed to detect, and discriminate AGs of *R. solani* from field, soil, and viruses along with their strains from *Brassica* species. High degree of genetic diversity has been recorded in major pathogens of crucifers in the form of pathotypes/races/strains differing in virulence, and host range on *Brassica* species and genotypes. The evolution of new virulence's of pathogens is more common in areas where crops with major gene resistance coupled with genetic uniformity have been sown. The genomes of *P. brassicae* pathotypes Pb2, Pb5, and Pb8 have been sequenced to gain insight into genome variations and its correlations with host specificity. Phylogenetics of pathotypes has been assessed. The changes in pathotypes structure under field conditions have been analyzed through whole genome DNA similarity sequences. The infection of pathogens in *Brassica* species activates host metabolism to regulate carbohydrates, respiration, lipid profile, enzymes, toxins, H_2O_2, OH, phenols, hormones, nucleic acid, proteins, electrolytes, GSL, ROS, and other metabolites which affects crucifer's physiology, biochemistry, and molecular events leading to pathogenesis. Several genes are differentially expressed during host–pathogen interaction for virulence at different stages of host infection and disease development. The genomics of crucifers host-pathosystem has been studies with simple, reproducible, and standardized methodology to elaborate genome sequencing of host as well as pathogen, events of host and pathogen interaction, signaling pathways, expression of different genes, analysis of transcriptome, biochemical changes, pathogenic variability, molecular markers, transition period from biotrophy to necrotrophy, detection, and identification of pathogens, pathotypes, and genes during the process of infection and pathogenesis of crucifer cops.

Keywords

Crucifer's · Pathogens · Genomics · Genome sequencing · Events of host–pathogen interaction · Omics technologies · Signaling pathways · QRT-PCR · Molecular functions · Choice of reference genes · Pathogens genome assembly · Host–pathogen recognition · Pathogenomics · Differential proteins/genes expression · Host-transcript regulation · Transition from biotrophy to necrotrophy · Hormonal regulation · Gene expression · Spliceosome genes · Evolutionary genomics · Molecular detection and identification of pathogens · Pathotypes and genes · Molecular markers · Genomic prediction tools · Correct diagnosis · Selection pressure · Race virulence · Host differentials · Proteomics · Aggressiveness · Hypovirulence · Metabolomics · Toxins · Host-biochemistry · Altered metabolism · Protocols · Methodology · Techniques

7.1 Introduction, Importance, and Scope of Crucifers' Host-Pathosystem

Crucifer's belong to the family Cruciferae containing over 3660 species which includes very important crops of human and animal needs yielding quality edible oil, industrial oilseeds, vegetables, and fodder crops. The largest group of crops contains 39 species of Brassicaceae grown all over the world including *Arabidopsis thaliana* which has been used widely for dissecting molecular mechanisms of crucifer's host-pathosystem through multiomics approaches. Under natural and cultivated conditions, crucifers are challenged by several abiotic, and biotic stresses. At global level, crucifers pathogens viz., *Albugo* (White rust), *Alternaria* (Alternaria blight), *Colletotrichum* (Anthracnose), *Erysiphe* (Powdery mildew), *Fusarium* (Fusarium wilt), *Hyaloperonospora* (Downy mildew), *Leptosphaeria* (Blackleg), *Plasmodiophora* (Clubroot), *Pseudocercosporella* (White leaf spot), *Pyrenopeziza* (Light leaf spot), *Sclerotinia* (Stem rot), Turnip mosaic virus (*TuMV*), *Verticillium* (Verticillium wilt), *Xanthomonas* (Black rot), and *Heterodera* (Cyst nematode) have major effects on crop production and yield. These pathogens have been used as model host-pathosystem to reveal genomics of crucifer's host-pathosystem. The genomics of plants was initiated after the sequencing of *A. thaliana* genome for the first time in 1990. The genome size of *A. thaliana* is 125 Mbp containing 25,498 genes encoding proteins from 11,000 families. Now, the genome of all the six *Brassica* species has been sequenced. The sequence analysis has revealed genome size of *B. carinata* 642 Mb, *B. juncea* 922 Mb, *B. napus* 925 Mb, *B. nigra* 591 Mb, *B. oleracea* 584.60 Mb, and *B. rapa* 485 Mb. The genome of major pathogens of crucifers has also been sequenced. The availability of genome sequence analysis of both host, and pathogen has allowed rapid identification of candidate genes during different events of host–pathogen interaction to understand genes governing pathogenesis, and host resistance. The application of omics technologies like NGS, Pangenomics, SNP, in silico, BSA, Ren Seq, Effectoromics, Transcriptomics, Proteomics, Secretomics, Interactomics, and Metabolomics has benefited greatly in revealing the complex biological, genetical, and molecular mechanisms of crucifer's host-pathosystem. Crucifers have developed molecular mechanisms in response to multiple stresses which are activated through complex signaling pathways for the expression of genes to overcome these stresses. Several genes are differentially expressed in response to multiple stresses. At global level, 16 pathogens causing crucifers diseases are considered as of major consequences based on their geographical distribution, host range, losses caused, and resources spent to manage them. Using crucifer's host-pathosystem with these pathogens, several genes have been identified, mapped on *Brassica* chromosomes with functional characterization, isolation, and cloning to produce resistant cultivars of *Brassica.* In recent years, QRT-PCR, a powerful and efficient technique has become the first choice in quantitative gene expression in crucifer's host-pathosystem for various biological, and molecular functions. However, data normalization is essential for reliable output of QRT-PCR assays to avoid unsuitable choice of reference genes (Saharan et al. 2005, 2014, 2016, 2017, 2019, 2021a; Herlihy et al. 2019; Näpflin et al. 2019).

7.2 Characterization of *Brassica* Species Genome

The genomes of diploid *Brassica* species are three to five times the size of the *Arabidopsis* genome, ranging from 0.97 pg/2C (468 Mb/1C; where C is haploid DNA per nucleus) for *B. nigra* to 1.37 pg/2C (662 Mb/1C) for *B. oleracea.* The amphidiploid genomes, which have two sets of chromosomes from each parent species, range from 2.29 pg/2C (1105 Mb/1C) to 2.56 pg/2C (1235 Mb/1C). Relationships among the three diploid species—*B. rapa* (syn. Rapa; 2n = 20, genome AA), *B. nigra* (2n = 16, genome BB), and *B. oleracea* (2n = 18, genome CC)—and three amphidiploids *B. napus* (2n = 38, genome AACC), *B. juncea* (2n = 36, genome AABB), and *B. carinata* (2n = 34, genome BBCC) are well known. Rapid-cycling strains of *B. oleracea* have been developed that have a life cycle as short as that of *Arabidopsis thaliana*, making them easier to study (UN 1935; Williams and Hill 1986; Paterson et al. 2001).

Neutral DNA polymorphisms are common among con-specific genotypes of most *Brassica* species, and at least 15 molecular maps have been produced, using at least 900 different publicly available *Brassica* and *Arabidopsis* DNA probes (which are each used with similar efficacy). Virtually all the probes hybridize to two or more loci, even in diploid *Brassica* species; however, only a subset of loci segregates for allelic variation in any one mapping population. The subset of genetically mapped loci can, therefore, be used to infer the locations of many additional sequence-tagged sites (STSs) (Lan et al. 2000). The comparative organization of the chromosomes of *Brassica* and *Arabidopsis* is well studied. On the basis of 186 corresponding loci, about 19 chromosome structural rearrangements differentiate *B. oleracea* and *A. thaliana* orthologs, suggesting that chromosomal tracts of about 20–25 cM remain largely co-linear in the two taxa (Lan et al. 2000). Microsynteny studies involving comparative genetic and physical mapping of specific chromosome segments have shown largely conserved gene order in *Arabidopsis* and *Brassica*, but some disruption in gene content by deletions or insertions. Furthermore, duplicated partial gene clusters are commonly found in both *Arabidopsis* and *Brassica* species. Comparative sequencing has permitted orthology assignment of some duplicated segments in *Arabidopsis* and *Brassica* species (Wroblewski et al. 2000).

As the members of the crucifer family are too diverse, the genomic structures are very complicated. In general, the cultivated *Brassica* species as per the triangle of U including diploid as well as all tetraploids species, the diploid possesses the most variability while there is less variability, as compared to the diploid, in the amphidiploid species as it is derived from the diploid species with the few chance crosses followed by the amphidiploidization. Also the selection pressure on the amphidiploid species is more due to its nucleo-cytoplasm and environmental instability. But still a considerable genetic diversity is present within these three amphidiploid species (Song et al. 1995). Based upon studies of genetic diversity, *B. napus* may be considered as the most ancient amphidiploid, succeeded by *B. juncea* and *B. carinata*. Two major factors are responsible for general diversity within amphidiploids: multiple hybridizations with different diploid parents and genome modifications following polyploidization. Early comparative studies conducted at the level of genetic linkage maps revealed extensive duplication within *Brassica* genomes (Lagercrantz and Lydiate 1996) and tracts of co-linearity disrupted by multiple rearrangements between the genomes of *B. nigra* and *A. thaliana* (Lagercrantz 1998). The genome of different U-triangle species along with *Arabidopsis thaliana* has been sequenced. Detailed description is given in Chap. 1 at Sects. 1.4.1–1.4.7.

7.3 Impact of Pathogen Genome Sequencing

The development of genomics, proteomics, transcriptomics, molecular, and biological techniques has revealed functional and evolutionary relationship of pathogen with each other including host-pathosystem of crucifers. Genomics, sequencing, and analysis of crucifer's pathogenesis genome have given deeper insight into the evolutionary history, close relationship between necrotrophy and biotrophy, consequences of biotrophy, identification of defense suppressing and their host targets, host specificity genes, gain and loss of genes, phylogenetic relationship with other organism's species, races, and strains. Genome sequencing has revealed clonal reproduction, genetic exchange between races, genome wide similarity between races, levels of divergence between races, identification of virulence genes to understand population biology of pathogens. Genome sequencing and comparative analysis help to identify the shared and species-specific pathogenicity factors/genes in closely related species. It contributes immensely to clarify taxonomy of pathogens at species level. It also helps to determine secondary metabolites encoding gene clusters and effector coding genes. Absence, and presence of species-specific regions on the chromosomes of the pathogen can be identified. Genome sequencing, and functional genomics approaches have increased insight into molecular plant–pathogen interaction which help to assess genetic transformation efficiency using homologous recombination. It is possible to estimate divergence time, and evolutionary status of related pathogenic species. A hybrid mapping, and genome assembly approaches help to investigate gene content variation in different isolates of pathogens. The studies on genome reveal its organization, function of genes, gene inactivation, modification, and expression regulation. Genome sequences also facilitate comparative transcriptomic and secretomic analysis of pathogenic species, clades, and isolates. Comparative genomics is an important tool to mine pathogenicity determinants, host specificity factors, and particularly type III effectors/gene in bacterial pathogens.

7.4 Pathogens Genome Assembly and Characteristics

The information on genome sequences, analysis, and assembly of limited pathogens of crucifers has been generated and made available. It has given deeper insight into the molecular and biological functions of host–pathogen interactions in different ecological situations in nature and directed in the modern agricultural operations. The *Albugo candida* genome size is 45.3 Mb which is significantly smaller than a comparable biotrophs, *Hyaloperonospora arabidopsidis*. The *A. candida* genome is deficient in enzymes required for nitrate, and sulfate assimilation and has a reduced number of cell wall degrading enzymes. It has low number (26) of RXLR effector-like genes. *Alternaria brassicae* genome assembly consists of 17 contigs of 34.14 Mb sequences. The contigs ranges in length from 66.7 to 7.01 Mb. Of the 17 contigs, one represents the mitochondrial genome. The assembly contains nine complete chromosomes. The *Colletotrichum higginsianum* genome is 53.4 Mb in size and distributed among 12 chromosomes including two mini-chromosomes less than 1 Mb. The fungi causing powdery mildew disease has about four times larger genome size than those of other ascomycetes. The genome of *G. orontii* is 160 Mb in size. The fungi *Fusarium oxysporum* genome is 57.3 Mb in size with 583 scaffolds. The G:C content is 47.4%. The genome analysis of *Hyaloperonospora* has revealed phylogeny of *H. parasitica* with other species infecting cruciferous genera, and species along with major species clusters in it. The *Leptosphaeria maculans* genome is 45 Mb in size, and organized in an unused patchwork fashion with G:C rich blocks known as isochore structure. The *Plasmodiophora brassicae* genome is compact and small in size of 24.2–25.5 Mb with high gene density. Thirteen chitin syntheses gene have been identified. There is reduction in intergenic space with multiple overlapping un-translated regions (UTRS) in *L. maculans*. The white stem rot pathogen *Sclerotinia sclerotiorum* genome is 38.3 Mb in size and contains 679 sequence contigs giving 8× sequencing coverage of the genome. A complete new assembly of *S. sclerotiorum* genome contains an additional 80 s, 146 bp. It produces a number of extracellular pectin–degrading enzymes, and biosyntheses genes of phytotoxins and other secondary metabolites. Bacterium *Xanthomonas campestris* comprises circular chromosomes of approximately 5,000,000 base pairs (bp), have a high G + C content, and do not carry plasmids. Genome wide comparisons have revealed three secretion system (type I, III, IV) associated with pathogenesis. The genome of *Turnip Mosaic Virus* (TuMV) isolates are 9798 nt in length, few are one to three nucleotide shorter in the terminal UTRs. Genome sequencing, comparative genome, transcriptome, and secretome analysis of pathogenic species, clades, and isolates facilitates determination of phylogeny, host specificity, pathogenicity factors/genes, evolution period, clonal reproduction, genetic exchange, and molecular genetical mechanisms and events in all host-pathosystem.

7.5 Genomic Regulation of Crucifers' Host–Pathogen Interaction

The genomics of crucifer's host–pathogen interaction is very complex and intriguing owing to be challenged by a large range of pathogen from necrotrophs (*Alternaria*, *Leptosphaeria*) to biotrophs (*Albugo*, *Hyaloperonospora*) and hemibiotrophs (*Sclerotinia*, *Colletotrichum*). Some pathogens are strictly foliar (*Alternaria*, *Colletotrichum*) to root invaders (*Fusarium*, *Verticillium*) and ectoparasites (*Erysiphe*) to endoparasites (*Plasmodiophora*). The mode of host entry is through natural openings (stomata, hydathodes), injury, and direct through appressorial formation and enzymatic action after softening of hard tissues. Mode of acquisition of nutrition is from absorption to holophytic. Mixed infection and co-infection of pathogens have synergistic effects on host response. Host range is very wide to narrow and exhibiting host specificity. The effect on the host tissues is from specific (foliage, stem) to systemic and vascular system. Each of these factors has great bearing on the host–pathogen recognition process during pre- and post-host–pathogen interaction. Immediately after establishment of recognition process, the pathogens respond through initiation and expression of virulence genes to suppress immune system of the host for successful pathogenesis. The expression level of virulence genes differs in host–pathogen interaction to manipulate host defense in favor of pathogens to feed them for infection, development, and survival to continue the life cycle.

The evaluation of published literature on host–pathogen systems reveals the expected recent increase in whole genome datasets for a broad range of host–pathogen studies, and also what is missing when addressing complex systems on ecological and spatiotemporal scales. This highlights a gap that future efforts might be specifically focused on in order to be able to project and test whether the identified underlying genetics of host–pathogen interactions scale-up from simple systems, such as a laboratory study, to ecological and spatiotemporal complex situations in the wild. The quantity and complexity of the sequence data generated in many recent and ongoing studies of host–pathogen interactions present a unique opportunity for further investigations addressing as yet unexplored aspects of the data. Perhaps most critically, however, it is suspected that this massive volume of complex data poses an increasing challenge for comparisons across studies. The lack of comprehensive cross-taxon comparative databases of host–pathogen interactions likely impedes the synthesis of individual host–pathogen studies and translation of new knowledge into solutions for real world problems. In the real-world, (a) pathogens attack hosts in the context of changing host environments, (b) these environments are increasingly impacted by anthropogenic forces such as climate change, and (c) are usually characterized by diverse communities of pathogenic and non-pathogenic organisms.

Studies of host–pathogen interactions suggest that these complexities are rarely captured in a single study. Thus, it will be essential to conduct comparative studies and perform meta-analyses of existing data across systems in order to achieve a comprehensive synthesis of how genomics can address host–pathogen interactions at different scales.

7.6 Genomic Mechanism of Crucifers' Host–Pathogen Interaction

Genomics of host–pathogen interactions have greatly advanced in perception, and understanding of molecular, and genetical mechanisms of host–pathogen recognition system and pathogenesis. Next-generation sequencing, and omics technologies have enabled the sequencing of entire host and pathogen genome or transcriptome more efficiently for better comprehension of host-pathosystem. Pathogenomics have revealed how pathogens adopt a specific host, subvert innate immune responses and change host range, and develop new virulent races/strains/pathotypes. In *Brassica–Albugo* interactions, the host–*R* genes control pathogen infection and development in compatible (susceptible) and incompatible (resistant/tolerant) cultivars. Differentially expressed proteins regulate defense reactions in the resistant varieties. The boost in cysteine and GSH levels is crucial for incompatible interaction between *A. candida* and *B. juncea*. In *A. candida–B. juncea* pathosystem, the timing of the induction of PR-5 in *R* and *S* varieties is crucial for mounting of an effective defense to the pathogen. During infection and pathogenesis of *Albugo* in *Brassica* genotypes differing in level of resistance (R, PR, MS, S), *R* genes regulate functions, and molecular mechanisms. In *Brassica–Alternaria* interactions species-specific genes of *Alternaria* with putative roles in virulence have been identified. Transcriptomic analysis has revealed six genes strongly associated with *A. brassicicola* pathogenesis. The *AbPf2* gene activates pathogenicity in *A. brassicicola*. The *AbNPS2* gene plays an important role in disease development and virulence. Among several genes identified as pathogenicity factors, *Amr1* gene is important for success of *A. brassicicola* as competitive saprophyte, and plant parasite. Microarray analysis of *Brassica* infected with *Alternaria* at 48 hpi revealed differential expression of 87 genes mainly involved in cell response signaling, cell wall degradation, protein degradation, auxins-induced metabolism, being downregulated. The change in host cell ultra-structure and transcriptome reprogramming has been observed in *Alternaria*-infected *Brassica*. Defense-related genes are activated also in *Brassica–Alternaria* susceptible interactions. Mitogen-activated protein (MAP) kinase and CAMP/PKA signaling pathways are involved in fungal growth, conidial formation, stress tolerance, and pathogenesis in *Arabidopsis* by *Colletotrichum higginsianum*. During pathogenesis, host peroxisomes, microtubules, and Golgi are novel targets for fungal effector genes. The Fus3/Ksi1-related MAPK chMK1 plays a significant role in cell wall integrity. In *Brassica-Erysiphe* interactions, three closely related genes, *MLO2*, *MLO6*, and *MLO12*, play a role in powdery mildew pathogenesis through suppression of basal defense responses. Powdery mildew effector genes promote pathogenesis through host transcriptional regulation. The WRKY 18/40 regulate induced expression of genes whose products are required for pathogen growth and reproduction. In response to powdery mildew infection, the expression of genes is regulated for different biological functions by transcriptional genes. The powdery mildew pathogenesis and disease development is governed by the genes *AtLFG1* and *AtLFG2* in *A. thaliana*. Several genes viz., *RSP1*, *RSP2*, *DMR1*, *LRR-RLK*, *DMR6*, and *DLO1* govern pathogenesis of *Brassica* by *Hyaloperonospora* enabling susceptibility process like attraction and attachment of pathogen to host cells, accommodation of specialized infection and feeding structures inside host cells, as well as nutrient production and transport from host to pathogen. Haustorium formation is initiated when pathogen secrete hydrolyses to locally degrade the host cell wall, and establish an entry point. Haustorial penetration is accompanied by reorganization of GFP-tagged cytoskeleton, and organelles in *Arabidopsis* mesophyll cells.

In *Brassica-Leptosphaeria* pathosystem, *BnMPK9-AvrLm1* interaction enhances cell death caused by BnMPk9. The virulence mechanisms are the transition period from biotrophy to necrotrophy to facilitate the acquisition of host nutrients by the pathogen. The AvrLM1 and AvrLM4-7 contribute to the aggressiveness of *L. maculans* analysis during leaf infection. Genome and transcriptome analysis during *Brassica* pathogenesis by *Leptosphaeria* has revealed several genes for 20 major functional categories. The GFP is expressed by the Cht promoter in stem lesions of *Brassica* during infection of *Leptosphaeria*. The gene *Pro1* is involved in clubroot pathogenesis by stimulating resting spore germination through its proteolytic activity. The genes *PbPP2A* and *PbHMG* are expressed in the early phases of infection in *Brassica rapa* roots. The BjN1t1 expression in *B. juncea* during infection promotes over production of auxins to form galls in infected roots. The role of different genes has been revealed through analysis of *P. brassicae* genome and life stage specific transcriptomic. Several effectors (genes) involved in the infection process (life cycle) of *P. brassicae* and subsequent clubroot development in its host have been identified. The *Plasmodiophora* pathogenesis is influenced by lipids in resting spores, glucosinolates content of the host, and host resistance level. Genes expression levels of primary and secondary zoospores govern different mechanisms utilized by the pathogen in causing primary and secondary infections of *Plasmodiophora*. Differential expressions of up- and downregulated genes and functional groups have been determined after 4, 7, and 10 DAI during life cycle stages of *Plasmodiophora*. In *Brassica* during *Plasmodiophora* infections, there is differential expression of several genes in host shoots, and roots regulating different biological, biochemical, and hormonal responses. In *Arabidopsis* roots infected with *Plasmodiophora*, hypoxia and sweet genes are expressed to form galls. Host species-specific genes are expressed during *Brassica–Plasmodiophora* interactions. The genes *Pb-YPT*, *Pb-Brep9*, and *Pb-PSA* regulate *P. brassicae* infection. During clubroot development, there is expression of MiRNA and immunophilins. *Plasmodiophora* manipulates host defense through methyltransferase by inactivating SA to Me SA. The GPCR signal transduction pathways play important roles in the growth, development, and pathogenicity of *P. brassicae* to crucifers

(Saharan et al. 2021a). During *Sclerotinia* pathogenesis of *Brassica*, four endo-polygalacturonase genes (*sspg1d*, *sspg3*, *sspg5*, *sspg6*) and two exo-PG genes (*ssxpg1*, *ssxpg2*) are expressed. Molecular analysis of *S. sclerotiorum* has revealed its biology, mycelial growth, virulence, pathogenesis, and signal events for sclerotia formation. Genes like *Mats*, *FoxE2*, *CAMP*, and Nsd1 control apothecial, and ascospore development. The *Sclerotinia* infection and pathogenesis is regulated by toxins, CWDEs, ROS, OA, Ssvp1, V263, Svf1, PKs13, and Nacα. During the course of *Brassica* infections by *Sclerotinia*, transcriptomic analysis revealed genes for functional groups like hydrolytic enzymes, CAZymes, secondary metabolites, detoxification, signaling, development, secreted effectors, oxalic acid, and reactive oxygen species production. The genes *SsSvf1* is involved in coping with ROS for pathogenesis of *Sclerotinia*. There is differential expression of genes during *Sclerotinia* interaction with *R* and *S* genotypes of *Brassica*. Integrated mRNA, sRNA, and degradome sequencing revealed complex responses to *Sclerotinia* infection in *Brassica*. The gene *SSITL* is an effector which suppresses resistance in the host at early stage of *Sclerotinia* infection. Oxaloacetate acetyl-hydrolase gene is involved in *Sclerotinia* pathogenesis. Three *Sclerotinia* genes (*SsCTR1*, *Ss CCS*, *Ss ATX1*) are upregulated during infection of *B. oleracea*. Several spliceosome genes are differentially altered during *Sclerotinia* infections. The P3 proteins are important in the infection cycle of *TuMV* and involved in virus replication, accumulation, symptomatology, resistance breeding, and cell to cell movement. The *BnSGS3* gene in *Brassica* has differentiated effects on the pathogenesis of viruses. Nitric oxide signaling and calcium protein kinases can modulate NADPH oxidase activity in modulating *A. thaliana-TuMV* pathosystem. Differential expression of proteins has been identified which are activated in response to *V. longisporum* in the leaf apoplast and xylem sap of *Brassica* in relation to disease development, photosynthetic electron transport, gas exchange, and nutrient status. The secretion system of *Xanthomonas* genome regulates pathogenesis of black rot in crucifers. The features and functions of Type iii effector genes of *Xanthomonas* in pathogenesis of crucifers have been characterized. The effectors/genes of *Xanthomonas* governing virulence and pathogenicity have been identified. Key pathogenicity determinants and virulence genes of *Xanthomonas* have been characterized and exported by Type ii and Type iii secretion system (T3S). The hormonal signaling in the pathways of *Xanthomonas* infection and pathogenesis of *Brassica* has been revealed.

The use of evolutionary genomics on the host–pathogen interaction and its impact can clear roadmap to combat the continuous threat of pathogens in a changing world. New tools will facilitate to resolve challenges of genomics of host with multiple pathogens interactions. The whole genome analysis approaches offer the potential to reveal new unbiased insight into the genetic basis of host-pathosystem of crucifers'.

7.7 Importance and Scope of Molecular Detection and Identification

Development of methods for the rapid identification of new and already known pathogens and races is imperative for sustainable disease management. With the advent of several molecular and DNA sequencing technologies, researchers can now generate high-resolution genotypic data from any organism. However, the quality of the data depends on the quality and purity of the DNA sample, which means that the pathogen under study needs to be in excess compared to the host plant or cultivated axenically, which is not possible for the biotroph like *A. candida*. The field pathogenomics of wheat yellow rust has been investigated using cDNA sequencing from infected leaves in the field, resulting in c. 37% of reads being mapped to the pathogen reference genome. In addition, complexity reduction methods have been developed for sequencing of specific subfractions of plant genomes (MutChrom Seq and exome capture in wheat). Jouet et al. (2019) reported a complexity reduction method (Pathogen Enrichment Sequencing or "Pen Seq") that enables capture and sequencing of plant-associated microbial sequences directly from field samples. It can explore race diversity in *A. candida* and the impact of *A. candida* on the host leaf microbiome.

To control disease outbreaks in crops, it is imperative that pathogen and pathogen population variability in the field can be quickly identified. This can be achieved by having a full set of genetic markers for different pathogens and pathotypes, not only contained within the local fields but universally. The initiatives for *P. brassicae* showed that *Cr 811* was a good candidate gene to differentiate pathotype 5 from the rest of the pathotypes (Feng et al. 2016). Molecular characterization of pathogen effectors is gaining momentum, with better genome assemblies and higher gene prediction accuracies. This apparent have better insights into the structures and functions of effector proteins, which allows more efficient screening and identification of *R* genes (Gibriel et al. 2016; Neik et al. 2017).

The accurate and rapid identification of the organism(s) that cause plant disease is essential for effective disease control. It enables more informed decisions to be made about cultivar choice and how and when chemicals can be used most effectively to control disease epidemics. Pathogen detection and identification is important for preventing the spread of disease by screening seeds and other planting material and is needed to implement quarantine regulations. It is also important more generally in many areas of plant pathology research, for example in: studies of disease epidemiology, studies of yield loss disease relationships, and in designing new strategies for disease control. Pathogen or disease diagnosis is, therefore, fundamental to virtually all aspects of plant pathology.

Traditional approaches to disease diagnosis generally first involve the interpretation of visual symptoms. This may be followed by laboratory identification, using for example selective media and microscopy, to confirm the diagnosis. In some cases, these methods are still the cheapest, simplest, and most appropriate. Conventional methods do, however, have a number of drawbacks, which has prompted the search for alternative diagnostic techniques. Traditional methods generally require skilled and specialized microbiological expertise, which often takes many years to acquire. There is a need to use more generic techniques that can be taught quickly and easily to relatively unskilled staff. Methods that involve culturing can often take days or weeks to complete and this is not acceptable when rapid, high throughput diagnosis is required. The results are not always conclusive, e.g., where similar symptoms can be caused by different pathogens or physiological conditions. Closely related organisms may be difficult to discriminate on the basis of morphological characters alone. It may also be necessary to discriminate between populations of the same pathogen that have specific properties, fungicide resistance, toxin production, or differences in virulence. Traditional methods may not be sensitive enough (e.g., where the detection of presymptomatic infection is needed) and as such much effort has been devoted to the development of novel methods for detecting and identifying plant pathogens over the last decade. The greatest effort has been on the development of diagnostics based on antibodies and nucleic acid technologies (Duncan and Torrance 1992; Schots et al. 1994; Ward 1994; Martin et al. 2000; McCartney et al. 2003). The details of protocols, methods, procedures, etc. have been covered in the chapter on techniques of crucifer's host-pathosystem.

Compared to plants and animals, pathogen has simple body plans with often morphologically and ecologically obscure structures. This poses challenges for accurate and precise identifications. The identification of pathogen encouraging the approach of integrative (polyphasic) taxonomy for species delimitation, i.e., the combination of genealogy (phylogeny), phenotype (including autocology), and reproductive biology (when feasible) allows objective evaluation of diagnostic characters, either phenotypic or molecular or both. Verification of identifications is crucial but often neglected. Because of clade-specific evolutionary histories, there is currently no single tool for the identification of pathogen, although DNA barcoding using the internal transcribed spacer (ITS) remains a first diagnosis, particularly in meta-barcoding studies. Secondary DNA barcodes are increasingly implemented for groups where ITS does not provide sufficient precision. Issues of pair wise sequence similarity-based identifications and OTU clustering, and multiple sequence alignment-based phylogenetic approaches with subsequent verification are recommended as more accurate alternatives. In meta-barcoding approaches, the trade-off between speed and accuracy and precision of molecular identifications must be carefully considered. Intragenomic variation of the ITS and other barcoding markers should be properly documented, as phylotype diversity is not necessarily a proxy of species richness. Important strategies to improve molecular identification of pathogen are: (1) broadly document intraspecific and intragenomic variation of barcoding markers; (2) substantially expand sequence repositories, focusing on under sampled clades and missing taxa; (3) improve curation of sequence labels in primary repositories and substantially increase the number of sequences based on verified material; and (4) link sequence data to digital information of voucher specimens including imagery. In parallel, technological improvements to genome sequencing offer promising alternatives to DNA barcoding in the future. Despite the prevalence of DNA-based pathogens taxonomy, phenotype-based approaches remain an important strategy to catalog the global diversity of microbes and establish initial species hypotheses (Lücking et al. 2020).

7.8 Detection and Identification of Pathogens, Pathotypes, and Effector Genes with Molecular Approach

The accurate, rapid detection and identification of the pathogens is essential for effective management of crucifers' diseases. The traditional approaches are time consuming, expansive, laborious, and less in accuracy in comparison to molecular approaches like barcoding markers, qRT-PCR, cDNA sequencing, pathogen enrichment sequencing (Pen seq), and use of sequence repositories. Pen Seq approach has been used to detect *Albugo* from infected leaves and to identify variations in pathogen genotypes. The distinct lineages that had diverged from each other were also identified. Many putative recombinant regions were identified between lineages. Pen Seq can identify *Albugo* race structure with phylogenetic lineage. The *Albugo candida* ITS1 DNA has been detected in symptomatic and asymptomatic host tissues. Molecular markers like RAPD and ESTs have been used for identification, detection, and analysis of genetic variability in *Alternaria* species infecting crucifers. In the genome of *Colletotrichum*, two effectors (genes), ChELP1 and ChELP2, homologs of LysM have been characterized. Molecular quantification of *Erysiphe* infection and development in host tissues have been assessed using qPCR assessing host cell entry, area covered by hyphae, conidia, and conidiophores. The expression of genes has been detected in response to powdery mildew infection with a broad range of transcriptional changes conferring susceptibility, resistance, and bimolecular regulations. The phylogenetic relationship of powdery mildew of crucifers with other powdery mildew species and isolates variations have been determined through sequence analysis of ITS rDNA and sequence comparison of DNA encoding 5.85 rDNA, ITS1, and ITS2. DNA (RAPD) finger print approach has been used to distinguish isolates of *H. parasitica* infecting different *Brassica* species. The combination of LMD and RNA sequencing of *Brassica* improves *Leptosphaeria* detection at the early stage of infection. Sequencing of gene fragments of β-tubulin, actin, and ITS region of rDNA from *L. biglobosa* isolates from different countries could show closer taxonomic similarities between different clades. Genomic prediction tools are valuable approaches in *Brassica* to assess *Leptosphaeria* species. *Plasmodiophora* virulence genes expressed during early stages of colonization of *Brassica* roots have been identified using SSH and EST analysis. AqPCR-based *Plasmodiophora* detection and spread in the soils has been developed for country-wide assessment. ITS sequencing and phylogenetic analysis of *Pseudocercosporella* isolates collected from *Brassica* species revealed a high degree of genetic similarity. Real-time PCR protocols have been developed to detect and discriminate anastomosis groups (AGs) of *R. solani* using ribosomal internal transcribed spacer (ITS) regions or β-Tubulin sequences of pathogen from field soils. Phenotypic pipeline has been developed to assess kinetics of disease symptoms caused by *Sclerotinia* in *A. thaliana* conferring susceptibility by NLR genes. A simple, rapid, sensitive, and cost-effective mRT-PCR protocol has been developed for detection and differentiation of *Brassica* viruses. Molecular approaches like DAS-ELISA, Rt-PCR, and ELISA can detect, identify, and confirm phylogenetic relationship of *Brassica* viruses and their strains. Simplex and multiplex PCR assays based on DNA sequence from rITS region and coding and noncoding regions of actin, elongation factors 1-α, glyceraldehydes-3-phosphate dehydrogenase, and tryptophan synthase genes are valuable approaches for identification, detection, and phylogenetic analysis of *Verticillium* species. *Verticillium longisporum* isolates have been characterized on the basis of a group-1 Intron in SSU-rRNA gene of pathogen. Isolates from two different species, i.e., *V. dahliae* and *V. albo-atrum* have been characterized based on AFLP analysis of rRNA and DNA from mitochondria. Several virulence genes of *Xanthomonas* have been identified (*pig, pig B, DF, EPSs, DSF, cyclic-di-GMP, rpfl, RpfF, RpfG, WspR, xantham, LPs*) which regulate production of enzymes for infection and development of black rot in *Brassica* by multiplication and movement of bacteria in host tissues. Several novel genes for resistance to *Xanthomonas* in *Brassica* have been identified by SSH–PCR-based techniques. Molecular identification of different pathotypes, pathogens, elicitors/effectors have been revealed using *R* genes in the form of isogenic lines of *Brassica*. The accuracy of qRT-PCR analysis data expression of target genes is affected by factors such as RNA quality, purity, and PCR amplification efficiency, technical, and biological variations. To control these variables and avoid bias in qRT-PCR, selection of appropriate RGs is crucial under multiple stresses. The interplay between multiple pathogens and associated microbiomes requires disentangling of hidden histories in genes and genomics of both hosts and pathogens. *R* genes homologues DNA fragments in *A. thaliana* have been identified to use them for cloning and in breeding for resistance against major pathogens of crucifers. The molecular approaches are valuable tools and their efficiency and accuracy depends on proper sampling and cares in using them for correct diagnosis (Lücking et al. 2020; Saharan et al. 2021a, b, c).

7.9 Significance of Pathogenic Variability

Pathogenic variability in the pathogens of crucifers has been determined in the form of races, pathotypes (fungal pathogens), pathovar (bacterial pathogens), strains (viral pathogens), and genetical variants (All pathogens). The standard procedure to detect pathogenic variability is to use isogenic lines of the host differentials and record differential interactions of different isolates of the pathogen on host genotypes. It is based on gene for gene interaction of host genotypes with pathogen isolates. The pathogen avirulent gene recognizes a *R*-gene in the host. On the basis of this differential interaction of host–pathogen genes, a race or pathotype is designated to identify pathogenic variability. These races and pathotypes differ in their virulence and aggressiveness to cause severe epidemics in different crucifer crops depending on host resistance genes in a cultivar and favorable environmental conditions for the development of the pathogen. Under natural conditions and modern agricultural settings, there is evolution of pathogenic variability in the form of new and virulent races through selection pressure of *R* genes in the genotypes/varieties of crucifer's crops during different stages of host–pathogen interactions. The brighter side of pathogenic variability is that a race or pathotype is excellent tool to identify *R* genes loci and *R* genes alleles on a locus of *Brassica* genotypes. It facilitates development of multigene-resistant cultivars to counteract so-called Boom and Bust cycle of resistant cultivars on the evolution of new virulent races of the pathogens. The propensity of a pathogen to evolve virulence depends on features of its biology (mode of reproduction) and its genome (amount of repetitive DNA). Avirulence effector genes in the pathogen are key pathogenic genes that are complementary to *R* genes of the host.

The plant disease resistance genes (*R* genes) that have been molecularly characterized so far can be grouped into several classes based on similarities in the function or amino acid sequence of the proteins they encode. The class containing the majority of cloned *R* genes is characterized by the presence of an N-terminal nucleotide binding site (NBS) and a C-terminal stretch of leucine-rich repeats (LRRs). A number of genes belonging to this class have been cloned, such as the *RPS2* (Bent et al. 1994; Mindrinos et al. 1994) and *RPM1* (Grant et al. 1995) genes (against *Pseudomonas syringae*) and the *RPP5* gene (against *Hyaloperonospora parasitica*) (Parker et al. 1997) from *Arabidopsis thaliana*, the *N* gene from tobacco (against tobacco mosaic virus) (Whitham et al. 1994), the *PRF* gene from tomato (against *Pseudomonas syringae*) (Salmeron et al. 1996), and the *L6* gene from flax (against *Melampsora lini*) (Lawrence et al. 1995). Although these *R* genes are found in diverse plant species and are active against a range of pathogens, the conserved NBS-LRR features suggest a common function in the defense response against pathogen attack, probably as part of the signal transduction pathway (Staskawicz et al. 1995).

The isolation and characterization of more of these genes is important because they may provide clues about the complex mechanisms of resistance, the interactions involved in pathogen recognition, and the evolution of the *R* genes. Furthermore, cloned genes can be transferred to other species (Rommens et al. 1995; Thilmony et al. 1995) to study the resistance mechanism in a completely different genetic background. It is known from hybridization experiments with the cloned NBS-LRR type of *R* genes that several related sequences can be found, both in the same species and in other species (Lawrence et al. 1995; Mindrinos et al. 1994; Parker et al. 1997; Salmeron et al. 1996; Whitham et al. 1994). Isolation of these sequences by polymerase chain reaction (PCR) may be attainable. Despite the general lack in DNA sequence conservation between *R* genes, there are a few conserved protein motifs present in the NBS. With degenerate primers based on these homologous regions, it is possible to amplify several resistance-gene-like (RGL) DNA fragments, as has been shown for soybean and potato (Kanazin et al. 1996; Leister et al. 1996; Yu et al. 1996). Some of these DNA fragments have been mapped in the vicinity of known disease resistance loci. A similar approach in *A. thaliana* has been applied with a degenerate primer based on the conserved amino acid sequence of the so-called P-loop motif and another degenerate primer downstream based on the sequence of the conserved domain 5 (Aarts et al. 1998; Lawrence et al. 1995).

A. thaliana is a good model species to test the general applicability of a PCR-based isolation method, as many of the plant–pathogen interactions in this species have been studied (Crute et al. 1994), the genetic positions of many resistance loci have been determined (Kunkel 1996; Holub 1997), and some resistance gene homologues are represented in the extensive EST data base (Aarts et al. 1998).

Much of current knowledge of plant infecting fungi is derived from narrow host range pathosystems, which often exhibit particular characteristics resulting from a pronounced evolutionary arms race between host and pathogen. This arms race often results in pathogens evolving virulence genes that are counteracted by corresponding resistance loci in the host, which are themselves counteracted by new or variant virulence genes in the pathogen, ad infinitum. This is termed a "gene-for-gene" interaction. Plant pathogenic fungi that co-evolve with their hosts in a gene-for-gene manner often exhibit compartmentalized genomes that contain alternating no repetitive, gene-rich regions and repetitive, gene sparse regions. Virulence-associated genes are often clustered in repetitive, gene sparse regions, and it is thought that this clustering and compartmentalization allows for their rapid evolution through transposable element (TE)-mediated duplication and mutation, without affecting housekeeping genes. This theory is known as the "two-speed-genome" hypothesis (Faino et al. 2016).

Enhanced mutation rates in repeat-rich regions in fungi may arise through a process termed repeat-induced point mutation (RIP). This process is a genomic defense against excessive TE activity, which could be deleterious if left uninhibited. It causes point mutations in duplicated sequences through 50-methylation of cytosine followed by domination to thymine. As a result, RIP-affected genomic regions often exhibit pronounced A/T content concomitant with a decrease in G/C content (Smith et al. 2012). Virulence-associated proteins are often found to have been affected by this process, which may be important for their rapid adaptation (Rouxel et al. 2011). Many of the virulence-associated genes of plant pathogenic fungi studied to date encode small secreted proteins, whose sole purpose is to manipulate plant defenses to advance fungal infection (Lo Presti et al. 2015). These proteins, termed "effectors," have been discovered in multiple plant pathogenic fungi and exhibit numerous different functions depending on fungal lifestyle. For example, necrotrophic fungi, which require dead tissue on which to feed, often produce effectors that promote cell death, whereas biotrophic fungi, which require living tissue, produce effectors that prevent cell death (Lyu et al. 2016). Hemibiotrophic fungi, which require both living and dead tissue at different life cycle stages, may produce different effectors at different time points during infection (Rudd et al. 2015). Many effector genes studied are species or even isolate specific and interact directly or indirectly with the products generated by particular intraspecifically polymorphic loci in the host, thus conforming to the gene-for-gene hypothesis (Bourras et al. 2015, 2016). However, several effectors are also conserved across various fungal species and have seemingly similar roles in pathogenesis on plants, interacting with loci conserved within a species and/or between species (Sanz-Martín et al. 2016; Redkar et al. 2015; Hemetsberger et al. 2015; Derbyshire et al. 2017).

7.10 Genomics of Pathogenic Variability

Under natural conditions and modern agricultural settings, the evolution of pathogenic variability in the form of new and virulent races is through selection pressure of *R* genes in crucifers' genotypes at different stages of host–pathogen interactions. Polyploidy in *Albugo candida* contributes to race diversification and evolution by selection pressure imposed from cultivated *Brassica* species. Molecular studies have revealed high degree of genetic diversity within *Albugo* pathogenic to *Brassica* species with wide host range and six new specialized species along with virulent races. The ITS and COX2 approach of molecular phylogenetic analysis and morphological variations of oospores have divided *A. candida* into three groups (1, 2, and 3) infecting different cruciferous with different races. Several races/pathotypes of *A. candida* have been identified at global level infecting different crucifers. The virulence of *A. candida* races varies from very narrow to wide infecting different *Brassica* species and genotypes. Phylogenetic analysis and nucleotide sequences have distinguished races of *A. candida* infecting different crucifers in different countries of the world. *A. candida* isolates from different crucifers have been characterized on the basis of virulence genes specificity to their hosts in Australia. A single dominant gene *AvrAC1* controls avirulence in race 2 of *A. candida* infecting *B. rapa* cv. Torch. The genetic variation among geographically distinct isolates of *A. brassicae* has been assessed using RAPD-PCR markers. The genetic structure of *A. brassicicola* population indicates the occurrence of sexual recombination's with a cryptic sexual stage. Mutation of *Amr1* gene in *A. brassicicola* causes increased virulence. Three pathotypes of *A. brassicae* viz., RM1, RM2, and V3 have been identified using eight commonly cultivated *Brassica* species. The isolates of *A. brassicae* collected from different geographical areas are different in conidial morphology, cultural characteristics, and symptom production on different hosts. *A brassicae* isolates from different locations show high degree of genetic similarity. There is differential protein expression by the isolates of *A. brassicae*. The isolates of *A. brassicae* show variability in biochemical contents sensitivity to different fungicides and plant extracts. The variation in relative levels of virulence in the form of disease incidence, severity, and defoliation within and between *Alternaria* spp. causing leaf spot disease on rapeseed-mustard has been recorded. Several pathotypes of *Alternaria* species have been identified with different designations using differential determinant attributes. The function of effectors/genes, *ChELP1* and *ChELP2* homologs of Lysm proteins has been characterized in *C. higginsianum.* In India, powdery mildew pathogen produces both asexual and sexual stages on *B. juncea*, so there is every possibility for *E. cruciferarum* to express pathogenic variability. The isolates of *H. parasitica* collected from different countries, locations, and host species have been characterized in the form of pathotypes showing host specificity and differential virulence. The isolates of *H. parasitica* from different host species and genotypes have been molecularly identified as pathotypes with their virulence potential using DNA (RAPD) finger printing. The isolates of *H. parasitica* collected and characterized on different crucifers as race and pathotypes cannot be compared in the absence of use of standardized host differentials (isogenic lines) and uniform designation/nomenclature system at international level. Hybridization of *Hyaloperonospora* isolates has been observed. The evolution of virulent races in *Leptosphaeria* has been observed in the areas where crops with major gene resistance coupled with genetic uniformity have been sown. The avirulence genes of *L. maculans* races have been identified from Europe. *L. maculans* field populations have high level of gene diversity and genotypic diversity in France. An avirulent-mutated *AvrLm* gene of *L. maculans* has been identified to detect new *R* genes for breeding blackleg-resistant cultivars of *Brassica*. The genome wide DNA variants and SNP haplotypes of *L. maculans* pathotypes have been identified.

Pathotypes of *Plasmodiophora brassicae* has been identified from various countries using different differential host sets and designating pathotypes in their own way without following any sound uniform system of nomenclature. The genomes of *P. brassicae* pathotypes Pb2, Pb5, and Pb8 have been sequenced to gain insight into genome variations and its correlation with host specificity. Proteomics of *P. brassicae* has revealed potential effectors/genes of pathotypes. The isolates of *P. brassicae* from different countries have been phylogenetically analyzed for similarities and differences of pathotypes. Polymorphism in *P. brassicae* in case of Korean isolates collected from field has been recorded. *Cr811* gene in *P. brassicae* pathotypes P5 plays an important role as molecular markers for identification of race P5, and other new virulences. The different geographical isolates of *P. brassicae* have been identified by rDNA sequence polymorphism. Pathogenic and genetic variations in *P. brassicae* pathotypes infecting a weed *C. flexuosa* and crucifers have been analyzed using RAPDs in Japan. The virulence of *P. brassicae* changes during interaction with CR genes in the host cultivars. The changes in *P. brassicae* pathotypes structure under field conditions have been analyzed through whole genome DNA similarity sequences. The application of RAPD seq has revealed distinct *P. brassicae* populations under Canadian conditions. ITS sequencing and phylogenetic analysis of *P. capsellae* isolates from *Brassica* hosts clearly differentiated them in virulence factors/races.

High level of genetic diversity and recombination's in *Sclerotinia* population in tropical countries has been identified with the groups of isolates as MCGs and MLHs infecting crucifers. The aggressiveness of *Sclerotinia* isolates on *Brassica* has been measured using AUDPC as one of the parameters. The sequence variation of ITS region of *S. sclerotiorum* isolates revealed higher heterogeneity and genetic variability showing presence of clonal and sexual progenies of the pathogen in *B. juncea* growing areas of India. The genetic diversity is higher in *Sclerotinia* populations from Canola and Sunflower in the NCUS. Distribution of microsatellite haplotypes regions is often on multiple crops. The genetic diversity and populations structure of *S. sclerotiorum* has been analyzed with DNA (RAPD) sequences from different countries of the world. The virulence factors of *Sclerotinia* has been identified on the basis of pathogenicity as virulent pathotypes (aggressiveness), genomic factors (dsRNA) controlled Hypovirulence, MCGs, cloned variables, haplotypes, genetic diversity of isolates, and population biology studies with markers such as MCGs, DNA finger printing or microsatellites.

The *TuMV* isolates collected from different hosts could not be characterized on *Brassica* species for virulence. The molecular characterization of isolates from different species of *Verticillium* indicated that isolates infecting crucifers are long spored (amphihaploids or allodiploids) as a result of *Verticillium* interspecific hybridization events. Black rot of crucifers' bacterium has been characterized into pathovars of *X. campestris* on different species of crucifers. The *X. campestris* isolates from oilseed rape are more genetically diverse with specialization to their nonhost than their *Brassica* hosts. The virulent effectors/genes of pathogens have different evolutionary mechanisms to counteract the defense mechanisms of the host. The identification, classification, and utilization of crucifer's pathogen pathotypes from different countries is most challenging task since scientists have not followed standardized uniform system and procedures in selecting host differentials, designation and nomenclature of pathotypes, pure isolates (single spore), and scoring of infection phenotypes. The expression of high genetic diversity by crucifers' pathogens under natural as well as modern agriculture settings is another challenge of variation in pathogenic populations (Saharan and Mehta 2008; Saharan et al. 2014, 2016, 2017, 2019, 2021a).

7.11 Metabolic Events of Crucifers' Host-Pathosystem

The Colletotrichum higginsianum is a hemibiotrophic ascomycetous fungus that causes economically important anthracnose diseases on numerous monocot and dicot crops worldwide. As a model pathosystem, the *Colletotrichum–Arabidopsis* interaction has the significant advantage that both organisms can be manipulated genetically. The genome sequence of *C. higginsianum* has provided insights into how genome structure and pathogen genetic variability has been shaped by transposable elements, and allows systematic approaches to longstanding areas of investigation, including infection structure differentiation and fungal–plant interactions. The *Arabidopsis–Colletotrichum* pathosystem provides an integrated system, with extensive information on the host plant and availability of genomes for both partners, to illustrate many of the important concepts governing fungal–plant interactions, and to serve as an excellent starting point for broad perspectives into issues in plant pathology.

Plants usually defend against microbial pathogens by activating both localized and systemic resistance responses. These responses include hypersensitive response, cell–wall fortification, synthesis of phytoalexins, and production of other antimicrobial secondary metabolites or pathogenesis-related proteins (PR proteins). Signaling molecules implicated in these inducible defense systems include salicylic acid (SA), jasmonic acid (JA), ethylene (ET), abscisic acid (ABA), auxin, gibberellins (GAs), cytokinins (CKs), brassinosteroids (BRs), and reactive oxygen species (ROS). These phytohormones can induce defense responses individually, and also interact synergistically or antagonistically to further orchestrate downstream signaling. A large number of fungal and oomycete pathogens have been reported to infect the model plant *A. thaliana*, either naturally or in the laboratory. As a typical hemibiotrophic fungus, *C. higginsianum* develops a series of specialized infection structures. In particular, the intracellular BH of *C. higginsianum* are equivalent to haustoria, and hence this pathosystem can provide insights into the molecular basis of biotrophy in obligatory parasitic organisms, such as rusts, powdery mildews, and downy mildews, all of which are not readily culturable or genetically manipulated (Shigenaga and Argueso 2016; Yan et al. 2018).

7.12 Crucifers' Host-Pathosystem Metabolomics

During *Albugo–Brassica* interaction, host metabolism is activated to regulate photosynthesis, carbohydrate metabolism, and respiration. The accumulation of soluble carbohydrate increase invertase and trigger sugar signal transduction leading to photosynthetic gene expression. The induction of hyperplasia and hypertrophy of leaf, stem, and floral parts is regulated in the *Albugo*-invaded tissues due to production of IAA. The lipid profile in the form of fatty acid composition differs in *Albugo* species. The enzymes like Alfa amylase, invertase, peroxidase, IAA-oxidase, and ascorbic acid are differentially produced in healthy and *Albugo*-infected *B. juncea* leaves and inflorescence. NRPS-type SM gene clusters in *A. brassicae* produce toxins (Destruxin B, HC toxin, siderophore) for virulence. An aspartate Dt × S4 is required for the production of β alanine from aspartic acid which is essential for production of the backbone structure of toxin Destruxin. The ROS production in *Brassica* infected host is very complex. H_2O_2 and OH production is high and low respectively in moderately resistant host. Enzymes PPO, POX, and CAT increases in resistant host. Alternaria leaf infection in mustard influences production of carbohydrate (sugars), phenols, and chlorophyll contents. *Alternaria–Arabidopsis* pathosystem has revealed regulation of *COL1* gene linked to amount of jasmonic acid and its analogs for functional characterization. The ROS, cell death, and callose accumulation show a spatial and temporal correlation with host susceptibility. *Amr1* gene synthesizes melanin in hyphal cells of *Alternaria* and reduces virulence due to lifestyle switch from pathogenesis to reproduction. *Alternaria* produces several metabolites/toxins which have effects on crucifers, physiology, biochemistry, and molecular process. Host-specific toxin has been used as tool to screen *Brassica* germplasm for sources of resistance to *Alternaria*. *A. brassicae* toxins destruxin B play role in signal transduction leading to programmed cell death, and suppression of defense system in *Brassica*. Several genes/proteins are differentially expressed in toxin-treated *Brassica* hosts for different molecular processes. The enzymes like cutinases and lipases contribute in the infection process of *Alternaria*. The *Brassica–Alternaria* interaction induces extreme alterations in host biochemistry to reduce photosynthetic area, defoliation, accelerated senescence, and low yield in susceptible genotypes. Arginine biosynthesis in conidia of *Colletotrichum* is essential for prepenetration process. Host metabolism supports the growth of *Erysiphe* with an increased level of ploidy of mesophyll cells in compatible interactions. *Hyaloperonospora* infection in *Brassica* host induces changes in metabolic processes like respiration, photosynthesis, nucleic acid, protein synthesis, phenol metabolism, endogenous growth substances electrolyte leakage, and enzymes. *Plasmodiophora brassicae* infection of *B. napus* increases aliphatic, indole, and aromatic GSLs. Aliphatic GSLs induced by JA in *B. napus* play a role during secondary infection stage of *Plasmodiophora* which is regulated by a gene *BnMYB* 28.1. *P. brassicae* infection induces increase in sugar (glucose, fructose) translocation in roots of *Brassica*. Several Br Sweets belonging to clade I and III are upregulated in CS-NIL on *P. brassicae* infection. The metabolism of auxins, cytokinin, and brassinosteroids hormones is altered in *P. brassicae*-infected plants. The *PbBSMT* gene is expressed during complete infection process of *Brassica–Plasmodiophora* host-pathosystem and to minimize chitin-triggered defense responses of the host. The gall formation in the infected roots of *Brassica* is due to altered metabolism, physiology, and modifications of the host cell walls induced by the *Plasmodiophora* through CWDEs genes. The genes like expansins, *XTHs*, *SHV3*, *SVL1*, *WAK*, and enzymes modification play a role in clubroot development during infection process. The genes encoding nutrient acquisition pathways have been identified during *Plasmodiophora* development in the host tissues. Polyamine biosynthesis (putrescine, spermidine, spermine) is vital in *Brassica* for the survival and pathogenesis of *P. brassicae*. For the clubroot development, auxin signaling pathway is activated by TIR family and *ABP1* genes. Metabolites produced in *Sclerotinia* mycelium play a role in sclerotium formation, and pathogenesis. The organic acids and enzymes are key metabolites of *Sclerotinia*. Enzymes like pectinolytic, proteases, cellulases, and gluco-amylases play significant role in pathogenesis by *Sclerotinia*. Several genes are differentially expressed by *Brassica–Sclerotinia* host-pathosystem for virulence at different stages of pathogenesis. *Sclerotinia* uses *Bna NPR1* gene to regulate specific physiological processes of *B. napus* to suppress defense response. The alteration of SA/JA ratio during infection of *Xanthomonas* determines a negative regulator of redox status and phenylpropanoid in the *Brassica–Xanthomonas* pathosystem.

7.13 Genomics Application to Develop Resistant and Genetically Modified (GM) Crops Against Multiple Stresses

The directed agricultural ecosystem with vast monoculture of genetically uniform crops at global level has provided ideal conditions, for exegesis vulnerability of crops to multiple stresses. The green revolution package of improved varieties with nonjudicious use of fertilizers, irrigation, and pesticides has made the crops more vulnerable to attack by multiple stresses. The genomics approaches have revolutionized the biological sciences like biotechnology, molecular biology, molecular genetics, molecular plant pathology, molecular breeding, and genetic engineering to manage the multiple stresses through host resistance and genetically modified (GM) crops. Genomics application during multiple stresses of crops have greatly advanced in perception and understanding in the multiomics era. The use of next-generation sequencing (NGS) technology has enabled the sequencing of entire genomes or transcriptomes of host and pathogen more efficiently for better comprehension of host-pathosystem to apply right approach at right time for breeding stress resistant crops. Crucifers' researchers have exploited multiple approaches including traditional and modern omics era technology to reveal host–pathogen interactions for better management of biological stresses as follows:

7.13.1 High-Quality Genome Assembles

The NGS technology has been applied to get high-quality genome sequences of major *Brassica* crops viz., *R. rapa*, *B. oleracea*, *B. nigra*, *B. napus*, and *B. juncea*. Similarly, high-quality genome assembles for the major pathogens of *Brassica* are now available. The availability of these high-quality genome assembles of both host and pathogens have greatly facilitated the discovery of candidates' genes for molecular mechanisms of host resistance and their functional operations during biological complex mechanisms of pathogens infection and pathogenesis to tackle them more efficiently. The genomics approaches can rapidly identify QTL's, *R* genes, and pathogenicity effector genes for breeding superior *Brassica* crops that are resistant to major pathogens.

7.13.2 Genomics of Host-Pathosystem

The effectors genes for infection, pathogenesis, virulence variability, production of metabolites, toxins, hormones, and induction of signal molecules have been identified from recognition of pathogen by the host to the completion of disease/pathogens life cycle during crucifer's host–pathogen interactions. Rapid and correct molecular detection and identification of pathogens, pathotypes, and their regularity genes technology has been applied.

7.13.3 Genetical and Molecular Mechanisms of Host Resistance

The genetical and molecular mechanisms have been revealed through traditional and omics approaches by identification of *R* genes, QTLs, on different chromosomes of hosts including *R* genes mapping, cloning, isolation, inheritance mechanism, and introgression approaches to breed resistant cultivars. Host resistance signaling network system to multiple stresses has been developed.

7.13.4 Genetically Modified (GM) Crops to Manage Biotic Stresses

The application of genetic engineering enables management of pathogen infection by:

1. Introducing genes from other plants or bacteria encoding enzymes like chitinase or glucanase. These enzymes are helpful in breakdown of chitin or glucan which are essential components of fungal cell walls.
2. Introgression of *R* genes to enhance innate host defense mechanisms, i.e., by activating phytoalexins, proteinase inhibitors, or toxic protein against pathogens.
3. Invoking hypersensitivity to restrict pathogen at site of infection.
4. The use of viral coat protein-mediated resistance against virus pathogens.
5. Improving host defenses to nematodes with the use of proteinase inhibitors (PIs).
6. Production of virus-free plants through tissue culture.
7. Use of gene silencing/RNAi approach.

7.14 Prospects of Indian Mustard Genetically Modified (GM) Crop

Development of transgenic Indian mustard plants with *chitinase* gene to incorporate resistance against *A. brassicae* was attempted by Mondal et al. (2003). The appearance of symptoms of the disease was delayed by a period of 10–15 days in transgenic plants than the control plants. Taj et al. (2004) transformed *B. juncea* with *osmotin* gene and reported that the transformed calli exhibited resistance to the purified *A. brassicae* toxin. It was found that introgression of *osmotin* gene in transformed calli expressed resistance by perturbation of cell cycle machinery. The transformed calli suppressed the expression of P53, thus inhibiting programmed cell death, while caspase 1 was not at all affected. Mondal et al. (2006) incorporated a PR protein-encoding *glucanase* from tomato into Indian mustard plants to have enhanced resistance toward *A. brassicae*. Glucanase kills the invading fungal pathogens by hydrolyzing glucan, a major component of their cell wall. Transgenic plants arrested growth of *Alternaria* hyphae by 15–54% during in vitro bioassay. Under greenhouse challenged conditions, reduced number, size, and spread of lesion/black spots were observed in transgenic plants along with the delay in symptom appearance time in comparison to the control plants. Borhan et al. (2010) transferred *A. thaliana*-derived *WRR4* gene into Indian mustard and the transgenic plants conferred significant level of white rust resistance. Chhikara et al. (2012) co-expressed a barley antifungal class II *chitinase* gene and type I ribosome inactivating protein (*RIP*) gene in Indian mustard to combat *Alternaria* blight disease. During in vitro bioassays, transgenic mustard plants exhibited 44% reduction in *A. brassicae* hyphal growth than the control plants. During green-house screening, when transgenic events were sprayed with fungal spores, delayed onset of the disease along with lesser number of lesions with reduced size spread was reported. Bangash et al. (2013) also optimized an efficient genetic transformation protocol in Indian mustard cv. NIFA RAYA with *WD* (Wasabi defensin) gene using *Agrobacterium*-mediated gene transfer technique. Antimicrobial peptides (AMP) have now become the target of genetic engineering of plants for incorporating disease-resistance traits. Cationic antimicrobial peptides (CAPs) and their synthetic versions are found to express resistance against a broad spectrum of phytopathogens. Rustagi et al. (2014) transformed *B. juncea* (cv. Varuna) with *MSRA1* gene to incorporate resistance against *Alternaria* blight and stem rot diseases. MSRA1 is a synthetic CAP having a fusion of cecropin A and mellitin CAPs with antimicrobial properties. In in vitro bioassays with *A. brassicae*, transgenic lines inhibited *Alternaria* hyphae growth by 44–62% and delayed and restricted infection protection varying from 69% to 85%. For stem rot disease, a protection rate of 56–71.5% was exhibited by the transgenic lines with less number and smaller size of sclerotia formation than the control plants. Ahmad et al. (2015) also transferred *chitinase* gene isolated from *Streptomyces griseus* HUT6037 into Indian mustard cv. Raya Anmol for incorporation of fungal tolerance.

Lectins are ubiquitous carbohydrate-binding proteins, which are found as secondary metabolites in plants and are found in conferring resistance to various biotic and abiotic stresses besides protecting plants against aphids. Kumar et al. (2015) transferred chickpea *lectin* gene into Indian mustard cv. Varuna to impart resistance against *A. brassicae* in transgenic lines as observed by the number of lesions and time taken for appearance of leaf necrosis, with the protection varying in the range of 36–60% in different transgenic plants as compared to the nontransformed plants. The transgenic plants also exhibited tolerance to salt stress (NaCl) and mannitol-induced water stress. Ali et al. (2017) reported that the constitutive overexpression of *NPR1* isolated from *B. juncea* in transgenic *B. juncea* cv. Varuna exhibited an enhanced resistance to both *Alternaria brassicae* and *Erysiphe cruciferarum* pathogens causing Alternria blight and powdery mildew, respectively. Tasleem et al. (2017) transformed *B. juncea* with *MPK3* gene and analyzed its role in imparting tolerance against *A. brassicae*. The estimation of activity of antioxidative enzyme was determined in transgenic plants. It was found that both ascorbate peroxidase (APX) and guaiacol peroxidase activity as well as proline content were higher in transgenic plants, which led to scavenging of ROS in transgenic plants produced as a result of *Alternaria* infection, which is responsible for inducing resistance trait in transgenic mustard plants. Arora et al. (2019) identified and isolated a CC-NB-LRR protein-coding *BjuWRR1* gene from Donskaja-IV, an exotic line of Indian mustard and transferred it into Indian mustard by *Agrobacterium*-mediated gene transfer technique to validate its role in providing resistance against white rust pathogen, *A. candida*. Transgenic mustard plants exhibited resistance against six isolates of white rust fungus collected from different Indian mustard growing zones of India (Table 7.1, Thakur et al. 2020).

7.15 Crucifers' Host-Pathosystem Genomics Methodology

Crucifers' host-pathosystem has become a favorable model system of biologists to get deeper insight into various biological functions operating and events induced by different gene expressions. The genomics of crucifers' host-pathosystem has been strengthened with different omics approaches/techniques used to reveal genome sequences of major *Brassica* crops and their pathogens. Some of the selected techniques/protocols developed and used by different scientists have been included in this chapter for the benefit of *Brassica* teachers, researchers, and students. The

Table 7.1 Genetic transformation in *Brassica juncea* for disease resistance Thakur et al. (2020)

Sr. no.	Genotype	Gene transferred	Target trait	References
1.	*B. juncea* cv. Pusa Jai Kisan	Chitinase	Resistance against *A. brassicae*	Mondal et al. (2003)
2.	*B. juncea*	*Osmotin*	Resistance against *A. brassicae*	Taj et al. (2004)
3.	*B. juncea* var. BD-3	*Chitinase* and *BAR*	Resistance against *A. brassicae*	Longdou et al. (2005)
4.	*B. juncea* cv. RLM	Tomato *glucanase*	Resistance against *A. brassicae*	Mondal et al. (2006)
5.	*B. juncea* cv. Pusa Jaikisan	Osmotin-ferritin	Resistance against *A. brassicae*	Nirupa et al. (2007)
6.	*B. juncea*	*WRR4*	Resistance against *A. candida*, white rust fungus	Borhan et al. (2010)
7.	*B. juncea*	A barley antifungal class II *chitinase* gene and type I ribosome inactivating protein (*RIP*)	Resistance against *A. brassicae*	Chhikara et al. (2012)
8.	*B. juncea* cv. Varuna	*MSRA1*	Resistance against Alternaria blight and stem rot fungi	Rustagi et al. (2014)
9.	*B. juncea* cv. Raya Anmol	*Chitinase*	Resistance against *A. brassicae*	Ahmad et al. (2015)
10.	*B. juncea* cv. Varuna	Chickpea *lectin*	Resistance against *A. brassicae*	Kumar et al. (2015)
11.	*B. juncea* cv. Varuna	*NPR1*	Resistance against *A. brassicae* and *Erysiphe cruciferarum*	Ali et al. (2017)
12.	*B. juncea* cv. Varuna	*BjuWRR1*	Resistance against *A. candida*, white rust fungus	Arora et al. (2019)

isolates of *A. candida* collected from fields have been characterized with the use of DNA sequence approaches. A PCR test specific only to the presence of *A. candida* DNA has been developed which allows separation of host DNA and detection of pathogen from both symptomatic and asymptomatic leaf materials as well as DNA from plant species for which *Albugo* infection is unknown. The protocol for genome sequencing and assembly of 16 *Alternaria* species with mRNA sequencing and transcript construction has been developed. The real-time PCR-based method has been used for quantitative estimation of seed infection of *Brassica* by *A. brassicae*. *Brassica–Alternaria* interactions have been analyzed with microarray approach. A procedure has been detailed for molecular and biochemical changes using *Brassica–Alternaria* host-pathosystem.

A detailed method has been developed for assessing virulence variability of *Alternaria* species infecting crucifer's. With the *Brassica–Colletotrichum* host-pathosystem, expression of MAPK genes has been characterized. A molecular identification technique for *Erysiphe* at anamorphic stage in conjunction with morphological and host data is very valuable for quick results. A procedure has been used to identify powdery mildew isolates through DNA sequence analysis and DNA markers analysis. Histological assessment of *Erysiphe* growth in host tissues has been revealed. A fast molecular marker identification of *Hyaloperonospora* method based on multiplex PCR (rDNA) analysis is useful for diagnosis of crucifers' downy mildew. DNA finger printing with RAPD analysis can distinguish *H. parasitica* isolates from *Brassica* species. A method for DNA isolation from infected hosts by *H. parasitica* has been developed. A method has been developed to characterize *Leptosphaeria* isolates for virulence on different *Brassica* species. The *L. maculans* field populations have been analyzed for genetic diversity. Phylogenetic relationship of R-loci in *Brassica* species to *Leptosphaeria* has been determined through DNA-PCR analysis. A protocol has been developed for transcriptome analysis of *Brassica–Leptosphaeria* pathosystem to reveal molecular mechanisms. The PCR-based protocols have been developed to detect *P. brassicae* from soil and plant along with its quantification. Bait plants have been used to validate the real-time PCR assay method. *Plasmodiophora* pathotype-specific genes have been identified with PCR and qPCR approach. The qPCR approach has been used for *P. brassicae* quantification from soil to assess response of time-based fertilizers and other control measures along with its spread. RNA seq analysis method has been developed for assessing genes expression during *Plasmodiophora* interactions with their host plants. *P. brassicae* genome has been sequenced along with transcriptomic analysis during *B. napus* infection. The approach of transcriptomic analysis of R and S genotypes of *Brassica* during *Brassica–Sclerotinia* interaction has revealed genes expression controlling resistance and susceptibility of host plants.

A serological test has been developed that allows the early detection of rapeseed petal infection by *Sclerotinia* inoculum under field conditions and has potential to visualize risk of infection. The infection of different viruses in *Brassica* has been detected with multiplex RT-PCR approach. The qRT-PCR assay has been used to detect expression levels of *BnSGS3* gene in different tissues of *Brassica* nontransgenic, single transgenic, and virus-inoculated plants. The presence of viruses on crucifers has been detected with both biological and serological methods. The isolates of *Verticillium* collected from different hosts and locations have been characterized with molecular PCR technique. Protocol has been developed to determine defensive metabolites in the phenylpropanoid pathways during *Brassica–Xanthomonas* interactions. A molecular method has been developed for genotyping of *Xanthomonas campestris* pv. *campestris* isolates. Multispectral imaging approach has been developed for presymptomatic identification of crucifers' pathogens/disease. RAPD-PCR approach has been used for the diagnosis of mixed infection in *Brassica* crops.

References

Aarts N, Metz M, Holub E, Staskawicz BJ, Daniels MJ, Parker JE (1998) Different requirements for *EDS1* and *NDR1* by disease resistance genes define at least two R gene-mediated signaling pathways in *Arabidopsis*. Proc Natl Acad Sci U S A 95:10306–10311

Ahmad B, Ambreen KSH, Khan MS, Haiser A, Khan I (2015) *Agrobacterium* mediated transformation of *Brassica juncea* (L) Czern & Coss. with chitinase gene conferring resistance against fungal infections. Pak J Bot 47(1):211–216

Ali S, Mir ZA, Tyagi A, Mehari H, Meena RP, Bhat JA, Yadav P, Papalou P, Rawat S, Grover A (2017) Overexpression of *NPR1* in *Brassica juncea* confers broad spectrum resistance to fungal pathogens. Front Plant Sci 8:1693. https://doi.org/10.3389/fpls.2017.01693

Arora H, Padmaja KL, Paritosh K, Mukhi N, Tewari A, Mukhopadhyay A, Gupta V, Pradhan AK, Pental D (2019) *BjuWRR1*, a CC-NB-LRR gene identified in *Brassica juncea*, confers resistance to white rust caused by *Albugo candida*. Theor Appl Genet 132:2223–2236. https://doi.org/10.1007/s00122-019-03350-z

Bangash SAK, Khan MS, Ambreen KSH, Siddique AN (2013) Genetic transformation of *Brassica juncea* with an antimicrobial Wasabi defensin gene. Pak J Bot 45(3):993–998

Bent AF, Kunkel BN, Dahlbeck D, Brown KL, Schmidt R, Giraudat J, Leung J, Staskawicz BJ (1994) *RPS2* of *Arabidopsis thaliana*: a leucine-rich repeat class of plant disease resistance. Science 265:1856–1860

Borhan MH, Holub EB, Kindrachuk C, Omidi M, Bozorgmanesh Frad G, Rimmer SR (2010) WRR4, a broad-spectrum TIR-NBLRR gene from *Arabidopsis thaliana* that confers white rust resistance in transgenic oilseed Brassica crops. Mol Plant Pathol 11:283–291. https://doi.org/10.1111/j.1364-3703.2009.00599.x

Bourras S, McNally KE, Ben-David R, Parlange F, Roffler S, Praz CR, Oberhaensli S, Menardo F, Stirnweis D, Frenkel Z, Schaefer LK, Flückiger S, Treier G, Herren G, Korol AB, Wicker T, Keller B (2015) Multiple avirulence loci and allele-specific effector recognition control the pm3 race-specific resistance of wheat to Powdery mildew. Plant Cell 27(10):2991–3012. https://doi.org/10.1105/tpc.15.00171

Bourras S, McNally KE, Müller MC, Wicker T, Keller B (2016) Avirulence genes in cereal powdery mildews: the gene-for-gene hypothesis 2.0. Front Plant Sci 7:241. https://doi.org/10.3389/fpls.2016.00241

Chhikara S, Chaudhury D, Dhankher OP, Jaiwal PK (2012) Combined expression of a barley class II chitinase and type I ribosome inactivating protein in transgenic *Brassica juncea* provides protection against *Alternaria brassicae*. Plant Cell Tissue Organ Cult 108:83–89

Crute IR, Beynon J, Dangl J, Holub E, Mauch-Mani B, Slusarenko A, Staskawicz B, Ausubel F (1994) Microbial pathogenesis of *Arabidopsis*. In: Meyerowitz EM, Somerville CR (eds) Arabidopsis. Cold Spring Harbor Laboratory, Cold Spring Harbor, NY, pp 705–747

Derbyshire M, Denton-Giles M, Hegedus D, Seifbarghy S, Rollins J, van Kan J, Seidl MF, Faino L, Mbengue M, Navaud O, Raffaele S, Hammond-Kosack K, Heard S, Oliver R (2017) The complete genome sequence of the phytopathogenic fungus *Sclerotinia sclerotiorum* reveals insights into the genome architecture of broad host range pathogens. Genome Biol Evol 9:593–618

Duncan JM, Torrance L (1992) In: Duncan JM, Torrance L (eds) Techniques for the rapid detection of plant pathogens. Blackwell Scientific, Oxford, pp 47–62

Faino L, Seidi MF, Shi-Kunne X, Pauper M, van den Berg GCM, Wittenberg AHJ, Thomma BPHJ (2016) Transposons passively and actively contribute to evolution of the two-speed genome of a fungal pathogen. Genome Res 26:1091–1100

Feng J, Jiang J, Feindel D, Strelkov SE, Hwang SF (2016) The gene *Cr811* is present exclusively in pathotype 5 and new emerged pathotypes of the clubroot pathogen *Plasmodiophora brassicae*. Eur J Plant Pathol 145:615–620. https://doi.org/10.1007/s10658-016-0903-0

Gibriel HA, Thomma BP, Seidl MF (2016) The age of effectors: genome-based discovery and applications. Phytopathology 106:1206–1212. https://doi.org/10.1094/PHYTO-02-16-0110-FI

Grant MR, Godiard L, Straube E, Ashfield T, Lewald J, Sattler A, Innes RW, Dangl JL (1995) Structure of the *Arabidopsis RPM1* gene enabling dual specificity disease resistance. Science 269:843–846

Hemetsberger C, Mueller AN, Matei A, Herrberger C, Hensel G, Kumlehn J, Mishra B, Sharma R, Thines M, Hückelhoven R, Doehlemann G (2015) The fungal core effector *Pep1* is conserved across smuts of dicots and monocots. New Phytol 206:1116–1126

Herlihy J, Ludwig NR, van den Ackerveken G, McDowell JM (2019) Oomycetes used in *Arabidopsis* research. Arabidopsis Book 17:e0188. https://doi.org/10.1199/tab.0188

Holub EB (1997) Organization of resistance genes in *Arabidopsis*. In: Crute IR, Holub EB, Burdon JJ (eds) The gene-for-gene relationship in plant-parasite interaction. CAB International, Wallingford, pp 5–26

Jouet A, Saunders DGO, McMullan M, Ward B, Furzer O, Jupe F, Cevik V, Hein I, Thilliez GJA, Holub E, van Oosterhout C, Jones JDG (2019) *Albugo candida* race diversity, ploidy and host-associated microbes revealed using DNA sequence capture on diseased plants in the field. New Phytol 221(3):1529–1543. https://doi.org/10.1111/nph.15417

Kanazin V, Marek LF, Shoemaker RC (1996) Resistance gene analogs are conserved and clustered in soybean. Proc Natl Acad Sci U S A 93:11746–11750

Kumar A, Banga SS, Meena PD, Kumar PR (2015) *Brassica* oilseeds: breeding and management. CABI Publishing, Oxfordshire, p 261

Kunkel BN (1996) A useful weed put to work: genetic analysis of disease resistance in *Arabidopsis thaliana*. Trends Genet 12:63–69

Lagercrantz U (1998) Comparative mapping between *Arabidopsis thaliana* and *Brassica nigra* indicates that *Brassica* genomes have evolved through extensive genome replication accompanied by chromosome fusions and frequent rearrangements. Genetics 150:1217–1228

Lagercrantz U, Lydiate DJ (1996) Comparative genome mapping in *Brassica*. Genetics 144(4): 1903–1910

Lan HY, Wang CH, Zhang LH, Liu GZ, Wan LL, Chen ZH, Tian YC (2000) Studies on transgenic oilseed rape (*Brassica napus*) plants transformed with beta-1, 3-glucanase and chitinase genes and its resistance to *Sclerotinia sclerotiorum*. Chin J Biotechnol 16:142–146

Lawrence GJ, Finnegan EJ, Ayliffe MA, Ellis JG (1995) The *L6* gene for flax rust resistance is related to the *Arabidopsis* bacterial resistance gene *RPS2* and the tobacco viral resistance gene N. Plant Cell 7:1195–1206

Leister D, Ballvora A, Salamini F, Gebhardt C (1996) A PCR based approach for isolating pathogen resistance genes from potato with potential for wide application in plants. Nat Genet 14:421–429

Lo Presti L, Lanver D, Schweizer G, Tanaka S, Liang L, Tollot M, Zuccaro A, Reissmann S, Kahmann R (2015) Fungal effectors and plant susceptibility. Annu Rev Plant Biol 66:513–545. https://doi.org/10.1146/annurev-arplant-043014-114623

Longdou L, Jun GW, Jingxue W, Hongying D, Ruili L (2005) Expression of chitinase gene in transgenic rape plants. Genet Si Biol Molecul Tom 6(1):167–170

Lücking R, Aime MC, Robbertse B, Miller AN, Ariyawansa HA, Aoki T, Cardinali G, Crous PW, Druzhinina IS, Geiser DM, Hawksworth DL, Hyde KD, Irinyi L, Jeewon R, Johnston PR, Kirk PM, Malosso E, May TW, Meyer W, Öpik M, Robert V, Stadler M, Thines M, Vu D, Yurkov AM, Zhang N, Schoch CL (2020) Unambiguous identification of fungi: where do we stand and how accurate and precise is fungal DNA barcoding? IMA Fungus 11:14. https://doi.org/10.1186/s43008-020-00033-z

Lyu X, Shen C, Fu Y, Xie J, Jiang D, Li G, Cheng J (2016) A small secreted virulence-related protein is essential for the necrotrophic interactions of *Sclerotinia sclerotiorum* with its host plants. PLoS Pathog 12:e1005435

Martin RR, James D, Levesque CA (2000) Impacts of molecular diagnostic technologies on plant disease management. Annu Rev Phytopathol 38:207–239

McCartney HA, Foster SJ, Fraaije BA, Ward E (2003) Molecular diagnostics for fungal plant pathogens. Pest Manag Sci 59(2):129–142. https://doi.org/10.1002/ps.575

Mindrinos M, Katagiri F, Yu GL, Ausubel FM (1994) The *A. thaliana* disease resistance gene *RPS2* encodes a protein containing a nucleotide-binding site and leucine-rich repeats. Cell 78:1089–1099

Mondal KK, Chatterjee SC, Viswakarma N, Bhattacharya RC, Grover A (2003) Chitinase mediated inhibitory activity of *Brassicas* transgenic on growth of *Alternaria brassicae*. Curr Microbiol 47:171–173. https://doi.org/10.1007/s00284-002-3980-6

Mondal KK, Bhattacharya RC, Koundal KR, Chatterjee SC (2006) Transgenic Indian mustard (*Brassica juncea*) expressing tomato glucanase leads to arrested growth of *Alternaria brassicae*. Plant Cell Rep 26:247–252

Näpflin K, O'Connor EA, Becks L, Bensch S, Ellis VA, Hafer-Hahmann N, Harding KC, Lindén SK, Olsen MT, Roved J, Sackton TB, Shultz AJ, Venkatakrishnan V, Videvall E, Westerdahl H, Winternitz JC, Edwards SV (2019) Genomics of host-pathogen interactions: challenges and opportunities across ecological and spatiotemporal scales. Peer J 7:e8013. https://doi.org/10.7717/peerj.8013

Neik TX, Barbetti MJ, Batley J (2017) Current status and challenges in identifying disease resistance genes in *Brassica napus*. Front Plant Sci 8:1788. https://doi.org/10.3389/fpls.2017.01788

Nirupa N, Prasad MNV, Jami SK, Kirti PB (2007) Optimization of Agrobacterium-mediated overexpression of osmotin-ferritin genes in *Brassica juncea*. Transgenic Plant J 1(2):384–392

Parker JE, Coleman MJ, Szabo V, Frost LN, Schmidt R, van der Biezen E, Moores T, Dean C, Daniels MJ, Jones JDG (1997) The Arabidopsis downy mildew resistance gene *Rpp5* shares similarity to the Toll and interleukin-1 receptors with N and L6. Plant Cell 9:879–894

Paterson AH, Lan T, Amasino R, Osborn TC, Quiros C (2001) *Brassica* genomics: a complement to, and early beneficiary of, the *Arabidopsis* sequence. Genome Biol 2(3):1011.1–1011.4

Redkar A, Bonequi MV, Doehlemann G (2015) Conservation of the *Ustilago maydis* effector See1 in related smuts. Plant Signal Behav 10:e1086855

Rommens CMT, Salmeron JM, Oldroyd GED, Staskawicz BJ (1995) Intergeneric transfer and functional expression of the tomato disease resistance gene *Pto*. Plant Cell 7:1537–1544

Rouxel T, Grandaubert J, Hane JK, Hoede C, van de Wouw AP, Couloux A, Dominguez V, Anthouard V, Bally P, Bourras S, Cozijnsen AJ, Ciuffetti LM, Degrave A, Dilmaghani A, Duret L, Fudal I, Goodwin SB, Gout L, Glaser N, Linglin J, Kema GH, Lapalu N, Lawrence CB, May K, Meyer M, Ollivier B, Poulain J, Schoch CL, Simon A, Spatafora JW, Stachowiak A, Turgeon BG, Tyler BM, Vincent D, Weissenbach J, Amselem J, Quesneville H, Oliver RP, Wincker P, Balesdent MH, Howlett BJ (2011) Effector diversification within compartments of the *Leptosphaeria maculans* genome affected by repeat-induced point mutations. Nat Commun 2:202. https://doi.org/10.1038/ncomms1189

Rudd JJ, Kanyuka K, Hassani-Pak K, Derbyshire M, Andongabo A, Devonshire J, Lysenko A, Saqi M, Desai NM, Powers SJ, Hooper J, Ambroso L, Bharti A, Farmer A, Hammond-Kosack KE, Dietrich RA, Courbot M (2015) Transcriptome and metabolite profiling of the infection cycle of *Zymoseptoria tritici* on wheat reveals a biphasic interaction with plant immunity involving differential pathogen chromosomal contributions and a variation on the hemibiotrophic lifestyle definition. Plant Physiol 167(3):1158–1185. https://doi.org/10.1104/pp.114.255927

Rustagi A, Kumar D, Shekhar S, Yusuf MA, Misra S, Sarin NB (2014) Transgenic *Brassica juncea* plants expressing *MsrA1*, a synthetic cationic antimicrobial peptide, exhibit resistance to fungal phytopathogens. Mol Biotechnol 56:535–545. https://doi.org/10.1007/s12033-013-9727-8

Saharan GS, Mehta NK (2008) *Sclerotinia* diseases of crop plants: biology, ecology and disease management. Springer, Delft, p 486

Saharan GS, Mehta NK, Sangwan MS (2005) Development of disease resistance in rapeseed-mustard. In: Saharan GS, Mehta NK, Sangwan MS (eds) Diseases of oilseed crops. Indus Pub Co., New Delhi, pp 561–617

Saharan GS, Verma PR, Meena PD, Kumar A (2014) White rust of crucifers: biology, ecology and disease management. Springer, New Delhi, p 244. https://doi.org/10.1007/978-81-322-1792-3

Saharan GS, Naresh NK, Meena PD (2016) Alternaria blight of crucifers: biology, ecology and disease management. Springer, Singapore, p 326. https://doi.org/10.1007/978-981-10-0021-8

Saharan GS, Mehta NK, Meena PD (2017) Downy mildew disease of crucifers: biology, ecology and disease management. Springer Nature, Singapore, p 357. https://doi.org/10.1007/978-981-10-7500-1

Saharan GS, Mehta NK, Meena PD (2019) Powdery mildew disease of crucifers: biology, ecology and disease management. Springer Nature, Singapore, p 362. https://doi.org/10.1007/978-981-13-9853-7

Saharan GS, Mehta NK, Meena PD (2021a) Clubroot disease of crucifers: biology, ecology and disease management. Springer Nature, Singapore, p 757. https://doi.org/10.1007/978-981-16-2133-8

Saharan GS, Mehta NK, Meena PD (2021b) Molecular mechanism of crucifer's host-resistance. Springer Nature, Singapore, p 835. https://doi.org/10.1007/978-981-16-1974-8

Saharan GS, Mehta NK, Meena PD (2021c) Genomics of crucifer's host-resistance. Springer Nature, Singapore, p 784. https://doi.org/10.1007/978-981-16-0862-9

Salmeron JM, Oldroyd GED, Rommens CMT, Scofield SR, Kim HS, Lavelle DT, Dahlbeck D, Staskawicz BJ (1996) Tomato *Prf* is a member of the leucine-rich repeat class of plant disease resistance genes and lies embedded within the *Pto* kinase gene cluster. Cell 86:123–133

Sanz-Martín JM, Pacheco-Arjona JR, Bello-Rico V, Vargas WA, Monod M, Díaz-Mínguez JM, Thon MR, Sukno SA (2016) A highly conserved metalloprotease effector enhances virulence in the maize anthracnose fungus *Colletotrichum graminicola*. Mol Plant Pathol 17(7):1048–1062. https://doi.org/10.1111/mpp.12347

Schots A, Dewey FM, Oliver R (1994) Modern assays for plant pathogenic fungi: identification, detection and quantification. CAB International, Wallingford

Shigenaga AM, Argueso CT (2016) No hormone to rule them all: interactions of plant hormones during the responses of plants to pathogens. Semin Cell Dev Biol 56:174–189. https://doi.org/10.1016/j.semcdb.2016.06.005

Smith KM, Galazka JM, Phatale PA, Connolly LR, Freitag M (2012) Centromeres of filamentous fungi. Chromosom Res 20:635–656

Song K, Slocum MK, Osborn TC (1995) Molecular marker analysis of genes controlling morphological variation in *Brassica rapa* (syn. *campestris*). Theor Appl Genet 90:1–10

Staskawicz BJ, Ausubel FM, Baker BJ, Ellis JG, Jones JDG (1995) Molecular genetics of plant disease resistance. Science 268:661–667

Taj G, Kumar A, Bansal KC, Garg GK (2004) Introgression of osmotin gene for creation of resistance against Alternaria blight by perturbation of cell cycle machinery. Indian J Biotechol 3:291–298

Tasleem M, Baunthiyal M, Kumar A, Taj G (2017) Determination of antioxidant activity in overexpressed *MPK3* transgenic *Brassica juncea* for induction of defense against *Alternaria* blight disease. J Pharma Phytochem 6(6):2579–2582

Thakur AK, Parmar N, Singh KH, Nanjundan J (2020) Current achievements and future prospects of genetic engineering in Indian mustard [*Brassica juncea* (L) Czern & Coss.]. Planta 252(4):56. https://doi.org/10.1007/s00425-020-03461-8

Thilmony RL, Chen Z, Bressan RA, Martin GB (1995) Expression of the tomato *Pto* gene in tobacco enhances resistance to *Pseudomonas syringae* pv *tabaci* expressing *avrPto*. Plant Cell 7:1529–1536

UN (1935) Genomic analysis in Brassica with special reference to the experimental formation of *B. napus* and peculiar mode of fertilization. Jpn J Bot 7:389–452

Ward E (1994) Use of the polymerase chain reaction for identifying plant pathogens. In: Blackman JP, Williamson B (eds) Ecology of plant pathogens. CAB International, Wallingford, pp 143–160

Whitham S, Dinesh-Kumar SP, Choi D, Hehl R, Corr C, Baker B (1994) The product of the tobacco mosaic virus resistance gene N: similarity to toll and the interleukin-1 receptor. Cell 78:1101–1115

Williams PH, Hill CB (1986) Rapid cycling populations of *Brassica*. Science 232:1385–1389

Wroblewski T, Coulibaly S, Sadowski J, Quiros CF (2000) Variation and phylogenetic utility of the *Arabidopsis thaliana Rps2* homolog in various species of the tribe Brassiceae. Mol Phylogenet Evol 16:440–448

Yan Y, Yuan Q, Tang J, Huang J, Hsiang T, Wei Y, Zheng L (2018) *Colletotrichum higginsianum* as a model for understanding host–pathogen interactions: a review. Int J Mol Sci 19(7):2142. https://doi.org/10.3390/ijms19072142

Yu YG, Buss GR, Maroof MAS (1996) Isolation of a superfamily of candidate disease-resistance genes in soybean based on a conserved nucleotide-binding site. Proc Natl Acad Sci U S A 93: 11751–11756

Protocols to Study Host-Pathosystems 8

Abstract

Crucifers' host-pathosystem has become a favorable model system of biologists to get deeper insight into various biological functions operating by induction of differential gene expressions. The genomics of crucifers' host-pathosystem has been strengthened with different omics approaches/techniques used to reveal genome sequences of major *Brassica* crops and their pathogens. Some of the selected techniques/protocols developed and used by different scientists have been included in this chapter for the benefit of *Brassica* teachers, researchers, and students. The isolates of *A. candida* collected from field have been characterized with the use of DNA sequence approaches. A PCR test specific only to the presence of *A. candida* DNA has been developed which allows separation of host DNA and detection of pathogen from both symptomatic and asymptomatic leaf material as well as DNA from plant species for which *Albugo* infection is unknown. The protocol for genome sequencing and assembly of 16 *Alternaria* species with mRNA sequencing and transcript construction has been developed. The real time PCR based method has been used for quantitative estimation of seed infection of *Brassica* by *A. brassicae*. The *Brassica–Alternaria* interactions have been analyzed with microarray approach. A procedure has been detailed for molecular and biochemical changes using *Brassica–Alternaria* host-pathosystem. A detailed method has been developed for assessing virulence variability of *Alternaria* species infecting crucifer's. With the *Brassica–Colletotrichum* host-pathosystem, expression of MAPK genes has been characterized. A molecular identification technique for *Erysiphe* at anamorphic stage in conjunction with morphological and host data is very valuable for quick results. A procedure has been used to identify powdery mildew isolates through DNA sequence analysis and DNA markers analysis. Histological assessment of *Erysiphe* growth in host tissues has been revealed. A fast molecular marker identification of *Hyaloperonospora* method based on multiplex PCR

G. Singh Saharan et al., *Genomics of Crucifer's Host-Pathosystem*,
https://doi.org/10.1007/978-981-19-3812-2_8

(rDNA) analysis is useful for diagnosis of crucifers' downy mildew. The DNA finger printing with RAPD analysis can distinguish *H. parasitica* isolates from *Brassica* species. A method for DNA isolation from infected hosts by *H. parasitica* has been developed. A method has been developed to characterize *Leptosphaeria* isolates for virulence on different *Brassica* species. The *Leptosphaeria maculans* field populations have been analyzed for genetic diversity. Phylogenetic relationship of R-loci in *Brassica* species to *Leptosphaeria* has been determined through DNA-PCR analysis. A protocol has been developed for transcriptome analysis of *Brassica–Leptosphaeria* pathosystem to reveal molecular mechanisms. The PCR based protocols have been developed to detect *P. brassicae* from soil, and plant along with its quantification. Bait plants have been used to validate the real time PCR assay method. The *Plasmodiophora* pathotype-specific genes have been identified with PCR and qPCR approach. The qPCR approach has been used for *P. brassicae* quantification from soil to assess response of time based fertilizers and other control measures along with its spread. The RNA seq analysis method has been developed for assessing genes expression during *Plasmodiophora* interactions with their host plants. The *Plasmodiophora brassicae* genome has been sequenced along with transcriptomic analysis during *B. napus* infection. The approach of transcriptomic analysis of R and S genotypes of *Brassica* during *Brassica–Sclerotinia* interaction has revealed genes expression controlling resistance and susceptibility of host plants. A serological test has been developed that allows the early detection of rapeseed petal infection by *Sclerotinia* inoculum under field conditions which has potential to visualize risk of infection. The infection of different viruses in *Brassica* has been detected with multiplex RT-PCR approach. The qRT-PCR assay has been used to detect expression levels of *BnSGS3* gene in different tissues of *Brassica* nontransgenic, single transgenic, and virus inoculated plants. The presence of viruses on crucifers has been detected with both biological and serological methods. The isolates of *Verticillium* collected from different hosts and locations have been characterized with molecular PCR technique. Protocol has been developed to determine defensive metabolites in the phenylpropanoid pathways during *Brassica–Xanthomonas* interactions. A molecular method has been developed for genotyping of *Xanthomonas campestris* pv. *campestris* isolates. Multispectral imaging approach has been developed for presymptomatic identification of crucifers' pathogens/disease. The RAPD-PCR approach has been used for the diagnosis of mixed infection in *Brassica* crops.

Keywords

Protocols for host-pathosystems studies · *Albugo* techniques · Use of DNA sequence approach · PCR assay for detection · *Alternaria* techniques · Genome sequencing · PCR-based assay for detection · Microarray analysis · Molecular and biochemical analysis · Virulence variability · *Colletotrichum* techniques · Characterization of MAPK gene · *Erysiphe* techniques · Molecular identification · DNA sequence analysis · DNA marker analysis · Histological assessment ·

Hyaloperonospora techniques · Molecular marker for identification · DNA finger printing · DNA isolation · *Leptosphaeria* techniques · Characterization of isolates · Genetic variability · Phylogenetic relationship of R-loci · Transcriptome analysis · *Plasmodiophora* techniques · PCR detection · PCR technique to identify pathogen · Real time PCR assay · Identification of pathotype-specific genes · Quantification and assessment with qPCR · PCR and quantitative PCR analysis · Genome sequencing and transcriptomic analysis · *Sclerotinia* techniques · Transcriptome analysis · Immunoassay for early detection · *TuMV* techniques · Detection by multiplex reverse transcription polymerase chain reaction · RT-PCR assay for detection of *BnSGS3* gene · *Verticillium* techniques · Molecular characterization · *Xanthomonas* techniques · Determination of defensive metabolites · Genotyping of *Xanthomonas* · Mixed infection study techniques · Multispectral imaging · RAPD-based molecular diagnosis

8.1 Introduction

Research and teaching of life sciences is based on sound, reliable, repeatable, and standardized protocols, procedures, methods, and techniques. The research results validation and acceptability requires a very sound repeatable technique to get correct results. The genomics of crucifers' host-pathosystem research has taken momentum after the sequencing of Brassica host and pathogens genome. Although several techniques have been developed with the use of *Brassica* host-pathosystem but the present chapter includes only selected following techniques for the benefit of the students, researchers and teachers to use for their future persuasion.

8.2 *Brassica–Albugo* Pathogenesis Protocols

8.2.1 Use of DNA Sequence Approach Under Field Conditions

The protocols was designed with 120mer biotinylated RNA baits based on both mitochondrial genes and nuclear "housekeeping" genes of 49 microbial species, comprising two rhizaria, five oomycetes, 12 fungi, and 30 bacteria with the aim of identifying these species from environmental samples. In addition, Jouet et al. (2018) designed baits to capture putative RxLR (Rehmany et al. 2005) and CHxC class effectors (Kemen et al. 2011; Links et al. 2011), and other secreted proteins, from oomycetes. In *Albugo candida*, baits also were designed both to 32 neutrally evolving loci and to a contig of ~400 kb with the aim of investigating the genetic diversity of wild pathogen populations. Neutrally evolving loci were selected using three independent neutrality tests performed on whole-genome data from seven laboratory isolates: Tajima's D (Tajima 1989), Fu's Fs (Fu, 1997), and dN/dS (Kimura 1977) whereas the ~400 kb contig was the longest contiguous sequence that could be assembled from *A. candida* at the time of the bait library design, using reads from isolate AcNc2 (Race 4, "contig 1" (McMullan et al. 2015), 246 genes including 11 with a predicted secretion signal, no putative effectors). AcNc2 was one of three races analyzed in a previous study that reported on the genetic diversity of five *A. candida* isolates. They also targeted the nuclear internal transcribed spacer (ITS) sequence in plants to facilitate host species identification. In total, 18,348,120 mer baits were synthesized that anneal specifically to targeted sequences (>2 Mb) without overlap. These baits were used to capture DNA from a total of 115 samples, including 91 *A. candida* isolates. DNA samples were barcoded and sequenced on three Hiseq lanes with 150 bp paired-end reads.

8.2.1.1 Preparation of Samples for Reconstruction Experiments

First set up reconstruction experiments to verify that DNA from targeted loci was enriched and sequenced in control samples, and assess the relationship between read depth and DNA abundance. To do this, two sets of controls are prepared. In samples #18 to 22, DNA from *Erysiphe cruciferarum*, *Hyaloperonospora arabidopsidis*, *Phytophthora infestans*, *Albugo laibachii*, *Pseudomonas syringae,* and *A. candida* was combined in various ways. These DNA samples were isolated either from infected leaves, or, for *P. syringae*, grown on medium before DNA extraction, included a noninfected *A. thaliana* sample as a negative control. Some samples varying relative molar amounts of DNA from *P. syringae* and *A. candida* were mixed so that the amount of DNA from *P. syringae* increased, and from *A. candida* decreased. Seven additional samples collected on various host species were used to investigate whether the ITS is sufficient for host species identification. Reads were mapped to all targets and average read depth as well as the breadth of coverage (the percentage of targeted base pairs with at least 109 read depth) was computed for each control organism.

8.2.1.2 Read Alignment and Consensus Sequences Calling

Reads were aligned to targets using the BWA-MEM algorithm of the Burrow–Wheelers Aligner BWA v.0.7.4 (Li and Durbin 2009). Read duplicates which may have arisen from amplification of adapter-ligated DNA and post-capture enrichment were discarded using SAMTOOLS v.0.1.19 (rmdup command). BAM files were further processed using ngsutils-0.5.7 to remove reads with >25% clipped bases and with the highest edit distance to the reference and lowest alignment score in case of multiple alignments (NM and AS tags). Read depth, the number of reads per base, as well as the breadth of coverage, the percentage of bases covered by at least 10 reads, was estimated using the depth command in SAMTOOLS. These statistics were averaged across loci or organisms depending on the analysis performed. Consensus sequences of *A. candida* loci were generated before nucleotide divergence and phylogenetic analyses. To do this, variants (>109) were called from the bam file produced by BWA using the MPILEUP command of SAMTOOLS and stored in BCF format. Conversion from BCF to VCF and then FASTQ format was performed using the "BCFTOOLS VIEW" and "VCFUTILS. PL VCF2FQ" commands in SAMTOOLS. A short shell script was written in Linux to convert FASTQ files into FASTA files. Also independently fed aforementioned bam and vcf files to

SHAPEIT2 v.2.20 (O'Connell et al. 2014) to obtain phased vcf files that were used to generate neutral and putative effector haplotypes using BCFTOOLS v.1.3.1 (Jouet et al. 2018). Sequences were finally aligned using the multiple alignment program MAFFT v7.127 (Katoh and Standley 2013). The reference *A. candida* isolate used in this study was Nc2 (Race 4; McMullan et al. 2015), except for some effectors that are absent from that particular isolate. There was a total of 6493, 24,249, 202,390 and 398,508 base positions in the final datasets of conserved loci, neutrally evolving loci, putative effectors and the ~400 kb contig, respectively.

8.2.1.3 Phylogenetic, Nucleotide Diversity, and Recombination Analyses

Phylogenies were inferred using RAXML v.7.7.3 (Stamatakis 2014) with the Generalized Time Reversible (GTR) model of nucleotide substitution, gamma distributed rate variation among sites (GTRGAMMA) and 100 bootstrap replicates. Pair wise nucleotide divergence was evaluated using MEGA v.6.06 (Tamura et al. 2013). It was computed as the number of base differences per site from averaging over all sequence pairs between or within lineages. Ambiguous positions were ignored in each sequence pair. The split network was inferred with SPLITSTREE v.4.14.2 (Huson and Bryant 2006) using the UNCORRECTEDP and the NEIGHBORNET methods and 500 bootstrap replicates. A more in-depth genetic diversity and selection analysis was also performed using DNASP v.5.10.01 (Librado and Rozas 2009) on neutral and putative effector haplotypes to compute six genetic diversity estimates (number of segregating sites, of haplotypes, of pair wise differences, heterozygosity, pi and theta) and three population genetic statistics (Tajima 1989; Fu and Li 1993; Fu 1997). Finally, recombination analysis was performed using the software HYBRID CHECK (Ward and van Oosterhout 2016). Using a sliding window approach, this software scans for sudden changes in nucleotide divergence between sequences and identifies potential recombination events when nucleotide identity is significantly increased. Recombination blocks were then dated using the same software and assuming a strict molecular clock with a mutation rate of 10×9 mutations/site/generation.

8.2.1.4 Evaluation of *Albugo candida* Heterozygosity and Ploidy Level

Heterozygosity was expressed as the proportion of heterozygous sites within individuals. Before estimating heterozygosity in *A. candida* isolates, sites where read depth was <509 and/or with a Phred-scaled quality (QUAL) score <100 were first removed from vcf files prepared above using vcflib (https://github.com/vcflib/vcflib#vcflib) and VCFTOOLS v.0.1.10 (Danecek et al. 2011) and isolates where $>10\%$ sites were removed were discarded. Heterozygosity was then estimated for the remaining isolates as the proportion of heterozygous sites at the ~400 kb contig. The mean and standard deviation of the mean percentage of heterozygous sites per *A. candida* lineage are reported. In addition, the ploidy level of *A. candida* was evaluated using the ~400 kb contig as in Yoshida et al. (2013). The distribution of the read proportion of each single nucleotide polymorphism (SNP) at heterozygous sites was graphed for each isolate using R v.3.1.2. This analysis was repeated using whole-genome data from laboratory isolates *AcNc2, AcEm2, AcEx1, Ac2v, Ac7v, AcBoT,* and *AcBoL.* In this case, reads were aligned to *AcNc2* (McMullan et al. 2015) and to *Ac2v* (Links et al. 2011) contigs. In all ploidy analyses, heterozygous sites where read depth were higher than 1.5 the average read depth was removed. This is to avoid biases due to mis-mapping of reads from paralogous sequences.

8.2.1.5 Microbiome Analysis

Albugo candida can infect its host both symptomatically and asymptomatically. Therefore, to compare the microbiome of *A. candida*-infected and no infected samples, Jouet et al. (2018) used plant samples from which *A. candida* locus Ev1786 could not be amplified and symptomatically infected plants. In total, 82 infected samples were used with high read depth at the 400 kb contig and 12 no infected plants. Reads generated for the above samples were fed into KRAKEN (v.0.10.5). First, the proportion of reads classified as Fungi or as Bacteria was averaged and compared between both groups of samples. In addition, the number of operational taxonomic units (OTUs) detected by KRAKEN was compared between the infected and no infected plants using a Mann–Whitney U-test. Common and rare microbial OTUs on the *A. candida*-infected plants were identified using a binomial test. In order to correct for an inflated type I error rate due to multiple testing, the critical value was divided by the total number of OTUs identified by KRAKEN, resulting in a Bonferroni corrected $\chi^2 = 1.92 \times 10^{-5}$ (Jouet et al. 2018).

8.2.2 A PCR Assay for Detection of *Albugo candida* DNA

8.2.2.1 DNA Extraction, Sequencing, and Primer Design

Genomic DNA was extracted using standard CTAB methods (Doyle and Doyle 1990). The universal primers ITS 1 (5′-TCCGTAGGTGAACCTGCGG-3′) and ITS 2 (5′-GCTGCGTTCTTCATCGATGC-3′) were initially used to amplify the internal transcribed spacer region 1 (ITS 1) between the 18S and 5.8S ribosomal genes (White et al. 1990). DNA extracted from inflorescences of *Lepidium oleraceum* that were heavily infected with *Albugo candida* typically yielded two PCR products, one derived from the host genome and the other putatively from the genome of the pathogen. These copies were of different sizes (~269 base pairs (bp) and197 bp, respectively), allowing them to be separated via agarose gel electrophoresis and sequenced individually. Following initial amplification, separated bands corresponding to the plant and pathogen copies of ITS 1 were manually excised and purified using Concert® Gel purification columns prior to re-amplification and direct sequencing (Big Dye Terminator method) using an ABI 310 automated sequencer at Land Care Research, Mt. Albert. Recovered sequences were then aligned and compared using the program Sequencer (Applied Biosystems). The identities of the plant and oomycete sequences were verified using previously generated *Lepidium* sequences as well as *Albugo* spp. sequences lodged on Genbank (BLAST/NCBI: www.ncbi.nlm.nih.gov).

This verification procedure confirmed that the pathogen was *A. candida*. Four separate *A. candida* sequences were generated, which were derived from different host plants growing. Once aligned together with the sequences from *L. oleraceum*, regions that differed between *A. candida* and the plant sequence, but were identical within the four sequences of *A. candida*, were identified at the opposite ends of the molecule. Based on this observation, internal primers that were specific for *A. candida* were designed for the ITS 1 region using the unique nucleotide differences apparent between host and pathogen sequences present across the border of ITS 1 and the 5.8S ribosomal gene. The objective was to develop a PCR test that was specific only to the presence of *A. candida* DNA. To do this, a 20-mer oligonucleotide primer was designed and called "ACAN-1" (5-′-CCACTGCTGAAAGTTTGTG-3′). This spanned the border of ITS 1 and 5.8S. Internal to the border of 18S ribosomal gene and ITS 1, another 20-mer primer (ACAN-2) was designed (5′-CTATCCACGTGAACCACTTG-3′).

8.2.2.2 Primer Effectiveness and Specificity Trials

The efficacy and template specificity of these primers was assessed by conducting a wide PCR assay of both symptomatic and asymptomatic leaf material of *L. oleraceum*, as well as of DNA from other plant species for which *Albugo* infection is unknown. This assay of symptomatic, asymptomatic and fungicidally treated material also served as a test for the presence of *A. candida* as a latent asymptomatic systemic infection. Assays were also performed on juvenile plants from the fitness trials and surface-sterilized seeds obtained from ARBG (DNA from surface-sterilized seeds was extracted individually using the above methods). In each case, random subsets of 20 plants from each of the growth experiment treatments were assayed for the presence of the pathogen. The PCR assay procedure involved DNA extraction from the target tissue sample (usually leaves) followed by PCR amplification using the primers specific for *A. candida*. PCR products were run out via agarose gel electrophoresis using TAE buffer at a constant voltage of 4 V/cm. PCR products could then be visualized under UV light after staining with ethidium bromide. A 1-kb (kilobase) DNA ladder (NEB) was run out alongside each PCR product to facilitate product size determination (Armstrong 2007).

8.3 *Brassica–Alternaria* Pathogenesis Protocols

8.3.1 *Alternaria* Genome Sequencing

8.3.1.1 Genome Sequencing and Assembly

Two isolates of *A. alternata* which were found to be co-infecting *B. juncea* leaves along with *A. brassicae* in experimental field station at Delhi, India (PN1 and PN2) were isolated and purified by single spore culture. High molecular-weight genomic DNA was extracted from mycelia of 5-day old cultures of *A. alternata* isolates using a method described earlier (Rajarammohan et al. 2019). Two micrograms of the high molecular weight genomic DNA was used for Nanopore library preparation using

LSK-108 ONT ligation protocol. The libraries were then run on R9.4 Spot ON MinION flow cells for 24 h. live base calling was enabled for all the runs. The MinION runs produced 4,14,210 and 2,68,910 reads amounting to 2.36 GB and 1.98 GB of data for *A. alternata* PN1 and PN2, respectively. The genomes were assembled de novo using the Canu assembler (version 1.6). Nanopolish (version 0.7.1) were used to compute an improved consensus sequence using the signal-level raw data for the assemblies.

8.3.1.2 mRNA Sequencing and Transcript Reconstruction

Total RNA was isolated from 15-day old fungal mycelia of *A. brassicae*, grown on Potato Dextrose Agar (PDA) and Radish Root Sucrose Agar (RRSA), using Qiagen RNeasy Plant Mini Kit following the manufacturer's instructions. 100 bp sequencing libraries were prepared using the TruSeq Stranded mRNA Library Prep Kit (Illumina). Paired-end sequencing was carried out using Illumina SBS v4 chemistry on an Illumina Hiseq 2500. The Hiseq run generated 50 million PE reads per sample. The raw reads were trimmed using Trimmomatic, and the trimmed reads were then mapped to the de novo genome assembly using STAR (version 2.5.0). Transcripts were reconstructed using Cufflinks (version 2.2.1) and likely coding regions were identified using Trans Decoder) (Trapnell et al. 2012; Haas et al. 2013).

8.3.1.3 Gene Prediction and Annotation

Repeat Modeler (version 1.0.11) was used for de novo repeat family identification. The de novo identified repeat library was used for masking the genome using Repeat Masker (version 4.0.7). The repeat-masked genome was used for gene predictions. For gene prediction, multiple lines of gene evidence were integrated using EVM (Evidence Modeler) (Haas et al. 2008). Two ab initio gene callers were used viz. AUGUSTUS (version 3.3.1) (Stanke et al. 2008) and Gene Mark-HMM-ES (version 2.0) (Ter-Hovhannisyan et al. 2008). Gene Mark-HMM-ES was self-trained on the repeat-masked genome whereas AUGUSTUS was trained on the genome and cDNA hints from *A. alternata*. RNA-Seq evidence in the form of coding regions identified by Trans Decoder was also used. Additionally, homology-based gene prediction was carried out using GeMoMa (version 1.4.2) (Keilwagen et al. 2016) with protein-coding genes of *A. longipes, A. arborescens,* and *A. alternata*. Each of these lines of evidence was presented to EVM as separate tracks. In EVM, weights were assigned to each evidence as follows: AUGUSTUS 1, Gene Mark- HMM-ES 1, GeMoMa 1, and RNA-Seq evidence 5. The genes predicted by EVM were used for all the downstream analyses. Genes were then annotated using BLAST (version 2.7.1+) against UniProt, SWISS-PROT, CAZy, MEROPS, and PHI-BASE. The fungal version of anti-SMASH (version 4.0) (Blin et al. 2017) was used to identify secondary metabolite gene clusters in the genomes. Candidate effector proteins were predicted using the following pipeline: (a) Signal P (version 4.1) (Petersen et al. 2017) and Phobius to identify secreted proteins, (b) TMHMM (version 2.0) (Krogh et al. 2001) to remove proteins with transmembrane domains, (c) predGPI to filter out proteins that harbored a GPI membrane-anchoring domain, and (d) Effector

P (version 2.0) (Sperschneider et al. 2016) to predict potential effectors from the above filtered protein set.

8.3.1.4 Orthology and Whole-Genome Phylogenetic Analysis

The genomes of 16 *Alternaria* species were included in the analysis with *Stemphylium lycopersici* as an out-group. The proteomes of the fungi were clustered using the Orthofinder (version 2.2.6) (Emms and Kelly 2015) pipeline with default parameters. The clusters were further analyzed with Mirlo (https://github.com/mthon/mirlo) to identify phylogenetically informative single copy gene families. These families were then concatenated into one large alignment and used for phylogenetic analysis. A phylogenetic tree was then inferred from the concatenated alignment by the neighbor joining method and Jukes Cantor substitution model using 1000 replicates. Orthologs were also identified for 13 other pathogenic Dothideomycetes for comparative analyses using the same pipeline as above.

8.3.1.5 Relationship of TEs and Repeat-Rich Regions to Genes and Gene Clusters

The overlap of repeat-rich regions and transposable elements (TEs) with (1) genes encoding secreted proteins, (2) effectors, and (3) secondary metabolite gene clusters were analyzed using the regions R package (Gel et al. 2016) in R. The analysis involved comparison of the overlap of each the above gene sets with transposable elements and repeat rich regions with a random set of genes selected from the genome. A distribution of means was established by running 10,000 permutation tests, which was then used to calculate a p-value (Rajarammohan et al. 2019).

8.3.2 The PCR-Based Assay for Detecting *Alternaria* in Seed

Current detection methods, based on culture and morphological identification of the fungus are time consuming, laborious, and not always reliable. Therefore, a polymerase chain reaction (PCR)-based assay was developed with *A. brassicae* specific primers designed on the basis of sequence of two clustered genes potentially involved in pathogenicity by Guillemette et al. (2004). Two sets of primers were selected for conventional and real time PCR, respectively. In both cases, *A. brassicae* is specifically detected using DNA extracted from seed. The real time PCR based method can be automated easily, and preliminary results indicate that it is efficient for quantitative estimation of seed infection (Saharan et al. 2016).

8.3.2.1 Preparation of Seed Samples

Seed samples are prepared from contaminated seed of radish and cabbage as described by Iacomi Vasilescu et al. (2002). For artificial contaminations, seed are disinfected with 1% sodium hypochlorite, rinsed with sterilized water, soaked for 1 h in a calibrated spore suspension (10^6 spores/mL), and dried at room temperature on sterilized filter paper. Disinfection and inoculation accuracy is confirmed by placing 20 seed on the surface of malt agar plates for 1 week at 25 °C. Seed batches showing

A. brassicae 0% (disinfected seed), or 100% *A. brassicae* grown (inoculated seed) are selected and mixed together to prepare seed samples with different levels of contamination (0, 5, 10, 50, and 100%). Samples are produced by placing 20 seed (artificially contaminated) or 100 seed (naturally contaminated) in a sterile tube and covering them with 0.5, and 3 mL, respectively, of liquid culture malt extract-dextrose-peptone (2% malt extract, 2% dextrose, 0.1% peptone) (MDP) culture medium. The tubes are incubated at 25 °C for 48 h with occasional shaking. This incubation step is an enrichment phase that allows an optimal increase of the fungal biomass from seed. The tubes are then vortexed briefly to separate mycelia and conidia from seed. The seeds are discarded, and the fungal structures collected on micro centrifuge filter (pore size 0.45 μm; Ultra free-Mc Millipore) by centrifugation at 5000 *g* for 30 min.

8.3.2.2 The DNA Manipulation

DNA is extracted from fungal mycelia and conidia by scraping the surface of Petri plates cultures (extraction from fungal culture), or collected on microcentrifuge filters (extraction from seed samples). Nucleic acids are isolated according to microwave mini-prep procedure described by Goodwin and Lee (1993). Alternatively, DNA is extracted using the nucleospin food kit according to the manufacturer's instructions.

8.3.2.3 The PCR-Based Assay

Specific oligonucleotides AB Csens (5′-CTGGTGAAAAGGTT GCGATCGT-3′) and AB Crev (5′-GTGACTTCATGAAATGACATTGATG-3′), complementary to 3′ end of the open reading frame (ORF) 2 and 115 sens (5′-A ACCCTATAGACC CACGTCGACTA-3′), and 115 rev (5′-GATGGTACGCAA GGCTTGGT-3′), complementary to a portion of ORFI are designed for the standard and real time PCR assays, respectively, using the appropriate software (Primer 3 online and ABI PRISM Primer express). The two ORFs are deduced from the nucleotide sequence of an *A. brassicae* genomic DNA fragment identified after screening a cosmid library with a probe corresponding to a portion of the nonribosomal peptide synthase (NRPS) gene. In order to test the specificity of the primer pairs against *A. brassicae* isolates, amplifications are performed using DNA extracted from pure fungal cultures from a range of *Alternaria* spp., and other fungi isolated from seed. Then the PCR procedure is applied to detect seed contamination. The universal primer pair ITS1/ITS2 is used as a positive control to assess the quality of the extracted DNA. Conventional PCR is performed using 2 μL of undiluted DNA preparation under the following conditions; Tris HCL, pH 9.0, 20 mM $(HN_4)_2SO_4$, 0.01% (w/v) Tween 20, 1.5 mM $MgCl_2$, 200 μM each deoxyribonucleotide triphosphate and 1 unit of thermo-stable DNA polymerase. A thermo jet thermo cycler is used with an initial step of 3 min at 95 °C; followed by 35 cycles of 30 s at 95 °C, 50 s at 60 °C, and 1 min at 72 °C, and a final step of 10 min at 72 °C. The amplification results are visualized after electrophoresis of an aliquot (10 μL) of the reaction mixture on a 1.2% agarose gel. The real time PCR reactions are carried out in a 25 μL final volume containing 2.5 μL of DNA, 12.5 μL of SYBR Green PCR

Master Mix 2X, 300 nM forward and reverse primers, and water up to 25 μL. An ABI PRISM 700 sequence Detection System (Applied Biosystems) is used with the following steps: 2 min at 50 °C, 10 min at 95 °C, and 40 cycles consisting in one step at 95 °C for 15 s followed by one step at 60 °C for 1 min. Nucleic acids are quantified in unknown samples by direct comparison to a standard. The DNA used as standard is extracted from fungal cultures, with the concentration measured by fluorometric assay before dilution (Saharan et al. 2016).

8.3.3 Microarray Analysis of *Brassica–Alternaria* Interactions

The time-course pattern of gene expression was examined in the second leaves of *B. oleracea* sprayed with *A. brassicicola* conidial suspension by cDNA hybridization to 29 k *Arabidopsis* Oligonucleotide Microarrays (University of Arizona). Each sample treated as a single biological replicate was collected from 3 control or 3 infected plants. For each time point (0, 12, 24, and 48 hpi), 2 biological replicates were analyzed and a common reference design was applied, where each sample was labeled with Cy5 and hybridized against the Cy3-labeled pool of the control plants. Each microarray probe sequence (~70-mer) was used as a query in a BLASTN search against *Brassica* nucleotide sequences available in the Gen Bank database. The best hit was recorded for each probe. Probes with high homology to *Brassica* targets were selected by applying a Bit score cutoff value of 40. For 66% of them, probe: target alignments were >60 nt in length and displayed >75% sequence identity. The total RNA was extracted with RNeasy Plant Mini Kit and samples were DNase-digested with a TURBO DNA-free kit. The sample quality was analyzed with a Nanodrop 1000 spectrophotometer and 2100 Bio-analyzer. For each labeling reaction, 15 μg of high-quality RNA (A_{260}/A280 and $A_{260}/A_{230} \geq 2$, RIN ≥ 9) was reverse-transcribed using the Super Script Indirect cDNA Labeling System (In vitro gen) according to the manufacturer's protocol. Aminoallyl-modified cDNA was coupled with 5 μL of Cy3 or Cy5 ester dye (Amersham) and unbound dye was removed with a MinElute Reaction Cleanup Kit. The efficiency of the labeling was monitored with Nanodrop measurements. Corresponding Cy3 and Cy5-cDNA were mixed, the volume was adjusted to 10 μL and the samples were heat-denatured at 95 °C for 10 min and added to 115 μL prewarmed (68 °C) hybridization buffer SlideHyb #3 (Ambion). Hybridization was carried out in a Hyb Array 12 workstation at 42 °C for 18 h. The microarrays were washed with 2 × SSC (Ambion) supplemented with 0.1% SDS at 42 °C, 0.5× SSC at 30 °C and 0.005× SSC at 25 °C each for 5 cycles (20 s flow, 40 s hold) and dried by centrifuging. Images were collected with a Scan Array Express scanner (PerkinElmer) at 100% laser intensity and variable PMT settings. The images were analyzed with GenePix 6.1 software (Molecular Devices) using the morphological opening background subtraction method with default parameters. The data were further analyzed in an R/Bioconductor environment using the limma package. Foreground and background intensities were read with "genepix. Custom" function. After background subtraction and print tip-loss normalization within arrays, followed by scaling, the log-ratios to have the same median-absolute-deviation (MAD) across arrays, a linear model was constructed for the contrasts of interest. Differential expression of genes for which the probes passed homology-based filtering criteria (see above) was assessed by moderated *t*-statistics and correcting for multiple testing using the false discovery rate (FDR). The adjusted *P*-value 0.1 was chosen as a significance threshold. For functional analysis and gene annotation, *A. thaliana* gene IDs originally assigned to the microarray probes were used. Only 8014 genes with high homology to *Brassica* were taken into consideration and used as a background in the Gene Ontology (GO) and MapMan analyses. A GO enrichment analysis of differentially regulated genes was performed separately for each time point against the list of all genes represented on the microarray using the ThaleMine v.4.1.1 online analysis tool. The p-value was calculated using a hyper-geometric distribution test and adjusted with the Holm–Bonferroni multiple test correction, and the significance threshold was set at a P-value < 0.05. The visualization of significantly regulated GO terms was performed using REVIGO, which reduces long lists of Gene Ontology terms by summarizing and removing redundant GO terms. The input data were prepared as follows. From the list of all microarray results, genes differentially expressed at 48 hpi (FDR < 0.1) were selected. The GO_cellular_process terms and the log fold change expression values of these genes were used as an input for REVIGO. Genes without assigned GO_cellular_process terms as well as genes for which GO: 0008150 (biological_process_unknown) was the only assigned GO term, were excluded from the analysis, resulting in an input list of 275 genes. The output list of REVIGO-summarized GO terms is visualized in Fig. 3.2 (Chap. 3). The MapMan (v.3.5.0) analysis was performed based on the probe mapping files for *A. thaliana* and pathways downloaded from the Map Man repository.

8.3.3.1 Photosynthetic Pigment Content

Chlorophyll was extracted in 100% methanol and measured using a spectrophotometer Power Wave XP. The content of chlorophyll a, chlorophyll b, and carotenoids was calculated according to Wellburn (1994). The samples were harvested from 3 control and 3 infected plants per time point, and each experiment was repeated independently 3 times ($n = 3$).

8.3.3.2 Chlorophyll a Fluorescence

The control and drop-inoculated infected plants were dark-adapted for 30 min, then each second leaf was detached and immediately subjected to analysis of the kinetics of chlorophyll a fluorescence quenching using a Handy FluorCam 1000-H System according to the manufacturer's built-in protocol. The measurements were performed on the selected area (diameter about 1.5 cm) of control untreated leaves, as well as infected ones containing necrosis and adjacent tissue. The duration of the reading procedure per one leaf was 4 min, and parameters such as F_0 (minimum fluorescence), F_m (maximum fluorescence), F_v (variable fluorescence) as well as F_p (peak fluorescence during the initial phase of Kautsky effect) were measured accordingly during 4 light adaptation (L1–L4), steady-state in light (Lss) and/or 3 dark relaxation periods (D1–D3). Based on the measured chlorophyll a fluorescence parameters, automatic calculation of over 45 parameters such as QYmax (maximum photo system II quantum yield, F_V/F_M), F_v/F_m (photo system II quantum yield), F_q/F_v (photo system II efficiency factor), Rfd (fluorescence decline ratio) and qL (fraction of photo system II open centers) in light adaptation periods and steady-state in light was performed and NPQ (nonphotochemical quenching) and qP (photochemical quenching of variable fluorescence) were also calculated in the dark relaxation periods. In each experiment, 3–4 control and 3–4 infected plants were used per time point. The experiment was repeated independently 3 times ($n = 3$).

8.3.3.3 Statistical Analysis

The statistical analysis of all the data obtained in this work, except the microarray data, was performed using analysis of variance (ANOVA) and post-hoc Duncan's test ($P < 0.05$) using STATISTICA v.13.3. (Macioszek et al. 2020).

8.3.4 Biochemical and Molecular Analysis for Mechanism of Resistance in *Brassica* to *Alternaria*

8.3.4.1 Plant Growth, Testing, and Sampling

Twenty seedlings per each tested cultigen (*Sinapis alba*, "Kamienna Głowa," PGH05I, and PGH09K) were grown from seeds for 45 days, and then inoculated with suspension of mixed *A. brassicicola* isolates and tested as per the optimized protocol. Each day, samples were collected from two plants per cultigen, at the same time (~11 a.m.). At 3 dpi, Kamienna Glowa started showing mild Alternaria dark spot symptoms, which exacerbated at 4 dpi; hence, sampling was stopped at this time. Collected samples were snap frozen in liquid nitrogen and stored at −80 °C until extracted.

8.3.4.2 Biochemical Assays

The homogenized samples were processed as in Nowicki et al. (2016). Extraction of soluble proteins (supernatant) was done with 5× (v: w) 50 mM potassium phosphate, pH 7.2, whereas the reactive oxygen species (ROS) and free phenols were assayed from the methanolic extracts (18,000 × *g* for 15 min at 4 °C each, with 10:1 (v: w) of 80% methanol in water), supernatants combined, and pellets were used for lignin assessment.

8.3.4.3 Assessment of ROS Generation (H_2O_2, HO*, and O_2^-)

Generation of either ROS species followed the published methods (Loreto et al. 2001; Loreto and Velikova 2001; Habu and Ibeh 2015; Moukette et al. 2015). Methanolic extracts were assayed at A390, A560, and A530 for H_2O_2, HO*, and O_2^-, respectively. The H_2O_2 as one of the standards; in the HO* and O_2^- assays, the absorbance readout was recalculated per mg of fresh tissue.

8.3.4.4 Assessment of Plant Redox Enzymic Activities

To analyze the dynamics of polyphenol oxidase (PPO), peroxidase (POX), superoxide dismutase (SOD), polygalacturonase (PGA), and catalase (CAT), standard methodology published elsewhere were used (Korgan et al. 2011; Moukette et al. 2015; Sheehy et al. 1988). Buffer extracted samples (as above) were assayed for protein using the BCA kit, and analyzed for linear assay range on a randomly chosen sample (0 dpi). Proper assays were run in microtiter plates for 3–5 min, respectively, whereas absorbance readouts were recorded every minute in the Epoch2T reader. Protein activity was calculated as change in absorbance per minute and per mg of protein (PPO) or metabolized substrate/min/mg protein (POX, SOD, PGA, CAT).

8.3.4.5 Accumulation of Free Phenols and Deposition of Lignins

Levels of lignins deposition and free phenolics were assessed with published spectrophotometric methods (Campbell and Ellis 1992; Cvikrova et al. 1992). Methanolic extracts and pellets were used, and the assays used technical lignins and gallic acid as the respective standards (Nowakowaska et al. 2019).

8.3.5 Virulence Variability of *Alternaria* species Infecting Crucifers' species

8.3.5.1 *Alternaria* spp. Isolates and Preparation of Inoculum

The hyphal tip isolates of 10 *Alternaria* spp. (*A. alternata*, *A. brassicae*, *A. conjuncta*, *A. ethzedia*, *A. eureka*, *A. hordeicola*, *A. infectoria*, *A. japonica*, *A. metachromatica*, and *A. tenuissima*) were used. The isolates were originally obtained from leaf samples taken during surveys of rapeseed crops across southern Australia (including Western Australia, South Australia, New South Wales, and Victoria). Representative isolates of *Alternaria* spp. had previously been phylogenetically identified (Al-lami et al. 2019a, 2020) using sequencing of two loci, major allergen Alta 1 (Hong et al. 2005) and plasma membrane ATPase (Lawrence et al. 2014), together with morphological (Simmons 1986, 2007) and phenotypical identification. The majority of isolates were selected from Western Australia, as it recorded the highest number of *Alternaria* spp. isolates. The number of isolates used for each species was: nine for *A. alternata*, 2 for *A. brassicae*, 8 for *A. conjuncta*, 19 for *A. ethzedia*, 11 for *A. eureka*, 10 for *A. hordeicola*, 14 for *A. infectoria*, 4 for *A. japonica*, 18 for *A. metachromatica*, and 6 for *A. tenuissima*. Isolates were grouped and named based on original location and host whereas all other isolates were collected from rapeseed. Two isolates each of *A. brassicae* (SA-2 and WA-7) and *A. japonica* (WA-46 and WA-47) were included for comparison checks with earlier studies conducted on both *B. napus* (cv. Thunder TT) and *B. juncea* (cv. Dune) (Al-lami et al. 2019a).

For production of hyphal inoculum, isolates were preserved in lyophilized ampoules and subsequently subcultured onto potato dextrose agar for 7 days at 23 ± 2 °C under cool white fluorescent light. Five agar plugs of 4 mm^2 containing growing mycelia were transferred from PDA into 150 mL of liquid growth medium

(24 g/L potato dextrose broth and 10 g/L peptone in water; Uloth et al. 2015). Cultures were placed on a rotary shaker at 150 rpm at 25 °C. After 5 days, actively growing mycelium was harvested by straining the broth through four layers of cheesecloth and washing well with sterile deionized water (SDW). The mycelium was macerated in SDW using a Mega Blender for 5 min; the initial inoculum concentration was estimated using a hemocytometer counting chamber and then adjusted to obtain mycelial fragments of 90 ± 10 μm length at a concentration of 4×10^6 fragments/mL (Gunasinghe et al. 2017). The hyphal inoculum was used immediately in virulence tests.

8.3.5.2 Virulence Tests

A single susceptible genotype, *B. napus* "Thunder TT," was used for virulence tests (Al-lami et al. 2019b, 2020). Ten replicate pots of each cultivar were used for each isolate, each pot sown with four seeds (pots 6 × 6 × 9 cm) and the subsequent seedlings were thinned to one plant per pot. Plants were maintained in a naturally lit greenhouse at 18 ± 4 °C, watered daily with deionized water and allowed to drain to field capacity. Plants were fertilized weekly with Thrive (Yates), a complete nutrient solution dissolved in water. At the 3-leaf stage, plants were placed into clear plastic boxes with lids, measuring 770 mm (L), 570 mm (W), 475 mm (H); both cotyledons and fully expanded leaves were spray-inoculated with 6 mL per plant of a hyphal suspension of each isolate with a concentration of 4×10^6 fragments/mL (or SDW as a control check) using a hand-held and operated aerosol sprayer. Plastic boxes were internally misted with SDW and lids closed for 4-days post-inoculation (dpi) to maintain high humidity. After 24-h postinoculation (hpi), the development of disease symptoms was examined at 6-h intervals (Al-lami et al. 2019b, 2020).

8.3.5.3 Disease Assessment

The reactions of both cotyledons and fully expanded leaves were recorded at 7 and 10 dpi (Hong and Fitt 1995). Plants were assessed for disease reaction in three ways: (a) leaf disease incidence (by assessing the percentage of leaves showing disease symptoms) on a 0–10 disease incidence scale (Barbetti 1987) where 0 = nil disease, 1 = 1%–10%, 2 = 11%–20%, 3 = 21%–30%, 4 = 31%–40%, 5 = 41%–50%, 6 = 51%–60%, 7 = 61%–70%, 8 = 71%–80%, 9 = 81%–90%, and 10 = 91%–100% of leaves diseased; (b) the percentage of leaf area diseased, by calculating the leaf area diseased across all leaves of each plant as a percentage of the total leaf area using the same scale (0–10) but for percentage of leaf area affected instead of disease incidence (Eshraghi et al. 2007); and (c) the degree of leaf senescence, by assessing the number of senescent leaves on plants diseased with Alternaria leaf spot compared to control plants (the number of senescent leaves on control plants was subtracted from the number on diseased plants), also on a 0–10 scale where 0 = no senescent leaves and 10 = >90% of leaves senesced from *Alternaria* spp. infection (Barbetti and Nichols 2005). Subsequently, each 0–10 score for leaf disease incidence, leaf area diseased, and leaf senescence was converted into percentage leaf disease incidence (% LDI), percentage leaf area diseased (% LAD), and percentage leaf senescence index (% LSI) index based on the method described by McKinney and Davis (1923), where:

$$\%\text{LDI}, \%\text{LAD or} \%\text{LSI} = \{[(a \times 0) + (b \times 1) + (c \times 2) + (d \times 3) + (e \times 4) + \ldots (k \times 10)] \times 100/[(a + b + c + d + e \ldots k) \times 10]\}$$

and where $a, b, c, d, e \ldots k$ are the number of plants with severity scores of 0, 1, 2, 3, 4, … 10, respectively. Subsequently, a distinct category of disease virulence (aggressiveness) was allocated to the isolates according to each of the three disease parameters. Virulences in relation to %LDI were classified as follows: L, low virulence (≤39%); M, moderate virulence (40%–59%); H, high virulence (≥60%). Virulences based on %LAD were classified as follows: L, low virulence (≤29%); M, moderate virulence (30%–49%); H, high virulence (≥50%). The % LSI was rated as low level of senescence for ≤10% of leaves senesced, intermediate level of senescence for 10%–20% of leaves senesced, and severe level of senescence where there was ≥20% of leaves senesced (Al-lami et al. 2019a, b).

8.3.5.4 Experimental Design and Statistical Analysis

The experiment was fully repeated twice. Each trial had 10 replicates arranged in randomized block design using the "Generate a Standard Design" function of GenStat 18th edition. Normality of data and homogeneity of variances from each experiment were tested before conducting analyses. A *t* test using GenStat was used to compare the disease indices (% LDI, % LAD, and % LSI) between the experiments. As no differences between the experiments were detected (i.e. $P > 0.05$), data for both trials were combined and re-analyzed as a single data set. The % LDI, % LAD, and % LSI were analyzed using one-way ANOVA function of GenStat. Fisher's least significant differences (LSDs) were used to separate significant differences between treatments ($P > 0.05$). Multiple comparison tests for means were performed using Bonferroni's method in GenStat. Bi-variate Pearson's correlation coefficients (*r* values) using SPSS Statistics for Windows v. 25.0 (IBM Corporation) were computed for aggregates to test the relationships between the disease assessment parameters across the *Alternaria* spp. overall and within each species (Al-Lami et al. 2019a, b).

8.4 *Brassica–Colletotrichum* Pathogenesis Protocols

8.4.1 Characterization of MAPK Gene During *Brassica–Colletotrichum* Interaction

8.4.1.1 Fungal Strains, Plants, and Culture Condition

The wild-type *C. higginsianum* IMI349061 (CH-1) was cultured on potato dextrose agar (PDA) at 25 °C and stored in PDA slants at 4 °C for further use. *Arabidopsis thaliana* wild-type Col-0 plants were used. Plants were grown in growth chambers at day and night temperatures of 20 ± 2 °C, with 12 h of light and 12 h of darkness.

Escherichia coli strain DH5a was used to propagate all plasmids and *Agrobacterium tumefaciens* strain EHA105 was used for fungal transformation.

8.4.1.2 Bioinformatics Data and Programs

The publicly available genomic sequence data base of *C. higginsianum* (http://genome.jgi.doe.gov/Colhi1/Colhi1.home.html) was used to characterize the CH063_08490 gene. The NCBI (http://www.ncbi.nlm.nih.gov/) and Uni Prot (http://www.*uniprot*.org/blast/) were used for Blastp analysis. The *Clustal* X program was used for amino acid alignments. MEGA program was used to produce the phylogenetic tree with un-rooted neighbor-joining method. The secondary structure prediction was completed with *Jnetpred* program.

8.4.1.3 The *ChMK1* Gene Replacement and Complementary

For characterizing the *ChMK1* gene, a ChMK1 replacement vector, pChMK1–3300, and a complementation vector, pChMK1-Com, were constructed. The replacement vector was constructed using the homologous recombination strategy. A0.9-kb fragment upstream of the ChMK1 ORF in the *C. higginsianum* genome was amplified with primers Rep-up F and Rep-up R and cloned in to the *Hind* III and *Sal* I sites on pMD18-*hph*, and the resulting construct was named pMD2-hph. The na 1.0-kb fragment downstream of *ChMK1* ORF was amplified with primers Rep-down F and Rep-down R and cloned between the *Xba* I and *Kpn* I sites inpMD2-*hph*, and the resulting construct was the *ChMK1* gene replacement vector, pChMK1–3300, which had the selective marker *hph* gene flanked by the ChMK1 ORF flanking sequences. The complementation vector was constructed that *ChMK1* cDNA was amplified by RT-PCR with primers Com F and Com R and cloned into the same sites of pCIT vector, which contained the constitutive P*trpC* promoter and terminator. Finally, the cDNA of *ChMK1* was digested with *Apa* I and cloned into pNeoP3300, resulting in *ChMK1* complementation vector, pChMK1-Com.

Agrobacterium-mediated transformation was performed as previously described (Liu et al. 2013) with some modification that plasmid-containing *A. tumefaciens* cultures (0.6 OD unit at 600 nm) were mixed 1:1 with *C. higginsianum* conidial suspension (10^6 spores/mL) in induction broth supplemented with 400 μM acetosyringone and cultured for 6 h at 25 °C 200 rpm, and then cultured on a cellophane membrane laid on co-induction medium supplemented with 400 μM acetosyringone at 22 °C for 48 h. The cellophane membrane were removed to new plates and overlaid with 15 mL PDA containing 500 μg/mL of cephalosporin and 50 μg/mL of hygromycin. After incubation at 25 °C for 72 h, transformants were transferred to PDA plates containing 50 μg/mL of hygromycin for a second round of selection. Transformants were confirmed primarily by RT-PCR with primers RTF and RTR and further confirmed by Southern blot. The *C. higginsianum β-tubulin* gene (CH063_04743) was used to normalize the RNA sample for each RT-PCR.

8.4.1.4 DNA Manipulation and Southern Blot Analysis

The genomic DNA of *C. higginsianum* wild-type strain and other derivative mutants was extracted according to the procedure (Sambrook and Russell 2001). Southern blot analysis was performed following the method described (Liu et al. 2013). For each sample, 15–20 μg genomic DNA was digested with Hind III, size-fractionated through a 0.8% agarose gel and mounted on positively charged nylon membrane. Then nylon membrane was then hybridized with a probe amplified by primers RT-F and RT-R and labeled with digoxigenin (DIG)-dUTP using the PCRDIG Probe Synthesis Kit in accordance with the manufacturer's instructions.

8.4.1.5 RNA Manipulation and qRT-PCR

Total RNA of fungal strains was isolated with TriZOL reagent (in vitro gen) according to manufacturer's instructions and stored at −80 °C for further study. The total RNA samples were treated with DNaseI (RNase Free) at 37 °C for 0.5 h and used to generate the first strand cDNA with Revert Aid TM First Strand cDNA Synthesis Kit (Fermentas) according to manufacturer's instructions. The first strand cDNA was stored at −20 °C for further study. Gene expression was analyzed by qRT-PCR using a Bio-Rad CFX96 (Bio-Rad) and SYBR Green PCR mix (Bio-Rad), according to the manufacturer's instructions. The *C. higginsianum* β-*tubulin* gene (CH063_04743) was used to normalize the RNA sample for each qRT-PCR. For each gene, qRT-PCR assays were repeated at least twice, with each repetition having three replicates. Primer pairs for qRT-PCR detections were designed across or flanking an intron.

8.4.1.6 Characterization of ChMK1 Deletion Transformant and Wild-Type Strain

The growth rates were assayed that ChMK1 deletion transformant and the wild-type strain were firstly cultivated on PDA for 7 days, then the mycelial agar discs were taken from the active colony edge and inoculated on the center of the PDA Petri dish at 25 °C before hyphal grow thaws examined. The colony morphology and conidiation of these strains were examined after being grown on PDA plate for 15 days at 25 °C. The virulence was assayed that *C. higginsianum* strains were cultured on PDA at 25 °C for 5 days, and the conidia were collected and washed with sterile distilled water twice. The conidial suspension (10^6conidia/mL) was spotted with 6 μL droplets on the leaf surface of 4-week-old *Arabidopsis.* Inoculated leaves were incubated in darkness at 25 °C, and the symptoms were observed at 6 days postinoculation (dpi). Conidial germination and appressorial formation were observed at 10 μL conidial suspension (10^6conidia/mL) was spotted on plastic cover slips and incubated at 25 °C for 24 h. Then, the infection-related morphogenesis was examined by Nikon Eclipse 80i microscope under bright-field model using 40× fold magnification.

8.4.1.7 Cell Wall Sensitivity Assay

The sensitivity of wild-type strain and the transformants to cell wall inhibitors were performed that mycelial plugs (4 mm in diameter) were respectively inoculated on to PDA containing 0.01% SDS, 300 μg/mL Calcofluor White (CFW), and 300 μg/mL Congo Red (CR) for 7 days as described (Zeng et al. 2012; Luo et al. 2014), and with PDA media as controls. Cell wall sensitivity to the compounds mentioned above was assayed by measuring the colony diameters as previously described (Fujioka et al. 2007; Valiante et al. 2009; Carbo and Perez-Martin 2010). The data assays were analyzed with Origin 7.5 using ANOVA (one-way, $P \leq 0.05$) (Wei et al. 2016).

8.5 *Brassica–Erysiphe* Pathogenesis Protocols

8.5.1 Molecular Identification of Anamorphic Powdery Mildews (Erysiphales)

The powdery mildew fungi (Erysiphales) are a common group of obligate plant pathogens that can be extremely difficult to identify, depending on the reproductive stage encountered. The ribosomal DNA Internal Transcribed Spacer (ITS) region was assessed for its usefulness to link anamorphic Erysiphales with their respective teleomorphs. PCR primers for the rDNA ITS region was designed, and found to have enhanced specificity for the 12 genera tested, even in the presence of contaminating fungi. Recent herbarium specimens of both anamorphic, and teleomorphic material yielded sufficient DNA for amplification using a Chelex-based extraction method. The ITS regions from 25 anamorphic specimens were sequenced, and compared with the ITS sequences of their suspected teleomorphs. In most cases, an ITS region sequence similarity of above 99% indicated that a correct match had been made. Although this technique will not always unambiguously identify an anamorphic specimen, it will provide valuable information to use in conjunction with morphological, and host range data to aid in the final identification (Cunnington et al. 2003).

8.5.1.1 Primer Design and Test Specimen Methods

Erysiphales specific primers for the ITS region was designed from 18S, and 28S sequences (Mori et al. 2000). The forward primer, PMITS1 (5′-TCGGACTGGCC(T/C)AGGGAGA-3′) was designed from 18S sequences of *Blumeria graminis* (Gen Bank Accession AB022361), *Erysiphe heraclei* (AB022390), *Golovinomyces cichoracearum* var. *cichoracearum* (AB022359), *Phyllactinia moricola* (AB0022400), *Podosphaera xanthii* (as *Sphaerotheca cucurbitae*, AB022409), and *Pleochaeta shiraiana* (AB022402), whereas the reverse primer PMITS2 (5-′-TCACTCGCCGTTACTGAGGT-3′) was designed from 28S sequences of *Erysiphe heraclei* (AB022391), *Erysiphe* (*Uncinula*) *mori* (AB022418), *Phyllactinia kakicola* (AB022372), *Golovinomyces cichoracearum* var. *cichoracearum* (AB022360), *Sawadaea polyfida* var. *japonica* (AB022364), and *Blumeria graminis* (AB022362). The primer sites were chosen by comparing these sequences with those from a range of other fungi available in Gen Bank. In particular, 26 ascomycete sequences from species in the Pezizales, Ophiostomatales, Microascales, Saccharomycetales, Chaetothyriales, Eurotiales, Hypocreales, Phyllachorales, Cyttariales, Rhytismatales, Dothidales, Sordariales, Onygenales, Heliotales, and Schizosaccharomycetales were used. To gain the specificity required, particular attention was paid to ensure that the 3′ end of both primers consisted of a sequence unique to the powdery mildews (Innis and Gelfand 1990; Kwok et al. 1990).

Fourteen test specimens were chosen to represent species from as many genera as possible, and also to include specimens with common contaminating fungi growing in the powdery mildew colony. Although most of these specimens were purely anamorphic, they were species that could be confidently identified by a combination of their characteristic anamorph morphology, and host. The only specimen not identified to species level was the *Sawadaea* species on *Acer*.

8.5.1.2 DNA Extraction and PCR

DNA was extracted using a protocol modified from Hirata and Takamatsu (1996). A small amount (<1 mm^3) of the fungus was scraped from the host plant, and placed into 50 μL of 5% Chelex 100 containing 0.01% Triton X-100. This was incubated at 56 °C for 2 h (vortexing briefly after 1 h), vortexed, incubated in boiling water for 8 min, vortexed, and spun in a bench-top centrifuge at 20,000 rpm for 5 min. The top 40 μL was removed, added to 40 μL isopropanol, mixed, left on ice for 10 min, and centrifuged for 10 min at 20,000 rpm. The supernatant was removed, and the sample was dried, and re-suspended in 40 μL of water.

The initial PCR was performed in 25 μL containing 5 μL DNA, 200 μM of each dNTP, 1.5 mM $MgCl_2$, 2.5 μL 10 × buffer, 4 ng each of primers PMITS1, and PMITS2, 6% Tween 20 (Lab chem.), 0.5 units of Amplitaq Gold, and 100 ng BSA. Reactions were over laid with 30 μL of light paraffin, and subjected to 10 min at 94 °C, 35 cycles of 1 min at 94 °C, 1 min at 65 °C, 2 min at 72 °C, and a final extension of 10 min at 72 °C. PCR products were detected by running 4 μL on a 1.4% agarose gel in TBE buffer. A nested PCR was performed in 50 μL as outlined above, but without Tween 20, using primers PMITS1, and ITS4 (White et al. 1990) at twice the previous concentration, and containing 1 μL of a 1:100 dilutions of the first round product, or if no product was visible then an undiluted 1 μL. Cycling times were the same, but with an annealing temperature of 60 °C. Nested PCR products were purified using a QIA quick PCR Purification Kit (Qiagen). To ensure that only the powdery mildew ITS region had been amplified, the products were sequenced directly using primers ITS5 (White et al. 1990), and ITS4, with an ABI PRISM BIGDYE Terminator Cycle Sequencing Kit (Perkin-Elmer). These sequences were used in similarity searches on the Gen bank database using the computer program Blast (Altschul et al. 1990) to confirm that they did belong to the Erysiphales.

8.5.1.3 Anamorph–Teleomorph Connection

The method described in the previous section was used to sequence the ITS region of several anamorphic specimens collected on a variety of Australian host plants. Additionally, some specimens from previous list were used. Cleistothecial material of their suspected teleomorphs was obtained from either Northern Hemisphere localities or from Australia, and the ITS region sequenced. Sequence data for several species were obtained from Gen Bank. In each case, to determine if a match had been made, a percentage DNA similarity was calculated by aligning the two sequences using Clustal W (Thompson et al. 1994), and using the program Homologies (Leunissen).

8.5.2 The DNA Sequence Analysis

DNA from the MGH isolate of *Erysiphe* was prepared by a slight modification of the method of Pfister (1997). Briefly, conidia (about 0.1 mL dry vol) were ground in a microcentrifuge tube with a plastic pestle. The ground conidia were resuspended in 0.4 mL lysis buffer (200 mM Tris/HCl, pH 8.5, 250 mM NaCl, 25 mM EDTA, 0.5% SDS), and the suspension was treated as described by Pfister (1997). One μL of the resulting DNA preparation was used for amplification by the polymerase chain reaction (PCR). The region of the ribosomal DNA containing the internal transcribed pacers (ITS) 1, and 2 was amplified using the primers ITS4, and NS7 (White et al. 1990), and the reaction conditions recommended by White et al. (1990) with an annealing temperature of 55 °C. The PCR products were sequenced without cloning utilizing ABI PRISM Dye Terminator chemistries, and ABI 377 DNA Sequencers. The primers ITS4, ITS2, NS8, NS7, ITS5, and ITS3 were used for DNA sequencing (White et al. 1990). Products from three independent PCR amplifications, one of which was performed on an independent DNA preparation, were sequenced to ensure that no PCR errors were obtained in the sequence. Analysis of the sequence data was carried out using Genetics Computer Group software, version 8 and the sequences of the ITS regions of the MGH isolate, and the *E. cichoracearum* UCSC isolate were compared using the GAOP algorithm (Plotnikova et al. 1998).

8.5.3 The DNA Marker Analysis

DNA for PCR was extracted from F_2 plants using a mini-prep procedure. Two or three young leaves were placed in a 1.5 mL minifuge tube, frozen in liquid nitrogen, and quickly ground to powder with a small pestle. The powder was thawed in 700 μL of extraction buffer containing 100 mM of Tris-HCI (pH 8.0), 500 mM of NaCI, 50 mM EDTA, 1.5% SDS, and 10 μM of 2-mercaptoethanol. The homogenate was mixed by vortex, and incubated at 65 °C for 10 min. Homogenates were mixed with 700 μL of phenol, and the upper phase after centrifugation for 5 min at 13,000 *g*, was transferred to a fresh tube, and extracted with 700 μL of phenol/chloroform (1:1). DNA in the aqueous phase was precipitated with 50 μL of 3 M NaAc (pH 5.2), and 450 μL of cold isopropanol, and collected by centrifugation. The pellets were dissolved in 100 μL of TE (10 mM Tris, 1 mM EDTA, pH 8.0), and stored at −20 °C. Larger scale DNA isolation for RFLP analysis of F_2 plants used 1–2 g of fresh leaves from 20 to 30 young F_3 plants, and the extraction was as for the mini-prep, scaled up proportionately. The DNA pellets were dissolved in 200–300 μL of TE. The primers of microsatellite markers (Bell and Ecker 1994), and CAPS markers (Konieczny and Ausubel 1993) were either obtained from Research Genetics or synthesized in house. Amplification reactions contained 0.3 μM of each primer,

0.2 μL of Taq polymerase (1 unit; Boehringer), 1 μL of DNA solution (about 5–10 ng), and 20 μL of PCR mixture containing 50 mM NH_4CI, 10 mM Tris, pH 8.0, 2–3 mM $MgCI_2$, 200 μM dNTPs.

Amplification was for 45 cycles: for microsatellite markers each cycle was 94 °C for 20 s, 55 °C for 20 s, and 72 °C for 40 s; for CAPS markers, each cycle was 94 °C for 30 s, 55 °C or higher, according to GC content, and primer length, for 30 s, and 72 °C for 60 s. Amplified products were resolved on a 1%–3% agarose (w/v) gel, depending upon the size of DNA fragments. New CAPS markers between La-*er,* and Ms-0 were developed from TSL, and PI using the following primers: TSL, forward primer 5′ CTGCCGCTGTCAAAGTCGTCAC 3′ reverse primer 5' CCGTCCTCCCA TGTCTCAGCTATG 3′; polymorphism revealed by Bcll; PI, forward primer 5′ GGGTAGAGGAAAGATCG AGATA 3′, reverse primer CACAAATAGCTTATCACAGAAT, polymorphism revealed by *Dral,* and *Alul.* The RFLP probe T01409 was amplified with the forward primer, 5′ CGGAGCTTATCAGATCCAGACA 3′, and reverse primer, 5′ CCTCAATTTTCCCCAAAGAAGA 3′. Other RFLP probes were amplified (BGL1), or cloned (pCITf16, obtained from Ohio DNA Stock Center, ABRC). After digestion, 1.5–2.0 μg of DNA was separated on a 1.0% agarose gel with 1 × TAE buffer, and then transferred to a nylon membrane (Hybond N, Amersham). Probes were labeled using the Pharmacia Oligo labeling kit, and hybridization was conducted according to manufacturer's instruction (Xiao et al. 1997).

8.5.4 Histological Assessment of *Erysiphe cruciferarum* Growth

Similar to other pathogenic powdery mildew and biotrophic fungi, *E. cruciferarum*, is noncultural on standard laboratory media, and reproduces and proliferates only on susceptible crucifer hosts. After inoculating the upper leaf surface with conidia of *E. cruciferarum*, samples were collected at different time periods for histological examination. Approximately four leaves from all infected plants of each genotype were examined for each time point. At 6 dpi, symptoms of infection were observed in white powdery mildew on the upper surface of B. napus leaves. Symptoms can be observed with the naked eye and increased with the progress of time (Plate 8.1d1, 3e1, 3f1). The observable signs of the pathogen included hyphae, conidia, conidiophores, and dead cells on the infected leaf surface, whereas visual symptoms were not observed in the *R. alboglabra* genotype (Plate 8.1g1–l1). At 1 dpi, the structures of pathogenic fungi appeared on the securities under microscopic examination, but lower compared with *B. napus* (Plate 8.1g). Microscopic examination showed the conidia of the pathogenic fungus produced germination tubes on the leaves of both genotypes, where conidia produced primary germ tube, aspersorium, and primary hyphae at 2 dpi, followed by hyphae in *B. napus* and to lesser extent in *R. alboglabra* (Plate 8.1b). The rapid growth and wide spread of the pathogen and hyphae were also observed. Several conidia were observed on the upper surface of the infected leaves of *B. napus*. The fungus growth increased with time (Plate 8.1c–f). On the contrary, the microscopic examination on *R. alboglabra* showed that most

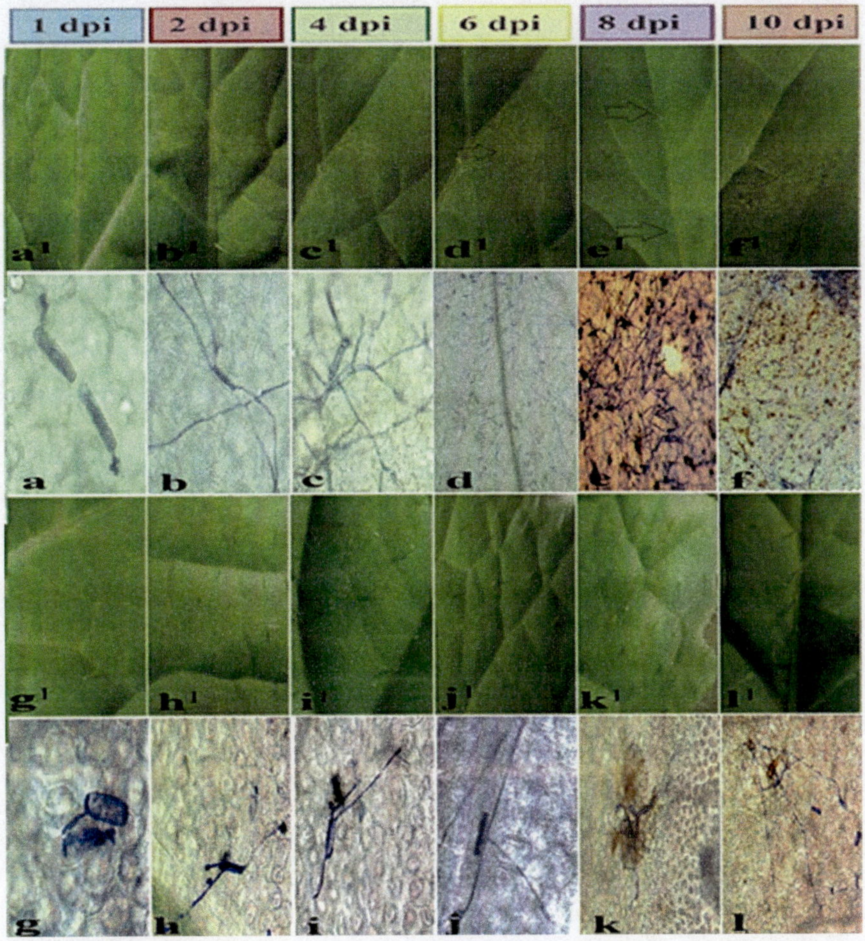

Plate 8.1 Symptoms of *Erysiphe cruciferarum* on leaves infected of *Brassica napus* of six time points. Symptom images were (a1, b1, c1, d1, e1, f1), and light micrograph were (a, b, c, d, e, f) for leaves of *R. alboglabra* infected by pathogen were (g1, h1, i1, j1, k1, l1), and light micrograph were (g, h, i, j, k, l) for leaves of RRCC infected by pathogen at 1, 2, 4, 6, 8, and 10-days post-inoculation (dpi) respectively. Stocks indicate to colonies and the growth of pathogenic fungus. Scale bars for light micrograph at 8 and 10 dpi are 25 μm (Alkooranee et al. 2015)

of the conidia produced germination tubes, but failed to format the hyphae. These conidia did not develop intensively and rapidly compared with *B. napus* during the same time period (Plate 8.1i–l). The results also showed the presence of dead cells in *B. napus*, and *R. alboglabra* increased. However, fungus growth was densely spread and increased with time. The microscopic examination clearly emerged after 6, 8, and 10 dpi (Plate 8.1f, 3i). The percentage of spores that successfully germinated on the leaf surface did not differ between the genotypes and time points. However, the further development of the fungus differed during cell penetration; development was lower for *R. alboglabra* (5%) compared with *B. napus* (75%) at 2 dpi. This continued penetration resistance reached 14% successful penetration at 4 dpi compared with 68% of B. napus (Plate 8.1g, h). Thus, no or very little colonies were formed on *R. alboglabra* at 4, 6, 8, and 10 dpi. Finally, the histological study confirmed that the *R. alboglabra* genotype was excluded from infection because it was able to prevent the pathogenic fungus *E. cruciferarum* from penetrating the cell wall of host successfully and inhibited the growth of the pathogen at different stages. This penetration resistance was activated at 1 dpi and allowed complete resistance under certain conditions (Alkooranee et al. 2015).

8.6 *Brassica–Hyaloperonospora* Pathogenesis Protocols

8.6.1 Molecular Marker for Identification of *Hyaloperonospora parasitica*

The PCR amplification of ribosomal RNA gene block (rDNA) spacers (ITS1 and ITS2) performed in 44 *H. parasitica* isolates from different *Brassica oleracea* cultivars and distinct geographic origins, revealed no length polymorphisms. ITS restriction analysis with three endonucleases, confirmed by sequencing, showed no fragment length polymorphisms among isolates. Furthermore, ITS amplification with DNA isolated from infected host tissues also allowed the detection of the fungus in incompatible interactions. The combination of the universal ITS4 and ITS5 primers, for amplification of full ITS, with a new specific forward internal primer for ITS2 (PpITS2F), originates a *H. parasitica*-specific amplicon, suitable for diagnosis. As ITS2 regions of *H. parasitica*, *B. oleracea*, other *B. oleracea* fungal pathogens and other *Hyaloperonospora* species are clearly distinct, a fast and reliable molecular identification method based on multiplex PCR amplification of full ITS and *H. parasitica* ITS2 is useful for the diagnosis of crucifer downy mildew. The method can be applied to diagnose the disease in the absence of fungal reproductive structures, thus being useful to detect nonsporulating interactions, early stages of infection on seedlings, and infected young leaves packed in sealed plastic bags. Screening of seed stocks in sanitary control is also a major application of this diagnostic method (Casimiro et al. 2004).

In fungal genomes, the highly conserved rRNA genes are separated by two less conserved internal transcribed regions, the internal transcribed spacers 1 and 2 (ITS1 and ITS2), which are therefore suitable for polymorphism studies among species or even at infra-specific level (Duncan et al. 1998; Mills et al. 1998). Amplified ribosomal DNA restriction analysis (ARDRA), using PCR primers based on conserved regions of the rRNA genes (White et al. 1990), followed by restriction with frequently cutting endonucleases, allows the easy assessment of sequence differences in ITS regions without length polymorphisms (Buscot et al. 1996; Lanfranco et al. 1998). Although fungal ITS1 has been shown to be more polymorphic at sequence level than ITS2 (Duncan et al. 1998), analysis of *H. parasitica* ITS1 sequences showed no differences among isolates collected from the same host species (*B. oleracea* and *A. thaliana*) and only 85% similarity between isolates from different hosts (Rehmany et al. 2000). In fact, host range must be associated with genetic differences of the isolates, classified as belonging to the same species, and these data point to the potential of ITS regions as molecular markers, both at species and *forma specialis* levels. There are, however, no available data on ITS2 variability and taxonomic relevance. Casimiro et al. (2004) characterize the ITS regions of *H. parasitica* isolates, from different *B. oleracea* crops and distinct geographic origins, in order to evaluate their potential as molecular markers for identification purposes.

8.6.1.1 *Hyaloperonospora parasitica* Isolates and Conidia Isolation

Forty-four *H. parasitica* isolates, collected from *B. oleracea* plants, were grown on seedlings of *B. oleracea* hosts "1" or "2." For each isolate, ca. 50 1-week-old seedlings were inoculated with two droplets of a conidia suspension (5×10^4 conidia mL^{-1}) per cotyledon and maintained in the dark, at 16 °C for 24 h. Then the inoculated seedlings were transferred to a growth room and maintained at 20 ± 1 °C, under a 20-h photoperiod. After 6 days of incubation, the seedlings were transferred to a dark room for 24 h, to induce sporulation. Cotyledons with sporulation were harvested and shaken in 50 mL of sterile distilled water to dislodge conidia. The conidial suspension was gauze filtered and centrifuged at 2600 *g* for 3 min. The pellet was re-suspended in 7.5 mL of sterile distilled water, aliquoted in 1.5-mL fractions and stored at −20 °C until use.

8.6.1.2 *Brassica oleracea* Fungal Pathogens Selection and Growth

Ten isolates of *B. oleracea* fungal pathogenic species or related species of the same genera were selected, namely *Fusarium culmorum*, *Trichoderma* sp., *Alternaria* sp., *Phoma* sp., *Phytophthora cinnamomi*, *Sordaria* sp., from collection, and from CECT (Collection Espanola de Cultivos Tipo) *F. oxysporum* (CECT 2154), *Sclerotinia sclerotiorum* (CECT 2882), *Mycosphaerella tassiana* (CECT 2665), and *Diaporthe phaseolorum* (CECT 2022). All fungi were grown on potato dextrose agar medium, with exception of *Phytophthora cinnamomi* which was grown on corn meal agar, at 28 °C for 7 days.

8.6.1.3 DNA Isolation

DNA was isolated using an adaptation of the Ferreira and Grattapaglia (1995) method. An aliquot of each *H. parasitica* conidial suspension was centrifuged at 6400 *g* for 3 min. The pellet, or 100 mg of each fungal mycelium (obtained by colony scraping), was macerated with 200 μL of glass beads (425–600 μm), and 500 μL of extraction buffer (CTAB 2%, 1.4 mol l^{-1} NaCl, 0.02 mol l^{-1} EDTA, 0.01 mol l^{-1} Tris-HCl pH 8.0, 1% PVP, 0.2% β-mercapto-ethanol, 0.1% Proteinase K), at 65 °C, were added. The suspension was incubated at 65 °C for 45 min, with mixing by inversion each 15 min. After cooling to room temperature, 500 μL of chloroform: isoamyl alcohol (24:1) were added, the tube was mixed by inversion and centrifuged at 16,700 *g* for 10 min. The upper aqueous phase was collected and the DNA was precipitated with 600 μL of isopropanol (−20 °C) for 1 h at −70 °C. After a 10-min centrifugation at 16,700 *g*, the pellet was washed with 500 μL of washing buffer (ethanol 70%, 0.15 mol l^{-1} NaCl) and centrifuged at 16,700 *g* for 5 min. The pellet was re-suspended in 25 μL of TE (0.01 mol l^{-1} Tris-HCl pH 8.0, 0.001 mol l^{-1} EDTA) and stored at 4 °C until utilization.

After maceration with liquid nitrogen and using the method above, DNA was extracted from 100 mg of short cycle *B. oleracea* CrGC3.1 and cabbage "Coracao-de-boi" seedling tissue, which were not infected with *H. parasitica*, and from 100 mg of Tronchuda cabbage "Algarvia" seedling tissue, either infected with the isolate P501 or uninfected.

8.6.1.4 ITS Amplification

To amplify ITS1, ITS2 and full ITS regions, either from the fungus or the hosts, the following primers were used (White et al. 1990): ITS2 and ITS5 to amplify ITS1 region; ITS3 and ITS4 to amplify ITS2 region; and ITS4 and ITS5 to amplify full ITS. Each reaction mixture contained 2 μL DNA, PCR buffer 1×, 0.0025 mol l^{-1} $MgCl_2$, 0.05% W1, 0.0002 mol l^{-1} of each dNTP, 0.001 mol l^{-1} of each primer and 2 U of Taq DNA Polymerase, in a final volume of 50 μL. To each PCR tube, ca. 50 μL of mineral oil were added and amplification occurred in a Robo Cycler 96 (Stratagene), according to the following amplification program: 4 min at 95 °C; 35 cycles of 1 min at 95 °C, 1 min at 56 °C and 2 min at 72 °C; 4 min at 72 °C. Each reaction sample was run on a 1.5% agarose gel, in 0.5 × TBE (0.05 mol l^{-1} Tris, 0.045 mol l^{-1} boric acid, 0.001 mol l^{-1} EDTA) at 90 V for 2 h 30 min, using 1 kb Plus standard (Gibco BRL) as molecular size marker. After ethidium bromide staining, the gels were analyzed with KODAK 1D 2.0 software.

For each isolate or host ITS regions, amplification was performed two to three times in order to assess the reproducibility of the method. The molecular sizes of *H. parasitica* ITS regions were estimated using the arithmetic average and standard error of the 44 isolates. Molecular sizes of host ITS regions were calculated as the average value of two replicates.

8.6.1.5 ITS Restriction Assay

To perform restriction digestion of amplified ITS regions, 10 μL samples of each PCR product, not purified, were digested with 5 U of each one of three restriction endo-nucleases, *Rsa*I (Nbl), *Hae* III, and *Sau* 3AI, in a final volume of 15 μL, according to manufacturer instructions. After a 3-h incubation period at 37 °C, 1.5 μL of bromophenol blue solution (0.25% bromophenol blue, 0.25% xylene cyanol, 15% Ficoll in water) was added to each sample to stop the reaction. Each reaction sample was run on a 1.5% agarose gel, in 0.5 × TBE at 90 V for 3 h, using a 100 bp standard as a molecular size marker. After ethidium bromide staining, the gels were analyzed with KODAK 1D 2Æ0 software. Reproducibility of the method was assessed with duplicate reactions. Molecular sizes of individual restriction fragments produced from *H. parasitica* ITS regions were estimated using the arithmetic average and standard error of the 44 isolates. For host ITS regions, molecular size estimations were based on two replications.

8.6.1.6 ITS2 Sequencing

In order to sequence the ITS2 region, 15 μL of the PCR reaction of isolate P524 were run in 1% agarose gel, in 0.5 × TBE at 90 V for 1 h 30 min. After ethidium bromide staining, the ITS2 band was extracted with a sterile scalpel and purified with the Concert Rapid Gel Extraction Systems kit. The purified product was cloned using the pGEM-T Easy Vector Systems kit, with the following adaptations: 3 μL of the purified PCR product in the ligation reaction; JM109 competent cells, after inoculation in TSS medium (1 × LB (1% tryptone, 0.5% yeast extract, 0.5% NaCl, pH 7.0), 10% PEG 6000, 5% DMSO, 0.05 mol L^{-1} $MgSO_4$, pH 6.5); SOC medium replaced by LB medium in JM109 transformation. The recombinant cells were plated in LB medium with 0.15 g L^{-1} ampicillin, 0.04 g L^{-1} IPTG and 0.04 g L^{-1} XGAL.

Screening of recombinant white colonies was performed after an overnight incubation of each colony in 2 mL of LB medium with 0.15 g L^{-1} ampicillin. Each cell suspension was centrifuged at 18,000 *g* for 1 min and the pellet was re-suspended in 150 μL TEG (0.05 mol L^{-1} glucose, 0.025 mol L^{-1} Tris-HCl, 0.01 mol L^{-1} EDTA, pH 8.0), followed by the addition of 200 μL 0.2 mol L^{-1} NaOH, 1% SDS. The suspension was mixed by inversion and chilled on ice, 200 μL 3 mol L^{-1} potassium acetate (pH 4.8) were added and the suspension was centrifuged at 18,000 *g* for 10 min. To the collected upper phase, 500 μL of isopropanol (−20 °C) were added. After a 30 min centrifugation at 18,000 *g*, the pellet was washed with 500 μL of 70% ethanol and centrifuged at 18,000 *g* for 5 min. The final pellet was re-suspended in 50 μL TE with RNase (0.05 g L^{-1}).

Restriction analysis of putative recombinants with the endonuclease *Pvu* II occurred for 2 h at 37 °C, according to the manufacturer instructions, in a final volume of 30 μL. Restriction products were resolved by electrophoresis in a 1% agarose gel, in 0.5 × TBE at 90 V for 1 h 30 min. The gel was stained with ethidium bromide and fragment molecular size was estimated with KODAK 1D 2.0 software. Recombinant colonies containing the insert were re-inoculated in LB medium with 0.15 g L^{-1} ampicillin and incubated overnight at 37 °C. Recombinant plasmid DNA was extracted with the Concert High Purity Plasmid Miniprep System kit. Sequencing was performed using the CEQ2000 Dye Terminator Cycle Sequencing kit and a capillary electrophoresis CEQ2000-XL sequencer, both directly from purified PCR product and from the cloned fragment. Both DNA strands were sequenced, with the primers T7 and SP6. BLASTN (Altschul et al. 1997) of the two sequences was performed in the GenBank database.

8.6.1.7 Internal Primer Design and Multiplex PCR

ITS sequences of *Hyaloperonospora* (26 from *H. parasitica* and 15 from other *Peronospora* spp.), *Albugo* (5), *Botrytis* (2), *Alternaria* (3), *Leptosphaeria* (2), *Plasmodiophora* (4), *Fusarium* (3), *Cladosporium* (2), *Trichoderma* (1), *Phoma* (1), *Diaporthe* (1), *Phytophthora* (1), *Sclerotinia* (1), *Mycosphaerella* (2), and *B. oleracea* (4) available in the GenBank database (http://www.ncbi.nlm.nih.gov) were aligned with hierarchical clustering (Corpet 1988) at INRA website (http://prodes.toulouse.fr/multialign/multialign.html). Based on this alignment, internal primers were designed for specific amplification of full ITS and ITS2 regions of

H. parasitica: PpITS1F (5′-CAAYTWTAATTGGGGG TCGTGATCTT-3′), PpITS2F (5′-AAGCGTGACG ATACTAATTTG-3′), and PpITS2R (5′-TGAAGTG CGGCCGAAGCTT-3′).

Three multiplex PCR amplifications were performed using the following combinations of primers: ITS3 and ITS4 (to amplify any ITS2 region) plus PpITS2F and PpITS2R (to specifically amplify *H. parasitica* ITS2 region); ITS5 and ITS4 (to amplify any full ITS region) and PpITS1F and PpITS2R (to specifically amplify *H. parasitica* full ITS region); and ITS5 and ITS4 plus PpITS2F to amplify any full ITS region and *H. parasitica* specific ITS2 region. Selectivity of internal primers was tested with samples corresponding to uninfected "Algarvia" cabbage; the same host infected with *H. parasitica*; the infected host DNA combined with *Alternaria* sp. and *Phytophthora cinnamomi* DNA; *H. parasitica* DNA free from host DNA contamination and each one of the 10 *B. oleracea* fungal pathogens.

Each reaction mixture contained 2 μL DNA, PCR buffer 1×, 0.0025 mol L^{-1} $MgCl_2$, 0.05% W1, 0.0004 mol L^{-1} of each dNTP, 0.001 mol L^{-1} of each primer and 2 U of Taq DNA Polymerase, in a final volume of 50 μL. To each PCR tube, ca. 50 μL of mineral oil were added and amplification occurred in a Robo Cycler 96 (Stratagene), according to the following amplification program: 4 min at 95 °C; 35 cycles of 1 min at 95 °C, 1 min at 54 °C and 2 min at 72 °C; 4 min at 72 °C. Each reaction sample was run on a 1.5% agarose gel, in 0.5 × TBE (0.05 mol L^{-1} Tris, 0.045 mol L^{-1} boric acid, 0.001 mol L^{-1} EDTA) at 90 V for 2 h 30 min, using 1 kb Plus standard as molecular size marker. After ethidium bromide staining, the gels were analyzed with KODAK 1D 2.0 software. For each sample, amplification was performed two to three times in order to assess the reproducibility of the method. Molecular sizes of ITS regions were calculated as the average value of two replications.

8.6.2 The DNA Finger Printing of *Hyaloperonospora parasitica*

RAPD analyses were carried out using genomic DNA from 16 isolates of *H. parasitica*. These included 7 field isolates from the *B. napus* pathotype and 4 field isolates from the *B. oleracea* pathotype (Sherriff and Lucas 1990) collected in different years from various sites in the UK. Within each pathotypes isolates were chosen on the basis of differential virulence to specific host cultivars, the oospore derived progeny P003 an P033 segregate for virulence to the oilseed rape (*B. napus* var. *oleifera*) cv. Cresor (Lucas et al. 1988), while P005 and P006 differ in virulence to the cauliflower (*B. oleracea* var. *cauliflora*) cv. Palermo Green (Moss et al. 1991). Other variable characters were mating type and differences in sensitivity to the acylalanine fungicide metalaxyl.

8.6.3 The DNA Isolation

8.6.3.1 Fungal Isolates

Genomic DNA was extracted from conidia of 4–5 days old *H. parasitica* infected cultivars maintained on host seedlings (*B. napus* cv. Mikado; *B. napus* cv. Cresor or *B. oleracea* cv. Offenham Compacta) in a growth room at a temperature of 15 ± 2 °C under a 14 h photoperiod at a light intensity of 77 $\mu Em^{-2}S^{-1}$. Conidia were dislodged from sporulating cotyledons by washing with sterile distilled water. The conidial suspension was centrifuged and the conidial pallet washed at least three times in sterile distilled water. Clean conidia were then vortexed with a mixture 1-and 6 mm diameter in a lysis buffer containing 100 mM TRIS HCL (ph 7.2), 100 *mM* EDTA, 10% (w/v) SDS and 2% (v/v) 2-mercaptoethanol. DNA was recovered from the suspension of broken conidia using the protocol described by Lee and Taylor (1990). The measurement of DNA concentration was done using a TKO 100 Dedicated Mini Fluorometer.

8.6.3.2 Isolates from Host Plants

Genomic DNA was extracted from cotyledons of uninfected 9 to 10 days old seedlings of maintenance hosts using the protocol described by Edwards et al. (1991).

8.6.3.3 Polymerase Chain Reaction (PCR) Materials

The 10 mers used random primers in the PCR with 10× concentration buffer. Amplification was carried out in a Model 60 Tempcycler and in a PHC-3 Dri-Block Thermal Cycler. Agarose (Ultrapure) was used.

8.6.3.4 Amplification Conditions

The amplification conditions were rigorously tested. A standard procedure was determined based on the protocol of Williams et al. (1990). The PCR volume was 25 μL and contained 0.2 μM of primer, 100 μM each of dATP, dCTP, dGTP, and dTTP, 25 ng template DNA and 0.5 U of Taq polymerase in 1 × PCR buffer (10 mM TRIS-HCL pH 8.3, 1.5 mM $MgC1_2$, 50 mM KCL, 0.1 mg/mL gelatine). During manipulations, the tubes were kept on ice. The reaction mixtures were overlaid with 25 μL mineral oil. Standard amplifications were performed in a Coy Model 60 Tempcycler programmed for 45 cycles of 1 min at 94 °C, 1 min at 36 °C and 2 min at 72 °C or in a Techne PHC-3 Dri-Block Thermal Cycler programmed for 45 cycles of 30 s at 93 °C with ramp time of 30 C/min, 40 s at 37 °C with a ramp time of 30 °C/min and 1 min 20 s at 72 °C with a ramp time of 20 °C/min. When programmed as above, both machines produced similar temperature profiles and amplification products. After the last cycle, the samples were kept at 72 °C for an additional 10 min and then cooled to 4 °C. Samples of 15 μL were analyzed by electrophoresis in a 1.5% (w/v) agarose gel containing 0.2 μg/mL ethidium bromide with 0.5 × TBE as buffer.

8.6.3.5 Hybridization Conditions

Selected amplification products obtained with template genomic DNA from isolate P003 were recovered from the gel and purified using Gene-Clean II. Random primed DNA labeling was carried out using $dCTP^{32}$. These fragments were used as probes in hybridization experiments with genomic DNA digested with *Eco*RI and *Hind*III. Genomic DNA (0.1 μg) was digested with 2 U of restriction enzyme, separated by electrophoresis and transferred to Gene Screen Plus membrane, according to the manufacturer's instructions. The filter was prehybridized for 4 h at 42 °C in a buffer containing 50% (v/v) form amide, 5× Denhardt's solution (Denhardt 1966), 3× SSPE, 0.5% (w/v) SDS. Hybridization was carried out overnight under the same conditions. The filter was washed once in 3× SSC with 0.1% (w/v) SDS, followed by another wash in 2× SSC containing 0.1% (w/v) SDS, each for 20 min at 65 °C. A final rinse in 2× SSC was carried out at room temperature, and the filter was exposed to X-ray film (Tham et al. 1994).

8.7 *Brassica–Leptosphaeria* Pathogenesis Protocols

8.7.1 Characterization of *Leptosphaeria* Isolates

8.7.1.1 Fungal Population to Identify Avirulence Gene Frequencies

A succession of trays of 10 days old seedlings of the susceptible variety Westar (no resistance genes) were exposed to the airborne inoculum. The *Leptosphaeria maculans* leaf spot population was sampled by collecting diseased cotyledons. From each cotyledon, one typical lesion was excised and placed on wet absorbent paper in a Petri dish. After allowing the lesion to sporulate for 24 h at room temperature, 30 single pycnidial isolates per plot were produced (one isolate per leaf spot). Spores oozing from single pycnidium were collected with a sterile needle under a magnifying lens and transferred onto malt-agar. Inoculum, consisting of suspensions of 107 pycnidiospores per mL, was obtained for each isolate. Specifically, V8 agar (V8 juice 160 mL l^{-1}, agar 20 g L^{-1}) was autoclaved for 20 min at 120 °C. After cooling, streptomycin concentrated solution 10% w/v in water was added to a final concentration of 0.1 g L^{-1}. For each isolate, agar plugs were transferred to V8 agar and grown for 10 days under near UV light. Pycnidiospores were dislodged from plates in sterile distilled water, filtered through muslin cloth. Spore concentrations were standardized to 107 spores per mL after counting aliquots with a Malassez cell under a magnifying lens, aliquoted and stored at −20 °C until tested for virulence.

8.7.1.2 Plants and Inoculation

A set of 17 oilseed rape lines or varieties with known resistance genes were used. After pregermination on wet filter paper, seeds of each variety were transplanted in trays with a 1:1:1 mix of sand, peat and compost, and grown in climate chamber under a 16-h photoperiod for 10–12 days, at temperatures of 18 °C night/20 °C day. A 10 μL drop of 10^7 conidia per mL suspension was placed on each lobe of prick-wounded cotyledons (4 inoculation sites per plant) for six plants of each variety, in two independent trays. A plastic cover was placed over the inoculated plants to create a 100% relative humidity (RH) atmosphere for 24 h in the dark.

8.7.1.3 Fungal Reference Isolates

At least two reference isolates with known phenotypes were included as controls in each test, so that both compatible (virulent) and incompatible (avirulent) reactions were generated and the expression of interactions was confirmed. Some prior information on isolates' genotypes was known from literature, and allowed expanding the information for genes not previously determined.

8.7.1.4 Scoring

Infection phenotypes were scored at 14 dpi and confirmed at 21 dpi as described in Chèvre et al. (2008). The level of cotyledon infection was scored on a 0 to 11 scale in which 0 = no disease and 11 = cotyledons collapsed from phoma stem canker. The scale was modified from the 0 to 9 scale described earlier in which 0 = no darkening around wound–typical response of noninoculated controls or controls inoculated with water only; 1 = limited blackening around wound and lesion diameter of 0.5–1.5 mm; 3 = dark necrotic lesion with diameter of 1.5–3.0 mm; 5 = nonsporulating, 3–6 mm in diameter, lesion sharply delimited by darkened necrotic tissue; 7 = grayish green tissue collapse, lesion of 3–5 mm, with sharply delimited, nondarkening margin; and 9 = c at approximately 10 days, accompanied by profuse sporulation in large lesion (greater than 5 mm), and lesions with diffuse, nondarkened margins. To this scale an additional category of 11 = cotyledon totally collapsed, very often dry was added (Chèvre et al. 2008) to categorize the most susceptible plants. Plants of classes 0–5 were considered as resistant as well as plants in classes 5–7 showing typical resistant plant reaction with blackening and limited symptoms that did not develop further between the two assessment dates at 14 and 21 days after inoculation. Plants of classes 9–11 were classified as susceptible, with symptoms appearing quickly, no blackening observed in lesions and with rapid disease progression between the two dates of assessment (14 and 21 days after inoculation). The scores were averaged over all infected plants (Bousset et al. 2020).

8.7.2 Genetic Variability in *Leptosphaeria*

8.7.2.1 Sampling of *Leptosphaeria maculans* Field Populations

A total of 401 isolates were obtained from pycnidiospores oozing from individual pycnidia of leaf lesions. The resulting cultures were sub-cultured as hyphal tips for further purification. All fungal cultures were maintained on V8 juice agar, and conidia were produced on V8 juice medium as previously described. Isolates of *L. maculans* were collected from four OSR fields in France as soon as the first leaf lesions were observed in Le Rheu (western France), Oucques (central France), and Grignon (near Paris) in autumn 2000 and in Versailles (near Paris) in autumn 2003. In each field, cultivar Drakkar, a cultivar that lacks the nine known resistance alleles (*Rlm* genes), was grown, and samples were taken from four arbitrarily selected plots

(2 by 1 m). From each 2-m^2 plot, 25 naturally infected plants were arbitrarily selected, and one leaf with phoma leaf lesions was collected from each plant. One lesion per leaf was arbitrarily selected, and the causal organism was isolated. In addition, isolations were made from each of 50 leaf lesions occurring on one leaf from Oucques and from every lesion occurring on five leaves (43 total lesions) collected at Versailles.

8.7.2.2 Pathogenicity Tests

Races of the *L. maculans* isolates were determined by inoculating the 401 isolates onto a *B. napus* differential set, comprising fixed lines or commercial cultivars with as few *Rlm* alleles as possible. Cotyledons of the fully susceptible cv. Westar (no known *Rlm* allele), the cvs. Columbus (*Rlm1* and *Rlm3*), Bristol (*Rlm2* and *Rlm9*), 22-1-1 (*Rlm3*), Jet Neuf (*Rlm4*), 23-2-1 (*Rlm7*), Falcon-MX (*Rlm4* and *Rlm6*), and Samourai-MX (*Rlm1* and *Rlm6*) or Darmor-MX (*Rlm6*) were inoculated according to established protocols. At 14–27 days after inoculation, symptoms were scored from 10 to 12 plants by using the IMASCORE rating scale comprising six infection classes (IC1–IC6), where IC1–IC3 corresponded to avirulent isolates (*AvrLm*) and IC4 to IC6 corresponded to virulent isolates (*avrLm*) (Balesdent et al. 2002).

8.7.2.3 Mating Type Determination

A multiplex PCR was developed to rapidly characterize the mating type of *L. maculans* isolates. It was used to determine the distribution of the *Mat1–1* and *Mat1–2* alleles within the 401 field isolates collected. The distribution of mating type alleles was analyzed with a χ^2 test.

8.7.2.4 Mini-Satellite Analysis

The evaluation of six polymorphic mini-satellite loci (*MinLm1* to *MinLm6*) was conducted, all on different chromosomes in *L. maculans*. New specific primers based on the published sequences (accession numbers AJ621802 and AJ621805, respectively) were designed for *MinLm3* and *MinLm5* by using Oligo v.5.0 software. Primers MinLm3-ULG (5-GGCTCGGTCGGTTAGTTA-3) and MinLm3-LLG (5-AATGATGTACAGGACGGGATTT-3) or primers MinLm5-ULG (5-GCCGCCCGCCGCCTTACC-3) and MinLm5-LLG (5-GAGCTCCTGCGCCACAGTG-3) hybridized to the flanks of the mini-satellite locus and allow amplification of the repeated motifs only. *MinLm2*, *MinLm4*, and *MinLm6* were amplified by using primers that had been previously described (Eckert et al. 2005).

For PCR, genomic DNA was extracted from pycnidiospores by using the DNeasy 96 plant kit (QIAGEN) as previously described (Attard et al. 2001). PCR amplification was done in a total volume of 15 μL containing 0.2 μM of each deoxynucleotide triphosphate, 1.2 μM of each primer, 0.6 U of *Taq* DNA polymerase (Qbiogen), 1.5 μL of a 10× reaction buffer supplied with the enzyme, and 10 to 30 ng of genomic DNA. PCR amplifications were done in an Eppendorf Master cycler EP Gradient thermocycler (Eppendorf), with 30 cycles of 94 °C for 30 s,

60 °C for 30 s, and 72 °C for 80 s, with a final extension at 72 °C for 5 min. The cycling conditions were the same for the amplification of all mini-satellite loci. The PCR products were separated by electrophoresis on a 2.5% SeaKem LE agarose gel (FMC) and visualized by staining with ethidium bromide and UV illumination. Allele sizes were determined by using Quantity One 1-D Analysis software (Bio-Rad) and comparing the bands for the alleles with those of the 1-kb ladder (In vitro gen) and with PCR products obtained after amplification of alleles of known size for each mini-satellite. Each allele identified at the six mini-satellite loci was previously sequenced from at least two isolates to determine the exact allele size and to confirm that the size polymorphism observed after electrophoresis was due to variability in the number of tandemly repeated core motifs. Sequencing of the same allele amplified from at least three different individuals also confirmed that sequences of the same size are homologous and size homoplasy did not occur. The data were scored as the number of repeat units for each mini-satellite (Gout et al. 2006).

8.7.2.5 Statistical Analysis

For statistical analyses, the 401 isolates were divided into a "leaf-scale group" comprising 50 isolates collected from one leaf in Oucques and 43 isolates collected from five leaves in Versailles and a "field-scale group" comprising 308 isolates. The field scale group was structured into four field populations, each of which was further subdivided into four subpopulations corresponding to four 2-m^2 plots. Genetic diversity, which was studied separately for each field population and for all fields sampled, was estimated from genotype and gene diversity. Genotypes of isolates in both the field- and leaf-scale groups were determined by combining information from the alleles at mini-satellite loci and at the mating type locus. Genotype diversity (G) was then estimated as the percentage of unique genotypes obtained in each sample unit, in each field population, and in the overall sample units. A discriminatory index (DI), based on Simpson's index of diversity, was used to estimate the discriminatory power of the six mini-satellites. DI measures the probability that two individuals randomly selected from a small sample will have the same haplotype. DI is determined by the number and relative frequencies of the different haplotypes and is calculated as follows:

$$\mathrm{DI} = \frac{1}{N(N-1)} \sum_{j-1}^{k} nj(nj-1)$$

where N is the total number of isolates, k is the total number of haplotypes, and nj is the number of isolates sharing the jth multilocus haplotype.

POPGENE version 1.31 software was used to compute allele frequency, Nei's gene diversity (h) for each population and Nei's measures of genetic identity (I) between populations over all loci. The mean gene diversity (H) was the mean of h over loci within each population. Gametic linkage disequilibria were computed by using GENEPOP version 3.3 software for each pair of loci within each of the four field populations and in the overall field samples. GENEPOP performs the Fisher exact test using a Markov chain to test for gametic disequilibrium between two loci. The null hypothesis H_0 was: "genotypes at one locus are independent from genotypes at the other locus." The significance of departure from linkage equilibrium was assessed by using Bonferroni adjusted *P* values. Linkage disequilibrium also was estimated by pair wise comparison of haplotypes using the index of multilocus gametic disequilibrium r_d, which tests to what extent individuals that are identical (or different) at one locus are more likely to be identical (or different) at the other loci. To determine whether an observation deviates significantly from the null hypothesis of random mating (no linkage disequilibrium among loci), the observed value was compared to the results of 500 randomized data sets. The program MULTILOCUS version 1.3 was used for both estimation and randomization (Gout et al. 2006).

Heterogeneity of allele frequencies among field populations was investigated for each locus with contingency tables analysis and the Fisher exact test as implemented in the GENEPOP program. The null hypothesis tested was H_0: "allele frequencies are homogeneous across populations." Genetic structure was analyzed with Nei's coefficient of population differentiation (*GST*), which indicates the proportion of the total genetic variation attributable to population differentiation. Analyses of molecular variance (AMOVA) were computed with Arlequin version 2.000 software and used to estimate how genetic variability is partitioned among fields, among plots within fields, and within plots. Statistics associated with these components of variance (CT, SC, and ST, respectively) were estimated and their significance tested by using a nonparametric per mutation approach (Excoffier et al. 1992).

8.7.3 Phylogenetic Relationship of R-Loci in *Brassica* to *Leptosphaeria*

8.7.3.1 Sample Collection, DNA Extraction, and PCR Amplification

Forty-five plant materials were used in this analysis. The samples comprised 15 accessions each of *B. rapa* and *B. oleracea* var. *virids* from the USDA germplasm collections, and 15 cultivars of *B. napus* from the National Winter Canola Variety Trials. More detailed information on the accessions is available at the USDA-ARS Germplasm Resources Information Network available online at (http://www.arsgrin.gov/npgs). Genomic DNA was extracted from leaves of 5 week old seedlings according to Edwards et al. (1991). Thirteen polymorphic RAPD primers: OPB01, OPB05, OPC08, OPE03, OPE11, OPE12, OPE14, OPE16, OPF02, OPF10, OPG02, OPT01 and OPI01 mapped to Chromosome 4 of *B. nigra* (Chevre et al. 1997), were selected for the study. The selected markers were obtained from MWG-Biotech PCR reactions were carried out in 25 μL volumes comprising the following: 0.4 mM of each dNTP, 1.5 mM $MgCl_2$, 1.25 units of gold Taq DNA polymerase, 0.4 pmol of each primer, 1x PCR buffer and 25 ng DNA template. The amplifications were performed on a Peltier Thermal Cycler. Amplifications comprised 45 cycles each having the following repeated sequence: Denaturation at 94 °C for 1 min, annealing at between 32 and 34 °C for 1 min depending on the marker, extension at 72 °C for

2 min. A final extension at 72 °C for 7 min was applied to all reactions. Amplified PCR products, mixed with a loading dye (0.25% bromophenol blue, 0.25% xylene cynol FF, 15% Ficoll) to increase its density and aid visualization, were loaded onto 1% agarose. A Bench-top 1 Kb DNA ladder was loaded alongside the PCR products for size calling. The fragments were separated under electric current for 4 h in a standard horizontal electrophoresis unit, followed by a 15 min ethidium bromide staining. Gels were de-stained and visualized under UV light. Gel images were captured using AlphaImager v5.5 2000.

8.7.3.2 Statistical Data Analysis

Amplified DNA fragments were evaluated for size in base pairs using Alpha Ease FC version 4.0.1. Polymorphic information content (PIC) values were calculated for each primer as demonstrated by Botstein et al. (1980) and also by using the online resource, Polymorphism Information Content Calculator (http://www.agri.huji.ac.il/~weller/Hayim/parent/PIC.htm). PIC compares the polymorphism levels across markers (Hongtrakul et al. 1977; Garcia-Mas et al. 2000) and is used to determine the usefulness of markers for specific studies. Polymorphic bands were scored as binary data (present = 1; absent = 0) across the genotypes to generate a binary data matrix. The matrix was then used to generate a Genetic similarity (GS) matrix based on Jaccard's coefficient of similarity (Jaccard 1908). In brief, GS $(ij) = 2a/(2a + b + c)$, where GS (ij) is the measure of genetic similarity between individuals i and j, a is the number of polymorphic bands that are shared by i and j, b is the number of bands present in i and absent in j, and c is the number of bands present in j and absent in i. The matrix of Genetic similarity was used to construct a dendrogram using the unweighted pair group method with arithmetic averages (UPGMA) algorithm (Ananga et al. 2008).

8.7.4 Transcriptome Analysis of *Brassica–Leptosphaeria* Pathosystem

8.7.4.1 Plant Growth and Inoculation

Susceptible Topas DH16516 and the *Rlm2* transgenic line NLA51–1 (Larkan et al. 2015) were grown in Sunshine mix #4 soil in controlled environments with a 16-h photoperiod (21 °C light, 16 °C dark). Seven-day old cotyledons were point inoculated with 10 μL of a *L. maculans* isolate 00–100 (*AvrLm2*; Ghanbarnia et al. 2015) pycnidia suspension (106) or sterile water (mock). All inoculations and tissue collections were performed between 11:00 AM and 2:00 PM to minimize time of day effects. To explore the role of calcium in resistance, 10 μL of 5 mM EGTA was added to the inoculation site as an additional treatment and allowed to dry before inoculation.

8.7.4.2 Tissue Processing

Inoculated cotyledons were processed for LMD at 1 and 3 dpi per the methods of Becker et al. (2017b). Inoculation sites were excised by making two cuts parallel to the cotyledon petiole-like structure. A minimum of 30 inoculation sites per biological replicate (3 biological replicates per treatment) were collected and fixed overnight in 3:1 (v:v) ethanol: acetic acid at 4 °C. Tissues were then dehydrated in a graded ethanol series (30 min each: 75, 85, 95, and 100%) followed by xylenes infiltration (1 h each: 3:1, 1:1, 1:3 ethanol: xylenes (v:v), 100% xylenes) at 4 °C. Ten paraffin chips were added to the xylenes and kept at 4 °C overnight. Following, samples were brought to room temperature before incubating at 42 °C for 30 min and then at 58 °C for 1 h. The mixture was replaced with 100% melted paraffin every hour for 4 h before embedding.

8.7.4.3 Sectioning and Laser Microdissection

Embedded tissues were sectioned at 10 μm using a Leica RM 2125 rotary microtome in RNase-free conditions and mounted on PEN membrane slides as per Millar et al. (2015). Slides were washed for 1 min in xylenes to deparaffinize before proceeding to LMD on a Leica LMD 6000. Sections of 100 μm proximal to the inoculation site were collected in 60 μL of lysis buffer. RNA was isolated from at least 20,000,000 μm^2 of collected tissue per biological replicate. RNA extractions were performed exactly as reported in Belmonte et al. (2013). Following, RNA quality was assessed using the Bioanalyzer, and were of sufficient quality for qRT-PCR and RNA-seq as described in Millar et al. (2015).

8.7.4.4 Library Construction

First and second strand cDNA synthesis was performed per the alternative HTR (C2) described in Kumar et al. (2012) with the following modifications: Input RNA decreased to 50–75 ng; no Poly(A) bead isolation was performed; anchored oligo (dT20) primers were used during first strand synthesis instead of random hexamers; half reaction volumes for all steps; cDNA sheared via sonication instead of enzymatic digestion. Libraries were prepared using the NEB Next Ultra II kit per the manufacturer's instructions. Quality and distribution of libraries was assessed with the Agilent 2100 Bio-analyzer, and libraries were quantified with the NEB Next Library Quant Kit for Illumina and the Nano Drop 3300 fluoro-spectrometer. Library size selection of the 300–500 bp region was performed with the E-gel size select system. Sequencing was performed on the HiSeq 2500 at Genome Prairie, SK, Canada.

8.7.4.5 Data Analysis

RNA sequencing of libraries produced ~423 million paired end reads, with an average of 17.6 million reads/sample (PRJNA477556). Adapter sequences, low-quality bases ($Q < 25$), and low-quality reads (avg $Q < 30$) were removed using the Trimmomatic v0.36 software (Bolger et al. 2014). Paired-end reads passing quality control were aligned to the *B. napus* genome v4.1 (Chalhoub et al. 2014) with HISAT2 in cufflinks compatibility mode, allowing no more than two mismatches. Output SAM files were converted to BAM and sorted with Samtools. Raw counts were obtained from BAM files using the HTSeq Python Framework (Anders et al. 2015). Dendrograms and similarity matrices were generated from raw counts using DESeq (Love et al. 2014). From BAM files, Cuffquant, Cuffnorm, and Cuffdiff (Trapnell et al. 2012) were used to generate normalized counts in FPKM and to identify significantly differentially expressed genes (DEGs; $P < 0.05$; false discovery rate = 0.05). Differentially expressed gene sets were input into SeqEnrich (Becker et al. 2017a) to identify enriched metabolic pathways, GO terms, promoter TF binding motifs, and predictive transcription factor networks. Networks were visualized in Cytoscape 3.6.0 (www.cytoscape.org). *L. maculans* transcriptome analyses were performed per Haddadi et al. (2016) and all reads were trimmed as above and mapped to the *L. maculans* isolate v23.1.3 genome (Rouxel et al. 2011). Transcript abundance was measured as Log2 RPKM (Reads Per Kilobase of exon per Million mapped reads). DEGs ($P < 0.05$; false discovery rate = 0.05) were identified with DESeq2 (Love et al. 2014). Identification of secretory protein signals was performed according to Haddadi et al. (2016).

8.7.4.6 Real-Time qPCR Experiments

RNA was converted to cDNA with the qScript cDNA Supermix (Quantabio). Following, qPCR was performed using the CFX Connect Real-Time System and PerfeCTa SYBR-Green Supermix as per manufacturer's instructions in a 15 μL reaction volume. The $\Delta\Delta C_t$ method (Rieu and Powers 2009) was used to compare relative mRNA abundance, normalized to the endogenous housekeeping gene *Actin* (GenBank AF111812.1). Statistics were performed as described in McLoughlin et al. (2018).

8.7.4.7 Jasmonic Acid Measurements

Inoculation sites (30 per bio-replicate, and five bio-replicates per treatment) were hand-dissected and immediately flash-frozen and pulverized in liquid nitrogen. Following, 5 mL of extraction solvent (2-propanol:H20:HCl, 2:1:0.002, pH 2.4) and 25 ng of dihydro jasmonic acid internal standard were added to each sample, which were then shaken at 4 °C for 30 min. Next, 7 mL of dichloromethane was added, and samples shaken again at 4 °C for 30 min. After centrifugation at 5300 RPM for 5 min, the lower organic phase was transferred and dried on ice in a nitrogen evaporator system. The dried sample was then dissolved in methanol, mixed aggressively, and filtered through 0.2 μM PTFE before ultraperformance liquid chromatography (UPLC). JA standards were prepared with internal standards to generate calibration curves and response factors. Samples were analyzed on a Waters Acquity UPLC/MS with Micro mass Quattro micro API, as described in Henriquez et al. (2016) and Becker et al. (2019).

8.8 *Brassica–Plasmodiophora* Pathogenesis Protocols

8.8.1 The PCR Detection of Clubroot Pathogen

A simple, one-step polymerase chain reaction (PCR) protocol was developed to detect the pathogen in plant, and soil samples at Canada (Cao et al. 2007). The primers TC1F and TC1R, based on a *P. brassicae* partial 18S ribosomal RNA (rRNA) gene sequence from GenBank, yielded a 548-bp product in the optimized PCR. A second pair of primers, TC2F, and TC2R, which amplified a fragment of the 18S and internal transcribed spacer (ITS) 1 regions of the rDNA repeat, also was tested and produced a 519-bp product. Neither set of primers amplified any DNA fragment from noninfected plant hosts, noninfested soil, or common soil fungi and bacteria tested in this study. Quantities of 100 fg or less of total *P. brassicae* DNA, or 1×10^3 resting spores per gram of soil, could be detected consistently using these primers and PCR protocol, corresponding to an index of disease of 11% or lower when the soil was bioassayed. The protocol also enabled detection of *P. brassicae* in symptomless root tissue 3 days after inoculation with the pathogen. Therefore, the PCR assay described by Cao et al. (2007) provide a reliable diagnosis for routine detection of *P. brassicae* in plant and soil materials in a specific and rapid manner.

8.8.2 The PCR Technique to Identify Pathogen

PCR amplification was conducted to confirm the pathogen. DNA was extracted from root galls and healthy roots (control) of Shepherd's purse (*Capsella bursa pastoris*) (SP). Two primers, TC2F (5′-AAACAACGAGTCAGCTTGAATGCTAGTGTG-3′), and TC2R (5′-CTTTAGTTGTGTTTCGGCTAGGATGGTTCG-3′) were used to detect *P. brassicae* (Cao et al. 2007). No PCR amplifications were observed with the control DNA as template. A fragment of the expected size (approximately 520 bp) was obtained when DNA was amplified from diseased roots of SP. These results suggest that the pathogen in the galled roots of SP is *P. brassicae*. Pathogenicity of *P. brassicae* in SP was tested on plants of both SP and Chinese cabbage (CC) (*B. campestris* ssp. *pekinensis*). A resting spore suspension prepared from naturally infected SP roots was mixed with a sterilized soil in two plastic pots, resulting in a final concentration of 5×10^6 spores/g soil. Soil treated with the same volume of sterile water was used as a control. Seeds of SP and CC were pregerminated on moist filter paper for 2 days (20 °C), and seeded into the infested and control pots. One seed per pot was planted for CC, and four seeds per pot for SP. The pots were placed in a chamber at 15 °C–25 °C under 12 h light and 12 h dark. Plants in each pot were uprooted after 4 weeks and the roots of each plant were washed under tap water and rated for clubroot disease. No disease symptoms were observed in the control treatments of SP or CC. Plants of both species showed symptoms of clubroot, with the disease incidence of 62.5%, and 100% on SP and CC, respectively. The pathogen was isolated from diseased roots of each plant and confirmed as *P. brassicae* based on morphological characteristics and PCR detection (Ren et al. 2014).

8.8.3 Detection of *Plasmodiophora brassicae* Using Loop-Mediated Isothermal DNA Amplification

Loop-mediated isothermal DNA amplification (LAMP) technique has been developed to detect *P. brassicae*. The set of three primer pairs was designed using LAMP software. The detection was performed with the GspSSD polymerase, isolated from bacteria *Geobacillus* sp., with strand displacement activity. DNA extraction from clubbed roots obtained from farmers' fields of oilseed rape infected by *P. brassicae* was done using a modified CTAB method. The reaction was performed for 60 min at 62 °C. The visual detection was done using CF × 96 Real Time PCR Detection System or Gerie II. The detection with LAMP proved its usefulness; it was easy, fast, and accurate and independent of plant age. The detection limit was 5 spores per 1 μL of the spore suspension, so LAMP was less sensitive than quantitative PCR tests. However, the method is cheap and simple, so, it is a good alternative, when it comes to practical use and the assessment of numerous samples (Kaczmarek et al. 2014).

8.8.4 A Real-Time PCR Assay to Detect *Plasmodiophora brassicae* Under Field Conditions

A protocol using real-time polymerase chain reaction (PCR) for the direct detection and quantification of *Plasmodiophora brassicae* in soil samples has been developed and used on naturally and artificially infested soil samples containing different concentrations of *P. brassicae* under Swedish soil samples collected from seven fields across three locations. Species-specific primers and a TaqMan fluorogenic probe were designed to amplify a small region of *P. brassicae* ribosomal DNA. Total genomic DNA was extracted and purified from soil samples using commercial kits. The amount of pathogen DNA was quantified using a standard curve generated by including reactions containing different amounts of a plasmid carrying the *P. brassicae* target sequence. The PCR assay was optimized to give high amplification efficiency and three to four copies of the target DNA sequence were detected. Regression analysis showed that the standard curve was linear over at least six orders of magnitude ($R^2 > 0.99$) and that the amplification efficiency was >92%. The detection limit in soil samples corresponded to 500 resting spores g^{-1} soil. The inter sample reproducibility was similar to, or higher than, that of assays for other pathogens quantified in soil samples. Bait plants were used to validate the real-time PCR assay. The protocol developed was used to investigate the spatial distribution of *P. brassicae* DNA in different fields and a significant difference was found between in-field sampling points. The reproducibility of soil sampling was evaluated and showed no significant differences for samples with low levels of inoculum, whereas at higher levels differences occurred. Indicator kriging was used for mapping the probability of detecting *P. brassicae* within a 2-ha area of a field. A threshold level of 5 fg plasmid DNA g^{-1} soil, corresponding to approximately 3×10^3 *P. brassicae* resting spores g^{-1} soil has been suggested for growing resistant cultivars. These results provide a robust and reliable technique for predicting the risk of disease development and for assessing the distribution of disease within fields (Wallenhammar et al. 2012).

8.8.4.1 In China

SYBR Green I real-time PCR assay has been developed to detect and quantify *Plasmodiophora brassicae* ribosomal DNA (rDNA) and internal transcribed spacer (ITS). A pair of primers PBF1/PBR1 was designed based on the conservative region of rDNA-ITS of *P. brassicae*. The positive plasmid pB12 was obtained and used as the template to create standard curve. The specificity, sensitivity, and reproducibility of real-time PCR were evaluated respectively. Naturally and artificially infested soil samples containing different concentrations of *P. brassicae* were detected. The results demonstrated that standard curve established by recombinant plasmid had shown a fine linear relationship between threshold cycle, and template concentration. The melting curve was specific with the correlation coefficient of 0.995, and the amplification efficiency was 93.8%. The detection limit of *P. brassicae* genomic DNA was approximately 40 copies per 25 μL. The sensitivity of the assay was at least 100-fold higher than conventional PCR. Only DNA from *P. brassicae* could be amplified and detected using this assay, suggesting the high specificity of this assay. The coefficient of variation was less than 3%, indicating that the PCR method revealed high reproducibility. The detection limit in soil samples corresponded to 1000 resting spores g^{-1} soil. Bait plants were used to validate the real-time PCR assay. This developed real-time PCR assay allows for fast and sensitive detection of *P. brassicae* in soil and should be useful in disease management and pathogen interception so as to prevent further spread of *P. brassicae* (Jin-ping et al. 2013).

8.8.5 Bioassay Technique of Wallenhammar et al. (2012)

Bait plants were used to validate the real-time PCR assay developed. The level of *P. brassicae* in naturally infested soil samples was assessed using a bioassay method previously developed by Wallenhammar (1996) and modified by Wallenhammar et al. (2000). The pots used were $90 \times 90 \times 95$ mm^3 (vefi). Infected plants were scored according to the following classes: 0 = no galls; 1 = enlarged lateral roots; 2 = enlarged tap root; 3 = enlarged napiform tap root; 4 = enlarged napiform tap root, lateral roots healthy; and 5 = enlarged napiform tap root, lateral roots infected. A disease severity index (DSI) was calculated according to the equation:

$$\mathrm{DSI} = \frac{\sum(N1 \times 1)}{(N \times 4)} \times 100\%$$

where *N*, class no. 1, no. of plants in each class; *N*, total no. of plants.

Samples of approximately 500 g were withdrawn for further DNA detection and dried stored at room temperature until extraction.

8.8.5.1 The DNA Extraction from Infected Roots, Fungal Material, and Plant Material

A modified E.Z.N.A. SP Plant DNA Miniprep Kit extraction method was used to extract DNA from plant and fungal material. A total of 30–130 mg of fresh mycelia, apothecia or infected root tissue was incubated with two to six stainless steel beads (2–3 mm in diameter) in 1.5-mL cetyltrimethyl ammonium bromide (CTAB) solution (20 g CTAB, 12.11 g Tris, 81.8 g NaCl and 7.4 g Na_2-ethylene diamine tetra-acetic acid (EDTA) L^{-1}; pH = 8.0) and 20 μL proteinase K (20 mg mL^{-1}, GeneScan) at 60 °C for 2 h after 30 s shaking at speed setting 4.5–5.5 using a Hybaid Ribo-lyser (Hybaid). The suspension was then centrifuged at >10,000 *g* for 5 min and 1 mL of the supernatant was mixed with 5 μL RNase A in a 1.5-mL tube and incubated at 60 °C for 15 min. After centrifugation at 20,000 *g* for 1 min, 600 μL supernatant were mixed with 210 μL SP2 buffer and incubated in a refrigerator for 5 min. The suspension was centrifuged at >10,000 *g* for 10 min, and a 400-μL portion of the supernatant was transferred to a 1.5-mL tube and mixed with 600 μL SP3 buffer. The supernatant was then transferred to a HiBind DNA Mini Column, washed and eluted in 100 μL elution buffer according to the manufacturer's protocol. All DNA solutions were stored at −20 °C.

8.8.5.2 The DNA Extraction from Soil

Soil samples were air-dried at room temperature (20 °C), plant material was removed by hand and the soil was pulverized using a mallet. DNA was extracted from 350-mg soil samples using the Fast-DNA SPIN Kit for Soil according to the manufacturer's instructions with the following modifications. After cell lysis for 30 s at speed setting 5.5, samples were centrifuged at 14,000 *g* for 5 min. At the tenth step, 600 μL of the supernatant were discarded and 600 μL of the resuspended mixture transferred to a SPIN Filter. DNA was eluted in 100 μL DES. Further purification of the eluate was performed using the Wizard DNA Clean-Up System and Illustra Micro Spin S-300 HR Columns according to the manufacturers' instructions with the following modifications. In the Promega kit, 900 μL of the resin was used and incubated for 5 min at 60 °C together with the DNA solution. DNA was eluted in 100 μL. The Micro Spin S-300 HR Columns were centrifuged for 2 min before adding the DNA eluate.

8.8.5.3 Primers and Probes

Primers and a probe with putative specificity to *P. brassicae* as mentioned below were designed using a ribosomal RNA gene sequence available in GenBank (accession number AF231027) and amplified a 103-bp fragment. The forward primer was based on the sequence of the forward primer used by Faggian et al. (1999) (Table 8.1). The 5′ terminal reporter dye of the probe was VIC and the 3′ quencher was TAMRA. Primer and probe sequences were checked for identity against other

Table 8.1 Primer and probe sequences used for the amplification and detection of the *Plasmodiophora brassicae* target sequence (Wallenhammar et al. 2012)

Primer/probe	Sequence[a]
Primer forward (PbF)	5′-AAA CAA CGA GTC AGC TTG AAT GC-3′
Primer reverse (PbR)	5′-TTC GCG CAC AAG CAC TTG-3′
Probe (PbP)	5′-CGC GCC ATG CGA CAC TGT TAA ATT G-3′

[a] Bases 4–23 of PbF are the same bases as in primer PbITS6 (Faggian et al. 1999)

known sequences using the BLASTN program. (http://www.ncbi.nlm.nih.gov/blast/Blast.cgi). Primers were purchased from MWG-Biotech.

8.8.5.4 Real-Time PCR

Real-time quantitative PCR was performed using the 7300 Real Time PCR System in a total volume of 25 μL. The reaction mixture included 1 × Taq Man Universal PCR Master Mix No AmpErase UNG, 0.12 μM PbF, 0.6 μM PbR, 0.2 μM PbP, and 5 μL template DNA. Soil DNA extracts was diluted fivefold before analysis and were analyzed in duplicate. The thermal cycling conditions consisted of an initial denaturation for 10 min at 95 °C followed by 45 cycles at 95 °C for 15 s and finally 59 °C for 60 s.

8.8.5.5 Quantification

Since *P. brassicae* is an obligate biotroph, it is difficult and laborious to extract pure genomic DNA from spores to use as a standard. Therefore, the amount of pathogen DNA was quantified using a standard curve generated by including reactions containing different amounts of a plasmid carrying the *P. brassicae* target sequence. The target DNA used to generate the calibration curves was a 607-bp sequence, amplified using a universal forward primer (NS7: 5′-to fg (10^{-15} g) plasmid DNA g^{-1} soil by taking into account that 350 mg soil were used for the DNA extraction and that 1 μL of the original DNA extract was analyzed. By analyzing the series of artificially infested soil samples containing different amounts of *P. brassicae* resting spores, it was possible to convert fg plasmid DNAg^{-1} soil to number of *P. brassicae* resting spores g^{-1} soil. Amplification efficiency (E) was calculated as a percentage using the equation:

$$E = 100 \times (10 - 1/k - 1)$$

where k is the slope of the equation describing the plot of C_t versus the logarithm of the DNA amount. GAGGCAATAACAGGTCTGTGATGC-3′) (White et al. 1990), and a *P. brassicae*-specific reverse primer (Pb3R: 5-′-ACCATACCCAGGGCGATTG-3′), which was cloned into pGEM-T vectors according to the manufacturer's instructions. Plasmid DNA was prepared using a QI Aprep Spin Mini prep Kit, and the DNA concentration was determined spectrophotometrically. A tenfold dilution series of the plasmid standard was analyzed in triplicate using the real-time PCR assay and C_t values were then plotted against the amount of plasmid DNA to create a standard curve. Soil samples with low levels of

P. brassicae DNA outside the range of the standard curve were quantified by extrapolating the standard curve to facilitate analysis of the results. The starting quantity of DNA in soil samples was transformed from ag (10^{-18} g) plasmid DNA.

8.8.5.6 Evaluation of Inhibition and Repeatability

Inhibition of the PCR reaction by contaminants in the DNA extracts was assessed for the soil samples analyzed in this study by analyzing a separate PCR reaction containing the soil DNA extract and 2 μL of the *P. brassicae* plasmid standard containing approximately 150 gene copies. The same amount of plasmid standard was added to a water control. The percentage inhibition effect (I) was calculated using the equation:

$$I = 100 \times \frac{\{(Q_{s+sp}) - (Q_s + Q_{sp})\}}{Q_s + Q_{sp}}$$

where Q_s is the quantity of the unknown sample, Q_{sp} is the quantity of the standard spiked in water and Q_{s+sp} is the quantity of the unknown sample spiked with the standard. The inter sample repeatability was evaluated on separate extractions of 10 soil subsamples quantified in a single real-time PCR assay.

8.8.5.7 Distribution of *Plasmodiophora brassicae* Inoculum and Risk Assessment

Indicator kriging was used for mapping the probability of detecting *P. brassicae* within a 2-ha area of field D at farm 2 (Gotene). This method is a nonlinear form of ordinary kriging in which the original data are transformed to a binary scale (0–1) (Burrough and McDonnell 1998). This approach, which has been applied frequently in soil mapping (Van Meirvenne and Goovaerts 2001; Lark and Ferguson 2004), is appropriate when the goal is to produce an estimate of the probability that a given threshold is exceeded. In this case, observations with a data value (z) of less than or equal to a specified threshold (T) (here T was set to 5.0 fg plasmid DNA g^{-1} soil) were assigned a value of 0. Other observations were given a value of 1. The transformed data were then used in variogram analysis and ordinary kriging (Webster and Oliver 1992; Burrough and McDonnell 1998). The map displays the probability of T being exceeded. The software GS+ (Gamma Design) was used for the geostatistical analysis and ARCGIS (ESRI) was used for mapping (Wallenhammar et al. 2012).

8.8.6 Bioassay Technique of Jin-ping et al. (2013)

Bait plants *Brassica pekinensis* (Lour.) Rupr. variety Zhongbai were used to validate the developed real-time PCR assay method. The level of *P. brassicae* in naturally infested soil samples was assessed using a bioassay method previously developed by Wallenhammar (1996) and modified by Wallenhammar et al. (2000). The pots used were 40 × 40 × 40 mm (vefi). Infected plants were scored according to the following classes: 0 = no galls; 1 = enlarged lateral roots; 2 = enlarged tap root; 3 = enlarged napi form tap root; 4 = enlarged napi form tap root, lateral roots healthy; and 5 = enlarged napi form tap root, lateral roots infected. A disease severity index (DSI) was calculated according to the equation:

$$\mathrm{DSI} = \frac{\sum(N1 \times 1)}{(N \times 4)} \times 100\%$$

where N, class no. 1, no. of plants in each class; N, total no. of plants.

8.8.6.1 The DNA Extraction from Infected Roots, Fungal Material, Bacterial Material, and Soil

All the mycelia, apothecia were incubated in solid media, and then transferred into liquid media for 7 days at 28 °C. A modified CTAB method was used to extract DNA from infected root tissue, fungal and bacterial material. Frozen mycelium was placed in 1.5 mL micro tubes, 600 μL of CTAB extraction buffer (2% CTAB, 0.3 mol L^{-1} NaCl, 1 mol L^{-1} Tris-HCl, 0.5 mol L^{-1} EDTA (pH 8.0)) was added, and the tubes were vortexed and then incubated at 65 °C for 30 min. Chloroform-isoamyl alcohol (24:1 mixture of chloroform and isoamyl alcohol) (1 volume) was added to each tube, and the tubes were mixed and centrifuged for 10 min at 13,000 *g*. The aqueous phase was transferred to a new tube, and the chloroform extraction was repeated. DNA was precipitated overnight at −20 °C after the addition of 0.1 volume of 3 mol L^{-1} sodium acetate (pH 8.0) and 2 volumes of cold 100% ethanol. The supernatant was discarded, and the pellets were washed with 70% ethanol and then dried by vacuum centrifugation. DNA was resuspended in 80 μL of TE (10 mmol L^{-1} Tris-HCl, 0.1 mmol L^{-1} EDTA (pH 8.0)) and then diluted 1:100 for use in PCRs in TE. The internal transcribed spacer (ITS) region of *P. brassicae* was based on GenBank (Y12831) and used for designing the primer PBF1 (5-′-GCTCCTGCGTGTCGCTGTTATT-3′) and PBR1 (5-′-CTGCGTTCTTCATCGTTGTGA-3′). Soil samples were air-dried at room temperature (20 ± 2 °C), plant material was removed by hand and the soil was pulverized using a mallet. DNA was extracted twice independently from 200 mg soil samples using the DNA SPIN Kit according to the manufacturer's instructions.

8.8.6.2 Real-Time PCR Assay

The DNA concentration was determined spectrophotometrically. A tenfold dilution series of the plasmid standard was analyzed in triplicate using the real-time PCR assay. Each reaction of 25 μL contained 10 ng of plasmid pB12, 0.5 μL of each 15 μmol L^{-1} primer (PBF1 and PBR1), 12.5 μL of 2 × SYBR Green master mix, and 0.4 μL of a 1000 × diluted reference dye. The PCR reaction was set up in triplicate for each DNA sample. Prior to amplification, an initial denaturation step was performed (95 °C for 3 min) ensuring complete denaturation of the DNA and activation of the *Taq* polymerase. The reactions were performed in a iCyeler iQ5 Sequence Detection System. This was followed by 40 cycles of 30 s at 95 °C, 30 s at

60 °C, and 35 s at 72 °C. Fluorescence was detected after each cycle. C_t values were then plotted against the amount of plasmid DNA to create a standard curve.

8.8.6.3 Sensitivity and Reproducibility of Real-Time PCR

The tenfold serial dilutions of the *P. brassicae* genomic DNA were made to measure the sensitivity of the assay. Every dilution repeated five times. Real-time PCR was performed using the optimal reaction system. Conventional PCR amplification was performed as control. The inter sample repeatability was evaluated on 10 separate soil DNA extracts of *P. brassicae* from one soil samples analyzed in the same real-time PCR. Standard deviation (SD) of the C_t values was determined based on the 10 replicates for each amount of target DNA.

8.8.6.4 The Detection of Artificially and Naturally Infested Soils

The artificially and naturally infested soil samples were detected using the optimal real-time PCR reaction system. The artificially infested soils sample that contained 1.0×10^8, 1.0×10^7, 1.0×10^6, 1.0×10^5, 1.0×10^4, 1.0×10^3, 5.0×10^2, 10, and 0 resting spores g^{-1} soil (DW) were measured to validate the sensitivity of the real-time PCR and determine the threshold level of clubroot. The total of 160 soil samples of wide type and the sterilized soil samples from lab were detected using real-time PCR and conventional PCR. The thermal cycling parameters of conventional PCR were initial denaturation at 94 °C for 3 min followed by 40 cycles consisting of denaturation at 94 °C for 1 min, annealing at 56 °C for 1 min, and extension at 72 °C for 1.5 min. A final extension at 72 °C for 7 min was done at the end of the amplification. Negative controls (no DNA template) were used in every experiment to test for the presence of contamination in reagents (Jin-ping et al. 2013).

8.8.7 Identification of Pathotype-Specific Genes by PCR and Quantitative PCR

Primers for the identification of the pathotypes-specific gene(s) were derived from Feng et al. (2013) with other primers used. The genomic locations of the primers are illustrated in Fig. 8.1. PCR analyses were conducted in an Eppendorf Mater cycler thermal cycler. Except in the genome walking analysis, all PCR analyses were run in a final volume of 20 μL in the Promega master mix with 0.25 μM of each primer and 100 ng in the template; for all primers, the annealing temperature was 52 °C, and/or 56 °C, all elongation time was 45 s, and the cycle number was 35. Quantitative PCR (qPCR) assays were conducted in a Step One Real Time PCR System using the GoTaq 2-step RT qPCR System (Promega) for RNA analysis or the power SYBR Green PCR Master Mix (Life Technologies) for DNA analysis. The *P. brassicae Actin 1* gene (GenBank accession number AAR88383) was used for normalization. Primer efficiency was verified using standard curves generated from the purified PCR product of the targeted fragments. The quantity of *Cr811* relative to *Actin1*, at the DNA or RNA level was calculated via the delta-delta method, with relative

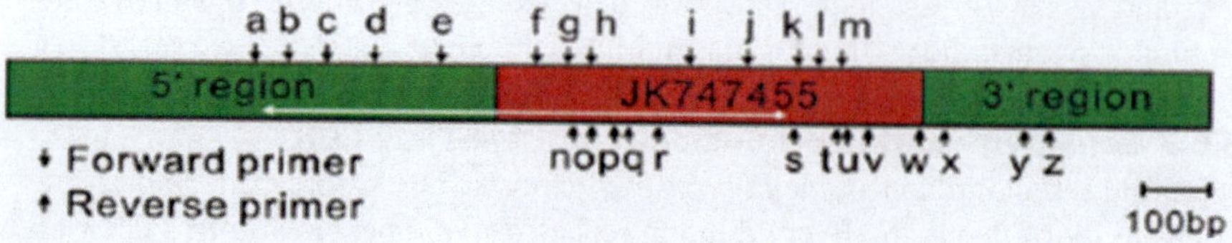

Fig. 8.1 Diagram of the genomic region of the *Cr 811* gene, JK 747455 GenBank accessions number of the expressed sequence tag that is a part of *Cr811* "5" region and "3" region. Sequences obtained by genome walking a to *z* primers whose sequences are linked. The white line denotes the sequenced cDNA (Zhang et al. 2015)

expression or amount $= 2^{-(\Delta CT\ \text{sample}\ -\ \Delta CT\ \text{control})}$. All qPCR amplification was performed in triplicate for each of three template preparations derived from different biological samples. Data from qPCR were analyzed for statistical significance using the general linear model (GLM) procedure of the Statistical Analysis System. All data were subjected to one-way analysis of variance and when appropriate, a multiple range test was conducted by Fisher's least significant difference (LSD) at $P \leq 0.01$ (Zhang et al. 2015).

8.8.8 Quantification and Control Assessment of Clubroot with qPCR

Conventionally the detection of *P. brassicae* is carried out by gall observations and bait tests. With the advent of molecular techniques, PCR amplification of DNA has been used to aid detection and study the diversity of *P. brassicae* (Buhariwalla et al. 1995; Buhariwalla and Mithen 1995; Cao et al. 2007; Faggian et al. 1999; Faggian and Strelkov 2009a, b; Ito et al. 1997, 1999; Wallenhammar and Ardwisson 2001). Methods of *P. brassicae* PCR detection are often based on ribosomal RNA (rRNA) gene internal transcribed spacer regions (ITS) (Ito et al. 1999). Detection and quantification is central to investigating the control of *P. brassicae*. Wakeham et al. (2017) quantified *P. brassicae* spores using real-time PCR (qPCR) and investigated the effect of Limex (LimeX70) and conventional lime-based products on clubroot control. Limex is a form of $CaCO_3$ produced as a byproduct of the sugar beet refining process (Harling 2007). Limex contains P_2O_5, MgO, and SO_3. The addition of calcium carbonate to clubroot contaminated land has been shown to improve control. However, large quantities of calcium carbonate are required to produce consistent results. This can raise the pH of the soil to levels which are inconsistent with the other crop production in the rotation (potatoes). The number of *P. brassicae* resting spores in soil was monitored in different control treatments and the subsequent effect on crop yield was ascertained.

8.8.9 Assessment of the Response of Lime-Based Fertilizers with qPCR

Determining *P. brassicae* infection within a field prior to crop planting has long posed a problem for choosing appropriate control treatments. The options for control of this pathogen are limited, and are based on adjusting pH with soil amendments. The quantitative polymerase chain reaction (qPCR) was investigated for measurement of this pathogen in different control treatments. The qPCR was capable of reliably quantifying *P. brassicae* at levels greater than and including 10^3resting spores/g soil. The assay was used to study the effect of lime-based products (LimeX) on the incidence of the clubroot pathogen in field trials with broccoli crops grown on contaminated land. The results showed that variation occurred in clubroot resting spore levels in treated and untreated plots during the crop growing period. In year one, there was a 96% decrease in spore load during the growth of the crop. Treatment with LimeX resulted in a greater marketable head weight of broccoli in 2 years of the field trials, and significantly reduced gall numbers on the roots in 1 year. The rate of lime (calcium carbonate) application was not found to have a significant effect, however, a greater reduction in clubroot was observed at higher LimeX concentrations (Wakeham et al. 2017).

8.8.10 The PCR and Quantitative PCR Analysis of Infested Soil Samples

Total genomic DNA's extracted from all soil samples collected from infested fields with a Fast DNA Spin kit for soil as per the manufacturer's instructions. An aliquot of 250 mg of soil is used per sample, with three replicates. Infestation by *P. brassicae* is detected by conventional PCR analysis (Cao et al. 2007). The amplification products are run on 1% agarose gels stained with SYBR Safe (Invitrogen), and visualized under UV light.

Quantitative PCR analysis is performed on diluted DNA (1/10 v/v) with sdH_2O to avoid inhibition of the reaction. The quantitative PCR amplification is conducted (Rennie et al. 2011). The qPCR amplifications are carried out in a Step One Plus Real Time PCR System in a 10 μL volume reaction (2.5 μL template DNA, 2.5 μL of 3.2 mM DR1F and DR1R primer mix, and 5 μL Dynamite qPCR Master mix. The reaction conditions consist of 95 °C for 2 min (initial denaturation step), followed by 35 cycles of 95 °C for 15 s (denaturation step) and 60 °C for 60 s (annealing, extension and fluorescence measurement step). A standard curve for the quantification of *P. brassicae* spores is generated with DNA isolated from known quantities of spores, which can be purified as per Castlebury et al. (1994) with modifications made by Cao et al. (2007). The 100 g of galled tissue, and 500 mL of water are blended, and the homogenate passed through a 450 μm metallic sieve. The resting spores are isolated by centrifugation in a 50% sucrose gradient, quantified with a hemocytometer (VWR), and adjusted to 1×10^8 resting spores/mL. Total DNA for the standard curve is extracted, and serially diluted with sdH_2O at tenfold intervals down to a dilution of 1×10^2 resting spores/mL. DNA from each dilution in the series is used as a template in the qPCR analysis according to Rennie et al. (2011). The correlation between the amount of pathogen DNA detected, and the soil pH is determined for the used samples (Ricarova et al. 2016).

8.8.11 Gene Expression (Transcriptomic) Analysis

RNASeq analysis was conducted for gene expression, RNA was prepared from hypocotyls or the upper 1 cm of control and infected plants (*A. thaliana* Cal 0) 16 and 26 DPI. Three independent biological replicates were used for the transcriptomic analysis and a second set of three biological replicates, grown at a different time, used for qRTPCR analysis. RNASeq was performed at the Applied Genomics and Analytical Technologies Dept., National Research Council of Canada using an Illumina Hi-Seq 2500 platform. Approximately 25 million reads were obtained from each sample. Paired end reads were mapped using Stampy (Lunter and Goodson 2011) to the *Arabidopsis* TAIR10 genome (https://www.arabidopsis.org/) and *P. brassicae* genome. Over 99% of the reads matched one genome or the other. For the 16 DPI samples ~13% (3.5 million) of the reads matched the *P. brassicae* genome in both hypocotyl and root samples. For the 26 DPI samples, 48% of the reads (12 million) matched the *P. brassicae* genome from hypocotyl tissue and 34% (nine million) reads from root tissue. Reads matching predicted transcripts were counted using HTSeq Count (Anders et al. 2015) using the default "union" method and differential gene expression analyzed using DESeq2 (Love et al. 2014). Phylogenetic trees were constructed using MEGA6 (Tamura et al. 2013). Fold-change data were overlaid on metabolic pathways using Mapman (Usadel et al. 2009). Quantitative RT-PCR was performed as described in Malinowski et al. (2012) using the following primers ARR4 (At1g10470) forward 5′-GCCTGGAATGACTGGTTATGAG-3′, reverse 5′-AGAAATCTTGAGCACCTTCCTC-3′; ARR5, (At3g48100), forward 5′-CTTCATTAGCATCACCGAAACTTC-3′, reverse 5′-AAGCTCTTGTCGCACTATCG-3′; ARR6, (At5g62920), forward 5′-GTCGGTTTTGAGGATTTGAAGG-3′, and reverse 5′-AGGCAAAATG-TTCTCGGAGG-3′, ARR7, (At1g19050), forward 5′-CGTAAAGTCATCGAGAGGTTGC-3′, reverse 5′-TCCGT-CACTATCAAATTCACCTTC-3′. All results were normalized against 18S rRNA expression.

A microarray analysis of control and infected ipt 1; 3; 5; 7 tissue was performed using control and infected root and hypocotyl tissue 16 DPI. RNA was prepared from duplicate samples and hybridized to Affymetrix Aragene-1-1-ST microarrays using the NASC International Affymetrix Service (http://affymetrix.arabidopsis.info/). Expression data contained in the CEL files provided from this service were normalized using the R package "oligo" with the RMA option (Carvalho and Irizarry 2010) and differential expression calculated using "limma" (Smyth 2005). Data are available from Array Express (www.ebi.ac.uk/arrayexpress) under the accessions

E-MTAB-4201 (microarray) and E-MTAB-4176 (RNA-Seq) (Malinowski et al. 2016).

8.8.12 The qPCR-Based Assessment of *Plasmodiophora* Spread in Soil

The DNA of *P. brassicae* was detected in 62% of the tested fields of Poland. The concentration of pathogen DNA measured using real-time qPCR technique reached 1.7×10^9 gene copies g^{-1} soil which is the equivalent of 7.7×10^8 spores. The spore loads in fields cultivated with Brassicaceae crops were mostly high and in 19.5% of the sampled fields, they exceeded 1×10^4 spores g^{-1} soil. The intensity of oilseed rape cultivation in these areas was very high, except for Podlasie, where soils are acidic, which favors the development of *P. brassicae*. Additionally, soils in areas with intensive cultivation of *Brassica* vegetables located in central Poland were also highly infested by the pathogen. Based on the concentration of *P. brassicae* DNA, recommendations for the farmers growing vegetable *Brassicas* and oilseed rape have been formulated, pointing out the areas where CR cultivars are indispensable. The soil sampling followed by real-time qPCR detection of *P. brassicae* can serve as a reliable tool for decision support for the farmers. The method should be implemented as a part of a toolbox used in the integrated management of clubroot worldwide.

8.8.12.1 Soil Sampling

Soil samples were collected between 2013 and 2019 from 431 fields across Poland under the cultivation of Brassicaceae crops, ca. 30 fields randomly sampled in each region, except for Podlasie (10 fields) and Swietokrzyskie (12 fields), where oilseed rape and vegetable *Brassicas* are less commonly grown. From each field the sample of approximately 2 L, consisting of ca. 40 subsamples (soil cores) was collected along W-transect. A soil auger with a diameter of 30 mm and a volume of 25 cm^3 was used to take subsamples representing the top 20 cm of the soil profile. The fields varied in size; in the case of vegetable *Brassica* cultivation they were usually 1 ha or below. Fields of oilseed rape ranged from approximately 10 ha to 50 ha. The number of subsamples was the same (40), regardless of the field size. The location of each tested field was determined.

8.8.12.2 DNA Extraction from Soil

Soil sub-samples were mixed by hand. The plant material and stones were removed. Then, the samples were air-dried at room temperature (20 °C). Then, soil samples were divided into two parts and ground with metal nuts in a paint shaker (2 replicates × 1 l), and then sieved through a 20 mm mesh. DNA extraction from soil was done according to the procedure described by Wallenhammar et al. (2012). Briefly, DNA was extracted from 350 mg soil samples (two biological replicates per sample to confirm inter-sample reproducibility of the assay) using the Fast DNA Spin Kit for Soil, and obtained eluate was then further purified using Wizard DNA Clean-Up System and Illustra MicroSpin S-300 HR Columns. To assess the quality of the isolated DNA, the concentration and purity of the DNA samples were measured on UV/Vis photometer. Only DNA samples with a ratio of the absorption values A260 to A280 between 1.8 and 2.0 was used. Forty randomly selected DNA samples (~10% of the studied lot) were subjected to electrophoresis on a 1.5% agarose gel and stained with ethidium bromide. All of the samples proved high purity of DNA; therefore, the spectrophotometric evaluation of the samples was considered as a sufficient method of DNA quality confirmation.

8.8.12.3 Real-Time qPCR

The sequences of primers and probes as well as qPCR conditions, were the same as described by Wallenhammar et al. (2012). The primers PbF 50-AAA CAA GTC AGC TTG AAT GC-30 and PbR 50-TTC GCG CAC AAG CAC TTG-30 were used to amplify a 103 bp fragment of ribosomal RNA gene (Gene Bank accession number AF231027). A probe PbP 50-CGC GCC ATG CAC TGT TAA ATT G-30 with VIC as the 50 terminal reporters dye and TAMRA as 30 quencher was used to improve qPCR specificity. Real-time quantitative PCR was performed using the 7300 Real-Time PCR System in a total volume of 25 μL. The qPCR reaction mixture included: 1 × TaqMan Universal PCR Master Mix No AmpErase UNG, 0.12 μm PbF, 0.6 μm PbR, 0.2 μm PbP and 5 μL DNA template diluted 1:5. The thermal cycling conditions were as followed: pre-PCR denaturation for 10 min at 95 °C, and then 45 cycles consisting of a 15 s denaturation step at 95 °C and 60 s annealing/extension step at 59 °C. All DNA samples were analyzed in duplicate, meaning that four measurements were performed for each soil sample to confirm intra-assay precision. Soil with a known amount of pathogen was used as a positive control in each qPCR run to confirm inter-assay precision and reproducibility. Moreover, a no template (negative) control was included in each qualitative assay. The amount of the pathogen was quantified using a standard curve obtained for a tenfold dilution series of the plasmid containing the synthetic gene of the *P. brassicae* target sequence.

8.8.12.4 Quantification of *Plasmodiophora brassicae* Spores

Spores of *P. brassicae* were extracted from Chinese cabbage cv. "Granaat" roots infected with five isolates originating from oilseed rape (*Brassica napus* L.) with visible symptoms of clubroot, collected at five regions of Poland. Roots (5 g) were macerated with MQ water (50 mL) using a blender. The suspension was then filtered through four layers of cheesecloth and diluted 100 times giving final spores concentration between 1.7×10^5 and 5×10^5. Spores were counted under the microscope Optika B-350 using a Fuchs-Rosenthal hemocytometer. DNA from spores was extracted using the same procedure as described previously and quantified using real-time qPCR as described above. Based on the frequency tabulation for $\log_{10}$ (No_spores) seven categories of spore loads were derived, containing samples with similar values. Each category has been attributed to the probable disease level and yield risk.

8.8.12.5 Cultivation Guidelines Based on qPCR Assay

Guidelines for farmers based on the levels of DNA copies g^{-1} soil have been adopted from Wallenhammar et al. (2012) and Almquist (2016). For DNA levels lower than 1300 DNA copies g^{-1} soil (3000 spores g^{-1} soil), the risk of yield loss in susceptible crops is regarded as lower than 10%. At DNA levels ranging from 1300 to 330,000 DNA copies g^{-1} soil (corresponding to 3000 and 750,000 spores g^{-1} soil, respectively), resistant cultivars are recommended. At levels above 330,000 DNA copies g^{-1} soil, the risk of multiplication of soil inoculum is considerable hence no cultivation of Brassicaceae crops is advised.

8.8.12.6 Statistical Analysis and Data Integration

Statistical significance of real-time qPCR results was studied using one-way ANOVA ($P = 0.05$) and Duncan's post hoc test (Statistica 9.0). The correlation of linear regression comparing microscopic and qPCR quantification of spores was confirmed by Pearson's correlation coefficient ($R = 0.974$) at $P < 0.05$. The results were analyzed for the percentage of oilseed rape and *Brassica* vegetables in relation to the cultivation area in individual provinces (Bienkowska et al. 2020).

8.8.13 The *Plasmodiophora brassicae* Genome Sequencing and Transcriptomic Analysis

The *P. brassicae* single-spore isolates SACAN-ss3 (classified as Pb2), SACAN-ss1 (Pb3), ORCA-ss4 (Pb5), AbotJE-ss1 (Pb6) and ORCA-ss2 (Pb8) [18] were used for genome sequencing and transcriptomic analysis during *B. napus* infection. The isolates were maintained as resting spores in mature root galls of the universally susceptible host Chinese cabbage (*B. rapa* var. *pekinensis*) "Granaat" at −80 °C. *B. napus* cv DH 112075 was derived from a cross between B. napus cultivars Cresor and Westar. Extraction of the resting spores and host inoculation were carried out as described by Strelkov et al. (2006) for frozen root tissue. The inoculated seedlings were grown in Sunshine #4 Mix/LA7 potting medium and kept in a greenhouse at 20 ± 2 °C with a 16 h photoperiod (light intensity of approximately 300 μmol/m^2/s). Galled roots were harvested 6 weeks after inoculation for the collection of resting spores. Resting spores to be used for the isolation of nucleic acids were purified further by using a continuous gradient of HS-40 colloidal silica as per the protocol of Castelbury et al. (1994). Additional transcriptomic analysis was performed on *A. thaliana* Col-0 inoculated with a Welsh pathotype of *P. brassicae* as described in (2012). This isolate was maintained on *B. rapa* var. *pekenensis* "Wong-Bok" and plants were inoculated with 1 mL of 10^6 spores mL^{-1}, 14 days after germination. Plants material (hypocotyls and the upper 1 cm of root) were harvested 16 and 26-days-post-inoculation.

8.8.13.1 The DNA and RNA Extraction

Purified mature spores of *P. brassicae* were collected from macerated clubs formed on roots of susceptible plants as described by Cao et al. (2007). Tissue (100–200 mg) was ground in liquid nitrogen using a mortar and pestle and then homogenized in Cell Lysis Solution-VF or CLS-Y using the Lysing Matrix A tube supplied with the kit and the single tube Mini-Bead beater. Homogenization was performed in 3 × 1 min cycles at 4800 rpm with 1 min of cooling on ice in between. The DNA was eluted, treated with RNAse A followed by Phenol: Chloroform: Isoamyl alcohol (P: C: I) extraction and C: I extraction, precipitated and resuspended in 10 mM Tris-HCl pH 8.5. RNA was extracted from the roots of infected plants. Tissues were ground in liquid nitrogen, and RNA was extracted with TRIzol and Ambion Pure Link RNA Mini kit according to the manufacturer's protocol. RNA quantity and quality were determined by Qubit (Life Technologies) and Bioanalyzer, respectively. RNA concentration for different samples was in the range of 57–100 ng μL^{-1}.

8.8.13.2 Genome Sequencing and Annotation

DNA from purified spores of Pb3 and Pb6 was used for shotgun and mate-paired sequencing using Roche 454 Titanium chemistry following the manufacturer's protocol (Roche). DNA sequencing was carried out. Raw sequencing reads were mapped to the genome of *B. napus* cv. DH12075 or *B. rapa* cv. Chiifu (Wang et al. 2011) to remove the possible plant DNA sequences in the Pb3 and Pb6 samples, respectively. *P. brassicae* genomes were assembled using Newbler assembly software. Repeat Masker and Repeat Modeler software (Kurtz et al. 2004) were used to analyze the repeats within the *P. brassicae* genomes.

Genome annotation was conducted using the ab initio gene prediction software Maker (v. 2.22) (Cantarel et al. 2008). The software was configured to use RNA sequence reads from infected roots and assembled 454 RNA reads generated from P. brassicae spores as well as Swiss Prot proteins as evidence. The gene prediction tools SNAP and Augustus trained on *P. brassicae* were used by Maker in an iterative fashion to predict *P. brassicae* genes. Putative transcript functions and GO analysis were performed using BLAST2GO. The entire annotation process was conducted for Pb3 and Pb6 independently, using only data from the one pathotype (i.e. no inter-pathotype data were used for training SNAP and Augustus).

A pipeline that is commonly used to predict secreted proteins from filamentous fungi (Guyon et al. 2014) was adapted for prediction of Pb3 secreted proteins. The criteria included in the search were presence of an N-terminal signal peptide, lack of transmembrane and membrane anchoring domains, and absence of subcellular localization signals using Signal P (v4.1), TMHMM (v2), Pred GPI and Target P (v1.1) software, respectively (Petersen et al. 2011). The fasta file of mature proteins generated by Signal P was used as input file for TMHMM. Proteins without a transmembrane helix that also lacked GPI and defined as secreted (S) by Target P (no RC threshold included) were considered as secreted proteins. SSPs with an RXLR motif were identified based on the criteria described previously using the RXLR motif finding tool provided by the Galaxy software (Cock et al. 2013).

8.8.13.3 The RNA-Sequencing and Transcriptomic Analysis

RNASeq was performed using the Illumina Hi-Seq 2500 system, a total of 1,075,036,260 sequence reads (about 89.6 million reads per time course) were generated from paired-end RNAseq libraries from *B. napus* root tissues 1, 2, 3, 4, 5, and 6 weeks after inoculation with *P. brassicae*, as well as libraries mock inoculated with water. All generated sequences were processed using the CASAVA algorithm prior to being aligned to the *P. brassicae* genome sequence using the RNA-Seq analysis application from CLC Genomics workbench (v. 7.0.4), where 89,317,981 sequences read was aligned to the *P. brassicae* genome. Under the assumption that no *P. brassicae* sequences should be present in the water-inoculated samples, the sequences detected in the water-inoculated libraries were subtracted from the values obtained in the *P. brassicae* inoculated libraries.

For the *A. thaliana* transcriptome analysis, paired-end RNA-seq libraries were prepared from the upper 1 cm root or hypocotyls 16 and 26 days after inoculation. Approximately 25 million reads were obtained per sample. Three independent biological replicates were analyzed for each time point with the appropriate uninfected controls. Paired end reads were mapped using Stampy (Lunter and Goodson 2011) to the *A. thaliana* genome TAIR10 (https://www.arabidopsis.org/) and *P. brassicae* Pb3 genome. More than 99% of the reads mapped to one genome or the other. Reads matching *P. brassicae* transcripts were counted using HTSeq (Anders et al. 2015) with the default "union" method. Differential gene expression, corrected for false discovery using the Benjamini–Hochberg method, was identified using DeSEQ2 (Love et al. 2014). For the *B. napus,* data comparisons were made between each week and for the *A. thaliana* data between each time point (16 and 26 DPI) in a tissue type (root or hypocotyl). For the production of heat maps, the output of the DESeq2 analysis was subject to regular log normalization and heat maps drawn using the R package heat map.2.

8.8.13.4 Hormonal Signature Analysis

The putative *A. thaliana* orthologues of differentially expressed genes (DEG) at different time point comparisons in *B. napus* were imported to HORMONOMETER (Volodarsky et al. 2009) to identify transcriptional similarities between DEG and *A. thaliana* genes associated with various plant hormones.

8.8.13.5 Detection of SNPs and Indels

Reads from each pathotype were aligned independently to the reference genome using CLC Genomics Workbench (v. 7.0.4). For each pathotype, a Binary Alignment Map (BAM) file was exported from the CLC Genomics Workbench. SNPs in all genotypes were called using the M pileup command from Samtools (v. 0.1.18) to produce a Variant Call Format (VCF) file. The VCF file was filtered to remove positions where the variant allele frequency was 100%. Additionally, if a single pathotype had a heterozygous genotype call, the position was removed. Due to the presence of homopolymer repeat errors in the 454 genome sequence, indels were found to be unreliable and were excluded from further analysis. The filtered VCF file was then processed using a custom Perl script to determine the number of haplotypes based on identical SNP patterns. Since the scaffolds of the reference genome are unordered, haplotype blocks were stopped when the end of a scaffold was reached. The recently published *P. brassicae* e3 isolate genome sequence (Schwelm et al.

2015) was aligned to the reference genome Pb3 using the Nucmer application within MUMmer (v 3.23) (Kurtz et al. 2004). SNPs were called from the aligned genome using the MUMmer application "showsnps." These SNPs were compared with the SNPs identified from the other pathotypes to define common SNPs (Rolfe et al. 2016).

8.9 *Brassica–Sclerotinia* Pathogenesis Protocols

8.9.1 Transcriptome Analysis of *Brassica–Sclerotinia* Pathosystem

8.9.1.1 Plant Material and Fungal Strain

Two *B. napus* doubled haploid lines differing in their susceptibility to *S. Sclerotiorum* were used. Both lines, NEP32 (susceptible) and NEP63 (resistant) were developed via microspore culture and were evaluated for their reaction to SSR. The susceptible line NEP32 was derived from spring type canola variety Helga (PI649136), and the resistant line NEP63 was developed from a cross between Helga and winter type canola accession PI458940. When inoculated using the petiole inoculation technique, the susceptible line develops lesions on the stem that expand to >4 cm in length with 100% girdling within 8-days-post-inoculation, eventually leading to wilting and death. Within the same period, the average lesion size on the resistant line is limited to <1 cm in length and <40% stem girdling, typically surrounded by purple margins suggesting accumulation of anthocyanins. The highly aggressive strain 1980 of *S. sclerotiorum* was used for inoculations and culture controls. Strain 1980 was chosen as it is the genome sequenced strain (Derbyshire et al. 2017; Kabbage et al. 2015).

8.9.1.2 Plant Growth and Inoculation

The seeds of NEP32 and NEP63 were surface sterilized by soaking them in 3% NaOCl for 1 min followed by immersion in 70% EtOH for 1 min and were rinsed three times in sterilized de-ionized water for 1 min. Seeds were planted in SunGro Sunshine mix #1 in 4 × 10 plastic plots with one seed per pot and plants were grown for 4 weeks in a growth chamber with a 16 h photoperiod, 21 °C/16 °C day/night temperature, and 60% relative humidity. Once germinated, seedlings were watered as necessary and fertilized once a week with 20-20-20 fertilizer. Four weeks after planting, plants were infected following an established petiole inoculation technique. The petiole of one leaf per plant was cut approximately 2.5 cm away from the stem and the stump was capped with PDA plugs containing hyphal tips of an actively growing 2 days-old, *S. sclerotiorum* colony using 1 mL pipette tip. Inoculated petioles (≈2.5 cm) were harvested at 8, 16, 24, and 48 h post inoculation (hpi), flash frozen in liquid nitrogen and stored at −80 °C until RNA extraction. Each treatment—time point consisted of three biological replicates; petioles collected from 5 individual plants were pooled to constitute one biological replicate (Zhao et al. 2004).

8.9.1.3 The RNA Extraction, Library Preparation, and Sequencing

Total RNA was extracted from inoculated petioles using RNeasy mini kit, and mRNA was isolated using Oligotex mRNA mini kit following manufacturers' instructions. cDNA libraries were constructed using NEB Next mRNA sample prep kit and NEB Next Multiplex Oligos kit. Similar procedure was followed to construct cDNA libraries from *S. sclerotiorum* mycelia grown on PDB in 10 cm Petri plates for 48 h under the same conditions described above. For sequencing, RNA from 8 and 16 hpi were pooled and referred as early infection stage (T1). Similarly, RNA from 24 and 48 hpi were pooled and referred as late infection stage (T2). The mRNA libraries were sequenced at the University of Minnesota Biomedical Genomics Center. Of the three biological replicates, two replicates per each treatment-time point were sequenced on Illumina GA-IIx platform (1× 76 bp Single Reads), while one replicate was sequenced on Illumina HiSeq 2500 platform (2× 50 bp Paired End).

8.9.1.4 Quality Control and Read Mapping to the Reference Genome

Quality checks for the raw fastq files were conducted through a pipeline consisting of FastQC and FastX toolkit (http://hannonlab.cshl.edu/fastxtoolkit/) to retain reads that were at least 50 bp long with a minimum quality score of 30. Reference genomes and the corresponding annotations for *S. sclerotiorum* and *B. napus* were downloaded from ensemble fungi (ftp://ftp.ensemblgenomes.org/pub/release-42/fungi/) and genoscope (http://www.genoscope.cns.fr/brassicanapus/data/), respectively. The most recent full genome sequence of *S. sclerotiorum* (*Sclerotinia sclerotiorum*_1980_uf_70_gca_001857865) was used. Indexes for each reference genome were built and quality filter-passed reads were mapped to the reference genomes using HiSat2. Raw read counts mapped to each gene from the HiSat2 generated alignments were obtained using the feature Counts command of the Sub-read package (Kim et al. 2015; Liao et al. 2013).

8.9.1.5 Differential Gene Expression Analysis

Differential gene expression (DGE) analysis was conducted using edge R package. The raw count data were normalized with trimmed mean of means (TMM) normalization method implemented in edge R. Principal component analysis (PCA) was conducted to determine relatedness of the biological replicates. PCA plots were generated using scatter plot 3d package in R. Statistical analysis was performed using negative binomial distribution extended to generalized linear models. Pair wise contrasts were performed following quasi-likelihood *F* tests. A false discovery rate (FDR) cutoff of 0.05 was applied to account for multiple testing corrections. A gene was considered as differentially expressed when the change in expression level was ≥twofold (absolute value of log twofold change (l2fc) ≥1), with an FDR-adjusted *P* value <0.05 (Lun et al. 2016).

8.9.1.6 Functional Classification and Enrichment Analyses of DEGs

Gene ontology (GO) enrichment analysis was performed in Blast2GO. GO terms were assigned to *S. sclerotiorum* total genes list in Blast2GO, which were used as the background list for enrichment analysis. Fishers' exact test implemented in Blast2GO was used to identify significantly enriched GO categories. A GO category was considered significantly enriched only when the p-value for that category was <0.05 after applying FDR correction (Conesa et al. 2005).

8.9.1.7 Differential Gene Expression Data Validation

The RNA-seq differential gene expression data was validated by performing qRT-PCR on six selected genes. Primers for qRT-PCR were designed using Primers-Blast. QuantiTect reverse transcription kit was used to synthesize cDNA from total RNA. Real-time quantification was performed using two technical replications in a BioRad CFX96 Real-Time system using iTaq Universal SYBR Green Super mix with the following cycling conditions: 95 °C for 30 s followed by 40 cycles of 95 °C for 10 s and 60 °C for 30 s. Expression levels of the DEGs were normalized against the *S. sclerotiorum* actin gene (sscle_14g099090) and the relative expression levels were calculated using the $2^{-\Delta\Delta Ct}$ method (Chittem et al. 2020).

8.9.2 Immunoassay for Early Detection of *Sclerotinia sclerotiorum*

A serological test has been developed that allows the early detection of infection of young rapeseed petals by *S. sclerotiorum.* Two steps are required to obtain an antiserum sufficiently specific for *S. sclerotiorum.* Soluble mycelial extracts of *S. sclerotiorum* are used to produce the first generation polyclonal antiserum. This is not specific for *S. sclerotiorum* in DAS-ELISA and allows the screening of cross-reacting species such as *Botrytis cinerea* a pathogen commonly present on rapeseed petals. Using a polyclonal *anti B. cinerea* serum enables the absorption by serial cycles of *S. sclerotiorum* antigens common to *B. cinerea.* Residual antigens are then used as immunogens for the production of 2s generation antisera (S_1 and S_2) which are then tested by DAS-ELISA. Cross-reactions with *B. cinerea* decrease with purification cycles of the immunogen whereas cross-reactions with some unrelated fungi slightly increase. *S. sclerotiorum* and *B. cinerea* are distinguishable using antiserum Sz (Jamaux and Spire 1994). However, Lefol and Morrall (1996) developed immunofluorescent staining of *Sclerotinia* ascospores on canola petals. The quantitative relationship among mean numbers of ascospores/petal, the percentage of petals carrying ascospores as judged by immunofluorescent staining and the percentage of petals carrying viable ascospores as judged by a planting technique is studied by regression analysis. The multiple infection transformation successfully linearizes the relationship between percent petals carrying ascospores and the mean numbers of ascospores per petal.

8.9.3 A Polymerase Chain Reaction (PCR) Assay for the Detection of Inoculum of *Sclerotinia sclerotiorum*

The development of a polymerase chain reaction (PCR) assay for the detection of inoculum of the plant pathogenic fungus *Sclerotinia sclerotiorum* is described. The PCR primers are designed using nuclear ribosomal DNA internal transcribed spacer sequences. Specific detection of DNA from *S. sclerotiorum* is possible even in the presence of a 40-fold excess of DNA from the closely related fungus *Botrytis cinerea*. PCR products are obtained from suspensions of untreated *S. sclerotiorum* ascospores alone, but DNA purification is required for detection in the presence of large numbers of *B. cinerea* conidiospores. Specific detection of inoculum of *S. sclerotiorum* is possible in field based air samples, using a Burkard spore trap and from inoculated oilseed rape petals. The assay has potential for incorporation into a risk management system for *S. sclerotiorum* in oilseed rape crops (Freeman et al. 2002).

8.10 *Brassica–Turnip Yellow Mosaic Virus* (*TuMV*) Pathogenesis Protocols

8.10.1 Detection of *Brassica* Viruses by Multiplex Reverse Transcription Polymerase Chain Reaction

8.10.1.1 Plant Material and Recombinant Plasmids

The *Nicotiana benthamiana* leaf tissues infected with *BrYV*-A, -B, C were used to standardize the multiplex RT-PCR. Each of these three genotypes infecting *N. benthamiana* was identified by RT-PCR and sequencing. Plasmids containing full-length genomic sequences of *BrYV*-A (Accession No. NC016038), *BrYV*-B (Accession No.HQ388351), *BrYV*-C (Accession No. KF015269), Turnip mosaic virus (*TuMV*) (Accession No. AF169561.2), Cucumber mosaic virus (CMV Fny strain) (Accession No. NC002034, NC002035, NC001440), Cucurbit aphid-borne yellows virus (*CABYV*) (Accession No. HQ439023), Beet western yellows virus Inner Mongolia isolate (*BWYV*-IM) (Accession No. EU636991), and plasmids containing partial genomic sequences of *TuYV* were used to test specificity of the multiplex PCR. The plasmid pMD19-*TuYV*-P0 contained the complete sequence of the ORF0 of *TuYV* (Accession No. NC003743) and the plasmid pMD19-56 #-1 contained the 4959–5163 nt sequence of *TuYV* which shared 96.4% nucleotide sequence identity with *TuYV* isolate Anhui (Accession No. KR706247.1).

8.10.1.2 RNA Extraction

Total RNA from 0.1 g of leaf tissue from infected plants was prepared by SDS-phenol/chloroform extraction. 600 μL of phenol: chloroform and 630 μL of extraction buffer (20 mM Tris–HCl, pH 7.8, 1% sodium dodecyl sulfate, 200 mM sodium chloride, and 5 mM EDTA) were added with continuous homogenizing. The thawed mixture was extracted two times with phenol: chloroform and separated by centrifugation. Total RNA in the supernatant was precipitated by equal volume of 4 M lithium chloride. Then it was washed two times with chilled 70% ethanol and one time with chilled 100% ethanol. The RNA was eluted in a final volume of 40 μL of diethylpyrocarbonate-treated (DEPC) water and stored at −20 °C for the following protocols.

8.10.1.3 Reverse Transcription and PCR

The revere transcription (RT) reaction was carried out using TaKaRa RNA PCR Kit (M-MLV) Ver. 3.0 (Takara) according to the manufacturer's protocol. The reaction was performed in a 20 μL PCR master mixture consisting of 4 μL 5 × RT Buffer, 1 μL BrY761R (10 μM), 1 μL TuY5163R (10 μM), 9 μL RNase Free double distilled water, 1 μL dNTP Mixture (each 2.5 mM), 0.5 μL RNase Inhibitor (40 U/μL), 0.5 μL MLV Reverse Transcriptase (200 U/μL) and 3 μL plant total RNA (~3000 ng). The RT reaction was incubated for 90 min at 37 °C. The products were stored at −20 °C for uniplex RT-PCR and multiplex PCR reactions. Uniplex PCR for BrYV was carried out in a 25 μL mixture containing 12.5 μL 2 × TSINGKE Master Mix (blue), 2 μL cDNA template, 0.5 μL of specific primers (10 μM) (Table 8.2), and 9.5 μL double distilled water. Multiplex PCR reactions was carried out in a 25 μL mixture containing 12.5 μL 2 × TSINGKE Master Mix (blue), 2 μL cDNA template, 0.5 μL of the six primers (10 μM) for each (Table 8.2), and 7.5 μL double distilled water. Amplified products (10 μL each) were electrophoresed in 1.5% agarose gels and stained with ethidium bromide to confirm the expected size of the fragments.

8.10.1.4 Cloning and Sequencing

Purified PCR products, amplified with primers BrYA484F/BrY761R, BrYB88F/BrY761R, BrYC257F/BrY761R from RT reaction products were inserted into pMD19-T and then transformed into competent cells of *Escherichia coli* MC1022. Recombinant clones were sequenced by Institute of Crop Science of CAAS.

8.10.1.5 Primer Design and Optimization of Annealing Temperatures

The full-length genomic sequences of *BrYV* and *TuYV* (*BrYV*-ABJ, NC016038; *BrYV*-AJS, HQ388350; *BrYV*-BBJ, HQ388349; *BrYV*-BJS, HQ388351; *BrYV*-CR, JN015068 and *BrYV*-CC, KF015269; *TuYV*, Nc003743; *TuYV* isolate Anhui, KR706247.1) were obtained from GenBank. The sequences of the designed primers used in this study are listed in Table 8.2. Gradient PCR was performed using

Table 8.2 List of primers used for developing multiplex RT-PCR (Zhao et al. 2016)

Primer	Primer sequence (5′ to 3′)	Positions
BrYA484F	5′-TACTTGGACTAGAGATGCTGAAAG-3′	BrYV-A 484–507 nt
BrYB88F	5′-CCTCCACCCAAAACAAGTAT-3′	BrYV-B 88–107 nt
BrYC257F	5′-CGAGTTTCCGTACTTGTTG-3′	BrYV-C 257–275 nt
BrY761R	5′-AGACCGAAGAGCTGAAAAGG-3′	BrYV-A 742–761 nt
TuY4959F	5′-AAGAGGCTTGCCCCTTCCTG-3′	TuYV 4959–4978 nt
TuY5163R	5′-AACCAAATCCGGTGTTGGAT-3′	TuYV 5144–5163 nt

different temperatures that were set randomly from 60 to 66 °C (60.0; 60.6; 61.4; 62.4; 63.8; 64.9; 65.5; 66.0 °C) by the PCR machine.

8.10.1.6 Sensitivity of Multiplex RT-PCR

The sensitivity was considered as the lowest concentration of viral RNA giving a strong positive signal in mRT-PCR. To determine this threshold, tenfold serially diluted DNA templates of the three genotypes of *BrYV* and *TuYV* were tested. The DNA samples range in quantity from 1 fg to 10 ng and the assay was carried out as described above.

8.10.1.7 Detection of Virus Genotypes in Field Samples by Multiplex RT-PCR

The samples showing symptoms such as yellowing and curl on leaves were collected from different fields. These samples were simultaneously used to test BrYV three genotypes by uniplex RT-PCR and multiplex RT-PCR (Zhao et al. 2016).

8.10.2 Quantitative RT-PCR Assay for Detection of *BnSGS3* Gene

Quantitative RT-PCR (qRT-PCR) was performed to detect the expression levels of BnSGS3 in different tissues of nontransgenic oilseed rape plants, single transgenic plants and the virus-inoculated plants. Primers Actin-F and Actin-R were used to amplify the reference β-Actin gene (*AF111812*), and primers Bn-F and Bn-R were used to amplify *BnSGS3* (all primers are listed in Table 8.3). The relative transcript levels of BnSGS3 $2^{-\Delta\Delta Ct}$ were obtained by calculating ΔC_t values [$\Delta C_t = C_t$ (*BnSGS3*) $- C_t$ (*Actin*)].

The qRT-PCR was also used to detect the accumulation of three viruses (by copy number) in the samples. Primers ORMV-F/R, CMV-F/R, and TuMV-F/R were used to amplify *CP* gene fragments of the three viruses via PCR. The PCR products were ligated to pMD-18T, which was transformed into *E. coli* DH5α, and plasmid DNA was subsequently extracted. The molecular weights of the *ORMV*, *TuMV* and *CMV* plasmids were 1,730,750.4 Da, 1,730,135 Da and 1,723,338.6 Da, respectively. The copy number of 1 μL plasmids was calculated according to Avogadro's formula "6.022 x 10^{23} molecules/moL," and the plasmids were diluted to 1.0×10^9 – 1.0×10^3 copies/μL concentration gradient, with three repeats per gradient, which were simultaneously subjected to qRT-PCR. The copy numbers of the three viruses in different samples were automatically calculated based on the sample C_t values and the standard curve.

Two independent biological replicates and three technical replicates were carried out for the inoculation experiment for each virus, taking the average value of two sets of data. The qRT-PCR was performed on the BioRad CF X 96 Real-Time System (C1000 Thermal Cycler). The reaction system included 10 μL SYBR Green PCR Master Mix, 0.5 μL of each upstream and downstream primer (10 μmol/L), 1 μL cDNA and 8 μL double distilled water. The reaction conditions were as follows:

Table 8.3 List of primers (Chen et al. 2015)

Primer	Sequence (5'–3')	T_m (C)	Product (bp)
	BnSGS3 cDNA amplification		
Bn-F1	TGAGGTTCTGGACAGGGATC		
Bn-R1	GCCGTCTTACTGAAAATGGA	55	1974
	Vector construction		
Bn-F2	GTACTCGAGGTTTGCTCTCTGTTTGGTT	55	1848
Bn-R2	CGTCTCGAGGCTAGTAGTCTTCTGTGTC		
BnRI-C	TATACTCGAGCAGTGGAAGGGTTTGGGTG	54	404
BnRI-KC	GCCCGAATTCTCCCTACTTGACAGTGTTG		
BnRI-A	CGAATCTAGACAGTGGAAGGGTTTGGGTG	54	404
BnRI-E	GCCGAAGCTTTCCCTACTTGACAGTGTTG		
	Detection of expression quantity		
Bn-F	TGGAAGGGTTTGGGTGAGGAG	62	181
Bn-R	GTGGACCATAGGAGTGGCGTG		
Actin-F	CTGGAATTGCTGACCGTATGAG	62	145
Actin-R	ATCTGTTGGAAAGTGCTGAGGG		
Bar-F	TGCCAGAAACCCACGTCAT	55	485
Bar-R	CTGCACCATCGTCAACCACTA		
	Detection of virus accumulation		
CMV-F	CCTCACCGGTACTGGTTTATC	62	107
CMV-R	CTTTCGCATGTCGCCAATATC		
TuMV-F	GGAAGTAAACGCTGGAACCT	62	96
TuMV-R	GCCACTCTTTGCTCGTATCT		
ORMV-F	CTGTGGCCATTAGGAGTCAA	62	108
ORMV-R	GCGCAGTAGTCCAAGGTAATA		

95 °C for 3 min, 95 °C for 10 s, 62 °C for 30 s, 72 °C for 15 s (40 cycles), and 65 °C for 5 s (Chen et al. 2015).

8.10.3 Biological and Serological Testing for the Presence of Cruciferous Viruses

Experimental plants known to have diagnostic value for the most important viruses of cruciferous plants (Rimmer et al. 2007; Plant Viruses Online, http://pvo.bio-mirror.cn/refs.htm) and references therein, namely *Nicotiana tabacum* cv. White Burley, *N. megalosiphon*, *N. glutinosa*, *N. clevelandii*, *N. benthamiana*, *Brassica rapa*, *Chenopodium amaranticolor*, *Ch. quinoa*, *Phaseolus vulgaris,* and *Pisum sativum* were mechanically inoculated with a homogenate prepared from the symptomatic field plants (Ohshima et al. 2002). The virus was maintained by serial passages in *B. rapa* and *N. benthamiana* hosts. For the preliminary virus diagnosis, double-radial immune diffusion tests were performed (Stefanac-Udbinac et al. 1963) using rabbit polyclonal antisera for the *TuMV*, Turnip yellow mosaic virus (*TYMV*) and Radish mosaic virus (*RaMV*).

8.10.3.1 The *Turnip Yellow Mosaic Virus (TuMV)* Full Genome Sequencing, Recombination, and Phylogenetic Analysis

The lyophilized material of *B. napus* from symptomatic field samples and experimental *N. benthamiana* plants was used to obtain single lesion The *TuMV* genomic material for the whole virus genome sequencing, as described by Tomimura et al. (2003). As recombination has been shown to be an important feature of *TuMV* evolution (Tomimura et al. 2003), the Croatian *TuMV* sequence (CRO184A) was checked for recombination events by RDP, Boot Scan, Max Chi, Chimaera and Si Scan methods implemented in RDP3 program (Martin et al. 2010). The obtained sequence was compared to 34 *TuMV* polyprotein sequences of nonrecombinant *TuMV* isolates representing all four phylogenetic lineages. Two genomic sequences of Japanese yam mosaic virus were used as out groups because BLAST searches had shown them to be the most closely and consistently related to those of *TuMV* (Tomimura et al. 2003). Sequences were aligned using CLUSTAL X, ver. 2.0 (Thompson et al. 1997) and analyzed using MEGA 5 (Tamura et al. 2011). Different evolutionary models were tested, but the obtained branching order was the same. The phylogenetic tree was generated using the neighbor-joining method applying Kimura 2-parameter evolutionary model as the common methodology in related papers. The tree topology was evaluated by bootstrap analysis based on 1000 repetitions.

8.10.3.2 Phytoplasma Identification by 16s rRNA Gene Analyses

For routine phytoplasma detection, nearly complete 16S rRNA, the spacer region and the beginning of 23S rRNA gene region were amplified by using universal phytoplasma primer pair P1/P7 (Deng and Hiruki 1991; Schneider et al. 1995) in a direct PCR assay, followed by nested PCR by using R16F2n/R2 primer pair (Gundersen and Lee 1996). All the procedures were performed as already described (Seruga et al. 2003). For phytoplasma classification, obtained amplicons were singly digested with restriction endonucleases Tru1I and Tsp509I according to the manufacturers' instructions and fragments were separated in 5% polyacrylamide gels in 1× TBE. DNA from AY phytoplasma reference strain HydB obtained from the phytoplasma collection maintained and PopP (Seruga et al. 2003) were used for comparisons.

In addition, P1/P7 PCR products of about 1800 bp were cloned into *Escherichia coli* INVaF' or XL1-Blue cells according to the instructions in the InsT/AcloneTM PCR Product Cloning Kit. PCR products of phytoplasma gene inserts from 3 to 6 selected clones obtained by employing universal plasmid primers M13pUCf/r were sequenced in both directions. In order to obtain full length sequences of P1/P7 amplicons, primer walking with R16R2n primer was performed. Classification of identified phytoplasma was additionally checked by using iPhyClassifier interactive online tool for in silico restriction fragment length polymorphism (RFLP) analyses (Zhao et al. 2009).

8.10.3.3 Phytoplasma Multigene Sequence Analysis

In order to have better insight into the rape phyllody phytoplasma strain, multigene sequence analysis was performed. The characterized genetic regions included the following: a major portion of the elongation factor Tu gene (*tufB*) primed by *fTufu/rTufu* (Schneider et al. 1997), a part of ribosomal protein operon (a fragment of rpsS gene, entire rplV gene, and a fragment of *rpsC* gene) primed by *rpF1/rpR1* (Lee et al. 1998), nearly complete phytoplasma translocation protein gene secY primed by the AYsecYF1/AYsecYR1 pair (Lee et al. 2006), groEL gene encoding chaperon GroEL (Mitrovic et al. 2011), and amp gene encoding variable antigenic membrane protein involved in interaction with insect hosts (Suzuki et al. 2006). In order to amplify the entire genes groEL and amp, new primers named groelAYF1 (50-GCTATTCTGGTAGCAAAGT- 30) and groelAYR1 (50-GTTCAAGAGCGTCTTTACA- 30) as well as groelAYend (50-GCTTC TATCTCTGCTTTG-30) and qns1 (50-CATTTCTTTYATTTGCTTCC-30) were designed, respectively. The PCR conditions for both groEL and amp gene amplification were as follows: a denaturation step of 4 min at 94 °C followed by 35 cycles consisting of 1 min at 94 °C, 2 min at 50 °C and 3 min at 68 °C and final extension step of 7 min at 68 °C. All gene designations are according to the NCBI accessions for the OY-M genome (Oshima et al. 2004). Restriction enzymes Tru1I and Tsp509I were used individually in the RFLP analyses of all amplified gene regions except groEL and amp. The amplification products primed by fTufu/rTufu, rpF1/rpR1, AYsecYF1/AYsecYR1, groelAYF1/R1 and amp of about 850, 1250, 1400, 2200 and 1350 bp, respectively, were sent for sequencing.

8.10.3.4 Phylogenetic Analysis of Phytoplasma Genes

The assembly of raw sequences and editing was done by using SEQUENCHER TM 4.7 software (http://www.genecodes.com/). Sequences were aligned (Thompson et al. 1997) and phylogenetic analyses were performed as described for viral sequences above (Tamura et al. 2011) by using different methods on different data sets depending on the analyzed genetic locus. Bootstrap analyses were performed (500 replicates) to estimate the stability of nodes and to support the inferred clades (Seruga-Music et al. 2014).

8.11 *Brassica–Verticillium* Pathogenesis Protocols

8.11.1 Molecular Characterization of *Verticillium* Isolates

8.11.1.1 Fungal Isolates

Thirty-six isolates of *V. longisporum* from different cruciferous hosts and locations, as well as two isolates from cauliflower, 14 isolates of *V. dahliae* from different genera of host plants, and at least one isolate of 19 other different phytopathogenic and nonphytopathogenic *Verticillium* species were used. All isolates were maintained as single-spored isolates on potato dextrose agar slopes (PDA) in the dark at 4 °C. When required, PDA plates were incubated in the dark, at 25 °C, and harvested accordingly.

8.11.1.2 The DNA Extraction

DNA was extracted by using extraction methods previously described (Typas et al. 1992). Buffers and electrophoresis techniques were according to standard protocols.

8.11.1.3 The PCR Amplification and Cloning

PCR amplification of the SSU-rRNA gene region (18S rDNA) from all *Verticillium* spp. isolates was performed with primers 18SVDF: GCGAAACTGCGAATGGCT and 18SVDR: GTAATGATCCCTCCGCTG on a PTC-100 Thermal Controller. An initial denaturation step of 3 min at 94 °C was followed by 32 repeated cycles of 1 min at 94 °C, 1 min at 60 °C, and 3 min at 72 °C, and completed with an extension step of 5 min at 72 °C. The reaction mixtures consisted of 50–100 ng genomic DNA as template, 2 mM $MgCl_2$, 1 unit Taq DNA Polymerase, 200 mM dNTPS, 13PCR buffer supplied with the enzyme (75 mM Tris-HCl, pH 9.0, 50 mM KCl, 20 mM $(NH_4)_2SO_4$, 0.001% BSA) and 50 pmoles of each primer. Primers 18SVDF and 18SVDR were designed from the sequenced data of the 18S gene region of one Greek isolate of *V. dahliae* spanning approximately 1650 bases. Thus, nearly the full-sized SSU rDNA was amplified, with the addition of 1 base at the 39 terminus end and with the exception of 79 bases at the 59 terminus end. Negative control reactions omitting DNA were included in all sets of amplifications to monitor the potential contamination by exogenous DNA. The reaction products were analyzed on 1.5% agarose gels in a 13TAE buffer (40 mM Tris acetate, 1 mm EDTA, pH 8.0) and were visualized under UV light. Individual PCR products were purified by using the JET-QUICK column (PCR Purification Spin Kit GENOMED) according to the manufacturer's instructions.

8.11.1.4 The DNA Sequencing and Data Analysis

PCR products (amplicons) were cloned in vector pAdvanTAge using the AdvanTageTM PCR Cloning Kit (CLONTECH). Amplicons were sequenced by using double stranded template (250 ng mL^{-1}) and 10 pmol of the forward and reverse PCR primers by the dideoxy chain termination method of Sanger (Sanger et al. 1977) with [^{35}S]-adenine and T7 DNA polymerase. Following the sequencing of a typical amplified product, five additional internal primers were designed to complete the sequencing in both directions. These were: VIF: AGGGACAGTCGG-GGGCA; KIF: GGAAGGGCACCACCAGGGGT; K2F: GAACCTCAAACAAGACCCACAAG for the 59 to 39 direction; and V2R: TGCCCCCGACTGTCCCT, K3R: GCTTGTGGGTCTTGTTTGAG for the 39 to 59 direction. Similarly, based on the extremities of the 839-bp intron, primers INT1: CCAGGGGTAAACG- GACTG and INT2: GGGCCATGCATTCGATTGC were designed to amplify the entire group-I intron from fungal isolates which contained such sequences. When PCR amplicons were cloned in vector pAdvanTAge, the forward and reverse universal primers were used for sequencing reactions. All sequencing reactions were run on standard 5% polyacrylamide gels in 13TBE buffer at 55 W for approximately 6.5 h. Gels were fixed in 10% acetic acid and exposed to Kodak X-OMAT AR film for the appropriate time. Sequence data entries, editing, and alignments were performed using the computer program DNA Star 6.8. Sequence pair distances were calculated using the Clustal V method with K-tuple size and gap penalty of 2 and 5 respectively. Computer database searches and sequence similarity were performed using BLAST (Altschul et al. 1990). A secondary structure model of the putative intron sequence was predicted according to the structural conventions for group-I introns (Burke et al. 1987) and secondary structures within loops were resolved with the energy minimization protocol (Jaeger et al. 1989, 1990) of the computer sub-optimal RNA folding program, Mulfold version 2.0, viewed with loop Dloop version 2.02 and finally modified in MacDrawPro (Gilbert 1990, 1994; Karapapa and Typas 2001).

8.12 *Brassica–Xanthomonas* Pathogenesis Protocols

8.12.1 Determination of Defensive Metabolites in the Phenylpropanoid Pathways

8.12.1.1 Total Phenolic and Flavonoid Contents

Total phenolic content in the leaves was determined by the Folin–Ciocalteu reagent assay (Lee et al. 2007). Total phenolic content was expressed as milligrams of gallic acid per gram fresh weight (FW). Total flavonoid content was measured by the aluminum chloride colorimetric assay (Zhishen et al. 1999) as expressed by milligrams of quercetin per gram FW.

8.12.1.2 Soluble and Insoluble Tannin Contents

Soluble and insoluble tannin contents were spectro-photometrically determined by the F–D method (Bubba et al. 2009). To determine the soluble and insoluble tannin content in 3.1 mL of ultrapure water, 0.1 mL of the extract and 300 mL of F–D reagent were added. After 3 min, 300 mL of a saturated aqueous solution of sodium carbonate was added. The results are expressed as milligrams of gallic acid per gram FW.

8.12.1.3 Proanthocyanidin and Total Hydroxycinnamic Acid (THA) Contents

The total proanthocyanidin content was measured using the 4-dimethylaminocinnamaldehyde (DMAC) assay (Prior et al. 2010). Total proanthocyanidins were quantified as catech in equivalents using a catech in standard curve. Determination of total hydroxycinnamic acid (THA) was performed by colorimetric methods using the chromogenic system of $HCl–NaNO_2–Na_2MoO_4–NaOH$ with chlorogenic acid standard (Štefan et al. 2014).

8.12.1.4 Glutathione and NADPH Redox

Total GSH and oxidized GSH (GSSG) were assayed using 5, 5-dithiobisnitrobenzoic acid (DTNB), according to the method of Rahman et al. (2006). For the determination of total GSH, the test solution was prepared by taking 2 mL of sample diluted with 31 mL of 100 mM K-PO_4 buffer containing 6.3 mM EDTA, then added with 140 mL of NADPH (0.248 mg/mL) and 20 mL of DTNB (6 mM), followed by GSH reductase (0.5 U). The absorbance was immediately recorded at 412 nm using a micro plate reader, and measurements were taken every 1 min for 4 min. The test solution for GSSG determination was prepared by taking 5 mL of sample diluted with 28 mL of 100 mM K-PO_4 buffer containing 6.3 mM EDTA, then added 140 mL of NADPH (0.248 mg/mL) and 20 mL of DTNB (6 mM), then GSH reductase (0.5 U) was added. The absorbance was immediately recorded at 412 nm using a micro plate reader, with measurements taken every 1 min for 4 min. Reduced GSH and oxidized GSSG were quantified using respective standard curves and expressed as nanomoles per gram FW. The resulting reduced/oxidized GSH/GSSG ratio was calculated.

For the determination of NADPH and NADPC concentration, fresh leaf samples (0.2 g) were immediately homogenized with 0.8 mL of 0.2 M NaOH for the NADPH assay and with 0.2 N HCl for the NADPC assay. The supernatants following a centrifugation 12,000 *g* for 10 min at 4 °C were heated at 95 °C for 1 min and stopped in ice-bath. The supernatant for NADPC determination was adjusted to pH 5–6 by 0.2 M NaOH or to pH 7–8 with 0.2 N HCl. The oxidized and reduced pyridine nucleotide contents were determined using the protocol of Queval and Noctor (2007).

8.12.1.5 Isolation of Total RNA and Quantitative Real-Time PCR

Total RNA was isolated from 200 mg leaf tissue using the SV Total RNA Isolation System (Promega). First-strand cDNAs were synthesized using the GoScript Reverse Transcription System. The gene expression level was quantified on a light cycler real-time PCR detection system (Bio-Rad) with SYBR Premix Ex TaqTM. All the quantifications were normalized to ACTIN. The qRT-PCR reactions were performed in triplicates for each of the three independent samples. Quantification of the relative transcript level was performed using the $2^{-\Delta\Delta Ct}$ method (Livak and Schmittgen 2001; Islam et al. 2017).

8.12.2 Genotyping of *Xanthomonas campestris* pv. *campestris*

8.12.2.1 Genomic DNA Extraction

For this purpose, genomic DNA from all 147 tested Xcc isolates and the reference strain NCPPB 1144 was extracted using the CTAB extraction method of Le Marrec et al. (2000) modified by Dimkic et al. (2013). Pure cultures of all isolates were grown on YDC agar at 26 °C for 48 h. Single bacterial colonies were suspended in 500 μL SDW and centrifuged at 10,000 *g* for 10 min. The pellet obtained was resuspended in TE buffer, and incubated with a mix of 10% (w/v) sodium dodecyl sulfate (SDS) and 20 mg mL^{-1} proteinase K in nuclease-free water at 37 °C for 30 min. Samples were then treated with 100 μL of 5 M NaCl and heated at 65 °C for 20 min after the addition of 3% hexadecyl-trimethyl ammonium bromide (CTAB, pH 8.0). The DNA was purified with 750 μL chloroform and centrifuged for 10 min at 10,000 *g*. The top (aqueous) phase was recovered by adding isopropanol and centrifugation at 10,000 *g* for 15 min. At the final phase, the obtained pellet was washed with 1 mL of 96% ice-cold ethanol, centrifuged at 10,000 *g* for 10 min and dried at room temperature for 30 min. The pelleted DNA was dissolved in 50 μL TE buffer (50 mM Tris, pH 8, 1 mM EDTA) and stored at −20 °C.

8.12.2.2 Repetitive Element PCR Fingerprinting

The method of repetitive element PCR (rep-PCR) fingerprinting was used to determine genetic diversity among the 147 Xcc isolates originating from six *Brassica* vegetables and winter oilseed rape collected from seven localities in two Serbian regions. The rep-PCR fingerprinting was performed using BOX (BOXA1R), ERIC (ERIC1R/ERIC2) and REP (REP1R-I/REP2-I) oligonucleotide primers (Louws et al. 1994; Versalovic et al. 1994). The DNA purity and concentrations for all tested isolates were measured with a Nanodrop spectrophotometer and equalized. All PCRs were performed according to a modified procedure of Louws et al. (1994) and Versalovic et al. (1994). A total of 25 μL PCR mix consisted of 1 μL of sample total DNA, 2.5 μL of 10 9 KAPA Taq Buffer B and 0.5 μL of 10 mM dNTP mixture; 2 μL (10 μM) of one of the three primer sets; 16.8 μL (ERIC and REP PCR) or 18.8 μL (BOX PCR) of ultrapure DNase/RNase-free water; and 0.2 μL of (5 UμL^{-1}) KAPA Taq polymerase.

Amplified PCR products (5 μL) mixed with 1.5 μL of Thermo Scientific 6 9 DNA-loading dye were separated using gel electrophoresis on 1% agarose gel with 0.5 9 TBE buffer (5 9 TBE: 1.1 M Tris; 900 mM borate; 25 mM EDTA; pH 8.3) and ethidium bromide (0.5 lg mL^{-1}). Gel electrophoresis was performed for 2.5 h under a constant voltage of 90 V and amperage of 300 mA. The surveyed distance of DNA fingerprints was checked in relation to a 100 bp DNA ladder. Differences in band position and the level of genetic similarity between the obtained profiles were computed using RESTDIST and NEIGHBOUR functions of the PHYLIP 3.69 program (Felsenstein 2005). The unweighted pair-group method with an arithmetic mean (UPGMA) was used to construct a phylogenetic tree with the NJPLOT tree-drawing program (Perriere and Gouy 1996).

8.12.2.3 The Multilocus Sequence Analysis

The Multilocus sequence analysis (MLSA) was performed with 18 representative *Xcc* isolates (Xc6, Xc9, Xc20, Xc23, Xc27, Xc30, Xc31, Xc32, Xc40, Xc48, Xc75, Xc110, Xc120, Xc150, Xc160, Xc170, Xc175, Xc178) selected on the basis of rep-PCR patterns obtained, and the reference strain NCPPB 1144. Two MLSA schemes have been used, with six housekeeping genes: *fusA, gap-1, gltA, gyrB1, lacF,* and *lepA* according to Almeida et al. (2010); and four genes *rpoD, dnaK, fyuA,* and *gyrB2* given by Young et al. (2008). The PCR amplification was performed in a total reaction volume of 25 μL, containing 12.5 μL of DreamTaq Green PCR Master

Mix, 1 μL of sample total DNA, 1 μL of each of the primers (10 μM) and 9.5 μL of ultrapure DNase/RNase-free water. Cycling programs for *fusA, gap-1, gltA, gyrB1, lacF* and *lepA* primers were performed according to Kyeon et al. (2016), starting with an initial denaturation step at 94 °C for 5 min; followed by 35 cycles of denaturation at 94 °C for 30 s, annealing at 58 °C for 30 s and polymerization at 72 °C for 30 s; with a final extension cycle at 72 °C for 7 min. The remaining four genes (*rpoD, dnaK, fyuA* and *gyrB2*) were amplified according to the method described by Young et al. (2008). The initial denaturation step was performed at 94 °C for 3 min; followed by 30 cycles of denaturation at 94 °C for 30 s, annealing at 54 °C for 30 s and polymerization at 72 °C for 1 min; with a final extension cycle at 72 °C for 10 min. The PCR products were purified using the QIAquick Gel Extraction kit and QIAquick/250 PCR Purification kit (QIAGEN) and sequenced in both directions by the Macrogen sequencing service (The Netherlands) using the same primers that were used for the amplification. The partial coding nucleotide sequences for all 10 genes were aligned using the automated CLUSTALW alignment feature of the BIOEDIT v. 7.0.5 program. All sequences were manually checked for quality and trimmed to the following lengths: *fusA* (663 nt), *gap1* (774 nt), *gltA* (547 nt), *gyrB1* (477 nt), *lacA* (507 nt), *lepA* (841 nt), *gyrB2* (774 nt), *rpoD* (778 nt), *fyuA* (631 nt) and *dnaK* (631 nt). The trimmed sequences were compared with those available in the NCBI (National Centre for Biotechnology Information) database using the nucleotide BLAST and deposited with accession numbers. Individual phylogenetic trees were constructed in MEGA 7 for each of the sequenced genes and rooted with *X. campestris* pv. *vesicatoria* (CP017190) from the NCBI database as an out group strain, using the neighbor-joining clustering method. Concatenated sequences with a total size of 6623 nt were also constructed to simultaneously compare all the selected isolates, the reference strain and an out group strain on the basis of all 10 used genes. In both cases, genetic distances were computed using the Kimura two-parameter nucleotide substitution model (Kimura 1980).

8.12.2.4 Pathogenicity

Pathogenicity tests were carried out by spraying a bacterial suspension on the surface of leaves (Popovic et al. 2013, 2014) of the seven tested *Brassica* plants–broccoli hybrid F_1 cv. Maraton (Sakata), and domestic cultivars of cabbage (Futoski), kale (Gvozdena Glava), cauliflower (Snezna Grudva), collard greens (unknown), kohlrabi (Backa Bela) and winter oilseed rape (Banacanka) obtained from the Institute of Field and Vegetable Crops (Novi Sad, Serbia). Plants were at the 2–3 true leaves stage (3-week-old plants). The seeds used in the tests were not chemically treated or disinfected, and so to avoid any potential contamination from seed borne pathogens (mostly fungi) that could affect seed germination and development, seeds were first sterilized by soaking in a 3% sodium hypochlorite (NaOCl) solution for 3–5 min, washed in tap water and then dried at room temperature. Potato dextrose agar (PDA) was used to test the success of disinfection. Plates with a few seeds were incubated at 25 ± 1 °C for 15 days in the dark to check disinfection success; untreated seeds were used as a control treatment. After 15 days, plates with developed seedlings were checked for their health status using a stereomicroscope. After checking seed sterility, young and healthy seedlings were sown in plastic boxes containing 66 wells (11 × 6), each measuring 4 × 5 cm in size, filled with a sterile growth substrate. The experiment was conducted in a completely randomized design in three replicates, with six host plants per replicate for each of the seven hosts (a total of 18 plants for each tested host). A total of 18 representative isolates, the same as used for molecular tests, and reference strain NCPPB 1144, were checked for their pathogenicity. Inoculation was made by spraying plants with the bacterial suspension (10^8 CFU mL^{-1} of a 48 h YDC culture in SDW). SDW was used as a negative control treatment, and plants treated with reference strain *Xcc* NCPPB 1144 were used as a positive control. After the inoculation, plants were kept under the plastic cover for the first 24 h to maintain RH at 90–100%. After 24 h and until the end of the experiment, the plastic cover was removed, and RH was reduced to ~50–60%. During the whole experiment, controlled conditions in the greenhouse were kept at 30 ± 2°C (day), and 20 ± 2°C (night), with a 12 h photoperiod. These conditions were similar to the usual conditions for *Brassica* cultivation in Serbia, which are also optimal for Xcc development. Irrigation was carried out once a day using a micro sprayer. To avoid splash transfer among isolates (cross contamination), treatments were separated with plastic barriers.

Disease was visually checked for symptom development 7 days after inoculation. Inoculated plants were scored on the 15th day after inoculation by rating five leaves per plant using the following scale: 1 = healthy leaves; 2 = < 10%; 3 = > 10%–25%; 4 = > 25%–50%; 5 = > 50%–75%; 6 = > 75%–100% infected leaf area.

8.12.2.5 Statistical Analysis

Basic statistical parameters were calculated and are presented in the tables. The Kolmogorov–Smirnov test of normality was used for analysis of variance, as was Levene's test for homogeneity of variance. The data obtained were subjected to analysis of variance (ANOVA), and separation of mean values of plant tissue necrosis in planta was accomplished by Tukey's HSD (honest significant difference) test. Values were considered significant at $P < 0.05$ for all tests. Statistical analyses were conducted by the general procedures of STATISTICA v. 7 (Stat Soft, Inc.) and SPSS STATISTICS v. 20 (SPSS, Inc.) (Popovic et al. 2019).

8.13 *Brassica*-Mixed Infection Diagnosis Protocols

8.13.1 Multispectral Imaging for Presymptomatic Identification of Cruciferous Diseases

Multispectral imaging (MSI) collect light reflected from the leaf surface. The spectrum of the reflected light is governed by both physical and biochemical interactions with the leaf, processes that will change during disease infection. Images are collected at different wave lengths in the optical range (350–1000 nm), where each pixel in the image represented by a vector, known as the spectral signature.

Analysis typically uses spectral information to study plant properties and conditions through spectral vegetation indices (SVIs), used commonly in remote sensing to describe vegetation health and density. Several SVIs have been developed to identify and detect plant disease. Using photometric stereo (PS) for analyzing plant tissue has attracted limited attention. Zhang et al. (2018) have highlighted the usefulness of the approach for extracting high resolution 3D imagery of leaves in order to aid vein extraction. It has been shown that different wavelengths used for light sources affect the validity of the assumption that the surface exhibits Lambertian reflectance. For example, in, reconstructions of human faces were shown to be more accurate under NIR than visible. This is thought to be due to the sub-surface scattering of NIR, which penetrates further into the skin, leading to a more diffuse reflectance. However, there is a compromise in using NIR for improving overall 3D reconstruction in that surface detail is lost. Whether the same issue affects plant tissue is not clear. Veys et al. (2019) developed a MSI system for disease diagnosis. This has been applied in a controlled environment for the analysis of disease progression and has the potential, with further development, to be used in the field. This MSI system uses narrowband light sources and a broadband silicon detector, in contrast to typical MSI and HSI systems where a broadband illumination source is used with a narrow band (typically diffraction grating based) detector. This selected spectral resolution allows the technology to be applied at a fraction of the cost of traditional systems, whilst improving signal-to-noise ratio (SNR) due to targeted reflectance peaks. The use of machine vision algorithms allows for early symptom detection (before symptoms can be detected by eye) allowing better quantification of disease progression with the potential, in field production scenarios, to allow earlier and hence more effective intervention (Veys et al. 2017).

The use of spectral imaging within the plant phenotyping and breeding community has been increasing due its utility as a noninvasive diagnostic tool. However, there is a lack of imaging systems targeted specifically at plant science duties, resulting in low precision for canopy-scale measurements. Veys et al. (2019) developed a prototype multispectral system designed specifically for plant studies and looks at its use as an early detection system for visually asymptomatic disease phases of *Pyrenopeziza brassicae* in *Brassica napus*. The analysis takes advantage of machine learning in the form of feature selection and novelty detection to facilitate the classification. An initial study into recording the morphology of the samples is also included to allow for further improvement to the system performance. The proposed method was able to detect light leaf spot infection with 92% accuracy when imaging entire oilseed rape plants from above, 12 days after inoculation and 13 days before the appearance of visible symptoms. False color mapping of spectral vegetation indices was used to quantify disease severity and its distribution within the plant canopy. In addition, the structure of the plant was recorded using photometric stereo, with the output influencing regions used for diagnosis. The shape of the plants was also recorded using photometric stereo, which allowed for reconstruction of the leaf angle and surface texture, although further work is needed to improve the fidelity due to uneven lighting distributions, to allow for reflectance compensation. The ability of active multispectral imaging has been demonstrated along with the improvement in time taken to detect light leaf spot at a high accuracy. The importance of capturing structural information is outlined, with its effect on reflectance and thus classification illustrated. The system could be used in plant breeding to enhance the selection of resistant cultivars, with its early and quantitative capability (Veys et al. 2019).

8.13.2 The RAPD-Based Molecular Diagnosis of Mixed Infections

Oilseed rape (*B. napus*) is attacked by parasitic fungi which often occur in mixed infections. Monitoring of these phytopathogens by morphological criteria is restricted due to their appearance especially in the later stages of disease development. Schneider et al. (1997) have developed molecular markers for a clear cut differentiation of a variety of rapeseed pathogenic fungi based on randomly amplified polymorphic DNA (RAPD). Twenty polymorphic fragments have been selected in southern hybridization experiments to test their taxon-specificity. Four amplification products give unspecific cross-hybridization patterns, one fragment corresponds to a genetic element common to three species within the genus *Alternaria* and 15 RAPD markers are highly specific for distinct fungal species. This report demonstrates the value of RAPD-PCR technique to amplify taxon-specific DNA fragments that can be used as hybridization probes for the diagnosis of a variety of rapeseed pathogens including *S. sclerotiorum.*

References

Alkooranee JT, Liu S, Aledan TR, Yin Y, Li M (2015) First report of powdery mildew caused by erysiphe cruciferarum on brassica napus in China. Plant Dis 99(11):1651

Al-Lami HFD, You MP, Barbetti MJ (2019a) Incidence, pathogenicity and diversity of *Alternaria* spp. associated with alternaria leaf spot of canola (*Brassica napus*) in Australia. Plant Pathol 68(3):492–503. https://doi.org/10.1111/ppa.12955

Al-Lami HFD, You MP, Mohammed AE, Barbetti MJ (2019b) Virulence variability across the *Alternaria* spp. population determines incidence and severity of alternaria leaf spot on rapeseed. Plant Pathol 69:506–517. https://doi.org/10.1111/ppa.13135

Al-Lami HFD, You MP, Barbetti MJ (2020) Relative host resistance to alternaria leaf spot in canola and mustard varieties is defined by *alternaria* species. Crop Pasture Sci 71(7):689–699. https://doi.org/10.1071/CP20024

Almeida NF, Yan S, Cai R, Clarke CR, Morris CE, Schaad NW, Schuenzel EL, Lacy GH, Sun X, Jones JB, Castillo JA, Bull CT, Leman S, Guttman DS, Setubal JC, Vinatzer BA (2010) PAMDB, a multilocus sequence typing and analysis database and website for plant-associated microbes. Phytopathology 100:208–215. https://doi.org/10.1094/PHYTO-100-3-0208

Almquist C (2016) Monitoring important soil-borne plant pathogens in Swedish crop production using real-time PCR. PhD Thesis. Swedish University of Agricultural Sciences, Uppsala, Sweden

Altschul SF, Gish W, Miller W, Myers EW, Lipman DJ (1990) Basic local alignment search tool. J Mol Biol 215:403–410

Altschul SF, Madden TL, Schäffer AA, Zhang J, Zhang Z, Miller W, Lipman DJ (1997) Gapped BLAST and PSI-BLAST: a new generation of protein database search programs. Nucleic Acids Res 25:3389–3402

Ananga AO, Cebert E, Soliman K, Kantety R, Konan K, Ochieng JW (2008) Phylogenetic relationships within and among *Brassica* species from RAPD loci associated with blackleg resistance. Afr J Biotechnol 7(9):1287–1293

Anders S, Pyl PT, Huber W (2015) *HTSeq*—a Python framework to work with high-throughput sequencing data. Bioinformatics 31(2):166–169. https://doi.org/10.1093/bioinformatics/btu638

Armstrong T (2007) Molecular detection and pathology of the oomycete *albugo candida* (white rust) in threatened coastal cresses. DOC Res Dev Series, 274, New Zealand Government, p. 1–18

Attard A, Gourgues M, Gout L, Schmit J, Roux J, Narcy JP, Balesdent MH, Rouxel T (2001) Molecular characterization and polymorphism of *MinLm1*, a mini-satellite from the phytopathogenic ascomycete *leptosphaeria maculans*. Curr Genet 40:54–64

Balesdent MH, Attard A, Kuhn ML, Rouxel T (2002) New avirulence genes in the phytopathogenic fungus *leptosphaeria maculans*. Phytopathology 92:1122–1133. https://doi.org/10.1094/PHYTO.2002.92.10.1122

Barbetti MJ (1987) Effects of three foliar diseases on biomass and seed yield for 11 cultivars of sub-terranean clover. Plant Dis 71:350–353

Barbetti MJ, Nichols PGH (2005) Field performance of subterranean clover germplasm in relation to severity of cercospora disease. Australasian Plant Pathol 34:197–201

Becker MG, Walker PL, Pulgar-Vidal NC, Belmonte MF (2017a) SeqEnrich: a tool to predict transcription factor networks from co-expressed *arabidopsis* and *brassica napus* gene sets. PLoS One 12(6):1–13. https://doi.org/10.1371/journal.pone.0178256

Becker MG, Zhang X, Walker PL, Wan JC, Millar JL, Khan D, Granger MJ, Cavers JD, Chan AC, Fernando DW, Belmonte MF (2017b) Transcriptome analysis of the *Brassica napus–leptosphaeria maculans* pathosystem identifies receptor, signalling and structural genes underlying plant resistance. Plant J 90(3):573–586. https://doi.org/10.1111/tpj.13514

Becker MG, Haddadi P, Wan J, Adam L, Walker P, Larkan NJ, Daayf F, Borhan MH, Belmonte MF (2019) Transcriptome analysis of *Rlm2*-mediated host immunity in the *brassica napus–leptosphaeria maculans* pathosystem. Mol Plant Microbe Interact 32(8):1001–1012. https://doi.org/10.1094/MPMI-01-19-0028-R

Bell CJ, Ecker JR (1994) Assignment of 30 microsatellite loci to the linkage map of *arabidopsis*. Genomics 19:137–144

Belmonte MF, Kirkbride RC, Stone SL, Pelletier JM, Bui AQ, Yeung EC, Hashimoto M, Fei J, Harada CM, Munoz MD, Le BH, Drews GN, Brady SM, Goldberg RB, Harada JJ (2013) Comprehensive developmental profiles of gene activity in regions and sub regions of the *arabidopsis* seed. Proc Natl Acad Sci U S A 110(5):E435–E444. https://doi.org/10.1073/pnas.1222061110

Bienkowska AC, Kaczmarek J, Schmidt KM, Nieróbca A, Czajka A, Jedryczka M (2020) Country-wide qPCR based assessment of *plasmodiophora brassicae* spread in agricultural soils and recommendations for the cultivation of brassicaceae crops in Poland. Pathogens 9:1070. https://doi.org/10.3390/pathogens9121070

Blin K, Medema MH, Kottmann R, Lee SY, Weber T (2017) The anti-SMASH database, a comprehensive database of microbial secondary metabolite biosynthetic gene clusters. Nucleic Acids Res 45:D555–D559

Bolger AM, Lohse M, Usadel B (2014) Trimmomatic: a flexible trimmer for illumina sequence data. Bioinformatics 30:2114–2120

Bousset L, Ermel M, Delourme R (2020) A *Leptosphaeria maculans* set of isolates characterized on all available differentials and used as control to identify virulence frequencies in a current French population. bioRxiv 2020-900167v1. https://doi.org/10.1101/2020.01.09.900167

Bubba MD, Giordani E, Pippucci L, Cincinelli A, Checchini L, Galvan P (2009) Changes in tannins, ascorbic acid and sugar content in astringent persimmons during on-tree growth and ripening and in response to different postharvest treatments. J Food Compos Anal 22(7): 668–677. https://doi.org/10.1016/j.jfca.2009.02.015

Buhariwalla S, Mithen R (1995) Cloning of a *Brassica* repetitive DNA element from resting spores of *Plasmodiophora brassicae*. Physiol Mol PIant Pathol 47:95–101

Buhariwalla S, Greaves S, Magrath R, Mithen R (1995) Development of specific PCR primers for the amplification of polymorphic DNA from the obligate root pathogen *plasmodiophora brassicae*. Physiol Mol PIant Pathol 47:83–94

Burke JM, Belfort M, Cech TR, Davies RW, Schweyen RJ, Shub DA, Szostak JW, Tabak HF (1987) Structural conventions for group I introns. Nucleic Acids Res 15(18):7217–7221. https://doi.org/10.1093/nar/15.18.7217

Burrough PA, McDonnell RA (1998) Principles of geographical information systems. Oxford University Press, Oxford

Buscot F, Wipf D, Di Battista C, Munch JC, Botton B, Martin F (1996) DNA polymorphism in morels: PCR/RFLP analysis of the ribosomal DNA spacers and microsatellite-primed PCR. Mycol Res 100:63–71

Campbell MM, Ellis BE (1992) Fungal elicitor-mediated responses in pine cell cultures: cell wall-bound phenolics. Phytochemistry 31:737–742

Cantarel BL, Korf I, Robb SMC, Parra G, Ross E, Moore B, Holt C, Alvarado AS, Yandell M (2008) MAKER: an easy-to-use annotation pipeline designed for emerging model organism genomes. Genome Res 18(1):188–196. https://doi.org/10.1101/gr.6743907

Cao T, Tewari J, Strelkov SE (2007) Molecular detection of *plasmodiophora brassicae*, causal agent of clubroot of crucifers, in plant and soil. Plant Dis 91:80–87

Carbo N, Perez-Martin J (2010) Activation of the cell wall integrity pathway promotes escape from G2 in the fungus *ustilago maydis*. PLoS Genet 6:e1001009. https://doi.org/10.1371/journal.pgen.1001009

Carvalho BS, Irizarry RA (2010) A framework for oligonucleotide microarray preprocessing. Bioinformatics 26(19):2363–2367

Casimiro S, Moura M, Ze-Ze L, Tenreiro R, Monteiro AA (2004) Internal transcribed spacer 2 amplicon as a molecular marker for identification of *peronospora parasitica* (crucifer downy mildew). J Appl Microbiol 96:579–587. https://doi.org/10.1111/j.1365-2672.2004.02193.x

Castlebury LA, Maddox JV, Glawe DA (1994) A technique for the extraction and purification of viable *plasmodiophora brassicae* resting spores from host root tissue. Mycologia 86:458–460

Chalhoub B, Denoeud F, Liu S, Parkin IA, Tang H, Wang X, Chiquet J, Belcram H, Tong C, Samans B, Corréa M, Da Silva C, Just J, Falentin C, Koh CS, Le Clainche I, Bernard M, Bento P, Noel B, Labadie K, Alberti A, Charles M, Arnaud D, Guo H, Daviaud C, Alamery S, Jabbari K, Zhao M, Edger PP, Chelaifa H, Tack D, Lassalle G, Mestiri I, Schnel N, Le Paslier MC, Fan G, Renault V, Bayer PE, Golicz AA, Manoli S, Lee TH, Thi VH, Chalabi S, Hu Q, Fan C, Tollenaere R, Lu Y, Battail C, Shen J, Sidebottom CH, Wang X, Canaguier A, Chauveau A, Bérard A, Deniot G, Guan M, Liu Z, Sun F, Lim YP, Lyons E, Town CD, Bancroft I, Wang X, Meng J, Ma J, Pires JC, King GJ, Brunel D, Delourme R, Renard M, Aury JM, Adams KL, Batley J, Snowdon RJ, Tost J, Edwards D, Zhou Y, Hua W, Sharpe AG, Paterson AH, Guan C, Wincker P (2014) Early allopolyploid evolution in the post-neolithic *brassica napus* oilseed genome. Science 345(6199):950–953

Chen H, Parimelalagan M, Lai YL, Lee KS, Koay ESC, Hapuarachchi HC, Ng LC, Ho PS, Chu JJH (2015) Development and evaluation of a SYBR green–based real-time multiplex RT-PCR assay for simultaneous detection and serotyping of dengue and chikungunya viruses. J Mol Diagn 17(6):722–728. https://doi.org/10.1016/j.jmoldx.2015.06.008

Chevre AM, Barret P, Eber F, Dupuy P, Brun H, Tanguy X, Renard M (1997) Selection of stable *Brassica napus-B. juncea* recombinant lines resistant to blackleg (*Leptosphaeria maculans*). 1. Identification of molecular markers, chromosomal and genomic origin of the introgression. Theor Appl Genet 95:1104–1111

Chèvre AM, Brun H, Eber F, Letanneur JC, Vallée P, Ermel M, Glais I, Li H, Sivasithamparam K, Barbetti MJ (2008) Stabilization of resistance to *leptosphaeria maculans* in *brassica napus–B. juncea* recombinant lines and its introgression into spring-type *brassica napus*. Plant Dis 92: 1208–1214

Chittem K, Yajima WR, Goswami RS, del Rio MLE (2020) Transcriptome analysis of the plant pathogen *Sclerotinia sclerotiorum* interaction with resistant and susceptible canola (*Brassica napus*) lines. PLoS One 15(3):e0229844. https://doi.org/10.1371/journal.pone.0229844

Cock PJA, Grüning BA, Paszkiewicz K, Pritchard L (2013) Galaxy tools and workflows for sequence analysis with applications in molecular plant pathology. Peer J 1:e167

Conesa A, Gotz S, Garcia-Gomez JM, Terol J, Talon M, Robles M (2005) Blast2GO: a universal tool for annotation, visualization and analysis in functional genomics research. Bioinformatics 21:3674–3676. https://doi.org/10.1093/bioinformatics/bti610

Cunnington JH, Takamatsu S, Lawrie A, Pascoe IG (2003) Molecular identification of anamorphic powdery mildews (Erysiphales). Australasian Plant Pathol 32:421–428

Cvikrova M, Binarova P, Eder J, Nedělník J (1992) Accumulation of phenolic acids in filtrate-treated alfalfa cell cultures derived from genotypes with different susceptibility to *Fusarium oxysporum*. J Plant Physiol 140:21–27

Danecek P, Auton A, Abecasis GR, Albers CA, Banks E, DePristo MA, Handsaker RE, Lunter G, Marth GT, Sherry ST, McVean G, Durbin R (2011) 1000 Genomics Project Analysis Group: the variant call format and VCF tools. Bioinformatics 27(15):2156–2158. https://doi.org/10.1093/bioinformatics/btr330

Deng S, Hiruki C (1991) Amplification of 16S rRNA genes from culturable and non-culturable mollicutes. J Microbiol Methods 14:53–61

Denhardt DT (1966) A membrane-filter technique for the detection of complementary DNA. Biochem Biophys Res Commun 23(5):641–646. https://doi.org/10.1016/0006-291x(66)90447-5

Derbyshire M, Denton-Giles M, Hegedus D, Seifbarghy S, Rollins J, van Kan J, Seidl MF, Faino L, Mbengue M, Navaud O, Raffaele S, Hammond-Kosack K, Heard S, Oliver R (2017) The complete genome sequence of the phytopathogenic fungus *sclerotinia sclerotiorum* reveals insights into the genome architecture of broad host range pathogens. Genome Biol Evol 9(3): 593–618

Dimkic I, Fira D, Stankovic S, Zivkovic S, Beric T, Gavrilovic V, Ivanovic Z (2013) Characterization and evaluation of two *Bacillus* strains, SS-12.6 and SS-13.1, as potential agents for the control of phytopathogenic bacteria and fungi. Biol Control 65:312–321

Doyle J, Doyle J (1990) Isolation of plant DNA from fresh tissue. Focus 12:13–15

Duncan JM, Cooke D, Brich P, Toth R (1998) Molecular variability in sexually reproducing fungal plant pathogens. In: Bridge PD, Couteaudier Y, Clarkson JM (eds) Molecular variability of fungal pathogens. CAB International, Wallingford, pp 19–39

Eckert M, Gout L, Rouxel T, Blaise F, Jedryczka M, Fitt BDL, Balesdent MH (2005) Identification and characterization of polymorphic mini-satellites in the phytopathogenic ascomycete *Leptosphaeria maculans*. Curr Genet 47:37–48

Edwards K, Johnstone C, Thompson C (1991) A simple and rapid method for the preparation of plant genomic DNA for PCR analysis. Nucleic Acids Res 19(6):1349. https://doi.org/10.1093/nar/19.6.1349

Emms DM, Kelly S (2015) OrthoFinder: solving fundamental biases in whole genome comparisons dramatically improves orthogroup inference accuracy. Genome Biol 16(1):157

Eshraghi L, Barbetti MJ, Li H, Danehloueipour N, Sivasithamparam K (2007) Resistance in oilseed rape (*Brassica napus*) and Indian mustard (*Brassica juncea*) to a mixture of *pseudocercosporella capsellae* isolates from Western Australia. Field Crop Res 101:37–43

Excoffier L, Smouse PE, Quattro JM (1992) Analysis of molecular variance inferred from metric distances among DNA haplotypes-application to human mitochondrial-DNA restriction data. Genetics 131:479–491

Faggian R, Strelkov SE (2009a) Detection and measurement of *Plasmodiophora brassicae*. J Plant Growth Regul 28:282–288

Faggian R, Strelkov SE (2009b) *Plasmodiophora brassicae* (clubroot)—a plant pathogen that alters host growth and productivity. J Plant Growth Regul 28:282–287

Faggian R, Bulman SR, Lawrie AC, Porter IJ (1999) Specific polymerase chain reaction primers for the detection of *plasmodiophora brassicae* in soil and water. Phytopathology 89:392–397

Felsenstein J (2005) PHYLIP (phylogeny inference package) version 3.6. Distributed by Author. Department of Genome Sciences, University of Washington, Seattle. http://evolution.genetics.washington.edu/phylip.html

Feng J, Hwang SF, Strelkov SE (2013) Assessment of gene expression profiles in primary and secondary zoospores of *Plasmodiophora brassicae* by dot-blot and real-time PCR. Microbiol Res 168:518–524

Ferreira ME, Grattapaglia D (1995) Introducao ao Uso de Marcadores RAPD e RFLP em Ana'lise Gene'tica. EMBRAPA-CENARGEN, Brasilia

Freeman J, Ward E, Calderon C, McCartney A (2002) A polymerase chain reaction (PCR) assay for the detection of inoculum of *Sclerotinia sclerotiorum*. Eur J Plant Pathol 108:877–886

Fu YX (1997) Statistical tests of neutrality of mutations against population growth, hitchhiking and background selection. Genetics 147:915–925

Fu YX, Li WH (1993) Statistical tests of neutrality of mutations. Genetics 133:693–709

Fujioka T, Mizutani O, Furukawa K, Sato N, Yoshimi A, Yamagata Y, Yamagata Y, Nakajima T, Abe K (2007) MpkA-dependent and independent cell wall integrity signaling in *Aspergillus nidulans*. Eukaryot Cell 6:1497–1510. https://doi.org/10.1128/EC.00281-06

Garcia-Mas J, Oliver M, Gomez-Paniagua H, de Vicente MC (2000) Comparing AFLP, RAPD and RFLP markers for measuring genetic diversity in melon. Theor Appl Genet 101:860–864

Gel B, Diez-Villanueva A, Serra E, Buschbeck M, Peinado MA, Malinverni R (2016) Region eR: an R/bioconductor package for the association analysis of genomic regions based on permutation tests. Bioinformatics 32(2):289–291

Ghanbarnia K, Fudal I, Larkan NJ, Links MG, Balesdent MH, Profotova B, Fernando WGDD, Rouxel T, Borhan MH (2015) Rapid identification of the *Leptosphaeria maculans* avirulence gene *AvrLm2* using an intraspecific comparative genomics approach. Mol Plant Pathol 16:699–709

Gilbert DG (1990) Mulfold, a Macintosh software program for prediction of RNA secondary structure by free energy minimization, version 2.0: sub-optimal folding with temperature dependence. Published electronically on the Internet, available via anonymous ftp to iubio.bio.indiana.edu

Gilbert D G (1994) Loop Dloop, a Macintosh software tool for drawing RNA secondary structures in molecular biology, version 1.2 a63. Published electronically on the Internet, available via anonymous ftp to iubio.bio.indiana.edu

Goodwin DC, Lee SB (1993) Microwave mini-prep of total genomic DNA from fungi, plants, protists and animals for PCR. BioTechniques 15:438–444

Gout L, Fudal I, Kuhn ML, Blaise F, Eckert M, Cattolico L, Balesdent MH, Rouxel T (2006) Lost in the middle of nowhere: the *AvrLm1* avirulence gene of the Dothideomycete *Leptosphaeria maculans*. Mol Microbiol 60:67–80

Guillemette T, Iacomi-Vasilescu B, Simoneau P (2004) Conventional and real-time PCR-based assay for detecting pathogenic *Alternaria brassicae* in cruciferous seed. Plant Dis 88:490–496

Gunasinghe N, You MY, Clode P, Cawthray GR, Barbetti MJ (2017) Unique infection structures produced by *Pseudocercosporella capsellae* on oilseed crops *Brassica carinata*, *B. juncea* and *B. napus* in Western Australia. Plant Pathol 66:304–315. https://doi.org/10.1111/ppa.12563

Gundersen DE, Lee IM (1996) Ultra-sensitive detection of phytoplasmas by nested-PCR assays using two universal primer pairs. Phytopathol Mediterr 35:114–151

Guyon K, Balague C, Roby D, Raffaele S (2014) Secretome analysis reveals effector candidates associated with broad host range necrotrophy in the fungal plant pathogen *Sclerotinia sclerotiorum*. BMC Genomics 15:336

Haas BJ, Salzberg SL, Zhu W, Pertea M, Allen JE, Orvis J, White O, Buell CR, Wortman JR (2008) Automated eukaryotic gene structure annotation using Evidence Modeler and the program to assemble spliced alignments. Genome Biol 9(1):R7

Haas BJ, Papanicolaou A, Yassour M, Grabherr M, Blood PD, Bowden J, Couger MB, Eccles D, Li B, Lieber M, MacManes MD, Ott M, Orvis J, Pochet N, Strozzi F, Weeks N, Westerman R, William T, Dewey CN, Henschel R, LeDuc RD, Friedman N, Regev A (2013) *De novo* transcript sequence reconstruction from RNA-seq using the trinity platform for reference generation and analysis. Nat Protoc 8(8):1494–1512

Habu JB, Ibeh BO (2015) *In vitro* antioxidant capacity and free radical scavenging evaluation of active metabolite constituents of Newbouldia laevis ethanolic leaf extract. Biol Res 48:16

Haddadi P, Ma L, Wang H, Borhan MH (2016) Genome-wide transcriptome analyses provides insights into the lifestyle transition and effector repertoire of *Leptosphaeria maculans* during colonization of *Brassica napus* seedlings. Mol Plant Pathol 17(8):1196–1210. https://doi.org/10.1111/mpp.12356

Harling R (2007) Clubroot control using novel and sustainable methods. Final report DEFRA project HH3227TFV

Henriquez MA, Soliman A, Li G, Hannoufa A, Ayele BT, Daayf F (2016) Molecular cloning, functional characterization and expression of potato (*Solanum tuberosum*) 1-deoxy-d-xylulose 5-phosphate synthase 1 (*StDXS1*) in response to *Phytophthora infestans*. Plant Sci 243:71–83

Hirata T, Takamatsu S (1996) Nucleotide sequence diversity of rDNA internal transcribed spacers extracted from conidia and cleistothecia of several powdery mildew fungi. Mycoscience 37: 283–288

Hong CX, Fitt BDL (1995) Effects of inoculum concentration, leaf age and wetness period on the development of dark leaf and pod spot (*Alternaria brassicae*) on oilseed rape (*Brassica napus*). Ann Appl Biol 127:283–295

Hong SG, Cramer RA, Lawrence CB, Pryor BM (2005) *Alt a 1* allergen homologs from *Alternaria* and related taxa: analysis of phylogenetic content and secondary structure. Fungal Genet Biol 42:119–129

Hongtrakul V, Huestis GM, Knapp SJ (1977) Amplified fragment length polymorphisms as a tool for DNA fingerprinting sunflower germplasm: genetic diversity among oilseed inbred lines. Theor Appl Genet 95:400–407

Huson DH, Bryant D (2006) Application of phylogenetic networks in evolutionary studies. Mol Biol Evol 23:254–267

Iacomi-Vasilescu B, Blancard D, Guénard M, Molinero-Demilly V, Laurent E, Simoneau P (2002) Development of a PCR-based diagnostic assay for detecting pathogenic *Alternaria* species in cruciferous seeds. Seed Sci Technol 30:87–95

Innis MA, Gelfand DH (1990) Optimization of PCRs. In: Innis MA, Gelfand DH, Sninsky JJ, White TJ (eds) PCR protocols: a guide to methods and applications. Academics Press, San Diego, CA, pp 3–12

Islam MT, Lee BR, Park SH, La VH, Bae DW, Kim TH (2017) Cultivar variation in hormonal balance is a significant determinant of disease susceptibility to *Xanthomonas campestris pv. campestris* in *Brassica napus*. Front Plant Sci 8:2121. https://doi.org/10.3389/fpls.2017.02121

Ito S, Maehara T, Tanaka S, Kameya-Iwaki M, Yano S, Kishi F (1997) Cloning of a single-copy DNA sequence unique to *Plasmodiophora brassicae*. Physiol Mol Plant Pathol 50:289–300

Ito S, Ichinose H, Yanagi C, Tanaka S, Kameya-Iwaki M, Kishi F (1999) Identification of an in planta-induced mRNA of *Plasmodiophora brassicae*. J Phytopathol 147:79–82

Jaccard P (1908) Nouvelles researches sur la distribution florale. Bull Soc Vaud Sci Nat 44:223–270

Jaeger JA, Turner DH, Zuker M (1989) Improved predictions of secondary structures for RNA. Proc Natl Acad Sci U S A 86:7706–7710

Jaeger JA, Turner DH, Zuker M (1990) Predicting optimal and suboptimal secondary structure for RNA. In: Doolittle RF (ed) Methods in enzymology. Academic Press, San Diego, CA, pp 281–306

Jamaux ID, Spire D (1994) Development of a polyclonal antibody-based immune assay for the early detection of *Sclerotinia sclerotiorum* in rapeseed petals. Plant Pathol 43(5):847–862. https://doi.org/10.1111/j.1365-3059.1994.tb01629.x

Jin-Ping L, Yan L, Xue-wen X, A-li C, Bao-ju L (2013) Development of a real-time PCR assay for *Plasmodiophora brassicae* and its detection in soil samples. J Integr Agric 12(10):1799–1806

Jouet A, Saunders DGO, McMullan M, Ward B, Furzer O, Jupe F, Cevik V, Hein I, Thilliez GJA, Holub E, van Oosterhout C, Jones JDG (2018) *Albugo candida* race diversity, ploidy and host-associated microbes revealed using DNA sequence capture on diseased plants in the field. New Phytol 221:1529–1543

Kabbage M, Yarden O, Dickman MB (2015) Pathogenic attributes of *Sclerotinia sclerotiorum*: switching from a biotrophic to necrotrophic lifestyle. Plant Sci 233:53–60

Kaczmarek J, Irzykowski W, Burzyński A, Jędryczka M (2014) The detection of *Plasmodiophora brassicae* using loop-mediated isothermal DNA amplification. Acta Agrobot 67(4):59–66. https://doi.org/10.5586/aa.2014.058

Karapapa VK, Typas MA (2001) Molecular characterization of the host-adapted pathogen *Verticillium longisporum* on the basis of a group-i intron found in the nuclear SSU-rRNA gene. Curr Microbiol 42:217–224. https://doi.org/10.1007/s002840010207

Katoh K, Standley DM (2013) MAFFT multiple sequence alignment software version 7: improvements in performance and usability. Mol Biol Evol 30:772–780

Keilwagen J, Wenk M, Erickson JL, Schattat MH, Grau J, Hartung F (2016) Using intron position conservation for homology-based gene prediction. Nucleic Acids Res 44(9):e89

Kemen E, Gardiner A, Schultz-Larsen T, Kemen AC, Balmuth AL, Robert-Seilaniantz A, Bailey K, Holub E, Studholme DJ, Maclean D, Jones JD (2011) Gene gain and loss during evolution of obligate parasitism in the white rust pathogen of *Arabidopsis thaliana*. PLoS Biol 9:e1001094

Kim D, Langmead B, Salzberg SL (2015) HISAT: A fast spliced aligner with low memory requirements. Nat Methods 12(4):357–360. pmid:25751142

Kimura M (1977) Preponderance of synonymous changes as evidence for the neutral theory of molecular evolution. Nature 267:275–276

Kimura M (1980) A simple method for estimating evolutionary rate of base substitutions through comparative studies of nucleotide sequences. J Mol Evol 16:111–120

Konieczny A, Ausubel FM (1993) A procedure for mapping *Arabidopsis* mutations using codominant ecotype-specific PCR-based markers. Plant J 4:403–410

Korgan S, Wolski EA, Cicore P, Suarez P, Capezio S, Huarte MA, Andreu AB (2011) *Solanum tarijense* reaction to *Phytophthora infestans* and the role of plant defence molecules. Plant Breed 130:231–236

Krogh A, Larsson B, von Heijne G, Sonnhammer EL (2001) Predicting transmembrane protein topology with a hidden Markov model: application to complete genomes. J Mol Biol 305(3): 567–580. https://doi.org/10.1006/jmbi.2000.4315

Kumar R, Ichihashi Y, Kimura S, Chitwood DH, Headland LR, Peng J, Maloof JN, Sinha NR (2012) A high-throughput method for illumina RNA-Seq library preparation. Front Plant Sci 3(8):202. https://doi.org/10.3389/fpls.2012.00202

Kurtz S, Phillippy A, Delcher AL, Smoot M, Shumway M, Antonescu C, Salzberg SL (2004) Versatile and open software for comparing large genomes. Genome Biol 5:R12

Kwok S, Kellogg DE, McKinnery N, Spasic D, Gada L, Levenson C, Sninsky JJ (1990) Effects of primer-template mismatches on the polymerase chain reaction: human immune deficiency virus type 1 model studies. Nucleic Acids Res 18:999–1005

Kyeon MS, Son SH, Noh YH, Kim YE, Lee HI, Cha JS (2016) *Xanthomonas euvesicatoria* causes bacterial spot disease on pepper plant in Korea. Plant Pathol J 32:431

Lanfranco L, Perotto S, Longato S, Mello A, Cometti V, Bonfante P (1998) Molecular approaches to investigate biodiversity in mycorrhizal fungi. In: Varma A (ed) Mycorrhiza manual. Springer-Verlag, Berlin, pp 353–372

Lark RM, Ferguson RB (2004) Mapping risk of soil nutrient deficiency or excess by disjunctive and indicator kriging. Geoderma 118:39–53

Larkan NJ, Ma L, Borhan MH (2015) The *Brassica napus* receptor-like protein RLM2 is encoded by a second allele of the LepR3/Rlm2 blackleg resistance locus. Plant Biotechnol J 13:983–992

Lawrence DP, Gannibal PB, Dugan FM, Pryor BM (2014) Characterization of *Alternaria* isolates from the infectoria species-group and a new taxon from Arrhenatherum. *Pseudoalternaria arrhenatheria* sp. nov. Mycol Prog 13:257–276

Le Marrec C, Hyronimus B, Bressollier P, Verneuil B, Urdaci MC (2000) Biochemical and genetic characterization of coagulin, a new antilisterial bacteriocin in the pediocin family of bacteriocins, produced by *Bacillus coagulans* I4. Appl Environ Microbiol 66:5213–5220

Lee SB, Taylor JW (1990) Isolation of DNA from fungal mycelia and single spores. In: Innis MA, Gelfand DH, Sninsky JJ, White TJ (eds) PCR protocols: a guide to methods and applications. Academic, San Diego, pp 282–287

Lee IM, Gundersen-Rindal DE, Davis RE, Bartoszyk IM (1998) Revised classification scheme of phytoplasmas based on RFLP analyses of 16S rRNA and ribosomal protein gene sequences. Int J Syst Bacteriol 48:1153–1169

Lee IM, Zhao Y, Bottner KD (2006) SecY gene sequence analysis for finer differentiation of diverse strains in the aster yellows phytoplasma group. Mol Cell Probes 20:87–91

Lee BR, Kim KY, Jung WJ, Avice JC, Ourry A, Kim TH (2007) Peroxidases and lignification in relation to the intensity of water-deficit stress in white clover (*Trifolium repens* L). J Exp Bot 58: 1271–1279. https://doi.org/10.1093/jxb/erl280

Lefol C, Morrall RAA (1996) Immunofluorescent staining of Sclerotinia ascospores on canola petals. Can J Plant Pathol 18(3):237–241

Li H, Durbin R (2009) Fast and accurate short read alignment with Burrows-Wheeler transform. Bioinformatics 25:1754–1760

Liao Y, Smyth G, Shi W (2013) Feature counts: an efficient general-purpose program for assigning sequence reads to genomic features. Bioinformatics 30(7):923–930. https://doi.org/10.1093/bioinformatics/btt656

Librado P, Rozas J (2009) DnaSP v5: a software for comprehensive analysis of DNA polymorphism data. Bioinformatics 25:1451–1452

Links MG, Holub E, Jiang RH, Sharpe AG, Hegedus D, Beynon E, Sillito D, Clarke WE, Uzuhashi S, Borhan MH (2011) *De novo* sequence assembly of *Albugo candida* reveals a small genome relative to other biotrophic oomycetes. BMC Genomics 12:503

Liu L, Zhao D, Zheng L, Hsiang T, Wei Y, Fu Y, Huang J (2013) Identification of virulence genes in the crucifer anthracnose fungus *Colletotrichum higginsianum* by insertional mutagenesis. Microbiol Pathol 64:6–17. https://doi.org/10.1016/j.micpath.2013.06.001

Livak KJ, Schmittgen TD (2001) Analysis of relative gene expression data using real-time quantitative PCR and the $2-\Delta\Delta CT$ method. Methods 25:402–408

Loreto F, Velikova V (2001) Isoprene produced by leaves protects the photosynthetic apparatus against ozone damage, quenches ozone products, and reduces lipid peroxidation of cellular membranes. Plant Physiol 127:1781–1787

Loreto F, Mannozzi M, Maris C, Nascetti P, Ferranti F, Pasqualini S (2001) Ozone quenching properties of isoprene and its antioxidant role in leaves. Plant Physiol 126:993–1000

Louws FJ, Fulbright DW, Stephens CT, de Bruijn FJ (1994) Specific genomic fingerprints of phytopathogenic *Xanthomonas* and *Pseudomonas* pathovars and strains generated with repetitive sequences and PCR. Appl Environ Microbiol 60:2286–2295

Love MI, Huber W, Anders S (2014) Moderated estimation of fold change and dispersion for RNA-Seq data with DESeq2. Genome Biol 15(12):550. https://doi.org/10.1101/002832

Lucas JA, Crute IR, Sherriff C, Gordon PL (1988) The identification of a gene for race-specific resistance to *Peronospora parasitica* (downy mildew) in *Brassica napus* var. *oleifera* (oilseed rape). Plant Pathol 37(4):538–545. https://doi.org/10.1111/j.1365-3059

Lun AT, Chen Y, Smyth GK (2016) It's DE-licious: a recipe for differential expression analyses of RNA-seq experiments using quasi-likelihood methods in edgeR. Methods Mol Biol 1418:391–416. PMID: 27008025

Lunter G, Goodson M (2011) Stampy: a statistical algorithm for sensitive and fast mapping of Illumina sequence reads. Genome Res 21(6):936–939

Luo Y, Zhang H, Qi L, Zhang S, Zhou X, Zhang Y, Xu JR (2014) FgKin1 kinase localizes to the septal pore and plays a role in hyphal growth, ascospore germination, pathogenesis, and localization of Tub1 beta-tubulins in *Fusarium graminearum*. New Phytol 204:943–954. https://doi.org/10.1111/nph.12953

Macioszek VK, Gapińska M, Zmienko A, Sobczak M, Skoczowski A, Oliwa J, Kononowicz AK (2020) Complexity of *Brassica oleracea-Alternaria brassicicola* susceptible interaction reveals downregulation of photosynthesis at ultrastructural, transcriptional, and physiological levels. Cell 9(10):2329. https://doi.org/10.3390/cells9102329

Malinowski R, Novak O, Borhan MH, Spichal L, Strnad M, Rolfe SA (2016) The role of cytokinins in clubroot disease. Eur J Plant Pathol 145:543–557. https://doi.org/10.1007/s10658-015-0845-y

Martin DP, Lemey P, Lott M, Moulton V, Posada D, Lefeuvre P (2010) *RDP3*: a flexible and fast computer program for analyzing recombination. Bioinformatics 26:2462–2463. https://doi.org/10.1093/bioinformatics/btq467

McKinney HH, Davis RJ (1923) Influence of soil temperature and moisture on infection of wheat seedlings by *Helminthosporium sativum*. Aust J Agric Res 26:195–218

McLoughlin AG, Wytinck N, Walker PL, Girnard IJ, Rashid KY, de Kievit T, Fernando WGD, Whyard S, Belmonte MF (2018) Identification and application of exogenous dsRNA confers plant protection against *Sclerotinia sclerotiorum* and *Botrytis cinerea*. Sci Rep 8:7320. https://doi.org/10.1038/s41598-018-25434-4

McMullan M, Gardiner A, Bailey K, Kemen E, Ward BJ, Cevik V, Robert-Seilaniantz A, Schultz-Larsen T, Balmuth A, Holub E, van Oosterhout C, Jones JDG (2015) Evidence for suppression of immunity as a driver for genomic introgressions and host range expansion in races of *Albugo candida*, a generalist parasite. eLife 4:e04550. https://doi.org/10.7554/eLife.04550

Millar JL, Becker MG, Belmonte MF (2015) Laser microdissection of plant tissues. In: Plant microtechniques and protocols. Springer International, Cham, pp 337–350. https://doi.org/10.1007/978-3-319-19944-3

Mills PR, Sreenivasaprasad S, Muthumeenakshi S (1998) Assessing diversity in *Colletotrichum* and *Trichoderma* species using molecular markers. In: Bridge PD, Couteaudier Y, Clarkson JM (eds) Molecular variability of fungal pathogens. CAB International, Wallingford, pp 105–120

Mitrovic J, Kakizawa S, Duduk B, Oshima K, Namba S, Bertaccini A (2011) The groEL gene as an additional marker for finer differentiation of '*Candidatus Phytoplasma asteris*'-related strains. Ann Appl Biol 159:41–48

Mori Y, Sato Y, Takamatsu S (2000) Evolutionary analysis of the powdery mildew fungi using nucleotide sequences of the nuclear ribosomal DNA. Mycologia 92:74–93

Moss NA, Lucas JA, Crute IR (1991) Evidence for differential response to isolates of *Peronospora parasitica* (downy mildew) in *Brassica rapa*. Test Agro Chem. cv. 12. Ann Appl Biol 118:96–97

Moukette BM, Pieme CA, Njimou JR, Biapa CPN, Marco B, Ngogang JY (2015) *In vitro* antioxidant properties, free radicals scavenging activities of extracts and polyphenol composition of a non-timber forest product used as spice: *Monodora myristica*. Biol Res 48:15

Nowakowska M, Wrzesińska M, Kamiński P, Szczechura W, Lichocka M, Tartanus M, Kozik EU, Nowicki M (2019) *Alternaria brassicicola*—Brassicaceae pathosystem: insights into the infection process and resistance mechanisms under optimized artificial bio-assay. Eur J Plant Pathol 153:131–151. https://doi.org/10.1007/s10658-018-1548-y

Nowicki M, Klosinska U, Szczechura W, Markiewicz M, Sowik I, Michalczuk L, Kozik EU (2016) Cellular and biochemical mechanisms of cucumber resistance against *Pseudoperonospora cubensis*. In: Proc XIth Eucarpia Meeting on Genetics and Breed of Cucurbitaceae 2016 at Warsaw, Poland

O'Connell J, Gurdasani D, Delaneau O, Pirastu N, Ulivi S, Cocca M, Traglia M, Huang J, Huffman JE, Rudan I, McQuillan R, Fraser RM, Campbell H, Polasek O, Asiki G, Ekoru K, Hayward C,

Wright AF, Vitart V, Navarro P, Zagury JF, Wilson JF, Toniolo D, Gasparini P, Soranzo N, Sandhu MS, Marchini J (2014) A general approach for haplotype phasing across the full spectrum of relatedness. PLoS Genet 10:e1004234

Ohshima K, Yamaguchi Y, Hirota R, Hamamoto T, Tomimura K, Tan Z, Sano T, Azuhata F, Walsh JA, Fletcher J, Chen J, Gera A, Gibbs A (2002) Molecular evolution of Turnip mosaic virus; evidence of host adaptation, genetic recombination and geographical spread. J Gen Virol 83: 1511–1521

Oshima K, Kakizawa S, Nishigawa H, Jung HY, Wei W, Suzuki S, Arashida R, Nakata D, Miyata S, Ugaki M, Namba S (2004) Reductive evolution suggested from the complete genome sequence of a plant pathogenic phytoplasma. Nat Genet 36:27–29

Perriere G, Gouy M (1996) WWW-query: an on-line retrieval system for biological sequence banks. Biochimie 78:364–369

Petersen TN, Brunak S, von Heijne G, Nielsen H (2011) SignalP 4.0: discriminating signal peptides from transmembrane regions. Nat Methods 8:785–786

Petersen TN, Lukjancenko O, Thomsen MCF, Maddalena Sperotto M, Lund O, Møller Aarestrup F, Sicheritz-Ponten T (2017) MGmapper: reference based mapping and taxonomy annotation of metagenomics sequence reads. PLoS One 12(5):e0176469. https://doi.org/10.1371/journal.pone.0176469

Pfister DH (1997) Castor, pollux, and life histories of fungi. Mycologia 89:1–23

Plotnikova JM, Reuber TL, Ausubel FM, Pfister DH (1998) Powdery mildew pathogenesis of *Arabidopsis thaliana*. Mycologia 90:1009–1016

Popovic T, Balaz J, Starovic M, Trkulja N, Ivanovic Z, Ignjatov M, Josic D (2013) First report of *Xanthomonas campestris* pv. *campestris* as the causal agent of black rot on oilseed rape (*Brassica napus*) in Serbia. Plant dis 97:418

Popovic T, Balaz J, Ignjatov M, Mitrovic P, Gavrilovic V, Josic D (2014) Identification and genetic characterisation of *Xanthomonas campestris* pv. *campestris* as an oilseed rape pathogen in Serbia. J Plant Pathol 96:553–560

Popovic T, Mitrovicb P, Jelusicc A, Dimkicd I, Marjanovic-Jeromelab A, Nikolicd I, Stankovic S (2019) Genetic diversity and virulence of *Xanthomonas campestris* pv. *campestris* isolates from *Brassica napus* and six *Brassica oleracea* crops in Serbia. Plant Pathol 68:1448–1457

Prior RL, Fan E, Ji H, Amy H, Nio C, Payne MJ, Reed J (2010) Multi-laboratory validation of a standard method for quantifying proanthocyanidins in cranberry powders. J Sci Food Agric 90: 1473–1478. https://doi.org/10.1002/jsfa.3966

Queval G, Noctor G (2007) A plate reader method for the measurement of NAD, NADP, glutathione, and ascorbate in tissue extracts: application to redox profiling during *Arabidopsis* rosette development. Anal Biochem 363:58–69. https://doi.org/10.1016/j.ab.2007.01.005

Rahman I, Kode A, Biswas SK (2006) Assay for quantitative determination of glutathione and glutathione disulfide levels using enzymatic recycling method. Nat Protoc 1:3159–3165. https://doi.org/10.1038/nprot.2006.378

Rajarammohan S, Pental D, Kaur J (2019) Near-complete genome assembly of *Alternaria brassicae*—a necrotrophic pathogen of *Brassica* crops. Mol Plant Microbe Interact 32(8): 928–930

Rehmany AP, Lynn JR, Tor M, Holub EB, Beynon JL (2000) A comparison of *Peronospora parasitica* (downy mildew) isolates from *Arabidopsis thaliana* and *Brassica oleracea* using amplified fragment length polymorphism and internal transcribed spacer 1 sequence analyses. Fungal Genet Biol 30:95–103. https://doi.org/10.1006/fgbi.2000.1216

Rehmany AP, Gordon A, Rose LE, Allen RL, Armstrong MR, Whisson SC, Kamoun S, Tyler BM, Birch PR, Beynon JL (2005) Differential recognition of highly divergent downy mildew avirulence gene alleles by *RPP1* resistance genes from two *Arabidopsis* lines. Plant Cell 17: 1839–1850

Ren L, Fang XP, Sun CC, Chen KR, Liu F, Li M, Xu L (2014) First report of clubroot on *Capsella bursa-pastoris* caused by *Plasmodiophora brassicae* in Sichuan Province of China. Plant Dis 98(5):687. https://doi.org/10.1094/PDIS-04-13-0395-PDN

Rennie DC, Manolii VP, Cao T, Hwang SF, Howard RJ, Strelkov SE (2011) Direct evidence of surface infestation of seeds and tubers by *Plasmodiophora brassicae* and quantification of spore loads. Plant Pathol 60:811–819

Ricarova V, Kazda J, Singh K, Ryšánek P (2016) Clubroot caused by *Plasmodiophora brassicae* Worn.: a review of emerging serious disease of oilseed rape in the Czech Republic. Plant Prot Sci 52:71–86. https://doi.org/10.17221/87/2015-PPS

Rieu I, Powers SJ (2009) Real-time quantitative RT-PCR: design, calculations, and statistics. Plant Cell Online 21(4):1031–1033. https://doi.org/10.1105/tpc.109.066001

Rimmer SR, Shattuck VI, Buchwaldt L (2007) Compendium of Brassica diseases. APS Press, St. Paul, MN

Rolfe SA, Strelkov SE, Links MG, Clarke WE, Robinson SJ, Djavaheri M, Malinowski R, Haddadi P, Kagale S, Parkin IAP, Taheri A, Borhan MH (2016) The compact genome of the plant pathogen *Plasmodiophora brassicae* is adapted to intracellular interactions with host *Brassica* spp. BMC Genomics 17:272. https://doi.org/10.1186/s12864-016-2597-2

Rouxel T, GrandaubertJ HJK, Hoede C, van de Wouw AP, Couloux A, Dominguez V, Anthouard V, Bally P, Bourras S, Cozijnsen AJ, Ciuffetti LM, Degrave A, Dilmaghani A, Duret L, Fudal I, Goodwin SB, Gout L, Glaser N, Linglin J, Kema GHJ, Lapalu N, Lawrence CB, May K, Meyer M, Ollivier B, Poulain J, Schoch CL, Simon A, Spatafora JW, Stachowiak A, Turgeon BG, Brett M, Tyler BM, Vincent D, Weissenbach J, Amselem J, Quesneville H, Oliver RP, Wincker P, Balesdent MH, Howlett BJ (2011) Effector diversification within compartments of the *Leptosphaeria maculans* genome affected by repeat-induced point mutations. Nat Commun 2:202. https://doi.org/10.1038/ncomms1189

Saharan G S, Mehta Naresh, Meena P D (2016) Alternaria blight of crucifers: biology, ecology and management. Springer Verlag, Singapore: p 326

Sambrook J, Russell DW (2001) Molecular cloning: a laboratory manual, 3rd edn. Cold Spring Harbor Laboratory Press, Cold Spring Harbor, NY

Sanger F, Nicklen S, Coulson AR (1977) DNA sequencing with chain-terminating inhibitors. Proc Natl Acad Sci U S A 74:5463–5467. https://doi.org/10.1073/pnas.74.12.5463

Schneider B, Seemuller E, Smart CD, Kirkpatrick BC (1995) Phylogenetic classification of plant pathogenic mycoplasma like organisms or phytoplasmas. In: Razin ES, Tully JG (eds) Molecular and diagnostic procedures in mycoplasmology. Academic Press, San Diego, CA, pp 369–380

Schneider B, Gibb KS, Seemuller E (1997) Sequence € and RFLP analysis of the elongation factor Tu gene used in differentiation and classification of phytoplasmas. Microbiologica 143:3381–3389

Schwelm A, Fogelqvist J, Knaust A, Jülke S, Lilja T, Bonilla-Rosso G, Karlsson M, Shevchenko A, Dhandapani V, Choi SR, Kim HG, Park JY, Lim YP, Ludwig-Müller J, Dixelius C (2015) The *Plasmodiophora brassicae* genome reveals insights in its life cycle and ancestry of chitin synthases. Sci Rep 5:11153. https://doi.org/10.1038/srep11153

Seruga Music M, Duc Nguyen H, Cerni S, Mamula D, Ohshima K, Skoric D (2014) Multilocus sequence analysis of 'Candidatus Phytoplasma asteris' strain and the genome analysis of *Turnip mosaic virus* co-infecting oilseed rape. J Appl Microbiol 117:774–785

Seruga M, Skoric D, Botti S, Paltrinieri S, Juretic N, Bertaccini AF (2003) Molecular characterization of a phytoplasma from the aster yellows (*16SrI*) group naturally infecting *Populus nigra* L. 'Italica' trees in Croatia. Forest Pathol 33:113–125

Sheehy RE, Kramer M, Hiatt WR (1988) Reduction of polygalacturonase activity in tomato fruit by antisense RNA. Proc Natl Acad Sci U S A 85:8805–8809

Sherriff C, Lucas JA (1990) The host range of isolates of downy mildew, *Peronospora parasitica* from *Brassica* crop species. Plant Pathol 39:77–91

Simmons EG (1986) Alternaria themes and variations. Mycotaxon 25:287–308

Simmons EG (2007) *Alternaria*: an identification manual. CBS Fungal Biodiversity Centre, Utrecht

Smyth GK (2005) Limma: linear models for microarray data. In: Gentleman R, Carey V, Huber W, Irizarry R, Dudoit S (eds) Bioinformatics and computational biology solutions using R and bioconductor. Springer, New York, pp 397–420

Sperschneider J, Gardiner DM, Dodds PN, Tini F, Covarelli L, Singh KB, Manners JM, Taylor JM (2016) Effector P: predicting fungal effector proteins from secretomes using machine learning. New Phytol 210(2):743–761

Stamatakis A (2014) RAxML version 8: a tool for phylogenetic analysis and post analysis of large phylogenies. Bioinformatics 30:1312–1313

Stanke M, Diekhans M, Baertsch R, Haussler D (2008) Using native and syntenically mapped cDNA alignments to improve *de novo* gene finding. Bioinformatics 24(5):637–644

Štefan MB, Rodríguez JV, Blažeković B, Kindl M, Vladimir-Knežević S (2014) Total hydroxycinnamic acids assay: prevalidation and application on Lamiaceae species. Food Anal Methods 7:326–336. https://doi.org/10.1007/s12161-013-9630-8

Stefanac-Udbinac Z, Milicic D, Zeljko M (1963) Wasserrüben- und Kohlrrübenmosaikvirus (*Turnip mosaic virus*) in Jugoslawien. Acta Bot Croat 22:107–117

Strelkov SE, Tewari JP, Smith-Degenhardt E (2006) Characterization of *Plasmodiophora brassicae* populations from Alberta, Canada. Can J Plant Pathol 28:467–474

Suzuki S, Oshima K, Kakizawa S, Arashida R, Jung HY, Yamaji Y, Nishigawa H, Ugaki M, Namba S (2006) Interaction between the membrane protein of a pathogen and insect microfilament complex determines insect-vector specificity. Proc Natl Acad Sci U S A 103(11):4252–4257. https://doi.org/10.1073/pnas.0508668103

Tajima F (1989) Statistical method for testing the neutral mutation hypothesis by DNA polymorphism. Genetics 123:585–595

Tamura K, Peterson D, Peterson N, Stecher G, Nei M, Kumar S (2011) *MEGA5*: molecular evolutionary genetics analysis using maximum likelihood, evolutionary distance, and maximum parsimony methods. Mol Biol Evol 28(10):2731–2739. https://doi.org/10.1093/molbev/msr121

Tamura K, Stecher G, Peterson D, Filipski A, Kumar S (2013) MEGA6: molecular evolutionary genetics analysis version 6.0. Mol Biol Evol 30:2725–2729

Ter-Hovhannisyan V, Lomsadze A, Chernoff YO, Borodovsky M (2008) Gene prediction in novel fungal genomes using an ab initio algorithm with unsupervised training. Genome Res 18(12): 1979–1990

Tham FY, Lucas JA, Wilson ZA (1994) DNA fingerprinting of *Peronospora parasitica*, a biotrophic fungal pathogen of crucifers. Theor Appl Genet 88:490–496

Thompson JD, Higgins DG, Gibson TJ, Clustal W (1994) Improving the sensitivity of progressive multiple sequence alignment through sequence weighting, position-specific gap penalties and weight matrix choice. Nucleic Acids Res 22(22):4673–4680. https://doi.org/10.1093/nar/22.22.4673

Thompson JD, Gibbson TJ, Plewniak F, Jeanmougin F, Higgins DG (1997) The ClustalX windows interface: flexible strategies for multiple sequence alignment aided by quality analysis tools. Nucleic Acids Res 25:4876–4882

Tomimura K, Gibbs AJ, Jenner CE, Walsh JA, Ohshima K (2003) The phylogeny of *Turnip mosaic virus*; comparisons of 38 genomic sequences reveal a Eurasian origin and a recent 'emergence' in East Asia. Mol Ecol 12:2099–2111

Trapnell C, Roberts A, Goff L, Pertea G, Kim D, Kelley DR, Pimentel H, Salzberg SL, Rinn JL, Pachter L (2012) Differential gene and transcript expression analysis of RNA-seq experiments with TopHat and Cufflinks. Nat Protoc 7(3):562–578. https://doi.org/10.1038/nprot.2012.016

Typas MA, Griffen AM, Bainbridge BW, Heale JB (1992) Restriction fragment length polymorphism in mitochondrial DNA and ribosomal RNA gene complexes as an aid to the characterisation of species and sub-species populations in the genus *Verticillium*. FEMS Microbiol Lett 95:157–162

Uloth MB, Clode PL, You MP, Barbetti MJ (2015) Attack modes and defence reactions in pathosystems involving *Sclerotinia sclerotiorum, Brassica carinata, B. juncea* and *B. napus*. Ann Bot 117(1):mcv150. https://doi.org/10.1093/aob/mcv150

Usadel B, Poree F, Nagel A, Lohse M, Czedik-Eysenberg A, Stitt M (2009) A guide to using MapMan to visualize and compare omics data in plants: a case study in the crop species. Maize Plant Cell Environ 32(9):1211–1229

Valiante V, Jain R, Heinekamp T, Brakhage AA (2009) The MpkA MAP kinase module regulates cell wall integrity signaling and pyomelanin formation in *Aspergillus fumigatus*. Fungal Genet Biol 46:909–918. https://doi.org/10.1016/j.fgb.2009.08.005

Van Meirvenne M, Goovaerts P (2001) Evaluating the probability of exceeding a site-specific soil cadmium contamination threshold. Geoderma 102:75–100

Versalovic J, Schneider M, de Bruijn FJ, Lupski FR (1994) Genomic fingerprinting of bacteria using repetitive sequence-based polymerase chain reaction. Methods Mol Cell Biol 5:25–40

Veys C, Hibbert J, Davis P, Grieve B (2017) An ultra-low-cost active multispectral crop diagnostics device. In: IEEE sensors. IEEE, Glasgow, pp 1–3. https://doi.org/10.1109/ICSENS.2017.8234211

Veys C, Chatziavgerinos F, AlSuwaidi A, Hibbert J, Hansen M, Bernotas G, Smith M, Yin H, Rolfe S, Grieve B (2019) Multispectral imaging for presymptomatic analysis of light leaf spot in oilseed rape. Plant Methods 15:4. https://doi.org/10.1186/s13007-019-0389-9

Volodarsky D, Leviatan N, Otcheretianski A, Fluhr R (2009) HORMONOMETER: a tool for discerning transcript signatures of hormone action in the *Arabidopsis* transcriptome. Plant Physiol 150:1796–1805. https://doi.org/10.1104/pp.109.138289

Wakeham AJ, Faggian R, Kennedy R (2017) Assessment of the response of *Plasmodiophora brassicae* in contaminated horticultural land, using lime based fertilizer concentrations. Food Energy Secur 6:122

Wallenhammar AC (1996) Prevalence of *Plasmodiophora brassicae* in a spring oilseed rape growing area in Central Sweden and factors influencing soil infestation levels. Plant Pathol 45:710–719

Wallenhammar AC, Arwidsson O (2001) Detection of *Plasmodiophora brassicae* by PCR in naturally infested soils. Eur J Plant Pathol 107:313–321

Wallenhammar AC, Johnson L, Gerhardson B (2000) Agronomic properties of partly clubroot-resistant spring oilseed turnip rape lines. J Phytopathol 148:495–499

Wallenhammar AC, Almquist C, Soderstrom M, Jonsson A (2012) In-field distribution of *Plasmodiophora brassicae* measured using quantitative real-time PCR. Plant Pathol 61:16–28

Wang TY, He F, Hu QW, Zhang Z (2011) A predicted protein-protein interaction network of the filamentous fungus *Neurospora crassa*. Mol BioSyst 7:2278–2285. https://doi.org/10.1039/c1mb05028a

Ward BJ, van Oosterhout C (2016) HybridCheck: software for the rapid detection, visualization and dating of recombinant regions in genome sequence data. Mol Ecol Resour 16:534–539

Webster R, Oliver MA (1992) Sample adequately to estimate variograms of soil properties. J Soil Sci 43:177–192

Wei L, Jian H, Lu K, Filardo F, Yin N, Liu L, Qu C, Li W, Du H, Li J (2016) Genome-wide association analysis and differential expression analysis of resistance to Sclerotinia stem rot in *Brassica napus*. Plant Biotechnol J 14:1368–1380. https://doi.org/10.1111/pbi.12501

Wellburn AR (1994) The spectral determination of chlorophylls a and b, as well as total carotenoids, using various solvents with spectrophotometers of different resolution. J Plant Physiol 144(3):307–313

White TJ, Bruns T, Lee S, Taylor JW (1990) Amplification and direct sequencing of fungal ribosomal RNA genes for phylogenetics. In: Innis MA, Gelfand DH, Sninsky JJ, White TJ (eds) PCR protocols. A guide to methods and application. Academic Press, London, pp 315–322

Williams JGK, Kubelik AR, Livak KJ, Rafalski JA, Tingey SV (1990) DNA polymorphisms amplified by arbitrary primers are useful as genetic markers. Nucleic Acids Res 18:6531–6535

Xiao SY, Ellwood S, Findlay K, Oliver RP, Turner JG (1997) Characterization of three loci controlling resistance of *Arabidopsis thaliana* accession Ms-0 to two powdery mildew diseases. Plant J 126(1):757–768

Yoshida K, Schuenemann VJ, Cano LM, Pais M, Mishra B, Sharma R, Lanz C, Martin FN, Kamoun S, Krause J et al (2013) The rise and fall of the *Phytophthora infestans* lineage that triggered the Irish potato famine. Elife 2:e0073

Young JM, Park DC, Shearman HM, Fargier E (2008) A multilocus sequence analysis of the genus *Xanthomonas*. Syst Appl Microbiol 31:366–377

Zeng F, Gong X, Hamid MI, Fu Y, Jiatao X, Cheng J, Jiang D (2012) A fungal cell wall integrity-associated MAP kinase cascade in *Coniothyrium minitans* is required for conidiation and mycoparasitism. Fungal Genet Biol 49:347–357. https://doi.org/10.1016/j.fgb.2012.02.008

Zhang H, Feng J, Zhang S, Zhang S, Li F, Strelkov SE, Sun R, Hwang SF (2015) Resistance to *Plasmodiophora brassicae* in *Brassica rapa* and *Brassica juncea* genotypes from China. Plant Dis 99:776–779. https://doi.org/10.1094/PDIS-08-14-0863-RE

Zhang L, Cai X, Wu J, Liu M, Grob S, Cheng F, Liang J, Cai C, Liu Z, Liu B, Wang F, Li S, Liu F, Li X, Cheng L, Yang W, Li MH, Grossniklaus U, Zheng H, Wang X (2018) Improved *Brassica rapa* reference genome by single-molecule sequencing and chromosome conformation capture technologies. Hortic Res 5:50. https://doi.org/10.1038/s41438-018-0071-9

Zhao J, Peltier AJ, Meng J, Osborn TC, Grau CR (2004) Evaluation of Sclerotinia stem rot resistance in oilseed *Brassica napus* using a inoculation technique under greenhouse conditions. Plant Dis 88:1033–1039

Zhao J, Buchwaldt L, Rimmer R, Sharpe A, McGregor L, Bekkaoui D, Hegedus D (2009) Patterns of differential gene expression in *Brassica napus* cultivars infected with *Sclerotinia sclerotiorum*. Mol Plant Pathol 10:635–649. https://doi.org/10.1111/j.1364-3703.2009.00558.x

Zhao J, Wu YX, Ho HH, Chen ZJ, Li XY, He YQ (2016) *PBT1*, a novel antimicrobial protein from the biocontrol agent *Bacillus subtilis* XF-1 against *Plasmodiophora brassicae*. Eur J Plant Pathol 145:583–590. https://doi.org/10.1007/s10658-016-0905-y

Zhishen J, Mengcheng T, Jianming W (1999) The determination of flavonoid contents in mulberry and their scavenging effects on superoxide radicals. Food Chem 64:555–559. https://doi.org/10.1016/S0308-8146(98)00102-2

Future Research Priorities of Crucifers' Host-Pathosystem

9

Abstract

During the last two decades of present century, significant progress has been made in comprehension of genomics of crucifers' host-pathosystem using multiomics technology by the *Brassica* scientists. However, there is always a scope for improvement in the technology to get better insight in this complex biological system being challenged by multiple stresses under natural and directed modern agricultural cultivation technology. It is imperative that a universal database set up of genomic studies kept updated. Cloning of more *R* genes from *Brassica* species against different stresses will strengthen better our management of host defense system. The information generation on induction of *R* genes and virulence genes at different stages of infection and pathogenesis will clear the complex life cycle of biotrophs and hemibiotrophs. More information on combining of race specific and non-race-specific *R* genes on durability of host resistance is required to get better insight on pathogen virulence evolution. There is a need to resolve controversial views on improved resistance by *R* genes pyramiding and its breakdown. Search of more *R* genes from wild relatives may provide better resources for some pathogens. More efforts are required to develop multiple disease resistance cultivars. The use of pangenomic approach can facilitate *R* genes mining from a wider plate form. The use of transcriptomics, proteomics, and metabolomics can provide new opportunities for mining *R* genes during prebreeding to promote better insight of host–pathogen interactions. The orthologous genes in *Brassica* should be investigated. Transgenic breeding against necrotrophs may be exploited. Effector genes and their effects during pathogenesis need in-depth studies. The development of cell biology and use of biometabolomics can offer better opportunities for solving many mysteries of the pathogens switching over from biotrophs to necrotrophs.

G. Singh Saharan et al., *Genomics of Crucifer's Host-Pathosystem*,
https://doi.org/10.1007/978-981-19-3812-2_9

Keywords

Genomics research priorities · Crucifers host-pathosystem · Genomics · Pangenomics · International collaborative research network · Universal data base set · Characterization of genes · Cloning of *R* genes · Induction of R and virulent genes · Gene mining · Molecular function · Identification of effectors/avirulence genes · Race-specific and non-race-specific · *R* genes pyramiding · *R* genes from wild relatives · Multiple disease resistance · Prebreeding · Orthologous genes · Transgenic breeding · Effector genes · Switching of biotrophs to necrotrophs · Mapping methods · Molecular host–pathogen interaction · Appressorium function · Biosynthetic pathways · IPM · Impact of climate change · Molecular host–pathogen interaction

9.1 Introduction

The crucifers host-pathosystem is a very complex biological system and genomics have dynamic influence on host–pathogen interactions. To comprehend this system, researchers have made every effort through multiomics approaches to reveal molecular mechanism involved in functional characterization of genes responsible for infection, virulence, and pathogenesis metabolic pathways. However, there is always a scope for refinement and improvement with generation protocols in every biological system. After scanning the information generated so far, some gaps have been identified for future priorities to get deeper insight in the genomics of crucifers' host-pathosystem as follows:

1. An international collaborative research network creating a world database for each pathogen of crucifers to resolve the nomenclature issue of *R* genes and pathotypes is established. Since dynamic changes can happen with host cultivars and pathogens in terms of cultivar resistance breakdowns and/or pathogens losing/gaining in terms of their pathogenicity/virulence over time, it is imperative that any universal database set up by regularly updated and kept current.
2. Except for the four *R* genes that have been cloned, the QTL for the major diseases in *Brassica* have been identified but the causal genes remain unknown. Cloning of *R* genes has historically been undertaken using traditional methods, such as map-based cloning. However, the recent genomics revolution has provided significant insights into the presence and absence of expressed genes in the *Brassica* A and C genomes using a pan-transcriptome approach. This, along with the recent pan-genome research readily, allows cloning of *R* genes in crops such as *Brassica* that have complex genomes. More *R* gene cloning work in *Brassica* will help us better understand how the plant defends itself against these pathogens and we can use this knowledge to exploit novel *R* genes in a quicker and more effective manner for breeding purposes.
3. The impact of climate change on crop diseases has been increasingly addressed over recent decades, making the characterization of new *R* genes more crucial

than ever. The MutRen Seq technology has been used to clone resistance genes. This technique allows more rapid and precise cloning. Such methodologies can now be similarly applied for precise alteration of disease resistance genes in *Brassica*.

4. Effective deployment of *R* genes in breeding *B. napus* resistant crops for enduring sustainability demands detailed knowledge about the ecology and life history of both host and pathogen and how their interactions may shape their evolution over time. Such knowledge can allow prediction of the host resistance at different stages of the infection processes for the pathogen, potentially allowing identification and targeting of the resistance gene (s) effective at the earliest infection stage in order to maximize impedance and/or even prevent the establishment of the pathogen and also curtail subsequent pathogen inoculum production and consequent secondary spread of the disease. While this approach could be particularly applicable for a disease such as Clubroot, there are still challenges as it is not yet possible to identify the co-relationship between root hair infection (primary) and root cortex infection (secondary). On the other hand, knowledge about controlling pathogen infection at different developmental stages of the host (e.g., seedling vs. adult) or even in different plant organs (e.g., cotyledons, leaves, stems, pods, and root) where the pathogen infection and colonization has the greatest adverse impact on the productivity of the plant (plant stems for *L. maculans* and *S. sclerotiorum*) have become more forthcoming for the major pathosystems in *B. napus*. Clearly, improving our understanding of the mechanism(s) at both life-stages of the pathogen infection and colonization processes, the development stages of the host, and the host–pathogen interactions should open up new opportunities to better understand and exploit approaches that maximize control of pathogen infection and spread to minimize consequent yield losses.
5. Sequencing of *Brassica* genomes will allow new discoveries and understandings regarding the genetic relationships of different *R* genes, within and between members of the Brassicaceae. Although hybridization between divergent groups in the Brassicaceae, which contains rich resources of *R* genes, is already possible, particularly in India, this is an area of great future potential that is likely to lead to identification of many new *R* genes as these methodologies become more widely adopted and as rates of success with wide-hybridization events improve. While scientists are able to make informed decisions about utilizing race-specific and/or non-race–specific resistance genes, further understanding of the benefits of different patterns of deployment for each type individually or rotation of both types, will identify better approaches toward maximizing the durability and effectiveness of both types of resistances, particularly *R* genes. Further, field deployment of race-specific and/or non-race–specific resistance genes will have a large impact on the dynamics within pathogen populations. Combining race-specific and race non-specific genes together in the one cultivar will also potentially increase durability of *R* genes deployed within *Brassica* species, and has already been demonstrated for *R* genes against *L. maculans*.

6. Improved resistance from gene pyramiding through marker-assisted selection was reported for the *P. brassicae-B. rapa* ssp. *pekinensis* pathosystem. However, this remains a controversial approach with some suggesting that such an approach could lead to an undesired outcome of breakdown of the multiple *R* genes pyramided. As we gain more knowledge about the molecular mechanisms of the plant immunity signaling pathway and the role of *R* genes, this will open up more effective and durable strategies to overcome pathogen attack.
7. For most of the *Brassica* species, the resistance resources for diseases are highly limited. In general, the A genome is rich in Turnip mosaic virus, Black rot, Downy mildew, and clubroot resistance, whereas the B genome possesses black rot and, Blackleg resistance, and the C genome harbors stem rot, Fusarium wilt, and downy mildew resistance. Fortunately, the six species in the *Brassica* genus and others, such as *B. incana*, *B. cretica* (C genome), and *B. fruticulosa* (B genome), as well as its close Brassicae relatives, could be used to facilitate resistance gene exchanges in breeding programs.
8. Interspecies crossing within the *Brassica* genus are widely adopted using embryo rescue, reciprocal crossing and MAS. There are few black rot-resistant resources in the C genome of *B. oleracea*, whereas high resistance is present in both the A and B genomes. Thus, interspecies hybridization has been used to transfer and utilize the resistance found in the A and B genomes. In addition, distant hybridization has been used by breeders to generate novel-resistant germplasms. In contrast, high-level resistance has been observed in a few wild species, such as *Erucastrum cardaminoides*, *Diplotaxis tenuisiliqua*, and *Sinapis arvensis*.
9. High levels of stem rot resistance from crosses between *B. napus*/*B. juncea* and *E. cardaminoides*/*D. tenuisiliqua*/*E. abyssinicum* have been obtained. The novel resistance germplasms generated in such studies provide valuable materials in future breeding programs for *Brassica* crops. However, there is still much work to be done for them to be practically applied in commercial cultivars. There are rich resources with clubroot resistance in *B. rapa* but very few in *B. oleracea*, and breeders have spent ~17 years introgressing resistance from Chinese cabbage (*B. rapa* subsp. *pekinensis*) cv. Parkin to different *B. oleracea* cultivars, which are now widely used.
10. Mapping methods based on high-throughput sequencing data have developed rapidly, especially SNP-based methods, such as KASP markers and microarrays, BSA/BSR, and GWAS. Based on whole genome level mutations and their association with trait values, GWAS enables fast and accurate target trait gene characterization using natural populations. In addition, KASP technology possesses high assay robustness and accuracy, and allows notable savings in terms of cost and time.
11. In many cases, there is one main disease in a specific production region and application of cultivars with resistance to that main disease is appropriate to realize the balance between defense and growth. However, for some continuous or high-intensity cropping regions, various pathogens may accumulate;

therefore, cultivars with multiple resistances to different diseases are in great need. Currently, MAS using abundant molecular markers enable us to realize the pyramiding of R alleles for different diseases and breed multiresistant cultivars. MAS combined with other methods, such as hybridization and microspore culture, has greatly shortened the breeding circle.

12. A single resistance gene is easily overcome because of pathogen variations and global climate changes. More durable resistance is urgently needed to ensure *Brassica* crop production. Complete race-specific resistance genes are very effective in a short period but are easily overcome by the pathogens; polygene-inherited resistance is thought to be more durable, but its effects might be unstable owing to variable environmental conditions. Thus, pyramiding qualitative genes with major quantitative loci in cultivars represents an ideal means to ensure the effectiveness and durability of resistance. Thus, combining quantitative resistance with single *R* genes is a promising strategy for resistance breeding. In addition, co-evolution between the host and the pathogen has been extensively studied; the mapped seven Blackleg resistance genes/alleles and their contrasting avirulence genes have indicated a rapid and fierce arms race between *Brassica* hosts, and *Leptosphaeria maculans* and the single use of one type of pyramiding could pose a high selection pressure on the pathogen, thus raise the risk of pathogen mutation and host resistance loss. So, the rationalized pyramiding of dissimilar sets of resistance genes/QTLs should be deployed in different cultivars to provide heterogeneity in the selection pressure on the pathogen population and may result in more durable resistance. In addition, pyramided resistance could be integrated with agricultural, physical, chemical, and biological controls to realize IPM, further maximize durability and guarantee stability.
13. Current molecular and omics methods, including transcriptomics, proteomics, and metabolics, provide new opportunities for mining genes in the resistance-regulating network, which could be either directly used in resistance breeding or used in indirect prebreeding studies to promote understanding of *Brassica*–pathogen interactions. In addition, the sequenced *Brassica* accessions do not contain all *R* genes due to variations between individuals, whereas the establishment of the pangenomes could facilitate gene mining from a wider platform.
14. Achieving expanded knowledge of the molecular host–pathogen interactions could provide evidence for disease control. Protein–protein interactions in the *Brassica–Turnip mosaic virus* system are a heavily researched topic. Previous studies have shown that the cytoplasmic inclusion protein determines the viral avirulence for TuRB01/01b/04, whereas P3 determines the avirulence of TuRB03/05.
15. In *Brassica* crops, although over ten *R* genes have been characterized to date, most avirulence or interaction genes in the pathogens have not yet been thoroughly characterized. Thus, the next emphasis should be on the establishment of global collections of pathogen isolates for the identification of important avirulence or interaction genes. In addition, for the host, the development of a

series of ILs through recurrent backcrossing to "Mendalize" the quantitative loci would be beneficial.

16. As many *R* genes have been cloned in *Arabidopsis*, the orthologous genes in *Brassica* can be investigated, which will greatly facilitate the cloning of these genes and the clarification of their molecular functions.
17. The next efforts should be more accurate identification of the *R* genes and clarification of their molecular mechanisms using emerging high-efficiency genomic, postgenomic, and omics methods, and more efficient application of the R resources through an integration of approaches, such as haploid culture, MAS, distant introgression, genome design, pyramiding, and transgenic breeding, to control the diseases and secure crucifers production.
18. Recent studies have shown the role of certain effector proteins, but most have still not been characterized and their molecular targets and function are still unknown. Most of the effector/avirulence genes in the postulated gene-for-gene model have not yet been identified molecularly.
19. With the advances in sequencing, additional sequences of different *Xcc* isolates, including multiple isolates from each race and additional sequences of isolates of closely related pathovars, should soon be available. The comparison of multiple sequences should then provide more clues on the important genes that determine pathogenicity, including effectors; the presence and absence of particular genes and the variability between isolates will provide clues on the role and evolution of these genes. Functional analysis of the genes can then confirm their role in the plant–pathogen interactions.
20. On the host side, a number of disease resistance genes in Brassicas and *A. thaliana* have been postulated, some have been mapped, but few have been cloned. The effort to identify more disease resistance genes should be continued, and some of the most important crops could be improved with the incorporation of disease resistance genes from closely related species via interspecific crosses and embryo rescue, or via transformation. It is also possible that the engineering of the promoters of known *R* genes to recognize multiple effectors could contribute to the development of broad-spectrum disease-resistant crops. The control of the disease could be improved by pyramiding disease resistance genes and/or by using multiple lines with different gene combinations that could be mixed and balanced to match the avirulence genes present in the bacterial population.
21. The application of functional genomics and proteomics to bacteria *in planta* to identify virulence factors, and the application of functional genomics and proteomics to both resistant and susceptible host plants inoculated with *Xcc*, will provide key information on the interaction between the bacteria and the hosts. Research on the diversity of *Xcc*, pathogenicity factors and evolution, together with host–pathogen interaction studies, should lead to improvements in the prevention and control of the black rot of crucifers.
22. The genome sequence of *C. higginsianum* has revealed a large number of putative effector proteins, but few effector proteins have been experimentally confirmed or characterized to date. The identification of virulence targets for the

hundreds of candidate effectors predicted from genome sequencing remains a major challenge, partly because protocols for high-throughput plant cellular assays are lacking.

23. Two effector proteins, ChELP1 and ChELP2, at the biotrophic stage in plants may be critical for suppressing chitin-triggered immune responses, while the basal expression levels in appressoria in vitro and in plants are required for efficient substrate penetration. Further work is needed to elucidate how these proteins contribute to appressorial function and the switch to invasive hyphal growth.
24. Lack of chromosome 11 leads to aborted infection during biotrophy, indicating that a number of potential genes from chromosome 11 have critical functions in manipulating plant host responses, and these can be selectively analyzed in future work to evaluate their possible function.
25. The comparative analyses of interactions between *Arabidopsis* and both nonadapted and adapted *Colletotrichum* species revealed that the adapted pathogen *C. higginsianum* induced papillary callose at a much lower frequency than non-adapted *Colletotrichum* species, indicating that this fungus may suppress pre-penetration resistance at the cell periphery. The mechanism of *C. higginsianum* suppression of prepenetration resistance remains to be elucidated by future experiments.
26. Many genes for synthesis of secondary metabolites are upregulated during plant infection, and many genes are specifically expressed during the biotrophic stage in *C. higginsianum*, indicating that appropriate gene expression during the biotrophic stage is a key for successful establishment of this fungus in host plants. So, the *C. higginsianum–Arabidopsis* pathosystem has tremendous potential for discovery of novel bioactive molecules, and identification of the corresponding biosynthetic pathways.
27. Due to the short biotrophic period, as well as switching to necrotrophic phase, identifying and assigning the temporal, spatial, and tissue-specific functions to virulence factors are real challenges for *Sclerotinia*. Future development of cell biology and biochemical tools tailored for *S. sclerotiorum* will offer opportunities for solving many mysteries of the fungus.

Index

G. Singh Saharan et al., *Genomics of Crucifer's Host-Pathosystem*,
https://doi.org/10.1007/978-981-19-3812-2

C

D

E

F

G

H

I

L

M

N

O